Perspectives on Geographical Marginality

Volume 3

Series editors

Walter Leimgruber, Fribourg, Switzerland
Etienne Nel, Dunedin, New Zealand
Stanko Pelc, Koper-Capodistria, Slovenia

This book series *Perspectives on Geographical Marginality* comprehensively overviews research, on areas and communities impacted by processes of marginalization as a result of globalization, economic, environmental, political and social change. This series seeks to discuss and determine what is geographical marginality by inviting leading international experts to publish theoretical and applied work. It also seeks to rigorously debate the degree to which local areas and communities are responding to these process of change and with what success.

The series stems from the International Geographical Union's (IGU), 'Commission on Globalization, Marginalization, and Regional and Local Response' (C12.29). As is suggested by its name, the commission researches the problem of geographical marginality offering a leading forum from which this series will be led. Marginality cannot be defined without putting it into a certain perspective: economic, political and social (including cultural). Marginality has to be clearly distinguished from peripherality. Marginal areas may be a part of periphery or even the centre, but "cannot really be attributed to them".

Proposed themes which will be covered include:

- Mountainous regions and globalization;
- Regional development and policy/or: Globalization and its impact on local and regional development;
- Theory of marginalization;
- Transformation of rural areas from the viewpoint of globalization and marginalization;
- Drivers of marginalization in border and peripheral areas.

More information about this series at http://www.springer.com/series/15046

Stanko Pelc · Miha Koderman
Editors

Nature, Tourism and Ethnicity as Drivers of (De)Marginalization

Insights to Marginality from Perspective of Sustainability and Development

Springer

Editors
Stanko Pelc
Faculty of Education
University of Primorska
Koper
Slovenia

Miha Koderman
Faculty of Humanities
University of Primorska
Koper
Slovenia

This book has been fully reviewed by external experts through a blinded review process. The final manuscript is a direct result of this review and revision process. For more information about this review process please contact the publisher directly.

ISSN 2367-0002 ISSN 2367-0010 (electronic)
Perspectives on Geographical Marginality
ISBN 978-3-319-59001-1 ISBN 978-3-319-59002-8 (eBook)
DOI 10.1007/978-3-319-59002-8

Library of Congress Control Number: 2017943097

Printed on acid-free paper

This Springer imprint is published by Springer Nature
The registered company is Springer International Publishing AG
The registered company address is: Gewerbestrasse 11, 6330 Cham, Switzerland

Preface

Geographical marginality as a conceptual framework for the research of multifaceted problems evolving from human impact on nature and vice versa is relatively new and its relevance needs to be further discussed in future. Sustainability, or, to be precise, the absence of it, has many links with marginality and marginalization. Both terms are to certain extent ambiguous as they are very widely used. That makes the topic of relationship between marginality and sustainability even more relevant. Future development should in all areas only be sustainable. Actually, if the development is not sustainable, it should not be named development because this term by the definition involves not just quantitative component (growth), but includes qualitative aspects, too. There cannot be any improvement in terms of quality if the growth in production neglects the impact of this development on the quality of life of future generations. Following the path of short-term profit maximization on the account of future generations means marginalization of our descendants and limitation of their prospects for long-term survival. However, unsustainability is not only a driver of long-term marginalization. Unsustainable economic practices are most common in those areas of the world where the awareness of environmental issues is still low and the need for any kind of sources of income is high. Multinational companies are always moving their operations in the direction of the cheapest labour and the lowest environmental restrictions, thus marginalizing the countries where they open their often environmentally offensive branches and employ their low paid workers that are working hard and have long working hours. Employees in the countries that used to host their branches before they moved them elsewhere are marginalized as well, as they become jobless. Therefore even better developed countries with a much higher public awareness of environmental issues and declared orientation towards sustainable development often implement dubious developmental solutions. As permanent growth of the GDP is so extremely important issue for politicians that work on four years re-election cycles, they are often giving priority to the solutions that result in short-term effects rather than those that will give the results sometime in the far future.

Unfortunately, systematic studies of marginality are rare and to come to some conclusion we have to use the insights given by many different partial studies that often use different approaches and deal with many different topics. What we try to find in the variety of these studies are different drivers of marginalization. In this third book of the series Perspectives on Geographical Marginality we are trying to reveal different aspects and different drivers of marginalization. The authors prepared their chapters on the basis of their papers presented at the last two conferences of the International Geographical Union's commission C12.29 Marginalization, Globalization and Regional and Local Responses, held in 2015 in Agri (Turkey) and in 2016 in Ljubljana, Koper and Maribor (Slovenia). The last conference focused on sustainability and this is the reason why the book starts with more nature oriented topics, followed by the insights into different roles that tourism plays in (de)marginalization process, drivers related to ethnicity and other predominantly economic topics.

We would like to express our appreciation to all the chapter authors who have contributed a wide variety of interesting topics that enable us to get better insight to marginality from many different perspectives.

Koper, Slovenia

Stanko Pelc
Miha Koderman

Contents

Contributors

Jamalunlaili Abdullah Faculty of Architecture, Planning and Surveying (FAPS), Department of Town and Regional Planning, Universiti Teknologi MARA Malaysia, Shah Alam, Malaysia

Elena Belonovskaya Institute of Geography of Russian Academy of Sciences, Moscow, Russia

Hugo Capellà i Miternique Department Geografia, Ed. Guillem Colom, Universitat Illes Balears, Palma de Mallorca, Spain

Marin Cvitanović Faculty of Science and Technology, Department of Life and Environmental Sciences, Bournemouth University, Poole, UK

Armand Faganel Faculty of Management, University of Primorska, Koper, Slovenia

Borna Fuerst-Bjeliš Faculty of Science, Department of Geography, University of Zagreb, Zagreb, Croatia

Raisa Gracheva Institute of Geography of Russian Academy of Sciences, Moscow, Russia

Tomáš Havlíček Faculty of Science, Department of Social Geography and Regional Development, Charles University, Prague 2, Czechia

Alenka Janko Spreizer Faculty of Humanities, Department of Anthropology and Cultural Studies, University of Primorska, Koper, Slovenia

Ruth Kark The Hebrew University of Jerusalem, Jerusalem, Israel

Miha Koderman Faculty of Humanities, Department of Geography, University of Primorska, Koper, Slovenia

Daichi Kohmoto Nara University of Education, Nara, Takabatake-cho, Japan

Walter Leimgruber Department of Geosciences, Geography, University of Fribourg, Fribourg, Switzerland

Avinoam Meir Department of Geography and Environmental Development, Ben Gurion University of the Negev, Beersheba, Israel

Warwick E. Murray Victoria University of Wellington, SGEES, Wellington, New Zealand

Firuza Begham Mustafa Faculty of Arts and Social Sciences, Department of Geography, University of Malaya, Kuala Lumpur, Malaysia

Janez Nared Anton Melik Geographical Institute of Research Centre of the Slovenian Academy of Sciences and Arts, Ljubljana, Slovenia

John Overton Victoria University of Wellington, SGEES, Wellington, New Zealand

Stanko Pelc Faculty of Education, University of Primorska, Koper, Slovenia

Noam Perry San José State University, San Jose, CA, USA

Urška Trček Miklavž na Dravskem polju, Slovenia

Anita Trnavčevič Faculty of Management, University of Primorska, Koper, Slovenia

Vera Vinogradova Institute of Geography of Russian Academy of Sciences, Moscow, Russia

Katja Vintar Mally Faculty of Arts, Department of Geography, University of Ljubljana, Ljubljana, Slovenia

Havatzelet Yahel Ben-Gurion Institute for the Study of Israel & Zionism, Ben-Gurion University of the Negev, Sde-Boker, Israel

Matija Zorn Anton Melik Geographical Institute, Research Centre of the Slovenian Academy of Sciences and Arts, Ljubljana, Slovenia

Igor Žiberna Faculty of Arts, Department of Geography, University of Maribor, Maribor, Slovenia

Jernej Zupančič Faculty of Arts, Department of Geography, University of Ljubljana, Ljubljana, Slovenia

List of Figures

List of Tables

Part I
Introduction

Chapter 1
Drivers of Marginalization from Different Perspectives

Stanko Pelc

1.1 Setting the Frame

The first two books in the series "Perspectives on Geographical Marginality" covered several topics that have been presented and discussed at the conference of the IGU Commission on Marginalization, Globalization and Regional and Local Responses, held in Nainital in the Indian Himalayan state of Uttarakhand. In such a mountainous setting the geometric view of marginality was clearly predominant and the mountains and the problems of mountain dwellers were the prevailing topics. In 2015 and 2016 the annual conferences of the commission were held in Ağri in eastern Anatolia (Turkey) and in Slovenia. At these conferences a wide range of topics has been presented, and this book contains a selection of those that are in one way or another linked to the problem of marginality and (de)marginalization.

The conference held in Turkey has failed to attract geographers from the host country, which clearly shows the marginality of the location where the conference has been held. The area populated by Kurds with a completely new university has been making promising developmental steps towards demarginalization, but unfortunately the peace process terminated in 2015 and the marginalization of territories settled by Kurds will obviously increase. This could be a very interesting topic, that would deserve to figure in our book, but unfortunately no author could write about it, considering that southeastern Turkey and academics living there are experiencing a very difficult situation at present.

The 2016 conference held in Slovenia on the other hand attracted several Slovenian authors that presented some interesting case studies from their own country. That is the reason why several chapters are dealing with Slovenia, a country that can hardly be an example of a typical marginal region. On the contrary, it is more a representative of the better developed part of the world. However, even

S. Pelc (✉)
Faculty of Education, University of Primorska, Cankarjeva 5, 6000 Koper, Slovenia
e-mail: stanko.pelc@pef.upr.si

S. Pelc and M. Koderman (eds.), *Nature, Tourism and Ethnicity as Drivers of (De)Marginalization*, Perspectives on Geographical Marginality 3,
DOI 10.1007/978-3-319-59002-8_1

in the most developed parts of the world there are individuals, groups and areas that are pushed to a marginal position from a certain point of view and by a certain process that is harmful for their economic, social or cultural situation. From that point of view, case studies from Slovenia can show many similarities with the majority of European nations.

1.2 Marginalization and Demarginalization

The simplest definition of marginalization may be that this is *the process in which something or somebody is pushed into a marginal position* (dictionary explanation: the social process of becoming or being made marginal (especially as a group within the larger society; (WordNet 3.0, Farlex clipart collection 2003–2008). Marginalization is always social, but it can be caused by many different factors. These can be defined as drivers of marginalization. Some of them may be physical by nature while others are social. In the last case, all the factors that are the result of human activities are considered as social. Whatever humans do is done within a social context. All activities have some kind of impact on other members of the society. When we are trying to observe the process of marginalization we intend to examine those activities and circumstances that have a certain effect on social structure with special regard to the changes in the position of certain social groups or entire societies. The result of the process of marginalization is a certain degree of marginality from a certain perspective. Two basic diametrically opposite perspectives are "self-perspective" and "other-perspective". A person or a social group may consider its situation as marginal and feel marginalized, or others consider a person (social group) as marginal. The only way to get insight into the perception of marginality of an individual or social group is the use of qualitative research. However, it is much more common that researchers start their research by considering marginality as an extremely unfavorable position from their perspective. Quantitative or qualitative indicators or a combination of both can be engaged in determining when and where this unfavorable position occurs. As in many other cases the crucial problem is the threshold. In the research of geographical marginality, we have not yet managed to come to widely accepted indicators of marginality that could be used to indicate the marginality point—the point from which the individuals, social groups or societies can be defined as marginal. Déry, Leimgruber and Zsilincsar raised this question too, referring to "Gino Germani (1980)" who accepted that there may be some "partial marginality", and the above authors went even a step further to "point to the fact that marginality *is always* partial, somewhere in-between, since no one can be fully marginal or fully not-marginal … marginality is always 'grey'. Hence, there should be a point where one crosses the fence; then again, there should be some measure of marginality or of non-marginality, when one is closer or further from the 'fence'—otherwise, we are condemned to work with constant 'blurriness'." (Déry et al. 2012, p. 8; original emphasis). They consider this dilemma as "similar to the one around the urban-rural dichotomy (and of threshold values in general): where do towns and villages begin? Statistics alone (number of inhabitants) do not help." (ibid.)

The non-existence of a clear "marginality point" is no longer a problem when we change the focus from marginality as such to the process of marginalization. In this case, we need to start with a social group (individual, society) in an unfavorable situation (spatial, economic, social) and to observe whether their situation improves (demarginalization) or worsens (marginalization). The above cited authors (Déry et al. 2012, p. 13) also concluded that "marginality is continuously changing (augmenting or diminishing)" and consider that this "left open the door for more research in this field and try to figure how to 'see' these changes."

We can say that the authors of the chapters in this book are presenting many different case studies mainly from this point of view even though they are not clearly pointing that out.

1.3 Nature as Socially Induced Factor of Marginalization

Humans are a part of nature, but we live within highly structured social contexts that enable us to overcome natural constraints on the one hand and build social constraints on the other. Societies of hunters and gatherers manage(d) to survive in very harsh natural conditions. We could say that there were no or only very few areas where people could not live (survive). It was always just the balance between available resources (food and shelter) and the number of people inhabiting certain area. Large areas with scarce resources allowed survival to few people. It all radically changed after the Neolithic revolution with the beginning of agriculture. In agricultural societies, the areas with appropriate natural conditions for growing crops became cultivated while those with worse conditions or those that were even inappropriate for growing crops did not attract human settlements and became uncultivated. Their significance for human survival became less relevant or even irrelevant. We can say that the phenomenon of marginal land appeared with the beginning of agriculture. This had nothing to do with profit but with energy input. The land where more energy was needed for cultivating than the energy got from the crops, was beyond the margin of cultivated land. Marginal land where the input and output of energy was equal was the land that could serve only for bare survival and was only cultivated in case that certain groups of people had no other option to survive than to cultivate this marginal land.

The extent and the location of marginal land in the above sense varies according to the development of technology and social organization. Deserts can be changed into fertile lands, new techniques of growing crops, new machines, new seeds, fertilizers, herbicides, pesticides etc. cause changes in the extent of cultivated land and shift the position of marginal land. Of course, the changes are not unidirectional. McCann (1995) in his book "People of the Plow—An Agricultural History of Ethiopia, 1800–1990" devoted one chapter (From Royal Fields to Marginal Lands) to describe the complex of political, social and economic changes that caused a dramatic shift from a once flourishing agricultural region into a region where in about hundred years agriculture deteriorated so much that the region like the rest of northern Ethiopia experienced a deadly famine as a consequence of the 1984–85 drought in combination with civil war and other socio-economic and demographic factors. The reason for such deterioration was that region lost its political importance (royal seat moved from Ankober to Addis Abeba), market conditions changed, incoming migrants used agricultural practices from the place of their origin inappropriate for the natural conditions in the region etc. A new use of the land (technology, ownership relations) transformed formerly valuable land into a marginal area.

The above example is taken from a feudal context. However, we must consider that in the capitalist system marginal land is no longer just the land that can guarantee barely enough crops (food) for people to survive from its yield. It is also the land that does not give any profit—the land where the invested capital in all forms on the long term equals the incomes of sold crops—where we have to deal with a 'no profit situation'.

Of course, the incomes depend on the market situation that is constantly changing (the prices of the crops going up and down). With the revolution in transportation every single farm has to compete with the rest of the world. It is therefore the global market that sets the limits where natural conditions for certain crops are appropriate or not for market oriented agriculture. Kang et al. (2013) in their article about marginal lands, quoting several authors such as Peterson and Galbraith, Strijker, Schroers, Heimlich, Hart etc. state that there are three different terms for marginal lands:

- physical marginal lands,
- production marginal lands, and
- economic marginal land.

According to Schroers (2006a, b, cited in Kang et al. 2013, p. 130) an economically marginal land is "an area where a cost-effective production is not possible, under given site conditions, cultivation techniques, agricultural policies as well as macro-economic and legal conditions".

In the physical and production sense, marginal lands "generally refer to the areas not only with low production, but also with limitations that make them unsuitable for agricultural practices and ecosystem function" (Heimlich 1989 and Hart 2001 cited in Kang et al. 2013, p. 130). Within agricultural production terminology marginal lands (marginal cropland, marginal farming land) is the opposite of prime farmland with the best combination of physical and chemical characteristics that make it the most productive.

Peterson and Galbraith (1932) when describing the situation on the westward slopes of Sierras descending from the top down, have written that:

> At some point on the slope the grade of land becomes such that it is possible under the conditions of the moment to cultivate the land and fulfill the requirements of the hitherto outlined 'necessary remuneration.' Presumably forestry and perhaps grazing have been

> carried out prior to this point on land of less satisfactory grade. Since cultivation must be defined more or less arbitrary it will be used here in the sense of actual manipulation of the soil for the purpose of producing crops (p. 299).

This fact is far from being the same for all crops, so it would be right to presume that this should be the point where any crop or combination of crops yields enough for the "necessary remuneration".

"The point where it is no longer possible to cultivate remuneratively a lower grade of land—the uppermost rim of remunerative cultivation on the hitherto mentioned slope—is the margin of cultivation. Subject to some further qualification that land may be considered truly marginal." (ibid.) When they present the factors determining the position of marginal land they first expose the human factor with its two main dimensions. The first one being the "capacity for remunerative cultivation" and the second one the "standard of living which he [the operator; S.P.] is willing or able to accept". From the present point of view we have to consider that agriculture has to compete with many other activities that usually offer higher remuneration and a higher standard of living on the one hand and that there are different kinds of state interventions that are "improving" the remuneration in order to control the extent of marginal lands on the other. Considering that other factors are the size of the farm unit, price, costs as well as the "relation of the margin of cultivation to the opportunity margin for non-agricultural use" (such as forestry, grazing or even recreation), Peterson and Galbraith (ibid.) state that "the margin of cultivation in abstract theory at least is highly mobile or dynamic concept."

From our point of view it is important to find and observe the main factors that influence the 'movements' of marginal land in a certain area under certain circumstances. Actually, these movements are the process of marginalization and demarginalization of agricultural land. It is obvious that natural conditions, no matter how important they may be for the productive function of agricultural land, are only the basic pre-condition for certain units of land to be marginal from an agricultural point of view. The rest are socio-economic factors therefore it is possible to consider the process of (de)marginalization of agricultural land as socially induced. Short term effects of socio-economic factors are predominant while the natural conditions are basically stable (long term changes of fertility, climate conditions etc.) even though some exceptional natural disasters may also have important short term effects on yields and annual remuneration.

Beside the above described approach to (de)marginalization (of agricultural land) we also must consider the wider aspect of (de)marginalization of nature in general. Enormous advances in the development of technology and the persistent need for the maximization of profit and for permanent economic growth brought humanity into the position 'man against nature'. For some time, it seemed that humanity overruled the nature and that the only purpose of nature is to serve humanity. Nature has been reduced to a resource provider. We can consider this reduction as the marginalization of nature. Nature strictly 'in service' of mankind is nevertheless utopian and this kind of anthropogenic view of nature is false. Facing the problems that almost all societies experienced in the sixties of the past century certain people concluded that humanity should know what it can expect if its attitude towards nature and natural resources does not change. The emergence of the Club of Rome and its 1972 book "The Limits to Growth" (Meadows et al. 1972) is a major step towards the changes in the attitude towards nature especially in connection with economic growth and development. The book presents the results of an expert team research that "examined five basic factors that determine, and therefore, ultimately limit, growth on this planet—population, agricultural production, natural resources, industrial production and pollution" (ibid. pp. 11–12). Two decades later the so-called Rio process started focusing on climate change as the major problem caused by human activities. It took another quarter of a century to get together 195 nations at the conference in Paris that agreed that the rise of the air temperature should be kept well below 2 °C. In the meantime, the Bruntland commission prepared the report 'Our Common Future' (World Commission on Environment and Development 1987) and since then the terms sustainable development and sustainability are very frequently used, although their true meaning may still be ambiguous.

The first part of this book opens with the chapter where the concept of marginality is debated from the viewpoint of its links to sustainability. Marginalization of nature and environment are rarely discussed (in geography). In comparison with geometrical marginality based on an extended center-periphery model (marginality from a socio-economic point of view), 'ecological marginality' shows the upside-down picture of marginal areas. What is considered central from economic perspective is marginal from environmental (ecological) point of view. The most modern Central Business Districts (CBDs) of today certainly are the financial and managing centers of economic activities. However, from an environmental point of view, CBDs are wastelands made of glass, concrete and asphalt. The vegetation plays purely a decorative role, being placed on the roofs of the buildings or at the edges of the streets. The centrality of natural wealth is on the contrary where the best-preserved areas of the planet are. Those that are far away from human greed and from activities such as mining, industry and even tourism.

However, when we observe the importance given to nature (environment) in decision making processes, we may see how often it played a minor, unimportant (marginal) role and in many cases, it still does. Unsustainable development still dominates worldwide. Huge areas of rainforests are cut down or burnt to give way to palm oil plantations or other forms of 'development'. There are certain links between marginalization of nature in the process of economic growth and marginalization of societies (social groups) living in the areas with unsustainable development that the opening chapter of the book's first part is devoted to.

In the following chapter Žiberna discusses the land use changes in the region of Svečinske gorice, a hilly area south of the Austrian-Slovenian border. The author considers the area as considerably marginal especially when compared with the region on the Austrian side of the border. This is in a certain contradiction with the natural conditions for vineyards. The vineyard areas of Svečinske gorice are among the most vital in northeastern Slovenia. However, the process of agricultural abandonment took place in the region, and subsequently the natural succession processes are able to progress. Land use changes show the growing share of forests while the share of arable land is decreasing. All that may result in further marginalization of the region. The author presents the results of the analysis of land-use changes between 2000 and 2015 in relation to selected physiographic features. In his research, he mainly focused on cultivated areas with a special regard on vineyards.

The following chapter by Zorn shifts attention from micro- to macro-scale and from land use to natural disasters. It tries to reveal another perspective of marginality induced by natural processes. Natural disasters frequently occur throughout the world. Nature is not choosing between developed and less developed countries. Earthquakes, forest fires, hurricanes etc. occur in both. However, many examples show that certain countries are much more vulnerable than others. The number of lives lost or people affected by natural disasters is much higher in less developed countries. The reasons are not only natural conditions. There are many social factors that Zorn discusses and points out weak social safety network, lack of early warning systems etc.

The last chapter in the first part of the book is dealing with relatively marginal part of the world—Northern Caucasus. Vinogradova, Gracheva and Belonovskaya present the changes in climate and land use. They believe that the conventional schemes of development of the spatial distribution of grassland ecosystems in intermountain basins of the Central Caucasus must be corrected due to historical changes in land use.

The intermountain basins of North Ossetia have been changed from natural ecosystems to cropland more than 3000 years ago. The current slope topography clearly reveals the long-term past human impact. The authors of the chapter have used a multidisciplinary approach to analyze the changes in land cover during the last sixty to sixty-five years. Especially in the last two decades the changes have been extremely profound. They estimated the changes in heat and humidity for the territory of the North Caucasus and are presenting the changes at different height levels after year 2000. The chapter informs us which vegetation types are recovering and succeeding former agricultural vegetation on former agricultural terraces. The process is explained by climate change and reduced human impact. The creation of natural parks, trails and tourist routes is considered as an option for sustainable development and demarginalization of the region.

1.4 Tourism as a Developmental Opportunity of Marginal Regions

Tourism as an economic activity is often considered to be very promising and with great potential for growth. A quick look on the UN World Tourist Organization website (UNWTO 2016) reveals that international tourist arrivals recently grew considerably. In 2015, a total of 1184 million international tourist arrivals were recorded worldwide. The growth of 4% or more, which is higher than the long term average, lasts since 2010. Tourism as such seems to be very promising and often imposes big expectations in areas with developmental problems. Especially in backward rural areas with relatively unspoiled nature and poor accessibility to industrial, commercial and residential centers, tourism is often seen as a redeeming developmental activity which may be highly overrated. Tourism cannot be a substitute for uncompetitive agriculture or missing industry. There are only few special places where the majority of the local population can live from the revenues from tourism. Not every single rural area can expect such an outcome from tourism development. However, tourism can play an important role in regional development when it is combined with the development of other sectors.

There are countless rural areas with more or less characteristics indicating marginality. Usually they are remote, agriculture is less viable, nonagricultural working places are not easily accessible. Young people go to school in relatively distant centers and eventually never come back or they even do not go to school—they just move in search of employment. The average age of population is growing, as is the share of farmers that are too old to work on their farms. Agricultural land is less and less intensively used or abandoned. The low standard of living provided by agriculture is no longer acceptable for young people, and this is the reason for their decision to leave and try to avoid inevitable marginality in case they stay and continue the practice of their parents. They may change one marginal situation with another in the city if they get a low paid job, cannot afford a proper apartment or even don't get a job and must live in a slum. In such cases emigration from an often relatively attractive rural area seems to be a mistake. However, the viewpoint of visitors is often in severe contradiction with the perspective of local inhabitants. While visitors usually only see the beauty of nature, astonishing sceneries, tranquility etc., local inhabitants work hard all-day long during many hours, living under the threat of natural hazards such as hail, frost, avalanches etc. that can ruin all their labourious efforts in a single moment. It is often hard for them to understand that there is this other view and that it can be an opportunity for additional income that may pave the way out of their marginal situation.

In mid 1990s I have been involved in an integral rural development project in a relatively remote rural area (Vojsko) near an old mining center Idrija in central part of Slovenia. The village Vojsko consists of many scattered houses (single farms) covering large area on a small karst plateau with several cultural and natural attractions that are worth visiting. As one of the outcomes of the first phase of the project, we prepared a promotional booklet presenting the mentioned attractions and the attractiveness of the location itself. To convince potential visitors that the location is worth visiting as it differs so much from urban and suburban areas I described it as "the place where you can still observe two hedgehogs fighting in the middle of the road in the moonlight of an early summer night". That was something that I experienced once when I was leaving the village after one of the late evening meetings with local inhabitants involved in the project. After the booklet was already published and in circulation the municipal development officer informed me that certain people from the village did not like the part of the booklet that was trying to present their village as a romantic, idyllic place—they would prefer that it would be presented as tough, harsh environment with unbearable living conditions expecting that this could provoke sympathy and eventually some developmental support. It was a bit of a disappointment, because we had several workshops and one would expect that people involved in the process of preparation of the 'developmental vision and goals' for their village would gain the awareness about what can attract tourist and thus enable the development of rural tourism. However, they obviously could not separate themselves from deeply rooted sense of their marginality that did not let them see what is their comparative advantage.

One of the important motives for people to travel and visit certain places is the search for something different, to meet people that differ from those that they meet in their everyday life, to change the environment that they constantly live in with the one completely different, often even leaving their 'comfort zone' in a search of an adventure.

Vojsko karst plateau certainly is something different and for some people even enchanting. However, it is only an hour and a half car drive from Slovenian capital Ljubljana. The inhabitants are Slovenians as the vast majority of population in Slovenia. The residents of the capital may look up on them as a bit backward, but they are not that much different and above all, there are plenty of similar places and people all over Slovenia. To attract visitors you need to persuade them that you are able to offer them something special. That is not an easy task nowadays as you need to compete with the global tourism market.

To see, explore and experience something much different than rural areas of one's own country, something that not so long ago one could only see in TV documentaries, means to travel to more distant places. Within the globalized society equipped with efficient modes of transportation this became possible and accessible to considerable share of 'westerners' and especially in recent years to ever growing numbers of 'easterners' (at first Japanese, Koreans, Taiwanese etc. and recently more and more Chinese). For many material culture of faraway places and the material evidence of old cultures that have vanished long time ago are still a dominant motive of their visit. However, we must not forget the interest for still existing cultures 'belonging to the past'. Cultures that may not differ that much from the one of visitor's recent or even long time ago ancestors.

In 2003 I attended an 'IGU Marginality Conference' in Kathmandu. After the conference I made a tourist visit to Chitwan National Park. At that time there were very few tourists in Nepal, because of the fear of the 'Maoists'! The number of tourist arrivals in 2003 was the lowest since the beginning of 1990s (338,000 compared to 363,000 in 1990 and 790,000 in 2014) (World Bank 2016). Nevertheless, the offer was similar then as it is today. Besides natural attractions such as riding on the elephant and searching for the rhinos, observing birds and other animal life while walking through the jungle or from a boat, watching elephant bathing and visiting Elephant Breeding Centre, they also offer the visit of the village and a Tharu dance. Tharu are the natives living in the Tarai, the southern part of Nepal where Chitwan National park is situated. During my visit of the village as a part of the 'tourist package' I felt a bit awkward even though we were only three intruders: the guide, me and my 17 years old son who was accompanying me. The guide was explaining about the life in the village, showed us around between and trough the huts of the people living in the village. It was like watching a 3D documentary where the three of us were completely separated from the observed persons even though we were moving among them. They were doing their ordinary daily tasks as usual not paying any attention to us, so it seemed as if we were in an open air museum with live people as the main exhibits. It was a clear sign of marginalization of local population and an example of how marginality can be used for tourist attraction. Of course, an average tourist usually does not reflect much of what he sees, besides he does not get the whole picture of what the Tharu people of Chitwan had to go through so that tourists from around the world can now enjoy the beauties and the adventures of the national park. McLean (1999) spent 20 months among Tharu people investigating the consequences that they suffered because of the establishment of the national park. Their traditional life has dramatically changed because of the park. Their deep bound with nature and woods where they were practicing their shifting cultivation has been broken as they lost their resources by being forced to leave the area. The relocation was conducted with force as McLean (1999, p. 40) was told:

> My informants told me how soldiers forcibly removed all the villages located inside the designated boundary of the park. Houses were burned down, fields and houses were trampled by elephants, men, women and children were threatened sometimes at gun point. Those who lived inside the boundaries tried to fight for their land, but lost and as a result became landless for life.

Similar is the case of Maasai in Africa (Azarya 2004). Those living near or in the areas that became protected as game or national parks also experienced several restrictions that limited their herding and had severe influences on their traditional way of life not to mention that in some cases they also had to evacuate protected area and to move elsewhere. Cited author in his article points out close tie between globalization and marginality as a tourist attraction. He tried "to show how tourism feeds on and further promotes an increasing attraction to the peripheral, even to the miserable, that are exhibited as objects of curiosity" (Azarya 2004, p. 950). The reason is of course the 'curious traveler' in his search for adventure and exploration that has money, time and means to visit places at the periphery and wants to see with his own eyes how 'exotic tribes live their primitive life'. The number of such curious explorers is considerable and is still growing. Similar to idyllic places of unspoiled nature where tourists are seeking peace and relaxation, the periphery with population living traditional way too, cannot accept masses of tourists, because masses destroy the core attractiveness of such places. The feeling of being "curious explorer" would be lost. There is no real adventure in mass consumerism. However, majority of tourists that are motivated by all kind of adventures still want all the amenities that they enjoy at home. Therefore, this kind of tourism has some benefits for local population too. As Azarya (2004, p. 953) states:

different cultures or experiencing different struggles such as suffering persecution or discrimination" (Rustenbach 2010). The analysis showed that this is relatively unimportant in explaining negative attitudes towards immigrants. The societal integration showed the strongest significance considering the connection with negative attitude. 'Natives' are afraid of immigrants because they bring too many unknown elements (customs, culture, language etc.). What consequences that may have is an uncertainty that they can hardly overcome, especially if their level of interpersonal trust is low and that may be one of the reasons for the development of a negative (anti-immigrant) attitude (ibid). At present, the millions of immigrants intensify the fear of 'the others', and the widespread anti-immigrant attitude strengthen (extreme) right wing parties. Public discourse is becoming increasingly hostile for immigrants. For those that just arrived illegally and live in all kind of 'receiving centers' (migrant camps) the degree of marginalization could hardly be higher. Those that have managed to immigrate legally and live in ghettos of big cities in poverty and social deprivation may be even more affected by negative attitudes of a considerable part of local populations.

The situation for migrants that managed to realize their dreams—study abroad, get a diploma and a relatively well paid job—is much different. For many of those who came from parts of the world where kinship is still very strong those dreams were realized with the help of their families. Therefore, they are obliged to return the favor by sending money back home and support the family. At the individual level these are the 'escapes' from marginality that have the reverse impact on the marginality at a larger scale (demarginalizing the region that these individuals have emigrated from). For some countries, such revenues coming from their diaspora may represent a considerable share of total revenue and may have at least some demarginalization effects.

A Common denominator of another group of drivers of marginalization is local and regional development. The first subgroup is related to the political transition from socialism to capitalism and to the impacts of this process on development at local or regional levels. From a political perspective, the major change was from a one-party to a pluralistic democratic system, and economically from a centrally planned to market economy and from state to private ownership. These changes had huge but diverse impacts on the different regions of former socialist countries. In many cases, this transition was simultaneously also the process of the disintegration of multinational states (Soviet Union, Yugoslavia, Czechoslovakia). Post-socialist territories became politically unstable and civil wars of different intensity and duration broke out (Bosnia and Herzegovina, Croatia, Chechnya etc.) The loss of lives, the destruction of cities and villages, and the displacement of thousands of people were the major factors of marginalization of regions where the armed conflicts took place. These territories were labeled as 'dangerous', 'conflict areas', 'mistrustful' etc., which also had economic consequences even for larger areas than the one that suffered from armed conflicts. I had to cancel a conference about planning issues for marginal areas in 1999 due to many cancellations that followed after NATO started bombing attacks on the Federal Republic of Yugoslavia (at that time consisting of Serbia including Kosovo and Montenegro). Even though the majority of participants would be geographers knowing that Slovenia is physically closer to Germany than to Serbia, many of them were afraid to travel to the 'conflict zone'. Potential tourists or investors probably have less knowledge about the geography of the region and were therefore even more discouraged to travel or do business with the region that belonged to the 'zone of conflict'. This is a clear example of marginalization of a wider territory because of political instability on post-socialist territories that escalated into military operations. It could be an interesting topic for a wider marginality study, but we lack such a case study in this book. The cases that we have are related to ownership issues and to social and economic changes at the local level.

The change from state owned land, infrastructure and means of production to private ownership was probably the one that affected post-socialist societies and economies even more than the change of the political system. Privatization and introduction of market economy led to a massive increase of inequality, not only among citizens, but also among the regions. Socialist Yugoslavia for example had a multilayered system of development support for underdeveloped regions. Many bigger companies had their branches in these regions, not to mention small local factories that were established as part of a planned development or because of a local initiative that was supported with the developmental program. After the collapse of the Yugoslav market and the invasion of foreign competitors on national markets, many of these economic subjects in underdeveloped regions went bankrupt, the workplaces were lost and workers had to find new jobs outside the region. That meant longer commuting times or even outmigration, not to mention the rise of the share of unemployed. All that was even more depressing for a population that had in former times experienced 'full employment' (if not in the region than at least elsewhere within the country, although some people preferred to go abroad to get a better paid job in Germany, Switzerland, Sweden etc.). The beginning of the transition was therefore the beginning of marginalization of the above-mentioned regions. However, when the process of the accession to the EU started, attention was increasingly devoted to their development. Slovenia was the first country to join, followed by Croatia, while the rest of former Yugoslavia is still engaged in the process of accession. The transition can therefore be regarded as a process of socio-economic changes that pushed many (underdeveloped) regions of the post-socialist world deeper into marginality, but has afterwards also brought changes in the attitude towards these regions resulting in different kinds of 'developmental' programs provided for them. Sometimes the success of such programs at the local level may depend on the necessary conditions such as the availability of land for basic infrastructure that is needed for the development of production or service zones. Unsolved property issues related to de-nationalization may be a reason for the marginalization of a local community that finds itself in uncertainty because the owner of the land that would be appropriate for development is not legally confirmed.

Another important factor in the process of marginalization related to local development is local level administrative division. The appropriate size of local (self-governing) units is of crucial importance for the successful functioning of these units. Defining what is the right size is unfortunately not an easy task. It is certain that it needs to be small so that people have more influence on the decision-making process and are more familiar with the issues that a local unit has to resolve. However, it must not be too small, because such units may be economically too weak and may also suffer a lack of human capital. These are the factors that are pushing such local units into marginality.

At the regional level, the post-socialist transition brought a lot of comparative advantages to the regions for the country capitals. Centralization caused growth of regional disparities and marginalization of regions that were lagging behind, not only behind the central but also the major part of the other regions. Those regions which became a part of the EU can count on the support of the ERDF (European Regional Development Fund). It pays special attention to regions that are marginal from a physical point of view (remote areas, mountainous regions) and therefore represents an important mechanism for demarginalization, but regional disparities continue to be an important issue to be addressed by the policy of regional development.

In some cases, the situation of at least one part of the population of an underdeveloped region may be improved (or worsened) by some specific economic activity. We have already devoted quite a lot of space to tourism, but there are some niche agricultural practices that may also play an important role in demarginalization at local or regional level. This part of the book deals with tea growing and aquaculture. Such economic activities can have demarginalizing effects on marginalized farmers if they bring additional income to their households or if their income grows because of the prosperity of the activity. However, the process may also take the opposite direction in case of a major change on the market. It may have disastrous consequences especially when the existence of a large share of the population depends on the activity that has suddenly become economically unviable.

In the chapter that opens the last part of the book the topic of migration, education and marginality attracted Overton and Murray's attention. They present networks and strategies of migrants in the Pacific Islands. They understand marginality in this case as an outcome of the location that weakens the connection of island states to the global economy and reduces the opportunities for their sustainable economic development.

The main topic of their research were education and migration strategies of a local population that needs to overcome the constraints that they meet in their home region and therefore have to find the opportunities elsewhere. There is a whole range of diverse strategies across the region varying in form, occurrence and political implications.

The strategies can only be expanded because of the existence of kin networks (Pacific people's diaspora) that may be further forged by states. The authors are discussing the possibilities for facilitation of those strategies that can strengthen the demarginalizing process trough migration and education.

The chapter that follows deals with unresolved problems with Church property that can have marginalizing effects. In it Havlíček argues that property settlement between the church and the state attracts little media attention in Czechia no matter how important it may be for regional development. The longer that this process lasts, the more difficult it is to reach an acceptable solution. Unresolved problems with land property that used to be the property of the church and was nationalized during the communist period have negative effects on local communities and sometimes regions as well. They could not use the ‘blocked’ land for development, because of the unresolved proprietary issues. The author presents the consequnces of such situation that lasted for two decades and reveals the differences among more and less affected regions according to the share of municipalitiy’s land that has been a subject of the restitution.

The Czech case is followed by two chapters presenting self-government changes and regional disparities in Slovenia. In his chapter Nared is discussing the past two reforms of the self-government system in Slovenia and their consequences for the polycentric urban system and regional development. He particularly exposes the cases of less accessible remote areas with developmental problems. The first reform had been introduced in the 1970s and the second in the 1990s. The result of the first reform was a socialist communal system with strong emphasis on polycentrism while the second brought further diffusion of power to small municipalities (whose number grew from 147 to 212 at present). Slovenia has no intermediate level units (between national and local level of government there is no regional one), therefore all power is divided between national and municipal authorities. The author presents his findings about the positive and negative effects of these reforms on local development and relates them to demarginalization in peripheral areas.

The findings of Vintar Mally to a certain extent confirm the findings of Nared. She presents regional disparities as the path to sustainability. As mentioned above Slovenia has no regional level political units. Therefore, statistical regions are spatial units that are usually used for different kinds of regional analysis. Slovenia is divided into 12 statistical regions. Vintar Mally assessed the economic development of these regions with a set of 32 indicators of sustainable development. The results showed considerably large and persistent disparities. These are largest in the field of economy and smallest in the field of environmental issues. The author discusses regional differences and tries to present her findings from the perspective of marginalization.

In the next chapter, we move from Slovenia to Sri Lanka and in a way also from macro- to micro-economic scale. Kohmoto presents the example of marginalization of tea estates in the changing global and national context. Sri Lanka is the world’s largest tea exporter, but is losing its competitiveness on the global market. Tea is still produced on large plantations from the colonial era, but on the other hand Sri Lanka was a pioneer in introducing organic tea production and thus benefitted from being first. The issues discussed in the chapter involve the relationships between organic tea growing on one side and sustainability, costs and prices as well as workforce involved on the other.

In her case study, Begham Mustafa examines the aquaculture system practiced by small farmers in Khulna in Bangladesh. These farmers prefer to grow shrimps (giant freshwater prawn—Macrobachium rosenbergii) instead of paddy. They have adopted modern farming technology that is very sustainable and environmentally friendly, and they are satisfied with their output. Their ponds are relatively small, but they earn enough to avoid poverty. Of course, they need certain skills to be successful (water management, pond preparation, maintenance, selection of species, seeds, equipment, material, feeding schedule etc.). Begham Mustafa believes that the understanding of management practice of farmers in Khulna can be a model for aquaculture management elsewhere and a means for demarginalizing farmers living under the poverty threshold.

The last chapter is a case study from Croatia. In it Cvitanović and Fuerst-Bjeliš are discussing the post-socialist transition as a driver of geographic marginalization. They are stating that geographic marginalization may also be a consequence of globalization, deindustrialization or economic transition. Some areas may benefit from these processes while others may be marginalized. Changes in society and economy that occurred during the post-socialist transition period (1991–2011) in the less developed region of Hrvatsko Zagorje in northern Croatia were the focus of their research. The region is traditionally agricultural and located on the Slovenian-Croatian border. Settlement patterns and changes in agriculture were analyzed. The research methods were a combination of remote sensing, household questionnaire surveys and regression analysis. The region experienced population aging, a change of educational structure, population density and employment. Consequently, settlement patterns and patterns of land use have changed. The authors are interpreting differences in spatial distribution of changes from the viewpoint of accessibility to centers, border location and natural conditions.

1.7 Conclusion

This book offers a variety of topics that can be this way or another linked to marginality and marginalization. This introductory chapter is an attempt to put them into a common framework. That is certainly not an easy task as the authors often differ in their understanding of marginality and marginalization. In many cases they are primarily following their research interest about a certain topic that is linked to marginalization, but their main focus was not on the analysis of this link. The conceptualization of marginality and marginalization from the viewpoint of their specific research foci has therefore been one of the aims of this chapter.

Even if the reader, who is reading about all the presented cases, bears in mind his own conception of geographical marginality and marginalization, these case studies give further insight into the complex topic of marginality and marginalization.

I hope that the conceptual issues discussed in this introductory chapter, will help the reader to concentrate on the marginality and marginalization point of view when reading about all the different case studies included in this book.

References

Azarya, V. (2004). Globalization and international tourism in developing countries: Marginality as a commercial commodity. *Current Sociology, 52*(6), 949–967.

Déry, S., Leimgruber, W., & Zsilincsar, W. (2012). Understanding marginality: Recent insights from a geographical perspective. *Hrvatski geografski glasnik, 74*(1), 5–18.

Erbentraut, J. (2016). *Here's what you should know about the dakota pipeline protest.* Retrieved November 5, 2016, from T Huffington Post: http://www.huffingtonpost.com/entry/dakota-access-protests-facts-facebook_us_5818baece4b064e1b4b4e3cf.

European Union agency for Fundamental rights. (2009). *EU-MIDIS European union minorities and discrimination survey.* Retrieved November 6, 2016, from European Commission: http://ec.europa.eu/justice/discrimination/files/roma_midis_survey_en.pdf.

Gomes, J., Cunha, L., & Figueiredo, R. (2007). Environmental marginality and natural disasters; A case study at Régua (Portugal), in the Douro wine region. In G. Jones, W. Leimgruber, & E. Nel (Eds.), *Issues in geographical marginality.* Grahamstown: Rhodes University.

Gosselin, P., Hrudey, S. E., Naeth, A., Plourde, A., Terrien, R., Van Der Kraak, G., & Xu, Z. (2010). *e Royal society of Canada expert panel: Environmental and health impacts of Canada's oil sands industry.* The Royal Society of Canada/La Société royale du Canada. Ottawa: The Royal Society of Canada/La Société royale du Canada.

Hart, J. F. (2001). Half a century of cropland change. *Geographical Review 91*, 525–543.

Heimlich, R. E. (1989). *Productivity and erodibility of U.S. cropland.* Economic Research Service, U.S. Dept. of Agriculture.

Kang, S., Post, W. M., Nichols, J. A., Wang, D., West, T. O., Bandaru, V., & Izaurralde, R. C. (2013). Marginal lands: Concept, assessment and management. *Journal of Agricultural Science, 5*(5).

Leimgruber, W. (2004). *Between global and local. Marginality and marginal regions in the context of globalization and deregulation.* Aldershot: Ashgate.

McCann, J. C. (1995). *People of the plow.* Retrieved October 28, 2016, from ProQuest Ebook Central: https://ebookcentral.proquest.com/lib/uprsi-ebooks/detail.action?docID=3445009.

McLean, J. (1999). Conservation and the impact of relocation on the Tharus of Chitwan, Nepal. *Himalaya, 19*(2), 38–44.

Meadows, D. H., Meadows, D. L., Randers, J., & Behrens, W. W., III. (1972). *The limits to growth.* New York: Universe Books.

Midwest Alliance for Infrastructure Now (MAIN). (2016). *Checking the facts once again.* Retrieved November 5, 2016, from Standing Rock Fact Checker: http://standingrockfactchecker.org/fact-checking-srst-claims-9-7-16/.

Park, R. E. (1928). Human migration and the marginal man. *American Journal of Sociology, 33*(6), 881–893. Retrieved from http://www.jstor.org/stable/2765982.

Peterson, G. M., & Galbraith, J. K. (1932). The concept of marginal land. *Journal of Farm Economics, 14*(2), 295–310.

Rustenbach, E. (2010). Sources of negative attitudes toward immigrants in Europe: A multi-level analysis. *The International Migration Review, 44*(1), 53–77.

Schroers, J. O. (2006a). *Physical marginal lands, production marginal lands, and economic marginal land.* Germany: University of Giessen.

Schroers, J. O. (2006b). *Towards the development of marginal land use depending on the framework of agricultural market, policy and production techniques.* Germany: University of Giessen.

Spiegel staff. (2014). *Europe's unwanted people.* Retrieved November 6, 2016, from Spiegel Online: http://www.spiegel.de/international/europe/europe-failing-to-protect-roma-from-discrimination-and-poverty-a-942057.html.

UNCHR. (2016). *Figures at glance.* Retrieved November 6, 2016, from UNCHR The UN Refugee Agency.

UNWTO. (2016). *Tourism market trends UNWTO*. Retrieved from World Tourism Organization UNWTO: http://mkt.unwto.org.

Weisberger, A. (1992). Marginality and its directions. *Sociological Forum, 7*(3), 425–446.

WordNet 3.0, Farlex clipart collection. (2003–2008). *Dictionary*. (Princeton University) Retrieved 10 9, 2016, from The Free Dictionary by Farlex: http://www.thefreedictionary.com/marginalization.

World Bank. (2016). *World development indicators| world databank*. Retrieved November 26, 2016, from http://databank.worldbank.org/data/reports.aspx?source=2&series=ST.INT.ARVL&country=NPL.

World Commission on Environment and Development. (1987). *Report of the world commission on environment and development: Our common future*. Retrieved November 3, 2016, from http://www.un-documents.net/our-common-future.pdf.

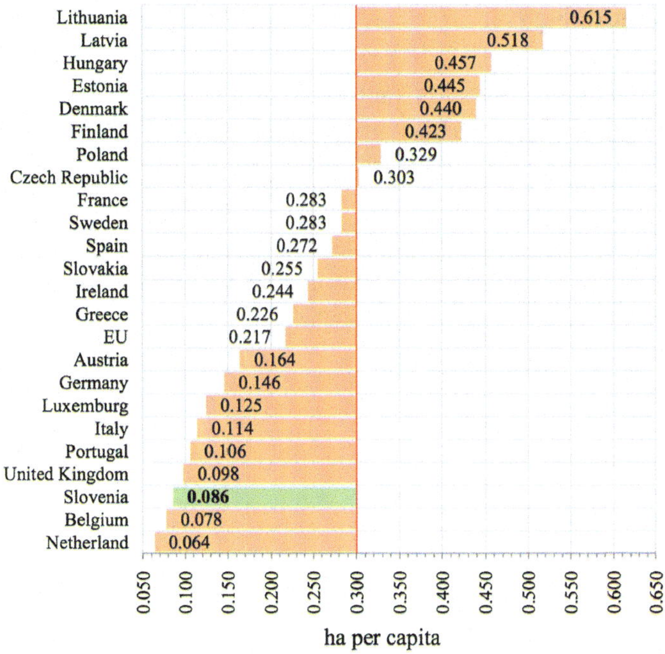

Fig. 3.1 Arable land per capita in selected countries of the European Union in 2009 (ha/per capita) (*Source* Eurostat regional yearbook 2012)

Svečinske gorice are one of the regions in north-eastern Slovenia, which can be denoted as the area with unfavourable socio-geographical characteristics. The area is characterized by a high ageing index: in the year 2016 most of the settlement have ageing index higher than 130, individual settlements, even higher than 160 (the average for Slovenia is 124). The average gross earnings for the area in the year 2016 was 1305 eur (average for Slovenia 1525). The area also have above-average unemployment (16.1% in 2016, the average for Slovenia was 11.8%) (http://gis.stat.si/#). Despite the great density of the road network and the nearnes of the second largest Slovenian city (Maribor), Svečinske gorice in the last 25 years remained on the periphery. Similar marginal status—not only physical, but also socio-geographical, have some regions, in which we can observe the alike condition (Haloze and Goričko). One of the forms of manifestation of marginalisation is also the reduction of cultivated areas and land-use changes in the direction of the overgrowing of arable areas and afforestation, which leads to a substantially lower utilisation of the natural potential. At a time when food is becoming a strategic component, this is a very unfavorable process.

3.2 Methodology

The article discusses changes in land use in the period from 2000 to 2015. The information about land cover/use has been taken from the Ministry of Agriculture, Forestry and Food (MAFF), which annually publishes reports on agricultural land use in 'shp' format (Ministry of Agriculture, Forestry and Food, 2015). For each category, vector data was converted to a raster image with a cell size of 5 m × 5 m. The methods of data capture have changed since 2000, when all forms of land use were classified in 21 categories, whereas in 2015 the number of categories rose to 26. By combining classes, we have designated eleven categories of land use:

- cultivated fields and gardens
- vineyards
- orchards
- other permanent crops
- meadows
- overgrown land
- mixed land use
- forest
- built-up and other artificial areas
- water areas,
- other.

Land use data has been compared to various geographic location data (altitude, relative elevation, slope inclination, and exposition) and to global solar exposure. The geographic location data was obtained using a digital model of elevations with a cell size of 5 m × 5 m (Archive of The Surveying and Mapping Authority of RS, 2011). Land use data for 2000 and 2015 were compared to determine the course of the changes in land use, as well as their intensity and connection to geographic location. Special attention was given to more intensive forms of land use, especially to cultivated fields and gardens, and vineyards and orchards.

To analyse the utilisation of the vine-growing potential in Svečinske gorice, the area was also classified with respect to the topoclimatic suitability for winegrowing. Two main criteria were used, specifically relative elevation and global solar radiation. Relative elevation is known to be an important factor in wine production; in locations with higher relative elevation, frosts and freezing temperatures are less likely to occur, while insolation (duration of solar exposure) is usually greater due to less frequent fog. Greater global solar radiation means a greater amount of energy, which is favourable for winegrowing (Žiberna 1992). In addition to astronomical factors (geographic latitude, position of the sun, which depends on the day of the year and time of day), relief characteristics were also considered (inclination and exposition of slopes) (Žiberna 2011, pp. 49–50) in the construction of the global solar exposure model (GSE). Relative elevation is more important in nighttime and global solar irradiation in daytime (Žiberna 1992, pp. 129–130). When defining the topoclimatic land capability of winegrowing locations, the

Table 3.1 Methods and types of topoclimatic land capability

Relative height (m)	Global solar radiation (kWh/m^2)		
	>1200	1000–1200	<1200
>50	1. class	1. class	2. class
25–50	2. class	2. class	3. class
<25	3. class	3. class	4. class

locations were divided into four classes, with first-class positions representing the areas of highest quality for winegrowing. The methods and types of topoclimatic land capability are shown in Table 3.1.

In the analysis of land use change, our primary aim was to identify the differences in area size across individual classes. Further steps also included an analysis of the direction of land use changes within the discussed 11 categories, which was then summarised according to the following categories:

- change in land use within arable land (e.g. cultivated field into vineyard or vineyard into orchard)
- change in land use within non-arable land (e.g. meadow into built-up area or overgrown area to forest)
- change in land use from non-arable into arable land or intensification (e.g. meadow into vineyard or overgrown land to cultivated field)
- change in land use from arable into non-arable land or extensification (e.g. cultivated field into overgrown land or vineyard into meadow).

Since viticulture is one of the most important economic activities in Svečinske gorice, the changes in land use were also considered in a wider context. To this end, our analysis also incorporated the existing information on wine-growing areas for the year 1824. Wine-growing data for 1824 was obtained from maps scanned from the Franciscean Cadastre, published on the website of the Archives of the Republic of Slovenia (http://arsq.gov.si/Query/suchinfo.aspx). The maps of all cadastral municipalities in Svečinske gorice were combined, compared, and geo-referenced. Our next step was the digitalisation of all wine-growing areas, with which a new class was formed; this was then used in follow-up spatial analyses, especially in comparison of wine-growing areas in relation to their natural-geographic factors. Land use data at the level of cadastral municipalities are not directly comparable, as the cadastral boundaries have been redrawn many times since 1824. However, the data clearly point to the differences in the size of specific land categories.

3.3 A Brief Geographical Description of Svečinske Gorice

Svečinske gorice is a micro-region of the western Slovenske gorice region. Its landscape is typical for sub-Pannonian north-eastern Slovenia—in other words, characterised by lowland and hilly sections—and it bears all the characteristics of the periphery of the Pannonian Basin. Having a specific climate and relief, as well as vegetation and soil, the area has excellent conditions for the development of agriculture, which has significantly marked the social-geographical traits of the area. Sub-Pannonian north-eastern Slovenia is a lowland, a former seabed, and later a bay of the Pannonian Sea in which tertiary rocks were deposited; these were covered by large quantities of sand and gravel in the area of Plio-Quaternary depressions. In terms of climate, this is a continental area, characterised by warm summers, cold winters, and usually sunny autumns. Agriculture is the traditional economic activity in sub-Pannonian Slovenia: arable farming and livestock breeding in the lowlands and fruit production and wine growing in the hilly terrains.

Svečinske gorice is the westernmost part of the Slovenske gorice region (Fig. 3.2). It is known to be composed of older Miocene rocks, largely of marine marl, sandstone, and clay, which in sporadic patches are also mixed with Litotamnia limestone (Kert 1994). Due to its resilient rock, the Svečinske gorice area differs from the Slovenske gorice area primarily in terms of greater absolute and relative elevation. This is especially important for the development of wine production, as surface share in the thermal belt is increased.

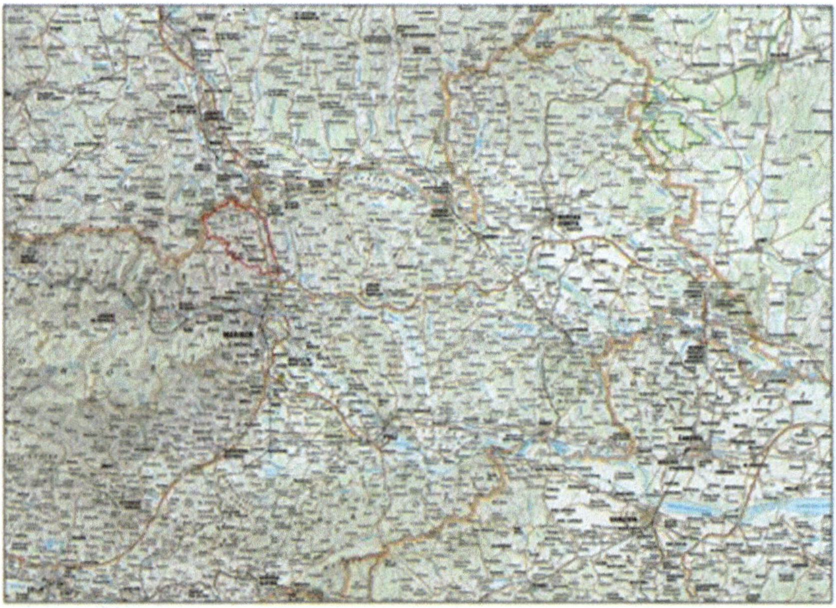

Fig. 3.2 The position of Svečinske gorice on the area of the northeastern Slovenia (*Source* DTK250, GURS 2016)

3.4 Changes in Land Use in Svečinske Gorice in the Period from 2000 to 2015

The area of Svečinske gorice, measuring 5626.9 ha in total, was dominated in 2000 by meadows (1837.3 ha or 32.7% of the area), followed by forests (1584.2 or 28.2% of the area), cultivated fields and gardens (688.3 ha or 12.2%), built-up areas (483.9 ha or 8.6%), orchards (476.5 ha or 8.5%), and vineyards (364.7 ha or 6.5%). In 2000, overgrown areas covered 81.9 ha or 1.5% of the area. In 2015, the land use structure was slightly different: the area is dominated by forest (1804.6 ha or 32.1%), followed by meadows (1768.8 ha or 31.4%), cultivated fields and gardens (572.5 ha or 10.2%), built-up areas (432.8 ha or 7.7%), overgrown areas (358.1 ha or 6.4%), orchards (355.2 ha or 6.3%), and vineyards (317.4 ha or 5.6%). In absolute terms, the most dynamic increase was recorded for overgrown land (for 276.2 ha or 4.9 percentage points), which was at the same time also the biggest change recorded in both absolute and relative terms. A significant increase was also detected in woodland surfaces (for 220.4 ha or 3.9 percentage points). A decrease was recorded for all types of agricultural land: the largest concerned orchards (by 121.3 ha or 2.2 percentage points), followed by cultivated fields and gardens (by 115.7 ha or 2.1 percentage points), and vineyards (by 47.4 ha or 0.8 percentage points) (Fig. 3.3).

Cultivated fields were reduced primarily at the end of the Pesnica River valleys and its tributaries: Svečina, Slatina, Plač, and Špičnik. A reduction was also observed in the Dobrenje brook valley, which is a consequence of the newly built highway. Several larger interconnected fields also disappeared at the end of the Cirknica valley, in the south-eastern part of the area (Fig. 3.4). The reduction of cultivated fields and gardens is most conspicuous on gradient slopes suitable for

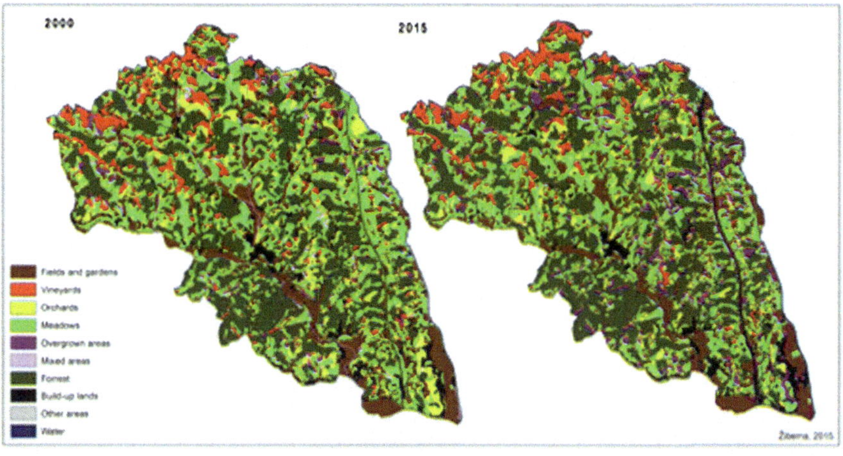

Fig. 3.3 Land use in Svečinske gorice in 2000 and 2015 (*Source* Ministry of Agriculture, Forestry and Food 2015)

farming: of the total 115.7 ha, 70.1 ha or 60.6% of fields disappeared on slopes with less than a 10° inclination and 90.8 ha or 78.5% on slopes between 10° and 15° gradient. In these cases, the reduction of cultivated fields and gardens is therefore not a consequence of rationalisation or less favourable cultivation conditions. As much as 32.8% of all surfaces used as cultivated fields are situated on slopes with a gradient lower than 2°, which is a defining feature of lowland areas. This reduction of cultivated fields and gardens is also hard to justify in terms of fertility, as the most drastic reductions can be detected in areas with eutric brown soil on Miocene rock sediments, which besides rigosoil (the anthropogenic type of soil found in vineyards, i.e. on steep slopes), is the most fertile soil in the Svečinske gorice area. As much as 57.2% of all cultivated fields and gardens that were situated on these soils have disappeared.

Fruit-growing areas were reduced in the lowland with a relative elevation of up to 25 m (18.2 ha or 15.0%) mainly in the belt between lowland and thermal belt (64.4 ha or 53.1%). Unfortunately, as much as 75.5% of all fruit-growing areas are situated on surfaces with favourable exposition (south, south-west, south-east). As much as 93.1% of all fruit-growing areas were reduced on slopes with the greatest solar exposure, i.e. with over 1000 kWh/m^2 of annual global solar radiation.

In the period considered, arable land has diminished by a total of 5.1% (to 283.9 ha). Areas that used to be agricultural land are evenly distributed throughout the territory, with a slightly higher concentration in the northern area of Svečinske gorice and on the slopes between brooks Cirknica and Dobrenje. In both cases, these surfaces used to be largely devoted to fruit-growing (Fig. 3.4).

An increase in overgrown land in Svečinske gorice can mainly be detected in the northern and western border area, and in the southern, south-eastern, and eastern sections. Newly emerging patches of overgrown land can also be traced along the new highway in the Dobrenje brook valley, where discontinuation of farming took

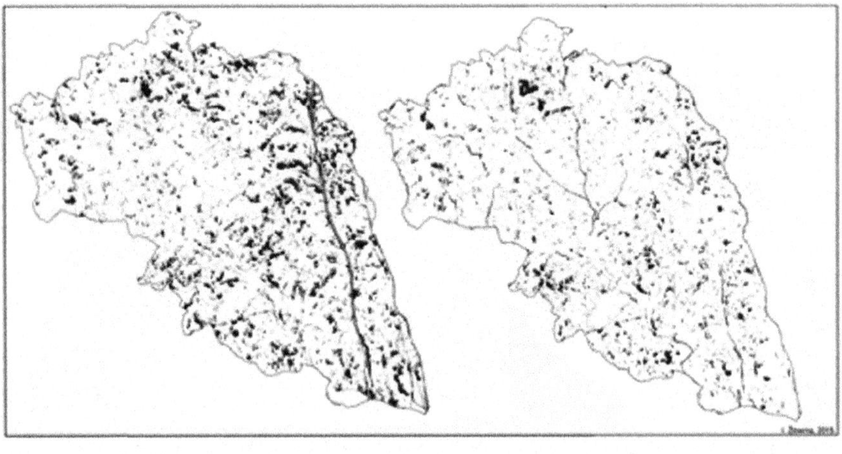

Fig. 3.4 The area of the reduction of fields and gardens (*left*) and arable land (*right*) in the area of Svečinske gorice between 2000 and 2015 (*Source* Own calculations)

place due to land fragmentation (Fig. 3.5). The increase is most noticeable in the lowland areas with a relative elevation of up to 25 m. In 2000, only 6.8 ha of land in this belt was overgrown, while by 2015, this surface area had grown over 13 times larger, i.e. 91.3 ha.

Forest areas also increased throughout the entire area of Svečinske gorice, most noticeably in the area between brooks Svečina and Plač and in the southern and eastern part of the area in question (Fig. 3.5). In 2000, as much as 128.7 ha (7.1%) of land presently covered with forest was used as meadows, 51.0 ha (2.8%) as mixed land use, 38.7 ha (2.1%) as overgrown land, and 37.8 ha (2.0%) as fruit-growing areas. Forest areas developed most quickly in low relative elevations: as much as 76.6% (168.4 ha) of all newly grown forest areas are situated on relative elevations up to 75 m.

Newly built-up areas (Fig. 3.6) are predominantly a consequence of the suburbanisation process. They emerged mainly in more densely populated areas in the south-eastern and southern part of the territory (in Pesniška dolina in Pesnica near Maribor, between Dolnja Počehova and Gradiška, and in Zgornja Kungota), in Svečinski potok valley (Svečina), and between Svečina and Ciringa. Built-up areas increased also as a result of the new highway and its associated infrastructure in the Dobrenjski potok valley.

The most common changes in land use were: cultivated field to meadow (on 243.1 ha or on 4.3% of the entire area discussed), meadow to overgrown land (212.5 ha or 3.8%), orchard to meadow (138.6 ha or 2.5%), meadow to cultivated field (137.6 ha or 2.4%), and meadow to forest (128.6 ha or 2.3%). Frequent changes of land use were: meadow to orchard (83.1 ha), vineyard to meadow (59.9 ha), meadow to built-up area (59.0 ha), mixed land use to forest (50.7 ha), and orchard to overgrown land (49.7 ha).

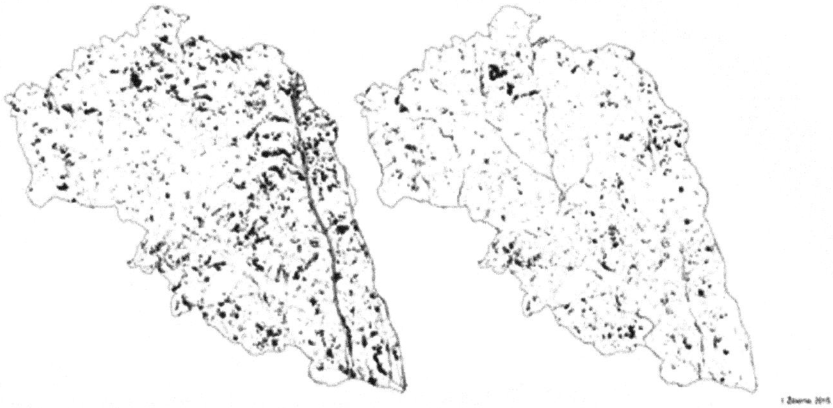

Fig. 3.5 Newly-created overgrowing areas (*left*) and newly-created forest areas (*right*) in the area of Svečinske gorice between 2000 and 2015 (*Source* Own calculations)

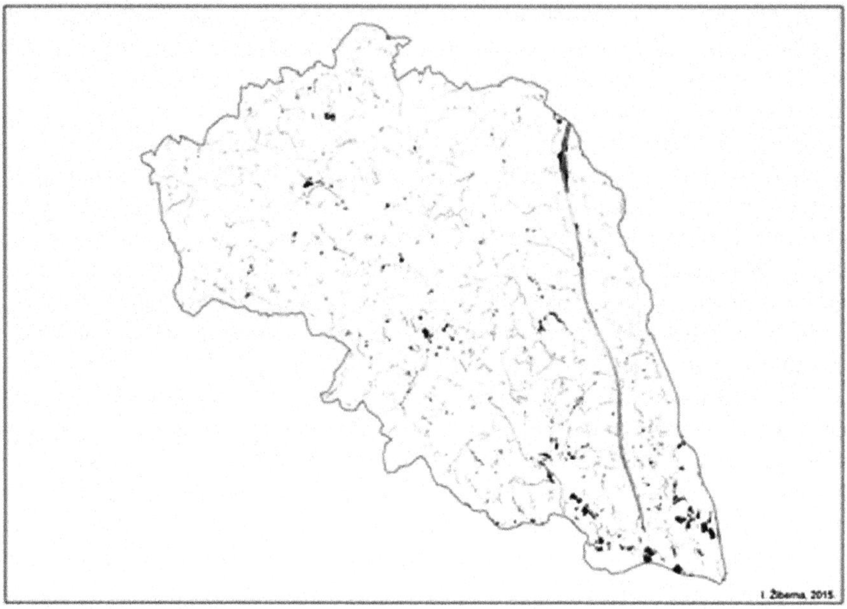

Fig. 3.6 Newly-created built-up areas in the area of Svečinske gorice between 2000 and 2015 (*Source* Own calculations)

In the period from 2000 to 2015, the existing forms of land use were preserved on 3872.2 ha or on 68.8% of the area under study. Changes in land use have thus occurred on approximately one third of the territory. Changes within arable land occurred on 73.3 ha (1.3% of the area), whereas changes within non-arable land occurred on as much as 715.3 ha (12.7% of the area); the process of conversion followed a pattern of meadow to overgrown area to forest. The processes of intensification (conversion of non-arable into arable land) occurred on 340.0 ha (6.0%), while the processes of extensification (conversion of arable into non-arable land) occurred on 623.9 ha (11.1% of the area). The ratio between intensification and extensification in Svečinske gorice was thus 1:1.83—for every hectare of new arable land there was 1.83 ha of non-arable land.

3.5 Changes in Wine-Growing Areas in Svečinske Gorice

Svečinske gorice has long been known as a typical wine-producing region (Zupanič 1969; Šilih 1979; Kert 1994). Because of this, the changes in winegrowing areas have been dealt with separately, in our analysis of land use changes covering the period from 1824 to the present-day.

and that the renewal of vineyards was carried out using American rootstock, which was ecologically more demanding and also frost-sensitive (Zupanič 1969). Vineyards were thus maintained in areas with more favourable topoclimatic conditions. This is confirmed also by the fact that in 1824 as much as 13.3% of vineyards were situated on relative elevations under 50 m (presently only 4.8%). In 1824, 68.8% of all vineyards were located on first-class sites (presently 77.5%). Nevertheless, wine-growing on steep terrains underwent a gradual reduction: in 1824, 58.3% of all vineyards were located on inclinations under 20°, whereas in 2015 this number rose to 76.8%. This shift in winegrowing to gentler slopes also coincided with the shift from manual to predominantly mechanised methods of winegrowing. Machining on steeper slopes can be relatively costly, and safety is also an issue. It can thus be said that despite a decrease in vine-growing land, vineyards are concentrated on more favourable locations, which are, however, still under used in Svečinske gorice.

3.6 Conclusion

Since 2000, there has been a substantial reduction of arable land in Svečinske gorice. From 2000 to 2015, arable land was reduced from 1529.5 ha (or 27.2% of the entire area) to 1245.6 ha (or 22.1% of the entire area), in other words, by 283.5 ha (or 5.1 percentage points) or annually by 17.7 ha. Arable land was reduced primarily on the account of fruit-growing areas (by 121.3 ha or 2.2 percentage points) and cultivated fields and gardens (by 115.7 ha or 2.1 percentage points). Reductions have also been observed in wine-growing areas, the region's hallmark, giving the landscape its distinctive appearance. From 2000 to 2015, wine-growing areas decreased by 47.4 ha or 0.8 percentage points. On the other hand, the areas that grew most markedly were primarily overgrown areas (by 276.2 ha or 4.9 percentage points) and forest areas (by 220.4 ha or 3.9 percentage points), which proves that the area is undergoing the distinct processes of overgrowing and afforestation. The ratio between intensification and extensification is 1:1.83, which is above the average recorded for north-eastern Slovenia, where the ratio is 1:1.60. Acknowledging the great potential for winegrowing in the area, it should be noted that only one fifth of all vineyards are located on first-class sites.

Svečinske gorice is one of the regions in north-eastern Slovenia, which can be denoted as the area with unfavourable socio-geographical characteristics. The area is characterized by a high ageing index: in the year 2016 most of the settlements had ageing index higher than 130, individual settlements, even higher than 160. The average gross earnings for the area in the year 2016 was 1305 eur (average for Slovenia 1525). The area also have above-average unemployment. Despite the great density of the road network and the nearness of the second largest Slovenian city (Maribor), Svečinske gorice in the last 25 years remained on the periphery. One of the forms of manifestation of marginalisation is also the reduction of cultivated areas and land-use changes in the direction of the overgrowing of arable areas and afforestation, which leads to a substantially lower utilisation of the natural potential. At a time when food is becoming a strategic component, this is a very unfavorable process. It can be seen that as far as land use is concerned, Svečinske gorice is treading a similar path to other marginalised areas of north-eastern Slovenia, such as north-eastern Goričko and eastern Haloze.

References

Eurostat regional yearbook. (2012). Land cover and land use.

Gabrovec, M., & Kladnik, D. (1997). Some new aspects of land use in Slovenia. *Acta Geographica, XXXVII*, 11–63.

Kert, B. (1994). Zahodne Slovenske gorice. Regionalna geografska monografija Slovenije. Elaborat. Raziskovalni inštitut Pedagoške fakultete. Maribor.

Kladnik, D. (1999). *Leksikon geografije podeželja*. Ljubljana: Inštitut za geografijo.

LUCAS (Land Use/Cover Area frame statistical Survey). (2009). Technical reference. Document C-3: Land use and Land Cover: Nomenclature. Directorate E: Agriculture and environment statistics; Statistical co-operation. Unit E-1: Farms, Agro-Environment and Rural Development. European Commission.

Perpar, A., & Kovačič, M. (2006). Prostorski vidiki razvoja kmetij. *Dela* (Vol. 25, pp. 61–71). Oddelek za geografijo. Filozofska fakulteta. Univerza v Ljubljani.

Plut, D. (2012). Prehranska varnost in Slovenija. *Dela* (Vol. 38, pp. 5–23). Oddelek za geografijo. Filozofska fakulteta. Univerza v Ljubljani.

Poročilo o stanju kmetijstva, živilstva, gozdarstva in ribištva v letu 2011, Ministrstvo za kmetijstvo in okolje. Ljubljana. 2012.

Pravilnik o evidenci dejanske rabe kmetijskih in gozdnih zemljišč. UL RS 122/2008.

Pravilnik o razvrstitvi kmetijskih gospodarstev v območja z omejenimi možnostmi za kmetijsko dejavnost. UL RS 25/2010.

Program razvoja podeželja Republike Slovenije za obdobje 2007–2013. Ministrstvo za kmetijstvo, gozdarstvo in prehrano. Ljubljana. 2007.

Šilih, F. (1979). Nekateri elementi geografskega razvoja Svečinskih goric. Mariborsko Podravje: zbornik 11. zborovanja slovenskih geografov. (115–125).

Vrabič Kek, B. (2012). Kakovost življenja. SURS. Ljubljana.

Žiberna, I. (1992). Vpliv klime na lego in razširjenost vinogradov na primeru Srednjih Slovenskih goric. *Geografski zbornik* (Vol. 32, pp. 51–139). SAZU. Ljubljana.

Žiberna, I. (2011). Izbrane naravnogeografske značilnosti občine Radlje. *Revija za geografijo* (Vol. 11, pp. 47–60). Oddelek za geografijo, Filozofska fakulteta, Maribor.

Zupanič, I. (1969). Zgodovina vinogradništva Slovenskih goric, Založba Obzorja. Maribor.

http://kazalci.arso.gov.si/?data=indicator&ind_id=87 (4.4.2013).

http://rkg.gov.si/GERK/ (2.2.2013).

http://arsq.gov.si/Query/suchinfo.aspx (20.1.2015).

http://gis.stat.si/# (15.8.2016).

Table 4.2 The greatest natural disasters in terms of absolute number of fatalities and in terms of damage for each year around the globe between 1995 and 2015 (adapted from Munich RE 2013, 2014, 2015; Zorn and Komac 2011; Zorn et al. 2012)

Year	No. of major natural disasters[a]	Greatest natural disasters in terms of absolute no. of fatalities	Greatest natural disasters in terms of damage
1995	615		Japan earthquake (Kobe): $131 bn; 6430 fatalities
1996	600		China flood (Yangtze River): $26 bn; 2700 fatalities
1997	530	Iran earthquakes: 2300 fatalities	Central Europe flood (Poland, Czech Republic): $5.3 bn
1998	702	Tropical cyclone in India (Gujarat; Jun.): $1.7 bn; 10,000 fatalities	China floods (May–Sept.): $30 bn; 3656 fatalities
1999	700	Turkey earthquake (İzmit); $24.5 bn; 17,800 fatalities	
2000	890	India floods (Aug.–Oct.): $1.2 bn; 1450 fatalities	Floods and landslides in the Alps (Oct.): $8.5 bn; 38 fatalities
2001	720	India earthquake (Gujarat; Jan.); $4.5 bn; 14,000 fatalities	Topical storm Allison (US, Texas; Jun.): $6 bn; 25 fatalities
2002	700	Afghanistan earthquake (Mar.): 2000 fatalities	Europe floods (Aug.): $18.5 bn; 230 fatalities
2003	700	Iran earthquake (Bam; Dec.): >22,000 fatalities	Europe heat wave and drought: $13 bn; >20,000 fatalities
2004	650	Southeast Asia earthquake and tsunami (Dec.): >200,000 fatalities	Japan earthquake (Honshu, Tokyo, Nagaoka; Oct.): $28 bn; 760 fatalities
2005	670	Pakistan and India earthquake (Oct.): 87,000 fatalities	Hurricane Katrina (US, Louisiana; Aug.): $125 bn; 1300 fatalities
2006	850	Indonesia earthquake (Yogyakarta, Java; May): $3.1 bn; 5750 fatalities	
2007	960	Cyclone Sidr (Bangladesh, India; Nov.): 3300 fatalities	Japan earthquake (Jul.): $12 bn; 11 fatalities
2008	750	Cyclone Nargis (Myanmar; May): 135,000 fatalities	China earthquake (Sichuan; May): $85 bn; 70,000 fatalities
2009	900	Indonesia earthquake (Sept.–Oct.): 1195 fatalities	Winter storm Klaus (France, Spain; Jan.): $5.1 bn; 26 fatalities
2010	970	Haiti earthquake (Jan.): $14 bn; 222,570 fatalities[b]	Chile earthquake (Feb.): $30 bn; 520 fatalities
2011	820	Japan earthquake and tsunami (March): $210 bn; 15,840 fatalities	
2012	920	Tropical cyclone (Philippines; Dec.): >1000 fatalities	Hurricane Sandy (US, Caribbean; Oct.): $50 bn; 210 fatalities

(continued)

Table 4.2 (continued)

Year	No. of major natural disasters[a]	Greatest natural disasters in terms of absolute no. of fatalities	Greatest natural disasters in terms of damage
2013	920	Tropical cyclone (Philippines, Vietnam, China; Nov.): $10 bn; 6095 fatalities	Floods (Central Europe): $15.2 bn; 25 fatalities
2014	980	Floods (India, Pakistan; Sept.): $5.1 bn; 665 fatalities	Tropical cyclone (India): $7 bn; 84 fatalities
2015	1060	Nepal earthquake (Apr.): $4.8 bn; 9000 fatalities	
Total	16,607		
Average	790.8		

[a]Based on the UN definition of great natural catastrophes
[b]Cavallo and Noy (2010, p. 31) report that in terms of the number of fatalities in relation to the total population of a country this was "the most catastrophic natural disaster in modern records" because about 3% of the population died

in the short run, and by 0.5% in the long run (they also cite another study with a 3.8% drop in income in the long run). The Munich Reinsurance Company (Munich RE), which has one of the largest climatic databases on natural disasters, also draws attention to the impact of climate change, but on the other hand also highlights the increased vulnerability of society (Munich RE 1998; 1999; 2010):

- Increase in the number and density of population, and urbanization (in the 1950s, less than 30% of the world population lived in urban settlements, whereas today this percentage has increased to over 50%; Fig. 4.2);
- Increase in the value of property, buildings, and infrastructure;
- Development in dangerous zones;
- Greater vulnerability of industrial society to natural disasters; and
- Destruction of the natural environment.

However, one must also be cautious in reporting the increase in damage because, as Guha-Sapir et al. (2004, p. 39) state, for many disasters in the past the damage was not even recorded. This is also connected with their claim that damage caused by disasters was "usually considerably underestimated." In addition, collecting data on damage is not systematic and also not methodologically uniform; data on direct damage predominate. On the other hand, developing countries in particular tend to inflate the actual costs of damage in order to obtain more international aid (Raschky 2008).

Table 4.2 shows that the natural disasters with the largest number of fatalities are not necessarily the most "expensive." The former are primarily common in developing countries (Guha-Sapir et al. 2004; Ibarrarán et al. 2009), where a large

Pipan, P. (2011a). Primerjava popotresne obnove v Italiji in Sloveniji po potresih v Zgornjem Posočju in Furlaniji (Comparing post-earthquake recovery in Italy and Slovenia after the earthquakes in the Soča Valley and Friuli). Ph.D. Theses. Koper: University of Primorska.

Pipan, P. (2011b). Sodelovanje javnosti v obnovi po naravnih nesrečah na primeru potresov v Furlaniji in Zgornjem Posočju v letih 1976, 1998 in 2004 (Public participation at the recovery after natural disasters on the example of earthquakes in Friuli and Upper Soča Valley in the years 1976, 1998 and 2004). In M. Zorn, B. Komac, R. Ciglič, & M. Pavšek (Eds.), *Neodgovorna odgovornost. Naravne nesreče* (Vol. 2, pp. 21–29). Ljubljana: Založba ZRC.

Pipan, P., & Zorn, M. (2013). Public participation in recovery after earthquakes in Friuli (NE Italy) and the Upper Soca Valley (NW Slovenia) in 1976, 1998, and 2004. *Natural Hazards and Earth System Sciences Discussions, 1*, 2231–2253. doi:10.5194/nhessd-1-2231-2013.

Raddatz, C. (2009). The wrath of god: Macroeconomic costs of natural disasters. *Policy Research Working Paper, 5039*. http://www-wds.worldbank.org/external/default/WDSContentServer/IW3P/IB/2009/11/05/000158349_20091105181816/Rendered/PDF/WPS5039.pdf. Retrieved August 19, 2011.

Raschky, P. A. (2008). Institutions and the losses from natural disasters. *Natural Hazards and Earth System Sciences, 8*, 627–634. doi:10.5194/nhess-8-627-2008.

Riebeek, H. (2005). The rising costs of natural hazards. *NASA Earth Observatory*, 28 March 2005. http://earthobservatory.nasa.gov/Features/RisingCost/printall.php. Retrieved November 17, 2010.

Sadowski, N. C., & Sutter, D. (2005). Hurricane fatalities and hurricane damages: Are safer hurricanes more damaging? *Southern Economic Journal, 72*(2), 422–432. doi:10.2307/20062119.

Sawada, X., & Takasaki, Y. (2017). Natural disaster, poverty, and development: An introduction. *World Development, 94*, 2–15. doi:10.1016/j.worlddev.2016.12.035.

Schmidt, A., Bloemertz, L., & Macamo, E. (Eds.). (2005). *Linking poverty reduction and disaster risk management*. Eschborn: Deutsche Gesellschaft für Technische Zusammenarbeit.

Schumacher, I., & Strobl, E. (2011). Economic development and losses due to natural disasters: The role of hazard exposure. *Ecological Economics, 72*, 97–105. doi:10.1016/j.ecolecon.2011.09.002.

Shepherd, A., Mitchell, T., Lewis, K., Lenhardt, A., Jones, L., Scott, L., & Muir-Wood, R. (2013). *The geography of poverty, disasters and climate extremes in 2030*. London: ODI.

Shi, P., Yang, X., Xu, W. & Wang, J. (2016). Mapping global mortality and affected population risks for multiple natural hazards. *International Journal of Disaster Risk Science, 7*(1), 54–62. doi:10.1007/s13753-016-0079-4.

Silbert, M., & Useche, M. (2012). Repeated natural disasters and poverty in Island nations: A decade of evidence from Indonesia. *PURC Working Paper*. University of Florida.

Sivakumar, M. V. K. (2008). Natural disasters and their mitigation for sustainable agricultural development. In R. Stefanski, & P. Pasteris (Eds.), *Management of natural and environmental resources for sustainable agricultural development* (pp. 172–191). Geneva: World Meteorological Organization.

Snel, M. (2002). Natural disasters, "At the whim of nature". *Environment & Poverty Times, 1*(1), 12.

The cycle of poverty and natural disasters. (2010). https://www.allianz.com/en/about_us/open-knowledge/topics/environment/articles/100316-the-cycle-of-poverty-and-natural-disasters.html/. Retrieved August 29, 2016.

The World Bank. (2016). New country classifications by income level. http://blogs.worldbank.org/opendata/new-country-classifications-2016. Retrieved August 19, 2016.

The World Bank: Disaster Risk Management Overview. http://www.worldbank.org/en/topic/disasterriskmanagement/overview. Retrieved August 29, 2016.

The World Factbook. (2011). https://www.cia.gov/library/publications/the-world-factbook/index.html. Retrieved August 19, 2011.

UNISDR Terminology on Disaster Risk Reduction. (2009). Geneva: UNISDR.

van Ness, J. (2005). Poverty and "Natural" Disasters. https://21stcenturychallenges.org/who-are-most-vulnerable-to-natural-hazards/. Retrieved August 29, 2016.

Who are most vulnerable to natural hazards? 21st Century Challenges: Informed Discussion from the Royal Geographical Society, 2016. https://21stcenturychallenges.org/who-are-most-vulnerable-to-natural-hazards/. Retrieved August 29, 2016.

Zorn, M., & Komac, B. (2011). Damage caused by natural disasters in Slovenia and globally between 1995 and 2010. *Acta geographica Slovenica, 51*(1), 7–41. doi:10.3986/AGS51101.

Zorn, M., & Hrvatin, M. (2015). Damage caused by natural disasters in Slovenia between 1991 and 2008. *Acta Geobalcanica, 1*(1), 33–43. doi:10.18509/AGB.2015.04.

Zorn, M., Ciglič, R., & Komac, B. (2012). Škoda in pokrajinski učinki večjih naravnih nesreč (Damage and landscape effects of major natural disasters). *Ujma, 26,* 70–79.

Chapter 5
Climate Change Effects on Mountain Regions Marginalized by Socio-Economic Transformation—The Case of North Caucasus

Vera Vinogradova, Raisa Gracheva and Elena Belonovskaya

5.1 Introduction

The North Caucasus, stretching from the Caspian Sea to the Black Sea, after the collapse of the Soviet Union became borderland between Russian Federation and neighbouring countries, Azerbaijan and Georgia. Mountain regions of the North Caucasus are traditional rural areas. Over the past 25 years, socio-economic situation of mountains has changed significantly as a result of the destruction of the planned economy, the transformation of land use and population outflow. In the geographical context, the mountain regions of the North Caucasus can be called as a marginal area. The current abandonment and underuse of the semi-natural grasslands, which are the most valuable landscapes of the rural mountain areas, complicate marginality and make it as many-sided problem. The same aspects of marginality are characteristic for remote regions of many countries (de Almeida 2012; de Castro 2012; Gonzales De La Rocha 2004; von Braun and Gatzweiler 2014 and others).

The North Caucasus is usually divided into Eastern, Central and Western Caucasus. At present, all parts of the North Caucasus have many similar problems relating to the border area, socio-economic and environment issues.

V. Vinogradova (✉) · R. Gracheva · E. Belonovskaya
Institute of Geography of Russian Academy of Sciences, Staromonetny Per. 29, 119017 Moscow, Russia
e-mail: vvvinog@yandex.ru; vinograd25vv@gmail.com

R. Gracheva
e-mail: gracheva@igras.ru; gracheva04@list.ru

E. Belonovskaya
e-mail: belena53@mail.ru

S. Pelc and M. Koderman (eds.), *Nature, Tourism and Ethnicity as Drivers of (De)Marginalization*, Perspectives on Geographical Marginality 3,
DOI 10.1007/978-3-319-59002-8_5

Our study area is Republic of North Ossetia-Alania situated in the Central Caucasus where all these issues of marginality are particularly evident. The question arises what are current climate change effects on the mountain area? Does climate change lead to deterioration of environment and marginality increasing or to improvement of landscape conditions and life quality?

5.2 Study Area

Central Caucasus is located between Mount Elbrus in the west and Mount Kazbek in the east. Its sub-parallel ridges alternate with wide uplifted intermountain basins, which have been the main areas of population settlement and economic activity in the past millennia. Due to the mountainous barriers, slopes of intermountain basins have contrasting climatic conditions (Vagin 2002). On the leeward southern and eastern slopes, high temperatures and lack of moisture in summer and cold winters with thin snow cover are characteristic. Those conditions decline position of woody vegetation; the dry steppes, typical steppes and meadow steppes replace each other with the altitude. The northern and western slopes are much more cool and wet. The elfin birch and mixed woodlands and post forest mesophilous meadows are spread there (Shiffers 1953; Belonovskaya 1995; Gracheva and Belonovskaya 2010). On the height of 2300–2400 m above the sea level mountain steppe ecosystems give place to the subalpine meadows.

Areas at altitudes of 1750–2560 m a. s. l. were studied in the wide intermontane basin between Bokovoi Ridge (Side Ridge) and Skalistyi Ridge (Rocky Ridge) in North Ossetia-Alania (Fig. 5.1).

The conventional classic scheme of spatial distribution of ecosystems must be corrected taking into account the population dynamics and land use history.

5.3 Population Dynamics, History and Changes of Land Use in the Mountains of North Ossetia

Outmigration from mountain regions can be documented using censuses data for the last 120 years. The reasons of population outflow are typical for many mountain regions of the world: dependence on natural processes, heavy manual labour and its low profitability, low living standard; in the second half of the 20th century, industry development, urban lifestyle and education became the main drivers of outmigration of young people.

The proportion of the mountain population within the Republic dropped from 20% in 1900 to 1% in 2010 (Fig. 5.2), and at present is about 0.6%. Due to the overall small size of the mountain population, there is danger of the extinction of mountain communities (Gracheva et al. 2012).

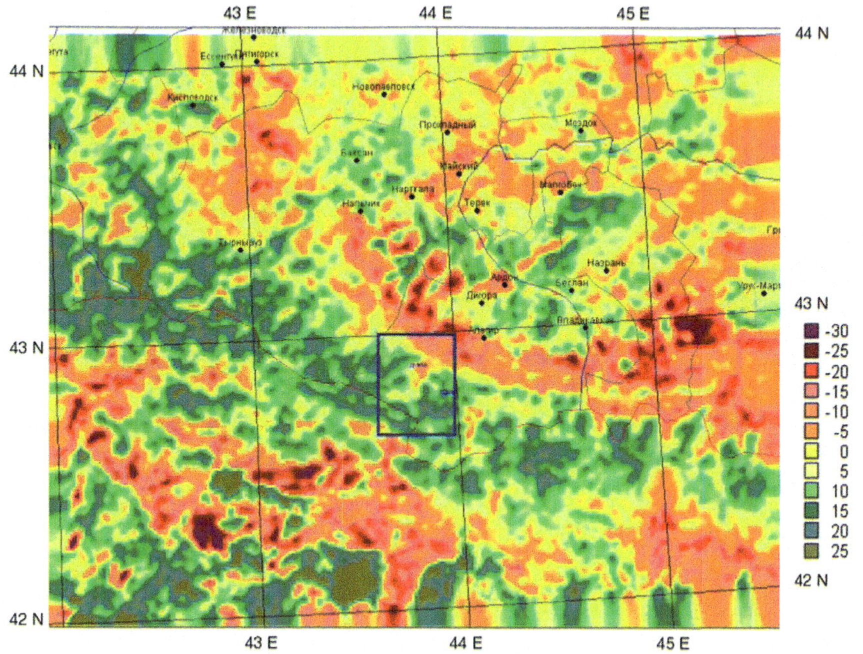

Fig. 5.4 Map of vegetation condition index (VCI). Changes over the period 2007–2014 compared to 2000–2006

The satellite climate extreme index (SCEI) showed the growth in the second period at the middle and high altitudinal areas where dry conditions converted to the neutral and moist ones. The index increased by 0.5–1.5 points. In the other parts of territory, especially on the plains and in the foothills, the aridity grew and the index decreased by 0.5–1.5 points (Vinogradova et al. 2015).

Linear trends of vegetation indices were calculated for different altitudinal belts. In the foothills and middle altitudinal areas the NDVI trend was negative, however at the higher levels its absolute values decreased and at the elevation of 1800–2000 m above sea level it became positive and grew up to maximal values in the alpine and subalpine belts at altitudes of 2800–3000 m a. s. l. (Fig. 5.6).

Similar to the NDVI, the VCI trend grew when the altitude increased. VCI trend became positive at the altitudes of approximately 1400 m a. s. l. It reached maximum in the alpine belt (2800–3000 m a. s. l). Thus, the trend of climate change during the last decades may have improved the conditions for vegetation growth and productivity.

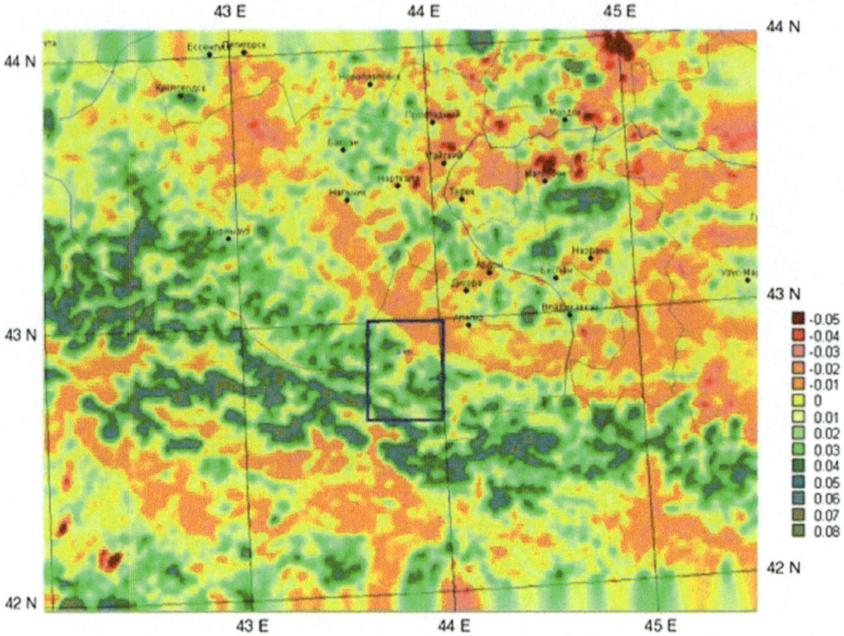

Fig. 5.5 The map of the vegetation index (NDVI). Changes over the period 2007–2014 compared to 2000–2006

Fig. 5.6 Dependence of NDVI linear trend on height in 2000–2014

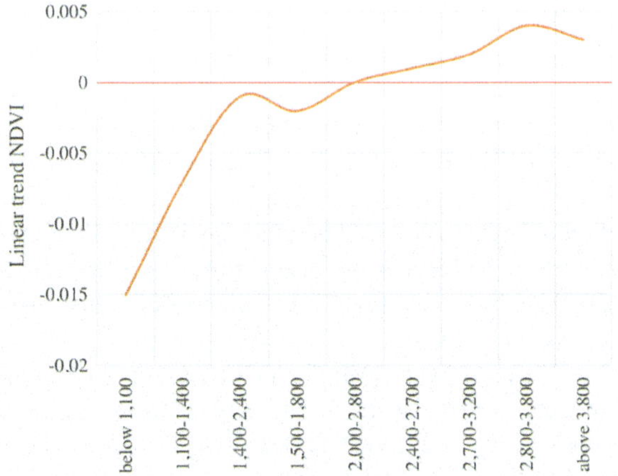

5.6 Meteorological Parameters

During climate warming, the sum of active temperatures on the considered territory was growing by 100–150 °C both for the flatland stations and for the stations located at the middle altitudes (Table 5.1). In the early 21st century, the sums of

for tourism purposes. They offer products with high value added (Plut et al. 2008, p. 30).

Another project, worth mentioning, is an international project, which was created in collaboration of the Austrian Alpine Association with the Austrian Federal Ministry of Agriculture, Forestry, Environment and Water Management. The project is entitled Mountaineering Villages and is carried out in the frames of Alpine Convention in places characterized by mountain tradition. The project aims to develop sustainable forms of tourism in accordance with its natural resources, culture and competencies in the field of mountaineering activities. The Austrian and Italian regions, where the project is already underway, recorded an increase in visits from walkers, hikers, climbers, ski mountaineers and others, which do not need an extensive tourist infrastructure for their activities (Ministrstvo za okolje in prostor 2015). Project financing has been carried out under European Agricultural Fund for Rural Development and the Austrian Federal Ministry of Agriculture, Forestry, Environment and Water Management (Planinska zveza Slovenije 2015). There were some initiatives for the implementation of this project in Slovenia. It could be carried out by Ministry of Environment and Spatial Planning and Alpine Association of Slovenia, but it is not yet clear which villages could be included (presumably the village of Zgornje Jezersko). The project will be launched in case that there will be some interest shown from the local authorities of alpine villages (Ministrstvo za okolje in prostor 2015).

6.3 Municipality of Bovec and Its Potential for the Development of Sustainable Tourism

Municipality of Bovec is the fourth largest municipality in Slovenia by its area. It is characterized by a border location in the alpine space. It borders on three Italian municipalities (Tarvisio, Chiusaforte and Resia) and is in close proximity with the Republic of Austria. Due to its location in the Julian Alps, almost 88% of the area of the municipality is hilly and 79% of the municipal territory is a part of the only national park in Slovenia—the Triglav National Park (Soča Valley Development Centre 2013, p. 20). As Pelc (2010, pp. 105–106) stated, border and poor accessibility play an important role in this region and are considered to be an obstacle for normal communication and logistics, which strongly influences the emigration (and, consequently, the depopulation).

Due to the natural conditions (relief, share of forest areas and erosion), farming restrictions and difficult transport accessibility, the municipality (together with other municipalities in the Upper Soča Valley region) is considered an area with limited potentials for the development of economic activities. As a result, the population is economically weak and demographic trends are unfavorable for decades (Lokalna akcijska skupina za razvoj Posočja 2008). The region is also one of the seismically most endangered in Slovenia and has had an intensive post-earthquake reconstruction in the past. At the same time, the area has the most intensive emigration of the population within the Triglav National Park. In the past hundred years, the number of population in municipality has decreased by three quarters (Rejec Brancelj and Smrekar 2000, p. 45). Along with the process of depopulation, most of the thirteen settlements in Municipality of Bovec face home abandonment and overgrowing of once cultivated agricultural land (Perko and Orožen Adamič 1999, p. 65). In comparison with the year 2000, the proportion of barren land has increased for 21 hectares or 105%, the share of utilized agricultural area has decreased for 4% (Statistical Office RS 2016). As Pelc estimates, marginalization of this region came from the loss of economic potential in the transition between the 19th and 20th century and can today be seen as a result of intensive depopulation. The area is at the same time located near some of the most recognizable tourist points in Slovenia (such as Bled, Bohinj, Kranjska Gora), which in regard to marginalization, make the area a bit more central. Tourism, as well as organic farming, is thus spontaneous response to the threat of marginalization (Pelc 2010, p. 112).

The population density in the municipality has been constantly decreasing in the last 15 years. In 2015, the municipality had an average of 8.5 inhabitants per km^2 (Statistical Office RS 2016). The number of inhabitants in 2015 decreased for 16.5% compared to 1971. Besides the loss of population, the area also witnesses the aging of the population. The aging index (Fig. 6.2) in the Municipality of Bovec grew faster than the national average (2000: 139.6; 2015: 190.4), while the average age of the population has also increased significantly from 2011 to 2015 (from 42.0 to 47.1 years).

Another significant factor, which has an impact on the development of the municipality, is Triglav National park that covers almost four-fifths of the

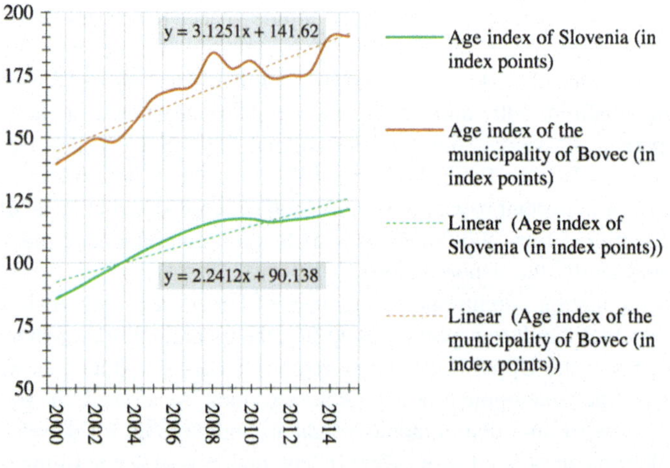

Fig. 6.2 Aging index in the Municipality of Bovec and Slovenia in the period between 2000 and 2015 (*Source* Statistical Office RS 2016)

Municipality of Bovec. In theory, this fact makes the idea of sustainability an unavoidable solution for the area. The importance of it was also emphasized by the Bovec Cooperative Development Society (2013); meanwhile Local Tourism Organization stressed out the meaning of green tourism for the region (Humar 2015). A civil initiative has also emerged in the municipality, which calls for the further consideration of the concept of sustainable development and strengthening it within the area of Municipality of Bovec. Nevertheless, in the scientific and technical literature, the municipality hasn't been recognized as a unit, which fully introduced the idea of sustainability in its development concept. According to the survey, which Plut et al. (2008) conducted among local residents in Triglav National Park (their research sample included more than 80% of all residents in the park), respondents see nature conservation as a crucial value, but consider standards of conservation regimes too strict (Groznik Zeiler 2011, p. 28). The Triglav National Park Act from 2010 provides development guidelines that promote the sustainable development: organic farming together with complementary farm activities, traditional building construction, and supply of products with designation of origin, geographical indication, brand development, sustainable energy consumption and development of eco-tourism (Triglav National Park Act 2010, Article 10).

From the perspective of tourist arrivals on a general level, Triglav National Park recorded a 32% increase between 2000 and 2009, meanwhile the Upper Soča Valley region (part of park where Municipality of Bovec is located), recorded decrease. This could be linked to region's greater dependence on the weather. Among other development problems of the tourism inside Triglav National Park, unfavorable demographic structure, lack of entrepreneurial initiative, seasonality of tourism, inadequate infrastructure, poorly developed public transport as well as problems with special planning have been identified. The solutions that were presented stress the importance of maintaining primeval features of area, integrity and diversity of natural and cultural heritage, and the need for development of specific tourist offer, which will increase the responsibility and attitude of tourists and visitors towards nature. The development of tourist products should also be based on the strengthening of coexistence of the local population with the nature, and should derive from the cooperation between tourism providers and Triglav National Park conservation policies (Mlekuž and Zupan 2011, pp. 22–33).

6.4 Tourist Activities and Opportunities for Sustainable Development in the Municipality of Bovec

Between 75,000 and 80,000 tourists annually visit the Municipality of Bovec. They generate about 200,000 overnight stays. Analysis of the tourism characteristics has revealed that foreign tourists are far most numerous guests, their numbers range between 50,000 and 63,000 visits annually (Fig. 6.3). The ratio between foreign and domestic guests was 78:22. The decline in the number of Slovenian tourists

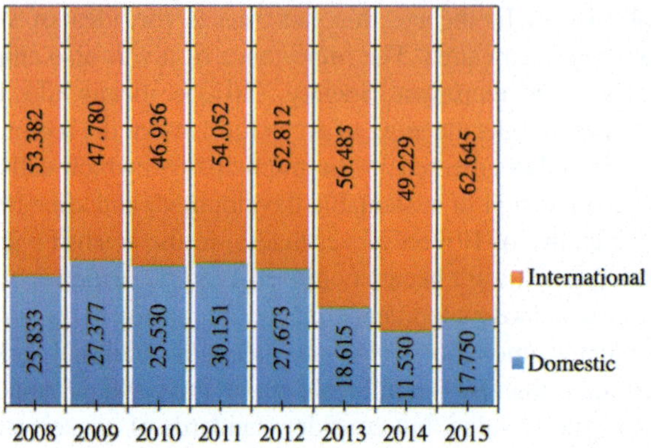

Fig. 6.3 The number of overnight stays in the Municipality of Bovec between 2008 and 2015 (*Source* Statistical Office RS 2016)

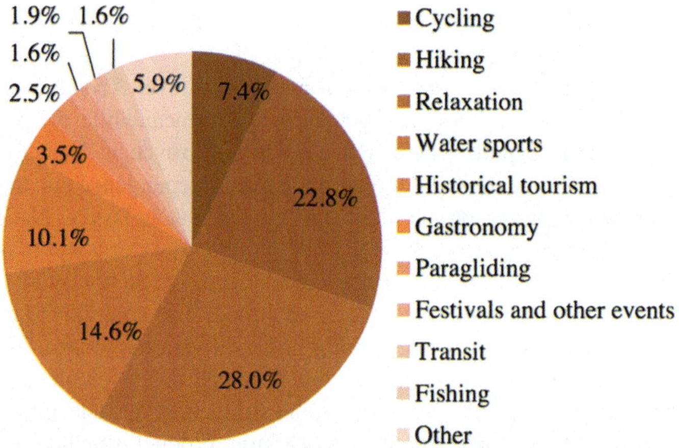

Fig. 6.4 Motives for visiting the upper Soča river valley according to a survey conducted by the Soča Valley Development Centre; N = 878 (*Source* Soča Valley Development Centre 2014)

between the years 2013 and 2016 can be interpreted with the fact that Kanin ski resort and cable car was closed (cable car started to operate again in the 2016/2017 season). Foreign guests tend to stay in municipality longer than domestic—in 2015, the average length of stay was 2.6 days for the foreign tourist and 2.0 days for domestic one. In 2015, 77% of foreign visitors generated 82% of all overnight stays (Statistical Office RS 2016).

Among the reasons for visiting the Municipality of Bovec, top four are as follows:

- a desire for relaxation and rest,
- hiking,
- water sports activities and
- historical tourism (Fig. 6.4).

open to foreigners without leading to a loss of its own customs (Grimalt and Prats I Serra 2001).

The arrival of the hippies to the island with their proposals and ways of thinking fit perfectly with the locals, creating a curious cohabitation without any type of conflict, despite the enormous differences between the two groups: a group of young people from the European and North American elite and the insular society that had maintained its traditions unaltered for centuries. Far from the logical clash that one might imagine, both parties knew how to respect each other and even developed a reciprocal admiration centred on the love that they shared for the island.

7.2.2 *The Hippy Seed*

The tourist boom that shook a large part of the Mediterranean coast also similarly impacted the Balearic Islands beginning in the mid-1960s (Ramón 2001). The number of tourists and visits—both national and international—transformed the economy and the infrastructure of the archipelago. Nevertheless, it is important to note that the impact was first observed in the Island of Mallorca and, to a lesser degree, in the Island of Menorca and the Pityusic Islands (Cirer-Costa 2001).

In this manner, Ibiza was not included in the process, but it attracted an international hippy movement that was amazed by the island's beauty. Starting in the 1970s, Ibiza, which had always been a very peripheral place, became a libertarian and tolerant exception under the somewhat-condescending watch of the Spanish authorities of the time (the dictatorship) to give the impression of openness and to incentivize tourists whose money was very welcome (Planells 2002).

In this manner, Ibiza became the flag bearer of an alternative lifestyle model that was in keeping with the values of the hippy movement and that came together with the values of the island societies that had stagnated in tradition in a very interesting manner. Despite the great cultural differences between locals and the new residents, everyone shared a deep love for the islands and their nature, becoming joint defenders of their values (Simard 2000).

7.2.3 *Just One More Mass Tourism Destination (1980s to 1990s)*

Development in the Balearic Islands led to widespread tourism development. With the growing arrival of tourists under the so-called sun and beach tourism model, the initial claim by the hippy movement gave way to democratization. The speed at which this process occurred forced the island to respond to a massive demand for tourism (Lalive D'Épinay 1982), that left little space for planning and was the result of improvisation performed by private capital with the backing of the public authorities. The impact on the fragile equilibrium of these islands came swiftly with the initial occupation of the coastline and the deep transformations in infrastructure to allow for the flood of tourists who came during a brief period in the year, based on the holidays and the favourable climate of the summer season (Ramón and Serra 2013).

After two long, out-of-control decades, investors themselves saw the limits and the long-term impact of tourism on resources and began to propose new strategies to diversify the tourism industry while maintaining the existing flows (Hall and Page 1998).

7.3 Music as Redemption: Ibiza's Global Reference (2000s)

Until the year 2000, the process of tourism development of Ibiza in places such as San Antonio, Santa Eulària, and the City of Ibiza itself appeared very similar to that of the rest of Baleares as a result of widespread tourism. However, in this section, we observe how Ibiza took a different course, rediscovering its beginnings. After observing the impact of widespread tourism on the island, the first response was to diversify tourism with actions such as extending the season, seeking different tourist groups, and developing different tourism activities such as sport and entertainment activities (Costa 1989). Nevertheless, the success of these strategies was limited, in large part because they were the same approaches that were implemented in other tourism destinations in the Mediterranean that were also saturated and, more concretely, they were in a context of competing at a disadvantage compared to other islands in the archipelago, particularly Mallorca. In comparison, Ibiza offered fewer choices, and its dependence on Mallorca was detrimental.

Nevertheless, Ibiza was able to develop an authentic strategy from its own entrepreneurs that would shape the future of the island and have an unparalleled impact around the world. Returning to its hippy heritage and libertarian image, there was a strategy that incentivized a diversity in terms of entertainment (Corbin 1995) and, in particular, nightlife through nightclubs and music (Augé 1992). The parties in Ibiza began to acquire a reputation for their diverse character and began to transform tourism in the island from a sun and beach family- friendly model to a demand for nightlife centred on a younger and more elitist public. This transformation emerged spontaneously, but over time, it acquired an almost collective business strategy dimension, with the more or less implicit support of the authorities (within a policy of lax laws).

7.3.1 The Impact of Global Nightlife in Ibiza: Youth Tourism

The promotion of the island's image through its parties with the release of record labels, elaborate shows (Bensa and Fassin 2002), and the transformation of buildings into large entertainment businesses (Rosa 2002), made it possible to return to the already known name of Ibiza and to catapult it as a symbol of entertainment and partying (Allemand 2010). This development was something that occurred not only at the national and international scale; rather, it was something that transformed the island into a global icon (Amirou 2000). The process of reinterpreting the hippy alternative image in a contemporary global context (Jauréguiberry 1994), came from local businessmen (from the island and from elsewhere). They were the actors who understood the authenticity of the island to convert it into a unique product that went beyond tourism. Given the party reputation of the island, the traditional sun and beach tourists began to stay away, which made it possible to attract visitors who were willing to spend large sums of money to experience the nightlife in Ibiza. There was a shift from tourism to the experience of shows and culture.

In this manner, by the mid-2000s, Ibiza had become internationally renowned in the electronic music industry (its own record labels and invited DJs) and for its nightclubs (Goulding and Shankar 2011) (such as Space, Privilege, Amnesia, Café del Mar, Pacha, Ushuaia, and Esparadis); it has also become one of the most sought-after stages during the summer in show business (with parties organized by the Matinée group, La Troya, Supermarxé, and Circuit). Simultaneously, from the cultural perspective, the construction of the image of the island had a minimal physical impact and, on the contrary, made it possible to emphasize the natural Mediterranean environment of the island and the libertarian heritage of the reinterpreted hippy movement (Perrot 1997).

With its network of nightclubs (Sönmez et al. 2013), Ibiza began to generate an entertainment niche that attracted record labels and international DJs, which transformed Ibiza into not only one of the destinations in the global circuit but also a destination with international renown, almost exclusively during the summer period (Fontaine and Fontana 1996). The dissemination of the reputation of its parties crossed borders, attracting people from all over the world to Ibiza. Simultaneously, it became a world-renowned destination through the dissemination of its parties through video, the Ibiza-based record labels, and the lyrics to songs that include Ibiza as an icon and that have been emulated in other parts of the world, such as "Ibiza's Party". This highlights the island's own style, which is defined by freewheeling parties and a relaxed environment that goes beyond labels or conventionalism. We find ourselves with the case of one of the few non-urban destinations that has managed to concentrate and attract nightlife, even going so far as to displace some of the international record labels and entertainment events in the main cities around the world (Nora 1974). A couple examples are London with the Ministry of Sound show and Paris Hilton's parties, which were associated with the Supermarxé parties.

Combining all of the efforts of individual businessmen, the nightlife industry is presented as an enormous entertainment platform with a volume of shows every night (during the summer season) that is found in an international city. Music and parties have incentivized a licit and illicit industry (Malborn 1999).

The result of this distinctive and authentic strategy, which was crafted by local actors but had a global impact, has led to a radical transformation of the tourism industry in the island in a process that has increased the elitism that had only been observed in certain exclusive Mediterranean destinations that were already well-established on the French Riviera (Saint-Tropez, Nice, Cannes, and Monaco itself).

Currently, the exclusivity of the island is beginning to generate a new transformation, slowly displacing the Ibiza nightlife in favor of luxury tourism, with yet-unquantifiable consequences.

7.3.2 From Popular to Exclusive: Pursuing Familiar Tourism

Even 10 years ago, Ibiza would have never imaged that, after the parties, it would become one of the most coveted tourist destinations. Today, the island is one of the largest consumers of luxury products, such as French champagne (according to data, Moët Chandon champagne). The trend towards the elitism of the island has caused a significant process of property speculation with multimillion-dollar mansions for rent and the implementation of luxury services and infrastructure, such as polo fields and the construction of a luxury sport seaport.

The island has become a symbol of success, whose label is sought after and controlled by the people of the island themselves, given that it has become synonymous with success and luxury.

The origin of all of these transformations was music and the music industry, which made it possible to disseminate the image of the island throughout the world through song lyrics and videos of the island (Bouillin-Dartevelle et al. 1991). In the first phase, this situation caused traditional sun and beach family tourists to be scared away, leaving out of fear of such sin; and in the second phase, the name recognition of Ibiza was exploited to transform it into an icon of entertainment and fun and, finally, owing to its success, towards luxury.

7.4 An Amazing Island: An Unreachable Destination (2010s)

By 2010, Ibiza had become its own brand around the world, independent of other powerful images such as the Balearic Islands or Spain. Ibiza went down its chosen road, a result of its success. Presently, however, it finds itself at a crossroads, between reconverting itself once again into the cutting edge of music, art, and the culture scene through entertainment or yielding to the emerging industry of luxury services, becoming part of the club of the most exclusive international destinations. Today, Ibiza is becoming a dreamlike tourist destination that is ever more inaccessible to ordinary tourists or spectators. It has become more focused on appealing to a select group of celebrities and rich businessmen. This current process entails the risk of moving away from the initial successful image of the hippy movement. The island could become a victim of its own success and renounce its alternative character, in the process losing part of its uniqueness and charm. It seems that the authorities are attempting a middle ground, considering luxury as a way of avoiding the impact of widespread tourism while conserving—even if, to a certain degree, it is false—its unique character based on a false simplicity and anonymity. Perhaps the authenticity of the process could be judged, although if we analyse its initial hippy origin story, it was also based on a utopia that was in large part sustained by many of the children of the global elite.

7.4.1 Ibiza as a Global Reference

The Ibiza concept has been successful at fitting perfectly with the existential and experiential character of contemporary societies. Faced with the loss of traditional collective symbols and faced with a fascinating universe of multicultural societies, individuals have begun to increasingly relativize that which has been rationally established and to value once again that which appeals to what is experienced through the senses. Music and nightlife have become catalysts towards an unparalleled space of freedom that the island has been able to understand from its previous hippy experience (Tutenges 2012). However, success did not merely consist of having a unique characteristic that could be of interest to others. Simultaneously, the island was able to find the appropriate channels with access to new information technology and the telecommunications boom (Missonnier et al. 2003), a unique method of disseminating its message in the freest manner possible. Entertainment makes it possible to reach people informally and experientially, making them feel a global message as something individual. The experience of Ibiza is lived first-hand; simultaneously, however, there has been a way to disseminate it to everyone on an equal basis internationally. What is interesting in this case is how the message that is transmitted through entertainment and music has become universal, despite cultural and historical differences. For this reason, we speak of a cultural process, given that it concerns an artistic message that can be interpreted by each one of us. Ibiza is more than an image or a brand, and it is viewed as the construction of one of the first collective global myths. For this reason, it would most likely be a mistake to confine it to a concrete image that would reduce its potential in the long term, which is what seems to be occurring now with the trend towards luxury.

7.4.2 Ibiza's Process of Exclusiveness

As a paradise, Ibiza seems to have become an exclusive space that is only accessible to those who can afford it. As a benefit, the luxury services development strategy mitigates ever-growing pressure of tourism on the northern Mediterranean region. Nevertheless, it is moving away from its original libertarian and egalitarian character, and it could fall into a middle-term vision, as noted above.

Ibiza could become another luxury destination, such as Monaco, Marbella, or one of the Emirates. However, it can continue to be one of the few places in the world that is centred on the concept of freedom and entertainment (Thornton 1995), a position most likely only shared with a few places such as Las Vegas. Ibiza is the only island in the world that, from a cultural and artistic perspective, has allowed for the construction of a dream, something that is only possible at certain times in some of the most important cities around the world such as Paris, New York, and Tokyo (McVeigh 2004). Ibiza is the only island that has been able to create a space of freedom and imagination within the contemporary Western world. Frequently disparaged and under the moralizing gaze, Ibiza's nightlife can be understood, on the contrary, as a new creative stage in the arts (music and show business) as well as a reconceptualization of entertainment from Western frameworks through new technologies (St John 2004). The reconceptualization of entertainment from Ibiza not only contributes economic incentives but also once again presents the foundations of entertainment themselves beyond tourism. It is the result of the reinvention of the island itself (Yonnet 1999), which is only recently being understood by certain academic fields (in the field of branding or marketing) and in terms of institutional recognition (by local and regional authorities).

We have presented an incredible case of how a marginalized island from the Mediterranean, very forgotten, became thanks to music and nightlife, one of the most well-known and exclusive spots in the world.

References

Allemand, R. (2010). De la mystification des pratiques à la négation du réel? Ethnographie de discothèques montpelliéraines. *Deviance et Societe, 34*(1), 29–48.

Amalvi, C. (1997). Le 14-juillet. Du Dies Irae à Jour de fête. In P. Nora (Ed.), *Les lieux de mémoires* (pp. 383–423). Paris, Gallimard, coll. Quarto, t. 1.

Amirou, R. (2000). *Imaginaire du tourisme cultural*. Paris: PUF.

Anderson, B. (1991). *Imagined communities*. Londres: Verso.

Augé, M. (1992). *Non-lieux: Introduction à une anthropologie de la surmodernité*. Paris: Seuil.

Bensa, A., & Fassin, E. (2002). Les sciences sociales face à l'événement. *Terrain, 38*, 5–20.

Bonnet, J. (1995). Les vitrines identitaires, une nouvelle source pour l'étude des recompositions identitaires. In C.-O. Carbonell (Ed.), *De l'Europe, identités et identité* (pp. 183–190). Toulouse: Presses universitaires des sciences sociales de Toulouse.

Bouillin-Dartevelle, R., Thoveron, G., & Noël, F. (1991). *Temps libre et pratiques culturelles*. Liège: Mardaga, coll. Création et communication.

Campo, D., & Ryan, B. D. (2008). The entertainment zone: Unplanned nightlife and the revitalization of the American downtown. *Journal of Urban Design, 13*(3), 291–315.

Cirer-Costa, J. C. (2001). Evolució de l'oferta de places turístiques a Eivissa i Formentera. 1950–2000. *Estudis sobre el turisme a Eivissa i Formentera, 2*, 73–93.

Corbin, A. dir. (1995). *L'Avènement des loisirs (1850–1960)*. Paris: Aubier.

Costa, N. (1989). *Sociologia del turismo: Interazioni e identità nel tempo libero*. Milan: Cooerativa Libreria I.U.L.M.

Dumazedier, J. (1962). *Vers une civilisation du loisir?*. Paris: Seuil.

Fontaine, A., & Fontana C. (1996). *Raver*. Paris: Anthropos.

Goulding, C. & Shankar, A. (2011). Club culture, neotribalism and ritualised behaviour. *Annals of Tourism Research, 38*(4), 14351453. doi:http://dx.doi.org/10.1016/j.annals.2011.03.013.

Grimalt Gelabert, M., & Prats i Serra, J. A. (Eds.). 2001. *Les Pitiüses: Terres i gent. Patronat de l'Escola Municipal de Mallorquí*. Manacor: Història.

Hall, C. M., & Page, S. J. (1998). *The geography of tourism and recreation: Environment, place and space*. London: Routledge.

Jauréguiberry, F. (1994). De l'appel au local comme effet inattendu de l'ubiquité médiatique, Espaces et sociétés, n (pp. 74–75).

Knox, D. (2009). Mobile practice and youth tourism. Culture of mass tourism. In Pons, M. Crang, & P. Travlou (Eds.), *Cultures of mass tourism: Doing the Mediterranean in the age of banal mobilities* (pp. 143–156). Farnham, Surrey, UK: Ashgate.

Lalive, D'Épinay, C., Bassand, M., Christe, E., & Gros, D. (1982). Temps libre; Culture de masse et cultures de classes aujourd'hui, Lausanne, Pierre-Marcel Favre. coll. Regards sociologiques.

Malborn, B. (1999). *Clubbing: Dancing, ecstasy and vitality*. London: Routledge.

McVeigh, M. (2004). *Nightclub design: The significance of performance in 21st century culture*. Manitoba: M.I.D., University of Manitoba.

Missonnier, S., & Lisandre, H. (Eds.). (2003). *Le virtuel: La présence de l'absent*. Paris: EDK éditions.

Mitchell, D. (2000). *Cultural geography. A critical introduction*. Oxford: Blackwell Publish.

Nash, C. (2000). Performativity in practice: Some recent work in cultural geography. *Progress in Human Geography, 24*(4), 653–664.

Nora, P. (1974). Le retour de l'événement. In J. Le Goff et P. Nora (dir.), Faire de l'histoire (Vol. 1). Paris: Gallimard.

Perrot, M. -D. (1997). Du vrai-faux au virtuel: Mondialisation culturelle et néo-utopie, Nouveaux cahiers de l'Institut Universitaire du Développement, n 6: La mondialisation des anti-sociétés. Espaces rêvés et lieux communs (pp. 41–63).

Planells, M. (2002). *El nacimiento de Babel -Ibiza años 60-*. Eivissa: José Ferrer y Vicent Guillamó.

Ramón, E. (2001). *Historia del turismo en Ibiza y Formentera 1900–2000*. Eivissa: Genial Ediciones Culturals.
Ramón, J., & Serra, A. (2013). Tolerancia y desarrollo turístico: Sant Antoni en los años cincuenta. *CULTUR-Revista de Cultura e Turismo, 7*(2), 132–156.
Rodaway, P. (1994). *Sensuous geographies*. London: Routledge.
Rosa, M. C. (Ed.). (2002). *Festa, lazer e cultura*. Saõ Paulo: Papirus.
Simard, M. (2000). Communautés locales et espace-monde. Les processus identitaires de la post-modernité. *Géographie et cultures, 36*, 3–20.
Sönmez, S., Apostolopoulos, Y., Theocharous, A., & Massengale, K. (2013). Bar crawls, foam parties, and clubbing networks: Mapping the risk environment of a Mediterranean nightlife resort. *Tourism Management Perspectives., 8*, 49–59.
St John, G. (2004). *Rave culture and religion*. London: Routledge.
Thornton, S. (1995). *Club cultures: Music, media, and subcultural capital*. Cambridge: Polity Press.
Touraine, A. (1969). *La société post-industrielle*. Paris: Denoël.
Tutenges, S. (2012). Nightlife tourism: A mixed methods study of young tourists at an international nightlife resort. *Tourist Studies, 12*(2), 132–150.
Yonnet, P. (1999). *Temps libre et lien social*. Paris: Gallimard.

Chapter 8
Cuban Tourism—The Marginalized Communist Country's First Step Towards Sustainable Development

Armand Faganel and Anita Trnavčevič

8.1 Introduction

Caribbean tourism market is flourishing, with over 28 million tourists in 2015 and an estimated expenditure of 34 billion USD (Caribbean Tourism Organization 2016). McLeod (2015) states that the Caribbean transformation of "sugar-based economies to tourist destinations has been remarkable". The island of Cuba represented a historically important tourist location before the Cuban Revolution took place. Tourism industry in Cuba almost disappeared after the revolution. Soviet Union financially supported Cuban economy for decades because of its anti-American feelings, sending them cheap oil and buying sugar with higher prices as found on world markets. After the disintegration of Soviet Union in the 1980s Russian help ceased and Cuban economy precipitated. In the early 1990s the Cuban economy decreased by 35%, Cuban export decreased by 80% while import decreased by 75% (Pérez-López 2002, pp. 508–509). Because of difficult economic times the government found tourism industry as a fair possibility to help the Cuban's economy.

Mass tourism slowly began to increase and problems started to arise. Responsible tourism is being developed in such a manner that it maintains vital in a tourist area for an unlimited time, without changing its social and economic characteristics, respecting the environment and culture. Tosun (2001) analyzed the challenges to sustainable tourism development in developing countries and stressed that despite the obvious positive consequences, brought by tourism development; it is not an easy task to implement such principles, especially if we take into account

A. Faganel (✉) · A. Trnavčevič
Faculty of Management, University of Primorska, Cankarjeva 5, 6000 Koper, Slovenia
e-mail: armand.faganel@fm-kp.si

A. Trnavčevič
e-mail: anita.trnavcevic@fm-kp.si

S. Pelc and M. Koderman (eds.), *Nature, Tourism and Ethnicity as Drivers of (De)Marginalization*, Perspectives on Geographical Marginality 3,
DOI 10.1007/978-3-319-59002-8_8

the socio-economic and political conditions in the developing world. Political and economic choices and trade-offs that have to be taken are difficult.

Leimgruber (2010) stressed that the concept of marginality cannot be limited only to the economic aspects, there are also social and cultural consequences of economically based marginalization to be taken into consideration, for example. Conradson and Pawson (2009) agree that regional marginality is besides economic also "an imagined and discursively constructed condition". Geographical marginality can be studied as a situation observed and measured from outside, but Pelc (2010) is arguing about the 'sensation of marginality' as a concept which is present among the population of certain area. Until 2008 some of Cuban hotels and other facilities were opened only for foreign tourists—they were accused of tourism apartheid (Roig-Franzia 2008; Mazzei 2012).

Due to decades long economic and political embargo of the island of Cuba, the marginality is seen both from inside and outside the country. Mazzei (2012) identifies tourists on Cuba as the privileged group and Cuban population as a marginalized group. But even the population is split between those who have the access to the tourism industry and US dollars and those who don't. The later face a realistic danger of poverty; so from dual economy a social duel might arise. Due to the return of this material inequality first tendencies of Cuba's future marginalization can be traced. Also the once-balanced city-country relations of Cuba are giving way to a growing disparity. The migration to the capital has increased since 1990 and begins to disintegrate the previous integrity of the territorial state (Klopfer and Mertins 2001). Burchardt (2002) is warning that colored citizens are the next possible target of marginalization after the Castro era. Darlington (2009) confirms this issue:

> Until 1959, Cuba had beaches, clubs and barbershops reserved for whites. One of the first things Fidel Castro did after his socialist revolution triumphed was declaring racial discrimination illegal. The revolution did deal an institutional blow to racism, but also incorrectly declared a centuries-old problem solved with just a decree or a law …

So we have present a multifaceted example of a marginalized Cuba, Caribbean region, underdeveloped, geographically isolated, where the GDP per capita reached 6790 USD in 2013, while translated into purchasing power (PPP) it equaled 20,649 USD (The World Bank 2016). A steady growth can be seen since 1996, when the GDP per capita was 2284 USD and PPP 7346 USD. After the USA–Cuba detant the number of tourists grew, 3.5 million visited Cuba in 2015—capacities were pushed to limits (Trotta 2016), while in 1995 there were 742,000 (The World Bank 2015). And it is interesting that Americans can only enter Cuba on officially sanctioned trips »such as ones organized by educational groups, or via countries with connecting flights, including Canada, Mexico or Panama« while »travel for tourism remains officially off-limits due to the economic embargo« (Fieser and Gayol 2015).

Inhabitants are very much aware of their own prosperity level, as there are many emigrants that left Cuba—mostly to Florida, USA; they are sending money to their relatives regularly. And there are many tourists, coming to the country, as the popularity of country as a tourist destination increases. We shouldn't forget that incoming tourists bring changes with them for local people. And we should keep in mind the influence of modern communication and information channels, cruise shipping industry etc. After the recent death of revolutionary leader Fidel Castro the first cautious steps of opening the country will be strengthened. Cuban government, although against the free market economy, understands the growing demand for tourism services as an opportunity to improve the economic situation of the Treasury, citizens are seizing the chance to better their financial status, but these changes bring some possible threats, too. The main question is how to create opportunities upon all of these challenges. Newly elected USA president Trump will probably not hesitate to find a common language with Cuban government as he showed to be quite pragmatic—interests of construction and gambling industries are free of ideology.

8.2 Literature Review and Analysis

The establishment of NAFTA, EU and other free trade agreements influenced the export of goods (mostly agricultural) from Cuba, making the tourism sector even more important. The loss of financial assistance from Soviet Union in early 1990's changed the approach to tourism industry (Lindeman et al. 2003), which is supposed to finance Cuba's social programs, such as free education and accessible health care. Pattullo (2005) documents the move away from a socialist model of state owned tourism infrastructure in 1987, introducing the joint venture ownership—hotel rooms tripled in ten years, from 1990 to 2000, reaching the number of 33,000 international standards fitting rooms in 200 hotels. The existing 63,000 rooms are filled up quickly, so the opening to USA helped private sector offering rooms in family homes, too (Hamre 2016). Despite the difficulties that American tourists are still facing because of restrictions imposed by their government, they enter Cuba through neighbouring countries, like Canada and Mexico; Cancun i.e. accounts the constant growth of flight passengers from USA while tourist arrivals remain flat (Acevedo et al. 2016).

Hingtgen et al. (2015) suggest that Cuba should enable a more entrepreneurial climate for a stronger private sector, and entrepreneurs should show more interest to develop tourism products which would diminish the importance of mass tourism. Entrepreneurial changes and their relevance in a collectivist society have been already examined by Koh (2002), Kline and Milburn (2010) and Hingtgen et al. (2015). While tourism promoters are trying to satisfy the growing demand and to help the struggling Cuban economy, planned development and increased numbers of visitors, mostly in Havana and Varadero, put in danger the cultural, social and natural environment. Majority of Cuban hotels—71% are situated on the beach, 2% in nature parks and the rest are in cities (Thomas et al. 2015). Especially the coastal regions are threatened, as new hotels are being built on the beachfront, endangering the traditional social environment, as much as the flora and fauna diversity.

Sustainable development of the tourism sector is already a big issue, Caribbean countries pursue the vision of a viable and resilient industry that provides better quality of life to their citizens, which should participate in the planning of future tourism development, countries should cooperate regionally (CTO 2008).

Other authors, like Bailey (2008) argue that governments under the pressure of global tourism trends change the policies to form more welcoming climate for international investment. This is a real tragedy, as in the run for a competitive offer they might destroy the social structure and natural richness of places. Gonzalez et al. (2014) warn that headless tourism-based city planning process leads to »an unequal, polarized, segregated city, where severely impoverished territories exist alongside intensely gentrified areas. These cities are socially fractured and fragmented in an urban development sense«. These processes create disabled »conurbations« very different from classical city concept. And it is quite clear that there exists no such money that would allow renovation of the impoverished two million Havana city, as the real owners flew from Cuba and nobody tried to restore their houses for 60 years—nowadays the real estate are pretty much in ruins.

Authors like Amendola (2000) and Soja (2008) haven't discussed or tackled the inequalities or segregation which tourism industry creates but they accorded that huge concentration of hotels and other tourist infrastructure becomes alienated from "normal" cities. Weaver (1988) introduced the term of "tourist plantation" economies, replacing former colonial plantation economies. Monreal (2005) has developed this argument for Cuba. Before that Ash and Turner (1976) wrote about undeveloped coastal areas turning into "pleasure peripheries" of Western cities. Wilkinson (2008) expressed the fear that in the post Castro era ecology in Cuba might become undesired obstacle for tourism plans (Gonzalez et al. 2014).

Perez et al. (2013) classified Cuba among ten most noteworthy islands regarding biodiversity, thanks to complex geology and tropical climate, 22% of surface is representing six Ramsar wetlands and ten National Parks. Smith (2009) noted growth of interest in the demand of local cultural tourism: if managed sustainably this might help to revitalize the cultural traditions. And yet, tourist offer has been focused on the classical tourism, based on beach and sun. Gonzalez et al. (2014) proposed an indicator system capable to assess the sustainability of nature-based destinations on Cuba, measuring social indicators, too, i.e. local population perception of public services improvement because of tourism, than the improvement in transportation infrastructure because of tourism etc. We mustn't forget that tourism brings also some negative effects to the lifestyle at tourist destination. Contrary to these tourism business helps prevent young people leaving the country, brings an increase in the number of local employees and similar.

Anfuso et al. (2014) benchmarked Cuban coastal scenic assessment, categorizing and classifying beach sites; La Habana and Varadero were marked as attractive, while many others received low scores for natural parameters. Authors suggest that in the latter the improvement is possible only addressing human parameters, e.g. removing litter and chaotic protective structures. But in Italy Concu and Atzeni (2012) found significant differences among the tourists depending if the beach type is urban or rural.

In the Cuban case the development of tourism brought not only the economic recovery but also social divisions (Espino 2000). Spencer (2016) is talking about geopolitical inequalities that emerge from tourists' mobility versus the "supposed place-boundedness" of Cubans. Most tourists, despite their willingness to discover other countries and cultures, don't want to change their usual lifestyle. Azarya (2004) writes that such tourists are being followed by certain amenities and so these come in the contact with local people that are servicing them, while Judd (2003) investigated Cuban tourist enclaves and points out that they "change consumption, suppress local culture and replace it with Disney culture".

People from Europe, Canada and South America are coming to Cuba also to use their health care (Wagner 2006) and this trend might turn into a major potential problem if doctors and hospitals will focus more on the tourist and neglect local citizens (Garret 2010). Iordache and Ciochina (2014) assessed the growth of medical tourism internationally by one fifth per year, representing a yearly income of one hundred billion dollars.

Canally and Carmichael (2011) discuss the manipulation of:

> … the political economy of tourist destination image, which traces the influence of inter-governmental and extra-governmental power structures that manipulate the image of a potential tourist destination (Cuba), to manufacture a discourse that aligns with the ideologies of the political elites in the US.

There is a global trend to revolutionize today's life model; Liu (2016) i.e. gives us a heads up, when she argues that "top-line growth doesn't ensure bottom-line prosperity". And she proposes that leaders should grab for the chance of remaking the country's economic development, using a wider vision.

Nevertheless, not every tourist wants the same kind of leisure experience. De la Pena et al. (2016) study on hedonic theory demonstrates that tourists expect price differentiation based on "membership of international hotel chains, high quality offers, diversified rooms, and adaptability to specific needs of each client". McLeod et al. (2016) identified a gap between the development of Caribbean islands and success of tourist industry; they advise reassessment of Caribbean tourism and redirection for sustainable benefits.

8.3 Discussion

Tourism is only one of the initiatives being promoted from Cuban government to improve the economic condition of the country. Of course it is beyond the scope of this chapter to discuss the influence of all these sectors; we are concentrating on tourism industry only and its effects on marginalization. There are two ways to see the tourism contribution in underdeveloped regions:

(a) the fight with poverty through tourism will be successful and sustainable only if it is accompanied with the network of conscious tourist operators and consumers; or

(b) it is seen as an aid to a poor community, without giving them the empowerment and needed knowledge.

Financial investors, construction lobbies and big tourism industry will mostly build on mass tourism, which gives to the local community precarious working places with low salaries. Competition within the world tourism industry and huge profit appetites direct capital owners to promote the number of tourists, competing with low prices. While living costs will increase, local people have to cope through the month with modest wages; from this point it is not very far to the appearance of alcohol problems, prostitution, drugs, crime.

Tourists' expectations about how to spend their leisure time are moving towards alternative tourism more and more. Some tourists prefer to spend holidays engaged in direct interactions with local people, their heritage and nature, growing personally and educating themselves. Cuba has a lot to show in this regard, several locations are inscribed in the World Heritage List (UNESCO 2016a): Archaeological landscape of the first coffee plantations in the South-East of Cuba; historic centre of Camagüey; old Havana and its fortification system; San Pedro de la Roca castle, Santiago de Cuba; Trinidad and the Valley de Los Ingenios; urban historic centre of Cienfuegos; Viñales valley; and two natural parks: Alejandro de Humboldt National park, and Desembarco del Granma National Park. Besides these sites there is also a tentative list waiting for consideration: National Schools of Art, Cubanacán, Ciénaga de Zapata National Park, and the Reef System in the Cuban Caribbean. The First coffee plantation i.e. offers remaining physical evidence of pioneering form of agriculture on a demanding terrain and demonstrates the Caribbean history from a social, economic and technological standpoint. While Trinidad, a historic and homogeneous city from 16th century, with Andalusian, Moorish and European neoclassical buildings, grew on the basis of flourishing sugar cane industry from nearby Valley de los Ingenios, where a living museum has been arranged with sugar mills, barracks, plantation houses, and more.

And then there is the Cuba's intangible cultural heritage, listed at UNESCO (2016b): unique La Tumba Francesca (French drums); while Rumba in Cuba, a festive combination of music and dances and all the practices associated, and Punto guajiro or Punto Cubano, a song genre of improvised poetic music art from 17th century are at the moment on-going nominations. The Cuban authorities and regional bodies are organizing capacity-building workshops on safeguarding of the intangible cultural heritage, where stakeholders like bearers of intangible heritage, national committees and institutions receive intensive training, offering a platform where participants reflect collectively on experiences and challenges in safeguarding intangible cultural heritage within the wider context of sustainable development (UNESCO 2014). Baldacchino (2007) also warned that we won't be able to understand island people and society "if we treat them as microcosmos of larger continents"; so it would be easier to engage and acknowledge "singularities and idiosyncrasies of their environmental and social landscapes". And Brohman (1996) already knew that alternative forms of tourism in rural areas might contribute to reverse the "historic trends of inequality and dependency".

Table 9.1 Languages (principal or 'mother' tongue) in Switzerland, 2014

Language	% of residents
German	63.3
French	22.7
Italian	8.1
Romansh	0.5
Other languages	20.9

Source Swiss Statistical Office (the total is 115.5% because many people indicated two or more languages as main/mother language)

2. A small minority group is divided into two or three groups by one or more state boundaries (Kurds, Catalans, Basques etc.).
3. A small minority living inside a state territory is unique and has no other ethnic, linguistic or cultural relatives (e.g. the Romansh in Switzerland, the Sorbs in Germany, etc.).
4. Diaspora designates a group that has been fragmented and lives scattered around the World. The Jews have been frequently quoted, but many Diasporas have emerged over time as a consequence of international migration.

Minorities (and thus diversity) are the result of historical processes. The present-day situation need not be final and is likely to change. Also, migrations as well as cultural policies are often the root of minority problems (Bennett 2001, p. 28).

"Diversity is the characteristic of nature and the basis of ecological stability" (Shiva 1993, p. 65). A high biological diversity is a sign for a sound ecosystem, both globally and regionally. This includes a high variety of plants and animals coexisting in a dynamic equilibrium. In agriculture, crop diversity is an important strategy to maintain soil fertility, defend the crops against pests, and ensure a varied food supply (Rattanavong 2008, p. 271). This is the basis of organic farming, where "biodiversity is both instrument and aim. Natural ecological balance, below and above ground, is a key to its success" (Scialabba 2003). Most people would agree on this basic principle, even if most current conventional agricultural practices oppose diversity.

What is important for the natural world also applies to human social systems. We are all different from each other, and although this difference may pose practical problems, it is vital in the long-term. While the importance of biodiversity has been gradually recognized, many people have not yet understood the richness of cultural diversity and its significance.

Diversity is the expression of the many ways nature and human societies organize, express and adapt themselves to equally varying conditions of life. "Diverse ecosystems give rise to diverse life forms, and to diverse cultures.… Cultural diversity and biological diversity go hand in hand" (Shiva 1993, p. 65). Proof is the fact that societies across the globe have developed different strategies to cope with similar conditions of life. Southeast Asia is a good example for diversity (Kumar and Siddique 2008).

The diversity in human systems differs from ecosystem diversity because value judgements play a central role in our thinking. Diversity will be seen as good and beneficial by those who are open-minded, but as bad and detrimental by people with a narrow outlook. Attitudes towards diversity in all fields (including nature) are therefore related to people's worldviews. Agricultural monocultures are an expression of short-term profit-oriented quantitative and linear thinking, whereas organic farming represents the very opposite (long-term orientation, quality and organic or systemic thinking). Linear thinking and uniformity are detrimental to both ecosystem and society, uniformity is the enemy of progress and an obstacle to development. The great inventors and thinkers were always marginal people, because their ideas deviated from the mainstream, but they took mankind forward.

This takes us to a fundamental principle to be applied with minorities: renouncing value-judgements along to the 'black-white' or 'good-bad' dividing line.[2] We are all human beings with strengths and weaknesses, and everyone possesses his and her particular gifts—on this level we are all equal. Differences occur on a lower level, and they are usually highlighted when it comes to deal with groups outside the mainstream. Politics often promotes a homogenous society, ostensibly in order to simplify life and administration, but homogeneity is an instrument of domination. Language is a manifestation of power (Raffestin 1980, p. 91) and occupies a particular place in politics (ibid. p. 97). Linguistic diversity, therefore, prevents the political power from totally and easily controlling its population. Minorities are obviously an obstacle to domination.

From an ecological perspective, homogeneity is negative. Monoculture farming leads to rapid environmental degradation: excessive synthetic fertilizer application results in soil depletion, and an increase in input will result in diminishing returns (Aubert 1996). It simplifies the technical side of agriculture, but ultimately annihilates its efforts. Heterogeneity is achieved through crop diversity (intercropping) and rotation, promoting diversity and maintaining soil fertility in the long run.

Our life is marked by short-term economic thinking, but little attention has been paid to the benefits diversity can offer the economy in the long run. In a seminal paper, the biologists Peter J. Edwards and Cyrus Abivardi emphasize the economic potential of biodiversity: "over-exploitation of resources, loss of genetic diversity and damage to ecological processes and life support systems have greatly reduced the planet's capacity to support people in both developed and developing countries" (Edwards and Abivardi 1998, p. 244). Maintaining biological diversity is in our interest, mainly because "we do not know how important species diversity is for the long-term stability of ecosystems; …" (ibid.).

The arguments are different in the human world, but the two diversities are related. Diversity in society is a stimulant to the development of social life. The variety of values, ideas and opinions drives human thinking, promotes innovations, teaches us to reflect, and helps to improve our quality of life. The former director of the United Nations Environmental Programme, Klaus Toepfer, summed up the

[2]Epitomized by the sheep on Animal Farm, bleating "Four legs good, two legs bad" (Orwell 1945).

analogy to biodiversity: "Respect for biological diversity implies respect for human diversity" (UNEP 1999, p. xi). In this volume on cultural and spiritual values of biodiversity, scientific analyses and source texts testify to the key role diversity plays in people's lives on all continents. Difference is regularly described as an asset, not as a drawback. Indigenous peoples, in particular, hold a lot of knowledge that would be useful for the future. "Indigenous peoples represent 95% of human cultural diversity, and, I daresay, 95% of humanity's breadth of knowledge for living sustainably on this Earth" (Joji Carino before the UN; General Assembly 1997, p. 32). In other words, there are economic benefits to be gained from cultural diversity. Biological and cultural diversity are comparable because of existing and potential benefits and because they help to integrate the human society into the ecosystem; they are in fact complementary (Table 9.2). Both include use values and non-use values, the former being of immediate or future economic importance, the latter acting on the human mind and soul. Both are equally important for the future of mankind.

Our western society has largely forgotten the contribution diversity in both domains has furnished to humanity in the past, is furnishing at present, and is able to furnish in the future. By emphasizing external (secular) values over spiritual ones, we have destroyed not only the environment but also human solidarity. Most traditional societies lived in harmony with nature, and there exists a "spiritual attachment to land and to the biodiversity that is found within that land or near the river" (Morea Veratua from Papua New Guinea, UNEP 1999, p. 154). "Traditional or indigenous knowledge systems of nature, environment, and the various ecosystems within it, represent the most valuable survival kit in rural areas" (ibid.). Spiritual (sacred) values are needed to preserve biodiversity, such as holy hills (the Dai in the Xishuangbanna region of Yunan; ibid., p. 381 f.) or sacred groves (Western Ghats; ibid., p. 382 ff.). This is also the message conveyed years ago by Schumacher (1999, in particular p. 24).

For centuries Europeans saw themselves as superior to nature and other cultures (Said 1978). They exported their values to other continents and influenced the transformation of traditional practices anchored in local knowledge about diversity. Modernization became the key term to improve the living conditions of supposedly primitive (non-European) societies. The consequences are dramatic, both for ecosystem and society, as an example by Jiang (2008) demonstrates. Her case study of a Khmu village that has moved from the traditional slash-and-burn agriculture to modern rubber plantations illustrates major recent transformations in the Mekong region. Rubber trees replaced the rainforest, which "preserves three times as much water as a rubber plantation" (Jiang 2008, p. 250). Also, "[r]ain washes away six times as much topsoil in rubber plantations than in tropical rainforests" (ibid., p. 248). As a result, the regional climate has become drier and hotter (ibid., p. 250). The Khmu used to live in close relationship with their environment. Their simple techniques (slash-and-burn agriculture, limited burning, using sowing sticks, and leaving tree stumps for new buds to grow) had a modest impact on the ecosystem. They lived on a variety of foodstuffs and survived if one crop would fail. Diversity ensured their survival. Large rubber plantations have replaced almost all

Table 9.2 Comparing biological and cultural diversity (Leimgruber 2014, p. 72, based on Edwards and Abivardi 1998 and Leimgruber 2004a)

	Use value			Non-use value	
	Direct use value	Ecological/cultural function value	Option value	Existence value	Bequest value
Biological diversity	Food (animal, vegetal) Raw materials (timber, fibres) Tourism	Flood control Photosynthesis Nutrient cycles Waste assimilation	Future medicines Genes for plant breeding New technology Resource substitute	Satisfaction that resources exist Psychological balance Re-creation	Altruism Between-generation equity Respect
Cultural diversity	Learning Alternatives Problem solving strategies	Exchange	Innovation Alternatives Experience	Satisfaction that variety exists Psychological balance Re-creation Intellectual challenge	Altruism Between-generation equity Experience

agricultural surfaces and created a high dependency on the world market for the sale of rubber and on external food sources. While their life was in a way secure because of nutritional diversity, they now depend on a monoculture and the whims of the world rubber market and its price fluctuations. Swidden farming has been condemned as detrimental to the environment, but a considerate practice with sufficiently long planting cycles allows the vegetation to recover (Miescher 1996, pp. 82–84).

Along with numerous authors (Anderson 1996; Attfield 1999; Carson 1962; Edwards and Abivardi 1998; Shiva 1993; Van Dieren 1995 and others), I insist that diversity is fundamental to the future of this planet. Species extinction robs us of a genetic resource pool, the extinction of languages reduces the range of ways of thinking. Thinking in diversity is more complex than linear thinking, but the only way of tackling a complex reality, because it stimulates mental flexibility. Shiva (1993) rightly criticizes western thinking as monocultures of the mind, as reductionist and unsustainable—similar to agricultural monocultures. Uniformity kills, both in agriculture and human society. "Yet what is 'unproductive' and 'waste' in the commercial context of the Green Revolution is now emerging as productive in the ecological context and as the only route to sustainable agriculture" (ibid. p. 58).

Diversity is an indicator of difference, and this is what geography is about, "the uneven distribution of phenomena over the earth surface" (Forsberg 1998, p. 19). All processes take place under different cultural, social, economic, and political circumstances; there is no uniformity. A 'geography of difference' (Miller 1998, p. 262) builds on the variety inherent in human thinking and acting. We must see difference as "a positive quality" (Leimgruber 2004c, p. 278). The political corollary is to recognize that unevenness (difference) must not lead to discrimination. The goal for development politics should be "to eliminate undesired (economic and social) disparities by maintaining the desirable (cultural) diversity" (ibid.).

9.3 The Negative Perception of Minorities

The minority 'issue' is only superficially a numerical question; it has much more to do with psychology, with perception, with "understanding ourselves" (Tuan 1974, p. 1). Many people have a problem with minorities, considering them a threat instead of an enrichment. Minorities may be a challenge to the social system, but they only become a threat if we consider them as such. Many groups suffer from such negative attitudes: foreigners, strangers, refugees, asylum-seekers, drug addicts, the poor, organic farmers,[3] members of religious sects, youngsters, the

[3]They are now recognized as a most valuable group in modern society since they work for, not against the ecosystem. The organic movement is almost 100 years old (Leimgruber et al. 1997), but its true merits have only been recognized since the 1980s.

elderly, the gay community, etc. Such negative perceptions of minority groups oscillate between indifference and violence.

Overemphasis of the 'We' against the 'Other' can result in skewed perceptions and lead to conflicts, as is the case in Myanmar. "The Burman-dominated authoritarian state saw diversity as a threat, and gave little priority to developing the ethnic borderlands; successive regimes also restricted ethnic political, cultural and social expression. These grievances run particularly deep in Rakhine State, where there is a strongly-held sense of separate identity, in part because it was historically never integrated into the Myanmar state." (International Crisis Group 2014, p. 7). Such a condescending attitude is the basis of discrimination, which fosters violence as the only answer.

Stereotypes usually arise out of individual cases but are not justified as generalizations. Judgements are often made following single incidents, and entire groups will eventually suffer. Knowledge and experience are the keys to avoid this:

> Of course, the fact that all actors move in situated contexts within larger totalities limits the knowledge they have of other contexts which they do not directly experience. All social actors know a great deal more than they ever directly live through, as a result of sedimentation of experience in language. But agents whose lives are spent in one type of milieu may be more or less ignorant of what goes on in others (Giddens 1984, p. 91; original emphasis).

To know a person or a country is likely to increase a constructive attitude and promote the understanding of people, countries or groups.

Prejudices result from a number of elements that influence the way minorities are perceived, treated and often discriminated: origin, religion, language, personal names.

The origin of immigrants is significant because they arrive from a different cultural background and socio-cultural system, which marks them as outsiders. This was the case with the Italian workers in Switzerland in the 1950s and 1960s, although they spoke one of our national languages. Their presence was felt as a burden on our society, and the political reaction became virulent. Years after, when other linguistic minorities arrived (from Spain, Portugal, Ex-Yugoslavia), the Italians vanished from the xenophobic discussion. We realize nowadays that they, as well as other immigrants, have considerably enriched our culture (e.g. in the field of food and eating habits).

Although religion has lost a lot of its significance in our society it is still an important cultural element. Switzerland is a Christian country with Protestants and Catholics as major groups, a difference still felt today (e.g. with public holidays). With the Tibetan refugees (1960s) we obtained a Buddhist minority, which integrated well with our society and managed to maintain its traditions. Initially, Muslims did not create problems, but the rise of Islamist terrorism has produced generalized fears of Islam and a widespread islamophobia.

Language is an important vehicle of communication and allows to identify a speaker's origin. In German-speaking Switzerland, local and regional dialects dominate conversations, and non-German speakers manifest considerable difficulties when it comes to learn German. An additional problem for them is that we do not speak the written German except under specific circumstances. Communications between employer and worker, between doctor and patient etc. are better conducted in dialect. This problem is unknown in the French and Italian speaking parts of the country.

Personal names are related to language. A strange name or even suffix (such as the Slav endings of –ic) may suffice in Switzerland to stigmatize a person, particularly in the German-speaking part. And Switzerland is by no means unique. "A Thai surname carries with it preferential treatment regarding various opportunities in Thai society, such as getting into the army or police or being promoted to higher positions in government. Many ethnic minorities therefore change their names in order to avoid discrimination" (McCaskill et al. 2008a, p. 17; McCaskill 2008, p. 312). The personal name is part of one's identity; by adopting a new surname, an individual may improve his/her chances in the mainstream society but risk discrimination and marginalization within his/her own group. But a change of name is not possible everywhere (in Switzerland, women can adopt their husband's, and men their wife's name).

Minorities are often seen as an obstacle or unwelcome element in a nation, they are "problematic" (Jackson-Preece 2004, p. 7) and viewed as not being helpful in state building. The origin of this idea lies probably with John Stuart Mill who argued that the existence of different languages within one state prevents the working of a representative government (ibid., p. 11). However, many states function well despite (or because) of minorities. Nationalism and chauvinism are often the source of tensions between dominant and minority groups, as in the Carpathian Basin (Romania) where Hungarian minorities were considered as "the main supporters of Hungarian irredentism and revanchism" (Kocsis and Kocsis-Hodosi 1998, p. 13).

9.4 Minorities as an Asset

The negative attitude towards minorities overlooks the role they play for social diversity. As pointed out above, minorities are beneficial because they remind us of the differences that exist within our societies and the diverse nature of mankind. However, the current globality paradigm is opposed to diversity and discards minorities as quantité négligeable. Our western civilization is characterized by a 'chess-board mentality' where everything has its orderly place in a 'mental square', there is no space for flexibility. This will eventually result in social and spatial marginalization.

Why are minorities an asset? The answer can be found in Table 9.2. The diversity due to minorities not only offers direct use values but also long-term option values. Every group has its own ways of thinking and acting, of responding to social and natural challenges, of securing its livelihood.

Minority studies focus on ethnic, linguistic and religious groups, leaving aside other possible categories. But even so this emphasis shows various difficulties in daily coexistence that can be traced back to our perception of people and should be corrected and overcome.

Among ethnic minorities, particularly indigenous peoples are still considered to be backward and not modern (McElwee 2008, p. 63). However, modern is a very relative term and refers mainly to technological achievements. From an ecosystems perspective, backwardness is a positive value. Ethnic minorities are often traditional societies that cherish sacred values, biodiversity, and the ecosystem (UNEP 1999). They take care of the environment in ways the western world has largely forgotten. By declaring mountains, forests, and watercourses as sacred and inhabited by nymphs, dwarfs, and other invisible creatures, they protected them from destruction. The taboo (Anderson 1996, p. 11) used to be a powerful instrument to this effect. We tend to discard the spiritual side as superstition because it does not fit with our secular worldview, and by discarding this important aspect, we ruin the ecosystem (and at the same time, with similar arguments, also our societies).

Linguistic minorities stand for a diversity of ways of thinking and problem solving, the opposite of linguistic uniformity, which "suppresses ways of thinking and expression" (Leimgruber 2004b, p. 189). To promote uniformity in languages as a political programme is an assault on this diversity. To be able to express one's ideas in one's own language is a human right: nobody must be discriminated because of language: "The non-discrimination principle is a fundamental rule of international law" (Ayton-Shenker 1995). On the international stage politicians prefer to express their views and ideas in their own language (in the UN or during press conferences). Article 27 of the International Covenant on Civil and Political Rights of 1966 (entry into force in 1976) is explicit:

> In those States in which ethnic, religious or linguistic minorities exist, persons belonging to such minorities shall not be denied the right, in community with the other members of their group, to enjoy their own culture, to profess and practise their own religion, or to use their own language.

The European Court of Human Rights follows the UN practice (Council of Europe 2011, p. 13):

> 20. According to the UN Committee on Economic, Social and Cultural Rights, the right of everyone to take part in cultural life, enshrined in Article 15 § 1 (a) of the International Covenant on Economic, Social and Cultural Rights, includes the right to express oneself in the language of one's choice. This may be particularly important for persons belonging to minorities, who have the right to preserve, promote and develop their own culture, including their language.

This holds good even if the individual countries can develop their own regulations, in particular in "communications with public authorities" or "to receive information in a language of one's choice" (ibid.).

Switzerland has solved this problem by declaring German, French and Italian as national and official languages, and Romansh as a national language. All federal legal texts must be in the three official languages and are equally binding. The

Chapter 10
Resettlement of Orang Asli (Aborigines) in Malaysia—Marginalization or Demarginalization of an Ethnic Group

Jamalunlaili Abdullah

10.1 Introduction

The Orang Asli, literally means "original people," are the indigenous minority of Peninsular Malaysia. However, they are a very small minority of the population, accounting for 149,723 (0.6%) of the total Malaysian population of 30 million people (JKOA 2006).

The Orang Asli tend to live in the interior of the peninsular, shying away from the rest of the population. They could be found living within or adjacent to Malaysian rainforests due to their shy nature and great dependence on agriculture and forest produce.

Due to their remote locations, the Orang Asli lag behind other ethnic groups in Peninsular Malaysia in terms of physical and socio-economic development. While Malaysia is considered as an upper middle income nation, most of Orang Asli are classified as extremely poor. Their settlements tend to consist of basic wooden houses which are of substandard quality compared to average Malaysian houses. In addition, due to their very remote locations, they lack access to quality healthcare, education and basic amenities which most Malaysians take for granted.

Due to this situation, the Malaysian government has embarked on various programs to improve the livelihood of the Orang Asli and perhaps to try to assimilate them into the mainstream Malaysian society. The two main programs are the resettlement of Orang Asli in new location (ex-situ development) and the redevelopment or improvement of existing settlements (in-situ development). The main objectives of these programs are to improve their settlement physical condition through provision of better houses and infrastructure, provide better access to

J. Abdullah (✉)
Faculty of Architecture, Planning and Surveying (FAPS),
Department of Town and Regional Planning, Universiti Teknologi MARA Malaysia,
Shah Alam, Malaysia
e-mail: jamal858@salam.uitm.edu.my

S. Pelc and M. Koderman (eds.), *Nature, Tourism and Ethnicity as Drivers of (De)Marginalization*, Perspectives on Geographical Marginality 3,
DOI 10.1007/978-3-319-59002-8_10

government services such as health and education facilities and to improve their socio-economic conditions.

The ultimate aim of the resettlement and redevelopment programs is to demarginalize the most marginalized segment of the Malaysian society. It is hoped that by relocating the community to areas closer to main population centre (ex-situ development) or by improving their existing settlement (in-situ development), the marginalization of Orang Asli, geographically and socio-economically, can be reduced significantly.

While the intention is noble, not all programs are able to achieve the objective of reduced marginalization. While some Orang Asli have become less marginalized due to the programs, some others may perhaps become more marginalized due to their inability to cope with different socio-economic environment in a different setting. This chapter presents the evaluation of the process of demarginalization and marginalization of Orang Asli in Peninsular Malaysia due to the resettlement and redevelopment scheme implemented by the government.

10.2 The Orang Asli in Malaysia

The Orang Asli are considered the earliest inhabitants of peninsular Malaysia, who are believed to have settled as early as 11,000 BC (Bellwood 2011). This very small minority consists of three main groups of which are Negrito, Senoi and Proto Malay which are divided into 18 sub-ethnic groups. Different ethnic, language, mode of livelihood and social organization pattern differentiate the Orang Asli. Orang Laut, Orang Seletar, and Mah Meri, who live close to the coast, are mainly fishermen while some Temuan, Jakun and Semai sub-ethnic groups have taken to permanent agriculture. The other tribes engage themselves in hill rice cultivation, and hunting and gathering of forest products to trade with local communities. A very small number of Negrito are still semi-nomadic where their livelihood is depending on seasonal forest. Although most of Orang Asli groups are forest dwellers or live in rural areas, a fair number live in or near urban areas where they are involved in waged and salaried jobs (Haron et al. 2009).

Living mostly in rural areas, within and adjacent to the tropical rainforests, their dominant economic activities are small scale agriculture and hunting and gatherings. Due to this geographic and socio-economic situation, Orang Asli is considered as the most marginalized group in Peninsular Malaysia. Orang Asli tend to practice communal land activities whereas Malaysia practices Torrent Land Law whereby private land ownership is recognized through land title issued by the land office which the Orang Asli do not possess with. Despite roaming the forests or working on the lands for many generations, the Orang Asli may not have the legal rights on the land which provide hardship to them should the forests are logged or the land taken by the state or individuals for economic development.

From the four types of processual marginality cited by Leimgruber (2004), namely the contingent, systemic, collateral, and leveraged, the Orang Asli in

Malaysia are considered as a result of the first two processual marginality. Systemic marginality which is produced by hegemonic forces within the Malaysian political and capitalist economic system generates inequities through the distribution of social, political and economic benefits. Being very small in number which translates into insignificant political force, and living in remote areas, the Orang Asli are unable to reap economic benefits of Malaysia's rapid economic growth and activities. Despite working on the lands for generation, without own individual titles, the Orang Asli face difficulties in staking ownership claims on the land. In addition, the Orang Asli are considered the least educated ethnic group and with practically no capitals make them a result of contingent marginality. Their uncompetitive characters in Malaysian capitalist economy make them further marginalized in the society.

Due to their special status as the earliest inhabitants and their marginalized setting, the Aboriginal Peoples Act 1954 or Act 134 (Government of Malaysia 2010) was passed to provide special protection to the Orang Asli. The Act has provided the provision under S6 and S7 for any state authority to gazette any area on the state land to be declared as Aboriginal Areas or Aboriginal Reserves for the occupancy of aboriginal peoples. In addition, under S8 the state authority may also grant rights of occupancy of any state land for the aboriginal peoples, but only as a tenant at will. Settlement areas for aboriginal people in Malaysia cover about 50,563 ha of land. From that area, about 19,701 ha are gazetted under the Aboriginal Peoples Act 1954 (Act 134) and another 30,489 ha were already approved by the state governments as settlement areas but yet to be gazetted under the Act. On the aspect of ownership, only 0.03% of the lands were owned individually by the aboriginal people (Idrus 2011). Most of the lands are occupied without proper registration by this community, which in the case of any land dealings or acquisitions, the rights of these aboriginal peoples are being denied.

The Orang Asli are known to be very attached to their cultures and lands. They live in their own community and have their own administration system led by a hereditary headman. Their use of land may differ from one tribe to another. There are at least 20 types of land usage related to their ways of life which include settlement area, sacred place, hunting area, area for staple planting, plant area for ritual/traditional events, and inherited orchards (Aliran 2010).

The land usage might be different according to their living area or way of life, but the Orang Asli have traditionally followed the system for a very long time. Although the Orang Asli traditionally live a nomadic lifestyle, some have established their living area and stay within it for generations. After the adoption of the Aboriginal People Act 1954 (Act 134), many settlement areas for Orang Asli have been gazetted which become proper and permanent Orang Asli villages. However, there are many other settlements which are not gazetted and may be taken over for other developments or abandoned due to the governments' programs of Orang Asli resettlement.

10.3 Resettlements and Redevelopment of Orang Asli Communities

Resettlement and redevelopment of orang Asli community have been conducted since the British colonial period due to the communist insurgency threats in the jungle. The early stage involved relocating the community away from areas under the Communist influence. In addition, the government felt that Orang Asli should not be marginalized from the rapid economic development experienced by the country. However, in line with the establishment of Department of Orang Asli (JHEOA) in 1954 under a new law called Aboriginal Peoples Ordinance, the focus of development has shifted to protect Orang Asli and their way of life from rapid economic development, as well as providing them education and health facilities and suitable development, especially in settled agriculture programs (Redzuan 2008).

The policy continued after Malaysia gained her independence in 1957. Various transformation programs were implemented by the government in order to improve the socio-economic status of Orang Asli and to ensure they will not continue to be further marginalized. These consist of the provision of basic amenities, education, and skills training in cooperation with state governments. Three major development programs were established; structured settlement programs, economic development, and social development (Redzuan 2008).

Structured settlement programs aim to provide better accessibility for the Orang Asli and to improve their socio-economic conditions. The programs can be ex-situ, involving relocating the Orang Asli into a totally new settlement, regardless the distance; or in-situ, improvement of their existing settlements. Specific programs are Regrouping Plan (RPS), Village Restructuring Project (PSK), New Village Plan (RKB), Natural Disasters Project and Orang Asli Settlement Customary Land Survey and Land Acquisition. These programs provide new or better housing and specific physical amenities such as piped water supply, electricity, roads, and education and health facilities to the Orang Asli communities.

In addition to structured settlement programs, various socio-economic development programs such as entrepreneur mentoring program, retail space development program, income generation program, and Social Welfare Development Scheme (SPKR) have been implemented by the government to enhance the socio-economic status of the Orang Asli. Facilities and programs to improve their health and education are also part of the whole government efforts.

10.4 Impacts of Resettlement and Redevelopment on Orang Asli

Many resettlement and redevelopment projects for Orang Asli have been implemented by the government throughout the country over the years. The aim is the same; to provide better accessibility to the community, to improve their access to health and education facilities and services, and to improve their socio-economic conditions (Idrus 2011). It is a strategy to integrate them into the mainstream economic development and the wider national community (Abdullah et al. 2016). While the objectives are noble and some resettled and redeveloped communities have achieved these benefits, certain communities may not achieve the desired effects. This is perhaps due to unfavourable locations of the new settlements, especially with regards to the traditional source of living from agriculture or jungle produce, to the social and psychological impacts of adapting to a new and modern environment.

The resettlements of Orang Asli have brought mixed results, judging by various studies evaluating their impacts in Malaysia. Lim (1997) argues that regrouping exercise of Jakun families in Keratung and Bukit Seruk 3 in the state of Pahang was a success due to higher income from permanent agriculture and wage earning. Improved housing and physical facilities, as well as the better confidence of the community were also evident. Other studies have also found that many Orang Asli communities were satisfied with their new settlement.

On the other hand, Ab Hadi et al. (2013) argue that the resettlement of Orang Asli has a negative impact on the environment, negating the aim of conservation initially envisaged. Forum Asia contends that it leads to increased poverty and deterioration of health among some villagers. Karim and Hashim (2012) in their study of resettled Orang Asli in urban Bukit Lanjan, find that many Orang Asli there had difficulties in adapting to new life in metropolitan settings.

The negative impacts of resettlements of Orang Asli cited in some literature in Malaysia follow the trends in some other countries around the world. In Australia, some of the impacts of resettlement include poverty, low self-esteem, a poor sense of emotional dependency, homelessness, high levels of family violence, suicide and substance abuse (Memmott et al. 2000). SPERI (2012) argues that resettlement of the highlander H'momg tribe in Luang Prbang, Laos, has led to land conflict and land dispute with Khmu community which was already there, as well loss of bio-cultural diversity and traditional knowledge. This chapter summarizes the findings of three case studies of Orang Asli communities' resettlement and redevelopment of which the author and his colleagues and students were directly involved in evaluating the impacts. For more details analyses of these cases, readers can refer to Abdullah et al. (2015, 2016, 2017).

The first case is on resettlements of 183 Orang Asli families who were resettled through Remote Area Integrated Program (Program Bersepadu Daerah Terpencil—PROSDET), Pantos in Pahang state. The 183 families were resettled from six villages upstream of Telom river, namely Kuala Suar, Lanai Baru, Perangkap, Tisut, Nyentil and Harong into a new settlement downstream. Each family was given a house together with six acre of palm oil plantation for their livelihood. The main reason for resettlement of the six villages were to provide better education and health services to the six river village communities in an easily accessible settlement, PROSDET Pantos.

The second case study is the resettlement of the Temuan tribe whose 247 ha of land which they traditionally used for agriculture and forest roaming activities were taken over in the mid 1990s by a developer for high density urban development. It is located only 17 km away from Kuala Lumpur City Centre. To ensure the Orang Asli will be able to continue their livelihood in the area, the developer and the Selangor state government allocated 18 ha of land for their new settlements within the area. It consists of 147 units of bungalows, given to each household, with community facilities such as multipurpose hall, shop houses, small mosques, school and Orang Asli mini museum. Besides, 130 units of apartment were given to single children above 16 years old. Some lands were provided to the Orang Asli for farming while the developer provided for trust funds and scholarships to the community.

The third case study is a redevelopment of Orang Asli village of Ulu Kuang, Selangor. It is considered a suburban location located about 30 min' drive north of Kuala Lumpur. The settlement had been gazetted as Orang Asli Reserve by the state government, covering 140.4 ha (347 acres). It is occupied by 470 people, consisting of 120 families (Department of Orang Asli Development—JKOA 2013).

The aim of all three case study resettlement or redevelopment projects is basically the same, to demarginalize the Orang Asli communities and to integrate them into the mainstream economic activities. Results of the author team's observations and surveys of almost all households in the communities reveal interesting findings on the impacts of these programs on the marginalization of the communities.

It was found that the third case study, Ulu Kuang, was the most successful program and had the most positive impacts on the Orang Asli. This was perhaps due to the fact that the nature of the program was redevelopment of existing settlement whereby the houses were rebuilt and rearranged in a more systematic manner on the same site. In addition, almost 400 acres of land in the settlement had been gazetted as Orang Asli reserve, which provides readily available land for agriculture activities for the community. In addition, being long-time residents of the community which is close to major employment centers have prepared these Orang Asli for employment in various industries near the community. Compared to some Orang Asli communities, the percentage who was still schooling or had completed at least primary school was much higher at 78%.

In terms of satisfaction with the housing facilities provided, for the three variables questioned, namely housing arrangement, house space and interior comfort, the mean on the Likert scale had changed from very unsatisfied to satisfied or very satisfied. The same findings were valid for all variables under the categories of commercial and services, infrastructure and utilities and public facilities. Most importantly, in terms of the three variables for socio-economic indicators, namely employment, income and property ownership, the mean responses had changed from very unsatisfied before the redevelopment to very satisfied after the development. The only improvements that the Orang Asli hoped to be implemented were reliable water supply and solid waste disposal system.

It was found that in-situ development which provides basic physical and infrastructure facilities have improved the livelihood of these Orang Asli community. This is rather different compared to some resettlements programs of Orang Asli in other areas. Perhaps this is due to the fact that these Orang Asli were not moved to new locations, thus providing continuity in their life. The fact that the land has been gazetted as Orang Asli reserve has played a big role in their success. In addition, being provided with basic education to take up employments in the urban Klang Valley has helped their socio-economic conditions as well.

On the other hand, the case study of the Temuan tribe in Bukit Lanjan was rather unique. They were given rather generous compensations, despite losing a big chunk of forested areas to new urban development. Although they had worked on the lands for generations, they did not have titles to the land. Compared to Ulu Kuang, Bukit Lanjan is located right in the heart of a very urbanized area of Kuala Lumpur metropolitan area. The bungalows and apartments given to them are worth millions now. The trust funds and scholarships given to the community also were supposed to provide them with better prospects of living in urban areas.

Although the tribe had moved a few times earlier which had caused some changes to their lives, the relocation to Desa Temuan of Bukit Lanjan was most profound in that they had to move from traditional settlement into a modern, planned housing estate surrounded by high rise modern development (Karim and Hashim 2012).

A survey and observation of the Desa Temuan twenty years after their resettlement have found a rather high number of vacant houses and houses rented to non-Orang Asli. Meanwhile most of the 130 apartment units were rented out to non-Orang Asli, the children given the apartments moving back to their parents' houses. It was also found that the museum no longer functions as an information centre. While the neighbourhood still looks decent, it was evident that non-Orang Asli have moved into the village development built exclusively for this tribe.

The high number of vacant houses was rather surprising considering that landed properties are very much sought after in the area. The Orang Asli are not allowed to sell the houses given to them for free. It was further revealed that occupants of the vacant houses felt uncomfortable living in the modern environment and have moved out to live with other Orang Asli tribes elsewhere in the Klang Valley.

A survey of 45 households who agreed to participate found that the average household size was 12.6, a much higher number than the Malaysian average of 4.5. This is because these Orang Asli live as extended family whereby married children with kids tend to live with their parents due to economic necessities. These children would rent out their apartments to earn extra income.

Forty percent of the respondents worked as labourers such as gardener, construction workers, and guards followed by self–employed (27%) and unemployed (17%). As in other Orang Asli communities generally, the Orang Asli do not hold high paying or professional jobs. The high unemployment rate, much higher than the Malaysian average of less than 4%, was rather worrying. In terms of income, 77% of the respondents earned less than RM1000 per month (RM4.4 = USD$1), the minimum wage in Malaysia. Thus, despite living in nice bungalows, the Orang Asli in Bukit Lanjan were considered relatively very poor, living in the affluent society of Damansara Perdana.

In education, the biggest percentage (37%) held a Malaysian Education certificate (SPM—equivalent to eleven years of schooling) with only 4.6% without any formal education. Those with tertiary education constituted 5.1% of the population, below than the Malaysian average of 15%. However, this education level of Orang Asli here was much better than other Orang Asli settlements, perhaps due to its location in urban areas where education facilities were more accessible staffed with very qualified teachers.

The results of the survey showed that only one third of the respondents felt comfortable in their new settlement; a majority would rather live in their original place which was a forest. This was not surprising since these respondents, who were head of households are used to their traditional lifestyle and had difficulty coping in a new and modern environment. In terms of standard of living, the highest percentage (38%) felt that theirs have gotten worse while 33% felt that theirs have improved. The former understood that their lack of education and low wage jobs have made it difficult to support their big families in high cost urban areas.

Majority of respondents complained that their biggest problems were those related to socio-economy, namely limited job opportunities and low education level. Social problems such as drinking, gambling and illegal racing were rather serious, while almost half felt that the physical setting surrounded by urban areas and away from forests made life unbearable. Although 40% of respondents indicated they prefer to move out, they have to stay put in the new settlement since they have no other place to call home.

MK Land, the developer, had provided some assistance including free apartments, land for farming, education scholarship and Malaysian Government trust funds. However, the Orang Asli had not been able to capitalize on these assistance perhaps due to their low education level, high cost of living, and inability to adjust to new urban setting.

The Temuan tribe of Bukit Lanjan is rather ironic, being much better off than their cousins in the countryside, yet being very marginalized in their own neighbourhoods, in which they have lived for generations. Perhaps, this is parallel to the case of marginalization of Italian speaking Swiss that Leimgruber (2004) had done extensive study of.

PROSDET Pantos, in a very rural area in the state of Pahang, provides a geographical setting opposite that of urban Bukit Lanjan. The 183 families from 6 villages upstream of the Telom River were resettled on cleared lands downstream. The main reason for resettlement exercise was to provide better health and education services to the Orang Asli since the old villages along the river could only be accessed by boats which would take almost 2 hrs. During the monsoon season, these villagers were virtually unreachable due to high water table and flooding.

In the old villages, the Orang Asli used to live in their individual wooden houses within their roaming areas where farming and hunting were their daily activities and main source of income. Some have cultivated their land with rubber and tropical fruits. At PROSDET Pantos, which was accessible by tarred road, these Orang Asli are resettled in brick detached houses provided free by the government. In addition, each household was given 2.4 ha (six acres) of land planted with oil palms to provide them with steady source of income.

Survey and focus groups discussion among the Orang Asli indicate in terms of satisfaction in the new settlement, the number is split in the middle. While 50% are unsatisfied with life in the new settlement, another 45% are satisfied. The main reasons given for dissatisfaction are lack of suitable land to garden/farm, the new houses are not suitable for their extended family, loss of income from cash crops and inadequate basic amenities.

It is rather interesting that some of the Orang Asli consider the new settlement as a transit home as their children attend primary school in the area while others consider it as their permanent homes. The first group are those from traditional villages who can still go back to their homes and orchards due to the short distance, while the latter's traditional home are much farther away which require high cost for river boats transportation.

In terms of aspirations, the majority wish for a secure and better standard of living and for their children to pursue higher education as well as to work in the public sector. Only 29% wish to continue family tradition of farming. Thus, it seems that despite some resistance to the resettlement by a small but vocal group within the community, the majority sees the new community as a necessary step to reduce their marginalization in the socio-economic development of the nation. This is perhaps due to the fact that while they received new home with 2.4 ha (six acres) of oil palm land, they can still choose to return to their traditional villages if they wish to, especially those whose traditional villages and orchards are still easily accessible.

10.5 Conclusion: Marginalization or Demarginalization of Orang Asli as a Result of Resettlement?

The redevelopment and resettlement programs of Orang Asli throughout Malaysia are implemented to demarginalize this ethnic group so that they would be able to reap the reward of socio-economic development enjoyed by the mainstream Malaysian society. However, as the three case studies have illustrated, in certain situations, the resettlements have caused certain segments of Orang Asli to be more marginalized as a result of these programs. These are the people who due to their low education level, lack of skills and confidence are unable to cope with life in the new resettlement areas. They tend to be older people who are used to living in the rural areas relying on resources of the forests and agriculture practices.

The types of programs implemented have direct impacts on the results. Redevelopment of Orang Asli settlement within their areas, especially in reserve or gazetted area, as evidenced in the case of Ulu Kuang, would have lower negative impacts on this ethnic group. The in-situ development has caused the least negative impacts to the community and provides greater mobility in terms of education and employment. This is perhaps the best example of demarginalization of the Orang Asli in Peninsular Malaysia.

On the other hand, resettlement of Orang Asli especially in urban setting as illustrated in the Bukit Lanjan case, has caused a significant number of the ethnic community to be more marginalized. The inability to cope in the new urban environment together with high cost of living has forced some families to abandon their new homes. Some of the olds have returned to rural or forested areas elsewhere while the young ones are grappling with social problems including drinking, gambling and illegal racing. The handsome property and monetary rewards provided have not been able to alleviate the sufferings experienced by some segments of the community.

In the case of PROSDET Pantos in Pahang, the result of redevelopment had caused some to be marginalized while others have become demarginalized. The difference is due to their access to orchards in traditional villages upstream, those whose land is rather substantial and close by are much well of after the resettlement while those whose holdings are much farther upstream have found it very hard to earn a living since the new settlement does not provide a steady source of income and the new houses require the payment of monthly utility bills. Even though they are provided with palm oil plantation, the return has been minimal due to poor soil condition and management issue of the plantation. However, one main positive development of this resettlement is the people's hope and confident that the new settlement can provide secure and better standard of living and greater opportunity for their children to pursue higher education.

The three cases have vividly shown that in addition to improved physical infrastructure which is rather easy to achieve, the government need to implement better strategies related to the socio-economic conditions of this marginalized ethnic group. These include better education and skill training opportunities to allow them to compete better in the modern economy. Despite being geographically and socio-economically marginalized throughout the history of the nation, the Orang Asli of Malaysia are always hopefully looking forward to the days when they will no longer be marginalized.

References

Ab Hadi, M. Y., Roddin, R., Razzaq, A., Mustafa, M. Z. & Baser, J. A. (2013). Poverty eradication through vocational education (tourism) among indigenenous people communities in Malaysia: Pro poor Tourism Approach (PPT). *Procedia—Social and Behavioral Sciences, 93*, 1840–1844.

Abdullah, J., Borhan, M. A., & Ahmad, C. B. (2015). Orang asli resettlement in urban environment at Bukit Lanjan, Selangor, Malaysia. *Procedia—Social and Behavioral Sciences, 201,* 71–79.

Abdullah, J., Mohamed Sayuti, N., Mohamad Arshad, A. A., & Embong, M. R. (2016). Living conditions in orang asli resettlement project (PROSDET) of Pantos, Pahang, Malaysia. *Procedia—Social and Behavioral Sciences, 222,* 143–150.

Abdullah, J., Alwi, M. A. & Ahmad, C. B. (2017). The impact of redevelopment towards physical and socioeconomy of orang asli in Ulu Kuang village, Rawang. In *Proceedings of 5th AMER International Conference On Quality Of Life, Chulalongkorn University, Bangkok, Thailand, 25–27 February 2017.*

Aliran (2010). *Bantahan dasar pemberimilikan tanah Orang Asli.* Aliran.

Bellwood, P. (2011). *Prehistory of the Indo-Malaysian archipelago.* Canberra: ANU E Press.

Government of Malaysia (2010). Aboriginal People Act 1954 (Act 134), Pub. L. No Act 134 (1954). Kuala Lumpur: Government Printer.

Forum Asia (2013). Indigenous people in ASEAN. Retrieved July 13, 2016, from http://forum-asia.org.

Haron, S., Nordin, N. M., A Wahab, R. & Hayati, F. (2009). *Lifelong learning among the Malaysian Orang Asli bateq community.* Malaysia: UiTM Shah Alam.

Idrus, R. (2011). The discourse of protection and the Orang Asli of peninsular Malaysia. *Akademika, 35,* 75–86.

JKOA, official website of jabatan kemajuan Orang Asli (2013). Ministry of rural and regional development, Malaysia. Retrieved August 15, 2016, from http://www.jakoa.gov.my.

Karim, H. A., & Hashim, A. H. (2012). The effects of a resettlement scheme on the socio-cultural changes of the Temuan community. *Procedia—Social and Behavioral Sciences, 42,* 362–373.

Leimgrueber, W. (2004). *Between global and local. Marginality and marginal regions in the context of globalization and regulation.* Aldershot: Ashgate.

Lim, Hin Fui (1997). Orang Asli, forest and development, Malaysian Forest Record, No 43.

Memmoot, P., Stacy, R., Chambers, C. & Keys, C. (2000). *Violence in indigenous communities.* Canberra: Report to National Crime Prevention, Attorney-General's Department.

SPERI (2012). Customary law-based forestland conflict resolution case study: Nhakhaluang—Longngau areas of Densavang and Phonsavat villages, Luang Prabang district, Luang Prabang province, Lao PDR.

Redzuan, M. (2008). *Isu, transformasi dan cabaran.* Serdang, Selangor, Malaysia: Universiti Putra Malaysia.

Chapter 11
Roma, Social Exclusion and Romani Settlements as Marginalized Place: The Case of Loke

Alenka Janko Spreizer

11.1 Introduction

Yes, Mem, ghettos exist. They create ghettos. How? Is this not a ghetto? We are pushed into one common settlement, where all inhabitants belong to the same nationality. We are not allowed to separate from and leave behind the [Romani] settlement, and they claim that there are no funds available for such spatial managements (Interview with a Roma resident from the settlement of Loke, August 16, 2016.)[1]

July 7th 2016, local newspaper Dolenjski list reported that several Romani representatives gathered in Dolenjska, (SE Slovenia), where Roma[2] are considered an "autochthonous ethnic community", appealing to solidarity and common interest of all Romani representatives, who addressed the state for immediate access of Roma to basic infrastructure (water, sanitation, electricity). They identified the urgent need for starting the process of legalization of Romani settlements and arrange appropriate content, funds and control of tenders, aiming at improvement of infrastructure in Romani settlements. Romani Associations expressed their legitimate expectations that the state and local communities should allocate funds for Romani residential issues in the state budget, increasing the funds for infrastructure. The president of the Romani Union of Revival, Darko Rudaš, proposed immediate

[1]All informants are anonymous and their names are coded for the reason of confidentiality.
[2]The term Roma is used here for communities traditionally known as Gypsies. Roma is a self-designation, but this term may also include Romanichels, Sinti, Ludar, and Gypsy-Travellers. Following the leading academic journal, the Romani Studies, the noun Roma and the adjective Romani is used.

A. Janko Spreizer (✉)
Department of Anthropology and Cultural Studies, Faculty of Humanities,
University of Primorska, Titov trg 5, 6000 Koper, Slovenia
e-mail: alenka.janko.spreizer@fhs.upr.si

S. Pelc and M. Koderman (eds.), *Nature, Tourism and Ethnicity as Drivers of (De)Marginalization*, Perspectives on Geographical Marginality 3,
DOI 10.1007/978-3-319-59002-8_11

installation of agricultural water pipes and electric off-road construction lockers for informal Roma settlements without drinking water and electricity.

Two weeks later, July 19th 2016, the leading national newspaper Delo featured another article with the title: *The state measures push Roma into ghettos: The state ignores the wishes from the field and does not co-finance the resettlement of families from illegal settlements.*

The article delivered the latest news on the informal Romani settlement in Loke, in the Municipality of Krško, and the joint efforts of municipality, the local community of Loke and Romani inhabitants to find a solution to the Romani housing problem and spatial management. The journalist reported about the answer of the Ministry of Economic Development and Technology (MEDT), after the local community of Krško applied for funds for the improvement of the Romani settlement infrastructure. MEDT rejected the application for funds, which was to be invested in the building of the house for a Romani family that agreed to leave Loke, and was able to find their location in the nearby village Gržeča vas outside the Romani settlements (Zore 2016).

Spatial and educational segregation of Roma is in fact as widespread in Slovenia as elsewhere in Europe. When Roma municipality councilors reflected on the Romani housing problem in previous years, they strongly rejected the racialized discourse of the regional civil initiative, which stigmatized their settlements in terms of "isolate or reserve" (Janko Spreizer 2011–2012), when similarly referred to socio-spatial seclusion (Wacquant 2010). Compared to the past, representing Romani settlements as ghettos was a major discourse turn. Rejecting the ethnic homogenization of settlements and segregation of Roma in Krško in one place, Moran Jurkovič, Romani councilor at this Municipality, labeled another Romani settlement, Kerinov Grm, as a ghetto, when critically reflecting on the efforts of the state, municipality and "ethnic community" of Roma to find a solution to their housing problem.

Spatial management as a part of social integration of Roma is in discourse represented as one of the most conflicting issues in the Dolenjska region. The rhetoric of Romani representatives implicitly reflects criticism when they address the slow and bureaucratic approaches of the state and local communities to the "solving of existential and residential problems" of Roma, which were claimed as one of the priorities of the Government Programs for Roma.[3]

[3]Slovenian government has accepted the Program of Measures for Helping Roma in the Republic of Slovenia in 1995, Equal Employment opportunities for Roma in 2000, The National Program of the Fight Against Poverty and Social Exclusion in 2000, and National action program for Roma 2010-2015. Romani housing and the urgent need for legalization of Romani settlements was addressed several times.

11.2 Socio-Spatial Seclusion of Roma and "Gypsy Ghettos": Theoretical Framework

We see them as people with special needs ... They need us, the majority of people in terms of thinking. Should I walk 20 steps backwards so that I can live with people who live in the Stone age? (Interview with a Non-Roma resident from the settlement of Loke, August 18, 2016.)

Locally present but spatially separated and confined in the woods or a remote agricultural land, imagined as morally and intellectually inferior or backward, described as socially and temporally distant from civilians,[4] Roma are represented as Others in Krško.

As elsewhere, Roma are in discourse constructed as "internal outsiders" (Bancroft 2005, Gay y Blasco 2016, p. 447). Placed on moral and political margins as in different areas of Europe, Roma are perceived as nomads by the state and settled populations and as outsider groups that seem to devalue space and transgress or disturb the meaning of place. In the pan-European institutional discourse on solving "the Roma issue" and in the imagination of local populations, Roma are often rejected and are the target of hostility from states and other residents. They are often victims of hatred, violence or different forms of discrimination in education, health, labour market and housing. Stigmatized, they are often spatially segregated in marginal and distant places. Analyzing the processes of spatial management and segregation, which are partially played in the arena of law at the level of national and local spaces and which trigger anti-Roma sentiments, Bancroft (2005, p. 4) showed how "space is an important factor in the interaction between outsider groups and the institutions and practices of modernity, nation and state".

Scholars, who discussed mobility, nomadism and space of Roma, Gypsies and Travellers (Kendall 1999; Theodosiou 2003; Levinson and Sparkes 2004; Sigona 2005) have showed that (imagined) nomadism is used to legitimize segregation and sedentarization policies (Levinson and Sparkes 2004; cf. Kendall 1999). Marginality of Roma is reproduced at the local, national and European level through a multitude of social, educational and spatial segregations (Gay y Blasco 2016: 447). Despite the international programs and initiatives Roma for social inclusion of Roma, they still live in spatially incarcerated areas, and their situation may be seen as catch 22 (Marushiakova and Popov, 2015).

This chapter was theoretically inspired by recent studies on Roma, spatialization, racialization and social exclusion that explore the social and cultural formation of "Gypsy ghettos" (cf. Vincze and Rat 2013; Powell 2013) in contemporary Europe. As Wacquant explained, socio-spatial seclusion is manifested in many configurations as a generic process that may concern institutions, activities and populations, such as elite, migrants or Roma. Socio-spatial seclusion may occur in the countryside or in an urban environment, around a city, in a milieu that is characterized as

[4]*Civili* or civilians is the term used both by Roma and non-Roma to refer the latter.

"spatial accumulation and intense accretion of various forms of capital (economic, cultural, social, and symbolic), fostered by administrative machinery" (Wacquant 2010, p. 165).

It was suggested that the "theoretical concept of the ghetto can elucidate the ways in which the spatial marginality of sites serves as a weapon of 'confinement and control' for the dominant and 'an integrative and protective' device for the stigmatized Gypsy-Traveller population" (Powell 2013) or underprivileged Roma of Eastern Europe (Vincze and Rat 2013).

Roma in Slovenia have the status of an ethnic community under the Constitution and the Roma Community Act (2007).[5] Over the last 25 years, the state has adopted ethnic legislation and several administrative measures regulating the living conditions of marginalized Roma. After 1995, the state has developed several programs for the Roma community,[6] but housing, residential problems and legalization of settlements started in 2002 in Krško. Roma are organized in several associations, unions, Councils and forums, but none of these institutions have political power: those institutions provide elaborates or recommendations to government. The government was not successful in implementing the national action program for Roma in previous years; many interlocutors see the main obstacle in the reason that jurisdiction is divided between the local communities, the municipalities and the state.

As a social anthropologist, I will present my ethnographic fieldwork research in the Romani settlements in the municipality of Krško. My ethnographic data were collected sporadically during my short visits in 2000, 2010, 2012 and in August 2016 in Loke at Leskovec pri Krškem. I carried out desktop research on the existing legislation and national documents on the spatial planning of Romani settlements, municipal planning documents of Krško on Romani issues (2002, 2008) and Detailed Municipal Spatial Plan for Gmajna 3 Housing (2015). I interviewed Roma and non-Roma [*civili*, civilians], the municipal councilors of Krško, the vice-mayor of Krško and the representatives of NGOs as well as the Amnesty International. I reconstructed the case, among other things, by way of media discourse on spatial management of Romani settlement, which highlighted Roma ghettoization and marginality.

11.3 Municipality of Krško, Spatial Management of Roma

In the 1980 s, they [the Roma] settled with us. Despite the presence of Roma, the municipality did not want to talk about them. Because if you are not talking about a problem, it is not detected. /…/ At that time [around 1993], the idea was to build a

[5]The Roma Community Act in the Republic of Slovenia—ZRomS-1 is a system act defining the role of state bodies and self-governing local communities' bodies in realizing special rights of the Roma Community; it also regulates the organization of the Roma Community at national and local level, as well as its financing. Roma community. Government of the Republic of Slovenia office for National Minorities 2016.

[6]Roma community. Government of the Republic of Slovenia office for National Minorities 2016.

joint [Romani settlement], where all Roma would be resettled from the municipality of Krško and Brežice. The location was a former rocket base in the barracks of Cerklje, which was state property. Ours [Roma] were on Loke. There was the family of Felix, others were from Rimš. His father was from there. Ours. He was civilian. (Interview with a Krško municipality officer, August 18, 2016.)

The municipality of Krško was one of the municipalities which stated the spatial management of Romani settlement in 2000, when they hosted round tables on legalization of Romani settlement. Media have reported that the municipality has purchased usurped farmland where the construction of the biggest Romany community started after 1995. Then, the municipality started with the legalization. First, they turned the farming land into a building site, and in later years they produced the location plan and urbanization of the plots. The parcels of land were sold at non-market price. Until 2007, the Krško municipality has already spent 150 million SIT or 625.939 EUR, and the state has recovered more than 70 million SIT or 292.105 EUR from the tenders. In 2009, the legalization of Kerinov Grm was finished, but the situation of other Romani settlements was still in liminal state (Brajović Hajdenkumer 2007).

Situated in south-eastern Slovenia, where the Sava valley expands into the vast plains of Krško polje, the Krško municipality is 100 km away from Slovenia's capital Ljubljana and 50 km away from Croatia's capital Zagreb. Krško is located just beside important traffic infrastructure, near the tenth railway corridor and near the A2 motorway, which is a part of an important pan-European corridor (Wikipedia 2017). When I drove down the highway to Leskovec pri Krškem, I was surprised to see huge halls of international corporations, such as Greichaber Logistik and Bosch, and a number of hypermarkets under a massive electric transmission line. The process of urbanization and the presence of the economic capital of multinational corporations, such as Deichman, Lidl, Müller and New Yorker are apparent at the very first glance.

Present-day Krško is the center of the Posavje region and a municipal center. Municipality has a population of 25,884 (SURS 2016), among whom were 352 Roma with permanent residence and a number of unregistered residents until 2007.[7]

There are several Romani settlements in the Krško area, the largest being the legalized settlement of Kerinov Grm. Both Kerinov Grm and the nearby smaller settlement of Drnovo were included in the formal spatial planning strategy (2002), whereas two other settlements Rimš and Loke near Leskovec, were labeled as illegal by the decision of municipality. The plot where the Rimš is located was returned to the Catholic church and Roma do not have access to water or electricity.

[7]Romska problematika. Občina Krško. [Romani issues. The Municipality of Krško].

11.4 Protective Confinement to Woods and Mutual Distancing in the Past: The Case of Loke

Previously, there was a forest company. Our ancestors worked there. Then they got the location. They said: you can stay. There was a tent and nothing else. Latter they brought the boards, because locals allowed us to build that wooden shag (Interview with a Roma resident from the settlement of Loke, August 16, 2016).

Roma came to us, and then they went into the state forest to prevent their removal. When the men were imprisoned, women with children remained in the place. Mom was their godmother when children were born. Mom wrote them letters and read them when a letter came from a man in prison (Interview with a municipality representative of Krško, August 18, 2016).

When I was looking for the Jurkovič family in GPS I lost the signal. After a phone call I came to the woods on the edge of Loke, where Roma built houses in the middle of a forest. I was looking for their address on the Google Maps and Google Street View, and I noticed that the camera stops in the bushes and then turns to the fence of the first house near the narrow local street. Why did it not enter into the settlement? Was it because this settlement was considered "illegal?" Or was it perceived as a dangerous zone into which the Big Brother did not dare to enter?

In this ethnically homogenous settlement Roma have lived for several generations; the Romani settlement is confined to the woods where Roma seem to have felt safe in what resembles the former socialist regime. Compared to the past when Roma of Krško were socially constructed as those "who have always been around here" or those who are still looking for their place under the sun and/or are seen as newcomers rather than local inhabitants (Janko Spreizer 2016, p. 174, p. 180), Roma in Loke were seen as those who came to the villagers of Loke or those who belonged to the place. The locality and belonging of Roma was narrated through the informal settlement of the family that emerged in the time of SFR Yugoslavia on nationalized land which was formerly the property of the local non-Roma families. At the time, the land belonged to the state and the plot was used by the local Forest Company. The company had certain economic benefits for not employing Roma. In exchange for their work (cleaning the forest and planting trees) without employment contracts, Roma were tolerated and symbolically they were given the right to stay there. With the local and state consent, they first set up tents 70 years ago and then they started building shacks. They kept to themselves and did not disturb other locals, since their movement was under the watchful eye of their neighbours. Perceived as "more civilized due to racial mixing with civilians", their presence in the woods was tolerated by the majority as they stayed confined from the village. They lived without water, electricity and other infrastructure. When men were imprisoned due to small thievery, women were allowed to stay in the woods and from time to time they came to locals asking for food, dress and at the birth of new-born babies Roma also asked neighbours civilians to be their godfathers or godmothers. The narrations revealed the asymmetric power relations between Roma who were asking for a place to stay, food, support and help. Because they were confined to the forest and socio-spatially separated from the village, they were not harassed by police or treated as a menace.

11.5 Romani Settlement Legalization Act of Krško, and Violation of Human Rights

Loke are not Roma settlement, because the municipality did not identify it as Roma settlement. This is a duality between formalism and reality. We do not know the interests of the mayor. Amnesty International constantly reminded us that the settlement is illegal, but it has been designated as illegal by the municipality. Krško applied to the tender for displacement of families, because they are aware of the problem, but then the municipality's application was rejected by the government, because this [new location] is not a Roma settlement. (Interview with Amnesty International representative, August 22, 2016).

After the disintegration of SFR Yugoslavia in 1991, the state land of the forest company in Loke was denationalized and given back to its owners, the Trenc family. The latter did not aim to drive the Roma away, but wanted to sell 8 hectares of the forested area to the municipality. The Trenc family was aware that Roma, poor and unemployed as they were, would not be able to get a loan for purchasing the plot, so they offered their property to be sold to the municipality. The municipality was not ready to buy the entire area and the Trenc family, reluctant to sell only one portion of it, settled with the Roma. Meanwhile, the Roma of Loke were able to connect the settlement to the water mains, but obtaining electricity was not possible, because the older members of the Trenc family had died and their heirs would not give their permission.

In 2002, the Municipality of Krško adopted the Romani Settlement Legalization Act and Loke 20[8] was not included in spatial management plan.

Discussion with local inhabitants [Roma and "civilians"] revealed that the settlement remained separated from the rest of the majority population. The settlement has no other infrastructure, such as sewage, but they pay the sewage disposal fee. They do not have access to public transport, but their children who attend school have the van transport at their disposal. On the other hand, the children who could attend a kindergarten could not be transported with a school van due to bureaucratic obstacles, since they are not students of the mandatory school. The inhabitants of Loke have bad infrastructure, but they obtain the one house number and in the past they could all be registered to the permanent address under the same house number. Until recently, the registration was not possible anymore, due to the owner dispute,

[8]The settlement of Loke has the same name as the location Loke 20, which is also the home and a permanent address of the Roma residents in the local community of Loke.

which is why newborn children are registered at the Center for social work, where they have their permanent residential address.

In 2008, 64 Roma lived at the permanent address Loke 20; the area was intended for building (Romska problematika 2008). The municipality intended to start the spatial managements in the location settled by the Roma, underlining that this location was actually a part of Leskovec. On the other hand, no provisions were made as to where the Roma would be resettled. With the administrative act Loke was proclaimed "illegal", since the municipality legalized only 2 settlements, Drnovo and Kerinov Grm, aiming at confining Roma to that location.

In 2009, the private company Si Investicije d.o.o. [SI Investments] bought the land from Trenc's heirs. In 2012 Roma heard that the land was intended for the building of an elite residential neighbourhood. With the support of local NGOs they organized a round table where they informed the local community about the possible dislodgment to a new location. A few years later, the private company agreed to sell the land to Roma, but the Romani dwellers could not afford to buy it at the market price. For the municipality, the price of the land was too high but in 2012 the municipality did not have an alternate solution for Loke (R. R. 2012). Since then, 8 Roma families have lived under the threat that their 5 houses would be bulldozed and "cleared" from the area. Several NGOs and Romani representatives underlined that basic human rights and special rights of Roma were dismissed to serve the interest of economic capital in the process of urbanization and transformation of the former socialist state to a neoliberal one (Interview with a Roma resident from the settlement of Loke, August 16, 2016).

Amnesty International (AI) has appealed several times that the state and the local communities do not respect international law on basic human rights, such as access to drinking water and lack of the legal security of tenure (2011, p. 15) in the case of Roma, but their appeals were ignored or rejected at the Courts. Families in Loke live in permanent fear of what will happen to their houses and they live under the threat of being forcefully evicted. In 2014, the UN Committee on Economic, Social and Cultural Rights (2014, E/C.12/SVN/CO/2) recommended that the State party should: "(a) Prioritize the legalization of Roma settlements, or work out other solutions in genuine consultation with the Roma communities. The state should also strive "not to carry out forced evictions of Roma, and enact legislation governing forced evictions which complies with the international standards and is in line with the Committee's general comment No. 7 (1997) on forced evictions. For those Roma who live in informal settlements, the state should provide "access to basic services, such as water, electricity and sanitation, also in accordance with the recommendations set forth by the Governmental Commission on Roma in 2011 that requested municipalities to provide water to Roma". It was also appealed that the state should "take effective measures to end segregation of Roma communities, and prevent acts of discrimination against Roma attempting to buy or rent housing outside their segregated areas" It was recommended that the state should also "facilitate access to social housing for Roma".

The latest spatial management plans of Krško (OPN UL RS, 61/2015) and the detailed local residential plan Gmajna 3 (OPPN, UL RS 13/2016) did not include the spatial management of the Romani settlement Loke. Before the plan Gmajna 3 was adopted in 2016, several local inhabitants tirelessly demanded to solve the problem of Romani housing. Some residents and municipality councilors even mentioned that this territory should belong to Roma. The paragraph 5 of Article 5 stated that at the location Gmajna 3 B are Romani houses on the private property and that the owner should "solve the question of their resettlement" (OPPN, UL RS 13/2016). Administrative machinery generated several documents of spatial management where Roma were excluded and the municipality did not plan any solution for resettlement of Roma settlement of Loke. Instead of planning the financial resources and time schedule for spatial management and resettlement of those who wanted to move outside Loke, the state and the municipality has conferred relocation and housing of Roma on the owner of the land.

After 2012 when the company SI Investicije revealed their future plans, Moran Jurkovič, the Romani municipality councilor, and the Municipality of Krško cooperatively continued with the activities for Roma housing. Jurkovič found a location with a house in Gržeča vas and the owner of a house agreed to sell the house for 45,000 EUR. However, Jurkovič was unable to apply for a loan because he was unemployed. His family lived on social support and they invested their funds in building the house in Loke.

The vice-mayor of Krško reported that the municipality applied for 15,000 EUR at the MEDT and their tender for the development of Romani infrastructure. The municipality would contribute the same amount of money and an equal share would be paid by Jurkovič's family in future years. In addition, a certain share would be given by unknown sponsors. The house would be the first property of the municipality and once the family was able to pay the whole amount, it would pass into its ownership (Interview with the vice-mayor of Krško, August 18 2016).

My interlocutors criticized the fact that the funds were available only for the resettling to ethnically segregated settlements and not for the resettlement outside Romani settlements. In July 2016 The MEDT explained to municipality of Krško and NGOs that the funds could only accommodate projects for the basic infrastructure in the Romani settlements and rejected their application for funding. Actors at the Ministry claimed that it was not possible to ignore the regional development legislation pursuant to which funding was available to accelerate the development of the settlement, where members of the Romani community live. The municipality was advised to apply for funds at the Ministry of Labour, Family, Social Affairs and Equal Opportunities to solve the housing or social problems of Roma.

The Head of the Office for Minorities, Stane Baluh, denied the intentional ghettoization of Roma imposed by the state and its ministries, but interlocutors and articles in newspapers reported that the process of ethnic segregation of Roma in Kerinov Grm already had negative consequences. The biggest Romani settlement of Krško, Kerinov Grm, was rather perceived as ghetto, a site of ethnically homogenized and criminalized group, which was confined to a location separated from the majority population (Zore 2016).

Krško's Municipality interlocutors insisted that the state should also provide funds to invest in housing for Romani families outside Romani settlements, if the state seriously intended to solve "Romani Issues".

When I visited this settlement known as a case of good practice (Amnesty International 2011) and talked to the inhabitants, I heard several stories which could not convince me that in this remote and legalized settlement the process of social inclusion was a smooth one. Roma who did not buy land plots there were imagining of leaving the place, because "it was not good to live in a segregated space". Representatives of Roma criticized the fact that there should be one big Romani settlement in Krško, and appealed for other options such as resettlement to non-Romani settlements.

My interlocutors at Amnesty International told me that after several years, there was still no strategy for Roma resettlement or a financial compensation for the so-called illegal settlements—or any other agreement that would eliminate the threatening possibility to become forcibly evicted and consequently homeless.

11.6 Conclusion

This chapter analyses socio-spatial seclusion that started in the former socialist state and which is still persistent in the countryside or in urban environments of contemporary Slovenia. Administrative machinery (the state with its minority legislation) has provided the legal framework, which regulates the special situation of Roma as an ethnic minority within the broader EU legislation on Roma. Despite several national plans on legalization of Romani settlements and spatial plans and regulations of the places of Roma at the municipality level, the Jurkovič family has not been able to regulate and legalize their dwellings in negotiations with the last owner, the local community or the state in the last 25 years.

The locals, NGOs, Roma representatives and media issued reports on spatial seclusion, underlining the claims that the state pushed Roma in ghettos, constructed Kerinov Grm as a separate space, populated by a homogenous ethnicity and labeled a poor community with a group of weak social capital: as a group, Roma are constructed as those with distinctive cultural habits, because they allow early marriages of their youths. The state and responsible ministries ignored the proposals of local inhabitants searching for displacement of Roma in Loke from their Romani settlement and emplacement in nearby /a/ farming house/s/ in a case if Roma are willing to leave Romani settlement.

In the interviews with my interlocutors, Krško was perceived as a wealthy local municipality[9] with the only nuclear power plant in Slovenia, two additional water power plants and the storage of nuclear waste from the power plant. The NGOs affirmed that the rich municipality could also have a big symbolic capital in solving

[9]As local community they receive 8,5 million EUR from the state budget as a compensation, and they distribute this money amongst their local inhabitants.

the Romani housing, but it seems that there is no such political agenda at the state level. Although the Municipality of Krško could invest its funds into the resettling of Roma from Loke, it was ready to invest only a part of municipality's funds into the legalization of Romani settlements: the vice-mayor insisted that the state with its ministries should also contribute, because the responsibility for legalization of Romani housing is divided between the local community and the state.

The state and the municipality remain trapped in the administrative machinery, claiming that Loke or Gmajna 3, section B, is not considered a Romani settlement. The new spatial plan has been drawn and it was accepted at the municipality meeting after the negotiation with some councilors. The OPPN Gmajna 3 was approved as if Romani buildings did not exist at the location. In narrations on the urbanization plans for Loke and Gmajna 3 some interlocutors were guessing that this place would be urbanized for the elite professional workers of the nuclear power plant. In accordance with their opinion, the place would be transferred to location for the elites and would become another kind of a place of social seclusion: a place of residents with symbolic, cultural and economic capital who stay enclosed in their rich neighbourhoods to which Roma do not belong. The sociospatial seclusion of Roma to Kerinov Grm and their social exclusion from the spatial management plan bears a clear message: people with a lesser cultural, economic and social capital cannot be seen as a possible part of an elite neighbourhood. Not protected by the state or the municipality and by not being included in the spatial management plan, they are symbolically erased and live under a serious threat that the owner will remove their houses one day, leaving them homeless.

Acknowledgements This research was partially supported by the Slovenian Research agency under the project "Ethnographies of Land and Water Routes: A Comparative Approach to (Im)mobility" (J6—6839). I thank to Nataša Rogelja and Nataša Gregorič Bon (ZRC SAZU) for comments that greatly improved my manuscript. I am deeply grateful to editors of this book, Stanko Pelc and Miha Koderman, for their suggestions and comments of my chapter. I also thank my interlocutors for sharing their stories. This research would not have been possible without the support of my husband Mitja Spreizer. Last, but not least I thank Manca Gašperšič and Gregor Pobežin for proofreading this work.

References

Amnesty International. (2011). *Parallel lives. Roma denied right to housing and water in Slovenia*. London: Amnesty international Ltd. General Secretariat.

Amnesty International. (2014). Romi. Retrieved January 10, 2017, http://www.amnesty.si/romi.

Bancroft, A. (2005). *Roma and Gypsy-travellers: Modernity, race, space, and exclusion*. Hants, Burlington: Ashgate Publishing Limited.

Brajović Hajdenkumer, S. (2007). Jutro nove misli. Retrieved November 22, 2017, http://tvslo.si/predvajaj/dosje-jutro-nove-misli/ava2.2254862/00:00:36/0:0:0/ .

Gay y Blasco, P. (2016). 'It's the best place for them': Normalising Roma segregation in Madrid. *Social Anthropology*, *24*(4), 446–461.

Janko Spreizer, A. (2011–12). Emotional geographies of exclusion in a selected roma settlement. *Etudes Tsiganes. Nouvelle Série*, (44–55), 174–189.

Janko Spreizer, A. (2016). From a tent to a house, from nomads to settlers: Constructions of space and place through Romani Narratives. In N. Gregorič Bon & J. Repič (Eds.), *Moving places. Relations, return and belonging* (Vol. 29, pp. 172–193). New York, Oxford: Berghnan, Easa Series.

Kendall, S. (1999). Sites of Resistance: Places on the Margin—The Traveller "Homeplace"'. In T. Acton (Ed.), *Gypsy politics and traveller identity* (pp. 70–89). Hatfield: University of Hertfordshire Press.

Levinson, M. P., & Sparkes, A. C. (2004). Gypsy Identity and orientations to space. *Journal of Contemporary, Ethnography, 33*(6), 704–734.

Marushiakova, E., & Popov, V. (2015). European policies for social inclusion of Roma: Catch 22? *Social Inclusion, 3,* 19–31.

MM. (2016). Romi stopili skupaj in bodo od države zahtevali konkretne izboljšave. *Dolenjski list*, 7 July. Retrieved January 10, 2017, from http://www.dolenjskilist.si/2016/07/07/158817/novice/dolenjska/Romi_stopili_skupaj_in_bodo_od_drzave_zahtevali_konkretne_izboljsave/.

Odlok o spremembah in dopolnitvah dolgoročnega plana Občin Krško (2002) UL RS 90/2002 (pp. 9915–9918). Retrieved January 10, 2017.

OPN. Odlok o Občinskem prostorskem načrtu občine Krško. 2015. UL RS, 61/2015. Retrieved January 10, 2017, from https://www.uradni-list.si/1/content?id=122934.

OPPN Gmajna 3. 2016. Odlok o občinskem podrobnem prostorskem načrtu stanovanjska gradnja Gmajna 3, Stran 1523. UL RS 13/2016, pp. 1523. Retrieved January 10, 2017, from https://www.uradni-list.si/1/content?id=125455#!/Odlok-o-obcinskem-podrobnem-prostorskem-nacrtu-stanovanjska-gradnja-Gmajna-3.

Powell, R. (2013). Loïc Wacquant's 'Ghetto'and Ethnic Minority Segregation in the UK: The neglected case of gypsy-travellers. *International Journal of Urban and Regional Research, 37* (1), 115–134.

Roma Community. (2017). Government of the Republic of Slovenia office for National Minorities. Constitutional and Legal Position. Retrieved January 10, 2017, from http://www.un.gov.si/en/minorities/roma_community/constitutional_and_legal_position/.

Romska problematika. Občina Krško. 2008. Retrieved January 10, 2017, from www.krsko.si/uploads/seje/57/ooci Y_Romska_problematika.doc.

R. R. (2012). Prihodnost romskega naselja Loke še vedno negotova, *Posavski obzornik*, May 29. Retrieved January 10, 2017, from https://www.posavskiobzornik.si/novice/prihodnost-romskega-naselja-e-vedno-negotova .

Sigona, N. (2005). Locating "the Gypsy Problem". The Roma in Italy. Stereotyping, labelling and "Nomad Camps". *Journal of Ethnic and Migration Studies, 31*(4), 741–756.

SURS. (2016). Prebivalstvo po naseljih, podrobni podatki, Slovenija, 1. januar 2016". Statistični urad Republike Slovenije. 6. junij 2016.

Theodosiou, A. (2004). Be-longing in a doubly occupied place: The Parakalamos Gypsy musicians. *Romani Studies, 14*(1), 25–58.

UN Committee on Economic, Social and Cultural Rights. (2014). (E/C.12/SVN/CO/2). Concluding observations on the 2nd periodic report of Slovenia: Committee on Economic, Social and Cultural Rights UN. Committee on Economic, Social and Cultural Rights (53rd sess.: 2014: Geneva) (2014-12-15).

Wacquant, L. (2010). Designing urban seclusion in the twenty-first century: The 2009 Roth-Symonds lecture. *Perspecta, 43,* 164–175.

Občina Krško. (2016). Retrieved January 10, 2017, from https://sl.wikipedia.org/wiki/Ob%C4%8Dina_Kr%C5%A1ko.

Vincze, E., & Rat, C. (2013). Spatialization and racialization of social exclusion. The social and cultural formation of "gypsy ghettos" in Romania in a European context. *Studia Universitatis Babes-Bolyai-Sociologia, 58*(2), 5–21. Retrieved January 12, 2017, from http://www.studia.ubbcluj.ro/download/pdf/816.pdf [Accessed 12 January 2017].

Zore, J. (2016). Država s spodbudami Rome stiska v gete. *Delo*, 19. July. Retrieved January 12, 2017, from http://www.delo.si/novice/slovenija/drzava-s-spodbudami-rome-stiska-v-gete.html.

Chapter 12
The Planned Process of Spatial Integration of Roma Settlements in Slovenia—The Way to Deghettoization and Demarginalization?

Jernej Zupančič

12.1 Introduction

In the last two decades, interest in Roma "issues" has been on a rise, namely from different perspectives (Bešter and Medvešek 2013). Issues of living in specific Roma settlements stand out greatly. Due to the separate development and special social circumstances, these settlements developed into rural and urban slums. Unorganized settlements are undoubtedly the cause of both personal and collective (local community, neighbourhood) marginalization and exclusion in almost all European countries (Macura and Macura-Vuksanović 2006). We can reasonably talk about the process of ghettoization and settlements as ghettoized spaces (Pirc 2013, p. 13). In some Roma settlements, the high density of buildings is not only a problem of low living standards, but also causes disputes and even conflicts between neighbours. On one hand, poor living conditions reflect the social marginalization and are also the most important reason for persisting in above mentioned marginalization on the other (Klopčič 2010). However, solving the residential problems of Roma is quite often founded on an illusionary expectation that it is possible, using primarily the instruments of social interventionism on one hand and systematic development of Romani settlements with technical or community infrastructure on the other—i.e. electrical installation, water distribution system, road surfaces with connections to the public road network, municipal cleansing service, and sewer system—to simultaneously achieve social indications of integrations, such as higher enrolment in educational institutions, educational achievements, and indirectly more successful competition on the labour market. We now have a decade's worth of more intensive interventions in Roma settlements with the goal of technical renovation and legalization. There is data, plans, and

J. Zupančič (✉)
Department of Geography, Faculty of Arts, University of Ljubljana,
Aškerčeva Cesta 2, 1000 Ljubljana, Slovenia
e-mail: jernej.zupancic@ff.uni-lj.si

S. Pelc and M. Koderman (eds.), *Nature, Tourism and Ethnicity as Drivers of (De)Marginalization*, Perspectives on Geographical Marginality 3,
DOI 10.1007/978-3-319-59002-8_12

analyses of situations and good practices, providing sufficient material for a performance review of various approaches and measures. In the last decade, community infrastructure has been constructed in many Roma settlements; however, the goals of integration are far from achieved (Zupančič 2015, p. 7, 22).

Currently, Roma community is experiencing rapid social and demographic transformation, which is reflected primarily in changes of the way of life and their spatial conduct. The transformation of Roma settlements is primarily a professional challenge, from analysis to planning and management (Zupančič 2013, p. 43). Roma settlements need to be intentionally and systematically demarginalized. Below we discuss the question of normative and functional integration of Roma settlements into the Slovenian settlement system. Roma settlements are thereby considered as spatial units of locally continuous Roma settlement, whether their residents identify themselves ethnically as Roma or not. The comprehensive discussion of Roma settlements considers their internal diversity, as well as spatial, material, and cultural potentials and actual characteristics of communication between the Roma and their neighbours.

12.2 Roma Settlements in Slovenia: Brief Overview

The belief that a high majority of Roma live in ethnically homogeneous "Roma-settlements" is not quite correct. Newer studies confirm an increased heterogeneity of these settlements and a high share and number of ethnic Roma who live dispersed across Slovenia. The number of settlements varies due to different understanding of the concept "Roma settlement". According to the records, there are 107 spatial units, considered as settlements where Roma represent the only residents or they are in great majority (Zupančič 2012, pp. 90–91). Those "settlements" are not recognized statistically as settlements due to the illegality and unsolved property issues. Later, studies found an even greater spatial dispersion of Roma; they also lived in many smaller settlements, so that the number of such settlement units climbs to about 130. Additionally, quite a few Roma left their environment and live in mixed marriages in urban environment, and new, in-depth studies show that many Roma settlement are actually ethnically mixed. Compared to over 6,000 "statistical" settlements in Slovenia, this represent just a bit under 2%, while Roma represent only 0.5% of total Slovenian population. Urban populations in larger cities are a completely new and another group. They live primarily in Ljubljana and Maribor; the latter are mostly immigrants from the areas of former Yugoslav federal republics (now independent countries), who came to Slovenia within the last three decades (Zupančič 2014, pp. 47–50).

Roma settlements are mostly newly formed, as permanent settlement became predominant only in the 1970s and later. Permanent settlement attempts were already undertaken in the 18th century in the territory of the former Habsburg Monarchy during the ruling period of Maria Theresa, but they proved unsuccessful (Štrukelj 1979). Later settlements, associated with quite forceful measures, also failed to have a permanent effect (Štrukelj 2004, p. 39). However, Roma were at least partly "territorialized"; they were registered in areas of certain municipalities, even though they did not have permanent residence in one place, but were moving in order to make a living (various "travelling" trades, also begging). Still, due to the aforementioned "territorialization", a segment of the Roma population settled down. Especially Roma in Slovenian eastern region Prekmurje predominantly settled down between World War I and II, and later. With them, nomadism gradually faded into folklore. The process of permanent settlement was a generation late in Dolenjska, Bela krajina, and Posavje region (Žugel 1991). Permanent settlement actually succeeded only after significant social changes and modernization processes of Roma society, and not due to intentional efforts by the government. With permanent settlement, some conveniences of urbanized and modernized society became at least partially available to them. However, permanent settlement quickly proved to be a continuation of the marginal social status, even in new circumstances. Permanent settlement was, in spatial terms, most often limited to a single family. Later, due to marriages and high fertility, it expanded significantly, and immediate family and distant relatives often moved there as well, so that Roma settlements obtained certain structural and physiognomic characteristics. Because of the lack of real estate ownership and legal regulations, financial assets, as well as a system of values, constructions differed greatly, and were mostly of poor quality. Roma settlements became open rural (and as well as urban) "slums" (Zupančič 2007, pp. 23–26).

Formation of Roma settlements is related to the characteristics of their culture and social processes. Stabilization started in the phase of intensive social restructuring of Slovenian society, from rural to semi-rural and industrial. Some families separated from previously familial communities, either because of the means of survival or because of various disputes (which were not a rare occurrence). We must also consider the process of dissolution of traditional Romani society in this context, which during modernization faced challenges of both economic and psychological nature (Zupančič 2009, p. 22). Because of familial connectedness, Roma settlements acted (and still do) socially very homogenously (Škraba 2007). Moving away often represented a kind of intentional ethnic assimilation.

Larger Roma settlements, such as Pušča (Rudaš 2011), Vanča vas or Sotina in the Prekmurje region (Komac 2005), Kerinov grm near Krško, Hudeje, Brezje—Žabjak in the Dolenjska region, Lokve in Bela krajina, and similar were formed by repopulation by residents who were not related. In these settlements, a new Roma community began forming in the phase of social modernization, which had stronger territorial and weaker familial ties. Real estate (location), which was not very important in the semi-nomadic phase (see more in Štrukelj 1991, pp. 38–42), began taking on new significance. Roma cultural identity could no longer persist purely on tradition and a way of life, but began to identify increasingly with the territory of permanent settlement—the location of permanent settlement was becoming the premise for identity. Permanent settlement became the venue for sociability and the starting point for livelihood and other functions (Klopčič 2007). The new dimension of spatial mobility was represented by motorization, as new means of transport allowed a wide area of activity to be established. This territorialization was dramatic, even shocking in a way, and often understood as coercion (Zupančič 2006). It was understood neither by Roma, who experienced it, nor the majority population, who was often a kind of victim as well, as Roma took over their real estate (Žugel 1991). The absence of concepts, which would accompany and direct the process of permanent settlement, was most likely a problem. For this reason, the formation of current permanent Roma settlements was seen from the perspective of the majority as uncontrollable. In reality, locations of most current Roma settlements were thought out: in places with the highest possibility for ensuring livelihood and where the opposition of real estate owners was still bearable. Most locations were probably only temporary at first; however, since they proved suitable, they became permanent settlements (Zupančič 2014).

12.3 The Concepts of Spatial Integration Are Concepts of Demarginalization

In 2007, under the auspices of a special expert group, a framework development plan for the Roma settlements in Slovenia was prepared, which stated that most Romani settlements could be developed and modernized in existing locations by establishing measures of legality, organizing ownership relationships, and establishing at least minimal technical and construction standards; the debated settlements would then be included in the Slovenian settlement system (Zupančič 2007). Only a few Roma settlements do not have these options and would require replacement locations or individual housing solutions to be found. Legality is one of the fundamental goals in resolving the housing issues of Roma, but this can be achieved in multiple ways, from the previously proposed legislative and systematic solutions to intervention solutions with inexpensive pre-fabricated constructions. Improvised construction solutions were especially common during rapid demographic growth, which, due to limited space, led to (excessive) density of buildings and related issues between neighbours. In this context, we must also mention the idea of the intervention legislation (1993) and the ad hoc legalization aimed at quickly establishing formal legality. Experience of this legal interventionism is poor —problems only became exacerbated. Intervention solutions increase the aspirations for governmental interventionism, reduce the willingness of population to use legal forms of spatial conduct, and strengthen the belief that creation of good living conditions is the task (including financial task) of society or government, and not an individual's potential (owner, user). In the long term, it leads towards a distinctly inert, self-oriented development, and away from integrative efforts. The position of gradual structural adaptation and simultaneous legalization became dominant. Roma settlement are to be gradually modernized, equipped with community infrastructure equivalent to that in other settlements in Slovenia, while legalization would be achieved as a legal completion of these processes. The idea is based also

functions, in addition to residential: work, education, relaxation, cultural life, etc. They thus enable increased functionality of Roma settlements, their path outwards, and therefore their establishment in the wider local environment (Zupančič 2015, p. 20)

12.6 Conclusion

The achieved forms of technical modernization and implemented legality in over half Romani settlements in Slovenia are otherwise encouraging results of quite substantial efforts into demarginalization of Roma residential areas. However, implementation of access to electricity, water, community installations, road network, and other technical goods do not by themselves bring what is most desired: integration. This is a more difficult and above all complex process that requires a stronger and more intensive interaction between participants in the process of spatial planning of Roma settlements. The obvious progress of some Roma settlements is proof of the effectiveness of presented construction. Demarginalization of Roma settlements will require additional attention in the future, especially in the segment of internal cultural and social instances of spatial conduct. On the other hand, the endeavours, done from time to time in the resolution of "issues" of Roma settlement, with its non-uniformly defined systems of norms, incentives, and sanctions, is guilty of producing, in its planning effectiveness, generally poorer results. This brings the entire complex of redevelopment and legalization of Roma settlements back to the starting point, to the developmental paradigm of Roma settlement: the manner, approaches, and ultimately key goals of integration efforts. Is the approach of raising the scope of social interventionism more suitable (effective and permanent in the long term), or, as opposed to now, is it the model of competitiveness, whereby social resolution has the function of also achieving other goals and is not just a value in itself?

It is better to lean towards the latter, since, using the former approach, leaves Roma more or less permanently stuck in a sort of controlled refuge as permanent beneficiaries of social transfers, pittance, and as subjects in tests of neighbourly tolerance. It is better to trust the developmental strengths of Roma, despite the errors and side tracks that happened, are happening, and will happen. Interventionism is necessary, but primarily as an instrument of slow but more reliable, more permanent, and more economical participative development model of marginal areas, especially if the politics of employment, education, and social security are more harmonized and subordinate to the principle of competitiveness and empowerment of members of this community in the labour market. Finally, Roma settlement areas can be reasonably considered in terms of spatial and cultural capital (largely unutilized), which can realize its goals in the paradigm of sustainable development, shaking off its existing economic and administrative tutorship.

References

Bešter, R., & Medvešek, M. (Eds.). (2013). *Dvig socialnega in kulturnega kapitala v okoljih, kjer živijo predstavniki romske skupnosti*. Ljubljana: Inštitut za narodnostna vprašanja.

Klopčič, V. (2007). *Položaj Romov v Sloveniji*. Inštitut za narodnostna vprašanja, Ljubljana: Romi in Gadže.

Klopčič, V. (2010). Romi v Evropi. Ljudstvo brez doma in brez groba. *Ars & Humanitas 4*(1/2), 182–202.

Komac, M. (2005). Izvlečki iz slovenske romologije". *Treatises and documents/Razprave in gradivo, 47,* 134–159.

Macura, V., & Macura-Vuksanović, Z. (2006). *Stanovanje i naselja Roma u Jugoistočnoj Europi: prikaz stanja i napretka u Srbiji*. Beograd: Društvo za unapređivanje romskih naselja.

Pirc, J. (2013). *Od skritosti do točke na zemljevidu*. Inštitut za narodnostna vprašanja, Ljubljana: Socialno-prostorski razvoj izbranih romskih naselij v Sloveniji.

Rudaš, D. (Ed.). (2011). *100 let naselja Pušča*. Murska Sobota: Krajevna skupnost Pušča.

Škraba, U. (2007). *Integracija Romov v občini Grosuplje*. Fakulteta za družbene vede, Ljubljana: Magistrsko delo.

Štrukelj, P. (1979). *Romi na Slovenskem*. Ljubljana: Cankarjeva založba.

Štrukelj, P. (1991). Etnološke raziskave romske populacije v Sloveniji. *Treatises and documents/Razprave in gradivo, 25,* 36–55.

Štrukelj, P. (2004). *Tisočletne podobe nemirnih nomadov*. Ljubljana: Družina.

Varga, R., & Friškič, S. (Ed.). (2013). *Strategija razvoja romske skupnosti v Pomurju do leta 2020. Pregled stanja, vizija razvoja, programi in projekti za razvoj romske skupnosti v Pomurju*. RRA Mura, Murska Sobota.

Zupančič, J. (2006). *Razmere kličejo po resni strokovni presoji: problematika romskih naselij*. Delo 48 (255), Sobotna priloga (Vol. 4, pp. 17–18).

Zupančič, J. (2007). Romska naselja kot posebni del naselbinskega sistema v Sloveniji. *Dela, 27,* 21–52.

Zupančič, J. (2009). Etnična in politična podoba ozemlja med Muro in Rabo. In T. Kikec (Ed.), *Pomurje: geografski pogledi na pokrajino ob Muri. Zveza geografov Slovenije, Ljubljana; Društvo geografov Pomurja, Murska Sobota* (pp. 19–34).

Zupančič, J. (2012). *Vključevanje romskih naselij v slovenski naselbinski sistem. In: Š. Humljan Urh (ed.) Država želi, da ostanemo cigani!: teoretske refleksije in prakse izključevanja/vključevanja Romov v Sloveniji. Fakulteta za socialno delo, Ljubljana* (pp. 89–101).

Zupančič, J. (2013). Romi in romska naselja v Sloveniji v procesih modernizacije. *Geografski obzornik, 60*(1/2), 41–53.

Zupančič, J. (2014). *Romi in romska naselja v Sloveniji*. Ljubljana: Založba FF.

Zupančič, J. (2015). O integraciji romskih naselij v slovenski naselbinski sistem. *Treatises and documents/Razprave in gradivo, 75,* 5–26.

Žugel, J. (1991). Integracija romske skupnosti v slovensko družbo. *Treatises and documents/Razprave in gradivo, 25,* 113–121.

Chapter 13
Multiculturalism and Ethnographic Museums in Israel: The Case of a Regional Bedouin Museum

Havatzelet Yahel, Ruth Kark and Noam Perry

13.1 Introduction

This chapter considers multiculturalism in Israeli society, and its expression in museums. It showcases one regional museum that aimed to present two different cultures in the Negev in Southern Israel.

The study is based on archival documentation, data collection, and interviews of people involved in the establishment of the center and museum, as well as current employees, guides, and visitors. We begin by examining ethnocentric museums in Israel, with special emphasis on minority ethnographic museums, posing the question whether Israel represents a multicultural society. Then we look at the existing world nomadic and Bedouin museums. We then present our case study—a museum of the Bedouin, part of the Joe Alon Center for Regional Studies which aims to reflect multiethnic and multicultural societies within the Negev. We explore how the Bedouin museum and the Center evolved through the years, look at the Negev Bedouin society, an ethnic minority group in Israel, and elaborate on the challenges to present differing identities and narratives in one regional center.

H. Yahel (✉)
Ben-Gurion Institute for the Study of Israel & Zionism,
Ben-Gurion University of the Negev, Sde-Boker, Israel
e-mail: Havatzelety@gmail.com

R. Kark
The Hebrew University of Jerusalem, Mount Scopus, 91905 Jerusalem, Israel
e-mail: ruthkark@mail.huji.ac.il

N. Perry
San José State University, San Jose, CA 95192, USA
e-mail: noam.perry@gmail.com

S. Pelc and M. Koderman (eds.), *Nature, Tourism and Ethnicity as Drivers of (De)Marginalization*, Perspectives on Geographical Marginality 3,
DOI 10.1007/978-3-319-59002-8_13

13.2 Museums of Nomads and Bedouin in the Middle East

For the sake of comparison we mention several museums that present nomadic and Bedouin cultures in the Middle East. Some of these were established by individual foreign and indigenous collectors and others by local or state authorities. We will not touch here upon topics of Orientalism (Said 1979), nationalism (Anderson 1983), and post-colonialism (Bhabha 2000).

Among the nomadic and Bedouin museums in the Middle East are:

In Turkey, Enver Yalçın recently opened the museum of nomads in Fethiye. The museum focuses on the life of the Yörük people in the southwestern district's Kargı village. It contains 2500 artifacts which include many ancient agriculture tools from different parts of Turkey.

Among the Museums in Egypt we can find several Bedouin museums in the Farafra Oasis in the Western Desert of Egypt.

An interesting museum is the El-Arish Bedouin Heritage Museum which focuses mainly on the Egyptian Bedouin lifestyle in the Sinai Peninsula in the last 100 years. This museum, established by the association for the preservation of the Sinai Bedouin heritage, contains artifacts such as tents, costumes, food utensils, jewelry, traditional medicine, and medicinal plants (*Al-Shark Al-Awsat* 2009).

In Jordan there are several museums. The "al-Hayat al-Shaabia," the Jordan Folklore and Popular Traditions Museums, opened in Amman in 1975. It displays nineteenth- and early twentieth-century nomadic desert Bedouin, village, and urban everyday lifestyle and culture (Guide2jordan 2017; Visit Jordan, Amman. 2017).

Another museum, also called "al-Hayat al-Shaabia" was established in 1987 in as-Salt. It aims to preserve the local heritage of the district of al-Balqa. Its collections illustrate Jordanian culture, showing how people faced the challenges of nature and how they adapted to their environment. It compares the lifestyles, houses and tents, weapons, and utensils of the villagers and nomads (Visit Jordan, As Salt 2017).

There is also the Ammarin museum at the Bedouin Camp in Beidha, Jordan (Nadinetoukan 2017).

In Jeddah, Saudi Arabia there are two relevant museums. The Jeddah Regional Museum of Archaeology and Ethnography is a major museum in the historic Khuzam Palace in the Al-Nuzlah al-Yamaniyah quarter in the south of Jeddah. The building was constructed as a royal palace from 1928 to 1932 under the supervision of Muhammed bin Laden. One collection contains archeological and historical artefacts ranging from the Stone Age to pre-Islamic and Islamic cultures. A second collection holds ethnographic items portraying the recent culture of the region. There are also two royal halls containing private historical items, documents, and photographs of King Abdul Aziz and King Saud (Cuddihy 2001; Trip adviser, Jeddah 2017).

Another museum in Jeddah is a historical museum in Tayebat International City, belonging to Abd al Rauf Khaleel. The museum complex contains 300 rooms, part of which displays Bedouin culture and costumes.

In 2014 as part of a program to promote the country's culture, Saudi Arabia planned to spend 1.7 billion USD to construct 230 museums. (News, Cultural Tourism 2014; Google 2017).

In Dubai, the Dubai Bedouin museum located in the historic neighbourhood of al-Fahidi, presents exhibitions on its Bedouin heritage, the history of the emirate, which is part of the United Arab Emirates, and its growth from the sands to a modern country (Visit Dubai 2017).

The National Museum of Qatar is currently under construction in the capital, Doha. The building was designed by architect Jean Nouvel, and was scheduled to open in 2016. The museum's mission is to celebrate the culture, heritage and future of Qatar and its people, embodying the pride and traditions of Qataris while offering international visitors insights on the natural history of the desert and the Persian Gulf, artifacts from Bedouin culture, historical exhibitions on the tribal wars and the establishment of the Qatari state, and finally, on the discovery of oil to the present.

13.3 Multiculturalism and Ethnographic Museums in Israel

Multiculturalism in Israeli society plays a role in the phenomenon of museums. There is a high ratio of museums to population size in Israel (over 200 existing museums) with a significant growth of ethnographic museums in the last two decades (representing over 10% of the total). This may reflect stages in the evolution of multiculturalism in the Israeli context: following the initial Jewish melting pot concept of integration of a diversity of immigrants, ethnic re-awakening took place in the country with ethnocentricism being expressed also in the form of museums (Perry and Kark 2017).

These local ethnographic museums function both to distinguish and to integrate the cultural groups they represent in Israeli society. Museums of Jewish communities in general reproduce Zionist narratives, while at the same time expanding the museum mission to include life in the Jewish Diaspora. Some museums of non-Jewish cultural groups function to reclaim their local, enduring, predominantly Arab-Palestinian narratives, which often conflict with the Jewish Zionist narrative. Museums in Israel face challenges in fulfilling the differing, and sometimes conflicting, needs of multiculturalism. They may thus serve the roles of both reflecting and contributing to the evolution of Israeli society with its component ethnic and social groupings (Kark and Perry 2012).

The ethnic re-awakening in Israel today includes the establishment dozens of ethnocentric, separatist new museums, which challenge Zionist and statist formative narratives. These museums serve their communities, and the entire Israeli society, by giving each community a space of its own, to reclaim and recount its individual story. Sometimes while glorifying to an exaggerated extent the contribution of certain groups at the expense of the epic narrative of Israeli statehood. Museums in

Israel can lead, contribute, and assume a very important role in nurturing multiculturalism, by giving groups individual and mutual respect; adequately emphasizing, while maintaining historic accuracy, the cultural idiosyncrasies of each ethnic group; and highlighting the roles and contribution to the development of Israeli society, in all of its groups and ethnic components (including women, Mizrahi Jews, Palestinians/Arabs, migrant workers, etc.).

At the same time, one must consider the dangers in comprehensive adoption of multiculturalism, which include regressing to racist discourse and emboldening un-enlightened un-democratic cultures. A blind adoption of multiculturalism could also contribute to an unjust disregard of the past contribution of the kibbutzim and other agricultural settlements, especially as the backdrop to today's struggles over real estate, or it could lead to backlash and the strengthening of political parties and groups that oppose multiculturalism all together. The display in museums brings forth the difficulties in bridging the gap between nationality-citizenship-uniformity and nurturing multiculturalism-difference-diversity of each and every group, as well as the extent to which each approach should be represented. It may be that the Israeli-Palestinian conflict follows a similar pattern. Every resolution of the conflict must go hand in hand with mutual change in perspectives over the conflicting national ethos of each group inhabiting this country. Here lies another leadership challenge for multicultural ethnographic museums (Perry and Kark 2017).

13.4 Bedouin Museums in Israel

In Israel we have the Museum of Bedouin Culture at Lahav in Southern Israel in the Negev (discussed in detail below), the Bedouin Heritage Center in Shibli in the Galilee, and the Bedouin Heritage Museum at Rahat, one of the Bedouin towns in the Negev, set up by the local residents to tell their own story, but which did not persist (see Fig. 13.1).

The Bedouin Heritage Center-Shibli was set up in 1897 for a short term by the local council and Diyab Shibli, who in 2000 re-opened it as a tourism center and a potential economic support for the village women. The village of Shibli, located at the foot of Mount Tabor, is populated by the remnants of the Arab a-Sabih Bedouin tribe, most of which scattered in Jordan during the 1948 war. In 1987, a Bedouin heritage week was held in the village; as part of it, an exhibit was made in which many Bedouin schools throughout Galilee participated. The local council continued to foster the site at which the exhibit was mounted. Afterward, the private collection of Diyab Shibli was added, and he was appointed director of the museum (Shibli studied museology at Tel Aviv University). Later the council decided to close the

Fig. 13.1 A Bedouin "orientalist" painting of nomadic life in the cultural center of the largest Bedouin town in the Negev, Rahat. Artist: Zuhir Al-Huzayl (by R. Kark, 2008)

exhibit, and Diyab Shibli chose to take it under his own auspices. He moved the items to a building that he rented next to his home and reopened the heritage center in 2000 (Shibli 1992).[1]

At the entrance, a movie on the Galilee Bedouins is screened, and a few objects are hung on the walls. The small exhibit is arranged in one large room, and contains traditional garb, housewares, vessels for making coffee, cooking and baking utensils, rugs embroidered in various stages of preparation, many agricultural implements (see Fig. 13.2), and personal items (jewelry, prayer beads, and the like). The exhibit does not have written explanations, and Shibli himself guides the visitors. Outside the building, there is a Bedouin hospitality tent as well as a garden of herbs and medicinal plants, which in Shibli's opinion complete the exhibit. Shibli hopes to develop the place. Currently the local council together with Wexner Foundation

[1]Shibli, Diyab. Director of the Bedouin Heritage Center, Shibli, Interview with Noam Perry, 17 June 2006.

Fig. 13.2 Display hall, Bedouin Heritage Center, Shibli (by N. Perry, 2006)

Alumni are promoting an extensive development plan for the village in order to make proper use of the village potential to attracts tourists, especially the Christian pilgrims visiting Mount Tabor.

13.5 The Joe Alon Center for Regional Studies

13.5.1 The History of the Center

The center was initiated by members of Kibbutz Lahav in the 1970s, who wished to establish within the kibbutz area a research center specializing in the Shfela and Negev regions. The center agenda was to facilitate three major tasks - an archeological collection, a center for environmental study including a display of flora and fauna, and a Bedouin folklore collection.[2] The triple agenda with a strong reliance on archeology is reflected in the symbol of the center: three stylized clay pots (see Fig. 13.3).

[2]Navon Avi interviewed by Havatzelet Yahel, Lahav, 4 Nov. 2015.

Fig. 13.3 The Joe Alon Center's symbol (Joe Alon Center Archive)

Joe Alon Center

The archeological display is intended to present nearby findings from several expeditions.[3] This includes relics from the chalcolithic period (around 3500 BC) through the Canaanite and the Jewish kingdom, the period of the second Jewish Temple, the Mishnaic and Talmudic eras (2nd–4th centuries C.E.). A second museum would hold the many Bedouin objects purchased from the Bedouin, or given as gifts by Bedouin to the neighbouring kibbutz. These objects were previously displayed in one of the Kibbutz warehouses.

The predominant founders and first directors were Avi Navon and Uzi Halamish, both members of Kibbutz Lahav; Halamish had previously served as head of the Benei Shimon Regional Council municipality, within whose boundaries Kibbutz Lahav is located.

Initial funding for the 60-m^2 building of the center was given as a loan by Mifal HaPayis, the national lottery of Israel with additional early support from the Joe Alon Association.[4] In July 1973, Colonel Yosef (Joe) Alon, a former air force pilot, and one of the founders of the Israeli air force, was assassinated in Maryland while serving as Israel's air attaché to the United States. His friends established a non-profit association in his name. The "Center for Regional and Folklore Studies" was dedicated to his memory in 1973, with the inauguration of the first building in 1975.

After the opening, the Keren Kayemet Le-Israel (Jewish National Fund-JNF-KKL) provided a substantial amount to enable it to build a second floor and observatory. The KKL is the primary Zionist movement involved with Jewish land purchases before the establishment of the State of Israel in 1948. It received a space for exhibiting its own important contribution to Jewish settlement in the Negev (Fig. 13.4). The rooftop observatory provides a wonderful view of the surroundings, and even has a telescope for gazing at the stars.

In the 1980s, the Center directors initiated an ambitious plan (Lissar 1980) (Fig. 13.5). It aimed to include five new buildings to facilitate the study of the history of the area and its characteristics, including the following:

- Collecting and displaying the archeological findings.
- Collecting and exhibiting local Bedouin ethnology and folklore.
- Collecting and exhibiting the local flora and fauna, a geological display, and samples of the local agriculture.

---

[3]See the list of the many archeological expeditions in Israeli Antiquity Authority website: http://www.hadashot-esi.org.il/search.aspx (accessed 5 Jan. 2017).

[4]Uzi Halamish interviewed by Havatzelet Yahel, Lahav, 8 Oct. 2015.

Fig. 13.4 KKL exhibition in Joe Alon Center (by H. Yahel 2015)

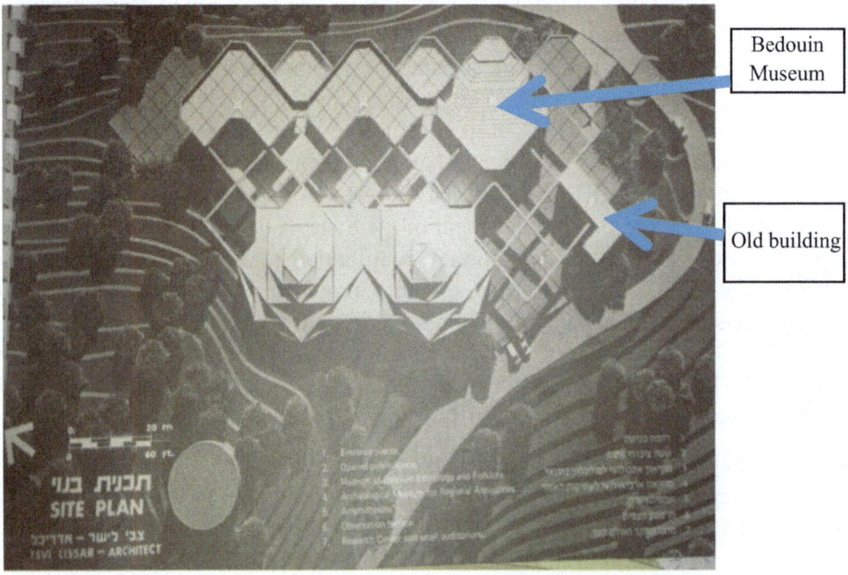

Fig. 13.5 Proposed site plan, architect Zvi Lissar, Program for Joe Alon Center, 1980 (Joe Alon Center Archive)

Fig. 13.6 Aerial photo of Joe Alon Center, around 1985 (Joe Alon Center Archive)

- Guiding pupils and visitors at the local historical site.
- Encouraging and developing tourism in the countryside.
- Assisting research delegations and local researchers.
- Organizing meetings and symposia.

However only the Museum of Bedouin Culture materialized (Fig. 13.6).

In that period, a large private collection of objects representing Bedouin tribes of the Sinai Peninsula was created by Orna Goren, an Israeli who lived with her family in central Sinai from 1975, where a substantive number of Bedouin live. As an archeologist, she was aware of the importance of preservation. Goren established a museum for Bedouin culture in Sinai. However, in 1982 as part of the peace treaty with Egypt, Israel was obliged to withdraw from Sinai and Goren needed a place for her exhibits. Navon offered to display her collection along with the existing Bedouin objects in the newly-built space. Goren became the Bedouin museum's first curator, a task she carried on until 2006 when Chavi Feingold, her assistant, took over the position.[5]

Halamish, Navon, and Goren traveled to the United States and Mexico to gain impressions of museums devoted to native tribes. In 1985, with the help of the Joe Alon Association, the KKL, Israel's government, and some private contributors, the

[5]Navon Avi interviewed by Havatzelet Yahel, Lahav, 4 Nov. 2015; Feingold Chavi, interviewed by Havatzelet Yahel, Lahav Forest, 9 Feb. 2017.

museum was reopened, combining the two collections. Since that time it has been managed by the Bnei Shimon Regional Council (Navon 1988).

Shortly before the reopening, many of the archeological items were stolen, and the Israel Antiquities Authority refused to grant the center suitable antiquities to replace those that had been stolen.

Despite the attention devoted to Bedouin culture, Halamish wished to maintain the museum as an interdisciplinary study center about the place and its entire region, and not only for the field of Bedouin culture. Halamish has said that it had been suggested to call the place the "National Museum for Bedouin Culture" and to refrain from delving into other subjects, but he was opposed to this idea, and therefore promoted the establishment of two additional wings.

13.5.2 The Jewish Heritage Wings

In the 1990s, Halamish began development of two new wings focused on Jewish history: the Bar Kochba Museum and the Museum of Jewish Pioneering Settlement in the Negev.

Bar Kochba led a revolt against the Roman Empire in 132 C.E. He established an independent Hasmonean Jewish state which he ruled for three years. Under his leadership, the Jewish community dug a complex system of underground caves and channels in which to hide themselves from the Romans. Several of these caves were discovered near the Center. The display included several replicas of connected caves, where visitors could see and sense the underground life of the Jewish warriors. It also had a film and several objects in glass cases with explanations. This museum was open only for a short while, due to safety issues (see Fig. 13.7).

The second new museum wing focused on Jewish settlement in the Negev in modern times. "Daroma," an impressive and costly audiovisual display was a central part of the exhibit, telling the history of Jewish settlement, mainly that of the 20th-century Zionist movement. This wing was also closed, unable to continue the expensive audiovisual display.

Fig. 13.7 A replica of underground Bar Kochba caves in the museum (by H. Yahel 2015)

When Halamish left his position as manager in 2010, his first successor, Eyal Zamir, put all the focus on the Bedouin museum. However, in 2013, another shift occurred. Ofir Libstein who volunteered to be the head of the center, explained:

> Until a few years ago, the Museum of Bedouin Heritage was the core, but today it is becoming the center for multicultural settlement in the Negev. We have two central wings, one that preserves the Bedouin heritage, and the other representing the progress of Jewish settlement in the Negev. We provide a meeting point between the two populations. When I say multiculturalism, I mean that you can find in this region, for example, Bedouin from the cities as well as recognized and unrecognized villages, along with people from Hashomer Hatzair kibbutzim, evacuees from Gush Katif who are instructors in the center, and religious and secular Jews who live together in Kibbutz Kramim.[6]

Barkai Saar, who took over the management after Libstein in 2015, emphasized the relations between Jewish life in the olden days and the nomadic Bedouin culture.[7]

13.6 The Museum of Bedouin Culture

13.6.1 Who Are the Negev Bedouin?

In the Negev desert, in the southern part of the State of Israel, live around 1 million people, of whom 230,000 are Muslim Bedouin and the rest Jewish. The Negev is a peripheral and marginal region.

The Bedouin are the poorest sector in Israel and suffer from high unemployment and a high birth rate (Annual Bureau of Statistics 2015). As a traditional community, the women are subject to various restrictions in the public sphere. One-third of the Bedouin men are married to more than one wife, although illegal under Israeli law (Lapidot-Firilla and Elhadad 2006; Abu-Rabia et al. 2008). The Negev Bedouin are not a homogenous community, comprising at least three different groups. The “genuine” or “true” Bedouin were once nomadic tribes living as herders in the rough conditions of the desert. For their survival they developed a unique way of life with traditions, strict laws, and an enforcement system. These Bedouin tribes roamed between the Arabian deserts, and arrived in the Negev mainly within the last 200 years. The fellahin—farmers who immigrated from the Nile valley in Egypt seeking available arable lands-joined the genuine Bedouin tribes and took upon themselves many of the Bedouin traditions. The third group consists of former slaves, who had been bought by sheikhs. In 1948 War and immediately thereafter, the Bedouin went through significant changes, with more than three-quarters of them fleeing from the Negev to neighbouring countries (Yahel and Kark 2015). The 11,000 Bedouin that remain within Israeli borders have

---

[6]*Hazman Ha-Yarok*, 28 Mar. 2013.

[7]Saar Barkai, interviewed by Havatzelet Yahel, Lahav Forest, 29 Sep. 2015.

gradually begun to settle in towns and villages, leaving their nomadic or semi-nomadic way of life. Currently, 60% of the Bedouin live in planned towns, while the rest are widely scattered in unauthorized settlements, were they suffer from a lack of infrastructure. The current problematic situation is strongly related to land disputes between the state and around 15% of them, in an area of about 600,000 dunams (Yahel 2017). Over the years since 1948 many state committees have proposed various solutions to the disputes, but with very limited success.

13.6.2 The Bedouin Culture Exhibitions

The permanent exhibition of the Bedouin culture-one of the largest of its kind-was opened in 1985. It includes around 500 heritage objects of the more than 3000 that the museum holds. About 100 of the displayed objects were collected by Goren in Sinai. The 828-m^2 exhibition was designed by Dodik Gal of the Israel Museum. The museum received official recognition from the state (Feingold personal interview 2017). One of the museum's brochures states:

> "[The museum] was established in 1985 when it became apparent that the formerly nomadic Bedouin tribes of Sinai and the Negev were in a process of transition to a more settled way of life, moving to established towns and villages. The Bedouin have recently acquired a modern way of life, reflected in cultural, social and economic changes as well as changes in housing, dress etc. As a result, many of the former Bedouin ways of life are disappearing, and with them the objects and traditions of Bedouin society, prompting a need to collect, preserve and record them before they completely disappear."[8]

The displays include mainly traditional objects: Bedouin arts, weaving, weapons, and artifacts (see Figs. 13.8, 13.9 and 13.10). It also presents a sheik's grave and the unique traditions around it. Household items, jewelry, and artwork were donated by Bedouin families. The brochure describes the setting:

> "The visitor walks around the display in a circle, resembling the wanderings of the Bedouin and the cycle of the seasons. Displayed here are different types of dwelling, articles of clothing, examples of crafts and traditional economies, pieces of jewelry, toys, etc. A display of authentic woven carpets hangs down from the ceiling into the central hall. The visitor continues down a flight of stairs to the lower floor, devoted to more spiritual aspects of traditional Bedouin life: history, society, the cycle of life, traditional medicine, and the customs of the Bedouin market."[9]

The Bedouin heritage is also presented in the traditional host tent outside. Visitors can hear stories and folktales, while the Bedouin host tends an open fire. Visitors are offered bitter coffee, sweetened tea, and baked pita straight from the *taboon* oven. In the outdoor setting, there are also live animals, such as camels,

---

[8]See in the museum website: http://www.joealon.org.il/?page=category&cat=63 (accessed 5 Jan. 2017).

[9]Ibid.

Fig. 13.8 Weaving loom, the Bedouin museum collection (by H. Yahel 2015)

Fig. 13.9 Bedouin weapons and other artifacts, the Bedouin museum collection (by T. Regev 2013)

Fig. 13.10 Camel saddle, the Bedouin museum collection (by H. Yahel 2015)

goats, and horses (Fig. 13.11). Outside the museum buildings a herb garden, including a section for herbal medicine was cultivated.

Bedouin objects are also presented, in the Satellite museum (mobile exhibits) initiated with the cooperation of educators from Rahat, a nearby Bedouin town.[10] "The Color of the Desert" exhibit was presented in Bedouin towns and villages as well as Jewish schools throughout the country.

13.6.3 Guides

The primary guides for the museum have always been Bedouin, offering tours in Hebrew and Arabic (Fig. 13.12). Most of the guides are Negev residents. Several, including head guide Auda Abu Ka'ud, are from the 'Arab a-Ramdin tribe, which prior to 1948 had lived on the lands where the museum is now located and had then gone to dwell in the Hebron hills, now under the Palestinian Authority. After 1967, these employees received permission to work in Israel.

[10]Halamish Uzi interviewed by Noam Perry, Lahav, 16 Aug. 2007.

Fig. 13.11 A camel in the museums yard (by H. Yahel 2015)

13.6.4 Management

Paid executives of the museum are all Jewish, although Bedouin representatives have been part of the governing board. An important aim of the museum was that it should operate in full cooperation with Bedouin community leaders.[11]

13.6.5 Visitors

At its peak, when all Israeli third-grade classes visited the museum as part of the curriculum, there were more than 50,000 visitors a year. When the Ministry of Education abolished the practice, the number of visitors dropped.[12] In 2012, there were about 15,000 visitors, with an increase in 2016 to about 35,000.[13] The internal distribution of visitors was inconsistent during the last years, however the number of Bedouin pupils have doubled in the last 4 years to about 10% of the visitors. However, according to interviews, there is an increased interest in the museum by

[11]Uzi Halamish interviewed by Havatzelet Yahel, Lahav, 8 Oct. 2015.

[12]Ofir Libstein, *Hazman Ha-Yarok*, 28 Mar. 2014 http://www.kibbutz.org.il/itonut/2013/dafyarok/130328_jo_alon.htm (accessed 9 Jan. 2017).

[13]Feingold Chavi, interviewed by Havatzelet Yahel, Lahav Forest, 9 Feb. 2017.

Bedouin individuals and leaders.[14] Halil Abu Shriki residence of Lakiya, a nearby Bedouin village brought his family to visit the museum, explaining:

> "I come here at least once a year and enjoy seeing the items and explaining our tradition to my children and grandchildren. I lived in a tent but they live in a house."[15]

His wife Na'ama wanted to have a picture with her daughter and grandchild in the small tent on the lower floor of the museum, noting they have nothing like that in their village (see Fig. 13.13).

13.6.6 Limitations of the Bedouin Exhibition

Regarding technical aspects, the exhibition has not changed since 1985, and is outdated. There is an urgent need to invest in conservation of the exhibits and displays.

As for its content, despite its clearly-stated aims and its cooperation with the Bedouin community, the museum has been criticized for not reflecting the many changes that have occurred in Bedouin society since the 1980s, including the accelerated urbanization and current social problems. The lack of connection to contemporary Bedouin society even led to the resignation of Orna Goren.[16] Almost all the displays present past traditions. Current Bedouin urban life, and the story of the new Bedouin villages and towns are hardly mentioned. In 2006, when Goren left her position at the museum she noted that her suggestions for changes had been rejected due to budgetary restraints. She further explained:

> "A museum should develop, to be connected to the Bedouin population and demonstrate what is happening within it today.... I think that I also represent the Bedouin ... who wish to see the change..."[17]

Feingold shares this view and thinks that the physical aspects of modern Bedouin life should be included in the display.[18]

But not everyone agrees that the museum should address current issues. Ali Abu Qrinat, a Bedouin museum guide for fifteen years believes that:

> "We [Bedouin] should present our past heritage, there is nothing worth presenting in the present."[19]

These contradictory views remain as an open question. The Bedouin community has gone through many changes in the last 100 years, from traditional nomadic to a

---

[14]Salah Abu-Habi, *Hazman Ha-Yarok,* 28 Mar. 2014.

[15]Halil Abu Shriki, interviewed by Havatzelet Yahel, Lahav Forest, 29 May 2015.

[16]Uzi Halamish interviewed by Havatzelet Yahel, Lahav, 8 Oct. 2015.

[17]*Haaretz*, 4 Sept. 2006.

[18]Feingold Chavi, interviewed by Havatzelet Yahel, Lahav Forest, 9 Feb. 2017.

[19]Ali Abu Qrinat, interviewed by Havatzelet Yahel, Lahav Forest, 29 May 2015.

Fig. 13.12 Ibrahim Abu Khaf during guided tour in the museum (by H. Yahel 2015)

Fig. 13.13 Naama Abu Shriki with her daughter and grandchild in a small tent (by H. Yahel 2015)

Fig. 13.14 Part of the wall hanging in the bedouin Museum (by H. Yahel 2015)

modern, western-like, permanent-dwelling way of life. Modernity brought, among other things, decrasing respect for the elderly and the traditional leadership; the empowerment of women, many of whom want to be more independent and to be able to study and work outside the village. It also gave power to the movement against polygamy. Other current challenges relate to ongoing land disputes, the question of unrecognized Bedouin settlements, and the poor quality of Bedouin public services. Whether they like it or not, new western perceptions are entering through the back door. One interesting example is the large embroidered wall hanging (more than 5.5 m^2, see Fig. 13.14). This work was a project of a workshop, the "embroidery of women's dreams." The stories and symbols depicted are not representative of desert life, but rather those of sedentary communities with many green trees.

13.6.7 The Bedouin—Joe Alon Center Relationship

The relationship between the Center and the Bedouin community has been one of respect and appreciation, but over the years it has become more complex. Four factors are notable: first, the museum was not established by the Bedouin community which is the subject of the exhibit; second, the museum is located outside of the many Bedouin settlements; third, it is managed by non-Bedouin Jews; fourth, the center does not represent a "single-ethnic community," but puts the Jewish and

Bedouin narratives side by side in the context of a place, since it was initially created as a regional museum for the Negev (Perry and Kark 2017).

These issues are not necessarily negative, especially the first two. Some Bedouin express their appreciation for the conservation of their objects, some of which are rare and unique, and would otherwise likely have vanished forever.

As for the location of the Center, it is in a "neutral" territory outside of inter-tribal disputes, since the Lahav Forest region was not part of any current Negev tribes' territory. There are no Bedouin tribes living in the Lahav forest, so there are no claims of discrimination raised by one or another tribe. Its neutral location assuages the fear that the museum might be "taken over" by one tribe at the expense of others, to present only their specific tribal legacy.[20]

But the two other issues are more problematic. Regarding the issue of Jewish management, the museum director has requested that Bedouin leaders become more involved, and over the years, the board of directors included Bedouin leaders. The center is seeking a Bedouin curator to work together with the Bedouin guides. According to Barkai and Feingold, until now no Negev Bedouin has expressed any interest in studying the field of museology.

As for the question of dual narratives, the current position of the museum board is in favor of change.[21] It was decided not to re-open the Jewish wings and to focus only on the Bedouin exhibition. They also drafted a new vision to the Center which states:

> "… to create a comfort zone for acquaintance between Jewish and Bedouin groups, since these meetings rarely happen elsewhere in the region."[22]

The aim is to promote the center as a bridge builder between Jews and Bedouin, while the Bedouin minority should feel empowered and respected.

In September 2016 the Bnei-Shimon Regional Council, decided to take over the old building that holds some of the center offices as well as the KKL and Jewish exhibitions. There, they opened high school classes for the Desert Stars initiative. Desert Stars is a nonprofit organization working to foster a young Bedouin leadership through a long intensive program.[23] This seems to be the final negative response to the debate over dual narratives under one roof in a regional museum in southern Israel.

---

[20] Ali Abu Qrinat interviewed by Havatzelet Yahel, Lahav Forest, 29 May 2015; Ibrahim Abu Kaf interviewed by Havatzelet Yahel, Lahav Forest, 29 Sep. 2015; Navon Avi interviewed by Havatzelet Yahel, Lahav, 4 Nov. 2015.

[21] Barkai Saar, interviewed by Havatzelet Yahel, Lahav Forest, 29 Sep. 2015; Eimi Chopin Ein-Gedi interviewed by Havatzelet Yahel, Lahav Forest, 29 Sep. 2015; Yarona Ben Shalom Richardson, interviewed by Havatzelet Yahel, Shoket, 1 Feb. 2016.

[22] Barkai Saar, interviewed by Havatzelet Yahel, Lahav Forest, 29 Sep. 2015.

[23] See Desert Stars homepage: http://www.desertstars.org.il/home (accessed 7 Jan. 2017).

13.6.8 The Future of the Center

The future of the regional center as a whole, and the Bedouin museum in particular is now uncertain. It is clear that the only museology activity will be that of the Bedouin museum. The high school management hopes to motivate the pupils to be involved in the Bedouin museum programs, but the museum staff have expressed concerns that in the future, the museum will become merely the "back yard" of the high school.

13.7 Conclusion

We have presented a case study of one regional center in southern Israel where an attempt was made to present two different narratives of local communities: the Jewish majority and the Bedouin minority. One of the questions raised was: is it possible to operate a regional multicultural ethnic museum in Israel? According to the experience of the Joe Alon Center, displaying two ethnic cultures with differing and contradicting histories and narratives under the same roof proved to be difficult and problematic. This outcome can be connected to poor management, and too many changes over the years, but our conclusion is that today in Israel, with so many challenges facing the communities, each one needs a place of its own in order to be empowered by ethnographic museums and centers.

The Joe Alon Center now prefers to focus on the Bedouin, and to put aside the Jewish history. This preference is both interesting and important, due to the fact that the Bedouin are a non-dominant minority group in Israel. This route provides a better chance to encourage the Bedouin to participate and get involved in the preservation of their traditions. The museum could play an important role in Bedouin empowerment, especially for the younger generation. We hope that all members involved in the municipality, in the high school, and in the center will collaborate in order to strengthen the positive potential and possible impact of the museum.

Some museums can function as bridges between different parts of Israeli society, and thus may contribute to its future as a multicultural state, as in the Joe Alon Museum's initial mission. As a part of a "supposedly multicultural national policy," Israel should embrace these museums, support them, and encourage additional under-represented cultural groups to establish museums.

References

Abu-Rabia, A., Elbedour, S., & Scham, S. (2008). Polygyny and post-nomadism among the Bedouin in Israel. *Anthropology Middle East, 3*(2), 20–37.

Al-Shark Al-Awsat. (2009). Sheet 11531. 26 Dec 2009. http://archive.aawsat.com/details.asp?issueno=10992&article=550108#.WISDIFw1N2A (in Arabic).

Anderson, B. R. (1983). *Imagined communities: Reflections on the origin and spread of nationalism*. London: Verso.

Annual Bureau of Statistics. (2015). http://www.cbs.gov.il/reader/shnaton/templ_shnaton.html?num_tab=st02_15x&CYear=2016. Retrieved December 28, 2016.

Bhabha, H. K. (2000). *The location of culture*. London: Routledge.

Cuddihy, K. (2001). *An A to Z of places and things Saudi* (p. 155). London: Stancey International.

Google, Saudi Arabia. (2017). https://www.google.co.il/search?sourceid=navclient&aq=&oq=%22SAUDI+ARABIA+Bedouin+museum%22&ie=UTF-8&rlz=1T4GGHP_enIL444IL444&q=%22SAUDI+ARABIA+Bedouin+museum%22&gs_l=hp….0.0.11.11249………..0.zWXPcPE_S3c. Retrieved December 28, 2016.

Guide2jordan. (2017). http://guide2jordan.com/jordan-folklore-and-popular-traditions-museums.

Kark, R., & Perry, N. (2012). Museums and Multiculturalism in Israel. In Y. Maos & Y. Charney (Eds.), *Geographers in Israel: A Collection of Papers, Prepared for the International Union of Geographers Conference, Germany*. Haifa: University of Haifa.

Lapidot-Firilla, A., & Elhadad, R. (2006). *Forbidden yet practiced: Polygamy and the cyclical making of Israeli policy*. Jerusalem: Center for Strategic & Policy Studies.

Lissar, Z. (1980). Site Plan.

Nadinetoukan, wanderer@largr. (2017). http://nadinetoukan.tumblr.com/post/34407888876/ammarin-museum. Retrieved February 8, 2017.

Navon, A. (1988). The center named for Joe Alon [in Hebrew]. In E. Schiller (Ed.), *Guide to Kibbutz museums* (pp. 105–108). Jerusalem: Ariel.

News, Cultural Tourism, 26.05.2014. http://www.ne-mo.org/news/article/nemo/saudi-arabia-to-spend-17-billion-usd-to-build-new-museums/363.html.

Perry, N., & Kark, R. (2017). *Ethnographic museums in Israel*. New York: Israel Academic Press.

Said, E. W. (1979). *Orientalism*. New York: Vintage.

Shibli, D. (1992). *Galilee Bedouins [in Hebrew]*. Tel Aviv: Integrated Pedagogical Center.

Trip Advisor, Jeddah. (2017). http://www.tripadvisor.com/LocationPhotoDirectLink-g295419-d4090222-i105834580-Al_Tayebat_International_City-Jeddah_Makkah_Province.html; http://www.visitjeddah.com/index. Retrieved February 8, 2017.

Visit Dubai. (2017). http://visitdubai.com/ar/see-thrill-play/heritage-and-culture/bedouin-life.

Visit Jordan, Amman 2017 (in Arabic). Retrieved February 8, 2017 http://ar.visitjordan.com/Wheretogo/amman/Museums.aspx.

Visit Jordan, As Salt (2017) Arabic]http://international.visitjordan.com/Wheretogo/AsSalt/HistoryCulture.aspx; http://culture.gov.jo/new. Retrieved February 8, 2017.

Yahel, H. (2017). The dispute between the state of Israel and the Bedouin over Negev lands (in Hebrew). *Forest, 17,* 54–66.

Yahel, H. (2015). *The policy of the Israeli government and state authorities regarding the Negev Bedouin, 1947–1989*. Ph.D. diss. Jerusalem: The Hebrew University of Jerusalem.

Yahel, H., & Kark, R. (2015). Israel Bedouin during the 1948 war: Departure and return. *Israel Affairs, 21*(1), 48–97.

Chapter 14
Recognition Policy of Bedouin Villages in Israel, Marginalization and the Ethic of Bio-cultural Diversity

Avinoam Meir

14.1 Introduction

Since the birth of the State of Israel in 1948 Bedouin settlement in the Negev desert has attracted the attention of countless scholars studying cultural, social, economic and political issues concerning the far-reaching changes in this indigenous minority. Countless recent studies were conducted on state policy of Bedouin settlement, with similar issues, cast within indigenous rights and social justice (e.g. Roded and Tzfadia 2013; Yiftachel and Roded 2014). The reason for such great attention is perhaps rooted in this group's (population 2016—240,000) extreme marginality in Israel. Bedouin marginality is multi-faceted: indigenous tribal traditional agro-pastoral people versus a settler modern urban industrial society; a Middle Eastern orthodox Muslim culture versus a Western Jewish secular culture; and an Israeli Arab-Palestinian identity versus Zionist Jewish identity. Their marginality is accentuated even within the Arab minority of Israel (about 1.2 million) as the most economically backward group (Abu-Saad and Lithwick 2000).

While these issues are important, environmental issues have been almost entirely missing from research discourses with only few studies (Meallem and Garb 2008; Kissinger and Karplus 2015) mostly limited in scope. This is striking given the extreme change in this close-to-nature culture and the recent growth of international interest in the impact of global climate change and resource exploitation on cultures and habitats of indigenous peoples (e.g. Salick and Byg 2007).

This lacuna is even more striking given that the policy of settlement of the Bedouin concerns a major environmental resource—that of about 1000 km^2 of territory (about 5% of national area) which they presently inhabit. This resource is contested by both the Bedouin for pursuing their traditional livelihood and by the

A. Meir (✉)
Department of Geography and Environmental Development,
Ben Gurion University of the Negev, Beersheba, Israel
e-mail: ameir@bgu.ac.il

S. Pelc and M. Koderman (eds.), *Nature, Tourism and Ethnicity as Drivers of (De)Marginalization*, Perspectives on Geographical Marginality 3,
DOI 10.1007/978-3-319-59002-8_14

state for implementing its national development projects. The conflict revolves around land ownership within what the Bedouin regard as their traditional tribal territories (Meir 2009). Yet little theoretical or empirical research has been conducted about *environmental aspects* of state Bedouin settlement policy and the benefits that may accrue to all involved—the Bedouin, the state and the environment if such issues are integrated into policy.

This situation may be termed environmental *marginalization*. The people inhabiting this desert environment are marginalized as their environment is driven to the backstage of state-indigenous politics. Adopting Leimgruber's (2004) quaternary classification of marginality, our concern is with a systemic marginality that is produced by hegemonic forces within the Israeli political and economic system. However, rather than generating inequities through the maldistribution of social, political and economic benefits, in our case we focus on the maldistribution of environmental benefits.

This chapter presents precisely this issue. Our purpose was to explore it from a conceptual-theoretical perspective such that it provides a wicket for a further empirical study. Since at macro-scale Bedouin detachment from their major subsistence resource of livestock grazing is primarily of a cultural nature, there is by necessity a close association between culture and environment. We submit that this linkage and the environmental shortcomings of state policy are best understood through the encompassing concept of biocultural diversity. However, based on our interpretations of this concept with regard to the Bedouin, we also explored briefly its interactions with the broader fields of environmental ethics and justice in terms of rights not only of the people but also of the eco-systemic environment involved.

14.2 Bedouin, Space, Environment and State

Our point of departure is an understanding of this society in its most fundamental historical condition as a close-to-nature grazing-cultivation pastoralist culture. Its social system is anchored in tribal organization (Marx 1974; Ben-David 2004) with an internal indigenous customary law (Bailey 2009; Stewart 2006) that leans upon historical accumulation of social knowledge. From the spatial and environmental perspectives (e.g., Stewart 1986; Perevolotsky 1987) this system covers delineation of territorial tribal boundaries and informal management regulations of space and eco-systemic resources that concern usage, cyclical movement and location of fixed and mobile abodes.

To best grasp the Bedouin-space-environment-State nexus we begin by the recently proposed concept of consummate space (Meir and Karplus 2015). This concept is particularly applicable to a unique small cultural group, such as an indigenous one, which constitutes distinctive spatial enclave within space of the greater society. We suggest that within the greater space, which has its grand mode of production (Lefebvre 1991), these cultural groups seek their distinctly optimal production of space. The search integrates land uses, modes of spatial and environmental management and 'regulation', and ideals, symbols and experiences constituting group-based customary patterns and practices. Consummate space implies perfectness in terms of congruence, stability and mutual harmony of the components vital for the fully-realized existence of this cultural group (see e.g. Harvey 1985; Karplus and Meir 2013). As an imagined entity it is manifested in intimate familiarity with and acceptance of its tangible components and their self-produced rules and regulations concerning the permitted and forbidden. Its stability is reflected in high individual and collective economic, environmental, social and political security and assuredness in their space, facilitating socially, culturally, politically and environmentally sustained resilience in the face of pressures applied by external processes of space production.

Bedouin space until mid-20th century may be described as relatively consummate whereby their livestock grazing and dry farming economy continued almost uninterrupted. This enabled a long period of accumulating environmental knowledge (Marx 1973; Bailey 1976) through intimate relationships with first nature (landforms) and second nature (biological and zoological entities). This vernacular knowledge has transformed into Bedouin's unique culture as their third nature (Karplus and Meir 2014) and a supporting pillar of their consummate space.

Similar to other pastoral societies in the 20th Century (Khazanov 1994), and indigenous peoples in general, since the early 1950s the State of Israel has been interfering with Bedouin delicate space-environment nexus. Broadly, state attitude towards the Bedouin was driven by security and geo-political motives. The resultant settlement policy involved considerable relocation of Bedouin groups from their traditional territories (Meir 1997; Ben-David 2004; Yahel and Kark 2014) and dispossession of their land rights (Yiftachel et al. 2012). Spatially, since the mid-1960s this policy intended to relocate all Bedouin population (then ~ 20,000) into state-planned towns. Until the late 1980s, seven such towns were established and about half of the Bedouin population was resettled. These new town-dwellers, belonging largely to a landless subgroup, were traditionally sub-ordinate to the real Bedouin landlords (Ben-David and Gonen 2001). As such they were not part of the land conflict. Relocation to towns was perceived a social upgrading as the state allotted them with small land plots for residential purposes in exchange for evacuating their spaces. The real Bedouin, constituting the second half of the population, have been refusing to wave their land rights and relocate to towns. They settled independently within their territories in dozens compounds of squatter villages which have become unrecognized by the state. The struggle for recognition involves demands for provision of municipal and state social services and infrastructures, and eligibility for municipal independence and elections. In the recent decade the state has recognized 12 villages (population ~ 60,000). However, the land conflict with the state is not resolved and this recognition has been "on paper" only without actual resettlement. The remainder of the population in the unrecognized villages, numbering about 30–40,000, has not even been included in this 'recognition' scheme. In fact, all unrecognized villages (population ~ 100,000), are still not regulated spatially.

Bedouin economy has been shifting toward integration with the secondary and tertiary sectors in the urban labour market. Yet, in many respects the Bedouin in these villages still resemble their traditional agro-pastoral subsistence economy. This group is the focus of our theoretical engagement with environmental aspects of state policy of recognition and settlement.

14.3 The Bedouin and Cultural Diversity

State policy towards Bedouin settlement desires their evacuation from traditional territories under the disguise of social modernization through the Western urbanization model (Meir 1997). Implementation of this policy involves functional detachment from subsistence grazing and dry farming economy. This policy, reflecting cultural transformation, completely disregards Bedouin indigenous historical attachment to their places and their significant material emotional and symbolic meanings. It constitutes thus a unidimensional and linear view of Bedouin culture premised on an assumption that this society has been a socially, economically and culturally homogenous pastoral-nomadic group. This assumption entailed therefore a vision of the desirable future for these people in which they are destined to linearly follow the urbanization trajectory of development.

We contend that, from the perspective of cultural diversity, these views of Bedouin culture are highly simplistic. In a globalized world cultural diversity that is, the quality of coexistence of diverse cultures within the same political realm, is viewed currently as highly vital for the survival of humanity in the long-term, calling thus for protection of unique sub-cultures (see UNESCO 2002). In examining pre-urbanization Bedouin society we suggest it is capable of contributing considerably to cultural diversity nationally and regionally. Those multiple dimensions of Bedouin marginality highlighted above are customarily viewed from marginality's negative perspective that requires adopting equity values in closing development gaps towards the majority culture. However, by adopting an alternative perspective of marginality we may view it precisely as a resource (Meir 1999) such that these dimensions may carry significant positive meanings. We may look at the Bedouin from the perspective of liminality in its interpretation as an edge culture. This interpretation implies the Bedouin practice a significantly different mode of living with modes of knowledge normally unavailable to the host culture. This knowledge constitutes a cultural capital and an important resource for the benefit of the wider society in various areas of life. This notion of marginality as an advantage is commensurate with Santos' (2007) post-abyssal thinking and the role of 'other' knowledge ecologies in shifting the status of marginal regions and peoples into center stage as equally legitimate and valuable to those of 'core' regions and cultures.

While the Israeli government ignores this local potential contribution to cultural diversity, the Bedouin were quite cognizant of it. In the late 1990s they launched a public campaign for state recognition and development in their villages. It was directed against the regional master plan prepared by the state in 1995 which ignored these villages altogether. The Bedouin prepared an alternative plan with one of its sections dealing specifically with their culture (RCBUV 1999) which we interpret here from the perspective of cultural diversity. In this document they argue that as part of the Arab minority sharing the same space with the Jewish majority they cannot be excluded from this regional bi-cultural reality.

However, the major notion in this plan was that of a multi-faceted Bedouin society. First, following the seven-town urbanization program, this society is now divided between urban culture and a rural culture, the latter still greatly oriented towards grazing and cultivation. The latter tier contains itself further sub-divisions, with groups that rely on grazing and those with greater inclination towards cultivation, both as cultural extensions of their semi-nomadic agro-pastoral heritage. Furthermore, historically geographically and tribally this subdivision is diverse in its origins in the Levant, northern Arabia and northern Africa. Finally, even within the unrecognized villages there are still some peasant landless subgroups whose ancestors became annexed by the Bedouin in late 19th Century for cultivating their land as sharecroppers or as slaves but did not settle in towns yet (Karplus 2010). As shown also by linguists Rosenhouse (1984) and Henkin-Roitfarb (2011) Bedouin dialects vary locally as well as nationally within the Israeli Palestinian Arab minority.

All these constitute an internal and external mosaic of considerable cultural diversity within Bedouin community. In Bedouin view (Meir 2005) their culture is capable of generating interbreeding with the Jewish majority culture, enriching thus regional cultural diversity and become an active partner in the regional multi-cultural system rather than a passive trailer of a unitary urban culture to which the state destines them.

14.4 Biocultural Diversity

The concept of cultural diversity and its crucial role in human existence draws upon its predecessor—biological diversity. For our purposes we adopt a basic common and encompassing definition that regards biodiversity as the "totality of genes, species and ecosystems of a region" (Larson 2001, p. 178). Biodiversity is regarded a vital resource, and is thought to be essential for the long-term survival of life on earth. Contrary to past ethics, in recent decades it has become conventionally accepted that all life forms including humans are equally sharing the same earth ecosystem. Therefore the two concepts should be viewed integrally in what has become accepted as *biocultural diversity* (Maffi 2001, 2007). Its basic notion is that by preserving the multitude of cultures, however small and local, global as well as local, biodiversity can be preserved with considerable benefits for the wellbeing of humanity also biodiversity.

The biocultural diversity discourse refers particularly to those liminal cultures such as indigenous peoples. These groups tend to inhabit marginal edge or transitional environments that are most often remote and characterized by extreme conditions avoided by core global and national cultures. These environments contain eco-systemic assets in flora and fauna highly valuable to the local cultures but not less to the core culture yet unfamiliar to its members. Turner et al. have shown that such edges of ecosystems often exhibit high levels of species richness. Similarly, cultural transitional areas, where cultures interact, are rich and diverse in cultural traits of many small groups who inhabit them. This results in availability of a wider overall range of traditional ecological knowledge since their indigenous inhabitants take action to maintain their ecological sustainability (Turner et al. 2003).

These notions are highly related to the concept of biocultural diversity. In communities that rely directly on local natural resources for subsistence the social and cultural structure may be viewed as reflective of a latent vernacular system of management and preservation of these resources. This system, as elaborated by Kideghesho (2008), is the reservoir of all beliefs, customs and taboos inherent in their heritage. In the absence of importable external resources, the system is dependent upon and tied to the local natural resources through a process of long and uninterrupted evolutionary and symbiotic local relationships. It carries thus the entire accumulated cultural knowledge produced in this symbiosis. This is embodied in the local group's language which is the reservoir of local concepts, terms and names that by necessity reflect closely the local biocultural diversity as its major source (Smith 2001).

Biocultural diversity as a theoretical concept has not escaped criticism. One of the major criticisms has been that the power of biocultural diversity is diminished when local indigenous communities suffer the impact of western economic and development powers. In such a reality, argues Holt (2005), the cultural diversity gradually disappears, to be followed then by the environmental values cherished traditionally within practice and language. But we wish to reverse this proposition. We maintain that if such cultural systems are given the opportunity to preserve their consummate space such that it remains uninterrupted, there is quite a significant potential of preserving their culture, their particular ecosystem (which no other culture can or wishes to inhabit), and thus its biodiversity. That is, policies of development should not highlight economic benefits solely but rather become 'culturalized' by aiming at cultural benefits to the local group too, particularly among indigenous peoples.

14.5 Bedouin Unrecognized Villages and Biocultural Diversity

Despite various constraints imposed by the state on Bedouin grazing and cultivation, such as limited grazing spaces and water allocation and despite non-profitability of this enterprise, the Bedouin continue to maintain their livestock (sheep and goats) estimated at about 200,000 heads (Landau et al. 2015) in both towns and the unrecognized villages. Many households still raise animals but at very small flock size. While men have entered the waged labour market, their women take a major role in this activity. For some households, with large flocks, this is their prime source of income. But for most others this is a cultural symbol and a vehicle for maintaining Bedouin tradition while also providing some extra income under conditions of permanent economic uncertainty, high unemployment rate and lowest income levels (Ginguld et al. 1997; Kressel 2003; Degen and el Meccawi 2009; Wachs and Tal 2009).

The last 2-3 decades have witnessed a lively research on the ecological value of Bedouin livestock grazing, attesting to the significance of the issue (Perevolotsky and Seligman 1998; Perevolotsky 2006; Wachs and Tal 2009; Golodets and Boeken 2006; Boeken 2012). The major conclusion by these studies is that, contrary to reduced biodiversity and increased land degradation effects of Bedouin grazing as previously argued (Society for Protection of Nature 2008), in Middle Eastern ecologies biodiversity can peak and land degradation reduced precisely through livestock grazing, provided that it is practiced at moderate levels that avoid blanket overgrazing and maximized livestock production value. As shown the Bedouin have the ecological know-how to maintain the quality of their grazing ecosystems under uninterrupted conditions (Bailey 1976) that is, politically uninterrupted consummate space, as revealed also by Meir and Tsoar (1996) regarding vegetation density, soil stability and desertification effects of overgrazing.

We may now relate this to the notion of biocultural diversity. Only about 7% of the Israeli urban Jewish population (the core cultural group) actually inhabits the Negev desert which constitutes about 50% of national territory. The Bedouin, in contrast, view this region as their natural habitat. Livestock grazing carries thus a potential ecological value in preservation of biodiversity. In other words it is Bedouin culture that best fits the ecological conditions of this ecosystem, possessing the know-how for merging between these two components of biocultural diversity. And yet, despite attempts by the state, Bedouin livestock grazing has been largely an unregulated enterprise, practiced particularly in the unrecognized villages. The risk of overgrazing and soil degradation at presently unrecognized spaces, where their consummate space has been interrupted severely, is significant.

Interpreting this from state Bedouin settlement policy, recognition of Bedouin villages may serve as a vehicle for achieving considerably greater regulation of livestock grazing. This can be reached through enhanced planning and implementation, including land ownership arrangements, of those villages that have recently already received formal recognition. Eco-systemic regulation of livestock grazing will thus become part of the overall regulation of Bedouin space in these villages as a vehicle toward reaching moderate grazing capable of increasing biodiversity and reducing land degradation. As shown by the above-reviewed studies, sufficient scientific knowledge has been accumulated concerning various aspects of flora, fauna, soils, and hydrological and geomorphological processes related to grazing in this region to facilitate establishment of criteria for moderate livestock grazing in terms of flock size, composition and spatio-temporal distribution. The vernacular knowledge accumulated culturally by this indigenous group will complement this rational scientific knowledge for the benefit of the local ecosystem.

14.6 Environmental Ethics Implications for Policy

Environmental benefits accrued from Bedouin grazing have already been accounted for (e.g. Wachs and Tal 2009). To this we may now add an environmental ethics perspective that is itself viewed from the perspective of marginality. The marginality discourse mostly highlights issues related to social and distributional justice. However, based on the concepts of consummate space and advantageous marginality proposed above, and following Karplus and Meir (2010), we may suggest that any social group (indigenous peoples in particular) has the cultural right for its unique spatiality that is, to produce its own consummate space within an ecosystem based on its cultural richness. Accepting the biocultural diversity paradigm, in which human cultures and their ecosystems are integrated at equal weights, it is insufficient to highlight human rights without equally highlighting ecological rights. We submit that as much as humans have the right to enjoy biodiversity as an ecosystem service for improving their livelihood and cultural sustainability, the ecosystem itself has an equal right to enjoy the cultural service generated by the cultural diversity of the people who inhabit it so that it can maintain its own biodiversity and become thus sustainable itself. This corresponds with the spirit of the deep ecology ethic (Naess 1989). The bi-directional ethic of biocultural diversity is an inherent ecological and cultural value, particularly among marginal indigenous peoples practicing tight relationships with nature.

This ethic is not only an abstract concept but one that carries practical implications. State policy of Bedouin settlement in the Negev may be interpreted as an attempt to reduce their internal and external contribution to cultural diversity through driving them towards the modern urban Western core culture. Consequently their right for spatiality, based on their unique pastoralist indigenous culture, has been denied, but concomitantly such is also their contribution to their ecosystem's sustainability, entailing thus its declining biodiversity. Linguistic studies have shown that Bedouin dialects, in terms of semantics of prepositions and cognitive aspects, have been changing inter-generationally following sedentarization (Rosenhouse 1984) including space-time terminology (Henkin and Cerqueglini forthcoming). As their forced settlement (sédentarisation) is associated with detachment from shepherding, which relied closely upon the natural environment, this linguistic change carries implications also for their cultural relations with the natural environment in terms of losing valuable vernacular indigenous knowledge of ecosystem management.

14.7 Conclusion

Following forced settlement (sédentarisation), forced relocation to towns and integration into the labour market economy, Bedouin traditional capacity to contribute to biocultural diversity has declined considerably. This is a lose-lose game for all entities involved—the Bedouin, the ecosystem, and the state. The alternative is an in situ recognition of all Bedouin villages in the Negev. This will allow the state to regularize livestock grazing at the desired moderate levels. Such policy change can contribute to the preservation and even increase of both biological diversity and cultural diversity through preserving significant components of traditional Bedouin pastoralist culture. It will facilitate materialization of the joint right for bioculturality by both the Bedouin and the ecological system and thus the right for social, spatial and environmental justice for this marginal group. This may become a win-win solution, working for the mutual benefit of all parties involved in this nexus of Bedouin, space, environment and state.

References

Abu-Saad, I., & Lithwick, H. (2000). *A way ahead: A development plan for the Bedouin towns in the Negev*. Beer Sheva: Negev Center for Regional Development.

Bailey, Y. (1976). Components of nature preservation in Bedouin culture. *Notes on the Bedouin* (Vol. 7, pp. 54–69). (Hebrew).

Bailey, C. (2009). *Justice without Government: Bedouin Law from Sinai and the Negev*. New Haven: Yale University Press.

Ben-David, Y. (2004). *The Bedouin in Israel—Social and Land Perspectives*. Jerusalem: Jeruslaem Institute for Israel Studies and The Institute for Research on Land Use Policy. (Hebrew).

Ben-David, Y., & Gonen, A. (2001) Bedouin and Fellaheen-Bedouin in the Urbanization Process in the Negev Jerusalem: Floresheimer Institute for Policy Studies. (Hebrew).

Boeken, B. (2012). *Grazing in Negev shrublands, paper presented at departmental seminar*. Department of Geography and Environmental Development: Ben Gurion University of the Negev, Beer Sheva, Israel (Hebrew).

Degen, A., & El-Meccawi, Sh. (2009). Livestock production among urban Negev Bedouin. *Outlook on Agriculture, 38*(4), 327–335.

Ginguld, M., Perevolotsky, A., & Ungar, E. (1997). Living on the margins: Livelihood strategies of Bedouin herd-owners in the northern Negev, Israel. *Human Ecology, 25,* 567–591.

Golodets, C., & Boeken, B. (2006). Moderate sheep grazing in semiarid shrubland alters small-scale soil surface structure and patch properties. *CATENA, 65,* 285–291.

Harvey, D. (1985). *The urbanization of capital*. Oxford: Blackwell.

Henkin-Roitfarb, R. (2011). Hebrew and Arabic in asymmetric contact in Israel. *Lodz Papers in Pragmatics, 7*(1), 61–100.

Henkin, R., & Cerqueglini, L. (Forthcoming). Cardinal directions in traditional Negev Arabic language and culture, *Anthropological Linguistics*.

Holt, L. (2005). The catch-22 of conservation: Indigenous peoples, biologist and cultural change. *Human Ecology, 33*(2), 199–214.

Karplus, Y. (2010). *The dynamics of construction of Bedouin Space in the Negev*. Ph.D. Dissertation, Ben Gurion University of the Negev, Beer Sheva. (Hebrew).

Karplus, Y., & Meir, A. (2010). The right for endogenic spatiality: The case of indigenous peoples. A discussion paper presented in Minerva Conference for Human Rights in Israel, Neve Ilan. (Hebrew).

Karplus, Y., & Meir, A. (2013). The production of space: A neglected aspect perspective in pastoral research. *Environment and Planning D: Society and Space, 31,* 23–42.

Karplus, Y., & Meir, A. (2014). From congruent to non-congruent spaces: dynamics of Bedouin production of space in Israel. *Geoforum, 52,* 180–192.

Kideghesho, J. (2008). Co-existence between the traditional societies and wild life in western Serengeti, Tanzania: Its relevancy in contemporary wildlife conservation efforts. *Biodiversity Conservation, 17,* 1861–1881.

Kissinger, M., & Karplus, Y. (2015). IPAT and the analysis of local human–environment impact processes: the case of indigenous Bedouin towns in Israel. *Environment, Development and Sustainability, 17*(1), 101–121.

Kressel, G. (2003). *Let shepherding endure: Applied anthropology and the preservation of a cultural tradition in Israel and the Middle East.* Albany: State University of New York Press.

Landau, Y., Evlagon, D., Abu-Rabia, A., & Abu-Siam, S. (2015). Seasonal grazing of Bedouin livestock in KKL forests. *Ya'ar, 15,* 30–39. (Hebrew).

Lefebvre, H. (1991 [1974]). *The production of space.* Blackwell: Oxford.

Larsson, Tor-Björn. (2001). *Biodiversity evaluation tools for European forests.* New York: Wiley-Blackwell.

Leimgruber, W. (2004). *Between global and local. Marginality and marginal regions in the context of globalization and deregulation.* Aldershot: Ashgate.

Maffi, L. (2001). Introduction: On the interdependence of biological and cultural diversity. In L. Maffi (Ed.), *On biocultural diversity* (pp. 1–53). Washington & London: Smithsonian Institution Press.

Maffi, L. (2007). Bicultural diversity and sustainability. In J. Pretty, A. Ballet, T. Benton, J. Guivant, D. Lee, D. Orr, et al. (Eds.), *Sage handbook on environment and society* (pp. 267–277). London: Sage Publications.

Marx, E. (1973). The ecological basis of the Bedouin society. *Notes on the Bedouin* (Vol. 4, pp. 33–58). (Hebrew).

Marx, E. (1974). *The Bedouin in the Negev.* Reshafim (Hebrew): Tel Aviv.

Meallem, I., & Garb, Y. (2008). The exposure of Bedouin women to waste related hazards, women and environments. *Fall/ Winter,* 44–48.

Meir, A. (1997). *As Nomadism ends: The Israeli Bedouin of the Negev.* Boulder Co: Westview.

Meir, A. (1999). Local government among marginalized ex-nomads: The Israeli Bedouin and the State. In H. Jussila, R. Majoral, & C. C. Mutambirwa (Eds.), *Marginality in space-past present and future* (pp. 101–119). Aldershot: Ashgate.

Meir, A. (2005). Bedouin, the Israeli state and insurgent planning: Globalization, localization or glocalization? *Cities, 22,* 201–215.

Meir, A. (2009). Contemporary state discourse and historical pastoral spatiality: Contradictions in the land conflict between the Israeli Bedouin and the state. *Ethnic and Racial Studies, 32*(5), 823–843.

Meir, A., & Karplus, Y. (2015). From consummate space to spatial intensity: Intercultural dynamics of production of space and its political significance. *HaReshet HaGeografit, 8,* 40–60 (Hebrew).

Meir, A., & Tzoar, H. (1996). International borders and range ecology: The case of Bedouin transborder grazing. *Human Ecology, 24*(1), 39–64.

Næss, A. (1989). *Ecology.* Community and Lifestyle: Outline of an Ecosophy. Cambridge University Press.

Perevolotsky, A. (1987). Territoriality and resource sharing among the Bedouin of Southem Sinai: A Socio-ecological interpretation. *Journal of Arid Environments, 13,* 153–161.

Perevolotsky, A., & Seligman, N. (1998). Role of grazing in Mediterranean rangeland ecosystems. *BioScience, 48*(12), 1007–1017.

Perevolotsky, A. (2006). Livestock grazing and biodiversity conservation in Mediterranean environments: the Israeli experience. *Options Méditerranéennes a, 67,* 51–56.
RCBUV (Regional Council for Unrecognized Bedouin Villages). (1999). *A Plan for Development of a Municipal Authority for the Unrecognized Bedouin-Arab Villages in the Negev.* Rahat. (Hebrew).
Roded, B., & Tzfadia, E. (2013). Indigenous land rights recognition: A comparative observation of the Arab-Bedouins in the Negev. *The Public Space, 7,* 66–99. (Hebrew).
Rosenhouse, J. (1984). Towards a classification of Bedouin dialects in Israel. *Bulletin of the School of Oriental and African Studies, 47*(3), 508–522.
Salick J. & Byg, A. (2007). *Indigenous peoples and climate change.* Report of Symposium 12–13 April 2007, Environmental Change Institute, Oxford.
Santos, Boaventura de Sousa. (2007). Beyond abyssal thinking: From global lines to ecologies of knowledges, *Review (Fernand Braudel Center), 30*(1), 45–89.
Society for Protection of Nature. (2008). Distinguishing between open spaces and development: Proposal for a policy of regulating Bedouin settlement in the Negev. Tel Aviv. (Hebrew).
Smith, A. S. (2001). On the coevolution of cultural, linguistic, and biological diversity, In: Maffi, L. (Ed.), *On biocultural diversity* (pp. 95–117). Washington & London: Smithsonian Institution Press.
Stewart, F. H. (1986). *Bedouin Boundaries in Central Sinai and the Southern Negev: A Document from the Ahaywat Tribe.* Wiesbaden: Otto Harasowitz.
Stewart, F. (2006). Customary law among the Bedouin of the Middle East and North Africa. In D. Chatty (Ed.), *Nomadic Societies in the Middle East and North Africa: Entering the 21st Century* (pp. 239–279). Leiden: Brill.
Turner, N. J., Davidson-Hunt, I. J., & O'Flaherty, M. (2003). Living on the edge: Ecological and cultural edges as sources of diversity for social–ecological resilience. *Human Ecology, 31*(3), 439–461.
UNESCO. (2002). *UNESCO Universal declaration on cultural diversity. UNESCO Universal UNESCO.*
Wachs, E., & Tal, A. (2009). Herd no more: Livestock husbandry policies and the environment in Israel. *Journal of Agricultural and Environmental Ethics, 22,* 401–422.
Yahel, H., & Kark, R. (2014). Israel Negev Bedouin during the 1948 War: Departure and return. *Israel Affairs.* doi:10.1080/13537121.2014.984421.
Yiftachel, O., Kedar, A., & Amara, A. (2012). A new perspective on the 'dead Negev' rule: property rights in Bedouin Space. *Law and Governance, 14*(1–2), 7–147. (Hebrew).
Yiftachel, O., & Roded, B. (2014). Between rights and denials: Bedouin indigeneity in the Negev/Naqab. *Public Space, 9,* 31–67. (Hebrew).

Part V
Possible Drivers of Marginalization from Different Perspectives

Chapter 15
Migration, Education and Marginality: Networks and Strategies in the Pacific Islands

John Overton and Warwick E. Murray

15.1 Introduction

Marginality is manifested and expressed in many different forms across the world. As a consequence, society and its constituent communities have evolved a diverse range of strategies to cope with and lessen, its effects. For many, migration is a common response to limited resources and opportunities. The search for higher education is also seen as way to build capability to respond, both within marginal areas and also in terms of its effect of facilitating migration. In this chapter we examine the case of the Pacific Islands in relation to migration and education. We first explore the context of marginality in the region and its multiple dimensions. We then focus on the role of education in expanding the options available for the regional population before analysing the strategies adopted by states, kinship networks and individuals to exploit linkages with external economies and polities in ways which ensure a flow of resources back to the Pacific. We conclude by suggesting that these strategies amount to processes of 'demarginalization': active, conscious and astute actions of communities to engage with the outside world that preserve hearths and cultures as valuable resources rather than, as is often portrayed, constraints to achieving wellbeing and sustainability.

J. Overton (✉) · W.E. Murray
Victoria University of Wellington, SGEES, Kelburn Parade, Wellington 6012, New Zealand
e-mail: john.overton@vuw.ac.nz

W.E. Murray
e-mail: warwick.murray@vuw.ac.nz

S. Pelc and M. Koderman (eds.), *Nature, Tourism and Ethnicity as Drivers of (De)Marginalization*, Perspectives on Geographical Marginality 3,
DOI 10.1007/978-3-319-59002-8_15

15.1.1 Marginality in Oceania

The Pacific Island region covers a vast area of the Earth's surface and includes a wide range of environments, populations, cultures and economies (Fig. 15.1). The concept of marginality seems to be particularly appropriate for this region and many authors have identified several dimensions to the vulnerability of people and countries in the region. We examine several of these in this section before offering a counter argument: that such negative views tend to portray the region as passive and fragile and thus limit the discourses of development available to people and states (Fry 1997). Instead, we draw on the work of Pacific scholars to open the debate to the possibilities of active and innovative forms of progressive development that link the region to the outside world.

15.1.2 Geographic

A simple reading of the geography of the Pacific Islands region stresses its geographic marginality. Firstly, there is the issue of isolation. Although we might see the Pacific islands as being located in the middle of one of the most dynamic and prosperous regions in the global economy—the so-called Asia Pacific region including North and South America, East and Southeast Asia and Australasia—the

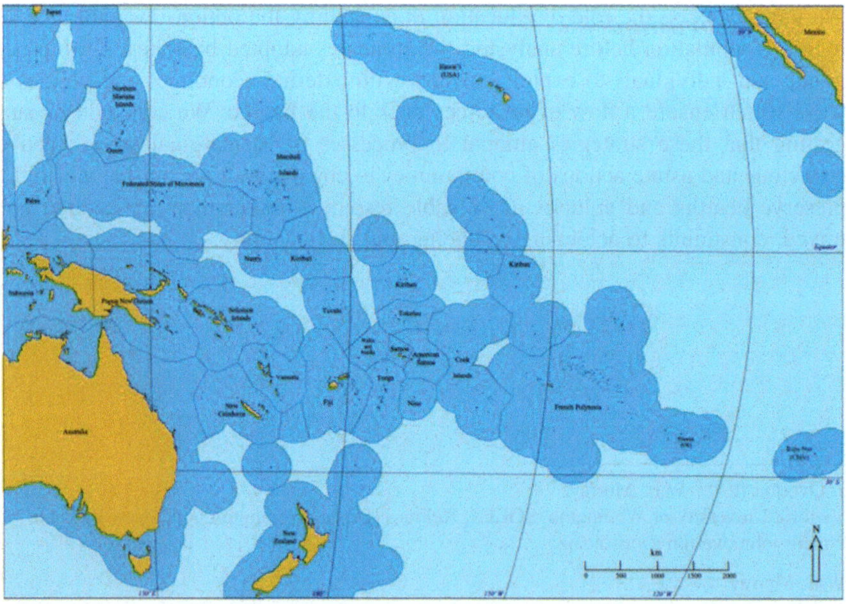

Fig. 15.1 Map of the Pacific Islands

reality is that, in terms of transport and communications, most of the Pacific Islands are not well connected to main trade routes and transport costs are high. This means that the islands are not closely integrated into the global economy. Isolation is compounded by fragmentation. Many island states are characterised not by a single concentrated land mass and population but by many small islands. The country of Fiji, for example, has a population of around 850,000 people and over 300 islands, 100 of which are inhabited. These islands and their communities need to be linked together with services and trade and the costs of air and sea transport to integrate these places illustrate the costs and difficulty of linking these countries to the global economy. Such 'islandness' is an important dimension of marginality.

There is also the issue of size. With the notable exception of Papua New Guinea, no Pacific Island country has a population in excess of a million people (Table 15.1). Many have fewer than 100,000 people and some are true microstates. The country of Tuvalu, for example is fully independent with its own government, legislature, bureaucracy, international airport, diplomatic representation, currency and flag. Yet its total population is just over 10,000 people. The per-capita costs of being a microstate and supporting this state infrastructure are very high indeed. Other similar sized or even smaller states and territories exist but often they are in a political relationship with a metropolitan state—a theme we examine more below. Smallness, then, imposes costs and this can be a significant economic burden. It also means that, whilst independent, small island states are sometimes seen as having a little presence in international political diplomacy.

15.1.3 Environmental

Geographic marginality is paralleled by environmental constraints. Although some of the larger Pacific Island Melanesian states (Papua New Guinea, Fiji, Solomon Islands, New Caledonia) have relatively substantial land masses, most of the island states of Polynesia are comprised of low lying coral atolls or the remnants of volcanic cones. The islands lie in an Ocean that experiences periodic cyclones, droughts and tsunamis. Furthermore, their often limited and fragile marine and terrestrial resources are subject to environmental degradation through pollution of lagoon waters, deforestation and coastal erosion.

However, although these contemporary environmental issues are serious and pressing, they have paled compared to recent concerns with climate change. The prospect—and many in the region would argue, the reality—of global warming and sea level rise has put the small low-lying islands of the Pacific at the centre of global concerns about climate change. Because many islands are coral atolls and often are no more than two or three metres above sea level at the highest, they are particularly susceptible. At its worst, predicted sea level rise may lead to many such islands—and in the case of Tokelau, Kiribati and Tuvalu, whole countries—disappearing beneath a rising sea, the effects of sea level are already being manifested in the form of coastal erosion and inundation and mounting saline intrusion of fresh water

Table 15.1 Populations and areas of Pacific Island States and territories

	Population (estim 2013)	Land area (km^2)	ODA per capita ($US)
American Samoa	56,500	199	n.a.
Cook Islands	15,200	237	1,006
Federated States of Micronesia	103,000	701	1,390
Fiji	859,200	18,333	106
French Polynesia	261,400	3,521	n.a.
Guam	174,900	541	n.a.
Kiribati	108,800	811	594
Marshall Islands	54,200	18	1,733
Nauru	10,500	21	2,741
New Caledonia	259,000	18,576	n.a.
Niue	1,500	259	12,200
Norfolk Island	1,895	35	n.a.
Northern Mariana Islands	55,700	457	n.a.
Palau	17,800	444	1,992
Papua New Guinea	7,398,500	462,840	89
Pitcairn	45	47	n.a.
Rapa Nui	5,700	164	n.a.
Samoa	187,400	2,934	631
Solomon Islands	610,800	28,000	472
Tokelau	1,200	12	20,050
Tonga	103,300	749	786
Tuvalu	10,900	26	2,459
Vanuatu	264,700	12,281	343
Wallis and Futuna	12,200	142	8,650

Source Pacific Community 2016

lenses, making habitation already problematic. Added to the real threat of sea level rise, global climate change also poses the threat of potential increases in the incidence of storm events and droughts and, through warming, threats to vital processes such as coral growth and key subsistence crops.

15.1.4 Economic

Allied to geographic marginality, and largely a result of it, is the economic marginalization of the region. Isolation and fragmentation as we have seen lead to high transport costs. It can cost as much if not more to ship products to an outer island from the country's main port as it costs to transport the same goods from a far distant metropolitan country to the main port. And shipping services may be far more irregular and infrequent. Smallness adds to these economic constraints for economies of scale cannot be achieved. These factors combine to mean that economic strategies based on export-led growth through exporting commodities for the global economy are simply not viable with small volumes and high transport costs. Indeed, global trade liberalization has severely affected former industries in the Pacific which enjoyed a degree of metropolitan protection and subsidy—including sugar, tropical fruit, rice and garments for example.

Some Pacific Island countries, such as Papua New Guinea, Fiji and New Caledonia enjoy rich natural resources: minerals, oil and gas, forests and fisheries. Yet these, again, are the larger states of Melanesia. For the other small island states, especially those with atoll environments, mineral and soil resources may be virtually non-existent. Large maritime exclusive economic zones may offer opportunities to exploit fisheries but this requires financial and regulatory resources usually beyond the capabilities of small populations.

15.1.5 Cultural

Finally we note that some authors have suggested that marginality and relative poverty in the Pacific islands has resulted from cultural factors. In the past, authors such as Watters (1969) have pointed to the way the persistence of traditional communal societies has limited the emergence of capitalism, enterprise and modernity. Communal land tenure and forms of exchange may promote social cohesion and subsistence but they also inhibit the ability of individuals to trade and exploit their lands and marine resources. Communal land cannot be readily used to gain credit—a key potential investment resource—and the individual accumulation of wealth is inhibited by cultural norms which encourage sharing and the conspicuous dispersal of wealth amongst one's kin. Thus, some have argued, the persistence of traditionalism and communal land tenure has further marginalized the economies of the Pacific Islands from an expanding and increasingly integrated global market economy.

15.1.6 Questioning the Discourse of Marginality

Yet we question these frames of reference for marginality in the region. It is true that these processes—the geographic, economic and environmental constraints in particular—do impose costs and do limit the ability of people in the region to compete in conventional forms of engagement with the global economy. However, the picture is far from one-sided. Pacific scholars in particular have rejected these views of the region and its people; views which stress vulnerability, fragility, smallness and passivity (Hau'ofa 1995; Thaman 2003; Teaiwa 2006). The Ocean is seen not as a constraint and a place of emptiness and poverty but rather as a productive and interconnected realm where people move, adapt, trade, and prosper. Culture is not a barrier but instead promotes an ethos of shared welfare and mutual responsibility—a worthy alternative to the individualism, violence and social isolation of modernism and a much more sustainable model than global capitalism (Batibasaqa et al. 1999).

Thus we approach the issues of marginalization in the Pacific Islands with this much more positive frame of local adaptation, resilience and agency. In the following sections we analyse how Pacific Island people have sought opportunities by actively interacting with the wider world and securing their welfare.

15.2 Education and Migration Strategies

Education lies at the centre of the ways many in putative marginal areas forge new futures for themselves and their children. Given the constraints and opportunities facing people of the Pacific Islands we see that this forms a central strategy in this region also.

15.2.1 Education in the Pacific

There is much emphasis on education in the Pacific region. For some countries with high fertility rates, there are young populations facing limited economic opportunities. Governments throughout the region have invested heavily in education, often with the assistance of aid donors (Murray and Overton 2011). The Millennium Development Goals spanning the 2000–2015 period did much to highlight the importance of universal access to primary and secondary education (Tolley 2008). Within the Pacific, many children living in remote areas and outer islands were not able to attend local primary schools, let alone progress to secondary education. The issue is not just about building more schools and training more teachers: roads and bus services are needed to get children to and from schools and accommodation is needed for children from small outer islands who have to attend secondary schools elsewhere. In some parts of the region though, the record was good: in Samoa, Tonga, Fiji, Niue, Cook Islands, and the French and American territories of the Pacific, high literacy rates were the result of many years of investment in primary education.

The focus on universal education and the considerable efforts to ensure this, however, has raised concerns about the appropriateness of education in the region. Several educationalists in the region have questioned the appropriateness of curricula which follow exclusively Western epistemologies, values and content (Gegeo 1998; Gegeo and Gegeo-Watson 2001; Coxon and Munce 2008; Sanga and Taufe'ulungaki 2005; Helu Thaman 1993). They have argued persuasively that not only do such curricula train students for urban employment often far removed from the realities and needs of the rural existence that most of their parents lead but also, they undervalue and demean indigenous knowledge and epistemologies. It could be questioned whether spending on such Western-styled education is merely guiding young people to get on a plane and go and train, live and work overseas (Gamlen et al., 2017). In response, there have been some important initiatives to redress the balance. At one level, there have been several rural training centres and local curriculum projects which have sought to utilise and strengthen local knowledge and skills (Fleming and Palomino-Schalscha 2016) and at the other, there has been a growth in tertiary training in the region, through institutions such as the pan-regional University of the South Pacific as well as various national universities and technical training colleges.

Yet despite these efforts and a considerable growth in the past three decades in the number of children attending schools and universities within the region, there is still a considerable flow of students overseas for education.

15.2.2 Education for Migration

Education is a major driver of migration. This occurs at different levels. For the most remote and marginal areas without suitable schools, children move from remote villages and islands often to capital cities on the main islands to attend secondary schools or tertiary institutes. Many others go overseas for what is perceived as better quality secondary schooling or university study. Government scholarships support some pupils but it is often family resources that invest in such movement and education, a sign of the value that is attached to education.

Migration for education raises a key issue for the long-term development of the Pacific. Overseas education equips students for life and work overseas but it often means they become less suited (in terms of skills and aspirations) for living at home in the islands. The skills and knowledge they attain overseas are more appropriate for the metropole and there may be little demand for specialist skills at home. Furthermore, even when the skills are applicable, employment in the big cities on the Pacific Rim (Sydney, Auckland, Los Angeles, Vancouver, Hong Kong) may appear more attractive and lucrative than a return to a poorly-paid civil service, nursing or teaching job at home.

15.2.3 Education for Development?

Therefore some key questions are raised regarding the role of education in development. It is universally acknowledged that education is good for development, whether from a neoliberal human capital development perspective or a more radical view of consciousness raising and political awareness (Mayo 2008; Colclough

2012). We could suggest that education is also a way of combatting marginality: it both develops skills to provide a way out for those living in marginal areas and raises awareness of the problems (and solutions) that lead to marginalization.

However, in relation to marginalization, education may contribute to the very process of marginalization. The 'brain drain' phenomenon (Adams and Page 2005; Docquier and Rapoport 2012; Gamlen 2014) involves the loss, in effect, of the youngest, brightest and best from marginal areas and they take their skills and incomes to more core and prosperous areas, never to return. Education may leave behind the elderly, low skilled and unemployed and this can be seen in some parts of the Pacific where depopulation is occurring (Gibson and Mackenzie 2011, 2012; Overton and Murray 2014).

On the other hand, a 'brain drain' may not be an entirely negative process and some authors have suggested a more optimistic view (Gamlen 2010; Jacobs 2016). Better educated expatriates may have left the Pacific Islands to work overseas but this does not mean they are either permanently removed from the region or distanced from its welfare. Pacific Island people living and working outside the region contribute greatly in the form of remittances back to their family members, churches and, ultimately, businesses in the islands. Such flows of remittances form substantial components of Pacific economies—sometimes up to a quarter or a third in countries such as Tonga and Samoa (see Connell and Brown 2005). Furthermore, it seems as if these remittance flows do endure over long periods of time, even inter-generationally (Macpherson 1992; Poirine 2006). And migrants also return home. They come back to look after relatives, enjoy a more peaceful life, retire or start a business. They bring back their savings, skills, experience and connections.

Therefore, we can suggest that education and migration are playing important but complex roles in transforming living standards and cultures in the Pacific Islands. It may be that education is undermining local knowledge systems and cultures and depleting local economies of skilled young people; yet it could also be simultaneously providing viable alternatives for otherwise underused human capital, contributing new sources of material wealth, and also, effectively, helping support and preserve long standing ways of life and social structures.

15.2.4 Uneven Opportunities and Impact

A further complication as we try to understand the links between education, migration and marginality in the Pacific, is that these trends are very uneven across the region. Some territories face easy access to the metropole and have high education levels and high migration. This is the case for many of the smaller Pacific countries linked politically and socially to France, USA and New Zealand in particular. Often their citizens have full citizenship rights to enter, live, attend school and work in the metropole. They can move freely. The inhabitants of other countries—the larger independent states—have fewer migration opportunities and, often as a result of much lower historic levels of foreign development assistance, are also

Table 15.2 Typology of education strategies

Education strategy	Education level/type	Migration implications	Development implication
Local customary	Non-formal, local, locally controlled	Restricted	Localised, conservative, traditional/customary, social status quo with some social and political challenges
Rural training centres	Locally based, wider skills	Low	Rural-based development based on local skills and resources, self-sufficiency, resilience
Universal primary education	National curriculum, basic literacy and numeracy skills	Limited	Universal participation, gender implications, basic education as a basis for national development goals. Possibilities for social transformation?
Secondary education	National curriculum, higher literacy etc. skills	Some movement to schools, likely to encourage future urbanisation for employment	A more literate national workforce, raised expectations for urban employment, knowledge of outside world
Vocational: national	Technical skills training—e.g. trades, nursing, religious	Movement to cities for training, some skills for international (e.g. nursing)	National development—skilled workforce.
Tertiary education	'National university'—builds national capabilities at higher level—general education and some professions	Movement to cities, some transferable skills/qualifications for overseas (but lower status)	National development—human capital-based. Builds skills in public service, teachers etc.
Vocational: global orientation	Targeted training schemes for overseas work (maritime etc.)—focus on international qualifications	Movement within country to centres but limits need to travel overseas for training	Global-oriented development with niche labour markets—remittances and saved incomes
Overseas tertiary: scholarships	Targeted degrees overseas—higher level skills and professions	Migration overseas to university but may be bonded to return home	National focus on higher skills (medicine, engineering etc.)—some opportunity for later migration overseas and remittances
Overseas education	Open choice to travel overseas for education. Privately funded or open government scholarships (not bonded)	Migration overseas often with intention to stay overseas	Globalisation—remittances and migration

Source Gamlen et al. forthcoming

struggling to provide basic primary and secondary education. Papua New Guinea and Solomon Islands, for example, have the lowest rates of literacy and school attendance in the wider region. In these cases, improved educational provision and access may lead to more children in school but it will not guarantee that they have jobs to go to when they graduate; nor will they easily be able to take their skills overseas to work.

Therefore, the relationship between education, migration and marginality is uneven. Better education, migration and the free flow of labour, remittances and ideas may contribute to some forms, and awareness, of marginalization, yet it can substantially mitigate and reverse many other marginalizing processes. For others, better education—faced with constrained migration—can exacerbate marginalization. It can provide inappropriate skills for a limited local labour markets, ignore and undermine local knowledge and endogenous development values and strategies and build a mounting population of urban-dwelling, educated but unemployed and dissatisfied people marginalized from mainstream, economy and society.

In order to begin to understand some of this complexity, we have begun to examine more closely the various educational options that might be open to governments and communities in the Pacific (Gamlen et al., 2017—see Table 15.2). Here we can see a wide range of educational strategies, from small-scale locally-based learning to international tertiary education. What we suggest is that there is no single educational strategy that is best or that will lessen the marginality and poverty that is sometimes experienced in the region. Rather, different states, families and individuals will perceive opportunities and costs which may or not match their particular situations and capabilities. Some resources can be provided by the state; some by communities; and some are only available to those who can call on considerable family financial resources. It is to these varied resources and networks that we now turn.

15.3 Networks and Relationships

To understand these complex relationships, we need to look at the way Pacific Island peoples have developed particular strategies in order to build social capital and exploit wider networks and relationships. In some ways the 'marginality' of the Pacific and its populations is an externally constructed and imposed framework. In a historic sense the peoples of the Pacific were adept at developing and maintaining networks and interactions at great distances given the sea-faring prowess of especially the Polynesians (Hau'ofa 2000). It is only with the advent of colonialism and the imposition of western notions of scale, distance and proximity that the concept of marginality has been applied. Furthermore, the arrival of the demarcation of territory in the form of borders within and between territories and (eventually) countries based on Westphalian notions of sovereignty also sits uncomfortably with the indigenous forms of globalisation that were long in place by the advent of the colonial period. In this sense then, marginality is a neo-colonial construct in the

Pacific. This is not to suggest that the real challenges of marginality in the context of surviving in the contemporary capitalist economy are not acutely felt; they are and the design of appropriate educational policy is crucial in overcoming this. But marginality is a constructed consequence of the particular political economic history of the region and the kinds of dependency that have evolved.

What are the kinds of relationship and networks both in history and in the contemporary scene that have evolved that have subverted this imposed view of marginality and how might this relate to the types of education and migration integration that we seek?

15.3.1 From 'Transnational Corporations of Kin' to 'Diasporic Social Networks of Kin'

Historically, kinship networks have been one of key processes used to subvert and challenge marginality both imagined and real. Pacific people, and this is especially the case in Polynesia, have always travelled great distances, fanning out across enormous swathes of ocean and maintaining social, economic and cultural networks over many centuries. In doing so a diaspora of Pacific peoples was long existent before the concept of diaspora was employed to describe it. To early Pacific peoples the concept of distance and isolation was quite different to how those things are perceived in the West. The ocean was seen a unifying entity not something which divides.

The strong sense of family that imbues Pacific life was the essential lifeblood of this indigenous globalisation. Communalism and collective ownership of resources which characterises traditional and many contemporary Pacific societies in the islands themselves has been contemporised and still drives these networks. As Pacific peoples in the contemporary world, due to political-economic arrangements with the rim countries—especially New Zealand and Australia—retain relatively free movement the diaspora has expanded territorially (see Barcham et al. 2009). Pacific Islanders who may be second or third generation born in New Zealand or Australia still feel a very keen sense that their home is on their island of origin. In an early paper on this Bertram and Watters referred such networks as transnational corporations of kin (1985; see also Munro 1990). Whilst this was useful in terms of identifying the important role of diasporas in allocating and optimising resource flows, it was essentially Western in terms of its construction. It can be argued that the frame of reference for networks of kin are not bordered nation-states and territories but diaspora which have long existed. Furthermore, the use of the term corporations is also misleading—Pacific Island people are not profit-maximisers; rather they tend to use networks to optimise the allocation of resources and ensure their equitable and sustainable flow. The West could learn a lot from the Pacific diaspora in terms of optimising globalisation. A better term to describe these entities might be diasporic social networks of kin. In the reality of the regional political economy there are of course real physical borders, but the particular formation of these networks allows relatively unbounded de-marginalisation to take place.

The above explains why remittances per capita in the Pacific diaspora are amongst the highest in the world (Connell and Brown 2005).[1] As transport technologies have improved there has been an increasing circular migration across the region where young people leave the island to study and work in the Rim and later return. Sometimes family members may not return for generations, but maintain the links through financial flows and flows in kind. This is why aunties and uncles in Auckland or Brisbane for example will host young Pacific islanders attending educational institutions. In all of this the original unbounded sense of place that Pacific peoples had before the arrival of Western imposed national boundaries is a driving factor. Notwithstanding this, it is sobering to note that many Pacific peoples occupy the lower socio-economic strata of the societies they live in on the Rim, and this has been and remains a significant brake on mobility. In some ways, Pacific peoples living in the former metropoles are more marginalised in social space than their ancestors who live on the islands. However, as this situation improves and the welfare of Pacific people rises, mobility is likely to increase further. It is such movements that offer the prospect of making the most of un-bounded space that might be promoted as development models for the region. De-marginalisation in this sense is something which Islanders have always practised although the technologies, velocities and spatialities have evolved. Education and work offer the prospect of income for the Pacific homelands—although this is also a potentially vulnerable strategy and imbalanced outflow raise the threat of unsustainable de-population especially in the smaller territories such as Niue, Tokelau and Wallis and Futuna for example.

15.3.2 Metropolitan Links and Subsidies

This inherent tendency of Pacific populations to seek unbounded space for the expansion of diasporic networks of kin has resulted in various strategies to facilitate this in the context of contemporary international and postcolonial relations. Unlike other parts of the colonised world, independence did not come rapidly after the Second World War exposed the weakness of many of the former empires. As we know, in Africa and Asia the fall of empire was rapid. This was not the case in the Pacific and it was not until the mid 1960s that the first island territories gained independence. Compared to Africa and Asia the push internally for independence was not as marked. This is not to say that were not violent struggles in the

[1]Although data are patchy and variable, the World Bank estimates that for 2012, remittances accounted for 22% of Samoa's GDP, 25% of Tonga's and 5% of Fiji's, though only 0.1% of Papua New Guinea's GDP. The world average for remittances is about 0.7% of GDP (see http://data.worldbank.org/indicator/BX.TRF.PWKR.DT.GD.ZS, accessed 3 March 2017).

Pacific– in New Caledonia (ultimately unsuccessful), and Samoa—there was conflict and this continues to the present in West Papua for example (Wesley-Smith 2007). After WW2 Western nations too were keen to maintain close links and in particular in the context of the Cold War as the Pacific Island held a geopolitical comparative advantage.

As a consequence of these particular historical processes only around half of the territories gained independence, some explicitly rejected it—such as Tokelau—and others obtained relationships with the former metropole that allowed the pursuit of a mixed and layered form of sovereignty. In Niue and the Cook Islands for example the concept of 'self-governing territories in free association' with New Zealand was used to replace colonial incorporation. This arrangement allows free movement of Niueans and Cook Islanders to and from New Zealand, grants them citizenship in both, allows the free-flow of labour and financial resources, and places foreign policy and defence largely in the hands of New Zealand (Connell 2008). This allows the Polynesian networks of kin to expand and circulate in ways which mirrors the way they have done for millennia. There is a range of other arrangements that have defined the 'sovereignty' of the Pacific territories including 'compact' agreements in the case the US territories and various forms of incorporation in the case of the French territories. These have evolved in part as a consequence of the interaction of the geopolitical needs of the metropole and territories—though it must be said that the former gains the most—especially in the case of the French and US territories. As such, Pacific sovereignty can be conceptualised on a continuum and it is this that allows relatively unbounded movement Pacific peoples have employed as de-marginalising strategies historically. This set of factors allows for easier migration access, especially to universities and secondary schools overseas. It also leads to significant flows of labour migration. As a consequence many more people from Wallis and Futuna, Cook Islands, Niue, Tokelau, American Samoa, for example live 'overseas' than at 'home' (though these concepts blur in the Pacific geographic imagination). Regional labour migration schemes have been established in Australia and New Zealand in the recent past and this allows for circular and temporary movement from places that do not have association agreements or compact such as Solomon Islands, Vanuatu, Fiji, and Tonga. A final consequence of the various close agreements that have been forged is the high levels of per capita aid that characterise the region. A material consequence is that small societies in the islands themselves can often sustain infrastructure—such as primary and secondary schools—that would not be possible in communities of similar sizes in New Zealand, France or the USA for example.

15.3.3 MIRAB Economies

Attempts have been made in the literature over the last four decades or so to summarise the relationships discussed above and provide a theoretical framework for their interpretation and evaluation. The most widely known in this regard is the

MIRAB concept (migration, aid remittances bureaucracy) (Bertram and Watters 1985; Bertram 1993, 2006; also Evans 1999; Oberst and McElroy 2007). In essence these authors argue that MIRAB is a sustainable strategy for small island states to sustain higher living standards than might be predicted by domestic income levels given few opportunities for production/export-led economic growth (Bertram 1986). Implicit in this is the idea that marginality (location, size, environment) is coped with or mediated through wider engagement with the metropole in terms of aid flows and remittances. In this small territories 'trade-off' sovereignty for resources and thereby adopt a range varied but similarly close relationships with former and current metropole (McElroy and Pearce 2006; McElroy and Parry 2012; Wesley-Smith 2007; Baldacchino 2006a, b). Various territories and countries of the Pacific correspond to the MIRAB model to varying degrees and over time its relevance has waxed and waned. There have been pushes in recent times to reduce aid for example, especially following the end of the Cold War, the rise of anti-immigration discourse in more recent times also threatens to undermine the sustainability of the model in countries and territories where this is not guaranteed in constitutions. Notwithstanding this, the role of remittances through kinship networks as discussed above, as well as levels of aid that remain high per capita, continue to support high government expenditure and employment in a number of Pacific societies. Aid is used to build educational networks and fund certain strategies and therefore remains very important. Most recently scholarships to institutions in the metropole have formed a very important and growing component of aid relationships with important consequences for the appropriateness or otherwise of development strategies encouraged by the donors.

But MIRAB can be criticised in three main ways. Firstly, it frames Pacific marginality from a Western perspective, seeing the Islands as dependent on the metropole. It could be argued that the relationship is in fact a result of witting strategy on the part of the Pacific and that dependency actually goes both ways. Secondly, and linked to this point, MIRAB overplays the importance of economic resource allocation in this process—and the role of communal and cultural processes are seen as a means to an economic end—the reality is far broader, more complex and less economically deterministic. Finally, MIRAB fails to recognise that facing marginality in the Pacific has a long history predating aid relationships established after the 1960s—and MIRAB is only one expression of that.

15.4 'Demarginalization' Through Unbounded Educational Spaces

What can we say about the role of education and migration in terms of marginality? In essence we suggest that they are actively used in combination as strategies to 'de-marginalize'. We see clear evidence for this across the Pacific where diasporic networks of kin that were formed before colonialism are contemporised and perpetuated through such flows. Unbounded spaces where such networks are encouraged can have de-marginalising effects, even when in a territorial sense marginality appears self-evident. In order to make sense of this we conclude with some thoughts under three headings; education and globalization, education as revolution or reaction, and education through state structures or individual agency.

15.4.1 Education and Globalization

In some ways what we see in this case study is a process of alternative globalisation. Not achieved through conventional neoliberal approaches involving free trade and export-led growth, which are completely unsuitable in the context of the Pacific region (Firth 2000) but rather the engagement of the resources of the metropole (aid, scholarships, labour migration) to facilitate preparation for and participation in the global labour markets (Connell 2010). This process builds on a much earlier movement of people that we have termed above 'indigenous globalization'.

As we have seen Pacific Island populations are highly mobile but also strongly connected back to their hearth—indigenous globalization is not a matter of a race to the bottom. Rather it involves a careful and selective engagement with the outside world whilst maintaining kin and home through diasporic flows of people, goods, money and knowledge that transcend borders. This type of progressive globalization is to be encouraged—others could learn a good deal from it. However, it is not without its risks and the potential for de-population where educational strategies are not appropriate hovers ever-present. The challenge is to design educational forms that allow locally-oriented progressive globalisation. This has not happened to date and much more research is required in this area.

15.4.2 Revolutionary or Reactionary?

The role of education in development and social progress is far from simple or uncontroversial—there is no linear connection between education and development and there are many examples of inappropriate strategies that may actually worsen progressive development prospects. The advantages of standardised universal education are accepted in a rather undifferentiated way in current policy—a silver bullet for development. We require a much more complex understanding in this regard—a more geographically differentiated reading. In the case of the Pacific we see a situation whereby education could either further marginalise or de-marginalise depending on the nature of the system that is designed and the process of its implementation. We need more sensitive research in order to build the evidence in the search for more geographically appropriate educational and migration strategies that for small territories and states and work in that area is on-going in Wallis and

Futuna, Papua New Guinea and a number of other case studies In the case of the Pacific, the role of migration for education as currently pursued can be seen as revolutionary—building skills for the global workforce, allowing migration and the pursuit of livelihoods elsewhere. It also leads to the questioning long-held customs and beliefs that may oppress certain disenfranchised groups including the young in general. Yet in practice it is simultaneously reactionary—those customs and kinship networks which define it retain a degree of control over resource and people flows. Education rarely challenges the status quo at home as we have seen in various guises in places such as Samoa, Vanuatu and Fiji among other places. To progress is to maintain the best of tradition whilst allowing for progressive change. In some ways in today's globalised world it is that combination which is required in order to facilitate de-marginalisation.

15.4.3 State Subsidies or Individual Agency?

Finally, we see in the literature that education is conventionally analysed through a rather Western-centric lens of the state. This viewpoint underpins the debate concerning which policies, what resources are to build schools, and what concepts to develop curricula are required among other things. Such an approach makes the assumption that education should be standardised and provided universally (if we can afford it). Whilst we certainly agree with the latter, the former is very problematic. Viewing the role of education in development from this frame emphasises bounded spaces and places national and territorial interest above all else.

It is in this context (what we term the retro liberal turn elsewhere—Murray and Overton 2016) that we see an increasing role of aid donors from the metropole moving away from MDG-inspired universal basic education to targeted skills training for their own labour markets (especially in New Zealand and Australia). Such a process can perpetuate marginalisation and has been evolved to benefit the metropole explicitly—it is essentially neo-colonial in our opinion.

In many Pacific Islands it is the agency of the family that is more important than the role of the state—and this agency transcends borders. Families find ways to by-pass state systems at home to access international resources, decisions about which children get education and which stay at home. It is this agency that offers the greatest promise in terms of access to appropriate education and the de-marginalisation of Pacific territories. This is not to romanticise the past and play-down the significant threats that exist but rather to point towards the importance of traditional forms of dealing with marginality as potential ways forward. The potential for agency varies widely across the region, according to socio-economic status, environment, geopolitical situation and many other factors. As noted previously some of the most marginalised components of Pacific Island society are found in settler populations in the metropole—a geographic conundrum in some ways. In facilitating the progress of these populations and those of the islands there is therefore still a critical role for the state providing universal primary education and a range of secondary and vocational education facilities which allow for choices at the individual level to made eventually. The provision of tertiary scholarships which brings benefits to the metropolitan universities, whilst laudable as part of a long-term complete strategy in theory, do not deserve in our opinion to be prioritised as they currently are as they marginalise rather than the opposite. States also need to intervene to allow diaspora to grow and evolve by maintaining pathways and reducing border controls as much as possible. It is this combination of enlightened state regulation that facilitates diasporic expansion and welfare optimisation through networks of education and migration that offers the most promise for the de-marginalisation of the Pacific Islands.

References

Adams, R. H., & Page, J. (2005). Do international migration and remittances reduce poverty in developing countries? *World Development, 33*(10), 1645–1669.

Baldacchino, G. (2006a). Managing the hinterland beyond: Two ideal-type strategies of economic development for small island territories. *Asia Pacific Viewpoint, 47*(1), 45–60.

Baldacchino, G. (2006b). Innovative development strategies from non-sovereign island jurisdictions? A global review of economic policy and governance practices. *World Development, 34* (5), 852–867.

Barcham, M., Scheyvens, R., & Overton, J. (2009). New Polynesian triangle: Rethinking Polynesian migration and development in the Pacific. *Asia Pacific Viewpoint, 50*(3), 322–337.

Batibasaqa, K., Overton, J., & Horsley, P. (1999). Vanua: Land, people and culture in Fiji. In J. Overton & R. Scheyvens (Eds.), *Strategies for sustainable development: experiences from the Pacific* (pp. 100–108). London and New York: Zed Books.

Bertram, I. G., & Watters, R. F. (1985). The MIRAB economy in South Pacific microstates. *Pacific Viewpoint, 26*(3), 497–519.

Bertram, I. G. (1986). "Sustainable development" in Pacific micro-economies. *World Development, 14*(7), 809–822.

Bertram, I. G. (1993). Sustainability, aid, and material welfare in small South Pacific Island economies, 1900–1990. *World Development, 21*(2), 247–258.

Bertram, I. G. (2006). Introduction: The MIRAB model in the twenty-first century. *Asia Pacific Viewpoint, 47*(1), 1–13.

Colclough, C. (2012). Education, poverty and development—mapping their interconnections. *Comparative Education, 48*(2), 135–148.

Connell, J. (2008). Niue: Embracing a culture of migration. *Journal of Ethnic and Migration Studies, 34*(6), 1021–1040.

Connell, J. (2010). Pacific islands in the global economy: Paradoxes of migration and culture. *Singapore Journal of Tropical Geography, 31*(1), 115–129.

Connell, J., & Brown, R. (2005). *Remittances in the Pacific: An overview*. Manila: Asian Development Bank.

Coxon, E., & Munce, K. (2008). The global education agenda and the delivery of aid to Pacific education. *Comparative Education, 44*(2), 147–165.

Docquier, F., & Rapoport, H. (2012). Globalization, brain drain, and development. *Journal of Economic Literature, 50*(3), 681–730.

Evans, M. (1999). Is Tonga's MIRAB economy sustainable? A view from the village and a view without it. *Pacific Studies, 22*(3), 137–166.

Fleming, K., & Palomino-Schalscha, M. (2016). Re-envisaging Education from a Diverse Economies Perspective: The Role of Rural Training Centers in the Solomon Islands.

Development, Education and Migration in Oceania Working Paper no. 2, School of Geography, Environment and Earth Sciences, Victoria University of Wellington, Wellington.

Firth, S. (2000). The Pacific islands and the globalization agenda. *The Contemporary Pacific, 12* (1), 178–192.

Fry, G. (1997). Framing the Islands: Knowledge and power in changing Australian images of the South Pacific. *The Contemporary Pacific, 9*(2), 305–344.

Gamlen, A. (2010). The new migration and development optimism. *Global Governance, 16*(3), 415–422.

Gamlen, A. (2014). The new migration and development pessimism. *Progress in Human Geography, 38*(4), 581–597.

Gamlen, A., Murray, W. E., & Overton, J. (2017). Investigating Education, Migration and Development—Moving Triangles in the Pacific. *New Zealand Geographer, 73*(1), 57–58.

Gegeo, D. W. (1998). Indigenous knowledge and empowerment: Rural development examined from within. *The Contemporary Pacific, 10*(2), 289–315.

Gegeo, D. W., & Watson-Gegeo, K. A. (2001). "How we know": Kwara'ae rural villagers doing indigenous epistemology. *The Contemporary Pacific, 13*(1), 55–88.

Gibson, J., & McKenzie, D. (2011). The microeconomic determinants of emigration and return migration of the best and brightest: Evidence from the Pacific. *Journal of Development Economics, 95*(1), 18–29.

Gibson, J., & Mckenzie, D. (2012). The economic consequences of 'brain drain'of the best and brightest: Microeconomic evidence from five countries. *The Economic Journal, 122*(560), 339–375.

Hau'ofa, E. (1995). Our sea of islands. In E. Waddell, V. Naidu & E. Hau'ofa (Eds.), *A New Oceania: Rediscovering our Sea of Islands* (pp. 38–48). Suva: University of the South Pacific.

Hau'ofa, E. (2000). The Ocean is us. In A. Hooper (Ed.), *Culture and sustainable development in the Pacific* (pp. 32–43). Canberra: Asia Pacific Press/Australian National University.

Helu Thaman, K. (1993). Culture and the curriculum in the South Pacific. *Comparative Education, 29*(3), 249–260.

Jacobs, A. C. (2016). *Exploring the role of education in a MIRAB economy: Brain drain or brain gain? The case of Wallis and Futuna.* Unpublished MDevStuds thesis. Wellington: Victoria University of Wellington.

Macpherson, C. (1992). Economic and political restructuring and the sustainability of migrant remittances: The case of Western Samoa. *The Contemporary Pacific, 4*(1), 109–135.

Mayo, P. (2008). Comparative and international perspectives on education in small states. *Comparative Education, 44*(2), 121–124.

McElroy, J., & Parry, C. (2012). The long-term propensity for political affiliation in island microstates. *Commonwealth and Comparative Politics, 50*(4), 403–421.

McElroy, J., & Pearce, K. (2006). The advantages of political affiliation: Dependent and independent small-island profiles. *Round Table, 386,* 529–539.

Munro, D. (1990). Transnational corporations of kin and the MIRAB system: The case of Tuvalu. *Pacific Viewpoint, 31*(1), 63–66.

Murray, W. E., & Overton, J. (2011). The inverse sovereignty effect: Aid, scale and neostructuralism in Oceania. *Asia Pacific Viewpoint, 52*(3), 272–284.

Murray, W. E., & Overton, J. (2016). Retroliberalism and the new aid regime of the 2010s. *Progress in Development Studies, 16*(3), 1–17.

Oberst, A., & McElroy, J. (2007). Contrasting socio-economic and demographic profiles of two, small island, economic species: MIRAB versus PROFIT/SITE. *Island Studies Journal, 2*(2), 163–176.

Overton, J., & Murray, W. E. (2014). Sovereignty for sale? Coping with marginality in the South Pacific—the example of Niue. *Hrvatski Geograkski Glasnik (Croatian Geographical Bulletin), 76*, 15–25.

Pacific Community. (2016). Pacific Island populations. Retrieved 22 August, 2016, from http://sdd.spc.int/en/news/latest-news/124-population-projections-by-pacific-island-countries-and-territories.

Poirine, B. (2006). Remittances sent by a growing altruistic diaspora: How do they grow over time? *Asia Pacific Viewpoint, 47*(1), 93–108.

Sanga, K., & Taufe'ulungaki, A. (Eds.). (2005). *International aid impacts on pacific education*. Institute for Research and Development in Maori and Pacific Education, Victoria University of Wellington, Wellington.

Teaiwa, T. K. (2006). On analogies: Rethinking the Pacific in a global context. *The Contemporary Pacific, 18*(1), 71–88.

Thaman, K. H. (2003). Decolonizing Pacific studies: Indigenous perspectives, knowledge, and wisdom in higher education. *The Contemporary Pacific, 15*(1), 1–17.

Tolley, H. C. (2008). The rise of basic education: A local issue or a construct of the global agenda?: A review of education priorities of key donors in the Pacific region. *The International Education Journal: Comparative Perspectives, 9*(2), 18–30.

Watters, R. F. (1969). *Koro: Economic development and social change in Fiji*. London: Oxford University Press.

Wesley-Smith, T. (2007). Self-determination in Oceania. *Race and Class, 48*(3), 29–46.

Chapter 16
Demarginalization and Church Property: The Case of Czechia

Tomáš Havlíček

16.1 Introduction

The presented study, examining the role of property reacquired by churches, especially the Roman Catholic Church while overcoming selected framework conditions in peripheral regions (Havlíček et al. 2008) of the prevailingly secular Czechia develops amongst other things the previous work (Hupková et al. 2015) which analysed the process in which the present democratic state returns the church property nationalized by the Communist regime after 1948. This represents a unique situation which poses certain risks (for example, present-day churches do not have sufficient experience with administering such property), but also brings significant opportunities that the following text intends to present using selected examples from peripheral regions of Czechia.

Recovered church property was blocked until 2012. It was not permitted to sell such property, to change its purpose regarding the land-use planning and to develop land falling in this category. This situation placed significant limits on further development of the affected, mostly peripheral, regions and often further increased their marginalisation. This state of affairs has been described in some detail on the example of the town of Červená Řečice in the Vysočina Region (Hupková et al. 2015). Opportunities for further development of the town were very limited because nearly all properties surrounding developed land were once owned by the Roman Catholic Church and the utilisation of such land was disabled by the "blocking provision" (Act No. 229/1991). This development presents a symbol of the negative repercussions of the unresolved property issues between the state and churches at the municipal level. More than 80% of the municipality's land is subject to the block on formerly church-owned property. Not only does the block on church

T. Havlíček (✉)
Faculty of Science, Department of Social Geography and Regional Development, Charles University, Albertov 6, 128 43 Prague 2, Czechia
e-mail: tomhav@natur.cuni.cz

S. Pelc and M. Koderman (eds.), *Nature, Tourism and Ethnicity as Drivers of (De)Marginalization*, Perspectives on Geographical Marginality 3,
DOI 10.1007/978-3-319-59002-8_16

properties result in stagnation in the development of infrastructure and new construction, but it also brings a series of associated effects which prevent or hinder the natural socio-economic development of a given territory. The municipalities miss out, in large degree, on the possibility of new residents moving in. Consequently, the influx of young families with children was limited, which in turn led to increases in the proportion of older residents. In connection with the low number of children in the municipalities, the local elementary school was often closed down (Havlíček et al. 2005). This blockage of properties also results in the devastation of the cultural and spiritual values of a municipality that is affected (Hupková et al. 2015). Architecturally, culturally and historically valuable buildings situated on blocked properties are frequently left to fall into disrepair and eventual ruin. The further postponement of an eventual solution will lead to irreparable damage to a number of valuable structures and, as such, to the loss of part of the local cultural-historical heritage. The primary concern for the municipalities is that contested properties cease to be blocked and that the new owners of these properties have complete control over them.

The study assumes that the returning of land to the original owners (i.e. the church) can open up new development potential for the peripheral regions in question. The objective of the presented study was, in particular, to describe and analyse the implications of the process of church restitution and the potential development of mainly peripheral regions as a result of the change in ownership and the related new activities.

16.2 Historical Aspects of Church-State Relations in Czechia

Throughout Czech history sacred and profane powers have been closely connected. This mutual connectivity and influence has led to many conflicts. Religion—more specifically, Christianity—played a decisive role in the initial process of creating a political state on the territory of present-day Czechia. For some time, beginning in the 10th century, Czechia found itself situated along the ideological divide between Eastern Orthodox and Western, Roman Catholic influences (Hupková et al. 2015). During the period of broader European reformation, the Czech lands fell increasingly under the influence of Protestant movements and churches that were connected with regional political forces. Protestant churches' attempts to take control of political power in the Czech lands were, in the end, suppressed by the Hapsburgs' strong anti-Reformation movement, following the Battle of Bílá Hora (White Mountain; in 1620). Religions other than Catholicism were forbidden by the Habsburg rulers and, eventually, non-Catholic religious communities were made illegal by the state. This step led to the emigration of Protestants or to their Catholicization during the Baroque (Counter-Reformation) period, which, incidentally, also involved the construction of numerous sacred symbols in the landscape (wayside shrines, chapels and crosses).

The creation of Czechoslovakia in 1918 brought about the separation of state and church (primarily Roman Catholic) power and yet, at the same time, a "state" Czechoslovak Hussite Church was established after the pattern of the Anglican Church in England. By the beginning of World War II, members of this church comprised more than 10% of Czechia's population. After this point, there was a gradual decrease in the number of adherents of the Czechoslovak Hussite Church and, at the present time, only 1% of the population claims membership of this "state" church (Havlíček 2006).

During the 20th century, two separate political dictatorships (Nazism and communism) led to significant state restrictions limiting the operations and freedoms of church institutions. State power even assumed certain aspects of religious thinking and behaviour in its ideologically charged discourses. So-called political religion (Mayr 1995) emerged, founded on ideologies taken from existing religious structures, which were, at the same time, strictly suppressed (Hupková et al. 2015). During the period of communist rule, the state limited and regulated church operations. The fall of communism in Europe and in Czechoslovakia brought a renewal of religious freedom and renewed separation of church and state powers. Under the communist regime, the state confiscated a large number of church-owned properties (3% of Czechia's land area) and part of this property was returned to the churches (Hupková et al. 2015).

The long and drawn-out nature of the talks concerning the eventual form of the property settlement between the state and the churches indicates just how complicated the topic is. Within Czechia's secularized society, a large majority of the general public is opposed to the idea of returning property to churches and religious societies. According to one church representative, 70% of Czechia's inhabitants do not agree with the return of church property to the churches (Hupková et al. 2015). Property settlement between the state and the churches that are affected is not a simple issue confined only to the subjects involved. Local self-governing authorities, i.e. municipalities, and to a lesser degree regions and private entities are important stakeholders in that they either use—or would like to use—blocked properties. The churches can be viewed as passive stakeholders due to the fact that they have never made an official proposal or request for settlement of the outstanding property issues. Instead, the churches have waited for the state to initiate negotiations. In contrast, the municipalities that are affected can be designated as active stakeholders which initiate and actively promote measures leading to the eventual settlement of the matter. This long process (20 years) of the restitution of church property is also the result of the secular society (Havlíček 2005) and its reflection in the decisions of the political system in Czechia.

16.3 Church-State Relations and Church Property in Czechia

The interplay between state and religious (church) powers has an important role in the development of every society and its territory. This is a relationship undergoing numerous transformations depending on the type of the secular power and religions involved. The separation of Church and State we saw in many European countries in the past has been taking place only in the recent years in Czechia (Knippenberg 2005). During their existence, religious institutions acquired mainly immovable assets; these were used for the purposes of worship and self-sustainable everyday operation of the church. With the increasing secularisation of European, and Czech societies, the influence of religious institutions diminished hand in hand with their property and importance as symbols in the religious landscape (Hupková et al. 2015). Many religious buildings and other immovable property were transformed into secular establishments in the past 50 years (Knipperberg 2005). Political changes, especially the onset of authoritative (primarily communist) regimes, were another reason for the change in the purpose of church property. Dictatorial regimes viewed religious institutions as an obstacle to success in controlling and manipulating society. They endeavoured to hamper freedom of speech and assembly while controlling or even seizing church property in order to weaken the role of churches in society (Havlíček et al. 2017).

During its rule, the communist regime in Czechoslovakia seized extensive property (buildings, forests, agricultural land, etc.) from churches and religious societies between 1948 and 1989 (most of the confiscations took place at the beginning of this period). The state made an attempt to remedy the property injustice of the communist dictatorship after 1989; however, only selected buildings (especially large churches, monasteries and some monastic buildings) were returned on the basis of the 1990 "list law" while a general property settlement between the state and churches/religious societies was yet to be accomplished (Hupková et al. 2015). This historical injustice was repaired only 22 years later through Act No. 428/2012 Coll., on property settlement with churches and religious societies. Forty percent of these former church assets (today owned by municipalities and regions) will not be returned; instead the state will pay to churches financial compensation worth CZK 59 billion (EUR 2.2 billion). Consequently, 56% of property held by the state will be returned to churches. 98% of these assets belong to the Roman Catholic Church; until recently they were administered mainly by the Land Fund of the Czech Republic (mostly agricultural land) and Lesy ČR (Forests of the Czech Republic), a state enterprise managing forest land. A large part of this property was returned to churches, mainly in 2014 and 2015. Only a small part of church property still remains to be returned. Thus, over 30,000 hectares of

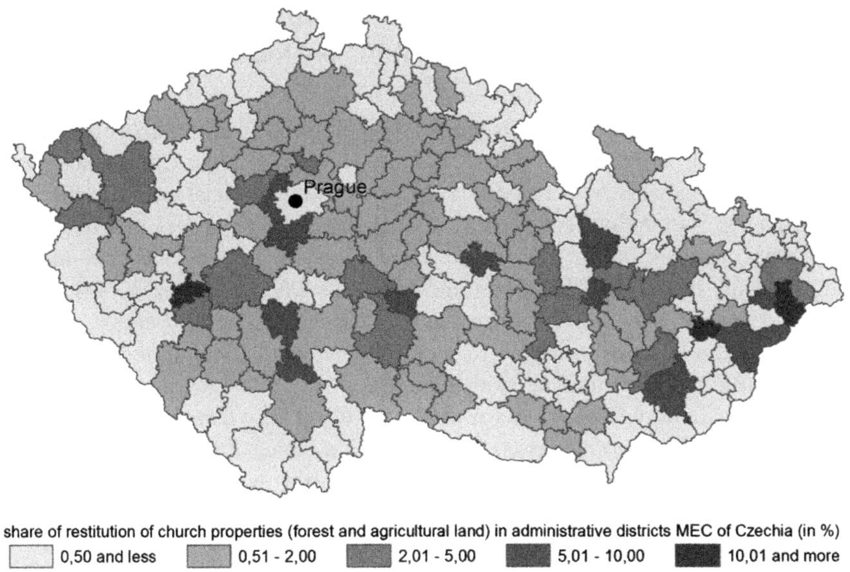

Fig. 16.1 Regional differentiation of restituted church properties in administrative districts of municipalities with extended competence (MEC) in Czechia (2016) *Datasource* Lesy (2016), Statní pozemkový úřad (2016)

agricultural land (Státní pozemkový úřad 2016) and around 80,000 hectares of forest land (Lesy 2016) were returned to churches by April 2016, which together accounts for approximately 1.4% of the total area of Czechia (Havlíček et al. 2017).

However, church restitution did not take place evenly in the territory of Czechia and the regional differentiation of this process shows several characteristics (Fig. 16.1). It is necessary to bear in mind in this respect that not all restitution claims of churches have been resolved to date, and the analysis presented is hence only a preliminary survey of the differentiation, although the resulting state of affairs is not likely to differ very significantly. The returning of immovable property to churches to date shows relatively large regional differences, which are much more significant in the eastern part of the country.

Significantly above-average figures (more than 2% of the area of the administrative district) are present especially in the MEC administrative districts near the border with Slovakia and the districts to the south and southeast of Prague.

Figures below average are shown by the northern regions near the border with Poland, with an above-average proportion of evangelic and evangelical churches, as well as in the southeast, near the border with Austria, with a very low proportion of forest land in usable land.

16.4 Restitution of Church Properties to the Roman Catholic Church as an Example of Demarginalization

Although the returning of church property by the state involves 16 churches and religious societies (Hupková et al. 2015), the Roman Catholic Church absolutely dominates the process as it receives 80% of the total financial compensation and more than 95% of the real estate returned to churches by the state. Consequently, in the initial analysis of the role of church real estate in the local and regional development of peripheral regions, the present study concentrated only on the Roman Catholic Church and its relevant activities. As mentioned above, it can be assumed that investment activities triggered by the church restitution can help stop marginalisation or even initiate gradual development, i.e. de-marginalisation of the relevant areas and regions. In this respect, it is also likely that the church will use the financial and immovable assets acquired within the restitution after 2012 for the benefit of the territories in need (peripheral territories) where the returned land is located and will not transfer much of these funds to centres.

The total investments of the Roman Catholic Church in Czechia induced by restitution amounted to nearly EUR 50 million in 2015 and were, and still are, directed to four major sectors:

(a) agriculture and forestry,
(b) administration, restoration and construction of buildings,
(c) purely financial investments,
(d) entrepreneurial projects with an innovative regional character such as mini-breweries owned by the Roman Catholic Church.

In addition to purely financial operations, the remaining three investment areas have an effect on the local and regional development of the regions concerned. In peripheral regions, investments of the Roman Catholic Church in agriculture and forestry primarily increase the number of jobs in areas with above-average unemployment, especially in the border regions of Czechia (Havlíček et al. 2008) where a large part of the real estate returned to the church are located, especially forests, arable land and pastures (Fig. 16.1). Other regions with above-average proportions of immovable property returned to churches are located in the east of the country near the border with Slovakia as well as in the northeast at the border with Saxony and in peripheral regions south of Prague where the unemployment rate is often twice as high as the national average.

The property returned includes mainly assets serving other than "ritual" purposes such as forests, meadows and farmland, as well as farm buildings within monasteries and castles. The church gradually adapts such buildings to mainly social and economic purposes. In the area of social welfare, the church sets up hospices, medical facilities specialising in aging-associated diseases (for example, Alzheimer centres) and retirement homes.

The economic purpose of such real estate includes mainly forestry and agriculture. In this way, several agricultural farms were established and deal with ecological farming (bio products) and specialised agricultural products (for example, gluten free food, special beer sorts, etc.).

Thanks to the returned property, churches were able to accomplish projects supporting their own operation as well as numerous endeavours in areas including charity, Christian ministry, education, medicine and cultural heritage, which often serve as a platform for dialogue between churches and the public. The church also lets many such spaces (those it cannot use for its own ends) for lease and, rarely, sells them to public and private entities.

It is too early yet to express the benefits of these activities e.g. in terms of decreasing unemployment or increasing proportion of young people. So far, the activities performed in the process of recovery of church property are building a platform for enhancing the social capital in peripheral areas and regions (Jančák et al. 2010) and hence a greater likelihood of further development, or de-marginalisation.

16.5 Conclusions

The process of church restitution, i.e. returning of property by the state, constitutes a new challenge for local and regional development in Czechia and, especially in peripheral regions, new opportunities for further development, in spite of the fact that Czech society is prevailingly secular and continues to oppose the returning process. During its rule between 1948 and 1989, the communist regime in Czechoslovakia confiscated extensive assets from the property of churches and religious societies. Although the state attempted to remedy the property injustice from the communist dictatorship after 1989, 22 more years had to pass before the matter was finally resolved (Hupková et al. 2015).

This chapter has sought to analyse the process of church restitution in general and its implications in local and regional development. It is too early to present a more thorough analysis of the effects of "newly" (after 2012) acquired church property on regional development of the relevant territories on the example of the Roman Catholic Church, but some basic initial trends can already be summarised: First, analyses show a significant regional differentiation of the church restitution process, which could bring benefits to peripheral regions with a high proportion of returned church property. In this respect, regional priorities of the Roman Catholic Church will be important. Second, it is obvious that the initial projects of the church (Arcibiskupství pražské 2016) were directed more to large cities (establishing community centres, church schools and care facilities) and, for the time being, less to selected peripheral territories (such as the northeast of Czechia). Third, the initial investments in peripheral, mainly rural areas are differentiated into the following three main categories:

- Projects related to care for ill and elderly people,
- Projects focusing on regional, mainly food products, and
- Projects related to forestry and agriculture.

Probably the next ten years will show the direct effects of church restitution and financial compensations on the local and regional development of peripheral regions.

Acknowledgements This chapter was prepared within the grant project Nr. 17-08370S of the Grant Agency of the Czech Republic: Post-secularism in Czechia: trends and regional differentiation. Thanks also to my colleague Dr. Jakub Lysák regarding the construction of the map.

References

Act No. 229/1991 Coll., o úpravě vlastnických vztahů k půdě a jinému zemědělskému majetku.

Act No. 428/2012 Coll., o majetkovém vyrovnání s církvemi a náboženskými společnostmi a o změně některých zákonů.

Arcibiskupství pražské, (Ed.). (2016). Aktuální projekty. Retrieved November 30, 2016, from http://www.apha.cz/aktualni-projekty.

Havlíček T. (2005). Czechia: Secularisation of the religious landscape. In Knippenberg, (Ed.), *The changing religious landscape of Europe* (pp. 189–200). Amsterdam: Het Spinhuis.

Havlíček, T., Jančák, V., Chromý, P., & Marada, M. (2005). Vybrané teoreticko-metodologické aspekty a trendy geografického výzkumu periferních oblastí. In M. Novotná (Ed.), *Problémy periferních oblastí* (pp. 6–24). KSGRR PřF: UK Praha.

Havlíček, T. (2006). Church-state relations in Czechia. *GeoJournal 67*(4), 331–340.

Havlíček, T., Chromý, P., Jančák, V., & Marada, M. (2008). Innere und äußere Peripherie am Beispiel Tschechiens. *Mitteilungen der Österreichischen Geographischen Gesellschaft, 150,* 299–316.

Havlíček, T., Klingorová, K., & Lysák, J. (2017). Atlas náboženství Česka. Praha: Karolinum. (in print).

Hupková, M., Havlíček, T., & Reeves, D. (2015). Secularization and church property: The case of Czechia. In S. Brunn (Ed.), *The changing world religion map. Sacred places, identities, practices and politics* (Vol. IV, Chap. 115, (9.5), pp. 2181–2200). Springer.

Jančák, V., Chromý, P., Marada, M, Havlíček, T., & Vondráčková, P. (2010). Sociální kapitál jako faktor rozvoje periferních oblastí: Analýza vybraných složek sociálního kapitálu v typově odlišných periferiích Česka. *Geografie, 115*(2), 207–222.

Knippenberg, H. (Ed.). (2005). *The changing religious landscape of Europe*. Amsterdam: Het Spinhuis.

Lesy, ČR (Ed.). (2016). Církevní restituce. [Czech forests (Ed.), (2016): Church restitution]. http://www.lesycr.cz/o-nas/cirkevni-restituce/Stranky/default.aspx .

Mayr, H. (1995). *Politische Religionen*. Freiburg im Breisgau: Herder.

Státní pozemkový úřad (Ed.). (2016). Majetkové vyrovnání s církvemi a náboženskými společnostmi. [The State Land Office (Ed.), (2016) Property settlement with churches and religious institutions]. http://www.spucr.cz/restituce/cirkevni-restituce/majetkove-vyrovnani-s-cirkvemi-a-nabozenskymi-spolecnostmi.html.

Chapter 17
Local Self-government Reforms in Slovenia: Discourse on Centrality and Peripherality

Janez Nared

17.1 Introduction

Despite its small size, Slovenia is remarkably rich in its natural, social, and cultural diversity, which characterizes its landscape and the settlement system (Marot et al. 2013; Ciglič and Perko 2013; Perko et al. 2015; Urbanc et al. 2015; Smrekar et al. 2016a, b). With more than six thousand settlements, the population is rather dispersed, and therefore providing services of general interest and services of general economic interest must be well defined to adequately cover the needs of the population throughout the county. To support balanced development of the country's territory, the Slovenian Ministry of the Environment and Spatial Planning reworked the Spatial Development Strategy of Slovenia (2004) and studies to support this process. One of these was a study of the polycentric settlement system in Slovenia (Nared et al. 2016), which clearly indicated the importance of municipal centers in the settlement system and providing services of general interest and services of general economic interest. Because the study uses the central place theory to explain the polycentric system, it automatically suggests the perception of the territory as a polygon of central places and peripheries as their counterparts.

The discussion of centrality and peripherality follows Christaller's (1933) theory of central places, Friedmann's (1966) core-periphery model, and Leimgruber's (1994) notion of marginality in geometric and economic terms, which all support the perception of a territory as a network of centers with adjacent peripheries. Although Friedmann's concept foresees that the urban system would become fully integrated with reduced spatial inequalities, the unequal distribution of economic activities, specialization, and a division of labour remain important factors in the contemporary territorial structure and explain different levels of centrality, as

J. Nared (✉)
Anton Melik Geographical Institute of Research Centre of the Slovenian Academy of Sciences and Arts, Gosposka ulica 13, 1000 Ljubljana, Slovenia
e-mail: janez.nared@zrc-sazu.si

S. Pelc and M. Koderman (eds.), *Nature, Tourism and Ethnicity as Drivers of (De)Marginalization*, Perspectives on Geographical Marginality 3,
DOI 10.1007/978-3-319-59002-8_17

suggested by Christaller's theory. This could also be explained by contingent and systemic marginality, which depend on market competition and hegemonic forces within the political and economic system that generate inequities through the distribution of social, political, and economic benefits (Leimgruber 2004, 61–62).

In Slovenia, the central place theory has a long tradition. Studies on central places started in the 1960s (Vrišer 1967) and were periodically carried out in the 1970s (Kokole 1971; Vrišer 1978), 1980s (Vrišer 1988), 2000s (Cigale 2002; Drozg 2005; Benkovič Krašovec 2006), and 2010s (Nared et al. 2016a, b; Nared and Razpotnik Viskovič 2016; Nared et al. 2017). Although the methodology changes from one study to another, the central focus is on providing services of general interest and services of general economic interest as well as market-oriented services to the population. The interest in the central place theory as a theoretical basis for the Slovenian settlement system is rooted in the concept of polycentric development in Slovenia, which has marked Slovenian territorial development since the 1960s. At least in theory, if not in fact, polycentrism has been the leading principle of Slovenian spatial development and a guiding force in defining the settlement system, local government, and regional policy (Nared 2003).

This chapter presents the impact that local government reforms have had on the development of Slovenian territory, particularly from the perspective of less-developed (peripheral) areas. In this regard, two main reforms are considered—one from 1960s/70s and one from the 1990s—by comparing their role in the development of the polycentric settlement system.

To determine the role that the last two local government reforms have had on territorial development in Slovenia, I examined central places in Slovenia in 2016 guided by the following research question: "How have the last two local government reforms in Slovenia influenced its settlement system?"

17.1.1 Local Government Reforms in Slovenia

The territorial organization of what is now Slovenia has undergone several transformations, mostly connected to major political processes such as the establishment of new states (the Austrian Monarchy, Austro-Hungarian Monarchy, State of Slovenes, Croats, and Serbs, Kingdom of Serbs, Croats, and Slovenes, Kingdom of Yugoslavia, Federal People's Republic of Yugoslavia, Socialist Federal Republic of Yugoslavia, and Republic of Slovenia) and constitutional changes defining the level of (de)centralization. Before the Second World War, there were 469 municipalities in what is now Slovenia and, following the establishment of a communist system after the war, municipalities played a central role in the decentralization process. They gradually gained more authority and at the same time their number decreased. In 1955 there were 130 municipalities; their number decreased to eighty-three by 1960 and to sixty-two by 1964 (Haček 2005). This was a one-tier system that combined local government functions and the functions of the central government. Responsibilities related to central government were at the forefront, corresponding to 85% of the functions (Šmidovnik 1994; Kopač 2005).

In 1963, the municipalities (or "communes") became a constitutional category, and thus central territorial units focused on providing basic services for their residents (Haček 2005) and on economic self-sufficiency (Šmidovnik 1994). The role of the municipalities increased again with the new constitution in 1974, in which, based on the Yugoslav self-management system, municipalities turned into the main territorial units for exercising self-management. Although it was expected that their role in terms of local government would increase, they remained the lowest level of the central government (Haček 2005).

Despite the fact that the municipalities were a kind of branch of the central government and that the local self-government found its application only in the local communities, municipalities became an important factor of development (Haček 2005). In particular, their role has increased with the concept of the polycentric settlement system, which was intended to develop regional centers as growth poles (Drozg and Premzl 1999). Instead of strengthening regional centers, their growth has been limited (Verlič Christiansen 1999) and they devolved to a local level. Municipal centers have become important supply centers and have experienced faster economic development. Pursuing the principles of the neo-classical growth theory, the regional policy of that time also helped redistribute economic power to less-developed areas (Nared 2003). By building new factories, the number of settlements with jobs rose from fewer than 2,000 in 1971 to 4,225 in 1991 (Drozg and Premzl 1999).

The local government reform from the 1960s and 1970s proved to be an important factor for spatial development in Slovenia, where its main effects were due to synergies of territorial reform, the newly established regional policy, and the concept of polycentric development of the settlement system, which altogether have shaped former municipal areas as important territorial units that still persist today in the form of administrative units as branches of the state administrative system at the local level and in the form of closer cooperation of the newly established municipalities derived from former larger municipalities (Rus et al. 2018).

The second major local government reform took place in 1994. At that time, the former communal system was dismantled with a thorough administrative reform that resulted in a two-tiered government system. The former "communes" were transformed into administrative units and represent a decentralized administrative organization of the state fully responsible for functions related to the central government, whereas functions related to local self-government were transferred to the newly established municipalities. The reform resulted in 147 municipalities, and their number grew over the following years: to 192 in 1998, 193 in 2002, 210 in 2006, and 212 in 2011.

The legal preconditions for establishing a new municipality were its population (which was never truly respected) and providing services of general interest and services of general economic interest that are in the domain of local self-government, such as primary school education, basic healthcare, water supply, waste management, and so on. Consequently, some small towns that sought to become municipal centers enhanced their service provision with new health clinics, full nine-year primary schools, and other services covered by the local level. Thus local government reform was also reflected in the settlement system when appraised on the basis of services of general interest and services of general economic interest.

17.2 Methods

The conceptual framework of this study comprised the following steps:

- A definition of central settlements in Slovenia;
- An analysis of the supply of central settlements with services of general interest in relation to the population size of that settlement (under- or oversupplied settlements);
- An exploration of the supply level of the respective settlement in relation to population growth and the settlement's role in the local government system (a former municipal center from the 1960s/70 s, a new municipal center from the 1990s/2000s, or a center without administrative functions).

17.2.1 Definition of Central Settlements in Slovenia

In defining central settlements, examples were followed from previous studies (Vrišer 1967, 1988; Kokole 1971; Cigale 2002; Drozg 2005; Benkovič Krašovec 2006) and six levels of centrality were defined (Table 17.1), ranging from centers of international importance as the most important level in the country to centers of rural importance at the other end of the spectrum (Nared et al. 2017). To rank the settlements in different centrality levels, types of services were defined that are expected at the specific level of centrality on the one hand (Table 17.1, Column 3) and centrality according to the population size on the other (Table 17.1, Column 2). Because many services of general interest coincide, it was decided to use only four main functions: public administration, education, healthcare, and the judiciary. In the end, both principles were merged into a combined index of level of centrality (l_{cen}). The index equally weighted the average level of centrality from the four functions ($\frac{\sum_1^4 f}{4}$) and the level of centrality based on population (l_{pop}).

$$l_{cen} = \frac{\frac{\sum_1^4 f}{4} + l_{pop}}{2} \qquad (1)$$

Table 17.1 Level of centrality for settlements and criteria for individual levels (Nared et al. 2017)

Level of centrality	Population	Expected functions	Centrality index value
1. Center of international importance	$\geq$ 50,001	• Public university • University medical center • Higher court	$\leq$ 1.50
2. Center of national importance	20,001–50,000	• College, university faculty, or academy • Large general hospital	1.51–2.50
3. Center of regional importance	10,001–20,000	• District court • Junior college • Hospital • High school	2.51–3.50
4. Center of inter-municipal importance	3,001–10,000	• Health center • Local government office • Local court	3.51–4.50
5. Center of local importance	1,501–3,000	• Nine-year primary school • Health clinic • Municipal headquarters	4.51–5.50
6. Center of rural importance	501–1,500	• Branch primary school	$\geq$ 5.51
6a. Center of rural importance with fewer than five hundred people	500 or less		Fewer than five hundred people and at least two functions

Although it was first intended to disregard settlements with a population under five hundred, many of them are well supplied, and so it was decided to add a new category of central settlements with a population below five hundred but that had to contain at least two of the four functions examined (Nared et al. 2017).

17.2.2 Defining Level of Supply

It was presumed that settlements of the same size should have a similar level of supply. In this regard, I compared the centrality level of each settlement according to its population with the centrality level of the settlement based on its supply with services of general interest. When the centrality based on population was lower than the centrality from the four functions, the settlement was oversupplied, and it was undersupplied when the centrality from the four functions was lower than that based on the population size.

17.2.3 *Defining the Supply Level of Settlements in Relation to Population Growth and the Settlements' Role in the Local Government System*

In the last step, I compared the level of supply a settlement has with the population change from 1971 to 2015. Thus I tested whether the under- or oversupply of an individual settlement was due to population change during the last four decades or whether the reasons for this should be sought in the local government reforms that took place at the same time.

17.3 Results

Based on the combined index of centrality, there are 396 central settlements in Slovenia's urban system (6.5% of all populated Slovenian settlements), of which there are (Fig. 17.1):

- 2 centers of international importance (Ljubljana and Maribor);
- 4 centers of national importance (Celje, Koper, Novo Mesto, and Kranj);
- 13 centers of regional importance;

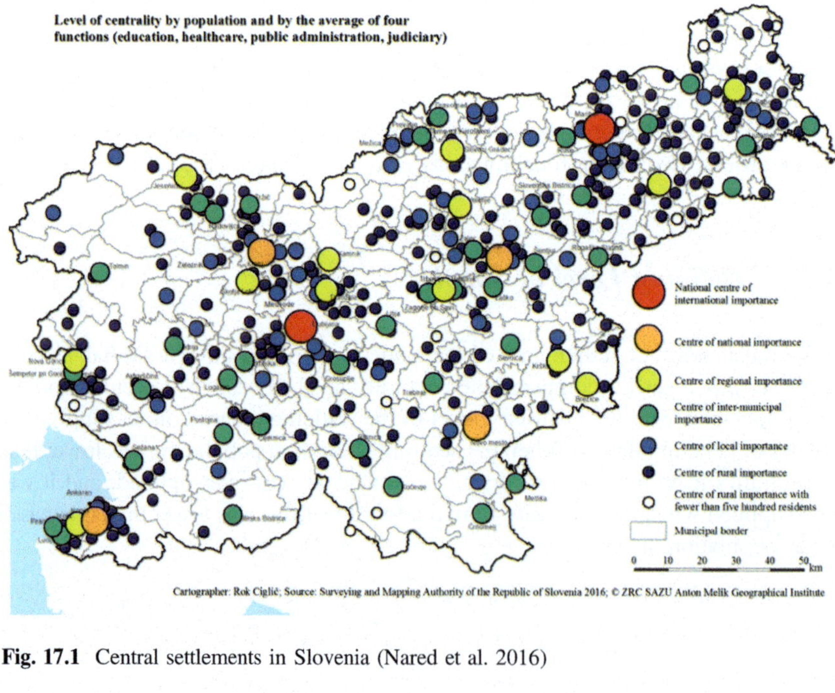

Fig. 17.1 Central settlements in Slovenia (Nared et al. 2016)

- 37 centers of inter-municipal importance;
- 58 centers of local importance;
- 282 centers of rural importance.

The second analytical step (Fig. 17.2) studied the difference between the level of centrality by the four functions and the level of centrality by population size (Table 17.1, Column 2) for each individual settlement. The results showed that the undersupplied settlements are mostly located near important national or regional centers, whereas the oversupplied settlements are mostly located in less-developed and sparsely populated areas. The undersupplied settlements predominate in suburban areas and are expected to be undersupplied due to intense population growth. On the other hand, the oversupplied settlements may have experienced a population decrease in the past four decades or, alternatively, increased their supply with new services (such as becoming a new municipal center).

To test this assumption, data on oversupplied and undersupplied settlements were compared with data on population growth during the last forty years; the results showed that the change in population only partly explains the results. I therefore created a matrix in which the x axis represents the population growth and the y axis the level of supply, and additionally compared the four categories of settlements with data on their administrative character (Fig. 17.3).

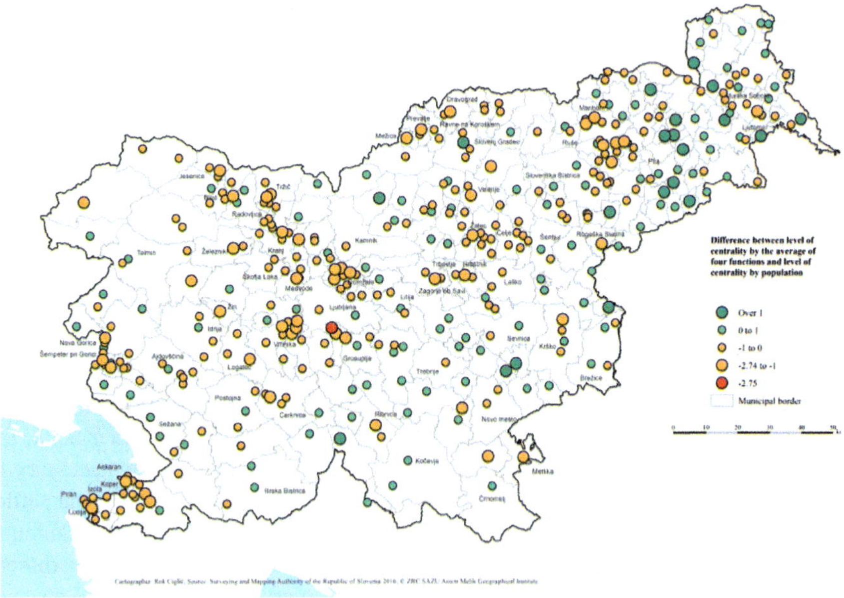

Fig. 17.2 Difference between level of centrality by the four functions and level of centrality by population size (Nared et al. 2016)

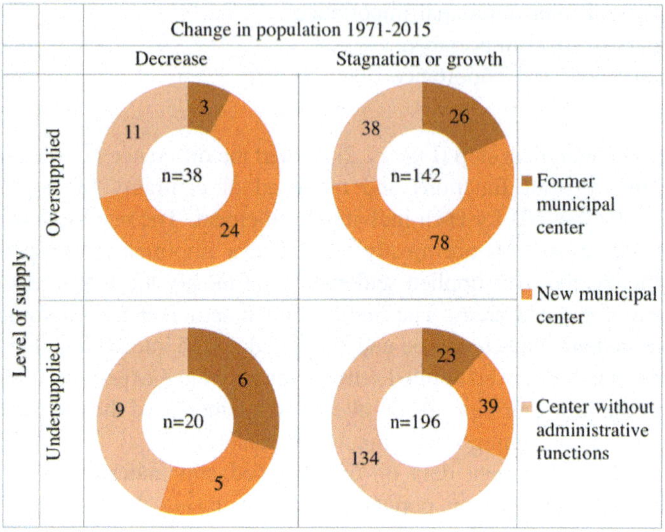

Fig. 17.3 Level of settlements' supply in relation to population growth over the last four decades

As evident from Fig. 17.3, there are only twenty undersupplied central settlements that experienced a decrease in population. Among the undersupplied central settlements whose population has grown (196), the great majority are centers without any administrative functions (134). These are followed by new municipal centers and former municipal centers.

Due to the general population increase in Slovenia between 1971 and 2015, the number of oversupplied central settlements with a decrease in population is also rather low. Among thirty-eight such central settlements, twenty-four are new municipal centers, three former municipal centers and eleven centers without administrative functions.

The group of oversupplied central settlements that have experienced population stagnation or growth is rather numerous. Out of 142 such central settlements, seventy-eight are new municipal centers, twenty-six former municipal centers, and thirty-eight centers without administrative functions.

More than 90% of the undersupplied central settlements have experienced population stagnation or growth and thus it is possible to confirm that the two categories have a strong negative correlation ($r_s = -0.32$). On the other hand, the same is not valid for the oversupplied central settlements, and so it is necessary to search for other reasons. One of them is surely the structure of the central settlements in term of their administrative status. As can be seen in Fig. 17.3, municipal centers play an important role, and so it can be concluded that local government reforms play an important role in the Slovenian settlement structure.

17.4 Discussion

The Slovenian settlement system is very dispersed. Not only there are many settlements (more than six thousand), but also at the city level the cities are numerous and only a few of them have a population over ten thousand. This is also visible in the system of central settlements, with almost four hundred central settlements that have their own functional areas competing for resources and many of them striving for decision-making power. This fact was clearly evident in the latest local government reform, in which there were many attempts to establish new municipalities, resulting in the current 212 municipalities. Being on their own, they were allowed to steer their own development, which was limited to basic infrastructure in many cases.

Because providing services of general interest and services of general economic interest is mandatory for each municipality, the areas that would like to become a new municipality enhanced their institutions. Thus some branch primary schools became full nine-year primary schools and some of the new municipal centers obtained their own health clinics. By becoming the seat of a municipality, they also provided better accessibility to services in the local domain (e.g., an administrative center, libraries, water treatment, waste management, and social care). Thus the centers became better supplied as well as responsible for many territorial management decisions. Although they are weak in terms of human and economic potential, they have important jurisdictions as well as access to tax revenues. In the absence of administration at the regional level, alongside the national level they are the only decision-making power regardless of their area and population.

This process went hand in hand with structural changes in recent decades that in spatial terms were most evident in demographic change and intense suburbanization processes (Ravbar 1997, 2005; Bole 2004). Areas surrounding centers of regional, national, and international importance have faced rapid economic and spatial development, enhanced by the new freeway network and increased car use (Bole 2004; Bole et al. 2012; Gabrovec and Razpotnik Visković 2012), whereas the developing suburban cities became increasingly residential areas with a very limited level of services and economic activities (Topole et al. 2006; Bole 2011). As shown in Fig. 17.2, this process is evident in the Ljubljana Basin (including Kranj and Jesenice), in the surroundings of Maribor, Celje, Velenje, and Nova Gorica, and in the coastal conurbation (Koper, Izola, and Piran). On the other hand, in less-developed and sparsely populated regions in southeastern and eastern Slovenia there are wider areas where centers are mainly oversupplied.

Although some areas with oversupplied centers have faced outmigration as a result of the postwar abandonment of farming (Ravbar 1995), this is not the main reason they have become oversupplied. In contrast, Kokole (1971) argued that centers in southeastern Slovenia are undersupplied due to the general developmental level of the area.

As indicated in Fig. 17.3, the oversupplied centers are mostly those that are municipal centers (72.8%), and among them the new municipal centers dominate, with more than 56.6% of the total oversupplied centers. This clearly shows that being a municipal center guarantees better access to services and thus provides better living conditions for the residents. Furthermore, it demonstrates the role that both local government reforms have had on the Slovenian settlement and spatial structure.

The impact that both reforms have had on the spatial development in Slovenia was strengthened by regional policy in both cases. In the 1970s there was an initial attempt to improve the development level of less-developed areas by introducing a traditional regional policy that strove to provide comparable living conditions for people no matter where they live (Nared 2003). This fostered the development of infrastructure and created new factories, and development was further strengthened by the concept of polycentric development (Ravbar 1995; Drozg and Premzl 1999). Importantly, polycentric development was not implemented at the regional level, as originally planned, but instead devolved to the local level, enhancing the development of the former municipal centers (Nared 2003). The local government reform starting in 1994 was similarly strengthened by a localized regional policy, especially in the last ten to fifteen years (Nared et al. 2017). In the absence of decision-makers at the regional level, the deciding power again devolved to the local level. Thus the main decision-making body at the regional level was a regional council represented by all of the mayors in the region (Rus et al. 2018), who put local interests before regional ones. The situation clearly reflected in contemporary regional development in Slovenia is visible in the centralization within Ljubljana and the Ljubljana Urban Region, and also in quickly developing municipal centers that are gaining in importance in relation to regional centers (Rus et al. 2013).

Despite the favorable results suggesting that the central settlements in peripheral regions are well supplied compared to those in suburban areas, this situation might blur the problems that peripheral areas have recently faced. One reason for this is methodological because the methodology used importantly depends on the four selected functions (public administration, education, healthcare, and the judiciary). Thus being a municipal center per se offers an advantage to the settlement when the network of central places is defined. In addition, small municipal centers do not have the authority and resources to fully use the potential that the status of an independent territorial body offers. Furthermore, the demographic decay of some very small municipalities has resulted in a situation without any children to attend the school, and it is also difficult to get teachers or physicians to work in more remote areas. Matters are even more complicated in the case of services of general economic interest and marketable services. Post offices, for example, are frequently commissioned to some local entrepreneurs (e.g., shops, bars, etc.) or limited to the lowest degree possible by providing services limited to certain days and hours (Nared et al. 2016). One of the important issues is also providing services of general interest in suburban areas. The vicinity of larger centers and their residential character hinders the development of services, which is often supported by the residents' behavior. Specifically, they use the services in areas where they work, which are often centers of regional, national, and international importance. Because of this, purchasing in residential areas is very limited and consequently services cannot develop. In the short term this is probably not a problem, but it could become one when the suburban population ages. The current dependence on cars, which is a consequence of increased mobility and spatial planning that has not taken the need for public transport into account, could cause gentrification and marginalization of contemporary residential suburban areas in the long term (Bole et al. 2012; Gabrovec and Razpotnik Visković 2012; Nared et al. 2012; Nared and Razpotnik Visković 2012).

Another important issue in studying the provision of services of general interest and services of general economic interest is the relation between cohesion and competitiveness. Contemporary studies of the polycentric settlement system favor the competitiveness of cities, mostly studying higher levels of centrality and not considering central settlements at the local level (Kloosterman and Musterd 2001; Parr 2004; Burdack and Bontje 2005; Green 2007; Meiers 2007, 2008; Burger and Meiers 2012). In contrast, for peripheral areas the crucial role of the centers is providing basic services, which places the discussion in the cohesion context. Thus, the creation of the settlement system may not only provide the foundations for a competitive economy but also pave the way to more balanced and cohesive development of the country's entire territory.

17.5 Conclusion

This chapter discussed the impacts that the last two local government reforms have had on territorial development in Slovenia. In this regard, it examined the central places in Slovenia in 2016, analyzing centrality based on population size, centrality based on the provision of services of general interest (public administration, education, healthcare, and the judiciary), and a combined index of centrality that included both of these aspects (population and services). By comparing the size of a settlement with the services it offers, it was possible to distinguish between oversupplied and undersupplied central settlements. Those were put in the context of population growth/decline and were additionally studied from the administrative perspective (i.e., formal municipal center, new municipal center, or center without administrative functions).

The results show that the provision of basic services in less-developed areas often surpasses the supply available in the same-sized suburban centers. This can be explained by the immense population growth in suburban areas, where the development of services and jobs has not followed apace. On the other hand, the oversupply of central settlements is not related to population decline, but instead to local government reforms in Slovenia over the last fifty years and additional jurisdictions and services that the new municipal centers have gained as local administrative headquarters.

In this regard, local government reforms and the newly established municipal centers play an important role in ensuring cohesion of the country's territory, particularly from the perspective of less-developed and peripheral areas. This role has even been strengthened by side effects of the concept of polycentric development in the last decades of the twentieth century, in which the initially planned promotion of regional centers devolved to municipal centers and a localized regional policy took root in Slovenia, particularly in the last fifteen years. In both cases, local centers have gained new authority and reduced their deficit in supply with services of general interest in comparison to the regional centers.

The results indicate that the provision of services of general interest is relatively rigid and inflexible, and that it adapts to the population's needs only in the long term, which has a positive effect especially on less-developed and marginal areas. Despite this, thorough reflection is needed to secure the provision of services in the longer term. Namely, adapting functions to population size would have a severe impact on less-developed and sparsely populated peripheral areas.

The results presented show that Christaller's theory is still relevant in spatial development. The functions that an individual settlement has make it a pull area for its surroundings. By devolving services to the local level during the last two local government reforms, settlement system became polycentric, whereas the increased mobility and persistent distribution of jobs in some limited centers on the one hand and the decentralization of the population into suburban residential areas on the other had enhanced the linkages among the centers, which confirms Friedmann's core-periphery model, suggesting that the settlement system is becoming increasingly complex and interconnected. At the same time, geometric marginality is not necessarily a disadvantage. As shown in this chapter, many geometrically peripheral settlements are better supplied with services of general interest in comparison to settlements in the suburban belts of large cities. This is due to the intentional and unintentional effects of local government reforms, which suggest that marginality could be overcome through prudent administrative measures.

Acknowledgements The author acknowledges the financial support of the Slovenian Research Agency (research core funding no. P6-0101) and from the project The Polycentric Network of Settlements and Access to Services of General Interest and Services of General Economic Interest, funded by the Slovenian Ministry of the Environment and Spatial Planning.

References

Benkovič Krašovec, M. 2006. *Vloga centralnih naselij prve in druge stopnje pri razvoju slovenskega podeželja*. Doktorska disertacija, Filozofska fakulteta Univerze v Ljubljani, Ljubljana.

Bole, D. (2004). Daily mobility of workers in Slovenia. *Acta geographica Slovenica, 44*(1), 25–45. doi:10.3986/AGS44102.

Bole, D. (2011). Changes in employee commuting: A comparative analysis of employee commuting to major slovenian employment centers from 2000 to 2009. *Acta geographica Slovenica, 51*(1), 89–108. doi:10.3986/AGS51104.

Bole, D., Gabrovec, M., Nared, J., & Razpotnik Visković, N. (2012). Integrated planning of public passenger transport between the city and the region: The case of Ljubljana. *Acta geographica Slovenica, 52*(1), 142–163. doi:10.3986/AGS52106.

Burdack, J., & Bontje, M. (2005). *Economic poles in the European metropolitan periphery and sustainable development*. Leipzig: Leibniz-Institut für Länderkunde.

Burger, M., & Meijers, E. (2012). Form follows function? Linking morphological and functional polycentricity. *Urban Studies*, *49*(5), 1127–1149. doi:10.1177/0042098011407095.

Christaller, W. (1933). *Die zentralen Orte in Süddeutschland*. Darmstadt: Wissenschaftliche Buchgesellschaft.

Cigale, D. (2002). Centralna naselja v Sloveniji in njihova vplivna območja v letu 1999. *Geografski Vestnik, 74*(1), 43–56.

Ciglič, R., & Perko, D. (2013). Europe's landscape hotspots. *Acta geographica Slovenica, 53*(1), 117–139. doi:10.3986/AGS53106.

Drozg, V. (2005). Koncepti policentrične ureditve Slovenije. *Dela, 24,* 147–158.

Drozg, V., & Premzl, V. (1999). Instrumenti regionalne politike v Sloveniji. *IB Revija, 33*(2/3), 76–83.

Friedmann, J. (1966). *Regional development policy: A case study of venezuela*. Cambridge: MIT Press.

Gabrovec, M., & Razpotnik Visković, N. (2012). Ustreznost omrežja javnega potniškega prometa v Ljubljanski urbani regiji z vidika razpršenosti poselitve. *Geografski Vestnik, 84*(2), 63–72.

Green, N. (2007). Functional polycentricity: A formal definition in terms of social network analysis. *Urban Studies*, *44*(11), 2077–2103. doi:10.1080/00420980701518941.

Haček, M. (2005). Zgodovinski kontekst nastajanja občin na Slovenskem. In M. Brezovšek & M. Haček (Eds.), *Lokalna demokracija II. Uresničevanje lokalne samouprave v Sloveniji* (pp. 43–67). Ljubljana: Fakulteta za družbene vede.

Kloosterman, R. C., & Musterd, S. (2001). The polycentric urban region: Towards a research agenda. *Urban Studies*, *38*(4), 623–633. doi:10.1080/00420980120035259.

Kokole, V. (1971). Centralni kraji v SR Sloveniji, problemi njihovega omrežja in njihovih gravitacijskih območij. *Geografski Zbornik, 12,* 112–131.

Kopač, J. (2005). Kraji in občine kot upravnoteritorialne enote 1945–1955. *Arhivi, 28*(2), 205–212.

Leimgruber, W. (1994). Marginality and marginal regions: Problems of definition. In *Marginality and development issues in marginal regions. Proceedings of the IGU study group development issues in marginal regions* (pp 1–18). Taipei: National Taiwan University.

Leimgruber, W. (2004). Between global and local: Marginality and marginal regions in the context of globalization and deregulation. Aldershot: Ashgate.

Marot, N., Kolarič, Š., & Golobič, M. (2013). Slovenia as the natural park of Europe? Territorial impact assessment in the case of natura 2000. *Acta geographica Slovenica, 53*(1), 91–116. doi:10.3986/AGS53105.

Meijers, E. (2007). From central place to network model: Theory and evidence of a paradigm change. *Tijdschrift voor economische en sociale geografie, 98*(2), 245–259. doi:10.1111/j.1467-9663.2007.00394.x.

Meijers, E. (2008). Measuring polycentricity and its promises. *European Planning Studies, 16*(9), 1313–1323. doi:10.1080/09654310802401805.

Nared, J. (2003). Legislation in the field of regional policy in Slovenia and an analysis of its spatial impact. *Acta geographica Slovenica, 43*(1), 85–110. doi:10.3986/AGS43103.

Nared, J., & Razpotnik Visković, N. (ed.) (2012). *Moving people: Towards sustainable mobility in European metropolitan regions*. Potsdam: Joint Spatial Planning Department Berlin-Brandenburg.

Nared, J., Bole, D., Gabrovec, M., Geršič, M., Goluža, M., Razpotnik Visković, N., & Rus, P. (2012). *Celostno načrtovanje javnega potniškega prometa v Ljubljanski urbani regiji*. Ljubljana: Založba ZRC.

Nared, J., & Razpotnik Visković, N. (2016). Somestja v Sloveniji. *Geografski vestnik, 88*(2), 51–68. doi:10.3986/GV88203.

Nared, J., Bole, D., & Ciglič, R. (2016a). Določanje stičnih naselij za vrednotenje opremljenosti naselij s storitvami splošnega in splošnega gospodarskega pomena. In R. Ciglič, M. Geršič, D. Perko, & M. Zorn (Eds.), *Prostorski podatki. GIS v Sloveniji 13* (pp. 193–202). Ljubljana.

Nared, J., Bole, D., Breg Valjavec, M., Ciglič, R., Černič Istenič, M., Goluža, M., et al. (2016b). *Policentrično omrežje središč in dostopnost prebivalstva do storitev splošnega in splošnega gospodarskega pomena. Končno poročilo*. Ljubljana: Geografski inštitut Antona Melika ZRC SAZU.

Nared, J., Bole, D., Breg Valjavec, M., Ciglič, R., Goluža, M., Kozina, J., et al. (2017). Centralna naselja v Sloveniji leta 2016. *Acta geographica Slovenica, 57*(2). doi:10.3986/AGS.4606.

Parr, J. (2004). The polycentric urban region: A closer inspection. *Regional Studies, 38*(3), 231–240. doi:10.1080/0034340042000211114.

Perko, D., Hrvatin, M., & Ciglič, R. (2015). A methodology for natural landscape typification of Slovenia. *Acta geographica Slovenica, 55*(2), 235–270. doi:10.3986/AGS.1938.

Ravbar, M. (1995). Zasnova poselitve v luči regionalnega razvoja Slovenije. *IB Revija, 29*(11/12), 21–38.

Ravbar, M. (1997). Slovene cities and suburbs in transformation. *Geografski Zbornik, 37*, 65–109.

Ravbar, M. (2005). » Urban sprawl «: Popačena slika (sub)urbanizacije v Sloveniji? *Geografski vestnik, 77*(1).

Rus, P., Razpotnik Visković, N., & Nared, J. (2013). Upravljanje območij z vidika sprememb funkcijskih zaledij centralnih krajev: primer Gorenjske. In I. Mrak, I. Potočnik Slavič, B. Rogelj (Eds.), *Gorenjska v obdobju glokalizacije* (pp. 253–268). Bled.

Rus, P., Nared, J., & Bojnec, Š. (2018). Oblike, področja in prostorske značilnosti medobčinskega sodelovanja v Ljubljanski urbani regiji. *Acta geographica Slovenica, 58*(1). doi:10.3986/AGS.4830.

Šmidovnik, J. (1994). Občina—temeljna samoupravna lokalna skupnost. *Teorija in praksa, 31* (1/2), 14–23.

Smrekar, A., Polajnar Horvat, K., & Erhartič, B. (2016). The beauty of landforms. *Acta geographica Slovenica, 56*(2), 321–335. doi:10.3986/AGS.3039.

Smrekar, A., Zorn, M., & Komac, B. (2016). Heritage protection through a geomorphologist's eyes: From recording to awareness raising. *Acta geographica Slovenica, 56*(1), 123–127. doi:10.3986/AGS.3348.

Strategija prostorskega razvoja Slovenije. (2004). Ministrstvo za okolje in prostor. Retrieved March 24, 2016, from http://www.mop.gov.si/fileadmin/mop.gov.si/pageuploads/publikacije/sprs_slo.pdf.

Topole, M., Bole, D., Petek, F., & Repolusk, P. (2006). Spatial and functional changes in built-up areas in selected slovene rural settlements after 1991. *Acta geographica Slovenica, 46*(2), 189–249. doi:10.3986/AGS.46203.

Urbanc, M., Gašperič, P., & Kozina, J. (2015). Geographical imagination of landscapes: Analysis of the book of photographs Slovenian landscapes. *Acta geographica Slovenica, 55*(1), 99–125. doi:10.3986/AGS.836.

Verlič Christiansen, B. (1999). Vpliv stanovanjske politike na trg delovne sile ter regionalno prostorska in okoljska dimenzija. *IB Revija, 33*(4), 75–86.

Vrišer, I. (1967). O centralnih naseljih. *Geografski Vestnik, 39*, 143–154.

Vrišer, I. (1978). Slovensko urbano omrežje in policentrični razvoj. In *Zbornik II. Slovensko-slovaškega geografskega simpozija* (pp. 65–76). Maribor.

Vrišer, I. (1988). The central places in the Socialist Republic of Slovenia in the year 1987. *Geografski zbornik, 28*, 123–151.

Chapter 18
Regional Disparities in Sustainable Development of Slovenia—Cohesion or Marginalization?

Katja Vintar Mally

18.1 Introduction

Sustainable development as a contemporary development paradigm imposes the requirement that progress in the economic, social, and environmental spheres be balanced. Sustainable development in close concert with environmental ethics strives to eliminate any kind of marginalization of populations or parts of the world. In investigating regional disparities we therefore proceed from the assumption that those regions which fail to follow the requirements of sustainable development will over time become even more marginalized.

Contemporary understanding of the process of marginalization focuses on geometric, ecological, economic and social types of marginality (Leimgruber 2007; Leimgruber 2010), in which marginalization is understood mainly as weak, inadequate integration into predominant structures and processes or as a lower level of development. Both are reflected in different dimensions of marginalization, but usually its socio-economic aspect is given greatest emphasis (Schmidt 2007). In order to better understand how marginalization occurs or emerges, it is important to know how systems operate (Déry et al. 2012), which is also in keeping with the approach used to study sustainable development. The crucial requirement of sustainable development is in fact this balancing of different development goals, in which right from the start there is a clash between socio-economic and environmental goals, since in the past the pursuit of the former as a rule occurred to the detriment of achieving the later. Pelc (2010) concludes that areas that are marginal based on social and economic criteria, and areas that are marginal based on environmental criteria, are usually mutually exclusive. Socio-economically more developed regions (i.e. central regions) are frequently more degraded and thus more

K. Vintar Mally (✉)
Faculty of Arts, Department of Geography, University of Ljubljana,
Aškerčeva cesta 2, 1000 Ljubljana, Slovenia
e-mail: katja.vintar@ff.uni-lj.si

S. Pelc and M. Koderman (eds.), *Nature, Tourism and Ethnicity as Drivers of (De)Marginalization*, Perspectives on Geographical Marginality 3,
DOI 10.1007/978-3-319-59002-8_18

marginalized from the environmental standpoint due to the environmental pressures of various human activities, while regions with a well-preserved natural environment (with high quality water resources, relatively unpolluted air and soil, greater biodiversity) have preserved their valuable ecosystem services precisely because of their social, economic, and geographical marginality.

In Slovenia, balancing regional development and guiding it towards the goals of sustainability in past decades has proven to be a difficult challenge. If we judge the country based on the extent of the use of natural resources and the pressure on the environment's ability to absorb the load, as is for example taken into consideration by calculations of the ecological footprint, its development model to date has been highly unsustainable. In the time since Slovenia's independence (1991) up until the global financial and economic crisis, a time of considerable economic growth, the ecological footprint of the average Slovenian inhabitant more than doubled, though in recent years, due mainly to the effects of the crisis, it has been somewhat reduced. But the Slovenian ecological footprint (5.8 global ha per capita in 2012) is still more than twice the biocapacity of the country's territory (2.4 global ha per capita) (Global Footprint Network 2016). The appropriation of natural resources and ecosystem services, which is above average on the European as well as the global scale, however, is not reflected in a proportionate socio-economic advancement of the country. Slovenia did raise its human development index from 0.766 to 0.880 in the period 1990–2014, but not its relative position, remaining in 25th place among all the countries of the world (UNDP 2015). Here it is important to stress that the number of inhabitants in the country has remained essentially unchanged and that many countries have achieved a similar level of prosperity with lower pressures on the environment and use of natural resources.

Slovenia is a member of the European Union, where sustainable development is given priority in various sector policies, the umbrella strategy Europe 2020 (European Commission 2010), and the special Sustainable Development Strategy (Council of the European Union 2006). In Slovenia, sustainable development was strongly emphasized in Slovenia's Development Strategy (IMAD 2005), which as the umbrella development strategy in the country "based on the principles of sustainable development and the integration of development policies" (IMAD 2005, p. 15) also emphasizes balanced regional development and the reduction of developmental lags in the least developed regions in the country as being among important development goals. The whole of the cohesion policy of the European Union (European Commission 2016) also strives towards this same goal.

At the regional level development trends in Slovenia are monitored only through examining the data from the so-called statistical regions, since an intermediate administrative level between the municipalities and the country as a whole does not exist. The next section of this chapter presents a comparison of the results of two studies of the development trends and current state in Slovenian regions that were carried out using indicators of sustainable development in the periods 1996–2002 (Vintar 2003) and 2010–2014 (Vintar Mally in press). The results of this evaluation of sustainability are compared with the findings of other development evaluations used by the government to monitor and encourage development in the direction of greater cohesion.

18.2 Economic, Social and Environmental Development Disparities in the Light of Sustainable Development

Guiding regions towards a more sustainable development path requires ongoing "measurement" of their progress. This is usually conducted using sustainable development indicators. Although the concept of sustainable development is often criticized for having too broad and vague a definition, it is in fact these indicators for monitoring it that define it most clearly. Thus, particular dimensions of sustainability or similar indicators appear repeatedly in different frameworks of indicators (see for example United Nations 2007; European Union 2015), and there is generally no dilemma as to whether for an increase in the level of sustainability it is desired that we "measure" increasingly higher or increasingly lower values.

In studies at the national or even more detailed spatial level, the set of indicators must be additionally adjusted to the specific features of the country and the availability of data. For the evaluation of development disparities among Slovenian regions we thus selected 32 indicators for the period 1996–2002, which we then also calculated for the period 2010–2014, though some of the original indicators were adjusted or replaced due to improvements in the collection of statistical data. These changes limit directly comparable results between the two periods but within a particular period they are entirely comparable. Both studies included 6 economic, 12 social, and 14 environmental indicators, which were selected as being representative of particular dimensions of sustainable development and by means of which we attempted to evaluate to what extent particular regions approach or diverge from the general goals of sustainable development.

For all indicators the calculation of the value for each region was followed by standardization, in order to eliminate differences due to units of measurements of variables. For each indicator we assigned scores to regions in a range from −2 (− −) for a negative contribution to sustainable development to +2 (++) for a positive contribution to sustainable development. As a basis for assigning scores we used the arithmetic mean of regions to divide positive and negative scores and the standard deviation to determine the threshold for doubling negative or positive scores. Thus if the value of the indicator for a particular region showed an above average successful approach towards the goal and was less than one standard deviation from the arithmetic mean of regions, we assigned a score of +1 (+), and for a larger deviation then +2 (++). Negative scores were assigned based on the same principle to regions with below average achievements. The assigning of scores was based on the prior decision as to whether it is desirable for a particular indicator from the standpoint of the goals of sustainable development to have a higher value (for example, increase in the extent of protected areas, spending on research and development, gross domestic product, the share of people with higher education, etc.) or as low a value as possible (for example, the share of unemployed, index of aging, water consumption per capita, production of municipal waste per capita, etc.). The overall evaluation is designed from the standpoint of an economically developed country, in which the majority of the population have their material needs satisfied and which strives towards the reduction of excessive pressures on the environment, dematerialization, and the elimination of interregional disparities in the welfare of inhabitants. It is important to stress that the same evaluation in developing countries would need to take into account reverse starting points in particular indicators with respect to desirable or undesirable phenomena. In such countries an increase in the aging index of the population, for example, would be evaluated as a positive contribution to sustainable development, since it would indicate a slowdown in the fertility rate and population growth as a precondition for faster socio-economic advancement, while in countries with an overly aged population a further increase in the index would only increase the threat to the social security system.

Based on the scores of regions for individual indicators, an average score for the economic, social, and environmental development of the region was calculated in which each indicator included carried equal weight in the final value. Based on the same principle the arithmetic mean was calculated for all three areas, which we call the indicator of sustainable regional development (ISRD).

Slovenia ranks among high income countries of the world, since according to gross national income per capita in 2014 (30,360 USD per capita, PPP) it occupied 36th place in the world and 17th place in Europe (World Bank 2016). During the economic crisis Slovenia's lag behind the average for the European Union increased sharply and in 2014 it achieved only 83% of average GDP per capita (PPP) for the European Union (IMAD 2016). However, based on the criteria of income we nevertheless cannot claim that the country or its individual regions are marginal. But certainly there are large economic disparities among Slovenian regions, as also indicated by the ratio between the region with the lowest (the Pomurska region) and the highest (Osrednjeslovenska region) GDP per capita, which over the period 1995–2014 increased from 1:1.8 to 1:2.1 (Statistical Office of the Republic of Slovenia 2016). For the evaluation of the successfulness of economic development of Slovenian regions in pursuing sustainable goals the following six indicators were used:

- Gross domestic product (€/capita),
- Gross value added (€/capita),
- Expenditure on fixed assets (€/capita),
- Average Research and Development expenditure (% of GDP),
- Disposable income (€/capita),
- Service sector employees (%).

Table 18.1 Average scores of Slovenian statistical regions in the main development spheres, 2010–2014

Statistical region	Economic indicators—average	Social indicators—average	Environmental indicators—average	ISRD
Pomurska	−1.17	−1.08	−0.07	−0.77
Podravska	−0.83	−0.67	−0.79	−0.76
Koroška	−0.83	−0.08	0.50	−0.14
Savinjska	0.17	−0.67	0.00	−0.17
Zasavska	−1.17	−0.92	0.00	−0.70
Spodnjeposavska	−0.67	−0.58	−0.36	−0.54
Jugovzhodna Slovenija	0.67	0.08	0.57	0.44
Osrednjeslovenska	2.00	0.92	−0.36	0.85
Gorenjska	0.00	1.17	−0.21	0.32
Notranjsko-kraška	−0.67	0.92	0.57	0.27
Goriška	0.50	0.92	−0.07	0.45
Obalno-kraška	0.50	0.25	0.14	0.30

For all the selected indicators, a higher value is desired for achieving greater economic sustainability, with the purpose of achieving greater material well-being in the regions, increasing the economic power of the population and the economy, and improving the employment structure and competitiveness of the economy. The average score of the economic sphere for the period 2010–2014 (Table 18.1) was highest in the Osrednjeslovenska region, which has a score that is higher by 3.17 than that of the economically weakest Pomurska region. Regions of the eastern part of Slovenia (especially Pomurska, Podravska, Koroška, and Zasavska) are pushed most to the economic periphery of the country (Fig. 18.1), and this situation was also shown by these regions in a study at the end of the 20th century. In general the rankings of the regions did not change essentially between the two periods, which indicates that these economic disparities are deeply rooted.

Similar findings also apply for disparities achieving social goals of sustainable development, which are based on ensuring the highest quality possible education, health care, and housing conditions for the population. With respect to preventing marginalization, in addition to access to public services the prevention of poverty and social exclusion is also very important, especially for more vulnerable groups (for example, unemployed women and less well educated people), as is maintaining the vitality of the population and reliance on endogenous human resources in the region. The aspects of social sustainability mentioned were studied using twelve indicators:

- Unemployed with uncompleted or completed primary school (%),
- Share of unemployed women (%),
- Population density (inhabitants/km^2),
- Population growth index,

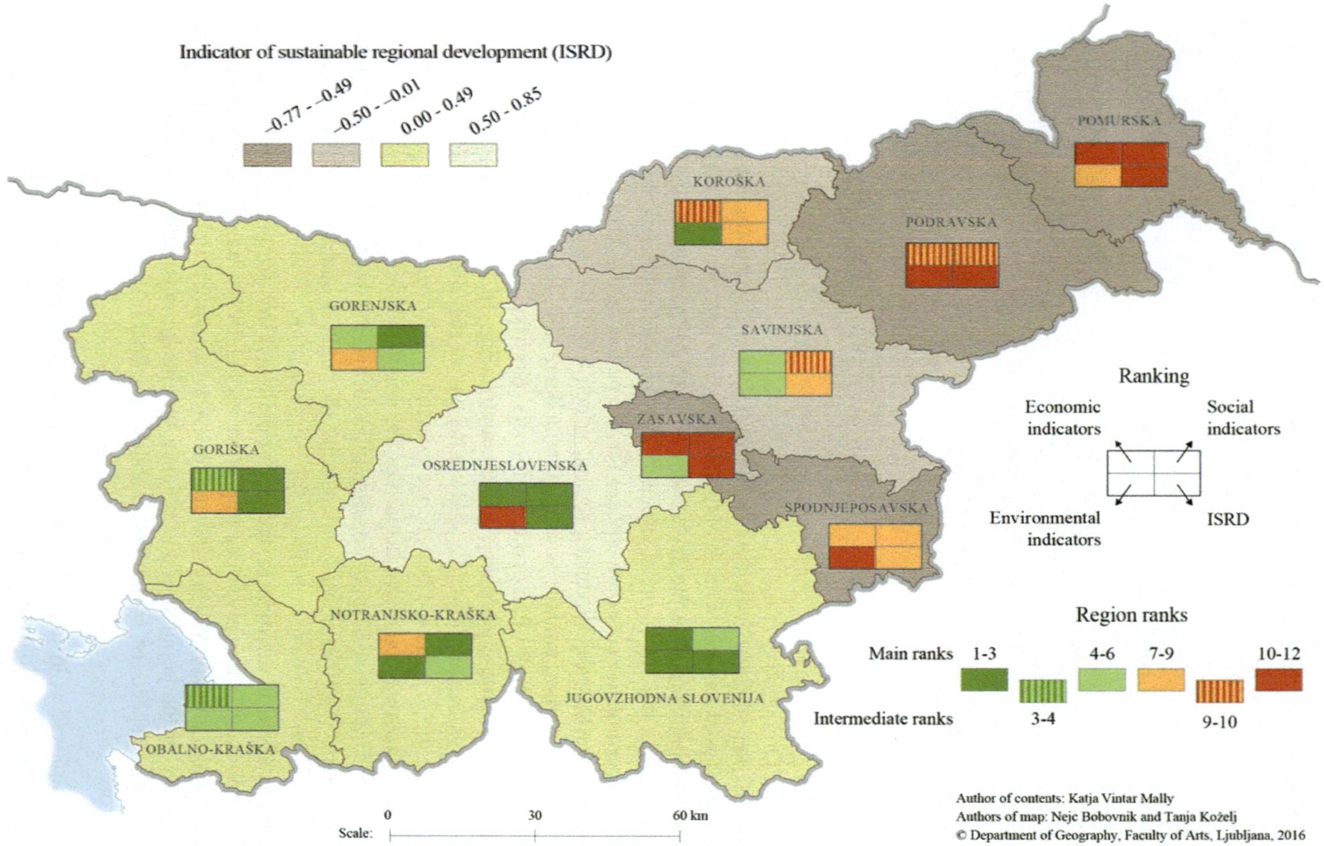

Fig. 18.1 Indicator of sustainable regional development and rankings of regions in the main development fields, 2010–2014

- Aging index,
- Average age at death (years),
- Recipients of social assistance benefits in cash (no. of recipients/1,000 inhabitants),
- Usable floor area (m^2/capita),
- Registered unemployment rate (%),
- Number of students per 1,000 inhabitants,
- Internet users (index),
- College degree holders (25–64 years) (%).

Social indicators monitor progress in increasing the quality of life, which improved according to some criteria but worsened according to others in Slovenian regions between the two periods studied. The population of Slovenia is very slowly increasing: over the period 1996–2014 it grew from 1,990,266 to 2,061,085 (Statistical Office of the Republic of Slovenia 2016), but at the same time some regions recorded a decline in population (especially Zasavska, Pomurska, and Koroška region), and calculations and comparisons of the indicators also point to a rapid aging of the population in all regions. In general the educational level of the population, housing conditions, and infrastructure improved, and life expectancy increased. On the other hand, the financial-economic crisis after 2008 was reflected in higher unemployment and significantly increased the number of people subjected to poverty and social exclusion.

Based on the average scores of regions in the social sphere it can be seen that the top six places (Fig. 18.1 and Table 18.1) are occupied by regions from the western half of the country, with the best conditions shown by the Gorenjska region, followed by Goriška, Notranjsko-kraška, and Osrednjeslovenska region. The absolute range of disparities in social scores (2.25) is smaller than that for economic scores but still greater than the range in environmental scores (1.36). Regions in the eastern part of the country show an above average aging of the population, and some also show a drop in the number of inhabitants; they have a poorer education structure and more difficulties in providing employment, and an increased risk of social exclusion for inhabitants.

A comprehensive evaluation of the current state and trends in the environmental sphere was most difficult since there is a lack of synthetic indicators and it is therefore necessary to use a larger number of indicators in order to capture at least the most important issues. With the combination of environmental indicators included we tried to cover all environmental elements (air, water, soil, biodiversity, space), either through evaluating their current state, the pressures they face from economic activities and population, or the responses of society to existing environmental problems. The final selection of fourteen key environmental indicators comprised the following:

- Organically farmed land (%),
- Wooded areas (m^2/capita),
- Road freight transport growth index,
- Intensively farmed land (m^2/capita),
- Quality of air (assessment),
- Municipal waste (kg/capita),
- Natura 2000 sites (%),
- Water consumption (m^3/capita),
- Average expenditure on environmental protection (% GDP),
- Built-up areas (%),
- Treated wastewater (m^3/capita),
- Housing with district heating in place (%),
- Motorization rate (cars/1,000 people),
- Livestock density index (LSU/ha).

In both periods studied the analysis of environmental indicators showed that interregional disparities in this sphere were the smallest, and scores in comparison to social and economic ones were more or less evenly distributed. At the same time the order of the ranking of the regions changed most between the two periods with respect to the average score for the environmental sphere. We can thus conclude that through a sound policy in the environmental sphere we could most easily achieve cohesion, as supported also by the fact that in most of the country the environment is still in a well preserved state.

The most favorable score in the environmental sphere in the period 2010–2014 was shared by the regions of Jugovzhodna Slovenija and Notranjsko-kraška, followed by Koroška (Fig. 18.1 and Table 18.1). The regions of Podravska, Osrednjeslovenska, and Spodnjeposavska most lagged behind the achievements of other regions. Regardless of the existing disparities among the regions it should be emphasized that since the end of the 20th century great progress has been achieved in the country in reducing air pollution and in monitoring pollutants, which can be seen in particular in the improved quality of water resources and air (replacement of energy sources, installation of desulphurization units and filters on the biggest air polluters, systematic construction of wastewater treatment plants, improved waste management, etc.). Likewise, there has been a growing trend towards environmentally friendly agricultural production (especially the growth of organic farming) and greater attention given to protected natural areas (national protected areas, the Natura 2000 European network of areas). On the other hand there have been some negative trends of increased traffic and the extent of material-energy flows, and an increasingly greater concentration of pressures from settlement, economic activities, and infrastructure in lowland areas, especially in the vicinity of regional centers.

18.3 Regional Disparities in Slovenia According to Different Measures of Development

A comparison of Slovenian regions according to the value of the synthetic indicator of sustainable regional development (ISRD) in both periods studied shows that the six westernmost regions have maintained above average (positive) values and the other six regions have below average (negative) values (Fig. 18.1). The ranking of regions has not essentially changed, since the rank of a particular region changed between the two periods by at most two places (in the second period there was an improvement in position by two places for Jugovzhodna Slovenija and a drop by two places for the Gorenjska and Podravska regions). The range of values in the period 2010–2014 was somewhat smaller but due to changes in the indicators included and the method of assigning scores we cannot attribute this directly to a reduction in disparities among regions. Due to greater development disparities between the eastern and western halves in the social and economic spheres there also appears a sharp dividing line between the two parts of the country in the ISRD (Table 18.2).

A comparison of the ISRD with the so-called development risk index (DRI), which is used at the national level for ranking regions according to level of development in particular program periods, shows greater differences in the rankings of regions. The latest available calculations of DRI are for the current program

Table 18.2 Changes in the development risk index and the indicator of sustainable regional development since the end of the 20th century

Statistical region	DRI[a]				ISRD			
	2001-2005[b]		2010-2014[c]		1996–2002		2010–2014	
	Score	Ranking	Score	Ranking	Score	Ranking	Score	Ranking
Osrednjeslovenska	8.7	1	35.5	1	0.74	1	0.85	1
Obalno-kraška	82.4	2	81.4	4	0.42	4	0.30	5
Gorenjska	83.1	3	66.6	3	0.67	2	0.32	4
Savinjska	92.3	4	92.6	5	-0.19	7	-0.17	8
Goriška	93.8	5	100.4	6	0.57	3	0.45	2
Jugovzhodna Slovenija	101.7	6	64.7	2	0.28	5	0.44	3
Koroška	103.9	7	121.6	8	-0.30	8	-0.14	7
Zasavska	113.9	8	125.1	11	-0.82	11	-0.70	10
Spodnjeposavska	116.8	9	101.5	7	-0.70	10	-0.54	9
Podravska	116.8	10	123.9	9	-0.70	9	-0.76	11
Notranjsko-kraška	127.0	11	124.8	10	0.07	6	0.27	6
Pomurska	159.5	12	161.8	12	-1.12	12	-0.77	12

Sources Vintar (2003), Pečar and Kavaš (2006), Pravilnik o razvrstitvi (2014); own calculations

[a]A higher value of the index means greater risk to the development of the region

[b]Period of data by means of which the development risk index was calculated for the program period 2007–2013

[c]Period of data by means of which the development risk index was calculated for the program period 2014–2020

period 2014–2020 and are based on data from the last few years before the start of this period (Table 18.2). In comparison with ISRD, DRI includes a smaller number of indicators, and the selection also changed in part based on the availability of data. The most recent set encompasses 14 indicators: GDP per capita, gross value added per employee, gross fixed capital formation in GDP, unemployment rate for young people (15–29 years), employment rate (20–64 years), proportion of the population with tertiary education (25–64 years), the share of gross domestic expenditure on research and development in GDP, the proportion of treated wastewater with at least secondary treatment, the proportion of protected land areas in the region, the share of the estimated damage caused by natural disasters in GDP, the unemployment rate, the index of aging, disposable income per capita, and population density (Pravilnik o razvrstitvi 2014).

The results of both synthetic indicators are not directly comparable due to different methods of standardization but we can compare the relative position of regions with respect to the selected method of evaluation. Due to the predominance of economic and social indicators in DRI (only three indicators could be characterized as environmental) the rankings of the regions differ significantly between DRI and ISRD: for example, two regions are ranked four places higher (Goriška and Notranjsko-kraška), and one region (Savinjska) is three places lower. With respect to DRI the Osrednjeslovenska region has the lowest development risk and the Pomurska region has the highest (Table 18.2). The governmental Institute of Macroeconomic Analysis and Development found that during the time of the last crisis, regional disparities in level of economic development and risk of poverty became smaller, but this was a result of the greater drop in economic activity or in other words the greater economic decline of regions that before 2008 had a more advantageous position (IMAD 2016). This thus does not mean a reduction in the marginalization of the least developed regions of the country. A more detailed insight into the disparities in the level of development of particular parts of the country is given by the coefficient of development of municipalities, which is used by the government to determine the level of co-financing of investments in Slovenian municipalities. The index is calculated from ten indicators that cover economic strength of the population and of the economy (income tax base per capita, gross value added of companies per employee), demographic characteristics (index of aging, density of settlement), employment opportunities (number of jobs for the size of the economically active population, unemployment rate, employment rate), environmental protection (share of Natura 2000 areas, proportion of the population with a connection to the public sewer system) and cultural heritage (cultural monuments and infrastructure) (Ministry of Finance of the Republic of Slovenia 2015).

Strikingly high values for the coefficient of development of municipalities are shown, as expected, for the central part of the country, especially the municipalities in the Osrednjeslovenska region, while low values point to the marginalization of particular areas along the border, especially those that are also in high mountain or forested Dinaric karst areas, or have poor transport connections with national and/or regional centers. Greater frequency of municipalities with a low coefficient of development is also noticeable in regions that lag behind economically (Fig. 18.2).

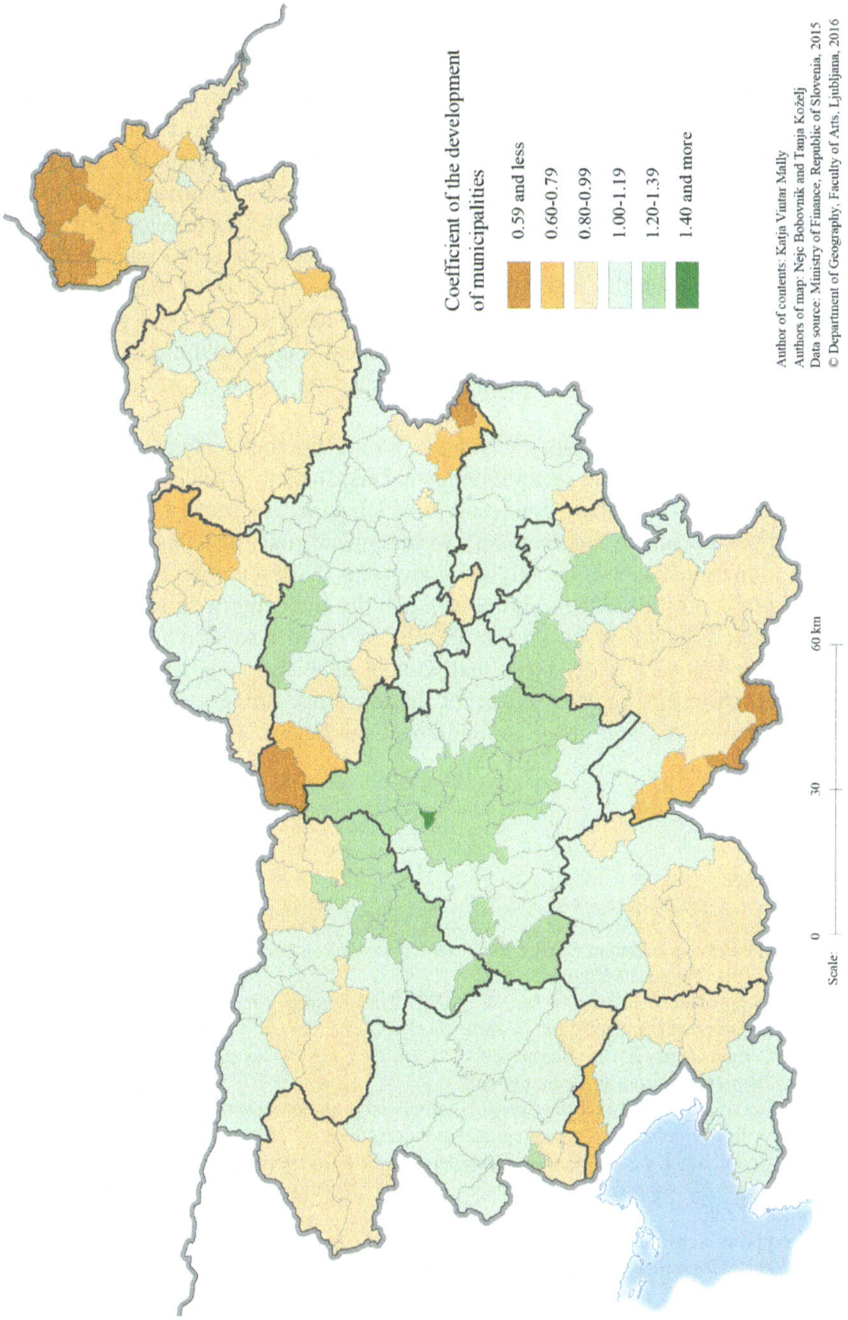

Fig. 18.2 Coefficient of the development of Slovenian municipalities in 2015

18.4 Conclusion

Quality of life for the 21st century population is no longer represented merely in terms of socio-economic well-being, such as that enabled by classical centers of power (for example, urban areas or regional and national centers), but also in terms of a high quality living environment, of which an indispensable part is an unspoiled and unpolluted natural environment offering numerous ecosystem services, which are valued also from aesthetic, recreational and other nonmaterial aspects. Sustainable development guides us toward balancing different development goals and the creation of the living conditions described. In general we cannot say of Slovenian regions that the disparities among them are growing, or that cohesion is increasing, since very diverse trends are recorded in different areas. But we can with certainty conclude that the prevailing development model in the country is still unsustainable, and that economic and social disparities are the most deeply rooted within it.

Evaluation using indicators can have a strong communicative value for the broader public, experts, and sectoral policymakers, since it indicates weaknesses in the functioning of the system, as a result of which particular parts of the system lag behind in the pursuit of contemporary development goals and thus become susceptible to marginalization. Usually these are areas that even now are faced with a range of problems, hence their ability or flexibility in coping with new challenges such as those brought on by climate change or broader social changes is questionable. Experience to date shows that without deliberate long-term measures in regional development we cannot expect cohesion, since certain regional disparities are very deep-rooted and create an additional obstacle to following a more sustainable path.

References

Council of the European Union. (2006). *Renewed EU sustainable development strategy*. Brussels: Council of the European Union.

Déry, S., Leimgruber, W., & Zsilincsar, W. (2012). Understanding marginality: Recent insights from a geographical perspective. *Croatian Geographical Bulletin, 74*(1), 5–18.

European Commission. (2010). Europe 2020, *A strategy for smart, sustainable and inclusive growth*. Brussels: European Commission.

European Commission. (2016). Cohesion policy frequently asked questions. Retrieved November 10, 2016, from http://ec.europa.eu/regional_policy/en/faq/.

European Union. (2015). *Sustainable development in the European Union: 2015 monitoring report of the EU sustainable development strategy*. Luxembourg: Publications Office of the European Union.

Global Footprint Network. (2016). National footprint accounts, 2016 Edition.

IMAD. (2005). *Slovenia's development strategy*. Ljubljana: Institute of Macroeconomic Analysis and Development.

IMAD. (2016). *Development Report 2016*. Ljubljana: Institute of Macroeconomic Analysis and Development. Retrieved November 10, 2016, from http://www.umar.gov.si/fileadmin/user_upload/publikacije/pr/2016/APoR_2016.pdf.

Leimgruber, M. (2007). Geographical marginality—past and new challenges. In G. Jones, W. Leimgruber, & E. Nel (Eds.), *Issues in geographical marginality* (pp. 2–11). Grahamstown: Rhodes University.

Leimgruber, M. (2010). Marginal regions. In B. Warf (Ed.), *Encyclopedia of geography* (pp. 1851–1853). Thousand Oaks, CA: SAGE Publications.

Ministry of Finance of the Republic of Slovenia. (2015). Določitev koeficientov razvitosti občin za leto 2015 z dne 5. 1. 2015. Retrieved November 13, 2016, from http://www.mf.gov.si/si/delovna_podrocja/lokalne_skupnosti/izracuni/dolocitev_koeficientov_razvitosti_obcin/za_leto_2015/.

Pelc, S. (2010). Environmental marginality: Reality or myth? In W. Leimgruber et al. (Eds.), *Geographical marginality as a global issue* (Vol. 1, pp. 50–58). General, Theoretical and Methodological Aspects. Dunedin: University of Otago.

Pečar, J., & Kavaš, D. (2006). *Metodologija izračuna indeksa razvojne ogroženosti za obdobje od 2007 do 2013*. Ljubljana: Institute of Macroeconomic Analysis and Development.

Pravilnik o razvrstitvi razvojnih regij po stopnji razvitosti za programsko obdobje 2014–2020. (2014). Ljubljana: Uradni list Republike Slovenije.

Schmidt, M. (2007). Some reflections on the definition and delimitation of geographical marginality. In G. Jones, W. Leimgruber, & E. Nel (Eds.), *Issues in geographical marginality* (pp. 34–43). Grahamstown: Rhodes University.

Statistical Office of the Republic of Slovenia. (2016). SI-Stat data portal. Retrieved November 5, 2016, from http://pxweb.stat.si/pxweb/dialog/statfile1.asp.

UNDP. (2015). *Human development report 2015*. New York: United Nations Development Programme.

United Nations. (2007). *Indicators of sustainable development: Guidelines and methodologies* (3rd ed.). New York: United Nations.

Vintar, K. (2003). *Okoljevarstveni vidiki sonaravnega regionalnega razvoja slovenije*. Ljubljana: Oddelek za geografijo Filozofske fakultete Univerze v Ljubljani.

Vintar Mally, K. (In press). Regional differences in Slovenia from the viewpoint of achieving Europe's sustainable development. *Acta geographica Slovenica*. doi:10.3986/AGS.3309.

World Bank. (2016). *World development indicators 2016*. Washington, DC: World Bank. doi:10.1596/978-1-4648-0683-4.

Chapter 19
Marginalization of Tea Estates in Sri Lanka in the Changing Global and National Context

Daichi Kohmoto

19.1 Introduction

Many developing countries around the world still depend on the export of a few agricultural products and/or mineral resources. Commercial plantation-based agriculture which expanded under colonial economics is a typical example in case.

Plantation agriculture can be seen as an expression of productivism whose objective is to produce food on a large scale through specialization and productivity improvements. As a reflection of this characteristic, plantations are still often cited for the various historical problems that they have not been able to rectify, including monoculture farming, structures that exploit and domineer over their workers, and environmental issues (See Beckford 2000; Harteminsk 2005 etc.). However, countries and regions where plantation agriculture is still practiced are beginning to feel the effects of the new liberalism-inspired globalization of food systems. In these countries, studies have confirmed economic, social and political instability (Nuhn 2004), and shifts to manufacture-based and service-based economies (Lincoln 2006).

This study focuses on the tea industry in Sri Lanka. Tea production in this country has been conducted by export-oriented commercial plantations inherited from the time of British colonial rule, and these plantations have had a major impact in defining the socioeconomic conditions in Sri Lanka.

Interviewing surveys were conducted at head offices of plantations and managing departments of tea estates, as well as with tea estate workers and residents in 2005. Information was also gathered from a variety of papers and other sources. Much of this study is based on my previous paper (Kohmoto 2008) written in Japanese.

D. Kohmoto (✉)
Nara University of Education, Nara, Takabatake-cho 630-8528, Japan
e-mail: daichizu@gmail.com

S. Pelc and M. Koderman (eds.), *Nature, Tourism and Ethnicity as Drivers of (De)Marginalization*, Perspectives on Geographical Marginality 3,
DOI 10.1007/978-3-319-59002-8_19

19.2 Global Tea Production and Distribution, and Where Sri Lanka Stands

Black tea produced in Sri Lanka is well-known as Ceylon Tea. However, the Sri Lankan tea industry, particularly the estate sector, is burdened with structural problems for sustainability. Tea production is growing around the world as a result of improvements in land productivity. Global consumption is also on the rise, primarily in developing countries. Consumption in advanced countries, however, is declining and this has caused a prolonged decline in global tea prices. Sri Lanka, the world's largest exporter, has continued to produce tea, primarily on plantation estates that were built during the colonial era.

Tea production in the four leading tea producing countries shows that while India and China have continued to expand production since 1960s, and Kenya has established its position as a new producer, Sri Lanka's production remained low until around 1990. Kenya had surpassed Sri Lanka in production amount after 2005. Both countries export most of the black tea they produce. Uganda, Tanzania and Malawi are up-and-coming producers of teas for export. In China, production of teas other than black teas is growing partly as a result of diversifying preferences in markets such as Japan.

At the time of 1980, top 10 destinations of Ceylon Tea were Iraq, UK, Egypt, USA, Pakistan, Saudi Arabia, Iran, Kuwait, Libya, Syria, and Canada. In 2014, they were Turkey, Russia, Iran, Iraq, UAE, Syria, Libya, Kuwait, Azerbaijan, Japan, and Chile. This change shows that while Sri Lanka does export to a wide range of countries, a relatively large proportion of their tea is exported to Russia and the Middle East. Kenya, on the other hand, exports to a small number of specific countries including the UK, Pakistan and Egypt now, resulting in market segregation between Sri Lankan exports. India's major markets include the former Soviet Union and UK, China's major markets (for black teas) include neighbouring countries such as Japan and Hong Kong, and major importers of Malawi's teas include the UK and neighbouring South Africa. The majority of these exports are believed to be black teas. Green tea exports to western Africa and other regions where green tea is consumed in large amounts primarily come from China.

Black tea consumption is high in the former Soviet Union, UK and other developed countries, and Pakistan. Tea consumption has been declining in developed countries including Germany, France, Australia and the UK, where an increasing number of consumers are shifting to other soft drinks such as mineral water and juices (Grigg 2003). In contrast, tea consumption in developing countries and the former Soviet Union is rising in all income levels mainly because teas are less expensive than coffee and are better thirst quenchers (Weinberg and Bealer 2001; Grigg 2003). While global production and consumption levels are both on an increasing trend, these consumption trends have put global tea prices on a declining or stagnating trend.

19.3 Sri Lanka's Tea Production and Distribution Structures

Even after 1948 when Sri Lanka gained independence from Britain, its economy was largely driven by the export of plantation crops (teas, coconut, and rubber). In 1977, the country introduced market economy policies to diversify its economy, and by 1990, clothing exports and workers working abroad had increased dramatically. Meanwhile, tea exports continued to grow with large volumes exported to Russia and the Middle East, both regions where tea consumption continues to grow, unlike Europe and North America.

However, things are not going as smoothly as they may appear. Particularly, the nationalization of tea plantations that occurred between 1972 and 1975 caused a decline in Sri Lanka's competitiveness in the international market. According to Manikam (1995) and Shanmugaratnam (1997), there was considerable discontent among the people of Sri Lanka (then called Ceylon), primarily among small-scale Sinhalese farmers who had had their land expropriated by British plantation operators, and this had been the backdrop for the nationalization movement. Private corporations who owned tea farms were threatened by nationalization and reduced business investments in the late 1950s; in particular, much of the British capital left Ceylon and moved to locations such as Kenya where they could pursue the "economy of scale". After all the plantations were requisitioned and nationalized between 1972 and 1975, various problems arose including corruption surrounding the hiring of plantation managers, lack of business acumen, and excessive political intervention in decision making processes. Furthermore, government policies that prioritized rice farming, export taxes, and a drop in international tea prices caused land and labour productivity to stagnate.

Sri Lanka's tea industry has been slow to introduce mechanization and new production methods. For example, tea factories around the world began employing a new and efficient method called the CTC manufacturing process in the 1980s. As of 2004, 97% of tea factories in Kenya used CTC whereas only 5% currently do so in Sri Lanka (based on information from these country's respective tea boards). Because introduction of this technique was delayed nationwide, European buyers, particularly the British buyers, came to see Sri Lanka as a "country that used orthodox methods" (according to an interview with the Sri Lanka Tea Board). Tainted by this image, many Sri Lankan tea factories that actually implemented CTC have not been able to generate profits from their sales, and the rate of CTC implementation has been falling since it peaked at 8% in 1995.

Why then has production been increasing since the 1990s? This has to do with Sri Lanka's regional disparities in tea production. There are three general categories of tea growing areas in Sri Lanka: lowlands (elevation of 600 m and lower), middle lands, and highlands (1200 m and higher; Fig. 19.1). All of these areas are located predominantly in the south-western region, which I will discuss in detail later.

Fig. 19.1 View of a typical landscape of tea estates in the highlands

Fig. 19.2 Tea production levels over the years by elevation (Compiled from the FAO Statistical Databases-www.fao.org)

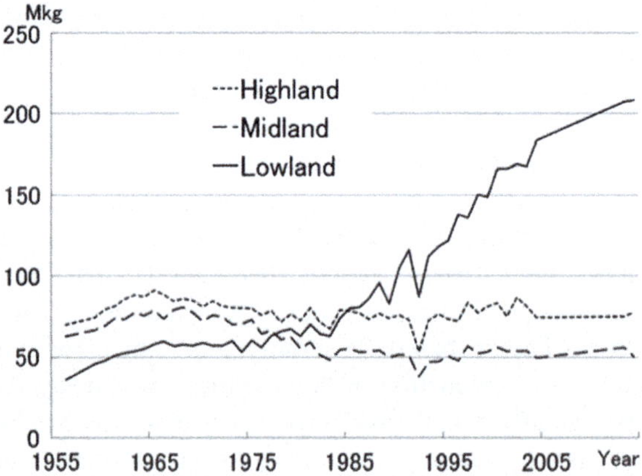

In terms of the lowlands, 69.9% of the tea growing acreage there (according to the 2002 agricultural census) are farmed by individuals growing tea on small plots (smaller than 4 ha; hereafter referred to as "small holdings") and not by the plantation sector. Growing area and productivity among small holdings improved dramatically in the 1980s thanks to increased government support (Fig. 19.2). Meanwhile, the plantation sector which mainly operates in the highlands and middle lands (making up 80.1% and 79.3% of the growing areas, respectively) has been undergoing privatization since 1992. The national government, while maintaining the state-owned status of land, established 22 regional plantation companies (RPCs) and separated the majority of plantations from the two national companies that had been controlling them. Since then, control of these RPCs have been gradually transferred to private corporations through a tender process, with active participation from foreign capital and domestic non-agricultural interests. These steps proved to be effective and resulted in a slight growing trend in production in the plantation sector.

According to the Sri Lanka Tea Board (2004), 95.5% of Sri Lankan tea are packed at tea farms and neighbouring factories before they are sent to Colombo for the weekly auction. Eight Sri Lankan brokers perform tastings on samples shipped from each factory and auction these teas to buyers. After the auction, they are shipped from the port of Colombo under the "Ceylon Tea" brand.

19.4 Problems in the Plantation Sector Stemming from Structural Changes

The plantation sector in the Sri Lankan tea industry is facing a variety of problems stemming from the structural changes described above. In the following discussion, I have categorized these issues into three broad categories:

(a) land use and the environment,
(b) sales price, and
(c) labour.

19.4.1 Land Use and the Environment

Humbel (1990, 1991) elaborated on the shift in tea farm land use in Sri Lanka by comparing aerial photos taken in 1956 and the 1980s. Figure 19.3 shows the results of his study as percentages of land used for tea production by DS division (a unit of land that loosely corresponds to local municipalities). In 1956, tea farms were mainly found in the middle—to highland provinces of Nuwara Eliya, Kandy and Badulla where plantations predominated. In the 1980s, tea farming acreage declined considerably in Kandy, which lies at the northern edge of the core tea farming region, and in the neighbouring province of Kegalle to the west. According to the agricultural census of 2002, tea farming acreage declined steeply in Nuwara Eliya and Badulla, with contrasting growth in the provinces of Ratnapura, Galle and Matara where small holdings predominated. These outcomes resulted from tea plantations in the middle—to highland areas going out of business and abandoning their land. In turn, this has resulted in a variety of issues including labour and

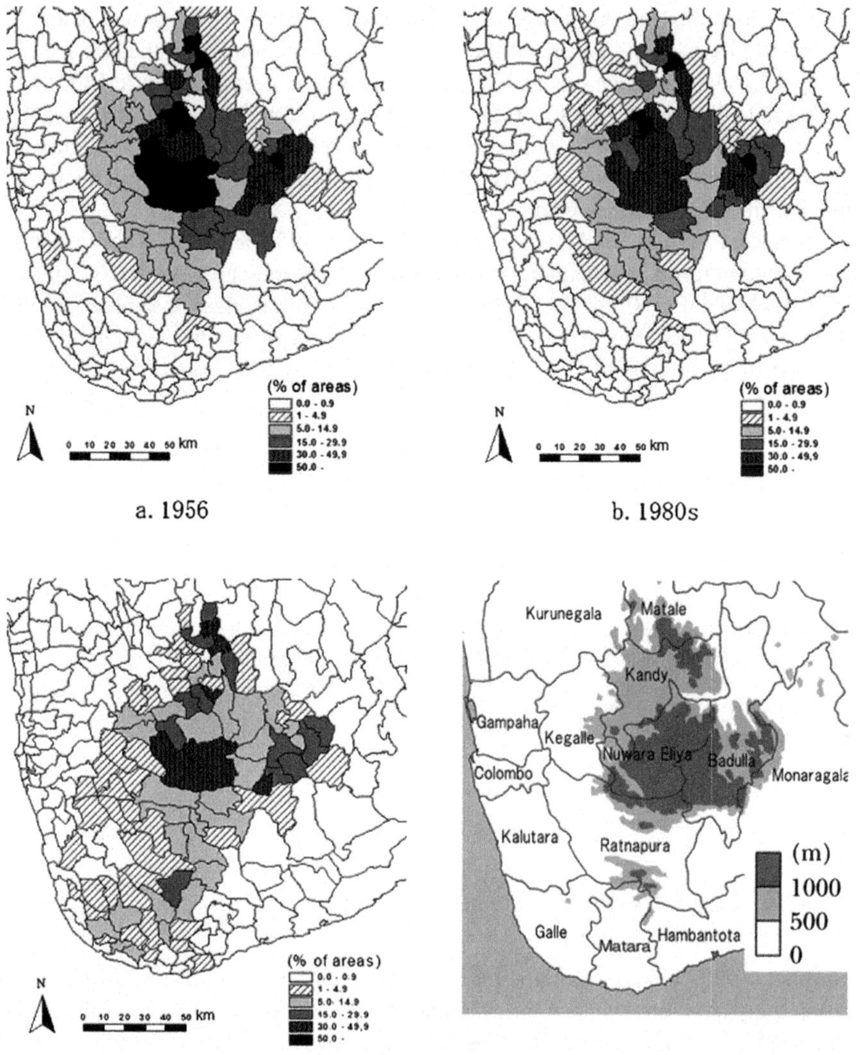

Fig. 19.3 Changes in distribution of tea lands by DS divisions

employment issues for tea farm workers, soil erosion and landslides, and water depletion (Humbel 1990 among others).

There have been considerable delays in the replanting and supplementary planting of tea plants in the plantation sector, even in tea fields that are being maintained; delays that were primarily due to inertia from the prolonged period of state ownership (Fig. 19.4). Of the entire harvested area in the middle—to highland regions where estates predominate, less than half are being planted with high-yield tea plants (vegetative propagated) and, in many cases, seedlings planted during the

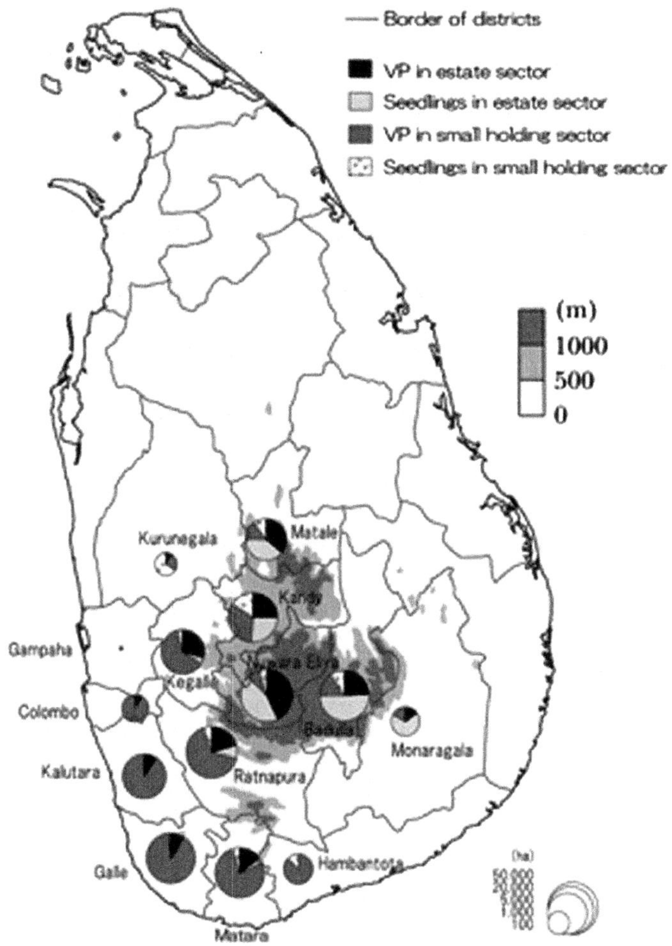

Fig. 19.4 Management agencies of tea estates and replanting situation of tea bushes by districts (*VP* = vegetative propagated; "Sri Lanka census of agriculture—2002: Report on estate Sector" and "Census of tea small holdings in Sri Lanka 2005", Department of Census and Statistics)

colonial era are still being picked (Fig. 19.5). This is in stark contrast to the lowlands where small holdings predominate. Tea plants begin to lose their vitality over the years and are said to have a life span of about 100 years. Plantation companies who aimed to gain short-term profits by increasing yield relied on chemically synthesized agricultural chemicals and chemical fertilizers that cost less compared to replantings or supplementary plantings. This has had a detrimental effect on sustainability, in terms of both the environment and business operations (Illukpitiya 2003; Illukpitiya et al. 2004).

Fig. 19.5 Difference in appearance of VPs (vegetative propagated) and seedlings (VPs are at the *bottom right* and seedlings are at the *upper left*)

19.4.2 Prices

Figure 19.4 shows the variations in the average price of 1 kg of tea during the periods 1987 through 1989, and 2002 through 2004 for each sub-district (also referred to as agro-climatic districts) as defined by the Sri Lanka Tea Board. This shows that the price of tea from the northern parts of the mountainous and hilly regions of Kandy, Matale and northern Nuwara Eliya have remained generally low since the 1987–89 period, and has seen a sustained relative drop leading up to the 2002–04 period. In terms of elevation, many of the tea farms in this region belong to the middle land category, and their low prices are mainly due to the fact that these farms are unable to produce teas with distinct characteristics such as those that characterize highland or lowland teas.

Teas from the southern part of Nuwara Eliya province, which lies in the southern part of the mountainous and hilly region (made up of tea producing areas called Western or Dimbulla, and Nuwara Eliya) where many of the tea fields are in high elevations, and the region centered around Badulla (tea producing area called Uva), but particularly the former, command high prices. This is because the characteristics of teas produced in these regions, particularly their refreshing aroma and bright color, match the tastes of consumers in countries such as Britain, and also because brands from these areas have established a favorable reputation. However, most sub-districts experienced a relative drop in their prices because the nationalization of plantations among other factors had made buyers leery. Meanwhile, teas from Koslanda/Haldumulla on the southern edge of Badulla, and Haputale were exceptions whose tea prices rose dramatically. I will discuss this phenomenon later.

Meanwhile, the price of teas from southern provinces, namely Kakutara, Galle, Matara and Ratnapura were second after Nuwara Eliya as of 1987–89, and continued to rise during 2002–2004. We believe this was due to the several factors: small holdings are predominant in these regions, demand increased in the middle east and Russia where teas with smokey aroma and rich flavor are preferred, these regions are suited for large volume production throughout the year due to their hot and humid climate, and teas from these regions are widely used in blends. Obiously, the trends in price for lowland teas is in stark contrast to those for teas produced in mountain and hilly regions where estates predominate.

19.4.3 Labour

The plantation sector has been plagued with labour drains and shortages, as well as ongoing increases in labour costs, and consequently, relative declines in productivity (Shanmugaratnam 1997; Wijeratne 2003). The majority of tea plantation workers are descendants of people who were forcibly relocated from southern India during Britain's colonial rule and are known as Indian Tamils. They make up 95% of the workers in the middle and highland tea farms (Fernando 2000). Indian Tamils were not granted nationality until 1974. In 1977, public schools that taught in Tamilese were built on and around plantations and this contributed to improving the level of education among Tamils (Little 1999).

Union movements gained momentum and wages have been increasing steadily. These developments, however, proved to be a major setback for plantation companies. Particularly, as education became widespread, younger people increasingly avoided working (Fig. 19.6) or living (Fig. 19.7) on tea farms, looking for employment opportunities in nearby towns or relocating to Colombo for work.

Additionally, social systems surrounding plantation workers have attracted much criticism. Research has been conducted on the oppression of female plantation workers even before privatization (Samarasinghe 1993), but recent studies have been more critical, based on gender theory perspectives that originated in developed industrial nations (Philips 2003, 2005). These studies prove that the social status of plantation workers after privatization has not fundamentally improved or has

Fig. 19.6 Tea estate workers picking tea

Fig. 19.7 Residents' houses called "lines" in a tea estate

worsened in some cases, and bring into question the validity of the current state of affairs.

Sri Lanka's tea industry, especially the plantation sector, is burdened with structural problems for achieving sustainability in the areas of the environment, economy and social justice.

19.5 The Implementation of Organic Farming: History and Overview

Sri Lanka is a pioneer with regard to the introduction of organic farming, including the world's first certified organic tea farm which was developed in response to demand for organic farm products from developed countries (Kohmoto 2006). Since 1986, there has been a rise in the number of Sri Lankan companies that hold organic certification (which were defined in developed countries) and export organically grown tea. In the plantation sector, six companies and seven tea farms have been certified organic. As mentioned in the previous chapter, case B in Table 19.1 (Figs. 19.8 and 19.9) was the first tea farm in the world to acquire organic certification. This was primarily due to heightened awareness for organic foods in western European countries and the environmental awareness within the mother company, the Stassen Group. In 1984, Stassen Natural Foods was established which implemented biodynamic farming in response to demands from German buyers, and was certified by Naturland, a German organic certification organization, in 1986. Twelve of the neighbouring tea farms (as of the current date) are jointly operated on a project basis. In 1991, the Needwood Organic Tea Estate (case D) broke away and is now under independent management.

Case C is one of the 20 tea farms that the RPC (regional plantation company) converted to an organic farm in 1992. In 1999, the company that operates case B

Table 19.1 Overview of certified organic tea estates

	Estate name	Company name	Established year of the company	Elevation	Area (ha)	Harvested area (ha)	VP ratio (%)	Number of workers
A	Venture bio tea garden	Stassen Natural Food (Pvt) Ltd.	1984	High	403	287	40.9	860
B	Idulgashinna bio tea garden	Stassen Natural Food (Pvt) Ltd.	1984	High	357	278	4.8	985
C	Koslanda organic tea garden	Maskeliya Plantations (Pvt) Ltd.	1992	Mid	1029	214	13.8	432
D	Needwood organic tea estate	Needwood Emmag (Pvt) Ltd.	1966	High	153	94	16.0	350
E	Thottulagara organic tea estate	Greenfield Bio Plantations (Pvt) Ltd.	1997	High	110	75	18.7	226
F	Nilmini estate	Nilmini Estate (Pvt) Ltd.		Low	39	14	100.0	
G	Maussawa estate	Guayapi Lanka (Pvt) Ltd.	1999	High	20	8	0.0	9

Based on interviews and various other information. Data from 2003 through 2005 was used
Case D separated from case B and become independent in 1991
VP ratio denotes the percentage of VPs or vegetative propagated plants planted per harvested area

Fig. 19.8 View of certified organic tea estate case B

Fig. 19.9 Organic tea grown on case B

converted one of the Stassen group's conventional tea fields to organic. This eventually became case A. The company's aim was to acquire a field in a region with different climatic conditions from case B to ensure a consistent supply of organically grown tea. The same year, Lanka Organics, a company that markets a broad range of Sri Lankan organic agricultural products, acquired a part of the tea field owned by the RPC (regional plantation company) and this became case E. Case F grows organic tea on a plot which was part of a deserted rubber plantation at an elevation of 300–400 m. Case G is run by Guayapi Tropical of France which acquired a mostly abandoned tea farm. The company began growing organic tea on this land in addition to providing ecotourism services.

The majority of farms that have been converted to organic farming are situated at high elevations, and are concentrated in the Haldumulla and Haputale regions (Fig. 19.10). We believe the main reasons for this is that all of these tea fields are on

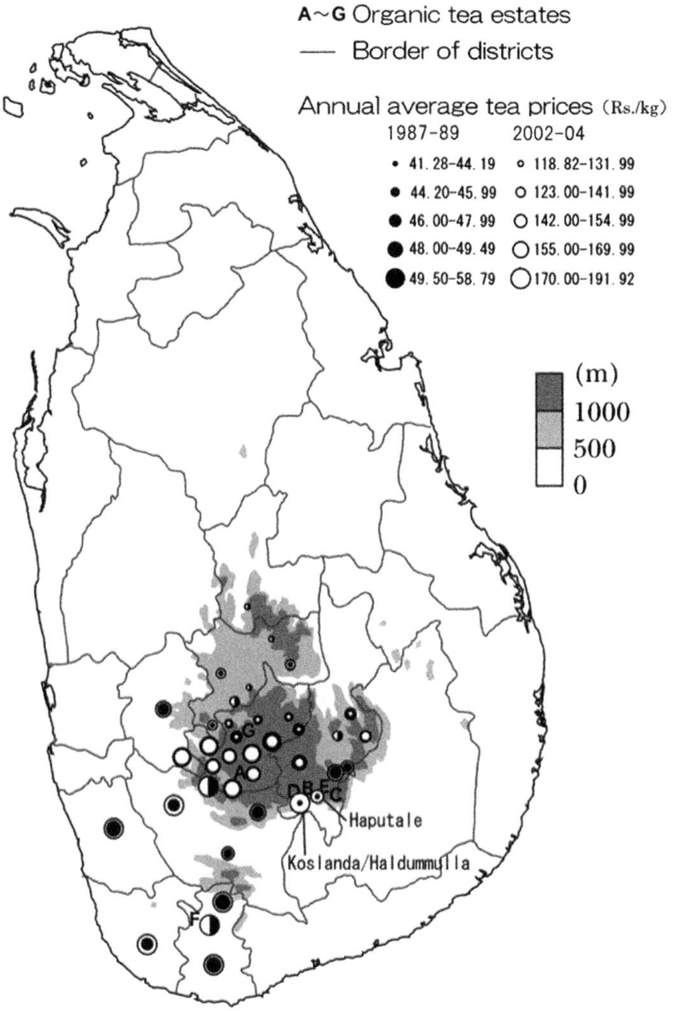

Fig. 19.10 Changes in tea prices by districts (After the statistics of Sri Lanka tea board. Annual average prices of tea are shown as a relative change, because the change of monetary value is large. The *circles* with a *white center* show the districts which the price fell relatively from 1987–89 to 2002–04. The *circles* with a *black center* show the districts which the price increased relatively. The cases both years are same size of the *circles* are shown as *semicircles* of *white* and *black*)

steep hills scattered with exposed rock (Fig. 19.8 shows case B) which makes it difficult to achieve large volumes and productivity improvements, and that consumers in developed countries tend to see tea grown in highlands as being better in quality.

A characteristic of organic teas is that they are not sold at weekly auctions in Colombo in the form of dried tea leaves, but are purchased directly by buyers who are primarily from developed countries. Distribution channels differ by company. For example, cases A and B transport dried tea leaves from their farm to their factory in Colombo, package them (see Fig. 19.8 for case B), and sell these packages to buyers from developed countries at their head offices next door. Case C completes packaging in their factory on the farm, transports the packages to their head office in Colombo, and exports them through buyers from developed countries as well as through domestic buyers. Case D has their head office on the farm so they complete the entire process on site and sell to buyers from developed countries.

19.6 Tea Farm Land Use and Its Effects on the Environment

According to Humbel (1990), Haldumulla region where cases A, C and D are located experienced one of the most dramatic rates of decline in tea farming acreage between 1956 and 1980 (38.0% decline)—a decline that still continues. By 2002, the tea farming acreage in this region had shrunk by 73.4% from 1956 levels. This was due to the region's conditions that make it difficult to improve productivity or increase yields. For example, case C harvested 307 kg/ha on its organic farm which is considerably lower than the average yield of 1576 kg/ha for this company which operates 18 conventional tea farms. What is more, this discrepancy had actually widened from 1994 levels, which were 348 and 1447 kg/ha, respectively. The VP ratio (percentage of vegetative propagated plants planted per harvested area) was a mere 13.8%. In fact, it was precisely these unfavorable conditions that had led the company to choose case C as a site for organic farming. One can say that organic farming contributes to maintaining these low-yield farms that would otherwise have been abandoned. Similar observations can be made of other certified organic tea farms. In fact, most yield only meager harvests of 300–400 kg/ha, extremely low even by Sri Lankan standards, and most have VP ratios of less than 20%.

Organic farming has also had an impact on environmental conservation. Organic farming uses no agricultural chemicals or chemical fertilizers, and instead uses nature's system of material circulation and its ecosystem. Therefore, it can have a major impact on environmental conservation. Of significance here in terms of tea farming is shade tree planting, which places tea plants in the shade (Fig. 19.9). The difference between organic and conventional farming is exemplified in the scenery that they create. In conventional farming, shade trees are usually removed to reduce the spacings between tea plants to increase productivity. Meanwhile, in organic farming, shade trees are planted to create a natural forest with tree layer, sub-tree layer, lower tree layer and shrub layer where tea plants make up the shrub layer. This setting contributes to better nitrogen fixation and moisture retention in tea plants, monoculture is avoided, and soil erosion is also prevented. As a result of avoiding monoculture, the plants become less susceptible to diseases and this allows operators to reduce the use of agricultural chemicals and chemical fertilizers. Additionally, this mode of planting ensures biological diversity as represented by microorganisms in the soil. The effects of organic farming on environmental conservation have been confirmed by many existing studies (Sivaram and Ganewatte 1999; Mohotti 2002, 2004 among others). In fact, the longer a farm has been certified organic the more progress there has been in terms of shade tree planting and their growth. While the total organic tea farming acreage amounts to only a small percentage of Sri Lanka's total tea farming acreage, they are all located in mountainous areas and near water sources. Therefore, these efforts are meaningful to conserve the environment.

19.7 Impact on the Tea Farming Business

Operators can expect a rise in tea prices by converting to organic farming. Figure 19.10 shows the change in annual prices for 1 kg of tea. It also shows that regions producing highly priced teas have shifted to low lying areas, and that, in contrast to the price of teas from the high- and middle lands that have been declining, teas from Koslanda/Haldumulla where cases C and D are located, and Haputale where cases B and E are located continue to increase in price. The former is a particular case in point. While the region posted the fourth lowest average annual price among sub-districts during the 1987–89 period, it went on to produce the highest priced teas for the 2000–04 period. To a large degree, this rise in price has been due to the "organic premium" conferred on these teas. According to per factory production figures for 2004 compiled by the Sri Lanka Tea Board, while production at Koslanda/Hadumulla (2 of 4 factories) and Haputale (2 of 10 factories), respectively made up only 10.0 and 14.4% of total production, a comparison of average prices shows that teas from Koslanda/Hadumulla and Haputale were respectively priced 2.7 and 1.6 times higher than conventional teas, confirming the effects of the organic premium.

However, the issue from the perspective of tea farm operators is whether this organic premium is enough to compensate for the extra production costs involved in certified organic tea farming. Production costs at certified organic tea farms were 293.78 (Rs./ha) for cases B and D (2001 figures from Jayakody and Shyamalie 2002) and 467.12 (2004 figures compiled by the author) for case C. Meanwhile, production costs at two neighbouring conventional tea farms operated by the same company, Ampitiakande/Mahakande tea farm (Ampitiakande/Mahakande Group; hereafter referred to as "A/M tea farm") and Poonagara tea farm (Poonagala Estate; hereafter referred to as "PE tea farm") were 148.88 and 148.41 (2004 figures compiled by the author), respectively; considerably lower than costs incurred at certified organic farms.

A breakdown of costs shows that tea picking makes up the largest percentage of costs. On organic farms, picking made up 19.9% (58.46 Rs./ha) for cases B and D

Fig. 19.11 Tree pruning and compost making on case B

combined, and 28.9% (135.00 Rs./ha) for case C. On conventional farms, tea picking accounted for a larger percentage of costs at 38.0% (56.60 Rs./ha) on the A/M tea farm and 28.3% (56.88 Rs./ha) on the PE tea farm. When we look at the percentage of costs spent in compost making (Fig. 19.11)—an important feature of organic farming—costs in cases B and D were 24.8% (72.86 Rs./ha), case C 17.3% (80.94 Rs./ha), A/M tea farm 0.3% (0.48 Rs./ha), and PE tea farm 0.0% (0.00 Rs./ ha), indicating the large percentage that compost making comprises in the overall cost of organic farming. The percentage of weed removal costs was also larger on organic farms at 4.6% (13.63 Rs./ha), 9.4% (44.11 Rs./ha), 2.6% (3.83 Rs./ha) and 2.5% (3.75 Rs./ha), respectively. And while costs for chemical fertilizers appear only for conventional farms at 0.0% (0.00 Rs./ha), 0.0% (0.00 Rs./ha), 9.7% (14.40 Rs./ha) and 9.5% (14.05 Rs.ha) respectively, these make up a smaller percentage compared to the cost of compost making in organic farming. As you can see, organic farms are generally more expensive to operate, with considerable costs going towards compost making and weed removal. This is because, in organic farming, the use of chemical fertilizers or agricultural chemicals is not allowed, operations tend to be labour intensive, and yields are often small. Costs incurred by case C are higher than cases B and D because, unlike other certified organic farms, they have a policy of producing their own organic fertilizer on-site. To date, they have not been able to generate revenues commensurate with these costs.

To overcome these obstacles, organic farms must find buyers who are willing to purchase organic teas at high prices. The company that operates cases A and B exports to buyers in Germany (126 t), Italy (72 t), France (54 t), Australia (54 t), US (36 t) and Japan (1 t). The company that runs case C exports to the US (4.3 t), Germany (1.0 t), France (1.0 t) and Japan (0.7 t) (export figures as of 2004).

Organic teas are exported exclusively to developed countries, which is in contrast to conventional teas that are exported to the Middle East and Russia. Organic operators make new buyer contacts primarily through their web sites, international trade shows for organic and natural foods, referrals from certifying organizations, and eco-tourism operations.

Meanwhile, competition with other organic tea producing countries has been intensifying. About two-thirds of the global production of organic black teas come from India (Kilcher et al. 2002). In Darjeeling, 20 of the 87 tea plantations have converted to organic since 1989 due to stagnating yield increases and various other reasons (Vossenaar and Wynen 2004; OECD 2005). In 1996, the Chinese Academy of Agricultural Sciences set up their Organic Tea Research and Development Center (OTRDC) to create their own organic certification, and this has brought about a rapid increase in the production of organically grown green and black teas (Vossenaar and Wynen 2004). In addition, Kenya, Tanzania, Myanmar and Seychelles also export organically grown teas (Harris et al. 2001). Meanwhile, market growth for organic products has been slowing in developed countries, creating a glut of organically grown teas (Wai 2006). Therefore, while Sri Lankan organic tea plantations have been the pioneers of the field, they are faced with difficulties in attracting buyers from developed countries and are unable to expect purchase prices to increase in the future.

In fact, the company that operates cases A and B has reportedly been seeing profits dwindle over the past few years compared to the first few years immediately after acquiring organic certification. Case C has not been able to find enough buyers who will pay prices commensurate with their high production costs, and currently distribute half of their organic tea production through the Colombo auction alongside their conventional teas. This company actually makes up for the losses incurred in their organic operations with profits from their conventional teas. An "emotional attachment" to the hard work that was put into these farms is the primary reason for this situation being tolerated within the company, and the company is reportedly discussing whether or not to discontinue their organic operations. Other companies have also attempted to implement organic farming in the past but were forced to bow out. For example, Bogawantalawa Plantations, an RPC that operates 11 tea farms, introduced organic farming on a portion (24.1 ha) of their tea farm in Dikoya (a high-elevation area in Nuwara Eliya) between 2001 and 2003, but discontinued this project because they were unable to project a profit. Similarly, Watawala Plantations, an RPC that operates 18 tea farms, introduced organic farming on a 127.5 ha area on their tea farm in Elgin (a high-elevation area in Nuwara Eliya) in 1998. However, because they were not able to turn a profit and also because of frequent labour disputes stemming from the labour intensive nature of organic farming, the company abandoned organic methods and resumed the use of chemical fertilizers and agricultural chemicals in 2003.

19.8 Impact on Tea Farm Workers and Residents

An important factor in plantation management, aside from the financial concerns we have seen in the examples above, is the management of workers and residents. According to a study of cases B, C and other farms by Guruge et al. (2005), workers on certified organic farms had lower concentrations of fluorinated organic compounds (FOCs) in their blood streams compared to city dwellers and workers working on conventional farms, indicating that workers on organic farms were healthier. Interviews with management have revealed that almost 100% of the workers and residents on cases A through E were Indian Tamil, which is essentially the same ethnic make up found in the middle and highland regions. However, since further details were not available, we conducted interviews on school education and job choices with individual households working at or living on cases B and E, and households on conventional tea farms that neighboured cases B and E, with the number of households interviewed being 10, (47 individuals), 24 (79 individuals), 14 (50 individuals) and 11 (55 individuals), respectively.

In Sri Lanka, education is mandatory for children aged 5 years or older for a period of 9 years, and the government strongly encourages its citizens to take up senior secondary education for an additional 2 years. Tuition is basically free in public schools. Table 19.2 shows that while the number of non-schooling individuals is smaller and the number of withdrawals before grade 6 (primary level education) is larger among worker or resident families of certified organic farms than residents of conventional farms, these two groups have several factors in common:

(a) a large number of individuals in their 30's or older had withdrawn from primary school;
(b) although the individual had enrolled after education became mandatory, people currently aged 17 or older, primarily women, had withdrawn without completing their mandatory education; and
(c) almost all of the students currently aged 16 go on to take senior secondary education. Another notable trait not shown in the table is that an increasing number of students, mainly males, are enrolled in Tamil schools in nearby towns instead of Tamil schools on their tea farms.

A study of occupations and places of employment by the author shows that individuals aged 45 or older were either tea farm workers or retirees. Residents who had completed school up to at least grade 10 fell into one of three categories:

(a) those who had no fixed employment and were searching for a job while continuing their studies or domestic work at home, or grew vegetables and sold them in nearby markets (if they were allowed to grow vegetables on the farm);
(b) those who commuted to jobs in neighbouring cities and other locations outside the tea farm; and

Table 19.2 School education background of tea estate residents

Organic tea estates													
(s)	Male						Female						Total
	No	Sc	Pr	Jn	Sn	Cl	No	Sc	Pr	Jn	Sn	Cl	
80							1						1
70				1			1						2
60			6						7	1			14
50			9	1	2				7	1			20
40			2	3	3				7	4			19
30			6	3	2		1		6				18
20			2	2	2		1	1	6	1	5	1	21
10		9		2	1			3		2	2		19
0		5					4	3					12
Total	0	14	25	12	10	0	8	7	33	9	7	1	126
Other tea estates													
(s)	No	Sc	Pr	Jn	Sn	Cl	No	Sc	Pr	Jn	Sn	Cl	
80													0
70			1				1						2
60	1		2	2			2						7
50	1		2						5		1		9
40			2	1	1		3			1			8
30	1		6	1	1				5	4		1	19
20	1		2	4	2	1	1			4	6	1	22
10		4		1	1			10		1			17
0	5	4					6	6					21
Total	9	8	15	9	5	1	13	16	10	10	7	2	105

No no schooling; *Sc* currently receiving schooling; *Pr* primary (grades 1–5); *Jr* junior secondary (grades 6–9); *Sn* senior secondary (grades 10–11); and *Cl* college (grades 12–13)
This table shows the number of people who withdrew before completing these stages

(c) those who worked in Colombo or overseas.

As the level of education among tea farm residents improved, the number of tea farm workers declined, with less than half of the people in their 20's working on tea farms. In case B however, only a few workers worked away from home, and this is largely due to the social welfare policies implemented by management. The company that operates cases A and B embodies the social egalitarian policies embraced by organic agriculture, and provides community-based social development programs rooted in equal opportunity for workers of both genders, and carries out renovations and repairs on worker housing (most of which are tenements called "lines" built in the colonial era) based on their motto "Applying the concept of organics to social development." Case C uses the premium that they earn from their fair-trade certification to purchase sewing machines and for other means to offer income opportunities to their workers, particularly women. In contrast, in case E where the previous management had neglected social welfare aspects of their operation due to their heavy focus on profit, workers went on strikes and economic gaps formed between the plantation's residents (Fig. 19.12).

While the decision to introduce organic farming to plantations is primarily made by the management with no participation from the workers or residents, the workers are forced to carry out labour-intensive work such as compost making and weed removal. Organic farming requires more labour than conventional farming in other areas as well. Therefore, certified organic farms must improve their social welfare programs beyond that which conventional farms provide and implement programs for empowering their workers (skill development and the granting of certain rights). They must clearly communicate, particularly to younger workers, the merits of

Fig. 19.12 Residents' houses called "lines" on case E tea estate (Of special note are the additions made to homes occupied by households with people working away from home)

working on an organic farm and ensure their dignity as tea farm workers. These approaches also coincide with policies embraced by fair-trade certification, which most farms have acquired, and also speak to consumer sensitivities. However, these measures are not easy to implement without sufficient profit to back things up. These findings show that Sri Lankan tea plantations implementing organic farming face corporate management problems and a range of other issues for their sustainability.

19.9 Conclusion: Marginality and Its Future

Tea estates in Sri Lanka are surely in the process of marginalization, and they have three characters. Firstly, their marginality is structural due to global and national socio-economic changes. Secondly, most of the workers are descendants of migrants, and they are ethnic minorities. Thirdly, they are living and working in marginal locations.

Under these conditions, organic farming on tea plantations is characterized by a variety of post-productivist elements, making it distinct from practices implemented on traditional plantations where productivism was the imperative.

However, tea plantations that have introduced organic farming face high production costs primarily from compost making and weed removal. Organic tea plantations are strongly impacted by general wage increases in the Sri Lankan tea industry due to their labour-intensive nature, but are not able to earn profits that are commensurate with these rising wages. The fact that attracting buyers from developed countries has not been as easy as initially envisioned, and that prices have not risen as anticipated are primary factors of this problem. If the current situation persists, not only is the number of organic tea farms unlikely to grow, but the survival of existing certified organic tea farms may be at risk.

To enhance its value, story of the farmers' life and social welfare, ecotourism namely product-oriented tours for limited number of consumers with special interest, and learning activities especially education for sustainable development for both producers and consumers are necessary.

References

Beckford, G. (2000). *Persistent poverty: Underdevelopment in plantation economies of the third world*. Kingston, Jamaica: University of the West Indies Press.

Fernando, M. (2000). *The story of Ceylon Tea*. Colombo: Mlesna (Ceylon).

Grigg, D. (2003). The worlds of tea and coffee: Patterns of consumption. *GeoJournal, 57,* 283–294.

Guruge, K. M., Taniyasu, S., Yamashita, N., Wijeratna, S., Mohotti, K. M., Seneviratne, H. R., et al. (2005). Perfluorinated organic compounds in human blood serum and seminal plasma: A study of urban and rural tea worker populations in Sri Lanka. *Journal of Environmental Monitoring, 7,* 371–377.

Harris, P. J. C., Browne, A. W., Barett, H. R., & Cadoret, K. (2001): *Facilitating the inclusion of the resource-poor in organic production and trade: Opportunities and constraints posed by certification*. Coventry: Henry Doubleday Research Association (HDRA).

Harteminsk, A. E. (2005). Plantation agriculture in the tropics: Environmental issues. *Outlook on Agriculture, 34,* 11–21.

Humbel, R. (1990). Decreasing extent of tea plantations: Chance for agricultural diversification or ecological threat? *Sri Lanka Journal of Tea Sciences, 59*(2), 65–81.

Humbel, R. (1991). *Tea area changes in Sri Lanka: Analysis of regional distribution, processes, mechanisms, and correlating factors of changes in the area cultivated with tea since 1956.* Zurich: University of Zurich.

Illukpitiya, I. M. P. M. P. (2003). Resource degradation in high altitude agro-ecosystems: An overview of the ecological capital management in tea plantations in the highlands of Uva in Sri Lanka. *Sabaragamuwa University Journal, 3,* 33–44.

Illukpitiya, P., Shanmugaratnam, N., & Kjosavik, D. J. (2004). Tea Agroecosystems in the Uva highlands of Sri Lanka: An evaluation of land management activities. *Mountain Research and Development, 24,* 52–59.

Kilcher, L., Schaefer, M., Richter, T.,van den Berge, P., Milz, J., Foppen, R., & Theunissen, A. (2002). *Organic coffee, cocoa and tea: Market, certification and production information for producers and international trading companies.* Research Institute of Organic Agriculture (FiBL), Frick, Switzerland, Naturland, Weinsberg, Germany, and Swiss Import Promotion Programme, Zurich.

Kohmoto, D. (2006). Development of organic farming and its mechanism in Sri Lanka. *Geographical Review of Japan, 79,* 373–397. (Japanese with English abstract).

Kohmoto, D. (2008). Structural changes in the tea industry and the introduction of organic farming methods into Sri Lanka with special attention to the estate sector. *Chigaku Zasshi (Jounal of Geography), 117,* 617–636. (Japanese with English abstract).

Lincoln, D. (2006). Beyond the plantation: Mauritius in the global division of labour. *Journal of Modern African Studies, 44,* 59–78.

Little, A. W. (1999). *Labouring to learn: Towards a political economy of plantations.* People and Education in Sri Lanka: MacMillan Press.

Manikam, P. P. (1995). *Tea plantations in crisis: An overview.* Colombo: Social Scientist's Association.

Mohotti, K. (2002). Bio prospecting of organic farmlands: Evidences from Sri Lankan organic tea soils. In *The Fifth IFOAM-Asia Scientific Conference: Organic Agriculture and Sustainable Development, IFOAM-Asia,* (pp. 163–167). China: Hanzhou.

Mohotti, A. J. (2004). Shade in tea: Is it beneficial? In A. K. N. Zoysa & M. T. Z. Mohamed (Eds.), *Proceedings of the First Symposium on Plantation Crop Research: Current Trends and Future Challenges, 21–31.* Talawakelle: The Tea Research Institute of Sri Lanka.

Nuhn, H. (2004). Coffee boom and coffee crisis: Open markets, global competition and consequences for developing countries. *Die Erde, 135,* 1–30.

OECD (2005). *Environmental requirements and market access.* Paris: OECD Publishing.

Philips, A. (2003). Rethinking culture and development: Marriage and gender among the Tea plantation workers in Sri Lanka. *Gender and Development, 11–2,* 20–29.

Philips, A. (2005). The kinship, marriage and gender experiences of tamil women in Sri Lanka's tea plantations. *Contributions to Indian Sociology, 39,* 107–142.

Samarasinghe, V. (1993). Puppets on a string: women's wage work and empowerment among female tea plantation workers of Sri Lanka. *Journal of Developing Areas, 27,* 329–339.

Shanmugaratnam, N. (1997). *Privatisation of tea plantations: The challenge of reforming production relations in Sri Lanka: An institutional historical perspective.* Colombo: Social Scientist's Association.

Sivaram, B., & Ganewatte, G. (1999). Cost-benefit of shade management in tea. *Tea Bulletin, 16,* 16–22.

Sri Lanka Tea Board. (2004). *Sri Lanka tea board annual report 2003*. Colombo: Sri Lanka Tea Board.

Vossenaar, R., & Wynen, E. (2004). *Trading opportunities for organic food products from developing countries*. New York: United Nations.

Wai, O. K. (2006). Organic farming in Asia. In H. Willer & M. Yussefi (Eds.), *The World of Organic Agriculture: Statistics and Emerging Trends 2006* (pp. 108–117). Frick, Switzerland: International Federation of Organic Agriculture Movements (IFOAM), Bonn, and Research Institute of Organic Agriculture (FiBL).

Weinberg, B. A., & Bealer, B. K. (2001). *The world of caffeine: The science and culture of the world's most popular drug*. Oxford: Routledge.

Wijeratne, M. A. (2003). Twin problems of the tea industry: Causes and a way-out. *Tea Bulletin, 18,* 5–9.

Chapter 20
Pond Management System Among Small Farmers in Khulna, Bangladesh

Firuza Begham Mustafa

20.1 Introduction

Aquaculture industry is growing tremendously in Bangladesh and the country is listed as top 5 world producer after China, India, Vietnam and Indonesia (FAO 2013). Aquaculture plays an important role in nutrition, income, employment and foreign exchange in Bangladesh. Fish contribute approximately 80% of the animal protein intake of the Bangladeshi population, 3.5% of GDP and 9% of export earnings (FAO 2012). Bangladesh aquaculture products' major market are European and Middle Eastern countries (FAO 2012).

The giant freshwater prawn, *Macrobrachium rosenbergii* is gaining popularity as a key aquaculture species and the global production currently exceed 220,000 tons (Rasid 2015). Since last 30 years many researches started on *Macrobrachium rosenbergii* (giant freshwater prawn) in diverse subjects. The earlier studies are focusing on enhancement of production, innovation in the technology, disease outbreak and prevention, soil investigation, water quality, feeding (Tacon and De Silva 1997), fry and brood stock selection and genetic studies (Hai Nguyen Thanh et al. 2015). Among social perspective research is included ethics in farming (Filipe et al. 2011), social implications, gender affiliation in aquaculture (Weeratunge 2010), labour issues, wages, marginalization local consumer, poverty, certification, organic farming (Paul and Vogl 2013), environmental impacts, land use changes and best management practice.

Pond management is one of the crucial aspects in rearing aquatic animals. It sometimes covers different aspects and it varies in perspectives. Mushtaq et al. (2007) state that pond management refers to managing the water distribution, accessibility, fees and political power in the management—which is more related to

F.B. Mustafa (✉)
Department of Geography, Faculty of Arts and Social Sciences, University of Malaya, 50603 Kuala Lumpur, Malaysia
e-mail: firuza@um.edu.my

S. Pelc and M. Koderman (eds.), *Nature, Tourism and Ethnicity as Drivers of (De)Marginalization*, Perspectives on Geographical Marginality 3,
DOI 10.1007/978-3-319-59002-8_20

irrigation pond management. While Brune and Drapcho, 1991 as cited in Bosma and Verdegem (2011) look at the balancing of water ecosystem as pond management (balancing production and decomposition processes).

Pond management actually is more about human involvement in the farming activities (Firuza 2005). In this study, pond management in aquaculture refers to pond preparation, water management, feeding, stocking density, genetic selection, growth and production. Best practised pond management plays an important role in determining the survival rate, yield per square meter, production capacity and cost factor for disease outbreak.

Paddy-fish culture has grown significantly in the 1980s in Southeast Asian countries. This type of farming has been first recorded in China as early as 2500 years ago and has been noted for its low input, high production, and efficacy in nutrient recycling (Liu and Cai 1998). The integration of fish into rice farming provides invaluable protein, especially for subsistence farmers who manage rain-fed agricultural systems. Rice fields provide shade and organic matter for fish, which in turn oxygenate soil and water, eat rice pests and favor nutrient recycling. These farming systems, as a part of an integrated ecosystem in line with the local cultural, environmental and economic conditions, are composed of complementary sub-agricultural ecosystems; these play important ecological service roles, such as bio-control and nitrogen fixation. The traditional low input rice–fish farming systems have a significant function in protecting the global environment and maintaining its biodiversity (Lu and Li 2006).

The paddy and aquatic culture started in 1970s in Thailand, Indonesia and Philippines with high challenges in the sustainability, water resources and environmental issues, low survival rate and yield (Little et al. 1996). Paddy-fish is very common despite much conflict in terms of sustainability and environmental impacts; but paddy-prawn farming especially *Macrobrachium rosenbergii* is a rare practice. The research presented in this chapter was conducted in order to reveal and understand the pond management practice among the *Macrobrachium rosenbergii* small farmers (less than 5 acre/2 ha) in Khulna Bangladesh. The fundamental objectives of the study were derived from questions how the small farmers in Khulna are producing *Macrobrachium rosenbergii* successfully with high survival rate in sustainable manner? The special focus has been on:

(i) farming system,
(ii) pond preparation,
(iii) seedling/brood stock
(iv) feeding management,
(v) fertilizer and chemicals,
(vi) water management
(vii) stocking and harvesting.

The description and analysis of pond management applied by *Macrobrachium rosenbergii* farmers can serve as a source of inspiration/idea for how this could be developed further, especially in terms of survival rate, yield production and sustainability.

The term small farmer refers to a traditional and semi-traditional farmers with special characteristic in their farming intensity practice, sometimes it is referred as homestead pond culture[1] and "quasi-peasant" in contrast with "quasi capitalist" (Belton and Azad 2012).

20.2 Methodology and Study Area

20.2.1 Method

Data was collected in November and December 2012 using combination of quantitative and qualitative methods. Open ended questionnaire was prepared to gather information regarding the characteristics of the farmers, farm management, assets, production, input, feeding, water management, practice and routine. In-depth interviews were conducted with five selected farmers from a few villages and different pattern of farming practices. Non-structured focus group discussions were used to gather information from the farmers and associated people such as processing company, agent/middle man, fisheries department and workers. The average duration of discussion was 2 hours.

Observation and field notes were done using transect walks. Transect walk is walking with or by local people through an area, observing, asking, listening, discussing, identifying (Chambers 1994). This method was used for the field survey. This method was very crucial for understanding of the farming environment and to record in situ observation. This direct interaction with the farmers generated on-the-spot questions that gleaned informal information on resource-use patterns which helped in understanding the farming practices and daily livelihood activities of the farmers (Paul and Vogl 2013).

A field work survey was done in selected farms in Khulna District (Southwest of Dhaka), Bangladesh. In-depth interview technique was used to gather information and data from the following five farmers Mr. Habibul Rahman (EPZ commercial farming area), Mr. Iqbal Hawladar (Mathardia, Mongla), Mr. Mahbubur Rahman (Rajibpur Village), Mr. Habib Sarwar Titu (Rajibpur Village) and Mr. Khukan Dhali (Rajibpur Village). The farmer's ponds size varied from 8 to 60 ares.

[1]Homestead pond culture is a small component of the household farm enterprise, used for multiple purposes including bathing, washing, watering livestock and irrigation for dyke or homestead vegetable production, whereas the fish produced by homestead aquaculture is generally marketed locally in small quantities at sub-district or union level markets by marketing intermediaries with limited buying power (Belton and Azad 2012).

20.2.2 Study Area

Bangladesh came into existence in 1971 when Bengali East Pakistan seceded from its union with West Pakistan. It is located between Burma and India in the Southern Asia with total area of 147,610 square kilometers out of which water covers 10,090 km^2. Major rivers that are crossing the country are Brahmaputra, Ganges (Padma) and Jamuna (Fig. 20.1). The Ganges and Brahmaputra rivers rise in the Himalayas (collectively with catchments in five countries: China, Nepal, India, Bhutan, Bangladesh) and ultimately deposit their sediments in the Bay of Bengal (Wilson and Goodbred 2015). Climate is tropical with mild winter (October to March), hot, humid summer (March to June) and humid, warm rainy monsoon (June to October). The country is frequently hit by cyclone and hurricane. About a third of this extremely poor country is flooded annually during the monsoon rainy season. The entire country is mostly flat alluvial plain with hilly southeastern part where the higher point is Mt. Keokradong with 1230 meters. Land use is 63.11% arable land, 3.07% permanent crops and other 34.5% (Bangladesh Country Index 2013).

Fish contribute about 63% of the available protein in the diet and the rest 37% protein comes from livestock and poultry. It indicates the importance of fish in contributing to the level of nutrition of the people of Bangladesh (DOF 2008). Field data was collected from farms in Mathardia, Mongla and Rajibpur villages in Khulna Division, southwest of Dhaka (Fig. 20.1).

The water resources of Bangladesh are vast (Table 20.1), covering 4.3 million hectares of inland water. A sub-tropical climate, suitable soil and water conditions create a very high potential for aquaculture development. Pond aquaculture is an integral component of the farming system in Bangladesh. Estimates show that the country has 0.3 million ha of ponds, offering huge potential for aquaculture (DOF, 2008 as cited in Murshed-E-Jahan and Pemsl 2011).

20.3 Discussion

20.3.1 *Overview of* Macrobrachium Rosenbergii *(Golda) Farming in Khulna*

Macrobrachium rosenbergii production in Bangladesh is mainly practiced by small farmers (small size of ponds) operating in almost every paddy field in Khulna division. Output from the small farms has turned Khulna to become the major producer of *Macrobrachium rosenbergii* in the country (Table 20.2). Almost 80% of the land area in Khulna is dominated by *Macrobrachium rosenbergii* co-existing with paddy (IAA—Integration Agriculture and Aquaculture). Before 1980s, Khulna's main production was paddy, but once the *Macrobrachium rosenbergii*

Fig. 20.1 Map of Bangladesh, note location of Khulna in southwest (Khulna location: 22° 48′ 51″ N, 89° 60’ 97″ E) Source: UN Department of Peacekeeping Operations, Cartographic Section, Map No. 3711, Rev. 2 http://www.un.org/Depts/Cartographic/map/profile/banglade.pdf

Table 20.1 Physical dimension of water resources in Bangladesh

Item	Amount
Coastal line (km)	710
Seawater area (sq. km)	68,480
Annual renewable surface water (billion m^3)	1160
Annual average rainfall (mm)	2300
Daily average delta discharge (million m^3)	3000
River discharge (million m^3 per second peak time)	1.5
Annual renewable surface water (billion m^3)	1160

Sources Combination data from DOF (2008), Ahmad et al. 2001 (in Chowdhury 2010)

Table 20.2 Shrimp and prawn production in Bangladesh in 2010–2011

Division	Bagda	Golda	Harina	Chaka	Others	Fish	Total
Dhaka	0.00	1270.4	48.48	0.00	16.16	276.8	1611.83
Khulna	38,451	35,165.1	17,488.9	2331.9	3497.8	47,702.3	144,636.8
Barisal	858.4	3283.75	239.53	31.93	47.91	1192.1	5653.58
Rangpur	0.00	12.11	0.00	0.00	5.28	7.60	24.99
Rajshahi	0.00	10.44	0.00	0.00	2.03	3.24	15.71
Chittagong	17,259.9	0.00	122.0	0.00	4501.8	11,105.8	32,989.45
Shyhet	0.0	4.1	0.00	0.00	0.00	2.3	6.4
Total	56,569.2	39,868.0	17,776.9	2363. 8	8070.9	60,290.0	184,938.7
Percent	30.6	21.6	9.6	1.3	4.4	32.6	100.0

Bagda: *Penaeus Monodon* Horina (Metapenaeus monoceros)
Golda: *Macrobrachium rosenbergii* Chaka (Fenneropenaeus indicus)
Source Department of Fisheries Bangladesh 2012

aquaculture of gher[2] started and started to boom, Khulna turned to a major producer of *Macrobrachium rosenbergii* in whole Bangladesh (Table 20.3). It contributes to almost 88% of *Macrobrachium rosenbergii* production in Bangladesh in 2010–2011, followed by Barisal (8.2%) and Dhaka (3%) (Fig. 20.2).

Khulna is a paradise for *Macrobrachium rosenbergii* farming, it has conducive environment and suitable landscape for the farming due to:

(1) abundance of freshwater around the year,
(2) suitable fertile soil,
(3) prefect climate-sunny and plenty of rain,

[2]Gher is a Bangla word used for the physical construction made for growing freshwater prawns in an impounded environment. These constructions, generally built in paddy fields, are used for growing both paddy and prawns. The ghers usually have a large dike with canals and an area dedicated for paddy cultivation within the gher (Nurazzaman 2012). Gher is therefore a modified rice field or pond located beside canal or river (Paul and Vogl 2013).

Table 20.3 Production of shrimp in 2010–2011 in Khulna District, Bangladesh

	Area of shrimp farm in ha			Production in metric tons				
District	Bagda	Golda	Total	Bagda	Golda	Others	Fish	Total
Bagerhat	47,900.00	18,556.06	66,456.06	13,177.00	12,828.90	8177.62	16,034.10	50,217.62
Chuadanga	0	0	0	0	0	0	0	0
Jessore	825.00	14,479.00	15,304.00	207.00	7525.20	2272.40	2,747.60	12,752.20
Jhenaidah	0.00	2.43	2.43	0.00	1.25	0.00	0.50	1.75
Khulna	36,557.18	13,960.42	50,517.60	9980.00	8409.14	785.00	16,601.76	35,775.90
Kushtia	0	0	0	0	0	0	0	0
Magura	0.00	7.31	7.31	0.00	4.80	1.00	1.32	7.12
Meherpur	0	0	0	0	0	0	0	0
Narail	0.00	2277.25	2277.25	0.00	1165.20	185.00	393.00	1743.20
Satkhira	60,348.00	7664.00	68,012.00	15,087.00	5230.56	11,897.45	11,924.00	44,139.01
Khulna division	*145,630.18*	*56,946.47*	*202,576.65*	*38,451.00*	*35,165.05*	*23,318.47*	*47,702.28*	*144,636.80*

Bagda: *Penaeus Monodon* Golda: *Macrobrachium rosenbergii*
Source Department of Fisheries Bangladesh 2012

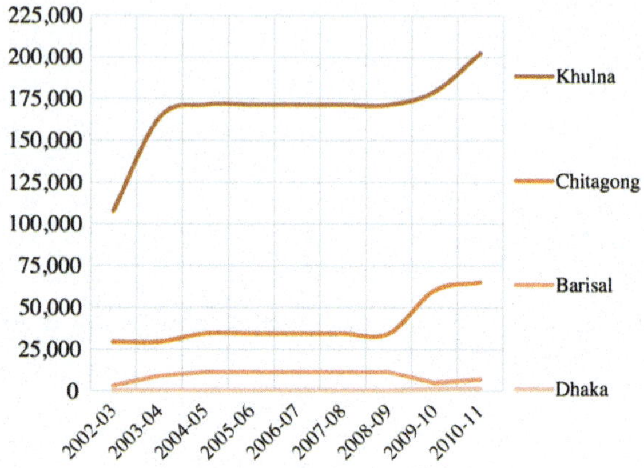

Fig. 20.2 Prawn and shrimp farming in Bangladesh division from 2002–2011 *Source* Department of Fisheries Bangladesh 2012

(4) physical surface: landscape, topography
(5) similarity of pond conditions with wild environment,
(6) natural fry from giant rivers,
(7) water logging (Shah 2006) and
(8) perfect salinity.

Other social factors supporting the production are:

(1) indigenous innovation in farm operation,
(2) availability of local/natural feed for prawn,
(3) partial organic practice,
(4) cheap labour, and
(5) existing paddy ponds.

20.3.2 Macrobrachium Rosenbergii *Farming Practice*

Farming System/Prawn Production System
Macrobrachium rosenbergii farmers in Khulna practice unique farming system, which incorporates paddy, prawn, vegetables and fruits in the farming areas. The farmers plant paddy during the dry season (summer) and rearing giant freshwater prawn during the rainy season (winter) and producing vegetables and fruits all year around. Table 20.4 shows the thematic schedule/calendar for *Macrobrachium rosenbergii* farming, paddy and other crop with climate index in Khulna division Bangladesh.

In the 1980s farmers produced paddy for whole year and catched aquatic animal from the riverine during winter (rainy season). Slowly, once the aquaculture activities entered the area, the farmers started integrating aquaculture in their paddy field. Almost 80–90% of the farmers in Khulna practice this method. The farmers involved in the activity are generally very small farmers with land area being less than 0.3 ha. Almost all the farmer display similar background small size of pond

Table 20.4 A thematic schedule/calendar for Macrobrachium rosenbergii farming, paddy and other crop with climate index in Khulna division Bangladesh

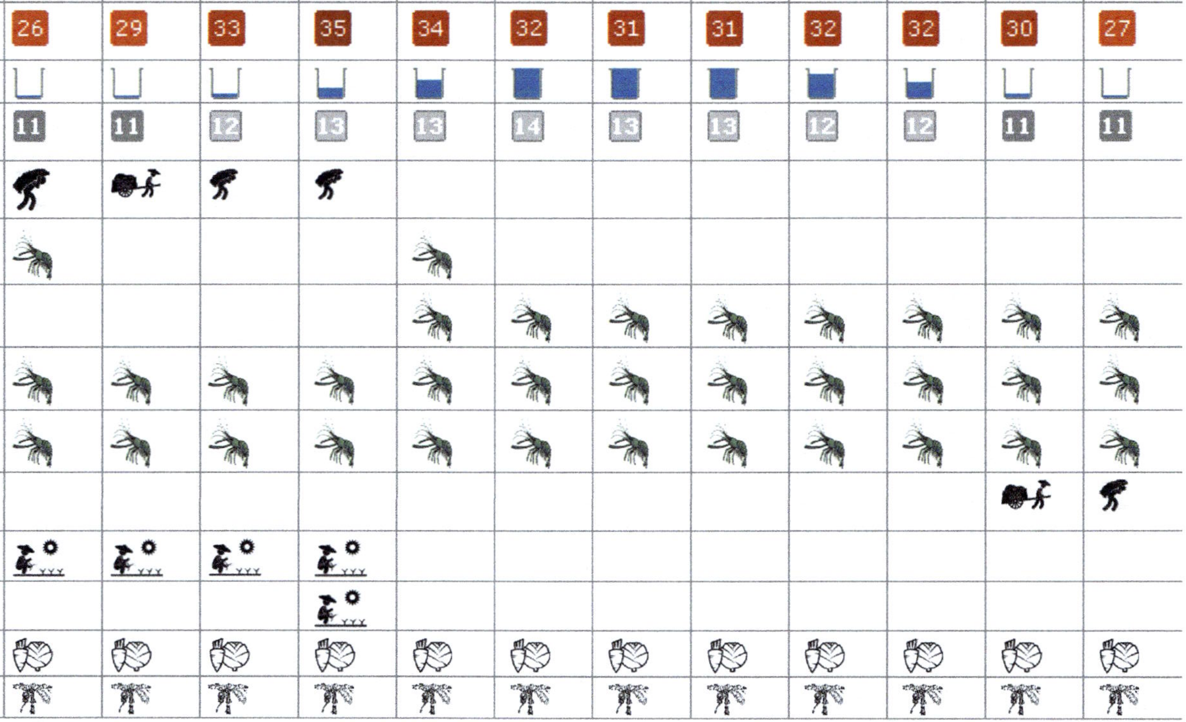

Descriptions	Jan	Feb	Mar	Apr	May	June	Jul	Aug	Sept	Oct	Nov	Dec
Max daytime temperature (°C)	26	29	33	35	34	32	31	31	32	32	30	27
Monthly rainfall (mm)												
Hours of daylight (Daily)	11	11	12	13	13	14	13	13	12	12	11	11
Pond preparation for aquaculture farming												
Fry stocking												
Prawn rearing entire pond												
Prawn rearing on pond canal												
Harvesting—continues												
Pond preparation for paddy cultivation												
Paddy planting												
Paddy harvest												
Vegetables												
Fruits[a]												

[a]coconut, mango, guava, jackfruit, banana, papaya

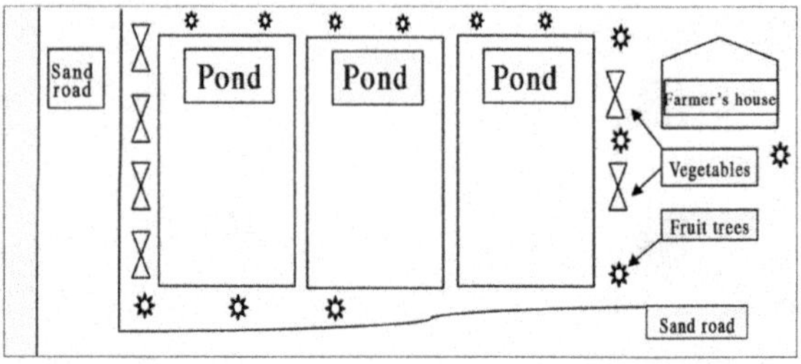

Fig. 20.3 Typical Macrobrachium rosenbergii farm layout in Khulna division Bangladesh

and labour source is mainly from family members. Seasonal labour contracted on a daily basis is only hired during pond preparation and harvesting. It is estimated that more than 90% of the 570 million farms worldwide are managed by an individual or a family and that they mostly rely on family labour (United Nations Conference on Trade and Development 2015). Family members are not paid any salary/wages. Family labour is not recognized as having monetary value by the prevailing social structure in Bangladesh (Paul and Vog 2013). Among the common activities involving family members are releasing post-larvae, feeding, pond guard, harvesting and marketing. There is difference between male and female member contributions.

Multi-usage of the pond areas is very common, farmer try utilizing maximum of the every space of the farm compound for production. Rearing animals like cows and sheep in the ponds areas which are located next to farmer's house. Most of ponds share same compound with farmer's house. Some perennial fruit crop are grown in the surrounding areas. Figure 20.3 shows the typical layout of the farm in Khulna area.

Farmers used dikes to grow vegetables and this spur additional income and food resources for the farmers. Among the vegetables planted in the pond areas are beans, gourds, eggplants, lady fingers, pumpkins, beet, bitter gourds, tomatoes and pepper. Farmers utilized the bench/bunds areas to grow fruit trees such as mango, banana, coconut, jackfruit guava, papaya and sugarcane.

This practice can be categorized as Integrated Agriculture and Aquaculture (IAA) referring to combining agricultural activities with aquacultural farming in the same areas at the same time. This system is very practical for small farmer. The major purpose of this practice is to earn as much as possible and to utilize all the available inputs.

In an IAA system the pond may provide additional benefits such as water storage capacity and improved soil fertility in addition to the returns from fish culture. Ponds can be used to process many forms of agricultural by products, including livestock and poultry manure and convert this manure into high-grade fish protein. Ponds and crops can be integrated using crops and crop residues as feeds and fertilizers for fish and pond sediments and water can be used as crop fertilizers and

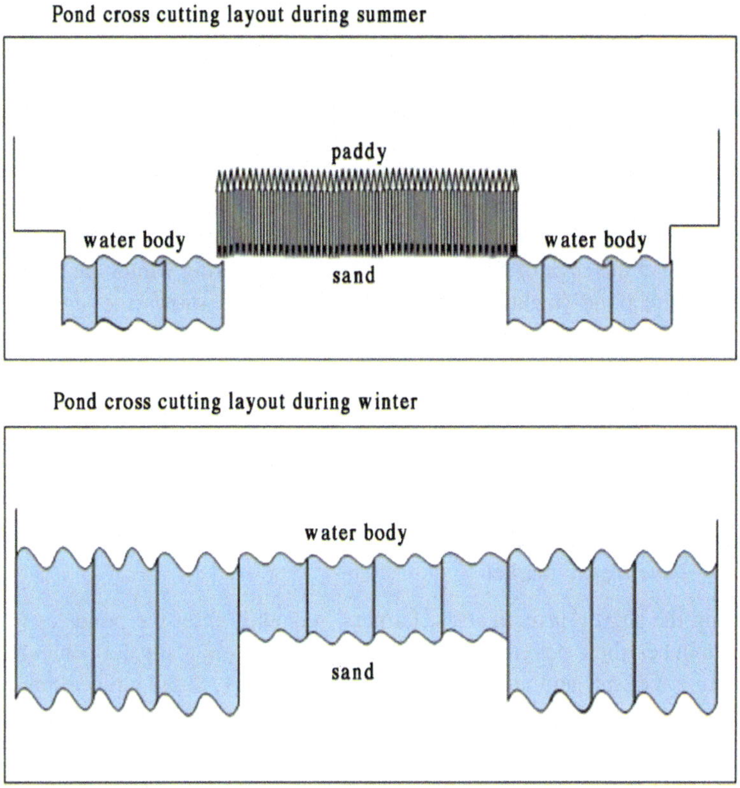

Fig. 20.4 Pond cross cutting layout in Khulna Bangladesh *Source* Fieldwork 2012

irrigation water, respectively (Little and Muir 1987 as cited in Murshed-E-Jahan and Pemsl 2011).

Pond preparation

The field investigation found there is a new concept of pond layout and system used by farmers with the following basic principal:

(i) pond bottom has two floors, higher in the middle and lower toward the edges (average one meter lower). During summer season- higher ground is used for paddy planting and lower ground used for aquaculture
(ii) during winter (raining season) the entire pond will be filled up (both lower and upper floors) and *Macrobrachium rosenbergii* will take place in the entire ponds.

Figure 20.4 shows pond cross cutting layout of the ponds in the study area. Similar concept for fish farming in paddy field is practiced in Shaanxi China (Wu 1994 in Liu and Cai 1998).

The farmers use same plot areas to plant paddy when there is less rain and utilize maximum the pond bottom which is highly nutrient to grow paddy. In rainy season, the farmers utilize the rain water to fill the pond and start their giant freshwater prawn farming. The farmers conduct their farming operations in tune with the weather. In many countries, especially Malaysia and Thailand; cleaning pond bottom is the major cause of environmental degradation as it pollutes the nearby riverine. Many farmers in this area change their pond layout arrangement to utilize all the fertile top soil to grow paddy (Ahmed 2001).

The most important feature is that the farmers are rearing *Macrobrachium rosenbergii* whole year in their pond where in summer *Macrobrachium rosenbergii* occupy canal at the edge of pond. On the other hand, during winter the rearing takes place in entire pond (including the paddy plots). This shows the significant role played by *Macrobrachium rosenbergii* in this farming system.

This aquatic species overshadow paddy in all areas, monopolize farming time, and are reared around the year. *Macrobrachium rosenbergii* merges as a personal preference for farmers due to:

(1) high ability to change it into cash anytime,
(2) high market appreciation (higher price compared to paddy),
(3) easy maintenance, and
(4) ready buyer/ready market

During the survey investigation, farmers frequently refer the ponds as a "bank" as it eliminates their poverty. The presented scenario is a special achievement in aquaculture industry and an example of a success story. Small aquaculture farmers in other developing country are suffering from poverty, insufficient income, marginalization, unsustainability and environmental issues.

Less than 50% of the pond area in general was used for canals and rest of the area for growing paddy. This practice, which started as a small venture in Bagherhat district, has spread to almost all parts of Bangladesh although still most of the activity is concentrated in the Southwestern part of the country. The farming area is reported to be growing at about 10% per year (Nandeesha 2003).

Seedling/fry/brood stock

Farmers collect the brood stock/fry from the wild or get supply from middle man which collects the wild fry from the nearby rivers. The farmers prefer to rear the local breed because:

(1) the survival rate is higher (Ahmed 2001);
(2) the immunity to disease;
(3) better adaptation to local ponds, perhaps because of the locality and natural characteristics of the local environment.

The farmers prefer wild fry because of its size, but are also facing challenges of limited supply. Meanwhile there is lower demand for hatchery fry because of its vulnerability and lower survival rate. Many farmers prefer to choose male fry (monotonous) as this will produce huge or large giant prawn and will lead to higher production.

Table 20.5 Pond management by farmers in Khulna District, Bangladesh

Description	Pond 1	Pond 2	Pond 3	Pond 4	Pond 5
Village name	Mongla	Mathardia	Rajibpur	Rajibpur	Rajibpur
Location	22° 48′ 51″N 89° 60’ 97″ E	22° 59′ 40″N 89° 65’ 10″ E	22° 77′ 984″N89° 43’ 69″ E	22°77′ 984″N 89° 43’ 698″ E	22° 77′ 97″N 89° 43’ 75″ E
Started/Age of aqua	1 year	4 years	8 years	3 years	4 years
Size of pond	4 acre/16,187 m^2	3 acre/12,140 m^2	3 acre/12,140 m^2	1.5 acre/6070 m^2	4 acre/16,187 m^2
Depth (meters)	1.0–1.5	1.0–2.0	1.1	1.1–1.2	1.0
Material	Sand/earth	Sand/earth	Sand/Earth	Earth	Earth
Stock density	12,000 male/acre 300 male/are	4000 fry and 4 kg fish 100 fry/are	8000 per acre 200 per are	6–8,000 per acre 150–200 per are	6000 per acre 150 per are
Cycle	Continues	Continues	Single stocking/continues harvesting	Single stocking/continues harvesting	Single stocking/continues harvesting
Pellet	Yes, pellet and natural resources	Yes, pellet and natural resources	Yes, pellet and natural resources-snails	Yes, pellet and natural resources-snails	Yes, pellet and natural resources-snails, corn
Drainage	Natural river	Natural river	n/a	n/a	n/a
Water source	River	River	Rain water	Rain water	Rain water
Fry sources	River	River	River	River	River
Survival rate	95%	80–85%	80%	80%	60%
Productivity (Kg/ha/season)	1200 kg/ha	320 kg/ha	213 kg/acre (for ha ×2.2)	213 kg/acre (for ha ×2.2)	200 kg/acre
Water Colour	Green	Light Green	Light green	Light green	Light green
Vegetation in ponds	Absent	Absent	Absent	Absent	Absent
Pond repairmen	Whenever needed	After raining season	Dry season	Dry season	Dry season

(continued)

Table 20.5 (continued)

Description	Pond 1	Pond 2	Pond 3	Pond 4	Pond 5
Village name	Mongla	Mathardia	Rajibpur	Rajibpur	Rajibpur
Removal of bottom	Yes, dry season (Jan-until April)	Yes, dry season (Dec, Jan, Feb March)	No removal-plough for paddy farming	No removal-plough for paddy farming	No removal-plough for paddy farming
Liming	Yes	Yes	Yes	Yes	Yes
Fertilizer	no, maybe	Yes, cow dung + phosphate	Yes, urea	Yes, urea + cow dung + TSP	Yes, urea
Harvesting	Continues	Continues	Continues	Continues	Continues
Water management	When needed, dry season	When needed-every 2 times per months	Depend on raining	Depend on raining	Depend on raining
Food management	2 times: 9 am in the morning and 5 pm in the evening	2 times: 8 am in the morning and 5 pm in the evening	2 times-morning and evening	2 times—5–6 am morning and 5 pm evening	3 times—5 am morning, 1 pm and 5 pm evening
Pond environment	Located in commercial areas (EPZ) Export Processing Zone-	Located in the middle of the village- at the surrounding pond —coconut, cow shelter, vegetables and farmer house is located close to farm–50 m	Surrounded by vegetable garden-lady finger, tomato, beans, bittergoud, banana, and other kitchen essential.	Surrounded by vegetable garden-lady finger, tomato, beans, bittergoud, banana, and other kitchen essential.	Surrounded by vegetable garden-lady finger, tomato, beans, bittergoud, banana, and other kitchen essential.

Pond 1 Mr. Habibul Rahman, *Pond 2* Mr. Iqbal Hawladar, *Pond 3* Mr. Mahbubur Rahman, *Pond 4* Mr. Habib Sarwar Titu, *Pond 5* Mr. Khukan Dhali

Feeding Management

Farmers use very minimal commercial inputs in their ponds. Very limited pellet or processed feed is given to the *Macrobrachium* rosenbergii owing to;

(1) higher purchasing cost,
(2) availability and
(3) accessibility.

The farmers get the feed supply from middle man and local distributer. They feed the prawn with additional natural food such as raw and boiled maize, crushed snail and soya. Farmers use local and natural inputs as feeding materials which are of lower cost and easily available. This is winning point for the small aquaculture farmers in Khulna as they minimize the operational cost. Most of the aquaculture farmers in Southeast Asia, especially in Malaysia, suffer from higher operational cost being dependent on processed feed, and are also using imported feed from Thailand. This is a high burden to farmers as feed represent 60–70% of the total cost of production.

The snail supply in mass quantity is done by poor marginalized people who earn livelihood by collecting sea snail. This value-chain supply help the marginal group to up-lift their life. Furthermore, the shell of the snails is used as a source for the production of lime which is very useful and important for liming pond bottom surface during the pond preparation process. Almost all the farmers practice feeding twice a day, usually at morning and afternoon. Interesting to note is that the farmers allow the aquatic species to find own food in the ponds which are rich in algae and other aquatic life. This practice is an important element that makes the whole farming system or production a semi-organic and semi intensive.

Fertilizers and chemicals

Farmers are using local and natural fertilizers in their pond. It is a common practice for farmers to utilize cow dungs as fertilizer, with the aim to increase the anaerobic activity in the pond. Besides, farmers also use urea and phosphate in ponds. Lime is used widely during pond preparation.

Water Management

Giant freshwater prawn farming in Khulna mostly uses water from river and rain. Bangladesh has huge freshwater sources; Ganga River and Madhumati River are running across Khulna. Plenty of rain almost through all the year with higher fall during the May-September period is ideal for the farming activities. Water quality is determined by farmers just by observation and experience. Farmers have no special equipment or instrument for the water testing. There is no single pond in the area that would have aeration installed.

Stocking and harvesting

The farmers have no strict guidelines in releasing fries into the ponds, the stocking varies from pond to pond. It was found the farmers release 6000–12,000 fries to each pond (Table 20.5). Some farmers release only male fries and some other releasse fish fingerlings for together for their own consumption. A range of carp species is cultured with the prawns, but harvested throughout the year mainly for household consumption. Farmers stocked Common carp, Cyprinuscarpio L., Silver carp, Hypophthalmichthysmolitrix, Indian major carp, Catlacatla, Labeorohita, Cirrhinusmrigala and Grass carp Ctenopharyngodonidellus (Ahmed 2001).

Farmers practice single stocking and continues harvesting—where the farmer does prawn size selection—when a prawn reaches certain preferable size (usually 250–600 g per piece) the farmer starts collecting the desired/marketable product and sell the produce to the middle man. Using this practice farmers get instant and relatively high income due to favourable price of prawn. These help them to redeem the products on the spot each time they need cash and keep them above the poverty line. Farmers idealize the pond considering it as a "bank" and their prawn is real "cash crop". This is very different from the situation in Malaysia where farmers are always "sour face" and traumatized with loss of production due to disease outbreaks or low survival rate.

20.3.3 Macrobrachium Rosenbergii *Farming Knowledge/Training*

Most of the *Macrobrachium rosenbergii* farmers in Khulna never had any formal training, vocational practice or education course. They learn the practice from their neighbour farmers, relatives and friends. This can be classified as knowledge gained through oral lesson and the knowledge spread from "mouth to mouth". Any new management practices will spread fast through oral discussion, Moreover, many farmers are illiterate only some with a primary education (Table 20.4), and all with peasant background. Some farmers mentioned that neighbours and friends who received training are the main source of technical assistance (Ahmed 2001).

20.4 Conclusion

The paddy-prawn integrated farming in Khulna, Bangladesh is an indigenous knowledge which improved from each cycle of farming and its spread through verbal communication. The paddy-prawn farming in Khulna is not to supply cheap animal protein to farmers, but is important income generator. The farmers involvement in paddy-prawn farming helped them to get rid of poverty and marginalization.

However, the farmers prefer to rear *Macrobrachium rosenbergii* instead of paddy farming only because of the cash value and ready market. The farmers have improved their living standards and manage to satisfy basic needs since they started the paddy-prawn farming.

The key success of farmers in Khulna, Bangladesh is being less dependent on processed commercial pellet and utilized farm waste including animals, maize and soya grain, while many marginal farmers in Malaysia, Indonesia, Thailand and Philippine are suffering from high production costs because of their dependence on processed feed and foreign labour.

Acknowledgements This work is supported by the High Impact Research (HIR) project funding of the Ministry of Higher Education, Malaysia. The author would like to thank the Ministry of Higher Education Malaysia for the required financial support and University Malaya HIR Office for their support in facilitating the field investigation trip.

References

Ahmad, Q. K., Biswas, A. K., Rangachari, R., & Sainju, M. M. (Eds.). (2001). *In Chowdhary. A Framework for Sustainable Development*. Brahmaputra-Meghna Region: University Press Limited, Dhaka, Bangladesh.

Ahmed, N. (2001). Socio-economic aspects of freshwater prawn culture development in Bangladesh. Ph.D thesis (unpublish). University of Stirling. Scotland.

About.com. (2013). Bangladesh Country Index. Retrieved July 17, 2014, from http://geography.about.com/library/cia/blcbangladesh.htm.

Belton, B., & Azad, A. (2012). The characteristics and status of pond aquaculture in Bangladesh. *Aquaculture, 358–359*(2012), 196–204.

Bosma, R. H., & Verdegem, M. C. J. (2011). Sustainable aquaculture in ponds: Principles, practices and limits. *Livestock Science, 139,* 58–68.

Chambers, R. (1994). The origins and practice of participatory rural appraisal. *World Development, 22*(7), 953–969.

Chowdhury, N. T. (2010). Water Management in Bangladesh: An Analytical Review.

Department of Fisheries 2010. (2012). Bangladesh Annual Report 2010. 2012. Fisheries Statistical Yearbook of Bangladesh 2010–2011. Volume 28. *Water Policy, 12,* 32–51.

FAO Fisheries and Aquaculture Department. (2013). Global Aquaculture Production Statistics for the year 2011.

FAO. 2012.

Felipe, J. A., Ferreira, M. A. M. & Coelho, M. (2011). An ethical issues in anti-commons management in aquaculture case in Portugal. *International Journal of Academic Research, 3* (1). Part I.

Firuza B. M. (2005). Impak Alam Sekitar Penternakan Udang Harimau di Kawasan Paya Bakau Bagan Tengkorak, Tanjung Karang, Selangor. Phd Thesis (Unpublished). Universiti Kebangsaan Malaysia. Malaysia.

Fisheries and Aquaculture Department. (2013). FAO Fisheries and Aquaculture Department has published the Global Aquaculture Production Statistics for the year 2011. Retrieved Mac 21, 2014, from ftp://ftp.fao.org/FI/news/GlobalAquacultureProductionStatistics2011.pdf.

Hai, N. T., Liu, Q., Zhao, L., Zhang, H., Liu, J., & Dang, N. H. (2015). Genetic diversity of cultured populations of giant freshwater prawn (*Macrobrachium rosenbergii*) in China using mtDNA COI and 16S rDNA markers. *Biochemical Systematics and Ecology, 62,* 261–269.

Little, D. C., Surintaraseree, P., & Taylor, N. I. (1996). Fish culture in rainfed rice fields of northeast Thailand. *Aquaculture, 140,* 295–321.

Liu, J., & Cai, Q. (1998). Integrated aquaculture in Chinese lakes and paddy fields. *Ecological Engineering, 11,* 49–59.

Lu, J., & Li, X. (2006). Review of rice–fish-farming systems in China—One of the Globally Important Ingenious Agricultural Heritage Systems (GIAHS). *Aquaculture, 260,* 106–113.

Murshed-E-Jahan, K., & Pemsl, D. E. (2011). The impact of integrated aquaculture–agriculture on small-scale farm sustainability and farmers' livelihoods: Experience from Bangladesh. *Agricultural Systems, 104,* 392–402.

Mushtaq, S., Dawe, D., Lin, H., & Moya, P. (2007). An assessment of collective action for pond management in Zhanghe Irrigation System (ZIS), China. *Agricultural Systems, 92,* 140–156.

Nandeesha, M. C. (2003). The "Gher Revolution" Innovations in freshwater prawn farming by Bangladesh farmers. *Aquaculture Asia, VIII*(4), 31–36.

Paul, B. G., & Vogl, C. R. (2013). Organic shrimp aquaculture for sustainable household livelihoods in Bangladesh. *Ocean and Coastal Management, 71,* 1–12.

Rasid, R. (2015). Dietary phytic acid and its effects on *Macrobrachium rosenbergii* (De Man, 1879). Ph.D thesis. University of Stirling. Scotland.

Shah, M. S. & Ahmed, M. K. (2006).Socio-economy and livelihoods of stakeholders at different levels of production, processing and marketing and value chain of freshwater prawn (*Macrobrachium rosenbergii*) in greater Khulna and Noakhali areas of Bangladesh. Bangladesh Fisheries Research Forum (BFRF).Value Chain and Market Assessment of Coastal and Marine Aquatic Products of Bangladesh.

Tacon, A. G. J., & De Silva, S. S. (1997). Feed preparation and feed management strategies within semi-intensive fish farming systems in the pics. *Aquaculture, 15*(1), 379.

United Nations Conference on Trade and Development. (2015). Commodities and Development Report 2015—Smallholder Farmers and Sustainable Commodity Development.

Weeratunge, N., Snyder, K. A., & Sze, C. P. (2010). Gleaner, fisher, trader, processor: Understanding gendered employment in fisheries and aquaculture. *Fish and Fisheries, 11,* 405–420.

Wilson, C. A., & Goodbred, S. L. (2015). Construction and maintenance of the Ganges-Brahmaputra-Meghna Delta: linking process, morphology and stratigraphy. *Annual Review of Marine Science, 7,* 67–88.

Wu, Y. X. (1994). Speed up the development of paddy fish culture and make new contribution to the promotion of rural economy. Paddy Field Fish Culture Has Bright Prospects. Fishery Bureau of Ministry of Agriculture, Beijing (pp. 2–10, in Chinese).

Chapter 21
Marginalization Between Border and Metropolis: Drivers of Socio-Spatial Change in Post-socialist Croatia

Marin Cvitanović and Borna Fuerst-Bjeliš

21.1 Introduction: Theoretical Background and Research Context

Marginalization as a process and marginality as a concept are extremely complex and can be defined in a number of ways. Basically, marginalization is the process of becoming marginal, or being pushed into a marginal position (Pelc 2016), and therefore, should be an intrinsically relational process and/or concept. Some authors see marginalization as a process of decline from a previous state (Déry et al. 2012). As a relational process and concept it cannot be static in space and time. It is constantly changing and is deeply dependent on wider socio-economic changes, such as globalization, deindustrialization or economic transition, which can cause profound and spatially heterogeneous changes, benefiting certain areas while creating disadvantages for others (Fuerst-Bjeliš and Cvitanović 2016). Marginalization is always a social process, although it can be triggered by a number of various factors.

Marginalization as a process is anchored in the concept of marginal, or marginality, and is an extremely broad and perhaps ambiguous term. What is certain is that it always has a negative connotation, such as being at the margins, or to be of minor importance or influence (Pelc 2016). But, of course, it is also a matter of perspective. Inner and outer perspectives may substantially differ, because they are of different scales and they do not refer to the same system (Déry et al. 2012).

M. Cvitanović (✉)
Faculty of Science and Technology, Department of Life and Environmental Sciences, Bournemouth University, Fern Barrow, Poole, UK
e-mail: mcvitanovic@bournemouth.ac.uk

B. Fuerst-Bjeliš
Faculty of Science, Department of Geography, University of Zagreb, Marulićev trg 19/II, 10000 Zagreb, Croatia
e-mail: bornafb@geog.pmf.hr

S. Pelc and M. Koderman (eds.), *Nature, Tourism and Ethnicity as Drivers of (De)Marginalization*, Perspectives on Geographical Marginality 3,
DOI 10.1007/978-3-319-59002-8_21

That is why most researchers include both qualitative and quantitative methods of measuring marginality.

In their review of the recent insights to marginality, Déry et al. (2012) identified a common denominator of the marginalization process. They found that in each case studied, there was a significant change in power relations (whether it was connected to relations within the system, or change in the case of integration into new (bigger) systems), followed by a lack of adequate adaptation.

In this view it follows that (one of) the most powerful and perhaps the most common triggers of marginalization is the integration of a local system into a new (mostly) larger one, when the local system, failing to adequately adapt to a new situation, is losing its relevance as well as becoming increasingly marginalized. So, as Déry, Leimgruber and Zsilincsar have put it: “paradoxically, integration may signify marginalization” (Déry et al. 2012).

Discussing the forms of marginality within the context of “processual” marginality, Leimgruber (2004) points specifically to the change in the system of production, market forces, and competition; which marginalizes people, regions, or countries.

Loss of importance in this context is usually characteristic for rural areas where the economic structure is dependent on primary resources such as agriculture, and where relative proximity to larger centers is found to be of crucial importance. As the review of the analysis by Déry et al. (2012) has shown so far, the process of marginalization, as a social process, usually manifests with a sharp decline in the number of jobs, population, increase in out-migration, as well as in economic dependency in general. Such indicators are therefore mostly used by researchers to quantitatively measure the marginality of the region.

Hrvatsko Zagorje (the northwestern-most Croatian border region, and the study area) displays most of the previously-discussed marginality issues. The region is traditionally a rural and agricultural area which has experienced a change in power relations since the beginning of the post-socialist transition period, i.e. since the beginning of the 1990s. According to Leimgruber’s typology (2004), the “processual” marginalization took place here, as it was triggered by a change in the system of production, market forces, and competition, deepening the pre-existing marginal character of the region. The region has since been integrated as a local system into a larger open market system that replaced the former system of a state-planned economy. The lack of adequate adaptation to the new conditions is reflected in deindustrialization followed by increasing unemployment, population outflow, and ageing.

The second important element of fostering marginalization, observed in previous research, is proximity to a larger urban center, in this case, the strongest and most-populated economic and political center in Croatia—the Zagreb metropolis (Fig. 21.1).

Hrvatsko Zagorje is also a border region, and as such, it is also experiencing what Leimgruber (2004) defines as “geometrical” marginality that may be close to the concept of periferality. Certainly, with its location in the northwest of Croatia on the border with a correspondingly marginalized part of Slovenia (Vintar Mally

Fig. 21.1 The location of the study area

2016), this region has been additionally burdened by the change of its border character after the dissolution of Socialist Federal Republic of Yugoslavia. What was formerly a largely administrative border within the same country suddenly became an international political border.

In the post-socialist period, the region of Hrvatsko Zagorje has been under the influence of its specific position between the border and a metropolis; between a kind of "dead end" and an extreme pulling force. At this point, we can define two key aspects of marginalization in the context of the post-socialist transition. Social aspects comprise processes of decline in the number of jobs, increasing unemployment, population outflow, and ageing; all followed by land use and land cover change reflected in the environment. Territorial aspects indicate the further "borderization" of the region, with the geometrical marginality or the peripheral character of the region on one side, and the strength of the major pulling force at the other.

The study of geographic marginalization during periods of rapid societal, economic and institutional changes such as post-socialist transitions can offer new insights into the characteristic of geographic marginality. Such analyses can be seen as a sort of a "natural experiment" with two political systems which occupied the same environment over a relatively short period of time, emphasizing the role of socio-economic changes that can drive marginalization processes in certain areas (Fuerst-Bjeliš and Cvitanovi 2016). The geography of marginalization tends to find answers about the causes and consequences of marginalization (Pelc 2016). In this regard, the main research questions of this study were:

- What are the spatial manifestations of marginalization?
- How do spatial manifestations of marginalization affect the development of the region?
- How do global trends act on local and traditional environments?
- How are peoples' attitudes and values driving changes in the environment?
- Do these factors foster marginalization and in what way?

To reach the answers, the research was based on a mixed methodology, combining a quantitative approach via remote sensing and regression modelling, and both qualitative and quantitative approaches using a household survey. Land use and land cover changes, as well as demographic and socio-economic variables, were employed in the context of acquiring the outer perspective of marginalization of the region, while the household-based approach was used to reach the inner perspective of marginalization.

21.2 Krapina-Zagorje County: The Study Area

Krapina-Zagorje County was formed as a regional administrative unit within the newly independent Republic of Croatia in 1992. It encompasses the southern part of the historical and traditional region of Hrvatsko Zagorje in Northern Croatia. With an area surface of 1,224 km^2 and a population of a little over 130,000, it is one of the most densely populated counties in the country. Characterized by small dispersed settlements and traditional subsistence agriculture, it is a typical rural area of Croatia. The average settlement size is less than 400 people and only 24% of the total population lives in urban areas. The area is characterized by a heterogeneous topography ranging from the alluvial plains of the Krapina river (around 150 m a.s. l.) to the peaks of Mt. Medvednica and Mt. Ivanščica reaching altitudes over 1000 m (Cvitanović 2014).

According to the earliest reliable documents, the first permanent settlements in Krapina-Zagorje County were located in hilly areas in altitudes of 300–500 m, away from the seasonally-flooded and swampy lowlands. Stronger colonization of the low-lying alluvial plains in the County started only in the 19th century, as a consequence of migration of population from the surrounding hills to more accessible river valleys in the area. The colonization of lower altitudes was facilitated by the agricultural and economic revolution, which was caused by the abolishment of serfdom in 1848. New economic development, mostly coal mines and the textile industry, occurred alongside the building of the railroad in the river valleys in the late 19th century; this attracted a large number of people in search of new employment opportunities (Crkvenčić 1957).

These broad-scale social and economic changes from the 19th century created the typical smallholder agriculture with highly fragmented estates in the county. The fragmentation of land is a result of a dynamic relief, but also of the aforementioned agricultural and economic revolution in the 19th century, when the previously communally-owned land (organized into cooperatives) was divided and distributed among the population. Already in 1895, three quarters of all estates in the central part of the County were smaller than 5 ha in surface area each (Crkvenčić 1958). A new cultural landscape was formed in the region, further strengthened after the Second World War when legal provisions prevented any substantial re-parcellation of land or increase in land holdings (Plaščak 2009). This traditional smallholder agriculture has remained the most important and most enduring element of the cultural landscape and cultural identity of the region to this day.

The period after the Second World War saw the formation of a state-planned economy within the communist and socialist framework in the Socialist Federative Republic of Yugoslavia, of which Croatia and Hrvatsko Zagorje were a part. The processes observed in Hrvatsko Zagorje were characteristic for the entire county, and for other republics of the Yugoslav Federation. They included accelerated economic development and a decrease in agricultural employment, with an increase in education and employment levels, particularly amongst women. Deagrarianization and migration to cities together with industrialization affected settlement patterns, especially around major cities and alongside major roads. Rural settlements located a greater distance from major cities were characterized by depopulation due to migration of population in search of employment. Those who continued to live in remote rural settlements created a new class of society in Yugoslavia, the so-called "worker-peasants". Worker-peasants commuted to cities on a daily or weekly basis (mostly due to lower education levels and lower income as a consequence), but also continued with their agricultural practices as a livelihood strategy in their settlements of origin (Spevec 2009).

After the collapse of socialism in 1991, the obsolete smallholder agriculture did not go through a process of modernization, and the total number of cattle was halved in a period of one decade (Croatian Ministry of Agriculture 2013). The processes of population ageing and migration to cities continued, causing agricultural abandonment in remote rural areas and also around major urban settlements. These processes are characteristic not only for traditional rural areas of Croatia, but of the broader region, and Krapina-Zagorje County was chosen as the best possible case study for the analysis of the interplay between the post-socialist transition and marginalization in traditional rural areas.

21.3 Data and Methods

This chapter is methodologically divided into three parts. The analysis of land cover change in the period of 1978–2011 was based on spectral analysis and classification of Landsat images. The focus has been placed on changes in agricultural land and settlement patterns as the best proxy for changes indirectly caused or mediated by the post-socialist transition. The analysis of causes of the observed land cover change was investigated through multiple regression analysis and further elaborated through a detailed questionnaire survey covering 262 households in the region.

The causes of the observed land cover changes were explained through the interaction between the environment and people. Human activities (e.g. agriculture, housing construction) were seen as drivers of environmental change, while simultaneously being influenced by the environment itself—as well as by both societal and economic changes. Under the assumption that land cover changes and different societal and economic changes can be identified and quantified, it was possible to create a model based on statistical correlations between these two types of variables. The importance of such a model is in creating a framework within which it is possible to evaluate the importance of different variables causing the observed changes, as well as for the possibility of predicting future changes (Woods and Skole 1998; Morita et al. 2007).

Apart from the environmental variables gathered from remote sensing images, the variables used in this research were based on the Population Censuses from 1991 and 2011, as well as on auxiliary information such as orto-photographs and statistical yearbooks. The multiple regression model was based on the fact that 98% of agricultural land in Krapina-Zagorje County is privately owned and that 89% of households still actively practice some form of agriculture.

21.3.1 Remote Sensing

Landsat TM and Landsat ETM+ satellite images were used in the remote sensing of Krapina-Zagorje County. The images were classified using K-means and Fuzzy-C unsupervised classification methods, and the anniversary years were 1978, 1991, and 2011. The achieved accuracy for agricultural land was between 75% and 85%, and around 70% for settlements. Therefore, with the help of auxiliary data (geo-referenced ortho-photographs) and several fieldworks, a substantial number of misclassified pixels were manually corrected, resulting in a higher level of accuracy for the final land cover maps. The land cover change maps were created by a post-classification overlay, and the observed trends in this study were further validated by similar findings from the Corine Land Cover Maps and through the aforementioned questionnaire survey.

21.3.2 Regression Analysis

The explanation of human-induced land cover changes depends largely on the ability to explain socio-geographic factors connected to observed changes (Woods and Skole 1998). In geographic research, different models of regression analysis are often used for such purposes (Rutherford et al. 2007; Zampella et al. 2007; Lo and Quattrochi 2003). The dependent variable (y), in this case the change in land cover, is a function of a series of independent variables $x_1,\ldots, x_p$ where $p \geq 1$. However, the variable y doesn't depend only on the variables $x_1 \ldots x_p$, but also on a certain number of variables which are not available (Reimann et al. 2008). Multiple regression analysis facilitates in creating a simplified link between an independent variable and a series of dependent variables, and is also a tool of evaluating their importance. In this research, the evaluation and reduction of a number of independent variables was achieved through backward stepwise regression. After the backward stepwise regression procedure, the only remaining variables in the final model were those which would significantly decrease the explanatory power of the model if they were eliminated.

The most often used (independent) variables in land cover modelling, apart from physical-geographic ones, are those connected to the education level of the population, population ageing, and population density. Previous research included a variety of variables, such as an increase of school-level children and literacy rates (Asia Pacific Network 2001), employment levels in different branches of industry, the number of cars per household or land prices (Morita et al. 2007), or different distance factors (Serneels and Lambin 2001). The variables used in this research were based on previous research and on expert knowledge of the area. They are: changes in population density; changes in higher education levels; changes in employment levels; and physical-geographic variables categorized into nine altitude and slope categories. They were calculated for each municipal unit (32 in total) because these are the smallest administrative units in Croatia for which we had available socio-economic data.

21.3.3 Questionnaire Survey

Multiple regression analysis enabled us to establish a correlation between e.g. changes in employment and agricultural abandonment, but did not establish causality or give insight into household management practices related to the observed changes. Therefore, a questionnaire survey on 262 households in the county was conducted as the third part of this research. This survey gave us additional insight during the interpretation of remote sensing data and multiple regression analysis, and also provided us with some new data which would have been unavailable otherwise.

The questionnaire consisted of 46 questions about the socio-economic status of the respondents, their livelihood strategies, the size and type of land in their possession, their plans regarding agriculture and possible resettlement, etc. A part of the questionnaire consisted of 14 Likert-type questions regarding cultural values about the land, living in rural/urban areas, and work. The questionnaire was analyzed using SPSS 20 software, and a Pearson and Spearman correlation coefficient; chi-square and t-test were used, depending on the nature of the analyzed variables.

21.4 Results

21.4.1 *Changes in the Land Cover of Krapina-Zagorje County from 1978 to 2011*

According to the results from the remote sensing analysis, the general land cover categories did not change significantly over the studied period, with over two-thirds of all land cover remaining unchanged during 34 years of socialism and post-socialist transition (Fig. 21.2). Forests have been the dominant and the most stable land cover in the region, covering 51% of the surface area in 1978, 46% in 1991, and 44% in 2011. They are mostly located on higher altitudes and outside zones of immediate human influence, and have been changing due to secondary succession (increase) or sourcing of wood for fuel (decrease) (Cvitanović et al. 2016).

The majority of observed land changes were in the grassland and agricultural land categories. While the socialist period (1978–1991) was characterized by a decrease in agricultural land, the post-socialist transition has witnessed a slight increase in agricultural land mostly at the expense of grasslands (Cvitanovic et al. 2017). A little over half of these changes occurred in altitudes up to 200 m.

The built-up areas covered 1% of the surface area in 1978, 2% in 1991, and 3% in 2011. The overwhelming majority of the observed increase in built-up areas occurred in altitudes up to 200 m.

21.4.2 *Multiple Regression Modelling*

The backward multiple regression analysis reduced the initial 13 variables (from the first step) to seven variables, explaining a statistically significant amount of

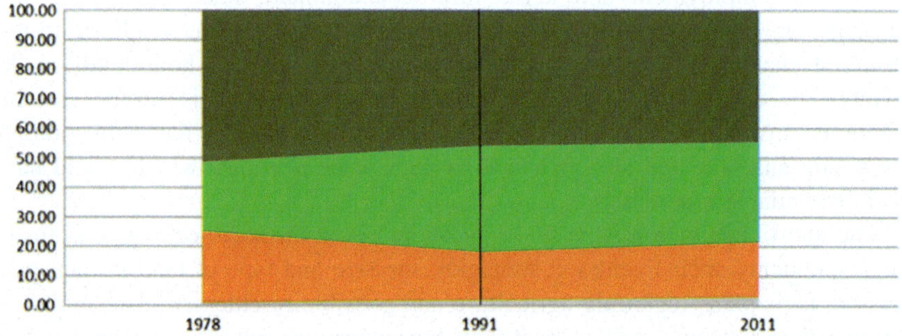

	1978	1991	2011
forest	51.59	46.04	44.65
grassland	23.33	35.86	33.80
arable land	24.06	16.04	18.43
built-up areas	1.02	2.06	3.12

Fig. 21.2 Changes in major land use/cover categories in the 1978–2011 period

variability of the transition between **built-up land** and **other categories**. The model accounted for 79.7% of the variance. Lower altitudes and flatlands were positively correlated with an increase in built-up areas, while higher altitudes and steeper slopes demonstrated a negative correlation. Regarding socio-economic variables—population density, higher levels of employment, and higher education levels were positively correlated with an increase in built-up areas.

Four variables remained in the model, explaining the **arable land to grassland** conversion that accounted for 73% of the variance. Increase in grasslands was negatively correlated with population density but positively correlated with an increase in education levels. This was also characteristic for higher altitudes and steeper slopes (Cvitanović et al. 2017).

Regarding the **grassland to arable land conversion**, the model managed to account for only 29% of the observed variance, and only the physical-geographic variables remained in the model. The increase of the share in flatland areas was positively connected with this process, while an increase in altitudes of 300–400 m demonstrated a negative correlation (Cvitanović et al. 2017).

21.4.3 Questionnaire Survey

One person per household (identified as the person making most of the decisions regarding agriculture and livelihood strategies in the household) was surveyed. The youngest respondent was 21 and the oldest 90. The average age of the respondents was 43, which is in accordance with the average age in the County; 59% of the respondents were women. The average size of the household was 3.9 and 19% of the households had only members older than 50 years of age.

Over 60% of all household dwellings of respondents for the questionnaire survey were reportedly built prior to 1978. During the studied period, 18.7% of respondents built a house, a further 2% built a barn, and 13% both a house and a barn. The most important reason for building a house was marriage and/or family expansion (53.5%), while 39% stated that the lack of space or inadequate quality of living areas was the main reason. Over 90% of the respondents did not plan on building a house or a barn in the future.

Regarding agricultural practices, around three-fourths of the land in the county is currently being cultivated. The main stated reasons for practicing agriculture were: tradition and way of life (33%); having an important source of food (28%); and cultivating high-quality organic products (26%). The reasons for not cultivating the land were: advanced age or illness (31%); economic reasons (25%); and other reasons such as lack of time, distance, etc. (Cvitanović et al. 2017). Around half of the surveyed households owned and used agricultural machinery; with 54% of the machinery having been bought prior to 1991 and a further 15% prior to 1978, emphasizing the low level of mechanization and its advanced age characteristic for traditional smallholder agriculture.

Less than 10% of the surveyed households owned cattle and less than 30% owned pigs. During the studied period, 61% of households reportedly reduced or completely abandoned cattle herding, with the main reasons being economic or health reasons (Cvitanović et al. 2017). In general, agriculture was seen as an important part of local tradition, and the land was seen as intrinsically very valuable. Such claims scored the highest level of agreement on the Likert scale in our questionnaire (Cvitanović et al. 2016).

The analysis regarding the activities of house and barn building demonstrated a statistically-significant difference related to the age of the respondents. Around 30% of the younger respondents (up to 39 years of age) built a house during the studied period, but only 19% of the middle-aged respondents (40 to 59 years of age) and 8.7% of respondents over 60 years of age did the same ($\chi^2 = 24{,}748$, df = 8, $p < 0{,}01$). There was also a statistically-significant difference between the age of the respondents and the reasons for (not) practicing agriculture ($\chi^2 = 61{,}706$, df = 16, $p < 0{,}01$). As much as 34% of the younger respondents had never practiced agriculture, while for the oldest group of respondents the share falls to 20%.

21.5 Discussion

Agricultural lands in Krapina-Zagorje County recorded relatively strong rates of land cover change during the later stages of socialism and the early part of the post-socialist transition. They include changes in forest cover, agricultural land, and built-up land. During the later stages of socialism, a recorded net loss in arable land was detected, and a consequential increase in grasslands. Such trends were noted earlier in other parts of Croatia as well, mostly connected to the processes of deagrarianization and migration to cities in search of employment in industry (Crkvenčić 1981; Vresk 1972). It was a part of a wider trend in socialist Yugoslavia comprising the deruralization and transformation of the countryside, and the transformation of the rural population into industrial workers (Slavuj Borčić et al. 2016; Seferagić 2006). Abandoned land was often termed "social fallow" (Hartke 1953), signifying the importance of the social character of land cover changes. Similar results have been recorded in neighbouring counties as well. The study by Horvat (2013) documented a loss of agricultural land amounting to 8% and an increase in built-up land of 3% in Međimurje County during the 1978–1992 period.

However, there was a noted change in trends during the post-socialist transition. Despite negative demographic and economic trends (ageing of population, employment outside agriculture, lack of interest in traditional agricultural practices), there was a slight increase (1.5%) in agricultural land in Krapina-Zagorje county in the 1991–2011 period, mostly at the expense of grasslands. The same process has been noted in the neighbouring Međimurje County (Horvat 2013).

21.5.1 Increase in Agriculture in More Marginalized Regions

The results from the multiple regression modelling and questionnaire survey pointed to the two-sided character of changes in agricultural land use/cover following the collapse of socialism. Areas close to the border with Slovenia and further away from main roads and local urban centers were characterized by more negative demographic and economic trends, but at the same time not necessarily by an overall decrease in agriculture. Such households are typical partially deagrarianized households practicing polyculture on small land parcels. The unfavorable education level and age structure of the population together with higher unemployment suggested that these areas were the least successful in adapting to new conditions of integration into the open market system, and transforming to market economy. So, while the system operates locally, its integration has been unsuccessful and the region is undergoing the processual marginalization. Agriculture was shown to be an important source of food for the household that also generates a small income from selling extra produce to friends and neighbours (Cvitanović et al. 2017). The unfavorable peripheral location (of the county) close to an international border and away from larger urban settlements indicated that the state border was playing a role in unfavorable economic development and negative socio-economic trends in the area. However, such marginalized areas can also be found in the northeastern part of the county, away from the international border, main traffic roads, and Zagreb as well, pointing to the additional existence of inner peripheries. The marginalization of these areas is seen as processual in its negative socio-economic trends, but the cause is, undoubtedly, at least partially geometric in nature.

21.5.2 Increase in Agriculture in Less Marginalized Areas

On the other hand, a certain number of settlements alongside important traffic routes and closer to Zagreb and/or local urban centers (characterized by comparatively favorable demographic and socio-economic trends) witnessed an increase in agriculture. The reason for this increase is probably different—as much as one quarter of respondents claimed that agriculture is their favorite pastime or hobby, and that the reason that they engaged in agriculture was to produce good-quality organic food (Cvitanović et al. 2017). Such answers were more common amongst younger respondents with comparatively higher income, but also amongst a number of retired respondents, many of them so-called "weekend-tourists". This shift from food production to a preference for natural and cultural amenities has been previously documented in post-industrial countries in the West—with agricultural land being used for recreation and enjoyment as well as to experience the rural lifestyle (Sorice et al. 2014; Johnson 2008). The reason for comparatively fewer negative

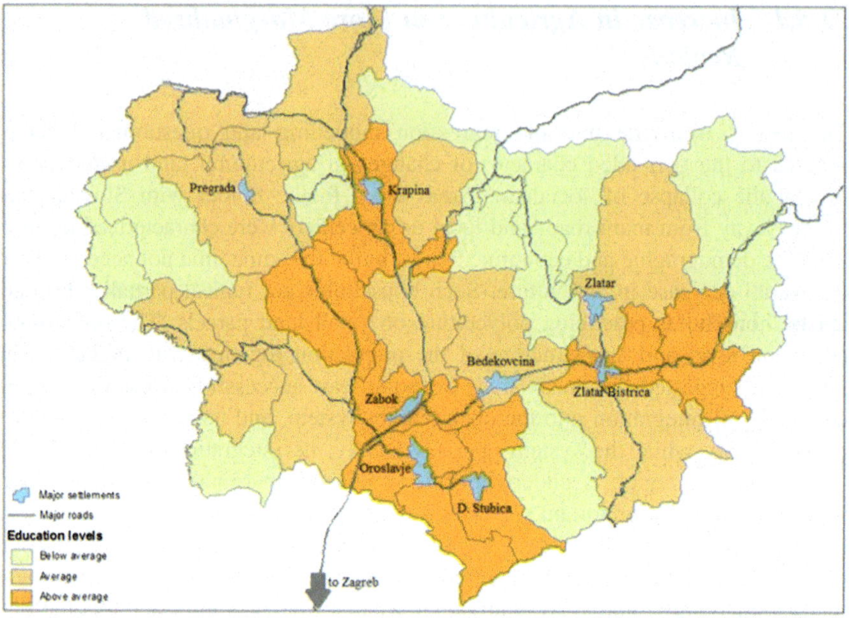

Fig. 21.3 Changes in education levels during the post-socialist period in Krapina-Zagorje County demonstrate the correlation between increased education levels and the proximity to major roads and larger urban settlements

socio-economic trends in this area is again, undoubtedly, partially geometric—proximity to Zagreb and/or local urban centers enables easier access to the employment market and educational, recreational, and health facilities (Fig. 21.3).

21.5.3 Increase in Built-up Areas

Further insight into land cover trends following the post-socialist transition was gained by analyzing the changes in the built-up category. New houses were being built mostly in lowlands and flatland areas in the county—areas with better traffic connections to Zagreb and to local urban centers. Additionally, the process of built-up expansion was negatively correlated with altitudes of 300 to 400 m and slopes with an inclination above 12°. The socio-economic variables which were statistically significant in this study were population density and higher education—areas with comparatively favorable trends in population and education dynamics recorded an increase in built-up land. The results indicated that the process of migration from hills in higher altitudes to lowland areas with better traffic connections (which began in the 19th century) has not yet finished. This was also characteristic during the socialist era, but post-socialist changes have since accelerated the process. Other research has also demonstrated that the migration from rural areas to municipal centers is still an ongoing process, characteristic for younger and better-educated segments of the population, while the older, less educated and less employable remain in rural areas (Njegač 1995; Spevec 2009). This was further corroborated by results from the questionnaire, where the younger population built houses and barns more often, on a statistically significant level.

21.6 Conclusion

The collapse of many subsidized industries in former socialist countries in Eastern Europe has resulted in higher unemployment levels. It can be seen as a change in power relations in places where a planned economic system has been integrated into a wider, global open market system with different characteristics. Subsidized industries were usually located in already marginalized areas, acting as a state-sponsored developmental tool (Vodopivec 1994). With the aforementioned changes and exposure to market forces, areas which were economically and demographically stronger managed to adapt to new circumstances, while less successful areas have been characterized by an accelerated decline. Such processes in a heterogeneous area have manifested themselves heterogeneously, both in economic and socio-demographic terms. Less favorable areas are turning again to self-subsistence agriculture as a household livelihood strategy, especially in households with aged members. These households are usually located in hilly and mountainous areas close to the border with Slovenia or far away from urban centers and/or important traffic routes. This indicates the importance of spatiality in the analysis of marginalization within the context of change in power relations. Conversely, areas with comparatively better socio-economic trends have been characterized by an increase in life-style agriculture which is seen as a cultural amenity or physical exercise. This is characteristic for areas with relatively larger local urban centers or those located on important traffic routes to Zagreb, enabling daily or weekly commutes.

Regarding settlement patterns, again the geometric part of the process of marginalization played an important role. Areas located in lowlands, closer to major traffic routes, have been characterized with an increase in built-up areas, while areas located in higher altitudes, far away from urban centers and closer to an international border have been characterized by out-migration.

These changes in the extent of agriculture and its character, as well as settlement patterns, are just symptoms of major changes caused by broad-scale political change and subsequent market reforms. More successful areas have been characterized with a variety of livelihood strategies which can include daily or weekly migration to urban centers, development of agro-tourism, and local industry. This variety of livelihood strategies has been unsuccessful in preventing the observed negative trends from occurring, but it has been able to weaken or slow them. On the other hand, areas with limited economic possibilities have been characterized with a sharp increase in unemployment and are turning again to traditional agriculture as a livelihood strategy. The location of these areas close to Slovenia indicates the role that the border plays in marginalization; but such areas can also be located away from the international border, pointing to the existence of inner peripheries. Such a process of economic decline and further marginalization would have probably continued without the political and economic changes of the post-socialist transition, but they have undoubtedly been accelerated by it. Agriculture can be a good indicator of such changes, but it is important to analyze its concomitant relationship to other potential indicators.

References

Asia Pacific Network for Global Change Research (2001). Land use and land cover change in Southeast Asia, a synthesis report. www.apn-gcr.org. Accessed 3 June 2013.

Crkvenčić, I. (1957). *Prigorje planinskog niza Ivančice. Geografski glasnik, 19*(1), 9–56.

Crkvenčić, I. (1958). *Prigorje planinskog niza Ivančice. Geografski glasnik, 20*(1), 1–48.

Crkvenčić, I. (1981). Socijalnogeografski aspekti pojave ugara, odnosno neobrađenih oranica. *Geografski glasnik, 43*(1), 95–106.

Croatian Ministry of Agriculture (2013). Strategija ruralnog razvoja RH 2008.-2013. http://www.mps.hr/default.aspx?id=3652. Accessed 11 Oct 2014.

Cvitanović, M., Blackburn, G.A., & Rudbeck Jepsen M. (2016). Characteristics and drivers of forest cover change in the post-socialist era in Croatia: evidence from a mixed-methods approach. *Regional Environmental Change,* 16, 1751–1763 (2016). doi:10.1007/s10113-016-0928-0.

Cvitanović, M., Lučev, I., Fuerst-Bjeliš, B., Slavuj Borčić, L., Horvat, S., & Valožić, L. (2017). Analyzing post-socialist grassland conversion in a traditional agricultural landscape—case study Croatia. *Journal of Rural Studies, 51,* 53–63. doi:10.1016/j.jrurstud.2017.01.008.

Déry, S., Leimgruber, W., & Zsilincsar, W. (2012). Understanding Marginality: Recent Insights from a Geographical Perspective. *Hrvatski geografski glasnik, 74*(1), 5–18.

Fuerst-Bjeliš, B., & Cvitanović, M. (2016). Post-socialist transition as a driver of geographic marginalization. In S. Pelc & M. Koderman (Eds.), *2016 Conference of IGU commission C12.29 globalization, marginalization, and regional and local responses. Book of abstracts and program* (pp. 36–37). Koper: University of Primorska Press.

Hartke, W. (1953). Die soziale Differenzierung der Agrar-Landschaft im Rhein-Main-Gebiet. *Erdkunde, 7*(1), 11–27.

Horvat, Z. (2013). Using landsat satellite imagery to determine land use/land cover changes in međimurje county. *Croatia. Hrvatski geografski glasnik, 75*(2), 5–28.

Johnson, B. E. (2008). Nature, affordability, and privacy as motivations for exurban living. *Urban Geography,* 29(7), 705–723 (2008). http://doi.org/10.2747/0272–3638.29.7.705.

Leimgruber, W. (2004). *Between Global and Local. Marginality and Marginal Regions in the Context of Globalization and Deregulation*. Aldershot: Ashgate.

Lo, C. P., & Quattrochi, D. A. (2003). Land use and land cover change, urban heat island phenomenon, and health implications: A remote sensing approach. *Photogrammetric Engineering and Remote Sensing, 69,* 1053–1063.

Morita, H., Hoshino, S., Kagatsume, M., & Mizuno, K. (2007). An Application of the Land Use Change Model for the Japan Case Study Area. Interim report IR-07-065/November. International Institute for Applied System Analysis, Laxenburg, Austria. http://citeseerx.ist.psu.edu. Accessed 28 Feb 2013.

Njegač, D. (1995). Oblici prostorne mobilnosti stanovništva Hrvatskog zagorja. *Geografski glasnik, 57,* 93–109.

Pelc, S. (2016). Marginality and sustainability. In S. Pelc & M. Koderman (Eds.), *2016 conference of IGU commission C12.29 globalization, marginalization, and regional and local responses. book of abstracts and program* (pp. 37–38). Koper: University of Primorska Press.

Plaščak, E. (2009). Kretanje posjeda poljoprivrednog zemljišta na području naselja Osijeka od proglašenja slobodnog kraljevskog grada do danas. *Anali Zavoda za znanstveni i umjetnički rad u Osijeku, 25,* 11–22.

Reimann, C., Filzmoser, P., Garrett, R., & Dutter, R. (2008). *Statistical Data Analysis Explained.* Chichester: Wiley.

Rutherford, G. N., Guisan, A., & Zimmermann, N. E. (2007). Evaluating sampling strategies and logistic regression methods for modelling complex land cover changes. *Journal of Applied Ecology, 44,* 414–424.

Serneels, S., & Lambin, E. F. (2001). Proximate causes of land-use change in Narok District, Kenya: a spatial statistical model. *Agriculture, Ecosystems & Environment, 85,* 65–81.

Slavuj Borčić, L., Cvitanović, M., & Lukić, A. (2016). Cultivating alternative spaces—Zagreb's community gardens in transition: From socialist to post-socialist perspective. *Geoforum, 77,* 51–60.

Seferagić, D. (2006). The pyramidal network of settlements in Croatia. *Sociologija sela, 44*(2–3), 355–368.

Sorice, M. G., Kreuter, U. P., Wilcox, B. P., & Fox III, W. E. (2014). Changing landowners, changing ecosystem? Land-ownership motivations as drivers of land management practices. *Journal of Environmental Management*, 133, 144–152 (2014). http://doi.org/10.1016/j.jenvman.2013.11.029.

Spevec, D. (2009). Populacijski potencijal Krapinsko-zagorske županije. *Hrvatski geografski-glasnik, 71*(2), 43–63.

Vintar Mally, K., 2016. Regional disparities and the path to sustainability: the case of Slovenia. In S. Pelc & M. Koderman (Eds.), *2016 conference of IGU commission C12.29 globalization, marginalization, and regional and local responses. Book of abstracts and program* (pp. 27–28). Koper: University of Primorska Press.

Vodopivec, M. (1994). Apropriability of returns in the Yugoslav firm. *Eastern Economic Journal, 20*(3), 337–348.

Vresk, M. (1972). Socijalni ugar i drugi oblici napuštanja agrarne aktivnosti kao posljedica emigracije i socijalnog diferenciranja stanovništva. *Geografski glasnik, 33–34*(1), 79–89.

Wood, C. H. and Skole D., 1998. Linking satellite, census, and survey data to study deforestation in the Brazilian Amazon. In D. Liverman, E. F. Moran, R. R. Rindfuss R. R. & Stern P. C. (Eds), *People and pixels: Linking remote sensing and social science* (pp. 70–93). Washington: National Academy Press.

Zampella, R. A., Dow, C. L., Lathrop, R. G., Bognar, J. A., Bunnell, J. F., Laidig, K. J. & Procopio, N. A. (2007). *Using multiple regression to quantify the effect of land use on surface water quality and aquatic communities in the New Jersey Pinelands.* Final report submitted to the U. S. Environmental Protection Agency EPA Grant Agreement NO. CD992770-01-0.

Part VI
Conclusions

Chapter 22
(De)Marginalization, Development and Sustainability: Brief Reflection of the Variety of Findings and Views Presented

Stanko Pelc and Miha Koderman

22.1 Introduction

The chapters of this book cover such a diversity of topics that at first sight it seems that they could hardly have anything in common. The intention was to tie them together with a common focus on geographical marginalization and sustainability thinking. However, there are still large gaps between the flourishing tourist destination of Ibiza and the problems that resettled Orang Aslis in Malaysia have to deal with, or between land use in hilly hinterland of Maribor in Slovenia and the role of Bedouin museum for desired multiculturality in Israel. Nevertheless, life itself is diverse and to comprehend its meaning one needs to go through many different experiences. To get better insights into marginality, marginalization and what drives or reverses it, one needs to be acquainted with several case studies focusing on different topics and from different perspectives as well.

Marginality as a "common denominator" or the research approach to diverse issues that geographers pay attention to, because of their spatial relevance, may be engaged with for several reasons. It attaches geography to sociology and other social sciences since marginality can be defined as both a spatial and as a social conceptual framework. Further it enables one to interpret changes that occur in society and space considering the direction of the process towards greater or lesser marginality. It already encompasses spatial, economic, social, cultural, political and other dimensions—so it may be considered as a kind of umbrella conceptual

S. Pelc (✉)
Faculty of Education, University of Primorska, Cankarjeva 5,
6000 Koper, Slovenia
e-mail: stanko.pelc@pef.upr.si

M. Koderman
Faculty of Humanities, Department of Geography, University of Primorska,
Titov trg 5, 6000 Koper, Slovenia
e-mail: miha.koderman@fhs.upr.si

S. Pelc and M. Koderman (eds.), *Nature, Tourism and Ethnicity as Drivers of (De)Marginalization*, Perspectives on Geographical Marginality 3,
DOI 10.1007/978-3-319-59002-8_22

framework for the interpretation of complex processes. Similarly von Braun and Gatzweiler "use the marginality lens to understand and explore the links between poverty, exclusion and ecology" (von Braun and Gatzweiler 2014, p. vi). Whenever geographical research focuses on phenomena that are closely related to human-environment relations and which are strongly influenced by either socio-economic/political or natural conditions (or both) that are causing worsening or improving individuals/groups wellbeing by pushing him/it to/away from sources of "wealth, power and prestige", the conceptual framework of marginality can be engaged with.

As sustainability has a strong emphasis on human-nature relations at least when we are considering it from the perspective of economic development the first part of the book deals with nature, both from the perspective of marginalization of nature and from the perspective of nature as being a driver of marginalization and in this concluding chapter some basic thoughts about this issue are given.

The second part of the book offers only a few case studies that focus on very diverse regions, but it is exactly this diversity that allows us to see how wide the range of geographical marginality studies can be and this is one of the most important issues discussed below.

Ethnicity as a driver of marginalization links geographical research of marginality with sociology and ethnology. Leimgruber considers ethnic mixture and the presence of all kind of minorities in society as the diversity that should be recognized by society as valuable in a sense as bio-diversity is while Meir goes even a step further and claims that cultural and bio-diversity may be strongly linked as in case of Beduins in Negev from his case study. However, the minorities are often marginalized because of the fears that the majority nurtures because of the lack of the knowledge about the culture and the character of minorities they share the same living space with. Or, as in a case of Slovenian Roma minority, they even don't want to share the same living space, thus pushing them into "ghetto" settlements.

Finally, we link this themes together and present common conclusions from case studies that range from kinship based educational migration in the area of Southern Pacific to post socialist changes in a rural hinterland of Zagreb in the region Hrvatsko Zagorje located at Croatian-Slovenian border. Two other case studies are closer to the last mentioned as they are focused on regional development in Slovenia while the other two are dealing with agricultural practices that may be considered as potential drivers of demarginalization—organic tea production in Sri Lanka and shrimp farming in Bangladesh.

22.2 Nature 'Marginalizing' or 'Marginalized'

According to Pelc's stance (Chap. 2) nature is both marginalizing and marginalized. It is marginalized within the decision-making processes that are deliberately ignoring long term effects of pollution and resource depletion thus pushing nature (environment) towards the margin of the value system putting short term benefits of social elites in its center. However, there are also short term survival strategies that force poor marginalized people to cause harm to the nature thus 'cutting the branch they are sitting on'.

The effects of natural disasters in areas with land degradation because of overpopulation (exceeded carrying capacity) are much austerer or may even be caused by the degradation. Zorn (Chap. 4) presented a whole range of data supporting the hypothesis that marginality from the global scale plays an important role in consequences that natural disasters cause, especially in terms of fatalities and from relative points of view in terms of damage too. From the perspective of social class it is obvious that people from low income group, especially those below poverty line, are most gravely affected by whole range of natural disasters. They live in more endangered areas, because these are the only areas they can afford to live in, the housing is usually of low quality and cannot guarantee proper safety. Besides, local and national authorities are often too incompetent and too corrupt, not to mention their lack of finances, to establish any kind of early warning system. Those marginalized socially or/and spatially can expect to get less (adequate) help and will usually have to wait longer to be helped. Thus, it may be concluded that life in marginal areas and at the margin of society is life at a much greater risk of a grave damage and potentially fatal event caused by a natural disaster. The effects of natural events become disasters within social context and within that context marginality plays an important role.

Clearly different perspective of marginalization may be obtained from the case study of land-use changes in Svečinske gorice (Chap. 3). It is a border area but far from being spatially marginal as it is located in the hinterland of an important Slovenian regional center Maribor (20–30 min by car). However, the vicinity of Maribor may be one of the reasons for the reduction of cultivated areas. Land-use changes such as the overgrowing of arable areas and afforestation resulted in lower utilization of the natural potential. It was easier for farmers to augment their income by engaging in non-agricultural activities in nearby Maribor than to invest into their small farms and the development of highly productive vineyards although there are almost perfect natural conditions for that as shown by Žiberna's maps. It may be stated that marginalization in this case is manifesting in the fact that local population is pushed in relatively unfavourable socio-economic situation in which it is not possible to utilize advantageous natural conditions of the area it lives in.

The last case study from Northern Caucasus (Chap. 5) reveals marginality from both a global and a regional scale. The study area is a part of global periphery and also marginal within the regional context due to the fact that it is mountainous and as such went through a process of transformation. The result is the restoration of mountain meadow steppe and sub-alpine meadows. Besides, agricultural lands including agricultural terraces are overgrown by pine and elfin birch woods. The authors of the chapter mainly focused on the climate changes that also influence this transformation. The overall result is the new type of landscape that may offer new opportunities for the remaining population. However, the idea for natural parks and tourist development should be further researched to evaluate what are the chances for such development as it strongly depends on the socio-economic situation within wider regional context and political readiness for appropriate action as well.

22.3 Does Tourism Have the Potential to Reverse the Process of Marginalization

Tourism represents an economic sector that is usually perceived as one of the most important developmental opportunities for geographically marginalized areas and regions. Poorly developed local communities and their authorities strive to promote the natural and cultural phenomena in their areas, as they expect economic benefits from potential visitors and investors. They anticipate that the development of tourism can significantly contribute to the local development in general, as this sector is known for its multiplier effects which encourage the growth in the primary and secondary sectors as well. According to Rusu (2011, p. 71), the term "Multiplier" means "that the value of expenditure is multiplied by some estimated factor in order to determine the total economic impact." Multiplier effects are therefore often referred to as secondary effects of tourism spending and can take many forms, including spending on goods and services by tourists visiting the area, investment by external sources, government spending on infrastructure and export of goods stimulated by tourism. In this regard, tourism can have the potential to alleviate and, in the long term, slowly reverse the process of marginalization. However, tourism development must also be addressed from the perspective of sustainability, as this activity is closely connected with nature and specific cultural environments, where thoughtless actions and strictly profit-oriented investments without the cooperation and inclusion of the local community can have devastating and irreversible consequences.

The first one of the three case studies dealing with tourism paid attention to the perspectives of sustainable tourism development in the area of the Upper Soča Valley in north-western Slovenia (Chap. 6). This peripheral border area may be regarded as geometrically marginal, but it is much more appropriate to state that, in some regards, the region experiences a certain degree of marginalization that results in emigration and depopulation, abandonment of arable land, as well as an above average unemployment rates. Due to its location in the mountainous terrain of the Julian Alps, tourism here is currently mainly propelled by outdoor sport and hiking leisure activities and is thus characterized by seasonality. As the authors pointed out, there is a lot of room for improvement, especially from the aspect of sustainability. As almost 80% of the municipal area is included into the national park and considering the astonishing potential of the natural beauty, sustainable tourism is the only acceptable option. Authors propose the developmental concepts such as culinary, cultural and geo-tourism as possible niche activities in the fragile mountainous environment. Their adoption could play an important role in preventing/reversing the process of marginalization.

Tourism and its multiplier effects have largely contributed to the development of the Balearic island of Ibiza, world famous and exclusive tourist destination in the Mediterranean (Chap. 7). The island of Ibiza experienced several different phases of tourism development and lost its appeal and attractiveness in the last decades of the 20th century, when new tourist destinations started emerging worldwide. In that period, the island was, in a way, marginalized by stiff competition. The process of marginalization was reversed in 2000s, when private and public investors in tourism sector started building a new image of the island, based on the club culture phenomenon with Ibiza's nightlife as the main attraction. This gradually led to a transformation of the once mass-tourism holiday resort into an exclusive, luxury tourism destination, which is only accessible to those who can afford it. The luxury tourism development has alleviated the growing pressure of tourism activities in the sensitive coastal environment, as the number of tourist arrivals on the island is nowadays much smaller than previously in the time of popular mass tourism. This may well make a contribution to the sustainability and nature preservation of the area, but the long-term effects of the elite-oriented tourism should also be devoted appropriate concern. Implementation of similar spatial and social segregation policies has in certain cases significantly contributed to the increase of the multi-layered gap between the local communities and tourists. Elitism has always been closely connected with the process of marginalization and the case of tourism development of the island of Ibiza may once again be put to the test.

The third tourism case study is focused on the tourism development of Cuba, also an island, only much bigger than Ibiza and undoubtedly marginal from several aspects (Chap. 8). At global scale Cuba belongs to the group of 'developing countries of the Third World',[1] one of the few remaining with a socialist regime and suffering from decades long economic sanctions and political isolation put on them by their powerful neighbour, USA. On the other hand "paradise" is a suitable word to describe the island's tropical weather, numerous beaches and nature conservation areas. This is the reason for the fact that mass tourism is increasing on yearly basis. Due to the country's poor economy, their ambiguous goals for the improvement of tourism, infrastructure development will depend on extensive foreign investments that will significantly affect the natural environment and social structure of the population, thus threatening to create new forms of marginalization. Cuba may already be on its way to becoming one more country where centrally governed decisions (mostly regarding the privatization of state owned land) have subsequently led to the rise of neoliberal considerations and populism alliances. The lessons learned from similar situations in certain countries of Eastern Europe which have already completed the process of transition from communist regime to

[1]It should be emphasized that this group of countries is very diverse and Cuba according to Human development index is close to Serbia and above Turkey, Bosnia and Herzegovina and China as well. The World Bank already stopped using the distinction between developed and developing countries.

democracy, could be valuable in terms of preventing the marginalization of the changing Cuban society in post-Castro period.

To answer the question in the heading: yes, tourism has the power to reverse the process of marginalization as shown in the case of Ibiza, while the other two cases are rather pointing out what could potentially work to achieve this, rather than giving firm evidence of such development. However, even as far as Ibiza is concerned, we must not neglect the fact, that it was still far from being marginal, although it was caught in the process of marginalization that could eventually have pushed the island's economy all the way to the 'edge'. Therefore, it was actually the change in tourism offerings, the innovation, that put Ibiza back on the 'path of success' and not tourism as such.

22.4 Ethnicity Issues—The Central Topic of Marginalization in Society and in Space as Well

The remaining population of 'natives' living in reservations is probably one of most common associations affiliated with the term marginal. When thinking about marginality in the European context, one of the first thoughts that would cross one's mind would probably be: 'Roma'. Ethnic as well as other minorities are very often the subject of marginalization. The majority's feeling of endangerment by otherness that usually characterizes minorities is the major driving force of the marginalization of these groups with prejudice as the fuel. However, it should not be so. Minorities should be valued as an important wealth that society possesses and ethnic diversity should be valued the same way biodiversity already is (Chap. 9). The majority should not be afraid of the positive discrimination of minorities. Giving them the freedom to choose how much to conform to the mainstream lifestyle and being able to accept their otherness without fear is the first step towards their demarginalization. There are many more steps which need to be taken in order to achieve the full acceptance of all 'the other' and the recognition of the diversity that they bring as an 'asset' for society.

The case studies presenting the situation of ethnic groups in our book cover Orang Aslis from Malaysia (Chap. 10), Slovenian Roma (Chaps. 11 and 12) and Bedouins from Israel (Chaps. 13 and 14). The first case study reveals marginalization of indigenous population of Malay Peninsula. They were resettled as the government decided that their traditional way of life was no longer possible due to their remoteness, poverty and thus poor accessibility to health care. They were certainly observed as marginal from perspective of authorities and the mainstream majority in Malaysia. However, the decision taken from above, based on the assumptions and made by government officials that probably do not have the empathy to understand and feel how native people assess their situation and how they value their way of life and their material and mental situation, certainly is an act of marginalization of this ethnic group even though it was meant to be for their benefit. As the field work interviews showed, the resettlement more often caused further marginalization rather than the opposite. The results were better in the cases where the settlement was rebuilt at the same location where people lived before and the worst in the case where the native population has been settled into urban environment, getting new houses with very good housing conditions, but with a total lack of opportunities for income generation. Their former strategies were not appropriate for their new socio-economic context and most of them were unable to adapt to new circumstances. The 'noble' action of authorities with the intention of improving the life conditions in majority of individual cases resulted in a worsening of life situation and thus in even severer marginalization of the affected individuals. The case study clearly showed that not giving certain group the right to have a major influence on the decisions concerning their life is an act of marginalization, often putting some or even all of members of such a group into more unfavourable situation than before. To reduce marginalization the voice of marginalized minority groups should not be ignored.

The problem of settlement is also present in two cases from the other side of the world and in much different circumstances, but with some similarities as well. The ethnic group in these two cases are not a part of native population, but also used to live nomadic way of life and had to settle permanent in recent decades. The problems concerning this settlement are segregation (Roma living in separate Roma settlements), property issues (settlements built on the land that they do not own), lack of basic infrastructure (water, electricity, sewage), often but not always very poor housing, etc. The first presented case (Chap. 11) shows how Slovenia in two and a half decades of its sovereignty did not manage to solve the afore-mentioned problems at a legislative level in such a way that the local authorities would have the capacity and means to deal with Roma integration into local society. Even though the Roma minority is represented by a Roma counsellor in every municipality with a Roma population, their voice may easily be overheard as shown in the presented case, their problems remain unsolved and their situation uncertain. The interests of majority prevail and the Roma remain marginalized. The second case study revealed the plan that should be followed in order to legalize and integrate the Roma settlements into Slovenian settlement system (Chap. 12). While the previous case study focused primarily on an individual family thus presenting what the core problems are and how are they perceived by the affected members of the minority group, this case presents the overall situation and the ideas for action needed to overcome the problems. It has been acknowledged by the author that there have been some positive results already achieved in the field of reconstruction of certain Roma settlements, but he emphasized the fact that the reconstruction and equipping the settlements with appropriate infrastructure is just the first necessary condition for the integration of Roma settlements into Slovenian settlement system. However, to achieve the integration of Roma population into Slovenian society is a much more complex problem and will take much more time and effort on the side of the majority population and on the side of Roma as well, not to mention political 'playmakers' on local as well as on national level. Nevertheless, if the Roma population will remain to live in separate Roma settlements, no matter how well equipped and built, they will still be secluded, marginalized and the integration even more difficult.

The last two cases are from Israel and they both deal with the Bedouins. Their marginalization is analyzed in Chap. 14, where it is examined from the perspective of (bio-)cultural diversity. The situation is very much like the one in the case of Orang Asli ethnic group in Malaysia. The resettlement of Bedouin to reduce their marginality most probably means that they are being marginalized even more. However, this resettlement may also be regarded as a possible act of diminishing biodiversity. Because many tribes were detached from pastoral life and relocated to towns by the Israeli government, their traditional activity, grazing of livestock, consequently became unprofitable, but is still maintained on a smaller scale as an important cultural symbol. Apart for being one of the prime sources of income for Bedouins that are continuing to live in unrecognized settlements in the Negev, sheep and goat grazing in the desert represents an important ecological value in the preservation of its biodiversity. One could therefore claim that the marginalization (in terms of relocation and the resultant spatial segregation) of the Bedouins has as a side effect manifested on a completely unimagined level—in the marginalization of Negev's nature and biodiversity. This case clearly shows how thoughtless political decisions can have complex and far reaching consequences that can threaten natural ecosystems. The solution for the demarginalization of Negev Bedouins could be not only through recognition of their settlements, but also through recognition of their role in maintaining ecosystem services and biodiversity. Financial support would be beneficial and could probably follow the example of the EU and its Common Agricultural Policy aiming to achieve higher biodiversity and to maintain provision of ecosystem services.

The other chapter devoted to Bedouins (Chap. 13) is devoted to the topics of multiculturalism and ethnographic museums. It analyses the case of the Regional Bedouin Museum in the Negev desert. As institutions, museums are an important part of the country's educational system, as their primary functions are the conservation of objects of cultural, historical or artistic importance, and their presentation for the public. In Israel, museums face significant challenges in fulfilling the differing, and sometimes conflicting, needs of multiculturalism. Here, one of the main problems is related to the objective, non-biased presentation of non-Jewish cultural groups with predominantly Arab-Palestinian narratives, which often conflicts with the Jewish Zionist narrative. Similar issues can be found in many other countries where the culture and politics of the dominant majority often prevail over minorities or non-dominant ethnic groups. The latter are usually marginalized by the dominant groups in several ways, as already discussed in the several chapters of this book, however such subtle marginalization can also be found in the museums' exhibition collections and accompanying explanatory notes, as well as information boards. To avoid this aspect of marginalization, the society's mainstream must step away from its majority-centred and sometimes nationalistic policies and include the relevant representatives of the non-dominant ethnic groups with active participation in the management of the museums and its collections. In this way, the existence of multiple cultural traditions within a single country could be properly addressed, while promoting the maintenance of cultural diversity.

22.5 Migration, Regional Development, Post Socialist Transition and Niche Agricultural Activities in the Same 'Marginality Basket'

In the last part of the book a large diversity of case studies is presented. The starting point are the Pacific islands—undoubtedly marginal areas from a global perspective (Chap. 15). The authors come to some conclusions regarding the possibilities for demarginalization through education. The first one is that migration based education may be pushing island societies in both directions – towards as well as from the margin, depending on what kind of strategies are implemented. They also recommend that the best of tradition should be maintained while at the same time the progressive change should be allowed to foster demarginalization. Furthermore, they consider different kinds of aid from the metropolis that is fostering education focused on skills being mainly needed for the labour market in the metropolis as neo-colonial and thus being a driver of marginalization, draining the periphery of its best resources. The family, on the other hand, still plays the most important role in decisions taken considering who and how can get the resources available for the appropriate education. According to the authors the role of the authorities is to guarantee basic education at local and secondary at regional level. They do not approve scholarships for tertiary education as they bring more benefits to metropolitan educational institutions (Australia, New Zealand) than to students from periphery (Pacific Islands). More effective, according to them, would be to foster the diaspora and let it grow by maintaining pathways and reducing border controls as much as possible. Close ties between diaspora and their families at home can be a powerful driver of demarginalization through networks of education and migration.

The other two case studies from 'Third World' (Sri Lanka and Bangladesh) revealed how certain agricultural activities may benefit some marginal (local) societies and improve their economic as well as social situation thus pushing them out of poverty and away from the margin of wider society. Although the focus in both cases is on the economic activities that are presented, the final emphasis is on the importance of these sustainable agricultural practices in the demarginalizing process. However, the case study about tea estates (Chap. 19) shows how complex is the situation that economies of the developing countries have to cope with in the context of globalization and worldwide competition. This is the reason for the marginalization of traditional tea production in Sri Lanka. Being the first to start with organic production pushed the process into the opposite direction, but still without any guarantee that organic tea producers can be competitive enough to maintain much more sustainable, but also costlier organic production. The case study from Bangladesh reveals how proper pond management with the use of sustainable practices of growing prawns can change the wellbeing of farmers in a marginal region (Chap. 20). It is exactly the use of sustainable methods of farming that makes the difference in this case and the author's stance is that this could be an example of good practice that could help farmers in other regions and other countries to change their unsustainable practices and escape the dependence on multinational companies that, in combination with low survival rates of their prawns keep them in marginality and poverty.

The remaining four case studies are from the Central European context (Slovenia, Croatia, Czechia) which from global scale are far from being marginal. Even from a wider regional scale (EU) these three countries could hardly be defined as marginal. However, at a micro-regional scale we may notice that there are processes that may be driving certain regions towards the margin when observed within the national context.

One of the Slovenian cases is showing how the reforms of local government system influenced the local situation through the dispersion of services and enhancement of the polycentric system (Chap. 17). Large numbers of new municipal centres managed to get new services thus improving living conditions for the population of peripheral areas far from higher rank regional centres. Furthermore, the process of devolvement of government at the local level to numerous small municipalities did contribute to higher levels of influence of local population on the decisions taken that directly influence their every day's life. At present, according to the author's stance the analysed process and the observed changes have had a more demarginalizing than marginalizing effects. However, small peripheral municipalities are economically weak and often also lack human capital, and it may therefore be questionable, how long can the described situation last.

The other case study from Slovenia is about regional disparities and sustainable development (Chap. 18). The observed development is not yet sustainable and developmental differences between regions are deep rooted and it will be hard to overcome them. That might be a potential danger for these regions to be caught by a process of marginalization instead of the opposite.

The last two case studies both examine the post-socialist transition. The one from Czechia focused on the process of the restitution of Church property. This remained unsettled for more than two decades thus blocking an important part of the land suitable for local development and therefore causing negative developmental effect in those administrative units that had an important share of such land. The situation may be defined as marginalization of those local units that could have develop, but were blocked by unsettled property issues between the state and the Church.

Finally, the Croatian case study examined the post socialist changes in a border region with Slovenia north-east from Zagreb (Chap. 21). Hrvatsko Zagorje is a traditional underdeveloped peripheral hilly region within wider commuting range of Zagreb, the capital of Croatia (about one hour car drive). The differences in land use changes and the overall economic, demographic etc. situation in the region is strongly related to accessibility to regional centres and especially to Zagreb. Higher and less accessible hilly areas are showing more characteristics that can be defined as the effects of marginalization such as the return to self-subsistent agriculture as a livelihood strategy by the older population or combined involvement in agricultural and non-agricultural activities (commuters—part time farmers). The transition from socialism to capitalism, according to the authors only accelerated the observed changes. Even in the heart of Europe it is difficult to avoid marginalization of less accessible areas within the context of 'market economy' and globalization.

22.6 Concluding Remark

As mentioned above marginality is a complex phenomenon and very diversely understood. Marginalization on the other hand, when defined as a process in which something or somebody is being pushed into a less favourable position and if this position is closer to the margin than to the average, can be traced and observed almost everywhere, not just at the extreme periphery. Driving forces of these changes can be defined as drivers of marginalization. In search of an understanding of these issues we have examined what factors play an active role in marginalization. Knowing and understanding is the key for appropriate intervention aimed at reversing the process of marginalization, enabling it to become a process of demarginalization. Case studies presented in this book reveal just selected small bits of a complex mosaic of marginality and marginalization. Much more still remains to be 'unveiled'.

References

Rusu, Sergiu. (2011). Tourism multiplier effect. *Journal of Economics and Business Research, XVII*(1), 70–76.

von Braun, J. & Gatzweiler, F. W. (Eds.). (2014). *Marginality: Addressing the Nexus of Poverty, Exclusion and Ecology*. Dordrecht, Heidelberg, New York, London: Springer. http://franzgatzweiler.org/wp-content/uploads/2013/10/Nexus-of-Poverty.pdf.

The manufacturer's authorised representative in the EU is Springer Nature Customer Service Centre GmbH, Europaplatz 3, 69115 Heidelberg, Germany. If you have any concerns regarding our products, please contact ProductSafety@springernature.com

Printed and bound by CPI Group (UK) Ltd, Croydon, CR0 4YY
01/10/2026
02225871-0006

Evelyn Kozak

Zur Notwendigkeit eines arbeitsrechtlichen und haftungsrechtlichen Whistleblowerschutzes

PL ACADEMIC RESEARCH

Bibliografische Information der Deutschen Nationalbibliothek
Die Deutsche Nationalbibliothek verzeichnet diese Publikation in der Deutschen Nationalbibliografie; detaillierte bibliografische Daten sind im Internet über http://dnb.d-nb.de abrufbar.

Zugl.: Berlin, Freie Univ., Diss., 2015

D 188
ISSN 2363-684X
ISBN 978-3-631-67207-5 (Print)
E-ISBN 978-3-653-06708-8 (E-Book)
DOI 10.3726/978-3-653-06708-8

PL Academic Research ist ein Imprint der Peter Lang GmbH.

Peter Lang – Frankfurt am Main · Bern · Bruxelles · New York · Oxford · Warszawa · Wien

Diese Publikation wurde begutachtet.

www.peterlang.com

Danksagung

Großer Dank gebührt meinem Doktorvater em. Univ.-Prof. Dr. iur. Dr. rer. pol. Dres. h.c. Franz Jürgen Säcker, der mir mit seinem Fachwissen zur Seite stand, viele seiner freien Stunden für mich opferte und mir nicht zuletzt auch durch private Gespräche zu einem wertvollen und unermüdlichen Wegbegleiter wurde. Prof. Säcker brachte mir viel Geduld entgegen und sorgte mit wertvollen Ratschlägen für das Gelingen der Arbeit. Er gab mir mit Seinem fundierten Fachwissen viele Anregungen für meine wissenschaftliche Arbeit.

Ich danke Univ.-Prof. Dr. Gerhard Seher für die Übernahme des Zweitgutachtens und die Auseinandersetzung mit dem gebietsübergreifenden Thema.

Ein ganz besonderer Dank geht an meine Familie, insbesondere an meine Eltern, die mir das Studium der Rechtswissenschaft ermöglichten und mir auch während der Anfertigung der Doktorarbeit immerzu unterstützend zur Seite standen. Ganz besonderen Dank dafür, dass sie immer das Beste für meine Schwester und mich tun und alles Erdenkliche bereit sind, dafür zu geben.

Ich möchte auch meiner Schwester Isabella dafür danken, dass sie mich immer motiviert hat und auch in schwierigen Stunden den Optimismus in mir geweckt hat. Die schwesterliche Verbundenheit bedeutet mir unbeschreiblich viel. Der Dank gilt auch ihrem Ehemann und meinem Schwager Daniel, weil er auch immer an mich geglaubt hat.

Allen meinen lieben Freunden danke ich für die Ausdauer, das Verständnis und den Rat. Ich möchte mich bei allen bedanken, die mich nicht nur tatkräftig unterstützt haben, sondern mich stets aufbauten und für die erforderliche Abwechslung sorgten.

Inhaltsverzeichnis

Teil 1: Einleitung

A. Hintergründe

Sowohl die Demokratie als auch der Rechtsstaat sind grundlegende Prinzipien, die die Ordnung und Einhaltung der Gesetze regeln, die der Gesetzgeber als Rechtsordnung vorgibt. Den Bürger trifft die Pflicht, sich gesetzeskonform zu verhalten und sein Handeln an der Rechtsordnung zu orientieren. Begeht er einen Gesetzesverstoß, muss er mit rechtlichen Konsequenzen rechnen.

Auch Unternehmen haben sich an Recht und Gesetz zu halten. Die Einhaltung der durch die Rechtsordnung geschützten und verankerten Rechte der einzelnen Bürger gehört zum Pflichtenkanon eines jeden Unternehmens. Wird gegen Recht und Gesetz verstoßen, drohen dem Rechtsverletzer Strafen und Sanktionen.

Was aber gilt für die Personen, die diese Gesetzesverstöße entdecken und dagegen vorgehen wollen?

Darf es in einem demokratischen Rechtsstaat Regelungen geben, die denjenigen sanktionieren, der Rechtsverstöße im Unternehmen aufdeckt und offenbart?

Die Frage wird in letzter Zeit häufig gestellt und geht mit der Überlegung einher, dass es schwer begründbar ist, einen Bürger arbeitsrechtlichen, strafrechtlichen/ordnungswidrigkeitenrechtlichen und haftungsrechtlichen Sanktionen auszusetzen, obwohl er sich mit der Offenbarung in den Dienst der Rechtsordnung gestellt hat.

Schutzwürdig sollte in einem demokratischen Verfassungsstaat die unternehmerische Verhaltensweise nur sein, wenn das Unternehmensgeheimnis mit dem geltenden Recht im Einklang steht. Das Recht darf nicht das Unrecht schützen. Die derzeitige Realität sieht jedoch anders aus. Weil es der Gesetzgeber von Anfang an unterlassen hat, den von ihm in das einfache Gesetzesrecht eingeführten Begriff des Betriebs- und Geschäftsgeheimnisses zu definieren, muss er von Rechtsprechung und Literatur näher konkretisiert werden. Die Rechtsprechungspraxis, insbesondere zu Whistleblower-Aktivitäten, zeigt, dass ohne Bedenken davon ausgegangen wird, das berichtenswerte Verhalten – mithin das gesetzeswidrige Verhalten – falle unter den Geheimnisschutz und sei durch arbeitsrechtliche Vorschriften und Gepflogenheiten sowie von Art. 12 GG und § 17 UWG geschützt.

In der Presse ist häufig über schwerwiegende Fehlverhalten und grobe Missstände in Unternehmen oder Organisationen zu lesen. Der Berichterstattung kann entnommen werden, dass eine innerbetriebliche Behebung des Fehlverhaltens nicht erfolgversprechend war – meistens liegt das daran, dass der Arbeitgeber selbst in die (illegalen) Machenschaften involviert war oder ist. Viele fragen sich in diesen Fällen, wie es dazu kommen konnte, und vor allem, warum nicht schon früher jemand dagegen vorgegangen ist. Aber wer will handeln, wenn er mit arbeitsrechtlichen, strafrechtlichen oder sogar haftungsrechtlichen Konsequenzen rechnen muss? Die

Bekanntmachung von Missständen bzw. Gesetzesverstößen im Rahmen eines Arbeitsverhältnisses wird als *Whistleblowing* bezeichnet.

Ein Whistleblower, der Kartellrechtswidrigkeiten aufdeckt, wird *Kronzeuge* genannt. Ausgangspunkt dafür ist ein Kartellverstoß. Der Kronzeuge meldet maßgebliche Kartelle und Verstöße gegen das Wettbewerbsrecht mit dem Ziel, einer Geldbuße der Wettbewerbsbehörden zu entkommen. Zudem bedient er sich Regelungen, die ihm Handlungsanweisungen vorgeben. Kronzeugenregelungen im Kartellrecht sind ein Werkzeug zur Ermittlung von Kartellen und zur Durchsetzung des Kartellverbots, wobei der Kronzeuge zur Erlangung der Straffreiheit bzw. Strafmilderung mit der Kartellbehörde „zusammenarbeiten" und somit zur Aufklärung des Sachverhalts beitragen muss. Die Besonderheit des Kronzeugen als Whistleblower ist, dass er oder zumindest sein Rechtsvorgänger im Unternehmen in das Geschehen involviert war und somit als Täter oder Mittäter zum Zeitpunkt der Tat zu qualifizieren ist. In dem Moment aber, in dem der Kronzeuge sich entschließt, das gesetzeswidrige Verhalten für sich zu beenden und offenzulegen, steht er in keiner rechtlich missbilligten Beziehung mehr zu den Missständen im Unternehmen.

Sowohl dem zunächst involvierten Whistleblower als auch dem unbeteiligten Whistleblower ist gemein, dass sie über besondere Informationen verfügen und bei Veröffentlichung dieser Betriebsinterna mit Repressalien rechnen müssen.

In letzter Zeit wird der Ruf nach rechtlichem Schutz für diese Personen lauter. Demjenigen, der ein Verhalten beenden will, dass gegen geltende Gesetze verstößt, um den Einklang der Rechtsordnung wiederherzustellen, sollte Immunität oder zumindest Amnestie zukommen.

B. Begriff und Bedeutung des Whistleblowings

Eine einheitliche, allgemein anerkannte Definition des Begriffs Whistleblowing hat sich im deutschen Rechtsraum bis jetzt nicht durchgesetzt. In vielen wissenschaftlichen Schriften wird aber folgende Definition bevorzugt: So spricht man häufig dann von Whistleblowing, wenn ein ehemaliges oder aktuelles Mitglied einer Organisation Kenntnis von illegalen, unmoralischen oder illegitimen Verhaltensweisen hat, die im Verantwortungsbereich der „Organisationsführung" liegen, und dieses Fehlverhalten (mangels Möglichkeit, es selbst zu beheben) gegenüber Personen oder Organisationen aufdeckt, die Handlungsmöglichkeiten besitzen.[1] Im weitesten Sinne sind darunter auch kritische Äußerungen, Beschwerden oder Anzeigen von abhängig Beschäftigten über Missstände oder Fehlverhalten in ihrem Unternehmen bzw.

---

1 *Miceli/Near*, Journal of Business Ethics, 1985, Vol. 4, Issue 1, S.4, „Whistleblowing is the disclosure by organization members (former or current) of illegal, immoral or illegitimate practices under the control of their employers, to persons or organizations that may be able to effect action."; *Briegel*, Whistleblowing-Systeme, S. 14 f; *Schneider*, Implementierung, S. 59; *Schulz*, Ethikrichtlinien, S. 142; *Schulz*, BB 2011, 629, 630; *Schemmel/Ruhmannseder/Witzigmann*, Hinweisgebersysteme, S. 3; *Donato*, Whistleblowing, S. 43.

gegenüber der zuständigen Behörde, staatlichen Stellen, der Presse oder sonstigen Dritten zu verstehen.[2] Der Begriff hat seinen Ursprung im angloamerikanischen Rechtskreis und leitet sich aus dem englischen "*to blow the whistle on somebody*" ab.[3] Dies bedeutet so viel wie „*etwas enthüllen*", „*jemanden verpfeifen*" oder „*über jemanden auspacken*".[4]

Von dieser Definition sind jedoch moralische Verhaltensweisen auszunehmen. Zwar ist ein unmoralisches Verhalten in gewissen Branchen sehr bedenklich, jedoch würde dessen Einbeziehung den Handlungsbereich des Whistleblowers zu sehr ausweiten. Whistleblower sind Diener der Rechtsordnung und keine Moral- oder Sittenwächter. Sie stellen sich in den objektiven Dienst der Rechtsordnung und sehen sich in der Pflicht, rechtswidrige Verhaltensweisen anzuzeigen (intern oder extern) und zukünftiges rechtswidriges Verhalten zu unterbinden. Ein Whistleblower ist somit jede Person, die zum Wohle der Allgemeinheit Rechtsverstöße bzw. rechtswidrige Verhaltensweisen (Missstände), die im Verantwortungsbereich der Unternehmensleitung liegen, aufdeckt und öffentlich macht. Um das positive Element verstärkt zum Ausdruck zu bringen, wird im deutschen Rechtsraum diese Person auch als Hinweisgeber qualifiziert.

Whistleblower dringen auf Abhilfe und verweigern unter Umständen auch ihre weitere Mitwirkung an der kritisierten Praxis. Haben sie damit keinen Erfolg, tragen sie gegebenenfalls ihre Kritik nach außen in die Öffentlichkeit.

Anlässlich der Verleihung des ersten Whistleblower-Preises[5] im Jahre 1999 an den russischen Ex-Marine-Kapitän und Wissenschaftler *Alexander Nikitin* aus St. Petersburg sagte der damalige Richter am Bundesverfassungsgericht Dr. Jürgen Kühling[6]:

> *„Auch das gesellschaftliche Umfeld des ‚Whistleblowers' ist gewöhnlich nicht auf seiner Seite. Sein Verhalten wird als Verrat eingestuft, gilt als illoyal. Ein tief verwurzeltes Ethos der Gefolgschaftstreue überlagert die Grundsätze einer aufgeklärten Ethik, die sein Verhalten gutheißt. Zustimmung erfährt er, wenn überhaupt, gewöhnlich von weither. Von Freunden gemieden, vom Recht verfolgt – das ist das gewöhnliche Schicksal dessen, der sich im Interesse von Frieden, Umwelt oder anderen höchstrangigen Rechtsgütern zum Bruch der Verschwiegenheit entschließt."*

Je nach Standpunkt des Betrachters mag man derartige Anzeigen als „Denunziantentum" ablehnen oder als Ausdruck von Verantwortungsgefühl preisen. In den USA

2 *Bussmann/Matschke*, CCZ 2009, 132, 136; *Schulz*, BB 2011, 629, 630.
3 Vgl. vertiefend *Fahrig*, Verhaltenskodex, S. 28; *Schemmel/Ruhmannseder/Witzigmann*, Hinweisgebersysteme, S. 5.
4 Vgl. Online-Wörterbuch, abrufbar unter: www.ponds.eu, (abgerufen am 20.10.2013).
5 Der Preis wird von der Vereinigung Deutscher Wissenschaftler (VDW) und der Deutschen Sektion der International Association of Lawyers Against Nuclear Arms (IALANA) gestiftet.
6 Zitat von Jürgen Kühling, aus: *Deiseroth/Falter*, Whistleblower in Altenpflege und Infektionsforschung, Vorbemerkung.

(und in vielen anderen *Common Law*-Staaten) sind Whistleblower gesellschaftlich allgemein anerkannt und werden rechtlich geschützt.[7] Aber auch in Deutschland ist dieses Thema zunehmend aktuell und Gegenstand gerichtlicher Entscheidungen. Im Übrigen existieren private Stiftungen und Einrichtungen, die solch ein von Zivilcourage getragenes Verhalten mit Preisen ehren.[8] Diese Sichtweise beruht auf der Erkenntnis, dass sich mangelnde Sicherheitsvorkehrungen oder finanzielle Missstände – um nur einige Problemfelder zu nennen – durch den Whistleblower am ehesten aufdecken lassen, da er als „Insider" über die entsprechenden Kenntnisse verfügt.[9]

I. Erscheinungsformen des Whistleblowings

Anders als in den USA, in denen der Begriff Whistleblowing durch gesetzliche Bestimmungen, insbesondere den *Whistleblower Protection Act*,[10] den *Sarbanes-Oxley Act*[11] und den *Dodd-Frank-Act*[12] geregelt wird,[13] existieren im nationalen Recht bisher keine entsprechenden Bestimmungen, die sich mit der Materie des Whistleblowings auseinandersetzen.[14] Meist fehlt bereits eine klare Abgrenzung zwischen den verschiedenen Erscheinungsformen des Whistleblowings. Die Literatur unterscheidet zwischen externem und internem Whistleblowing, was erhebliche Auswirkungen auf den Schutz des Hinweisgebers hat.[15] Denn je nachdem, welche der beiden Formen der Hinweisgeber nutzt, ergeben sich unterschiedliche Voraussetzungen und Folgen hinsichtlich eines arbeitsgerichtlichen Verfahrens und des

7 Vgl. *Müller*, NZA 2002, 424, 425. Schutz erfährt der Whistleblower unter anderem durch den *Sarbanes-Oxley Act*, der nach dem *Whistleblower Protection Act* von 1989 die wichtigste bundesrechtliche Erweiterung zum Whistleblowingrecht darstellt.

8 U. a. Vereinigung Deutscher Wissenschaftler e.V., der regelmäßig Whistleblower-Preise verleiht.

9 *Müller*, NZA 2002, 424, 425; z. B. die Challenger-Katastrophe: Am 28. Januar 1986, 73 Sekunden nach dem Start der Mission STS-51-L, brach die Raumfähre in etwa 15 Kilometer Höhe auseinander. Dabei starben alle sieben Astronauten. Als Grund wurde das Versagen eines oder mehrerer Dichtungsringe in einer der seitlichen Feststoffraketen ermittelt. Zuvor hatte ein Mitarbeiter intern auf die Möglichkeit des Versagens der Dichtungsringe bei Kälte hingewiesen und vor diesem Szenario gewarnt.

10 Whistleblower Protection Act of 1989, Public Law 101–12 – April 10, 1989, abrufbar unter: http://www.gpo.gov/fdsys/pkg/STATUTE-103/pdf/STATUTE-103-Pg16.pdf, (abgerufen am 22.11.2013).

11 Sarbanes-Oxley Act („SOX"), Public Law 107–204 – July 30, 2002, abrufbar unter: https://www.sec.gov/about/laws/soa2002.pdf, (abgerufen am 22.11.2013).

12 Dodd–Frank Wall Street Reform and Consumer Protection Act („Dodd-Frank Act"), Public Law 111–203 – July 21, 2010, abrufbar unter: http://www.gpo.gov/fdsys/pkg/PLAW-111publ203/pdf/PLAW-111publ203.pdf, (abgerufen am 22.11.2013).

13 Siehe auch *Mengel*, CCZ 2012, 146, 152.

14 *V. Busekist/Fahrig*, BB 2013, 119, 120.

15 U. a. *Fahrig*, NJOZ 2010, 975, 976.

arbeitsrechtlichen Schutzes. Wendet sich der Arbeitnehmer an seinen Arbeitgeber, an eine betriebsinterne Beschwerdestelle oder an eine dafür ausdrücklich eingerichtete *Whistleblower-Hotline*,[16] handelt es sich um internes Whistleblowing.[17] Hierdurch wird Mitarbeitern die Möglichkeit gegeben, diskret und vertraulich Fehlverhalten zu melden und dadurch innerbetrieblich Abhilfemaßnahmen zu erreichen. Auf der anderen Seite erhält der Arbeitgeber durch einen intern offengelegten Hinweis die Chance, die Angelegenheit schneller, kostengünstiger und ohne Aufsehen der Öffentlichkeit zu korrigieren.[18] Hilfreich ist daher der Verweis an einen *Compliance-Beauftragten*,[19] aber auch eine Telefonschaltung zu einem externen Rechtsanwalt mag sich anbieten, wenn anzunehmen ist, dass Hinweisgeber diesem wegen dessen beruflicher Verschwiegenheitsverpflichtung mehr Vertrauen entgegenbringen.[20] Logischerweise setzt dies jedoch ein integriertes und funktionierendes *Compliance-System* im Unternehmen und vor allem die gute Organisation und Durchführung eines solchen Systems durch einen Compliance-Beauftragten voraus.[21]

Wird im Unternehmen nicht oder nicht in angemessener Form auf das angezeigte Fehlverhalten reagiert oder sind die Verantwortlichen nicht willens zu handeln, bleibt dem Whistleblower nur der externe Weg.

Von externem Whistleblowing spricht man, wenn sich der Hinweisgeber nach außen, das heißt an staatliche Behörden, Interessenverbände oder unmittelbar an die öffentliche Presse wendet.[22] Oft wird darin der falsche Weg gesehen, weil durch die öffentliche Preisgabe die Schäden materieller und immaterieller Art (negative Schlagzeilen, Rufschädigung) für das Unternehmen unkalkulierbar sind und die Funktionsunfähigkeit des Betriebs droht.[23] In diesen Fällen entwickelt sich ein Spannungsverhältnis zwischen Zivilcourage und Denunziantentum. Während Aufsichtsbehörden in der Regel eine Aufklärung unter Ausschluss der Öffentlichkeit anstreben, geht es der Presse vorrangig um die öffentliche Aufmerksamkeit. Deswegen wird der Gang an die Presse als Form des externen Whistleblowings am kritischsten gesehen.

---

16 *Mengel*, Compliance und Arbeitsrecht, Kap. 5, Rn. 66; *Lampert*, in: Hauschka, Corporate Compliance, § 9, Rn. 35.

17 *Fahrig*, NJOZ 2010, 975, 976.

18 *Groneberg*, Whistleblowing, S. 36.

19 Zu den Aufgaben und zur Stellung eines Compliance-Beauftragten im Unternehmen *Meier*, NZA 2011, 779.

20 *Lampert*, in: Hauschka, Corporate Compliance, § 9, Rn. 35.

21 *Lampert*, in: Hauschka, Corporate Compliance, § 9, Rn. 34.

22 *Schmolke*, RIW 2012, 224, 226.

23 *Donato*, Whistleblowing, S. 14.

II. Whistleblowing im öffentlichen Sektor

Aufgrund des thematischen Zusammenhangs soll in gebotener Kürze die Rechtslage zum Whistleblowing im öffentlichen Sektor aufgezeigt werden. Eine vertiefte Auseinandersetzung findet nicht statt, da für Whistleblowing von Angehörigen des öffentlichen Dienstes ein anderer Maßstab anzuwenden ist. Grund dafür ist das besondere Vertrauensverhältnis. Dabei sind einige Pflichten, die dem Kern der hergebrachten Grundsätze des Berufsbeamtentums angehören, besonders relevant: die allgemeine Treuepflicht, welche die Verfassungstreuepflicht, die Gehorsamspflicht sowie die Neutralitätspflicht beinhaltet, das Dienstweggebot, die Verschwiegenheitspflicht sowie die Pflicht zu achtungs- und vertrauenswürdigem Verhalten. Deshalb ist ein Vergleich mit einem Whistleblower in der Privatwirtschaft nicht angebracht.

Durch das Urteil des EGMR[24] im Fall der Altenpflegerin *Brigitte Heinisch* ist die Debatte über die Zulässigkeit von Whistleblowing außerhalb des rein privatwirtschaftlichen Bereiches angeheizt worden. Der Fall betraf die Kündigung einer Arbeitnehmerin, deren beklagter Arbeitgeber im Mehrheitsbesitz des Landes Berlin stand.

Im öffentlichen Sektor sind Angestellte des öffentlichen Dienstes und Beamte zu unterscheiden. Beide Bereiche sind solche des öffentlichen Dienstes, jedoch gelten unterschiedliche Gesetze und Maßstäbe.[25] Für Angestellte des öffentlichen Dienstes ist die Verschwiegenheitspflicht in § 3 Abs. 1 TVöD/§ 3 Abs. 2 TV-L tarifvertraglich geregelt: Verschwiegenheit ist zu wahren über Angelegenheiten, deren Geheimhaltung gesetzlich vorgesehen oder vom Arbeitgeber angeordnet worden ist. An die Verschwiegenheitspflicht eines Beschäftigten des öffentlichen Dienstes sind höhere Maßstäbe anzulegen als bei Beschäftigten der Privatwirtschaft; insbesondere sind die dem Staat zur Verfügung gestellten Daten und Erkenntnisse wegen des Datenschutzes zugunsten des Bürgers zu schützen.[26] § 3 Abs. 2 TV-L statuiert sogar eine gesetzliche Verschwiegenheitspflicht über die Beendigung des Arbeitsverhältnisses hinaus, was die besondere Bedeutung dieser Pflicht hervorhebt. Zwar besteht im privatrechtlichen Arbeitsrecht die Möglichkeit, eine (nachvertragliche) Verschwiegenheitspflicht vertraglich zu vereinbaren. Jedoch ist dann die Grenze des billigen Ermessens nach § 315 BGB zu beachten, sowie eventuell eine AGB-Prüfung vorzunehmen.

Für Beamte ergibt sich die Rechtslage aus dem Bundesbeamtengesetz oder dem Beamtenstatusgesetz. Beamte stehen aufgrund von Art. 33 Abs. 5 GG (hergebrachte Grundsätze des Berufsbeamtentums) in einem besonderen Pflichten- und Treueverhältnis, das ihnen wesentlich engere Grenzen im Hinblick auf ihr Verhalten gegenüber ihrem Dienstherrn vorgibt, als dies bei den übrigen Arbeitnehmern der

---

24 EGMR, Urt. v. 21.7.2011 – 28274/08, NJW 2011, 3501 (Heinisch/Deutschland).

25 *Brock*, öAT 2011, 243, 244; zu Verhaltenspflichten im öffentlichen Sektor *Groneberg*, Whistleblowing, S. 210.

26 *Stier*, in: Bepler/Böhle/Meerkamp/Russ, TVöD, § 3, Rn. 3.

Fall ist.[27] Hinsichtlich des Whistleblowings weisen bestimmte Regelungen auf besondere Dienstpflichten hin, z. B.:

- öffentlich-rechtliches Treueverhältnis gem. § 4 BBG/§ 3 BeamtStG,
- Eintreten für die freiheitlich demokratische Grundordnung gem. § 60 BBG/§ 33 BeamtStG,
- achtungs- und vertrauensvolles Verhalten gem. § 61 BBG/§ 34 BeamtStG,
- Einhaltung des Dienstweges gem. § 125 BBG/§ 36 BeamtStG und
- die Verschwiegenheitspflicht gem. § 67 BBG/§ 37 BeamtStG.

Neben der gesetzlich begründeten Pflicht, Straftaten anzuzeigen (§ 138 StGB) und für den Erhalt der freiheitlich demokratischen Grundordnung einzutreten, sind § 67 Abs. 2 BBG bzw. § 37 Abs. 2 BeamtStG Spezialregelungen für die Offenlegung von Korruptionsdelikten durch Beamte. Zur Umsetzung von Art. 9 des Zivilrechtsübereinkommens über Korruption des Europarats vom 4. November 1999[28] wurde mit § 67 Abs. 2 S. 1 Nr. 3 BBG bzw. § 37 Abs. 2 S. 1 Nr. 3 BeamtStG für externes Whistleblowing eine Ausnahme von der Verschwiegenheitspflicht geschaffen.[29] Die Pflicht zur Amtsverschwiegenheit gilt demnach nicht, wenn Beamte einen durch Tatsachen begründeten Verdacht einer Korruptionsstraftat nach den §§ 331 bis 337 StGB gegenüber einer zuständigen Stelle anzeigen. Die Vorschriften legen nahe, dass dabei sowohl interne (zuständige oberste Behörde) als auch externe Instanzen (Strafverfolgungsbehörden) angerufen werden können. Der Beamte ist daher grundsätzlich frei in der Wahl der Hinweisstelle. Zudem ist seine Motivlage irrelevant, da in § 67 Abs. 2 S. 1 Nr. 3 BBG bzw. § 37 Abs. 2 S. 1 Nr. 3 BeamtStG keine Gründe für die Offenlegung thematisiert werden.[30] Damit ist der deutsche Gesetzgeber einen Schritt weiter gegangen als Art. 9 des Zivilrechtsübereinkommens des Europarats über Korruption, der nur auf den in redlicher Absicht handelnden Beschäftigten abstellt.[31]

Eine weitere beamtenspezifische Regelung, die Bezüge zum Whistleblowing aufweist, ist das Remonstrationsverfahren nach § 63 Abs. 2 BBG bzw. § 36 Abs. 2 BeamtStG. Ausgangspunkt ist das Spannungsverhältnis, das dadurch entstehen kann, dass Beamte auf der einen Seite Anordnungen von Vorgesetzten ausführen müssen (Gehorsamspflicht) und auf der anderen Seite für die Rechtmäßigkeit ihrer

---

27 BVerfG, Beschl. v. 22.5.1975 – 2 BvL 13/73, JZ 1975, 561; *Müller*, NZA 2002, 424, 426.

28 Zivilrechtsübereinkommens des Europarats über Korruption, GRECO (99), vom 4. November 1999 (Straßburg), abrufbar unter: http://www.coe.int/t/dghl/monitoring/greco/documents/1999/Greco%281999%291_EN.pdf, (abgerufen am 5.3.2014).

29 Die Vorschrift stellt zudem sicher, dass Beamte die einen Korruptionsverdacht mitteilen, keinerlei Nachteile zu befürchten haben, siehe *Abraham*, ZRP 2012, 11, 12.

30 *Herold*, ZBR 2013, 8, 13; vgl. auch *Király*, DÖV 2010, 894, 896.

31 Vgl. Art. 9 des Zivilrechtsübereinkommens: „Jede Vertragspartei sieht in ihrem innerstaatlichen Recht vor, dass Beschäftigte, die den zuständigen Personen oder Behörden in redlicher Absicht einen begründeten Korruptionsverdacht mitteilen, angemessen vor ungerechtfertigten Nachteilen geschützt werden."

dienstlichen Handlungen die volle persönliche Verantwortung tragen. Das Remonstrationsverfahren sieht vor, dass Beamte Bedenken gegen die Rechtmäßigkeit von Anordnungen gegenüber ihren Vorgesetzten geltend machen.[32] Wird die Anordnung aufrechterhalten und hilft auch die nächsthöhere Stelle nicht ab, müssen Beamte diese trotzdem ausführen, sind aber von der eigenen Verantwortung befreit.[33] Da bei einer Remonstration allerdings regelmäßig nichts „enthüllt" wird, sondern lediglich eine andere rechtliche Bewertung bekannter Tatsachen dargelegt werden darf, führt diese nicht zur gewünschten Offenlegung von rechtswidrigen Verhaltensweisen und Missständen. Ferner können Beamte lediglich gegen Anordnungen remonstrieren, die ihnen aufgetragen wurden. Somit können Bedenken gegen die Rechtmäßigkeit von Handlungen der jeweiligen Behörde, die Kollegen oder Vorgesetzte betreffen, nicht mithilfe des Remonstrationsverfahrens vorgebracht werden.

Neben dem gesetzlichen Antrags- und Beschwerderecht nach § 17 Abs. 1 ArbSchG i. V. m. § 125 BBG und dem gesetzlich geregelten Anzeigerecht des § 17 Abs. 2 ArbSchG, dass für Angestellte aufgrund eines privatrechtlichen Arbeitsverhältnisses gilt, ergibt sich die Zulässigkeit von Whistleblowing aus den allgemeinen verwaltungsrechtlichen Vorschriften und deren Konkretisierungen durch die Rechtsprechung. Wie beim privatwirtschaftlichen Arbeitsrecht besteht der Grundkonflikt in den widerstreitenden Interessen von Dienstherr und Beamten. Insbesondere der Grundsatz des Berufsbeamtentums nach Art. 33 Abs. 5 GG, aus dem die besondere Verschwiegenheitspflicht hergeleitet werden kann, genießt höchste Priorität und erschwert es einem Beamten, seine Grundrechte wahrzunehmen.[34]

Die Güterabwägung in Form der praktischen Konkordanz kann ergeben, dass das Recht auf freie Meinungsäußerung zurücktreten muss. Die Rechtsprechung hat dabei als Maßstab zur Abwägung die aus dem Zivilrecht bekannte „Stufentheorie" herangezogen, nach der interne Abhilfeversuche externen Meldungen vorgehen.[35] Was den Hinweisgehalt angeht, können ebenfalls Vergleiche zum Arbeitsrecht in der Privatwirtschaft gezogen werden. Wissentliche Falschaussagen, Leichtfertigkeit im Falle schwerwiegender Anschuldigungen sowie haltlose Vorwürfe sind nicht geschützt und stellen einen Pflichtverstoß dar. Auch die Entscheidung des EGMR[36] zum Fall *Heinisch* führt nicht zu einer Abweichung von den bisherigen, in der Rechtsprechung angelegten Grundsätzen.[37] Besondere beamtenrechtliche Relevanz kommt dem expliziten Verweis auf das Urteil im Fall *Guja/Moldawien*[38] zu, das sich erstmalig

32 *Battis*, BBG, § 63, Rn. 4; gleiches gilt für Beamte nach dem BeamtStG.

33 Vgl. *Battis*, BBG, § 63, Rn. 4 mit Verweis auf BT-Drs. 16/7074, S. 219 und BT-Drs. 16/4027, S. 31.

34 So auch *Bauschke*, öAT 2012, 271, 274.

35 So schon BVerfG, Beschl. v. 28.4.1970 – 1 BvR 690/65, NJW 1970, 1498, 1501.

36 EGMR, Urt. v. 21.7.2011 – 28274/08, NJW 2011, 3501 (Heinisch/Deutschland).

37 Vgl. *Reufels/Molle*, KSzW 2012, 3, 8; *Brock*, öAT 2011, 243, 247; *Sagan*, DB 2012, Beilage zu Heft 6, 11, 12.

38 EGMR, Urt. v. 12.2.2008 – 14277/04 (Guja/Moldavia), abrufbar unter: http://echr.ketse.com/doc/14277.04-en-20080212/view/, (abgerufen am 5.3.2014).

mit der Weitergabe von internen Informationen an die Medien durch einen Beamten auseinanderzusetzen hatte.[39] Laut EGMR ist gerade bei Beamten (und Angestellten des öffentlichen Dienstes) die Loyalitäts- und Zurückhaltungspflicht besonders stark ausgeprägt.[40] Daher haben sie grundsätzlich Genauigkeit und Zuverlässigkeit der Informationen zu prüfen und aufgrund ihrer Verschwiegenheitspflicht zunächst den Vorgesetzten zu informieren.[41] Nur wenn dies eindeutig zu keinem Erfolg führen kann, können sie sich an die Öffentlichkeit wenden, wobei stets zu prüfen ist, ob ihnen nicht doch ein milderes und gleich effektives Mittel zur Verfügung steht, um den Missstand zu beheben.

Für Angestellte des öffentlichen Dienstes, deren Arbeitsverhältnis auf einem privatrechtlichen Vertrag beruht, sind tarifvertraglich keine Bestimmungen zum Schutz von Whistleblowing getroffen worden. Insofern ist in diesen Fällen ein Schutzdefizit zu verzeichnen.

Der beamtenrechtliche Schutz von Whistleblowern ergibt sich aus der Rechtsfolge der Ausnahmeregelung zur Verschwiegenheitspflicht. Denn mit einer zulässigen Anzeige nach § 67 Abs. 2 S. 1 Nr. 3 BBG bzw. § 37 Abs. 2 S. 1 Nr. 3 BeamtStG begeht der Beamte kein Dienstvergehen. Die Anzeige darf kein Anlass für Disziplinar- oder Personalmaßnahmen sein.[42] Zudem sieht Art. 9 des Zivilrechtsübereinkommens vor, dass Beschäftigte, die einen begründeten Korruptionsverdacht mitteilen, angemessen vor ungerechtfertigten Nachteilen geschützt werden sollen.[43]

Insgesamt ist festzuhalten, dass Whistleblowing im öffentlichen Sektor Rechtsprobleme und Interessenkollisionen aufweist, die im zivilrechtlichen Pendant ebenfalls bekannt sind, wie die weiteren Ausführungen zeigen werden. Grundsätzlich hat jeder Beschäftigte im öffentlichen Dienst den internen Dienstweg zu durchlaufen. Nur für die Korruptionstatbestände nach §§ 331–337 StGB ist ein externes beamtenrechtliches Anzeigerecht vorgesehen. Dennoch bleiben in allen anderen Fällen praktische Unsicherheiten. Ein einheitliches Gesamtkonzept für das Whistleblowing fehlt.

Trotz gewisser Ähnlichkeiten zum privaten Sektor beschäftigt sich diese Arbeit nicht weiter mit Angehörigen des öffentlichen Dienstes, da bei diesen Beschäftigten aufgrund der gesteigerten Treuepflicht die rechtspolitische Problematik anders beurteilt werden muss als bei Arbeitnehmern aus der Privatwirtschaft.

---

39 EGMR, Urt. v. 21.7.2011 – 28274/08 (Rn. 63), NJW 2011, 3501, 3503 (Heinisch/Deutschland).

40 EGMR, Urt. v. 21.7.2011 – 28274/08 (Rn. 64), NJW 2011, 3501, 3503 (Heinisch/Deutschland).

41 EGMR, Urt. v. 21.7.2011 – 28274/08 (Rn. 65), NJW 2011, 3501, 3503 (Heinisch/Deutschland), nähere Ausführungen zum Urteil siehe Teil 4, A, IV.

42 Ähnlich *Király*, ZRP 2011, 146, 146.

43 *Herold*, ZBR 2013, 8, 13.

C. Motive des Whistleblowers und ihre Relevanz

"Our lives begin to end the day we become silent about things that matter."
(Martin Luther King)

Personen, die nicht wegschauen, sondern genau hinsehen und notfalls auch laut darüber reden, opfern sich selbstlos zum Nutzen einer bedeutenden Sache auf. Die einen verehren sie als „Heilige" einer säkularen Kultur.[44] Für andere handelt es sich um Verräter oder Profilneurotiker, die sich wichtigmachen und aufspielen: *Whistleblower.*

Der Preis, den Whistleblower zahlen müssen, ist meist sehr hoch: Sie werden als Lügner, geistig Verwirrte, inkompetente Wichtigtuer oder ähnlich herabwürdigend bezeichnet.[45] Ein Bericht über eine US-amerikanische Studie zeigt, dass die Konsequenzen für Whistleblower dramatische Dimensionen erreichen können:[46]

- 90 Prozent verloren ihren Job oder wurden degradiert
- 27 Prozent wurden juristisch verfolgt
- 27 Prozent brauchten psychiatrische oder medizinische Hilfe
- 17 Prozent verloren ihr Haus
- 15 Prozent ließen sich scheiden
- 10 Prozent versuchten Selbstmord zu begehen.

I. Motive des Whistleblowers

Auf die Frage, warum es Menschen gibt, die viel riskieren, um einen Missstand zu beseitigen, während andere nichts tun, gibt es keine eindeutige Antwort. Entscheidend ist wohl vor allem die Persönlichkeit: Der Whistleblower richtet sein Verhalten an Uneigennützigkeit, Gemeinwohlinteresse[47] und Zivilcourage aus. Im Zusammenhang mit der Zivilcourage ist das staatsbürgerliche Verantwortungsbewusstsein[48] hervorzuheben: Der Whistleblower sieht sich in der Verantwortung, durch Bekanntmachung der Missstände eine gesellschaftliche Diskussion anzuregen und somit die Gesellschaft für bestimmte Themen zu sensibilisieren. Er fühlt sich zudem

---

44 *Grant,* Journal of Business Ethics, Vol. 39 (2002), 391, 392.

45 *Schemmel/Ruhmannseder/Witzigmann,* Hinweisgebersysteme, S. 87 m. w. N.; *Graser,* Whistleblowing, S. 6 f.; *Groneberg,* Whistleblowing, S. 40; *Deiseroth/Derleder,* ZRP 2008, 248, 250; *Rohde-Liebenau/Peter,* AiB 2004, 615, 616.

46 Vgl. *Leisinger,* Whistleblowing und Corporate Reputation Management, S. 4 f. mit Verweis auf *Grace/Cohen,* Business Ethics: Australian Problems and Cases, S. 148. Obwohl die Studie länderbezogen ist und deswegen keine allgemeingültige Geltung erlangen kann, soll sie beispielhaft zeigen, mit welchen materiellen und immateriellen Folgen ein Whistleblower zu rechnen hat.

47 *Graser,* Whistleblowing, S. 105; *Groneberg,* Whistleblowing, S. 42, die den Whistleblower als selbstlosen Verteidiger des Allgemeinwohls darstellen.

48 *Schemmel/Ruhmannseder/Witzigmann,* Hinweisgebersysteme, S. 20.

verpflichtet, die Gesellschaft vor gesetzeswidrigen Verhaltensweisen[49] zu schützen, die enorme Folgen für die Umwelt, für Leib und Leben, für das Eigentum und nicht zuletzt für die nachfolgenden Generationen haben können. Da der Whistleblower auch Teil der Gesellschaft und Träger der grundrechtlich geschützten Rechte ist, will er sich selbst und seine Güter schützen; insbesondere wenn die Wahrung der Menschenrechte auf dem Spiel steht. Deshalb geschieht Whistleblowing in Fällen mit Folgen für die ganze Menschheit nicht nur aus rein altruistischen Gründen. Trotz alledem ist das Handeln vom Bewusstsein für Gerechtigkeit[50] getragen und dient dem Schutz der Allgemeinheit. Oft ist das Handeln mit dem Gedanken verbunden, die Gesellschaft auf den Straftäter aufmerksam zu machen, um ihr Verhalten ihm gegenüber zu überdenken. Damit geht einher, dass die Unternehmen oder Personen, die mit dem „Straftäter" kooperieren und zusammenarbeiten, ihre Zusammenarbeit mit ihm beenden und die Kontakte vorerst, bis zur Besserung der Lage, einfrieren.

Auch jobspezifische Motive spielen eine gewichtige Rolle. Arbeitnehmer besitzen eine hohe Loyalität gegenüber ihrem Unternehmen und möchten als Hinweisgeber auch den Arbeitgeber, der nicht am Missstand beteiligt ist, vor Schaden bewahren. Als Arbeitnehmer haben Whistleblower zudem ein Interesse daran, sich beruflich zu profilieren und ihre Zuverlässigkeit in betrieblichen Abläufen zum Ausdruck zu bringen.[51]

Neben dem Bild des altruistisch handelnden Whistleblowers darf nicht übersehen werden, dass es auch Whistleblower gibt, die aus rein egoistischen Motiven handeln. Dabei sind die Motive der Whistleblower u. a. Rache und Opportunismus.[52] Der Opportunist strebt vorrangig nicht die Durchsetzung seiner Ideale an, sondern die Sicherung seiner beruflichen Zukunft.[53] Er wird zum Whistleblower, um den Schutz auszunutzen, den er in einigen Ländern gesetzlich erhält. Der Whistleblower, der aus Rache agiert, ist meist schon aus dem Unternehmen ausgeschieden oder steht in keiner guten Beziehung zu seinem Arbeitgeber. Der Wunsch nach Rache wird nicht selten von der Gier nach finanziellen Vorteilen begleitet. Obwohl die Vorwürfe gegen den Arbeitgeber stimmen mögen, ist die Intention eine andere.

Die Motive bzw. die Interessen des Whistleblowers im Kartellrecht haben einen anderen Schwerpunkt. Der Kronzeuge konzentriert sich hier darauf, das wettbewerbswidrige Verhalten für die Zukunft zu beenden und dem Bußgeld der Kartellbehörde zu entgehen. Der Whistleblower ist oftmals in das gesetzeswidrige Verhalten involviert gewesen, bis er sich entschieden hat, sich selbst und die anderen

49 Gesetzeswidrige Verhaltensweisen sind: Verstöße gegen arbeitsrechtliche Vorschriften, Verstöße gegen das Wirtschafts- und Umweltstrafrecht aber auch im weitesten Sinne Delikte gegen Leib und Leben. Unter Wirtschaftskriminalität werden illegale Handlungen wie z. B. Diebstahl, Unterschlagung, Geldwäsche, Betrug, Korruption, Industriespionage, Computerkriminalität und Falschbilanzierung gefasst.

50 *Leisinger*, Whistleblowing und Corporate Reputation Management, S. 77.

51 *Donato*, Whistleblowing, S. 20; *Jos/Tompkins/Hays*, PAR 1989, Vol. 49, No. 6, 552, 557.

52 Siehe ausführlich *Donato*, Whistleblowing, S. 21 m. w. N.

53 *Donato*, Whistleblowing, S. 21.

beteiligten Kartellanten bei der Kartellbehörde anzuzeigen. Das Motiv, selbst von der Strafe verschont zu bleiben,[54] hat Priorität.

Setzt man sich also mit dem Wesen des Whistleblowers auseinander, wird klar, dass es einen typischen Whistleblower wohl nicht gibt. Abgesehen davon, dass es besonders schwierig ist, das wahre Motiv des Whistleblowers zu ermitteln, scheint eine genaue Abgrenzung zwischen den selbstlosen und egoistischen Beweggründen fast unmöglich. Dies ist nicht zuletzt der subjektiven Wahrnehmung der Menschen geschuldet. Je nachdem, wie der Whistleblower selbst zu Gerechtigkeit und Verantwortungsbewusstsein steht, ergeben sich die Grundlagen für seine Entscheidung.

II. Relevanz der Motive

Bei allen Versuchen, die Beweggründe zu konkretisieren, muss bedacht werden, dass Motiv und Resultat nicht immer konvergieren: tugendhafte und ehrenwerte Motive bescheren nicht immer das gewünschte positive Resultat; umgekehrt können Handlungen, die auf eigennützigen oder rachsüchtigen Motiven beruhen, im Ergebnis zum Nutzen der Allgemeinheit sein und eine unerwartet positive Folge nach sich ziehen. Positive Folgen des Whistleblowings ergeben sich in der Weise, dass das Insiderwissen an (organisationsinterne oder -externe) Dritte weitergegeben wird und dadurch Informationen über gesetzeswidrige Verhaltensweisen ans Licht kommen, die andernfalls – wenn überhaupt – erst sehr viel später aufgedeckt worden wären.[55]

Das Motiv für das Handeln ist jedoch rechtlich irrelevant. Ein Staat, der vom Rechtsstaatsprinzip geprägt ist, darf die Strafbarkeit oder Straflosigkeit von gesetzeswidrigem Handeln nicht vom Motiv des Whistleblowers abhängig machen. Gesetzesverstöße dürfen nicht geduldet werden. Erst recht dürfen diejenigen, die Gesetzesverstöße publik machen, keinen rechtlichen Konsequenzen ausgesetzt werden. Ein Erfordernis, ein Recht zu schützen (Recht auf Wahrung des Betriebsgeheimnisses), dessen Ausübung nach geltender Rechtslage Unrecht wäre, ist nicht ersichtlich. Die Motive des Whistleblowers dürfen demnach keine Rolle spielen. Es darf der Rechtsordnung grundsätzlich nicht darauf ankommen, aus welchem Grund jemand einen Rechtsverstoß eines anderen meldet, sofern dieser sachlich richtig ist oder zumindest ein berechtigter und nicht haltloser Verdacht auf einen Rechtsverstoß vorlag.[56] Der Hinweis dient nicht nur dem Schutz vor weiteren Gefahren, sondern eröffnet die Möglichkeit, die entsprechenden Sachverhalte aufzuklären und die Verantwortlichen juristisch und/oder außer-juristisch zu verfolgen.[57] Dieser positive Effekt ergibt sich unabhängig davon, ob der Whistleblower aus Rache oder aus rein

---

54 Siehe Bekanntmachung des Bundeskartellamtes Nr. 9/2006 über den Erlass und die Reduktion von Geldbußen und Kartellsachen – Bonusregelung -, Abschnitt B.

55 Vgl. *Göpfert/Landauer*, NZA-Beil. 1/2011, 16, 21; *Graser*, Whistleblowing, S. 8; *Reuter*, ZRP 2011, 104, 107.

56 So ähnlich *Kort*, FS Kreutz, S. 247, 261; *Thüsing/Forst*, in: Thüsing, Beschäftigtendatenschutz und Compliance, § 6, Rn. 36.

57 So ähnlich *Schulz*, Ethikrichtlinien, S. 27.

altruistischen Motiven handelt. Aber auch die Verhütung von Fehlentwicklungen ist ein gesamtgesellschaftliches Interesse von besonders hohem Rang. Daher ist ist die sich herauskristallisierende positive Bewertung des Phänomens Whistleblowing durchaus gerechtfertigt.

D. Problemaufriss und Gegenstand der Untersuchung

Ein Blick auf die veröffentlichte Rechtsprechung zeigt, dass der gesetzliche Schutz von Whistleblowern bedeutender denn je ist.

Der Ausgang des Falles *Heinisch vs. Bundesrepublik Deutschland* vor dem Europäischen Gerichtshof für Menschenrechte hat deutlich gemacht, in welche Richtung sich die deutsche Rechtsprechung zu bewegen hat. Die Interessenabwägung zwischen der Meinungsfreiheit des Whistleblowers und der Verletzung der Rücksichtnahmepflichten aus dem Arbeitsverhältnis ist von größter Bedeutung und lässt eine einheitliche Entscheidung aller Fälle nicht zu. Ein einheitlicher Katalog, der als Maßstab für die Gewichtung der jeweiligen Interessen der Beteiligten dient, besteht nicht, wäre aber wünschenswert. Es gibt auch kein Gesetz, das die Besonderheiten des Whistleblowings berücksichtigt. Der Whistleblower, der sich in den Dienst der Rechtsordnung stellt, gesetzeswidriges Verhalten offenbart und zu dessen Beendigung beiträgt, ist somit rechtlichen Konsequenzen ausgesetzt. Denn durch das Whistleblowing ergeben sich nicht nur arbeitsrechtliche, sondern auch strafrechtliche/ordnungswidrigkeitenrechtliche und haftungsrechtliche Folgen, die aufgrund fehlender Regelungen und der einhergehenden Rechtsunsicherheit für den Whistleblower unkalkulierbar sind. Die Angst vor Benachteiligungen oder aussichtslosen Gerichtsverfahren ist groß und stellt ein Hemmnis für potenzielle Hinweisgeber dar.

Zudem darf die Motivation keine Rolle spielen und dem Whistleblower den – noch zu kodifizierenden – rechtlichen Schutz verwehren.

Der Whistleblower, der Kartellrechtswidrigkeiten offenbart, wird zwar als Kronzeuge bezeichnet, jedoch sind bei ihm die gleichen Aspekte zu berücksichtigen. Er hat verstärkt mit haftungsrechtlichen Konsequenzen zu rechnen, da er anfangs an der offengelegten Tat beteiligt war. Insbesondere ist der Whistleblower in diesen Fällen meist nicht ein einzelner Arbeitnehmer, sondern das Unternehmen bzw. die Unternehmensleitung. Das Handeln des Unternehmens als Whistleblower ist daher im strafrechtlichen/ordnungs-widrigkeitenrechtlichen und haftungsrechtlichen Teil der Arbeit verstärkt zu berücksichtigen.

Es ist an der Zeit, die ungeklärten Fragen zum Schutz von Whistleblowern und die sich daraus ergebenden Folgeprobleme zu beantworten und vertretbare Lösungsansätze zu finden. Denn offensichtlich fehlt es an rechtlichem Schutz von Personen, die sich und ihre Karriere zur Aufrechterhaltung der Rechtsordnung aufopfern.

Ziel dieser Arbeit ist es daher, die Notwendigkeit eines gesetzlichen Whistleblowerschutzes aufzuzeigen.

Dazu wird der Schutz des Whistleblowers nach geltender Rechtslage (de lege lata) aufgezeigt. Darauffolgend werden mögliche zukünftige Regelungen (de lege ferenda) in Form von Gesetzesentwürfen untersucht und deren Effektivität und Praktikabilität für den Ausbau des Whistleblowerschutzes beurteilt. Dabei wird die nationale sowie europäische Rechtsprechung einbezogen.

Der erste Teil führt in die Materie des Whistleblowings ein. Hierdurch soll ein Gefühl für die Problematik und für die Dringlichkeit einer Lösung vermittelt werden. In Teil zwei wird vertieft auf den Whistleblower im Kartellrecht eingegangen. Im dritten Teil werden praxisbezogene Fälle dargestellt. Die Teile vier und fünf stellen den Whistleblowerschutz *de lege lata* und *de lege ferenda* dar. Schließlich wird im sechsten Teil eine Bewertung abgegeben und ein Ausblick auf zukünftige Regelungen gewagt.

Insbesondere werden drei Bereiche des Whistleblowerschutzes untersucht. Neben dem Arbeitsrecht werden in gleicher Weise das strafrechtliche bzw. ordnungswidrigkeitenrechtliche und das haftungsrechtliche Gebiet beleuchtet. Der Begriff Hinweisgeber wird als Synonym für den Begriff Whistleblower verwendet.

Zur Erstellung dieser Arbeit werden alle notwendigen juristischen Werke sowie nationale und europäische Gesetzesmaterialien verwendet, die bis Mai 2015 zur Verfügung standen.

Teil 2: Der kartellrechtliche Kronzeuge als Whistleblower

Der Begriff Whistleblowing dient als Oberbegriff für ein Verhalten, das der Aufdeckung von Gesetzeswidrigkeiten dient. Im Kartellrecht wird der Whistleblower als Kronzeuge bezeichnet. Die besondere Bezeichnung rührt daher, dass der am Kartell beteiligte Whistleblower bei der Offenlegung des Missstandes besondere Voraussetzungen zu erfüllen hat, um eine für ihn vorteilhafte „Gegenleistung" für seine Offenbarung zu erhalten. Eine weitere Besonderheit ergibt sich daraus, dass der Whistleblower im Kartellrecht sowohl ein Unternehmen als auch eine natürliche Person sein kann, wobei letztlich nur das Unternehmen haftbar gemacht werden kann. Bevor näher auf das Kronzeugenprogramm eingegangen wird, erfolgt eine kurze historische und rechtliche Einordnung des Kronzeugen.

A. Historische und rechtliche Einordnung

Der Begriff des Kronzeugen stammt ursprünglich aus dem Strafverfahrensrecht. Eine Vorreiterrolle nahm England ein. Der Begriff leitet sich aus dem englischen Ausdruck *„to give evidence for the crown*" ab.[58] Damit war jedermann gemeint, der für die Krone aussagte, da in der britischen Monarchie der Staatsanwalt im Namen des Staatsoberhaupts – durch die Krone symbolisiert – seine Beschuldigungen vor Gericht vortrug. Mittlerweile hat sich die Bedeutung des Begriffes gewandelt: Aufgrund der wichtigen Bedeutung eines Kronzeugen und der Verwendung von Zeugen, die selbst Straftäter sind oder sogar als Mittäter aussagen, wird heute insbesondere in der deutschen Sprache unter einem Kronzeugen ein vorbelasteter Zeuge verstanden. Stellt der Kronzeuge sich der Staatsanwaltschaft zur Verfügung und führt seine Aussage zur Verurteilung des Angeklagten, so sichert die Staatsanwaltschaft dem Kronzeugen im Gegenzug gem. § 46 b StGB[59] Strafmilderung bis

58 „Kronzeuge (m.) 'Hauptbelastungszeuge' (19. Jh.), Wiedergabe von engl. *king's evidence*, im englischen Strafrecht ein Mittäter, den die Krone (vertreten durch den Staatsanwalt) gegen das Versprechen der Strafmilderung oder Straffreiheit als Belastungszeugen benutzt", *Berlin-Brandenburgische Akademie der Wissenschaften* (Hrsg.), Etymologisches Wörterbuch – Digitales Wörterbuch der deutschen Sprache, Stand 2014.

59 Zuletzt geändert durch das 46. Strafrechtsänderungsgesetz – Beschränkung der Möglichkeit zur Strafmilderung bei Aufklärungs- und Präventionshilfe – v. 10.6.2013, BGBl I-2013, S. 1497 (Datum des Inkrafttretens: 1.8.2013); vertiefend dazu *Peglau*, NJW 2013, 1910.

hin zum völligen Straferlass zu und leitet unter Umständen auch Maßnahmen des Zeugenschutzes ein.[60]

Die bestehende Kronzeugenregelung im Kartellrecht verfolgt das gleiche Ziel. Grundsätzlich sollte die Sanktion, die für einen Verstoß gegen das Kartellverbot verhängt wird, nach dem Handlungsunrecht und dem verursachten Schaden bemessen werden. Sind mehrere Unternehmen als Kartellbeteiligte zu qualifizieren, müssten ihnen in gleichem Maße Sanktionen drohen. Im Kartellrecht wird der Kronzeuge jedoch anders behandelt, denn das „Nachtatverhalten" wird mitberücksichtigt.[61] So kann es zu unterschiedlichen Sanktionshöhen oder sogar zum vollständigen Erlass der Sanktion gegenüber einem Unternehmen kommen, das in gleicher Weise gehandelt hat wie das andere beteiligte Unternehmen.

Der Begriff Whistleblowing hat sich auch im Kartellrecht etabliert und beschreibt dabei die Personen oder die Unternehmen, die von der Kronzeugenregelung Gebrauch machen.[62] Diese Whistleblower sind aber von den „typischerweise" unbeteiligten Whistleblowern abzugrenzen, und zwar insoweit, als sie Zeuge der Anklage und im Gegensatz zu normalen Zeugen in den Fall involviert sind. Zwar gibt es auch Fälle, in denen der Kronzeuge an den kartellrechtswidrigen Machenschaften nicht beteiligt war, sondern der Rechtsvorgänger. Der Kronzeuge, der die Kronzeugenregelung in Anspruch nimmt, hat dann für Taten des Rechtsvorgängers einzustehen. Nach der Neuerung durch die 8. GWB-Novelle[63], die am 30. Juni 2013 in Kraft getreten ist, kann eine Geldbuße nach § 30 Abs. 1 OWiG im Fall der Gesamtrechtsnachfolge oder einer teilweisen Gesamtrechtsnachfolge durch Aufspaltung (§ 123 Abs. 1 UmwG) auch gegen den oder die Rechtsnachfolger verhängt werden.[64] Jedoch spielen diese Kronzeugen eine untergeordnete Rolle, sodass im Weiteren nicht näher auf sie eingegangen wird.

Die Kronzeugenregelung beruht auf einem gegenseitigen Geschäft:[65] Während der Kronzeuge mit seiner Aussage andere Beteiligte belastet, wird ihm ein Erlass oder die Reduzierung der Kartellgeldbuße in Aussicht gestellt. Wie der Kronzeuge verfügt der Whistleblower auch über Insiderwissen, das zur Aufdeckung von Missständen führt.

Vorbild für eine kartellrechtliche Kronzeugenregelung war das Leniency-Programm des US-Justizministeriums aus dem Jahr 1978.[66] Durch die Kronzeugenregelung sollten die Unternehmen verstärkt dazu bewogen werden, mit den

---

60 *Streng*, in: Kindhäuser/Neumann/Paeffgen, Kommentar StGB, § 46 b StGB, Rn. 11 f.; *Eisenberg*, Beweisrecht der StPO, Rn. 942, 948.

61 So ähnlich auch *Lutz*, BB 2000, 677, 679.

62 Vgl. *Bürkle*, DB 2004, 2158, 2159 f.

63 Achtes Gesetz zur Änderung des Gesetzes gegen Wettbewerbsbeschränkungen, v. 29.6.2013, BGBl. I-2013, S. 1738.

64 § 30 Abs. 2a OWiG enthält die Neuerung zur Rechtsnachfolge in die Bußgeldverantwortlichkeit; vertiefend *Bosch*, NJW 2014, 1714, 1715.

65 *Groneberg*, Whistleblowing, S. 35.

66 Vgl. *Lutz*, BB 2000, 677, 678 f.

Kartellbehörden zu kooperieren. Den Unternehmen sollte ein Anreiz geliefert werden, die Beteiligung und wettbewerbswidrige Handlungen einzustellen und zur Aufklärung beizutragen.[67] 1996 veröffentlichte die Europäische Kommission ihre „eigene" *Leniency Notice*[68], die überwiegend den Regelungen des US-amerikanischen Leniency-Programms glich.[69] Doch aufgrund mangelnder Transparenz wurde die Regelung durch die *Leniency Notice* von 2002 ersetzt. Nach weiteren Überarbeitungen wurde schließlich eine neue *Leniency Notice* im Amtsblatt im Jahre 2006 veröffentlicht, die jetzt als Grundlage für die Bemessung der Geldbußen auf europäischer Ebene dient.[70]

Mit Inkrafttreten der Verordnung Nr. 1/2003 (EG)[71] wurde der Vorrang der europäischen Wettbewerbsregeln vor dem jeweiligen nationalen Kartellrecht der Mitgliedstaaten festgelegt.[72] Der deutsche Gesetzgeber implementierte die erforderlichen Regelungen erstmals in der siebten GWB-Novelle vom 7. Juli 2005.[73] Das nationale Recht ist anzuwenden, wenn es keinen grenzüberschreitenden Sachverhalt gibt. Die Bestimmungen zur Durchsetzung des Kartellverbots sind in den §§ 48, 54 f., 81 GWB zu finden.

Als Pendant zur *Leniency Notice* der Europäischen Kommission hat das Bundeskartellamt die (inhaltsgleiche) Bonusregelung[74] eingeführt, in der die Richtlinien für die Festsetzung von Geldbußen aufgeführt sind. Eine ausdrückliche Ermächtigungsgrundlage für den Erlass einer solchen Bonusregelung und für den Erlass von Bußgeldbemessungsleitlinien ist in § 81 Abs. 7 GWB zu finden.[75] Die Bekanntmachung des Bundeskartellamtes stellt die aktuelle Grundlage für den Erlass und die Reduktion von Geldbußen in Kartellsachen dar. Die Bonusregelung betrifft das verwaltungsgerichtliche Verfahren der Behörden, ein zivilrechtliches Verfahren wird nicht berücksichtigt. Nach dieser Regelung wird den „Kartellsündern" ihre Strafe in Form von Geldbußen ganz oder teilweise erlassen.

---

67 So auch *Böttcher*, Die Pflicht des Vorstandes, S. 31.

68 Mitteilung der Kommission über die Nichtfestsetzung oder die niedrigere Festsetzung von Geldbußen in Kartellsachen, ABl. C 207 vom 18.7.1996.

69 Zur Kronzeugenmitteilung siehe *Arhold*, EuZW 1999, 165, 170.

70 Mitteilung der Kommission über den Erlass und die Ermäßigung von Geldbußen in Kartellsachen, ABl. C 298/17 vom 8.12.2006.

71 Verordnung Nr. 1/2003 (EG) vom 16.12.2002 zur Durchführung der in den Artikel 81 und 82 des Vertrages niedergelegten Wettbewerbsregeln, ABl. EG 2003, Nr. L 1, S. 1.

72 Art. 3 VO 1/2003 (EG).

73 Siebtes Gesetz zur Änderung des Gesetzes gegen Wettbewerbsbeschränkungen vom 7.7.2005, BGBl. I-2005, S. 1954.

74 Bekanntmachung des Bundeskartellamtes Nr. 9/2006 über den Erlass und die Reduktion von Geldbußen in Kartellsachen – Bonusregelung – vom 7.3.2006, abrufbar unter: http://www.bundeskartellamt.de/SharedDocs/Publikation/DE/Bekanntmachungen/Bekanntmachung%20-%20Bonusregelung.html?nn=4136556, (abgerufen am 10.12.2013).

75 *Dannecker/Biermann*, in: Immenga/Mestmäcker, Wettbewerbsrecht: GWB, § 81 GWB, Rn. 469.

B. Inhalt und Wesen eines Kronzeugenprogramms

I. Repression und Prävention

Ebenso wie das Kronzeugenprogramm der EU-Kommission bezweckt die Bonusregelung des Bundeskartellamts, Kartelle von innen heraus aufzudecken.[76] Eine typische Eigenschaft aller Kronzeugenregelungen ist, dass kooperierende Unternehmen, Unternehmensvereinigungen und beteiligte natürliche Personen eine bevorzugte Behandlung bei der Festlegung der Sanktion erhalten. Der sanktionsrechtliche Anreiz bestimmter Rechtsvorteile bei der Kooperation mit den Behörden zur Aufklärung des wettbewerbswidrigen Verhaltens ist Inhalt einer solchen Kronzeugenregelung.[77] Auf diese Weise erhalten die staatlichen Stellen Informationen über die jeweilige Kartellrechtsverletzung.

Die aus der Kronzeugenregelung resultierende verminderte Geldbuße belohnt den Beteiligten dafür, dass er sich auf die Seite des „Guten" geschlagen hat und gibt Anreize zum Ausstieg aus der kartellrechtlich verbotenen Handlung.

Zudem wirkt sie darauf hin, solche wettbewerbswidrigen Absprachen nicht zu wiederholen. Nicht jeder Beteiligte kann die Kronzeugenregelung in Anspruch nehmen und somit in den Genuss von Bußgeldfreiheit kommen.[78] Denn dieses Verfahren unterliegt strengen Regelungen und Voraussetzungen, die auch in zeitlicher Hinsicht eine gewichtige Rolle spielen. So kann es dazu kommen, dass nur einem Kartellbeteiligten die Geldbuße erlassen wird, dem anderen jedoch aufgrund verspäteter Stellung des Antrags keine kooperative Hilfe zur Aufdeckung des Kartells beigemessen werden kann. Der Anreiz zum Ausstieg trägt also auch Unsicherheit in die Reihen der Täter und destabilisiert ihr gegenseitiges Vertrauen, das eine unverzichtbare Basis für verbotene Verhaltensweisen ist. Da im Falle eines funktionierenden Kronzeugenprogramms jederzeit mit kooperationsbereiten Beteiligten gerechnet werden muss, wird eine Kultur des Misstrauens geschaffen.[79]

Das Kronzeugenprogramm dient insofern nicht nur der Repression von illegalen Absprachen, sondern auch der Prävention neuer rechtswidriger Vereinbarungen.

---

76 *Dannecker/Biermann*, in: Immenga/Mestmäcker, Wettbewerbsrecht: GWB, § 81 GWB, Rn. 417.

77 *Dannecker/Biermann*, in: Immenga/Mestmäcker, Wettbewerbsrecht: GWB, § 81 GWB, Rn. 416 f.

78 Siehe dazu *Dannecker/Biermann*, in: Immenga/Mestmäcker, Wettbewerbsrecht: GWB, § 81 GWB, Rn. 423, 430. Es kommt darauf an, wer sich als erster an das Bundeskartellamt wendet und/oder wie die Qualität der Informationen zu werten ist.

79 *Denoth*, Kronzeugenregelung, S. 106; *Schneider*, Kronzeugenregelung, S. 21.

II. Public und Private Enforcement

Die Durchsetzung und Anwendung von Kartellrecht geschieht auf zwei Wegen: zum einen durch Behörden, zum anderen durch private Personen oder Unternehmen.[80]

Die hoheitliche Durchsetzung des Kartellrechts in der Europäischen Union zeichnet sich durch ein paralleles Ineinanderwirken[81] zweier Rechtsordnungen[82] aus: Einerseits erfolgt sie durch die EU-Kommission und andererseits durch die Kartellbehörden und Gerichte der Mitgliedstaaten der Europäischen Union. Um die Einhaltung der Art. 101 und 102 AEUV gewährleisten zu können, hat die Kommission die Durchsetzung der in den Artikeln niedergelegten Grundsätze zu überwachen. Zum Zwecke der effektiven Gestaltung wird der EU-Kommission durch die Verordnung Nr. 1/2003 (EG) die Befugnis erteilt, Ermittlungsverfahren durchzuführen und Strafen zu verhängen.[83] Für die Ermittlungen ist jede anerkannte Form der Untersuchung erlaubt:[84] von Befragungen der Beteiligten über Auskunftsrechte bis hin zur Durchsuchung der Räumlichkeiten und Durchsicht der gefundenen Dokumente. Darüber hinaus ist die EU-Kommission befugt, Verbots- und Verpflichtungsentscheidungen zu treffen sowie Geldbußen und Zwangsgelder festzusetzen.[85] Die Höhe der zu entrichtenden Geldbuße bemisst sich nach der Schwere der Zuwiderhandlung und deren Dauer. Die genauen Einzelheiten ergeben sich aus den Leitlinien für das Verfahren zur Festsetzung von Geldbußen[86] sowie aus der Kronzeugenmitteilung von 2006.[87]

Das Zusammenspiel von europäischem und deutschem Kartellrecht dient der Entlastung der Kommission – die EU-Kommission soll für Grundsatzfragen wettbewerbspolitischen Inhalts und die nationalen Kartellbehörden für die Großzahl der Verfahren zuständig sein.[88]

Die zivilrechtliche Durchsetzung hingegen kann nur auf der Ebene der Mitgliedstaaten erfolgen, da die Europäische Union keine eigene Zivilgerichtsbarkeit

---

80 Einen Einblick in das Verhältnis von public und private Enforcement bieten *Mundt/Meeßen*, in: Oberender, Private und öffentliche Kartellrechtsdurchsetzung, S. 39, 42.

81 Dieses Ineinanderwirken ist der Dezentralisierung geschuldet, die mit dem Erlass der Verordnung Nr. 1/2003 (EG) angestrebt wurde, siehe *Emmerich*, Kartellrecht, S. 177.

82 *Meessen*, in: Loewenheim/Meessen/Riesenkampff, Kartellrecht, Einführung, Rn. 82 f.

83 *Schmidt*, in: Immenga/Mestmäcker, EU-Wettbewerbsrecht, Kartellverfahrensverordnung 1/2003, Rn. 7.

84 Art. 17 Verordnung Nr. 1/2003 (EG); *Burrichter/Hennig*, in: Immenga/Mestmäcker, EU-Wettbewerbsrecht, Vorbemerkung zu Art. 17–22 VO Nr. 1/2003, Rn. 1, 2.

85 *Dannecker/Biermann*, in: Immenga/Mestmäcker, EU-Wettbewerbsrecht, Art. 24 VO 1/2003, Rn. 1.

86 Leitlinien für das Verfahren zur Festsetzung von Geldbußen gemäß Art. 23 Abs. 2 a) der VO Nr. 1/2003 (EG), ABl. C 210/2 vom 1.9.2006.

87 Mitteilung der Kommission über den Erlass und die Ermäßigung von Geldbußen in Kartellsachen, ABl. C 298/17 vom 8.12.2006.

88 *Zagrosek*, Kronzeugenregelungen, S. 84, 88.

unterhält und keine Kompetenz in der Zivilgerichtsbarkeit hat.[89] Daher ergeben sich die Ansprüche von Privatpersonen oder Unternehmen aus dem jeweiligen Recht der Mitgliedstaaten.[90] Schon früh hat der EuGH[91] zur privaten Rechtsdurchsetzung ausgeführt, dass die Wirksamkeit des Kartellverbots beeinträchtigt würde, wenn nicht jedermann die Möglichkeit habe, seinen durch wettbewerbswidriges Handeln entstandenen Schaden gerichtlich durchsetzen zu können. Die Einzelheiten der Durchsetzung und des Verfahrens blieben zwar Sache der nationalen Gerichte, jedoch müssten auch die vom EuGH aus Art. 4 Abs. 3 EUV hergeleiteten Grundsätze der Wirksamkeit und Gleichwertigkeit sowie die allgemeinen Grundsätze des Gemeinschaftsrechts beachtet werden.[92]

Problematisch ist in dem Zusammenhang nur, wie die privaten und öffentlich-rechtlichen Vollzugsebenen zu verzahnen sind.[93] Auf der einen Seite ist fraglich, ob bei der Festsetzung der Rechtsfolge beide Vollzugsebenen aufeinander abgestimmt werden müssen. Auf der anderen Seite – und das bildet den Schwerpunkt des wissenschaftlichen Diskurses – geht es darum, ob Aufgaben der privaten und öffentlich-rechtlichen Vollzugsebenen „vermischt/übernommen" werden können bzw. sollen, um die Durchsetzung des Kartellrechts nach europäischen Grundsätzen gewährleisten zu können. Denn es wird – wie sich noch zeigen wird – gerne mit der „Effektivität der Kronzeugenregelung" argumentiert.

Da die Bonusregelung Regelungen enthält, die denen des Kronzeugenprogramms weitestgehend gleichen, ergeben sich auch auf Bundesebene die gleichen Probleme, insbesondere der Konflikt zwischen der öffentlichen und privaten Kartellrechtsdurchsetzung. Zum einen soll möglichst effektiv und wirkungsvoll gegen Kartelle im Wege der behördlichen Untersuchung vorgegangen werden, zum anderen sollen die Anreize für die Kartellmitglieder, den Kartellbehörden freiwillig Informationen zu liefern und im Gegenzug dafür straffrei auszugehen, nicht zunichtegemacht werden. Der Anreiz für eine Anzeige und Kooperation mit den Behörden vermindert sich zunehmend, wenn das angezeigte Unternehmen Gefahr läuft, auf der Grundlage der von ihm gelieferten Informationen mit Schadensersatzklagen belastet zu werden.[94] Die in den angelegten Akten enthaltenen Informationen und Beweise könnten den Klägern zugänglich gemacht werden, und somit muss das Unternehmen neben

---

89 *Zagrosek*, Kronzeugenregelungen, S. 83 f.

90 Anstelle vieler *Frenz*, Handbuch Europarecht, Band 2, Rn. 1099 f.

91 Vgl. EuGH, Urt. v. 20.9.2001 – Rs. C-453/99, Slg. 2001, I-6297 (Rn. 26), EuZW 2001, 715, 716 (Courage/Crehan); sowie EuGH, Urt. v. 13.7.2006 – verb. Rs. C-295/04-C-298/04, Slg. 2006, I-6641 (Rn. 60), EuZW 2006, 529, 533 (Manfredi u. a.).

92 Der Grundsatz der Wirksamkeit verlangt, dass die mitgliedstaatlichen Gerichte in ihrer Durchsetzung der Rechtsansprüche in dergestalt verfahren, dass die Verwirklichung der Rechte, die das Gemeinschaftsrecht gewährt, nicht obsolet oder übermäßig erschwert wird.

93 Eingehend dazu *Thomas*, in: Oberender, Private und öffentliche Kartellrechtsdurchsetzung, S. 55, 69.

94 *Denoth*, Kronzeugenregelung, S. 4.

der Offenlegung von Unternehmensgeheimnissen mit einer Klage[95] rechnen, die es finanziell empfindlich trifft. Die durch die Kronzeugenregelung erreichte Bußgeldbefreiung verliert aus der Sicht des Unternehmens seine Wirkung, wenn die finanziellen Konsequenzen der Zivilrechtsklage schwerer ins Gewicht fallen.[96]

Es stellt sich die Frage, ob und wie den Whistleblowern, die zukünftig von der Kooperation mit den Behörden und einer Kronzeugenregelung Abstand nehmen könnten, das Risiko von Schadensersatzklagen genommen und Schutz gegeben werden kann.

C. Die Besonderheit des an der Tat beteiligten Whistleblowers

Neben den Fällen, in denen der Whistleblower als unbeteiligter Hinweisgeber tätig wird, gibt es auch Fälle, in denen der Whistleblower an dem Missstand oder der rechtswidrigen Tat mitgewirkt hat. Der Schwerpunkt wird hierbei auf den Kartellbeteiligten gelegt, der im Nachgang die Bonusregelung nutzt. Durch seine vorherige Beteiligung hat er den Missstand gefördert und dazu beigetragen, dass das gesetzeswidrige Handeln in der Form praktiziert werden konnte, in der es ohne sein Zutun nicht vorgenommen worden wäre.

I. Die Interessen des an der Tat beteiligten Whistleblowers

Was sind die Interessen und Motive eines (zunächst) an der Tat beteiligten Whistleblowers? In dem Augenblick, in dem sich der Whistleblower entschließt, die Missstände zu offenbaren, hat er sich auf die Seite „des Guten" geschlagen. Es findet also ein moralischer und emotionaler Wandel in der Person des Whistleblowers statt. Er hat sich bewusst dazu entschieden, den Missstand zu beenden und zukünftig keinen Gesetzesverstoß zu begehen. Dadurch ist das einst bestehende Gerüst der illegalen Handlungen zusammengefallen, womit die Mittäter nicht gerechnet haben.

Der Whistleblower, der sich durch die Mitteilung an die zuständige Kartellbehörde entscheidet, das Kartell zu beenden und die Behörde im Rahmen der notwendigen Zusammenarbeit über die Beteiligten zu informieren, ist eine wichtige Schlüsselfigur. Indem er Informationen zunächst in Unkenntnis der anderen Kartellanten der Behörde offenbart, erhöht sich die Wahrscheinlichkeit, die anderen Beteiligten bei der wettbewerbswidrigen Handlung aufzugreifen. Der Whistleblower leistet also einen nicht zu unterschätzenden Beitrag, kartellrechtliche Missstände aufzudecken und zu beenden. Dass sein Handeln nicht nur rein altruistischen Motiven entspringt,

---

95 So wurden dem Bundeskartellamt im Referenzzeitraum 2004 bis 2012 2055 neue Kartellzivilverfahren bekannt gegeben, darunter 480 kartellrechtliche Schadensersatzklagen allein für Deutschland, vgl. *Mundt/Meeßen*, in: Oberender, Private und öffentliche Kartellrechtsdurchsetzung, S. 39, 40.

96 *Sweeney*, Melbourne University Law Review (2006), Vol. 30, Issue 3, S. 873, 877.

sondern durchaus auch von Egoismus getragen ist, lässt den „Whistleblower-Status" nicht entfallen. Wie schon angemerkt, handelt der Whistleblower nicht selten aus einem Bündel von Motiven.

Neben der Frage, was einen zunächst an der Tat beteiligten Whistleblower dazu bewegt, sich auf die Seite der Rechtsordnung zu stellen und ungesetzliches Handeln zu beenden, stellt sich die Frage, wie das Handeln zu würdigen ist.

Einerseits kann man die Ansicht vertreten, dass der innere Wandel des Whistleblowers außer Acht zu lassen ist. Er hat an der Straftat bzw. einer Ordnungswidrigkeit mitgewirkt, sodass er auch die Konsequenzen in vollem Umfang zu tragen hat. Das heißt, dass ihm kein arbeitsrechtlicher, strafrechtlicher/ordnungswidrigkeitenrechtlicher und zivilrechtlicher Schutz gebührt.

Andererseits ist zu bedenken, dass trotz der anfänglichen Mitwirkung und des Umstandes, dass bereits Schaden materieller und immaterieller Art entstanden ist, der Gesetzesverstoß durch das Handeln des Whistleblowers für die Zukunft beendet wurde. Nicht nur die Gefahrenabwehr, sondern auch die Prävention von entsprechenden Fehlentwicklungen sind Interessen sowohl der Gesellschaft also auch des Whistleblowers. Ihm muss deshalb, wie dem an der Tat unbeteiligten Whistleblower, hinreichend Schutz vor den rechtlichen Folgen zukommen. Ob dieser so weit wie bei einem unbeteiligten Whistleblower reichen soll, ist im Laufe der Arbeit aufzuklären.

II. Das Unternehmen als „Whistleblower"

Die Bonusregelung bezieht sich nach deren Randnummer 1[97] nicht nur auf natürliche Personen, sondern auch auf juristische Personen und Personenvereinigungen (§ 30 OWiG).[98] Demnach können sich auch Unternehmen als gesellschaftsrechtliche Einheit dazu entscheiden, das Kartell für die Zukunft zu beenden. Da dies der Regelfall ist, wird auf das Unternehmen als Whistleblower hauptsächlich eingegangen. Die Interessen und Motive sind die gleichen wie bei einer natürlichen Person. Da hinter dem Unternehmen natürliche Personen stehen,[99] können die obigen Ausführungen im übertragenden Sinne herangezogen werden. Dementsprechend muss auch die Würdigung gleich ausfallen. Es macht einen besonderen Eindruck, wenn das ganze Unternehmen als Einheit sich entscheidet, gesetzeswidriges Verhalten zu beenden und zu unterbinden. Die Intention, Missstände zu beheben, Gefahren zu beseitigen und Rechtsverstöße zu ahnden, geht vom ganzen Unternehmen aus.

Der beteiligte Whistleblower ist im besonderen Maße haftungsrechtlichen Konsequenzen ausgesetzt, da er die Folge des begangenen Gesetzesverstoßes mitverursacht hat bzw. mit zu verantworten hat. Deshalb ist bei der Frage nach

---

97 Bekanntmachung des Bundeskartellamtes Nr. 9/2006 über den Erlass und die Reduktion von Geldbußen und Kartellsachen – Bonusregelung -, Abschnitt A, Rn. 1.

98 *Dannecker/Biermann*, in: Immenga/Mestmäcker, Wettbewerbsrecht: GWB, § 81 GWB, Rn. 419.

99 Siehe zur Handlungsfreiheit *Di Fabio*, in: Maunz/Dürig, GG-Kommentar, Art. 2 Abs.1 GG, Rn. 10 m. w. N.

dem haftungsrechtlichen Schutz gegenüber Ansprüchen Dritter besonders auf den beteiligten Whistleblower einzugehen. Bei Kartellrechtswidrigkeiten kann eine natürliche Person als Whistleblower, die keine leitende Position hat, im Gegensatz zum Unternehmen als Whistleblower nur begrenzt bzw. gar nicht haftbar gemacht werden. Dementsprechend ist in diesem Teil der Untersuchung das Unternehmen gemeint, wenn von einem Whistleblower gesprochen wird, und nicht der Arbeitnehmer.

D. Vergleichbare Bonusregelung in anderen Rechtsgebieten

Fraglich ist, ob sich auch in anderen Rechtsgebieten eine vergleichbare Bonusregelung finden lässt. Wie eben erwähnt, dient diese Regelung der finanziellen und rechtlichen Besserstellung des Kronzeugen. Durch die Inanspruchnahme der besonderen kartellrechtlichen Bonusregelung kann der Whistleblower der (vollen) Geldbuße wegen Verletzung einer Ordnungswidrigkeit entkommen.

Eine allgemeine staatliche Whistleblower-Prämie gibt es in der Bundesrepublik Deutschland nicht. Laut eines Berichts des Handelsblattes[100] sollen nach dem Willen der EU-Kommission die Mitgliedstaaten der Europäischen Union künftig finanzielle Anreize für Whistleblower ausloben, deren Hinweise zur Aufdeckung von Wirtschaftsstraftaten führen. Die angedachte EU-Verordnung soll Kapitalmarktverbrechen wie Insiderhandel und Marktmanipulation bekämpfen. In dem Vorschlag wurde angeregt, dass die Mitgliedstaaten „finanzielle Anreize für Personen" schaffen, „die besonders wichtige Informationen über mögliche Verstöße gegen diese Verordnung bereitstellen".[101] Doch aufgrund weitreichender Probleme in der praktischen Umsetzung ist (zumindest) in Deutschland bis dato keine Whistleblower-Prämie eingeführt worden. Zudem bezogen sich die Überlegungen nur auf Wirtschaftsstraftaten,[102] also im Kern auf Korruptionsdelikte, das Steuerstrafrecht, Untreue, Insolvenzdelikte, Betrug und Diebstahl von geistigem Eigentum. Ob dazu auch Umweltstraftaten im weitesten Sinne zählen, ist fraglich. Auch welchen rechtlichen Wert diese Zahlung haben soll, ist unklar. Es wird von einer Art „Fangprämie" gesprochen, die weder einen schadensersatzrechtlichen noch einen haftungsrechtlichen Charakter haben soll. Die Belohnung soll lediglich einen finanziellen Anreiz darstellen, Insiderwissen offen zu legen. Probleme ergeben sich zudem, weil der

---

100 Vgl. Handelsblatt, Artikel vom 25.10.2012, abrufbar unter: http://www.handelsblatt.com/politik/international/wirtschaftskriminalitaet-eu-kommission-will-whistleblower-praemie-einfuehren/7301992.html, (abgerufen am 29.11.2014).

101 Vgl. *Buchert*, CCZ 2013, 144, 144.

102 Wirtschaftsstrafrecht ist der Oberbegriff für alle Strafvorschriften, die im Bereich der Wirtschaft liegende Tatbestände unter Strafe stellen. Ob eine Strafvorschrift inhaltlich Wirtschaftsstrafrecht ist, ergibt sich aus deren Schutzzweck, vgl. *Grützner/Jakob*, Compliance von A-Z, Wirtschaftsstrafrecht.

Hinweisgeber durch die Auslobung einer Whistleblower-Prämie in Konflikt mit einem im Unternehmen integrierten Hinweisgebersystem geraten kann. Mit einem funktionierenden Hinweisgebersystem kann sichergestellt werden, dass empfindliche Informationen nicht nach außen gelangen. Mit der Einführung einer staatlichen Whistleblower-Prämie würde externes Whistleblowing attraktiv werden.

Übrig bliebe nur die Anwendung der strafrechtlichen Kronzeugenregelung nach § 46 b StGB,[103] bei der in strafrechtlicher Hinsicht ein „Bonus" gewährt wird. Eine Strafrahmenverschiebung nach § 46 b StGB kann sich der Kronzeuge nur „verdienen", wenn er entweder durch freiwilliges Offenbaren seines Wissens wesentlich dazu beigetragen hat, dass eine Straftat aus dem Katalog des § 100 a Abs. 2 StPO aufgedeckt wird (Aufklärungsvariante) oder eine solche Tat verhindert wird (Präventionsvariante). Indem der Kronzeuge die Tatbestände erfüllt, kann er Strafmilderung oder (unter bestimmten Voraussetzungen) Straffreiheit erlangen. Kronzeuge im Sinne von § 46 b StGB kann jedoch nur sein, wer Täter oder Teilnehmer einer Straftat ist, die mit einer im Mindestmaß erhöhten Freiheitsstrafe oder mit einer lebenslangen Freiheitsstrafe bedroht ist. Für eine im Mindestmaß erhöhte Freiheitsstrafe kommt es nach § 46 b Abs. 1 S. 2 StGB nicht nur auf den an § 38 Abs. 2 StGB zu messenden Grundstrafrahmen nebst etwaigen Qualifikationen an, sondern auch auf Schärfungen für besonders schwere Fälle. Damit sind vor allem Regelbeispiele und zwingende Beispielfälle erfasst.[104]

Seit dem Inkrafttreten der Neuregelung des § 46 b StGB im August 2013 kann der Kronzeuge nur noch Strafmilderung erlangen, wenn er sein Wissen im Rahmen sog. „Zusammenhangstaten" mitgeteilt hat.[105] Bis zu der Neuerung konnte der Kronzeuge auch dann in den Genuss der Strafmilderung kommen, wenn keinerlei Zusammenhang zwischen der begangenen Tat und der Tat, zu der er sein Wissen offenbarte, bestand. Nun ist § 46 b StGB nur anwendbar, wenn zwischen der Tat des Kronzeugen und der Tat, auf die sich seine Angaben beziehen, ein Zusammenhang besteht;[106] der Kronzeuge also beispielsweise eine Waffe gestohlen hat (Diebstahl) und Angaben zu Personen macht, die diese Waffe – mit seinem Wissen – bei einem Raub verwendet haben (schwerer Raub).

Es fallen also nur Personen in den Anwendungsbereich der strafrechtlichen Kronzeugenregelung, die in einem Zusammenhang mit der zu untersuchenden Tat stehen und deren Hinweis von Gegenständen handelt, die in § 100 a Abs. 2 StPO aufgezählt sind. Damit scheiden regelmäßig Fälle aus, die sich z. B. um das Thema Umweltschutz drehen, obwohl die Verstöße strafrechtlich geahndet werden. Entscheidet sich ein Whistleblower – involviert oder nicht – dazu, einen

103 Zuletzt geändert durch das 46. Strafrechtsänderungsgesetz – Beschränkung der Möglichkeit zur Strafmilderung bei Aufklärungs- und Präventionshilfe – v. 10.6.2013, BGBl I-2013, S. 1497.

104 *Dann*, CCZ 2010, 30 mit Beispielen.

105 BT-Drs. 17/9695, S. 1.

106 Vgl. dazu *Peglau*, NJW 2013, 1910, 1912.

umweltrechtlichen Verstoß zu melden, kann er keine rechtliche Besserstellung erwarten. Weder haftungsrechtlich noch strafrechtlich erhält er einen „Bonus" für die Offenbarung des Gesetzesverstoßes. Dabei ist auch die Vielfalt der umweltrechtlichen Gesetze nicht hilfreich: Weder im KrWG (Kreislaufwirtschaftsgesetz), BBodSchG (Bundesbodenschutzgesetz), BNatSchG (Bundesnaturschutzgesetz), BImSchG (Bundesimmissionsschutzgesetz) noch im WHG (Wasserhaushaltsgesetz), TEHG (Treibhausgas-Emissionshandelsgesetz) oder sonstigen umweltrechtlichen Spezialgesetzen ist von einer Bonusregelung die Rede.

Betrifft der Hinweis des Whistleblowers eine Steuerstraftat, ist zwischen (leichteren) Steuerordnungswidrigkeiten und (schwereren) Steuerstraftaten zu unterscheiden. Dabei werden unterschiedlich abgestufte Sanktionen (Bußgelder, Geldstrafen und Freiheitsstrafen) festgelegt. Steuerstraftaten sind u. a. in den §§ 369 ff. der Abgabenordnung (AO) geregelt. Danach sind Steuerstraftaten insbesondere alle Taten, die nach den Steuergesetzen strafbar sind. Zu den Steuerstraftaten gehören u. a.:

- Steuerhinterziehung, § 370 AO[107]
- Bannbruch, § 372 AO
- gewerbsmäßiger, gewaltsamer und bandenmäßiger Schmuggel, § 373 AO
- Steuerhehlerei, § 374 AO
- Steuerzeichenfälschung, § 369 Abs.1 Nr. 3 AO i.V.m. §§ 148–150 StGB.

Welche konkrete Strafe ein Gericht im Einzelfall verhängt, ist von verschiedenen Umständen abhängig. Die konkrete Strafe wird bei der Strafzumessung ermittelt. Eine Besonderheit im Steuerstrafrecht ist die Möglichkeit, trotz einer Steuerhinterziehung Straffreiheit durch eine rechtzeitige und ordnungsgemäße Selbstanzeige (§ 371 AO) zu erhalten. Die strafbefreiende Selbstanzeige nach § 371 AO regelt einen besonderen persönlichen Strafaufhebungsgrund, der neben die Vorschriften über den Rücktritt vom Versuch und die tätige Reue tritt.[108] Geknüpft an Korrektur und Wiedergutmachung, schützt die Selbstanzeige nachträglich vor Bestrafung wegen einer begangenen Steuerhinterziehung.[109] Gemäß § 371 Abs. 3 AO tritt diese nachträgliche strafbefreiende Wirkung für einen Tatbeteiligten (Täter oder Teilnehmer) aber nur dann ein, wenn er die Steuern binnen angemessener Frist nachzahlt.[110] Demnach erlangt der Täter/Teilnehmer keine vollständige, sondern nur eine strafrechtliche Amnestie. Mit der kartellrechtlichen Bonusregelung, nach der das Bußgeld erlassen bzw. ermäßigt wird, ist die Selbstanzeige nicht vergleichbar. Die Selbstanzeige schließt nur die Bestrafung wegen Steuerhinterziehung aus. Auf

107 Der gesetzliche Strafrahmen für die (einfache) Steuerhinterziehung beträgt bis zu 5 Jahren Freiheitsstrafe oder Geldstrafe. In besonders schweren Fällen beträgt die Freiheitsstrafe 6 Monate bis zu 10 Jahren.
108 *Schmitz*, in: MüKo StGB, § 369 AO, Rn. 24.
109 *Kohler*, in: MüKo StGB, § 371 AO, Rn. 1.
110 *Jäger*, in: Klein, AO, § 371 AO, Rn. 1.

andere (Steuer-)Straftaten ist die Vorschrift des § 371 AO – soweit nicht ausdrücklich auf sie verwiesen wird[111] – nicht analog anwendbar.[112]

Bei Steuerordnungswidrigkeiten handelt es sich um geringfügige Verstöße gegen steuerliche Normen. Im Unterschied zu Steuerstraftaten, welche vorsätzlich begangen sein müssen, lässt das Gesetz für die Steuerordnungswidrigkeiten Fahrlässigkeit genügen, oftmals in Form der Leichtfertigkeit. (Steuer-) Ordnungswidrigkeiten[113] werden mit einem Bußgeld geahndet.

Im Ergebnis ist festzuhalten, dass es nur im Kartellrecht eine Bonusregelung gibt, die dem Täter die (vollständige) Zahlung des Bußgeldes erlässt. Die Offenlegung durch den Whistleblower wird deshalb in vielen Fällen nicht aus finanziellen Gründen erfolgen. Umso schwieriger ist es für diese Personen, die Gesetzesverstöße zu melden. Ob die Schaffung weiterer Bonusregelungen in anderen Rechtsgebieten sinnvoll ist, lässt sich schwer beantworten.

---

111 Bei einer leichtfertigen Steuerverkürzung gem. § 378 sind § 371 Abs. 3 und 4 entsprechend anwendbar. Alle übrigen Steuerordnungswidrigkeiten, die weder einen ausdrücklichen Verweis auf § 371 noch eine § 378 Abs. 3 Satz 1 vergleichbare Regelung enthalten, können demgegenüber nicht, auch nicht analog, von der Selbstanzeigenorm profitieren, vgl. *Kohler*, in: MüKo StGB, § 371 AO, Rn. 34.

112 *Kohler*, in: MüKo StGB, § 371 AO, Rn. 237 m. w. N.

113 Ordnungswidrigkeiten sind die leichtfertige Steuerverkürzung (§ 378 AO), die Steuergefährdung (§ 379 AO), die Gefährdung von Abzugssteuern (§ 380 AO), die Verbrauchsteuergefährdung (§ 381 AO), sowie die Gefährdung von Eingangsabgaben (§ 382 AO).

Teil 3: Aktivismus gegen den Gesetzesverstoß

Wer sich mit dem Thema Whistleblowing beschäftigt, stößt häufig auf den Begriff „Zivilcourage". Zivilcourage ist gleichbedeutend mit sozialem Mut, also einem bestimmten Typus sozial verantwortlichen Handelns. Dieser Bürgermut und die Erkenntnis, dass rechtswidriges Handeln im Unternehmen nicht geduldet werden darf, bewegen den Whistleblower dazu, gegen Gesetzesverstöße im Unternehmen vorzugehen. Die folgenden Beispielfälle befassen sich mit Gesetzesverstößen in bestimmten Branchen und Bereichen (Gesundheit, Umwelt, Steuern und Wettbewerb) und sollen exemplarisch zeigen, wie gegen Missstände im Unternehmen vorgegangen wird und welche Schäden entstehen können, wenn die Anzeichen und Hinweise eines Whistleblowers nicht wahrgenommen werden. Auch wenn eingangs betont wurde, dass Whistleblowing im öffentlichen Sektor nicht Gegenstand der Untersuchungen sein wird, betrifft der Fall im Bereich Steuern den öffentlichen Dienst, um die Vielfältigkeit von Whistleblower-Aktivitäten zu verdeutlichen.

Einige Personen, um die es im Folgenden gehen wird, erhielten für ihre Zivilcourage und ihr Engagement den Whistleblower-Preis. Die deutsche IALANA hat zusammen mit der Vereinigung deutscher Wissenschaftler (VDW) diesen Preis gestiftet. Er wird an Persönlichkeiten vergeben, die – häufig unter Inkaufnahme beträchtlicher Risiken für Arbeitsplatz und Karriere – Missstände aufdecken und nach außen bekannt machen, die ihnen in ihrer dienstlichen oder amtlichen Tätigkeit bekannt geworden sind.[114]

A. Whistleblowing im Bereich Gesundheit

I. Beispielsfall: Der Fall Strecker

Der LKW-Fahrer Miroslaw Strecker übernahm im August 2007 in Hamburg 13 Paletten tiefgefrorene Schlachtabfälle, die nach Bayern sollten.[115] Die elf Tonnen

---

114 So unter: http://www.ialana.de/arbeitsfelder/whistleblowing/whistleblower-preis/allgemeines-zum-whistleblowerpreis-und-seiner-verleihung; IALANA e.V. steht für "International Association Of Lawyers Against Nuclear Arms", (abgerufen am 10.11.2013).

115 Die Informationsgewinnung erfolgte aus dem Artikel des BMI vom 14.10.2011, abrufbar unter http://www.bmi.bund.de/SharedDocs/Standardartikel/DE/Themen/PolitikGesellschaft/ohneMarginalspalte/xy_1_2011.html,(abgerufenam28.11.2014); dem Artikel des Tagesspiegels vom 25.8.2013, abrufbar unter http://www.tagesspiegel.de/politik/zwischen-helden-und-verraetertum-deutschland-unternimmt-wenig-um-whistleblower-zu-schuetzen/8686118–3.html, (abgerufen am 28.11.2014); dem Artikel aus der Badischen Zeitung vom 30.9.2013, abrufbar unter http://www.badische-zeitung.de/deutschland-1/wie-ergeht-es-dem-whistleblower-im-gammelfleisch-skandal—75711790.html, (abgerufen am 28.11.2014).

Fleisch waren laut den Papieren mit der offiziellen Kategorie „K3" klassifiziert. Das heißt, dass das Fleisch nicht für den menschlichen Verzehr geeignet ist, aber für Katzen- und Hundefutter verwendet werden darf.

Herr Strecker hatte gefrorene Rinderaugen, vier Jahre alte Hammelleber aus Neuseeland und verdorbene Fleischlappen geladen. Das Fleisch war höchstens noch für Tierfutter zu gebrauchen – doch in der Fabrik wurden Lebensmittel für Menschen verarbeitet. Beim Abladen bemerkte er, dass der Chef sämtliche Zettel, die das Fleisch als „K3"-Ware und „Nicht für den menschlichen Verzehr geeignet" kennzeichneten, abriss und verschwinden ließ.

Er rief bei der Polizei an, wurde aber zur Berufsgenossenschaft verwiesen. Beim dritten Anruf erreichte er das Veterinäramt, das noch am selben Tag den Betrieb überprüfte. Nachdem sich die Polizei doch der Sache annahm, reagierten auch die Behörden und deckten in der Folge einen riesigen Skandal auf: Das Gammelfleisch sollte umetikettiert und an einen Döner-Lieferanten in Berlin verkauft werden. Auf diese Weise waren im Vorfeld bereits 150 Tonnen Schlachtabfälle mit rund 150.000 Euro Umsatz in Umlauf gebracht worden.[116] Die „Wurst- und Fleischfabrik" wurde vom Staatsanwalt noch am selben Tag geschlossen.

Nach einiger Zeit wurde Herr Strecker das erste Mal krank. Man riet ihm zur Kündigung. Seine Chefs teilten ihn nach seiner Genesung künftig für die schweren Touren mit vielen Ladestopps ein. Die hoch gestapelten Paletten durfte er eigentlich nicht heben, weil ihm dies der Arzt nach einer Schulteroperation geraten hatte. Seine Vorgesetzten wussten das. Nach längerer erneuter Krankheit wurde Herr Strecker 2010 von seinem Arbeitgeber gekündigt – ein Zusammenhang mit seiner couragierten Tat wird vermutet, kann aber nicht bewiesen werden.[117] Er kämpfte vor Gericht und erstritt eine Abfindung.

Miroslaw Strecker deckte einen der größten Gammelfleisch-Skandale in Deutschland auf. 2007 wurde er für seine Zivilcourage vom damaligen Bundeslandwirtschaftsminister Horst Seehofer mit der erstmals verliehenen Goldenen Plakette des Bundeslandwirtschaftsministeriums ausgezeichnet.

II. Besonderes Interesse

Betrachtet man den Fall des Miroslaw Strecker, wird offensichtlich, welche Folgen das Nichtaufdecken eines Skandals haben kann. Ungenießbare Lebensmittel sind für die Gesundheit der Menschen gefährlich und können schwerwiegende Krankheiten hervorrufen. Im vorliegenden Fall waren hochrangige Güter gefährdet, nämlich die Gesundheit und körperliche Integrität. Diese Güter sind grundrechtlich geschützt.

---

116 Artikel des BMI vom 14.10.2011, abrufbar unter http://www.bmi.bund.de/SharedDocs/Standardartikel/DE/Themen/PolitikGesellschaft/ohneMarginalspalte/xy_1_2011.html, (abgerufen am 28.11.2014).

117 Artikel des BMI vom 14.10.2011, abrufbar unter http://www.bmi.bund.de/SharedDocs/Standardartikel/DE/Themen/PolitikGesellschaft/ohneMarginalspalte/xy_1_2011.html, (abgerufen am 28.11.2014).

Schon deshalb hat der Whistleblower ein besonderes Interesse daran, Fehlverhalten zu offenbaren, die sich gegen grundrechtlich geschützte Güter richten.

Zwar wurden die Haupttäter wegen gewerbsmäßigen Betrugs zu zwei Jahren Haft verurteilt; zwei Komplizen erhielten Bewährungsstrafen (2 Jahre bzw. 15 Monate) und Geldstrafen (10.000 Euro bzw. 2.200 Euro). Dennoch ist das Inverkehrbringen von verdorbenem Fleisch die schwerwiegendere Tat (Straftat gegen das Lebensmittel- und Futtermittelgesetzbuch)[118], weil durch sie die Verletzung der Gesundheit der Menschen billigend in Kauf genommen wurde und sie direkte gesundheitliche Auswirkungen auf die Bevölkerung hätte haben können. Hätte man einen Kausalzusammenhang herstellen können, dass eine Person ernsthaft aufgrund des Verzehrs verdorbenen Fleisches erkrankt ist, hätte sich der Betrüger zusätzlich wegen einer Körperverletzung verantworten müssen. Der Whistleblower Miroslaw Strecker hat die Gesetzesverstöße durch die Meldung an die Polizei und die Behörden aufgehalten.

Auch wenn die Folgen und Auswirkungen des „Umetikettierens" anders gewertet werden sollten, handelt es sich in dem Fall um eine gesetzeswidrige Verhaltensweise des Arbeitgebers, die nur durch das besondere Pflichtbewusstsein und Verantwortungsgefühl des Arbeitnehmers beendet werden konnte.

B. Whistleblowing im Bereich Umweltschutz/Naturschutz

I. Beispielsfall: Der Fall Moormann

Herr Dr. Rainer Moormann arbeitete seit 35 Jahren in der Kernforschungsanlage (KFA), dem heutigen Forschungszentrum (FZJ), in Jülich. Zu seinen wissenschaftlichen Arbeitsschwerpunkten zählte über lange Zeit die Sicherheit von Kugelhaufen-Reaktoren. Ein Versuchsreaktor dieses Typs (AVR) mit einer Kapazität von 15 Megawatt war in Jülich bis 1988 in Betrieb. Herr Dr. Moormann ist durch seine Untersuchungen zu dem Schluss gelangt, dass mit der Kugelhaufen-HTR-Technologie andere, nicht minder bedrohliche Störfallmöglichkeiten und Risiken mit katastrophalen Folgen für Mensch und Umwelt verbunden sind.[119]

Herr Dr. Moormann hat aufgedeckt, dass der 1988 endgültig stillgelegte Versuchsreaktor in Jülich im Normalbetrieb jahrelang unzureichend gegen überhöhte

118 Lebensmittel- und Futtermittelgesetzbuch (LFGB) in der Fassung der Bekanntmachung v. 3.6.2013, BGBl. I-2013, S. 1426, das durch Artikel 2 des Gesetzes vom 5.12.2014, BGBl. I-2013, S. 1975, geändert worden ist. Das LFGB ist durch das Gesetz zur Neuordnung des Lebensmittel- und des Futtermittelrechts (NeuordG) vom 1.9.2005, BGBl. I-2005, S. 2618, 3007 geschaffen worden.

119 Begründung zum Whistleblower-Preis, S. 1, abrufbar unter: http://www.vdw-ev.de/images/stories/vdwdokumente/aktuelles/preistrger%20moormann.pdf, (abgerufen am 10.11.2013).

Betriebstemperaturen im Reaktorkern gesichert war.[120] Hinweisen auf zu hohe Reaktortemperaturen wurde nicht hinreichend nachgegangen. Er hat Indizien dafür vorgelegt, dass der Betreiber-Gesellschaft mutmaßlich schon seit Ende der 1970er Jahre die Problematik überhöhter Betriebstemperaturen aufgefallen war. Die Aufsichtsbehörde gab sich mit der Vorlage von Modellrechnungen durch die Betreiber-Gesellschaft zufrieden. Durch die Untersuchungen von Herrn Dr. Moormann ist auch der begründete Verdacht aufgekommen, dass der AVR Jülich am 13. Mai 1978 nur knapp einem GAU mit den verheerenden Folgen einer weitflächigen radioaktiven Verseuchung der Umwelt entging.[121] Betreiber-Gesellschaft, Forschungszentrum Jülich und Aufsichtsbehörden in Land und Bund haben sich der Aufarbeitung dieser Gefahrensituation bisher nicht hinreichend gestellt. In der Folge des 1978 als normaler Störfall eingestuften Wassereinbruchs gelangte beim Abpumpen radioaktiv hoch kontaminiertes Wasser aus dem Reaktorbehälter ins Erdreich unter dem Reaktor und ins Grundwasser. Die konkreten Auswirkungen dieser Kontamination liegen bis heute im Dunkeln. Über damit verbundene Gesundheitsgefährdungen sowie einen möglichen Zusammenhang mit gehäuften Leukämieerkrankungen im Umland herrscht Ungewissheit.

Wie intern seit Langem bekannt, ist der Reaktordruckbehälter radioaktiv hoch kontaminiert. Die genauen Ursachen für die hohe Kontamination konnten bisher nicht ermittelt werden, zumal der Reaktor mit Beton ausgegossen wurde, um den radioaktiven Staub zu binden.

Herr Dr. Moormann hat teuer für seine Zivilcourage zahlen müssen. Er wurde intern und von der externen „Kugelhaufen-Community“ als Nestbeschmutzer diffamiert und als „verrückt“ verleumdet. Seine Arbeitsgruppe im FZ Jülich wurde aufgelöst. Er selbst wurde in eine andere Abteilung versetzt. Dort wurde er aufgefordert, seine „nuklearfeindlichen Aktivitäten“ einzustellen, da man auf Aufträge aus dem Nuklearbereich angewiesen sei.[122]

II. Exkurs: Störfälle in Hochrisikoanlagen

Das deutsche Atomhaftungsrecht ist in den §§ 25 AtG i. V. m. dem Pariser Atomhaftungsübereinkommen (PÜ) geregelt. Nach § 25 Abs. 1 AtG i. V. m. Art. 3, 6 PÜ haftet allein der Inhaber eines Kernkraftwerks für Schäden, die auf einem von seiner Anlage ausgehenden nuklearen Ereignis beruhen. Dabei handelt es sich um eine

---

120 Vgl. Informationen, abrufbar unter: http://www.heise.de/newsticker/meldung/Whistleblower-Preis-2011-geht-an-Kernforscher-1255724.html, (abgerufen am 11.11.2013).

121 Begründung zum Whistleblower-Preis, S. 1, abrufbar unter: http://www.vdw-ev.de/images/stories/vdwdokumente/aktuelles/preistrger%20moormann.pdf, (abgerufen am 11.11.2013).

122 Begründung zum Whistleblower-Preis, S. 2, abrufbar unter: http://www.vdw-ev.de/images/stories/vdwdokumente/aktuelles/preistrger%20moormann.pdf, (abgerufen am 11.11.2013).

Gefährdungshaftung, die weder Rechtswidrigkeit noch Verschulden voraussetzt. Zudem gibt es eine Störfall-Verordnung,[123] die in Betriebsbereichen zu beachten ist, in denen gefährliche Stoffe in bestimmten Mengen vorhanden sind. Die Betreiber solcher Anlagen haben bestimmte Vorkehrungen zu treffen, damit keine Störfälle entstehen, insbesondere haben sie Meldepflichten (§ 19 12. BImSchV) einzuhalten. Wer gegen diese Pflichten verstößt, handelt nach § 21 Abs. 1 Nr. 15 12. BImSchV ordnungswidrig.

Die wichtigsten strafrechtlichen Tatbestände in diesem Bereich sind die Tötungs-, Körperverletzungs-, Umwelt- und die Sprengstoffdelikte. Das deutsche Strafrecht kann aber nur eingeschränkt herangezogen werden, da sich juristische Personen nicht strafbar machen können.[124] Zur Strafverfolgung müssten einzelne Unternehmensangehörige herausgefiltert werden, was wegen der Aufgabenaufteilung und oftmals unübersichtlichen Verantwortungsstrukturen sehr schwierig ist.[125]

Im Umweltstrafrecht ist an § 328 StGB zu denken, insbesondere an Abs. 3, nach dessen Nr. 1 verschiedenes Verwenden (Lagern, Bearbeiten, Verarbeiten, sonstiges Verwenden) von radioaktiven Stoffen oder von gefährlichen Stoffen und Gemischen i. S. d. Art. 3 der Verordnung Nr. 1272/2008 (EG)[126] beim Betrieb einer Anlage mit Strafe bedroht ist, sofern der Täter verwaltungsrechtliche Pflichten verletzt und eine Gefahr für die Gesundheit eines anderen, fremde Sachen von bedeutendem Wert, Tiere oder Pflanzen, Gewässer, die Luft oder den Boden hervorruft.[127]

Als verwaltungsrechtliche Pflichten kommen dabei Pflichten aus dem Chemikalien- und Gefahrstoffrecht sowie dem Transportrecht in Betracht,[128] etwa gravierende Verstöße gegen Ladungs-, Informations- und Überwachungspflichten.[129] Die Vorschriften in den Gefahrgut-Verordnungen (GGVSEB, GGVSee, ADR-Übk., COTIF und GGAV)[130] legen die verwaltungsrechtlichen Pflichten fest, die bei den Beförderungsvorgängen der Güter zu beachten sind. Die einschlägigen

---

123 Störfall-Verordnung (12. Verordnung zur Durchführung des Bundes-Immissionsschutzgesetzes) vom 8.6.2005, BGBl. I-2005, S. 1598, die zuletzt durch Art. 1 der Verordnung vom 14.8.2013, BGBl. I-2013, S. 3230, geändert worden ist.

124 Vgl. aber § 30 OWiG, nach dem eine Geldbuße gegen Unternehmen festgesetzt werden kann.

125 Notfallschutz bei Kernkraft-, Chemie- und Sondermüllanlagen – Bericht zur wissenschaftlichen Fachtagung am 5.5.2011 in Berlin *Burbat*, ZUR 2011, 499, 501.

126 Verordnung (EG) Nr. 1272/2008 des Europäischen Parlamentes und des Rates vom 16. Dezember 2008 über die Einstufung, Kennzeichnung und Verpackung von Stoffen und Gemischen, zur Änderung und Aufhebung der Richtlinien 67/548/EWG und 1999/45/EG und zur Änderung der Verordnung (EG), 31.12.2008, ABl. L 353/1.

127 Inhalt des § 328 StGB in neuster Fassung nach dem Strafrechtsänderungsgesetz, 45. Strafrechtsänderungs-gesetz zur Umsetzung der Richtlinie des Europäischen Parlaments und des Rates über den strafrechtlichen Schutz der Umwelt vom 6.4.2011, BT-Drs. 17/5391; Beschlussempfehlung und Bericht des Rechtsausschusses (6. Ausschuss) zu dem Gesetzentwurf der Bundesregierung, BT-Drs. 17/7674.

128 OVG Lüneburg, Urt. v. 14.2.2003 – 7 ME 64/02, NuR 2003, 565, 566 f.

129 *Rengier*, Das moderne Umweltstrafrecht, S. 33.

130 Weitere Ausführungen siehe *Alt*, in: MüKo StGB, § 328 StGB, Rn. 43, 45.

verwaltungsrechtlichen Pflichten können sich aber auch aus den für Gewässer-, Luft- und Bodenschutz sowie aus für den Tier- und Pflanzenschutz relevanten Vorschriften ergeben.[131] Ferner ergeben sich Nachweispflichten aus §§ 50 ff. KrWG, gefahrgutspezifischen Vorschriften zur Regelung des Straßenverkehrs[132] sowie arbeitsschutzrechtlichen Vorschriften[133] gegen Gase, Nebel, Dämpfe und Stäube.

Täter im Sinne der Abs. 2 Nr. 1 und Abs. 3 kann aber nur derjenige sein, der verwaltungsrechtlich verpflichtet ist.[134] Strafrechtlich können die Pflichten insofern nach § 14 StGB übergehen. Amtsträger, die nicht selbst die Tathandlung ausführen, kommen nicht als Täter nach Abs. 1 und 3 in Betracht.[135] Sie können nach Abs. 2 Täter sein, beispielsweise, wenn sie entgegen einer Garantenpflicht die Herbeiführung einer nuklearen Explosion nicht verhindern.

III. Besonderes Interesse

Nach dem 2011 die Betreiber-Gesellschaft eine Expertengruppe mit der Aufarbeitung der Reaktorgeschichte beauftragt hatte, wurde im April 2014 der Abschlussbericht vorgelegt.[136] Demzufolge kam es über Jahrzehnte immer wieder zu Störfällen und wahrscheinlich mindestens zweimal zur Kontaminierung der Umwelt. Dem Bericht ist zudem zu entnehmen, dass man Notfallsysteme bewusst abgeschaltet hatte und die Betreiber bei Störfällen darauf verzichteten, die Störung zu melden. Es wurde offensichtlich gegen eine gesetzliche Betreiberpflicht verstoßen. Zwar war damals die heute geltende Störfall-Verordnung (12. BImSchV) noch nicht in Kraft, nach der zumindest gegen die Meldepflicht nach § 19 12. BImSchV verstoßen worden wäre. Nach dem Bericht bestand aber gleichwohl eine Pflicht gegenüber der Aufsichtsbehörde, Störfälle zu melden, zumindest aber die Pflicht zu einer Korrektur der „Störfall-Einstufung".[137]

Zwar geht es in dem oben geschilderten Fall ebenfalls um wichtige Güter wie Gesundheit und Leben eines Menschen (wenn die Umweltschäden sich auswirken), in erster Linie ist jedoch durch das Verhalten die Umwelt[138] schwer betroffen.

---

131 *Heine*, JR 1996, 300, 302 f.; *Möhrenschlager*, NStZ 1994, 566, 567; *Heine/Hecker*, in: Schönke/Schröder, Kommentar StGB, § 328 StGB, Rn. 21 f.

132 *Rengier*, Das moderne Umweltstrafrecht, S. 35 f.; *ders.*, FS Boujong, S. 791, 799; *Heine/Hecker*, in: Schönke/Schröder, Kommentar StGB, § 328 StGB, Rn. 22, 23.

133 *Alt*, in: MüKo StGB, § 328 StGB, Rn. 45, m. w. N.

134 *Ransiek*, in: Kindhäuser/Neumann/Paeffgen, Kommentar StGB, § 328, Rn. 15 mit Verweis auf Ausführungen zu § 324 a, Rn. 27.

135 *Ransiek*, in: Kindhäuser/Neumann/Paeffgen, Kommentar StGB, § 328, Rn. 15.

136 Der Bericht ist abrufbar unter: http://www.fz-juelich.de/portal/DE/UeberUns/selbstverstaendnis/verantwortung/avr/avr-expertengruppe/_node.html, (abgerufen am 6.12.2014).

137 Bericht der AVR-Expertengruppe (abrufbar siehe eben), Kurzfassung, S. 10 f.

138 Allg. zur Bedeutung des Umweltschutzes für das Arbeitsverhältnis *Schmitt-Schönenberg*, AuR 1994, 281.

Art. 20 a GG weist den Schutz der natürlichen Lebensgrundlagen als Staatsziel aus. Auch wenn der Begriff der „Staatszielbestimmung" konturlos erscheinen mag, so besteht doch Einigkeit, dass Staatsziele ihrer wesensmäßigen Struktur nach keine anspruchsbegründende Wirkung für den Einzelnen entfalten, sondern grundsätzlich nur objektiv-rechtlich zu verstehende Verfassungsgrundsätze darstellen.[139] Dennoch ist die Umwelt ein Gut, dessen Erhaltung im Interesse eines großen Teils der Bevölkerung steht. Die Beweggründe des Whistleblowers, in Fällen tätig zu werden, in denen die Umwelt betroffen ist, liegen hauptsächlich darin, dass Umweltschäden über längere Generationen anhalten und zu Folgeproblemen für das ganze Klima in der Welt führen können. Der Umweltschutz hat einen großen Wirkungsbereich: die Erhaltung und den Schutz der Umwelt sowie die Verbesserung ihrer Qualität, den Schutz der menschlichen Gesundheit, die umsichtige und rationelle Verwendung der natürlichen Ressourcen sowie die Förderung von Maßnahmen auf internationaler Ebene zur Bewältigung regionaler oder globaler Umweltprobleme und insbesondere die Bekämpfung des Klimawandels.[140]

C. Whistleblowing im Bereich Steuern

I. Beispielsfall: Der Fall Schmenger/Wehrheim

Die Herren Rudolf Schmenger und Frank Wehrheim waren über viele Jahre Steuerfahnder im sog. „Bankenteam" am Finanzamt Frankfurt.[141] Es war bekannt, dass Bankbedienstete entsprechend den Vorgaben und Erwartungen ihrer Vorgesetzten viele wichtige Kunden über Jahre hinweg dabei unterstützt haben, Gelder anonymisiert am deutschen Steuerfiskus vorbei nach Luxemburg, in die Schweiz oder andere Steueroasen zu transferieren und anzulegen.

Die Steuerfahndungsabteilung, in der Herr Schmenger und Herr Wehrheim mit steuerstrafrechtlichen Ermittlungsverfahren im Bankenbereich befasst waren, operierte seit Jahren sehr erfolgreich. Als sie erfuhren, dass vorgesetzte Dienststellen der hessischen Finanzverwaltung ihre Ermittlungsarbeit aus nicht nachvollziehbaren Gründen mittels einer so genannten „Amtsverfügung" vom 30. August 2001 zu beeinträchtigen suchten, entstand ein Konflikt.[142]

Rudolf Schmenger, Frank Wehrheim und ihre Kollegen wussten aus langjähriger Erfahrung, dass Gelder häufig in kleineren Tranchen von unter 300.000 DM ins Ausland verschoben werden.[143] Aufgrund der Amtsverfügung sollte künftig jedoch

139 *Scholz*, in: Maunz/Dürig, GG-Kommentar, Art. 20 a GG, Rn. 33.

140 So lauten auch die Ziele der unionalen Umweltpolitik (Art. 191 f. AUV), siehe *Terhechte*, in: Grabitz/Hilf/Nettesheim, Das Recht der EU, Art. 3 EUV, Rn. 55.

141 *Deiseroth/Falter*, Whistleblower in der Steuerfahndung, S. 9.

142 *Deiseroth/Falter*, Whistleblower in der Steuerfahndung, S. 9; Dokument Nr. 1 (Amtsverfügung 2001/08 vom 30.8.2001 vom Finanzamt Frankfurt a. M.), in: *Deiseroth/Falter*, Whistleblower in der Steuerfahndung, S. 107.

143 *Deiseroth/Falter*, Whistleblower in der Steuerfahndung, S. 10.

nur ein steuerstrafrechtlicher Anfangsverdacht angenommen werden, wenn es sich um Summen von über 300.000 DM handelte. Es wurde von ihnen verlangt, erst in Kenntnis aller erforderlichen Daten nach Beweisen zu suchen, wobei das Finanzamt sich auf die in einer Entscheidung des LG Freiburg aufgestellten allgemeinen Grundsätze stützte.[144] Sowohl Herr Schmenger und Herr Wehrheim als auch die hessische Generalstaatsanwaltschaft hielten „die Anknüpfung eines Tatverdachts im Sinne des § 152 Abs. 2 StPO“ an bestimmte, zum Teil mehr oder weniger willkürliche Kriterien für bedenklich. Mit diesen Kriterien ließe sich ein Anfangsverdacht weder nachvollziehbar begründen noch ablehnen.[145] Ferner sprach ihre Erfahrung dafür, dass Banken, die über Jahre Beihilfe zur Steuerhinterziehung geleistet haben, auch in den Folgejahren strafbare Finanzpraktiken betreiben würden.[146]

Die Herren Schmenger und Wehrheim sowie viele ihrer Kollegen sahen deshalb die Gefahr, dass erhebliche Steueransprüche der öffentlichen Haushalte gegen Steuersünder nicht durchgesetzt und verloren gehen würden, dadurch der „Gleichheitsgrundsatz“ der Besteuerung verletzt werde und sie sich letztlich der „Strafvereitelung im Amt“ strafbar machten.[147]

Die Begebenheiten brachten Rudolf Schmenger und Frank Wehrheim dazu, ihre Bedenken offen vorzutragen. Anstatt die Kritik seitens der Leitung des Finanzamtes aufzugreifen, wurde mit Repressalien, Mobbing und Einschüchterungsversuchen gegenüber den Steuerfahndern reagiert. Mehr als ein Dutzend kritischer Steuerfahnder wurden versetzt und zwei Steuerfahnder sowie Sachgebietsleiter mussten die Steuerfahndung ganz verlassen, darunter Herr Schmenger und Herr Wehrheim.[148] Rudolf Schmenger und andere Steuerfahnder wurden zum Amtsarzt geschickt und auf der Basis schwer nachvollziehbarer psychiatrischer Gutachten als angeblich dienstunfähig „zwangspensioniert“.[149]

II. Besonderes Interesse

Dieser Fall unterscheidet sich in zwei wichtigen Punkten von den vorher genannten Fällen. Zum einen geht es hier um Steuergerechtigkeit und nicht um den Schutz hochrangiger Güter wie Leben, Gesundheit und die Umwelt, die allesamt verfassungsrechtlich geschützt sind. Zum anderen betrifft dieser Fall den öffentlichen

---

144 Dokument Nr. 1 (Amtsverfügung 2001/08 vom 30.8.2001 vom Finanzamt Frankfurt a. M.), in: *Deiseroth/Falter*, Whistleblower in der Steuerfahndung, S. 107.

145 LT-Drs. 16/5360, S. 39 f.

146 *Deiseroth/Falter*, Whistleblower in der Steuerfahndung, S. 10.

147 *Deiseroth/Falter*, Whistleblower in der Steuerfahndung, S. 10.

148 Begründung zum Whistleblower-Preis, S. 2, abrufbar unter: http://www.vdw-ev.de/images/stories/vdwdokumente/aktuelles/jurybegruendung%20whistleblower-preis%202009.pdf, (abgerufen am 11.11.2013).

149 Begründung zum Whistleblower-Preis, S. 2, abrufbar unter: http://www.vdw-ev.de/images/stories/vdwdokumente/aktuelles/jurybegruendung%20whistleblower-preis%202009.pdf, mit Verweis auf „Stern“ 51/2008, S. 33, (abgerufen am 11.11.2013).

Sektor. Das Bundesverfassungsgericht[150] betont zwar, dass andere, etwa volkswirtschaftliche, sozialpolitische oder ökologische Zwecke die steuerliche (Un-) Gleichbehandlung rechtfertigen können. Aber Steuerungerechtigkeit aufgrund von kriminellen Machenschaften wird keinesfalls geduldet. Noch dazu sind in diesem Fall Bankbedienstete und die Landesfinanzverwaltung in die Steuerstraftaten involviert, sodass auch ein großer Vertrauensbruch vorliegt. In diesem Fall ist das Interesse des Whistleblowers auf Gerechtigkeit und die Unterbindung von gesetzeswidrigem Verhalten in der oberen Bedienstetenebene gerichtet. Steuerhinterziehung ist ein Verstoß gegen das Steuerstrafrecht gem. § 370 AO und kann, wie in der jüngsten Presse betont[151], nicht (mehr) als Kavaliersdelikt bezeichnet werden. Die Beihilfe zur Steuerhinterziehung wird ebenfalls strafrechtlich verfolgt.

D. Whistleblowing im Bereich Wettbewerb

I. Kartellstrafrecht im Allgemeinen

In bestimmten Fällen stellt sich die Frage nach Whistleblowing bei einer strafbaren Handlung im Kartellrecht. Der Gesetzgeber hat sich im Jahr 1997[152] entschlossen, § 298 StGB (Wettbewerbsbeschränkende Absprachen bei Ausschreibungen) als Straftatbestand gegen den Wettbewerb in den 26. Abschnitt des StGB einzufügen. Geschütztes Rechtsgut ist vorrangig das Allgemeininteresse an einem freien Wettbewerb; geschützt wird aber auch das Vermögen des Veranstalters einer Ausschreibung und der (möglichen) Mitbewerber.[153] Aufgrund des abweichenden Schutzzwecks von § 263 StGB, wird § 298 StGB nicht als *lex specialis* zu § 263 StGB angesehen.[154] Neben den weiterhin anwendbaren ordnungswidrigkeitenrechtlichen Bestimmungen (insbesondere § 81 Abs. 1 bis Abs. 3 GWB) können § 298 StGB und § 263 StGB parallel zur Anwendung kommen. In der Praxis scheint allerdings

---

150 BVerfG, Urt. v. 20.4.2004 – 1 BvR 1748/99, 905/00, NVwZ 2004, 846, 849; näher dazu *Kischel*, in: BeckOK GG, Art. 3 GG, Rn. 148.

151 Siehe Artikel vom 23.3.212, abrufbar unter: http://www.wiwo.de/finanzen/steuernrecht/rein-rechtlich-steuersuendern-droht-noch-schneller-gefaengnis-seite-all/6364082-all.html, (abgerufen am 12.11.2013).

152 Eingeführt durch Art. 1 Nr. 3 des Gesetzes zur Bekämpfung der Korruption (Korruptionsbekämpfungsgesetz– KorrBekG) v. 13.8.1997, vgl. BGBl. I-1997, S. 2038. Eingehend zur Entstehungsgeschichte des § 298 StGB *Wedlich*, Submissionsabsprachen, S. 21 f.

153 *Klusmann*, in: Wiedemann, Kartellrecht, § 56, Rn. 1 mit Verweis auf *Achenbach*, WuW 1997, 958, 959; *Kleinmann/Berg*, BB 1998, 277, 279; *Wunderlich*, Die Akzessorietät des § 298 StGB zum GWB, S. 38, 70; *Kühl*, in: Lackner/Kühl, StGB-Kommentar, § 298 StGB, Rn. 1; allgemein dazu *Pasewaldt*, ZIS 2008, 84.

154 *Klusmann*, in: Wiedemann, Kartellrecht, § 56, Rn. 1 mit Verweis auf *Achenbach*, WuW 1997, 958, 959; *Kleinmann/Berg*, BB 1998, 277, 281; *Dölling*, ZStW 112 (2000), 334, 348: Idealkonkurrenz; a. A. *Walter*, JZ 2002, 254, 256.

(aufgrund der leichteren Beweisbarkeit seiner Tatbestandsmerkmale) eher § 298 StGB zum Zug zu kommen.[155]

Die Vorschrift gilt für alle Ausschreibungen, also für offene und beschränkte Ausschreibungen (Abs. 1) wie für die freihändige Vergabe nach vorausgegangenem Teilnahmewettbewerb (Abs. 2). Prinzipiell ist seine Anwendung unabhängig davon, ob es sich um einen öffentlichen oder privaten Auftraggeber handelt. Die bloße Absprache zwischen Wettbewerbern genügt nicht – vielmehr muss ein Angebot abgegeben werden, das auf einer kartellrechtswidrigen Absprache beruht.[156] Gegenüber natürlichen Personen können Taten nach §§ 263 und 298 StGB mit Geldstrafe oder mit Freiheitsstrafe bis zu fünf Jahren geahndet werden. Soweit Ordnungswidrigkeiten verfolgt werden, gilt unverändert auch insoweit der Bußgeldrahmen des § 81 Abs. 4 GWB bzw. des § 17 Abs. 4 OWiG.[157]

Neben §§ 298 und 263 StGB (Submissionsbetrug) kommen noch Straftaten wie Bestechung/Vorteilsgewährung, Untreue, Erpressung und Nötigung in Betracht.

II. Beispielsfall: Das Bierkartell 2014[158]

Das Bundeskartellamt hat Geldbußen wegen verbotener Bierpreisabsprachen in Höhe von insgesamt 106,5 Mio. € gegen die Unternehmen Bitburger Braugruppe GmbH (Bitburger), Krombacher Brauerei Bernhard Schadeberg GmbH & Co. KG (Krombacher), C. & A. Veltins GmbH & Co. KG (Veltins), Warsteiner Brauerei Haus Cramer KG (Warsteiner) und Privat-Brauerei Ernst Barre GmbH (Barre) sowie gegen sieben persönlich Verantwortliche verhängt.

Ausgelöst hatte das Verfahren ein Bonusantrag der Anheuser-Busch InBev Germany Holding GmbH (AB InBev), gegen die wegen der Anwendung der Bonusregelung des Bundeskartellamtes keine Geldbuße[159] verhängt wurde. Die Ermittlungen gegen zwei weitere Brauereikonzerne waren zu dem Zeitpunkt noch nicht abgeschlossen. Gleiches gilt für vier regionale Brauereien aus Nordrhein-Westfalen sowie den entsprechenden Regionalverband wegen der Beteiligung an einem regionalen Absprachekreis.

---

155 *Klusmann*, in: Wiedemann, Kartellrecht, § 56, Rn. 1.

156 Ausführlich *Heine/Eisele*, in: Schönke/Schröder, Kommentar StGB, § 298 StGB, Rn. 1, 11 f.

157 *Klusmann*, in: Wiedemann, Kartellrecht, § 56, Rn. 4.

158 Alle folgenden Informationen ergeben sich aus der Pressemitteilung des Bundeskartellamtes vom 13.1.2014, abrufbar unter: http://www.bundeskartellamt.de/SharedDocs/Meldung/DE/Pressemitteilungen/2014/13_01_2014_Fernsehbiere.html, (abgerufen am 15.1.2014).

159 Nach der Bonusregelung („Kronzeugenregelung") des Bundeskartellamtes wird dem Kartellanten das Bußgeld erlassen, wenn er die Voraussetzungen nach Abschnitt B der Mitteilung erfüllt. Danach muss der Kartellant u. a. uneingeschränkt und ununterbrochen mit dem Kartellamt hinsichtlich der Aufklärung zusammenarbeiten.

In gemeinsamen Treffen und bilateralen Kontakten trafen zunächst die überregional tätigen Brauereien eine Vereinbarung über eine Preiserhöhung, über die betroffenen Gebinde (Fass- und/oder Flaschenbier) und über ihre Größenordnung. Anschließend stimmten sich einige der überregional tätigen Brauereien (AB InBev, Veltins und Warsteiner) mit in Nordrhein-Westfalen tätigen regionalen Brauereien (u. a. Barre) auf Sitzungen des regionalen Brauereiverbandes im Juni 2006 und September 2007 über diese Preiserhöhungen ab.

Im Laufe des Verfahrens haben neben AB InBev noch Bitburger, Krombacher, Veltins und Warsteiner auf Basis der Bonusregelung mit dem Bundeskartellamt kooperiert. Darüber hinaus konnte mit allen fünf mit einer Geldbuße bedachten Unternehmen eine einvernehmliche Verfahrensbeendigung (sog. Settlement[160]) erzielt werden.

Beides –Kooperation und Settlement– wurde zugunsten der jeweiligen Unternehmen bußgeldmindernd[161] berücksichtigt.

III. Besonderes Interesse

In diesem Fall war der Whistleblower in die Ordnungswidrigkeit, mithin den Gesetzesverstoß selbst involviert. Er hat als Beteiligter des Kartells an der Tat so lange mitgewirkt, bis er sich entschieden hat, gegen den Rechtsverstoß vorzugehen. Zwar können in Fällen von Kartellrechtswidrigkeiten auch unbeteiligte Whistleblower den Missstand offenbaren. Jedoch konzentriert sich diese Arbeit bei Kartelldelikten auf den an der Tat beteiligten Whistleblower. Durch das Offenbaren und die „Selbstanzeige" hat er dazu beigetragen, dass der Verstoß gegen das Kartellrecht beendet wird. Er sorgt zudem durch die Kooperation mit der zuständigen Behörde dafür, dass auch alle anderen Beteiligten das Fehlverhalten einstellen. Das dabei ein egoistisches Motiv im Vordergrund steht (Bußgeldbefreiung), ist irrelevant. Ebenfalls ist unbeachtlich, dass im vorliegenden Fall ein Unternehmen als Whistleblower agiert hat, weil sowohl Unternehmen als auch natürliche Personen im Kartellrecht Whistleblower sein können.

Zwar betrifft der Schaden, der durch die rechtswidrigen Preisabsprachen entstanden ist, nur das Vermögen im weitesten Sinne. Aber auch dieses Recht muss geschützt werden. Durch die Offenlegung des Missstandes (Kartell) hat der Whistleblower sich auf die Seite der Rechtsordnung gestellt, um künftigen Schaden am Gemeinwohl, insbesondere von unwissenden Verbrauchern, abzuwenden.

---

160 Als *Settlement* bezeichnet man die einvernehmliche Verfahrensbeendigung, siehe dazu das Merkblatt zu *Settlement*-Verfahren des Bundeskartellamtes in Bußgeldsachen vom 23.12.2013, abrufbar unter: http://www.bundeskartellamt.de/SharedDocs/Publikation/DE/Merkbl%C3%A4tter/Merkblatt-Settlement.html?nn=3591418, (abgerufen am 15.1.2014).

161 Nach der Bonusregelung („Kronzeugenregelung") des Bundeskartellamtes kann das Bußgeld reduziert werden, wenn der Kartellant die Voraussetzungen nach Abschnitt C der Mitteilung erfüllt. Danach muss der Kartellant u. a. uneingeschränkt und ununterbrochen mit dem Kartellamt hinsichtlich der Aufklärung zusammenarbeiten.

Teil 4: Der Schutz des Whistleblowers de lege lata

A. Arbeitsrechtlicher Schutz

Whistleblowing birgt Chancen und Risiken für Arbeitgeber als auch für Arbeitnehmer. Durch den Hinweis des Arbeitnehmers auf bestehende Missstände im Unternehmen bekommt der Arbeitgeber die Chance, diese ohne viel Aufsehen zu beseitigen und das gestörte Arbeitsklima wieder zu stärken. Die Chance für den Arbeitnehmer besteht darin, durch seinen Hinweis den Gesetzesverstoß zu beenden und die Wahrscheinlichkeit einer Wiederholung zu minimieren. Entscheidet sich der Arbeitnehmer für die Offenbarung eines Missstandes, muss er jedoch damit rechnen, dass der Arbeitgeber arbeitsrechtliche Maßnahmen gegen ihn ergreift, die ihn nachteilig treffen. Dem Arbeitgeber sind hinsichtlich seiner Reaktion aber Grenzen gesetzt. Im Folgenden werden die Handlungsmöglichkeiten des Arbeitgebers in einer Whistleblowing-Situation und der jeweilige Schutz des Whistleblowers gegen die einzelnen Reaktionen erläutert.

I. Derzeitige Rechtslage – der unbeteiligte Whistleblower

Die folgende Darstellung der gegenwärtigen Rechtslage soll aufzeigen, ob und in welchem Umfang der Whistleblower durch die Rechtsordnung arbeitsrechtlichen Schutz erlangt. Da im deutschen Rechtsraum keine besonderen Schutzbestimmungen[162] für den Whistleblower bestehen, bleibt nur der Rückgriff auf die allgemeinen Gesetze.

1. Das Kündigungsverbot

In arbeitsrechtlicher Hinsicht ist die naheliegendste Reaktion des Arbeitgebers eine Kündigung. Anknüpfungspunkt für arbeitsrechtliche Sanktionen ist eine Verletzung arbeitsvertraglicher Pflichten. Voraussetzung für die Wirksamkeit der Kündigung ist also das Vorliegen eines Kündigungsgrundes. Angenommen, das Whistleblowing ist „an sich“ rechtmäßig, kann keine vertragliche Pflichtverletzung bestehen; es fehlt somit an einem Kündigungsgrund. Ein Hinweis auf diesen speziellen Fall im Rahmen des Kündigungsrechts ist in den gesetzlichen Bestimmungen nicht zu finden.

---

162 Im Allgemeinen dazu *Becker*, DB 2011, 2202; *Simon/Schilling*, BB 2011, 2421; *Leuchten*, ZRP 2012, 142; Gegen eine Notwendigkeit eines besonderen Schutzes *Rudkowski*, CCZ 2013, 204, 208; *Simonet*, RdA 2013, 236, 238 f.

Nur die Literatur verweist ansatzweise darauf, wie im Falle des Whistleblowings die mutmaßliche Pflichtverletzung zu beurteilen ist.[163]

Denkbare Pflichtverletzungen können sich zum einen aus einer Verletzung der vertraglichen Hauptleistungspflicht ergeben, zum anderen aus der Verletzung vertraglicher Nebenpflichten. Nicht nur die fristlose außerordentliche Kündigung, sondern auch eine ordentliche Kündigung kann dann in Frage kommen.[164]

In Betracht kommt demnach zunächst eine Verletzung der Hauptleistungspflicht aufgrund des Hinweises – sei es unternehmensintern oder -extern – des Arbeitnehmers auf ein gesetzeswidriges Verhalten im Unternehmen. Gemäß § 611 Abs. 1 BGB besteht die Hauptleistungspflicht darin, die arbeitsvertraglich näher konkretisierte und geschuldete Arbeitsleistung zu erbringen.[165] Erfolgt der Hinweis oder die mit ihm einhergehenden Aktivitäten während der Arbeitszeit, deren Umfang in erster Linie durch den Arbeitsvertrag geregelt wird,[166] kann darin eine Pflichtverletzung wegen nicht- oder nicht wie vereinbart erbrachter Leistung liegen.[167] Indem der Hinweisgeber während seiner Arbeitszeit Vorbereitungen trifft, um Missstände publik zu machen, oder vielleicht sogar Beweise im Unternehmen sammelt, um seinen Hinweis zu belegen, ist er (schuldhaft) nicht für den Arbeitgeber verfügbar und kommt somit seiner Arbeitspflicht nicht nach.

Eine Pflichtverletzung kann auch vorliegen, wenn der Hinweisgeber die Erfüllung seiner Arbeitspflicht verweigert. Gründe dafür können unter anderem sein, dass der Arbeitnehmer befürchtet, durch sein (Weiter-)Handeln selbst gegen gesetzliche Regeln zu verstoßen, oder er glaubt, ihm drohten infolge unternehmensinterner Regelverstöße Gefahren für Gesundheit und Leib und Leben, wenn er am Arbeitsplatz weiterhin seine Arbeitspflicht erfüllen muss.

Es kann sich aber ein Leistungsverweigerungsrecht aus § 618 BGB ergeben. Erfüllt der Arbeitgeber seine Fürsorgepflicht nach § 618 BGB nicht, kann der Arbeitnehmer die Leistung verweigern.[168] Erforderlich ist eine *objektiv* bestehende Verletzung der Schutzpflichten aus § 618 BGB.[169] Einer *unmittelbar drohenden* Gefahr bedarf es dagegen nicht. Fraglich ist, wie es sich verhält, wenn der Arbeitgeber dem Arbeitnehmer eine Weisung erteilt, sich rechtswidrig zu verhalten. Denn in diesem Fall verletzt der Arbeitgeber nicht selbst eine Schutzpflicht im Sinne von § 618 BGB,

---

163 Mit Bezug auf die Rechtsprechung des EGMR *Küttner*, Personalbuch, Whistleblowing, Rn. 3 f.; *Grabenwarter*, in: Maunz/Düring, GG-Kommentar, Art. 5 GG, Rn. 185; *Simon/Schilling*, BB 2011, 2421, 2426; *Forst*, NJW 2011, 3477, 3481; *Bauschke*, öAT 2012, 271, 272 f.; *Hergenröder*, in: MüKo BGB, § 1 KSchG, Rn. 243 f.; *Müller-Glöge*, in: Erfurter Kommentar, § 626 BGB, Rn. 64 f.; *Rolfs*, in: BeckOK Arbeitsrecht, § 1 KSchG, Rn. 295.

164 *V. Busekist/Fahrig*, BB 2013, 119, 122.

165 *Müller-Glöge*, in: MüKo BGB, § 611 BGB, Rn. 1011.

166 *Preis*, in: Erfurter Kommentar, § 611 BGB, Rn. 652 f.

167 *Preis*, in: Erfurter Kommentar, § 611 BGB, Rn. 641, 706.

168 *Henssler*, in: MüKo BGB, § 618 BGB, Rn. 91 m.w.N.

169 *Henssler*, in: Müko BGB, § 618 BGB, Rn. 91.

sondern er weist seinen Arbeitnehmer an, eine Pflicht zu verletzen – er handelt also mittelbar. Weigert sich der Arbeitnehmer, die ihm obliegende Arbeitspflicht zu erbringen, so steht dem Arbeitgeber bei (beharrlicher) Arbeitsverweigerung in der Regel ein außerordentliches fristloses Kündigungsrecht nach § 626 BGB zu.[170] Leistungsverweigerungsrechte bei gleichzeitigem Erhalt des Entgeltanspruchs werden dem Arbeitnehmer jedoch zugebilligt, wenn der Arbeitgeber z. B. gegen Arbeitsschutzvorschriften verstößt oder dem Arbeitnehmer Weisungen erteilt, deren Ausführung gegen gesetzliche Verbote (§ 134 BGB), die guten Sitten (§ 138 BGB) oder die Gewissensfreiheit[171] (Art. 4 Abs. 2 GG) verstößt. So gibt es durchaus die Möglichkeit, dass der Arbeitnehmer berechtigterweise die Arbeitstätigkeit verweigert. Die Weigerung stellt dann keine arbeitsvertragliche Pflichtverletzung dar.

Das Leistungsverweigerungsrecht spielt bei Whistleblowing-Aktivitäten also eine große Rolle. Unter welchen gesetzlichen Voraussetzungen jedoch bei einem Gesetzesverstoß im Unternehmen eine Verweigerung der vertraglich geschuldeten Leistung möglich ist, bleibt genauso unklar wie die Frage nach dem Erhalt des Vergütungsanspruches. Denn § 134 BGB behandelt nur das Schicksal von Rechtsgeschäften – auf andere Rechtshandlungen und bloße Maßnahmen bezieht sich die Vorschrift nicht.[172]

Darüber hinaus kann ein Hinweis nur dann eine arbeitsvertragliche Pflichtverletzung sein, wenn es Teil der Hauptleistungspflicht des Arbeitnehmers ist, negative Äußerungen in Bezug auf das Unternehmen zu unterlassen. Diese besondere Pflicht kann aber nicht für sämtliche Beschäftigte gelten, sondern allenfalls für Arbeitnehmer mit einem bestimmten Beschäftigungsbild.[173]

Die Verletzung einer Hauptleistungspflicht durch das Whistleblowing lässt sich also nur ausnahmsweise bejahen. Jedoch kann durchaus die Verletzung einer vertraglichen Nebenpflicht Grund für eine Kündigung sein.

Hinsichtlich einer möglichen Kündigung wegen Verletzung einer arbeitsvertraglichen Nebenpflicht ist besonders auf die Verletzung der Rücksichtnahmepflicht bzw. Verschwiegenheitspflicht einzugehen. Nebenpflichten ergeben sich zumeist aus dem Arbeitsvertrag in Verbindung mit §§ 241 Abs. 2, 242 BGB.[174] Wichtigste

---

170 BAG, Urt. v. 21.11.1996 – 2 AZR 357/95, NZA 1997, 487, 488.

171 BAG, Urt. v. 24.2.2011 – 2 AZR 636/09, NZA 2011, 1087, 1089, 1091: „Ist der Arbeitnehmer auf Grund eines offenbarten beachtlichen Glaubenskonflikts teilweise außerstande, seine arbeitsvertraglichen Verpflichtungen zu erfüllen, berechtigt dies den Arbeitgeber gleichwohl nicht zur Kündigung, wenn er den Arbeitnehmer im Betrieb oder Unternehmen entweder innerhalb des vertraglich vereinbarten Leistungsspektrums oder aber zu geänderten Vertragsbedingungen unter Vermeidung des Konflikts sinnvoll weiterbeschäftigen kann."

172 *Armbrüster*, in: MüKo BGB, § 134 BGB, Rn. 24.

173 Siehe *Schemmel/Ruhmannseder/Witzigmann*, Hinweisgebersysteme, S. 113, die als Paradebeispiel den unternehmenseigenen Pressesprecher anbringen.

174 *Dütz/Thüsing*, Arbeitsrecht, Rn. 151.

Nebenpflicht des Arbeitnehmers ist seine allgemeine Pflicht zur Rücksichtnahme auf die Interessen und Rechtsgüter des Arbeitgebers (früher: Treuepflicht).[175]

Eine andere bedeutsame Nebenpflicht ist die Verschwiegenheitspflicht.[176] Der Arbeitnehmer darf nicht über geschäftliche Belange des Arbeitgebers berichten, soweit dessen Interessen dadurch betroffen werden, es sei denn, der Verschwiegenheitsbruch ist durch die Grundrechte des Arbeitnehmers gerechtfertigt.[177] Der Arbeitnehmer darf insbesondere keine kreditschädigenden Äußerungen über seinen Arbeitgeber in der Öffentlichkeit verbreiten (§ 824 BGB).[178] Neben Tatsachen und Vorgängen, die dem Arbeitnehmer im Zusammenhang mit seiner Stellung im Betrieb bekannt geworden sind, sind auch Betriebs- und Geschäftsgeheimnisse Gegenstand der allgemeinen Verschwiegenheitspflicht, angelehnt an § 17 UWG. Indem der Hinweisgeber die Missstände der Öffentlichkeit mitteilt und damit betriebsinterne Vorgänge aufdeckt, verletzt er seine Verschwiegenheitspflicht.[179]

Ferner bestehen besondere Verschwiegenheitspflichten, die auf bestimmte Berufsgruppen, wie Ärzte und Rechtsanwälte, zugeschnitten sind.

a. *Offenbarungspflichten*

Das Whistleblowing stellt keinen Verstoß gegen eine arbeitsvertragliche Pflicht dar, soweit der Whistleblower zur Offenbarung der entsprechenden Information verpflichtet oder berechtigt ist.

Wichtig sind hier besonders die gesetzlichen Offenbarungspflichten. Neben den Zeugnispflichten im Rahmen gerichtlicher und behördlicher Verfahren, die staatsbürgerliche Pflichten darstellen, spielt die Anzeigepflicht nach § 138 StGB und die innerbetriebliche Anzeigepflicht nach dem Arbeitsschutzgesetz eine entscheidende Rolle.

Zeugnispflichten ergeben sich sowohl im Strafverfahren (vgl. §§ 48 Abs. 1, 59 StPO) als auch im Zivilprozess (vgl. §§ 373, 380, 390 ZPO). Demnach ist der Zeuge grundsätzlich verpflichtet, zur Vernehmung zu erscheinen, wahrheitsgemäß auszusagen und seine Aussage auf Verlangen zu beeiden.[180] Die Pflicht besteht auch dann, wenn der Arbeitnehmer eine Aussage zu machen hat, die zuungunsten seines Arbeitgebers ausfallen oder die zum Verstoß gegen die arbeitsvertragliche

175 *Dütz/Thüsing*, Arbeitsrecht, Rn. 151.

176 *Schemmel/Ruhmannseder/Witzigmann*, Hinweisgebersysteme, S. 115; *Preis*, in: Erfurter Kommentar, § 611 BGB, Rn. 710 f.; *Müller-Glöge*, in: MüKo BGB, § 611 BGB, Rn. 1088 f.; *Dütz/Thüsing*, Arbeitsrecht, Rn. 156 f.

177 *Preis*, in: Erfurter Kommentar, § 611 BGB, Rn. 715.

178 *Dütz/Thüsing*, Arbeitsrecht, Rn. 156.

179 Zur Frage, ob auch „Geheimnisse mit illegalem Inhalt" geschützt sind, siehe Teil 4, B, I.

180 Ähnlich Vorschriften finden sich auch in anderen Verfahrensordnungen, z.B. AGG, FGO, SGG und VwGO; für die StPO: *Senge*, in: Karlsruher Kommentar StPO, § 48, Rn. 2; für die ZPO: *Huber*, in: Musielak, ZPO, § 737, Rn. 9.

Verschwiegenheitspflicht führen würde.[181] Denn Sinn des prozessualen Schweigerechts[182] ist es nicht, die Interessen des Arbeitgebers zu wahren, sondern die Konfliktsituation, in der sich der Zeuge befindet, zu lösen.[183]

§ 138 StGB bedroht die Nichtanzeige bestimmter, abschließend aufgezählter Delikte mit Strafe, sofern die Tathandlung oder der Erfolg durch eine Anzeige möglicherweise noch verhindert worden wäre. Dabei ist der Begriff der Anzeige nicht mit dem des § 158 StPO identisch, sodass es genügt, wenn der Pflichtige eine Warnung mit dem Ziel ausspricht, die Tatausführung oder den Eintritt des Taterfolgs zu verhindern.[184] Das bedeutet für den Whistleblower, dass die Anzeigeerstattung im Arbeitsrecht nicht als Pflichtverletzung gewertet werden kann, wenn die Nichtanzeige einer bevorstehenden Straftat nach Maßgabe des § 138 StGB mit Strafe bedroht ist. § 138 StGB spielt aber eine eher untergeordnete Rolle im Zusammenhang mit Whistleblowing, da die Missstände und rechtswidrigen Verhaltensweisen oft dem Bereich der Körperverletzung, dem Umweltstrafrecht und dem Wirtschaftsstrafrecht zuzuordnen sind, diese aber nicht im Katalog des § 138 StGB aufgeführt sind.

Daneben verpflichtet § 16 Abs. 1 ArbSchG die Beschäftigten, dem Arbeitgeber oder dem zuständigen Vorgesetzten jede von ihnen festgestellte unmittelbare erhebliche Gefahr für die Sicherheit und Gesundheit sowie jeden an dem Schutzsystem festgestellten Defekt unverzüglich, das heißt ohne schuldhaftes Zögern, zu melden. Der Kreis der zuständigen Vorgesetzten in diesem Sinne ist in § 13 ArbSchG festgelegt. Je nach den Umständen ist die Meldung also bei Vorstand, Geschäftsführung, Direktor, Betriebs- oder Abteilungsleiter, Meister, Polier, Vorarbeiter, Schichtführer oder Supervisor zu erstatten. Die Beschäftigten haben nach den oben angeführten Sachkundekriterien das Recht, nach pflichtgemäßem Ermessen als Adressat ihrer Meldung den Vorgesetzten auszuwählen, der nach ihrer Meinung schnell und sachkundig die notwendigen Maßnahmen veranlasst.[185] Eine unmittelbare erhebliche Gefahr liegt dann vor, wenn der Eintritt nicht mehr abgewendet werden kann oder sehr wahrscheinlich ist oder der zu befürchtende Schaden nach Art und Umfang sehr schwer wäre.[186] Im Gegensatz zu den oben genannten Offenbarungspflichten ist es dem Whistleblower hier, wenn man strikt nach dem Wortlaut geht, nur erlaubt, innerbetrieblich auf Missstände hinzuweisen, jedoch nicht außerbetrieblich. Der

---

181 So auch *Schemmel/Ruhmannseder/Witzigmann*, Hinweisgebersysteme, S. 125.

182 Besondere Zeugnisverweigerungsrecht bzw. Schweigerechte ergeben sich aus §§ 52, 53, 53a, 55 StPO und §§ 383, 384 ZPO.

183 Für § 52 StPO siehe *Senge*, in: Karlsruher Kommentar StPO, § 52, Rn. 1; für § 383 ZPO siehe *Huber*, in: Musielak, ZPO, § 383, Rn. 1.

184 *Hohmann*, in: MüKo StGB, § 138 StGB, Rn. 13 m. w. N.

185 *Butz*, in: Kollmer/Klindt, ArbSchG, § 16, Rn. 11.

186 *Butz*, in: Kollmer/Kindt, ArbSchG, § 15, Rn. 43. Nicht nur dann, wenn der Schaden wahrscheinlich eintritt (Gefahrenabwehr), sondern auch im Fall, dass sich ein bereits vorliegender Schaden wahrscheinlich gravierend verschlimmert (Schadensbegrenzung), ist vom Vorliegen einer unmittelbaren erheblichen Gefahr auszugehen.

Adressatenkreis der Meldung ist nur auf Personen zugeschnitten, die im Betrieb zuständig sind oder zum Betrieb gehören.

b. Anzeige- und Beschwerderechte

Ebenso wenig wie bei einer Offenbarungspflicht kann ein Hinweis auf Missstände im Unternehmen als Pflichtverletzung des Whistleblowers gewertet werden, wenn ihm ein Recht zur Anzeige oder Beschwerde zusteht. Eine gesetzliche Regelung, die es Arbeitnehmern erlaubt, auf Missstände im Unternehmen hinzuweisen, besteht nicht. Lediglich das allgemeine Recht zur Erstattung einer Strafanzeige gegenüber unternehmensexternen Stellen nach § 158 Abs. 1 StPO erlaubt jedermann, eine Anzeige zu erstatten, wenn nach Meinung des Anzeigenden Anlass für Ermittlungen der Strafverfolgungsbehörden besteht. Es ist unerheblich, ob der Anzeigeerstatter durch die angezeigte Straftat in irgendeiner Weise mittelbar oder unmittelbar selbst betroffen ist oder nicht.[187] Erforderlich ist aber die Mitteilung des Sachverhalts in der Art, das ihr zureichende tatsächliche Anhaltspunkte für eine Straftat (§ 152 Abs. 2 StPO) entnommen werden können.[188] Die bloße Wiedergabe einer angeblich verletzten Strafbestimmung ist für sich keine Anzeige.[189] Obwohl das allgemeine Anzeigerecht nach § 158 StPO keinerlei Einschränkungen vorsieht, besteht es dennoch nicht vorbehaltlos. Besteht ein Konflikt zwischen zwei vom Grundgesetz geschützten Rechtsgütern, so muss eine Gesamtabwägung der berührten Interessen unter Berücksichtigung der Umstände des Einzelfalles erfolgen.[190] Beim Whistleblowing ist beim Arbeitnehmer auf das Anzeigerecht als Ausfluss des in Art. 2 Abs. 1 GG verbürgten Rechts auf freie Entfaltung der Persönlichkeit sowie auf das Rechtsstaatsprinzip als Grundlage einzugehen.[191] Auf Seiten des Arbeitgebers steht die in Art. 12 Abs. 1 GG verbürgte Unternehmensfreiheit und die damit einhergehende Rücksichtnahmepflicht.[192] Zu dieser einzelfallbezogenen Interessenabwägung hat jüngst die europäische Rechtsprechung[193] eine wegweisende Entscheidung erlassen. Auf diese wird im nächsten Abschnitt vertieft eingegangen.

Ein besonderes Anzeigerecht ergibt sich aus § 17 Abs. 2 ArbSchG,[194] wonach die zuständige Behörde von den Beschäftigten des Betriebs zu verständigen ist, wenn

---

187 *Griesbaum*, in: Karlsruher Kommentar StPO, § 158, Rn. 4.
188 *Griesbaum*, in: Karlsruher Kommentar StPO, § 159, Rn. 15.
189 *Griesbaum*, in: Karlsruher Kommentar StPO, § 158, Rn. 15.
190 *Herbert/Oberrath*, NZA 2005, 193, 197, die von einer Einschränkung des Anzeigerechts nach § 158 StPO aufgrund der besonderen Beziehung zwischen Arbeitgeber und Arbeitnehmer sprechen.
191 BVerfG, Beschl. v. 25.2.1987 – 1 BvR 1086/85, NJW 1987, 1929, 1930.
192 BAG, Urt. v. 24.6.2004 – 2 AZR 63/03, NJW 2005, 619, 621.
193 EGMR, Urt. v. 21.07.2011 – 28274/08, NJW 2011, 3501 (Heinisch/Deutschland).
194 Bis 31.12.2004 galt § 21 Abs. 6 GefStoffV als lex specialis zu § 17 Abs. 2 ArbSchG. Danach konnten die Beschäftigten bei Erreichen oder Überschreiten von Grenzwerten für das Auftreten von Gefahrstoffen am Arbeitsplatz nach Ausschöpfung innerbetrieblicher Möglichkeiten, Beschwerde an die zuständige Stelle einreichen.

sie aufgrund konkreter Anhaltspunkte der Auffassung sind, dass die vom Arbeitgeber getroffenen Maßnahmen und bereitgestellten Mitteln nicht ausreichen, um die Sicherheit und den Gesundheitsschutz bei der Arbeit zu gewährleisten. Der Arbeitnehmer darf sich jedoch nicht unmittelbar an die externe Stelle (Aufsichtsbehörde) wenden, sondern muss zunächst den Arbeitgeber – erfolglos – zur Abhilfe aufgefordert haben.[195]

Zuletzt ist auf ein innerbetriebliches Beschwerderecht einzugehen: § 13 Abs. 1 AGG eröffnet Beschäftigten das Recht, sich bei den zuständigen Stellen des Betriebes, des Unternehmens oder der Dienststelle zu beschweren, wenn sie sich im Zusammenhang mit ihrem Beschäftigungsverhältnis vom Arbeitgeber, vom Vorgesetzten, anderen Beschäftigten oder Dritten benachteiligt fühlen.[196]

Zusammenfassend ist festzustellen, dass es durchaus Regelungen gibt, die das Handeln des Whistleblowers rechtfertigen und ihn in besonderen Fällen sogar dazu verpflichten. Greifen diese Rechte und Pflichten in einem Sachverhalt, kann ein Hinweis nicht als pflichtwidrig angesehen werden. Die Vorschriften haben jedoch gemeinsam, dass sie nur in besonderen Situationen gelten. Außerdem gibt es Vorschriften, die sich nur auf innerbetriebliche Hinweise beziehen, also nur das interne Whistleblowing „erlauben". Wer also unternehmensextern auf Missstände oder Regelverstöße aufmerksam macht, läuft Gefahr, dass sein Verhalten durch den Arbeitgeber oder im Rahmen einer (gerichtlichen) Überprüfung als Verletzung arbeitsvertraglicher Pflichten eingestuft wird.[197] Eine allgemeine gesetzliche Regelung, welche ein Anzeigerecht des Arbeitnehmers wegen Gesetzesverstößen im Unternehmen festschreibt, existiert in Deutschland nicht.

2. *Sonderfall Druckkündigung*

Bei der Druckkündigung handelt es sich um eine besondere Form der Kündigung, bei der vom Arbeitgeber auf Druck – unter Androhung von Nachteilen – von Dritten (Belegschaft, Kunden, Betriebsrat) die Aussprache einer Kündigung gegenüber dem Arbeitnehmer verlangt wird. Nachteile können u. a. die Eigenkündigung oder die Nichterteilung eines Auftrags sein.[198] Es wird unterschieden zwischen echter und

---

Die aktuelle Fassung der Verordnung enthält keine entsprechende Regelung mehr, GefStoffV v. 26.11.2010, BGBl. I-2010, 1643.

195 *Wank*, in: Erfurter Kommentar, § 17 ArbSchG, Rn. 2 mit Verweis auf die Rechtsprechung des EGMR vom 21.7.2011. Vorherige Abhilfe setzt aber voraus, dass der Arbeitgeber gewillt ist Abhilfe zu leisten – dies wird in den Fällen zu verneinen sein, in denen er selbst in die Tat oder den Missstand involviert ist.

196 Erfasst werden Benachteiligungen im Hinblick auf die Rasse, der ethnischen Herkunft, des Geschlechts, der Religion oder Weltanschauung, einer Behinderung, des Alters oder der sexuellen Identität (§ 1 AGG); ausführlich dazu *Oetker*, NZA 2008, 264.

197 So auch *Schemmel/Ruhmannseder/Witzigmann*, Hinweisgebersysteme, S. 153.

198 *Hergenröder*, in: MüKo BGB, § 1 KSchG, Rn. 245.

unechter Druckkündigung. Entscheidet sich der Arbeitgeber unabhängig von dem auf ihn ausgeübten Druck für eine Kündigung und liegt ein Kündigungsgrund vor, handelt es sich um eine unechte Druckkündigung. Der Arbeitgeber hat in dieser Situation die Entscheidungsfreiheit, ob er den Arbeitnehmer kündigt oder nicht.[199] Liegt hingegen kein objektiver Grund für eine Kündigung vor und wird dem Arbeitgeber mit Nachteilen gedroht, handelt es sich um eine echte Druckkündigung. Hier muss der Arbeitgeber zunächst versuchen, die Drohungen durch andere Maßnahmen abzuwenden – er muss sich schützend vor den Arbeitnehmer stellen.[200] Dem Arbeitgeber wird damit eine interne Handlungsgrenze gesetzt. Stellt sich der Arbeitgeber auf die Seite des Druckausübenden und spricht eine Kündigung aus, ist diese aufgrund eines Verstoßes gegen den *ultima-ratio*-Grundsatz unwirksam.[201] Das Handeln des Arbeitgebers kann nur in besonderen Ausnahmefällen gerechtfertigt sein. Das ist der Fall, wenn alle Abwendungsversuche ausgeschöpft sind und der Arbeitgeber einen unzumutbaren Schaden für den ganzen Betrieb in Kauf nehmen müsste.[202] Erhält die Belegschaft vom Whistleblowing eines Arbeitnehmers Kenntnis, ist dieser also nicht davor geschützt, „Opfer“ einer Druckkündigung zu werden. Er muss damit rechnen, dass Mitarbeiter sein Verhalten missbilligen und als illoyal einstufen. Je weniger Zuspruch und Akzeptanz für das Thema Whistleblowing besteht, desto stärker und schädigender das Druckmittel. Diejenigen, die gegen das ungeschriebene Gesetz der „Gemeinschaft im Betrieb“ verstoßen, werden ausgeschlossen.[203] Geleitet von Angst vor erheblichen Nachteilen und Schäden für das ganze Unternehmen wird der Arbeitgeber die Druckkündigung durchaus in Erwägung ziehen, ohne dass es einen objektiven Grund für eine Kündigung gäbe.

Rechtsdogmatisch ist die Einordnung einer Druckkündigung in das System der Kündigungsgründe unklar. Insbesondere ist in der Literatur umstritten,[204] ob eine Druckkündigung als betriebsbedingte Kündigung sozial gerechtfertigt sein kann.

---

199 Die vom Dritten geschaffene Drucksituation kann in solchen Fällen höchstens bei der Interessenabwägung eine Rolle spielen, vgl. *Hergenröder*, in: MüKo BGB, § 1 KSchG, Rn. 245 m. w. N.

200 BAG, Urt. v. 19.6.1986 – 2 AZR 563/85, NZA 1987, 21, 22; BAG, Urt. v. 4.10.1990 – 2 AZR 201/90, NZA 1991, 468, 469.

201 BAG, Urt. v. 18.9.1975 – 2 AZR 311/74, JZ 1976, 323, Leitsatz 1; *Oetker*, in: Erfurter Kommentar, § 1 KSchG, Rn. 185.

202 BAG, Urt. v. 18.7.2013 – 6 AZR 420/12, NZA 2014, 109, 112; BAG, Urt. v. 31.1.1996 – 2 AZR 158/95, NZA 1996, 581, 584; BAG, Urt. v. 19.6.1986 – 2 AZR 563/85, NZA 1987, 21, 22; BAG, Urt. v. 10.10.1957 – 2 AZR 32/56, AP Nr. 1 zu § 626 BGB, Leitsatz 1.

203 *Deiseroth/Derleder*, ZRP 2008, 248, 250.

204 Zustimmend *Griebeling*, in: GK-KSchG, § 1 KSchG, Rn. 586 a; *Rolfs*, in: BeckOK Arbeitsrecht, § 1 KSchG, Rn. 422; *Henssler*, in: MüKo BGB, § 626 BGB, Rn. 253, 255; Ablehnend *Kiel*, in: Ascheid/Preis/Schmidt, Kündigungsrecht, § 1 KSchG, Rn. 521; *Krause*, in: v. Hoyningen-Huene/Linck, KSchG, § 1 KSchG, Rn. 346; *Oetker*, in: Erfurter Kommentar, § 1 KSchG, Rn. 184.

Mit seiner Entscheidung vom 18. Juli 2013[205] hat das BAG die bisherige Rechtsprechung zur betriebsbedingten Druckkündigung bestätigt und gesagt, dass der objektiv unbegründete Druck eines Vertragspartners des Arbeitgebers eine betriebsbedingte Druckkündigung rechtfertigt. Das BAG kommt zu dem Schluss, dass der Auftragsverlust eine Drucksituation erzeuge, auf die der Arbeitgeber unternehmerisch reagieren müsse, um den Fortbestand des Betriebs zu sichern. Diese Situation sei mit einer echten Druckkündigung vergleichbar. Entstünden bei der Verwirklichung der Drohung schwere wirtschaftliche Nachteile, so müsse der Arbeitgeber aus wirtschaftlichen Gründen handeln.[206]

Auch wenn die Kündigung aus betriebsbedingten Gründen grundsätzlich nur gerechtfertigt ist, wenn eine unternehmerische Entscheidung den Wegfall von Beschäftigungsbedarf zur Folge hat, und die Argumentation des BAG in diesem Fall schwer nachvollziehbar ist, bestätigt sie doch die Zulässigkeit einer sog. „echten Druckkündigung". Allerdings hat sich der Arbeitgeber zunächst schützend vor den betroffenen Arbeitnehmer zu stellen.[207]

3. *Auflösungsantrag*

Nicht mit einer Druckkündigung wegen Whistleblowings, sondern mit dem Auflösungsantrag eines Arbeitgebers beschäftigte sich vor kurzem das LAG Schleswig-Holstein.[208] Der Sachverhalt sowie die Urteilsbegründung waren zweigeteilt. Nachdem das LAG zunächst betonte, dass es eine Druckkündigung grundsätzlich als zulässig erachte, verglich es zutreffenderweise die Drucksituation in diesem Fall mit der Konstellation beim Mobbing und sprach sich gegen die Zulässigkeit einer solchen Kündigung aus. Es konnte dahingestellt bleiben, ob die beiden Zeugen wirklich ernsthaft damit drohten, den Betrieb der Beklagten zu verlassen, wenn der Kläger bleibe. Entscheidend war, dass das Bemühen der Arbeitgeberin, die Drucksituation ohne Kündigung zu lösen, als unzureichend empfunden wurde.[209] Denn eine Kündigung sollte nach Ausschöpfen aller Mittel das letzte Mittel sein. Zu Recht hat das LAG im Ergebnis eine Drucksituation der Arbeitgeberin abgelehnt, wenn das Verhalten des Arbeitnehmers geraume Zeit – im vorliegenden Fall jahrelang – geduldet wird. Anschließend setzte sich das Gericht mit dem Thema des Whistleblowings auseinander und ließ den Auflösungsantrag (§ 9 Abs. 1 S. 2 KSchG) zu, den der Arbeitgeber nach (erfolgloser) Kündigung hilfsweise beantragte. Nach Ausspruch der Kündigung hatte der Arbeitnehmer gegenüber der Agentur für

205 BAG, Urt. v. 18.7.2013 – 6 AZR 420/12, NZA 2014, 109.
206 BAG, Urt. v. 18.7.2013 – 6 AZR 420/12, NZA 2014, 109, 112; siehe dazu *Hamacher*, NZA 2014, 134 f.
207 BAG, Urt. v. 18.7.2013 – 6 AZR 420/12, NZA 2014, 109, 112.
208 LAG Schleswig-Holstein, Urt. v. 20.3.2012 – 2 Sa 331/11, BB 2012, 1862.
209 LAG Schleswig-Holstein, Urt. v. 20.3.2012 – 2 Sa 331/11, BB 2012, 1862, 1864.

Arbeit eine sog. „Kurzarbeitsmissbrauchsanzeige“[210] abgegeben. Das Gericht sah in der Anzeige des Arbeitnehmers eine Rechtfertigung für einen Auflösungsantrag: Es könne nachvollziehbar die gedeihliche Zusammenarbeit der Arbeitsvertragsparteien stören, wenn die Arbeitgeberin nicht sicher sein könne, ob künftig ohne vorherige innerbetriebliche Klärung bei Meinungsverschiedenheiten Behörden eingeschaltet oder Strafanzeigen erstattet werden.[211]

In der Urteilsbegründung spielte das Motiv des Arbeitnehmers eine große Rolle. Denn dieser war offensichtlich wegen des laufenden Kündigungsverfahrens von Revanchegedanken bestimmt.[212] So stellt die Auflösung des Arbeitsverhältnisses eine Reaktionsmöglichkeit des Arbeitgebers dar, wenn das Verhalten des Whistleblowers eine den Betriebszwecken dienliche weitere Zusammenarbeit zwischen den Parteien nicht erwarten lässt.

4. Das Benachteiligungsverbot

Zudem ist der Arbeitgeber grundsätzlich berechtigt, eine Abmahnung auszusprechen. Daraufhin kann ein negativer Eintrag in die Personalakte erfolgen. Ebenfalls kommt eine Änderungskündigung oder „einfache“ Versetzung in Frage, um den Mitarbeiter des Arbeitsplatzes zu verweisen und ihn nachteilig zu versetzen.[213] Diese und ähnliche Maßnahmen stellen eine Benachteiligung des Arbeitnehmers dar. Eine solche kann auch darin bestehen, dass dem Arbeitnehmer Vorteile vorenthalten werden, die der Arbeitgeber anderen Arbeitnehmern gewährt, die entsprechende Rechte nicht ausgeübt haben.[214]

§ 612 a BGB statuiert ein allgemeines Benachteiligungsverbot. Nach dem Maßregelungsverbot dieser Vorschrift darf der Arbeitgeber einen Arbeitnehmer bei einer Vereinbarung oder einer Maßnahme nicht benachteiligen, weil dieser in zulässiger Weise seine Rechte ausübt. Eine Maßregelung können sowohl Vereinbarungen (rechtsgeschäftliche Abmachungen) als auch Maßnahmen in Form von einseitigen rechtsgeschäftlichen (z. B. Kündigung) oder tatsächlichen Schritten sein.[215] Ob eine Benachteiligung des Arbeitnehmers nach § 612 a BGB vorliegt, ist zunächst durch einen Vergleich der Situation des Arbeitnehmers vor und nach der Maßnahme

---

210 Der Kläger hat sich zwar nicht direkt an die Polizei oder Staatsanwaltschaft gewendet, sondern er hat eine Anzeige gegenüber der Bundesagentur für Arbeit erstattet. Die E-Mail enthielt laut Ermittlungsakten aber ausdrücklich das Wort „Anzeige“.

211 LAG Schleswig-Holstein, Urt. v. 20.3.2012 – 2 Sa 331/11, BB 2012, 1862, 1866.

212 Vgl. auch *Lelley*, BB 2012, 1862, 1867.

213 *Dendorfer*, in: Moll, Arbeitsrecht, § 35, Rn. 148.

214 BAG, Urt. v. 21.9.2011 – 7 AZR 150/10, NZA 2012, 317, 321 zum Fall eines Arbeitgebers, der einem befristet beschäftigten Arbeitnehmer keinen Folgevertrag anbot, weil er seine ihm zustehenden Rechte (hier Art. 5 GG) ausgeübt hat.

215 *Fuchs*, in: BeckOK BGB, § 612 a BGB, Rn. 4.

oder Vereinbarung zu beurteilen.[216] Doch geht das Tatbestandsmerkmal „Benachteiligung" darüber hinaus. Es ist nur erfüllt, wenn der Arbeitgeber zwischen verschiedenen, gleichermaßen rechtmäßigen Maßnahmen wählen konnte.[217] Hat er sein Verhalten an der Rechtsordnung orientiert, liegt keine Benachteiligung des Arbeitnehmers vor.[218] Dieses Verbot umfasst auch, dass ein Arbeitnehmer nicht wegen eines rechtmäßigen Hinweises benachteiligt werden darf, also bei Ausübung eines rechtmäßigen Whistleblowings.

Allerdings ist unklar, was unter einer rechtmäßigen Rechtsausübung zu verstehen ist,[219] insbesondere wann das Handeln des Whistleblowers rechtmäßig ist. Die Zulässigkeit der Rechtsausübung beurteilt sich nach der Gesamtrechtsordnung, worunter die Ausübung von Grundrechten fällt, soweit sie im Arbeitsverhältnis rechtserheblich sind.[220] Weitgehend sicher ist jedoch, wie das Tatbestandsmerkmal „Ausübung seiner Rechte" auszulegen ist: Die Rechtsausübung kann im Erheben eines Anspruchs oder von Einreden und Einwendungen liegen.[221] Sie kann sich aber auch aus einem tatsächlichen Verhalten ergeben. Unerheblich ist, ob der Arbeitnehmer seine Rechte schriftlich oder mündlich bzw. unmittelbar oder mittelbar über einen Anwalt ausübt.[222]

Es gibt jedoch, wie sich noch zeigen wird, verschiedene Formen und Abstufungen des rechtmäßigen Verhaltens beim Whistleblowing. Erst durch eine Gesamtschau wird deutlich, ob ein Whistleblower den Gesetzesverstoß offenbaren durfte. Die Zulässigkeit der Rechtsausübung des Whistleblowers ist abhängig davon, inwieweit Abhilfen auf anderen Wegen erfolgversprechend sind.

5. Schadensersatzanspruch des Whistleblowers bei Verstoß gegen das Benachteiligungsverbot

Wird gegen das Benachteiligungsverbot verstoßen, stellt sich die Frage nach einem Anspruch auf Entschädigung, Schadensersatz oder Unterlassen der rechtswidrigen Handlung.

In Betracht kommt der Schadensersatzanspruch nach § 280 BGB. Ein Arbeitsverhältnis ist ein Schuldverhältnis i. S. d. § 280 BGB. Weil grundsätzlich gem. § 280

216 *Müller-Glöge*, in: MüKo BGB, § 612 a BGB, Rn. 14.

217 *Müller-Glöge*, in: MüKo BGB, § 612 a BGB, Rn. 14.

218 BAG, Urt. v. 14.2.2007 – 7 AZR 95/06, NZA 2007, 803, 805 f.; *Faulenbach*, Maßregelungsverbot, S. 86 f., 103, geht davon aus, es liege schon keine Maßnahme des Arbeitgebers vor, wenn ihm für seine Reaktion kein Handlungsspielraum eröffnet sei.

219 *Linck*, in: Ascheid/Preis/Schmidt, Kündigungsrecht, § 612 a BGB, Rn. 7, der nur darauf hinweist, dass die Rechtsausübung objektiv rechtmäßig sein muss.

220 *Preis*, in: Erfurter Kommentar, § 612 a BGB, Rn. 6.

221 *Müller-Glöge*, in: MüKo BGB, § 612 a BGB, Rn. 8; *Linck*, in: Ascheid/Preis/Schmidt, Kündigungsrecht, § 612 a BGB, Rn. 7b.

222 *Linck*, in: Ascheid/Preis/Schmidt, Kündigungsrecht, § 612 a BGB, Rn. 7b.

Abs. 1 Satz 2 BGB vermutet wird, dass die Pflichtverletzung zu vertreten ist, wird dem Whistleblower die Durchsetzung von Schadensersatzansprüchen gegen seinen Arbeitgeber deutlich erleichtert, ohne dass dem Arbeitgeber die Möglichkeit genommen wäre, sich zu entlasten.[223] Ob der Verstoß gegen das Benachteiligungsverbot eine Pflichtverletzung im Sinne von § 280 Abs. 1 BGB darstellt, kann dahingestellt bleiben; zumindest ist eine Pflichtverletzung im Sinne von § 241 Abs. 2 BGB gegeben.[224] Inwieweit eine Entschädigung für Nichtvermögensschaden zu zahlen ist, kann nur durch Heranziehung allgemeiner Vorschriften geklärt werden. Fraglich ist auch, wie ein Verstoß durch einen Mitarbeiter schadensersatzrechtlich zu behandeln ist. Für einen Anspruch aus § 280 BGB bedarf es einer Pflichtverletzung im Rahmen eines besonderes Verhältnisses (z. B eines Schuldverhältnisses). Ein solches Verhältnis zwischen Arbeitnehmern zu konstruieren, welches auch noch durch gegenseitige Pflichten in Bezug auf eine Benachteiligung bzw. das Verbot einer Benachteiligung ausgestaltet ist, ist schwierig. Übrig bliebe jedoch § 823 BGB, wobei fraglich ist, worin die Rechtsgutsverletzung und die Verletzungshandlung liegen sollen. Allerdings besteht die Möglichkeit, dem Arbeitgeber das Verhalten seiner Mitarbeiter zuzurechnen oder in dem Nichteinschreiten bzw. Nichtbeheben der Benachteiligung eine Pflichtverletzung (durch den Arbeitgeber) zu sehen. Anspruchsgegner ist dann in der Regel der Arbeitgeber.

Ein Anspruch auf Unterlassung aufgrund Diskriminierung oder Verletzung des allgemeinen Persönlichkeitsrechts kann sich hingegen aus dem AGG[225] oder §§ 823, 1004 BGB ergeben (quasi-negatorischer Unterlassungsanspruch).[226]

6. Beweislastumkehr

Aus dem Umstand, dass der Whistleblower bei der Berufung auf das Benachteiligungsverbot nach § 612 a BGB gezwungen ist nachzuweisen, dass die von ihm gerügte Benachteiligung auf einem zulässigen Whistleblowing beruht, ergibt sich, dass es einer Beweislastregelung zugunsten des Whistleblowers bedarf.[227] Als Vorbild

---

223 So *Rudkowski*, CCZ 2013, 204, 208.

224 Vgl. *Richardi/Fischinger*, in: v. Staudinger, BGB-Kommentar, § 612 a BGB, Rn. 32 „Außerdem können Verstöße gegen das Maßregelungsverbot Unterlassungs-, Beseitigungs- und Schadensersatzansprüche (§§ 280 I i.V.m. 241 II, §§ 823 II i.V.m. 612a BGB) des Arbeitnehmers auslösen."

225 § 21 AGG sieht eine eigene Anspruchsgrundlage für Schadensersatz-, Beseitigungs- und Unterlassungsansprüche bei Verstoß gegen das Diskriminierungsverbot vor; außerdem steht dem Betroffenen nach § 15 Abs. 5 AGG das Recht zu, nach anderen Vorschriften bestehende Ansprüche geltend zu machen *Schlachter*, in: Erfurter Kommentar, § 15 AGG, Rn. 19.

226 *Reichold*, in: Münchener Handbuch zum Arbeitsrecht, § 86, Rn. 27; *Thüsing*, in: MüKo BGB, § 15 AGG, Rn. 51 f.

227 *Rudkowski*, CCZ 2013, 204, 209.

könnte die Regelung des § 22 AGG dienen.[228] Hierfür spräche das Gebot der europarechtskonformen Auslegung: der Wortlaut von Art. 4 RL 97/80/EG, Art. 8 RL 2000/43/EG und Art. 10 RL 2000/78/EG differenziert nicht nach der Anspruchsgrundlage. Auch bezieht sich die Beweislastregel nicht ausdrücklich auf Ansprüche aus diesem Gesetz, sondern bezieht alle Verfahren ein, in denen eine Benachteiligung geltend gemacht wird.[229] Gleiches müsste dann bei einem Verstoß gegen das Benachteiligungsverbot gelten. Eine Beweislastumkehr ließe sich also mit einer analogen Heranziehung dieser Regeln begründen. Voraussetzung dafür ist aber eine planwidrige Regelungslücke. Diese lässt sich kaum begründen, wenn man bedenkt, dass dem Gesetzgeber die Regelungslücke bekannt war und er in Gesetzesüberarbeitungen (Schuldrechtsreform und Einführung des AGG) keinen Handlungsbedarf sah.[230] Die Beweislastregelung des § 611 a BGB a. F. bzw. dessen Nachfolgeregelung § 22 AGG kann somit nicht übertragen werden.[231] Dem Arbeitnehmer kann aber im Einzelfall eine Beweiserleichterung durch einen Anscheinsbeweis hinsichtlich der subjektiven Tatbestandsmerkmale und der Kausalität zugutekommen bei offensichtlichem Zusammenhang zwischen Benachteiligung und Rechtsausübung, wenn er sich insofern (ausnahmsweise) auf anerkannte Erfahrungssätze berufen kann.[232] Dass diese Beweisvorlage mit besonderen Erschwernissen verbunden ist, ist offensichtlich.

Eine gesetzliche Regelung, die dem Whistleblower den Beweis einer Benachteiligung erleichtert, fehlt. Ihn auf komplizierte andere Regelungen zu verweisen, die nur in besonderen Ausnahmefällen greifen, ist unzureichend und wenig hilfreich.

---

228 *Rudkowski*, CCZ 2013, 204, 209.

229 *Thüsing*, in: MüKo BGB, § 22 AGG, Rn. 5.

230 *Benecke*, NZA 2011, 481, 484; *Rudkowski*, CCZ 2013, 204, 209, die darauf hinweist, dass es keinen Grund gebe, den Whistleblower gegenüber anderen Benachteiligten besser zu stellen.

231 Vgl. BAG, Urt. v. 2.4.1987 – 2 AZR 227/86, AP BGB § 612 a Nr. 1, unter II.1.cc; BAG, Urt. v. 25.11.1993 – 2 AZR 517/93, NZA 1994, 837, 838; *Preis*, in: Erfurter Kommentar, § 612 a BGB, Rn. 22; *Krause*, SAE 2003, 205, 210; *Däubler*, in: Kittner/Däubler/Zwanziger, Kündigungsschutzrecht, § 612 a BGB, Rn. 22; *Benecke*, NZA 2011, 481, 484; a. A.: *Linck*, in: Ascheid/Preis/Schmidt, Kündigungsrecht, § 612 a BGB, Rn. 22 (wegen gemeinschaftsrechtskonformer Auslegung); *Belling/v. Steinau-Steinrück*. DB 1993, 534, 536.

232 So *Müller-Glöge*, in: MüKo BGB, § 612 a BGB, Rn. 24, mit Verweis auf BAG, Urt. v. 21.11.2000 – 9 AZR 665/99, NZA 2001, 1093, 1094; weitergehend BAG, Urt. v. 23.4.2009 – 6 AZR 189/08, NZA 2009, 974, 975 (gestufte Darlegungslast: es genügt, wenn der Sachvortrag des Arbeitnehmers die Kausalität indiziert); *Preis*, in: Erfurter Kommentar, § 612 a BGB, Rn. 22; *Däubler*, in: Kittner/Däubler/Zwanziger, Kündigungsschutzrecht, § 612 a BGB, Rn. 23.

7. Schadensersatzansprüche des Arbeitgebers

Der Whistleblower hat nicht nur mit einer Kündigung oder einer Benachteiligung zu rechnen, sondern muss auch befürchten, dass der Arbeitgeber von ihm nach den oben genannten Anspruchsnormen Ersatz des durch die Offenlegung verursachten betrieblichen und finanziellen Schadens verlangt. Selbst wenn der Whistleblower an dem rechtswidrigen Verhalten in keiner Weise beteiligt war, können ihm durch das Whistleblowing zivilrechtliche Folgen drohen: bei einem nicht berechtigten (und/ oder unwahren) Hinweis kann ihn eine Schadensersatzpflicht nach der allgemeinen Regelung des §§ 280 Abs. 1 i. V. m. § 241 Abs. 2 BGB oder nach Maßgabe des § 9 UWG in Verbindung mit § 3 Abs. 1 sowie § 4 Nr. 8 und Nr. 11 UWG treffen.[233] Der Arbeitgeber wird durch die Veröffentlichung der Missstände regelmäßig einen Schaden im Sinne von §§ 280, 252 BGB erleiden, da entweder potenzielle Kunden abspringen oder die Repräsentation auf dem Markt nicht mehr den gewünschten (finanziellen) Erfolg mit sich bringt.[234] Die Pflichtverletzung besteht dabei in der Hinweiserteilung und in der damit einhergehenden Veröffentlichung der Verhaltensweisen und Betriebsgeheimnisse im Unternehmen.[235] Auch eine Haftung nach § 823 Abs. 1 BGB ist nicht ausgeschlossen. Denn die wahre, aber geschäftsschädigende Tatsachenbehauptung vermag, im Gegensatz zu unwahren Behauptungen – weil § 823 Abs. 2 BGB i. V. m. einem Schutzgesetz vorgeht – einen Eingriff in ein „sonstiges Recht" im Sinne des § 823 Abs. 1 BGB zu begründen.[236] Dagegen ist ein Gewinn nicht ersatzfähig, wenn dieser nur bei Ausübung einer rechtlich verbotenen Tätigkeit angefallen wäre.[237] Diese Konstellation wird regelmäßig bei Kartellrechtsfällen zu bejahen sein. Denn nach § 1 GWB sind Vereinbarungen zwischen Unternehmen, Beschlüsse von Unternehmensvereinigungen und aufeinander abgestimmte Verhaltensweisen, die

233 So auch *Schemmel/Ruhmannseder/Witzigmann*, Hinweisgebersysteme, S. 154, die als Pflichtverletzung, die den Schadensersatz begründet, aber die unberechtigte Weitergabe der Informationen meinen, die erfolgt, wenn man sich nicht an die Regeln eines im Unternehmen implementierten Hinweisgebersystems hält.

234 Zum Begriff des entgangenen Gewinns *Oetker*, in: MüKo BGB, § 252 BGB, Rn. 5, 6.

235 Ablehnend in einer Entscheidung des LAG Hamm, Urt. v. 21.7.2011 – 11 Sa 2248/10, öAT 2012, 19. Zwar sei das Ermittlungsverfahren und die dadurch einhergehende öffentliche Berichterstattung erst durch die anonymen Strafanzeigen ausgelöst worden, jedoch fehle es an den Voraussetzungen einer Schadensersatzpflicht des Anzeigeerstatters. Da ein Interesse der Allgemeinheit an der Aufklärung von Straftaten besteht, stellt die Erstattung einer Strafanzeige regelmäßig eine erlaubte und gebilligte Möglichkeit dar.

236 *Staudinger*, in: Schulze, BGB-Kommentar, § 823 BGB, Rn. 127.

237 BGH, Urt. v. 7.5.1974 – VI ZR 7/73, NJW 1974, 1374, 1376 f.; BGH, Urt. v. 6.7.1976 – VI ZR 122/75, NJW 1976, 1883, 1884; OLG Karlsruhe, Urt. v. 6.4.1993 – 18a U 138/92, NJW-RR 1993, 918, 919; *Bydlinski*, FS Deutsch, S. 63, 69 f.; *Stürner*, VersR 1976, 1012, 1013 f.; *Teichmann*, in: Jauernig, BGB, § 252 BGB, Rn. 1; *Grüneberg*, in: Palandt, BGB-Kommentar, § 252 BGB, Rn. 2; *Oetker*, in: MüKo BGB, § 252 BGB, Rn. 7.

eine Verhinderung, Einschränkung oder Verfälschung des Wettbewerbs bezwecken oder bewirken, verboten.

Eine speziell auf den Whistleblower zugeschnittene Schutzvorschrift ist in der deutschen Rechtsordnung nicht vorhanden. Stellt sich das Whistleblowing also als Pflichtverletzung dar – hauptsächlich liegt im Fall des unbeteiligten Whistleblowers eine Nebenpflichtverletzung vor –, weil keine Offenbarungspflicht oder kein Anzeigerecht die Pflichtverletzung ausschließt, ist der Tatbestand des §§ 280, 241 Abs. 2 BGB erfüllt. § 280 Abs. 1 BGB käme also grundsätzlich auch in Betracht, wenn der rechtmäßige Hinweis erteilt wurde, aber nicht durch die besonderen Rechte zur Offenlegung gerechtfertigt war.

a. Notwehr nach § 227 BGB?

Es stellt sich die Frage, ob der Whistleblower einem Schadensersatzanspruch des Arbeitgebers entgehen kann, weil er in Notwehr gehandelt hat. Lägen die Voraussetzungen des § 227 Abs. 2 BGB vor, entfiele die sonst anzunehmende Rechtswidrigkeit der Handlung.[238] Eine Schadensersatzpflicht nach den §§ 823 ff. BGB wäre damit ausgeschlossen. Der Unterschied zum strafrechtlichen Notwehrrecht liegt insbesondere in der Zielsetzung: Während das Straf- und Ordnungswidrigkeitenrecht auf die Sanktionierung besonders sozialschädlichen Verhaltens abzielt, statuiert das Zivilrecht ein Ausgleichssystem zwischen privaten Interessen, das vor allem auf Störungen der Güterzuordnung reagiert. Zwar gleicht § 227 BGB dem § 32 StGB, beide Normen schließen sich jedoch nicht aus. Ein wegen Notwehr nach § 32 StGB strafloses Verhalten kann in besonderen Konstellationen durchaus Schadensersatzpflichten aus §§ 823 f. BGB nach sich ziehen und umgekehrt.[239] § 227 BGB kommt jedoch keine Rechtfertigungswirkung im Straf- und Ordnungswidrigkeitenrecht zu.[240]

Das Whistleblowing könnte eine notwendige und gebotene Notwehrhandlung darstellen. Die Notwehrlage ist dabei in der gesetzeswidrigen Handlungsweise des Arbeitgebers zu sehen. Eine Angriffshandlung ist jedes positive Tun, das Rechte und Rechtsgüter eines anderen beeinträchtigt. Geschützt sind alle individuellen Rechte oder Rechtsgüter, z. B. die in § 823 Abs. 1 BGB genannten, aber auch das Recht am eigenen Bild[241] oder das seelische und körperliche Wohlbefinden.[242] Die Verteidigung der Rechte Dritter ist als Nothilfe von § 227 Abs. 2 BGB („von sich oder einem anderen") mit umfasst. Die Rechtsordnung als Ganzes dagegen ist ebenso wenig notwehrfähig wie Rechtsgüter der Allgemeinheit bzw. die öffentliche Sicherheit und Ordnung. Diese Rechtsgüter könnten nur bei § 34 StGB in Betracht kommen. Notwehrfähige Rechte, die der Whistleblower durch die Offenbarung (Verteidigungshandlung) schützt, sind also nicht nur Rechtsgüter, die das Leben

238 *Grothe*, in: MüKo BGB, § 227 BGB, Rn. 1.
239 *Grothe*, in: MüKo BGB, § 227 BGB, Rn. 2.
240 *Grothe*, in: MüKo BGB, § 227 BGB, Rn. 2.
241 OLG Hamburg, Beschl. v. 14.4.1972 – 1 Ws 84/72, NJW 1972, 1290.
242 *Dörner*, in: Schulze, BGB-Kommentar, § 227 BGB, Rn. 2.

oder die körperliche Unversehrtheit betreffen, sondern auch sonstige Rechte nach § 823 Abs. 1 BGB, wie z. B. dingliche Rechte oder auch Immaterialgüterrechte.[243]

Der Angriff ist rechtswidrig, wenn das Verhalten gegen ein rechtliches Ge- oder Verbot verstößt und nicht vom Gesetz ausnahmsweise gestattet wird. Begeht der Arbeitgeber einen Gesetzesverstoß, in dem er beispielsweise durch umweltverschmutzende Praktiken Grund und Boden eines anderen kontaminiert, ist eine Angriffshandlung gegeben. Fraglich ist nur, ob diese noch gegenwärtig ist. Der Angriff ist gegenwärtig, solange er unmittelbar bevorsteht, andauert oder noch nicht abgeschlossen ist.[244] In vielen Fällen, in denen der Whistleblower von der rechtswidrigen Handlung des Arbeitgebers Kenntnis erlangt, sind die „Angriffe" schon endgültig abgeschlossen, sodass es an der Gegenwärtigkeit und somit an der Notwehrlage fehlt.

In einigen Fällen dauert der rechtswidrige Angriff durch den Arbeitgeber aber noch an oder steht erst unmittelbar bevor. Dies gilt für jene Fallkonstellationen, in denen der Arbeitgeber augenscheinlich rechtswidrig die Rechtsgüter des Arbeitnehmers, also des Whistleblowers, oder eines Dritten verletzt. Dann ist eine Notwehrlage zu bejahen.

Für eine Rechtfertigung nach § 227 BGB muss die Notwehrhandlung des Whistleblowers geboten und erforderlich sein. Erforderlich ist jene Verteidigung, die zur Abwehr des Angriffs (zumindest teilweise) geeignet ist und zugleich das relativ mildeste Gegenmittel darstellt.[245] Dass Whistleblowing grundsätzlich geeignet ist, Gesetzesverstöße des Arbeitgebers zu unterbinden, steht außer Frage. Durch die Offenbarung wird der Arbeitgeber gezwungen, sein rechtswidriges Verhalten einzustellen. Ist das Whistleblowing aber das mildeste und gleichermaßen geeignetste Mittel? Es gibt verschiedene Erscheinungsformen des Whistleblowings: neben internem Whistleblowing gibt es das externe Whistleblowing gegenüber zuständigen Stellen (Behörden) oder Dritten, wie z. B. der Presse.

Deshalb ist zu unterscheiden: Grundsätzlich muss der Angegriffene unter mehreren zur Verfügung stehenden und gleich wirksamen Mitteln das für den Angreifer am wenigsten gefährlichste wählen. Eine Verhältnismäßigkeitsprüfung findet nicht statt.[246]

Dem Whistleblower stehen folgende Mittel zur Auswahl:

- Information an den Vorgesetzten oder andere Geheimnisträger innerhalb des Unternehmens (interne Stelle)
- Nutzung eines Hinweisgebersystems, sofern ein solches besteht (interne Stelle)

243 Vgl. dazu *Wagner*, in: MüKo BGB, § 823 BGB, Rn. 205 f.

244 *Dennhardt*, in: BeckOK BGB, § 227 BGB, Rn. 9.

245 BGH, Beschl. v. 21.11.2012 – 2 StR 311/12, NStZ-RR 2013, 105, 106. Zwar wird dort auf den § 32 StGB eingegangen – dieser ist jedoch bei den Ausführungen zum § 227 BGB heranzuziehen.

246 *Grothe*, in: MüKo BGB, § 227 BGB, Rn. 17.

- Hinweis an die Polizei, Staatsanwaltschaft oder sonstige Behörden (zuständige Stellen)
- Benachrichtigung der Allgemeinheit (Dritte).

Handelt es sich bei den verletzten Rechtsgütern um hochrangige Güter, wie Leben und Gesundheit des Whistleblowers oder eines Dritten, ist das geeignetste Mittel die Benachrichtigung der zuständigen Stelle – und in besonders akuten Fällen die Benachrichtigung der Allgemeinheit. Sind bei dem Gesetzesverstoß durch den Arbeitgeber weniger bedeutende Rechtsgüter betroffen, ist ein interner Hinweis geeignet, den Angriff abzuwehren. Schlägt der Hinweis an eine interne Stelle fehl, weil keine Abhilfe durch die verantwortlichen Personen geschaffen wird – und der Arbeitgeber in die Tat involviert ist –, hat der Whistleblower sich an eine zuständige Stelle zu wenden. Ungeachtet der Diskussion in der Literatur, ob § 227 BGB einen Verteidigungswillen voraussetzt, ist ein solcher bei dem Whistleblower gegeben, da er in Kenntnis der Notwehrlage handelt und bewusst den Angriff abwehren will.

Bei der Gebotenheit der Verteidigungshandlung gibt es zwar einige Einschränkungen[247] – diese greifen aber nicht im Falle des Whistleblowings.

Im Ergebnis ist festzustellen, dass Whistleblowing gerechtfertigt sein kann. Eine durch Notwehr/Nothilfe gedeckte Handlung ist nicht widerrechtlich. Sie begründet deshalb keine Schadensersatzpflicht aus unerlaubter Handlung (§ 823 BGB).

b. Schadensersatz und Rechtsstaatsprinzip

Differenzierter sieht das LAG Hamm das Problem des Schadensersatzes:[248] Es hat jüngst entschieden, dass es mit dem Rechtsstaatsprinzip nicht vereinbar ist, wenn derjenige, der in gutem Glauben und nicht leichtfertig eine Strafanzeige erstattet, Nachteile erleidet.

In dem Fall ging es um einen Schadensersatzanspruch des Arbeitgebers gegen eine ehemalige Arbeitnehmerin wegen Whistleblowings. Die Beklagten waren eine ehemalige Chefärztin und deren Lebensgefährte.[249] Im Jahr 2008 erhielten das klagende Klinikum sowie Patienten und Angehörige verstorbener Patienten mehrere, teils anonyme Schreiben, in denen von angeblichen Qualitätsmängeln und darauf zurückzuführenden Todesfällen die Rede war. Gegen mehrere Ärzte des Klinikums wurde wiederholt anonym Strafanzeige wegen vorsätzlicher bzw. fahrlässiger Tötungsdelikte erstattet. Es stellte sich im Nachhinein heraus, dass der Lebensgefährte der ehemaligen Chefärztin die anonymen Schreiben verfasst hatte. Nach Einstellung des Ermittlungsverfahrens gegen das Klinikum wegen „erwiesener Unschuld", klagte das Klinikum auf Schadensersatz.

---

247 Dazu zählen: Grobes Missverhältnis; Angriff schuldlos Handelnder, Irrender, Betrunkener; Unfugabwehr; Enge persönliche Beziehungen; Notwehrprovokation.

248 LAG Hamm, Urt. v. 21.7.2011 – 11 Sa 2248/10, ZWH 2012, 253, 255.

249 Zusammenfassung des Sachverhalts *Pacholski*, öAT 2012, 19, 19.

Das LAG Hamm bestätigte die Klageabweisung des Arbeitsgerichts Münster[250]. Die gesetzlichen Voraussetzungen eines Schadensanspruches seien nicht erfüllt. Da der Beklagte zu keinem Zeitpunkt in einem Arbeitsverhältnis zur Klägerin gestanden habe, könnten Schadensersatzansprüche gegen den Lebensgefährten nicht auf vertraglicher, sondern nur auf gesetzlicher Grundlage bestehen. Auch die ehemalige Chefärztin schulde dem Klinikum keinen Schadensersatz. Daran ändere auch der Umstand nichts, dass sie zu der Zeit in einem Arbeitsverhältnis zur Klägerin stand und deshalb auch ein vertraglicher Schadensersatzanspruch in Betracht zu ziehen sei.[251] Die Klägerin stützte sich auf § 823 Abs. 1 und Abs. 2 BGB (Verletzung eines absoluten Rechts/Verstoß gegen ein Schutzgesetz), § 824 BGB (Kreditgefährdung) und § 826 BGB (sittenwidrige vorsätzliche Schädigung). All diesen Anspruchsgrundlagen ist gemeinsam, dass der Schaden durch ein rechtswidriges und schuldhaftes Verhalten des Anspruchsgegners verursacht sein muss. Zwischen dem rechtswidrigen und schuldhaften Verhalten des Schädigers und der Rechtsbeeinträchtigung muss ein Ursachenzusammenhang bestehen.

In der Entscheidung des LAG Hamm wird auf eine Entscheidung des BVerfG[252] verwiesen, dass eine Handhabung des Schadensersatzrechts, die den gutgläubigen Strafanzeigeerstatter mit dem Risiko des Schadensersatzes für den Fall belastet, dass seine Anzeige nicht zum Beweis des behaupteten Vorwurfs führt, gegen Art. 2 Abs. 1 GG und das Rechtsstaatsprinzip verstößt. Da nicht festgestellt werden konnte, dass die Strafanzeigen bewusst wahrheitswidrig oder leichtfertig erstattet wurden, sind beide Beklagte nicht zum Schadensersatz verpflichtet.[253] Zudem besteht regelmäßig ein Interesse der Allgemeinheit an der Aufklärung von Straftaten.[254] Deshalb ist die Erstattung einer Strafanzeige regelmäßig erlaubt.

Die Entscheidung des LAG Hamm macht deutlich, dass Schadensersatzansprüche gegen den redlichen Whistleblower wegen einer Strafanzeige – unabhängig davon, ob er (noch) in einem Arbeitsverhältnis mit dem Arbeitgeber steht oder nicht – erfolglos bleiben. Zwar ist diese keine höchstrichterliche Entscheidung, jedoch zeigt sie eine gewisse Tendenz im Umgang mit solchen Schadensersatzklagen. Die Begründung des Gerichts ist untermauert mit zahlreichen Zitaten höchstrichterlicher Entscheidungen, insbesondere solchen des Bundesverfassungsgerichts. Demnach dürfte eine gegenteilige Entscheidung eines anderen Gerichts schwerlich zu begründen sein; sie ist jedoch nicht ausgeschlossen.

Es ist festzuhalten, dass eine gesetzliche Bestimmung, die den Schadensersatz in einer Whistleblower-Situation regelt, fehlt. Zwar kann sich die Rechtmäßigkeit des Whistleblowing an den dargestellten Maßstäben bei einer arbeitsrechtlichen Kündigung orientieren. Dennoch steht dies einem Schutzgesetz nicht gleich, da

250 ArbG Münster, Urt. v. 29.10.2010 – 4 Ca 272/10.
251 Rechtsprechungsticker, RDG 2012, 72, 73.
252 BVerfG, Beschl. v. 25.2.1987 – 1 BvR 1086/85, NJW 1987, 1929 f.
253 Rechtsprechungsticker, RDG 2012, 72, 73.
254 So schon BVerfG, Beschl. v. 2.7.2001 – 1 BvR 2049/00, NJW 2001, 3474, 3475.

der Arbeitnehmer nur unter erschwerten Bedingungen an die gerichtlichen Entscheidungen gelangt und er gegebenenfalls mit Interpretationsschwierigkeiten zu rechnen hat. Zudem ist keinesfalls gewährleistet, dass alle Gerichte, insbesondere die Gerichte unterer Instanzen, gleichermaßen urteilen werden. Insoweit könnte – abgesehen von der Frage nach der Gegenwärtigkeit des Angriffs durch den Arbeitgeber – in solchen Fällen nur der Rückgriff auf den Notwehrtatbestand des § 227 BGB Abhilfe schaffen.

8. Fazit

Die bestehende Rechtslage stellt durchaus Instrumente zur Verfügung, die die Position eines Whistleblowers stärken können – sie bieten jedoch keinen hinreichenden und umfassenden Schutz. Die Vorschriften gehen von einem (älteren) Verständnis der Pflichtverletzung und Kündigungsgründe aus und lassen wenig Raum für neue besondere Einzelfallkonstellationen. Auch eine Abstufung der einzelnen Verhaltensweisen des Whistleblowers und die Abgrenzung zwischen internem und zulässigem externem Whistleblowing sind an keiner Stelle kodifiziert oder durch Auslegung zu ermitteln.

Die Maßnahmen, mit denen gegen einen Whistleblower vorgegangen werden kann, richten sich nach den allgemeinen Regeln der – besonderen – Schuldverhältnisse. Es fehlt an einer differenzierten Betrachtung der allgemeinen Regeln. Eine Hervorhebung und Konkretisierung der Besonderheiten, die beim Whistleblowing zu beachten sind, würde zu Rechtssicherheit für beide Vertragsparteien führen. Es muss eindeutig festgelegt werden, wann Whistleblowing zulässig ist und wann nicht.

II. Rechtslage beim an der Tat beteiligten Whistleblower

Die bisherigen Ausführungen zum Schutz des Whistleblowers *de lege lata* beziehen sich nur auf den Hinweisgeber, der nicht an dem Missstand („der Tat“) beteiligt war. Fraglich ist, wie zu verfahren ist, wenn der Whistleblower sich zunächst an der rechtswidrigen Tat beteiligt hat und erst im Nachhinein von seinem Verhalten abgerückt ist.

Die Pflichtverletzung hat hier eine andere Qualität. Es geht also nicht „nur“ um das Whistleblowing an sich, sondern auch um die (Mit-)Täterschaft oder Teilnahme des Arbeitnehmers an dem offenbarten Missstand. Die Rechtsprechung und Literatur hat sich mit dieser besonderen Problematik noch nicht auseinandergesetzt. Fest steht: Die Begehung einer Straftat im Rahmen des Arbeitsverhältnisses ist grundsätzlich ein Kündigungsgrund.[255]

255 *Hergenröder*, in: MüKo BGB, § 1 KSchG, Rn. 278.

1. Die Kündigung

Neben den oben genannten Folgen, die jeden Whistleblower treffen können, tritt hinzu, dass der Arbeitnehmer sich wegen Beteiligung an einer Straftat strafbar gemacht haben kann.[256] Die Begehung einer Straftat im dienstlichen Bereich stellt grundsätzlich eine Verletzung des Arbeitsvertrages dar und kann eine verhaltensbedingte oder außerordentliche Kündigung rechtfertigen.[257] Straftaten, die *im Rahmen des Arbeitsverhältnisses* zum Nachteil des Arbeitgebers begangen werden, sind Arbeitsvertragspflichtverletzungen, die eine verhaltensbedingte Kündigung – regelmäßig sogar *ohne Abmahnung* – rechtfertigen.[258] Straftaten, die *außerhalb des Arbeitsverhältnisses* begangen werden, können die persönliche Eignung des Arbeitnehmers in Frage stellen. Die Begehung einer Straftat kann in besonderen Fällen sogar eine personenbedingte Kündigung nach sich ziehen.[259]

Demnach drohen einem an der Tat (oder dem Missstand) beteiligten Whistleblower aufgrund des Whistleblowings sowie der Beteiligung an der Tat arbeitsrechtliche Konsequenzen. Er ist gewissermaßen doppelt bestraft, obwohl die Hemmschwelle größer war, das (eigene) Verhalten zu offenbaren, da er selbst mit Konsequenzen rechnen muss.

Einen besonderen gesetzlichen Schutz gibt es in diesem Fall nicht. Es kann nur auf die allgemeinen Regeln zurückgegriffen werden. Denn ob ein die Kündigung rechtfertigender Grund gegeben ist oder nicht, bestimmt sich danach, ob eine konkrete Beeinträchtigung des Arbeitsverhältnisses gegeben ist. Das BAG zieht für die Entscheidung über die Sozialwidrigkeit das Kriterium der Zumutbarkeit heran. Eine Kündigung ist demnach sozial gerechtfertigt, wenn die Störung eine Fortsetzung des Arbeitsverhältnisses unzumutbar macht.[260] Kündigungsrechtlich relevant sind nur erhebliche Beeinträchtigungen der unternehmerischen Interessen; geringfügige

---

256 Hilft der Arbeitnehmer z. B. dem Arbeitgeber, verschmutzten und hochgiftigen Abfall falsch zu entsorgen, kann die Verwirklichung von Umweltdelikten in Betracht kommen. Behandelt ein Arbeitnehmer in einer Pflegeeinrichtung liegende Patienten nicht nach Vorschrift, kann eine Körperverletzung in Betracht kommen. Zwar macht sich der Arbeitgeber strafbar, wenn er solche Arbeitsanweisung erteilt; erkennt der Arbeitnehmer jedoch, dass durch die Praktiken gegen ein Gesetz verstoßen wird und handelt er den Anweisungen entsprechend bzw. unterstützt den Arbeitgeber, macht er sich ebenfalls strafbar.

257 *Dörner/Vossen*, in: Ascheid/Preis/Schmidt, Kündigungsrecht, § 1 KSchG, Rn. 256.

258 So schon BAG, Urt. v. 20.9.1984 – 2 AZR 633/82, AP Nr. 80 zu § 626 BGB; *Hergenröder*, in: MüKo BGB, § 626 BGB, Rn. 278.

259 Siehe zuletzt BAG, Urt. v. 20.6.2013 – 2 AZR 583/12, NZA 2013, 1345, 1346. Als Wachpolizist hatte der Kläger weitgehend die gleichen Befugnisse wie ein verbeamteter Polizist. Die Herstellung eines verbotenen Betäubungsmittels in nicht unerheblichem Umfang war mit dieser hoheitlichen Funktion unvereinbar und stellte nach der Interessenabwägung einen sozial gerechtfertigten Kündigungsgrund dar.

260 So z. B. BAG, Urt. v. 17.1.1991 – 2 AZR 375/90, NZA 1991, 557, 558.

Störungen sind hinzunehmen.[261] Eine Art „mildernder Umstand" wegen Beendigung des rechtswidrigen Verhaltens ist in keiner Vorschrift als Regelungsgegenstand zu erkennen. Auch wenn man das Prognoseprinzip[262] heranzieht oder eine Abwägung durchführt, ergibt sich kein anderes Ergebnis.

Wird eine Straftat im Rahmen des Arbeitsverhältnisses begangen, geschieht das regelmäßig zum Nachteil des Arbeitgebers und hat einen erheblichen Vertrauensverlust zur Folge. Eine Fortsetzung des Arbeitsverhältnisses ist dann oftmals nicht mehr zumutbar. Im Rahmen des Whistleblowings muss dies jedoch nicht immer der Fall sein. Der Whistleblower ist von seinem rechtswidrigen Verhalten abgerückt und hat sein Verhalten reflektiert. Dadurch ist für den Arbeitgeber ersichtlich, dass er den Vertrauensbruch beseitigen will bzw. seiner Person wieder Vertrauen geschenkt werden kann. Demnach muss der einmalige Vertrauensbruch nicht zwangsläufig zur Lösung des Arbeitsverhältnisses führen. Wenn der Arbeitnehmer sich im Rahmen des Whistleblowings dafür entscheidet, den Missstand zu beheben, muss dies positiv bewertet und geschätzt werden.

2. *Schadensersatz gegenüber dem Arbeitgeber*

Hat der Whistleblower an der Tat mitgewirkt, kann er wegen einer Pflichtverletzung haftbar gemacht werden. Anders als im Fall des unbeteiligten Whistleblowers, bei dem eine Verletzung der Verschwiegenheitspflicht im Raume steht (Vertrauensbereich), könnte hier eine schuldhafte Pflichtverletzung im Leistungsbereich oder Unternehmensbereich vorliegen.[263] Dieses Verhalten des Arbeitnehmers kann Schadensersatzansprüche nach sich ziehen. Da der Arbeitsvertrag ein gegenseitiger Vertrag im Sinne der §§ 320 ff. BGB ist, gelten für Leistungsstörungen im Arbeitsverhältnis grundsätzlich die allgemeinen schuldrechtlichen Vorschriften. Die arbeitsvertragliche Hauptpflicht ist die Arbeitsleistung des Arbeitnehmers. Eine Verletzung dieser Pflicht kann vorliegen, wenn der Arbeitnehmer die Arbeitsleistung nicht oder schlecht erbringt. Aber auch Verletzungen von Nebenpflichten können Schadensersatzansprüche auslösen.[264] Da die Begehung der Straftat oder die Beteiligung an dem Missstand regelmäßig nicht die Hauptleistungspflicht tangiert, kann sich ein Schadensersatzanspruch des Arbeitgebers auch hier nur aus einer Nebenpflichtverletzung ergeben. Anders wäre es nur, wenn die Begehung einer rechtswidrigen Tat eine Schlechtleistung darstellt.

---

261 *Dörner/Vossen*, in: Ascheid/Preis/Schmidt, Kündigungsrecht, § 1 KSchG, Rn. 64.

262 Auf der Basis des in der Vergangenheit liegenden vertragswidrigen Verhaltens ist zu beurteilen, ob das Vertragsverhältnis durch diese Vorkommnisse auch künftig beeinträchtigt werden wird, etwa weil mit Wiederholungen zu rechnen ist oder das Vertrauensverhältnis irreparabel zerstört ist *Stoffels*, in: BeckOK Arbeitsrecht, § 626 BGB, Rn. 72.

263 Zur Kündigung wegen Pflichtverletzung *Hergenröder*, in: MüKo BGB, § 1 KSchG, Rn. 191.

264 *Dütz/Thüsing*, Arbeitsrecht, Rn. 187.

Bei Nebenpflichtverletzungen des Arbeitnehmers kommen vertragliche Schadensersatzansprüche nach den allgemeinen Regeln in Betracht, insbesondere nach § 280 Abs. 1 BGB für die Verletzung von Nichtleistungspflichten.[265] Hat der Arbeitgeber durch die Begehung der Straftat also einen Schaden erlitten, kann er Ersatz fordern. Begeht der Arbeitnehmer eine Pflichtverletzung im Rahmen des Arbeitsverhältnisses, so wird nicht vermutet, dass der Arbeitnehmer diese Pflichtverletzung zu vertreten hat. Abweichend von der Regelung in § 280 Abs. 1 S. 2 BGB trägt bei Pflichtverletzungen der Arbeitgeber die Beweislast für das Vertretenmüssen des Arbeitnehmers, § 619 a BGB.[266] Zwar kommt grundsätzlich bei einer Schlechtleistung oder Nebenpflichtverletzung des Arbeitnehmers eine Haftungsbeschränkung[267] in Betracht. Doch wird hier davon ausgegangen, dass der Arbeitnehmer das Delikt vorsätzlich begeht; bei vorsätzlichem Handeln haftet er trotz der Haftungsbeschränkung in Form des innerbetrieblichen Schadensausgleichs voll.[268] Die Besonderheit, dass der Arbeitnehmer seine Tat durch die Offenlegung selbst anzeigt und damit eine Form der Reue zeigt, findet keine Berücksichtigung.

Daneben können Ansprüche aus §§ 823 ff. BGB gegeben sein, wenn ein geschütztes Rechtsgut schuldhaft verletzt wird.

3. *Fazit*

Mangels spezifischer Normen, muss – wie beim unbeteiligten Whistleblower – auf die allgemeinen Rechtsvorschriften zurückgegriffen werden. Dass der Arbeitnehmer als beteiligter Whistleblower freiwillig von seiner Tat abrückt, wird zu keinem Zeitpunkt berücksichtigt. Sein Verhalten wird als Straftat qualifiziert, bei der das nachträgliche Wohlverhalten ohne Bedeutung ist und zur Kündigung berechtigt. Die Aufgabe der Tat wird nicht bei der Zumutbarkeit der Kündigung in Erwägung gezogen. Es ist zwar zuzugeben, dass eine Berücksichtigung des Verhaltens des Whistleblowers nach den gegenwärtigen allgemeinen Regeln schwerlich zu begründen ist. Dennoch darf in einem demokratischen Rechtsstaat ein gesetzeswidriges Handeln nicht „geschützt", bzw. derjenige, der dieses Verhalten öffentlich macht, (arbeitsrechtlich) nicht bestraft werden. Zwar begeht der involvierte Whistleblower eine Pflichtverletzung, jedoch offenbart er Handlungspraktiken, die gegen das Gesetz verstoßen.

Zudem ist zu beachten, dass ohne den Whistleblower die Gesetzesverstöße in Zukunft weiter bestanden hätten und mit an Sicherheit grenzender Wahrscheinlichkeit mehr Schäden an gemeinschaftlichen Rechtsgütern, aber auch an solchen des Arbeitgebers entstanden wären.

---

265 *Dütz/Thüsing*, Arbeitsrecht, Rn. 209.

266 Vgl. *Schreiber*, in: Schulze, BGB-Kommentar, § 619 a, Rn. 1.

267 Siehe dazu BAG, Urt. v. 15.11.2012 – 8 AZR 705/11, NJOZ 2013, 709, 710, mit Verweis auf die vom Großen Senat entwickelten Grundsätze zur Arbeitnehmerhaftung BAG GS, Beschl. v. 27.9.1994 – GS 1/89 (A), NZA 1994, 1083, 1084.

268 Ausführlicher dazu *Schreiber*, in: Schulze, BGB-Kommentar, § 619 a, Rn. 4 f.

ausgesprochene ordentliche Kündigung unwirksam. Es liege weder ein wichtiger Kündigungsgrund im Sinne des § 626 Abs. 1 BGB vor, noch sei die Kündigung im Sinne des § 1 Abs. 1 KSchG sozial gerechtfertigt.[315]

Das Gericht führte aus, dass das Arbeitsverhältnis aus wichtigen verhaltensbedingten Gründen nur dann ohne Einhaltung einer Kündigungsfrist außerordentlich gekündigt werden könne, wenn das Verhalten des Arbeitnehmers und die andauernde Beeinträchtigung des Arbeitsverhältnisses derart schwerwiegend seien, dass dem Arbeitgeber selbst die Fortsetzung des Arbeitsverhältnisses bis zum Ablauf der ordentlichen Kündigungsfrist nicht zugemutet werden kann.[316]

Es stellte fest, das Vorbringen der Beklagten genüge nicht den Anforderungen einer verhaltensbedingten Kündigung:

> *„Es könne dabei dahinstehen, ob die Klägerin – wie von Beklagtenseite behauptet – Dienstpläne, die in der Folgezeit in der RBB-Sendung Abendschau gezeigt wurden, an Dritte weitergab und Strafanzeige gegen die Geschäftsführung der Beklagten wegen Betrugs und Nötigung erstattete, ob entsprechend der Verhaltensweise an sich gehalten sind, eine verhaltensbedingte Kündigung zu rechtfertigen und ob die von der Beklagten insoweit gerügten Pflichtverletzungen der Klägerin tatsächlich im für die gerichtliche Prüfung der Wirksamkeit der Kündigung vom 9. Februar 2005 allein maßgeblichen Zeitpunkt, dem des Zugangs dieses Kündigungsschreibens, bei der Klägerin bereits vorlagen, denn die aufgeführten Vorwürfe waren wie die Verteilung des zweiten Flugblattes nach Zugang der Kündigung vom 9. Februar 2005 nach Beklagtenvortrag ‚nicht Bestandteil der Kündigung'."*[317]

Das Gericht betonte zudem, dass die Klägerin ihre arbeitsvertraglichen Pflichten nicht verletzt habe, da die Äußerungen auf dem Flugblatt vom Grundrecht auf Meinungsfreiheit gedeckt seien.

Mit der überragenden Bedeutung des Grundrechts aus Art. 5 Abs. 1 GG wäre es unvereinbar, wenn das Grundrecht in der betrieblichen Arbeitswelt, welche die Lebensgrundlage zahlreicher Staatsbürger wesentlich bestimmt, gar nicht oder nur eingeschränkt anwendbar wäre. Unter Berücksichtigung des Art. 5 Abs. 1 GG, der für eine freiheitlich demokratische Staatsordnung schlechthin konstituierend sei, liege keine Verletzung der vertraglichen Rücksichtnahmepflicht und der im Grundgesetz geschützten Interessen der Beklagten vor, denn der Wortlaut des Flugblattes setze sich zwar polemisch mit der Beklagten auseinander, die Grenzen strafrechtlich relevanten Verhaltens würden hingegen nicht überschritten – zumal ein gewisser Personalmangel augenscheinlich auch objektiv vorliege.[318]

Die Klägerin habe mit dem Schreiben an die Geschäftsführung, das auf die unzumutbaren Umstände hinwies, vielmehr in Wahrnehmung berechtigter Interessen gehandelt. Sie habe verbleibende höherrangige unternehmensinterne

---

315 ArbG Berlin, Urt. v. 3.8.2005 – 39 Ca 4775/05, BeckRS 2011, 77275.
316 ArbG Berlin, Urt. v. 3.8.2005 – 39 Ca 4775/05, BeckRS 2011, 77275.
317 ArbG Berlin, Urt. v. 3.8.2005 – 39 Ca 4775/05, BeckRS 2011, 77275.
318 Vgl. ArbG Berlin, Urt. v. 3.8.2005 – 39 Ca 4775/05, BeckRS 2011, 77275.

Interventionsmöglichkeiten zur Bewirkung von ihr für notwendig erachteter Veränderungen der Betriebsorganisation genutzt, nachdem ihre Überlastanzeigen erfolglos geblieben waren und Handlungsmöglichkeiten auf näheren Ebenen der Betriebshierarchie ausgeschöpft schienen.[319] Dies rechtfertige somit auch keine Kündigung.

Nach Ansicht des Arbeitsgerichts Berlin war die Kündigung also unwirksam.

bb. Landesarbeitsgericht Berlin

Das Landesarbeitsgericht Berlin[320] hob in der Berufung das Urteil des Arbeitsgerichts Berlin auf und stellte fest, dass die Kündigung rechtmäßig gewesen sei. Die Strafanzeige der Beschwerdeführerin stelle einen wichtigen Grund im Sinne von § 626 BGB dar.

Im Rahmen des arbeitsrechtlichen Verhältnisses seien die vertraglichen Rücksichtnahmepflichten zu beachten und zu konkretisieren:

> *„Zunächst darf der Arbeitnehmer die Strafanzeige nicht auf wissentlich oder leichtfertig unwahre Tatsachen stützen. Abgesehen von diesen Fällen darf sich die Anzeige des Arbeitnehmers nicht als eine unverhältnismäßige Reaktion auf ein Verhalten des Arbeitgebers oder seines Repräsentanten darstellen. Dabei können als Indizien für eine unverhältnismäßige Reaktion des anzeigenden Arbeitnehmers sowohl die Berechtigung der Anzeige als auch die Motivation des Anzeigenden oder ein fehlender innerbetrieblicher Hinweis auf die angezeigten Missstände sprechen“.*[321]

Das Gericht betonte weiter, dass die Motive des Arbeitnehmers eine besondere Bedeutung verdienten und die Erstattung der Anzeige eine unverhältnismäßige Reaktion sein könne, wenn sie ausschließlich den Arbeitgeber schädigen bzw. „fertig machen“ wolle. Durch ein derartiges pflichtwidriges Verhalten nehme der Arbeitnehmer keine Verfassungsrechte wahr, sondern verhalte sich – jedenfalls gegenüber dem Arbeitgeber – rechtsmissbräuchlich.[322]

Das Gericht führte aus, dass einer innerbetrieblichen Klärung nicht generell der Vorrang gebühre. Es sei im Einzelfall zu bestimmen, wann dem Arbeitnehmer eine vorherige innerbetriebliche Anzeige ohne weiteres zumutbar sei und ein Unterlassen ein pflichtwidriges Verhalten darstelle:

> *„Eine vorherige innerbetriebliche Meldung und Klärung ist dem Arbeitnehmer unzumutbar, wenn er Kenntnis von Straftaten erhält, durch deren Nichtanzeige er sich selbst einer Strafverfolgung aussetzen würde. Entsprechendes gilt auch bei schwerwiegenden Straftaten und vom Arbeitgeber selbst begangenen Straftaten. Hier muss regelmäßig die*

319 Vgl. ArbG Berlin, Urt. v. 3.8.2005 – 39 Ca 4775/05, BeckRS 2011, 77275.
320 Vgl. LAG Berlin, Urt. v. 28.3.2006 – 7 Sa 1884/05, AuR 2007, 51.
321 Vgl. LAG Berlin, Urt. v. 28.3.2006 – 7 Sa 1884/05, AuR 20007, 51, 52; so schon BAG, Urt. v. 3.7.2003 – 2 AZR 235/02 – NZA 2004, 427, 430.
322 Vgl. LAG Berlin, Urt. v. 28.3.2006 – 7 Sa 1884/05, AuR 2007, 51, 52.

Pflicht des Arbeitnehmers zur Rücksichtnahme auf die Interessen des Arbeitgebers zurücktreten."

Das Gericht kommt danach jedoch zu dem Schluss, dass bei Anwendung dieses Maßstabes sich die von der Klägerin erstattete Strafanzeige als grober Verstoß gegen ihre arbeitsvertraglichen Rücksichtnahmepflichten erweise, mithin als wichtiger Grund für eine außerordentliche Kündigung an sich geeignet sei.[323] Die Klägerin habe nämlich ihre Anzeige leichtfertig auf Tatsachen gegründet, die im Prozess nicht dargelegt werde konnten und schon insoweit keine berechtigten Interessen wahrgenommen. Zudem stelle sich die gesamte Anzeige als unverhältnismäßige Reaktion auf die Weigerung der Beklagten dar, den von der Klägerin behaupteten Personalmangel anzuerkennen.[324] Das LAG argumentiert, dass die Klägerin mit dieser Anzeige und einer sich daraus ergebenden öffentlichen Diskussion eine Kampagne gegen die Beklagte zu eröffnen bezweckte. Die Klägerin habe für den von ihr erhobenen Vorwurf nicht ansatzweise einen erwiderungsfähigen und einem Beweis zugänglichen Sachverhalt dargestellt, vielmehr habe sie mit den von ihr erhobenen Vorwürfen wie dem Abrechnungsbetrug und der gewagten Konstruktion eines Betruges zu Lasten der Angehörigen und Bewohner Aufsehen erregen und Druck auf die Beklagte ausüben wollen.[325]

Unter Berücksichtigung aller Umstände des Einzelfalls und unter Abwägung der Interessen der Vertragsparteien sei es der Beklagten auch nicht zuzumuten, das Arbeitsverhältnis bis zum Ablauf der ordentlichen Kündigungsfrist fortzusetzen,[326] wodurch sich die streitgegenständliche Kündigung vom 9. Februar 2005 als rechtswirksam erweise.

cc. Bundesarbeitsgericht/Bundesverfassungsgericht

Das Landesarbeitsgericht hat die Klage abgewiesen und die Revision gegen das Urteil nicht zugelassen. Hiergegen wendete sich Frau Heinisch nun mit einer Nichtzulassungsbeschwerde, die sich auf Divergenz, grundsätzliche Bedeutung einer entscheidungserheblichen Rechtsfrage und Verletzung des Anspruchs auf rechtliches Gehör stützte.

Das Bundesarbeitsgericht wies die Nichtzulassungsbeschwerde mit Beschluss vom 6. Juni 2007 zurück. Als Grundsatzbeschwerde entspreche ihre Begründung nicht den gesetzlichen Anforderungen.[327] Dies gelte auch für die Rüge der Verletzung des Anspruchs auf rechtliches Gehör.[328] Unter dem Gesichtspunkt der Divergenz sei die Beschwerde ebenfalls unbegründet.[329]

---

323 Vgl. LAG Berlin, Urt. v. 28.3.2006 – 7 Sa 1884/05, AuR 2007, 51, 52.
324 Vgl. LAG Berlin, Urt. v. 28.3.2006 – 7 Sa 1884/05, AuR 2007, 51, 52.
325 Vgl. LAG Berlin, Urt. v. 28.3.2006 – 7 Sa 1884/05, AuR 2007, 51, 52.
326 Vgl. LAG Berlin, Urt. v. 28.3.2006 – 7 Sa 1884/05, AuR 2007, 51, 53.
327 BAG, Beschl. v. 6.6.2007 – 4 AZN 487/06, BeckRS 2011, 77276.
328 BAG, Beschl. v. 6.6.2007 – 4 AZN 487/06, BeckRS 2011, 77276.
329 BAG, Beschl. v. 6.6.2007 – 4 AZN 487/06, BeckRS 2011, 77276.

Die eingelegte Verfassungsbeschwerde wurde vom Bundesverfassungsgericht[330] ohne weitere Begründung nicht zur Entscheidung angenommen.

c. *Die Rechtsprechung des Europäischen Gerichtshofs für Menschenrechte*

Am 9. Juni 2008 hat die Beschwerdeführerin beim Europäischen Gerichtshof für Menschenrechte Beschwerde eingelegt und gerügt, die Kündigung habe ihr in Art. 10 EMRK garantiertes Recht auf Freiheit der Meinungsäußerung verletzt und außerdem sei das Verfahren vor dem Landesarbeitsgericht nicht fair gewesen. Am 27. Juli 2011 hat der Gerichtshof einstimmig die Beschwerde wegen der Verletzung des Anspruches auf ein faires Verfahren für unzulässig, im Übrigen für zulässig erklärt und festgestellt, dass Art. 10 EMRK verletzt sei sowie Deutschland nach Art. 41 EMRK verurteilt, binnen drei Monaten an die Beschwerdeführerin 10.000 Euro als Ersatz für einen Nichtvermögensschaden und 5000 Euro für Kosten und Auslagen zu zahlen.[331]

Oberflächlich betrachtet behandelt die Entscheidung des EGMR den Fall des Whistleblowings in einer neuartigen Tiefe und Breite. Bei genauerem Hinsehen aber zeigt sich, dass die Bedeutung dieser Entscheidung überschätzt wird. Denn der EGMR hat in der Entscheidung lediglich jene Kontrollmaßstäbe angewendet, die der deutschen Rechtsprechung schon bekannt waren. Dennoch ist der Gerichtshof zu einem anderen Ergebnis gekommen, was die Bedeutung einer vertieften Auseinandersetzung mit dem Thema Whistleblowing unterstreicht. Allein wegen der erstmaligen Konkretisierung der Verhaltensmaßstäbe und der Aufstellung von richtungsweisenden Abwägungskriterien kann die Entscheidung als „Paukenschlag“ bezeichnet werden.

Bevor auf die einzelnen Entscheidungspunkte des Gerichtshofs eingegangen wird, soll zunächst die Bedeutung der EMRK für das deutsche Recht erläutert werden.

aa. Die Bedeutung der EMRK und der Entscheidungen des EGMR im deutschen Recht

Die EMRK und ihre Zusatzprotokolle sind völkerrechtliche Verträge, denen der Bundesgesetzgeber gem. Art 59 Abs. 2 GG durch förmliches Gesetz zugestimmt hat.[332] Dadurch hat er sie in die innerstaatliche Rechtsordnung übernommen und ihnen rechtliche Geltung geschenkt. Die Europäische Menschenrechtskonvention ist 1950 von der Bundesrepublik ratifiziert worden[333] und besitzt den Rang einfachen Bundesrechts.[334] Die Vertragsstaaten der EMRK sind gem. Art. 56 Ziff. 3 EMRK

330 BVerfG, Beschl. v. 6.12.2007 – 1 BvR 1905/07.
331 EGMR, Urt. v. 21.7.2011 – 28274/08, NJW 2011, 3501 (Heinisch/Deutschland).
332 *Schlachter*, RdA 2012, 108, 109.
333 Bekanntmachung der aktuellen Fassung der EMRK in BGBl. II-2002, S. 1054.
334 BVerfG, Beschl. v. 14.10.2004 – 2 BvR 1481/04, NJW 2004, 3407, 3408, 3411.

verpflichtet, das Konventionsrecht in ihrem Hoheitsgebiet zu wahren. Alle staatlichen Stellen einschließlich der Gerichte haben das Konventionsrecht anzuwenden.[335]

Die nationalen Gerichte haben deshalb die Bestimmungen der Konvention bei der Interpretation des nationalen Rechts, auch der Grundrechte und der rechtsstaatlichen Garantien, zu berücksichtigen[336] und im Rahmen methodisch vertretbarer Auslegung zu beachten und anzuwenden. Da das Grundgesetz prinzipiell völkerrechtsfreundlich auszulegen ist, macht das BVerfG in seinen Entscheidungen deutlich,[337] dass die EMRK auch die Auslegung der Grundrechte beeinflusst. Die deutschen Grundrechte müssen daher selbst als Ausprägung der Menschenrechte verstanden werden, die sie in Form einer „Mindestausprägung" garantieren.[338] Die Auslegung nationalen Rechts am Maßstab des Konventionsrechts ist als Bestandteil der Bindung an Recht und Gesetz nach Art. 20 Abs. 3 GG zu verstehen. Die Reichweite der von den Rechtsanwendern geforderten Auslegungsanstrengungen bezieht sich sowohl auf die EMRK (als niedergeschriebenes Recht) als auch auf die Interpretation durch den EGMR. Das Bundesverfassungsgericht[339] betont, dass es keineswegs ausschließlich verpflichtet ist, diese Kooperation von nationalem und Konventionsrecht zu gewährleisten, sondern dass vielmehr alle Fachgerichte dazu verpflichtet sind, die EMRK und die Rechtsprechung des EGMR in ihren Entscheidungen ausreichend zu berücksichtigen. Mithin sind alle deutschen Gerichte verpflichtet, sich bei Entscheidungen mit Menschenrechtsbezug auch an den einschlägigen Urteilen des Gerichtshofs für Menschenrechte zu orientieren.

Eine unzureichende Berücksichtigung der EMRK verletzt den Grundrechtsträger in seinem Grundrecht in Verbindung mit dem Rechtsstaatsprinzip.[340]

bb. Wesentliche Eckpunkte der Heinisch-Entscheidung

Zunächst geht der Gerichtshof allgemein auf Art. 10 EMRK ein, indem er anhand von älteren Entscheidungen die ausschlaggebenden Kriterien aufzeigt.

Das Recht auf Meinungsäußerungsfreiheit nach Art. 10 EMRK schließt die Freiheit mit ein, Informationen und Ideen ohne behördliche Eingriffe empfangen und weitergeben zu können. Darunter fällt auch das Whistleblowing.[341]

---

335 EGMR, Urt. v. 3.2.2011 – 18136/02 (Rn. 38), NZA 2012, 199, 200 f. (Siebenhaar/Deutschland).

336 BVerfG, Urt. v. 29.5.1990 – 2 BvR 254/88, NJW 1990, 2741, 2743.

337 BVerfG, Urt. v. 14.10.2004 – 2 BvR 1481/04, NJW 2004, 3407, 3410 f. Es sind, wie das BVerfG es formuliert „grundsätzlich die vom Gerichtshof in seiner Abwägung berücksichtigten Aspekte auch in die verfassungsrechtliche Würdigung, namentlich die Verhältnismäßigkeit, einzubeziehen, und es hat eine Auseinandersetzung mit den vom Gerichtshof gefundenen Abwägungsergebnissen stattzufinden."; bestätigt durch BVerfG, Urt. v. 4.5.2011 – 2 BvR 2365/09, NJW 2011, 1931, 1934 f.

338 *Schlachter*, RdA 2012, 108, 109.

339 BVerfG, Urt. v. 21.7.2010 – 1 BvR 420/09, NJW 2010, 3008, 3015.

340 BVerfG, Urt. v. 14.10.2004 – 2 BvR 1481/04, NJW 2004, 3407, 3410 f.

341 So schon EGMR, Urt. v. 12.2.2008 – 14277/04 (Rn. 70), (Guja/Moldawien).

Der Gerichtshof hatte schon früher entschieden, dass die Freiheit der Meinungsäußerung auch am Arbeitsplatz gilt, auch dann, wenn auf die Beziehung zwischen Arbeitgeber und Arbeitnehmer Privatrecht anzuwenden sei.[342] Der Staat habe die Verpflichtung, die Rechte auch im Verhältnis von Privatpersonen zu schützen.[343]

In dieses Recht werde eingegriffen, wenn an seine Ausübung Nachteile wie die fristlose Kündigung geknüpft werden und staatliche Regelungen keinen Schutz vor solchen Nachteilen böten. Ebenfalls stelle es einen Eingriff dar, wenn deutsche Gerichte die Entlassung der Beschwerdeführerin in ihrem Urteil für rechtmäßig erachteten. Lasse der Staat tatbestandliche Eingriffe zu, müssten diese sich im Rahmen dessen halten, was nach dem Konventionsrecht zulässig sei.

Ein solcher Eingriff könne das hier in Frage stehende Recht aus Art. 10 EMRK verletzen, wenn er nicht „gesetzlich vorgesehen" sei, kein „berechtigtes Ziel" im Sinne von Art. 10 Abs. 2 EMRK verfolge und nicht „in einer demokratischen Gesellschaft notwendig" sei, um dieses Ziel zu erreichen.[344]

Das Prüfprogramm nach der EMRK gleicht der nationalen Prüfung nach dem Kündigungsrecht – es ist die Verhältnismäßigkeit zu prüfen. Abgestellt wird auf § 626 BGB. § 626 BGB ist die Grundlage für die fristlose Kündigung. Die Regelung, nach der jeder Vertragsteil berechtigt ist, dem anderen zu kündigen, verfolge auch ein legitimes Ziel, nämlich den Schutz des geschäftlichen Rufs und der Interessen des Arbeitgebers.[345] Der EGMR prüft dann im Rahmen einer Interessenabwägung, ob der Eingriff „in einer demokratischen Gesellschaft notwendig" ist. Der Gerichtshof richtet sich nach seiner ständigen Rechtsprechung[346] und stellt klar, dass Informationen eines Angestellten des öffentlichen Dienstes über rechtswidriges Verhalten oder Straftaten am Arbeitsplatz unter bestimmten Voraussetzungen vom Anwendungsbereich des Art. 10 EMRK erfasst würden.[347] Gleiches gelte für einen Arbeitnehmer der Privatwirtschaft.

Im Rahmen der Abwägung komme durch die Erstattung der Anzeige die Verletzung einer vertraglichen Rücksichtnahmepflicht nach § 241 Abs. 2 BGB in Betracht, die einen wichtigen Grund im Sinne des § 626 BGB darstellen könne.[348] Ob diese Pflicht tatsächlich verletzt wurde, sei im Rahmen der Abwägung der betroffenen, grundrechtlich geschützten Interessen der Parteien festzustellen.

---

342 EGMR, Urt. v. 26.2.2009 – 29492/05 (Rn. 85), (Kudeshkina/Russland).

343 EGMR, Urt. v. 29.2.2000 – 39293/98 (Rn. 38), (Fuentes Bobo/Spanien).

344 EGMR, Urt. v. 21.7.2011 – 28274/08 (Rn. 46), NJW 2011, 3501, 3502 (Heinisch/Deutschland).

345 EGMR, Urt. v. 21.7.2011 – 28274/08 (Rn. 48), NJW 2011, 3501, 3502 (Heinisch/Deutschland).

346 EGMR, Urt. v. 15.2.2005 – 68416/01 (Rn. 87), NJW 2006, 1255, 1258 f. (Steel and Morris/Vereinigtes Königreich).

347 EGMR, Urt. v. 21.7.2011 – 28274/08 (Rn. 62), NJW 2011, 3501, 3503 (Heinisch/Deutschland).

348 EGMR, Urt. v. 21.7.2011 – 28274/08 (Rn. 64), NJW 2011, 3501, 3503 (Heinisch/Deutschland).

Auf der Seite des Gekündigten stehe das sich aus dem Rechtsstaatsprinzip (Art. 20 Abs. 3 GG) ergebende Recht, sich an Strafverfolgungsbehörden zu wenden, und zwar auch dann, wenn sich (nur) der Verdacht auf eine Straftat gegen einen Vertragspartner des Anzeigenden richte. Daneben könne sich der Gekündigte auf das Recht auf Meinungsfreiheit berufen. Tatsachenbehauptungen fielen zumindest dann unter Art. 5 Abs. 1 GG bzw. Art. 10 EMRK, wenn sie mit Werturteilen verbunden oder für die Bildung von Meinungen relevant seien.[349] Bei der Prüfung, ob der Eingriff in die Freiheit der Meinungsäußerung verhältnismäßig war, müsse, neben dem grundsätzlichen Versuch betriebsinterner Abhilfe durch den Arbeitnehmer, berücksichtigt werden, ob dem Beschwerdeführer andere wirksame Mittel zur Verfügung standen, die Missstände abzustellen, die er offenbaren wollte (bzw. offenbart hat).[350] Ebenfalls müsse bei der Prüfung im Rahmen des Art. 10 Abs. 2 EMRK der Aspekt des „öffentlichen Interesses" besonders beachtet werden. Außerdem sei die Freiheit der Meinungsäußerung mit Pflichten und Verantwortung verbunden, sodass jeder, der Informationen weitergeben will, diese Umstände – soweit möglich – sorgfältig zu prüfen habe.[351] Die Weitergabe von Informationen wegen persönlichem Groll oder Feindschaft oder Erwartung finanzieller Vorteile, rechtfertige keinen starken Schutz – wichtig sei der gute Glaube, das öffentliche Interesse und die Überzeugung von der Wahrheit der Information.[352]

Problematisch könnte dieser Gesichtspunkt werden, wenn die Anzeige nicht anonym erstattet wurde, da die Argumentation dann fundierter auszufallen hat als bei anonymem Handeln.[353]

Auf Seiten des Kündigenden gehe es um den Schutz des guten Rufs des Unternehmens[354] sowie um die unternehmerische Freiheit nach Art. 12 Abs. 1 GG, die dem Arbeitgeber das Recht gebe, nur mit solchen Mitarbeitern zu arbeiten, die das Unternehmen fördern und es vor Schaden bewahren.

Daraus folgt, dass eine fristlose Kündigung des Hinweisgebers jedenfalls dann als unverhältnismäßig einzustufen ist, wenn ihn entweder eine Pflicht zum Handeln trifft (z. B. nach § 138 StGB), der Vorwurf auf Tatsachen beruht, die Anzeige nicht aus zu missbilligenden Motiven erfolgt und die innerbetriebliche Abhilfe im Vorfeld erfolglos geblieben ist.

---

349 BVerfG, Urt. v. 22.6.1982 – 1 BvR 1376/79, NJW 1983, 1415, 1416.

350 Vgl. EGMR, Urt. v. 21.7.2011 – 28274/08 (Rn. 65), NJW 2011, 3501, 3503 (Heinisch/Deutschland) mit Verweis auf EGMR, Urt. v. 12.2.2008 – 14277/04 (Rn. 73), (Guja/Moldawien).

351 Vgl. EGMR, Urt. v. 20.5.1999 – 21980/93 (Rn. 65), JuS 2000, 805, 806 (Bladet Tromsø/Norwegen).

352 EGMR, Urt. v. 21.7.2011 – 28274/08 (Rn. 69), NJW 2011, 3501, 3503 (Heinisch/Deutschland); EGMR, Urt. v. 12.2.2008 – 14277/04 (Rn. 77), (Guja/Moldawien).

353 *Schlachter*, RdA 2012, 108, 110.

354 EGMR, Urt. v. 21.7.2011 – 28274/08 (Rn. 64), NJW 2011, 3501, 3503 (Heinisch/Deutschland).

Diese grundsätzlichen Abwägungskriterien werden vom EGMR konkretisiert und auf den Fall *Heinisch* zugeschnitten.

(1) Das öffentliche Interesse an der Information

Die von der Beschwerdeführerin weitergegebenen Informationen müssen von öffentlichem Interesse sein. Dies war vorliegend der Fall.[355] Der Gerichtshof betonte die Besonderheit des Falles, indem er feststellte, dass es sich um alte, kranke Patienten in Pflegeheimen handelte, die häufig nicht in der Lage seien, selbst auf die Missstände in der Pflege aufmerksam zu machen. Die Weitergabe von Informationen über Mängel der Pflege sei daher durch Dritte dringend geboten, um Missbräuche zu verhindern.[356] Kontrollen durch besondere Dienste, wie den medizinischen Dienst der Krankenkassen, stellen kein ausreichendes Instrument dar, insbesondere weil diese Kontrollen nach vorheriger Ankündigung stattfinden.[357] Besonders offensichtlich ist das öffentliche Interesse in diesem Fall, weil es um eine Pflegeeinrichtung geht, die in Händen der öffentlichen Hand liegt. Es geht um das Vertrauen in die angemessene Erbringung von Leistungen im Rahmen der öffentlichen Daseinsvorsorge durch den Staat.[358]

(2) Andere Möglichkeiten der Aufdeckung

Des Weiteren ist bei der Interessenabwägung zu klären, ob eine andere Möglichkeit als die Erstattung der Anzeige gegen den Arbeitgeber bestand. Der EGMR verwies hier auf das Urteil des BAG[359] vom 3. Juli 2003, nach dem von einem Arbeitnehmer eine vorherige innerbetriebliche Klärung vernünftigerweise nicht erwartet werden kann, wenn der Arbeitgeber Kenntnis von der Straftat hat, eine Abhilfe keinen Erfolg verspricht oder der Arbeitnehmer sich durch die Nichtanzeige selbst der Strafverfolgung aussetzen würde.

Der Gerichtshof stellt klar: Habe der Arbeitnehmer den Arbeitgeber auf die gesetzeswidrige Praxis hingewiesen und sorge dieser nicht für Abhilfe, bestehe keine Loyalitätspflicht mehr gegenüber dem Arbeitgeber.[360]

Diese Erwägungen wurden im Fall *Heinisch* herangezogen. Da die Beschwerdeführerin mehrmals durch Beschwerden an den Arbeitgeber zu erkennen gegeben

---

355 EGMR, Urt. v. 21.7.2011 – 28274/08 (Rn. 71), NJW 2011, 3501, 3503 (Heinisch/Deutschland).

356 EGMR, Urt. v. 21.7.2011 – 28274/08 (Rn. 71), NJW 2011, 3501, 3503 (Heinisch/Deutschland).

357 *Schlachter*, RdA 2012, 108, 111.

358 EGMR, Urt. v. 21.7.2011 – 28274/08 (Rn. 71), NJW 2011, 3501, 3503 (Heinisch/Deutschland).

359 BAG, Urt. v. 3.7.2003 – 2 AZR 235/02, NZA 2004, 427, 430.

360 EGMR, Urt. v. 21.7.2011 – 28274/08 (Rn. 73), NJW 2011, 3501, 3504 (Heinisch/Deutschland).

hat, welche Umstände in der Einrichtung herrschten, hat sie ihre Loyalitätspflicht in ausreichendem Maße erfüllt.

(3) Wahrheit der offengelegten Informationen

Weiter hebt der EGMR hervor, dass die weitergegebenen Informationen „echt", also tatsächlich zutreffend sein müssen. Die Freiheit der Meinungsäußerung sei mit Pflichten und Verantwortung verbunden, und jeder, der Informationen weitergeben wolle, müsse, soweit ihm das nach den Umständen möglich sei, prüfen, ob sie genau und zuverlässig seien, besonders dann, wenn der Informant zur Vertraulichkeit und Loyalität gegenüber dem Arbeitgeber verpflichtet sei.[361] Der EGMR stellt fest, dass die Argumente des deutschen Gerichts, es bestünden Zweifel an der Richtigkeit der in der Strafanzeige aufgestellten Behauptungen, weil die Beschwerdeführerin sie im Ermittlungsverfahren nicht weiter spezifiziert und keine weiteren Zeugen benannt habe, nicht haltbar seien.[362] Sie habe objektive Umstände dargelegt, die nachweisbar der Wahrheit entsprachen. Zudem habe sie bei der Überlastungsanzeige tatsachenbasiert und gutgläubig gehandelt,[363] denn ihre krankheitsbedingten Fehlzeiten sprächen für das Vorliegen belastender Umstände.[364] Der Mangel an Beweisen könne zwar zur Folge haben, dass vorläufige Ermittlungsverfahren eingestellt werden – er rechtfertige aber nicht den Schluss, dass die Angaben in der Strafanzeige einer Tatsachengrundlage entbehrten oder von Anfang an leichtfertig gewesen wären.[365] Da es keine betriebsinterne Anlaufstelle gegeben habe, sei die Beschwerdeführerin gezwungen gewesen, betriebsexterne Unterstützung zu suchen, auch wenn dies mit einer Rufschädigung des Pflegeheims verbunden war. Dies müsse aber hinter dem Informationsbedürfnis der Öffentlichkeit zurücktreten.[366]

(4) Die subjektive Vermutung – „guter Glaube"

Die Beschwerdeführerin hatte vorgetragen, dass sie die Strafanzeige hauptsächlich wegen der möglichen Gefahren für die Gesundheit der Bewohner der Pflegeeinrichtung erstattet habe. Die nationalen Gerichte meinten dagegen, sie habe nur Druck auf ihren Arbeitgeber ausüben wollen, da sie keine Tatsachengrundlage

---

361 EGMR, Urt. v. 13.5.1992- 18957/91, Decisions and Reports, Bd. 73, S. 225, 230 f. (Haseldine/Vereinigtes Königreich).

362 EGMR, Urt. v. 21.7.2011 – 28274/08 (Rn. 81), NJW 2011, 3501, 3504 (Heinisch/ Deutschland).

363 Die Motivation der Erstattung der Anzeige bezieht auch das BAG in die Abwägung mit ein, vgl. BAG, Urt. v. 4.7.1991 – 2 AZR 80/91, Rn. 20.

364 Vgl. auch *Schlachter*, RdA 2011, 108, 111.

365 EGMR, Urt. v. 21.7.2011 – 28274/08 (Rn. 81), NJW 2011, 3501, 3504 (Heinisch/ Deutschland).

366 *Schlachter*, RdA 2011, 108, 111.

für den Abrechnungsbetrug habe vorbringen können, und daher nicht gutgläubig gehandelt habe.[367]

Der Gerichtshof lässt dieses Argument nicht gelten. Er habe keinen Zweifel daran, dass die Beschwerdeführerin in gutem Glauben gehandelt habe und der Überzeugung war, es liege im öffentlichen Interesse, die angeblichen Straftaten ihres Arbeitgebers den Strafverfolgungsbehörden anzuzeigen. Sie sei zudem überzeugt davon gewesen, dass ihr kein anderes, diskreteres Mittel zur Verbesserung der Lage zur Verfügung stünde.[368] Des Weiteren hätten schon frühere Berichte des medizinischen Dienstes der Krankenkassen den schlechten Zustand beschrieben, woraus die Beschwerdeführerin herleiten konnte, dass sich an der Situation nichts ändern werde. Sie sah die Strafanzeige als letztes Mittel zur Abhilfe der Mängel in der Altenpflege und nicht als grundlosen persönlichen Angriff auf ihren Arbeitgeber.[369]

Diese Erwägungen reichen nach Ansicht des Gerichtshofs[370] aus, um von einem guten Glauben der Beschwerdeführerin auszugehen, als sie die Strafanzeige gegen den Arbeitgeber erstattete.

(5) Leichtfertigkeit der Anzeige

Der EGMR weist zudem auf den Umstand hin, dass sich die Beschwerdeführerin vor der Anzeigeerstattung mit einem Anwalt beraten habe; mit der Einholung von Rechtsrat sei eine „Leichtfertigkeit" der Anschuldigungen ausgeschlossen.[371] Die Anzeige der Beschwerdeführerin habe einen tatsächlichen Hintergrund und sei nicht wissentlich oder leichtfertig falsch gewesen. Nach ihren Erfahrungen mit vielen ergebnislosen betriebsinternen Beschwerden habe sie annehmen können, dass die Strafanzeige das letzte Mittel zur Verbesserung der Pflegesituation sei.[372] Der EGMR betont – anders als das LAG –, dass alleine die mangelnde Beweisbarkeit der erhobenen Vorwürfe nicht zur Annahme einer leichtfertigen Anzeige führen könne.[373]

---

367 *Schlachter*, RdA 2011, 108, 111.

368 EGMR, Urt. v. 21.7.2011 – 28274/08 (Rn. 83), NJW 2011, 3501, 3504 (Heinisch/Deutschland).

369 EGMR, Urt. v. 21.7.2011 – 28274/08 (Rn. 84), NJW 2011, 3501, 3505 (Heinisch/Deutschland).

370 EGMR, Urt. v. 21.7.2011 – 28274/08 (Rn. 87), NJW 2011, 3501, 3505 (Heinisch/Deutschland).

371 EGMR, Urt. v. 21.7.2011 – 28274/08 (Rn. 86), NJW 2011, 3501, 3505 (Heinisch/Deutschland).

372 So schon zum Punkt „Wahrheit der Informationen" und „guter Glaube", EGMR, Urt. v. 21.7.2011 – 28274/08 (Rn. 84), NJW 2011, 3501, 3504, 3505 (Heinisch/Deutschland).

373 EGMR, Urt. v. 21.7.2011 – 28274/08 (Rn. 79, 80), NJW 2011, 3501, 3504 (Heinisch/Deutschland).

cc. Die Bindungswirkung der Entscheidung

Die Entscheidung des Europäischen Gerichtshofs für Menschenrechte hat nach den Art. 42, 44 EMRK für die Parteien nur formelle Rechtskraft.[374] Daneben kommt dem Tenor nach Maßgabe des Art. 46 EMRK auch materielle Rechtskraft zu.[375] Zwar hat die Europäische Menschenrechtskonvention in Deutschland den Rang eines einfachen Bundesgesetzes, allerdings kann sie nach Ansicht des BVerfG nicht durch einfaches Gesetz abbedungen werden. Es ist viel mehr zu beachten, dass das Konventionsrecht die Auslegung des Bundesrechts, insbesondere der Grundrechte, beeinflusst.[376] Deshalb ist die Frage schwierig zu beantworten, ob die Entscheidung des EMRK eine Vorbildfunktion für parallel gelagerte Fälle hat. Nach Ansicht des BVerfG besteht keine absolute Bindung deutscher Gerichte an die Entscheidungen des Gerichtshofs.[377] Es gehört aber zur Bindung an Gesetz und Recht gem. Art. 20 Abs. 3 GG, die EMRK und die Entscheidungen des EGMR im Rahmen methodisch vertretbarer Gesetzesauslegung zu berücksichtigen. Bei der Berücksichtigung von Entscheidungen des Gerichtshofs haben die staatlichen Organe die Auswirkungen auf die nationale Rechtsordnung und ihre Rechtsanwendung einzubeziehen.

Den deutschen Gerichten verbleibt also ein nicht unerheblicher Spielraum, von der Rechtsauffassung des Gerichtshofs im Einzelfall abzuweichen. Jedoch dürfen die geforderten Schutzstandards nicht unterschritten werden.

dd. Fazit

Zusammenfassend sind insbesondere zwei Gesichtspunkte hervorzuheben, die in der Entscheidung besonders betont wurden und ein abweichendes Ergebnis von der Entscheidung des LAG Berlin zulassen: das besondere öffentliche Interesse an der Aufklärung und Aufdeckung von Missständen im Pflegebereich (genauer gesagt: das öffentliche Vertrauen in die Qualitätsstandards der Altenpflege ausgeführt durch eine staatlich gehaltene Gesellschaft) und die Ablehnung der beweisrechtlichen Argumentation für das Vorliegen einer „leichtfertigen" Falschanzeige. Deshalb kommt der Gerichtshof unter Abwägung aller betroffenen Interessen zu dem Ergebnis, dass der Eingriff in das Recht der Beschwerdeführerin auf Freiheit der Meinungsäußerung, insbesondere in ihr Recht auf Weitergabe von Informationen, „in einer demokratischen Gesellschaft [nicht] notwendig" war.[378] Das LAG Berlin hat in seinem ausführlich begründeten Urteil vom 28. März 2006 die entsprechenden Rechtssätze angewandt, jedoch die beweisrechtliche Situation anders gewertet. Denn nach den Unterlagen des LAG zum Zeitpunkt des Verfahrens konnte der vermeintliche

---

374 Vgl. *Forst*, NJW 2011, 3477, 3479.

375 Vgl. *Forst*, NJW 2011, 3477, 3479.

376 U. a. BVerfG, Urt. v. 14.10.2004 – 2 BvR 1481/04, NJW 2004, 3407, 3408.

377 BVerfG, Urt. v. 14.10.2004 – 2 BvR 1481/04, NJW 2004, 3407, 3408, 3409; wesentlich bestätigt durch BVerfG, Urt. v. 4.5.2011 – 2 BvR 2365/09, NJW 2011, 1931, 1945.

378 EGMR, Urt. v. 21.7.2011 – 28274/08 (Rn. 93), NJW 2011, 3501, 3505 (Heinisch/ Deutschland).

Abrechnungsbetrug weder im Rahmen des Strafverfahrens noch im Rahmen des Kündigungsschutzprozesses näher dargelegt und bewiesen werden, worin das LAG eine „polemische Diktion der Strafanzeige“ sah.[379] Der EGMR hat demgegenüber befunden, dass die Arbeitnehmerin bereits vor der Strafanzeige die für den Vorwurf des Betrugs relevanten Tatsachen mitgeteilt habe.[380] Im Nachhinein lässt sich schwer beurteilen, wie es zu dieser Subsumtion gekommen ist und ob dieses Ergebnis auf eine zukünftige Sonderrechtsprechung des EGMR zum Whistleblowing für öffentliche Unternehmen hinweist.[381] Fest steht, dass eine Pflichtverletzung nur gegeben ist, wenn falsche Informationen wissentlich und/oder leichtfertig weitergegeben werden – diesen Tatbestand sah der Gerichtshof nicht erfüllt. Die Abwägung hat deshalb ergeben, dass in einer demokratischen Gesellschaft das öffentliche Interesse an Informationen über Mängel in der Heimpflege für Senioren in einer staatlichen Einrichtung so wichtig ist, dass es das Interesse des Unternehmens am Schutz des guten Rufs im Geschäftsverkehr und seiner wirtschaftlichen Interessen überwiegt.[382] Die fristlose Kündigung war deshalb unverhältnismäßig hart.

Zwar war die Arbeitgeberin ein Unternehmen in staatlicher Hand (das Land Berlin war Anteilseigner), trotzdem hat der Gerichtshof in seiner Argumentation zu Art. 10 EMRK nicht zwischen Arbeitgebern aus der Privatwirtschaft und Arbeitgebern aus dem öffentlichen Sektor unterschieden.[383] Der Gerichtshof musste auch nicht differenzieren, denn das Recht aus Art. 10 EMRK ist in zweifacher Hinsicht zu gewährleisten: der Staat ist verpflichtet die Meinungsfreiheit gegenüber Eingriffen des Staates (Abwehrrecht) und im Rahmen der Drittwirkung gegenüber Eingriffen privater Dritter zu schützen.[384] Demnach sind die aufgestellten Kriterien in jedem Arbeitsverhältnis anzuwenden. In der *Heinisch*-Entscheidung hat der Gerichtshof die für Art. 10 EMRK relevante Frage, ob ein Unternehmen privatrechtlich oder öffentlich-rechtlich organisiert ist, nicht weiter spezifiziert. Allerdings steht fest,

---

379 Vgl. *Mengel*, CCZ 2011, 229, 230.

380 „Es stimmt zwar, dass die rechtliche Einordnung des Verhaltens der Arbeitgeberin als schwerer Betrug erstmals in der vom Rechtsanwalt der Beschwerdeführerin verfassten Strafanzeige vom 7. Dezember 2004 Erwähnung fand, der Gerichtshof stellt aber fest, dass die Beschwerdeführerin in ihren früheren Beschwerden gegenüber ihrer Arbeitgeberin durchaus auf den Sachverhalt, auf dem ihre anschließende Strafanzeige basierte, und zwar einschließlich der Tatsache, dass Leistungen nicht ordnungsgemäß dokumentiert worden seien, hingewiesen hatte.“, EGMR, Urt. v. 21.7.2011 – 28274/08 (Rn. 72), NJW 2011, 3501, 3503 f. (Heinisch/Deutschland).

381 So *Mengel*, CCZ 2011, 229, 231.

382 So im Ergebnis EGMR, Urt. v. 21.7.2011 – 28274/08 (Rn. 93, 94), NJW 2011, 3501, 3505 (Heinisch/Deutschland).

383 Vgl. Ausführungen EGMR, Urt. v. 21.7.2011 – 28274/08 (Rn. 64), NJW 2011, 3501, 3503 (Heinisch/Deutschland).

384 So auch *Schubert*, German Yearbook of international Law 2011(54), S. 753, 758.

dass der Staat sich seiner Verpflichtung nicht entledigen kann, wenn er seine Aufgaben an ein privates Unternehmen delegiert.[385]

Die nationalen Gerichte sind nun angehalten, Wege zu finden, die Ansätze des EGMR in ihre Rechtsprechungspraxis zu integrieren. Diese Grundsätze sind dem BAG nämlich nicht fremd, auch wenn der EGMR vertieft auf die Dogmatik der Meinungsfreiheit eingegangen ist. Die Abwägungskriterien sind nun für die Überprüfung von Whistleblower-Fällen entscheidend[386] und können als Richtschnur für alle Arbeitsverhältnisse dienen. Der Gerichtshof hat nicht eindeutig festgehalten, das Recht auf Meinungsfreiheit bei einem Arbeitgeber aus dem privaten Sektor weiter einzuschränken. Auch wenn der Staat nicht als Arbeitgeber tätig ist, und somit kein *besonderes* öffentliches Interesse an den Qualitätsstandards in öffentlich-rechtlichen Pflegeeinrichtungen betroffen ist, so hat er doch den Schutz desjenigen Bürgers zu gewährleisten, der zum Schutz der durch die Konvention garantierten Rechte der anderen Bürger Missstände seines (privaten) Arbeitgebers öffentlich macht.[387]

Neben den besonderen gesetzlichen Vorschriften, die dem Whistleblower das Recht geben, Sachverhalte zu offenbaren oder anzuzeigen, kann sich ein solches Recht auch aus den Grundrechten des Anzeigenden ergeben. Der arbeitsrechtliche Schutz des Whistleblowers ist aber trotzdem sehr kritisch zu sehen: Er ist nur in besonderen Fällen gegeben. Der arbeitsrechtliche Schutz des Whistleblowers ist *de lege lata*, auch im Hinblick auf andere Repressalien und deren Folgen, nicht ausreichend. Denn es besteht die Gefahr versteckter Sanktionen oder Benachteiligungen, gegen die sich der Arbeitnehmer nur schwer zur Wehr setzen kann, da sich regelmäßig kaum nachweisen lässt, dass die Maßnahmen eine Reaktion auf den erfolgten Hinweis darstellen.[388] Zudem genügt es nicht, dass nur Gerichte die Kriterien anzuwenden haben, sondern auch der Arbeitgeber muss im Vorhinein wissen, wie er sich in bestimmten Situationen zu verhalten hat, um Kündigungen zu verhüten. Die Entscheidung bezieht weder Stellung dazu, wie der Arbeitgeber sich zu verhalten hat, noch, welche anderen Arten des Whistleblowings sowie Hinweisgegenstände zulässig sind.

Im Ergebnis hat die *Heinisch*-Entscheidung das deutsche Recht nicht wie erwartet erneuert. Trotz allem sollte eine Kodifizierung der Kriterien und Handlungsweisen erwogen werden. Der Gerichtshof hat in seiner Folgeentscheidung *Sosinowska vs. Polen*[389] nahezu die gleichen Abwägungskriterien angewandt, obwohl die

385 *Meyer-Ladewig*, EMRK, Art. 1, Rn. 9 m.w.N.

386 *Schubert*, German Yearbook of international Law 2011(54), S. 753, 759.

387 *Schubert*, German Yearbook of international Law 2011(54), S. 753, 761.

388 So auch *Graser*, Whistleblowing, S. 167 f.; *Schmitt*, Whistleblowing, S. 95.

389 EGMR, Urt. v. 18.10.2011 – 10247/09 (Sosinowska/Polen). Hier hatte sich eine Lungenspezialistin unter Berufung auf Art. 10 EMRK gegen eine disziplinarrechtliche Strafe gewandt, die ihr auferlegt worden war, nachdem sie das Fehlverhalten und die mangelnde Sachkompetenz ihrer Vorgesetzten gegenüber dem Klinikdirektor, Kollegen und externen Experten – unter Bezugnahme auf einzelne konkrete diagnostische und therapeutische Entscheidungen – gerügt hatte.

Heinisch-Entscheidung zu diesem Zeitpunkt noch nicht rechtskräftig war. Dies spricht für Kontinuität und Beständigkeit der Kriterien. In der Folgeentscheidung wird der Ansatz, das öffentliche Interesse an der Offenbarung eines Missstandes als wichtigen Faktor in Rechnung zu stellen, weitergeführt.

3. *Whistleblowing gegenüber sonstigen Dritten*

Von den bereits genannten Fällen ist derjenige zu unterscheiden, der die direkte Informationsweitergabe an Dritte betrifft, also etwa an die Presse, Nicht-Regierungsorganisationen oder die unmittelbare Öffentlichkeit. Die „zuständigen Behörden" sind dabei nicht „sonstige Dritte". Sowohl die Dritten als auch die zuständigen Behörden sind Adressaten von externem Whistleblowing. Jedoch ist das Ergebnis der Interessenabwägung bei der Hinweiserteilung an Dritte ein anderes.[390] Die Weitergabe der Informationen an die unmittelbare Öffentlichkeit (Dritte), ohne den Versuch zu unternehmen, den Sachverhalt zunächst an eine zuständige Behörde zu melden, muss im Regelfall ausgeschlossen werden oder zumindest als „ultima ratio" angesehen werden. Fraglich ist demnach, wie der Schutz des Whistleblowers ausgeprägt ist, wenn er die Hinweise nach „Außen" gibt.

Bei der Abwägung sind die Interessen anders zu gewichten als im oben genannten Fall. Den Interessen des Arbeitnehmers kann nicht mehr so viel Gewicht beigemessen werden. Denn zur Ausübung des staatsbürgerlichen Rechts gem. Art. 2 Abs. 1 GG in Verbindung mit Art. 20 Abs. 3 GG gehört nicht die Einschaltung einer nicht-staatlichen Stelle, auch wenn im Ergebnis ein staatliches Verfahren in Gang kommen sollte.[391] Folgt aufgrund eines externen Whistleblowings eine arbeitsrechtliche Maßnahme (Kündigung), fällt die Abwägung regelmäßig zugunsten des Arbeitgebers aus. Dem ist zuzustimmen, wenn die geforderte und zumutbare innerbetrieblich Abhilfe nicht versucht wird. Anders sind die Fälle zu behandeln, in denen eine externe Meldung den einzigen (Aus)-Weg darstellt. Allein die Meinungsfreiheit nach Art. 5 Abs. 1 GG könnte greifen – bei anonymen Whistleblowern ist dies bereits fraglich. Denn nach der Rechtsprechung des BAG im Jahr 2004 fällt eine solche anonyme Anzeige nicht in den Schutzbereich des Art. 5 Abs. 1 GG – ihr fehlt es gerade an dem konstituierenden Element der Subjektivität.[392] Ohne die deutlich erkennbare persönliche Zuordnung kann eine anonyme Äußerung nicht an der geistigen Auseinandersetzung teilnehmen.[393] Anders entschied der BGH 2009: Eine Beschränkung der Meinungsäußerungsfreiheit auf Äußerungen, die einem bestimmten Individuum zugeordnet werden können, ist mit Art. 5 Abs. 1 S. 1 GG

---

390 So auch *Rudkowski*, CCZ 2013, 204, 207.

391 Etwas überspitzt *Rudkowski*, CCZ 2013, 204, 207: „Durch die Wendung an nichtstaatliche Stellen erkennt der Arbeitnehmer weder das staatliche Gewaltmonopol noch das Rechtsstaatsprinzip allgemein an: Er misstraut den Behörden und entscheidet sich gegen gesetzlich festgelegte Sanktionen."

392 BAG, Urt. v. 3.7.2003 – 2 AZR 235/02, NZA 2004, 427, 430.

393 BAG, Urt. v. 3.7.2003 – 2 AZR 235/02, NZA 2004, 427, 430.

nicht vereinbar.[394] Die Verpflichtung, sich namentlich zu einer bestimmten Meinung zu bekennen, könnte dazu führen, dass der Einzelne aus Furcht vor Repressalien oder sonstigen negativen Auswirkungen seine Meinung nicht äußert. Dieser Gefahr der Selbstzensur soll durch das Grundrecht auf freie Meinungsäußerung entgegengewirkt werden.[395]

Es ist festzuhalten, dass ein Rückgriff auf Art. 5 Abs. 1 GG möglich bleibt. Dies ist insofern von Bedeutung, als der EGMR in seinem Urteil besonders auf die Meinungsfreiheit des Arbeitnehmers einging und diesem Recht besondere Aufmerksamkeit beigemessen hat. In Art. 5 Abs. 1 GG einen besonderen Rechtfertigungsgrund für Whistleblowing an nicht-staatliche Stellen zu sehen, liegt daher nahe.

Zudem ist an Sachverhalte zu denken, an denen die Öffentlichkeit ein besonderes Interesse hat, weil es um hochrangige Güter eines jeden Einzelnen geht. Ist ein Rechtsgut akut gefährdet, ist der Gang zur Behörde nicht zielführend. Während sich die Behörde mit dem Sachverhalt befasst und eventuell Maßnahmen einleitet, kann es schon zu spät sein. Es bedarf in diesen Situationen einer schnellen Reaktion und Warnung. Diese kann nur erreicht werden, wenn die relevanten Informationen nach außen gelangen, z. B. mit Hilfe von Nicht-Regierungsorganisationen oder besonderen Meldestellen. Sie können sich durch Pressmitteilungen, Internetauftritte oder sonstigen Informationsveranstaltungen direkt an die Öffentlichkeit wenden und z. B. Warnungen vor gefährlichen Produkten, verschmutzten/belasteten Lebensmitteln oder akuten umweltschädlichen Auswirkungen aussprechen. Deshalb ist den Interessen des Whistleblowers auch in diesem Fall ein besonderes Gewicht beizumessen.

Es ist jedoch zu bedenken, dass Rechte und Interessen des Arbeitgebers umso mehr betroffen sind, da durch die Einschaltung sonstiger Dritter die herrschenden Zustände im Unternehmen in der Öffentlichkeit bekannt gemacht werden. Rufschädigungen oder finanzielle Schäden sind die logische Folge. Erweisen sich die Anschuldigungen als unwahr, könnte der Ruf des Arbeitgebers unbegründet beschädigt worden sein.[396] Eine Behebung des schon eingetretenen Schadens ist dann kaum möglich. Deswegen ist grundsätzlich der Gang zur Behörde dem Gang an die Presse bzw. sonstigen Dritten vorzuziehen– soweit keine akute und unmittelbare Gefährdung droht. Aufgrund der gravierenden Gefährdung der Arbeitgeberinteressen muss dem Arbeitgeber wenigstens die Gelegenheit gegeben werden, vom Missstand oder Rechtsverstoß Kenntnis zu erlangen und gegebenenfalls einzuschreiten. Diese Möglichkeit erhält er, wenn der Whistleblower den Gang zur Behörde wählt, denn die Behörde wird sich im Rahmen der Untersuchung mit dem Arbeitgeber

394 BGH, Urt. v. 23.6.2009 – VI ZR 196/08, NJW 2009, 2888, 2892; *Grabenwarter*, in: Maunz/Düring, GG-Kommentar, Art. 5 Abs. 1 GG, Rn. 86; *Starck*, in: v. Mangoldt/Klein/Starck, Kommentar zum GG, Art. 5 GG, Rn. 32; *Fechner*. in, Stern/Becker, GG-Kommentar, Art. 5 GG, Rn. 83 m. w. N.

395 BGH, Urt. v. 23.6.2009 – VI ZR 196/08, NJW 2009, 2888, 2892.

396 So auch *Rudkowski*, CCZ 2013, 204, 208.

auseinandersetzen, wohingegen die Presse meist vorrangig die Befriedigung der Neugier der Leserschaft im Blick hat.

Vor diesem Hintergrund kann lediglich in Ausnahmefällen die Einschaltung sonstiger Dritter zulässig sein. Ein solcher Fall könnte vorliegen, wenn durch die Rechtsverstöße Leben und Gesundheit von Menschen oder die Sicherheit und Umwelt gefährdet würden.[397] Die Weitergabe von Informationen an die Öffentlichkeit muss jedenfalls gestattet und gerechtfertigt sein, wenn der Arbeitnehmer rechtzeitige und hinreichende Hilfe auch von staatlicher Seite nicht erwarten kann.

Auch bei der Offenlegung von Gesetzesverstößen gegenüber Dritten ist der Whistleblower also unzureichend geschützt. Weder gesetzliche Regelungen noch die Rechtsprechung zeigen einem Whistleblower geschützte Verhaltensweisen auf. Zwar wurde gerichtlich festgestellt, dass vorrangig betriebliche Abhilfe zu ersuchen ist; wann aber externes Whistleblowing an Dritte zulässig ist, ist unklar. Dies führt zur Unsicherheit des Whistleblowers. Der Whistleblower kann daher meist nur schwer einschätzen, ob und wann er sich direkt an eine unternehmensexterne Stelle wenden darf, weil alle anderen Maßnahmen unzumutbar sind. Die Festlegung eines Stufenverhältnisses zwischen internem und externem Whistleblowing gegenüber Behörden sowie sonstigen Dritten könnte Abhilfe schaffen. Wer unternehmensextern auf Gesetzesverstöße aufmerksam machen will, läuft derzeit Gefahr, dass dieses Verhalten als Verletzung arbeitsvertraglicher Pflichten eingestuft wird.

B. Strafrechtlicher und ordnungswidrigkeitenrechtlicher Schutz

Es stellt sich auch die Frage nach dem strafrechtlichen und dem ordnungswidrigkeitenrechtlichen Schutz eines Whistleblowers. Von zentraler Bedeutung für die Strafbarkeit des Whistleblowers sind die Rechtfertigungstatbestände: Neben einer (mutmaßlichen) Einwilligung kommt der rechtfertigende Notstand nach § 34 StGB in Betracht. Die Notwehr nach § 32 StGB scheidet regelmäßig aus, weil es sich bei dem Rechtsverstoß des Arbeitgebers nicht um einen akuten und unmittelbaren Angriff[398] auf die Rechtsgüter des Whistleblowers oder des Dritten handelt. § 32 StGB schützt nur Individualinteressen – nicht Interessen der Allgemeinheit – und geht von einem anderen Prinzip aus,[399] das für eine Interessenabwägung keinen Raum lässt. Zudem beruht die Notwehrhandlung nach § 32 StGB nicht auf dem Prinzip der Solidarität mit dem Opfer, sondern auf dem Selbstschutz des Täters

---

397 So auch der Gesetzentwurf zum Hinweisgeberschutzgesetz, § 6 Abs. 2 HinwGebSchG-E, BT-Drs. 17/8567; ähnlich *Leuchten*, ZRP 2012, 142, 144.

398 Zwar wird in Abschnitt A dieses Teils der Arbeit angedeutet, dass ausnahmsweise ein gegenwärtiger Angriff im Sinne von § 227 BGB vorliegen kann – hier steht jedoch § 34 StGB als passendere Vorschrift zur Verfügung.

399 *Erb*, in: MüKo StGB, § 34 StGB, Rn. 26.

vor dem Angreifer.[400] § 227 BGB kommt keine Rechtfertigungswirkung im Straf- und Ordnungswidrigkeitenrecht zu;[401] §§ 228, 904 BGB sind schon tatbestandlich ausgeschlossen.

§ 34 StGB schützt jedes beliebige Rechtsgut, unabhängig davon, in welcher Form das Rechtsgut den Schutz durch die Rechtsordnung erfahren hat, also nicht nur strafrechtlich geschützte Rechtsgüter, sondern jedes anerkannte Rechtsgut, das einer Gefährdung ausgesetzt sein kann.[402]

I. Schutz des Whistleblowers als Tatunbeteiligter

Publikationen, die sich mit dem Thema Whistleblowing und dem strafrechtlichen Schutz beschäftigen, weisen auf Eines deutlich hin: Die Einführung eines Hinweisgebersystems minimiert das Strafbarkeitsrisiko des Whistleblowers.[403] Ein funktionierendes und wirksames, innerbetriebliches Hinweisgebersystem kann dazu führen, dass grundsätzlich strafbare Handlungen, wie z. B. der Geheimnisverrat, nicht strafbar sind, weil der Hinweis von der Unternehmensleitung gedeckt sein wird.[404] Daraus ergibt sich ein strafrechtlicher Rechtfertigungstatbestand. Auch im Übrigen wäre die Gefahr der Strafverfolgung durch die Einführung eines Hinweisgebersystems äußerst gering, da in vielen Fällen schon der objektive Tatbestand des § 203 StGB bzw. § 17 UWG nicht erfüllt sein würde.[405]

Diese Intention ist nachvollziehbar, jedoch besteht momentan keine gesetzliche Pflicht zur Implementierung eines innerbetrieblichen Hinweisgebersystems oder eines Compliance-Programms mit einer integrierten Whistleblower-Hotline.[406] Sollte ein Hinweisgebersystem in einem Unternehmen implementiert werden, ist dies zu begrüßen. Dennoch ist nicht klar, wie dieses System ausgestaltet ist und ob dadurch dem Whistleblower ausreichend Schutz zukommt. Daher ist auch ausreichender Schutz auf gesetzlicher Ebene zu gewährleisten.

---

400 *Neumann*, in: Kindhäuser/Neumann/Paeffgen, Kommentar StGB, § 34, Rn. 13.

401 Siehe *Grothe*, in: MüKo BGB, § 227 BGB, Rn. 2.

402 *Erb*, in: MüKo StGB, § 34 StGB, Rn. 55.

403 Instruktiv dazu *Zimmermann*, ArbRAktuell 2012, 58, 61; *Donato*, Whistleblowing, S. 137 f.; *Schulz*, Ethikrichtlinien, S. 175 f. („Whistleblower-Klausel"); *Lutterbach*, Whistleblowing, S. 48 f.; *Moosmayer*, Compliance, Abschnitt D, S. 56 f.; *Mengel*, CCZ 2008, 85, 88 f.; *Schemmel/Ruhmannseder/Witzigmann*, Hinweisgebersysteme, S. 89 f.; *Zimmer/Seebacher*, CCZ 2013, 31, 33.

404 *Schemmel/Ruhmannseder/Witzigmann*, Hinweisgebersysteme, S. 111.

405 *Schemmel/Ruhmannseder/Witzigmann*, Hinweisgebersysteme, S. 112.

406 Zwar wurde von der Regierungskommission und dem Bundesjustizamt ein Regelwerk „Deutscher Corporate Governance Kodex" veröffentlicht, der jährlich der nationalen und internationalen Entwicklung im Unternehmensbereich angepasst wird. Ein rechtlicher Zwang, den Empfehlungen des Kodex zu folgen, besteht im Ergebnis jedoch nicht. Näheres dazu in Teil 5, A.

1. *Strafbarkeit nach dem StGB*

Offenbart der unbeteiligte Whistleblower den Gesetzesverstoß, ist er strafrechtlichen Konsequenzen ausgesetzt, die sich unter anderem aus dem Strafgesetzbuch ergeben.

Neben dem Tatbestand des § 17 UWG, dessen persönlicher Anwendungsbereich sämtliche Unternehmensbeschäftigte[407] mit einschließt, drohen auch andere besondere Vorschriften dem Verrat von Geheimnissen mit Strafe.

Einzelnen Berufsgruppen ist aufgrund eines berufsspezifischen Vertrauensverhältnisses eine besondere Geheimhaltungspflicht auferlegt, deren Verletzung § 203 StGB unter Strafe stellt. Danach macht sich strafbar, wer ein fremdes Geheimnis, namentlich ein zum persönlichen Lebensbereich gehörendes Geheimnis oder ein Betriebs- oder Geschäftsgeheimnis offenbart, das dem Täter in einer besonderen beruflichen Funktion anvertraut worden oder sonst bekannt geworden ist.[408] Zu den angesprochenen Personen gehören unter anderem Ärzte, Anwälte, Steuerberater und auch das Klinikpersonal. Der Kreis tauglicher Täter ist relativ eng gefasst: Zum einen muss der Täter aus dem im Katalog genannten Berufsträgerkreis stammen. Zum anderen werden nur Geheimnisse erfasst, welche diese Personen gerade in dieser Eigenschaft (und nicht etwa als Privatperson) erlangt haben.[409] Demnach macht sich ein Whistleblower nicht strafbar, wenn er nicht zu dem genannten Täterkreis gehört oder von den Informationen in sonstiger Weise Kenntnis erlangt hat. § 17 UWG ist *lex specialis* zu § 203 StGB.[410] Tateinheit kann aber vorliegen, wenn das Unternehmensgeheimnis zugleich Geheimnis einer Privatperson ist.[411]

In besonderen Fällen kann noch eine Strafbarkeit nach § 164 StGB (falsche Verdächtigung) gegeben sein. Denn in gewissen Fällen ist es dem Whistleblower nicht möglich oder gar unzumutbar, die Stichhaltigkeit seines Hinweises abschließend zu prüfen. Die Anschuldigungen können sich also im Nachhinein in einigen Punkten als falsch erweisen. Neben dem Merkmal des Verdächtigens als Tathandlung muss das betreffende Verhalten des Whistleblowers auch dazu geeignet sein, ein behördliches Verfahren in Gang zu setzen oder eine andere behördliche Maßnahme herbeizuführen.[412] Dies ist beim (externen) Whistleblowing regelmäßig der Fall, sodass eine Strafbarkeit wegen falscher Verdächtigung in Betracht kommt.

Ein Whistleblower berichtet über ein strafrechtlich relevantes Fehlverhalten von Personen aus dem Unternehmen gegenüber internen Stellen, Ermittlungsbehörden oder der Öffentlichkeit. Dadurch werden die Persönlichkeitsrechte der betroffenen Personen unmittelbar berührt. In Frage kommen also auch die Ehrschutzdelikte

---

407 *Diemer*, in: Erbs/Kohlhaas, Strafrechtliche Nebengesetze, § 17 UWG, Rn. 18.
408 *Lenckner/Eisele*, in: Schönke/Schröder, Kommentar StGB, § 203 StGB, Rn. 4.
409 *Lenckner/Eisele*, in Schönke/Schröder, Kommentar StGB, § 203 StGB, Rn. 12.
410 *Cierniak/Pohlit*, MüKo StGB, § 203 StGB, Rn. 143.
411 *Cierniak/Pohlit*, MüKo StGB, § 203 StGB, Rn. 143.
412 *Zopfs*, in: MüKo StGB, § 164 StGB, Rn. 23.

nach §§ 185 ff. StGB, wie Beleidigung, üble Nachrede und Verleumdung.[413] Bei diesen gibt es einen besonderen, auf die Meinungsfreiheit zugeschnittenen Rechtfertigungsgrund. Nach § 193 StGB sind Äußerungen, die zur Ausführung oder Verteidigung von Rechten oder zur Wahrnehmung berechtigter Interessen gemacht werden, nur insofern strafbar, als das Vorhandensein einer Beleidigung aus der Form der Äußerung oder aus den Umständen, unter welchen sie geschah, hervorgeht.[414] Ein „berechtigtes Interesse" ist sowohl das eigene Interesse als auch das Interesse der Allgemeinheit an der Verfolgung verübter Straftaten.[415] Es genügt, dass beim Whistleblower tatsächliche Anhaltspunkte für die Begehung einer Straftat durch einen anderen vorliegen, sodass auch strafprozessual ein Anfangsverdacht angenommen werden kann.[416] Stellt sich dann bei der Ermittlung heraus, dass der Verdacht unbegründet ist, ist die vorher getätigte Anzeige des Whistleblowers trotzdem nach § 193 StGB gerechtfertigt.[417] Dies steht auch im Einklang mit der Wertung des Bundesarbeitsgerichts.[418]

Wie gerade angedeutet, können den Whistleblower, je „nachdem", welchen Weg er einschlägt, Konsequenzen treffen. Zwar scheitert die Strafbarkeit nach dem StGB letztlich meist an den Voraussetzungen des subjektiven Tatbestands oder dem Eingreifen eines Rechtfertigungsgrundes.[419] Dennoch ist die Einleitung eines Ermittlungsverfahrens gegen den Whistleblower nicht ausgeschlossen. Auch wenn das Verfahren nach § 170 Abs. 2 StPO (Einstellung des Verfahrens) oder §§ 153 ff. StPO (Einstellung des Verfahrens/Absehen von Strafverfolgung) eingestellt wird,[420] besteht das Risiko, ins Visier der Strafverfolgungsbehörden zu geraten und damit verbundenen psychischen sowie physischen Belastungen ausgesetzt zu sein.

Hauptsächlich steht jedoch die Strafbarkeit wegen Verletzung des Geschäfts- und Betriebsgeheimnisses im Raum, die gesondert im folgenden Abschnitt angesprochen wird.

---

413 *Zimmermann*, ArbRAktuell 2012, 58, 60.

414 *Lenckner/Eisele*, in: Schönke/Schröder, Kommentar StGB, § 193 StGB, Rn. 1.

415 So z. B. BVerfG, Beschl. v. 25.1.1961 – 1 BvR 9/57, NJW 1961, 819, 821; BGH, Urt. v. 22.12.1959 – VI ZR 175/58, NJW 1960, 476, 477; *Kühl*, in: Lackner/Kühl, StGB-Kommentar, § 193 StGB, Rn. 7; andeutend *Lenckner/Eisele*, in: Schönke/Schröder, Kommentar StGB, § 193 StGB, Rn. 13.

416 *Lenckner/Eisele*, in: Schönke/Schröder, Kommentar StGB, § 193 StGB, Rn. 20.

417 *Zimmermann*, ArbRAktuell 2012, 58, 61.

418 Vgl. BAG, Urt. v. 3.7.2003 – 2 AZR 235/02 , NZA 2004, 427, 429; *Zimmermann*, ArbRAktuell 2012, 58, 61.

419 So auch *Schemmel/Ruhmannseder/Witzigmann*, Hinweisgebersysteme, S. 111.

420 Vgl. *Schemmel/Ruhmannseder/Witzigmann*, Hinweisgebersysteme, S. 111.

2. *Strafbarkeit nach dem UWG*

Aus der dienstvertraglichen Loyalitätspflicht folgt regelmäßig eine Geheimhaltungs- bzw. Verschwiegenheitspflicht über interne Vorgänge und Tatsachen.[421] Das größte strafrechtliche Risiko droht den an der Tat unbeteiligten Hinweisgebern im Hinblick auf Vorschriften, die den Verrat von Betriebs- und Geschäftsgeheimnissen[422] unter Strafe stellen. Der Whistleblower macht sich nach § 17 UWG strafbar, wenn er etwa zur Untermauerung seiner Behauptungen Originale oder Kopien von Geschäftsunterlagen an Dritte weitergibt[423] oder jemandem zu Zwecken des Wettbewerbs, aus Eigennutz, zugunsten eines Dritten oder in der Absicht, dem Inhaber des Unternehmens Schaden zuzufügen, solche Geheimnisse mitteilt. Die Vorschrift bezweckt in erster Linie, Unternehmen vor Verletzungen ihrer Geschäfts- und Betriebsgeheimnisse zu schützen (Integritätsinteresse). Bei der Ermittlung des Geheimnisses kommt es darauf an, ob der Betriebsinhaber ein berechtigtes wirtschaftliches Interesse an der Geheimhaltung hat. Eine Legaldefinition des Geheimnisbegriffes gibt es jedoch nicht.[424] Entwickelt aus der Norm des § 17 UWG sind nach gefestigter Rechtsprechung und herrschender Lehre unter Betriebs- und Geschäftsgeheimnissen alle auf ein Unternehmen bezogenen Tatsachen, Umstände und Vorgänge, die nicht offenkundig, sondern nur einem begrenzten Personenkreis zugänglich sind, und an deren Nichtverbreitung der Rechtsträger ein berechtigtes Interesse hat, zu verstehen.[425]

Während früher noch ein Teil der Literatur[426] „illegalen Geheimnissen"[427] den Schutz des § 17 UWG von vornherein versagte, ist sie heute überwiegend der

421 *Linck*, in: Schaub, Arbeitsrechtshandbuch, § 53, Rn. 51.

422 Einig ist man sich darüber, dass dieses Gut verfassungsrechtlich geschützt ist. Uneinig ist man sich aber über die Grundlage: für die Einschlägigkeit des Art. 12 Abs. 1 GG u. a BVerfG, Beschl. v. 14.3.2006 – 1 BvR 2087/03 u.a. NVwZ 2006, 1041, 1042; *Wolff*, NJW 1997, 98, 99; für die Einschlägigkeit des Art. 14 Abs. 1 GG *Siekmann*, ZIP 1994, 651, 652; für die Einschlägigkeit des Art. 2 Abs. 1 GG BGH, Urt. v. 8.2.1994 – VI ZR 286/93, NJW 1994, 1281, 1282.

423 *Zimmermann*, ArbRAktuell 2012, 58, 59.

424 *Beyerbach*, Unternehmensinformation, S. 89, 90. Lediglich § 93 StGB enthält eine Legaldefinition des Staatsgeheimnisses.

425 BGH, Beschl. v. 21.1.2014 – EnVR 12/12, RdE 2014, 276, 282; BGH, Urt. v. 4.9.2013 – 5 StR 152/13, NStZ 2014, 325, Leitsatz 1; BAG, Urt. v. 16.3.1982 – 3 AZR 83/79, NJW 1983, 134, Leitsatz 2; *Köhler*, in: Köhler/Bornkamm, UWG, § 17, Rn. 4; *Henssler*, in: MüKo BGB, § 626 BGB, Rn. 194; *Többens*, NStZ 2000, 505, 506; *Beyerbach*,Untern ehmensinformation, S. 90 m. w. N.

426 So z. B. *Rützel*, GRUR 1995, 557, 560; *Kotthoff/Gabel*, in: HK-WettbR, § 17 UWG, Rn. 8; *Richters/Wodtke*, NZA-RR 2003, 281, 282.

427 Der Sprachgebrauch „illegales Geheimnis" ist so gesehen falsch, denn nicht das Geheimnis ist illegal, sondern sein Gegenstand. Trotzdem wird der Begriff „illegales Geheimnis" zwecks Vereinheitlichung genutzt.

Meinung,[428] dass es grundsätzlich bedeutungslos ist, ob das Geheimnis einen sitten- oder gesetzeswidrigen Inhalt hat, also ein „illegales Geheimnis" ist (z. B. geheim gehaltene Schmiergeldpraxis; Produktion unter Verstoß gegen Umweltschutzvorschriften; Kartellbeteiligung). Die Strafbarkeit entfällt somit – wie es früher der Fall gewesen wäre[429] – nicht, wenn der Whistleblower etwa Kartellabsprachen oder Steuerhinterziehungen offenlegt.[430] Es sind also auch „illegale Geheimnisse" von § 17 UWG umfasst.[431] Für diese Ansicht sprechen hauptsächlich drei Gründe: Erstens würde eine Nicht-Einbeziehung von illegalen Geheimnissen im Widerspruch zur Rechtsordnung stehen. Sowohl in §§ 93, 97 a StGB[432] – mit Ausnahme der Klassifizierungen in § 93 Abs. 2 StGB[433] – als auch in § 203 StGB[434] können Gegenstand von Geheimnissen Illegalitäten sein. Das berechtigte schutzwürdige Interesse einer Person an der Geheimhaltung einer Tatsache besteht unabhängig davon, welchen Inhalt das Geheimnis hat – ein berechtigtes schutzwürdiges Interesse kann demnach im Rahmen des § 203 StGB auch an der Wahrung eines „illegalen Geheimnisses" bestehen.[435] Da grundsätzlich vom Standpunkt des Geheimnisgeschützten auszugehen ist, spielt es objektiv keine Rolle, ob das Rechtsgut rechtlich missbilligt wird, es sich also um einen Gesetzesverstoß handelt. Der Tatbestand des § 203 StGB erfasst also auch eine Straftat oder deren Planung.[436] Die Bewertung des Geheimnisses wird dem

---

428 *Köhler*, in: Köhler/Bornkamm, UWG, § 17, Rn. 9; *Rengier*, in: Fezer, UWG, § 17, Rn. 21; *Dittrich*, in: Müller-Gugenberger/Bieneck, Wirtschaftsstrafrecht, § 33, Rn. 51; *Harte-Bavendamm*, in: Harte-Bavendamm/Henning-Bodewig, UWG, § 17, Rn. 5 f.; *v. Pelchrzim*, CCZ 2009, 25, 26; a. A. *Diemer*, in: Erbs/Kohlhaas, Strafrechtliche Nebengesetze, § 17 UWG, Rn. 16; *Möhrenschlager*, in: Wabnitz/Janovsky, Handbuch WStR, Kap. 13, Rn. 10; *Rützel*, GRUR 1995, 557 f.; *Engländer/Zimmermann*, NZWiSt 2012, 328, 330 f.

429 Vgl. *Koch*, ZIS 2008, 500, 503.

430 Es kann nämlich durchaus sein, dass auch „illegale" Tatsachen wirtschaftlich betrachtet einen Wert für das Unternehmen darstellen können. Deswegen sollte ihnen der Geheimhaltungscharakter nicht sofort abgesprochen werden, vgl. *Herbert/Oberrath*, NZA 2005, 193, 196; *Brammsen*, in: MüKo UWG, § 17, Rn. 22; *Ohly*, in: Ohly/Sosnitza, UWG, § 17, Rn. 12; *Mayer*, GRUR 2011, 884, 887.

431 *Köhler*, in: Köhler/Bornkamm, UWG, § 17, Rn. 9; *Ohly*, in: Ohly/Sosnitza, UWG, § 17, Rn. 12; *Mayer*, GRUR 2011, 884, 887; *Rengier*, in: Fezer, UWG, § 17, Rn. 21; *Koch*, ZIS 2008, 500, 503.

432 *Paeffgen*, in: Kindhäuser/Neumann/Paeffgen, Kommentar StGB, § 93 StGB, Rn. 32.

433 Das „illegale Geheimnis" fällt trotz der normativen Einschränkung in § 93 Abs. 2 StGB grundsätzlich unter Abs. 1 und dementsprechend auch unter dessen Schutz. Das bedeutet, dass alle Gegenstände, die nicht unter Abs. 2 fallen und illegal sind, den Schutz des Abs. 1 StGB genießen.

434 *Kargl*, in: Kindhäuser/Neumann/Paeffgen, Kommentar StGB, § 203 StGB, Rn. 8 mit Verweis auf BVerfG, Beschl. v. 24.5.1977 – 2 BvR 988/75, NJW 1977, 1489, 1491.

435 *Cierniak/Pohlit*, in: MüKo StGB, § 203 StGB, Rn. 21.

436 *Lenckner/Eisele*, in: Schönke/Schröder, Kommentar StGB, § 203 StGB, Rn. 7; *Kargl*, in: Kindhäuser/Neumann/Paeffgen, Kommentar StGB, § 203 StGB, Rn. 8.

Betroffenen überlassen. Dass illegale Geheimnisse strafrechtlich geschützt sind, ist demnach keinesfalls ungewöhnlich.[437] Dies lässt sich auf Gemeinwohlerwägungen zurückführen und dient insbesondere der Verschwiegenheit der in § 203 StGB aufgeführten Berufsgruppen.[438]

Zweitens darf der Arbeitnehmer nicht als Kontrollorgan[439] des Arbeitgebers „rekrutiert" werden. Ihm könne nicht die Befugnis erteilt werden, die Gesetzmäßigkeit von innerbetrieblichen Vorgängen zu beurteilen und zu entscheiden, ob dieser Vorgang offenzulegen sei oder nicht. Würden die Geheimnisse illegalen Inhalts also nicht in den Schutzbereich des § 17 UWG fallen, hätte der Arbeitnehmer die Möglichkeit, selbst zu entscheiden, welche Geheimnisse illegal sind oder nicht. Eine falsche Entscheidung ist dabei nicht ausgeschlossen.

Der dritte Grund entspringt der strafrechtlichen Dogmatik. Ohne die Einbeziehung von „illegalen Geheimnissen" in den Schutzbereich des § 17 UWG entsteht eine Strafbarkeitslücke in der Verwertungsalternative des § 17 Abs. 2 Nr. 2 UWG. Denn gerade mit illegalen Geheimnissen wird am „Markt" gehandelt. Ein Profitinteresse des Arbeitnehmers ist nämlich auch bei Weitergabe von „illegalen Geheimnissen" gegeben, wenn die Informationen an Dritte, insbesondere an Wettbewerber des Arbeitgebers, „verkauft" werden.[440] Der Konkurrent kann dann – durch Aufdeckung der illegalen Informationen und der sich ergebenden behördlichen Untersuchung – eine bestehende Wettbewerbssituation für sich zum Vorteil verändern und positiven Nutzen ziehen. Deswegen ist bei Umständen, die die Wettbewerbsfähigkeit beeinflussen können, anzunehmen, dass der Betriebsinhaber auch an ihnen, also den „illegalen Geheimnissen", ein Geheimhaltungsinteresse hat.[441]

Die genannten Gründe führen zum Ergebnis, „illegale Geheimnisse" mit in den Schutzbereich des § 17 UWG aufzunehmen und in der Offenlegung dieser Geheimnisse eine Verletzung von Betriebs- und Geschäftsgeheimnissen zu sehen.[442] Ein Geheimhaltungsinteresse des Unternehmens besteht auch an Informationen über rechtswidrige Sachverhalte.

---

437 Siehe auch *Koch*, ZIS 2008, 500, 503 m. w. N; a. A. *Engländer/Zimmermann*, NZWiSt 2012, 328, 333, die darauf hinweisen, dass aufgrund des anders gelagerten Schutzzweckes eine andere Interpretation des Begriffs des Unternehmensgeheimnisses gerechtfertigt ist.

438 *Lenckner/Eisele*, in: Schönke/Schröder, Kommentar StGB, § 203 StGB, Rn. 3.

439 Der Arbeitnehmer ist weder „Sittenrichter noch Kontrollorgan gegenüber dem Arbeitgeber", *Otto*, wistra 1988, 125, 126 m. w. N.; *Többens*, NStZ 2000, 505, 506; *Mayer*, GRUR 2011, 884, 887.

440 So ähnlich *Rützel*, GRUR 1995, 557, 561, der aber zu dem Ergebnis kommt, dass die Einbeziehung illegaler Geheimnisse in den Begriff des Unternehmensgeheimnisses sich u. a. nicht durch das Bestehen einer Strafbarkeitslücke rechtfertige.

441 Auch *Többens*, NStZ 2000, 505, 506.

442 Eine Alternative wäre die Sachverhalte nicht zu schützen, die gegen die guten Sitten oder die öffentliche Ordnung verstoßen oder eine Straftat von erheblicher Bedeutung verwirklichen vgl. *Beyerbach*, Unternehmensinformation, S. 101.

Folgt man der Meinung,[443] dass die Norm nicht dem Interesse der Allgemeinheit an einem unverfälschten Wettbewerb dient, sondern nur dem Geheimhaltungsbedürfnis des betroffenen Unternehmens, kommt man zu keinem anderen Ergebnis. Nur wenn man § 17 UWG als Vermögensdelikt versteht und den juristisch-ökonomischen Vermögensbegriff[444] vertritt, müsste den illegalen Geheimnissen die Vermögensqualität abgesprochen werden.[445]

Entscheidend ist neben dem Mitteilungsgegenstand die fehlende Befugnis zur Weitergabe. Unbefugt handelt jener, der bei der Offenlegung nicht durch ein gesetzliches Anzeigerecht geschützt ist oder einer Auskunftspflicht unterliegt.[446] Beim Whistleblowing wird es sich meist um Fälle handeln, in denen der objektive Tatbestand erfüllt ist. Bei der Weitergabe der Informationen handelt es sich um eine unbefugte – weil nicht vom geschützten Anzeigegenstand des § 138 StGB umfasste – Bekanntmachung.[447] Denn nicht selten betreffen die offengelegten Sachverhalte Straftaten, die nicht im Katalog des § 138 StGB aufgezählt sind, wie z. B. Umweltdelikte.

In subjektiver Hinsicht ist der Tatbestand nur erfüllt, wenn neben einer Vorsatzform auch ein unrechtssteigernder Beweggrund hinzukommt (überschießende Innentendenz).[448] Oft ist das Handeln der Whistleblower von einer moralischen Pflicht zum Handeln für die Gemeinschaft getragen. Aufgrund der Vielfältigkeit der Motive ist jedoch nicht auszuschließen, dass alle vier in § 17 UWG genannten Beweggründe[449] in Betracht kommen.

3. *Die Rechtfertigung nach § 34 StGB*

Fraglich ist, ob der Whistleblower vor den beschriebenen möglichen Konsequenzen durch die Rechtsordnung geschützt wird. Neben der Strafbarkeit nach dem StGB steht auch die Strafbarkeit nach dem UWG im Raum. Bei Vorliegen des Tatbestandes

---

443 BGH, Urt. v. 24.1.2006 – XI ZR 384/03, NJW 2006, 830, 838; *Beater*, Unlauterer Wettbewerb, Rn. 1876; *Brammsen*, MüKo UWG, § 17, Rn. 4; *Mayer*, GRUR 2011, 884, 885; a. A. *Diemer*, in: Erbs/Kohlhaas, Strafrechtliche Nebengesetze, § 17 UWG, Rn. 2; *Ohly*, in: Ohly/Sosnitza, UWG, § 17, Rn. 1; *Köhler*, in: Köhler/Bornkamm, UWG, § 17, Rn. 2.

444 Danach zählen zum Vermögen nur die geldwerten Güter der Person, die nicht rechtlich missbilligt werden.

445 *Engländer/Zimmermann*, NZWiSt 2012, 328, 333.

446 *Ohly*, in: Ohly/Sosnitza, UWG, § 17, Rn. 28.

447 Zu den einzelnen Punkten siehe *Schemmel/Ruhmannseder/Witzigmann*, Hinweisgebersysteme, S. 95 f.

448 *Janssen/Maluga*, in: MüKo StGB, § 17 UWG, Rn. 58.

449 Nach § 17 Abs. 1 Satz 1 UWG sind die Beweggründe (1.) zu Zwecken des Wettbewerbs, (2.) aus Eigennutz, (3.) zugunsten eines Dritten oder (4.) in der Absicht, dem Inhaber des Unternehmens Schaden zuzufügen.

ist an eine mögliche Rechtfertigung nach § 34 StGB zu denken.[450] Zwar ist das UWG ein strafrechtliches Nebengesetz, jedoch ist § 34 StGB nach gleichen Maßstäben zu prüfen.

Dem Notstand im Sinne des § 34 StGB liegt, soweit es um die erforderliche Güterabwägung zwischen Erhaltungs- und Eingriffsgut geht, das Prinzip des überwiegenden Interesses zugrunde.[451] Wesentliche Elemente des rechtfertigenden Notstands sind die Notstandslage, die Notstandshandlung, das subjektive Rechtfertigungselement des Notstandtäters und die Interessenabwägung.

Die Notstandslage besteht in einer gegenwärtigen Gefahr für Leben, Leib, Freiheit, Ehre, Eigentum oder ein anderes Rechtsgut, die nicht anders abgewendet werden kann als durch Einwirkung auf gleichermaßen anerkannte rechtliche Interessen.

a. Gefahr für ein notstandsfähiges Rechtsgut

Geschützt ist jedes Rechtsgut des Einzelnen und der Allgemeinheit,[452] soweit es in der konkreten Situation schutzwürdig und schutzbedürftig ist. Neben den beispielhaft aufgezählten Gütern Leben, Leib, Freiheit, Ehre und Eigentum kann als anderes Rechtsgut jedes rechtlich geschützte Interesse in Frage kommen, gleichgültig von welchem Teil der Rechtsordnung es diesen Schutz erfährt.[453] Dazu zählen auch zivilrechtlich geschützte Rechtsgüter nach § 823 BGB, wie z. B. der Persönlichkeitsschutz. Im Wirtschaftsstrafrecht kommen als Erhaltungsrechtsgüter insbesondere das (Gläubiger-) Vermögen (§§ 399 ff. AktG, §§ 263, 263a, 265b, 266b, 305a StGB, § 148 GenG etc.), das Eigentum (§ 305a StGB etc.), das öffentliche Vertrauen (§ 399 AktG, §§ 264 a, 331 ff. StGB, §§ 147 GenG, §§ 82, 84, 85 GmbHG etc.) und die Lauterkeit des Wettbewerbs (§§ 298,[454] 299 StGB etc.) in Betracht.[455]

§ 34 StGB setzt weiterhin eine Gefahr für das Rechtsgut voraus. Eine Gefahr liegt vor, wenn ein Zustand gegeben ist, in dem aufgrund tatsächlicher Umstände

---

450 § 34 StGB liegt insbesondere dann vor, wenn es um die Wahrnehmung des Strafverfolgungsinteresses geht, *Ebert-Weidenfeller*, in: Achenbach/Ransiek, Wirtschaftsstrafrecht, S. 327.

451 *Kühl*, in: Lackner/Kühl, StGB-Kommentar, § 34 StGB, Rn. 1.

452 Eine Beschränkung aus Individualinteressen wie bei § 32 StGB würde dem Wesen des Notstandes nicht entsprechen. Einer Rechtfertigung ist dann jedoch unter dem Gesichtspunkt der Erforderlichkeit enge Grenzen gesetzt: Zulässig ist danach privates Handeln zum Schutz der Allgemeinheit wegen der primären Zuständigkeit staatlicher Organe nur im äußersten Notfall *Perron*, in: Schönke/Schröder, Kommentar StGB, § 34 StGB, Rn. 10 m. w. N.

453 *Perron*, in: Schönke/Schröder, Kommentar StGB, § 34 StGB, Rn. 9.

454 Nur wenn speziellen Rechtfertigungsgründe nach §§ 2, 3 GWB nicht vorliegen, ist eine Rechtfertigung nach den strafrechtlichen Rechtfertigungsgründen in Betracht zu ziehen, *Dannecker*, in: Kindhäuser/Neumann/Paeffgen, Kommentar StGB, § 298 StGB, Rn. 84.

455 *Lutterbach*, Whistleblowing, S. 94.

die Wahrscheinlichkeit des Eintritts eines schädigenden Ereignisses besteht.[456] Wahrscheinlich ist der Eintritt, wenn die Möglichkeit naheliegt oder begründete Besorgnis besteht – eine bloß allgemeine Möglichkeit hingegen ist nicht ausreichend.[457] Dieses Wahrscheinlichkeitsurteil knüpft an die im Handlungszeitpunkt vorliegenden Umstände an und prognostiziert die zu erwartende Gesamtentwicklung.[458] Maßgeblich ist dabei nach überwiegender Ansicht der Literatur die Sicht eines objektiven Beobachters, der in Kenntnis der Umstände die Situation *ex ante* beurteilt.[459] Damit ist der Standpunkt eines Beobachters gemeint, dem die im kritischen Augenblick wesentlichen Umstände bekannt sind.

Hinsichtlich des Wahrscheinlichkeitsgrads dieses objektiven, sachkundigen *ex ante*-Urteils genügt eine nur ganz entfernte Möglichkeit ebenso wenig wie die nur „gedankliche Möglichkeit".[460] Vielmehr reicht es aus, wenn die Wahrscheinlichkeit einen über die allgemeinen Lebensrisiken hinausgehenden Grad erreicht hat, von dem an man sich vernünftigerweise auf die Möglichkeit des schädigenden Ereignisses einzustellen pflegt.[461] Als Orientierungshilfe für den Whistleblower wird in Betracht gezogen,[462] den prozessualen Begriff des Anfangsverdachts nach § 152 Abs. 2 StPO heranzuziehen. Ein Anfangsverdacht besteht, wenn es nach zureichenden tatsächlichen Anhaltspunkten möglich erscheint, dass eine verfolgbare Straftat vorliegt.[463]

b. Gegenwärtigkeit

Problematischer ist die Gegenwärtigkeit der Gefahr. Dies ergibt sich daraus, dass die Sachverhalte, die der Whistleblower offenbaren möchte, oft vergangene, das heißt beendete Straftaten umfassen oder der Täter schon aus dem Unternehmen ausgeschieden ist. Gegenwärtig ist eine Gefahr nur, wenn bei natürlicher Weiterentwicklung der Dinge der Eintritt des Schadens sicher oder höchstwahrscheinlich ist, falls nicht alsbald Abwehrmaßnahmen ergriffen werden. Der ungewöhnliche Zustand kann nach menschlicher Erfahrung und natürlicher Weiterentwicklung

---

456 BGH, Urt. v. 25.3.2003 – 1 StR 483/02, NJW 2003, 2464, 2466. Nach h. M. ist der Gefahrbegriff bei § 34 StGB nicht anders als sonst im StGB zu verstehen *Kühl*, in: Lackner/Kühl, StGB-Kommentar, § 34 StGB, Rn. 2.

457 *Perron*, in: Schönke/Schröder, Kommentar StGB, § 34 StGB, Rn. 12.

458 *Lackner*, in: Lackner/Kühl, StGB-Kommentar, § 34 StGB, Rn. 2.

459 BGH, Urt. v. 25.3.2003 – 1 StR 483/02, NJW 2003, 2464, 2466; *Perron*, in: Schönke/Schröder, Kommentar StGB, § 34 StGB, Rn. 12, der das ex ante-Urteil nur auf die in der Handlungssituation gegebenen Umstände stützen will; *Kühl*, in: Lackner/Kühl, StGB-Kommentar, § 34 StGB, Rn. 2; *Momsen*, in: BeckOK StGB, § 34 StGB, Rn. 4; *Kretschmer*, Jura 2005, 662, 663; *Roxin*, AT I § 15, Rn. 12; *Wessels/Beulke/Satzger*, StrafR AT, Rn. 304.

460 *Perron*, in: Schönke/Schröder, Kommentar StGB, § 34 StGB, Rn. 15 m. w. N.

461 *Perron*, in: Schönke/Schröder, Kommentar StGB, § 34 StGB, Rn. 15.

462 *Lutterbach*, Whistleblowing, S. 95.

463 *Pfeiffer*, Kommentar StPO, § 152 StPO, Rn. 1a.

jederzeit in einen Schaden umschlagen.[464] Auch die Dauergefahr fällt unter den Begriff der Gefahr nach § 34 StGB.[465] Dem Whistleblower ist also noch eine zusätzliche Möglichkeit gegeben, die er positiv nutzen kann. Eine Dauergefahr liegt vor, wenn nicht ausgeschlossen ist, dass der Eintritt des Schadens noch eine Weile auf sich warten lässt, sich aber jederzeit aktualisieren kann.[466] Gegenwärtig ist eine solche Dauergefahr, wenn sie so dringend ist, dass sie nur durch unverzügliches Handeln wirksam abgewendet werden kann.[467]

Für den Whistleblower ergibt sich hinsichtlich der zu schützenden Rechtsgüter folgende Lage: entweder es handelt sich bei den Gegenständen seiner Offenbarung um dauerhaft gefährdete Rechtsgüter oder es muss eine hinreichend konkrete Wiederholungsgefahr[468] bestehen, um von einer gegenwärtigen Gefahr sprechen zu können. Typische illegale Schmiergeldpraktiken[469] oder die dauerhafte und sich immer wiederholende Umwelt- und Gesundheitsschädigung durch unsachgemäße Entsorgung von Abfällen könnten daher grundsätzlich als gegenwärtige Gefahr erfasst werden. Eine Notstandslage für den Whistleblower besteht somit nur nicht in den Fällen, in denen nur ein vager Hinweis auf eine Gefahr besteht oder sich diese als nicht (mehr) gegenwärtig erweist.

c. Nicht anders abwendbar

Nach § 34 StGB darf die Gefahr nicht anders abwendbar sein als durch die Begehung der Tat. Inhaltlich entspricht dieses Merkmal der Erforderlichkeit bei der Notwehr nach § 32 StGB. Erforderlich ist demgemäß, was zur Abwendung der Gefahr geeignet ist und unter Berücksichtigung aller aus der *ex ante*-Sicht eines sachkundigen objektiven Beobachters erkennbaren Umstände als der sicherste Weg zur Erhaltung des gefährdeten Gutes erscheint.[470] Unter mehreren geeigneten Mitteln ist das relativ mildeste zu wählen.[471] Besteht eine Ausweichmöglichkeit oder ist staatliche Hilfe zu erlangen, so ist davon Gebrauch zu machen. Ist das mildeste Mittel zugleich das aussichtsreichste, so ist dieses auch das allein erforderliche. Das Gleiche gilt für das geeignetste unter mehreren gleich oder annähernd gleich schweren Mitteln und umgekehrt für die am wenigsten gravierende unter mehreren (annähernd) gleich geeigneten Maßnahmen.[472] Vor allem aber ist nach der grundsätzlichen Möglichkeit,

464 BGH, Urt. v. 25.3.2003 – 1 StR 483/02, NJW 2003, 2464, 2466.
465 BGH, Urt. v. 15.5.1979 – 1 StR 74/79, NJW 1979, 2053, 2054.
466 BGH, Urt. v. 15.5.1979 – 1 StR 74/79, NJW 1979, 2053, 2054.
467 BGH, Urt. v. 25.3.2003 – 1 StR 483/02, NJW 2003, 2464, 2466.
468 Eine Wiederholungsgefahr kann dann angenommen werden, wenn die Begehung weiterer deliktischer Handlungen höchst wahrscheinlich ist und naheliegt *Lutterbach*, Whistleblowing, S. 97.
469 *Lutterbach*, Whistleblowing, S. 97.
470 BGH, Urt. v. 25.3.1952 – 1 StR 172/51, BGHSt 2, 242.
471 OLG Düsseldorf, Urt. v. 25.10.2005 – III-5 Ss 63/05- 33/05 I, NJW 2006, 630, 631.
472 *Perron*, in: Schönke/Schröder, Kommentar StGB, § 34 StGB, Rn. 20a.

den Erfolg abzuwenden, zu fragen.[473] Ausreichend ist es, wenn die Schadensabwendung nicht ganz unwahrscheinlich ist.[474]

Erteilt der Whistleblower einen Hinweis, so hat dieses Verhalten zunächst nur mittelbaren Einfluss auf die Notstandslage, da er selbst nicht unmittelbar eingreift. Sein Handeln kann aber trotzdem zur Gefahrenabwendung geeignet sein, und zwar immer in den Fällen, in denen der Adressat (oder der Adressatenkreis) des Hinweises Einflussmöglichkeiten auf die Auslöser der gegenwärtigen Gefahr besitzt. Gemäß der obigen Definition genügt es, wenn durch das Handeln des Whistleblowers zu dem Zeitpunkt eine nicht ganz unwahrscheinliche Chance besteht, die Gefahr der Rechtsgutsverletzung zu beseitigen.

Probleme können sich bei Prüfung der Erforderlichkeit ergeben. Der Whistleblower hat nämlich zwischen den ihm zur Verfügung stehenden Handlungsalternativen zu wählen. Jedoch kann der Whistleblower nur wählen, wenn er von seiner Wahlmöglichkeit Kenntnis hat und die mildere Handlungsalternative mindestens gleich effektiv ist.

Es stehen dem Whistleblower mehrere geeignete Mittel zur Verfügung:[475]

- Information des Vorgesetzten oder anderer Geheimnisträger innerhalb des Unternehmens, sofern diese nicht selbst Verantwortliche der Straftat sind
- Nutzung eines bestehenden Hinweisgebersystems
- Hinweis an die Polizei, Staatsanwaltschaft oder Finanzbehörde
- Benachrichtigung der Allgemeinheit.[476]

Die Benachrichtigung der Allgemeinheit stellt einen Sonderfall dar. In vielen Fällen wird ein Hinweis an die Öffentlichkeit nicht das mildeste Mittel sein, jedoch besonders effektiv, da er den größten Adressatenkreis erreicht. Da ein Hinweis an die staatlichen Behörden wohl fast gleich effektiv sein wird (nur der Eingriff in das Rechtsgut relativ schwer wiegt), ist die Frage nach dem besten, also geeignetsten und mildesten Mittel schwer zu beantworten. Eine Rechtfertigung nach § 34 StGB ist also im Rahmen einer Interessenabwägung festzustellen.[477]

d. Interessenabwägung

Nach § 34 S. 1 StGB ist die Tat nicht rechtswidrig, wenn bei Abwägung der widerstreitenden Interessen, namentlich der betroffenen Rechtsgüter und des Grades der ihnen drohenden Gefahren, das geschützte Interesse das beeinträchtigte wesentlich überwiegt.

---

473 *Michalowski*, ZStW 109 (1997), 519, 532.

474 *Perron*, in: Schönke/Schröder, Kommentar StGB, § 34 StGB, Rn. 19; *Erb*, in: MüKo StGB, § 34 StGB, Rn. 91 f.

475 Ausführlich dazu *Lutterbach*, Whistleblowing, S. 99.

476 Gemeint sind: Presse, Whistleblowing-Stiftungen und Vereinigungen, Internetblogs, Fernsehsender.

477 So auch *Lutterbach*, Whistleblowing, S. 100.

Bei der Prüfung des Rangverhältnisses der Güter ist eine Gesamtwürdigung aller Umstände erforderlich. Im Allgemeinen ist von einer Hierarchie auszugehen, nach der formalen Ordnungsbelangen eine geringere Bedeutung zukommt als „handfesten" Rechtsgütern. Unter letzteren sind Eigentum und Vermögen im Allgemeinen von geringerem Gewicht als Persönlichkeitswerte, von denen wiederum das Leben den höchsten Rang einnimmt, gefolgt von Gesundheit, persönlicher Freiheit und sonstigen Persönlichkeitswerten.[478]

Weiterhin ist der Grad der den Rechtsgütern drohenden Gefahren zu berücksichtigen, und zwar sowohl für das geschützte Rechtsgut als auch für das beeinträchtigte. Damit ist gemeint, dass die Gefahr abzuwägen ist, der die betroffenen Rechtsgüter mit oder ohne Vornahme der Notstandshandlung jeweils ausgesetzt sind.[479]

Abzuwägen sind die betroffenen Rechtsgüter bzw. Interessen mit den Erhaltungsrechtsgütern, welche durch das Whistleblowing geschützt werden sollen.

aa. Einzelne Rechtsgüter – Erhaltungsrechtsgüter –

Neben den vorrangig behandelten Wirtschaftsstraftaten darf nicht vergessen werden, dass die Gesetzesverstöße, die aufgedeckt werden sollen, auch im Bereich der Körperverletzung nach §§ 223 StGB angesiedelt sein können. Denn nicht selten haben die Missstände weitreichende Folgen für die körperliche Integrität und die Gesundheit; im schlimmsten Fall kann sogar eine Lebensgefährdung die Folge sein. Es kann beispielsweise in „gesundheitsorientierten" Betrieben die Gefahr bestehen, dass Arbeitnehmer oder Patienten direkt einer Körperverletzung ausgesetzt werden, aber auch außerhalb des Betriebs kann durch Nichtbeachtung von Schutz- oder Abfallvorschriften die breite Bevölkerung Gesundheitsschädigungen erleiden. Schutzgut der §§ 223 ff. StGB ist das körperliche Wohl, das sowohl durch körperliche als auch durch seelische Einwirkungen verletzt werden kann.[480]

Das Persönlichkeitsrecht kann auch ein Rechtsgut sein, dass der Whistleblower durch seine Offenbarung schützen will. Unter Persönlichkeitsrecht wird das Recht des Einzelnen auf Achtung seiner Menschenwürde und auf Entfaltung seiner individuellen Persönlichkeit verstanden.[481] Das betrifft die Rücksichtnahme auf das Wohl und die berechtigten Interessen des Arbeitnehmers, insbesondere auf Schutz vor Gesundheitsgefahren[482] und Einhaltung von menschenwürdigen Arbeitsbedingungen. Das Rechtsgut schützt auch den persönlichen Bereich vor unbefugtem Eindringen oder Ausforschen, wie z. B. dem Zusenden von SPAMs auf den privaten PC oder Telefonwerbung.[483] Ergreift der Arbeitgeber unerlaubterweise Maßnahmen,

478 So *Erb*, in: MüKo StGB, § 34 StGB, Rn. 112.
479 *Erb*, in: MüKo StGB, § 34 StGB, Rn. 129.
480 *Eser*, in: Schönke/Schröder, Kommentar StGB, § 223 StGB, Rn. 1.
481 BGH, Urt. v. 2.4.1957 – VI ZR 9/56, NJW 1957, 1146, 1147.
482 *Schaub/Koch*, in: Schaub, Arbeitsrecht von A-Z, Persönlichkeitsrecht.
483 Vgl. bzgl. der Telefonwerbung BGH, Urt. v. 6.10.1972 – I ZR 54/71, NJW 1973, 42.

die das Persönlichkeitsrecht seiner Arbeitnehmer oder Dritter verletzen, begeht er einen rechtswidrigen Eingriff.

In dem Zusammenhang sind auch die Delikte zu nennen, die Taten gegen die Umwelt unter Strafe stellen. Schutzgut der §§ 324 ff. StGB ist die Umwelt in ihrer Funktion als gegenwärtige und künftige Lebensgrundlage des Menschen.[484] Das Interesse am Erhalt der einzelnen Umweltmedien (Wasser, Boden, Luft) und Umweltgüter (Tiere und Pflanzen) ist auf den Menschen bezogen, was besonders deutlich wird bei den Tatbeständen, die wie §§ 325, 325 a, 330 a StGB auf den Schutz der körperlichen Integrität des Menschen abstellen.[485]

Das von § 298 StGB geschützte Rechtsgut ist das immer wieder bestätigte Vertrauen des Einzelnen in die Funktionsfähigkeit des freien und fairen Wettbewerbs, vor allem der redlichen Emittenten eines Angebots und der Veranstalter einer Ausschreibung in die Gewährleistung eines regelhaften, von untereinander abgesprochenen Angeboten unbeeinträchtigten Wettbewerbsverfahrens zur Beschaffung von Waren oder Dienstleistungen.[486] § 299 StGB schützt vorrangig das Allgemeininteresse an „lauteren", also fairen Wettbewerbsbedingungen.[487] Im Mittelpunkt steht die Sicherung des Wettbewerbs vor unlauteren Einflussnahmen. Daneben sind aber auch potenzielle Vermögensinteressen von Mitbewerbern und Geschäftsherrn geschützt.[488]

Geschütztes Rechtsgut des § 263 StGB ist ausschließlich das Vermögen,[489] nicht die Redlichkeit im Geschäftsverkehr und auch nicht die Dispositionsfreiheit als solche – geschützt ist auch ausländisches Staatsvermögen.[490] Wegen des großen Dunkelfeldes, aber auch wegen der relativ hohen Schadenssummen ist das Whistleblowing hier besonders relevant.[491]

bb. Einzelne Rechtsgüter – Eingriffsrechtsgüter –

§ 17 UWG schützt den Geheimbereich eines Unternehmens vor einem unredlichen Eingriff[492] – ohne Rücksicht darauf, aus welchem Grund Geschäftsgeheimnisse verraten oder erlangt werden. Aufgrund der weiten Auslegung sowie der Besonderheit in Abs. 5 ist geschütztes Rechtsgut auch das Betriebs- und Geschäftsgeheimnis als individuelles Rechtsgut sowie der Wettbewerb als Institution der

484 *Rengier*, StrafR BT II, § 47, Rn. 10.
485 *Ransiek*, in: Kindhäuser/Neumann/Paeffgen, Kommentar StGB, Vorb. zu §§ 324 f. StGB, Rn. 7.
486 *Hohmann*, in: MüKo StGB, § 298 StGB, Rn. 1.
487 *Heine/Eisele*, in: Schönke/Schröder, Kommentar StGB, § 299 StGB, Rn. 2.
488 *Heine/Eisele*, in: Schönke/Schröder, Kommentar StGB, § 299 StGB, Rn. 2 m. w. N.
489 Vgl. BGH, Beschl. v. 18.7.1961 – 1 StR 606/60, NJW 1961, 1876, 1876.
490 *Kühl*, in: Lackner/Kühl, StGB-Kommentar, § 263 StGB, Rn. 2 m. w. N.
491 *Lutterbach*, Whistleblowing, S. 102.
492 BGH, Urt. v. 24.11.1959 – 1 StR 439/59, NJW 1960, 207, 207.

Wirtschaftsordnung und damit als soziales Rechtsgut.[493] Durch die Einführung der Schutzzwecktrias in § 1 UWG hat der Gesetzgeber zudem klargestellt, dass die Marktteilnehmer, insbesondere die Verbraucher und die Mitbewerber, durch das UWG gleichermaßen und gleichrangig geschützt werden.[494] Als Betriebs- und Geschäftsgeheimnisse werden alle auf ein Unternehmen bezogene Tatsachen, Umstände und Vorgänge verstanden, die nicht offenkundig, sondern nur einem begrenzten Personenkreis zugänglich sind, und an deren Nichtverbreitung der Rechtsträger ein berechtigtes Interesse hat.[495] Betriebsgeheimnisse umfassen im Wesentlichen technisches Wissen im weitesten Sinne; Geschäftsgeheimnisse betreffen vornehmlich kaufmännisches Wissen. Wie schon im vorherigen Abschnitt erläutert, fallen auch „illegale Geheimnisse" unter § 17 UWG.

Für die Bestimmung des Rechtsguts des § 203 StGB muss auf gewichtige Einzelelemente abgestellt werden, die für dessen Bestimmung erhebliche Relevanz entfalten. Unter anderem werden Theorien zur Bestimmung zur Hilfe genommen,[496] bei der folgende Themen im Mittelpunkt stehen: persönliche Geheimsphäre des Einzelnen, materielle „Privatheit" und allgemeines Vertrauen in die Verschwiegenheit der Angehörigen bestimmter Berufe bzw. der Träger bestimmter Funktionen. In erster Linie ist jedoch das Individualinteresse an der Geheimhaltung bestimmter Tatsachen geschützt.[497] Im Vordergrund steht der Individualschutz des Bürgers – allenfalls mittelbar ist auch das Vertrauen der Allgemeinheit in die Verschwiegenheit von Amtsträgern und andere Bedienstete und damit die Funktionsfähigkeit der Verwaltung geschützt.[498]

cc. Rangverhältnis

Die durch die Notstandshandlung betroffenen Güter sind gegenüberzustellen, wobei sowohl die unmittelbar als auch die nur mittelbar betroffenen Rechtsgüter zu berücksichtigen sind. Das menschliche Leben darf dabei gegen ein anderes Leben aufgrund des absoluten Lebensschutzes nicht abgewägt werden.[499] Auch genießt die körperliche Integrität Vorrang gegenüber einem „materiellen" Rechtsgut.[500] Kann die Gefahr nur dadurch abgewendet werden, dass das Betriebsgeheimnis offenbart wird,

---

493 *Diemer*, in: Erbs/Kohlhaas, Strafrechtliche Nebengesetze, § 17 UWG, Rn. 1.

494 *Janssen/Maluga*, in: MüKo StGB, § 17 UWG, Rn. 9 f.; kritisch zur Einbeziehung des Verbraucherschutzes siehe *Otto*, GRUR 1982, 274, 279.

495 BVerfG, Beschl. v. 14.3.2006 – 1 BvR 2087/03 und 1 BvR 2111/03, MMR 2006, 375, 376.

496 *Cierniak/Pohlit*, in: MüKo StGB, § 203 StGB, Rn. 2, 3 m. w. N.

497 So schon BGH, Urt. v. 25.03.1993 – IX ZR 192/92, NJW 1993, 1638, 1638 wo es um die Individualsphäre eines Patienten ging.

498 *Lenckner/Eisele*, in: Schönke/Schröder, Kommentar StGB, § 203 StGB, Rn. 3.

499 *Neumann*, in: Kindhäuser/Neumann/Paeffgen, Kommentar StGB, § 34 StGB, Rn. 74 m. w. N.

500 *Erb*, in: MüKo StGB, § 34 StGB, Rn. 116; *Perron*, in: Schönke/Schröder, Kommentar StGB, § 34 StGB, Rn. 23.

ist der Eingriff trotz des daraus resultierenden Schadens gerechtfertigt, weil die Art und die Intensität der Gefahr für Rechtsgüter wie die körperliche Integrität höher zu bewerten ist. Alle anderen Rechtsgüter, vor allem die hier in Betracht kommenden wirtschaftlichen Rechtsgüter, sind jedoch einer Abwägung zugänglich. Ein eindeutiges Rangverhältnis der Rechtsgüter des § 17 UWG (Betriebs- und Geschäftsgeheimnis) wie auch der §§ 203, 263, 298, 299 StGB – sowohl der Eingriffsrechtsgüter als auch der Erhaltungsrechtsgüter – lässt sich allerdings nicht aufstellen. Höchstens im Hinblick auf das Vermögen als Inbegriff aller wirtschaftlichen Güter lässt sich wegen des niederen Rangs dieses Rechtsguts ein anderes Rangverhältnis herleiten.

dd. Grad der drohenden Gefahren

Weiterhin ist im Rahmen der Interessenabwägung der Grad der den Rechtsgütern drohenden Gefahren zu berücksichtigen, und zwar sowohl für das geschützte Rechtsgut als auch für das beeinträchtigte.[501] Als relevante Gesichtspunkte können Art, Ursprung, Intensität und Nähe der Gefahr sowie Art und Umfang der drohenden Werteinbußen eine Rolle spielen.[502] Je höher der Gefahrengrad für ein Rechtsgut im Verhältnis zur Wahrscheinlichkeit der Beseitigungsmöglichkeit, desto größer ist die Schutzbedürftigkeit.[503]

Je nachdem, an welchen Adressaten sich der Whistleblower mit dem Hinweis wendet, ist der Grad der drohenden Rechtsgutsverletzung höher oder niedriger. Deshalb ist zwischen der innerbetrieblichen Offenlegung, der Offenlegung gegenüber Ermittlungsbehörden und der Benachrichtigung der Allgemeinheit zu unterscheiden.[504]

Die höchste Wahrscheinlichkeit für einen Schaden an den Eingriffsrechtsgütern ist gegeben, wenn der Whistleblower an die Öffentlichkeit geht und die Missstände und Machenschaften der breiten Allgemeinheit bekannt macht. Zwar ist diese Art der Offenlegung am effektivsten, jedoch auch am unkontrollierbarsten hinsichtlich des Schadenumfangs.

Im Gegensatz dazu ist die Verletzung eines Eingriffsrechtsgutes bei einer innerbetrieblichen Meldung am geringsten. Das Geheimnis gelangt nicht an die Öffentlichkeit und bei Bestehen eines Hinweisgebersystems kann eine zeitnahe Behebung der Missstände erreicht werden. Jedoch ist die Implementierung eines Hinweisgebersystems oder Ähnlichem keine Pflicht in Deutschland und deswegen kann nur unter Vorbehalt darauf abgestellt werden. Des Weiteren geht die Effektivität eines solchen Whistleblower-Systems ganz verloren, wenn die verantwortlichen Personen selbst als Täter zu qualifizieren sind, da dann kaum mit Abhilfe zu rechnen ist.

---

501 *Erb*, in: MüKo StGB, § 34 StGB, Rn. 129.

502 *Erb*, in: MüKo StGB, § 34 StGB, Rn. 130 f.

503 *Perron*, in Schönke/Schröder, Kommentar StGB, § 34 StGB, Rn. 27; *Erb*, in: MüKo StGB, § 34 StGB, Rn. 130.

504 Ausführlich dazu *Lutterbach*, Whistleblowing, S. 104 f.

Dementsprechend würde der Erfolg des Whistleblowers vom Zufall abhängen – von der Person des Verantwortlichen.

Die Offenlegung der Hinweise gegenüber Ermittlungsbehörden liegt zwischen den beschriebenen Fällen: Zum einen werden die berechtigten Interessen des Unternehmens geschützt, weil die Geheimnisse nicht an die breite Öffentlichkeit getragen werden, zum anderen ist diese Art der Offenlegung besonders effektiv für die Erhaltung der Rechtsgüter des Whistleblowers und der Allgemeinheit, da die Behörden im Regelfall gut und schnell wirkungsvolle Maßnahmen ergreifen können. Der Grad der drohenden Gefahren ist im mittleren Bereich anzusiedeln.

Die Offenlegung von brisanten Hinweisen an die Ermittlungsbehörden stellt somit eine rechtfertigende Notstandshandlung dar.

Aus der Abwägung herausgenommen wurden jedoch jene Rechtsgüter, die das Leben und die körperliche Integrität zum Inhalt haben und allgemeinen Vorrang genießen. Aus diesem Umstand ist zu schließen, dass ein Hinweis, der Verstöße gegen diese Rechtsgüter zum Gegenstand hat, erlaubt ist. Legt der Whistleblower also eine Gefahr für das Leben oder die körperliche Integrität offen, ist dieser Hinweis aufgrund der erheblichen Bedeutung und des besonderen Vorrangs dieser Rechtsgüter zulässig. Eine rechtfertigende Notstandshandlung ist in dieser Konstellation zu bejahen.

ee. Bewertung

Ein wesentliches Überwiegen der Erhaltungsrechtsgüter gegenüber den Eingriffsrechtsgütern des § 17 UWG ist bei Nutzung eines Hinweisgebersystems und bei einer Mitteilung gegenüber Ermittlungsbehörden gegeben. Grundsätzlich ist die Benachrichtigung der Allgemeinheit und der Medien keine gerechtfertigte Notstandshandlung. Wendet sich der Whistleblower direkt an die Öffentlichkeit, steht dies grundsätzlich in keinem Verhältnis zu den drohenden Gefahren für die Rechtsgüter des § 17 UWG. Ausnahmen ergeben sich, wenn besonders bedeutende und schützenswerte Rechtsgüter im Spiel sind, wie etwa das Leben oder die körperliche Integrität.

§ 203 StGB und § 17 UWG sind nicht unmittelbar vergleichbar. Der stärkere Persönlichkeitsbezug bei § 203 StGB gebietet strengere Anforderungen an eine Rechtfertigung. Das bedeutet, dass ein Hinweisgeber, der in den „Täterkreis“ des § 203 StGB fällt, höhere Anforderungen erfüllen muss, um sein Handeln zu rechtfertigen. Zudem handelt es sich bei § 17 UWG um Nebenstrafrecht und die Geheimnisse des § 203 StGB sind durch ein prozessuales Zeugnisverweigerungsrecht in § 53 StPO zusätzlich geschützt. Daraus ergibt sich eine deutliche Abstufung zwischen den Geheimnisschutzvorschriften.[505] Im konkreten Einzelfall und in ihrem konkreten Ausmaß müssen sich die Erhaltungsrechtsgüter am Maßstab des § 138 StGB und

505 *Lutterbach*, Whistleblowing, S. 113.

des § 139 Abs. 3 StGB orientieren.[506] Nur bei bevorstehendem (wirtschaftlich) hohem Schaden kann eine gerechtfertigte Offenlegung von Hinweisgegenständen in Frage kommen.[507]

Ein anderes Ergebnis würde sich nur ergeben, wenn man die „illegalen Geheimnisse" aus dem Schutzbereich des § 17 UWG herausnähme. Und zwar nicht durch Auslegung, sondern durch eine Gesetzesänderung. Eine Kodifikation, die „gesetzeswidrige Geheimnisse" nicht unter den Tatbestand des Betriebs- und Geschäftsgeheimnisses fallen lässt, würde die derzeitige prekäre Lage auflösen. Im Hinblick auf den Whistleblowerschutz würde dies auch sinnvoll sein: Er hätte Gewissheit, dass er strafrechtlich nicht wegen der Offenlegung eines Gesetzesverstoßes verfolgt werden kann.

Notwendig ist dies jedoch nicht. Das Problem kann auch im Rahmen der Rechtfertigung gelöst werden, sofern die angesprochenen Maßstäbe angewendet werden. Denn auch wenn man „gesetzeswidrige Geheimnisse" aus dem Tatbestand des § 17 UWG ausschließen würde, ergäbe sich das Stufenverhältnis zwischen internem und externem Whistleblowing. Eine Reform des § 17 UWG wäre zwar wünschenswert und würde die Rechtslage vereinfachen. Ausschlaggebend ist, dass ein Ergebnis zugunsten des Whistleblowers erzielt wird. Prioritär ist die Straflosigkeit bei der Offenlegung von Gesetzesverstößen. Ob seine Handlung schon nicht tatbestandsmäßig oder (nur) gerechtfertigt ist, ist zweitrangig.

e. Subjektives Rechtfertigungselement

Subjektiv fordert § 34 StGB, das der Täter handelt, um die Gefahr von sich oder einem anderen Bedrohten abzuwenden, sog. Rettungswille. Dabei darf der Rettungswille nicht neben anderen Motiven völlig zurücktreten. Das heißt, dass auch andere Beweggründe eine Rolle spielen können. Beseitigt der Täter jedoch unwissend eine Gefahrenlage, so bleibt sein Verhalten rechtswidrig.[508] Dass Whistleblowing auch aus eigennützigen Motiven geschieht und nicht aus reinem Altruismus, ist deshalb nicht schädlich. Liegen die oben genannten Voraussetzungen vor, wird die Rechtfertigung im Regelfall nicht am subjektiven Element scheitern.

f. Gesamtresümee

Grundsätzlich kann der Whistleblower einer Strafbarkeit entkommen. Die Prüfung der Rechtfertigungskriterien und insbesondere die in jedem Fall individuell durchzuführende Abwägung kann jedoch nicht von einem Whistleblower erwartet werden. Der Whistleblower braucht klare gesetzliche Bestimmungen, die ihm

---

506 Im Ergebnis *Kargl*, in: Kindhäuser/Neumann/Paeffgen, Kommentar StGB, § 203 StGB, Rn. 66.

507 *Lutterbach*, Whistleblowing, S. 125. Die Aufdeckung einer langjährigen Schmiergeldpraxis oder einer sog. „schwarzen Kasse" erfüllt die Voraussetzungen, nicht aber Einzeltaten auf geringem monetärem Niveau.

508 *Kühl*, in: Lackner/Kühl, StGB-Kommentar, § 34 StGB, Rn. 5.

aufzeigen, wann Whistleblowing strafrechtlich bedenkenlos ist und welche Hinweisgegenstände zulässig sind. Insbesondere die Kriterien der Abwägung müssen in einer Regelung erkennbar gemacht werden, sodass er schon im Vorfeld entscheiden kann, ob das betreffende Rechtsgut überwiegt und eine interne oder externe Offenbarung angebracht ist. Eine gänzlich neue Vorschrift braucht es in der Hinsicht nicht – es genügt die Integrierung eines Hinweises auf die Rechtfertigungstatbestände und das Stufenverhältnis zwischen internem und externem Whistleblowing in einem Schutzgesetz, um die strafrechtliche Situation beim Whistleblowing zu verdeutlichen. Die Überlegung, § 17 UWG zu reformieren und den Gegenstand des Geheimnisses zu kodifizieren, kann eine Alternative sein. Würde gesetzlich vorgeschrieben, dass nur jenes Geheimnis unter den Geheimnisschutz des § 17 UWG fällt, das keinen Gesetzesverstoß zum Inhalt hat, wäre der Whistleblower straffrei. Die tatbestandliche Erfüllung von § 17 UWG wäre dann ausgeschlossen.

II. Schutz des Whistleblowers als Tatbeteiligter

In strafrechtlicher Hinsicht kommen bei einem zunächst an der Tat beteiligten Whistleblower neben dem oben genannten Straftatbeständen auch noch jene hinzu, die durch die Beteiligung verwirklicht sein können.

1. Strafbarkeit nach dem StGB

Wegen der Vielfalt können hier nicht alle in Betracht kommenden Delikte erörtert werden. Von der Körperverletzung bis hin zu Umweltstraftaten kann alles verwirklicht sein. Auch das Kartellstrafrecht und sein Sonderstraftatbestand § 298 StGB[509] ist zu beachten. Zur Verwirklichung des Tatbestandes bedarf es der Abgabe eines Angebots in einem Ausschreibungsverfahren über Waren oder gewerbliche Dienstleistungen, die auf einer rechtswidrigen Absprache beruht.[510] Die Absprache muss darauf abzielen, den Veranstalter des Vergabeverfahrens zur Annahme eines bestimmten Angebots zu veranlassen. Die wettbewerbsbeschränkende Absprache selbst bildet als Zuwiderhandlung gegen das Verbot wettbewerbsbeschränkender Vereinbarungen eine Kartellordnungswidrigkeit gem. § 81 GWB. Neben wettbewerbsbeschränkenden Absprachen bei Ausschreibungen (§ 298 StGB) ist die besondere Art des Betrugs in Form des Submissionsbetrugs nach § 263 StGB hervorzuheben.[511] Die Täuschung des Veranstalters eines Vergabeverfahrens besteht in der Angebotsabgabe durch den Veranstalter. Mit der Angebotsabgabe ist

509 § 298 StGB sieht vor, dass sich derjenige strafbar macht, der bei einer Ausschreibung über Waren oder gewerbliche Leistungen ein Angebot abgibt, das auf einer Absprache beruht, die den Ausschreibenden zur Annahme eines bestimmten Angebots veranlassen soll.

510 *Kühl*, in: Lackner/Kühl, StGB-Kommentar, § 298 StGB, Rn. 3 f.

511 Ausführlich dazu *Klusmann*, in: Wiedemann, Kartellrecht, § 56, Rn. 12 f.

die konkludente Erklärung verbunden, dass dieses Angebot ohne eine vorherige Absprache zwischen den Bietern zustande gekommen ist.[512]

Vorrangig ist aber auf die Besonderheiten der Täterschaft und Teilnahme einzugehen, da der Whistleblower oft in seiner Position als Arbeitnehmer eine strafrechtlich eher „untergeordnete Rolle“ einnimmt und selten als Haupttäter agiert.

Der Whistleblower muss zunächst ein geeigneter Täter sein. Problematisch ist dies bei sog. eigenhändigen Delikten[513] und bei Sonderdelikten.[514] Bei Sonderdelikten kommt der Whistleblower nur als Täter in Betracht, wenn er eine bestimmte Eigenschaft aufweist, also eine bestimmte Täterqualifikation hat. Diese ergibt sich in den meisten Fällen aus einer außerstrafrechtlichen Pflichtenstellung.[515] Die allgemeine Täterschaft ergibt sich unmittelbar aus der Tatbestandsverwirklichung. Wer die mit Strafe bedrohte Handlung also selbst begeht (soweit möglich) und in seiner Person sämtliche Merkmale des objektiven und subjektiven Unrechtstatbestandes erfüllt, ist Täter im Sinne von § 25 Abs. 1 Alt. 1 StGB.[516] Mittäterschaft ist hingegen die gemeinschaftliche Begehung einer Straftat durch bewusst gewolltes Zusammenwirken.[517] Bei der Teilnahme besteht die Besonderheit, dass der Whistleblower mit

---

512 BGH, Urt. v. 11.7.2001 – 1 StR 576/00, NJW 2001, 3718, 3719; worin die Vermögensverfügung zu sehen ist (Eingehungsbetrug vs. Erfüllungsbetrug) wird nicht einheitlich entschieden, vgl. *Achenbach*, in: Achenbach/Ransiek, Wirtschaftsstrafrecht, S. 338 f. m. w. N.

513 Bei den eigenhändigen Delikten ergibt nach herkömmlicher Auffassung die Auslegung des betreffenden Tatbestands, dass *nur die eigenhändige Vornahme der strafbaren Handlung* den Unwert des Delikts realisiert (z. B. § 154 Abs. 1 StGB, § 173 Abs. 1 StGB, § 174 Abs. 1 StGB, § 176 Abs. 1 StGB, § 179 Abs. 1 StGB, § 182 Abs. 1 StGB, § 316 Abs. 1 StGB, § 323a Abs. 1 StGB, § 339 StGB) *Heintschel-Heinegg*, in: BeckOK StGB, Lexikon, Deliktstypen, G, Rn. 15. Wer die Ausführungshandlung nicht persönlich vornimmt, kann nicht Täter, Mittäter oder mittelbarer Täter, sondern nur Teilnehmer sein *Wessels/Beulke/Satzger*, StrafR AT, Rn. 40.

514 *Echte Sonderdelikte* sind Straftaten, die überhaupt nur ein Sonderpflichtiger verwirklichen kann, z. B. § 331 Abs. 1 StGB, § 339 StGB. Die Sonderpflicht wirkt bei dieser Deliktsgruppe *strafbegründend*, vgl. § 28 Abs. 1 StGB. *Unechte Sonderdelikte* sind Allgemeindelikte, die für den Fall, dass sie durch einen Sonderpflichtigen verwirklicht werden, besondere Rechtsfolgen vorsehen. Zumeist wird der Strafrahmen erhöht. Die Sonderpflicht wirkt *strafschärfend*, vgl. § 28 Abs. 2 StGB, *Wessels/Beulke/Satzger*, StrafR AT, Rn. 39.

515 *Heintschel-Heinegg*, in: BeckOK StGB, Lexikon, Deliktstypen, Q, Rn. 37.

516 *Wessels/Beulke/Satzger*, StrafR AT, Rn. 508.

517 Jeder Beteiligte ist als „gleichberechtigter Partner“ Mitträger des gemeinsamen Tatentschlusses und der gemeinsamen Tatbestandsverwirklichung, so dass die einzelnen Tatbeiträge sich zu einem einheitlichen Ganzen vervollständigen und der Erfolg jedem voll zuzurechnen ist, *Wessels/Beulke/Satzger*, StrafR AT, Rn. 526 m. w. N.

einem sog. Doppelvorsatz[518] handeln muss. Demnach reicht es nicht aus, dass der Whistleblower dem Haupttäter bewusst nur bei dessen vorsätzlicher Tat behilflich ist. Er muss sowohl bei seiner Gehilfenhandlung als auch bezüglich der Haupttat mit Vorsatz handeln.[519] Nur dann kommt eine Teilnahme in Betracht.

Neben der Möglichkeit des Rücktritts sehen einzelne Vorschriften selbst bei vollendeter Straftat die Erlangung von Straffreiheit durch tätige Reue vor – andere Vorschriften eröffnen dem Gericht die Möglichkeit, die Strafe zu mildern oder ganz von Strafe abzusehen.[520] Abgesehen vom klassischen Rücktritt, der nur möglich ist, wenn der Versuch nicht fehlgeschlagen oder das Delikt nicht vollendet[521] ist,[522] stellt sich hier die Frage, ob schon das Handeln des Whistleblowers den Ausschluss der Strafbarkeit rechtfertigen kann. Aufgrund der gefestigten strafrechtlichen Dogmatik und der engen Voraussetzungen einer Rücktrittshandlung, aber auch aufgrund der „Neuheit" des Phänomens Whistleblowing im deutschen Rechtsraum gibt es darauf noch keine Antwort.

2. Schutz durch das StGB

Anders als bei der Prüfung der Rechtfertigung des unbeteiligten Whistleblowers ist hier vor allem zu berücksichtigen, dass der Whistleblower zunächst bewusst als (Mit-)Täter oder Teilnehmer eine (andere) Tat ausgeführt hat – ohne Rechtfertigungsgrund – und im Nachhinein Strafbefreiung durch das Whistleblowing erlangen will. Diese ist nur denkbar durch einen Rücktritt oder tätige Reue.

---

518 *Joecks*, in: MüKo StGB, § 26 StGB, Rn. 54; *Joecks*, in: MüKo StGB, § 27 StGB, Rn. 90, der in beiden Fällen der Meinung ist, dass die Begrifflichkeit „Doppelvorsatz" überflüssig ist.

519 Teilnehmer ist, wer einen anderen zu dessen vorsätzlich begangener rechtswidriger Tat vorsätzlich bestimmt (Anstiftung, § 26) oder ihm zu einer solchen Tat vorsätzlich Hilfe leistet (Beihilfe, § 27) *Wessels/Beulke/Satzger*, StrafR AT, Rn. 505.

520 Bsp. für tätige Reue: § 139 Abs. 4 § 264 Abs. 5, § 264a Abs. 2, § 265b Abs. 2, § 306e Abs. 2, § 314a Abs. 3, § 298 Abs. 3 StGB; Bsp. für Strafmilderung: § 83a, § 98 Abs. 2, § 142 Abs. 4, § 261 Abs. 10, § 306e Abs.1 StGB.

521 Eine Ausnahme bildet § 24 Abs. 2 Satz 2 Alt. 2 StGB, der diejenigen Fälle erfasst, in denen der Tatbeitrag des Zurücktretenden zwar für den Versuch, aber nicht mehr für die Vollendung kausal geworden ist: sei es, dass der Beteiligte seinen Beitrag im Versuchsstadium vollständig neutralisiert hat, oder dass der Beitrag von den übrigen Beteiligten nicht benutzt wird bzw. sich als überflüssig erweist, näher dazu *Eser/Bosch*, in: Schönke/Schröder, Kommentar StGB, § 24 StGB, Rn. 97 f.

522 *Wessels/Beulke/Satzger*, StrafR AT, Rn. 627, 628; die Rücktrittsregelung von § 24 Abs. 1 S. 1 StGB beruht auf der Grundunterscheidung von unbeendetem und beendetem Versuch mit jeweils unterschiedlichen und daher jeweils eindeutig festzustellenden Erfordernissen, vgl. *Eser/Bosch*, in: Schönke/Schröder, Kommentar StGB, § 24 StGB, Rn. 6.

a. Der Rücktritt

Strafbefreiung durch Rücktritt vom Versuch ist nach § 24 Abs. 1 StGB nur zu erlangen, wenn die Tat nicht vollendet, der objektive Unrechtstatbestand also nicht vollständig verwirklicht wird.

Kommt es trotz der Rücktrittshandlung bzw. der Rücktrittsbemühungen des Täters zum Eintritt des tatbestandsmäßigen Erfolges, kann eine Straflosigkeit nur in Betracht kommen, wenn die objektive Zurechenbarkeit des konkreten Erfolges zu verneinen ist.[523]

Sollte es doch dazu kommen, dass die Tat im Versuchsstadium steckenbleibt, weil noch nicht alles Erforderliche zur Tatvollendung getan wurde, kann eine Strafbefreiung in Frage kommen, wenn der Täter freiwillig die weitere Ausführung der Tat aufgibt oder deren Vollendung verhindert (§ 24 Abs. 1 S. 1) oder wenn er sich freiwillig und ernsthaft bemüht, die Vollendung zu verhindern (§ 24 Abs. 1 S. 2). Zu beachten ist hier aber wiederum, dass diese Art des Rücktritts nur möglich ist, wenn es sich um einen Alleintäter handelt. Sind mehrere an der Tat beteiligt, wie es beim Whistleblower als „unterstützender" Täter im Rahmen des Arbeitsverhältnisses regelmäßig der Fall sein wird, so wird der rücktrittswillige Beteiligte – Täter oder Teilnehmer – nach § 24 Abs. 2 StGB nicht bestraft, wenn er freiwillig die Vollendung der Tat, an deren Versuch er beteiligt war, verhindert oder wenn er sich ernsthaft um die Verhinderung der Vollendung bemüht, falls die Tat ohne sein Zutun nicht vollendet oder unabhängig von seinem früheren Tatbeitrag begangen wird. Die Verhinderung ist auch durch Unterlassen weiterer Mitarbeit möglich, wenn ohne Mitwirkung des Beteiligten die Vollendung der Tat unmöglich wird oder wenn nur der Zurücktretende die Vollendung noch herbeiführen könnte.[524]

Für den Whistleblower wäre also ein Rücktritt noch möglich, wenn der Rechtsverstoß ohne sein Zutun gar nicht verwirklicht werden könnte, z. B. die Umweltverschmutzung nicht stattfindet, weil der Whistleblower den von ihm vorbereiteten Abfall nach Anweisung seines Vorgesetzten kurz vor der Entsorgung doch nicht unsachgemäß entsorgt und im Zuge dessen lieber die Umweltbehörden informiert, um die Tat nicht durch einen anderen Mitarbeiter zur Vollendung kommen zu lassen. Aber wie sich zeigt, sind solche Fälle schwer zu konstruieren, weil die Tat des Whistleblowers selbst bereits das Versuchsstadium erreicht haben muss, um zurücktreten zu können.

Der Rücktritt im Vorbereitungsstadium ist nach den allgemeinen Zurechnungs- und Teilnahmegrundsätzen zu behandeln. Danach erlangt derjenige keine Strafbefreiung, der eine Straftat mitgeplant und bereits durch eigene Handlungen gefördert hat und sich noch im Vorbereitungsstadium vom gemeinsam Vorhaben lossagt, wenn die übrigen Beteiligten die geplante Tat ohne ihn ausführen und dabei aus

523 *Wessels/Beulke/Satzger*, StrafR AT, Rn. 627; *Rudolphi*, in: SK-StGB, § 24 StGB, Rn. 16, 28.

524 *Otto*, Jura 1992, 423, 430; *Kudlich*, JuS 1999, 449, 450; *Kühl*, in: Lackner/Kühl, StGB-Kommentar, § 24 StGB, Rn. 25.

seinen fortwirkenden Tatbeiträgen Nutzen ziehen.[525] Der Rücktrittswillige muss seinen Tatbeitrag vollständig rückgängig machen – denn durch das „Überbleibsel" seines Beitrags können die anderen Beteiligten immer noch die Tat zur Vollendung bringen. Eine Strafbarkeit wegen Beihilfe kann die Folge sein.[526]

Angenommen, der Whistleblower hat die Schwelle zum Versuch überschritten und kann wirksam zurücktreten, stellt sich die Frage nach der Freiwilligkeit seines Handelns. Die Freiwilligkeit stellt ein wichtiges Kriterium dar und ist nur gegeben, wenn der Rücktrittswillige nicht durch zwingende Hinderungsgründe veranlasst wird, sondern der Rücktritt eigenen autonomen Motiven des Täters entspringt.[527] Als Motive kommen dabei Gewissensbisse, Reue, Scham, Mitleid mit dem Opfer, seelische Erschütterung und Angst vor Strafe in Betracht.[528]

Da der Whistleblower regelmäßig handelt, um die Allgemeinheit zu informieren und sie vor Schäden materieller und immaterieller Art zu schützen, kann nicht davon ausgegangen werden, dass der Whistleblower nicht aus seiner eigenen Entscheidung heraus handelt und nicht „Herr seiner Entschlüsse"[529] ist.

Fraglich ist, ob durch die Offenlegung des Whistleblowers die Tat eigenständig verhindert werden muss oder ob es genügt, wenn erst durch das Eingreifen der Behörden, die den Hinweis erhalten haben, die Vollendung der Tat verhindert werden kann. Zwar ist der Rücktritt ein persönlicher Strafaufhebungsgrund, dennoch kann nicht vom Rücktrittswilligen verlangt werden, dass er ohne technische oder menschliche Hilfe die Tat verhindert.[530] Er verfügt meist gar nicht über die nötigen Mittel – das kann ihm jedoch nicht vorgeworfen werden. Erforderlich ist das ernsthafte Bemühen um die Vollendungsverhinderung oder das Unschädlichmachen des Tatbeitrages sowie die Freiwilligkeit.

Je nach Sachverhalt ist es durchaus möglich, dass der Whistleblower durch die Offenbarung der Missstände oder des Gesetzesverstoßes, an denen er auch beteiligt war, noch zurücktreten kann. In vielen bekannt gewordenen Fällen war der Whistleblower jedoch entweder Unbeteiligter oder die Tat wurde schon vollendet und der Whistleblower versuchte, durch die Offenlegung ein Wiederholen für die Zukunft zu verhindern.

---

525 *Zaczyk*, in: Kindhäuser/Neumann/Paeffgen, Kommentar StGB, § 24 StGB, Rn. 117.

526 BGH, Beschl. v. 11.3.1999 – 4 StR 56/99, NStZ 2000, 449, 450.

527 *Herzberg/Hoffmann-Holland*, in: MüKo StGB, § 24 StGB, Rn. 117.

528 *Zaczyk*, in: Kindhäuser/Neumann/Paeffgen, Kommentar StGB, § 24 StGB, Rn. 73 mit Beispielen und weiteren Nachweisen.

529 Siehe dazu BGH, Urt. v. 17.12.1992 – 4 StR 532/92, NStZ 1993, 279, 279.

530 Die Tatverhinderung schließt die Einbeziehung Dritter nicht aus, solange die Nichtvollendung dem Zurücktretenden zurechenbar ist *Beckemper*, in: BeckOK StGB, § 24 StGB, Rn. 64.

b. Die tätige Reue

Ist das Delikt vollendet, so ist § 24 StGB nicht anwendbar. Einzelne Vorschriften verlagern den Vollendungszeitpunkt sehr weit nach vorne, sehen aber dafür bei vollendeter Straftat die Erlangung von Straffreiheit durch sog. tätige Reue vor. In anderen Vorschriften wird dem Gericht in diesen Fällen die Möglichkeit eröffnet, die Strafe zu mildern oder von Strafe abzusehen. Tätige Reue ist nur möglich, wenn das Gesetz dies vorsieht.

Eine relevante Vorschrift ist dabei § 298 Abs. 3 StGB.[531] Straflos bleibt danach, wer freiwillig die Angebotsannahme durch den Veranstalter oder die spätere Leistungserbringung durch den Auftraggeber verhindert (§ 298 Abs. 3 S. 1) bzw. sich bei Ausbleiben der Annahme des Angebots oder der Erbringung der Leistung aus anderen Gründen freiwillig und ernsthaft bemüht, die Annahme des Angebots oder das Erbringen der Leistung des Ausschreibenden zu verhindern (§ 298 Abs. 3 S. 2).

Die tätige Reue hat einen sehr weiten Anwendungsbereich. Auch wenn die Schädigung des Wettbewerbs schon eingetreten ist und nicht mehr rückgängig gemacht werden kann, ist sie noch möglich, und zwar dann, wenn die Leistungserbringung in Form von Geldzahlung nach Durchführung des Auftrags verhindert wird.[532] Deshalb ist einschränkend zu verlangen, dass der Täter die Leistungserbringung durch den Veranstalter vollständig verhindert; eine bereits erfolgte Teilzahlung versperrt demzufolge den Weg des Täters zur Strafbefreiung.[533]

Im Hinblick auf wettbewerbsbeschränkende Absprachen nach § 298 StGB, die notwendigerweise mehrere Beteiligte voraussetzen, ist immer an die Regelung des § 24 Abs. 2 StGB zu denken, deren Grundsätze entsprechend auf die tätige Reue anzuwenden sind.[534] Die Definitionen der Merkmale „Freiwilligkeit", „Verhindern" und „ernsthaftes Bemühen" sind ebenfalls aus § 24 StGB abzuleiten. Die Regelung des § 298 Abs. 3 StGB gilt nur für Straftaten nach § 298 StGB und nicht für Betrug, Urkundenfälschung und andere Straftaten.[535]

Bei Submissionsabsprachen als Kartellrechtsverstoß kann daher trotz Vollendung Straffreiheit zu erlangen sein. Jedoch ist diese Möglichkeit nur auf Submissionsabsprachen begrenzt, andere wettbewerbswidrige Delikte sind nicht erfasst. Da Submissionskartelle regelmäßig dem Ziel dienen, den Preiswettbewerb außer Kraft zu setzen, liegt die Sozialschädlichkeit nicht primär in dem betrügerisch

---

531 Erläuterungen sind ähnlich wie bei den betrugsähnlichen Straftatbeständen §§ 264 Abs. 4, 264 a Abs. 3, 265 b Abs. 2 StGB (Subventions-, Kapitalanlage- und Kreditbetrug), die auch eine Regelung für tätige Reue enthalten.

532 *Dannecker*, in: Kindhäuser/Neumann/Paeffgen, Kommentar StGB, § 298 StGB, Rn. 85.

533 *Dannecker*, in: Kindhäuser/Neumann/Paeffgen, Kommentar StGB, § 298 StGB, Rn. 85 m. w. N.

534 *Hohmann*, in: MüKo StGB, § 298 StGB, Rn. 116; *Kühl*, in: Lackner/Kühl, StGB-Kommentar, § 298 StGB, Rn. 8.

535 *Dannecker*, in: Kindhäuser/Neumann/Paeffgen, Kommentar StGB, § 298 StGB, Rn. 87.

herbeigeführten Vermögensschaden des Ausschreibenden, sondern in den volkswirtschaftlichen Nachteilen solcher Absprachen.[536] Obwohl die „normalen" Preiskartelle ebenfalls ein hohes Risiko für die Gesellschaft bergen, sieht der Gesetzgeber hier keine Vergleichbarkeit, sonst wäre hierfür ein ähnlicher Tatbestand geschaffen worden. Auch eine analoge Anwendung wird nicht diskutiert. Demnach bleibt kein Raum für die Anwendung der Grundsätze der tätigen Reue auf andere Kartelltypen.

Der Whistleblower, der ehemals als Kartellant tätig war und durch die Offenlegung des Preiskartells dessen Beendigung anstrebt bzw. weiteres Handeln für die Zukunft unterbinden will, ist strafrechtlich nicht geschützt. Das könnte insbesondere daran liegen, dass die Begehung wettbewerbswidriger Handlungen in Form eines Kartells – außer in Form eines Submissionskartells – nicht strafbar im Sinne des Strafgesetzbuches ist. Neben der strafrechtlichen Verfolgung nach § 298 StGB besteht, soweit der Täter für ein Unternehmen oder eine Unternehmensvereinigung handelte, die Möglichkeit der Auferlegung einer durch die deutschen Kartellbehörden festzusetzenden Unternehmensgeldbuße nach § 81 GWB i. V. m. §§ 9, 30, 130 OWiG.

3. Ahndung nach dem OWiG

Hat der an dem Missstand beteiligte Whistleblower keine Straftat, sondern eine Ordnungswidrigkeit verübt, ist das OWiG heranzuziehen. Konsequenzen für den unbeteiligten Whistleblower folgen aus dem OWiG nicht. Die Unterscheidung zwischen Straftaten und Ordnungswidrigkeiten bestimmt sich danach, ob das Gesetz als Rechtsfolge Strafe oder Geldbuße androht bzw. die Tat als ordnungswidriges Verhalten bezeichnet. Wird Strafe angedroht, handelt es sich um eine Straftat. Wird eine Geldbuße angedroht, handelt es sich um eine Ordnungswidrigkeit.

Ist ein missbilligtes Verhalten von einem geringeren sozialethischen Gewicht, kann der Gesetzgeber dieses Verhalten als Ordnungswidrigkeit bezeichnen und in einem gesonderten Gesetz regeln.[537]

Daneben hält das Gesetz eine bedeutsame Zurechnungsvorschrift bereit. Diese ist insofern von besonderer Bedeutung, als nicht nur natürliche Personen, sondern auch Unternehmen Whistleblowing betreiben können. Insbesondere ist hierbei auf Unternehmen einzugehen, die an einem Kartell beteiligt sind. Obwohl eine juristische Person oder eine Personenvereinigung nicht selbst handeln kann, sondern nur mittels ihrer Organe oder Vertreter, und somit nicht als Rechtssubjekt[538] im Sinne des klassischen Strafrechts in Frage kommt, gestattet § 30 OWiG die Verhängung

536 *Dannecker*, in: Kindhäuser/Neumann/Paeffgen, Kommentar StGB, § 298 StGB, Rn. 7.

537 *Bohnert*, OWiG Kommentar, § 1, Rn. 4; der Gesetzgeber hat die „allgemeinen" Spielregeln zur Verfolgung und Ahndung von Ordnungswidrigkeiten im OWiG festgehalten. Besondere ordnungswidrigkeitenrechtliche Tatbestände finden sich in den jeweiligen Gesetzen, wie z. B. § 81 GWB.

538 Im Mittelpunkt des Rechtsgeschehens steht der Mensch als Rechtssubjekt. Anknüpfungspunkt im Strafrecht ist somit das mit sozialschädlichen Folgen verbundene menschliche Verhalten *Wessels/Beulke/Satzger*, StrafR AT, Rn. 81 f.

einer Geldbuße gegen diese.[539] Den betroffenen juristischen Personen und Personenvereinigungen wird das Handeln ihrer Organe und Vertreter zugerechnet.[540] Nach Abs. 1 Nr. 1 des § 30 OWiG, der auch den Täterkreis nennt, ist die Verletzung einer betriebsbezogenen Pflicht vonnöten. Als Bezugstat kommt nur eine Straftat oder Ordnungswidrigkeit in Betracht, durch die Pflichten verletzt werden, welche die juristische Person oder die Personenvereinigung trifft – insbesondere zählen dazu alle in § 81 GWB genannten Ge- und Verbote.

Eine vergleichbare Zurechnung des Handelns von Repräsentanten findet in § 9 OWiG statt. Die Vorschrift regelt in sachlicher Übereinstimmung mit § 14 StGB die sog. Organ- und Vertreterhaftung und führt zu einer Ausweitung der bußgeldrechtlichen Haftung auf Personen, die für den eigentlichen Normadressaten handeln.[541] Normadressaten des § 9 Abs. 1 Nr. 1 OWiG sind ausschließlich die vertretungsberechtigten Organe der juristischen Person sowie die Mitglieder eines solchen Organs. Im Gegensatz zum Strafrecht gilt im OWiG der Begriff des Einheitstäters, was bedeutet, dass nicht zwischen Täterschaft und Teilnahme unterschieden wird, sondern jeder bußgeldrechtlich belangt werden kann, auch wenn das besondere persönliche Merkmal (§ 9 OWiG) nur bei einem Beteiligten vorliegt.[542] Nach § 9 OWiG ist deshalb ein Gesetz, nach dem besondere persönliche Merkmale die Strafbarkeit begründen, auf bestimmte Vertreter und Beauftragte anzuwenden, auch wenn diese Merkmale nicht bei ihnen, aber bei dem Vertretenen bzw. dem Betriebsinhaber vorliegen.[543] Beteiligt sich jemand an einem Sonderdelikt, der selbst weder die von dem Tatbestand geforderte Rolle einnimmt, noch über die Merkmalszurechnung nach § 9 OWiG an ihm teilhat, kommt es darauf an, ob die Haupttat eine Ordnungswidrigkeit oder eine Straftat ist. Das Ordnungswidrigkeitenrecht schreibt eine umfassende Zurechnung unter den Beteiligten vor. Es kommt nur darauf an, dass überhaupt (unmittelbar oder über § 9 OWiG) ein Sonderpflichtiger an der Tat beteiligt ist.[544]

---

539 *Bohnert*, OWiG Kommentar, § 30, Rn. 1, 5. § 30 überträgt das Delikt einer natürlichen Person auf eine JP oder PV, als hätte diese das Delikt begangen. Die natürliche Person muss hierfür einen Deliktstatbestand erfüllt haben und Repräsentant der JP oder PV sein und eine betriebsbezogene Pflicht verletzt haben.

540 *Cramer/Pananis*, in: Loewenheim/Meessen/Riesenkampff, Kartellrecht, § 81 GWB, Rn. 35.

541 *Bohnert*, OWiG Kommentar, § 9, Rn. 1.

542 *Cramer/Pananis*, in: Loewenheim/Meessen/Riesenkampff, Kartellrecht, § 81 GWB, Rn. 13.

543 So z. B. bei der Vorenthaltung von Sozialversicherungsbeitragen gem. § 266 a Abs. 1 StGB: Das Tätermerkmal „Arbeitgeber" ist auf bestimmte Leitungspersonen eines Unternehmens anzuwenden, wenn eigentlich der Arbeitsvertrag mit dem Rechtsträger des Unternehmens geschlossen worden ist, wenn aber diese Leistungsperson in ihrer Leitungsfunktion die tatbestandsmäßige Handlung des Vorenthaltens ausgeführt hat, *Achenbach*, in: Achenbach/Ransiek, Wirtschaftsstrafrecht, S. 26.

544 *Achenbach*, in: Achenbach/Ransiek, Wirtschaftsstrafrecht, S. 31.

Das StGB enthält hingegen in § 28 StGB eine Regelung, die die Strafe des Teilnehmers nach Maßgabe des § 49 Abs. 1 StGB mildern kann, wenn strafbegründende besondere persönliche Merkmale des Täters beim Teilnehmer fehlen.[545]

Über § 130 OWiG kann darüber hinaus gegen den Inhaber eines Betriebs wegen einer vorsätzlich oder fahrlässig begangenen Aufsichtspflichtverletzung eine Geldbuße verhängt werden. § 130 OWiG ist also ein echtes Unterlassungsdelikt. Zwar lässt die Vorschrift weitgehend offen, was im Einzelnen von der Aufsichtspflicht umfasst ist, jedoch sind nach Abs. 1 Satz 2 dieser Vorschrift „auch" die Bestellung, sorgfältige Auswahl und Überwachung von Aufsichtspersonen gemeint. Eine Konkretisierung der Pflicht ist nur nach den Besonderheiten des Einzelfalls möglich.[546] Die Details der Aufsichtspflichtverletzung waren häufig Gegenstand der Rechtsprechung des BGH und der Oberlandesgerichte zum Kartellordnungswidrigkeitenrecht nach dem GWB.[547] Die Besonderheit dieser Norm liegt darin, dass die Haftung des Unternehmens nicht an die Vorschrift anknüpft, die im Unternehmen konkret verletzt wurde, sondern an das Unrecht, dass aus der Verletzung einer besonderen Aufsichtspflicht resultiert. Der objektive Tatbestand des § 130 OWiG setzt daher voraus, dass die zur Verhinderung von Zuwiderhandlungen gegen betriebs- und unternehmensbezogene Pflichten erforderlichen und zumutbaren Aufsichtsmaßnahmen[548] nicht vorgenommen werden. Der Vorsatz bzw. die Fahrlässigkeit muss sich dementsprechend nur darauf beziehen. Die Regelung fußt auf dem Gedanken, dass die Betriebsinhaber bzw. die vertretungsberechtigten Organe der juristischen Personen regelmäßig nicht alle Aufgaben im Unternehmen persönlich bewältigen können und daher Aufgaben an Mitarbeiter delegieren dürfen. Die verantwortlichen Personen müssen aber verpflichtet bleiben, angemessene Aufsichts- und Kontrollmaßnahmen zu schaffen, die die Pflichterfüllung sicherstellen.[549] Als Adressaten der Zurechnung des besonderen persönlichen Merkmals der Unternehmensinhaberschaft sind taugliche Täter auch die dort angeführten Organe, Organmitglieder, Vertreter und Beauftragten. Ziel der in § 130 OWiG statuierten Aufsicht muss es sein, Zuwiderhandlungen gegen solche, den Inhaber treffende Pflichten zu verhindern, deren Verletzung mit Strafe oder Geldbuße bedroht ist.[550]

---

545 *Achenbach* verweist in dem Zusammenhang auf den rechtswissenschaftlichen Streit über den Begriff des „besonderen persönlichen Merkmals", der sich nicht mit § 14 StGB deckt *Achenbach*, in: Achenbach/Ransiek, Wirtschaftsstrafrecht, S. 31.

546 Zu den in Betracht kommenden einzelnen Aufsichtsmaßnahmen *Achenbach*, in: Achenbach/Ransiek, Wirtschaftsstrafrecht, S. 39.

547 OLG Düsseldorf, Urt. v. 5.4.2006 – VI-2 Kart 5/05 OWi, VI-2 Kart 6/05 OWi, 2 Kart 5/05 OWi, 2 Kart 6/05 OWi, WUW/E DE-R 1893, 1896 f.; BGH, Urt. v. 24.3.1987 – KRB 8/86, WUW/E BGH 2394, 2397.

548 Siehe zu den Aufsichtspflichten *Cramer/Pananis*, in: Loewenheim/Meessen/Riesenkampff, Kartellrecht, § 81 GWB, Rn. 29 f.

549 *Cramer/Pananis*, in: Loewenheim/Meessen/Riesenkampff, Kartellrecht, § 81 GWB, Rn. 25 f.

550 *Wittig*, Wirtschaftsstrafrecht, S. 80.

Die Bemessung der Geldbuße bestimmt sich nach § 17 OWiG. § 17 Abs. 1 OWiG betont, dass grundsätzlich die im jeweiligen Gesetz angedrohte Geldbuße zu Grunde zu legen ist. Der Regelung des Verfalls nach § 29 a OWiG kommt lediglich subsidiärer Charakter zu, weil der wirtschaftliche Vorteil, den der Täter oder ein Dritter aus der Ordnungswidrigkeit gezogen hat, in der Regel bereits nach § 17 Abs. 4 bzw. § 30 Abs. 3 OWiG abgeschöpft wird.

Die Zurechnung von Zuwiderhandlungen in Unternehmen geschieht also auf drei Ebenen:[551]

- auf der ersten Ebene erfolgt die Sanktionierung der individuell handelnden Person,
- auf der zweiten Ebene wird die für die Tat des unmittelbar Handelnden verantwortliche Person in der Unternehmensleitung zur Rechenschaft gezogen,
- auf der dritten Ebene wird das Unternehmen selbst sanktioniert.

Neben den allgemeinen Regeln des Ordnungswidrigkeitenrechts (OWiG) gibt es Gesetze, die Ordnungswidrigkeiten für bestimmte Bereiche regeln. Da diese Arbeit auch den Schutz des kartellrechtlichen Whistleblowers untersucht, muss auf die ordnungswidrigkeitenrechtlichen Folgen nach dem GWB eingegangen werden. Die Bußgeldtatbestände des § 81 GWB enthalten überwiegend Sonderdelikte, bei denen die bußgeldbewehrten Pflichten von vornherein nur juristischen Personen und Personenvereinigungen auferlegt sind, die selbst nicht in sanktionsrechtlich relevanter Weise handeln können. Gem. § 81 Abs. 1 i. V. m Abs. 4 GWB wird ein Verhalten mit einer Kartellordnungswidrigkeit bedroht, das gegen Art. 101 oder Art. 102 AEUV[552] verstößt. Gegen die in diesen Vorschriften niedergelegten Grundsätze verstößt, wer als „Unternehmen" rechtswidrige Vereinbarungen trifft oder eine beherrschende Stellung missbräuchlich ausnutzt. Für das deutsche Kartellrecht gelten nahezu die gleichen Grundsätze. Die Tatbestände der Zuwiderhandlungen sind in § 81 Abs. 2 und 3 GWB zu finden. Neben der Zuwiderhandlung gegen das Kartellverbot und gegen das Verbot der Ausnutzung einer marktbeherrschenden Stellung zählt § 81 Abs. 2 Nr. 1 GWB weitere Tatbestände einer Zuwiderhandlung auf.[553]

Auch für Unternehmen besteht also ein Risiko, ordnungswidrigkeitenrechtlich belangt zu werden. Durch die Vielfalt der Zurechnungsvorschriften, die auch im Kartellordnungswidrigkeitenrecht gelten, kann jeder Beteiligte – gleich in welcher Form er gehandelt hat – bußgeldrechtlich zur Rechenschaft gezogen werden. Denn es darf nicht vergessen werden, dass auch ein Unternehmen oder Repräsentanten als Whistleblower agieren können. Bei maßgeblicher Tatbeteiligung kann das betroffe-

551 Vgl. *Achenbach*, JuS 1990, 601, 602.

552 In Art. 101 und 102 AEUV sind die europäischen Wettbewerbsgrundsätze (Kartellverbot, missbräuchliche Ausnutzung einer beherrschenden Stellung) verankert, die dem deutschen Recht gleichen.

553 Ausführlich dazu *Achenbach*, in: Achenbach/Ransiek, Wirtschaftsstrafrecht, S. 359 f; Zuwiderhandlungen gegen vollziehbare Anordnungen regelt Abs. 2 Nr. 2.

ne Unternehmen durch das Offenlegen des Sachverhalts selbst sanktioniert werden. Da das Unternehmen strafrechtlich nicht selbst handeln kann, kann es auch nicht bestraft werden. Es verbleibt nur die Verhängung eines Bußgeldes. Ob eine schärfere Strafandrohung zu einer veränderten Handlungspraxis innerhalb der Unternehmen führt oder ob ein Unternehmensstrafrecht nützlich und rechtspolitisch notwendig ist, wird diskutiert.[554]

4. *Ordnungswidrigkeitenrechtlicher Schutz*

Die Verfolgung und Ahndung nach dem OWiG betrifft sowohl die natürliche Person als Whistleblower als auch das Unternehmen, das wie ein Whistleblower agiert. Dies gilt natürlich nur, soweit der Whistleblower in irgendeiner Weise an den Gesetzesverstößen beteiligt war und eine Ordnungswidrigkeit begangen wurde. Fraglich ist, ob durch das Whistleblowing und damit das nachträgliche Wohlverhalten dem Whistleblower auch hier Amnestie zukommen kann.

a. *Allgemeines Ordnungswidrigkeitenrecht*

Nach § 1 OWiG ist eine Ordnungswidrigkeit eine rechtswidrige und vorwerfbare Handlung, die den Tatbestand eines Gesetzes verwirklicht, das die Ahndung mit einer Geldbuße zulässt. Wie das Strafrecht hält auch das Ordnungswidrigkeitenrecht besondere strafbefreiende Tatbestände bereit.

Nach § 15 OWiG ist eine durch Notwehr gebotene Handlung nicht rechtswidrig. § 15 Abs. 2 definiert Notwehr als die erforderliche Verteidigungshandlung, um einen gegenwärtigen rechtswidrigen Angriff von sich oder einem anderen abzuwenden. Der rechtfertigende Notstand nach § 16 OWiG stellt einen zweiten Rechtfertigungsgrund dar. Die Regelungen entsprechen den § 32 und § 34 StGB.[555] Es handelt hiernach nicht rechtswidrig, wer in einer gegenwärtigen, nicht anders abwendbaren Gefahr für ein geschütztes Rechtsgut oder rechtliches Interesse agiert, wenn bei Abwägung der widerstreitenden Interessen das verteidigte Interesse das beeinträchtigte Interesse wesentlich überwiegt. Daneben können gesetzliche Rechtfertigungsgründe aus allen Rechtsbereichen in das Ordnungswidrigkeitenrecht hineinwirken, auch ungeschriebene, sofern sie gewohnheitsrechtlich anerkannt sind.[556] Die Voraussetzungen der Rechtfertigung richten sich nach den Grundsätzen des StGB und können vergleichend herangezogen werden.[557] Unter die schützenswerten Interessen fallen neben den herkömmlichen Rechtsgütern fast alle individuellen Positionen tatsächlicher wie rechtlicher Art.[558] Rechtsgut im Sinne von § 16 OWiG

---

554 Vgl. Ausführungen dazu *Leipold*, ZRP 2013, 34, 35.
555 *Bohnert*, OWiG Kommentar, § 15, Rn. 1.
556 *Bohnert*, OWiG Kommentar, § 15, Rn. 1.
557 Siehe ausführlich dazu *Bohnert*, OWiG Kommentar, § 15, Rn. 3 f.
558 *Bohnert*, OWiG Kommentar, § 15, Rn. 8.

sind alle Güter, die einen Rechtswert haben[559] – Leben, Leib, Freiheit und Eigentum sind lediglich Beispiele.

Auch das Ordnungswidrigkeitenrecht sieht also eine Straffreiheit aufgrund eines gerechtfertigten Handelns vor. Ist der Whistleblower ein (ehemaliger) Beteiligter einer Ordnungswidrigkeit, könnte ein Rechtfertigungsgrund für sein Handeln sprechen, insbesondere weil es im Ordnungswidrigkeitenrecht die behördliche Erlaubnis als Rechtfertigungsgrund gibt. Viele Ahndungstatbestände enthalten nämlich Wendungen wie „ohne behördliche Erlaubnis" oder „unter Verstoß gegen ein behördliches Verbot".

Diese Konstellation greift aber nicht bei Kartelldelikten. Im Fall des Kartellanten liegt die Rechtsverletzung in der Verletzung eines wettbewerbsrechtlichen Ge- oder Verbots und nicht, wie beim unbeteiligten Whistleblower, im Whistleblowing an sich. Nur dort kommt eine Rechtfertigung des Handelns in Frage, weil der Betroffene erst durch das Whistleblowing einen (maßgeblichen) Rechtsverstoß begeht. Der Kartellant aber hat schon vor dem Zeitpunkt des Whistleblowings eine Norm verletzt. Zwar gibt es auch im Ordnungswidrigkeitenrecht eine Rücktrittsregelung (vgl. § 13 OWiG). Jedoch hilft im Falle eines ehemaligen Kartellanten – und zumeist auch bei Whistleblowern, die an einer anderen Ordnungswidrigkeit beteiligt waren –, dessen Tat schon über das Versuchsstadium hinaus ist, einer Rücktrittsregelung nicht weiter. Vorschriften über die tätige Reue enthält das Ordnungswidrigkeitenrecht, soweit ersichtlich, nur in § 378 Abs. 3 AO.

b. § 81 Abs. 7 GWB und die Bonusregelung

Zwar kommen für die „spezielle" Ordnungswidrigkeit des § 81 GWB (Bußgeldvorschrift im Kartellrecht) ergänzend die Vorschriften des OWiG in Betracht. Dies gilt insbesondere für die §§ 9, 14, 29 a, 30 und 130 OWiG. Jedoch hält das Kartellordnungswidrigkeitenrecht eine besondere Vorschrift bereit, die sich auf das Bußgeld auswirkt, das gegen den Kartellbeteiligten verhängt werden soll: Die Bonusregelung des Bundeskartellamtes.

Die Bekanntmachung des Bundeskartellamts Nr. 68/2000[560] vom 17. April 2000 war der erstmalige Versuch, eine Richtlinie für die Festsetzung von Bußgeldern gegenüber aufklärungsbereiten Kartellmitgliedern zu erarbeiten. Knapp sechs Jahre später wurde im März 2006 die „alte" Bonusregelung durch die Bekanntmachung Nr. 9/2006[561] ersetzt und gilt seitdem als aktuelle Grundlage für den Erlass und

---

559 *Bohnert*, OWiG Kommentar, § 16, Rn. 3 mit der Anmerkung, dass der Begriff des Rechtsguts zu eng sei, denn der Gesetzestext sieht jedes rechtlich geschützte Interesse als notstandsfähig an.

560 Bekanntmachung des Bundeskartellamtes Nr. 68/2000 vom 17.4.2000 über Richtlinien des Bundeskartellamtes für die Festsetzung von Geldbußen – Bonusregelung -, abrufbar unter http://www.bundeskartellamt.de, (abgerufen am 20.10.2013).

561 Bekanntmachung des Bundeskartellamtes Nr. 9/2006 vom 7.3.2006 über den Erlass und die Reduktion von Geldbußen in Kartellsachen – Bonusregelung -, abrufbar

die Reduktion von Geldbußen in Kartellsachen. Gründe waren unter anderem die hohe Intransparenz und die Rechtsunsicherheit der Beteiligten, ob ihnen tatsächlich eine Vergünstigung zuteilwürde.[562] Die neue Bonusregelung findet auf alle Anträge Anwendung, die nach März 2006 gestellt worden sind.[563] Ebenso wie das Kronzeugenprogramm der Europäischen Kommission bezweckt die Bonusregelung des Bundeskartellamtes, durch Anreize – in Form von Amnestie – die Kartellbeteiligten zu motivieren, aus Kartellen auszusteigen, diese zu beenden oder bei Beteiligung das Kartell zu melden.[564] Durch die Meldung an die Wettbewerbsbehörden und dem gleichzeitigen Ausstieg kann zukünftigen Kartellen effektiv begegnet werden.

Im Gegensatz zur alten Regelung geht die Bekanntmachung Nr. 9/2006 auf eine ausdrückliche Ermächtigungsgrundlage im GWB zurück, die mit der 7. GWB-Novelle[565] in das Gesetz aufgenommen wurde. In § 81 Abs. 7 heißt es:

> *„Das Bundeskartellamt kann allgemeine Verwaltungsgrundsätze über die Ausübung seines Ermessens bei der Bemessung der Geldbuße, insbesondere für die Feststellung der Bußgeldhöhe als auch für die Zusammenarbeit mit ausländischen Wettbewerbsbehörden, festlegen.“*

Damit ermächtigt § 81 Abs. 7 GWB das Bundeskartellamt ausdrücklich zum Erlass einer Bonusregelung, und gibt dieser eine gesetzliche Grundlage.[566] Soweit das Bundeskartellamt auf der Grundlage von § 81 Abs. 7 GWB einen Verwaltungsgrundsatz erlässt, ist es aufgrund der Selbstbindung an die Einhaltung von Recht und Gesetz gebunden.[567] Die Gerichte sind aber regelmäßig nicht an solche Verwaltungsgrundsätze gebunden.[568]

Die Bonusregelung, die nach ihrer Randnummer 1 sowohl für natürliche Personen als auch für Unternehmen gilt, unterscheidet zwischen zwei Fallgruppen: dem Erlass (Abschnitt B) und der Reduktion einer Geldbuße (Abschnitt C). Die Einstufung erfolgt danach, ob eine Selbstanzeige vor oder nach Einleitung des Ermittlungsverfahrens stattgefunden hat, sowie danach, welche Rolle der Betroffe-

---

unter http://www.bundeskartellamt.de/SharedDocs/Publikation/DE/Bekanntmachungen/Bekanntmachung%20-%20Bonusregelung.html?nn=4136556, (abgerufen am 20.10.2013).

562 Zu weiteren Gründen der Neufassung *Engelsing*, ZWeR 2006, 179, 183 f.

563 Rn. 25 der Bekanntmachung Nr. 9/2006.

564 *Dannecker/Biermann*, in: Immenga/Mestmäcker, Wettbewerbsrecht: GWB, § 81 GWB, Rn. 417.

565 Siebtes Gesetz zur Änderung des Gesetzes gegen Wettbewerbsbeschränkungen vom 7. Juli 2005, BGBl. I-2005, S. 1954.

566 *Dannecker/Biermann*, in: Immenga/Mestmäcker, Wettbewerbsrecht: GWB, § 81 GWB, Rn. 418, 469.

567 *Herzog/Grzeszick*, in: Maunz/Dürig, GG-Kommentar, Art. 20 GG, Rn. 59 f.

568 *Bechtold*, Kartellgesetz, § 81 GWB, Rn. 41; *Raum*, in: Langen/Bunte, Deutsches KartellR, § 81 GWB, Rn. 148; vgl. OLG Düsseldorf, Beschl. v. 19.1.2000 – 2b Ss (OWi) 290/99 – (OWi) 2/00 I, NStZ-RR 2000, 218, 219.

ne in dem Kartell und bei der Aufklärung des Kartells gespielt hat. In jedem Fall muss der Kartellbeteiligte allgemeinen Kooperationspflichten nachkommen, die in Abschnitt D der Bonusregelung näher beschrieben werden. Soweit alle Voraussetzungen[569] vorliegen, verzichtet das Bundeskartellamt generell auf die Festsetzung einer Geldbuße.

Abschnitt F Randnummer 24 stellt zudem klar, dass das Bundeskartellamt verpflichtet bleibt, Verfahren gegen natürliche Personen gemäß § 41 OWiG an die Staatsanwaltschaft abzugeben, wenn es sich bei der Tat um eine Straftat handelt.

Nach derzeitiger Rechtslage existiert somit ein Instrument, den Whistleblower ordnungswidrigkeitenrechtlich in Kartellsachen zu schützen. Indem der Whistleblower das Kartell gegenüber der Kartellbehörde offenbart, kann er in den Genuss der Bußgeldreduzierung oder gar des Bußgelderlasses kommen. Im Gegenzug für seine Selbstanzeige und „Auslieferung" wird ihm Schutz gewährt. Er ist vom Bußgeld (teilweise) befreit und hat – soweit keine Straftat im Spiel ist – keine weiteren strafrechtlichen oder ordnungswidrigkeitenrechtlichen Konsequenzen zu befürchten. Die Bonusregelung des Bundeskartellamtes stellt eine echte Kronzeugenregelung dar. Es wurde erkannt, dass Whistleblowing in Kartellsachen einen positiven Nebeneffekt hat: die Bekämpfung verbotener Absprachen zwischen Wettbewerbern über Preise, Absatzquoten und Marktaufteilungen. Die Bonusregelung „unterstützt" die Whistleblower in ihrem Handeln und erhöht die Bereitschaft eines Arbeitnehmers bzw. eines Unternehmens als Whistleblower zu agieren.

C. Haftungsrechtlicher Schutz gegenüber Ansprüchen Dritter

Im Folgenden wird untersucht, ob und inwieweit der Whistleblower vor zivilrechtlichen Schadensersatzansprüchen geschützt ist.

Bei der Beurteilung der Frage muss wiederum unterschieden werden, ob der Whistleblower sich an der offengelegten Tat beteiligt hat oder sonst involviert war, oder ob er nur aufgrund des Whistleblowings an sich in Anspruch genommen werden kann. Es werden die in Betracht kommenden Anspruchsgrundlagen aufgezeigt und geklärt, ob Schutzmechanismen bestehen. Besonderes Augenmerk wird dabei auf Ansprüche gegen den Whistleblower als ehemaliger Kartellant und die Problematik der Akteneinsicht des Anspruchsberechtigten in die Kronzeugenakten gelegt.

I. Schutz vor zivilrechtlichen Schadensersatzansprüchen gegen den an der Tat unbeteiligten Whistleblower

Hat der Arbeitgeber gegen ein Gesetz verstoßen, erleiden Dritte dadurch regelmäßig einen Schaden. Die Geltendmachung von Schadensersatzansprüchen, etwa wegen

569 Ausführlich *Danneker/Biermann*, in: Immenga/Mestmäcker, Wettbewerbsrecht: GWB, § 81 GWB, Rn. 513 f.

Verletzung von Leib und Gesundheit oder Vermögen, wird erfolglos sein, wenn der Whistleblower an der Tat nicht beteiligt war. Der unbeteiligte Whistleblower hat keinen Bezug zu der Tat und kann dafür auch nicht belangt werden. Jedoch könnte es eine Norm geben, die den Whistleblower als Arbeitnehmer dazu verpflichtet zu handeln, wenn er den Missstand erkennt, um die Allgemeinheit vor Schäden zu bewahren.

Es gibt keine allgemeine Pflicht einer einzelnen Person, die Allgemeinheit vor Schäden zu schützen. Niemand ist verpflichtet, sich oder seine Güter zu opfern, um die Allgemeinheit zu retten. Dies folgt schon aus den Vorschriften zum strafrechtlichen Unterlassen, nach denen grundsätzlich die Zumutbarkeit des Handelns zu prüfen ist.[570] Danach ist grundsätzlich eine Person nur zu Hilfe verpflichtet, wenn es ihr zumutbar ist. Aber auch eine Verkehrssicherungspflicht[571] als besondere Pflicht zum Schutz allgemeiner Güter besteht nicht,[572] da der unbeteiligte Whistleblower schon keine Gefahr verursacht hat.

Eine erfolgreiche Geltendmachung eines Schadensersatzanspruches durch einen Dritten gegenüber einem an der Tat unbeteiligten Whistleblower ist somit ausgeschlossen, eines besonderen Schutzes bedarf es also nicht.

II. Schutz vor zivilrechtlichen Schadensersatzansprüchen gegen den an der Tat beteiligten Whistleblower

Zivilrechtliche Schadensersatzansprüche kommen vermehrt bei dem involvierten Whistleblower in Betracht. Ist der Whistleblower an dem Missstand in irgendeiner Weise beteiligt gewesen oder hat er durch seine Beihilfe die Tat gefördert und ist dadurch ein Schaden entstanden – gleich welcher Art –, kann ein Dritter den Whistleblower haftbar machen.

---

570 *Heuchemer*, in: BeckOK StGB, § 13 StGB, Rn. 73; *Sternberg-Lieben/Hecker*, in: Schönke/Schröder, Kommentar StGB, § 323 c StGB, Rn. 18.

571 Fallgruppen, die eine Verkehrssicherungspflicht begründen können: Eröffnung eines Verkehrs; Eingriffe in den bereits bestehenden Verkehr oder die Teilnahme daran; Verantwortlichkeit für gefährliche Sachen; Übernahme einer Aufgabe (vgl. als gesetzliche Fälle die §§ 831 II, 832 II, 834, 838); Ausübung eines besonderen Berufes oder Gewerbes, ausführlicher *Staudinger*, in: Schulze, BGB-Kommentar, § 823, Rn. 64.

572 Die Verkehrssicherungspflicht lässt sich in ihrem Grundtatbestand als allg. Rechtspflicht umschreiben, wonach derjenige, der in seinem Verantwortungsbereich Gefahren schafft oder andauern lässt, alle geeigneten, erforderlichen und zumutbaren Vorkehrungen treffen muss, um Gefahren von Dritten abzuwenden. Eine solche Pflicht kann nur angenommen werden, soweit sich nach sachkundigem Urteil die naheliegende Möglichkeit der Schädigung von Rechtsgütern anderer ergibt, vgl. *Staudinger*, in: Schulze, BGB-Kommentar, § 823, Rn. 60.

1. *Ansprüche des Dritten gegen den beteiligten Whistleblower*

Kommt ein Dritter bei Ausführung einer betrieblichen Tätigkeit zu Schaden, haftet ihm der Arbeitnehmer im Außenverhältnis voll, insbesondere nach § 823 BGB. Dabei muss ein gesetzlich genanntes Rechtsgut oder ein sonstiges Recht des Dritten verletzt sein und der Arbeitnehmer muss in zurechenbarer Weise die Rechtsgutsverletzung zu verantworten haben.[573] Es können Verletzungen der Gesundheit nach § 823 Abs. 1 (z. B. bei umweltverschmutzenden Praktiken) eine Rolle spielen, aber auch die Verletzung des Vermögensinteresses,[574] das durch ein Schutzgesetz nach § 823 Abs. 2 geschützt sein kann (z. B. § 263 StGB als Schutzgesetz).

Nach den Grundsätzen des innerbetrieblichen Schadensausgleichs hat der Arbeitnehmer aber grundsätzlich gegenüber seinem Arbeitgeber einen Freistellungsanspruch, und zwar dann, wenn der Arbeitnehmer – unterstellt der Arbeitgeber wäre der Geschädigte – auch nicht oder nicht voll haften würde.[575] Dies gilt aber nur, wenn die Grundsätze des innerbetrieblichen Haftungsausgleichs greifen. Bei der vorsätzlichen Begehung einer Straftat und den sich daraus ergebenden Schäden greifen die Grundsätze nicht.[576] Allerdings muss auch der Schaden vom Vorsatz erfasst sein,[577] sodass ein „einfacher" vorsätzlicher Pflichtverstoß nicht genügt. Der geschädigte Dritte kann demnach gegen den Arbeitnehmer vorgehen, wenn aufgrund der Straftat eines der in § 823 BGB geschützten Rechtsgüter verletzt wurde.[578] Der Schwerpunkt wird dann wohl auf der Verletzung eines Schutzgesetzes im Sinne des § 823 Abs. 2 BGB liegen.

2. *Tätige Reue im Zivilrecht?*

Zwar wird meist in Fällen, in denen der Arbeitnehmer als Whistleblower agiert und an der Tat beteiligt ist, der Arbeitgeber als Mit- oder Haupttäter ebenfalls zivilrechtlichen Ansprüchen ausgesetzt sein. Dies ändert jedoch nichts daran, dass der Whistleblower in Anspruch genommen wird, obwohl er den Gesetzesverstoß beendet hat.

Wie sich nachträgliches Wohlverhalten auf den Rechtsnachteil – hier Schadensersatz an einen Dritten – auswirkt, wurde zwar vereinzelt im Schrifttum untersucht,[579] kann aber aufgrund der veralteten Rechtslage nicht mehr herangezogen werden.

---

573 *Teichmann*, in: Jauernig, BGB, § 823, Rn. 20.

574 So auch *Alexander*, Schadensersatz und Abschöpfung, S. 123.

575 *Schreiber*, in: Schulze, BGB-Kommentar, § 619 a, Rn. 8 m. w. N.

576 St. Rspr. BAG, Urt. v. 18.1.2007 – 8 AZR 250/06, NZA 2007, 1230, 1235; BAG, Urt. v. 18.4.2002 – 8 AZR 348/01, NZA 2003, 37, 39 f.; BAG, Urt. v. 23.1.1997 – 8 AZR 893/95, NZA 1998, 140 f.; *Henssler*, in: MüKo BGB, § 619 a BGB, Rn. 38; *Linck*, in: Schaub, Arbeitsrechtshandbuch, § 59, Rn. 24, 48.

577 Siehe BAG, Urt. v. 18.4.2002 – 8 AZR 348/01, NJW 2003, 377, 380 f.

578 *Wagner*, in: MüKo BGB, § 823 BGB, Rn. 115.

579 Siehe *Knütel*, Tätige Reue im Zivilrecht.

Die Frage, ob ein „nachträgliches Wohlverhalten“ eine bereits eingetretene Rechtsfolge ungeschehen machen kann oder ob dies dem Betroffenen die Rechte wieder entzieht/beschränkt, die ihm aufgrund des Fehlverhaltens schon erwachsen waren, ist unklar. Wieso sollte die Rückkehr zur Rechtstreue nicht belohnt werden? Der Whistleblower, der zunächst an der Tat mitgewirkt hat, ist spätestens durch das Whistleblowing wieder zu gesetzmäßigem Verhalten zurückgekehrt. Ihm keine haftungsrechtliche Entlastung zuzusprechen, ist unangemessen. Das Gesetz sieht jedoch eine solche „Belohnung“ nicht vor – auch die Rechtsprechung hält sich zurück.

Im Ergebnis erhält also der zunächst an der Tat beteiligte Whistleblower derzeit keinen haftungsrechtlichen Schutz durch die Rechtsordnung.

3. Ansprüche des Dritten gegen den am Kartell beteiligten Whistleblower

Die haftungsrechtliche Problematik des Whistleblowers im Kartellrecht stellt den Schwerpunkt dieses Abschnitts dar. Die Interessen des Whistleblowers (Kronzeugen), der Behörde und des Geschädigten sind dabei in Ausgleich zu bringen. Unter welchen Voraussetzungen kann ein geschädigter Dritter vom Whistleblower Schadensersatz verlangen und welche Probleme ergeben sich im Hinblick auf die besondere Stellung des Whistleblowers?

Wenn im Folgenden vom Whistleblower die Rede ist, ist damit das Unternehmen gemeint, weil der Kreis der Ersatzpflichtigen im Kartellrecht begrenzt ist. Zwar können auch einzelne Organe oder einzelne Gesellschafter bzw. die Geschäftsleitung einen Kartellrechtsverstoß begehen.[580] Jedoch wird das Verhalten der Organe dem Unternehmen, soweit es in der Rechtsform einer Handelsgesellschaft oder einer juristischen Person organisiert ist, nach § 31 BGB zugerechnet. Bei einem Verstoß eines Gesellschafters ergibt sich die Haftung aus § 31 BGB in Verbindung mit dem weiten Täterbegriff des § 9 OWiG. Im Übrigen haften die Handelnden aus der Geschäftsleitung nur, wenn sie selbst den jeweiligen Deliktstatbestand erfüllen können.[581] Werden andere Personen für ein Unternehmen in kartellrechtswidriger Weise tätig, bleibt nur der Rückgriff auf § 831 BGB.[582] Entscheidend ist dann, ob das Unternehmen bzw. die zuständigen Organe ihre Organisations- und Aufsichtspflicht schuldhaft verletzt haben, was in der Regel zu bejahen ist, wenn Arbeitnehmer gegen Gesetze verstoßen. Ersatzpflichtig ist also in erster Linie der Träger des Unternehmens. Eine direkte Inanspruchnahme des Arbeitnehmers, der keine leitende Position bekleidet, ist nicht möglich.

---

580 *Emmerich*, in: Immenga/Mestmäcker, Wettbewerbsrecht: GWB, § 33 GWB, Rn. 30, 31.

581 *Emmerich*, in: Immenga/Mestmäcker, Wettbewerbsrecht: GWB, § 33 GWB, Rn. 31.

582 *Wagner*, in: MüKo BGB, § 831 BGB, Rn. 14.

a. Grundlage kartellrechtlicher Schadensersatzklagen

Nach § 33 Abs. 3 GWB kann wegen Schadensersatz in Anspruch genommen werden, wer vorsätzlich oder fahrlässig einen Verstoß im Sinne von § 33 Abs. 1 GWB begeht, also eine kartellrechtliche Vorschrift oder eine kartellbehördliche Verfügung im Sinne von § 33 Abs. 1 GWB verletzt.[583] § 33 Abs. 1 GWB unterscheidet nicht mehr zwischen Verstößen gegen nationales und EG-Kartellrecht, wodurch ein Ausweichen auf § 823 Abs. 2 BGB bei Verstößen gegen Gemeinschaftskartellrecht obsolet wurde. Gegenüber § 823 Abs. 2 BGB bildet der kartellrechtliche Schadensersatzanspruch nunmehr die speziellere Anspruchsgrundlage.[584] Dagegen ist die Generalklausel des § 826 BGB neben § 33 GWB anwendbar. Für Kartellrechtsverstöße, die vor dem 1. Juli 2005 begangen wurden, gilt § 33 GWB aber nicht.[585] Für Altfälle greift § 823 Abs. 2 BGB in Verbindung mit § 1 GWB, Art. 101 und 102 AEUV. Das im GWB und AEUV verkörperte Kartellverbot zieht einen großen Kreis an Aktivlegitimierten (entsprechend dem § 33 Abs. 3 GWB) nach sich.[586]

§ 33 Abs. 3 Satz 1 i. V. m. § 33 Abs. 1 GWB ist seit der 7. GWB-Novelle also einheitliche Anspruchsgrundlage für Schadensersatzansprüche wegen schuldhaften Verstoßes gegen die Vorschriften des GWB, Art. 101 und Art. 102 AEUV sowie kartellbehördliche Verfügungen. Mangels spezialgesetzlicher Regelungen bestimmt sich der Inhalt und Umfang des Schadensersatzanspruchs geschädigter Dritter gegen das kartellbeteiligte Unternehmen aus § 33 Abs. 3 GWB i. V. m. §§ 249, 252 BGB. Der Anspruch ist der Sache nach ein deliktischer Anspruch, so dass die allgemeinen Vorschriften der §§ 823 ff., 249 ff. BGB anwendbar sind, sofern nicht die kartellrechtlichen Sonderbestimmungen aus § 33 Abs. 3 S. 2 bis 5 GWB eingreifen.[587] Dies gilt insbesondere für das Mitverschulden nach § 254 BGB, die Art des Schadensersatzes und den Schadensumfang.[588] Die Berechnung der konkreten Schadenshöhe erfolgt auf Grundlage der Differenzhypothese:[589] Der Geschädigte ist so zu stellen, wie er stünde, wenn es den Kartellverstoß nicht gegeben hätte. Der konkrete Schaden besteht aus der Differenz zwischen dem tatsächlich gezahlten Kartellpreis und dem hypothetischen Wettbewerbspreis zuzüglich eines entgangenen Gewinns[590] und

583 Die Bestimmungen des GWB unterliegen einem ganz eigenen Sanktionssystem, vgl. *Alexander*, Schadensersatz und Abschöpfung, S. 339.

584 *Emmerich*, in: Immenga/Mestmäcker, Wettbewerbsrecht: GWB, § 33 GWB, Rn. 113; *Rehbinder*, in: Loewenheim/Meessen/Riesenkampff, Kartellrecht, § 33 GWB, Rn. 56, der betont, dass auch § 823 Abs. 1 BGB unter dem Gesichtspunkt des Eingriffs in das Recht am eingerichteten und ausgeübten Gewerbebetrieb hinter § 33 GWB zurücktritt, da diese Regelung nach allgemeiner Meinung nur subsidiär gilt.

585 *Zimmer/Logemann*, WuW 2006, 982, 984 f.

586 BGH, Urt. v. 28.6.2011 – KZR 75/10, NJW 2012, 928, 928.

587 *Alexander*, Schadensersatz und Abschöpfung, S. 390.

588 *Rehbinder*, in: Loewenheim/Meessen/Riesenkampff, Kartellrecht, § 33 GWB, Rn. 31.

589 OLG Karlsruhe, Urt. v. 28.1.2004 – 6 U 183/03, NJW 2004, 2243, 2244.

590 Vgl. *Hellwig*, in: Basedow, Private Enforcement, S. 121 f.

der Zinsen in Höhe von 5 Prozent über dem Basiszinssatz (§§ 33 Abs. 2 Satz 4, 5 GWB i. V. m. § 288 BGB).[591] Da sich die hypothetischen Wettbewerbspreise nicht immer ermitteln lassen, steht es dem Richter nach § 33 Abs. 3 Satz 3 GWB frei, eine Schätzung des Schadens nach § 287 ZPO[592] vorzunehmen.

Gemäß §§ 830, 840 BGB haften dabei alle Kartellanten gesamtschuldnerisch[593] – auch der Kronzeuge ist in die Haftung mit eingeschlossen.[594] Deshalb kann grundsätzlich gegenüber jedem Kartellbeteiligten ein Schadensersatzanspruch geltend gemacht werden. Durch den gemeinsamen Vollzug der Kartellabsprache und das wettbewerbswidrige Verhalten trägt jeder einzelne zur Schadensentstehung bei, sodass ein Geschädigter nach §§ 840 Abs. 1, 421 f. BGB auch ausschließlich vom Kronzeugen den Ersatz des Schadens in voller Höhe verlangen kann.

Daneben kommen bereicherungsrechtliche Ansprüche wegen arglistiger Täuschung nach § 123 Abs. 1 BGB in Betracht, die bei der Rückabwicklung die Rechtsfolge des §§ 812, 818 BGB auslösen und bei Verletzung einer bestehenden Aufklärungspflicht einen Schadensersatzanspruch nach § 311 Abs. 2 BGB (c. i. c.) nach sich ziehen.[595] Die Darlegungs- und Beweislast für den Rechtsverstoß trägt der Schadensersatzkläger.[596]

Im Falle sog. *follow on*-Klagen[597] sind die Zivilrichter insoweit an die rechtskräftige Entscheidung des Bundeskartellamtes, der Europäischen Kommission sowie der Wettbewerbsbehörden anderer Mitgliedstaaten gebunden (§ 33 Abs. 4 GWB). Diesem Gedanken entspricht auch Art. 16 Abs. 1 der Verordnung Nr. 1/2003 (EG), wonach mitgliedstaatliche Gerichte keine Entscheidungen über Vereinbarungen, Beschlüsse oder Verhaltensweisen (im Sinne der Art. 101 oder 102 AEUV) erlassen dürfen, die bereits Gegenstand einer Entscheidung der Kommission sind und der Entscheidung der Kommission zuwiderlaufen würden.[598] Zwar befreit § 33 Abs. 4

---

591 EuGH, Urt. v. 13.7.2006 – verb. Rs. C-295/04-C-298/04, Slg. 2006, I-6641 (Rn. 95), EuZW 2006, 529, 535 (Manfredi u. a.).

592 § 287 ZPO bezieht sich dabei auf die haftungsausfüllende Kausalität und damit auf die Feststellung, ob ein Schaden eingetreten ist, welchen Umfang er hat und ob er auf dem entsprechenden Verhalten beruht *Meyer*, GRUR 2006, 27, 30.

593 BGH, Urt. v. 28.6.2011 – KZR 75/10, NJW 2012, 928, 935; zum Gesamtschuldnerausgleich i. S. der §§ 426 Abs. 1 S. 1, 254 Abs. 1 BGB *Dreher*, FS Möschel, S. 149, 156 f.; *Krüger*, Kartellregress, S.77 f.

594 *Zöttl/Schlepper*, EuZW 2012, 573, 573.

595 BGH, Beschl. v. 28.1.2010 – VII ZR 50/09, ZfBR 2010, 368, 369; *Bornkamm*, in: Langen/Bunte, Deutsches KartellR, § 33 GWB, Rn. 99; *Zöttl/Schlepper*, EuZW 2012, 573, 574; a. A. OLG Düsseldorf, Urt. v. 20.2.2009 – 22 U 135/08, NJW-Spezial 2011, 14, unter Verneinung des Anfechtungsgrundes.

596 BGH, Beschl. v. 28.1.2010 – VII ZR 50/09, ZfBR 2010, 368, 369.

597 *Follow on*-Klagen sind Klagen, bei denen sich der Kläger die Bindungswirkung der Bußgeldentscheidung einer Wettbewerbsbehörde für deutsche Gerichte bezüglich der Rechtsverletzung nach § 33 Abs. 4 GWB zunutze macht.

598 Zu den Grenzen der Wortlautauslegung in Bezug auf das Merkmal „Zuwiderlaufen“ *Imgrund*, Die Bindung der deutschen Zivilgerichte, S. 95 f.

GWB den Kläger nicht von seiner Pflicht, den Schaden sowie die Kausalität zwischen Verstoß und eingetretenem Schaden und das Verschulden nachzuweisen. Jedoch erübrigt die Tatbestandswirkung nach § 33 Abs. 4 GWB die grundlegende Feststellung eines Kartellverstoßes, die oftmals erhebliche Schwierigkeiten bereitet.[599] Außerdem wird durch die Bindungswirkung eine Vergeudung der knappen Ressourcen im Justizwesen durch doppelte Prüfungen derselben Sach- und Rechtsfragen vermieden.

Gemäß § 33 Abs. 1 Satz 3 GWB ist derjenige anspruchsberechtigt, der durch den Kartellrechtsverstoß als Mitbewerber oder sonstiger Marktteilnehmer „betroffen" ist. Betroffen ist der direkte Abnehmer der kartellbehafteten Produkte, also die unmittelbare Marktgegenseite, wie die unmittelbaren Lieferanten, Abnehmer und Endverbraucher. Lange Zeit war umstritten, ob auch der indirekte Abnehmer anspruchsberechtigt ist. Unter einem indirekten Abnehmer versteht man z. B. die Kunden der Kunden von Kartellmitgliedern und die Endverbraucher – also den letzten in einer Vertriebskette. Mit Blick auf die Rechtsprechung des EuGH und den *effet utile* hat der BGH[600] entschieden, dass indirekte Abnehmer mit in den Kreis der Anspruchsberechtigten einzubeziehen seien. Nach europäischem Recht müsse „jedermann" Ersatz des Schadens verlangen können, der ihm durch einen Verstoß gegen Art. 101 oder 102 AEUV entstehe.[601] Verhaltensweisen, die sich auf die Marktgegenseite insgesamt auswirken, von der Schadensersatzsanktion auszunehmen, wäre mit der Pflicht der nationalen Gerichte, dem Kartellverbot volle Wirksamkeit zu verleihen,[602] nicht vereinbar und würde zu einer zweckwidrigen Entlastung gerade solcher Kartelltäter führen, die Schäden mit großer Breitenwirkung verursachen.[603] Des Weiteren sind nach § 33 Abs. 2 Nr. 1 GWB auch rechtsfähige Verbände zur Förderung gewerblicher oder selbstständiger beruflicher Interessen anspruchsberechtigt, wenn ihnen eine erhebliche Zahl von betroffenen Unternehmen im Sinne des Absatzes 1 Satz 3 angehört und sie insbesondere nach ihrer personellen, sachlichen und finanziellen Ausstattung imstande sind, ihre satzungsmäßigen Aufgaben der Verfolgung gewerblicher oder selbstständiger beruflicher Interessen tatsächlich wahrzunehmen. Ansonsten sind Verbände grundsätzlich nicht anspruchsberechtigt.

599 *Berrisch/Burianski*, WuW 2005, 878, 881.

600 BGH, Urt. v. 28.6.2011 – KZR 75/10, NJW 2012, 928.

601 BGH, Urt. v. 28.6.2011 – KZR 75/10, NJW 2012, 928, 930, unter Verweis auf den EuGH, Urt. v. 13.7.2006 – verb. Rs. C-295/04-C-298/04, Slg. 2006, I-6619 (Rn. 60, 61), EuZW 2006, 529, 531 (Manfredi u. a.).

602 BGH, Urt. v. 28.6.2001 – KZR 75/10, NJW 2012, 928, 929 f. mit Verweis auf EuGH, Urt. v. 20.9.2001 – Rs. C-453/99, Slg. 2001, I-6297 (Rn. 26), EuZW 2001, 715 (Courage/Crehan) und EuGH, Urt. v. 13.7.2006 – verb. Rs. C-295/04-C-298/04, Slg. 2006, I-6619 (Rn. 60), EuZW 2006, 529 (Manfredi u. a.).

603 BGH, Urt. v. 28.6.2001 – KZR 75/10, NJW 2012, 928, 929 f.; vgl. *Bulst*, NJW 2004, 2201, 2202; *Berrisch/Burianski*, WuW 2005, 878, 881; *Emmerich*, in: Immenga/Mestmäcker, Wettbewerbsrecht: GWB, § 33 GWB, Rn. 16.

Umstritten war ebenfalls, wie mit dem *passing-on*-Einwand umzugehen ist. Im Rahmen des Vorteilsausgleichs[604] ist zu fragen, ob die Weitergabe der auf der Kartellvereinbarung beruhenden Preiserhöhung an die nachgelagerte Marktstufe angerechnet werden muss und dieser Schadensposten damit für die unmittelbare Marktgegenseite nicht mehr ersatzfähig ist. Mit Blick auf die erforderliche Kausalität zwischen Schädigung und Vorteil und einer wertenden Betrachtung spricht einiges für einen Ausschluss der *passing-on defense*.

Der Ausschluss kann jedoch dazu führen, dass der Schädiger mehrfach in Anspruch genommen[605] und der Direktabnehmer ungerechtfertigt bereichert wird.[606] § 33 Abs. 3 Satz 2 GWB lässt die Frage nach der Zulassung des Einwands offen. Die Rechtsprechung hat die Lücke gefüllt, der BGH erlaubt grundsätzlich den *passing-on*-Einwand.[607] Um eine mögliche Mehrfachbelastung des Beklagten in Grenzen zu halten, lässt der BGH den Einwand der Schadensabwälzung zu und verweist den Schädiger auf die Möglichkeit der Streitverkündung.[608]

Danach sind dem Geschädigten zufließende Vorteile aus dem Weiterverkauf der Kartellprodukte bzw. der Produkte, zu denen sie weiterverarbeitet wurden, auf seinen Schaden anzurechnen, wenn der Vorteil bei wertender Betrachtung adäquat kausal auf dem schädigenden Ereignis (Kartellabsprache) beruht, die Anrechnung dem Geschädigten zumutbar ist und den Schädiger nicht unangemessen entlastet.[609] Damit hat der BGH die Rechtsunsicherheit beseitigt, ob der Einwand der Vorteilsanrechnung zulässig ist.

Zur Haftung bei Erwerb von einem Kartellaußenseiter hat der Europäische Gerichtshof in einer Grundsatzentscheidung am 5. Juni 2014[610] entschieden, dass Kartellbeteiligte (auch) für einen Schaden haften, der durch die Verfälschung des Marktpreises infolge des Kartells entstanden ist. Damit hat der EuGH so genannte Preisschirmeffekte („Umbrella-Pricing") anerkannt und den Umfang der Schadensersatzhaftung von Kartellteilnehmern neu bestimmt.

Für das deutsche Recht ist die Frage, ob die Kartellbeteiligten für den Schaden haften, der einem Angehörigen der Marktgegenseite dadurch entstanden ist, dass er das Produkt nicht von einem Kartellbeteiligten selbst, sondern von einem Kartellaußenseiter erworben hat, dieser aber infolge eines Preisschimeffekts ebenfalls einen erhöhten Preis verlangt hat, höchstrichterlich nicht entschieden. Im Schrifttum[611]

604 Näher dazu *Bornkamm*, in: Langen/Bunte, Deutsches KartellR, § 33 GWB, Rn. 112; vertiefend *Soyez*, EuZW 2012, 100.

605 *Kleinlein/Schubert*, WuW 2012, 345, 345.

606 *Bulst*, ZWeR 2012, 70, 80.

607 BGH, Urt. v. 28.6.2011 – KZR 75/10, NJW 2012, 928, 932.

608 BGH, Urt. v. 28.6.2011 – KZR 75/10, NJW 2012, 928, 934.

609 BGH, Urt. v. 28.6.2011 – KZR 75/10, NJW 2012, 928, 934.

610 EuGH, Urt. v. 5.6.2014 – Rs. C-557/12, EuZW 2014, 586 (Kone); siehe auch Anmerkung *Schoene*, GRUR-Prax 2014, 293, 294.

611 *Alexander*, Schadensersatz und Abschöpfung, S. 388 f; *Bulst*, Schadensersatzansprüche, S. 255; *Meeßen*, Schadensersatz, S. 256 f; a. A. *Stöber*, EuZW 2014, 257, 258 f.

wird eine Haftung der Kartellbeteiligten mit der Begründung verneint, dass die Preisfestsetzung des Kartellaußenseiters auf dessen autonomer Entscheidung beruhe, weshalb sie den Kartellbeteiligten nicht zugerechnet werden könne. Diese Ansicht übersieht jedoch, dass in einem solchen Fall die Vorgaben des Europarechts zu berücksichtigen sind. Diese Erkenntnis hat im Aufzugskartell-Fall den österreichischen OGH im Jahr 2012 zu einem Vorabentscheidungsverfahren an den EuGH veranlasst.[612] Der EuGH antwortete auf die Vorlagefrage, dass Art. 101 AEUV dahin auszulegen sei, dass er einer Auslegung und Anwendung des innerstaatlichen Rechts eines Mitgliedstaates entgegenstehe, wonach es aus Rechtsgründen kategorisch ausgeschlossen sei, dass die an einem Kartell beteiligten Unternehmen zivilrechtlich für Schäden haften, die daraus resultieren, dass ein an diesem Kartell nicht beteiligtes Unternehmen in Anbetracht der Machenschaften des Kartells seine Preise höher festgesetzt habe, als es dies ohne das Kartell getan hätte.[613] In der Urteilsbegründung weist der Gerichtshof darauf hin, dass die praktische Wirksamkeit des Verbots wettbewerbswidriger Absprachen beeinträchtigt wäre, wenn der Einzelne nicht Ersatz des ihm durch einen Verstoß gegen die Wettbewerbsregeln entstandenen Schadens verlangen könne. Ein Kartell kann zur Folge haben, dass Unternehmen, die ihm nicht angehören, ihre Preise an den durch das Kartell entstandenen Marktpreis anpassen, was den Kartellbeteiligten nicht verborgen bleiben kann. Deshalb könne eine Entscheidung über die Festsetzung eines Angebotspreises, auch wenn sie als völlig autonome Entscheidung anzusehen ist, unter Heranziehung eines durch das Kartell verfälschten Marktpreises getroffen werden.[614] Ist also erwiesen, dass das Kartell nach den Umständen des konkreten Falles und insbesondere den Besonderheiten des betreffenden Marktes eine Erhöhung der geltenden Preise durch nicht am Kartell beteiligte Wettbewerber zur Folge haben kann, müssen die durch diese Preissteigerung Geschädigten von den Kartellbeteiligten den Ersatz des ihnen entstandenen Schadens verlangen können, auch wenn keine vertragliche Beziehung zwischen ihnen bestand.[615] Um den Kartellbeteiligten jedoch keine uferlose Haftung aufzuerlegen, ist für die Bestimmung, welche Schäden von der Haftpflicht umfasst sind, ein „ursächlicher Zusammenhang“[616] zu fordern. Die Anwendung dieses Begriffs ist grundsätzlich Aufgabe des innerstaatlichen Rechts des einzelnen

---

612 OGH Wien, Beschl. v. 17.10.2012 – 7 Ob 48/12b, EuZW 2013, 80, Leitsatz, WuW 2013, 453.

613 EuGH, Urt. v. 5.6.2014 – Rs. C-557/12 (Rn. 37), (Kone); so auch schon Ansicht der Generalanwältin *Kokott* in ihrem Schlussantrag vom 30.1.2014, NZKart 2014, 117 (Schlussantrag Kone).

614 EuGH, Urt. v. 5.6.2014 – Rs. C-557/12 (Rn. 29), (Kone).

615 Vgl. EuGH, Urt. v. 5.6.2014 – Rs. C-557/12 (Rn. 34), (Kone).

616 Generalanwältin *Kokott* zieht in ihrem Schlussantrag den Maßstab heran, den der EuGH in Bezug auf die außervertragliche Haftung der Unionsorgane nach Art. 340 Abs. 2 AEUV anlegt: Es müsse demnach ein „hinreichend unmittelbarer Kausalzusammenhang“ zwischen dem Fehlverhalten und dem Schaden bestehen, NZKart 2014, 117, Rn. 34 (Schlussantrag Kone).

Mitgliedstaates.[617] Für das deutsche Recht sind die §§ 249 ff. BGB und die Regeln des allgemeinen deutschen Schadensrechts maßgeblich,[618] die im Grundsatz auf § 33 GWB Anwendung finden.[619]

Auch wenn im Einzelfall zu entscheiden ist, ob die Festlegung überhöhter Preise durch Kartellaußenseiter adäquat-kausal auf einem Kartellverstoß beruht, besteht nun grundsätzlich die Möglichkeit, dass Geschädigte von Beteiligten eines Preiskartells nach § 33 Abs. 1 S. 1, Abs. 3 S.1 GWB auch für die auf Preisschirmeffekten beruhenden Schäden Ersatz erhalten können.

b. Änderungen nach der 8. GWB-Novelle vom 26. Juni 2013

Um die gesellschaftspolitische Anpassung und vor allem auch die Erfüllung europarechtlicher Vorgaben zu gewährleisten, ist eine regelmäßige Novellierung der gesetzlichen Lage angebracht. Eine aktuelle Neuerung im Wettbewerbsrecht ist die 8. GWB-Novelle vom 26. Juni 2013.[620]

Die 8. Novelle des GWB ist am 29. Juni 2013 im Bundesgesetzblatt verkündet worden. Die Novelle soll die wettbewerblichen Rahmenbedingungen modernisieren und optimieren sowie die Durchsetzung des GWB effizienter gestalten. Zudem sollen Unterschiede zwischen der deutschen und europäischen Fusionskontrolle verringert werden. Die Novelle führt auch zu einzelnen Änderungen im Verfahrens- und Sanktionsrecht. Der letztgenannte Bereich und die einhergehenden Änderungen werden im Folgenden näher skizziert.

aa. Verwaltungsverfahren

Die Novelle stellt in verwaltungsverfahrensrechtlicher Hinsicht klar, dass das Bundeskartellamt bei einem Verstoß gegen Kartellrecht auch Abhilfemaßnahmen erlassen kann, die in die Substanz des Unternehmens eingreifen (sog. strukturelle Maßnahmen). Darüber hinaus wird gesetzlich verankert, dass die Kartellbehörde im Rahmen einer Abstellungsverfügung auch eine Rückerstattung der infolge kartellrechtswidrigen Verhaltens erzielten Vorteile anordnen kann.

bb. Bußgeldrecht

Die Ahndung von Kartellrechtsverstößen erfolgt nach dem deutschen Ordnungswidrigkeitenrecht (OWiG). Bei Kartellrechtsverstößen kann das Bundeskartellamt eine

---

617 EuGH, Urt. v. 5.6.2014 – Rs. C-557/12 (Rn. 24, 32), (Kone).

618 Ausführlicher *Stöber*, EuZW 2014, 257, 259.

619 *Alexander*, Schadensersatz und Abschöpfung, S. 156 f., 390; *Lübbig*, in: MüKo GWB, § 33 GWB, Rn. 78.

620 Achtes Gesetz zur Änderung des Gesetzes gegen Wettbewerbsbeschränkungen vom 26.6.2013, BGBl. I-2013, S. 1738.

Geldbuße von bis zu 10% des gesamten Konzernumsatzes verhängen.[621] Nach einem Urteil des BGH[622] im Februar 2013 handelt es sich im Gegensatz zum europäischen Recht dabei nicht um eine Kappungsgrenze, sondern nach verfassungskonformer Auslegung um einen Bußgeldrahmen. In der Bekanntmachung des Bundeskartellamts vom 25. Juni 2013 sind nun Leitlinien[623] festgelegt, wie das Bundeskartellamt schwerwiegende Kartellrechtsverstöße zu ahnden hat und welche Kriterien für die Bußgeldzumessung herangezogen werden.

Praktische Bedeutung über das Kartellrecht hinaus hat die Änderung des § 30 OWiG. Nach § 30 Abs. 2 a OWiG hat der Gesamtrechtsnachfolger für Bußgelder zu haften, die der Rechtsvorgänger verwirkt hat. Der Gesamtrechtsnachfolger haftet also, wenn und weil der ursprünglich haftende Rechtsträger nicht mehr existiert. Entgegen dem Vorschlag des Bundesrates, ist in das Gesetz keine Regelung zur kartellrechtlichen Aufsichtspflicht der Muttergesellschaft über ihre Tochterunternehmen aufgenommen worden. Die Bundesregierung hat im Gesetzgebungsverfahren erklärt, dass die bestehenden Regelungen ausreichend sind.[624]

cc. Akteneinsichtsrecht

Das Akteneinsichtsrecht, das in einem Entwurf in § 81 b GWB Ref-E angedacht wurde, ist in der 8. GWB-Novelle nicht berücksichtigt worden. Es geht insbesondere um die Frage, wie und in welchem Umfang den Geschädigten ein Akteneinsichtsrecht in Bonusanträgen gewährt werden soll. Mit Hilfe dieses Rechts soll den Geschädigten der Weg zur Schadensersatzklage erleichtert werden. Diese Frage ist nun mangels gesetzlicher Regelung von der Rechtsprechung zu beantworten. Ein kurzer Überblick soll zeigen, was im Entwurf zum § 81 b GWB vorgesehen war.

(1) Entwurf eines § 81 b GWB

Der Referentenentwurf zur 8. GWB-Novelle sah einen § 81 b GWB vor, der die Vertraulichkeit von Aufklärungsbeiträgen regeln sollte.[625] Der Schutz von Informationen,

---

621 Grundlage ist Art. 23 II VO 1/2003 und § 81 Abs. 4 Satz 2 GWB; kritisch dazu *Buntscheck*, EuZW 2007, 423, 425 f.

622 BGH, Urt. v. 26.2.2013 – KRB 20/12, NJW 2013, 1972, 1973.

623 Bekanntmachung des Bundeskartellamts vom 25. Juni 2013 – Leitlinien für die Bußgeldzumessung in Kartellordnungswidrigkeitenverfahren, abrufbar unter: http://www.bundeskartellamt.de/SharedDocs/Publikation/DE/Leitlinien/Bekanntmachung%20-%20Bu%C3%9Fgeldleitlinien-Juni%202013.pdf?__blob=publicationFile&v=5, (abgerufen am 4.4.2014).

624 Siehe Gegenäußerung der Bundesregierung, BT-Drs. 17/9852 (Anlage 4), S. 49, unter 1 b).

625 Referentenentwurf für ein 8. Gesetz zur Änderung des Gesetzes gegen Wettbewerbsbeschränkungen, S. 17, abrufbar unter: http://www.bmwi.de/BMWi/Redaktion/PDF/G/gwb-8-aenderung-referentenentwurf,property=pdf,bereich=bmwi,sprache=de,rwb=true.pdf, (abgerufen am 11.11.2013).

die gegenüber den Kartellbehörden im Wege des Bonusverfahrens offenbart werden, ist aktuell kaum vorhanden. Schadenersatzkläger verlangen Akteneinsicht in die gefertigten Unterlagen, um ihren Schaden beweisen bzw. beziffern zu können.

Im Referentenentwurf war mit § 81 b GWB Ref-E eine Regelung enthalten, die Akteneinsicht in Bonusanträge generell ausschloss.[626] Darin hieß es in Absatz 1:

> *„Akteneinsicht in einen Antrag auf Erlass oder Reduktion einer Geldbuße und die dazu übermittelten Beweismittel nach § 406e der Strafprozessordnung, auch in Verbindung mit § 46 Absatz 1 des Gesetzes über Ordnungswidrigkeiten, findet nicht statt."*

Der Referentenentwurf hatte sich für eine allgemeine Regelung und nicht für eine Abwägung im Einzelfall entschieden.[627] Zur Sicherstellung der Effektivität von Bonusprogrammen sei eine solche Regelung in abstrakter Weise sinnvoll und auch möglich.[628]

(2) Fehlgeschlagene Implementierung

Im Regierungsentwurf und in der aktuellen Fassung des GWB ist weder von einem § 81 b GWB noch von einem Akteneinsichtsrecht die Rede.

Möglicherweise hielt man das Problem nach den Entscheidungen der Gerichte für erledigt. Die aktuelle Diskussion in der Wissenschaft verdeutlicht aber, dass dies nicht der Fall ist.

Zudem ist nicht sicher, ob eine Regelung wie die des § 81 b GWB Ref-E das „Dilemma" des Akteneinsichtsrechts lösen würde, da auch Wettbewerbsbehörden anderer Mitgliedstaaten Interesse an Informationen haben. § 81 b GWB Ref-E würde eine solche Weitergabe nicht unterbinden, da §§ 50 a bis 50 c GWB unberührt blieben.[629] Nach dem Urteil des EuGH im Fall *Donau Chemie*[630] kann auch bezweifelt werden, ob eine Regelung wie § 81 b GWB Ref-E im Anwendungsbereich des Art. 101 AEUV europarechtskonform gewesen wäre.[631]

In der Bekanntmachung der Gegenäußerung der Bundesregierung[632] auf die Vorschläge des Bundesrates wird auch zum Anspruch auf Akteneinsicht kurz Stellung

---

626 Referentenentwurf für ein 8. Gesetz zur Änderung des Gesetzes gegen Wettbewerbsbeschränkungen, S. 17, abrufbar unter: http://www.bmwi.de/BMWi/Redaktion/PDF/G/gwb-8-aenderung-referentenentwurf,property=pdf,bereich=bmwi,sprache=de,rwb=true.pdf, (abgerufen am 11.11.2013).

627 Siehe *Fritzsche*, DB 2012, 845, 852.

628 *Fritzsche*, DB 2012, 845, 852; Referentenentwurf für ein 8. Gesetz zur Änderung des Gesetzes gegen Wettbewerbsbeschränkungen, S. 51, abrufbar unter: http://www.bmwi.de/BMWi/Redaktion/PDF/G/gwb-8-aenderung-referentenentwurf,property=pdf,bereich=bmwi,sprache=de,rwb=true.pdf, (abgerufen am 11.11.2013); anders *Fornasier/Sanner*, WuW 2011, 1067, 1072.

629 So *Fritzsche*, DB 2012, 845, 852.

630 EuGH, Urt. v. 6.6.2013 – Rs. C-536/11, JZ 2013, 734 (Donau Chemie u. a.).

631 So auch *Bosch/Fritsche*, NJW 2013, 2225, 2229.

632 Gegenäußerung der Bundesregierung, BT-Drs. 17/9852, S. 49, 1 c).

genommen. Der Vorschlag des Bundesrats lief darauf hinaus, den Umfang der Akteneinsicht bei Bonusanträgen im Gesetz gegen Wettbewerbsbeschränkungen klarzustellen. Die Bundesregierung vermochte sich laut Gegenäußerung einem solchen Vorschlag nicht anzuschließen. Danach teile die Bundesregierung zwar die Einschätzung des Bundesrates, dass die Bonusregelung in Deutschland eine sehr wichtige Rolle bei der Kartellverfolgung spiele.[633] Jedoch habe das AG Bonn[634] in seiner Entscheidung das Recht der Akteneinsicht auf Grundlage des § 406 e StPO schon sehr weit eingeschränkt, wonach keine Bedürftigkeit für eine Regelung mehr bestehe.[635] Die Bundesregierung betont weiterhin, dass sie die Rechtsprechung der Gerichte zu Akteneinsichtsanträgen bei Kartellrechtsverfahren weiter verfolgen (bis März 2013) und entsprechende Maßnahmen ergreifen werde, sobald sich Bedarf an einer Regelung ergebe.[636]

Eine Einführung ist also vorläufig gescheitert.

dd. Anonymes Hinweisgebersystem

Das Bundeskartellamt hat der Verfolgung und Ahndung illegaler Absprachen, insbesondere der Verfolgung von „Hardcore-Kartellen", stets eine hohe Priorität eingeräumt. Um dieses Ziel weiterhin erreichen zu können und die Kartellverfolgung so effektiv wie möglich zu gestalten, wurden verschiedene Instrumente eingeführt: Eines ist das anonyme Hinweisgebersystem, das 2012 eingerichtet wurde[637] und in die Zeit der Novellierung fällt. Auf der Internetseite des Bundeskartellamtes wird auf die nötige Hilfe der Hinweisgeber hingewiesen und die Annahme der Information unter Nutzung eines Links angeboten. Die Information kann durch schriftlichen oder telefonischen Kontakt übermittelt werden. Überdies wird betont, dass die Weitergabe der Informationen auch anonym erfolgen kann. Das Bundeskartellamt betont, dass solche Insider-Kenntnisse sehr wertvoll für die Aufdeckung und Ahndung von Kartellen sein können und dabei helfen, weitere Schäden zu verhindern. Für diesen Fall hat das Bundeskartellamt ein standardisiertes Hinweisgebersystem eingerichtet – eine technische Rückverfolgung des Hinweises ist unmöglich. Das System ist von einem öffentlich bestellten und vereidigten Sachverständigen zertifiziert und wird auch in der Korruptionsbekämpfung von bekannten Unternehmen sowie bei Landeskriminalämtern und Polizeibehörden verwendet.[638]

---

633 Gegenäußerung der Bundesregierung, BT-Drs. 17/9852, S. 49, 1 c).

634 AG Bonn, Beschl. v. 18.1.2012 – 51 Gs 53/09, GRUR-RR 2012, 178.

635 Gegenäußerung der Bundesregierung, BT-Drs. 17/9852, S. 49, 1 c).

636 Gegenäußerung der Bundesregierung, BT-Drs. 17/9852, S. 49, 1 c).

637 Siehe http://www.bundeskartellamt.de/DE/Kartellverbot/Anonyme_Hinweise/anonymehinweise_artikel.html?nn=3591496, (abgerufen am 4.4.2014).

638 Siehe http://www.bundeskartellamt.de/DE/Kartellverbot/Anonyme_Hinweise/anonymehinweise_artikel.html?nn=3591496, der auf die Seite der Business Keeper AG verweist, die das Hinweisgebersystem entwickelt hat, (abgerufen am 4.4.2014).

Zusammenfassend kann festgestellt werden, dass das aktuelle Problem der Akteneinsicht in der Politik durchaus erkannt wurde. Eine Implementierung einer gesetzlichen Vorschrift ist nicht gelungen. Hilfe bei der Lösung des Problems können jedoch gerichtliche Entscheidungen bieten, die in jüngster Zeit vermehrt diese Thematik zum Gegenstand hatten.

4. Schutz des kartellrechtlichen Whistleblowers durch Begrenzung des Akteneinsichtsrechts

Die Bonusregelung des Bundeskartellamts sichert die Effektivität und Attraktivität der Kartellrechtsdurchsetzung. Indem die Antragsteller die Sicherheit bekommen, ganz oder zumindest teilweise von dem Bußgeld verschont zu bleiben, wird ihnen der Ausstieg aus der Beteiligung erleichtert. Doch die jüngste Rechtsprechung zum Akteneinsichtsrecht eines Dritten nach § 406 e StPO in Verbindung mit § 46 OWiG zwecks Vorbereitung einer Schadensersatzklage hat gezeigt, dass private und behördliche Kartellrechtsdurchsetzung sich gegenseitig behindern können. Je einfacher die Durchsetzung privater Kartellschadensersatzansprüche, desto unattraktiver sind Kronzeugenprogramme. Aber gerade die Kronzeugenprogramme helfen dem Whistleblower, sich zu offenbaren und den Missstand publik zu machen. Sie geben ihm die Sicherheit, trotz der vorherigen Teilnahme eine „Begünstigung" zu erlangen. Ist mit einer erhöhten Inanspruchnahme des Whistleblowers durch die Einsicht in die Akten zu rechnen, ist der Whistleblower haftungsrechtlich schutzlos. Denn trotz der Offenbarung des Gesetzesverstoßes ist er das Ziel vorzeitiger und umfänglicher Inanspruchnahme durch Dritte. In dem Zusammenhang stellt sich die Frage nach dem Verhältnis zwischen dem *private* und *public enforcement.*

a. Das Akteneinsichtsrecht im Rahmen von Kronzeugenverfahren

Das Akteneinsichtsrecht hat erheblichen Einfluss auf die Effektivität von Kronzeugenregelungen. Die Kartellbehörden haben den Konflikt aufzulösen, der sich aus den widerstreitenden Interessen des Klägers (möglichst viel Informationen für eine Schadensersatzklage erhalten) und des Whistleblowers (weitgehender Verschluss der an die Behörde gelieferten Informationen) ergibt. Kronzeugenerklärungen beinhalten detaillierte Informationen über die Struktur eines Kartells bzw. über den spezifischen „Tatbeitrag" des Kartellanten sowie die Mitwirkung des oder der übrigen Beteiligten.[639] Wird dem Kartellgeschädigten erlaubt, uneingeschränkte Einsicht in die Akten zu nehmen, besteht nicht nur die Gefahr, dass sensible Informationen, wie etwa Geschäftsgeheimnisse, an die Öffentlichkeit treten, sondern auch, dass der Kronzeuge als Whistleblower der erste Kartellbeteiligte ist, der Ziel von Schadensersatzklagen wird.[640] Die Folge ist eine erhebliche Abschreckungswirkung.

---

639 *Schroll*, Effektivität der Kronzeugenprogramme, S. 186.

640 *Schroll*, Effektivität der Kronzeugenprogramme, S. 163; *Böge*, in: Basedow, Private Enforcement, S. 217, 223; *Kersting*, ZWeR 2008, 252, 263; *Schroeder*, FS Bechtold,

Die Dokumente bzw. Akten, die im Rahmen eines Kronzeugenverfahrens angelegt werden, können sich in verschiedenen Händen befinden. Diese unterschiedliche Betrachtung ist wichtig, da abhängig von den individuellen Besitzverhältnissen unterschiedliche Anspruchsgrundlagen zu erwägen sind. Daher werden zunächst die grundsätzlichen Zugriffsmöglichkeiten auf nationaler Seite betrachtet, um dann das Weißbuch der EU-Kommission und die aktuelle Rechtsprechung zum Akteneinsichtsrecht darzustellen. Sind Akten im Besitz der EU-Kommission, ergibt sich ein besonderes Normenverhältnis im Hinblick auf die Einsichtsrechte Dritter, das gesondert erläutert wird.

aa. Zugriff auf Dokumente im Besitz des Bundeskartellamtes

Möchte der Kartellgeschädigte im Rahmen seiner privaten Schadensersatzklage Akten einsehen, gelingt ihm das nicht als Verfahrensbeteiligter. Als geschädigter Dritter hat er jedoch die Möglichkeit, nach §§ 406 e Abs. 1, 475 StPO i. V. m. § 46 Abs. 1 OWiG seinen Anwalt mit der Aufgabe zu betrauen, die dem Gericht vorliegenden Ermittlungsakten – u. a. Akten des Bundeskartellamtes – einzusehen. Dabei muss jedoch ein berechtigtes Interesse dargelegt werden. Dann ist eine Einsicht in die Akten während jedes Verfahrensstadiums, auch in einer frühen Phase des Ermittlungsverfahrens, möglich.[641] Auch wenn alle Voraussetzungen vorliegen,[642] wird das Recht auf Akteneinsicht nicht schrankenlos gewährt. Nach § 406 e Abs. 2 StPO kann das Einsichtsbegehren versagt werden, soweit überwiegende schutzwürdige Interessen des Beschuldigten oder anderer Personen entgegenstehen und/oder der Untersuchungszweck gefährdet erscheint bzw. durch die Akteneinsicht das Verfahren erheblich verzögert wird.

Gemäß § 1 Abs. 1 IFG[643] hat jeder, d. h. jede natürliche und juristische Person des Privatrechts, gegenüber den Behörden des Bundes einen voraussetzungslosen Anspruch auf Zugang zu amtlichen Informationen. Der Zweck des IFG besteht dabei in einer Steigerung der Transparenz des Verwaltungshandelns, um die effektive Wahrnehmung von Bürgerrechten zu gewährleisten.[644] Das IFG gewährt jedoch ausschließlich den Zugang zu amtlichen Informationen, sodass Informationen aus Ordnungswidrigkeitenverfahren – wie die Kronzeugendokumente – nicht über das

---

S. 437, 439; ebenso Kommentar des BDI, S. 4, in: Public consultation on amendment of the Leniency Notice – comments received, October 2006.

641 *Schaefer*, NJW-Spezial 2007, 327, 328; *Schroll*, Effektivität der Kronzeugenprogramme, S. 204.

642 Als Anspruchssteller muss die Person als „Verletzter" i. S. d. Norm zu qualifizieren sein. Ebenfalls muss darüber hinaus der Verletzte ein berechtigtes Interesse darlegen können: die Akteneinsicht muss gerade zur Interessenwahrnehmung des Verletzten erforderlich sein, vgl. *Kiethe*, wistra 2006, 50, 51.

643 Gesetz zur Regelung des Zugangs zu Informationen des Bundes (Informationsfreiheitsgesetz) vom 5.9.2005, BGBl. I-2005, S. 2722.

644 Gesetzentwurf Informationsfreiheitsgesetz, BT-Drs. 15/4493 v. 14.12.2004, Begründung A. II.

IFG eingesehen werden können, sondern höchstens auf Grundlage von spezialgesetzlichen Regelungen.[645]

bb. Akteneinsicht durch Auskunftsersuchen der Gerichte

Wird dem Geschädigten (über seinen Anwalt) keine Einsicht in die Akten der Kartellbehörde gewährt, besteht die Möglichkeit, Kronzeugendokumente, die sich im Besitz der Behörden befinden, durch das mit der Schadensersatzklage befasste Gericht nach den §§ 273 Abs. 2 Nr. 2 und 432 Abs. 1 ZPO in den Prozess einzuführen. Nach § 273 Abs. 2 Nr. 2 ZPO kann zur Vorbereitung jedes Termins der Vorsitzende oder ein von ihm bestimmtes Mitglied des Prozessgerichts Behörden oder Träger eines öffentlichen Amtes um Mitteilung von Urkunden oder um Erteilung amtlicher Auskünfte ersuchen. Gemäß § 432 Abs. 1 ZPO können die Schadensersatzkläger (Beweisführer) förmlichen Beweis durch den Antrag antreten, zu der Beweisfrage bei der Kartellbehörde eine amtliche Auskunft einzuholen oder die Kartellbehörden um Mitteilung einer Urkunde zu ersuchen. Daneben können die Zivilgerichte die Kartellbehörden auch von Amts wegen anfragen. Sowohl § 273 ZPO als auch § 432 ZPO setzen jedoch voraus, dass der Antragsteller die benötigte Urkunde so genau bezeichnet, dass die Behörde das Ersuchen ordnungsgemäß erledigen und die betreffende Urkunde auffinden kann.[646] Folglich muss der Schadensersatzkläger das beweiserhebliche Dokument möglichst konkret benennen. Ob ein bloßer Verweis des Geschädigten bzw. des Klägers auf die Kronzeugenerklärung bzw. den Antrag auf Kronzeugenbehandlung ausreicht, ist fraglich, dürfte aber im Hinblick auf die Durchsetzung der Rechte des Kartellgeschädigten und ein effektives *private enforcement* zu bejahen sein.[647]

cc. Exkurs: Datenübermittlung

Das OLG Hamm[648] entschied, dass Kronzeugenanträge und sonstige Kooperationsbeiträge in der staatsanwaltschaftlichen Akte dem Zivilgericht im Schadensersatzverfahren zugänglich sind. Die Gewährung von Einsicht in die bei der Staatsanwaltschaft geführten Ermittlungsakten richte sich nach den §§ 474 Abs. 1, 477 Abs. 2 und Abs. 4, 478 StPO. Die Tatbestandswirkung des § 33 Abs. 4 GWB macht nach Ansicht des Gerichts das Akteneinsichtsgesuch „nicht erforderlich“

---

645 Der Anspruch aus §§ 406 e, 475 StPO i.V.m. § 46 Abs. 1 OWiG ist subsidiär, vgl. *Jüntgen*, WuW 2007, 128, 131; *Schroll*, Effektivität der Kronzeugenprogramme, S. 239.

646 Für § 432 ZPO: *Reichold*, in: Thomas/Putzo, ZPO, § 432, Rn. 2; für § 273 ZPO: *Greger*, in: Zöller, ZPO, § 273, Rn. 3, 7.

647 So auch *Schroll*, Effektivität der Kronzeugenprogramme, S. 240.

648 OLG Hamm, Beschl. v. 26.11.2013 – 1 VAs 116/13 – 120/13 und 122/13, BB 2014, 526.

im Sinne des § 474 Abs. 1 StPO.[649] Sowohl der für den Zivilprozess maßgebliche Beibringungsgrundsatz als auch die Systematik der abgestuften Akteneinsicht in § 147, § 406e, § 474 und § 475 StPO stehen der Aktenübersendung an ein Zivilgericht nicht entgegen. Außerdem stellen Bonusanträge in Kronzeugenprogrammen trotz der den Kartellanten zugesicherten Vertraulichkeit keine ungewöhnliche Art von Daten im Sinne des § 477 Abs. 4 S. 2 StPO dar.[650]

Das BVerfG,[651] das sich mit einer hierauf eingelegten Verfassungsbeschwerde auseinandersetzen musste, entschied, dass ein Verstoß gegen den Schutz von Betriebs- und Geschäftsgeheimnissen aus Art. 12 Abs. 1 GG nicht vorliege. Die angegriffenen Entscheidungen würden das Grundrecht der Beschwerdeführerin nicht verkennen. Im zu entscheidenden Fall begegne das Zusammenspiel der §§ 474, 477 StPO und § 273 Abs. 2 Nr. 2 ZPO in der Auslegung des OLG, gegen die sich die Beschwerdeführerin wendete, keinen verfassungsrechtlichen Bedenken.[652]

dd. Zugriff auf Dokumente im Besitz des Kronzeugen

Neben der Möglichkeit, Zugriff auf Unterlagen zu erhalten, die sich im Besitz der Behörden befinden, können die Dokumente eingesehen werden, die im Besitz des Kronzeugen verblieben sind. Gemäß § 273 Abs. 2 Nr. 5 ZPO können deutsche Gerichte zur Vorbereitung des Termins Anordnungen gemäß §§ 142, 144 ZPO erlassen, Urkunden oder sonstige Unterlagen vorzulegen, die sich im Besitz einer Partei oder eines Dritten befinden. Jedoch kann der Richter wegen § 142 Abs. 1 ZPO Urkunden zu Beweiszwecken nicht unabhängig von einem schlüssigen Parteivortrag anfordern.[653] Es bedarf eines substantiierten Parteivortrags.[654] Angesichts der Ausführungen des Gerichtshofs[655] und der neuen Entwicklung zum Akteneinsichtsrecht ist aber davon auszugehen, dass ein mit einer Schadensersatzklage befasstes Gericht aufgrund des strukturellen Informationsdefizits und der Informationsasymmetrie der geschädigten Dritten keine allzu strengen Anforderungen an die Substantiierung des Parteivortrags stellen wird.[656]

Zudem stehen die Anordnungen nach §§ 142, 144 ZPO im Ermessen des Gerichts. Die Erwägung, dass Zivilgerichte zum Schutz der Effektivität von Kronzeugenprogrammen Anordnungen zur Vorlegung oder Inaugenscheinnahme von Kronzeugendokumenten im Rahmen des Ermessens ablehnen, ist aufgrund der aktuellen

649 OLG Hamm, Beschl. v. 26.11.2013 – 1 VAs 116/13 – 120/13 und 122/13, BB 2014, 526, 530.
650 OLG Hamm, Beschl. v. 26.11.2013 – 1 VAs 116/13 – 120/13 und 122/13, BB 2014, 526, Leitsatz 3.
651 BVerfG, Beschl. v. 6.3.2014 – 1 BvR 3541/13 u. a., NJW 2014, 1581.
652 BVerfG, Beschl. v. 6.3.2014 – 1 BvR 3541/13 u. a., NJW 2014, 1581, 1581.
653 BGH, Urt. v. 1.8.2006 – X ZR 114/03, NJW-RR 2007, 106 f.
654 Ausführlich dazu *Schroll*, Effektivität der Kronzeugenprogramme, S. 217 f.
655 BGH, Urt. v. 1.8.2006 – X ZR 114/03, NJW-RR 2007, 106, 107.
656 *Schroll*, Effektivität der Kronzeugenprogramme, S. 219.

Rechtsprechung nicht mehr haltbar.[657] Denn Zivilgerichte sind im Gegensatz zu Kartellbehörden nicht ausschließlich der objektiven Wirtschaftsaufsicht, sondern vor allem dem effektiven Individualrechtsschutz verpflichtet.

Sowohl die Rechtsprechung auf nationaler wie auf europäischer Ebene als auch die EU-Kommission haben sich eingehend mit der Problematik der privaten Rechtsdurchsetzung, insbesondere dem Akteneinsichtsrecht im Rahmen privater Schadensersatzklagen, beschäftigt.

Die EU-Kommission hat schon im Jahr 2008 versucht, Lösungsansätze zu liefern.

b. Weißbuch der Europäischen Kommission „Schadensersatzklagen wegen der Verletzung des EG-Wettbewerbsrechts"

Die EU-Kommission hatte im Dezember 2005 das Grünbuch über Schadensersatzklagen wegen Verletzung des EU-Wettbewerbsrechts vorgelegt. Darin wurden Änderungen für die nationalen Rechtsordnungen im Bereich der kartellrechtlichen Schadensersatzklagen empfohlen und begründet. Im April 2008 wurde schließlich auf der Basis des Grünbuches[658] das Weißbuch[659] über Schadensersatzklagen wegen Verletzung des EU-Wettbewerbsrechts (damals EG-Wettbewerbsrechts) publiziert.

Die Gründe für die Vorlage dieses Weißbuches liegen in den Prinzipien des Gemeinschaftsrechts. Danach muss jeder Bürger und jedes Unternehmen, die aufgrund einer Zuwiderhandlung gegen EU-Wettbewerbsvorschriften der Art. 101 und 102 AEUV (ex-Art. 81 und 82 EG-Vertrag) einen Schaden erleiden, vom Rechtsverletzer Schadensersatz fordern können.[660] Die Kommission kam schon 2005 in ihrem Grünbuch zu dem Schluss, dass der Missstand eines nicht geltend gemachten oder zugesprochenen Schadensersatzes auf diverse Hindernisse in den materiellen und verfahrensrechtlichen Vorschriften der Mitgliedstaaten zurückzuführen sei. Derartige Schadensersatzfälle wiesen eine Reihe von besonderen Merkmalen auf, denen das traditionelle Haftungs- und Verfahrensrecht nur unzureichend Rechnung trage – dadurch bestehe ein hohes Maß an Rechtsunsicherheit.[661]

Den ineffektiven Schadensersatzklagen im Wettbewerbsrecht kann dadurch begegnet werden, dass sowohl die Gemeinschaft als auch die Mitgliedstaaten Rahmenbedingungen schaffen, die einen Mindeststandard bei der Durchsetzung von Schadensersatzklagen im Zusammenhang mit den Artikel 101 und 102 AEUV (ex-Artikel 81 und 82 EG-Vertrag) gewährleisten. In dem Weißbuch werden konkrete

---

657 So schon *Jüntgen*, WuW 2007, 128, 133 f.

658 Grünbuch der Europäischen Kommission über Schadensersatzklagen wegen Verletzung des EU-Wettbewerbsrechts vom 19.12.2005, KOM (2005), 672 endgültig.

659 Weißbuch der Europäischen Kommission über Schadensersatzklagen wegen Verletzung des EG-Wettbewerbsrechts vom 2.4.2008, KOM (2008), 165 endgültig.

660 So auch EuGH, Urt. v. 20.9.2001 – Rs. C-453/99, Slg. 2001, I-6297 (Rn. 24, 26), EuZW 2001, 715, 716 (Courage/Crehan) und EuGH, Urt. v. 13.7.2006 – verb. Rs. C-295/04-C-298/04, Slg. 2006, I-6619 (Rn. 59, 61), EuZW 2006, 529 , 532 (Manfredi u. a.).

661 Weißbuch der EU-Kommission, KOM (2008), 165 endgültig, Abschnitt 1.1.

Vorschläge unterbreitet, die Hindernisse bei Durchsetzung von wettbewerbsrechtlichen Schadenersatzansprüchen zu beseitigen.

aa. Prinzipien und Ziele des Weißbuches

Vorrangiges Ziel des Weißbuches ist, die rechtlichen Rahmenbedingungen für die Geschädigten zu verbessern, damit sie ihre im Gemeinschaftsrecht verankerten Rechte besser durchsetzen können. Wichtiges Leitprinzip ist die vollständige Entschädigung der Geschädigten.[662] Ein bedeutender Mechanismus ist unter anderem, dass den Geschädigten die Kosten der Wettbewerbsverstöße abgenommen und durch die Rechtsverletzer getragen werden. Wirksame Instrumente zur privaten Rechtsdurchsetzung (*private enforcement*) erhöhen die Wahrscheinlichkeit, Rechtsverletzer vollständig zur Verantwortung zu ziehen. Des Weiteren schrecken sie wettbewerbswidrig handelnde Akteure ab, da sie damit rechnen müssen, bei Aufdeckung ihrer Machenschaften zusätzlich mit einer privaten Schadensersatzklage belastet zu werden. Diese Abschreckung wiederum könnte die Einhaltung der EG-Wettbewerbsvorschriften durch die potenziellen Rechtsverletzer fördern.

Die Kommission lässt sich zudem davon leiten, dass für eine wirksamere Durchsetzung von Schadensersatzansprüchen ein europäischer Rechtsrahmen erforderlich ist. Deshalb werden in dem Weißbuch angemessene Maßnahmen vorgeschlagen, die sich auf die europäische Rechtskultur und -tradition stützen.[663]

Ein weiteres Leitprinzip besteht darin, einen gerechten und ausgewogenen Ausgleich zwischen der behördlichen und der privaten Durchsetzung mittels Schadensersatzklagen zu schaffen, ohne dass das *public enforcement* ersetzt oder gefährdet wird.

Im letzten Absatz des Abschnitts 1.2 des Weißbuches bekräftigt die Kommission noch einmal, dass es jedem Opfer eines Wettbewerbsverstoßes prinzipiell möglich sein muss, seinen Schadensersatzanspruch wirksam geltend zu machen. Deshalb beziehen sich die im Weißbuch der Kommission vorgeschlagenen Maßnahmen grundsätzlich auf alle Gruppen von Geschädigten und alle Formen von Verstößen gegen Artikel 101 und 102 AEUV (ex-Artikel 81 und 82 EG-Vertrag) sowie auf alle Wirtschaftszweige. Die Kommission hält es zudem für angezeigt, sowohl Folgeklagen wie auch eigenständige Klagen zu berücksichtigen.

bb. Vorgeschlagene Maßnahmen und rechtspolitische Optionen

In Abschnitt 2 des Weißbuches werden konkrete Maßnahmen vorgeschlagen, die eine Verbesserung der Durchsetzung privater Schadensersatzklagen gewährleisten sollen. Allen Opfern soll ein Zugang zu wirksamen Rechtsschutzinstrumenten gegeben werden.

---

662 Weißbuch der EU-Kommission, KOM (2008), 165 endgültig, Abschnitt 1.2.
663 Weißbuch der EU-Kommission, KOM (2008), 165 endgültig, Abschnitt 1.2.

Im Folgenden werden nur Maßnahmen erläutert, die für das Verhältnis des Whistleblowerschutzes zum Akteneinsichtsrecht/Beweismittelzugang des Geschädigten relevant sind.

(1) Klagebefugnis

In der *Courage/Crehan*-Entscheidung des Europäischen Gerichtshofs wurde von den Richtern festgehalten, dass „jedermann", dem durch einen Kartellrechtsverstoß ein Schaden entstanden ist, vor einzelstaatlichen Gerichten Ersatz verlangen kann.[664] Dieser Grundsatz gilt auch für indirekte Abnehmer, also für diejenigen Abnehmer, die keinen direkten Kontakt mit dem Rechtsverletzer hatten, aber dennoch unter Umständen einen erheblichen Schaden erlitten haben, weil auf sie in der Vertriebskette rechtswidrige Preisaufschläge abgewälzt wurden.

Weiterhin empfiehlt die EU-Kommission Maßnahmen, um den kollektiven Rechtsschutz zu fördern. Es muss eine Bündelung der individuellen Schadensersatzforderungen ermöglicht werden. Insbesondere bei relativ geringwertigen Streuschäden schrecken einzelne Verbraucher, aber auch kleinere Unternehmen, häufig vor Individualklagen zurück angesichts der mit einem Gerichtsverfahren verbundenen Kosten, Verzögerungen, Unwägbarkeiten, Risiken und Belastungen.

Die Kommission hat zwei Lösungsvorschläge, um einen kollektiven Rechtsschutz zu gewährleisten:[665]

- Verbandsklagen, die von qualifizierten Einrichtungen für eine Gruppe bezeichneter oder, in eher begrenzten Fällen, identifizierbarer Einzelpersonen erhoben werden.[666]
- Opt-in-Gruppenklagen,[667] zu denen sich einzelne Opfer ausdrücklich zusammenschließen, um ihre jeweiligen Schadensersatzansprüche in einer einzigen Klage zusammenzufassen.

---

664 EuGH, Urt. v. 20.9.2001 – Rs. C-453/99, Slg. 2001, I-6297 (Rn. 24, 26), EuZW 2001, 715, 716 (Courage/Crehan), bestätigt durch Entscheidung des EuGH, Urt. v. 13.7.2006 – verb. Rs. C-295/04-C-298/04, Slg. 2006, I-6619 (Rn. 61), EuZW 2006, 529, 532 f. (Manfredi u. a.).

665 Weißbuch der EU-Kommission, KOM (2008), 165 endgültig, Abschnitt 2.1; vgl. auch *Heinemann*, in: Zäch/Heinemann/Kellerhals, The Development of Competition Law, S. 306 f.

666 Diese Einrichtungen werden entweder a) im Vorhinein offiziell von einem Mitgliedstaat benannt oder b) von einem Mitgliedstaat ad hoc für einen bestimmten Wettbewerbsverstoß ermächtigt, im Namen einiger oder aller ihrer Mitglieder Verbandsklagen einzureichen.

667 „*Opt-in*-Gruppenklagen" sind Klagen im Namen von mehreren namentlich erwähnten Geschädigten, denen sich weitere Geschädigte anschließen können. Im Gegensatz dazu sind „*Opt-out*-Klagen" Klagen, bei denen ein einzelner Geschädigter Klage im Namen der ganzen Gruppe erheben kann, während die übrigen Geschädigten dieser Gruppe noch nicht bestimmt sind.

Da qualifizierte Einrichtungen nicht immer in der Lage oder bereit sein werden, allen Schadensfällen mit Hilfe einer Verbandsklage nachzugehen, müssen Verbandsklagen durch Opt-in-Klagen ergänzt werden, um einen wirksamen Schutz gewährleisten zu können.[668]

(2) Zugang zu Beweismitteln

In Bezug auf die Sachverhaltsfeststellung und Bewertung sind Wettbewerbsfälle außergewöhnlich aufwendig. Viele zentrale Beweismittel, die für die Begründung eines wettbewerbsrechtlichen Schadensersatzanspruches erforderlich sind, werden häufig geheim gehalten und sind, da sie sich in der Sphäre der Beklagten oder Dritter befinden, den Klägern nicht hinreichend bekannt.[669]

Der Vorschlag der Kommission geht dahin, ein Mindestniveau für die Offenlegung von Beweismitteln zwischen den Streitparteien festzulegen. Um den in der Richtlinie 2004/48/EG zur Durchsetzung der Rechte des geistigen Eigentums zugrunde liegenden Ansatz fortzuführen und der strukturellen Informationsasymmetrie entgegenzuwirken, sollte der Zugang zu Beweismitteln nach Ansicht der Kommission auf der Grundlage eines Tatsachenvortrags gewährt werden. Ebenfalls sind die Plausibilität des Vortrags und die Verhältnismäßigkeit des Offenlegungsbegehrens gerichtlich zu prüfen.

Die Kommission schlägt daher Folgendes vor:[670]

- Die einzelstaatlichen Gerichte sollten unter bestimmten Voraussetzungen befugt sein, die Prozessparteien oder Dritte anzuweisen, genau bezeichnete Kategorien von relevanten Beweismitteln offenzulegen.
- Ein solcher Offenlegungsbeschluss sollte unter anderem voraussetzen, dass der Kläger:
    - alle Tatsachen und Beweismittel, die ihm vernünftigerweise zugänglich sind, dargelegt hat, und diese plausible Gründe für die Annahme enthalten, dass er durch einen Wettbewerbsverstoß des Beklagten Schaden erlitten hat;
    - zur Überzeugung des Gerichts nachgewiesen hat, dass es ihm unter vernünftigerweise zumutbaren Anstrengungen nicht möglich ist, die für eine Begründetheit der Klage weiterhin erforderlichen Beweise anderweitig beizubringen;
    - die verschiedenen Kategorien von Beweismitteln, für die er eine Offenlegung beantragt, genau genug bezeichnet hat und
    - das Gericht überzeugt hat, dass die beantragte Offenlegung sowohl für die Klage erheblich als auch erforderlich und verhältnismäßig ist.

---

668 Weißbuch der EU-Kommission, KOM (2008), 165 endgültig, Abschnitt 2.1, letzter Absatz.

669 Weißbuch der EU-Kommission, KOM (2008), 165 endgültig, Abschnitt 2.2.

670 Weißbuch der EU-Kommission, KOM (2008), 165 endgültig, Abschnitt 2.2. sowie Kapitel 3 des Arbeitspapiers der Kommissionsdienststellen.

Die Kommission fordert zudem, dass ein angemessener Schutz für Unternehmenserklärungen im Rahmen von Kronzeugenprogrammen und für die Untersuchungen von Wettbewerbsbehörden gewährleistet werden sollte. Um die Beweismittel vor Vernichtung zu schützen oder der Verweigerung der angeordneten Offenlegung entgegenzukommen, sollen die Gerichte befugt sein, hinreichend abschreckende Sanktionen zu verhängen.[671]

(3) Verhältnis zwischen Kronzeugenprogrammen und Schadensersatzklagen

Sowohl für die behördliche als auch für die private Durchsetzung des Wettbewerbsrechts ist es von Bedeutung, dass Kronzeugenprogramme ihre Attraktivität behalten.[672] Es muss daher ein angemessener Schutz vor Offenlegung von Informationen sichergestellt werden, damit der Antragsteller im Rahmen der Unternehmenserklärungen nicht schlechter gestellt wird als die anderen Beteiligten. Eine drohende Offenlegung schränkt die Effektivität und Qualität der Angaben des Antragstellers ein.

Der Vorschlag der Kommission sieht deshalb vor:[673]

- Der Schutz sollte für alle Unternehmenserklärungen gelten, in denen eine Kronzeugenbehandlung für einen Verstoß gegen Artikel 101 AEUV (ex-Artikel 81 EG-Vertrag) beantragt wurde.
- Der Schutz sollte unabhängig davon gelten, ob dem Antrag auf Kronzeugenbehandlung stattgegeben wird oder der Antrag abgewiesen wird oder die Wettbewerbsbehörde gar nicht über den Antrag entscheidet.

Eine weitere Maßnahme, um Kronzeugenprogramme attraktiv zu halten, könnte eine Begrenzung der zivilrechtlichen Haftung von Kronzeugen sein, denen die Geldbuße erlassen wird. Deshalb regt die Kommission eine Diskussion an,[674] die zivilrechtliche Haftung jener Kronzeugen, denen eine Geldbuße erlassen wurde, auf Schadensersatzansprüche ihrer direkten und indirekten Vertragspartner zu begrenzen.

cc. Umsetzung des Weißbuches

Das Weißbuch setzt keine verbindlichen Regeln fest und die EU-Kommission hat auch nicht verlauten lassen, wie die nächsten Entwicklungsschritte bei der privaten Rechtsdurchsetzung aussehen sollten. Lediglich zu einer Stellungnahme wurde auf

---

671 Darunter fällt auch die Möglichkeit, im Verfahren der Schadensersatzklage nachteilige Schlussfolgerungen für die betroffene Partei zu ziehen, Weißbuch der EU-Kommission, KOM (2008), 165 endgültig, Abschnitt 2.2, letzter Absatz sowie Kapitel 3 des Arbeitspapiers der Kommissionsdienststellen.

672 Weißbuch der EU-Kommission, KOM (2008), 165 endgültig, Abschnitt 2.9.

673 Weißbuch der EU-Kommission, KOM (2008), 165 endgültig, Abschnitt 2.9. sowie Kapitel 10 Abschnitt B.1 des Arbeitspapiers der Kommissionsdienststellen.

674 Weißbuch der EU-Kommission, KOM (2008), 165 endgültig, Abschnitt 2.9, letzter Absatz mit Verweis auf Kapitel 10 Abschnitt B.2 des Arbeitspapiers der Kommissionsdienststellen.

der letzten Seite des Weißbuches eingeladen. Die bei der EU-Kommission eingegangenen Stellungnahmen sind öffentlich zugänglich.[675] Ob eine Richtlinie zum kartellrechtlichen Schadensersatz die Vorschläge des Weißbuches hinreichend ersetzen oder sogar verbessern kann, wird in einem der nachfolgenden Abschnitte näher erläutert. Ein erster Schritt in Richtung einer rechtlich verbindlichen Regelung für kartellrechtliche Schadensersatzklagen ist somit getan.

c. *Der Fall Pfleiderer*

Das Verhältnis zwischen *private* und *public enforcement* ist nicht eindeutig geklärt.

Die Kommission erkennt an, dass ein Gleichgewicht zwischen den Formen der Aufklärung und Durchführung gefunden werden muss und betont, dass die private Durchsetzung mittels Schadensersatzklagen die behördliche Durchsetzung ergänzen, nicht aber ersetzen oder gefährden soll. Um aber private Kartellschadensersatzklagen erfolgreich auf den Weg bringen zu können, muss der Kläger genügend Informationen haben, um seinen Schaden beziffern zu können. Dies kann ihm nur gelingen, wenn er Einsicht in die Akten des Verfahrens erlangt. Es stellt sich also die Frage nach der Zulässigkeit der Akteneinsicht in Kronzeugenunterlagen, um eine Klage vorzubereiten. Im Folgenden werden Entscheidungen erläutert, die eine Antwort auf die Frage geben können. Begonnen wird mit dem Fall *Pfleiderer*.

aa. Die Entscheidung des Bundeskartellamtes

Das AG Bonn[676] hat mit seinem Beschluss vom 18. Januar 2012 in der Rechtssache „Pfleiderer-AG/BKartA“ eine Grundsatzentscheidung getroffen. Es entschied, dass Kartellgeschädigte zur Vorbereitung einer privaten Schadensersatzklage keine Einsicht in die Kronzeugenunterlagen erhalten sollen und das Bundeskartellamt in der Folge berechtigt ist, Kartellgeschädigten die Einsicht in die Unterlagen zu verweigern.[677]

Bis zum Beschluss des AG Bonn sind zwei Verfahrensstadien durchlaufen worden.

Das Bundeskartellamt hatte 2008 gegen die am Kartell beteiligten Unternehmen (Dekorpapierhersteller) Geldbußen in Höhe von insgesamt 62 Mio. Euro wegen Preis- und Kapazitätsstilllegungsabsprachen verhängt.[678] Als Kundin der am Kartell

---

675 Stellungnahmen zum Weißbuch sind verfügbar unter: http://ec.europa.eu/competition/antitrust/actionsdamages/white_paper_comments/iv_de.pdf, (abgerufen am 15.3.2014).

676 AG Bonn, Beschl. v. 18.1.2012 – 51 Gs 53/09, GRUR-RR 2012, 178 (Pfleiderer II).

677 Zur weiteren Information vgl. Pressemitteilung des BKartA vom 30.1.2012, abrufbar unter: http://www.bundeskartellamt.de/SharedDocs/Meldung/DE/Pressemit teilungen/2012/30_01_2012_Pfleiderer.html, (abgerufen am 20.11.2013).

678 Vgl. Pressemitteilung des BKartA vom 5.2.2008, abrufbar unter: http://www.bundes kartellamt.de/SharedDocs/Meldung/DE/Pressemitteilungen/2008/05_02_2008_Dekopapierhersteller.html, (abgerufen am 20.11.2013).

beteiligten Dekorpapierhersteller beantragte die Pfleiderer-AG am 26. Februar 2008 beim Bundeskartellamt Einsicht in die Verfahrensakten des Bundeskartellamtes nach § 406 e StPO i. V. m. § 46 OWiG, um die private Schadensersatzklage vorzubereiten. § 406 e Abs. 1 StPO erlaubt einem Rechtsanwalt oder einer Rechtsanwältin die Einsicht in Akten des Verfahrens, um zum Vorteil des Verletzten Beweise oder Informationen zu erlangen. Dies ist nur möglich, wenn ein berechtigtes Interesse dargelegt wird. Gemäß § 406 e Abs. 2 StPO kann dem Berechtigten die Akteneinsicht versagt werden, soweit überwiegende schutzwürdige Interessen des Beschuldigten oder anderer Personen entgegenstehen.

Im vorliegenden Fall versagte das Bundeskartellamt die Akteneinsicht in die Kronzeugenunterlagen, da andernfalls das Bonusprogramm an Wirksamkeit verlieren würde.[679] Das Bundeskartellamt lehnte (zumindest teilweise) zum Schutz des Kronzeugenprogramms den Antrag auf Akteneinsicht ab.[680]

Gegen diese Entscheidung wandte sich die Pfleiderer-AG am 27. November 2008 mit Antrag an das AG Bonn. Am 4. August 2009 setzte das AG Bonn das Verfahren aus und legte dem EuGH gem. Art. 267 AEUV die Frage zur Vorabentscheidung vor, ob kartellrechtliche Bestimmungen des Unionsrechts, insbesondere die Artikel 11 und 12 der Verordnung Nr. 1/2003 (EG) sowie Artikel 10 Abs. 2 in Verbindung mit Artikel 3 Abs. 1 lit. g EG dahingehend auszulegen seien, dass Geschädigte eines Kartells zur Geltendmachung zivilrechtlicher Ansprüche keine Akteneinsicht in Bonusanträge und von Bonusantragstellern in diesem Zusammenhang freiwillig herausgegebene Informationen und Unterlagen erhalten dürfen, die eine mitgliedstaatliche Behörde im Rahmen eines nationalen Bonusprogramms erhalten hat.[681]

bb. Die Entscheidung des EuGH

Gemäß dem Tenor des Urteils sind die kartellrechtlichen Bestimmungen der Union, insbesondere die Verordnung Nr. 1/2003 (EG) des Rates vom 16. Dezember 2002 zur Durchführung der in den Artikel 101 und 102 AEUV niedergelegten Wettbewerbsregeln dahingehend auszulegen, dass sie es nicht verbieten, dass eine durch einen Verstoß gegen das Wettbewerbsrecht der Union geschädigte und Schadensersatz fordernde Person Zugang zu Dokumenten eines Kronzeugenverfahrens erhält, die den Urheber dieses Verstoßes betreffen – es ist jedoch Sache der Gerichte der Mitgliedstaaten, auf der Grundlage des jeweiligen nationalen Rechts unter Abwägung

679 Vgl. *Mäger/Zimmer/Milde*, WuW 2009, 885, 893; *Seitz*, GRUR-RR 2012, 137, 138.

680 *Seitz*, GRUR-RR 2012, 137, 138; so sieht es auch der Präsident des BKartA, *Mundt*, FS Möschel, S. 427, 438, der aufgrund der höheren Effektivität für einen „Vorrang“ der behördlichen Kartellverfolgung vor privatrechtlichen Schadensersatzansprüchen plädiert.

681 AG Bonn, EuGH-Vorlagebeschl. v. 4.8.2009 – 51 Gs 53/09, NZG 2010, 60 f.

1 StPO.[709] Danach sei die Akteneinsicht zu versagen, „soweit überwiegende schutzwürdige Interessen des Beschuldigten oder anderer Personen entgegenstehen". Dies sei dann der Fall, wenn das Interesse des Betroffenen oder der Nebenbetroffenen an der Geheimhaltung bestimmter in den Akten enthaltenen und sie betreffenden Erkenntnisse gewichtiger sei als das berechtigte Interesse des Verletzten, den Akteninhalt insoweit einsehen zu können – erforderlich sei eine Abwägung aller maßgeblichen Umstände des Einzelfalls.[710]

Das Gericht unterteilte die Dokumente in zwei Kategorien: zum einen die Geschäftsgeheimnisse sowie die persönlichen Daten der Betroffenen in den Bußgeldbescheiden, zum anderen die "übrigen Aktenbestandteile", die sich aus dem Bonusantrag und den freiwillig übergebenen Dokumenten zusammensetzen. Für letztere versagte das Gericht die Akteneinsicht mit der Begründung, dass das Interesse an der Geheimhaltung der darin festgehaltenen Daten gewichtiger sei, als das Interesse der Antragsteller auf Akteneinsicht.[711] Die Einsicht sei insbesondere aus Gründen des Vertrauensschutzes zu verwehren. Der Kronzeuge habe auf die in Randnummer 22 der Bonusregelung niedergelegte Geheimhaltung vertraut, in der das Bundeskartellamt versichert, dass es privaten Dritten Akteneinsicht in die Bonusanträge und die dazu übermittelten Unterlagen grundsätzlich nicht gewähren wird.[712] Diese Zusage sei auf das öffentliche Interesse an einer wirksamen und effizienten Verfolgung von Kartellverstößen zurückzuführen.[713] Darüber hinaus sei eine Akteneinsicht nicht erforderlich, weil die Einlassung der Kartellanten ohnehin keine Informationen zur Schadenshöhe enthielt.[714]

Was die erste Kategorie betrifft, führt nach Ansicht des Gerichts die nach § 406 e Abs. 2 Satz 1 StPO gebotene Abwägung zu dem Ergebnis, dass lediglich in die Bußgeldbescheide Einsicht zu gewähren ist, in denen die Geschäftsgeheimnisse der Nebenbetroffenen und die persönlichen Daten der Betroffenen anonymisiert sind. Das Interesse an der Geheimhaltung über den Tatvorwurf hinausgehender

709 OLG Düsseldorf, Beschl. v. 22.8.2012 – V-4 Kart 5/11 (OWi), V-4 Kart 6/11 (OWi), (Kaffeeröster), WRP 2012, 1596, 1600.

710 OLG Düsseldorf, Beschl. v. 22.8.2012 – V-4 Kart 5/11 (OWi), V-4 Kart 6/11 (OWi), (Kaffeeröster), WRP 2012, 1596, 1599.

711 OLG Düsseldorf, Beschl. v. 22.8.2012 – V-4 Kart 5/11 (OWi), V-4 Kart 6/11 (OWi), (Kaffeeröster), WRP 2012, 1596, 1600.

712 OLG Düsseldorf, Beschl. v. 22.8.2012 – V-4 Kart 5/11 (OWi), V-4 Kart 6/11 (OWi), (Kaffeeröster), WRP 2012, 1596, 1600; siehe auch *Mäger/Zimmer/Milde*, WuW 2009, 885, 893.

713 OLG Düsseldorf, Beschl. v. 22.8.2012 – V-4 Kart 5/11 (OWi), V-4 Kart 6/11 (OWi), (Kaffeeröster), WRP 2012, 1596, 1600.

714 OLG Düsseldorf, Beschl. v. 22.8.2012 – V-4 Kart 5/11 (OWi), V-4 Kart 6/11 (OWi), (Kaffeeröster), WRP 2012, 1596, 1600.

Geschäftsgeheimnisse und der persönlichen Daten, wie Geburtsdaten, überwiege das Informationsinteresse der Antragstellerin.[715]

Aber auch eine Einsicht in die übrige, um Geschäftsgeheimnisse bereinigte Fassung der Verfahrensakte und in die um Geschäftsgeheimnisse bereinigten Asservate sei im konkreten Fall den Antragstellern nicht zu gewähren.[716] Das Gericht stützte sich hierbei auf § 406 e Abs. 2 Satz 2 StPO. Danach kann die Akteneinsicht verweigert werden, wenn durch sie das Verfahren erheblich verzögert würde. Dies sei gegeben, da die vorzunehmende Anonymisierung der Geschäftsgeheimnisse in den Akten zu einer erheblichen Verfahrensverzögerung führen würde.[717] Im Ergebnis ließ das Gericht lediglich die Einsichtnahme in die um Geschäftsgeheimnisse und persönliche Angaben geschwärzten Bußgeldbescheide sowie die Überlassung einer Liste der sichergestellten Asservate zu.

bb. Rechtliche Würdigung

Dem Beschluss ist insoweit zuzustimmen, als die Individualinteressen der Parteien – und nicht die abstrakt-öffentlichen Interessen – in einer Einzelabwägung als vorrangig angesehen werden. Die Orientierung an Individualinteressen führt nicht zu einer Stärkung der privaten Rechtsdurchsetzung, sondern zur Stärkung der privaten Rechte der Kronzeugen.[718]

Wie schon oben ausgeführt, dürfen die nationalen Gerichte die Durchsetzung und Verwirklichung des Unionsrechts nicht praktisch unmöglich machen oder erschweren.[719] Seit der *Pfleiderer*-Entscheidung ist klar, dass dem Geschädigten der Akteneinsichtsanspruch nicht gänzlich verwehrt werden darf. Es ist ein im Unionsrecht niedergelegter Grundsatz, dass jedermann Ersatz des Schadens verlangen kann, der ihm durch ein Verhalten entstanden ist, das den Wettbewerb beschränken oder verfälschen kann.[720] Diese bisherige in Randnummer 22 der Bonusregelung zum Ausdruck kommende Verwaltungspraxis des Bundeskartellamtes, nach der Akteneinsichtgesuche in Kronzeugendokumente stets abgelehnt wurden und der sich daraus entwickelnde Vertrauensschutz der Kartellanten auf Geheimhaltung,

---

715 OLG Düsseldorf, Beschl. v. 22.8.2012 – V-4 Kart 5/11 (OWi), V-4 Kart 6/11 (OWi), (Kaffeeröster), WRP 2012, 1596, 1600.

716 OLG Düsseldorf, Beschl. v. 22.8.2012 – V-4 Kart 5/11 (OWi), V-4 Kart 6/11 (OWi), (Kaffeeröster), WRP 2012, 1596, 1601.

717 OLG Düsseldorf, Beschl. v. 22.8.2012 – V-4 Kart 5/11 (OWi), V-4 Kart 6/11 (OWi), (Kaffeeröster), WRP 2012, 1596, 1601.

718 So auch *Dworschak/Maritzen*, WuW 2013, 829, 835.

719 Vgl. EuGH, Urt. v. 16.7.1998 – Rs. C-298/96, Slg. 1998, I-4767 (Rn. 23, 24), EuZW 1998, 603, 605 (Oelmühle).

720 Vgl. EuGH, Urt. v. 14.6.2011 – Rs. C-360/09, Slg. 2011, I-5161 (Rn. 28), EuZW 2011, 598, 599 (Pfleiderer-AG/BKartA); so auch schon EuGH, Urt. v. 13.7.2006 – verb. Rs. C-295/04 bis C-298/04, Slg. 2006, I-6619 (Rn. 59, 61), EuZW 2006, 529 (Manfredi u. a.); EuGH, Urt. v. 20.9.2001 – Rs. C-453/99, Slg. 2001, I-6297 (Rn. 24, 26), EuZW 2001, 715 (Courage/Crehan).

steht insoweit im Widerspruch zum Unionsrecht.[721] Zwar schließt diese Tatsache den Rückgriff auf Vertrauensgesichtspunkte nicht schlechthin aus,[722] jedoch tritt dem Vertrauensschutz im Hinblick auf eine europarechtswidrige Verwaltungspraxis das Interesse an der Wirksamkeit und Durchsetzbarkeit des Unionsrechts entgegen.[723] Aufgrund des zu beachtenden Grundsatzes des *effet utile* ist eine besondere Gewichtung des Vertrauensschutzes bedenklich.

e. Der Fall Donau Chemie – Modifikation der Pfleiderer-Entscheidung?

Hintergrund war ein beim OLG Wien als Kartellgericht anhängig gemachtes Verfahren wegen eines Antrags der Unternehmensvereinigung Verband Druck & Medientechnik auf Einsicht in die Akten des von der Wettbewerbsbehörde gegen die Donau Chemie AG (u. a.) betriebenen Gerichtsverfahrens.[724] Die Gesellschaften wurden aufgrund ihrer Teilnahme an einem gegen Art. 102 AEUV verstoßenden Kartell zur Zahlung einer Geldbuße in Höhe von 1,5 Mio. Euro verurteilt.[725] Mit Beschluss vom 4. Oktober 2010 bestätigte der Oberste Gerichtshof als Rekursgericht die Entscheidung des OLG Wien, die damit rechtskräftig wurde.[726] Die Ermittlungen der Bundeswettbewerbsbehörde (BWB) wurden aufgrund von Informationen zweier Kronzeugen (Donau Chemie und DC Druck-Chemie Süd GmbH & Co. KG) eingeleitet. *Donau Chemie* hatte in diesem Verfahren als erster Kronzeuge zwar einige Informationen und Beweismittel vorgelegt, den genauen Umfang der Beteiligung am Kartell und einen Zeitraum der Zuwiderhandlung von mehr als zwei Jahren jedoch verschwiegen. Deshalb hatte das Kartellgericht auf Antrag der BWB entschieden, dass *Donau Chemie* doch eine Geldbuße zahlen müsse.[727]

Die geforderte Einsicht in die Akten wurde von allen Parteien mit Ausnahme der Bundeswettbewerbsbehörde abgelehnt. Im Rahmen des Verfahrens hatte das

---

721 *Dworschak/Maritzen*, WuW 2013, 829, 836.

722 Der Grundsatz des Vertrauensschutzes ist auch auf europäischer Ebene anerkannt, siehe Vorabentscheidungsverfahren EuGH, Urt. v. 18.6.2013 – Rs. C-681/11 (Rn. 40), WuW 2013, 893, 896 f. (Schenker).

723 Zum Vertrauensschutz bei gemeinschaftsrechtswidriger Subventionsgewährung vgl. BVerwG, Urt. v. 17.2.1993 – 11 C 47/92, NJW 1993, 2764, 2765; *Sachs*, in: Stelkens/Bonk/Sachs, VwVfG, § 48, Rn. 268.

724 OLG Wien als Kartellgericht – KG, Urt. v. 26.3.2010 – 29 Kt 5/09.

725 Veröffentlichung der Bundeswettbewerbsbehörde, BWB/K-159 Druckchemikalien OGH, abrufbar unter: http://www.bwb.gv.at/KartelleUndMarkmachtmissbrauch/Entscheidungen/Seiten/BWBK159-Druckchemikalien.aspx, (abgerufen am 5.5.2014).

726 (Oberster Gerichtshof) KOG, Beschl. v. 4.10.2010 – 16 Ok 5/10 (Druckchemikalien), RdW 2011, 221.

727 Veröffentlichung der Bundeswettbewerbsbehörde, BWB/K-159 Druckchemikalien OGH, abrufbar unter: http://www.bwb.gv.at/KartelleUndMarkmachtmissbrauch/Entscheidungen/Seiten/BWBK159-Druckchemikalien.aspx, (abgerufen am 5.5.2014).

Kartellgericht Zweifel, ob § 39 Abs. 2 öKartG mit der Auslegung des Unionsrechts vereinbar ist. Angesichts der in der *Pfleiderer*-Entscheidung betonten Notwendigkeit einer Einzelfallabwägung durch den nationalen Richter legte das Kartellgericht am 12. Oktober 2011[728] dem EuGH die Frage nach der Vereinbarkeit von § 39 Abs. 2 öKartG[729] mit dem Unionsrecht vor. Zwar bezog sich die Vorlagefrage nicht nur auf die Kronzeugendokumente, sondern auch auf den Inhalt der Ermittlungsakten insgesamt, dennoch konnten die Ausführungen des EuGH zu Kronzeugendokumenten herangezogen werden.

aa. Die Entscheidung des EuGH

Der Gerichtshof wies darauf hin, dass das Unionsrecht dem Einzelnen Pflichten auferlege, aber auch Rechte verleihen könne und dass solche Rechte nicht nur entstünden, wenn die Verträge dies ausdrücklich bestimmten, sondern auch aufgrund von eindeutigen Verpflichtungen, die diese Verträge dem Einzelnen wie auch den Mitgliedstaaten und den Organen der Union auferlegten.[730] In diesem Zusammenhang habe der Gerichtshof schon festgestellt, dass die Grundsätze der Art. 101 und 102 AEUV und das in ihnen kodifizierte Verbot des wettbewerbswidrigen Verhaltens an Wirksamkeit verlieren würde, wenn dem Einzelnen der Schaden nicht ersetzt würde, der ihm durch einen wettbewerbswidrigen Vertrag entstanden sei.[731] Es sei ständige Rechtsprechung, dass die nationalen Gerichte dafür Sorge tragen müssten, die Rechte zu schützen, die das Gemeinschaftsrecht dem Einzelnen verleihe.[732]

Des Weiteren wird, wie schon in der Sache *Courage/Crehan* und *Manfredi*, ausgeführt, dass das Verlangen von Schadensersatz dazu geeignet sei, von Vereinbarungen oder Verhaltensweisen abzuhalten, die den Wettbewerb beschränken oder

---

728 OLG Wien als Kartellgericht (Vorabentscheidungsersuchen), Beschl. v. 12.10.2011 – 29 Kt 5/09–86 (Donau Chemie), WuW KRInt 2005, 547.

729 § 39 Abs. 2 des österreichischen Kartellgesetzes 2005 (im Folgenden: öKartG) bestimmt: „In die Akten des Kartellgerichts können am Verfahren nicht als Partei beteiligte Personen nur mit Zustimmung der Parteien Einsicht nehmen."

730 EuGH, Urt. v. 6.6.2013 – Rs. C-536/11 (Rn. 20), JZ 2013, 734, 735 (Donau Chemie u.a) mit Verweis auf EuGH, Urt. v. 19.11.1991 – Rs. C- 6/90 und C-9/90, Slg. 1991, I-5357 (Rn. 30), NJW 1992, 165, 166 f. (Francovich u. a.) sowie Urt. v. 20.9.2001 – Rs. C-453/99, Slg. 2001, I-6297 (Rn. 19), EuZW 2001, 715, 716 (Courage/Crehan) und die dort angeführte Rechtsprechung.

731 EuGH, Urt. v. 6.6.2013 – Rs. C-536/11 (Rn. 21), JZ 2013, 734, 735 (Donau Chemie u. a.) mit Verweis auf EuGH, Urt. v. 20.9.2001 – Rs. C-453/99, Slg. 2001, I-6297 (Rn. 26), EuZW 2001, 715, 716 (Courage/Crehan).

732 EuGH, Urt. v. 6.6.2013 – Rs. C-536/11 (Rn. 22), JZ 2013, 734, 735 (Donau Chemie u. a.); EuGH, Urt. v. 19.6.1990 – Rs. C-213/89, Slg. 1990, I-2433 (Rn. 19), NJW 1991, 2271, 2272 (Factortame u. a.); EuGH, Urt. v. 20.9.2001 – Rs. C-453/99, Slg. 2001, I-6297 (Rn. 25), EuZW 2001, 715, 716 (Courage/Crehan), sowie EuGH, Urt. v. 13.7.2006 – verb. Rs. C-295/04 bis C-298/04, Slg. 2006, I-6619 (Rn. 89), EuZW 2006, 529, 535 (Manfredi u. a.).

verfälschen können, und dass dieses Recht einen wirksamen Schutz gegen nachteilige Folgen biete. Mangels einschlägiger Normen im Unionsrecht sei es Sache der Mitgliedstaaten, nationale Vorschriften zu erlassen und anzuwenden, die das Recht der geschädigten Personen auf Zugang zu Dokumenten regeln.[733] Dabei sei von den Mitgliedstaaten der Äquivalenz- und Effektivitätsgrundsatz zu beachten.[734]

Mit der Betonung auf das in der *Pfleiderer*-Entscheidung etablierte Gebot der Einzelfallabwägung[735] beantwortet der Gerichtshof die Vorlagefrage schließlich dahingehend, dass § 39 Abs. 2 öKartG mit dem Unionsrecht unvereinbar sei. Es verbiete sich jede starre Regelung[736] im Hinblick auf die wirksame Anwendung der Art. 101, 102 AEUV, was eine Abwägung umso notwendiger mache.

In den nachfolgenden Abschnitten bemüht sich der Gerichtshof um eine Konkretisierung seines Gebots der Einzelfallabwägung und wendet sich bei der Strukturierung der Umstände den konkreten Individualinteressen zu – dem Interesse der Geschädigten an der Wiedergutmachung des erlittenen Schadens wie dem entgegenstehenden Interesse an Schutz des Geschäftsgeheimnisses und personenbezogener Daten.[737] Es sei eine Abwägung zwischen den Interessen, die eine Übermittlung der Informationen rechtfertigen, und denen, die deren Schutz rechtfertigen, von den nationalen Gerichten im Rahmen des nationalen Rechts nur im Einzelfall unter Berücksichtigung aller maßgeblichen Umstände vorzunehmen.[738]

Nach § 39 Abs. 2 öKartG wird nur Einsicht in die Akten des Kartellgerichts gewährt, wenn keine der Parteien des Verfahrens sich dem widersetzt. Nach Ansicht des Gerichtshofs bietet diese Regelung den Gerichten keine Möglichkeit, die unionsrechtlich geschützten Interessen gegeneinander abzuwägen.[739] Die Gefahr sei groß, dass in der Praxis jeder Antrag auf Einsichtsnahme systematisch abgelehnt werde, insbesondere wenn Schriftstücke betroffen seien, deren Weitergabe den Interessen der Parteien des Verfahrens zuwiderlaufe, einschließlich solcher Schriftstücke, die Beweise für den Schaden im Rahmen der Schadensersatzklage enthalten könnten.[740]

---

733 EuGH, Urt. v. 6.6.2013 – Rs. C-536/11 (Rn. 26), JZ 2013, 734, 735 (Donau Chemie u. a.) mit Verweis auf EuGH, Urt. v. 14.6.2011 – Rs. C-360/09, Slg. 2011, I-5161 (Rn. 23), EuZW 2011, 598, 599 (Pfleiderer-AG/BKartA).

734 EuGH, Urt. v. 6.6.2013 – Rs. C-536/11 (Rn. 27), JZ 2013, 734, 735 (Donau Chemie u. a.) mit Verweis auf die schon in den vorherigen Fußnoten genannten Urteile Courage/Crehan, Manfredi, Pfleiderer.

735 EuGH, Urt. v. 6.6.2013 – Rs. C-536/11 (Rn. 30), JZ 2013, 734, 735 (Donau Chemie u. a.).

736 EuGH, Urt. v. 6.6.2013 – Rs. C-536/11 (Rn. 31), JZ 2013, 734, 735 (Donau Chemie u. a.).

737 EuGH, Urt. v. 6.6.2013 – Rs. C-536/11 (Rn. 33), JZ 2013, 734, 735 (Donau Chemie u. a.).

738 EuGH, Urt. v. 6.6.2013 – Rs. C-536/11 (Rn. 34), JZ 2013, 734, 735 (Donau Chemie u. a.) und die Anmerkung in Rn. 35, dass das nationale Recht nicht so ausgestaltet werden darf, dass für die nationalen Gerichte jede Möglichkeit einer Abwägung im Einzelfall ausgeschlossen ist.

739 EuGH, Urt. v. 6.6.2013 – Rs. C-536/11 (Rn. 37), JZ 2013, 734, 735 (Donau Chemie u. a.).

740 EuGH, Urt. v. 6.6.2013 – Rs. C-536/11 (Rn. 38), JZ 2013, 734, 735 (Donau Chemie u. a.).

Da die Regelung es dem Geschädigten unmöglich mache, Zugang zu den betreffenden Dokumenten zu erhalten, ohne dass dem Umstand Rechnung getragen würde, dass dieser Zugang die einzige Möglichkeit für den Betroffenen darstellen könnte, sich die zur Begründung der Schadensersatzklage notwendigen Beweise zu verschaffen, sei sie geeignet, das von den Betroffenen aus dem Unionsrecht hergeleitete Recht auf Schadensersatz übermäßig zu erschweren.[741]

Nachdem der Gerichtshof auf das Interesse an der Attraktivität der Kronzeugenregelung und der Effektivität der öffentlich-rechtlichen Kartellrechtsdurchsetzung einging, betonte er nochmals, dass eine systematische Verweigerung des Dokumentenzugangs nicht gerechtfertigt sei.[742] Die maßgebliche Verweigerung des Zugangs müsse bei jedem einzelnen Dokument auf zwingende Gründe in Bezug auf den Schutz des geltend gemachten Interesses hin geprüft werden – nur wenn Gefahr bestehe, dass ein bestimmtes Schriftstück konkret das öffentliche Interesse an der Wirksamkeit des nationalen Kronzeugenprogramms beeinträchtigen könnte, könne die Nichtweitergabe dieses Schriftstückes gerechtfertigt sein.[743]

Nach alldem beantwortete der Gerichtshof die vorgelegte Frage wie folgt: Das Unionsrecht stehe einer Bestimmung des nationalen Rechts wie dem vorliegenden § 39 Abs. 2 öKartG entgegen.

bb. Rechtliche Würdigung

Der EuGH stützt seine Entscheidung auf die bereits in den Rechtssachen *Courage/Crehan* und *Manfredi* entwickelten Grundsätze der Äquivalenz und Effektivität. Der in der Entscheidung vornehmlich behandelte Effektivitätsgrundsatz verlangt, dass die Ausgestaltung des nationalen Rechts „die Ausübung der durch die Unionsrechtsordnung verliehenen Rechte nicht praktisch unmöglich machen oder übermäßig erschweren“ darf.[744] Indem der EuGH die Abwägungsaufgabe der nationalen Gerichte zu einer Abwägungspflicht machte,[745] entwickelte er die *Pfleiderer*-Entscheidung weiter. Ein generelles Verbot des Dokumentenzugangs ist nicht zulässig.

---

741 EuGH, Urt. v. 6.6.2013 – Rs. C-536/11 (Rn. 39), JZ 2013, 734, 735 (Donau Chemie u. a.).

742 EuGH, Urt. v. 6.6.2013 – Rs. C-536/11 (Rn. 43), JZ 2013, 734, 735 (Donau Chemie u. a.).

743 EuGH, Urt. v. 6.6.2013 – Rs. C-536/11 (Rn. 48), JZ 2013, 734, 735 (Donau Chemie u. a.).

744 EuGH, Urt. v. 6.6.2013 – Rs. C-536/11 (Rn. 27), JZ 2013, 734, 735 (Donau Chemie u. a.); EuGH, Urt. v. 20.9.2001 – Rs. C-453/99, Slg. 2001, I-6297 (Rn. 29), EuZW 2001, 715, 716 (Courage/Crehan); EuGH, Urt. v. 13.7.2006 – verb. Rs. C-295/04-C-298/04, Slg. 2006, I-6619 (Rn. 62), EuZW 2006, 529, 533 (Manfredi u. a.); EuGH, Urt. v. 14.6.2011 – Rs. C-360/09, Slg. 2011, I-5161 (Rn. 24), EuZW 2011, 598, 599 (Pfleiderer-AG/BKartA) mit Verweis auf ältere Rechtsprechung im Zusammenhang mit Beihilfenrückforderung EuGH, Urt. v. 16.7.1998 – Rs. C-298/96, Slg. 1998, I-4767 (Rn. 23, 24), EuZW 1998, 603 (Oelmühle).

745 So auch *Kersting*, JZ 2013, 737, 738.

Diese Grenzziehung durch den EuGH lenkt die nationale Gesetzgebung in ihrer Arbeit und verdeutlicht den Umfang der Gestaltungsfreiheit des nationalen Gesetzgebers.

Der EuGH versucht in seiner *Donau Chemie*-Entscheidung, das Einzelabwägungsgebot zu konkretisieren. Dies gelingt ihm aber nur insoweit, als die Unsicherheit über dessen Adressatenkreis ausgeräumt wird und die Abwägungspflicht dem nationalen Richter zugeschrieben wird und nicht der mitgliedstaatlichen Gesetzgebung. Im Übrigen überzeugen die Ausführungen des EuGH nicht, da er die Individualinteressen und öffentlich-abstrakten Interessen an der Wirksamkeit der Kartellrechtsdurchsetzung nicht hinreichend abwägt. Der EuGH konzentriert sich vorrangig auf die Individualinteressen der Parteien und misst dem Argument, dass der Zugang zu Dokumenten im Interesse der Vertraulichkeit von Kronzeugenunterlagen und des Schutzes der Attraktivität von Kronzeugenprogrammen verweigert werden müsse, nur geringe Bedeutung zu. Dadurch, dass der Gerichtshof eine Verweigerung nur zulassen möchte, wenn infolgedessen konkrete nachteilige Folgen entstehen könnten, grenzt er die Verweigerungsmöglichkeit zu sehr ein.[746] Was die konkreten nachteiligen Folgen sind und ab wann sie eintreten, wird nicht dargelegt. Für eine Versagung der Akteneinsicht besteht vor diesem Hintergrund praktisch kaum noch Raum.

Der EuGH räumt in dieser Entscheidung den Interessen der Geschädigten nahezu uneingeschränkten Vorrang ein, weil er die Gefahr sehr hoch einschätzt, dass Schadensersatzansprüche nicht durchgesetzt werden.

Die Privilegierung von Kronzeugen ist nicht mehr in jeder Hinsicht aufrechterhalten. Die Einsicht in die Akten führt zur übermäßigen Inanspruchnahme des kartellrechtlichen Whistleblowers – von ausreichendem haftungsrechtlichen Schutz kann deshalb nicht die Rede sein.

cc. Folgerungen für das nationale Recht

Folgende Konsequenzen ergeben sich für das nationale Recht:[747]

I. In erster Linie ist der nationale Richter in der Pflicht, eine Entscheidung im Einzelfall zu treffen und bei der Abwägung alle relevanten Gesichtspunkte in Betracht zu ziehen, insbesondere die Interessen der betroffenen Parteien hinreichend zu gewichten.

II. Der nationale Gesetzgeber darf keine Regelung schaffen, in der generell der Zugang zu Kronzeugendokumenten verweigert wird. Demnach wäre der im Referentenentwurf angedachte § 81 b GWB Ref-E in der 8. GWB-Novelle, der eine generelle Versagung der Akteneinsicht in Kronzeugendokumente regeln sollte, mit dem Unionsrecht nunmehr unvereinbar.

---

746 So auch *Kersting*, JZ 2013, 737, 738; *Dworschak/Maritzen*, WuW 2013, 829, 839.

747 Angelehnt an den Aufbau bei *Kersting*, JZ 2013, 737, 738.

III. Auch wenn dem Gesetzgeber Grenzen in der Gestaltungsfreiheit gesetzt sind, so kann er doch die Abwägungskriterien für die Gerichte vorzeichnen und strukturieren. Der Gesetzgeber sollte jedoch eine abschließende Aufzählung vermeiden und den nationalen Gerichten die letzte Entscheidungsbefugnis im Einzelfall belassen.
IV. Die Abwägung im Einzelfall kann zu unterschiedlichen Ergebnissen für die einzelnen Geschädigten führen. Der Gesetzgeber sollte dafür eine Lösung parat haben. Er könnte regeln, dass kein automatischer Anspruch für ähnlich gelagerte Fälle besteht, sondern dass dieser für jeden Fall bzw. in jedem Verfahren gesondert festgestellt werden muss. Erhält aber in demselben Verfahren ein Geschädigter Akteneinsicht, sollte diese den anderen Geschädigten auch gewährt werden. Der Gesetzgeber sollte deshalb eine Ausnahmeregelung kodifizieren, die es dem Gericht erlaubt, in besonders gelagerten Fällen anders zu entscheiden.
V. Das entscheidende Gericht hat jedes einzelne Dokument zu überprüfen und dann zu entscheiden, ob dem Geschädigten unter Berücksichtigung der Interessen des Schädigers Einsicht gewährt werden kann. Um allen Interessen gerecht zu werden, kann der Gesetzgeber beschließen, dass ein unbeteiligter Dritter bei der Auswahl der Akteninhalte zur Verfügung steht, der zur Verschwiegenheit verpflichtet ist.

f. Grenzüberschreitende Kartelle und die Transparenzverordnung

In diesem Zusammenhang stellt sich die Frage nach der Akteneinsicht in Kronzeugendokumente der EU-Kommission. Diese betrifft besonders die kartellverfahrensrechtliche Praxis auf der Ebene grenzüberschreitender Kartelle. Auf Grundlage von Art. 27 Abs. 2 der Verordnung Nr. 1/2003 (EG) in Verbindung mit Art. 15 der Verordnung Nr. 773/2004 (EG)[748] steht das Recht auf Akteneinsicht lediglich den Parteien, also den unmittelbar durch die abschließende Kommissionsentscheidung betroffenen Unternehmen,[749] nach Zustellung der Mitteilung der Beschwerdepunkte zu, wenn keine Geschäftsgeheimnisse betroffen sind. Dritte werden, sofern sie ein ausreichendes Interesse nachweisen, lediglich über Art und Gegenstand des Verfahrens unterrichtet, Art. 13 Abs. 1 Verordnung Nr. 773/2004 (EG). Für die Beschwerdeführer[750] besteht ein Akteneinsichtsrecht nur dann, wenn die Kommission die Beschwerde abzuweisen beabsichtigt, und auch dann nur im Hinblick auf eine um Geschäftsgeheimnisse und sonstige vertragliche Informationen bereinigte

748 Verordnung Nr. 773/2004 (EG) der Kommission v. 7.4.2004 über die Durchführung von Verfahren auf der Grundlage der Art. 81 und 82 EG-Vertrag durch die Kommission, ABl. L 123, vom 27.4.2004, S. 18.
749 *Kellerbauer*, WuW 2011, 688, 689.
750 Dabei sind Beschwerdeführer Unternehmen, die in einer Beschwerde Kartellverstöße anderer Unternehmen bei der Kommission zur Anzeige gebracht haben.

Fassung der Akte.[751] Die Akteneinsicht wird den Adressaten von Beschwerdepunkten also in vollem Umfang,[752] Beschwerdeführern und sonstigen Dritten nur in begrenztem Umfang gewährt.[753] Die eingesehenen Unterlagen dürfen nur für Gerichts- und Verwaltungsverfahren zur Anwendung der Art. 101 und 102 AEUV (und von nationalem Recht, soweit es parallel dazu angewandt wird)[754] verwendet werden.[755]

Die EU-Kommission ist, bevor sie eine Entscheidung gegen eine Partei erlässt, verpflichtet, den jeweiligen Personen und Unternehmen die Gelegenheit zu geben, zu den gegen sie vorgebrachten Beschwerdepunkten oder Einwendungen Stellung und Einsicht in die Akten der EU-Kommission zu nehmen, damit ihre Verteidigungsrechte während des Verfahrens vollständig gewahrt werden.[756] Interne Schriftstücke und sonstige vertrauliche Informationen sind von der Akteneinsicht ausgenommen,[757] worunter auch die Unternehmenserklärung und andere Kronzeugendokumente zu zählen sind.[758]

Potenzielle Zivilkläger können also im Ergebnis aus der Verordnung Nr. 773/2004 (EG) kein Einsichtsrecht ableiten und Informationen aus den Erklärungen erhalten.

---

751 Art. 8 Abs. 1 VO Nr. 773/2004 (EG).

752 Vgl. dazu Art. 10 und 15 der VO Nr. 773/2004 (EG).

753 Das Gericht erster Instanz hat festgestellt, dass Beschwerdeführer nicht über die gleichen Rechte und Garantien verfügen wie die Adressaten einer Mitteilung der Beschwerdepunkte. Beschwerdeführer können daher kein Recht auf Akteneinsicht geltend machen, wie dies den Adressaten einer Mitteilung der Beschwerdepunkte zukommt, vgl. dazu EuG, Urt. v. 15.7.1994 – Rs. T-17/93, Slg. 1994, II-595 (Rn. 34, 35), EuZW 1995, 115, 116 (Matra Hachette SA). Das Gericht führt des Weiteren aus, dass die Rechte Dritter auf das Recht beschränkt sind, sich am Verwaltungsverfahren zu beteiligen; vgl. zu den beschränkten Einsichtsrechten von Beschwerdeführern, Art. 8 Verordnung Nr. 773/2004 (EG); *Ritter*, in: Immenga/Mestmäcker, EU-Wettbewerbsrecht, Kartellverfahrensverordnung 1/2003, Kap. 8, Rn. 18.

754 Art. 3 Abs. 1 VO Nr. 1/2003 (EG).

755 Art. 15 Abs. 4 VO Nr. 773/2004 (EG).

756 Mitteilung der Kommission über die Regeln für die Einsicht in Kommissionsakten in Fällen einer Anwendung der Artikel 81 und 82 EG-Vertrag, Artikel 53, 54 und 57 des EWR-Abkommens und der Verordnung Nr. 139/2004 (EG), ABl. C 325/07, vom 22.12.2005, Rn. 1 und 10.

757 Vgl. u. a. Mitteilung der Kommission über die Regeln für die Einsicht in Kommissionsakten in Fällen einer Anwendung der Artikel 81 und 82 EG-Vertrag, Artikel 53, 54 und 57 des EWR-Abkommens und der Verordnung Nr. 139/2004 (EG), ABl. C 325/07, vom 22.12.2005, Rn. 10.

758 *Denoth*, Kronzeugenregelung, S. 287.

aa. Die Transparenzverordnung

Losgelöst vom kartellverfahrensrechtlichen Akteneinsichtsrecht ist der allgemein ausgerichtete, aus dem Demokratiegedanken[759] gespeiste Anspruch auf Dokumentenzugang nach Art. 2 Abs. 1 Transparenzverordnung.[760] Er soll entsprechend Art. 1 lit. a) Transparenz-VO einen größtmöglichen Zugang zu Dokumenten gewährleisten, sodass grundsätzlich alle Dokumente der Organe für die Öffentlichkeit zugänglich sind.

(1) Anwendbarkeit

Fraglich ist, wie das Verhältnis der Transparenz-VO zu speziellen Dokumentenzugangsregelungen im Allgemeinen und zum kartellverfahrensrechtlichen Akteneinsichtsrecht im Besonderen zu beurteilen ist. Voraussetzung für einen Zugang zu Kartellverfahrensakten nach der Transparenzverordnung ist, dass diese neben oder anstelle der Kartellverfahrensordnung in Verbindung mit der Verordnung Nr. 773/2004 (EG) anwendbar ist. Zwar ist letztere später erlassen und enthält speziellere Regelungen,[761] jedoch will die Transparenzverordnung gerade einen möglichst umfassenden Schutz des Dokumentenzugangs garantieren. Die damit einhergehende Geltung für alle Bereiche wäre mit einem Ausschluss der Kronzeugendokumente nicht vereinbar. Zum einen spricht für die uneingeschränkte Anwendbarkeit der Wortlaut des Art. 2 Abs. 3 der Transparenz-VO, der ausdrücklich alle Tätigkeiten der Union erfasst sehen will.[762] Zum anderen sollten nach dem 12. Erwägungsgrund der Transparenzverordnung alle Bestimmungen über den Zugang der Organe mit ihr in Einklang stehen,[763] also demnach auch die „engere" speziellere Vorschrift über den Zugang zu Akten in Kronzeugenverfahren.

Die im ursprünglichen Entwurf vorgesehene *lex specialis*-Regelung, nach der die Transparenz-VO nicht gelten sollte, „sofern es Sondervorschriften hinsichtlich des Zugang zu Dokumenten gibt",[764] wurde ersatzlos gestrichen.[765] Zwar gilt der Grundsatz „*lex specialis derogat legi generali*", der Normwidersprüche verhindern will und

759 EuGH, Urt. v. 21.7.2011 – Rs. C-506/08 P, Slg. 2011, I-6237 (Rn. 72), EuZW 2012, 22, 23 (Schweden/MyTravel).

760 Verordnung Nr. 1049/2001 (EG) des Europäischen Parlamentes und des Rates vom 30.5.2001 über den Zugang der Öffentlichkeit zu Dokumenten des Europäischen Parlamentes, des Rates und der Kommission, ABl. L 145/43, vom 31.5.2001.

761 Für eine partielle Verdrängungswirkung *Kleine*, ZWeR 2007, 303, 307; mit Nachweisen „Pro et contra" bei *Westhoff*, Zugang zu Beweismitteln bei Schadensersatzklagen im KartellR, S. 119 f.

762 Zuletzt siehe EuGH, Urt. v. 21.7.2011 – Rs. C-506/08 P, Slg. 2011, I-2637 (Rn. 88), EuZW 2012, 22, 23 (Schweden/MyTravel) m. Anm. Kellerbauer.

763 *Frenz*, EuZW 2013, 778, 779 mit Verweis auf *Milde*, Schutz der Kronzeugen, S. 242 f., 246 f.

764 KOM (2000) 30, Art. 2 Abs. 2; vgl. auch Vorschlag der Kommission zur Änderung der Transparenz-VO vom 30.4.2008, KOM (2008) 229.

765 *Kleine*, ZWeR 2007, 303, 306.

der Einheit der Rechtsordnung dient.[766] Jedoch gilt dieser Grundsatz und damit die zwingende Verdrängung der allgemeineren Norm nur, wenn die Rechtsfolgen sich gegenseitig ausschließen und es folglich um zwei verschiedene Intentionen geht.[767] Dies ist hier nicht der Fall. Wo im Einzelfall Diskrepanzen entstehen, können diese im Wege teleologischer oder systematischer Auslegung beseitigt werden.[768]

Der EuGH lehnte bezogen auf die Beihilfeverfahrensverordnung[769] ein Spezialitätsverhältnis zur Transparenzverordnung ab und verfolgte diesen Ansatz in mehreren Entscheidungen.[770] In der Entscheidung *Schweden/MyTravel* bezog der EuGH anstandslos die Transparenzverordnung in seine Entscheidungsfindung ein.[771]

Auch das EuG hat in jüngerer Rechtsprechung[772] die Transparenz-VO auf Dokumente aus Kartellverfahren angewendet.[773]

Für die kartellrechtliche Fragestellung ergibt sich damit folgendes Ergebnis: Gründe gegen eine kartellverfahrensrechtlich abgeleitete Beschränkung der Transparenz-VO sind nicht ersichtlich.

(2) Anwendungsbereich

Nach Art. 2 der Transparenz-VO hat jeder Unionsbürger sowie jede natürliche oder juristische Person mit Wohnsitz oder Sitz in einem Mitgliedstaat Zugang zu den Dokumenten der Organe. Der Zugang ist unter anderem zu verweigern, wenn Inspektions-, Untersuchungs- und Audittätigkeiten beeinträchtigt würden.[774] Darunter fasst die Kommission auch die Kronzeugenverfahren.[775] Art. 4 Abs. 2 a. E. der Transparenz-VO regelt die Rückausnahme, nach der der Zugang nicht

766 *Engisch*, Einführung in das juristische Denken, S. 275.

767 *Larenz*, Methodenlehre, S. 268.

768 *Larenz*, Methodenlehre, S. 268; *Tietje/Nowrot*, EWS 2006, 486, 487.

769 Verordnung Nr. 659/1999 (EG) des Rates v. 22.3.1999 über besonderer Vorschriften für die Anwendung von Art. 93 des EG-Vertrags, Abl. (EG) L 83/1, vom 27.3.1999.

770 EuGH, Urt. v. 28.6.2012 – Rs. C-404/10 P (Rn. 110), WuW 2012, 983, 991 (Éditions Odile Jacob); EuGH, Urt. v. 29.6.2010 – Rs. C-28/08 P, Slg. 2010, I- 6055 (Rn. 56), EuZW 2010, 617, 618 (Bavarian Lager); EuGH, Urt. v. 21.9.2010 – Rs. C-514/07 P u. a., Slg. 2010, I-8533 (Rn. 69), (Schweden/API).

771 Vgl. EuGH, Urt. v. 21.7.2011 – Rs. C-506/08 P, Slg. 2011, I-6237 (Rn. 72, 73), EuZW 2012, 22 (Schweden/MyTravel).

772 Zur grundsätzlichen Anwendbarkeit schon EuG, Urt. v. 13.4.2005 – Rs. T-2/03, Slg. 2005, II-1121 (Rn. 66 f.), EuZW 2005, 566 (Verein für Konsumenteninformation). Das Gericht ging in der Entscheidung nicht weiter auf die Frage der Anwendbarkeit der Verordnung Nr. 1049/2001 (EG) ein, sondern prüfte nur die Voraussetzungen; vgl. zuletzt EuG, Urt. v. 15.12.2011 – Rs. T-437/08, Slg. 2011, II-8251 (Rn. 24, 26), EuZW 2012, 315, 316 (CDC Hydrogene Peroxide).

773 In diesem Sinne *Palzer*, EuR 2012, 583, 589; *Dittrich*, WuW 2012, 133, 139.

774 Art. 4 Abs. 2 Transparenz-VO.

775 So die Kommission, vgl. dazu EuG, Urt. v. 9.9.2008 – Rs. T-403/05, Slg. 2008, II-2027 (Rn. 21), (MyTravel).

verweigert werden darf, wenn ein überwiegendes öffentliches Interesse an der Verbreitung besteht. Da die private Rechtsdurchsetzung nicht nur das Individualinteresse schützt, sondern auch dem Schutz des Wettbewerbs als solchem und damit dem öffentlichen Interesse dient, wäre eine Verweigerung, die sich auf die private Rechtsdurchsetzung stützt, dem Grunde nach nicht zulässig.[776] Die Kommission berücksichtigte dies bislang jedoch nicht, sondern lehnte die Akteneinsicht ohne Prüfung der Rückausnahme ab. Das Europäische Gericht hat sich daraufhin in jüngeren Entscheidungen gegen die Sichtweise der Kommission – und in Konkretisierung der Vorgaben des EuGH – gestellt.[777] Das EuG hatte in diesen Entscheidungen (CDC und EnBW) erstmals die Gelegenheit, über die Bedeutung der Kronzeugenregelungen bei der Auslegung der Ausnahmebestimmung des Art. 4 Abs. 2 Transparenz-VO zu urteilen.

bb. Die Entscheidungen des Europäischen Gerichts

Durch die Entscheidungen des EuG wurde ein neuer Maßstab zur Prüfung der Ausnahmetatbestände nach Art. 4 Transparenz-VO entwickelt. Dabei sind die relevanten privaten und öffentlichen Interessen adäquat einzubeziehen.[778]

(1) Schutz der geschäftlichen Interessen

Die grundrechtlich geschützten geschäftlichen Interessen einer natürlichen oder juristischen Person, einschließlich des geistigen Eigentums, dürfen durch Verbreitung der Dokumente nicht beeinträchtigt werden, es sei denn es besteht ein überwiegendes öffentliches Interesse an der Verbreitung.[779] Das EuG schränkt dabei die „geschäftlichen Interessen" dahingehend ein, dass nicht sämtliche Informationen über eine Gesellschaft und ihre Geschäftsbeziehungen darunter fallen. Andernfalls würde der Zweck der Transparenz-VO, der Öffentlichkeit einen größtmöglichen Zugang zu Dokumenten der Organe zu gewähren, vereitelt.[780] Vor diesem Hintergrund erkennt das Europäische Gericht keine berührten schutzwürdigen Interessen, wenn lediglich Einsicht in das Inhaltsverzeichnis gewünscht ist, das die in der Kommissionsakte enthaltenen Dokumente aufführt. Etwas anderes gilt, wenn sich daraus sensible Daten ergeben würden, etwa Informationen über Preise, Kostenstruktur, Marktanteile oder Ähnliches, und dies eine konkrete Zuordnung zu einem

---

776 So ähnlich auch *Koch*, JZ 2013, 390, 392.

777 Vgl. zuletzt daraufhin ergangene Entscheidungen EuG, Urt. v. 15.12.2011 – Rs. T-437/08, Slg. 2011, II-8251 (Rn. 36 f.), EuZW 2012, 315, 316 f. (CDC Hydrogene Peroxide) und EuG, Urt. v. 22.5.2012 – Rs. T-344/08 (Rn. 40 f.), WuW/E EU-R 2356, 2360 (EnBW).

778 Unmittelbar auch EuG, Urt. v. 15.12.2011 – Rs. T-437/08, Slg. 2011, II-8251 (Rn. 44, 58), EuZW 2012, 315, 316 (CDC Hydrogene Peroxide).

779 Art. 4 Abs. 2, 1. Spiegelstrich Transparenz-VO.

780 EuG, Urt. v. 15.12.2011 – Rs. T-437/08, Slg. 2011, II-8251 (Rn. 44), EuZW 2012, 315, 317 m.w.N. (CDC Hydrogene Peroxide).

Unternehmen ermöglichen könnte.[781] Allein ein Hinweis darauf, dass die erbetenen Informationen für Schadensersatzklagen verwendet werden, genügt nicht.[782] Auch das Interesse, Kosten zu vermeiden, die aus einer notwendigen Verteidigung erwachsen, stuft das EuG vor dem Hintergrund der *Courage/Crehan*-Entscheidung des EuGH nicht als schutzwürdiges geschäftliches Interesse ein.[783] Weiter geht es davon aus, dass eine Art Wechselwirkung zwischen dem Alter der fraglichen Information und der Intensität besteht, in der das betreffende Unternehmen durch eine Veröffentlichung in seine geschäftlichen Interessen berührt wird.[784] Die mit zunehmendem Alter abnehmende Schutzwürdigkeit der Informationen hätte die Kommission deshalb in der Entscheidung *EnBW* zu einer konkreten und individuellen Prüfung der jeweiligen Dokumente veranlassen sollen.[785]

(2) Schutz des Wettbewerbsverfahrens

Der Schutz des Zwecks der Untersuchung[786] betrifft auch das wettbewerbsrechtliche Verfahren und die damit verbundene Untersuchung. Die Verwendung von Dokumenten allein dafür entbindet aber nicht von einer konkreten und individuellen Prüfung.[787] Das EuG stellt zunächst klar, dass durch Art. 4 Abs. 2, 3. Spiegelstrich Transparenz-VO nicht die Untersuchung geschützt werden solle, sondern nur deren Zweck, der im Fall eines Verfahrens in Wettbewerbssachen darin bestehe zu überprüfen, ob ein Verstoß gegen das Kartell- oder das Missbrauchsverbot begangen worden sei und ggf. Sanktionen zu verhängen seien.[788] Dies gelte allerdings nur für die Dauer des Kartellverfahrens und nicht, wenn bereits eine endgültige Entscheidung getroffen wurde, mit der die Akte, zu der Zugang beantragt ist, geschlossen wurde[789] – unbeschadet der Möglichkeit, dass diese Entscheidung in der Folge von den Gerichten möglicherweise für nichtig erklärt werde.[790] Andernfalls

---

781 EuG, Urt. v. 15.12.2011 – Rs. T-437/08, Slg. 2011, II-8251 (Rn. 45), EuZW 2012, 315, 317 (CDC Hydrogene Peroxide).

782 EuG, Urt. v. 15.12.2011 – Rs. T-437/08, Slg. 2011, II-8251 (Rn. 48), EuZW 2012, 315, 317 (CDC Hydrogene Peroxide).

783 EuG, Urt. v. 15.12.2011 – Rs. T-437/08, Slg. 2011, II-8251 (Rn. 49), EuZW 2012, 315, 318 (CDC Hydrogene Peroxide).

784 EuG, Urt. v. 22.5.2012 – Rs. T-344/08, (Rn. 131 f.), WuW/E EU-R 2356, 2377 f. (EnBW): „Insoweit schwächen sich die nachteiligen Folgen, die sich aus der Verbreitung einer wirtschaftlich sensiblen Information ergeben können, mit zunehmendem Alter der Information ab“ (Rn. 139).

785 EuG, Urt. v. 22.5.2012 – Rs. T-344/08 (Rn. 142), WuW/E EU-R 2356, 2378 (EnBW).

786 So in Art. 4 Abs. 2, 3. Spiegelstrich Transparenz-VO.

787 EuG, Urt. v. 22.5.2012 – Rs. T-344/08 (Rn. 85), WuW/E EU-R 2356, 2369 (EnBW).

788 EuG, Urt. v. 15.12.2011 – Rs. T-437/08, Slg. 2011, II-8251 (Rn. 59), EuZW 2012, 315 (CDC Hydrogene Peroxide).

789 EuG, Urt. v. 22.5.2012 – Rs. T-344/08 (Rn. 116), WuW/E EU-R 2356, 2373 f. (EnBW).

790 EuG, Urt. v. 15.12.2011 – Rs. T-437/08, Slg. 2011, II-8251 (Rn. 62), EuZW 2012, 315 (CDC Hydrogene Peroxide).

liefe es darauf hinaus, den Zugang zu diesen Dokumenten von unvorhersehbaren Ereignissen abhängig zu machen, nämlich vom Ausgang dieses gerichtlichen Verfahrens und den Konsequenzen, die die Kommission daraus ziehen könnte. Dies würde im Widerspruch zu dem Ziel stehen, der Öffentlichkeit den größtmöglichen Zugang zu Dokumenten der Organe zu gewährleisten.[791]

Das jeweilige Dokument muss essentiell sein, um die wettbewerbliche Untersuchungstätigkeit der Kommission zu schützen, was einer engen Auslegung nicht entgegensteht.[792] Es ist zu beachten, dass das Kronzeugenprogramm, dessen Wirksamkeit die Kommission schützen möchte, nicht das einzige Mittel ist, um zu gewährleisten, dass die Wettbewerbsregeln der Union eingehalten werden. Schadensersatzklagen vor den nationalen Gerichten können ebenfalls wesentlich zur Aufrechterhaltung eines wirksamen Wettbewerbs in der Union beitragen.[793]

(3) Schlussfolgerung

Das EuG legt strenge Maßstäbe bei der Prüfung des Akteneinsichtsanspruchs an. Es entspricht der Konzeption der Transparenz-VO, den Zugang zu Unionsdokumenten – unabhängig vom Politikbereich – als Grundsatz festzuschreiben und die Verweigerung zur begründungsbedürftigen Ausnahme zu machen.[794] Dabei kommt dem Erfordernis einer konkreten Prüfung eines jeden Dokuments besondere Bedeutung zu.[795] Eine pauschale Zurückhaltung der gesamten Verfahrensakte ist also nicht möglich. Das EuG ist im Ergebnis zu der Auffassung gelangt, dass die Kommission ohne konkrete Analyse jedes einzelnen Dokuments nicht annehmen darf, dass alle begehrten Dokumente offenkundig unter Art. 4 Abs. 2, 3. Gedankenstrich der Verordnung Nr. 1049/2001 (EG) fallen. Soweit sich Kronzeugendokumente in bestimmten Grenzen unter die Ausnahmeregelung des Art. 4 Abs. 2, 3. Spiegelstrich Transparenz-VO subsumieren lassen, ist über die Zulässigkeit ihrer Zurückhaltung durch die Kommission noch keine endgültige Entscheidung getroffen. Überwiegt das öffentliche Interesse das Interesse an der Geheimhaltung, wären diese Dokumente gleichwohl nach Art. 4 Abs. 2, 3. Spiegelstrich Transparenz-VO zu offenbaren. Dazu jedoch hatte sich das Europäische Gericht weder in der Entscheidung *Hydrogen Peroxide* (CDC) noch in der Entscheidung *EnBW* äußern müssen. Vielmehr ging es zunächst um die Auslegung der Ausnahmeregelung in Art. 4 Abs. 2

---

791 EuG, Urt. v. 15.12.2011 – Rs. T-437/08, Slg. 2011, II-8251 (Rn. 64, 65), EuZW 2012, 315 (CDC Hydrogene Peroxide).

792 EuG, Urt. v. 22.5.2012 – Rs. T-344/08 (Rn. 126), WuW/E EU-R 2356, 2375 (EnBW).

793 EuG, Urt. v. 22.5.2012 – Rs. T-344/08 (Rn. 128), WuW/E EU-R 2356, 2375 (EnBW), mit Verweis auf die Courage-Entscheidung des EuGH; so auch schon EuG, Urt. v. 15.12.2011 – Rs. T-437/08, Slg. 2011, II-8251 (Rn. 77), EuZW 2012, 315, 320 (CDC Hydrogene Peroxide).

794 *Palzer*, EuR 2012, 583, 598, 599.

795 So auch der EuGH, Urt. v. 6.6.2013 – Rs. C-536/11 (Rn. 43), JZ 2013, 734, 735 (Donau Chemie u. a.).

Transparenz-VO – die Abwägungskriterien können jedoch anderen Entscheidungen entnommen werden. Danach lässt sich ein überwiegendes öffentliches Interesse nicht (allein) damit begründen, dass die Dokumente dem Antragsteller ermöglichen, seine Argumente im Rahmen der Schadensersatzklage besser geltend zu machen.[796] Vielmehr muss dieses Interesse objektiv und allgemein sein und darf nicht mit besonderen oder privaten Interessen, z. B. in Bezug auf eine Klage gegen die Gemeinschaftsorgane, verwechselt werden, da solche besonderen oder privaten Interessen kein erheblicher Gesichtspunkt sind für die Interessenabwägung nach Art. 4 Abs. 3 Unterabsatz 2 Transparenz-VO.[797] Gleiches gilt für die Frage, ob ein überwiegendes öffentliches Interesse nach Art. 4 Abs. 2 a. E. dieser Verordnung vorliegt.[798]

cc. Die Entscheidung des EuGH

Mit seinem Urteil vom 27. Februar 2014[799] hob der EuGH das Urteil des EuG auf und bestätigte weitgehend die Entscheidung, mit der die Kommission den Zugangsantrag abgelehnt hatte. Der EuGH ging dabei davon aus, dass sich die Kommission pauschal auf eine allgemeine Vermutung für das Vorliegen von Ausnahmen vom Zugangsrecht stützen durfte. Dem Leitsatz entsprechend, besteht für den Inhalt von Akten eines Kartellverfahrens die allgemeine Vermutung (zugunsten einer Verweigerung des Zugangs), dass die Verbreitung von Dokumenten grundsätzlich den Schutz eines der in Art. 4 Transparenz-VO aufgezählten Interessen beeinträchtigen würde, so dass die Kommission nicht angehalten ist, jedes einzelne Dokument der in Rede stehenden Akte, die ein nicht abgeschlossenes Kartellverfahren betrifft, konkret und individuell zu prüfen. Diese allgemeine Vermutung war im konkreten Fall laut EuGH nicht widerlegt worden.[800] Der Gerichtshof maß der kartellrechtlichen Schadensersatzklage, die der Antragsteller unter Verwendung der begehrten Informationen erheben wollte, im konkreten Fall kein öffentliches Interesse bei, das die durch die Ausnahmeregelung geschützten Interessen hätte überwiegen können.[801] Eine solche allgemeine Vermutung bestand bislang etwa für Verwaltungsakte eines Beihilfekontrollverfahrens.[802]

Der EuGH hat in seiner Entscheidung die bislang unbeantwortete Frage, ob kartellrechtliche Schadensersatzklagen im öffentlichen Interesse im Sinne von Art. 4

796 EuG, Urt. v. 9.9.2008 – Rs. T-403/05, Slg. 2008, II-2027 (Rn. 65) (MyTravel/Kommission); offen lassend EuGH, Urt. v. 21.7.2011 – Rs. C-506/08 P, Slg. 2011, I- 6237 (Rn. 120, 121) (Schweden/MyTravel).

797 EuG, Urt. v. 9.9.2008 – Rs. T-403/05, Slg. 2008, II-2027 (Rn. 65) (MyTravel/Kommission).

798 EuG, Urt. v. 9.9.2008 – Rs. T-403/05, Slg. 2008, II-2027 (Rn. 68) (MyTravel/Kommisison).

799 EuGH, Urt. v. 27.2.2014 – Rs. C-365/12, EuZW 2014, 311 (Kommission/EnBW).

800 Vgl. EuGH, Urt. v. 27.2.2014 – Rs. C-365/12 (Rn. 102), EuZW 2014, 311, 316 (Kommission/EnBW).

801 Vgl. Ausführungen bei *Hempel*, EuZW 2014, 297, 298 f.

802 EuGH, Urt. v. 29.6.2010 – Rs. C-139/07 P (Rn. 46), EuZW 2010, 624 (Kommission/TGI).

Abs. 2 Transparenz-VO stehen, mehr oder weniger bejaht. Es obliegt aber einem jeden, der Schadensersatz wegen eines Verstoßes gegen Art. 101 AEUV (ex-Art. 81 EG-Vertrag) begehrt, nachzuweisen, dass für ihn die Notwendigkeit des Zugangs zu einem Dokument der Kommissionsakte besteht, damit die Kommission die beiderseitigen Interessen der Beteiligten abwägen kann.[803] Ist eine solche Notwendigkeit nicht gegeben, kann das Interesse an Schadensersatz wegen eines Kartellrechtsverstoßes kein überwiegendes öffentliches Interesse im Sinne von Art. 4 Abs. 2 der Transparenz-VO darstellen.[804] In dem Zusammenhang hat der Gerichtshof auf seine Rechtsprechung in der Rechtssache *Donau Chemie* verwiesen und vereinheitlicht damit einerseits die EU-rechtlichen Vorgaben für den Zugang zu Kartellverfahrensakten der Kommission nach der Transparenz-VO und andererseits die Einsicht in die Kartellverfahrensakten der mitgliedstaatlichen Kartellbehörden nach deren mitgliedstaatlichem Recht.[805]

Die Transparenz-VO regelt den Zugang restriktiv und beschränkt diesen auf die Parteien und Beschwerdeführer, und zwar unter dem Vorbehalt, dass weder Geschäftsgeheimnisse oder vertrauliche Informationen noch interne Schriftstücke der Kommission verbreitet und die zugänglich gemachten Dokumente nur für Gerichts- und Verwaltungsverfahren zur Anwendung von Art. 101 AEUV verwendet werden. Daraus folgt, dass die Kommission in einem solchen Fall bei der Anwendung der Ausnahmen nach Art. 4 Abs. 2 Transparenz-VO ohne eine konkrete und individuelle Prüfung der Dokumente zu der Annahme berechtigt ist, dass durch die Verbreitung die Interessen nach Art. 4 Abs. 2 Transparenz-VO beeinträchtigt wären.[806]

Im Ergebnis reicht zur Widerlegung der bloße Hinweis auf das Recht jedes Unionsbürgers nicht aus, Ersatz des Schadens zu verlangen, der diesem durch einen Verstoß gegen Art. 101 AEUV entstanden ist.

dd. Ausblick

Whistleblowing in Kartellsachen sperrt nicht automatisch und pauschal den Dokumentenzugang.[807] Grundsätzlich können potenzielle Kartellgeschädigte Zugang zu den Kommissionsakten nach der Transparenzverordnung verlangen.[808] Der Ansatz der wettbewerbsrechtlichen Regelungen, der sich aus der Transparenzverordnung herauskristallisiert, gilt nicht uneingeschränkt. Dennoch bestätigt das EuG in

---

803 EuGH, Urt. v. 27.2.2014 – Rs. C-365/12 (Rn. 107), EuZW 2014, 311, 316 (Kommission/EnBW).

804 EuGH, Urt. v. 27.2.2014 – Rs. C-365/12 (Rn. 108), EuZW 2014, 311, 316 (Kommission/EnBW).

805 *Hempel*, EuZW 2014, 297, 299.

806 EuGH, Urt. v. 27.2.2014 – Rs. C-365/12 (Rn. 116), EuZW 2014, 311, 315 (Kommission/EnBW).

807 So auch *Frenz*, EuZW 2013, 778, 781.

808 Vgl. Prüfung nach EuG, Urt. v. 13.4.2005 – Rs. T-2/03, Slg. 2005, II-1121, EuZW 2005, 566 (Verein für Konsumenteninformation).

seinen Entscheidungen einen weitgehenden Zugang, ohne dass dieser unter Hinweis auf den Antragsgegenstand relativiert werden könnte. Insofern scheint die Transparenz-VO nicht mehr vollends mit der Ansicht der Kommission übereinzustimmen; dem Argument des wirksamen Schutzes des Kronzeugenprogramms wird weniger Bedeutung beigemessen. Die Transparenz-VO könnte mehr und mehr zum „Stolperstein“[809] werden. Das EuG betonte mehrmals, dass die Kommission von dem Argumentationsmuster viel zu sehr in pauschaler Weise Gebrauch mache und dabei nicht sehe, dass die Kronzeugenprogramme nicht das einzige Mittel zur Gewährleistung der Wettbewerbsregeln der Union seien, sondern auch Schadensersatzklagen vor den nationalen Gerichten wesentlich zu einem wirksamen Wettbewerbs beitrügen[810] und ihnen deshalb mehr Gewicht zukommen müsse. Schadensersatzkläger können trotz der aufgeführten Ausnahmetatbestände im Anschluss an eine Entscheidung auch gegen den Willen der EU-Kommission Zugang zu EU-Kommissionsakten – einschließlich der darin eventuell enthaltenen Kronzeugenerklärungen – erlangen. Die Transparenz-VO eröffnet dem geschädigten Dritten damit grundsätzlich den Zugriff auf die Dokumente; sogar über das kartellrechtliche Akteneinsichtsrecht hinaus.

Ein anderer Blickwinkel ergibt sich jedoch, wenn man die Entscheidung des EuGH heranzieht. Danach dürfte eher auf Seiten der Kartellschadensersatzkläger – anders als nach der Entscheidung des EuG – die Transparenz-VO an Attraktivität verloren haben. Denn die aus der Entscheidung gefolgerte allgemeine Vermutung, dass Ausnahmetatbestände vorliegen, stellt einen „Stolperstein“ für die Geschädigten dar, die ihren Schaden geltend machen wollen. Zwar ergibt sich aus dem Urteil des EuGH, dass der Zugang zu Kronzeugenanträgen nicht ausgeschlossen ist. Jedoch muss der Antragsteller (nun) zeigen können, dass er für die Erhebung seiner Schadensersatzklage auf Informationen aus dem Kronzeugenantrag angewiesen ist. Das wird ihm regelmäßig nicht gelingen, da er zum einen nicht abschätzen kann, welche Informationen der Kronzeugenantrag enthält und ob sie ihm weiter helfen, und zum anderen die Kronzeugen davon absehen, Umstände zu berichten, die für die Feststellung des Kartellrechtsverstoßes nicht benötigt, aber für die Schadensermittlung durch Geschädigte nützlich sein können.[811]

Ob das Interesse an der Geheimhaltung betrieblicher Geheimnisse bzw. der Nichtveröffentlichung der Details der dem Kartellbeteiligten zur Last gelegten Zuwiderhandlung einen erhöhten Schutz verdient, bleibt offen.

Insoweit kann sich aufgrund des unterschiedlich behandelten Einsichtsrechts durchaus ergeben, dass potenzielle Kronzeugenanträge auf Erlass oder Ermäßigung des Bußgeldes zurückgehen und kartellbehördliche Ermittlungserfolge seltener

---

809 *Palzer*, EuR 2012, 583, 606.

810 EuG, Urt. v. 15.12.2011 – Rs. T-437/08, Slg. 2011, II-8251 (Rn. 35), EuZW 2012, 315, 317 (CDC Hydrogene Peroxide) und EuG, Urt. v. 22.5.2012 – Rs. T-344/08 (Rn. 128), WuW/E, EU-R 2356, 2375 (EnBW).

811 *Hempel*, EuZW 2014, 297, 300.

werden. Das allgemeine, umfangreiche Einsichtsrecht nach der Transparenz-VO macht den Whistleblower in Kartellsachen besonders angreifbar und nimmt ihm die vom Kronzeugenprogramm zugesicherte Privilegierung.

Aus rechtspolitischer Sicht und aufgrund der kaum befriedigend zu lösenden Problematik im Bereich der Akteneinsicht wäre es gut vertretbar, die Lösung in einer zivilrechtlichen Privilegierung des Kronzeugen zu suchen. Dem würde ein weitergehendes Akteneinsichtsrecht ohne Belastung der Effektivität von Kronzeugenprogrammen fördernd beiseite stehen, denn dem Kronzeugen würde von anderer Seite eine Privilegierung zukommen.

5. *Europäische Richtlinie zu Schadensersatzklagen im Kartellrecht*

Am 11. Juni 2013 hatte die Europäische Kommission einen Richtlinienvorschlag[812] zu kartellrechtlichen Schadensersatzklagen in der EU veröffentlicht. Dieser wurde von einem Bericht zur Folgenabschätzung in Form eines Begleitpapiers[813] sowie von einer Mitteilung[814] zur Ermittlung des Schadensumfangs bei Schadensersatzklagen wegen Wettbewerbsverstößen (ergänzt durch einen „Praktischen Leitfaden"[815]) begleitet. Zusätzlich hatte die Kommission eine Empfehlung für Grundsätze für kollektive Unterlassungs- und Schadensersatzverfahren in den Mitgliedstaaten ausgesprochen.[816]

Die Kommission verfolgte damit zum einen das Ziel, das Zusammenspiel von privater und behördlicher Rechtsdurchsetzung (*private and public enforcement*) zu optimieren und zum anderen die Opfer von Zuwiderhandlungen gegen das EU-Wettbewerbsrecht vollständig zu entschädigen.

---

812 Vorschlag für eine Richtlinie des Europäischen Parlaments und des Rates über bestimmte Vorschriften für Schadensersatzklagen nach einzelstaatlichem Recht wegen Zuwiderhandlungen gegen wettbewerbsrechtliche Bestimmungen der Mitgliedstaaten und der Europäischen Union vom 11.6.2013, KOM (2013), 404 final.

813 Zusammenfassung der Folgenabschätzung, Begleitpapier zum Richtlinienvorschlag vom 11.6.2013, SWD (2013), 204 final.

814 Mitteilung der Kommission zur Ermittlung des Schadensumfangs bei Schadensersatzklagen wegen Zuwiderhandlungen gegen Art. 101 und 102 des Vertrags über die Arbeitsweise der Europäischen Union, 2013/C 167/07, ABl. EU Nr. C 167, vom 13. 6. 2013, S. 19.

815 Praktischer Leitfaden zur Ermittlung des Schadensumfangs bei Schadensersatzklagen im Zusammenhang mit Zuwiderhandlungen gegen Art. 101 oder 102 des Vertrags über die Arbeitsweise der Europäischen Union, abrufbar unter: http://ec.europa.eu/competition/antitrust/actionsdamages/quantification_guide_de.pdf, (abgerufen am 1.6.2014).

816 Empfehlung über Grundsätze für kollektive Unterlassungs- und Schadensersatzverfahren in den Mitgliedstaaten, ABl. 2013 Nr. L 201, vom 11.6.2013, S. 60.

Nach letzten Änderungen durch das Europäische Parlament[817] ist die Richtlinie am 25. Dezember 2014[818] in Kraft getreten.

a. Erwägungsgründe und Regelungsgegenstände

Für die Wirksamkeit der in Art. 101 und 102 AEUV festgelegten Verbote ist es erforderlich, dass jeder – seien es Einzelne, Verbraucher, Unternehmen oder Behörden – vor einzelstaatlichen Gerichten Ersatz des Schadens verlangen kann, der ihm durch eine Zuwiderhandlung gegen diese Bestimmungen entstanden ist.[819] Das Unionsrecht auf Ersatz des kartellrechtlichen Schadens muss daher voraussetzen, dass in jedem Mitgliedstaat Verfahrensvorschriften bestehen, die gewährleisten, dass dieses Recht wirksam geltend gemacht werden kann. Außerdem müssen die behördliche und private Rechtsdurchsetzung effizient koordiniert werden.[820] In den weiteren Erwägungsgründen zählt die Richtlinie neben Problemen, die sich aus dem Spannungsverhältnis zwischen *private* und *public enforcement* ergeben, Schwierigkeiten wegen der uneinheitlichen Regelung auf europäischer Ebene auf. Deshalb wird abschließend ausgeführt:

> *„Da die Ziele dieser Richtlinie, nämlich Vorschriften für Schadensersatzklagen wegen Verletzung des Wettbewerbsrechts der Union festzulegen, um die volle Wirkung der Artikel 101 und 102 AEUV und das reibungslose Funktionieren des Binnenmarkts für Unternehmen und Verbraucher zu gewährleisten, von den Mitgliedstaaten nicht ausreichend verwirklicht werden können, sondern vielmehr wegen der erforderlichen Wirksamkeit und Kohärenz der Anwendung der Artikel 101 und 102 AEUV auf Unionsebene besser zu verwirklichen sind, kann die Union im Einklang mit dem in Artikel 5 EUV verankerten Subsidiaritätsprinzip tätig werden."*[821]

Zwar kann die Richtlinie dem Whistleblower nicht unmittelbaren Schutz gewähren – aufgrund der Umsetzungspflicht der Mitgliedstaaten, die sich bei der Umsetzung an die Vorgaben der Richtlinie zu halten haben, kann jedoch jetzt schon erahnt

817 Legislative Entschließung des Europäischen Parlaments vom 17. April 2014 zu dem Vorschlag für eine Richtlinie des Europäischen Parlaments und des Rates über bestimmte Vorschriften für Schadensersatzklagen nach einzelstaatlichem Recht wegen Zuwiderhandlungen gegen wettbewerbsrechtliche Bestimmungen der Mitgliedstaaten und der Europäischen Union (COM(2013)0404 – C7-0170/2013 – 2013/0185(COD)) (Ordentliches Gesetzgebungsverfahren: erste Lesung), abrufbar unter: http://www.europarl.europa.eu/sides/getDoc.do?type=TA&reference=P7-TA-2014-0451&language=DE&ring=A7-2014-0089, (abgerufen am 20.6.2014).

818 Richtlinie über bestimmte Vorschriften für Schadensersatzklagen nach nationalem Recht wegen Zuwiderhandlungen gegen wettbewerbsrechtliche Bestimmungen der Mitgliedstaaten und der Europäischen Union („Schadensersatzrichtlinie") 2014/104/EU, ABL. L 349 vom 5.12.2014, S. 1–19.

819 Schadensersatzrichtlinie 2014/104/EU, Erwägungsgrund 3.

820 Schadensersatzrichtlinie 2014/104/EU, Erwägungsgrund 6.

821 Schadensersatzrichtlinie 2014/104/EU, Erwägungsgrund 54.

werden, welche Regelungen künftig in Deutschland gelten werden. Im Folgenden werden einige Regelungsgegenstände kurz dargestellt, wobei nur die Artikel behandelt werden, die die Offenlegung von Beweismitteln und die Privilegierung des Kronzeugen/Whistleblowers zum Gegenstand haben.

aa. Recht auf vollständigen Schadensersatz (Art. 3 RL)

Art. 3 der RL ist die Norm, die den Kreis der Anspruchsberechtigten festlegt und aufzeigt, welche einzelnen Schadensposten ersetzt werden können. Absatz 3 betont, dass der vollständige Ersatz im Rahmen dieser Richtlinie nicht zur Überkompensation führen darf, unabhängig davon, ob es sich dabei um Strafschadensersatz, Mehrfachentschädigungen oder andere Arten von Schadensersatz handelt.

bb. Offenlegung von Beweismitteln (Art. 5–7 RL)

Die Art. 5 bis 7 der RL regeln die Offenlegung von Dokumenten und die beschränkte Verwendung von allein durch Einsicht in die Akten einer Wettbewerbsbehörde erlangten Beweismitteln. Art. 8 RL legt fest, welche Sanktionen die einzelstaatlichen Gerichte verhängen können, wenn Offenlegungsanordnungen gegen die Parteien, Dritte und ihre rechtlichen Vertreter nicht befolgt werden.

(1) Offenlegung von Beweismitteln (Art. 5 RL)

Nach Art. 5 Abs. 1 RL haben die Mitgliedstaaten zu gewährleisten, dass in Schadensersatzverfahren in der Union auf Antrag eines Klägers die einzelstaatlichen Gerichte unter den Voraussetzungen dieses Kapitels die Offenlegung von *relevanten* Beweismitteln durch den Beklagten oder einen Dritten, die sich in deren Verfügungsgewalt befinden, anordnen können. Der Kläger muss eine *fundierte Begründung* vorgelegt haben, die die Plausibilität seines Schadensersatzanspruchs ausreichend stützt. Der Zugang zu den Beweismitteln darf nicht mit einem unzumutbaren Aufwand verbunden sein.

Nach Abs. 2 müssen die Mitgliedstaaten gewährleisten, dass die einzelstaatlichen Gerichte die Offenlegung von bestimmten einzelnen Beweismitteln oder relevanten Kategorien von Beweismitteln anordnen können. Dies hat so *genau und präzise abgegrenzt* zu erfolgen, wie dies auf der Grundlage der mit zumutbarem Aufwand zugänglichen Tatsachen in der *substantiierten Begründung* möglich ist. Es genügt demnach – soweit nicht anders möglich – die möglichst konkrete Bezeichnung offenzulegender Beweismittel der Kategorie nach.

In Abs. 3 ist der Grundsatz der Verhältnismäßigkeit festgeschrieben. Insbesondere haben die Mitgliedstaaten zu berücksichtigen, inwieweit der Anspruch oder der Einwand durch zugängliche Tatsachen und Beweismittel gestützt wird, die den Antrag auf Offenlegung von Beweismitteln rechtfertigen. In die Berücksichtigung haben Umfang und Kosten der Offenlegung einzufließen. Zudem ist das Kriterium, ob die offenzulegenden Beweismittel vertrauliche Informationen – insbesondere Dritte betreffende Informationen – enthalten, zu beachten.

Besondere Aufmerksamkeit ist Abs. 4 zu schenken. Darin ist bestimmt, dass die Mitgliedstaaten gewährleisten, dass die einzelstaatlichen Gerichte befugt sind, die Offenlegung von Beweismitteln zu veranlassen, die vertrauliche Informationen enthalten, wenn sie dies bei Schadensersatzklagen als *sachdienlich* erachten. Jedoch haben sie auch zu gewährleisten, dass die einzelstaatlichen Gerichte bei der Veranlassung der Offenlegung solcher Informationen über wirksame *Maßnahmen für deren Schutz* verfügen.

Danach ist der Zugang zu allen Dokumenten im Sinne von Art. 5 RL, gleich welcher Art, grundsätzlich zulässig.

In Abs. 5 wird klargestellt, dass das Interesse von Unternehmen, Schadensersatzklagen aufgrund von Zuwiderhandlungen gegen das Wettbewerbsrecht zu vermeiden, nicht schützenswert ist.

Überdies wird in Abs. 7 noch einmal hervorgehoben, dass die Mitgliedstaaten gewährleisten müssen, dass diejenigen, von denen die Offenlegung verlangt wird, *Gelegenheit zur Anhörung* erhalten, bevor ein nationales Gericht die Offenlegung aufgrund dieses Artikels anordnet.

Abs. 8 des Artikels erlaubt den Mitgliedstaaten, unbeschadet der Absätze 4 und 7 sowie des Art. 6 RL, Vorschriften beizubehalten oder einzuführen, die zu einer umfassenderen Offenlegung von Beweismitteln führen.

(2) Offenlegung von Beweismitteln, die in den Akten einer Wettbewerbsbehörde enthalten sind (Art. 6 RL)

Art. 6 RL regelt die Bestimmungen, die, neben denen des Art. 5, für die angeordnete Offenlegung von Beweismitteln, die in den Akten einer Wettbewerbsbehörde enthalten sind, bei Schadensersatzklagen relevant sind. Der Zugang der Öffentlichkeit zu Dokumenten nach der Verordnung Nr. 1049/2001 (EG) wird ebenso wenig berührt, wie die nach Unionsrecht oder nationalem Recht geltenden Vorschriften und geltende Anwendungspraxis im Bereich des Schutzes der internen Unterlagen von Wettbewerbsbehörden und des Schriftverkehrs zwischen Wettbewerbsbehörden (Abs. 3).

Abs. 4 stellt klare Maßstäbe für die Verhältnismäßigkeit auf. Bei einer Anordnung zur Offenlegung von Informationen

> *„berücksichtigen die nationalen Gerichte gemäß den Kriterien des Artikels 5 Absatz 3,*
>
> *a) ob der Antrag eigens hinsichtlich Art, Gegenstand oder Inhalt der der Wettbewerbsbehörde übermittelten oder in deren Akten enthaltenen Unterlagen formuliert wurde und nicht unspezifisch in Bezug auf die der Wettbewerbsbehörde übermittelten Unterlagen,*
>
> *b) ob die Partei, die die Offenlegung beantragt, diesen Antrag im Rahmen einer Schadensersatzklage vor einem nationalen Gericht stellt, und*
>
> *c) ob […] die Notwendigkeit besteht, die Wirksamkeit der öffentlichen Durchsetzung des Wettbewerbsrechts zu wahren.“*

Nach Abs. 5 dürfen die nationalen Gerichte die Offenlegung der folgenden Kategorien von Beweismitteln erst dann anordnen,

> *„wenn eine Wettbewerbsbehörde ihr Verfahren durch Erlass eines Beschlusses oder in anderer Weise beendet hat:*
>
> *a) Informationen, die von einer natürlichen oder juristischen Person eigens für das wettbewerbsbehördliche Verfahren erstellt wurden,*
>
> *b) Informationen, die die Wettbewerbsbehörden im Laufe ihres Verfahrens erstellt und den Parteien übermittelt haben, und*
>
> *c) Vergleichsausführungen, die zurückgezogen wurden."*

Art. 6 Abs. 6 der RL schreibt fest, dass die Offenlegung von Kronzeugenunternehmenserklärungen und Vergleichsausführungen durch eine Partei oder einen Dritten zum Zwecke von Schadenersatzklagen *zu keinem Zeitpunkt von nationalen Gerichten angeordnet* werden kann.

Jedoch kann ein Kläger nach Abs. 7 einen begründeten Antrag stellen, wonach ein einzelstaatliches Gericht die in Absatz 6 genannten Unterlagen einsehen darf, um zu gewährleisten, dass der Inhalt der Unterlagen den in Art. 4 (Begriffsbestimmung) Nr. 16 (Kronzeugenerklärungen) und Nr. 18 (Vergleichsausführungen) enthaltenen Begriffsbestimmungen entspricht. Bei dieser Beurteilung können die einzelstaatlichen Gerichte nur die zuständige Wettbewerbsbehörde um Unterstützung bitten. Die Verfasser der betreffenden Unterlagen können auch angehört werden. Das Gericht darf auf keinen Fall den anderen Parteien oder Dritten Zugang zu diesen Beweismitteln gewähren. Abs. 8 und 9 regeln die Fälle, in denen Unterlagen oder Teile von Unterlagen angefordert werden, die nicht unter Abs. 6 fallen. Die Offenlegung der Beweismittel kann dann angeordnet werden. In Abs. 10 des Artikels wird betont, dass die Mitgliedstaaten zu gewährleisten haben, dass die nationalen Gerichte die Offenlegung von Beweismitteln bei der Wettbewerbsbehörde beantragen können, wenn von einer Partei oder einem Dritten die angeforderten Beweismittel *nicht mit zumutbarem Aufwand* erlangt werden können.

Abs. 11 erlaubt der Wettbewerbsbehörde die Vorlage einer Bemerkung vor Gericht, wenn sie ihre Ansichten über die Verhältnismäßigkeit von Offenlegungsanträgen darlegen möchte.

(3) Beschränkungen für die Verwendung von allein durch Einsicht in die Akten einer Wettbewerbsbehörde erlangten Beweismittel (Art. 7 RL)

Nach Maßgabe des Art. 7 der RL sind Beweismittel, die unter eine der in Art. 6 Abs. 6 RL aufgeführten Kategorien fallen und von einer natürlichen oder juristischen Person allein durch Einsicht in die Akten einer Wettbewerbsbehörde erlangt wurden, in Verfahren auf Schadensersatz *unzulässig* oder müssen auf andere Weise nach dem anzuwendenden *nationalen Recht geschützt* werden. Das Gleiche gilt nach Abs. 2 für Beweismittel, die unter eine der in Art. 6 Abs. 5 aufgeführten Kategorien fallen. Damit soll sichergestellt sein, dass die Beschränkungen für die Verwendung von Beweismitteln ihre volle Wirkung entfalten, bis die Wettbewerbsbehörde ihr Verfahren durch Erlass eines Beschlusses oder in anderer Weise eingestellt hat. Nach Abs. 3 können Beweismittel, die von einer natürlichen oder juristischen Person allein durch Einsicht in die Akten einer Wettbewerbsbehörde erlangt wurden und die nicht unter Abs. 1 oder 2 fallen, in einem Schadensersatzverfahren nur von dieser Person oder von der natürlichen oder juristischen Person verwendet werden, die in ihre Rechte eintritt. Das gilt auch für die Person, die ihren Anspruch erworben hat.

cc. Gesamtschuldnerische Haftung (Art. 11 RL)

Art. 11 Abs. 1 der RL gibt den Grundsatz der Gesamtschuldnerhaftung wieder, wonach jedes der am Wettbewerbsverstoß beteiligten Unternehmen zum vollständigen Ersatz des Schadens verpflichtet ist. Der Geschädigte hat das Recht, von jedem Beteiligten vollständigen Schadensersatz zu verlangen, bis der Schaden vollständig ersetzt ist.

Eine Sonderstellung für bestimmte Unternehmen sieht Abs. 2 vor. Kleine oder mittlere Unternehmen gemäß der Definition in der Empfehlung 2003/361/EG der Kommission haften nur gegenüber ihren unmittelbaren und mittelbaren Abnehmern, wenn bestimmte Voraussetzungen vorliegen, die in a/b) und i/ii) aufgelistet sind. Diese Ausnahme gilt nach Abs. 3 nicht, wenn das Unternehmen die Zuwiderhandlung organisiert oder andere Unternehmen gezwungen hat, sich an der Zuwiderhandlung zu beteiligen oder ein früherer Wettbewerbsrechtsverstoß des Unternehmens festgestellt wurde.

Abs. 4 hat die Außenhaftung des Kronzeugen zum Gegenstand. Abweichend von Abs. 1 haftet ein Unternehmen, dem der Erlass der Geldbuße zuerkannt wurde, gesamtschuldnerisch

a) gegenüber seinen unmittelbaren oder mittelbaren Abnehmern oder Lieferanten und
b) gegenüber anderen Geschädigten nur dann, wenn von den anderen Unternehmen, die an derselben Zuwiderhandlung gegen das Wettbewerbsrecht beteiligt waren, kein vollständiger Schadensersatz erlangt werden kann.

Ein Rechtsverletzer kann von anderen Rechtsverletzern einen Ausgleichsbetrag verlangen. Dessen Höhe wird anhand ihrer relativen Verantwortung für den durch den Verstoß gegen das Wettbewerbsrecht verursachten Schaden bestimmt (Abs. 5). Der Ausgleichsbetrag eines Rechtsverletzers, dem im Rahmen eines Kronzeugen-Programms der Erlass der Geldbuße zuerkannt wurde, darf nicht höher sein als der Schaden, den er seinen eigenen unmittelbaren oder mittelbaren Abnehmern oder Lieferanten verursacht hat. Nach Abs. 6 haben die Mitgliedstaaten zu gewährleisten, dass, soweit durch die Zuwiderhandlung gegen das Wettbewerbsrecht anderen Geschädigten als den unmittelbaren oder mittelbaren Abnehmern oder Lieferanten der Rechtsverletzer ein Schaden entstanden ist, der Ausgleichsbetrag eines Kronzeugen gegenüber den anderen Rechtsverletzern anhand seiner relativen Verantwortung für diesen Schaden bestimmt wird.

dd. Wirkung von Vergleichen auf anschließende Schadensersatzklagen (Art. 19 RL)

Art. 19 Abs. 1 der RL ordnet an, dass sich der Schadensersatzanspruch des Geschädigten nach einem Vergleich proportional zu dem Schadensanteil verringert, der ihm durch den an der Einigung beteiligten Rechtsverletzer entstanden ist. Den verbleibenden Anspruch hat der Geschädigte dann nur gegenüber dem nicht am Vergleich beteiligten Rechtsverletzer geltend zu machen. Die übrigen zahlenden Schädiger können von dem an der Einigung beteiligten Rechtsverletzer keinen Ausgleichsbetrag verlangen (Abs. 2). Abweichend von dieser Vorschrift muss gemäß Abs. 3 der an dem Vergleich beteiligte Rechtsverletzer nur bei völliger Zahlungsunfähigkeit der anderen Rechtsverletzer für die Anteile aufkommen. Bei der Festlegung des Ausgleichsbetrags, den ein Rechtsverletzer von einem anderen Rechtsverletzer entsprechend seiner relativen Verantwortung für den durch die Zuwiderhandlung gegen das Wettbewerbsrecht verursachten Schaden verlangen kann, sind frühere Vergleiche, an denen der betreffende Rechtsverletzer beteiligt war, von den Gerichten zu berücksichtigen.

b. Rechtliche Würdigung

Kritikwürdig ist die „abgeschwächte Form" der Substantiierung des Parteivortrags zur Einsicht in die Kronzeugendokumente. Danach soll es ausreichen, dass bei der Substantiierung des Vortrags „nur" die Kategorie der Beweismittel bezeichnet wird. Zwar ist es grundsätzlich zu begrüßen, dass auf das Erfordernis der Substantiierung nicht gänzlich verzichtet wird. Auch wird damit dem Umstand Rechnung getragen, dass aufgrund des strukturellen Informationsdefizits des kartellrechtlichen Schadensersatzklägers dieser nicht über detaillierte Informationen verfügen kann. Jedoch führt die oben genannte Möglichkeit letztlich zu einer Aufweichung der Substantiierungspflicht. Eine Klarstellung, insbesondere was unter „Kategorie der Beweismittel" zu verstehen ist, ist sinnvoll.

Zudem ist fraglich, ob der in Art. 6 RL uneingeschränkt garantierte Schutz von Kronzeugendokumenten mit der Rechtsprechung des EuGH vereinbar ist. Wenn es dem nationalen Gesetzgeber nach dem Urteil *Donau Chemie* verwehrt ist, die Akteneinsicht in bestimmte Dokumente wie die Kronzeugenerklärungen generell zu untersagen, muss dies auch für den europäischen Gesetzgeber gelten. Zur Akteneinsicht der Geschädigten hat der Gerichtshof betont, dass Kronzeugenprogramme einer wirksamen Anwendung des Kartellverbots gem. Art. 101 AEUV dienten. Doch „jede starre Regel" zum Schutz dieser Programme beeinträchtige das Recht auf Schadensersatz. Insbesondere dürfe der Zugang zu Kronzeugenakten nicht „systematisch" verweigert werden. Insofern ist die in Art. 6 RL geregelte vollständige Verweigerung des Zugangs zu Kronzeugendokumenten mit der Rechtsprechung des EuGH nicht vereinbar.[822]

---

822 So auch *Kersting*, JZ 2013, 737, 739; a. A. *Fiedler*, BB 2013, 2179, 2184.

Die Privilegierung des Kronzeugen im Rahmen der gesamtschuldnerischen Haftung ist ein gutes Instrument.[823] Festgeschrieben ist eine materiell-rechtliche Privilegierung des Kronzeugen, der nur direkt von seinen unmittelbaren oder mittelbaren Abnehmern oder Lieferanten in Anspruch genommen werden soll. Ansonsten besteht lediglich eine subsidiäre Haftung. Die Regelung des Art. 11 RL gewährt dem Kronzeugen einen gewissen haftungsrechtlichen Schutz.

Eine Benachteiligung der Geschädigten besteht nur auf den ersten Blick. Die Rückausnahme, dass der Kronzeuge in Anspruch genommen werden darf, wenn die anderen Kartellanten für den Schaden nicht aufkommen bzw. den Schaden nicht kompensieren können, ist im Hinblick auf den hinreichend zu gewährenden Schutz der Geschädigten zu begrüßen. Art. 11 RL schlägt eine Möglichkeit vor, die zu einer echten Haftungsprivilegierung in Form der Ausfallhaftung führt und die Interessen der Geschädigten – anders als bei einer vollkommenen Haftungsbeschränkung – angemessen berücksichtigt. Zudem führt die Regelung zur gewünschten Optimierung des Verhältnisses privater und öffentlicher Kartellrechtsdurchsetzung.

Im Ergebnis muss sich ein Wandel vollziehen. Das Interesse der Wettbewerbsbehörden, eine Beeinträchtigung der Kronzeugenprogramme zu verhindern, um sie attraktiv zu halten, kann nicht mehr durch die völlige Verweigerung der Einsicht in die Akten verfolgt werden. Die Mitgliedstaaten sowie die Kommission und die Behörden müssen dazu übergehen, Kronzeugen auf anderem Wege zu privilegieren. Dies könnte durch eine materiell-rechtliche Besserstellung des Kronzeugen geschehen. Durch den Erlass der Richtlinie über kartellrechtliche Schadensersatzklagen sind erste Schritte in diese Richtung unternommen worden. Durch die Privilegierung des Kronzeugen bei der gesamtschuldnerischen Haftung schafft die Richtlinie eine Grundlage, das Verhältnis zwischen *private* und *public enforcement* zu verbessern und die Rechtsentwicklung im Europäischen Raum zu harmonisieren.

Die Schadensersatzrichtlinie 2014/104/EU soll es den Opfern von Kartellen erleichtern, Schadensersatzansprüche gegen die Kartellteilnehmer durchzusetzen. Wer in der Europäischen Union wegen eines Kartells überhöhte Preise gezahlt hat, wird es künftig leichter haben, vor Gericht Schadensersatz zu beanspruchen. Der Kronzeuge wird aber auch nicht vernachlässigt – ihn privilegieren besondere Regelungen zu Kronzeugenerklärungen sowie zum Gesamtschuldnerausgleich. Die Mitgliedstaaten und einzelstaatlichen Gerichte haben einen großen Anteil an dieser Entwicklung zu tragen. Die nationalen Vorschriften sind entsprechend den Vorgaben zu reformieren. Die Richtlinie beseitigt eine Reihe praktischer Hindernisse für Schadensersatzklagen. Sie entfaltet zwar keine unmittelbare horizontale Wirkung,

823 Der BDI hält die Regelung in Art. 11 grundsätzlich für begrüßenswert, sieht jedoch Bedürftigkeit zur Änderung im Bereich der Beweislast beim Kartellantenregress, vgl. Stellungnahme des BDI zum Richtlinienvorschlag, Dok.-Nr. D 0607, vom 31.8.2013, S. 6, abrufbar unter: http://www.bdi.eu/download_content/KonjunkturStandortUndWettbewerb/D_0607_-_BDI-STN_-_Schadensersatzklagen.pdf, (abgerufen am 10.6.2014).

allerdings können die Richtlinien-Vorgaben bereits jetzt gerichtlich berücksichtigt werden.

Opfer von Wettbewerbsverstößen haben nach der Richtlinie Anspruch auf die volle Kompensation ihres Schadens zuzüglich Zinsen. Hat der Verstoß eine Preiserhöhung verursacht und wird diese in der Vertriebskette „weitergegeben", steht der Schadenersatz (auch) dem Letztgeschädigten zu. Hingegen soll die Richtlinie nicht zu einer Über-Kompensation wie in den USA führen („*punitive damages*"). Die Kartellopfer haben – unter Beachtung des Beginns der Verjährungsfrist – mindestens fünf Jahre lang Zeit, um ihre Schadensersatzklage zu erheben. Daneben sind drei wichtige Aspekte hervorzuheben:

Erstens sieht die Richtlinie zur Erleichterung von Schadenersatzklagen vor, dass die nationalen Gerichte grundsätzlich die Offenlegung von Beweismitteln durch Unternehmen anordnen können. Dabei werden sie die Verhältnismäßigkeit dieser Anordnungen sicherstellen und für einen gebührenden Schutz vertraulicher Angaben sorgen. Ausgenommen von der Offenlegungspflicht bleiben jedoch Erklärungen von Kronzeugen und Vergleichsausführungen. Damit sollen Kartellteilnehmer angeregt werden, mit den Behörden zusammenzuarbeiten. Es soll verhindert werden, dass Gerichte die Offenlegung von Kronzeugenerklärungen und Vergleichsausführungen, die in den Akten der Wettbewerbsbehörde enthalten sind, anordnen können. Damit eine solche Vorgabe mit der Rechtsprechung des EuGH im Einklang steht, müssen die Mitgliedstaaten gewährleisten, dass die einzelstaatlichen Gerichte die Offenlegung aller Dokumente bei der Wettbewerbsbehörde beantragen können. Zwar wird innerhalb der Richtlinie in einigen Artikel oft darauf verwiesen, dass dem Kläger alle relevanten Beweismittel durch Anordnung offengelegt werden können. Ein Kläger kann zudem einen begründeten Antrag stellen, wonach ein einzelstaatliches Gericht die in Art. 6 Abs. 6 (Offenlegung von Beweismitteln, die in den Akten einer Wettbewerbsbehörde enthalten sind) genannten Unterlagen einsehen darf. So ist gewährleistet, dass der Inhalt der Unterlagen den in Art. 4 (Begriffsbestimmung) Nr. 16 (Kronzeugenerklärung) und 18 (Vergleichsausführung) enthaltenen Begriffsbestimmungen entspricht. Bei dieser Beurteilung können die einzelstaatlichen Gerichte nur die zuständige Wettbewerbsbehörde um Unterstützung bitten. Die Verfasser der betreffenden Unterlagen können überdies Gelegenheit zur Anhörung erhalten. Das Gericht darf aber auf keinen Fall den anderen Parteien oder Dritten Zugang zu diesen Unterlagen gewähren. Es wird also lediglich gestattet, die Akten einzusehen, um zu überprüfen, ob in den Unterlagen wirklich (nur) die Kronzeugenerklärung und die Vergleichsausführung enthalten sind und keine anderen – relevanten – Beweismittel.

Zweitens muss neben Antitrust-Entscheidungen der EU-Kommission auch ein Kartellrechtsverstoß durch eine nationale Wettbewerbsbehörde in Schadenersatzverfahren vor den Gerichten dieses Mitgliedstaats unwiderleglich festgestellt werden. Bestandskräftige Entscheidungen von einzelstaatlichen Gerichten sollen zumindest als *Prima-facie*-Beweis akzeptiert werden.

Drittens wird die gesamtschuldnerische Haftung neu geregelt. Neben der Sonderregelung für kleine oder mittlere Unternehmen haften im Regelfall die einzelnen

Teilnehmer eines Kartells gesamtschuldnerisch: Ein Opfer kann von jedem von ihnen den vollen Schadenersatz verlangen. Unternehmen, die eine Kronzeugenbehandlung erreicht haben, müssen allerdings nur für den von ihnen verursachten Schaden bei unmittelbaren und mittelbaren Abnehmern oder Lieferanten geradestehen, es sei denn, die anderen Kartellteilnehmer sind zahlungsunfähig. Dann sind Kronzeugen auch gegenüber den anderen Geschädigten ersatzpflichtig. Zudem wirken einvernehmliche Regelungen zwischen dem Geschädigten und einem Rechtsverletzer auf anschließende Schadensersatzklagen ein.

All diese Neuerungen lösen nicht das Problem, dass sich eine individuelle Klage für den Einzelnen oft nicht lohnt, wenn sich der Schaden auf viele Opfer verteilt. Deshalb hat die EU-Kommission im Sommer 2013[824] in einer rechtlich nicht bindenden Empfehlung den Mitgliedstaaten geraten, Verfahren für eine „kollektive Rechtsdurchsetzung" einzurichten.

Am Ende wird die Praxis entscheiden, ob sich durch die Umsetzung der Richtlinie die privaten Schadensersatzklagen und die öffentliche Durchsetzung des Kartellrechts durch die Wettbewerbsbehörden ergänzen. Die Grundsätze sind klar: Zum einen wird vollständiger Schadensersatz für die Opfer erreicht (einschließlich entgangener Gewinne und Zinsen). Zum anderen wird die zentrale Rolle der Wettbewerbsbehörden bei der Untersuchung und Ahndung von Verstößen gestärkt, und so eine abschreckende Wirkung erzielt. Unterstützt werden die Behörden im Rahmen der Untersuchung durch Kronzeugen, die dafür privilegiert werden sollen. Ob diese Ziele mit der Richtlinie erreicht werden können, wird sich erst im Laufe der Zeit zeigen.

Die Umsetzungspflicht der Mitgliedstaaten zieht einige Änderungen der bestehenden Vorschriften in Deutschland nach sich. Besonders im Zivilprozessrecht würde die Umsetzung der Richtlinie eine Veränderung der Denkweise erforderlich machen, um den Kartellbeteiligten eine weitreichende Offenlegung der betriebsinternen Dokumente, wie in Art. 5 RL gefordert, abverlangen zu können. Die Reichweite des § 142 ZPO, nach dem ein Prozessgericht die Herausgabe von Unterlagen vom Beklagten oder einem Dritten anordnen kann, ist dadurch eingeschränkt, dass der Kläger im Kartellverfahren außerhalb des Geschehens steht und regelmäßig die erforderlichen Dokumente nicht genau genug beschreiben kann. Die Offenlegungsregeln der §§ 142 bis 144 ZPO reichen dafür nicht aus, da die Richtlinie über § 142 ZPO hinausgeht.

Hinsichtlich der Einsicht in Kronzeugenerklärungen/Unternehmensvereinbarungen nach Art. 6 RL und der Wirkung einzelstaatlicher Entscheidungen ergeben sich keine gesetzlichen Änderungen durch die Umsetzung der Richtlinie. Die Richtlinie stellt letztendlich das Ergebnis einer Abwägung (zugunsten des Kronzeugen) dar, die für alle Kronzeugenunterlagen gelten soll.

824 Die EU-Kommission hat die Mitgliedstaaten aufgerufen, kollektive Rechtsschutzverfahren einzuführen (11.6.2013), abrufbar unter: http://ec.europa.eu/deutschland/press/pr_releases/11453_de.htm, (abgerufen am 10.6.2014).

Einer teilweisen Kodifizierung bedarf es im Bereich der gesamtschuldnerischen Haftung. Derzeit haftet in Deutschland der Kronzeuge wie alle Kartellbeteiligten gem. §§ 830 Abs. 1 S. 1, 840 BGB gesamtschuldnerisch.[825] Eine eingeschränkte Haftung im Innenverhältnis wird von der herrschenden Meinung in Deutschland bereits nach geltendem Recht anerkannt (§ 426 i. V. m. § 254 BGB analog). Die Sonderbestimmung für Kronzeugen in Form der Ausfallhaftung im Außenverhältnis stellt eine Neuheit dar. Da sich dieses Modell der Haftungsprivilegierung aus dem Gesetz nicht ergibt und ein materiell-rechtliches Haftungsprivileg des Kronzeugen gegenüber Kartellopfern bislang nicht anerkannt war, ist eine gesetzliche Kodifikation nötig.

Die Mitgliedstaaten haben nun bis zum 27. Dezember 2016 Zeit, um die Bestimmungen in innerstaatliches Recht umzusetzen. Das bedeutet für den haftungsrechtlichen Schutz des Whistleblowers eine Besserung. Durch die Neugestaltung der Haftungsbegrenzung zugunsten des Whistleblowers wird ein Schutzmechanismus geschaffen, der den kartellrechtlichen Kronzeugen vor ausufernder Inanspruchnahme durch Dritte schützt. Eine Integration eines haftungsrechtlichen Schutzes in ein deutsches Gesetzes führt zu Rechtssicherheit und Vertrauen in die Schutzinstrumente. Der Whistleblower hat Gewissheit, inwieweit ihn haftungsrechtliche Konsequenzen treffen. Er schreckt daher weniger vor der Offenlegung des Missstandes zurück. Nachdem die Bonusregelung ihm ordnungswidrigkeitenrechtlichen Schutz gebietet, kann den Whistleblower – soweit eine passende Vorschrift implementiert wird – auch (zumindest teilweise) haftungsrechtlicher Schutz zukommen. Aufgrund der schwer zu durchblickenden Rechtsprechung der nationalen und europäischen Gerichte erscheint dies unausweichlich. Ein gesetzlicher Schutz des Whistleblowers im Haftungsrecht ist zu befürworten.

6. Gesamtfazit

Die Ausführungen haben gezeigt, dass sowohl auf nationaler als auch auf europäischer Ebene verstärkt mit Schadensersatzklagen Dritter gerechnet werden muss. Die höchstrichterlichen Entscheidungen lassen es nicht zu, das Akteneinsichtsrecht den Geschädigten gänzlich zu verweigern – gleich ob dies auf gesetzlichem Weg geschieht oder durch Richterspruch. Was dies für den Whistleblower bedeutet, liegt auf der Hand. Der Whistleblower, der als Kartellant agierte, durch seine Offenlegung aber von seinem Verhalten abgerückt ist, muss mit gehäuften Schadensersatzforderungen rechnen. Der Whistleblower hat umso mehr zu befürchten, (auch frühzeitig) das Ziel von Schadensersatzklagen zu werden, weil er derjenige ist, der alle Informationen an die Behörde weitergeleitet hat, aus der die Akten zur Einsicht entstehen. Gerade diese Akten werden dann zur Vorbereitung eines Schadensersatzanspruches genutzt. Im Rahmen der Geltendmachung des Einsichtsrechts in Dokumente, die sich im Besitz des Bundeskartellamtes befinden, ist die Einsicht

825 BGH, Urt. v. 28.6.2011 – KZR 75/10, BB 2012, 75, Leitsatz.

auch schon in einer früheren Phase des Verfahrensstadiums möglich. In den Akten, die sehr umfangreich und undurchsichtig sein können, lässt sich der Kartellbeteiligte, der die Unternehmenserklärung abgegeben hat, am leichtesten ermitteln. Der Moment der Akteneinsicht bestimmt die Wahl, wer als erstes „angegriffen" wird. Es besteht die Gefahr, dass die besondere Stellung des Kronzeugen ausgenutzt wird und ihn überwiegend die Pflicht zur zivilgerichtlichen Auseinandersetzung trifft. Diese Gefahr aber kann ihm aufgrund der jüngst ergangenen höchstrichterlichen europäischen Entscheidungen zum Akteneinsichtsrecht nicht genommen werden.

Nur die Umsetzung der Schadensersatzrichtlinie könnte dazu führen, dass die Folgen der (vorranigen) Inanspruchnahme des Kronzeugen weniger einschneidend sind. Denn die Richtlinie regelt zum einen die Akteneinsicht bzw. die Offenlegung von Akteninhalten differenzierter und zum anderen besteht nach Art. 11 der RL die Möglichkeit, die Haftung des Kronzeugen gegenüber Dritten zu begrenzen.

Dem Whistleblower, der zwar nicht an einer Kartellrechtswidrigkeit beteiligt war, aber an einer sonstigen Tat mitgewirkt hat, gebührt momentan kein gesetzlicher Schutz vor den haftungsrechtlichen Folgen. Das nachträgliche Abrücken von der Tat muss jedoch honoriert werden. Dafür reicht auch eine Reduzierung der Haftung aus. Sie gibt dem Whistleblower das Signal, dass es sich lohnt, sich in den Dienst der Rechtsordnung zu stellen und Gesetzesverstöße im Unternehmen trotz eigener Beteiligung offenzulegen. Dass die tätige Reue bei gesetzeswidrigem Handeln berücksichtigt wird, ist der Rechtsordnung nicht fremd – aber nur im Strafrecht geläufig. Deshalb ist an einen Tatbestand der tätigen Reue im Zivilrecht zu denken. Der Arbeitnehmer sieht sich sonst gezwungen, Missstände im Unternehmen nicht zu offenbaren, weil er seinen finanziellen Ruin befürchten muss.

Teil 5: Der Schutz des Whistleblowers de lege ferenda

A. Arbeitsrechtlicher Schutz durch Kodifizierung eines Schutzgesetzes

In Deutschland erfolgt die rechtliche Bewertung von Whistleblowing bisher hauptsächlich durch Rückgriff auf die allgemeinen Gesetze, Grundsätze und Prinzipien. Dies führt meist zu Einzelfallentscheidungen auf der Grundlage von Abwägungen, die vorab kaum berechenbar sind. Zwar bietet die *Heinisch*-Entscheidung des EGMR nun Anhaltspunkte zur Bewertung des Whistleblower-Verhaltens. Jedoch fehlen klare Definitionen und auf den Fall zugeschnittene Auslegungsregeln, die zu einem ausreichenden Informantenschutz führen.

Vor diesem Hintergrund stellt sich die Frage, ob ein gesetzlicher Schutz das Spannungsverhältnis zwischen moralischer Pflicht zur Anzeige von Missständen einerseits und der Sanktionierung des Arbeitnehmers bei einer Anzeige andererseits auflösen kann. Ein gesetzlicher Whistleblowerschutz könnte die geschützten und zu schützenden Rechte und Pflichten des Whistleblowers, aber auch die des betroffenen Arbeitgebers kodifizieren. Bereits auf dem G20-Gipfel in Seoul am 11. und 12. November 2010 hatten sich die Teilnehmer, darunter die Bundeskanzlerin Angela Merkel als Vertreterin der Bundesrepublik Deutschland, darauf verständigt „bis Ende 2012 Regeln zum Whistleblowerschutz zu erlassen und umzusetzen, [...] um Hinweisgeber, die gutgläubig einen Verdacht auf Korruption melden, vor Diskriminierung und Vergeltungsmaßnahmen zu schützen".[826] Zwar zielte das geplante Vorhaben nur auf die Korruption und nicht auf andere wichtige Hinweisgegenstände ab, jedoch sieht die Bundesregierung auch in diesem Bereich bis dato keinen Handlungsbedarf.

Der 2013 geschlossene Koalitionsvertrag der großen Koalition von CDU/CSU und SPD spricht das Thema zwar unter der Überschrift „Informantenschutz im Arbeitsverhältnis" an, sagt jedoch lediglich, dass geprüft wird, ob die internationalen Vorgaben hinreichend umgesetzt sind.[827] Weiteres bleibt abzuwarten.

Bis vor kurzem stellten die nachfolgend betrachteten Gesetzesentwürfe, die in der vorherigen (17.) Legislaturperiode von Parteien eingebracht wurden, die

826 Abschlusserklärung vom 12.11.2010, Annex III: G 20 Anti-Corruption Action Plan, Point 7, vgl. *Deiseroth*, PersR 2011, 457, 458 oder abrufbar unter: http://www.oecd.org/daf/anti-bribery/G20_Anti-Corruption_Action_Plan.pdf, (abgerufen am 9.9.2014).

827 Koalitionsvertrag zwischen der CDU, CSU und SPD, 18. Legislaturperiode, vom 27.11.2013, S. 70.

einzigen Kodifikationsversuche dar. Im November 2014 wagte die Bundestagsfraktion Bündnis 90/Die Grünen einen neuen Versuch.

I. Entwurf eines Hinweisgeberschutzgesetzes

Mit einem Gesetzentwurf hatte die Oppositionsfraktion im Bundestag 2012 offensichtlich auf das Urteil des EGMR[828] vom 21. Juli 2011 reagiert. Der Gesetzentwurf ist, gemessen an dem Ziel, die momentane Whistleblower-Situation zu verbessern, aus Sicht der Unternehmenspraxis durchaus zu begrüßen. Zudem könnte eine gesetzliche Kodifizierung die Gerichte entlasten, da die Rechtsfindung dann durch einen Blick ins Gesetz erleichtert oder zumindest durch Auslegung kodifizierter Bestimmungen erreicht wird.

1. Hintergründe

Am 07. Februar 2012 brachte die SPD-Fraktion im Deutschen Bundestag den Entwurf[829] eines Gesetzes zum Schutz von Hinweisgebern – Whistleblowern (Hinweisgeberschutzgesetz – HinwGebSchG) ein. Bereits 2008 stellte die damalige Bundesregierung durch das Ministerium für Arbeit und Soziales, Ernährung, Landwirtschaft und Verbraucherschutz und das Ministerium für Justiz einen entsprechenden Gesetzentwurf vor.[830] Ziel war es, genau wie beim neuen Entwurf, eine klare und eindeutige Regelung im Bereich des Informantenschutzes zu schaffen um damit die Rechtssicherheit des Whistleblowers zu stärken.[831]

Der Gesetzentwurf vom Februar 2012 sollte den Schutz von Beschäftigten verbessern, die dazu beitragen, in ihren Organisationen und Unternehmen Missstände oder sogar Straftaten aufzudecken. Denn nach Ansicht der Fraktion sei dieser Schutz in Deutschland nur unzureichend geregelt.[832] Bisher berechtigen gesetzliche Anzeigerechte Beschäftigte nur ausnahmsweise dazu, innerbetriebliche Missstände extern anzuzeigen und häufig muss erst ein innerbetriebliches Beschwerdeverfahren durchlaufen werden.[833] Dies schafft Unsicherheit für die Betroffenen und verhindert die Offenlegung und Aufklärung von Missständen. Der von der Fraktion vorgeschlagene Gesetzentwurf eines Hinweisgeberschutzgesetzes sei daher nötig, um Beschäftigte vor arbeitsrechtlichen Nachteilen zu schützen. Gegen eine gesetzliche Regelung wurde eingewandt, dass die Einführung eines entsprechenden Hinweisgebersystems ein Klima der Denunziation fördere, wodurch das Betriebsklima nachhaltig Schaden nehme. Dem sollte im Entwurf von 2012 mit einem am Grundsatz der Verhältnismäßigkeit orientierten Anzeigerecht entgegengewirkt

828 EGMR, Urt. v. 21.7.2011 – 28274/08, NJW 2011, 3501 (Heinisch/Deutschland).
829 BT-Drs. 17/8567.
830 BT-Ausschussdrs. 16(10)849.
831 BT-Ausschussdrs. 16(10)849, S. 2.
832 BT-Drs. 17/8567, S. 5, allgemeine Begründung.
833 BT-Drs. 17/8567, S. 5, allgemeine Begründung.

werden. Zudem zeigten Erfahrungen aus anderen Staaten (z. B. *Whistleblower Protection Act, Sarbanes-Oxley Act*, USA; *Public Interest Disclosure Act*, Großbritannien), in denen bereits Hinweisgeberregelungen existieren, dass keineswegs ein „Klima der Denunziation" geschaffen wird.[834]

2. *Der Entwurf*

a. *Überblick*

Der vorgelegte Gesetzentwurf besteht insgesamt aus 15 Vorschriften. In den §§ 2 und 3 finden sich Aussagen zum Anwendungsbereich sowie Begriffsbestimmungen. In § 4 Abs. 1 wird festgelegt, dass Hinweisgeber wegen ergangener Hinweise nicht benachteiligt werden dürfen. Dem Arbeitgeber werden in § 5 Maßnahmen und Pflichten auferlegt, um eine Benachteiligung zu verhindern. Die entscheidende Vorschrift des Gesetzentwurfs ist § 6. Darin ist das Anzeigerecht des Hinweisgebers geregelt. Gemäß Abs. 1 kann sich der Hinweisgeber bei Missständen sowohl an den Arbeitgeber oder an innerbetrieblich zuständige Stellen (internes Whistleblowing) als auch an außerbetriebliche Stellen (externes Whistleblowing) wenden. Ein Hinweis an Dritte, die keiner Verschwiegenheitsverpflichtung unterliegen, oder an die Öffentlichkeit ist nach Abs. 2 eingeschränkt. In § 7 sind Leistungsverweigerungsrechte des hinweisgebenden Arbeitnehmers geregelt, während die §§ 8 und 9 Unterlassungs- und Schadensersatzansprüche bei rechtswidriger Benachteiligung des Hinweisgebers gewähren. § 10 regelt einen Schadensersatzanspruch des Arbeitgebers gegen den Hinweisgeber im Falle eines rechtswidrigen Hinweises. Um einem „Klima der Denunziation" vorzubeugen, soll § 10 die „Verantwortung der Hinweisgebenden als mündige Bürger"[835] unterstreichen. § 11 verknüpft den Entwurf mit dem BDSG und schafft für unternehmensinterne Whistleblowing-Programme einen eigenen Rechtfertigungsgrund gemäß § 4 Abs. 1 BDSG. § 12 sieht Regelungen zur Beweislast vor. Die §§ 13 und 14 normieren schließlich die Unabdingbarkeit der Vorschriften des Hinweisgeberschutzgesetzes sowie Sanktionsmöglichkeiten (§ 14) bei Verstößen gegen das Benachteiligungsverbot.

b. *Begründung*

Der das Gesetz einleitende Paragraph (§ 1) soll die vom Normgeber mit dem Gesetz verfolgten Absichten verdeutlichen. Es sollen Rahmenbedingungen geschaffen werden, um Benachteiligungen von Hinweisgebern zu verhindern.[836] Der nachfolgende Paragraph legt den Anwendungsbereich fest, wodurch der schutzbedürftige Personenkreis einen Mindestschutz erfahren soll. Die in § 3 aufgenommenen Begriffsbestimmungen haben eine klarstellende Funktion.[837] Die Begriffe

834 BT-Drs. 17/8567, S. 6, allgemeine Begründung.
835 BT-Drs. 17/8567, S. 8, Begründung zu § 10 HinwGebSchG-E.
836 BT-Drs. 17/8567, S. 7, Begründung zu § 1 HinwGebSchG-E.
837 BT-Drs. 17/8567, S. 7, Begründung zu § 3 HinwGebSchG-E.

„Benachteiligung", „Missstand", „Hinweis", „Arbeitgeber" und „zuständige externe Stellen" werden näher erläutert, um auch für juristische Laien verständlich zu sein. Auf diese Weise soll die Rechtsanwendung und Akzeptanz des Gesetzes in der Bevölkerung gefördert werden.[838] § 4 regelt das Benachteiligungsverbot, wonach Hinweisgeber wegen der Inanspruchnahme des Anzeigerechts und Leistungsverweigerungsrechts nicht gemaßregelt werden dürfen. Außerdem wird in Absatz 1 bestimmt, dass Kündigungen aufgrund rechtmäßiger Hinweise ausgeschlossen sind. Nach der Begründung des Entwurfs soll dies einen effektiven Schutz von Beschäftigten gewährleisten.[839] Absatz 2 des Paragraphen verdeutlicht noch, dass ein Verstoß gegen dieses Verbot durch den Arbeitgeber oder andere Beschäftigte eine Verletzung vertraglicher Pflichten darstellt. Damit ist der Weg für die tatbestandliche Erfüllung von Schadensersatznormen geebnet. Die Absätze 1 und 3 des § 5 regeln Maßnahmen und Pflichten der Arbeitgeber bzw. des Arbeitgebers in Bezug auf das Benachteiligungsverbot. § 5 verpflichtet dazu, präventive Maßnahmen gegen Benachteiligungen einzuleiten, um der Fürsorgepflicht des Arbeitgebers gegenüber seinen Beschäftigten gerecht zu werden. Mit der Norm sollen darüber hinaus durch die in Absatz 2 aufgezeigten Maßnahmen für die Beschäftigten eine offene, vorbeugende und Missstände erkennende Betriebsatmosphäre geschaffen werden.[840]

§ 6 ist, neben § 4, ein weiteres Kernelement zur Stärkung der Rechte von Hinweisgebern. Die Norm regelt das Anzeigerecht. Nach § 6 Abs. 1 des Entwurfs kann der Hinweisgeber sich an eine betriebsinterne Stelle wenden. Er hat nach Satz 3 auch das Recht, sich sofort an eine externe Stelle, also an diejenige Behörde, die für die Entgegennahme der Anzeige oder Beseitigung des jeweiligen Missstandes örtlich und sachlich zuständig ist,[841] zu wenden; eine vorherige Information des Arbeitgebers (innerbetriebliche Abhilfe) und ein Verlangen nach Abhilfe sind vor der Kontaktierung einer externen Stelle aber nicht erforderlich. Da Mitteilungen von betrieblichen Rechtsverstößen und Missständen an die Öffentlichkeit dazu geeignet sind, den Ruf des Arbeitgebers zu schädigen, wird durch Absatz 2 das externe Whistleblowing begrenzt. Danach dürfen nur Mitteilungen an die Öffentlichkeit oder Dritte weitergeleitet werden, wenn wegen des Missstandes das Leben oder die Gesundheit von Menschen oder die Umwelt gefährdet sind (Nr. 1) oder wenn die zuständige Behörde nicht angemessen auf den Hinweis reagiert hat (Nr. 2). Eine unangemessene Reaktion liegt nach Nr. 2 vor, wenn die Behörde den Eingang des Hinweises nicht unverzüglich bestätigt, sie nicht innerhalb einer angemessenen Frist eine inhaltliche Prüfung einleitet oder nicht innerhalb einer bestimmten Zeit (die im Einzelfall zu ermitteln ist) mittels gebührenden Sachstandsberichts den Hinweisgeber informiert. In der Begründung zum Entwurf wird klargestellt, dass es dem Hinweisgeber trotz des hohen Rangs der Meinungsäußerungsfreiheit zumutbar ist, vorrangig auf das

838 BT-Drs. 17/8567, S. 7, Begründung zu § 3 HinwGebSchG-E.
839 BT-Drs. 17/8567, S. 7, Begründung zu § 4 HinwGebSchG-E.
840 BT-Drs. 17/8567, S. 7, 8, Begründung zu § 5 HinwGebSchG-E.
841 BT-Drs. 17/8567, § 3 HinwGebSchG-E (Begriffsbestimmung).

„mildere" Mittel der Behördenanzeige verwiesen zu werden, und die Öffentlichkeit erst dann zu informieren, wenn die Voraussetzungen nach Nr. 1 und Nr. 2 des § 6 Abs. 2 gegeben sind.[842] Darüber hinaus erlaubt § 7 dem Hinweisgeber seine Tätigkeit ohne Verlust des Entgeltanspruchs einzustellen. Dies ist ihm jedoch nach Absatz 1 des § 7 nur in dem Fall gestattet, in dem der Hinweisgeber davon ausgehen darf, sich bei der Ausführung der Tätigkeit der Gefahr der Verfolgung einer Straftat oder der Ahndung einer nicht geringfügigen Ordnungswidrigkeit auszusetzen. Durch diese Regelung soll der Druck auf den Arbeitgeber erhöht werden, geeignete Maßnahmen zur Abhilfe zu treffen.[843] Die Formulierung in § 7 Abs. 1 soll dem Umstand Rechnung tragen, dass dem Hinweisgeber nicht nur die subjektive Prüfung der Sachlage obliegt, sondern er sich auch mit der objektiven Sachlage auseinandersetzen muss.[844] Eine abstrakte Gefahr soll dabei ausreichend sein.[845]

§ 8 und § 9 des Entwurfs regeln die Beseitigung und Unterlassung sowie die Entschädigung und den Schadensersatz. Sinn und Zweck dieser Normen ist es, dem Betroffenen einen Ausgleich für die erlittenen Einbußen zu gewähren und eine Genugtuung gegenüber dem Schädiger zu ermöglichen.[846] Gemäß § 9 Abs. 1 ist Adressat dieser Vorschrift der Benachteiligende, also nicht nur der Arbeitgeber, sondern auch ein Kollege bzw. Mitarbeiter. Auf diese Weise soll eine Abschreckung gegenüber allen potenziell Benachteiligenden erzielt werden.[847]

§ 10 des Entwurfs gibt dem Arbeitgeber oder sonstigen Beteiligten das Recht, vom Hinweisgeber Schadensersatz einzufordern, wenn er wegen eines unrechtmäßigen Hinweises im Sinne von § 2 Abs. 2 einen Schaden verursacht hat. Durch diese Norm wird klargestellt, dass der Hinweisgeber ebenfalls Pflichten unterliegt und bei unrechtmäßigem Handeln eine Sanktion zu befürchten hat, nämlich die Kompensation des durch den unrechtmäßigen Hinweis entstandenen Schadens.

§ 11 erlaubt Betrieben und Unternehmen, ein Hinweisgebersystem einzurichten. Auf diese Weise soll im Sinne von § 4 Abs. 1 des Bundesdatenschutzgesetzes eine ausdrückliche Rechtfertigung[848] für die Einführung von Whistleblowing-Programmen in Unternehmen kodifiziert werden.[849] Bisher bestand für jene Unternehmen nur die Möglichkeit, sich auf „andere Rechtsvorschriften" im Sinne des § 4 Abs. 1

842 BT-Drs. 17/8567, S. 8, Begründung zu § 6 HinwGebSchG-E.
843 BT-Drs. 17/8567, S. 8, Begründung zu § 7 HinwGebSchG-E.
844 BT-Drs. 17/8567, S. 8, Begründung zu § 7 HinwGebSchG-E.
845 BT-Drs. 17/8567, S. 8, Begründung zu § 7 HinwGebSchG-E.
846 BT-Drs. 17/8567, S. 8, Begründung zu § 9 HinwGebSchG-E.
847 BT-Drs. 17/8567, S. 8, Begründung zu § 9 HinwGebSchG-E.
848 Im Rahmen eines Whistleblowing-Systems werden personenbezogene Daten im Sinne von § 3 BDSG des Beschuldigten und ggf. auch des Hinweisgebers erhoben, übermittelt und gespeichert. Die Übermittlung liegt in der Meldung der angeblichen Verletzungshandlung an die zuständige Stelle. Diese Vorgänge bedürfen gemäß § 4 Abs. 1 BDSG einer Rechtfertigung, soweit nicht eine andere Vorschrift sie erlaubt, vgl. *Mengel*, Compliance und Arbeitsrecht, Kap. 7, Rn. 40.
849 BT-Drs. 17/8567, S. 8, Begründung zu § 11 HinwGebSchG-E.

BDSG zu berufen oder sich auf die Interessenabwägung des § 28 Abs. 1 BDSG (bzw. § 32 BDSG) zu stützen.[850] Diese Norm dient somit der Harmonisierung des internationalen Rechts- und Wirtschaftsverkehrs.[851]

Mit § 12 (Beweislast) soll die Rechtsposition des strukturell unterlegenen Hinweisgebers gestärkt werden. Dies dient dem gesetzgeberischen Ziel, die Kultur des Whistleblowings zu fördern.[852]

Die §§ 13 und 14 regeln die Schluss- und Bußgeldvorschriften. Mit der Klarstellung in § 14, dass ein Verstoß gegen § 4 als Ordnungswidrigkeit geahndet wird, soll Benachteiligungen vorbeugend begegnet werden.[853] Nach der Begründung führt diese Regelung zu einer persönlichen Haftung von Vorständen und Geschäftsführern, die ihre Unternehmensorganisation nicht in dem Maße gestalten oder überwachen, dass Benachteiligungen vermieden werden. Die in Abs. 2 festgelegte Höhe der Geldbuße auf bis zu 50.000 Euro soll eine hinreichende Abschreckung gewährleisten, erlaubt aber auch eine entsprechende Reduzierung im Einzelfall.[854]

3. *Kritische Würdigung*

Der Gesetzentwurf, der eine Vielzahl unbestimmter Rechtsbegriffe enthält, hat einige Schwächen und dürfte sich in dieser Form wohl auch in der laufenden Legislaturperiode nicht durchsetzen.[855] Die pauschalen und wenig präzisen Regelungen sind nicht förderlich für die Rechtssicherheit des Hinweisgebers. Einige Bestimmungen sind auch im Hinblick auf die nationale Verfassung und die Rechtsprechung des Europäischen Gerichtshofs für Menschenrechte sehr bedenklich.

Während § 1 HinwGebSchG-E als Nicht-Regelung lediglich die Ziele dieses Entwurfs wiedergibt, enthält § 3 HinwGebSchG-E bereits einige problematische Regelungen. So ist die Definition des „Missstandes" in Absatz 2 zu weit gefasst. Nach dem Entwurf soll bereits bei jeder Pflichtverletzung im Unternehmen ein Missstand vorliegen. Pflichtverletzungen des Arbeitgebers liegen aber auch schon in banalen Situationen vor, wie der Falschaufstellung der Urlaubstage.[856] Eine solche Bagatelle ist aber mit Sicherheit nicht gemeint gewesen, als der Begriff des Missstandes definiert wurde. Eine solche Ausweitung führt zur Praxisuntauglichkeit, jedoch nicht zum Schutz des Whistleblowers. Auch die Bezeichnung „im Umfeld einer unternehmerischen oder betrieblichen Tätigkeit" ist sehr vage. Es scheint, als wolle man den Bereich räumlich und zeitlich eingrenzen. Aber wie eng oder weit man in der Begrenzung gehen soll, steht nicht fest. Daher wäre es hilfreich, den Begriff im

---

850 BT-Drs. 17/8567, S. 9, Begründung zu § 11 HinwGebSchG-E.
851 BT-Drs. 17/8567, S. 9, Begründung zu § 11 HinwGebSchG-E.
852 BT-Drs. 17/8567, S. 9, Begründung zu § 12 HinwGebSchG-E.
853 BT-Drs. 17/8567, S. 9, Begründung zu § 14 HinwGebSchG-E.
854 BT-Drs. 17/8567, S. 9, Begründung zu § 14 HinwGebSchG-E.
855 So auch *v. Busekist/Fahrig*, BB 2013, 119, 124; *Mengel*, CCZ 2012, 146, 146; grundlegend auch *Rudkowski*, CCZ 2013, 204, 208 f.
856 So *Mengel*, CCZ 2012, 146, 146.

Zusammenhang mit einer katalogartigen Aufzählung von relevanten Pflichtverletzungen zu versehen. Nicht jede möglicherweise in Betracht kommende Verletzung kann in diesen Katalog Eingang finden, jedoch kann dieser als „Richtlinie“ dienen. Ähnliche Verletzungen können durch Auslegung unter die im Katalog genannte Pflichtverletzung subsumiert werden. Diese Art der Eingrenzung kann zur Feststellung des Missstandes beitragen und dem Whistleblower mehr Sicherheit geben.

Schließlich scheint § 3 Abs. 2 Satz 2 HinwGebSchG-E eine gewisse „Wesentlichkeitsschwelle“ vorzugeben, die durch den Zusatz „ein Missstand liegt auch vor, wenn eine Gefahr für Leben und Gesundheit von Menschen oder für die Umwelt droht“ betont wird. Nach dem Wortlaut des Absatzes 1 kann aber nicht davon ausgegangen werden, dass in Satz 1 eine ähnlich hohe Schwelle für den Begriff des Missstandes vorausgesetzt wird.[857] Um dies klarzustellen, sollte eine Ergänzung eingefügt werden, die gut mit dem oben genannten Problem gekoppelt werden kann. Sie sollte definieren, was einen Missstand (Gesetzesverstoß), und darauf aufbauend, was einen *wesentlichen* Missstand ausmacht.

In diesem Zusammenhang kommt die Frage auf, ob § 3 Abs. 2 Satz 2 HinwGebSchG-E dem Wortlaut nach Missstände mit einbezieht, die nicht der betrieblichen Sphäre zuzuordnen sind. Ob dieser Satz also so zu verstehen ist, dass damit ein Tatbestand für Whistleblowing zu öffentlichen Gefährdungslagen geschaffen werden soll. Andernfalls hätte das Wort „auch“ in Satz 2 keinen Sinn. Die Formulierung macht den Eindruck, als solle den Arbeitgeber (trotzdem) die Pflicht nach §§ 4, 5 HinwGebSchG-E treffen, obwohl er nicht der Verursacher ist.[858] Dies kann wohl nicht gemeint sein – eine Präzisierung des Wortlauts wäre also geboten. Eine Vorschrift mit dieser Formulierung könnte dazu führen, dass primär öffentlich-rechtlich zu bewältigende Aufgaben und Lasten auf private Unternehmen übertragen werden.

Während § 4 des Entwurfs den § 612 a BGB wiedergibt und die Kündigung wegen rechtmäßiger Hinweise ausschließt, verpflichtet § 5 HinwGebSchG-E den Arbeitgeber, vorbeugende Maßnahmen zu treffen. Jedoch benennt die Norm nicht, was genau mit vorbeugenden Maßnahmen gemeint ist. Eine reine Information der Beschäftigten über die Verhinderung von Benachteiligungen erscheint nicht ausreichend. Angemessen wäre zum Beispiel ein Angebot an Schulungen zum Thema Whistleblowing und Compliance. Außerdem könnte gesetzlich ein zeitlicher Rahmen festgelegt werden, wie oft eine solche Informationsveranstaltung durchzuführen ist. Die Bezeichnung „in geeigneter Art und Weise“ schafft weder Rechtssicherheit noch ist sie vorteilhaft für den Arbeitgeber, der im Zweifelsfall die ordnungsgemäße Erfüllung seiner Pflicht nach § 5 Abs. 2 HinwGebSchG-E darlegen muss.

Insbesondere § 6 des Entwurfs ist höchst bedenklich. Zentral ist die Festlegung in § 6 Abs. 1 Satz 3, dass ein Hinweisgeber sich „sofort“ an eine „externe Stelle“ wenden

857 *Mengel*, CCZ 2012, 146, 147.
858 Vgl. auch *Mengel*, CCZ 2012, 146, 147.

darf.[859] Zu beachten ist hierbei, dass § 3 Abs. 5 HinwGebSchG-E lediglich den Begriff der „zuständigen externen Stelle" definiert hat und § 6 Abs. 1 HinwGebSchG-E von einer „externen Stelle" spricht. Diese Ungenauigkeit lässt sich leicht berichtigen. Viel bedenklicher ist, dass die zum externen Whistleblowing ergangene Rechtsprechung, nach der Arbeitgeber grundsätzlich erwarten dürfen, dass Arbeitnehmer auf einen Missstand zunächst unternehmensintern aufmerksam machen, keine Beachtung findet.[860] Die vorrangige interne Abhilfe fußt auf den grundrechtlich verbürgten Interessen und Rechten des Arbeitgebers und ist dem Grundsatz der Verhältnismäßigkeit geschuldet. Die Verfasser des Entwurfs haben dem Kriterium der mehrpoligen Interessenabwägung zu wenig Beachtung geschenkt. Das *Heinisch*-Urteil des EGMR hat aufgezählt, nach welchen Kriterien eine Interessenabwägung stattzufinden hat. Dabei ist die vorherige innerbetriebliche Abhilfe grundsätzlich einzubeziehen. Nur in Ausnahmefällen darf eine solche Abhilfe unterlassen werden. Diese Abhilfe von vornherein lediglich im Wege einer „Kann-Vorschrift" zu deklarieren, schafft keine Rechtsklarheit und gefährdet den Betriebsfrieden erheblich. So wird die Chance auf eine zivilrechtliche Lösung ohne staatliche Beteiligung und öffentliche Anprangerung verspielt.

Zudem führt diese „Sofortmeldung" zur Überlastung des staatlichen Handlungsapparats. Schließlich kann jeder Missstand, der unter anderem auch zivilrechtliche oder arbeitsrechtliche Streitigkeiten umfassen kann, sofort der Polizei oder der Staatsanwaltschaft gemeldet werden.[861] Dem Begriff des Missstandes in § 3 HinwGebSchG-E zufolge könnte jede noch so banale Pflichtverletzung Gegenstand eines Hinweises sein. Folglich ist eine Vermischung öffentlich-rechtlicher und zivilrechtlicher Strukturen kaum zu vermeiden. Eine Parallele zur Rechtsprechung[862] zum Whistleblowing ist nicht zu erkennen.

Absatz 2 des § 6 HinwGebSchG-E enthält gute Differenzierungen hinsichtlich der Weitergabe von Informationen an externe Dritte. Doch die handwerkliche Technik überzeugt nicht: Die unbestimmten Begriffe „angemessen", „gebührend" und „unverzüglich" tragen nicht zur erwünschten Rechtssicherheit bei. Während „unverzüglich" noch unter Hinzuziehung zivilrechtlicher Normen bestimmt werden kann,[863] bedürften die anderen Begriffe der näheren Erläuterung. Dies ist besonders

---

859 Für die verfassungsrechtliche Bedenklichkeit dieser Regelung *v. Busekist/Fahrig*, BB 2013, 119, 124.

860 EGMR, Urt. v. 21.7.2011 – 28274/08, NJW 2011, 3501 (Heinisch/Deutschland); BAG, Urt. v. 3.7.2003 – 2 AZR 235/02, NZA 2004, 427 ff.; *Schmidt*, in: Erfurter Kommentar, Art. 5 GG, Rn. 37; *Simon/Schilling*, BB 2011, 2421, 2424.

861 Vgl. auch *Mengel*, CCZ 2012, 146, 147.

862 Vgl. BAG, Urt. v. 3.7.2003 – 2 AZR 235/02, NZA 2004, 427, 431; BAG, Urt. v. 7.12.2006 – 2 AZR 400/05, NZA 2007, 502, 504; *Müller-Glöge*, in: MüKo BGB, § 626 BGB, Rn. 64 f.; *Mengel*, Compliance und Arbeitsrecht, Kap. 5, Rn. 59 f.

863 *Valenthin*, in: BeckOK BGB, § 174 BGB, Rn. 9 m. w. N: Das Tatbestandsmerkmal der Unverzüglichkeit ist an § 121 BGB zu orientieren und bedeutet ohne schuldhaftes Zögern. Das bedeutet aber nicht, dass die Zurückweisung sofort erfolgen muss.

wichtig, da die Begriffe eine zeitliche Schnittstelle für den Whistleblower bedeuten, bei deren Eintritt er nach § 6 Abs. 2 HinwGebSchG-E die Erlaubnis hat, sich sofort an externe Dritte zu wenden.

Problematisch ist auch die Regelung des § 10 HinwGebSchG-E. Die Vorschrift normiert das Recht des Arbeitgebers, Schadensersatz vom Arbeitnehmer zu fordern, wenn durch die Hinweiserteilung ein Schaden entstanden ist. Zwar ist dies nur dann möglich, wenn der Hinweis rechtswidrig war. Wann ein Hinweis im Sinne von § 10 des Entwurfes unrechtmäßig sein soll, wird aber an keiner Stelle näher erläutert. Es wird lediglich in § 6 des Entwurfs klargestellt, dass ein Hinweis nicht leichtfertig offengelegt werden darf. Doch wie zu verfahren ist, wenn der Hinweis nicht leichtfertig veröffentlicht wurde, sich jedoch im Nachhinein im Rahmen der Untersuchung der Behörden als unwahr erweist, ist unklar. Auch die Schadenshöhe, die durch den Hinweisgeber kompensiert werden soll, ist nicht begrenzt, obwohl klar ist, dass der Arbeitnehmer regelmäßig nicht in der Lage sein wird, den gesamten Schaden zu ersetzen. Weiter stellt sich die Frage, wer mit „sonstigen Beteiligten" gemeint ist, deren Schaden der Hinweisgeber ebenfalls zu ersetzen hat.

§ 11 HinwGebSchG-E des Entwurfs ist eine sinnvolle Regelung. Es ist zu begrüßen, dass die Schaffung eines Hinweisgebersystems in Unternehmen angeregt wird. Warum aber ist diese weitestgehend in der Compliance-Community etablierte und akzeptierte Institution nicht verpflichtend vorgesehen? Der Entwurf enthält nur eine „Kann-Vorschrift", obwohl gerade die Verpflichtung zur Implementierung eines solchen Compliance-System viele Probleme lösen könnte.[864] Die Vorschrift regelt so gesehen nichts Neues, denn dass Unternehmen freiwillig ein Compliance-Programm einführen können, ist bereits geltende Rechtslage. Die Kodifikation einer Verpflichtung, zumindest in gewissem Umfang, ist wünschenswert. Ob die Rechte des Arbeitgebers auf freie Unternehmensführung und -gestaltung dabei beeinträchtigt werden, ist eine andere Frage, die sich im Wege einer interessengerechten Abwägung klären lässt.

4. *Ausblick*

Im Ergebnis ist festzuhalten, dass sich zwar durch die *Heinisch*-Entscheidungen eine weitestgehend klare Richtschnur für Kündigungen wegen Whistleblowing herausgebildet hat. Der Whistleblower ist deswegen keineswegs völlig schutzlos gestellt.[865] Eine Kodifizierung ist trotzdem notwendig. Denn Rechtssicherheit kann nur hergestellt werden, wenn jedem Bürger die geltende Rechtslage zugänglich und verständlich gemacht wird. Aus Sicht des Rechtssuchenden ist es einfacher, das Recht in einer niedergeschriebenen Norm zu finden, als es sich aus einer Vielzahl von Entscheidungen zusammenzusuchen.[866] Ein Whistleblower muss wissen,

---

864 Vertiefend zur Implementierung, Durchsetzung und Organisation von Compliance *Mengel*, Compliance und Arbeitsrecht.

865 So auch der Bund der Richterinnen und Richter der Arbeitsgerichtsbarkeit in ihrer Stellungnahme zum Hinweisgeberschutzgesetz, vgl. BT-Ausschussdrs. 17(11)808.

866 So auch *Kort*, FS Kreutz, S. 247, 252.

was erlaubt ist und was nicht. Außerdem kann ein Hinweisgeber nur Gewissheit erfahren, wenn er sich auf Rechtsnormen berufen kann, die von jedem Arbeitgeber einzuhalten sind. Ein Schutzgesetz, in dem Handlungsanweisungen sowie Ge- und Verbote verankert sind, gibt dem Whistleblower optimale Rechtssicherheit. Der vorgelegte Entwurf ist dafür nicht geeignet. Zum einen müssen unbestimmte Rechtsbegriffe vermieden werden. Zum anderen ist deutlicher zwischen internem und externem Whistleblowing zu unterscheiden. Hilfreich für eine Optimierung des HinwGebSchG wäre es, anerkannte Whistleblower-Organisationen hinzuzuziehen, die sich eingehend mit Whistleblowern und deren Fällen auseinandergesetzt haben und dadurch die „Schwachstellen" des Gesetzentwurfs besser erkennen könnten.

Eine andere Möglichkeit wäre eine Konkretisierung schon bestehender Regelungen, z. B. eine Änderung des § 612 a BGB und eine Implementierung eines § 612 b BGB.

II. Entwurf eines Whistleblower-Schutzgesetzes

Die Fraktion Bündnis 90/Die Grünen schlagen ein Gesetz zur Förderung von Transparenz und zum Diskriminierungsschutz von Hinweisgeberinnen und Hinweisgebern vor, durch das § 612 a BGB reformiert und ein § 612 b BGB implementiert werden soll.

Da der Entwurf einen anderen Weg vorschlägt als derjenige der SPD-Fraktion, wird er im Folgenden kurz skizziert. Die Darstellung konzentriert sich auf die vorgesehenen Änderungen des Bürgerlichen Gesetzbuches.

1. Der Entwurf 2012

Der Entwurf[867] sieht Änderungen des Bürgerlichen Gesetzbuches, des Berufsbildungsgesetzes, des Bundesbeamtengesetzes sowie des Beamtenstatusgesetzes vor. Diese Änderungen sollen Whistleblowern arbeits- bzw. dienstrechtlichen Diskriminierungsschutz gewähren und regeln, unter welchen Voraussetzungen sie sich an eine außerbetriebliche zuständige Stelle bzw. direkt an die Öffentlichkeit wenden dürfen.[868]

a. Regelungsgegenstände

§ 612 a BGB, der das Maßregelungsverbot zum Gegenstand hat, soll einen Absatz 2 erhalten. Der bisherige Wortlaut wird der „neue" Absatz 1. Der neu einzufügende Absatz 2 soll folgenden Inhalt haben:[869]

867 Gesetzentwurf zur Förderung von Transparenz und zum Diskriminierungsschutz von Hinweisgeberinnen und Hinweisgeber (Whistleblower-Schutzgesetz) vom 23.5.2012, BT-Drs. 17/9782.

868 BT-Drs. 17/9782, S. 2.

869 BT-Drs. 17/9782, S. 3, Artikel 1.

„(2) Sofern ein Arbeitnehmer Tatsachen glaubhaft macht, die eine Benachteiligung wegen zulässiger Ausübung seiner Rechte erkennbar werden lassen, trägt der Arbeitgeber die Beweislast dafür, dass kein Verstoß gegen Absatz 1 vorliegt.“

Die Ergänzung wird damit begründet, dass nur durch die Beweislastregelung negative Folgen für die Arbeitnehmer eingeschränkt werden können. Die in den neuen § 612 a Abs. 2 BGB aufgenommene Abstufung der Beweislast ist an § 22 AGG angelehnt und trägt dem Umstand Rechnung, dass dem Whistleblower der Nachweis der Kausalität zwischen der zulässigen Rechtsausübung und einer Benachteiligung kaum gelingen kann, da es sich um eine innere Tatsache handelt.[870]

Des Weiteren sieht der Entwurf vor, dass nach dem § 612 a BGB ein § 612 b BGB-E eingefügt wird, der die Überschrift „Anzeigerecht“ trägt und in 6 Absätze unterteilt ist.

Absatz 1 legt dem Arbeitnehmer die Pflicht auf, sich zuerst an den Arbeitgeber oder an eine innerbetriebliche zuständige Stelle zu wenden, wenn er konkrete Anhaltspunkte hat, dass im Zusammenhang mit der betrieblichen Tätigkeit rechtliche Pflichten verletzt werden oder eine solche Verletzung droht.

Durch diese Bestimmung wird die rechtliche Zulässigkeit des Handelns des Whistleblowers normiert. Das hat zur Folge, dass die Kontaktierung einer internen Stelle somit als Grundlage für eine verhaltensbedingte Kündigung, Abmahnung oder andere nachteilige Maßnahme ausscheidet. Der Entwurf sichert den Schutz des Whistleblowers auch insoweit, als bei der Beurteilung der Voraussetzungen des Absatzes 1 der subjektive Beurteilungshorizont des Arbeitnehmers heranzuziehen ist.[871] Dennoch sollen Kündigungen oder arbeitsrechtliche und deliktische Schadensersatzansprüche nicht unmöglich sein, wenn die Anschuldigungen, etwa bei Racheakten oder Ähnlichem, substanzlos sind. Der Entwurf sieht nur vor, dass sich der Schutz (auch) auf falsche Rückschlüsse erstreckt, solange der Hinweis „aufgrund konkreter Anhaltspunkte“ erfolgt.[872]

Nach Absatz 2 des § 612 b BGB-E hat der Arbeitnehmer das Recht, sich an eine zuständige außerbetriebliche Stelle zu wenden, wenn der Arbeitgeber dem Verlangen nach Abhilfe nicht binnen angemessener Frist oder nach Auffassung des Arbeitnehmers aufgrund konkreter Anhaltspunkte nicht oder nicht ausreichend nachkommt. Nach Satz 2 ist die vorherige innerbetriebliche Abhilfe nicht notwendig, wenn dies dem Arbeitnehmer nicht zumutbar ist. Unzumutbarkeit soll insbesondere dann vorliegen (Satz 3), wenn eine gegenwärtige Gefahr für Leben, Körper, Gesundheit, Persönlichkeitsrecht, Freiheit der Person, Stabilität des Finanzsystems oder Umwelt droht (Nr. 1), wenn im Zusammenhang mit der betrieblichen Tätigkeit eine Straftat begangen worden ist (Nr. 2), wenn eine Straftat geplant ist,

---

870 BT-Drs. 17/9782, S. 9, Begründung zu § 612 b BGB-E.

871 BT-Drs. 17/9782, S. 9, Begründung zu Absatz 1 von § 612 b BGB-E.

872 BT-Drs. 17/9782, S. 9, 10, Begründung zu Absatz 1 von § 612 b BGB-E; die Formulierung „aufgrund konkreter Anhaltspunkte“ entspricht derjenigen aus dem Anzeigerecht in § 17 Abs. 2 ArbSchG.

durch deren Nichtanzeige der Arbeitnehmer selbst der Strafverfolgung ausgesetzt würde (Nr. 3) oder wenn eine innerbetriebliche Abhilfe nicht oder nicht ausreichend erfolgen wird (Nr. 4). Im Fall der Nr. 2 trägt der Arbeitnehmer nach Satz 4 nicht die Beweislast dafür, dass er der Auffassung war, dass im Zusammenhang mit der betrieblichen Tätigkeit eine Straftat begangen worden ist.

Das Stufensystem zwischen den beiden Absätzen zeigt, dass grundsätzlich immer eine vorherige innerbetriebliche Abhilfe erforderlich ist. Der Begriff der Unzumutbarkeit wird durch den Katalog der Nummern 1 bis 4 konkretisiert. Laut Begründung dient Nr. 3 der Einheitlichkeit der Rechtsordnung. Es wird Bezug auf § 138 StGB genommen, welcher die Nichtanzeige bestimmter Straftaten unter Strafe stellt.[873]

Unter einer „zuständigen außerbetrieblichen Stelle" ist grundsätzlich eine Stelle zu verstehen, die zur Abhilfe bzw. zur Beseitigung des jeweiligen Missstandes oder behaupteten Rechtsverstoßes zuständig ist.[874] Je nachdem, um welchen Verstoß oder Missstand es sich handelt, können die Polizei, die Staatsanwaltschaft, die allgemeinen Ordnungsbehörden, die Gesundheitsämter oder die Umwelt- und Gewerbeaufsichtsbehörden zuständig sein. Zudem sind auch solche Stellen erfasst, die unter anderem Überwachungs- und Kontrollfunktionen ausüben, wie z. B. Datenschutzbeauftragte, der Medizinische Dienst der Krankenversicherungen, Ombudsstellen, die Finanzkontrolle Schwarzarbeit sowie die Berufskammern. Die Presse stellt laut Begründung keine „zuständige außerbetriebliche Stelle" im Sinne des Absatzes 2 dar.[875]

§ 612 b Abs. 2 Satz 4 BGB-E enthält eine Beweislastumkehr zugunsten des Arbeitnehmers. Dies entspricht laut Begründung der Rechtsprechung des BVerfG[876] und des BAG,[877] nach der eine Strafanzeige stets zulässig ist, es sei denn der Arbeitnehmer ist im Wissen oder hat leichtfertig verkannt, dass die Vorwürfe in der Anzeige nicht der Wahrheit entsprechen. Nach Satz 4 hat also der Arbeitgeber den Beweis zu erbringen, dass der Arbeitnehmer nicht im Recht ist.

Absatz 3 des § 612 b BGB-E erlaubt dem Arbeitnehmer, sich direkt an die Öffentlichkeit zu wenden, wenn das öffentliche Interesse an der Information das betriebliche Interesse an deren Geheimhaltung erheblich überwiegt. Nach Satz 2 ist ein solches öffentliches Interesse insbesondere dann gegeben, wenn der Arbeitnehmer aufgrund konkreter Anhaltspunkte annimmt, dass im Zusammenhang mit der betrieblichen Tätigkeit eine gegenwärtige erhebliche Gefahr für Leben, Körper, Gesundheit, Persönlichkeit oder Freiheit der Person, Stabilität des Finanzsystems oder Umwelt droht.

Absatz 3 enthält sozusagen eine Steigerung zu Absatz 2, indem er dem Arbeitnehmer erlaubt, sich sofort an die Öffentlichkeit zu wenden, wenn eine gegenwärtige

---

873 BT-Drs. 17/9782, S. 10, Begründung zu Absatz 2 von § 612 b BGB-E.
874 BT-Drs. 17/9782, S. 10, Begründung zu Absatz 2 von § 612 b BGB-E.
875 BT-Drs. 17/9782, S. 10, Begründung zu Absatz 2 von § 612 b BGB-E.
876 BVerfG, Beschl. v. 25.2.1987 – BvR 1086/85, NJW 1987, 1929, 1930; BVerfG, Beschl. v. 2.7.2001 – 1 BvR 2049/00, NZA 2001, 888, 889.
877 BAG, Urt. v. 7.12.2006 – 2 AZR 400/05, NZA 2007, 502, 504; BAG, Urt. v. 3.7.2003 – 2 AZR 235/02, NZA 2004, 427, 429.

erhebliche Gefahr für die aufgezählten Güter droht. Im Gegensatz zu Absatz 2 wird hier eine Gefahr erheblichen Ausmaßes verlangt, um die besondere Dringlichkeit hervorzuheben. Die Erheblichkeit der Gefahr bezieht sich laut Begründung sowohl auf die Wahrscheinlichkeit des Schadenseintritts als auch auf den Umfang des drohenden Schadens.[878] Mit dem Begriff „Öffentlichkeit" sind nicht nur die Presse und andere Medien erfasst, sondern auch Veröffentlichungen im Internet.[879]

Gemäß Absatz 4 des Entwurfs zu § 612 b BGB-E wird dem Arbeitnehmer das Recht eingeräumt, eine verkörperte Wiedergabe von betrieblichen Informationen herzustellen und weiterzuleiten, soweit dies erforderlich ist. Hierunter soll etwa das Anfertigen einer Kopie oder das Fotografieren von Dokumenten fallen.[880]

Das in Absatz 5 zu § 612 b BGB-E geregelte Abweichungsverbot soll sicherstellen, dass die Regelungen nicht umgangen werden oder abweichende nachteilige Vereinbarungen vertraglich geregelt werden.

§ 612 b Abs. 6 BGB-E sorgt dafür, dass Beschwerde- und Anzeigerechte sowie Äußerungsrechte der Arbeitnehmer nach anderen Vorschriften von der Neuregelung unberührt bleiben. Damit soll unter anderem den verfassungsrechtlichen Vorgaben von Art. 17 GG Rechnung getragen werden. Dieser sichert allen Bürgern ein Petitionsrecht zu, durch das sie sich mit Bitten und Beschwerden schriftlich an die zuständigen Stellen sowie an die Volksvertretung wenden können.

b. Rechtliche Würdigung

Entgegen dem von der SPD-Fraktion im Februar 2012 eingebrachten Vorschlag, der ein eigenes Gesetz zum Hinweisgeberschutz vorsieht, schlug die Bundestagsfraktion Bündnis 90/Die Grünen ein Artikelgesetz vor. Dieses beschränkte sich im Wesentlichen auf Änderungen einer bereits bestehenden Norm (des § 612 a BGB) und auf die Einführung eines § 612 b BGB. Ein klares Signal, Whistleblowing zu fördern, ist dies nicht. Die Etablierung eines eigenen Gesetzes zum Schutz von Whistleblowing wäre hingegen ein deutliches Zeichen. Außerdem sollte auf das Problem der Rechtmäßigkeit einer Kündigung besonders hingewiesen und nicht mit einer Benachteiligung im Sinne des § 612 a BGB gleichgesetzt werden. Eine besondere Hervorhebung des Kündigungsschutzes bei einem (rechtmäßigen) Hinweis wäre wünschenswert.

Auch andere Punkte sind zu kritisieren: Neben Regelungslücken bei untypischen Beschäftigungsverhältnissen gibt es Bedenken bei der konkreten Ausführung. Auch wenn in der Begründung geklärt wird, was mit „rechtlichen Pflichten" in § 612 b Abs. 1 BGB-E gemeint ist, besteht die Gefahr, dass der Hinweisgeber oder die Hinweisgeberin diesen Begriff in einer Weise für sich definiert, der zu umfassend ist und über das Ziel hinausschießt. So könnte leicht jede „Lappalie" den Hinweisgeber dazu bewegen, „Alarm" zu schlagen. Es wären damit auch Bagatellverletzungen, die allenfalls privatrechtlich bewehrt sind, erfasst.

---

878 BT-Drs. 17/9782, S. 11, Begründung zu Absatz 3 von § 612 b BGB-E.
879 BT-Drs. 17/9782, S. 11, Begründung zu Absatz 3 von § 612 b BGB-E.
880 BT-Drs. 17/9782, S. 11, Begründung zu Absatz 4 von § 612 b BGB-E.

Des Weiteren ist fraglich, wie der Abschnitt „nach Auffassung des Arbeitnehmers nicht oder nicht ausreichend nachkommt" in § 612 b Abs. 2 BGB-E interpretiert werden soll. Zwar ist zu begrüßen, dass der Entwurf einen subjektiven Beurteilungshorizont des Arbeitnehmers heranzieht. Dennoch besteht ein hohes Missbrauchsrisiko, wenn dem Arbeitnehmer überlassen bleibt zu entscheiden, wann keine hinreichende Abhilfe stattfindet. Die Voraussetzung ist zu unbestimmt. Eine Verbesserung könnte durch das Aufstellen einer Erheblichkeitsschwelle erreicht werden.

Was mit „angemessener Frist" gemeint sein soll, ist ebenfalls nicht erkennbar. So besteht kein Maßstab, nach dem sich ein Hinweisgeber richten kann. Eine ausreichende Abhilfe im Betrieb, eventuell durch Errichtung bestimmter Vorkehrungen und Einrichtungen, bedarf Zeit – ungeduldige und unerfahrene Arbeitnehmer können damit nicht umgehen. Die Gefahr einer vorschnellen Anzeigeerstattung ist groß.

Zu begrüßen ist, dass im Entwurf eine unterschiedliche Ausdrucksweise für die gegenwärtige Gefahr in Absatz 2 und die gegenwärtige erhebliche Gefahr in Absatz 3 gewählt wurde. Dies unterstreicht die besondere Differenzierung zwischen den beiden Hinweisadressaten beim externen Whistleblowing.

Problematisch erscheint § 612 b Abs. 2 Nr. 2 BGB-E, der nicht unterscheidet, durch wen die Straftat im Zusammenhang mit der betrieblichen Tätigkeit begangen worden sein muss. Sind auch andere Arbeitnehmer als Straftäter zu qualifizieren, würde dies dem Whistleblower im großen Umfang Tor und Tür öffnen, wenn externes Whistleblowing schon bei einer beliebigen, möglicherweise geringfügigen Straftat irgendeines Arbeitnehmers zulässig wäre. Eine Konkretisierung auf den Straftatbestand wäre hilfreich.

Kritisch zu sehen ist der Absatz 4 des § 612 b BGB-E, der es erlaubt, Kopien und Fotografien von Dokumenten des Arbeitgebers herzustellen, um so Informationen an Dritte weitergeben zu können. Bedenken bestehen angesichts einer möglichen Verletzung der Rechte des Arbeitgebers auf Wahrung des Betriebsgeheimnisses oder der Wirtschaftsfreiheit. Einem Arbeitnehmer das Recht einzuräumen, Vervielfältigungen von betriebsinternen Dokumenten zu erstellen, übersteigt die „Kompetenz" des einfachen Bürgers – schließlich gibt es Ordnungskräfte und die Polizei, die in zulässigerweise Zutritt zu den Betriebsräumen und Zugang zu den entsprechenden Dokumenten erlangen können.

Auch ist zu betonen, dass Regelungen zur schadensersatzrechtlichen Haftung fehlen. Das Risiko des Hinweisgebers, vom Arbeitgeber in Anspruch genommen zu werden, ist groß. Es muss geregelt werden, wann ein Anspruch geltend gemacht werden darf, und vor allem, in welcher Höhe er zu begrenzen ist. Schließlich gilt zwischen dem Arbeitgeber und dem Arbeitnehmer grundsätzlich das Prinzip des innerbetrieblichen Schadensausgleichs.

Zu begrüßen sind allerdings die Beweislastregeln zugunsten der Arbeitnehmer, da so der Hinweisgeber effektiv in seiner Offenlegung unterstützt wird.

2. Der Entwurf 2014

Die Bundestagsfraktion Bündnis 90/Die Grünen hat am 7. November 2014 ein Gesetzentwurf in den Bundestag eingebracht,[881] der den Whistleblower vor Repressalien des Arbeitgebers und Kündigungen schützen und Rechtssicherheit schaffen soll. Aufgebaut ist er auf dem Entwurf, der in der. 17. Legislaturperiode[882] eingebracht und bereits zuvor erläutert wurde. Der neue Gesetzentwurf zeigt, dass die Diskussion um einen Whistleblowerschutz noch aktuell ist. Gefordert wird von der Bundestagsfraktion ein Artikelgesetz, welches das Bürgerliche Gesetzbuch, Berufsbildungsgesetz, Bundesbeamtengesetz, Beamtenstatusgesetz und das Strafgesetzbuch ändert. Die Erwägungen haben sich zum vorherigen Entwurf kaum verändert; neuen Anstoß hat jedoch der Fall des *Edward Snowden* gegeben,[883] der weltweit durch die Veröffentlichung von Staatsgeheimnissen Schlagzeilen gemacht hat. Besonders wird auf die vorgeschlagene Änderung des Bürgerlichen Gesetzbuches eingegangen.

a. Regelungsgegenstände

Der aktuelle Entwurf ist prinzipiell wie der vorherige aufgebaut. § 612 a BGB soll einen Absatz 2 erhalten und es wird ein § 612 b BGB eingefügt. Grundsätzlich hat der Arbeitnehmer nach § 612 b BGB-E sich zuerst an den Arbeitgeber oder an eine zur innerbetrieblichen Klärung zuständige Stelle zu wenden.

Im Gegensatz zum Entwurf von 2012 soll nach § 612 b Abs. 2 BGB-E der Arbeitnehmer das Recht haben, sich sofort an eine außerbetriebliche Stelle zu wenden, wenn eine innerbetriebliche Stelle nicht besteht. Zusätzlich soll es dem Arbeitnehmer erlaubt sein, sich an eine zuständige außerbetriebliche Stelle zu wenden, wenn der Arbeitgeber dem Abhilfebegehren des Arbeitnehmers nicht ausreichend nachkommt oder eine vorherige Abhilfe dem Arbeitnehmer unzumutbar ist. Wann eine Unzumutbarkeit vorliegt, wird in den Nummern 1 bis 4 geklärt, die inhaltsgleich mit dem vorherigen Gesetzentwurf sind. Lediglich in Nr. 2, nach dem es dem Arbeitnehmer unzumutbar ist, interne Abhilfe abzuwarten, wenn es um eine Straftat geht, wurde der Begriff „erheblich“ eingefügt. Somit muss es sich bei der Straftat, die im Zusammenhang mit einer betrieblichen Tätigkeit steht, um eine erhebliche Straftat handeln. Zudem wurde der letzte Satz des Absatzes 2 verkürzt. So reicht es aus, wenn der Arbeitnehmer konkrete Anhaltspunkte in den Fällen der Nr. 2 vorträgt. Von „Beweislast“ ist nicht mehr die Rede.

---

881 Gesetzentwurf zur Förderung von Transparenz und zum Diskriminierungsschutz von Hinweisgeberinnen und Hinweisgebern (Whistleblower-Schutzgesetz) vom 4.11.2014, BT-Drs. 18/3039.

882 BT-Drs. 17/9782.

883 Alles Wichtige zum NSA-Skandal, abrufbar unter: http://www.zeit.de/digital/datenschutz/2013-10/hintergrund-nsa-skandal, (abgerufen am 7.12.2014).

In Absatz 3, der dem Arbeitnehmer das Recht geben soll, sich an die Öffentlichkeit zu wenden – wobei unter „Öffentlichkeit" alle Print- und elektronischen Medien, Veröffentlichungen im Internet sowie alle traditionellen Formen der Weitergabe fallen –, ist hinzugefügt worden, dass ein überwiegendes Interesse am Bekanntwerden der Information auch gegeben ist, wenn der Arbeitnehmer aufgrund konkreter Anhaltspunkte annimmt, dass im Zusammenhang mit der betrieblichen Tätigkeit eine Begehung von erheblichen Straftaten droht. Im Übrigen ist ein überwiegendes öffentliches Interesse gegeben, wenn eine erhebliche Gefahr für das Leben, die körperliche Unversehrtheit, die Gesundheit, das Persönlichkeitsrecht, die Freiheit der Person, die Stabilität des Finanzsystems oder die Umwelt droht.

§ 612 b Abs. 4 und Abs. 5 BGB-E sind unverändert geblieben.

In Absatz 6 wurde eine Änderung vorgenommen: Neben dem Hinweis, dass durch das Gesetz unter anderem die Beschwerderechte des Arbeitnehmers nach anderen Rechtsvorschriften unberührt bleiben, bleiben nun auch die Rechte der Arbeitnehmer nach anderen Verwaltungsvorschriften unberührt.

b. Rechtliche Würdigung

Die Verfasser des Gesetzentwurfs sehen die Notwendigkeit verschiedener gesetzlicher Änderungen. Nicht nur das Arbeitsrecht – sowohl das privatwirtschaftliche als auch das öffentlich-rechtliche Arbeitsrecht – soll angepasst werden, sondern auch das Strafgesetzbuch. Der Entwurf zum Strafgesetzbuch sieht eine Implementierung eines § 97 c und d StGB (Offenbarung von Staatsgeheimnissen) sowie die Wiedereinfügung des § 353 c StGB (Offenbaren von Geheimnissen im Rahmen eines Dienstverhältnisses) vor. Auf beide Normen wird aber in dieser Arbeit nicht näher eingegangen. Die umfängliche gesetzliche Verankerung des Whistleblowerschutzes vor arbeitsrechtlichen und strafrechtlichen Konsequenzen erfolgt nicht zuletzt aus den Erfahrungen mit vergangenen Ereignissen.

Der Entwurf und seine Intention sind grundsätzlich zu begrüßen. Da sich jedoch kaum bedeutende Änderungen zu dem Entwurf von 2012 ergeben, ist er rechtlich ähnlich zu würdigen. Es fehlt an begrifflichen Konkretisierungen wie an einer Regelung zur schadensersatzrechtlichen Haftung. Zudem ist nicht verständlich, warum in der Begründung zu § 612 b Abs. 1 BGB-E davon ausgegangen wird,[884] dass Whistleblowing aus anderen als rein altruistischen Motiven (hier Racheakt) mit einer leichtfertigen oder gar wissentlichen unwahren Offenbarung gleichzusetzen ist. Schließlich wird nur verlangt, dass Anzeigen auf konkreten Anhaltspunkten beruhen, der Arbeitnehmer von der Rechtswidrigkeit des Handelns überzeugt ist und nicht offensichtlich Unwahres behauptet. Dies kann aber auch der Fall sein, wenn ein Arbeitnehmer aus egoistischen Motiven handelt. Deshalb sollte insgesamt dem Motiv keine Beachtung geschenkt werden.

Positiv hervorzuheben ist, dass die Entwurfsverfasser nur Rechtsgüter von erheblicher Bedeutung erfassen, wenn es darum geht, sich mit einem Hinweis an eine

884 BT-Drs. 18/3039, S. 17, Begründung zu § 612 b Abs. 1 BGB-E.

betriebsexterne Stelle zu wenden. Dies verhindert eine zu weitreichende Regelung und verdeutlicht die Besonderheit der erlaubten externen Berichterstattung. Neu ist zudem, dass die Straftat „erheblich" sein muss. Damit soll verhindert werden, dass die Hinweiserteilung nach außen Bagatelldelikte umfasst. Positiv ist ebenfalls, dass in § 612 b BGB-E Abs. 2 von „zuständige[r] außerbetriebliche[r] Stelle" und in Abs. 3 von „Öffentlichkeit" die Rede ist. Das Stufenverhältnis zwischen internem Whistleblowing, externem Whistleblowing gegenüber zuständigen Stellen und externem Whistleblowing gegenüber Dritten wird dadurch sichtbar.

Trotz guter Ansätze ändert dies nichts an der Notwendigkeit, ein Gesetz zu schaffen, dass so verständlich und konkret ist, dass auch ohne Einsicht in die Begründung des Gesetzgebers deutlich wird, wie der Arbeitnehmer und der Arbeitgeber sich im Rahmen des Whistleblowings zu verhalten haben.

3. Gesamtresümee

Beide Gesetzesentwürfe – der SPD-Fraktion sowie der Bündnis90/Die Grünen-Fraktion – enthalten durchaus geeignete Bestimmungen, die die Position des Arbeitnehmers als Whistleblower stärken. Dennoch gehen sie nicht weit genug. Von einem „Randphänomen Whistleblowing" kann nicht mehr die Rede sein. Würden die Entwürfe in dieser Form erneut in den Bundestag eingebracht werden, hätten sie keine Aussicht auf Erfolg. Auch der aktuelle Entwurf der Bundestagsfraktion Bündnis 90/Die Grünen ist lückenhaft und unspezifisch.

Vorrangig muss definiert werden, welche Personen und Hinweisgegenstände besonderen Schutz genießen sollen. Des Weiteren darf der Motivation des Whistleblowers keine Bedeutung beigemessen werden. Motive sind in einem Rechtsstreit regelmäßig schwer nachzuweisen, sodass sie kaum justiziabel sind. Eine Regelung, die auf die Motivation des Whistleblowers abstellt, führt ferner zu einer verringerten Rechtssicherheit und senkt somit die Bereitschaft, sich als Hinweisgeber zu betätigen. Schließlich kann es der Rechtsordnung grundsätzlich egal sein, aus welchem Grund jemand einen Rechtsverstoß eines anderen meldet, sofern dieser nur sachlich richtig ist oder zumindest ein berechtigter Verdacht bestand.[885] Der Zweck des Whistleblowings liegt in der Offenbarung von Rechtsverstößen – für eine „sittliche" Überprüfung der Motivation des Whistleblowers bleibt da kein Raum.

Positiv hervorzuheben ist bei beiden Entwürfen die Möglichkeit des externen Whistleblowings; insbesondere die Betonung auf die vom öffentlichen Interesse getragenen Rechtsgüter, wie beispielsweise Leben, Gesundheit und Umwelt. Auch wenn die Ausführungen dazu differenziert ausfallen, wird eingesehen, dass bei der Gefährdung von höherrangigen Gütern eine andere Sichtweise geboten ist als bei „nur" vermögensschädigenden Missständen. Deshalb sollte im Ergebnis eine Meldung an die Öffentlichkeit (Presse, Medien) nur in äußerst schwerwiegenden Fällen und als *ultima ratio* in Betracht kommen. Da beide Entwürfe gute Ansätze

---

885 *Thüsing/Forst*, in: Thüsing, Beschäftigtendatenschutz und Compliance, § 6, Rn. 36.

enthalten, ist – vorbehaltlich der Ausbesserung einiger inhaltlicher Punkte – eine Kombination des Entwurfs der SPD-Fraktion aus dem Jahr 2012 und des Entwurfs der Fraktion Bündnis 90/Die Grünen von 2014 in Erwägung zu ziehen. Das Resultat sollte ein eigenes Gesetz zum Schutz von Hinweisgebern (Whistleblower) sein.

III. „Guidelines on Whistleblowing" – Europäische Kommission 2012

Im Dezember 2012 verfasste der Vizepräsident Šefčovič eine Mitteilung an die Europäische Kommission über Leitlinien für die Meldung von ernsten Missständen.[886]

Zwar sind die Regeln und Leitlinien der Kommission für die Meldung von Missständen nur auf Bedienstete, unabhängig von der jeweiligen dienstlichen Stellung, anwendbar.[887] Aber trotz des eingeschränkten persönlichen Anwendungsbereiches liefert die Mitteilung gute Ansätze für den Umgang mit Whistleblowern, die in bestimmten Fällen auf ein „normales" Arbeitsverhältnis übertragen werden können.

Gleich zu Beginn der Mitteilung wird betont, dass es wichtig für verantwortungsvolle Organisationen und die dort beschäftigen Personen sei, über Verfahren für die Mitteilung von Betrugsfällen, Korruption und sonstigem schwerwiegenden Verhalten zu verfügen. Doch ohne hinreichenden Schutz, falle es schwer, Bediensteten zur Meldung von Bedenken oder Missständen zu bewegen. Solange Bedienstete nicht darauf vertrauen könnten, dass ihnen durch die Meldung keine Nachteile entstünden, werden sie Stillschweigen bewahren. Deshalb sei es oberste Priorität, den Betroffenen allgemein akzeptierte und anerkannte Wege vorzuzeichnen, über die sich die Betroffenen mit ihren Bedenken an interne und auch externe Stellen wenden könnten. Werde ihnen eine Möglichkeit geboten, Bedenken frei zu äußern, ohne Gefahr zu laufen, durch Repressalien benachteiligt zu werden, ermutige sie das zum Handeln.

Im Einvernehmen mit dem OLAF[888] hat die Kommission daher Leitlinien erstellt, die im Folgenden kurz skizziert werden.

1. Die Leitlinien im Überblick

Gemäß den Leitlinien sind Bedienstete verpflichtet, schwerwiegende Unregelmäßigkeiten zu melden. Schwerwiegende Unregelmäßigkeiten stellen rechtswidrige Handlungen, einschließlich Betrug und Korruption, sowie schwerwiegendes

---

886 Communication to the Commission from Vice-President Šefčovič on Guidelines on Whistleblowing, 6.12.2012, SEC (2012) 679 final.

887 SEC (2012) 679 final, Abschnitt 1.3.

888 OLAF (European Anti-Fraud Office), Europäisches Amt für Betrugsbekämpfung. Das OLAF untersucht Fälle von Betrug zum Nachteil des EU-Haushalts, von Korruption sowie von schwerwiegendem Fehlverhalten innerhalb der Organe und Einrichtungen der EU und entwickelt eine Betrugsbekämpfungsstrategie für die Europäische Kommission.

Fehlverhalten im Beruf dar. Ausgenommen sind jedoch Meldungen über Informationen, die[889]

- bereits öffentlich zugänglich sind,
- unbegründete Gerüchte und Gerede darstellen,
- triviale Angelegenheiten zum Gegenstand haben,
- Meinungsverschiedenheiten über politische Ausrichtung und Maßnahmen beinhalten oder
- nicht mit der Aufgabenerfüllung einer Person im Zusammenhang stehen.

Diese Regeln sind ferner nicht auf Informationen anwendbar, für die dem Personal spezifische Verfahren zur Verfügung stehen oder die nicht als angemessen oder wahr angesehen werden können.[890]

Das Mitteilungsverfahren sieht vor, dass Bedienstete, die in Ausübung oder anlässlich der Ausübung ihrer Aufgaben davon Kenntnis erhalten, dass schwerwiegende Unregelmäßigkeiten vorliegen oder auftreten können, verpflichtet sind, ihren unmittelbaren Vorgesetzten, ihren Generaldirektor oder Dienststellenleiter unverzüglich schriftlich zu unterrichten.[891] Wird befürchtet, dass diese Mitteilung zu Repressalien führen könnte oder der vorgesehene Empfänger der Meldung von den schwerwiegenden Unregelmäßigkeiten persönlich betroffen ist, kann der direkte Weg der internen Meldung umgangen und die Meldung an den Generalsekretär oder direkt an das OLAF geleitet werden.[892]

Das OLAF oder die Kommission haben nach Erhalt der Information 60 Tage Zeit, den Zeitraum anzugeben, den sie für angemessen und erforderlich halten, um geeignete Maßnahmen zu ergreifen. Werden in diesem Zeitraum keine Maßnahmen ergriffen oder kann der Hinweisgeber nachweisen, dass der festgesetzte Zeitraum wegen der Umstände des Falls unangemessen ist, kann er die Missstände an externe Stellen (den Präsidenten des Rates, des Parlaments oder des Rechnungshofes oder den Europäischen Bürgerbeauftragten) melden.[893] Auch in diesem Fall gilt der Schutz des Hinweisgebers bei Vorliegen der oben genannten Voraussetzungen weiter.[894]

Der Schutz des Informanten spiegelt sich darin wider, dass derjenige, der gegen den Bediensteten eine Maßnahme ergriffen hat, die Beweislast dafür trägt, dass die Maßnahme aus anderen Gründen als der Meldung der Unregelmäßigkeit erfolgt ist. Von den Bediensteten wird nicht erwartet, dass sie den Nachweis für das Fehlverhalten erbringen. Sie verlieren den Schutz auch nicht deshalb, weil sich ihre Bedenken nachträglich als unbegründet erweisen.[895]

---

889 SEC (2012) 679 final, Abschnitt 1.4.
890 SEC (2012) 679 final, Abschnitt 1.4, der die bestimmten Verfahren aufzählt.
891 SEC (2012) 679 final, Abschnitt 2.
892 SEC (2012) 679 final, Abschnitt 2.
893 SEC (2012) 679 final, Abschnitt 2.
894 SEC (2012) 679 final, Abschnitt 2.
895 SEC (2012) 679 final, Abschnitt 3.

Des Weiteren gelten spezifische Schutzmaßnahmen, wie z. B. Vertraulichkeit der Identität, Mobilität, Anonymität und Sanktionen für Personen, die Repressalien ergreifen.[896]

2. Würdigung

Die einzelnen Ausführungen dieser Leitlinie sind nicht für eine Übertragung in andere Systeme geeignet, da sie speziell auf das Bedienstetensystem innerhalb der Organe und Einrichtungen der EU zugeschnitten sind. Das Grundgerüst des Mitteilungsverfahrens, nach dem vor der externen Informationsübermittlung eine interne Übermittlung stattzufinden hat, kann aber durchaus als Leitlinie für alle Arbeitsverhältnisse herangezogen werden. Insbesondere ist die Präzision der zeitlichen Abfolge im Mitteilungsverfahren hervorzuheben. Die genaue zeitliche Beschreibung dient zum einen der schnellen Bearbeitung der Fälle durch die Kommission bzw. das OLAF, zum anderen der Rechtssicherheit des Betroffenen, der dadurch genaue Vorstellungen über die „Wartezeit" hat und mit Ablauf der 60-Tage-Frist die nächsten Schritte überdenken kann.

IV. "Recommendation on Protecting Whistleblowers" – Europarat 2014

Der Europäische Ausschuss für justizielle Zusammenarbeit (CDCJ) hat im Juni 2013 in Straßburg einen Entwurf einer Empfehlung für den Umgang mit Whistleblowern veröffentlicht.[897] Der Entwurf beinhaltete Vorschläge, ergänzt durch Begriffsbestimmungen und Verfahrensmechanismen, mit welcher Vorgehensweise dem Whistleblower hinreichend Schutz gewährt werden kann. Auf seiner 88. Plenarsitzung hat der Ausschuss den Entwurf überarbeitet und freigegeben.[898] Am 30. April 2014 wurde der Entwurf[899] vom Ministerkomitee angenommen. Die folgende Analyse gibt einen Überblick über die wichtigsten Inhalte der Empfehlung.

---

896 SEC (2012) 679 final, Abschnitt 3.

897 Bureau of the European Committee on legal Co-Operation (CDCJ-BU), 93rd meeting, 19–21 June 2013 Strasbourg, "Draft Recommendation on Protecting Whistleblowers", CDCJ (2013) Misc7 final.

898 "Draft Recommendation on Protecting Whistleblowers" from the CDCJ at its 88th plenary meeting, abrufbar unter: http://www.coe.int/t/dghl/standardsetting/cdcj/Whistleblowers/CDCJ%20DRAFT%20RECOMMENDATION%20ON%20PROTECTING%20WHISTLEBLOWERS_en.pdf, (abgerufen am 20.10.2013).

899 Recommendation CM/Rec (2014)7 of the Committee of Ministers to member States on the protection of whistleblowers, 30. April 2014, abrufbar unter: https://wcd.coe.int/ViewDoc.jsp?id=2188855&Site=CM, (abgerufen am 1.6.2014).

1. *Die Empfehlung im Überblick*

Nachdem eingehend auf die Prinzipien und Ziele des Europarats eingegangen sowie auf die einschlägigen Menschenrechte, Abkommen und Resolutionen aufmerksam gemacht wurde, betont das Komitee die Wichtigkeit von Rahmenbedingungen zum Schutz von Whistleblowern. Es wird den Mitgliedstaaten angeraten, die Empfehlungen innerhalb der nationalen Systeme zum Schutz von Whistleblowern zu adaptieren.

Der Schutz des Whistleblowers soll so ausgerichtet werden, dass er „problemlos" Berichte und Angaben machen kann, die Informationen von öffentlichem Interesse enthalten. Unter Klarstellung, dass es Sache der Mitgliedstaaten sei zu bestimmen, was im öffentlichen Interesse liege, empfiehlt das Komitee, sowohl Verstöße gegen Menschenrechte als auch Risiken für die öffentliche Gesundheit, die Sicherheit und die Umwelt als vom öffentlichen Interesse getragen anzusehen.[900] Der persönliche Anwendungsbereich soll alle Personen umfassen, die entweder im öffentlichen oder privaten Sektor tätig sind, unabhängig von der Art ihrer Arbeitsbeziehung und unabhängig davon, ob sie bezahlt werden oder nicht. Gleichgültig ist auch, ob sie sich in der vorvertraglichen Verhandlungsphase befinden oder das Verhältnis bereits beendet ist.[901]

Der normative Rahmen soll einen umfassenden und kohärenten Ansatz liefern, um Berichterstattungen und Offenlegungen von Informationen von öffentlichem Interesse zu ermöglichen bzw. zu fördern. Jede Person, die durch die Veröffentlichung der Informationen, auch wenn sie ungenau oder irreführend sind, direkt oder indirekt benachteiligt wird, sollte den Schutz behalten und Rechtsmittel gegen dieses Verhalten nach den allgemeinen Regeln des Privatrechts einlegen können – eine Kündigung „aufgrund des Whistleblowings" soll ebenfalls ausgeschlossen sein.[902]

Die Mitgliedstaaten werden angehalten, geeignete Maßnahmen und Vorkehrungen zu treffen, um Whistleblowern den Weg zur Berichterstattung und Offenlegung möglichst leicht zu machen. Wege zur Offenlegung der Hinweise können sein:[903]

- Berichterstattung innerhalb einer Organisation oder eines Unternehmens,
- Berichterstattung an relevante öffentliche Aufsichtsbehörden, Strafverfolgungsbehörden und Aufsichtsgremien,
- Berichterstattung an die Öffentlichkeit, z. B. an Journalisten.

In Fällen der internen Berichterstattung sollen die Arbeitgeber dafür sorgen, dass geeignete Vorkehrungen getroffen und Einrichtungen an Ort und Stelle angeboten werden. Arbeitnehmer und ihre Vertreter sollen über Maßnahmen unterrichtet und bei der Einrichtung eines internen Berichterstattungssystems eingebunden

900 "Recommendation on Protecting Whistleblowers", CM/Rec(2014)7, I.1, 2.
901 "Recommendation on Protecting Whistleblowers", CM/Rec(2014)7, II. 3,4.
902 "Recommendation on Protecting Whistleblowers", CM/Rec(2014)7, III, 10,11.
903 "Recommendation on Protecting Whistleblowers", CM/Rec(2014)7, IV, 14.

werden.[904] Als allgemeine Regel soll gelten, dass interne Berichterstattung und die Meldung an relevante öffentliche Aufsichtsbehörden sowie Strafverfolgungsbehörden und Aufsichtsgremien gefördert werden.[905] Zudem soll die Identität der Whistleblower vertraulich behandelt werden.

Die Berichte und Veröffentlichungen von Informanten sollten unverzüglich und, wenn nötig, mit Hilfe der Arbeitgeber, der entsprechenden öffentlichen Behörde, der Strafverfolgungsbehörde oder des Aufsichtsorgans untersucht werden.[906]

Informanten sollen vor Vergeltungsmaßnahmen jeglicher Form, sei es direkt oder mittelbar durch Personen, die für oder im Auftrag des Arbeitgebers handeln, geschützt werden.[907] Auch wenn die Informationen schließlich nicht der Wahrheit entsprechen oder nicht vom öffentlichem Interesse getragen sind, soll der Schutz nicht aberkannt werden, wenn der Whistleblower berechtigte Gründe für die Annahme der Richtigkeit der Information hatte. Ist in einem Betrieb ein internes Berichtssystem eingeführt worden und hat der Whistleblower die Information sofort an die Öffentlichkeit weitergegeben, kann dies bei der Entscheidung über die Rechtsmittel oder über das Maß an Schutz in Betracht gezogen werden.[908] Findet aufgrund einer nachteiligen arbeitgeberseitigen Maßnahme ein Gerichtsverfahren statt, soll der Arbeitgeber die Beweislast dafür tragen, dass die arbeitsrechtliche Maßnahme nicht aufgrund des Whistleblowings eingeleitet wurde.[909]

2. *Würdigung*

Die Empfehlungen des Ministerkomitees sind zielführend. Auch wenn es den Anschein hat, dass viele Punkte nur rudimentär geregelt wurden, ist dennoch eine Grundlage geschaffen, die von den Mitgliedstaaten als Rahmenbedingung aufgenommen und ausgebaut werden kann. Insbesondere ist zu begrüßen, dass trotz nachträglich bestätigter (nicht beabsichtigter) Unwahrheit oder Nichtbestehen des öffentlichen Interesses dem Whistleblower Schutz zukommen soll. Positiv ist auch die Regelung zur Beweislastregelung, die ebenfalls den Schutz des Betroffenen stärkt. Nicht geklärt ist jedoch die Frage, ob es möglich ist, dass der Hinweisgeber sich sofort an externe Stellen wendet. Eine Ausnahmeregelung besteht nicht. Es wird aber geregelt, wie zu verfahren ist, wenn der Hinweisgeber ein im Betrieb bestehendes internes System nicht nutzt, obwohl dies möglich ist. Da die Regelungen allerdings nur eine Empfehlung an die Mitgliedstaaten sind, ihre nationalen Systeme

904 "Recommendation on Protecting Whistleblowers", CM/Rec(2014)7, IV, 16.
905 "Recommendation on Protecting Whistleblowers", CM/Rec(2014)7, IV, 17.
906 "Recommendation on Protecting Whistleblowers", CM/Rec(2014)7, VI, 19.
907 "Recommendation on Protecting Whistleblowers", CM/Rec(2014)7, VII, 21. Als Formen solcher Vergeltung werden genannt: Entlassung, Suspendierung, Rückstufung, Verlust der Beförderungsmöglichkeiten, Strafversetzungen, Abzüge vom Gehalt, Belästigung oder andere diskriminierende Behandlungen.
908 "Recommendation on Protecting Whistleblowers", CM/Rec(2014)7, VII, 24.
909 "Recommendation on Protecting Whistleblowers", CM/Rec(2014)7, VII, 25.

nach ihnen auszurichten, können nationale Bestimmungen die angesprochenen Lücken füllen.

Die Empfehlungen in dieser Veröffentlichung sind jedenfalls richtungsweisend.

V. Richtlinienentwurf zum Schutz vertraulicher Geschäftsinformationen

Im Zusammenhang mit der möglichen Integrierung eines Whistleblowerschutzes ist auch kurz auf den Entwurf der Europäischen Kommission über eine Richtlinie zum Schutz vertraulichen Know-hows und vertraulicher Geschäftsinformationen (Geschäftsgeheimnisse) vor rechtswidrigem Erwerb sowie rechtswidriger Nutzung und Offenlegung[910] einzugehen. Am 26. Mai 2014 haben die zuständigen Minister auf der Tagung des Rates eine allgemeine Ausrichtung zu diesem Vorschlag der Kommission verabschiedet.[911]

Der am 28. November 2013 erstellte Vorschlag soll einen einheitlichen europäischen Schutz vor unrechtmäßigem Zugriff Dritter auf geheime Unternehmensinformationen schaffen. Bislang differiert nicht nur das Verständnis, was ein Geschäfts- und Betriebsgeheimnis ausmacht, erheblich zwischen den einzelnen Mitgliedstaaten. Auch die gesetzlichen Bestimmungen zum Schutz und zur Durchsetzung solcher Geheimnisse sind länderabhängig äußerst unterschiedlich ausgestaltet. Gerade für grenzüberschreitend agierende Unternehmen stellt sich die Situation daher seit Langem unbefriedigend dar, da ein effektiver Geheimnisschutz kaum möglich ist.

Art. 4 Abs. 1 des Richtlinienentwurfs enthält eine Auflistung von Tatbeständen, bei denen von einem rechtmäßigen Erwerb bzw. einer rechtmäßigen Nutzung oder Offenlegung eines Geschäftsgeheimnisses auszugehen ist. Genannt werden etwa die unabhängige Entdeckung oder Beobachtung eines Produkts, das öffentlich verfügbar gemacht wurde, oder jede andere Vorgehensweise, die unter den gegebenen Umständen mit einer seriösen Geschäftspraxis vereinbar ist.[912]

Der Richtlinienentwurf thematisiert ausdrücklich auch die Offenlegung von Geschäftsgeheimnissen durch Whistleblower. Nach Art. 4 Abs. 2 des Richtlinienentwurfs sollen die Mitgliedstaaten sicherstellen, dass kein Recht auf Inanspruchnahme der in dieser Richtlinie vorgesehenen Maßnahmen, Verfahren oder Rechtsbehelfe besteht,[913] wenn der angebliche Erwerb bzw. die angebliche Nutzung oder Offenlegung des Geschäftsgeheimnisses in bestimmten Situationen erfolgt ist. So soll der Schutz

---

910 Vorschlag der Europäischen Kommission über eine Richtlinie über den Schutz vertraulichen Know-hows und vertraulicher Geschäftsinformationen (Geschäftsgeheimnisse) vor rechtswidrigem Erwerb sowie rechtswidriger Nutzung und Offenlegung, 28.11.2013, COM(2013) 813 final.

911 Mitteilung der Kommission, abrufbar unter: http://ec.europa.eu/internal_market/iprenforcement/trade_secrets/index_de.htm, (abgerufen am 10.8.2014).

912 Art. 4 Abs. 1 RL-E.

913 Umfasst sind Unterlassungs-, Schadensersatz-, Rückruf- und Vernichtungsansprüche (Art. 11–14 RL-E).

(des Trägers des Geheimnisses) nicht für Geschäftsgeheimnisse gelten, wenn dadurch eine gesetzliche Pflicht erfüllt, ein Recht auf freie Meinungsfreiheit oder Informationsfreiheit wahrgenommen wird oder deren Offenlegung insoweit dem öffentlichen Interesse dient, als damit Rechtsverstöße aufgedeckt werden, die strafrechtlich oder als Ordnungswidrigkeit sanktioniert werden.[914] Einhellig sind damit Situationen gemeint, die typischen Whistleblowing-Aktivitäten entsprechen. Nach dem 12. Erwägungsgrund des Richtlinienvorschlags ist es wichtig sicherzustellen, dass die Ziele dieser Richtlinien nicht uneingeschränkt verfolgt werden. Es muss gewährleistet werden, dass die vorgesehenen Maßnahmen und Rechtsbehelfe, die dem Träger des Geschäftsgeheimnisses und damit dem Antragsteller zugutekommen, nicht die Meinungsäußerungsfreiheit der offenbarenden Person (Whistleblower) einschränken und somit Whistleblower-Aktivitäten unmöglich machen. Daher soll sich der Schutz von Geschäftsgeheimnissen nicht auf Fälle erstrecken, in denen die Offenlegung eines Geschäftsgeheimnisses dem öffentlichen Interesse dient, weil ein ordnungswidriges Verhalten oder eine strafbare Handlung aufgedeckt wird.[915]

Zwar steht der Whistleblowerschutz nicht im Mittelpunkt dieses Richtlinienvorschlags. Dennoch wird an geeigneter Stelle auf die Offenlegung von Betriebsinterna durch einen Whistleblower eingegangen. Die Regelung stellt einen Grundbaustein für die Versagung des arbeitgeberseitigen Schadensersatzanspruches gegen den Whistleblower dar. Sie will den Whistleblower vor Inanspruchnahme durch den Arbeitgeber schützen, ihm also teilweise arbeitsrechtlichen Schutz gewähren. Dieser Aspekt könnte auch bei der Strafbarkeit des Whistleblowers nach § 17 UWG eine Rolle spielen. Allerdings sagt der Richtlinienvorschlag nichts darüber aus, ob ein „illegales Geheimnis" in den Anwendungsbereich einzubeziehen ist. Nach dem Sinn und Zweck kann sich jedoch nichts anderes ergeben. Sonst wäre die Erlaubnisvorschrift überflüssig. Außerdem haben auch „illegale Geheimnisse" einen kommerziellen Wert, was die Anwendbarkeit der Richtlinie durch Art. 2 Abs.1 lit. b) bestätigt.

Es kann daher Folgendes festgestellt werden: Die Versagung der Ansprüche des Geheimnisträgers gegen den Whistleblower ist zwar ein richtiger Schritt, führt aber nicht zu einem hinreichenden Whistleblowerschutz; insbesondere bietet der Entwurf keinen Schutz vor Kündigungen, Mobbing sowie strafrechtlichen/ordnungswidrigkeitenrechtlichen Konsequenzen oder haftungsrechtlicher Inanspruchnahme durch Dritte.

VI. Schutz durch Implementierung eines Compliance-Systems?

Compliance im gesellschaftsrechtlichen Sinn wird umschrieben als „eine präventiv wirkende und das Risiko der Haftung des Unternehmens bzw. der Unternehmensleitung minimierende Unternehmensorganisation".[916] Seit dem Bekanntwerden von

914 Vgl. Erwägungsgrund 12 des RL-Entwurfs.

915 Vgl. Erwägungsgrund 12 a. E. des RL-Entwurfs.

916 *Kiethe*, GmbHR 2007, 393, 393 f.

Rechtsverstößen namhafter Unternehmen[917] und besorgniserregenden Statistiken zu Wirtschaftsdelikten[918] wird der Ruf nach Einführung von Compliance-Regeln und Compliance-Organisationen in Unternehmen immer lauter. Dahinter verbirgt sich insbesondere das Bestreben der Unternehmensführung nach Risikoabwehr durch Risikomanagement und die Angst vor Repressalien.[919] Die Ausrichtung und das Handeln eines Unternehmens nach Gesetz und Ethik – „Corporate Compliance" – stellt eine wirtschaftliche Notwendigkeit dar, um in materieller Hinsicht Schäden in Millionenhöhe zu vermeiden und in immaterieller Hinsicht einer Rufschädigung oder dem Verlust von Kundenbeziehungen zu entgehen.[920]

Für Unternehmen bedeutet Compliance heute typischerweise die Sicherstellung der rechtlichen und organisatorischen Einhaltung aller für sie maßgeblichen Gesetze und Normen, die Beachtung öffentlich-rechtlicher und strafrechtlicher Regeln, insbesondere aufsichtsrechtlicher Natur, z. B. der Börsenaufsicht, der Banken- und Finanzdienstleistungsaufsicht, der Arbeitsschutzbehörden und sonstiger (allgemeiner) Aufsichtsbehörden.[921]

Compliance muss in Deutschland immer im Zusammenhang mit dem Deutschen Governance Kodex gesehen werden.[922] Dieser stellt einen sog. „*Code of Best Practice*" dar, der von einer vom Bundesministerium für Justiz einberufenen Regierungskommission erstmals im August 2002 veröffentlicht worden ist und seither regelmäßig aktualisiert wird. Er soll dazu beitragen, dass die in Deutschland geltenden Regeln für die Unternehmensleitung und -überwachung für nationale und internationale Investoren transparent gemacht werden und so das Vertrauen in die Unternehmensführung deutscher Gesellschaften und damit mittelbar in den deutschen Kapitalmarkt gestärkt wird.[923]

Die *ARAG/Garmenbeck*-Entscheidung[924] des Bundesgerichtshofs spielt eine wichtige Rolle bei der Förderung von Compliance. Der Bundesgerichtshof benennt die

---

917 Dazu *Behringer*, in: Behringer, Compliance kompakt – Best Practice im Compliance-Management, S. 33 f.

918 Vgl. *Nestler/Salvenmoser/Bussmann*, Studie Wirtschaftskriminalität 2013, einzusehen über www.pwc.de, (abgerufen am 10.1.2014). Zwar wird in der Studie von einem Rückgang der Kriminalität gesprochen – dennoch bewegt sich die Zahl auf einem hohen Niveau.

919 So auch *Lelley*, Compliance im Arbeitsrecht, S. 9 f.

920 *Maschmann*, in: Maschmann, Corporate Compliance und Arbeitsrecht, S. 7 m. w. N; *Wybitul*, BB 2009, 606, 611.

921 *Mengel*, Compliance und Arbeitsrecht, Einleitung. Zudem ist die Festlegung der Unternehmensphilosophie und -kultur sowie die – teils auch überobligatorische – Selbstverpflichtung von Unternehmen zu moralischem, wirtschaftlichem und auch ökologisch nachhaltigem Handeln immer wichtiger geworden.

922 Aktuelle Fassung des Deutschen Corporate Governance Kodex vom 13.5.2013, vgl. http://www.corporate-governance-code.de/ger/download/kodex_2013/D_CorGov_Endfassung_Mai_2013_markiert.pdf, (abgerufen am 15.12.2013).

923 *Kreitner*, in: Küttner, Personalbuch, Compliance, Rn. 2.

924 BGH, Urt. v. 21.4.1997 – II ZR 175/95, NJW 1997, 1926, 1927 f.

besonderen Pflichten, die durch die Vorstandsmitglieder einer Aktiengesellschaft einzuhalten sind und welche Konsequenzen drohen, wenn gegen diese Pflicht verstoßen wird. Zusammenfassend besteht die Organisationspflicht des Vorstandes in der Unternehmensplanung, der Unternehmenskoordinierung, der Unternehmenskontrolle und der Besetzung der obersten Führungsstellen im Unternehmen.[925] Die Überwachungspflichten des Vorstandes begleiten die Organisationspflicht, da die Kontrolle des Vorstandes die ordnungsgemäße Ausführung der Organisationsvorgaben sicherstellen soll. Der Vorstand kann zwar Überwachungsaufgaben an Personen unterhalb der Vorstandsebene delegieren, muss aber weiterhin die oberste Aufsicht führen.[926]

1. *Bedeutung im Arbeitsrecht*

Compliance wirkt sich nicht nur im Wirtschaftsrecht aus, sondern auch im Arbeitsrecht. Im Arbeitsrecht besteht gleichermaßen ein erhebliches Interesse an rechtskonformem Verhalten. Die arbeitsrechtliche Implementierung von Compliance- und Ethikrichtlinien ist notwendig, um das Verhalten aller Beteiligten im Unternehmen zu regulieren. Das Arbeitsrecht macht wesentliche inhaltliche Compliance-Vorgaben, da es eine Vielzahl von oftmals strafbewehrten Regelungen enthält.[927] So verpflichtet das Allgemeine Gleichbehandlungsgesetz (AGG)[928] den Arbeitgeber, jegliche Benachteiligung in Form von Diskriminierung im Sinne des § 1 AGG zu unterbinden (§ 12 Abs. 1 AGG) und dagegen vorzugehen, ob präventiv oder repressiv (§ 12 Abs. 3 und 4 AGG).[929] Bei Missachtung droht dem Arbeitgeber unter anderem die Zahlung von Schadensersatz an den Geschädigten. Um dieses Szenario zu vermeiden, scheint es für beide Seiten essentiell, ein Compliance-Programm, welches Regelungen zum Verbot von Benachteiligungen enthält, im Unternehmen zu implementieren und dessen Einhaltung zu sichern.

Das Arbeitsrecht setzt jedoch auch Grenzen für „überschießende" Compliance-Bestimmungen, indem es eine umfassende Interessenabwägung vornimmt.[930] Dabei sind insbesondere das grundrechtlich geschützte Persönlichkeitsrecht und das darin enthaltende Recht auf Datenschutz zu beachten. Wird dem Arbeitnehmer durch eine Compliance-Vorschrift vertraglich eine Pflicht auferlegt, die seine Nebenpflicht

925 Vgl. *Semler/Spindler*, in: MüKo AktG, Vor § 76 AktG, Rn. 59 f.

926 OLG Düsseldorf, Beschl. v. 12.11.1998 – 2 Ss (OWi) 385/98 – (OWi) 112/98 III, NStZ-RR 1999, 151, 151; *Fleischer*, AG 2003, 291, 295.

927 *Umnuß/Dworschak/Scheele*, in: Umnuß, Corporate Compliance Checklisten, Kap. 1 A, Rn. 2.

928 Allgemeines Gleichbehandlungsgesetz vom 14.8.2006, BGBl. I-2006, S. 1897, das zuletzt durch Artikel 8 des Gesetzes vom 3.4.2013, BGBl. I-2013, S. 610 geändert worden ist.

929 Vgl. *Mengel*, Compliance und Arbeitsrecht, Einleitung, C.

930 *Umnuß/Dworschak/Scheele*, in: Umnuß, Corporate Compliance Checklisten, Kap. 1 A, Rn. 4.

konkretisiert und damit über das normale Pflichtenprogramm hinausgeht, ist stets zu prüfen, ob das Recht auf persönliche Entfaltung des Arbeitnehmers eine zu große Einschränkung erfährt.[931]

Die infrage kommenden Compliance-Themen im Arbeitsrecht lassen sich grob in zwei Bereiche unterteilen: das individuelle und das kollektive Arbeitsrecht. In kollektiver Hinsicht können sich Probleme sowohl im Betriebsverfassungsrecht als auch im Tarifrecht ergeben. Im Individualarbeitsrecht ist, neben dem Persönlichkeitsschutz, besonderes Augenmerk auf die Themen Datenschutz, Arbeitsschutz und Arbeitssicherheit, Diskriminierungsschutz sowie besondere Auskunfts- und Aushangpflichten gegenüber den Arbeitnehmern zu legen.[932]

Rechtlich problematisch ist im Rahmen der persönlichen Bindung der Arbeitnehmer an Compliance-Bestimmungen die Art der Kodifizierung. Dem Arbeitgeber stehen die Instrumente des Direktionsrechts, der arbeitsvertraglichen Vereinbarung oder der Betriebsvereinbarung zur Auswahl. Entscheidet er sich dafür, die compliance-relevanten Vorschriften und Richtlinien durch Arbeitsvertragsvereinbarung festzulegen, muss eine Angemessenheitsprüfung nach den §§ 307 ff. BGB stattfinden.[933] Sog. Whistleblowing-Klauseln, die Arbeitnehmer vertraglich zur unaufgeforderten Mitteilung über Umstände im Unternehmen anhalten,[934] müssen dementsprechend auf ihre Angemessenheit hin überprüft werden. Zwar entspricht es der gefestigten Rechtsprechung des Bundesarbeitsgerichts, dass der Arbeitnehmer aufgrund arbeitsvertraglicher Nebenpflichten zur Rücksichtnahme verpflichtet ist und den Arbeitgeber über alle wesentlichen Vorkommnisse im Betrieb informieren muss, um so Schäden zu verhindern.[935] Eine generelle, auch unwesentliche Vorkommnisse betreffende Meldepflicht ist aber unverhältnismäßig.[936] Der Arbeitnehmer darf aufgrund des *nemo tenetur*-Grundsatzes zudem im Rahmen eines Whistleblowing-Systems nicht angehalten werden, sich selbst zu bezichtigen. Allgemeine Kronzeugenregelungen, die einen arbeitsrechtlichen Sanktionsverzicht ohne Bezug zu konkreten Aufklärungsmaßnahmen zusagen, sind unwirksam.[937]

---

931 Vgl. zum umfassenden Verbot von Liebesbeziehungen am Arbeitsplatz LAG Düsseldorf, Beschl. v. 14.11.2005 – 10 TaBV 46/05, NZA-RR 2006, 81, 87 f.

932 *Kreitner*, in: Küttner, Personalbuch, Compliance, A, Rn. 4 f.

933 Vgl. *Schreiber*, NZA-RR 2010, 617, 618; siehe dazu BAG, Beschl. v. 22.7.2008 – 1 ABR 40/07, NZA 2008, 1248, 1255.

934 So *Kreitner*, in: Küttner, Personalbuch, Compliance, A, Rn. 6.

935 BAG, Urt. v. 3.7.2003 – 2 AZR 235/02, NZA 2004, 427, 429; BAG, Urt. v. 7.12.2006 – 2 AZR 400/05, NZA 2007, 502, 503; *Preis*, in: Erfurter Kommentar, § 611 BGB, Rn. 744; siehe zur Vermögensschädigung *Stahlhacke/Preis/Vossen*, Kündigung und Kündigungsschutz, § 22, Rn. 692; *Gach/Rützel*, BB 1997, 1959, 1961.

936 Vgl. im Ergebnis *Schulz*, BB 2011, 629, 631 f.

937 Ausführlich dazu *Göpfert/Merten/Siegrist*, NJW 2008, 1703, 1704.

Als effektivstes Mittel zur Aufdeckung von Compliance-Verstößen hat sich inzwischen in vielen Unternehmen die Einführung von Whistleblowing-[938] oder Ombuds-Systemen[939] etabliert. Whistleblowing-Systeme ermöglichen es einem Arbeitnehmer in einem bestimmten Verfahren, gegebenenfalls auch anonym, Rechtsverstöße (insbesondere gegen die aufgestellten betriebsinternen Ethik-Richtlinien) Dritten oder direkt dem Unternehmen zu melden und Informationen über Verhaltensweisen dieser Mitarbeiter abzugeben.[940] Die Gestaltung eines solchen Systems birgt aber gewisse rechtliche Probleme, u. a. im Datenschutz. Um handfeste Eingriffe in die Rechte des Verdächtigten und, je nach Ausgestaltung, in die Rechte des Hinweisgebers zu vermeiden, sollten die Grenzen der zulässigen Hinweisgebung klar sein. Die Grenzziehung kann durch eine Konkretisierung des Hinweisgegenstandes erreicht werden. Bereits aus der Natur der Sache ergibt sich, dass in erster Linie unternehmensbezogene Vorgänge und Verhaltensweisen vom Hinweisgegenstand erfasst sein sollten.[941] Da das Compliance-Programm auf einer Unternehmensentscheidung fußt und die Unternehmensleitung mit der Implementierung das Ziel verfolgt, ordnungsgemäßes und gesetzestreues Verhalten im Unternehmen zu fördern, ist es nur logisch, dass Verhaltensweisen im Betrieb oder jedenfalls während der Arbeitszeit als Hinweisgegenstand anzusehen sind. Zur Erleichterung der innerbetrieblichen Meldung von Straftaten oder Pflichtwidrigkeiten ist die Einrichtung einer internen unabhängigen Beschwerdestelle als (notwendiger) Bestandteil eines Compliance-Systems also von besonderer Bedeutung.[942]

Die Hinweisgeber sollten einen Hinweis vertraulich oder anonym abgeben können, wobei fraglich ist, ob die Anonymität überhaupt eine gute Option für den Whistleblower ist. So ist z. B. bei kleinteiligen Arbeitszusammenhängen und hoch spezialisierten Arbeitsgruppen zumeist schon durch die zu übermittelnde Information ein Rückschluss auf einen begrenzten Personenkreis möglich, von dem ein Hinweis zu erwarten war.[943] Der Whistleblower ist trotz Anonymität nicht davor geschützt, dass seine Identität erraten wird. Zudem sind Whistleblower aufgrund der „Heimlichtuerei“ weniger glaubwürdig, was sich auch auf die Überprüfung des Hinweises auswirkt.

Das soziale Klima innerhalb des Unternehmens könnte schlechter werden, wenn die Arbeitnehmer damit rechnen müssten, jederzeit als Verdächtige in eine Hinweisüberprüfung zu geraten. Es ist außerdem leichter, den Schutz des Hinweisgebers vor Vergeltungsmaßnahmen zu organisieren – vor allem wenn dieser Schutz gesetzlich gesichert ist –, wenn die Anschuldigung der Beschwerdestelle offen vorgebracht

---

938 *Wisskirchen/Körber/Bissels*, BB 2006, 1567, 1567; zur Einrichtung *Schneider/Nowak*, FS Kreutz, S. 855, 864.

939 *Rosbach*, CCZ 2008, 101, 104; *Schulz*, BB 2011, 629, 629.

940 *Breinlinger/Krader*, RDV 2006, 60, 60; *Mahnhold*, NZA 2008, 737, 737.

941 *Schemmel/Ruhmannseder/Witzigmann*, Hinweisgebersysteme, S. 185.

942 So auch *Seel*, MDR 2012, 9, 12; gegen Notwendigkeit *Schneider/Nowak*, FS Kreutz, S. 855, 865.

943 So *Strack*, Compliance-Berater 4/2014, 113, 117.

wird.[944] Es muss deshalb ein klares Bekenntnis aller Verantwortlichen im Unternehmen geben, dass kein ehrlicher Whistleblower Nachteile erleiden wird. Nur wenn das Hinweisgebersystem auch in seiner Anwendungspraxis keine „Black-Box" ist, werden die potenziellen Nutzer sehen, dass sie mit einem Hinweis etwas erreichen können und Vertrauen in das System fassen.[945]

Für eine wirksame und effiziente Compliance-Struktur ist es unerlässlich, einen oder mehrere Compliance-Officer[946] als verantwortliche Personen zu ernennen. Der Compliance-Officer macht der Geschäftsleitung zunächst Vorschläge für die Einrichtung des Compliance-Systems, also für alle organisatorischen Maßnahmen, mit denen erreicht werden soll, dass sich das Unternehmen und die für das Unternehmen handelnden Personen rechtskonform verhalten.[947] Er ist zuständig für die Implementierung, Dokumentation und Weiterentwicklung des Systems, welche ebenso wie die Einführung von der Geschäftsleitung beschlossen werden.[948] Daneben hat der Compliance-Officer drei grundlegende praktische Funktionen zu erfüllen: Prävention, Aufdeckung eines Fehlverhaltens und Reaktion auf ein solches Verhalten. Dazu stehen ihm verschiedene Instrumente zur Verfügung,[949] wie z. B.:

- Erstellung und ständige Fortentwicklung der Compliance-Regeln
- Unterstützung der Unternehmensleitung bei der Durchsetzung der Regeln in Unternehmensprozessen
- Präventive Beratung und Training der zugeordneten Unternehmenseinheiten
- Einführung und Koordination von Kontrollen zur Überwachung der Regeln
- Unterstützung bei internen Untersuchungen von Fehlverhalten
- Sicherstellung von geeigneten disziplinarischen Maßnahmen nach Aufdeckung von Fehlverhalten
- Entgegennahme von (anonymen) Hinweisen über eine unternehmensinterne Whistleblower-Hotline[950].

Die Person des Compliance-Officers muss eine verantwortungsvolle und willensstarke Person sein, um das Compliance-Programm effektiv und konsequent in das Unternehmen zu implementieren und durchzuführen. Sie muss zudem besonders vertrauenswürdig sein, damit Whistleblower das System nutzen und den Hinweis ihm gegenüber offenbaren.

Die Einführung einer „Whistleblowing-Hotline" ist grundsätzlich zulässig.[951] Die Hotline soll dazu dienen, die Meldung von Pflichtverletzungen mit betrieblichem Bezug zu erleichtern. Diese Hotlines können intern installiert sein oder durch externe

---

944 *Neundorf*, in: Hauschka, Corporate Compliance, § 30, Rn. 32.
945 *Strack*, Compliance-Berater 4/2014, 113, 117.
946 *Illing/Umnuß*, CCZ 2009, 1 f.
947 *Bürkle*, in: Hauschka, Corporate Compliance, § 8, Rn. 24.
948 *Bürkle*, in: Hauschka, Corporate Compliance, § 8, Rn. 24.
949 Vgl. *Moosmayer*, Compliance, Abschnitt D, S. 39.
950 *Bürkle*, in: Hauschka, Corporate Compliance, § 8, Rn. 26 m.w.N.
951 BAG, Beschl. v. 22.7.2008 – 1 ABR 40/07, NZA 2008, 1248, 1253 f.

Anbieter bereitgestellt werden. Der externe Anbieter übernimmt dabei die telefonische Aufnahme des Hinweises, der namentlich oder anonym abgegeben werden kann, und übermittelt den ausgewerteten Bericht an die Compliance-Organisation des beauftragten Unternehmens. Auf diese Weise kann das Unternehmen gegen Missstände vorgehen, während der Whistleblower anonym bleibt, solange und soweit er das wünscht.[952] Bei der internen Installation einer „Telefonhotline" direkt bei der Compliance-Organisation im Unternehmen spielt der Ombudsmann eine wichtige Rolle. Er ist ein externer Anwalt, der vom Unternehmen beauftragt wird und Mitarbeitern als Anlaufstelle zur Verfügung steht.[953] Dabei legt der Ombudsmann nur die Informationen offen, für die er vom Whistleblower autorisiert wurde – dies gilt auch für die Identität.[954] Allen Hinweisgebersystemen ist gemein, dass sie einer sorgfältigen Überprüfung unterzogen werden müssen, und zwar in rechtlicher wie unternehmerischer Hinsicht. Die Einführung muss dem geltenden Recht entsprechen und sollte an die Organisation des Unternehmens angepasst werden. Für mittelständische Unternehmen wird ein Hinweisgebersystem in Form der Einsetzung eines Ombudsmannes regelmäßig die beste Variante darstellen.[955] Spezialisierte Anwaltskanzleien führen diese Tätigkeit mittlerweile für eine Vielzahl von Unternehmen aus, was sich günstig auf die Kosten der Unternehmen auswirkt. Eher abzuraten ist davon, den „Hausanwalt" des mittelständischen Unternehmens als Ombudsmann zu nutzen.[956] Er ist dem Unternehmen zu sehr als Berater verbunden, um die für die Tätigkeit als Ombudsmann erforderliche Distanz und Unabhängigkeit gegenüber seinem Auftraggeber zu wahren.

2. Rechtspflicht zur Einführung eines Compliance-Systems

Die Frage nach einer allgemeinen Rechtspflicht, ein Compliance-System in Unternehmen einzuführen, wird aktuell in der Literatur diskutiert. Ein Teil[957] der Literatur versucht die Existenz einer allgemeinen Rechtspflicht für alle Unternehmen mittelbar aus § 130 OWiG, § 91 Abs. 2 AktG und aus einer Reihe von spezialgesetzlichen Normen, wie § 52a Abs. 2 BImSchG, § 53 KrW/AbfG, Art. 11 EG-Geldwäsche Richtlinie und § 14 GeldwäscheG, abzuleiten. Im Wege einer Gesamtanalogie wird aus den Spezialvorschriften gefolgert, dass für alle Unternehmen eine Verpflichtung zur Einrichtung eines Compliance-Systems bestehe.[958] Dem wird mit dem Argument

---

952 *Moosmayer*, Compliance, Abschnitt D, S. 57.
953 *Moosmayer*, Compliance, Abschnitt D, S. 58.
954 Ausführlich *Buchert*, CCZ 2008, 148.
955 Vgl. *Moosmayer*, Compliance, Abschnitt D, S. 59.
956 So *Moosmayer*, Compliance, Abschnitt D, S. 59.
957 Vgl. *Schneider*, ZIP 2003, 645, 647 f.; *Nave/Bonenberger*, BB 2008, 734, 735; im Ergebnis auch *Schaupensteiner*, NZA-Beil. 1/2011, 8, 10; *Schneider/Nowak*, FS Kreutz, S. 855, 861 „gesellschaftsrechtliche Pflicht".
958 *Schneider*, ZIP 2003, 645, 648.

erwidert,[959] dass diese Spezialvorschriften nicht genügten, um eine Rechtspflicht zu konstruieren. Anders könne dies bei Aktiengesellschaften aussehen, wenn die Grundlage für die Pflicht zur Einführung eines Compliance-Systems aus dem Deutschen Corporate Governance Kodex[960](DCGK) entnommen würde.[961] Nach Ziffer 4.1.3 des Deutschen Corporate Governance Kodex hat der Vorstand für die Einhaltung der gesetzlichen Bestimmungen und der unternehmensinternen Richtlinien zu sorgen und darauf hinzuwirken, dass die Beachtung durch die Konzernunternehmen gewährleistet wird (Compliance). Die im Kodex enthaltenen Anregungen und Empfehlungen dienen dazu, den Führungsorganen einen Leitfaden zu geben, wie sie sich bei Erfüllung ihrer Aufgaben im Sinne einer effizienten Corporate Governance verhalten sollen, sog. *Code of Best Practice.*[962] Zu beachten ist jedoch, dass dem DCGK keine Gesetzeskraft zukommt. Vielmehr enthält er nur Empfehlungen für die Selbstverpflichtung von Unternehmen.[963] Aus dieser „Empfehlung" kann aber noch keine Rechtspflicht zur Einführung eines Compliance-Programms abgeleitet werden.[964] Zwar verpflichtet § 161 AktG den Vorstand und Aufsichtsrat, jährlich eine Erklärung abzugeben, inwiefern den Empfehlungen des Kodex entsprochen wurde. Jedoch ändert dies nichts an der Entscheidungsfreiheit des Unternehmens über die Ausgestaltung von Unternehmensregeln.[965] Auch die Ableitung einer Rechtspflicht aus § 91 Abs. 2 AktG wird überwiegend abgelehnt und stattdessen auf das Leitungsermessen des Vorstandes (*Business Judgement Rule*) nach § 93 Abs. 1 AktG verwiesen.[966] Zwar muss der Vorstand geeignete organisatorische Vorkehrungen

---

959 Unter anderem Nachweise bei *Liese*, BB Spezial 2008, Nr. 5, 17, 18 f.

960 Deutscher Corporate Governance Kodex (erstmals 2002 im Bundesanzeiger veröffentlicht; die Inhalte werden jährlich vor den Hintergrund von nationalen und internationalen Entwicklungen überprüft und angepasst), Ziffer 4.1.3. in der neuen Fassung vom 15. März 2013, abrufbar unter: http://www.corporate-governance-code.de/ger/download/kodex_2013/D_CorGov_Endfassung_Mai_2013.pdf, (abgerufen am 20.12.2013).

961 So z. B. *Bürkle*, BB 2007, 1797, 1800.

962 *Schemmel/Ruhmannseder/Witzigmann*, Hinweisgebersysteme, S. 15.

963 OLG München, Urt. v. 6.8.2008 – 7 U 5628/07, BB 2009, 232, 233; *Hauschka*, in: Hauschka, Corporate Compliance, § 1, Rn. 23, m. w. N; *Seibt*, AG 2002, 249, 250; im Ergebnis ebenfalls gegen eine gesetzliche Handlungspflicht *Thüsing*, in: Thüsing, Beschäftigtendatenschutz und Compliance, § 2, Rn. 16.

964 So aber *Bürkle*, BB 2007, 1797, 1800; vgl. gegen die Annahme einer aus dem DCGK abgeleiteten Verpflichtung des Vorstands zur Einführung eines Compliance-Systems *Bachmann*, in: Bachmann/Baums/Goette/Hauschka/Hirtz/Schäfer, Gesellschaftsrecht in der Diskussion, S. 65, 72; *Schemmel/Ruhmannseder/Witzigmann*, Hinweisgebersysteme, S. 15; *Thümmel*, CCZ 2008, 141, 141.

965 Ähnlich *Schemmel/Ruhmannseder/Witzigmann*, Hinweisgebersysteme, S. 15.

966 *Fleischer*, in: Spindler/Stilz, AktG, § 91, Rn. 43; *Kort*, in: GK-AktG, § 91, Rn. 65; *Spindler*, in: MüKo AktG, § 91, Rn. 36; *Bürkle*, BB 2005, 565, 568 f.; *Fleischer*, AG 2003, 291, 299 f.; *ders.*, NJW 2009, 2337, 2338; *Hauschka*, AG 2004, 461, 462; *ders.*, ZIP 2004, 877, 878; *Koch*, WM 2009, 1013, 1014 f.

treffen, um die ordnungsgemäße Aufgabenerfüllung laufend zu überwachen und auf diese Weise sicherzustellen, dass das geltende Recht unternehmensweit Beachtung findet. Bei der Entscheidung, welche konkreten Maßnahmen getroffen werden, steht dem Vorstand allerdings ein Spielraum zu.[967] Im Einzelfall kann jedoch durch die Dispositionsbefugnis des Unternehmens das Leitungsermessen des Vorstandes auf Null reduziert werden, wenn sich die Überwachungspflicht zu einer konkreten Pflicht des Vorstandes verdichtet, in der auf die Einführung eines Compliance-Systems bestanden wird.[968] Auch für börsennotierte Aktiengesellschaften ist der DCGK nicht verbindlich[969] – § 161 AktG verpflichtet jedes Unternehmen nur, Rechenschaft darüber abzulegen, ob und warum sie den DCGK befolgen oder nicht.[970] Aus § 161 AktG und Ziffer 4.1.3 des DCGK folgt demnach keine Pflicht zur Errichtung eines allgemeinen Compliance-Systems.

Fraglich ist, ob eine solche Pflicht für andere Gesellschaftsformen gilt. Eine analoge Anwendung der angesprochenen Bestimmungen, insbesondere auf die GmbH, wird zwar diskutiert,[971] ist aber mangels ausdrücklicher Regelung im GmbHG schwer zu begründen.

Es sei jedoch darauf hingewiesen, dass sich aus § 12 Abs. 5, § 13 AGG ergibt, dass der Arbeitgeber verpflichtet ist, eine Beschwerdestelle einzurichten. Da es sich um eine erforderliche Maßnahme nach § 12 Abs. AGG handelt, ist ein Ermessen ausgeschlossen.[972] Lediglich die weitere Ausgestaltung und Organisation bleibt dem Arbeitgeber überlassen. Unter die Organisationshoheit fällt unter anderem die Entscheidung über den Ort der einzurichtenden Beschwerdestelle, die Besetzung der Stelle und die Bearbeitung der eingehenden Beschwerden.[973] Aufgrund des mit dem AGG verfolgten Ziels sind die Beschwerden gegenständlich beschränkt. Es geht ausschließlich um die Entgegennahme von Beschwerden, die sich im Bereich

967 Grundlegend dazu BGH, Urt. v. 21.4.1997 – II ZR 175/95, NJW 1997, 1926; *Bürkle*, BB 2005, 565, 568 f.; anders *Schaefer/Baumann*, NJW 2011, 3601, 3602, die aus § 107 Abs. 3 Satz 2 AktG herleiten, dass zumindest bei börsennotierten Unternehmen der Gesetzgeber davon ausgeht, dass eine solche Gesellschaft eine Compliance-Organisation entsprechend internationalem Standard hat, weil die Norm von einem „internem Kontrollsystem" spricht.

968 *Fleischer*, in: Fleischer, Handbuch des Vorstandsrechts, § 8, Rn. 43 f.; *Spindler*, in: MüKo AktG, § 91 AktG, Rn. 36; *Maschmann*, in: Dölling, Handbuch der Korruptionsprävention, S. 179.

969 *Spindler*, in: MüKo AktG, § 91 AktG, Rn. 36.

970 *Thüsing*, in: Thüsing, Beschäftigtendatenschutz und Compliance, § 2, Rn. 16.

971 Z. B. *Spindler*, in: MüKo AktG, § 91 AktG, Rn. 42; *Lensdorf*, CR 2007, 413, 414 (IT-Compliance); *A. Kießling/E. Kießling*, WM 2003, 513, 523; *Schneider*, GmbHR 2010, 1313, 1316; *Kort*, GmbHR 2013, 566, 567; *Trips*, NVwZ 2003, 804, 810 f. fordert eine entsprechende Regelung für die öffentliche Verwaltung.

972 LArbG Nürnberg, Beschl. v. 19.2.2008 – 6 TaBV 80/07, DB 2009, 71, 72; *Oetker*, NZA 2008, 264, 266.

973 BAG, Beschl. v. 21.7.2009 – 1 ABR 42/08, NZA 2009, 1049, 1050; *Schlachter*, in: Erfurter Kommentar, § 13 AGG, Rn. 2; *Oetker*, NZA 2008, 264, 267 f.

der Benachteiligung (Diskriminierung) im Beschäftigungsverhältnis bewegen (vgl. § 13 Abs. 1 AGG). Deswegen kann aus dem AGG folglich auch keine allgemeine Pflicht begründet werden, organisatorische Maßnahmen zur Entgegennahme und Behandlung von Hinweisen auf Gesetzes- oder Regelverstöße zu treffen.[974]

3. *Würdigung*

In einer Zeit, in der das Thema Compliance immer mehr in den Vordergrund juristischer Diskurse rückt, muss an eine gesetzliche Pflicht zur Einführung eines Compliance-Systems gedacht werden. Im Zusammenhang mit der Schaffung eines Whistleblowerschutzgesetzes oder einer Änderung des § 612 a BGB könnte eine allgemein verbindliche Norm erarbeitet werden, die für alle Unternehmen, gleich welcher Gesellschaftsform, gilt. § 11 des Entwurfs zum Hinweisgeberschutzgesetz enthält eine ergänzende Vorschrift, in der von der Einrichtung eines Hinweisgebersystems in Unternehmen und Betrieben gesprochen wird. Zwar ist das Hinweisgebersystem nur ein Teil eines Compliance-Systems. Jedoch handelt es sich um einen wesentlichen Teil. § 11 des HinwGebSchG-E verpflichtet die Unternehmen aber nicht, es steht den Unternehmen stattdessen frei, ein solches System zu errichten.

Eine Implementierung hat nur Vorteile für das Unternehmen. Das Risiko von Schäden z. B. durch Rufschädigung minimiert sich und das System führt allgemein zu einer guten Wertestellung im Unternehmen – sowohl im Verhältnis der Mitarbeiter untereinander als auch in der Außendarstellung. Das Klima der Angst und des gegenseitigen Misstrauens wird abnehmen, wenn Mitarbeitern deutlich gemacht wird, dass das Unternehmen Hinweise begrüßt und darin keinesfalls eine Form der Illoyalität sieht. Dadurch wächst das Verständnis für das Thema Whistleblowing in der Belegschaft. Neben der Qualifizierung eines solchen Systems als „Frühwarnsystem"[975] überzeugt, dass ein solches System signifikant das Risiko von externem Whistleblowing minimiert. Je frühzeitiger Risiken erkannt werden, desto wahrscheinlicher ist es, potenziellen Gefahren und Schäden durch geeignete Maßnahmen entgegenzuwirken.

Enthält das Compliance-Programm eine zusätzliche Whistleblower-Regelung, die jegliche Repressalien gegenüber dem Whistleblower ausschließt, ist es eine geeignete Schutzregelung. Es besteht ein in sich geschlossener, funktionierender Mechanismus, der es dem Whistleblower erlaubt, sich (auch) anonym zu äußern, seinen Hinweis zu offenbaren und dabei von Repressalien verschont zu bleiben.

Soweit jedoch einer der Bestandteile eines Hinweisgeber-Systems (Hinweisgegenstand, Hinweismeldung, Anonymität, vertraulicher Compliance-Officer, Schutz vor Repressalien) nicht funktioniert, kann nicht erwartet werden, dass der Whistleblower sich schutzlos ausliefert. Es stellt sich daher die Frage, ob dem

974 So auch *Schemmel/Ruhmannseder/Witzigmann*, Hinweisgebersysteme, S. 74; *Sänger*, Whistleblowing, S. 216.

975 So etwa *Hofmann*, ZCG 2006, 121, 126; *Sänger*, Whistleblowing, S. 66; *Wrase/Fabritius*, CCZ 2011, 69, 70.

Whistleblower die Möglichkeit eröffnet werden muss, sich an die Behörden zu wenden, wenn dem Compliance-Officer kein Vertrauen geschenkt werden kann.

Ist im Betrieb kein Whistleblower-System eingerichtet und muss sich der Arbeitnehmer zunächst innerbetrieblich an den Arbeitgeber wenden, müsste bei Scheitern der innerbetrieblichen Hilfe das gleiche Ergebnis bestehen. Der Whistleblower muss bei Unzumutbarkeit oder Scheitern der internen Abhilfe durch den Arbeitgeber die Behörde einschalten dürfen. Dies müsste auch dann der Fall sein, wenn das innerbetrieblich eingerichtete System, dass die Berichterstattung an den Arbeitgeber „ersetzt", nicht funktioniert. Dagegen könnte aber sprechen, dass der Compliance-Officer nicht identisch mit dem Arbeitgeber ist. Er trifft eigene Entscheidungen und ist abgesehen von bestimmten arbeitgeberseitigen Weisungen[976] frei in seinem Handeln. Demnach kann es durchaus vorkommen, dass der Compliance-Officer sich anders verhält oder andere Entscheidungen trifft als der Arbeitgeber. Das würde dafür sprechen, dem Arbeitgeber die Chance zu geben, das Whistleblowing – entgegen der Ansicht des Compliance-Officers – selbst (neu) zu bewerten und weitere Schritte einzuleiten.

Ausschlaggebend ist jedoch Folgendes: Der Arbeitgeber hat den Compliance-Officer ausgewählt und dazu bestimmt, die Ansichten des Unternehmens im Hinblick auf regelkonformes Verhalten zu vertreten und (compliance-relevante) Entscheidungen zum Vorteil für das Unternehmen zu treffen. Den Arbeitgeber trifft zumindest das Delegations- bzw. Überwachungsverschulden,[977] wenn der Compliance-Officer nicht vertragsgemäß bzw. weisungsgemäß handelt. Er hat dafür einzustehen, wenn dem Whistleblower keine oder keine ausreichende Hilfe durch den Compliance-Officer angeboten wird.

Diese Situation ist also damit zu vergleichen, dass eine vorherige Abhilfe beim Arbeitgeber – ohne eingeführtes Compliance-System – fehlgeschlagen ist. Beiden Konstellationen ist gemein, dass der Rechtsverstoß mit Hilfe innerbetrieblicher Mittel nicht beseitigt werden kann und deswegen eine externe Offenbarung notwendig wird. In einem demokratischen Rechtssaat muss aber den Personen, die den Rechtsverstoß beenden möchten, die Möglichkeit gegeben werden, dies zu tun, ohne rechtlichen Konsequenzen ausgesetzt zu werden. Whistleblower gehen derzeit in Fällen, in denen sie annehmen, ein interner Ausgleich sei ihnen nicht zuzumuten, ein hohes Risiko ein, wenn sie sich direkt an Behörden, Presse, Dritte oder die Öffentlichkeit wenden.

Ein neuer Gesetzentwurf sollte daher differenziert gestaltet sein. Angemessen ist der Hinweis auf die vorrangige Nutzung eines innerbetrieblichen Hinweisgebersystems, aber auch auf eine mögliche behördliche Aufklärung. Daher sollte ein

---

976 *Bürkle*, in: Hauschka, Corporate Compliance, § 8, Rn. 33 m.w.N.

977 Mit der Delegation wandelt sich die Pflicht der Mitglieder des Leitungsorgans zu einer Pflicht zur ordnungsgemäßen Delegation um, welche die Sorgfalt bei der Auswahl des Delegierten, dessen Anweisung und seine Überwachung umfasst, vgl. *Lackhoff/Schulz*, CCZ 2010, 81, 84; ausführlich *Fleischer*, AG 2003, 291, 293 f.

Gesetzentwurf zum Whistleblowing zwar seinen Schwerpunkt auf unternehmensinterne Abhilfemöglichkeiten legen, aber nicht alle anderen Optionen ausschließen.

Ohne die Verpflichtung der Unternehmen, Compliance- und dementsprechend Whistleblowing-Systeme zu implementieren, bleibt dem Whistleblower oft keine große Wahl, an wen er sich wendet.

B. Strafrechtlicher und ordnungswidrigkeitenrechtlicher Schutz durch Strafmilderung

Im Zusammenhang mit der Darlegung des Schutzes *de lege ferenda* ist fraglich, inwieweit bei den Lösungsansätzen *de lege ferenda* die Problematik des strafrechtlichen bzw. ordnungswidrigkeitenrechtlichen Schutzes bedacht wurde. In den erörterten Gesetzesentwürfen, die sich hauptsächlich auf die arbeitsrechtliche Problematik des Whistleblowings konzentrieren, ist ein strafrechtlicher Schutz nicht erwähnt. Dies hängt größtenteils damit zusammen, dass die Strafbarkeit des Whistleblowers nicht eindeutig geklärt ist. In Betracht kommt nur § 17 UWG.[978] Eine grundsätzliche Einigung über den Geheimnisbegriff besteht dabei nicht. Überwiegend wird jedoch angenommen, dass auch „illegale" Geheimnisse in den Schutzbereich des § 17 UWG fallen. Demnach macht sich der Whistleblower nach § 17 UWG strafbar, wenn er innerbetriebliche Missstände, die durch illegale Verhaltensweisen entstehen, offenbart und anzeigt. Andere Straftatbestände kommen nur in Betracht, soweit der Whistleblower an dem Missstand beteiligt war.

Rechtswissenschaftliche Publikationen stützen sich beim strafrechtlichen Schutz vor § 17 UWG auf den Rechtfertigungstatbestand des § 34 StGB. Wie in Teil 4 B dieser Arbeit ausgeführt, überzeugt dieser Weg. Daher ist es vertretbar zu behaupten, dass es keines besonderen strafrechtlichen Schutzes des Whistleblowers bedarf. Ein Hinweis auf die strafrechtliche Lage und Abwägungskriterien in einem Hinweisgeberschutzgesetz wäre aber zu begrüßen.

Dasselbe gilt für den am Missstand beteiligten Whistleblower. Ihm steht in den meisten Fällen bezüglich der begangenen Tat kein Rechtfertigungstatbestand zur Seite und auch die Regelungen über den Rücktritt oder die tätige Reue greifen nur ausnahmsweise. Dennoch ist dieses Ergebnis gerecht. Auch wenn der Whistleblower sich nach seiner Beteiligung dazu entschieden hat, das rechtswidrige Handeln zu beenden, hat er doch gewichtig zur Verwirklichung des Straftatbestandes oder der Ordnungswidrigkeit beigetragen. Die Strafrechtsdogmatik lässt es nicht zu, den Whistleblower anders zu behandeln als andere Täter.

Vergleicht man den Whistleblower, der an einer sonstigen Tat beteiligt war mit dem am Kartell beteiligten Whistleblower, scheint es ungerecht, letzteren durch Erlass/Reduzierung des Bußgeldes zu privilegieren und ersterem kein Instrument

978 Andere Straftatbestände entfallen regelmäßig aufgrund der Nicht-Erfüllung des subjektiven Tatbestandes oder aufgrund des besonderen Rechtfertigungstatbestandes des § 193 StGB.

zur Verfügung zu stellen, dass ihn strafrechtlich/ordnungswidrigkeitenrechtlich vor den Folgen schützt. Die Differenzierung ist aber gerechtfertigt: Der am Kartell beteiligte Whistleblower hat eine Ordnungswidrigkeit begangen, die Verbraucher aufgrund überhöhter Preise um ihr Geld bringt. Der Schwerpunkt liegt also in einem materiellen Schaden. Der Whistleblower, der an einer sonstigen Tat beteiligt war, hat unter Umständen Straftaten begangen, die ein höherrangiges Rechtsgut verletzt haben als Vermögen oder Eigentum. Die unterschiedliche Wertigkeit der Rechtsgüter rechtfertigt also eine Differenzierung. Auch wenn der Whistleblower in Missstände verwickelt ist, die „nur" materielle Schäden zur Folge haben, ist diese unterschiedliche Betrachtungsweise gerechtfertigt. Nicht umsonst hat der Gesetzgeber das OWiG geschaffen, um im Gegensatz zum StGB ein missbilligtes Verhalten von einem geringeren sozialethischen Gewicht zu bestrafen.

Unterstützt der Whistleblower die Behörden bei der Identifizierung möglicher weiterer Beteiligten, steht ihm – wie bei der Bonusregelung im Kartellrecht – eine Norm zur Verfügung (§ 46 b StGB), nach der eine Strafmilderung oder ein Straferlass stattfindet.

Ist die maßgebliche Tat nicht im Katalog des § 100 a StPO aufgeführt, ist auf § 46 StGB zurückzugreifen, wonach aufgrund bestimmter beim Täter vorliegender Kriterien die Strafe individuell zugemessen werden kann. Insbesondere spielt beim Whistleblowing § 46 Abs. 2 StGB[979] eine besondere Rolle. Insofern besteht die Möglichkeit, dass im Rahmen der richterlichen Feststellung ein in der Person des Whistleblowers liegender Umstand zur begünstigenden Strafzumessung führt. Die Vorschrift bietet bei Whistleblower-freundlicher Auslegung und Anwendung ausreichend strafrechtlichen Schutz.

Hat der Whistleblower Ordnungswidrigkeiten begangen, kann § 17 OWiG zur Bußgeldmilderung führen. Danach kann die Höhe der Geldbuße nach Abs. 3 ebenfalls individuell festgelegt werden. Die Bemühungen des Täters um Wiedergutmachung nach Vollendung des Delikts werden durch mildere Bußgeldbemessung oder durch Einstellung des Verfahrens (§ 47 OWiG) honoriert.

Ist der Whistleblower ein ehemaliger Kartellbeteiligter, ist das besondere Ordnungswidrigkeitenrecht (§ 81 GWB) zu beachten und auf die Bonusregelung des Bundeskartellamtes einzugehen. Sie erlaubt den Erlass bzw. die Ermäßigung der Geldbuße, die der Kartellbeteiligte hätte zahlen müssen, wenn er nicht mit der Kartellbehörde im ausreichenden Maße kooperiert hätte. Dem Whistleblower im Kartellrecht steht also ein Instrument zur Verfügung, das ihn ordnungswidrigkeitenrechtlich vor der Verhängung einer Geldbuße schützt. Eine andere, den Whistleblower schützende Bonus- bzw. Kronzeugenregelung – abgesehen von § 46 b StGB – findet sich nicht.

---

979 Nach § 46 Abs. 2 StGB wägt das Gericht bei der Zumessung die Umstände, die für und gegen den Täter sprechen, gegeneinander ab. Dabei kommen u. a. sein Verhalten nach der Tat, besonders sein Bemühen, den Schaden wiedergutzumachen [...] in Betracht.

C. Zivilrechtlicher Schutz durch Haftungsbegrenzung

Hat durch den Gesetzesverstoß im Unternehmen ein außenstehender Dritter einen Schaden erlitten, wird dieser gegen den Schädiger vorgehen. Ist der Whistleblower an dem Missstand nicht beteiligt, hat die Geltendmachung eines Anspruchs keinen Erfolg. Demnach bedarf dieser auch keines haftungsrechtlichen Schutzes. Ist der Whistleblower jedoch involviert, kann ein Dritter Schadensersatz verlangen.

Grundsätzlich richtet sich der Anspruch gegen den Schädigenden, also gegen die natürliche Person selbst. Im Kartellrecht tritt meist das Unternehmen als Whistleblower auf, das letztendlich auch für die zivilrechtlichen Schäden haftbar gemacht wird. Ersatzpflichtiger ist immer das Unternehmen oder ein Repräsentant bzw. die Geschäftsleitung des Unternehmens, aber nicht der einzelne („einfache") Arbeitnehmer. Auch wenn der Arbeitnehmer also als Whistleblower agiert hat, treffen ihn die haftungsrechtlichen Konsequenzen nach dem GWB nicht, sondern nur das Unternehmen als organisatorische und wirtschaftliche Einheit.

Keine gesetzliche Regelung berücksichtigt den Umstand, dass der an der Tat involvierte Whistleblower nachträglich durch die Offenlegung des Gesetzesverstoßes den Missstand beendet. Deshalb muss überlegt werden, wie dem Whistleblower Schutz zukommen kann.

I. Schutz des beteiligten Whistleblowers gegenüber Ansprüchen geschädigter Dritter

Die Auswirkungen nachträglichen Wohlverhaltens bei einer Verletzung zivilrechtlicher Pflichten wurden in der Literatur angesprochen,[980] diese Erwägungen können aber aufgrund der veralteten Gesetzeslage nicht mehr aufgegriffen werden. Da das Phänomen Whistleblowing in Deutschland noch in den „Kinderschuhen" steckt, ist eine Diskussion über den haftungsrechtlichen Schutz des Whistleblowers noch nicht entfacht worden. Um den Whistleblower, der sich zunächst an der Tat beteiligt hat, vor zivilrechtlichen Ansprüchen Dritter zu schützen, sollte darüber nachgedacht werden, die Rechtsfigur der tätigen Reue in das Zivilrecht aufzunehmen. Sie könnte entweder im Rahmen des Verschuldens oder der Schadenszumessung Berücksichtigung finden. Die zivilrechtliche Haftung des Whistleblowers gegenüber dem Dritten wäre dann (zumindest) begrenzt. Der Dritte müsste bei der Geltendmachung des Anspruchs beachten, dass der Whistleblower sein schädigendes Verhalten durch nachträgliche Offenbarung und Einstellung wiedergutzumachen versucht. Dass die tätige Reue bei der Zumessung der Folge („Strafe") einbezogen wird, ist der deutschen Rechtsordnung nicht fremd. Eine Übertragung des Gedankens auf das Zivilrecht läge also nahe. Rechtstechnisch könnte dies durch Implementierung eines Absatzes oder einer Norm im Schadensersatzrecht des Bürgerlichen Gesetzbuches (§§ 823 ff. BGB) oder in einem Whistleblowerschutzgesetz umgesetzt werden.

980 Siehe *Knütel*, Tätige Reue im Zivilrecht.

II. Schutz des am Kartell beteiligten Whistleblowers gegenüber Ansprüchen geschädigter Dritter

Die besondere Problematik bei Whistleblowern im Kartellrecht rührt daher, dass das Kartellrecht eine spezielle schadensersatzrechtliche Norm bereithält. Zudem sind Grundsätze des europäischen Rechts einzubeziehen. Die Erkenntnis, dass sowohl das *private* als auch das *public enforcement* in nationalen und europäischen Normen sowie deren wettbewerbsrechtlichen Grundsätzen gleichermaßen verankert ist, wirft die Frage auf, wie das Spannungsverhältnis künftig aufzulösen ist.

Die obigen Ausführungen haben gezeigt, dass eine Lösung über das Akteneinsichtsrecht nicht befriedigend ist. Der Konflikt zwischen Kronzeugenregelung und Schadensersatzklagen kann aber vermindert werden, indem versucht wird, auf ganz anderem Wege den Kronzeugen zu „entlasten" und so das Recht auf Akteneinsicht zuzulassen; beispielsweise könnte die zivilrechtliche Haftung im Innen- sowie Außenverhältnis begrenzt oder die Existenz der Kronzeugendokumente auf ein Mindestmaß reduziert werden.

1. Gesetzliche Regelung zur Vorwegnahme des Abwägungsergebnisses oder der Abwägungskriterien

In Betracht kommt die vielfach geforderte „Vorwegnahme des Abwägungsergebnisses", durch die dem Dritten das Akteneinsichtsrecht gesetzlich versagt wird.[981] Die Versagung wäre nicht nur aufgrund der zunehmenden Forderung nach Rechtssicherheit begründet, sondern auch aufgrund der Effektivität der öffentlich-rechtlichen Kartellrechtsdurchsetzung, die durch den Zugang Dritter zu Kronzeugenakten abgenommen hat. Der Forderung nach einer gesetzlichen Regelung kann nach dem *Donau Chemie*-Urteil des EuGH aber nicht mehr nachgekommen werden. Mit dem aus dem Effektivitätsgebot hergeleiteten Gebot der Einzelfallabwägung wendet sich der EuGH gegen jede starre mitgliedsstaatliche Regelung der Akteneinsicht in Kronzeugendokumente.[982]

Damit ist weiterhin eine Entscheidung im Einzelfall notwendig. Zwar verlangt der EuGH eine konkrete Abwägung durch die Gerichte, ohne dass der Gesetzgeber ein Ergebnis vorschreibt. Eine gesetzliche Abwägungsgrundlage ist aber nicht ausgeschlossen.[983]

So bleibt eine gesetzliche Kodifikation von relevanten Abwägungskriterien grundsätzlich möglich. Eine entsprechende Regelung ließe sich etwa in das GWB integrieren. Die Kriterien können aus den Entscheidungen der nationalen und europäischen Gerichte entnommen oder im Wege der grundsatznahen Auslegung

981 *Ritter/Käseberg*, WuW 2012, 661, 667; *Fritsche*, DB 2012, 845, 852; *Kersting*, JZ 2012, 42, 44; *Kapp*, WuW 2012, 474, 477; *Koch*, JZ 2013, 390, 392 f.

982 EuGH, Urt. v. 6.6.2013 – Rs. C-536/11(Rn. 31), JZ 2013, 734, 735 (Donau Chemie u. a.).

983 So auch *Fritzsche*, DB 2012, 845, 852.

der wettbewerbsrechtlichen Grundsätze entwickelt werden. Sicherlich ist es für den Gesetzgeber schwierig, sachgerechte Kriterien zu bestimmen. Vollkommene Rechtssicherheit lässt sich damit nicht erzielen.[984]

Man könnte etwa darüber nachdenken, ob die Struktur des Absatzmarktes ein taugliches Abwägungskriterium ist. Denn gibt es nur wenige Abnehmer mit geringen Streuschäden, ist ohnehin nicht mit einer effektiven privaten Kartellrechtsdurchsetzung zu rechnen.[985] Des Weiteren könnte geprüft werden, ob die fraglichen Informationen anderweitig offengelegt werden können, etwa bei Hausdurchsuchungen oder ähnlichen Situationen.[986] Danach ist, unabhängig von einem Kronzeugenantrag, mit Bekanntwerden dieser Informationen zu rechnen. Die Beeinträchtigung der öffentlich-rechtlichen Kartellrechtsdurchsetzung steht somit nicht mehr im Raum.

2. Beschränkung von Kronzeugendokumenten

Es könnten auch Maßnahmen in Betracht gezogen werden, die bewirken, dass Kronzeugendokumente, insbesondere Unternehmenserklärungen bzw. Selbstanzeigen, gar nicht erst erstellt werden müssen.

Das Rechtssystem der USA verfolgt einen Ansatz, der dabei als Leitfaden dienen könnte. Unternehmen, die von einem Sanktionserlass profitieren wollen, müssen bei der Antitrust Division zunächst nur einen sog. „*proffer*" (Angebot zur Hilfe bzw. Kooperation) in mündlicher Form vorbringen.[987] Danach haben die Unternehmen diesen Antrag zu vervollständigen, indem sie die Fakten zu der kartellrechtlichen Zuwiderhandlung zusammenstellen und die dazugehörigen Informationen und Dokumente liefern. Die Dokumente stellen eine Art „*road map*" dar, die die Antitrust Division dazu bewegen kann, die Durchführung einer Untersuchung (*grand jury investigation*) gegen die übrigen Kartellmitglieder anzuordnen.[988]

Der große Unterschied zur europäischen Untersuchung ist, dass die Beweismittel gegen die übrigen Kartellmitglieder, auf denen die Verurteilung basiert, hauptsächlich aus der Untersuchung der Antitrust Division und nicht aus einer Erklärung des kooperierenden Unternehmens stammen. Diese Vorgehensweise führt dazu, dass die in den Verfahrensakten der Antitrust Division befindlichen Dokumente in erster Linie nicht das kooperierende Unternehmen belasten. Vielmehr klagen sie das Unternehmen an, bei dem die zur Verfolgung des wettbewerbswidrigen Verhaltens notwendigen Unterlagen gefunden wurden.[989] Dieser wesentliche Unterschied zur europäischen und damit auch zur deutschen Vorgehensweise bei der

---

984 So auch *Koch*, JZ 2013, 390, 392; *Kersting*, JZ 2013, 737, 738.

985 So z. B. *Dworschak/Maritzen*, WuW 2013, 829, 840.

986 Vgl. *Fornasier/Sanner*, WuW 2011, 1067, 1078.

987 *Denoth*, Kronzeugenregelung, S. 336.

988 *Logemann*, Der kartellrechtliche Schadensersatz, S. 169 mit Verweis auf *Hempel*, Privater Rechtsschutz im Kartellrecht, S. 18.

989 *Denoth*, Kronzeugenregelung, S. 337.

Zusammenführung der notwendigen Unterlagen liegt vor allem darin begründet, dass das US-amerikanische Recht unterschiedliche Voraussetzungen für die Gewährung einer Sanktionsbefreiung vorschreibt.[990]

Fraglich ist, ob es überhaupt möglich sein kann, diese Vorgehensweise in das deutsche System des *public enforcement* zu implementieren.

Zu betonen ist zunächst, dass auch im US-amerikanischen System ein Risiko besteht. Aufgrund des weiten Anwendungsbereiches der *discovery rules* müssen die Dokumente im Zuge eines *discovery*-Begehrens offengelegt werden.[991] Insofern sollte dieser Teil des US-amerikanischen Systems nicht vollständig übernommen werden. Die Informationen, welche der Behörde vom kooperierenden Unternehmen zugestellt werden, dürfen nicht als Grundlage für die Entscheidung und somit nicht als Beweismittel verwendet werden.[992] Diese Unterlagen würden dann nicht Bestandteil der Verfahrensakten, sondern lediglich Hilfsmittel, aufgrund derer die Wettbewerbsbehörde ihre Untersuchungen durchführen kann. Die Abgabe einer „belastenden" Unternehmenserklärung, welche die heutige Fassung des Kronzeugenprogramms vorschreibt, gäbe es dann nicht mehr. Die Behörden sind auf eigene Untersuchungsergebnisse angewiesen, nur diese werden Bestandteil der Akte.

Die Einführung einer veränderten Mitwirkungspflicht des kooperierenden Unternehmens am Untersuchungsverfahren ginge mit einer umfassenden Reform des Kronzeugenprogramms einher. Die jetzigen Voraussetzungen für den Erlass oder die Ermäßigung des Bußgeldes müssten geändert werden. Die bisher geforderte Informationsqualität müsste aufgegeben und die Mitwirkung entsprechend des US-amerikanischen Rechts lediglich auf vollständige und ununterbrochene Kooperation mit der Behörde begrenzt werden.

Schließlich stellt sich noch das Problem der notwendigen Ressourcen. Die Kartellbehörden müssten mehr Aufwand betreiben und benötigten mehr Mitarbeiter sowie eine neue gerichtliche Kompetenzaufteilung zwischen Zivilgerichten und Wettbewerbsbehörden. Obwohl eine stetige Verbesserung der Kartellverfolgung zu verzeichnen ist,[993] bestehen noch immer Probleme wegen Informationsflut und langer Verfahrensdauer infolge der Ressourcenknappheit.[994] Ob sich das Problem löst, wenn den Kartellbehörden noch mehr Untersuchungsaufwand aufgebürdet

---

990 Vgl. Ausführungen zur Kronzeugenregelung im US-amerikanischen Recht *Zagrosek*, Kronzeugenregelungen, S. 56 f.

991 *Logemann*, Der kartellrechtliche Schadensersatz, S. 199.

992 *Denoth*, Kronzeugenregelung, S. 337 m. w. N.

993 Pressemitteilung des Bundeskartellamtes zum Ausbau der Kartellverfolgung vom 14.7.2011, abrufbar unter: http://www.dawnraid.de/wp-content/uploads/2011/07/BkartA_PM-14-07-20111.pdf, (abgerufen am 1.6.2014).

994 Hintergrundpapier des Bundeskartellamtes zur Tagung des Arbeitskreises Kartellrecht am 4.10.2012, S. 3, abrufbar unter: http://www.bundeskartellamt.de/SharedDocs/Publikation/DE/Diskussions_Hintergrundpapier/Bundeskartellamt%20-%20Kartellbu%C3%9Fgeldverfahren%20zwischen%20deutschem%20Systemdenken%20und%20europ%C3%A4ischer%20Konvergenz.html, (abgerufen am 2.6.2014).

wird, weil die kooperierenden Unternehmen in ihrer Informationsweitergabe beschränkt sind, ist sehr fraglich.

3. *Beschränkung der zivilrechtlichen Haftung des Kronzeugen*

Eine weitere Maßnahme, Kronzeugenprogramme attraktiv zu machen, wäre die Begrenzung der zivilrechtlichen Haftung der Kronzeugen als Ausgleich für die Inanspruchnahme durch Dritte. Der Whistleblower könnte durch eine Haftungsbeschränkung sowohl im Innenverhältnis der Kartellbeteiligten als auch im Außenverhältnis gegenüber dem Geschädigten privilegiert werden.

a. Gesamtschuldnerische Haftung

Alle Kartellteilnehmer haften für den durch das Kartell verursachten Schaden nach §§ 830, 840 BGB als Gesamtschuldner. Das hat für den Geschädigten den Vorteil, dass er nach Belieben jeden Kartellanten in Anspruch nehmen kann (§ 421 BGB). Der Geschädigte ist insbesondere nicht darauf beschränkt, gegen seinen Vertragspartner vorzugehen.[995] Erst recht muss der Geschädigte nicht gegen alle Kartellbeteiligten vorgehen. Es genügt die Geltendmachung gegen einen Kartellanten, um vollen Schadensausgleich zu erlangen. §§ 830, 840 BGB stellen ausdrücklich klar, dass alle Beteiligten für den verursachten Schaden haften. Folglich braucht der Geschädigte auch keinen Nachweis der jeweiligen Tatbeiträge der einzelnen Schädiger erbringen.[996]

Die gesamtschuldnerische Haftung ohne jegliche Privilegierung könnte zu einer verstärkten Abschreckung führen. Dem Beteiligten wird bewusst werden, dass er im Extremfall ganz alleine haften muss. Deshalb werden hierfür Lösungsansätze konstruiert, die zur „Sonderbehandlung“ des Kronzeugen führen können.

Probleme ergeben sich insoweit, als eine haftungsbegrenzende Regelung im Widerspruch zur Rechtsprechung des EuGH stehen könnte. Bei der Privilegierung des Kronzeugen könnte das Risiko der Nichtleistung (z. B. im Falle der Insolvenz) durch die anderen Kartellanten den Geschädigten auferlegt werden.[997] Dies könnte dazu führen, dass die volle Haftung des Kronzeugen wieder auflebt und eine abschreckende Wirkung erzeugt wird.[998] Um das zu vermeiden, bedarf es *de lege ferenda* eines ergänzenden Mechanismus, der mittlerweile in der Literatur mit der Haftungsbeschränkung im Innenverhältnis gefunden wurde.

---

995 *Krüger*, in: Oberender, Private und öffentliche Kartellrechtsdurchsetzung, S. 79, 80; *Logemann*, Der kartellrechtliche Schadensersatz, S. 75.

996 *Krüger*, in: Oberender, Private und öffentliche Kartellrechtsdurchsetzung, S. 79, 81 m. w. N.

997 So auch schon *Kersting*, ZWeR 2008, 252, 268.

998 *Koch*, JZ 2013, 390, 394.

b. Gesamtschuldnerausgleich von Kartellmitgliedern im Innenverhältnis

Der Gesamtschuldner, an den sich der Gläubiger gehalten hat, soll nicht die volle Last tragen, sondern im Innenverhältnis Ansprüche gegen seine Mitschuldner auf Ausgleich von Leistung haben, die er über seine Quote hinaus erbracht hat.[999] Insbesondere vor dem Hintergrund, dass ein Kronzeuge für Schadensersatzkläger häufig das primäre Ziel ihrer Ansprüche darstellt,[1000] muss dem Kronzeugen gegenüber seinen Kartellbrüdern ein Regressanspruch zustehen. Auch die Gefahr des Insolvenzrisikos der Kartellbeteiligten, die zunächst nicht zahlen mussten, ist nicht zu unterschätzen. Zudem kann es durchaus vorkommen, dass der leistende Kronzeuge jahrelang keinen Regress nehmen kann, wenn (europäische) Gerichte nicht über die gegen die Bußgeldentscheidung gerichtete Klage entschieden haben[1001] und somit die Verhältnisse nicht geklärt sind.

De lege lata besteht bereits eine Grundlage, nach der eine Privilegierung des Kronzeugen erreicht werden kann. Im Rahmen des Gesamtschuldnerausgleichs zwischen den Kartellanten könnte der Kronzeuge im Innenverhältnis von der Haftung freigestellt werden bzw. der ihm anfallende Haftungsanteil um den Faktor der Bußgeldreduzierung gekürzt werden.[1002] Dieses Ergebnis kann dogmatisch durch einen modifizierenden Rückgriff auf § 426 BGB i. V. m. § 254 Abs. 2 BGB erzielt werden. In der Folge haftet der Kronzeuge zumindest mit einem geringeren Anteil. Nach § 426 Abs. 1 BGB erfolgt der Ausgleich grundsätzlich zu gleichen Anteilen, „soweit nicht ein anderes bestimmt ist". Ob eine solche andere „Bestimmung" in der Privilegierung des Kronzeugen durch die Kronzeugenregelung zu sehen ist, ist fraglich. Eine gesetzliche Regelung fehlt. Die Beurteilung, ob „ein anderes bestimmt ist" kann sich neben gesetzlichen oder vertraglichen Vorgaben auch aus sonstigen Wertungen ergeben. Eine Besserstellung des Kronzeugen kann durchaus damit begründet werden, dass sein Verhalten im öffentlichen Interesse der Aufklärung von Kartellrechtsverstößen liegt.[1003] Was die Höhe des verbleibenden Haftungsanteils angeht, könnte insoweit der Anteil des Kronzeugen am Gesamtschuldnerausgleich im Verhältnis zu der ihm gewährten Reduktion des Bußgeldes beschränkt werden.

Bei der Aufteilung der Schadenshaftung unter Gesamtschuldnern der Höhe nach ist der Rechtsgedanke des § 254 BGB heranzuziehen,[1004] der es ermöglicht, die jeweilige Haftung nach Verursachungsbeiträgen differenzierend zu bestimmen.

---

999 *Krüger*, in: Oberender, Private und öffentliche Kartellrechtsdurchsetzung, S. 79, 86.

1000 Vgl. *Schroll*, Effektivität der Kronzeugenprogramme, S. 170; *Böge*, in: Basedow, Private Enforcement, S. 223.

1001 *Schroll*, Effektivität der Kronzeugenprogramme, S. 170 mit Verweis auf *Jürgens/Seeliger*, EWS 2006, 337, 341.

1002 So bereits der Vorschlag im Weißbuch der EU-Kommission, KOM (2008), 165 endgültig, S. 12.

1003 So auch *Alexander*, Schadensersatz und Abschöpfung, S. 423.

1004 Vgl. BGH, Beschl. v. 9.7.2013 – KZR 15/12, BB 2013, 2639, 2640 m. w. N.; *Koch*, JZ 2013, 390, 394 mit Verweis auf *Kersting*, JZ 2012, 42, 45; *Krüger*, NZKart 2013, 483, 484.

Nach § 254 BGB hängt der Umfang des Schadensersatzanspruchs davon ab, inwieweit der Geschädigte den Schaden selbst verursacht hat. Die Aufdeckung des Kartells könnte als schadensmindernder Faktor auf Seiten des Kronzeugen zu berücksichtigen sein.[1005] Der Beitrag zur Schadensminderung[1006] fließt somit im Rahmen der Verteilung der Haftungsquote in die Bewertung mit ein. Genau genommen findet dadurch keine Privilegierung statt, sondern es wird nur die Haftung konsequent nach Verursachungsanteilen bestimmt. Dann ist es nur folgerichtig, dass die anderen Kartellanten, die keinen Beitrag zur Schadensminderung geleistet haben, mehr Verantwortung zu tragen haben. Die Mitkartellanten haften dann entsprechend ihrer Mitverursachungsbeiträge und nicht darüber hinaus.[1007]

Im Ergebnis findet bei der Feststellung der Verursachungsbeiträge eine Abwägung statt, in der zahlreiche Umstände Berücksichtigung finden können.[1008] Auch wenn die einzelnen Abwägungsfaktoren umstritten sind,[1009] so dürften doch die beherrschende Rolle eines Kartellmitgliedes (im Gegensatz zu der des Mitläufers), die Höhe der Marktanteile und Umsätze sowie der Umfang der Aufklärungsbeiträge im Vordergrund stehen.[1010]

Daraus ergibt sich die Folgefrage, welche Kronzeugen die „Haftungsprivilegierung" in Anspruch nehmen können. Neben dem ersten Kronzeugen, ohne den das Kartell nicht aufgedeckt worden wäre, bestehen noch Kronzeugen der „zweiten Stufe", die dem Bundeskartellamt weitere Informationen zum Nachweis des Kartells zugeleitet haben. Unter den Gesichtspunkten der Erforderlichkeit und Praktikabilität spricht einiges dafür, nur den ersten Kronzeugen, nicht aber die nachfolgenden Kartellmitglieder, die einen Antrag gestellt haben, im Gesamtschuldnerausgleich voll zu „privilegieren".[1011] Dieses Ergebnis kann auf den Rechtsgedanken des § 254 Abs. 2 BGB gestützt werden, der nur denjenigen belohnen will, der den Schaden mindert oder abwendet. Indem der erste Kronzeuge das Kartell öffentlich macht und die Behörden bei der Untersuchung unterstützt, wendet er weiteren Schaden

---

1005 *Kersting*, ZWeR 2008, 252, 267; *ders.* JZ 2012, 42, 45; *Koch*, JZ 2013, 390, 393; kritisch *Bien*, EuZW 2011, 889, 890.

1006 Auch wenn die Kooperation des Kronzeugen bis zur Aufdeckung den durch die wettbewerbswidrige Handlung entstandenen Schaden nicht (rückwirkend) verringert, so verhindert sie doch eine weitere Zunahme des Schadens.

1007 So auch *Dworschak/Maritzen*, WuW 2013, 829, 842.

1008 *Krüger*, in: Oberender, Private und öffentliche Kartellrechtsdurchsetzung, S. 79, 86.

1009 Vgl. zu den einzelnen Abwägungsfaktoren *Krüger*, Kartellregress, S. 63 f.; *Dreher*, FS Möschel, S. 149, 156 f.; *Meeßen*, Schadensersatz, S. 559 f; *Vollrath*, NZKart 2013, 434, 442.

1010 *Krüger*, in: Oberender, Private und öffentliche Kartellrechtsdurchsetzung, S. 79, 86 f; *Legner*, WRP 2014, 1163, 1164.

1011 Für die Privilegierung nur des ersten Kronzeugen, vgl. *Dworschak/Maritzen*, WuW 2013, 829, 842 f; für eine Freistellung entsprechend der jeweils gewährten Bußgeldreduktion, vgl. *Kersting*, JZ 2013, 737, 739; *Koch*, JZ 2013, 390, 393 f.; für richter liches Ermessen, vgl. *Alexander*, Schadensersatz und Abschöpfung, S. 422 f.

ab. Ab dann findet kein wettbewerbswidriges Verhalten mehr statt. Die nachfolgenden Beweismittel von Kronzeugen „der zweiten Stufe" mindern weiteren Schaden nicht, weil in den meisten Fällen das Kartell bereits beendet ist oder kurz vor der Beendigung steht.

Eine haftungsrechtliche Privilegierung des Kronzeugen im Innenverhältnis wird von der herrschenden Meinung im deutschen Recht bereits anerkannt (§ 426 i. V. m. § 254 Abs. 2 BGB analog). Der Kartellregress im Innenverhältnis ist für den Schutz des Whistleblowers im Kartellrecht aber nicht ausreichend, weil unkalkulierbar ist, wann und in welchem Umfang er mit dem Regress der anderen Beteiligten rechnen kann.

c. *Sonderstellung des Kronzeugen im Außenverhältnis*

Das eben erläuterte Modell bezieht sich nur auf die Haftungsverteilung im Innenverhältnis. Der Kronzeuge ist im Außenverhältnis gegenüber den Geschädigten weiterhin vollumfänglich verpflichtet, den Schaden zu kompensieren. Um ein effektives Kronzeugenprogramm zu erhalten, werden gesetzliche Haftungsbefreiungs- bzw. Haftungsbegrenzungstatbestände erwogen.[1012] Dies stößt jedoch auf Kritik.[1013] Abgesehen davon, dass dieser Ansatz *de lege lata* nicht zu konstruieren ist, würde eine Haftungsprivilegierung des Kronzeugen dazu führen, dass letztlich der Geschädigte belastet würde.[1014] Er würde der Möglichkeit beraubt, den ihm entstanden Schaden vollständig zu liquidieren, wenn der Kronzeuge dem Geschädigten entgegenhalten könnte, er hafte nicht in voller Höhe. Bedenken ergeben sich ebenfalls, wenn man die Haftungsprivilegierung damit erreichen wollte, den durch die Freistellung des Kronzeugen entfallenden Teil des Schadensersatzes auf die übrigen Kartellbeteiligten zu verteilen.[1015] Denn diese Variante käme einem Strafschadensersatz gleich. Die übrigen Kartellbeteiligten müssten eine höhere Schadenssumme aufbringen, als ihnen nach ihren Verursachungsbeiträgen tatsächlich zuzurechnen wäre.[1016] Abgesehen davon, hat der BGH zur Problematik des Strafschadensersatzes entschieden,[1017] dass dieser unter anderem nicht mit dem Verbot der ungerechtfertigten Bereicherung vereinbar sei. Zu berücksichtigen ist auch, dass durch den Verweis des Geschädigten auf andere Kartellbeteiligte gegebenenfalls dessen Ansprüche nicht durchsetzbar sind, weil er lediglich Informationen über den Kronzeugen besitzt. Diese wurden schließlich im Rahmen des behördlichen Verfahrens gewonnen.

1012 *Böge*, in: Basedow, Private Enforcement, S. 223 f.; *Glöckner*, WRP 2007, 490, 495.

1013 *Böge*, in: Basedow, Private Enforcement, S. 233; *Schroll*, Effektivität der Kronzeugenprogramme, S. 171; *Alexander*, Schadensersatz und Abschöpfung, S. 423.

1014 So auch *Alexander*, Schadensersatz und Abschöpfung, S. 421 mit dem abschließenden Hinweis, dass man nicht aus den Augen verlieren dürfe, wie es mit der Behandlung der Opfer im Vergleich zur Behandlung der Täter stehe.

1015 *Schroll*, Effektivität der Kronzeugenprogramme, S. 171

1016 *Schroll*, Effektivität der Kronzeugenprogramme, S. 171.

1017 BGH, Urt. v. 4.6.1992 – IX ZR 149/91, NJW 1992, 3096, 3103.

Auf belastendes Material gegen die übrigen Kartellbeteiligten kann er oftmals nicht zurückgreifen.[1018]

Die Überlegungen führen dazu, dass nach jetziger nationaler Rechtslage dem Kronzeugen keine Sonderstellung im Außenverhältnis zukommen kann. Mit diesem Modell wäre der Geschädigte zu sehr belastet.

4. *Schutz durch Compliance und Risikomanagement im Kartellrecht*

Auf die Bedeutung von Compliance-Programmen und deren Implementierung im Unternehmen wurde schon im Rahmen des arbeitsrechtlichen Schutzes (*de lege ferenda*) hingewiesen. Bei der Ausgestaltung der Maßnahmen zur Verhinderung und Aufdeckung von Wettbewerbsverstößen im Unternehmen sind allerdings Besonderheiten zu beachten. Bei der Einrichtung eines Compliance-Systems spielen die sonstigen Themen, wie z. B. die Überwachung der Buchhaltung oder die Annahme von Geschenken, eine geringere Rolle, da Absprachen oft auf mündlichen Verhandlungen beruhen und zwecks Geheimhaltung nicht dokumentiert werden. Das Hauptaugenmerk liegt bei der kartellrechtlichen Compliance auf der Vermeidung von Kartellabsprachen und Wettbewerbsverzerrung. Deshalb sind vorrangig Regelungen vonnöten, die Absprachen zwischen Unternehmen oder übermäßige Nutzung der Marktmacht verhindern.

Es gibt keinen allgemeingültigen Weg, um kartellrechtlichen Risiken zu begegnen. Die Einhaltung und Umsetzung der geltenden Vorschriften können die Unternehmen nach eigenem Ermessen bestimmen. Faktoren, die den Umfang und die Intensität der Umsetzung beeinflussen können, sind die Größe des Unternehmens, sein Geschäftsfeld und das jeweilige Risiko, in einen Verstoß gegen EU-Wettbewerbsrecht verwickelt zu sein. Wichtige Gründe für Unternehmen, die Wettbewerbsregeln einzuhalten, sind neben dem Ruf als ethisch verantwortungsvolles Unternehmen auch die Vermeidung hoher Kosten, die aus der Nichteinhaltung bestimmter Rechtspflichten resultieren können. Darunter ist auch die Zahlung von Schadensersatz zu fassen, den ein Dritter geltend macht, wenn sich ein Unternehmen wettbewerbswidrig verhält. Weiter könnte sich die Medienberichterstattung nicht nur nachteilig auf den Ruf des betreffenden Unternehmens auswirken, sondern auch Unverständnis bei den Kunden und Verbrauchern auslösen. Sie könnten sich betrogen fühlen und dadurch vom Kauf der Produkte oder von der Inanspruchnahme von Dienstleistungen Abstand nehmen.

a. *Standpunkt der Kommission*

Die EU-Kommission veröffentlichte im Jahr 2012 eine Broschüre über wettbewerbsrechtliche Compliance.[1019] Darin wird beschrieben, was Unternehmen tun können,

---

1018 *Schroll*, Effektivität der Kronzeugenprogramme, S. 172.

1019 EU-Kommission, Wettbewerbsrechtliche Compliance – Was Unternehmen tun können, um die EU-Wettbewerbsvorschriften besser einzuhalten, 2012 (Erstfas-

um die EU-Wettbewerbsvorschriften einzuhalten. Die als Leitfaden gedachte Broschüre enthält neben der Aufforderung zur Einhaltung des Wettbewerbsrechts auch eine Darstellung der Folgen, die bei einer Nichtbefolgung auf ein Unternehmen zukommen können. Der Hauptteil ist eine Anleitung für Unternehmen, wie sie Compliance sicherstellen. Zudem geht die EU-Kommission auf die Tätigkeiten der Wettbewerbsbehörden ein. Sie betont, dass die Wettbewerbsbehörden befugt sind, Maßnahmen zur Einhaltung der Wettbewerbsregeln zu ergreifen. Die von der Kommission veröffentlichten Berichte, Mitteilungen, Leitlinien und Gerichtsurteile stehen ihnen dabei unterstützend zur Seite.[1020]

Zunächst sollten die Unternehmen eine Strategie entwickeln, deren oberstes Ziel die Sensibilisierung für potenzielle wettbewerbsrechtliche Probleme und die Vermittlung von Verhinderungsmaßnahmen sein soll.[1021] Dabei sind alle Unternehmensebenen, also die Mitarbeiter wie auch die mittlere oder obere Führungsebene, zu beteiligen. Die EU-Kommission listet für die Ermittlung des Risikos für einen möglichen Wettbewerbsverstoß Faktoren auf, die für die Risikoanalyse von Belang sein können:[1022]

- Risikofaktor: Wirtschaftszweig – Wirtschaftszweige, in denen gehäuft Verstöße aufgetreten sind, bedürfen besonderer Aufmerksamkeit
- Risikofaktor: Häufigkeit und Grad der Zusammenarbeit mit Wettbewerbern – zum Beispiel im Rahmen von Branchentreffen oder innerhalb von Verbänden, aber auch im täglichen Geschäftsgebaren
- Risikofaktor: Merkmale des Marktes – zum Beispiel die Stellung des Unternehmens, seiner Wettbewerber und etwaige Markteintrittsschranken. Ein Unternehmen mit einer marktbeherrschenden Stellung muss andere Vorkehrungen treffen als ein Unternehmen in einem Bereich mit einem hohen Risiko der Kartellbildung.

Neben der Risikoanalyse empfiehlt die EU-Kommission, die Compliance-Strategie innerhalb des ganzen Unternehmens bekannt zu machen.[1023] Diese Bekanntmachung sollte verständlich, in schriftlicher Form und für alle zugänglich veröffentlicht werden, sodass insbesondere auch die möglichen Folgen bei Nichteinhaltung der gesetzten Regeln für die Belegschaft deutlich sichtbar sind. Die Veröffentlichung könnte

sung in englischer Sprache 2011), abrufbar unter: http://bookshop.europa.eu/de/wettbewerbsrechtliche-compliance-pbKD3211985/, (abgerufen am 22.3.2014).

1020 EU-Kommission, Wettbewerbsrechtliche Compliance – Was Unternehmen tun können, um die EU-Wettbewerbsvorschriften besser einzuhalten, 2012, S. 13.

1021 EU-Kommission, Wettbewerbsrechtliche Compliance – Was Unternehmen tun können, um die EU-Wettbewerbsvorschriften besser einzuhalten, 2012, S. 14.

1022 EU-Kommission, Wettbewerbsrechtliche Compliance – Was Unternehmen tun können, um die EU-Wettbewerbsvorschriften besser einzuhalten, 2012, S. 14.

1023 EU-Kommission, Wettbewerbsrechtliche Compliance – Was Unternehmen tun können, um die EU-Wettbewerbsvorschriften besser einzuhalten, 2012, S. 14.

z. B. in Form von Merkblättern erfolgen, die unter anderem die Themenbereiche[1024] horizontale Absprachen (insbesondere Preisabsprachen) und abgestimmte Verhaltensweisen, Verbandsaktivitäten, vertikale Beziehungen, marktbeherrschende Stellung eines Unternehmens, Umfang und Gestaltung von unternehmensinternen Untersuchungen veranschaulichen.

Neben der richtigen Strategie und der Unterrichtung aller Mitarbeiter ist die ausdrückliche Unterstützung der Unternehmensleitung von entscheidender Bedeutung.[1025] Dazu kann auch ein klares Bekenntnis der Unternehmensleitung zur Rechtstreue gezählt werden. Meist begehen Mitarbeiter einen Kartellrechtsverstoß nicht, um sich persönlich zu bereichern. Stattdessen sind sie der Meinung, ausschließlich im Interesse des Unternehmens gehandelt zu haben. Deshalb ist es angebracht, dass die Unternehmensführung in Form eines Compliance Commitments/Mission Statements oder eines vergleichbaren Dokuments deutlich macht, dass Kartellverstöße gerade nicht im Sinne des Unternehmens sind.[1026] Es soll eine erkennbare Compliance-Kultur geschaffen werden, die entsprechend der Größe und dem Risiko für das Unternehmen unterschiedlich ausfallen kann.

Um die Mitarbeiter zur Einhaltung der Compliance-Vorschriften zu motivieren, schlägt die EU-Kommission unter anderem vor, einen Anreiz für Mitarbeiter zu schaffen. Die vorbildliche Einhaltung von Compliance-Pflichten könnte zum Beispiel Bestandteil von Stellenanforderungen sein oder für eine Personalbeurteilung als Kriterium herangezogen werden.[1027]

Neben der rein inhaltlichen Wissensvermittlung an die Mitarbeiter ist es ratsam, zusätzlich Hinweise zu Verhaltensregeln zu geben. Dies ist insbesondere bei denjenigen Mitarbeitern angebracht, die in Kontakt zu Wettbewerbern stehen, etwa im Rahmen von Verbandsarbeit[1028] oder Messen.[1029] Es sollte aber nicht davor Halt gemacht werden, den Mitarbeitern die möglichen Konsequenzen in persönlicher Hinsicht, aber auch für das Unternehmen als ganze Einheit bewusst zu machen und hinreichend auf die Höhe der Geldstrafen im Fall der Zuwiderhandlung hinzuweisen.[1030]

Darüber hinaus betont die EU-Kommission in ihrer Broschüre, wie wichtig die ordnungsgemäße interne Berichterstattung sowie die ständige Beratung,

1024 *Lampert/Matthey*, in: Hauschka, Corporate Compliance, § 26, Rn. 54 f.

1025 EU-Kommission, Wettbewerbsrechtliche Compliance – Was Unternehmen tun können, um die EU-Wettbewerbsvorschriften besser einzuhalten, 2012, S. 15.

1026 *Lampert/Matthey*, in: Hauschka, Corporate Compliance, § 26, Rn. 53.

1027 EU-Kommission, Wettbewerbsrechtliche Compliance – Was Unternehmen tun können, um die EU-Wettbewerbsvorschriften besser einzuhalten, 2012, S. 15.

1028 Vertiefend zur Kartellrechts-Compliance in der Verbandsarbeit siehe *Kapp/Hummel*, CCZ 2013, 240 f.; vgl. zum Spannungsfeld Kartellrecht und Verbandstätigkeit *Brouwer*, CCZ 2009, 161, 162.

1029 *Moosmayer*, Compliance, Abschnitt E, S. 94.

1030 *Lampert/Matthey*, in: Hauschka, Corporate Compliance, § 26, Rn. 53, wo vorgeschlagen wird, die Konsequenzen gleich im Zusammenhang mit der Darstellung des *Compliance Commitments* zu verknüpfen.

Fortbildung[1031] und die Wirksamkeit des Audits und Monitorings sind.[1032] Nur durch Schulungen können die Mitarbeiter auf dem aktuellsten Stand gehalten werden und ihre Erinnerungen von vorherigen Zusammenkünften auffrischen. Neben den klassischen Präsenzschulungen können Schulungen auch über die Methode des E-Learnings stattfinden.[1033]

Überdies ist bei der kartellrechtlichen Compliance an die Einrichtung einer Whistleblower-Hotline zu denken. Der Schwerpunkt sollte auf der Einrichtung einer externen (von Dritten betriebenen) Meldestelle und auf der Möglichkeit der anonymen Hinweiserteilung liegen.

Die Kommission unterstützt alle Bemühungen der Unternehmen, die der Einhaltung der einschlägigen Wettbewerbsvorschriften dienen. Werden die oben genannten Regeln beachtet, sind Wettbewerbsverstöße des Unternehmens unwahrscheinlich. Das sollte das Ziel eines jeden Unternehmens sein: Wettbewerbsverstößen vorbeugen und nicht erst nach einem aufgedeckten Verstoß Compliance-Programme installieren.

Zwar gibt die EU-Kommission nützliche Informationen zur Einrichtung von Compliance-Programmen, aber sie betont auch, dass sie keine Programme genehmigt. Das bedeutet, dass sie sich nicht zu einzelnen Compliance-Programmen äußert oder sie bewilligt. Es sei nicht Aufgabe der Kommission, eine solche Entscheidung zu treffen, zumal die Unternehmen am besten wüssten, was für eine Compliance-Strategie sie benötigen.[1034] Zudem weist die Kommission ausdrücklich darauf hin, dass Compliance-Programme von Unternehmen bei einem aufgedeckten Verstoß eine Minderung der etwaigen Geldbußen nicht begründen können. Bei der Höhe der Geldbuße wird, je nachdem in welcher Form der Kooperationsbeitrag durch das Unternehmen geleistet wird, die spezifische Situation (Erlangung der Kronzeugenstellung) des betreffenden Unternehmens berücksichtigt.[1035] Nutzt das Unternehmen das Kronzeugenprogramm, wird ihm die Geldbuße erlassen oder ermäßigt. Die Bonusregelung bemisst die Geldbuße nach dem Grad der Kooperation. Dass es ein Compliance-Programm in dem betreffenden Unternehmen gibt, reicht weder aus, um Verstößen gegen Wettbewerbsregeln zu begegnen, noch wird dies

1031 So auch *Moosmayer*, Compliance, Abschnitt E, S. 93 f.

1032 EU-Kommission, Wettbewerbsrechtliche Compliance – Was Unternehmen tun können, um die EU-Wettbewerbsvorschriften besser einzuhalten, 2012, S. 16.

1033 *Lampert/Matthey*, in: Hauschka, Corporate Compliance, § 26, Rn. 76; *E-Learning* wird als Form von Lernen verstanden, bei der elektronische oder digitale Medien für die Präsentation und Distribution von Lernmaterialien und/oder zur Unterstützung zwischenmenschlicher Kommunikation zum Einsatz kommen.

1034 EU-Kommission, Wettbewerbsrechtliche Compliance – Was Unternehmen tun können, um die EU-Wettbewerbsvorschriften besser einzuhalten, 2012, S. 18.

1035 EU-Kommission, Wettbewerbsrechtliche Compliance – Was Unternehmen tun können, um die EU-Wettbewerbsvorschriften besser einzuhalten, 2012, S. 19.

als mildernder Umstand betrachtet.[1036] Dennoch raten die Wettbewerbsbehörden und die EU-Kommission, ein solches Programm einzuführen, um rechtswidrigem Verhalten vorzubeugen bzw. eine Wiederholung zu vermeiden. Zudem verhindert das Compliance-Programm mittelbar Schadensersatzklagen durch Dritte wegen Verstoß gegen das Kartellrecht.

b. Exkurs: Berücksichtigung eines Compliance-Programms bei der Bußgeldbemessung

Nach der früheren Praxis der EU-Kommission wurden eingerichtete Compliance-Programme in die Bußgeldbemessung einbezogen. Auch das Europäische Gericht hat sich bei seiner Entscheidungsfindung davon leiten lassen.[1037] Seitdem jedoch die EU-Kommission zu der Auffassung gelangt ist, Compliance-Programme seien außer Acht zu lassen, hat auch die Rechtsprechung diese Praxis abgesegnet.[1038] Zur Begründung führte das Gericht in der Sache *Degussa*[1039] an, die Tatsache, dass in früheren Verfahren Compliance-Programme mildernd berücksichtigt wurden, bedeute nicht, dass dies auch in anderen Fällen so beibehalten werden müsse. Die Installation eines Compliance-Programms ändere nichts daran, dass „die festgestellte Zuwiderhandlung tatsächlich begangen" worden sei.[1040] Auch der EuGH ist der Linie der EU-Kommission und des EuG mit der gleichen Begründung gefolgt.[1041]

Diese Praxis der Gerichte und der Kommission schafft jedoch keinen Anreiz für Unternehmen, ein Compliance-Programm einzuführen. Es bedarf eines Anreizsystems für die Unternehmen, um die Implementierung und Durchsetzung eines Compliance-Programms attraktiv zu machen mit der Folge einer optimalen Kartellprävention. Wenn Unternehmen nicht haften, weil sie nachweisen, dass sie durch Ausgestaltung und nachhaltige Durchsetzung ihrer Compliance-Organisation bestmögliche präventive Anstrengungen unternommen haben, gäbe ihnen dies Anreize, noch mehr in Compliance zu investieren.[1042]

1036 Das Programm muss vom Unternehmen und den Mitarbeitern auch umgesetzt werden, EuGH, Urt. v. 28.6.2005 – verb. Rs. C-189/02 P, C-202/02, C-205/02 P bis C-208/02 P und C-213/02, Slg. 2005, I-5425, (Rn. 373) (Dansk Rørindustri u. a.).

1037 Vgl. EuG, Urt. v. 14.7.1994 – Rs. T-77/92, Slg. 1994, II-549 (Rn. 93), EuZW 1994, 666 (Parker Pen).

1038 Vgl. EuG, Urt. v. 8.7.2008 – Rs. T-53/03, Slg. 2008, II-1333 (Rn. 423) (BPB); EuG, Urt. v. 26.4.2007 – Rs. T-109/02 u. a., Slg. 2007, II-947 (Rn. 653) (Bolloré); EuG, Urt. v. 20.3.2002 – Rs. T-16/99, Slg. 2002, II-1633 (Rn. 326) (LögstörRör).

1039 EuG, Urt. v. 5.4.2006 – Rs. T-279/02, Slg. 2006, II-897, EuZW 2006, 384 (Degussa).

1040 Vgl. EuG, Urt. v. 5.4.2006 – Rs. T-279/02, Slg. 2006, II-897 (Leitsatz), EuZW 2006, 384, 384 (Degussa).

1041 EuGH, Urt. v. 28.6.2005 – verb. Rs. C-189/02 P, C-202/02, C-205/02 P bis C-208/02 P und C-213/02 (Rn. 371 f.) (Dansk Rørindustri u. a.).

1042 *Gehring/Kasten/Mäger*, CCZ 2013, 1, 8.

aa. Compliance-Anreizsystem

Mögliche Gestaltungsansätze für ein Compliance-Anreizsystem können sich durch einen Blick in das ausländische Kartellrecht ergeben.

So gewährt beispielsweise das *UK Office of Fair Trade* (OFT) in Großbritannien in Einzelfällen eine Bußgeldreduktion von bis zu 10 % bei Unternehmen, die adäquate Schritte für ein sicheres Compliance-System unternommen haben.[1043] Im vergangenen Jahrzehnt hat das OFT in einer Vielzahl von Fällen die Angemessenheit von Compliance-Maßnahmen geprüft und auf dieser Grundlage niedrigere Sanktionen auferlegt.[1044]

Die französische *Autorité de la Concurrence* hat am 10. Februar 2012 Leitlinien zu kartellrechtlichen Compliance-Programmen erlassen, nach denen bei Einrichtung eines Compliance-Programms das Bußgeld um ebenfalls bis zu 10 % reduziert werden kann.[1045] Grundlage dieser Mitteilung ist die gesetzliche Regelung des Art. L. 464–2 III Code de Commerce, wonach eine künftige Verhaltensänderung des Unternehmens sanktionsmindernd berücksichtigt werden kann.

Ein bestehendes Compliance-Programm wirkt sich nach Auffassung der französischen Wettbewerbsbehörde dann (mittelbar) auf die Höhe des Bußgeldes aus, wenn der Kartellverstoß intern durch das Compliance-Programm aufgedeckt wird.[1046] Tatsächlich führt die Bußgeldminderung und nicht das Compliance-Programm „an sich" in diesen Fällen zur Entscheidung, den Wettbewerbsverstoß selbst zu melden und die Kronzeugenregelung in Anspruch zu nehmen.[1047] Die Wettbewerbsbehörde weist aber darauf hin, dass durch die Einführung oder Erweiterung eines Compliance-Programms unmittelbar eine Bußgeldreduzierung für zukünftiges positives Verhalten gem. Art. L. 464–2 III Code de Commerce erreicht werden kann. Darauf bezieht sich die oben genannte Minderung. Voraussetzung für die Reduktion ist nach den Leitlinien, anders als in der britischen Fassung, dass das betroffene Unternehmen sich zur Einrichtung bzw. Verbesserung eines Compliance-Programms im Rahmen einer Vereinbarung mit der Behörde verpflichtet.[1048] Es darf also noch kein Compliance-Programm bestanden haben. Es genügt, dass eine Einführung erst

---

1043 Dies hat die Behörde mehrfach ausdrücklich festgehalten, vgl. OFT's guidance as to the appropriate amount of a penalty, OFT 423, September 2012, abrufbar unter: http://www.oft.gov.uk/shared_oft/business_leaflets/ca98_guidelines/oft423.pdf, Seite 12, Rn. 2.15, (abgerufen am 10.5.2014).

1044 Nachweise zu zahlreichen entsprechenden OFT-Entscheidungen bei *Wils*, Journal of Antitrust Enforcement, 2013, Vol. 1, No. 1, pp. 52, 55, Fn. 11.

1045 Framework-Document of 10 February 2012 on Antitrust Compliance Programmes (Englische Fassung), S. 9, Fn. 31, abrufbar unter: http://www.autoritedelaconcurrence.fr/doc/framework_document_compliance_10february2012.pdf, (abgerufen am 10.5.2014); ausführlich dazu siehe *Aufdermauer*, CCZ 2012, 194.

1046 *Aufdermauer*, CCZ 2012, 194, 196.

1047 *Aufdermauer*, CCZ 2012, 194, 196.

1048 Framework-Document of 10 February 2012 on Antitrust Compliance Programmes (Englische Fassung), S. 1, 9.

im Nachhinein stattfindet. In Abschnitt 3 der Leitlinien werden die Anforderungen für ein effektives Compliance-Programm nach den Vorstellungen der Wettbewerbsbehörde aufgezählt. Hat ein Unternehmen ein Compliance-Programm nach den in den Leitlinien festgelegten Standards eingerichtet oder ein bestehendes Programm gemäß den Standards angepasst, ist die Wettbewerbsbehörde zu einer Reduzierung der Geldbuße um bis zu 10 % bereit. Zudem findet eine Kontrolle statt, ob das Compliance-Programm umgesetzt worden ist.

Im deutschen Kartellrecht besteht grundsätzlich ebenfalls die Möglichkeit der Enthaftung bei Zuwiderhandlungen durch die §§ 30, 130 OWiG. Danach werden bei ausreichenden Aufsichtsmaßnahmen zur Verhinderung von Kartellrechtverstößen dennoch begangene Zuwiderhandlungen nicht zugerechnet (§ 130 Abs. 1 OWiG). Dies wird im Grundsatz auch vom Bundeskartellamt anerkannt,[1049] hat aber in der veröffentlichten Entscheidungspraxis noch keinen Niederschlag gefunden.[1050] Das Bundeskartellamt begründet seine Ansicht, Compliance-Systeme nicht bußgeldmindernd zu berücksichtigen, damit, dass die Pflicht zur Aufsicht eine gesetzliche Obliegenheit nach § 130 OWiG darstelle, bei deren Erfüllung das Handeln nicht als mildernder Umstand gewürdigt werden könne.[1051] Die Pflicht des Unternehmens, seinen Aufsichtsaufgaben gerecht zu werden, dürfe keine Besonderheit darstellen, die durch einen „Bonus" gewürdigt werde. Sie müsse vielmehr als Selbstverständlichkeit angesehen werden.

Das deutsche Kartellrecht enthält aber keine ersichtlichen Hindernisse für eine Berücksichtigung von Compliance-Anstrengungen oder bereits implementierten Compliance-Programmen bei einem Verstoß.[1052]

Auch das europäische Kartellrecht verbietet eine Minderung bei ernsthaften Compliance-Bemühungen nicht. Ohne dass dafür normative Änderungen nötig wären, könnten im Rahmen des Art. 23 Abs. 3 der Verordnung Nr. 1/2003 (EG) bei der Schwere des Verstoßes und der Höhe der Geldbuße die Anstrengungen des Unternehmens zu Compliance-Programmen positiv berücksichtigt werden.[1053] Auch die Bußgeldleitlinien könnten so umgestaltet werden, dass die Bemühungen für ein effektives und nachhaltiges Compliance-Programm als mildernder Umstand anzusehen sind.[1054] Gerade die frühere Praxis der EU-Kommission hat gezeigt, dass Compliance-Programme durchaus mildernd berücksichtigt werden können. Somit wäre eine Neuausrichtung der Entscheidungspraxis für die Zukunft unter Rückkehr zu der seinerzeit vertretenen Auffassung ohne weiteres möglich.[1055]

1049 *Pampel*, BB 2007, 1636, 1638.

1050 *Kasten*, in: Mäger, Europäisches Kartellrecht, 2. Kap., Rn. 138.

1051 *Pampel*, BB 2007, 1636, 1638.

1052 So *Gehring/Kasten/Mäger*, CCZ 2013, 1, 10.

1053 Vgl. auch *Kasten*, in: Mäger, Europäisches Kartellrecht, 2. Kap., Rn. 141.

1054 Leitlinien für das Verfahren zur Festsetzung von Geldbußen gemäß Art. 23 Abs. 2 lit. a) der Verordnung (EG) Nr. 1/2003, ABl. 2006 C 210/02.

1055 *Gehring/Kasten/Mäger*, CCZ 2013, 1, 11 m. w. N.

Wie dargestellt, ist die Berücksichtigung eines vorhandenen Compliance-Programms also nicht unmöglich und das europäische Kartellrecht würde in der jetzigen Fassung bereits normative Ansatzpunkte für eine moderne, am Präventionsziel orientierte Sanktionsausrichtung bieten. Die Behörden können ein Compliance-Anreizsystem auch praktisch umsetzen. Neben repressivem Handeln könnten sie durch Maßnahmen zur Vermeidung von Kartellverstößen auch vorbeugend tätig werden. Die Förderung der präventiven Maßnahmen, etwa durch Belohnung, würde dem Sinn und Zweck des Kartellbußgelds entsprechen, das neben der Bestrafung auch spezialpräventive Ziele (vorbeugende Verhütung) verfolgt.[1056] Die Unternehmen werden so motiviert, Compliance-Programme zu implementieren und sie effektiv und effizient durchzusetzen.

Wie das französische Modell zeigt, kann die Berücksichtigung jedoch nur greifen, wenn das Compliance-Programm nach erfolgter Zuwiderhandlung eingeführt wurde.

bb. Best Practice Compliance-Programm

Wie bereits angedeutet, gehören die Aufsichtspflichten zu den Obliegenheiten der Unternehmen nach § 130 OWiG. § 130 OWiG verpflichtet den Unternehmensinhaber zu Aufsichtsmaßnahmen, die erforderlich sind, um Zuwiderhandlungen gegen betriebliche Pflichten zu unterbinden. Die Einführung eines Compliance-Programms zur Erfüllung der Aufsichtspflicht kann dann nicht als besonderer Umstand angesehen werden. Anders sieht es bei *Best Practice* Compliance-Programmen aus, die über die „normalen" Compliance-Programme und die vom Gesetzgeber angedachten Mindeststandards hinausgehen.[1057] Die Implementierung eines überobligatorischen Compliance-Programms ist mehr als die Erfüllung einer gesetzlichen Obliegenheit und verdient eine entsprechende Würdigung seitens der Kartellbehörde.

In der Praxis wird die Berücksichtigung von den Behörden weder explizit abgelehnt, noch halten gesetzliche Regelungen diese Möglichkeit offen. Denn das Verhalten des Betroffenen nach der Tat stellt nach Ansicht der Rechtsprechung einen Aspekt dar, der sich „im Rahmen der von § 17 Abs. 3 OWiG vorgesehenen Zumessungserwägungen hält".[1058] Aus diesem Nachtatverhalten lassen sich Rückschlüsse auf den Unrechtsgehalt der Tat und die innere Einstellung des Täters ziehen,[1059] die somit für die Beurteilung des „Vorwurfs" i. S. d. § 17 Abs. 3 OWiG (mit)bestimmend

1056 Vgl. auch *Krebs/Eufinger/Jung*, CCZ 2011, 213, 215.

1057 Vgl. *Krebs/Eufinger/Jung*, CCZ 2011, 213, 215.

1058 Vgl. OLG Düsseldorf, Urt. v. 27.3.2006 – VI-Kart 3/05 (OWi), WuW/E DE-R 1733, 1738, 1739.

1059 *Rau*, Compliance und Unternehmensverantwortlichkeit, S. 147; vgl. *Dannecker/Biermann*, in: Immenga/Mestmäcker, Wettbewerbsrecht: GWB, § 81 GWB, Rn. 381.

sind.[1060] Ferner wird in den Bußgeldleitlinien des Bundeskartellamtes das Nachtatverhalten des Unternehmens als mildernder Umstand angeführt.[1061]

Es ist sowohl in zeitlicher (Compliance-Programm vor oder nach der Zuwiderhandlung) als auch in qualitativer Hinsicht („normales“ oder *Best Practice* Compliance-Programm[1062]) zu unterscheiden, wann eine Berücksichtigung durch die Behörden stattfinden kann.

Eine bußgeldreduzierende Berücksichtigung von kartellrechtlichen Compliance-Programmen ist also nicht ausgeschlossen. Eine gesetzliche Grundlage oder eine Art Leitlinie, die präventive Maßnahmen der Unternehmen belohnend unterstützt, wäre wünschenswert.[1063] Die deutsche Zuordnung zum Ordnungswidrigkeitenrecht und dessen Anbindung an das Strafrecht gebieten die Berücksichtigung jedes *Best Practice* Compliance-Programms, d. h. eines jeden Compliance-Programms, das über den allgemeinen Pflichtstandard hinausgeht. Aus § 130 OWiG kann nicht die generelle Pflicht zur Schaffung eines *Best Practice* Compliance-Programmes abgeleitet werden. Gerade deshalb ist jedes Programm, das von seiner Konzeption und praktischen Handhabung in der Lage ist, einen Großteil der Kartellrechtsverstöße zu verhindern, bußgeldmindernd zu berücksichtigen.[1064]

c. *Ausblick*

Die aktuelle Auseinandersetzung bietet dem Gesetzgeber und den Kartellbehörden die Chance, in die Diskussion über verpflichtende Compliance-Programme einzusteigen und eine angemessene Lösung zu erarbeiten. Compliance zielt nicht nur auf die Einhaltung rechtlicher Regelungen, sondern generell auf die Vermeidung der mit Rechtsverletzungen verbundenen wirtschaftlichen Risiken. Die Unternehmen müssen einsehen, dass ein solches Programm nur Vorteile schafft. Denn dadurch entgeht der kartellrechtliche Whistleblower auch mittelbar den haftungsrechtlichen Konsequenzen. Deswegen ist im Zusammenhang mit der Forderung einer Compliance-Pflicht nicht nur an eine Bußgeldreduzierung zu denken, sondern ebenfalls an die Verhinderung von haftungsrechtlichen Ansprüchen durch Dritte.

1060 *Krebs/Eufinger/Jung*, CCZ 2011, 213, 215.

1061 Mitteilung des Bundeskartellamts, Bußgeldleitlinien, Rn. 17.

1062 Inwieweit überobligatorische Anstrengungen (Best Practice Compliance-Programme) gegeben sind, welche über die „normalen“ Maßnahmen hinausgehen – und damit im Rahmen der Bußgeldbemessung berücksichtigungsfähig sind –, hängt maßgeblich von der Kartellanfälligkeit der jeweiligen Branche sowie der Historie und Größe des Unternehmens ab. Insbesondere in Bezug auf Großkonzerne müsste der Standard in vielen Fällen deutlich nach oben angehoben werden. Die Bandbreite der zusätzlich zu ergreifenden Maßnahmen reicht dabei bis zu vollständig ausgebauten Compliance-Programmen, *Krebs/Eufinger/Jung*, CCZ 2011, 213, 217.

1063 So schon *Van Vormizeele*, CCZ 2009, 41, 48, der zu dem Entschluss kommt, dass es einer gesetzgeberischen Initiative bedürfe, um *de lege ferenda* die Präventivbemühungen von Unternehmen bei der Kartellbekämpfung stärker zu berücksichtigen.

1064 *Krebs/Eufinger/Jung*, CCZ 2011, 213, 217.

III. Schutz durch Umsetzung der Europäischen Richtlinie zu Schadensersatzklagen im Kartellrecht

Zwar ist die Richtlinie 2014/104/EU[1065] über bestimmte Vorschriften für Schadensersatzklagen nach nationalem Recht wegen Zuwiderhandlungen gegen wettbewerbsrechtliche Bestimmungen der Mitgliedstaaten und der Europäischen Union („Schadensersatzrichtlinie“) nun in Kraft getreten, es bedarf jedoch einer Umsetzung ins nationale Recht. Dies ist erforderlich, nicht nur um einer Vertragspflicht – gestützt auf dem Vertrag über die Arbeitsweise der EU – nachzukommen, sondern auch um auf horizontaler Ebene eine unmittelbare Wirkung der Richtlinienbestimmungen hervorzurufen. Entsprechend den Ausführungen zur Richtlinie im vorherigen Teil der Arbeit (Teil 4, C, II, 5) hat der nationale Gesetzgeber Anpassungen und Änderungen der nationalen Vorschriften vorzunehmen. Mit der Umsetzung wird der Richtlinieninhalt Teil der nationalen Rechtsordnung und gilt somit für alle, die vom Umsetzungsakt (z. B. ein Gesetz) betroffen sind. In welcher Form die Richtlinie umgesetzt wird, bleibt dem Mitgliedstaat überlassen, solange die nationalen Vorschriften den unionalen Bestimmungen entsprechen bzw. deren Gehalt in richtlinienkonformer Auslegung ermittelt werden kann.

1065 Richtlinie über bestimmte Vorschriften für Schadensersatzklagen nach nationalem Recht wegen Zuwiderhandlungen gegen wettbewerbsrechtliche Bestimmungen der Mitgliedstaaten und der Europäischen Union („Schadensersatzrichtlinie“) 2014/104/EU, ABL. L 349 vom 5.12.2014, S. 1–19.

Teil 6: Bewertung und Ausblick

A. Zusammenfassung und Bewertung

Gegenstand der Untersuchung war die Darstellung des Whistleblowerschutzes *de lege lata* und *de lege ferenda*. Dabei wurde der arbeitsrechtliche, strafrechtliche/ordnungswidrigkeitenrechtliche und haftungsrechtliche Schutz geprüft. Anlass dafür war die zunehmende Kritik in Deutschland an einem unzureichenden Schutz für Whistleblower, der erfolglose Versuch von (ehemaligen) Bundestagsparteien Gesetzesentwürfe auf den Weg zu bringen und nicht zuletzt der aufsehenerregende Fall des amerikanischen Whistleblowers *Edward Snowden*, der weltweit Schlagzeilen gemacht hat.

Es hat sich gezeigt, dass die Motive eines Whistleblowers vielfältig sind und es keinen „typischen" Whistleblower gibt. Dabei wurde herausgearbeitet, dass in einem demokratischen Rechtsstaat das Motiv keine Rolle spielen darf. Ist ein Rechtsverstoß im Unternehmen begangen worden, muss dem Whistleblower unabhängig von der Motivlage das Recht gegeben werden, den Verstoß zu melden. Selbst eine zunächst an der Tat beteiligte Person kann im Nachhinein als Whistleblower agieren. Dass diese Person nicht aus rein altruistischen Motiven handelt, ist irrelevant. Besonderes Augenmerk verdient dabei der Kartellbeteiligte – als Unternehmen -, der die Bonusregelung des Bundeskartellamtes in Anspruch nimmt, um einer Kartellbuße zu entgehen.

Im Ergebnis ist eine gesetzliche Kodifikation des Whistleblowerschutzes im Arbeitsrecht und im Haftungsrecht nötig. Die geltende Gesetzeslage in Deutschland ist nicht ausreichend.

Im Strafrecht kann das Handeln des unbeteiligten Whistleblowers gerechtfertigt und somit straffrei sein. Trotzdem ist ein Hinweis auf die straf- sowie ordnungswidrigkeitenrechtliche Rechtslage in einem Schutzgesetz angebracht. War der Whistleblower an der Tat beteiligt, muss er strafrechtlich bzw. ordnungswidrigkeitenrechtlich verfolgt werden – jedoch sollte unbedingt erwogen werden, ihn aufgrund des nachträglichen Wohlverhaltens bei der Strafzumessung zu begünstigen. Für den kartellrechtlichen Whistleblower gilt ohnehin die Bonusregelung.

Empfohlen wird ein eigenes Schutzgesetz, das alle drei angesprochenen Schutzbereiche beinhaltet, und sowohl dem Whistleblower als auch dem Arbeitgeber deutlich macht, welche Handlungen rechtmäßig sind. Den Arbeitnehmern, die auf Missstände hinweisen wollen, muss die Möglichkeit gegeben werden, die Konsequenzen ihres Handelns abzuschätzen. Als Vorlage bietet sich eine Kombination der in der Arbeit dargestellten Gesetzesentwürfe an, da alle gute Ansatzpunkte enthalten.

Besonders hervorzuheben sind der arbeitsrechtliche und haftungsrechtliche Schutz. Arbeitgeber sind derzeit noch nicht im Bilde, wie sie mit dem Phänomen des Whistleblowings umzugehen haben. Whistleblowing darf kein Kündigungsgrund sein. Erst recht, wenn gewisse Anzeige- und/oder Beschwerderechte bestehen – aber auch, wenn der Verrat von Betriebsinterna aus anderen Gründen gerechtfertigt ist, ist eine Nebenpflichtverletzung auszuschließen. Damit geht einher, dass der Arbeitgeber keinen Schadensersatz verlangen darf. Der zivilrechtliche Erlaubnistatbestand der Notwehr erlaubt es dem Whistleblower, gegen die Handlungen des Arbeitgebers vorzugehen. Solange berechtigte Zweifel an der Rechtmäßigkeit betrieblicher Vorgänge bestehen, muss der Whistleblower die Möglichkeit haben, seine Bedenken kundzutun.

Eine Verankerung der in der *Heinisch*-Entscheidung aufgestellten Grundsätze und weitergehender Regelungen in ein Schutzgesetz ist daher notwendig – nicht nur, weil das Richterrecht für Arbeitnehmer nicht so leicht zugänglich ist wie ein Gesetz. Eine allgemeingültige Kodifizierung ist auch geboten, weil die Einzelfallentscheidungen der Gerichte für Whistleblower keine rechtssichere Prognose zulassen.

Haftungsrechtlicher Schutz muss dem Whistleblower zudem vor Schadensersatzklagen geschädigter Dritter zukommen. Schadensersatzklagen gegen Kartellmitglieder befinden sich in der deutschen Rechtsprechung noch in den „Kinderschuhen", sodass in den nächsten Jahren vermehrt mit weiteren Lösungsansätzen zur Behandlung des *private enforcement* und der Privilegierung des Kronzeugen gerechnet werden kann. Ob sich das Konzept der Gesamtschuld, wie in der Richtlinie 2014/104/EU festgelegt, bewährt oder negative Auswirkungen auf die private sowie öffentliche Kartellrechtsdurchsetzung hat, wird sich zeigen. Betrachtet man die Richtlinie in seiner Gesamtheit, wird deutlich, dass durch verschiedene Instrumente versucht wird, einen angemessenen Ausgleich zwischen privater und öffentlicher Kartellrechtsdurchsetzung zu finden. Das Zusammenspiel der Regelungen zur Offenlegung von Beweismitteln einerseits und Verjährungsfristen andererseits gewährt dem privaten Schadensersatzkläger genügend Schutz.

Ob man für den haftungsrechtlichen Schutz des ehemaligen Kartellbeteiligten eine passende Vorschrift im BGB (Haftungsrecht §§ 830, 840 BGB) verortet oder in einem Whistleblower- bzw. Hinweisgeberschutzgesetz, sei dem Gesetzgeber überlassen. Ratsam ist die Einfügung einer passenden Regelung an beiden Stellen.

Auch an eine angemessene Vorschrift zur tätigen Reue ist zu denken, die den Whistleblower, der sich an einer sonstigen Tat im Unternehmen beteiligt hat, entlastet. Eine Bestimmung, die das nachträgliche Wohlverhalten und die Bemühungen zur Wiedergutmachung des Whistleblowers berücksichtigt, könnte im Rahmen der Schadensermittlung und -zumessung im Bereich der §§ 823 f. BGB implementiert werden und/oder im BGB und einem Whistleblower- bzw. Hinweisgeberschutzgesetz.

Ziel dieser Arbeit ist es, die Notwendigkeit eines Schutzgesetzes aufzuzeigen. Verschiedene Lösungsansätze in Literatur und Rechtsprechung können als Hilfestellung dienen, um ein Gesetz auf den Weg zu bringen. Dies würde nicht nur zur Entlastung der Gerichte führen, sondern auch zu einem friedvollen und respektvollen Umgang zwischen den Arbeitsvertragsparteien.

B. Thesen

1) Gegenstand des Whistleblowings sind Gesetzesverstöße, die die Verletzung von Rechtsgütern eines jeden Einzelnen und der Allgemeinheit zur Folge haben.
2) Um dem Whistleblower hinreichend Schutz zu gewähren, ist eine gesetzliche Regelung auf den Weg zu bringen, die nicht nach einzelnen Beschäftigungsformen differenziert und den persönlichen Anwendungsbereich klar definiert.
3) Das Schutzgesetz muss inhaltlich drei Bereiche beinhalten: den arbeitsrechtlichen, strafrechtlichen/ordnungswidrigkeitenrechtlichen und haftungsrechtlichen Schutz.
4) Im Arbeitsrecht muss gewährleistet werden, dass der Whistleblower vor Kündigungen und anderen Benachteiligungen geschützt ist. Whistleblowing bei berechtigtem Handeln darf keinen Kündigungsgrund darstellen. Auch eine Benachteiligung durch den Arbeitgeber und andere Arbeitnehmer muss unterbunden werden. Wird eine solche Benachteiligung festgestellt, muss der Arbeitgeber mit Schadensersatz rechnen. Die Darlegungs- und Beweislast trifft den Arbeitgeber. Er muss zeigen, dass keine Benachteiligung stattgefunden hat – insbesondere nicht aufgrund des Whistleblowings.
   Beim zunächst an der Tat beteiligten Whistleblower darf die Rechtslage nicht anders ausgestaltet sein. Auch ihm darf nicht wegen des Whistleblowings gekündigt werden. Ist es dem Arbeitgeber aufgrund der begangenen (Straf-)Tat nachweislich unzumutbar, den Whistleblower weiterzubeschäftigen, muss eine Kündigung nach den allgemeinen Vorschriften möglich sein, wobei insbesondere das Ultima-Ratio-Prinzip zu beachten und eine Abwägung vorzunehmen ist.
5) Der sachliche Anwendungsbereich sollte ein Regel-Ausnahme-Prinzip beinhalten. Aufgrund der vertraglichen Interessenwahrungs- und Schadensabwendungspflicht ist der Arbeitnehmer grundsätzlich gehalten, sich zunächst an den Arbeitgeber zu wenden. Das bedeutet, dass dem internen Whistleblowing der Vorrang gebühren muss.
6) Die Regel, die ausnahmsweise ein externes Whistleblowing erlaubt, muss inhaltlich konkret sein. Dies ist die kritischste Erscheinungsform des Whistleblowings, weil sie mit weitreichenden Folgen für ein Unternehmen verbunden ist. Die gesetzliche Regelung muss daher statuieren, wann nicht mehr mit einer ausreichenden und angemessenen Abhilfe des Arbeitgebers zu rechnen ist (zeitlich und inhaltlich) sowie welche Hinweisgegenstände vom externen Whistleblowing geschützt sein sollen. Zudem ist der „Öffentlichkeits"-begriff differenziert zu verwenden: zum einen soll die Möglichkeit bestehen, sich an externe zuständige Behörden zu wenden, zum anderen soll darauf eingegangen werden, wann ein Hinweis an Dritte, wie beispielsweise die Presse, zulässig ist. Wann ein sofortiger Hinweis an die Öffentlichkeit zulässig ist, sollte deshalb explizit geregelt werden – dabei sind jene Rechtsgüter aufzuzählen, die externes Whistleblowing zulassen. Darunter fallen: Leben, Körper, Gesundheit, Persönlichkeit und Freiheit einer Person, Stabilität des Finanzsystems oder

der Umweltschutz. Der Ausnahmecharakter dieser Regelung ist genauso klarzustellen wie die Wichtigkeit zu handeln, wenn eine unmittelbare, ernsthafte Gefahr für die oben genannten Rechtsgüter droht. Steht die Gefahr unmittelbar bevor und ist eine Abhilfe durch die Behörden nicht mehr ausreichend, soll der Whistleblower berechtigt sein, den Hinweis direkt an Dritte offenzulegen.

7) Das Motiv des Whistleblowers ist für den Schutz seiner durch das Schutzgesetz gesicherten Rechte unbeachtlich, solange damit der Einklang der Rechtsordnung wiederhergestellt wird.

8) Zudem ist der Whistleblower vor Schadensersatzklagen des Arbeitgebers zu schützen. Solchen Schadensersatzansprüchen darf der Whistleblower nicht ausgesetzt werden, da es mit den Grundgeboten des Rechtsstaates unvereinbar ist, wenn derjenige, der rechtmäßig eine Strafanzeige erstattet hat, Nachteile erleidet.

9) Im Strafrecht muss betont werden, welche Konsequenzen dem Whistleblower drohen. Eine separate Vorschrift im Schutzgesetz soll verdeutlichen, dass Whistleblowing durchaus den Tatbestand des § 17 UWG (und andere Geheimnistatbestände) erfüllen kann. Anschließend muss darauf hingewiesen werden, dass eine Rechtfertigung möglich ist und der Whistleblower Straffreiheit erlangen kann. Bedeutende Rechtsgüter, die bei der Abwägung den Geheimnisverrat rechtfertigen, sollten beispielhaft benannt werden. Dies trägt zur Rechtssicherheit bei.

   Was den Whistleblower betrifft, der sich zunächst an der Tat beteiligt hat, kann hinsichtlich der Strafbarkeit keine Änderung der Gesetzeslage stattfinden. Jedoch sollte in der Strafzumessung das nachträgliche Wohlverhalten einbezogen und wohlwollend berücksichtigt werden.

10) Dem Whistleblower muss haftungsrechtlicher Schutz vor Ansprüchen Dritter zukommen. Sowohl dem am Kartell beteiligten als auch dem an einer sonstigen Tat beteiligten Whistleblower muss geholfen werden, zivilrechtlich nicht übermäßig belastet zu werden.

11) Ist der Whistleblower an einer anderen Tat als an einer Kartellrechtswidrigkeit beteiligt, sollte darüber nachgedacht werden, die tätige Reue ins Zivilrecht aufzunehmen oder zumindest eine Vorschrift, die ein nachträgliches Verhalten berücksichtigt.

12) Die Wechselwirkung zwischen der öffentlich-rechtlichen und der privaten Vollzugsebene im Kartellrecht muss stärker ins Blickfeld rücken. Entweder muss bei der Verhängung des Bußgeldes von den Kartellämtern berücksichtigt werden, dass auch Schadensersatzklagen drohen, oder die Schadensersatzpflicht muss in irgendeiner Form begrenzt werden. Der Schutz des Whistleblowers kann entweder durch eine Begrenzung der vom Kronzeugen gegenüber der Behörde gemachten vertraulichen Angaben im Rahmen der zivilrechtlichen Beweiserhebung stattfinden oder im Rahmen der zivilrechtlichen Gesamtschuldnerhaftung sind die Haftungsanteile differenziert zu verteilen.

13) Die herrschende Meinung in Deutschland bejaht bereits nach geltendem Recht ein Haftungsprivileg des Kronzeugen beim Kartellregress im Innen-

verhältnis nach § 426 BGB i. V. m. § 254 BGB analog. Danach ist für den Gesamtschuldnerausgleich von Kartellmitgliedern entscheidend, in welchem Umfang diese an der Begründung oder der Höhe des auszugleichenden Schadens mitgewirkt haben. Die konkrete Ausgestaltung des Privilegs ist bislang noch unklar. Eine Haftungsbeschränkung im Außenverhältnis kann aus dem geltenden Recht nicht hergeleitet werden.

14) Das Haftungsmodell nach Artikel 11 der EU-Richtlinie 2014/104/EU enthält in Absatz 4 eine Sonderbestimmung, die den Kronzeugen von der gesamtschuldnerischen Haftung im Außenverhältnis zum Teil ausnimmt. Danach soll der Kronzeuge anderen Geschädigten *außer* seinen unmittelbaren oder mittelbaren Abnehmern *nur* haften, wenn diese nachweisen, dass sie von den anderen Kartellbeteiligten keinen vollständigen Schadensersatz erhalten können. Diese Sonderbestimmung soll allein für das Unternehmen gelten, dem das Bußgeld nach dem anwendbaren Kronzeugenprogramm (hier Bonusregelung) erlassen worden ist, also dem „ersten" Kronzeugen. Andere Unternehmen, die im Nachgang der Behörde Beweismaterial zukommen lassen und deshalb eine Reduzierung des Bußgeldes erfahren haben, sind von der Regelung ausgenommen. Diese Richtlinienbestimmung bietet ausreichend Schutz für den Kronzeugen und verfehlt damit nicht das Ziel, einerseits den Kartellgeschädigten zu einer Kompensation ihres Schadens zu verhelfen und andererseits die Attraktivität der behördlichen Kronzeugenprogramme aufrechtzuerhalten.

15) Die Implementierung einer solchen Regelung im deutschen Recht würde zu einer Haftungsprivilegierung des Whistleblowers im Kartellrecht führen. Da der Inhalt einer solchen Regelung mit den Grundsätzen der Gesamtschuld dem Grunde nach vereinbar ist und der Kronzeuge nicht einseitig zuungunsten der privaten Kartellrechtsdurchsetzung Schutz genießt, ist eine Übernahme ins deutsche Recht zu befürworten.

Literaturverzeichnis

Abraham, Jens: Whistleblowing – Neue Chance für eine Kurswende!?, ZRP 2012, 11–13 (Zitiert: Abraham, ZRP 2012)

Achenbach, Hans: Pönalisierung von Ausschreibungsabsprachen und Verselbständigung der Unternehmensgeldbuße durch das Korruptionsbekämpfungsgesetz 1997, WuW 1997, 958–964 (Zitiert: Achenbach, WuW 1997)

Achenbach, Hans: Die Sanktionen gegen die Unternehmensdelinquenz im Umbruch, JuS 1990, 601–608 (Zitiert: Achenbach, JuS 1990)

Achenbach, Hans/Ransiek, Andreas: Handbuch Wirtschaftsstrafrecht, 3. Auflage, 2012 (Zitiert: Bearbeiter, in: Achenbach/Ransiek, Wirtschaftsstrafrecht)

Alexander, Christian: Schadensersatz und Abschöpfung im Lauterkeits- und Kartellrecht, 2010 (Zitiert: Alexander, Schadensersatz und Abschöpfung)

Arhold, Christoph: Das Geldbußenregime nach der Kartell-Verordnung, EuZW 1999, 165–175 (Zitiert: Arhold, EuZW 1999)

Ascheid, Rainer/Preis, Ulrich/Schmidt, Ingrid: Kündigungsrecht, 4. Auflage, 2012 (Zitiert: Bearbeiter, in: Ascheid/Preis/Schmidt, Kündigungsrecht)

Aufdermauer, Christian: Die Compliance-Leitlinien der französischen Wettbewerbsbehörde, CCZ 2012, 194–197 (Zitiert: Aufdermauer, CCZ 2012)

Bachmann, Gregor: Compliance – Rechtsgrundlagen und offene Fragen, in: Bachmann, Gregor/Baums, Theodor/Goette, Wulf/Hauschka, Christoph/Hirtz, Bernd/Schäfer, Carsten, Gesellschaftsrecht in der Diskussion 2007, 2008 S. 65–102 (Zitiert: Bachmann, in: Bachmann/Baums/Goette/Hauschka/Hirtz/Schäfer, Gesellschaftsrecht in der Diskussion)

Bamberger, Heinz Georg/Roth, Herbert: Beck'scher Online-Kommentar zum Bürgerlichen Gesetzbuch, Edition 32, Stand: August 2014 (Zitiert: Bearbeiter, in: BeckOK BGB)

Battis, Ulrich: Kommentar zum Bundesbeamtengesetz, 4. Auflage, 2009 (Zitiert: Battis, BBG)

Bauschke, Hans-Joachim: Whistleblowing – ein Zwischenbericht, öAT 2012, 271–274 (Zitiert: Bauschke, öAT 2012)

Beater, Axel: Unlauterer Wettbewerb, 2011 (Zitiert: Beater, Unlauterer Wettbewerb)

Bechtold, Rainer: Kommentar zum Kartellgesetz, 7. Auflage, 2013 (Zitiert: Bechtold, Kartellgesetz)

Becker, Cornelius: Das Urteil des EGMR zum Whistleblowing – Neuer Lösungsweg auch für deutsche Arbeitsgerichte?, DB 2011, 2202–2204 (Zitiert: Becker, DB 2011)

Behringer, Stefan: Compliance kompakt – Best Practice im Compliance-Management, 2. Auflage, 2011 (Zitiert: Bearbeiter, in: Behringer, Compliance kompakt – Best Practice im Compliance-Management)

Belling, Detlev/v. Steinau-Steinrück, Robert: Freiwillige Leistungen des Arbeitgebers als Maßregelung streikender Arbeitnehmer?, DB 1993, 534–538 (Zitiert: Belling/v. Steinau-Steinrück, DB 1993)

Benecke, Martina: Umfang und Grenzen des Maßregelungsverbots und des Verbots der „Viktimisierung", NZA 2011, 481–486 (Zitiert: Benecke, NZA 2011)

Bepler, Klaus/Böhle, Thomas/Meerkamp, Achim/Russ, Willi: Kommentar zum Tarifrecht für den öffentlichen Dienst (TVöD), Edition 30, Stand: Juni 2014 (Zitiert: Bearbeiter, in Bepler/Böhle/Meerkamp/Russ, TVöD)

Berrisch, Georg/Burianski, Markus: Kartellrechtliche Schadensersatzansprüche nach der 7. GWB-Novelle, WuW 2005, 878–888 (Zitiert: Berrisch/Burianski, WuW 2005)

Beyerbach, Hannes: Die geheime Unternehmensinformation, Dissertation 2012 (Zitiert: Beyerbach, Unternehmensinformation)

Bien, Florian: Überlegungen zu einer haftungsrechtlichen Privilegierung des Kartellkronzeugen, EuZW 2011, 889–890 (Zitiert: Bien, EuZW 2011)

Böge, Ulf: Leniency Programs and the Private Enforcement of European Competition Law, in: Basedow, Jürgen, Private Enforcement in EC Competition Law, 2007, S. 217–228 (Zitiert: Böge, in: Basedow, Private Enforcement)

Bohnert, Joachim: Kommentar zum Ordnungswidrigkeitengesetz, 3. Auflage, 2010 (Zitiert: Bohnert, OWiG Kommentar)

Bosch, Wolfgang: Die Entwicklung des deutschen und europäischen Kartellrechts, NJW 2014, 1714–1720 (Zitiert: Bosch, NJW 2014)

Bosch, Wolfgang/Fritzsche, Alexander: Die 8. GWB-Novelle – Konvergenz und eigene wettbewerbspolitische Akzente, NJW 2013, 2225–2230 (Zitiert: Bosch/Fritzsche, NJW 2013)

Böttcher, Jan Tibor: Die Pflicht des Vorstandes der Aktiengesellschaft, kartellrechtliche Kronzeugenregelungen zu nutzen, um Schaden von der Gesellschaft abzuwenden, Dissertation 2009 (Zitiert: Böttcher, Die Pflicht des Vorstandes)

Breinlinger, Astrid/Krader, Gabriela: Whistleblowing – Chancen und Risiken bei der Umsetzung von anonym nutzbaren Hinweisgebersystemen im Rahmen des Compliance-Managements von Unternehmen, RDV 2006, 60–70 (Zitiert: Breinlinger/Krader, RDV 2006)

Briegel, Torsten: Einrichtung und Ausgestaltung betriebsinterner Whistleblowing-Systeme, Dissertation 2009 (Zitiert: Briegel, Whistleblowing-Systeme)

Brock, Martin: Neue Regeln für Whistleblower im öffentlichen Dienst – Folgen der Heinisch-Entscheidung des EGMR vom 21.07.2011, öAT 2011, 243–247 (Zitiert: Brock, öAT 2011)

Berücksichtigung des Entwurfs eines neuen § 612 a BGB, Dissertation 2011 (Zitiert: Groneberg, Whistleblowing)

Gruson, Michael/Kubicek, Matthias: Der Sarbanes-Oxley Act, Corporate Governance und das deutsche Aktienrecht (Teil I), AG 2003, 337–352 (Zitiert: Gruson/Kubicek, AG 2003)

Grützner, Thomas/Jakob, Alexander: Compliance von A-Z, 1. Auflage, 2010 (Zitiert: Grützner/Jakob, Compliance von A-Z)

Hamacher, Anno: Neues zur betriebsbedingten Druckkündigung – Bedrückend?, NZA 2014, 134–136 (Zitiert: Hamacher, NZA 2014)

Hannich, Rolf: Karlsruher Kommentar zur Strafprozessordnung mit GVG, EGGVG und EMRK, 7. Auflage, 2013 (Zitiert: Bearbeiter, in: Karlsruher Kommentar StPO)

Harte-Bavendamm, Henning/Henning-Bodewig, Frauke: Kommentar zum Gesetz gegen den unlauteren Wettbewerb (UWG) mit Preisangabenverordnung, 3. Auflage, 2013 (Zitiert: Bearbeiter, in: Harte-Bavendamm/Henning-Bodewig, UWG)

Hauger, Nils/Palzer, Christoph: Verweigerung der Einsichtnahme in Bonusanträge nach EuGH „Pfleiderer" – die Folgeentscheidung des AG Bonn vom 18.1.2012 – 51 Gs 53/09, EWS 2012, 124–130 (Zitiert: Hauger/Palzer, EWS 2012)

Hauschka, Christoph: Corporate Compliance – Unternehmensorganisatorische Ansätze zur Erfüllung der Pflichten von Vorständen und Geschäftsführern, AG 2004, 461–475 (Zitiert: Hauschka, AG 2004)

Hauschka, Christoph: Corporate Compliance – Handbuch der Haftungsvermeidung im Unternehmen, 2. Auflage, 2010 (Zitiert: Bearbeiter, in: Hauschka, Corporate Compliance)

Hauschka, Christoph: Compliance am Beispiel der Korruptionsbekämpfung, ZIP 2004, 877–883 (Zitiert: Hauschka, ZIP 2004)

Heermann, Peter/Schlingloff, Jochen: Münchener Kommentar zum Lauterkeitsrecht (UWG), Band 1 und 2, 2. Auflage, 2014 (Zitiert: Bearbeiter, in: MüKo UWG)

Heine, Günter: Zur Verwaltungsakzessorietät im Umweltstrafrecht, JR 1996, 300–304 (Zitiert: Heine, JR 1996)

Heinemann, Andreas: Private Enforcement in Europe, in: Zäch, Roger/Heinemann, Andreas/Kellerhals, Andreas, The Development of Competition Law, 2010 S. 306–325 (Zitiert: Heinemann, in: Zäch/Heinemann/Kellerhals, The Development of Competition Law)

Hellwig, Martin: Private Damage Claims and the Passing-On Defense in Horizontal Price-Fixing Cases – An Economist's Perspective, in: Basedow, Jürgen, Private Enforcement in EC Competition Law, 2007, S. 121–162 (Zitiert: Hellwig, in: Basedow, Private Enforcement)

Hempel, Rolf: Anmerkung zur Entscheidung des EuGH vom 06.06.2013 (C 536/11; EuZW 2013, 586) – Zur Frage des Akteneinsichtsrechts durch Schadenersatzkläger im kartellrechtlichen Verfahren, EuZW 2013, 589–590 (Zitiert: Hempel, EuZW 2013)

Hempel, Rolf: Einsicht in Kartellverfahrensakten nach der Transparenzverordnung- Neues aus Luxemburg, EuZW 2014, 297–300 (Zitiert: Hempel, EuZW 2014)

Hempel, Rolf: Privater Rechtsschutz im Kartellrecht, Dissertation 2002 (Zitiert: Hempel, Privater Rechtsschutz im Kartellrecht)

Herbert, Manfred/Oberrath, Jörg-Dieter: Schweigen ist Gold? Rechtliche Vorgaben für den Umgang des Arbeitnehmers mit seiner Kenntnis über Rechtsverstöße im Betrieb, NZA 2005, 193–199 (Zitiert: Herbert/Oberrath, NZA 2005)

Herold, Nico: Die beamtenrechtliche Zulässigkeit des „Whistleblowing", ZBR 2013, 8–14 (Zitiert: Herold, ZBR 2013)

Hofmann, Stefan: Whistleblowing: Heldentat oder Verrat? Hinweisgeber im Spannungsfeld zwischen Zivilcourage und Denunziantentum, ZCG 2006, 121–126 (Zitiert: Hofmann, ZCG 2006)

Hopt, Klaus/Wiedemann, Herbert: Großkommentar zum Aktiengesetz, 4. Auflage, 2012 (Zitiert: Bearbeiter, in: GK-AktG)

Imgrund, Jan: Die Bindung der deutschen Zivilgerichte an Beschlüsse von Kommission und Behörden der Europäischen Union im Kartellrecht, Dissertation 2011 (Zitiert: Imgrund, Die Bindung der deutschen Zivilgerichte)

Immenga, Ulrich/Mestmäcker, Ernst-Joachim: Wettbewerbsrecht – Kommentar zum Europäischen Kartellrecht, 5. Auflage, 2012 (Zitiert: Bearbeiter, in: Immenga/Mestmäcker, EU-Wettbewerbsrecht)

Immenga, Ulrich/Mestmäcker, Ernst-Joachim: Wettbewerbsrecht – Kommentar zum Deutschen Kartellrecht (GWB), 4. Auflage, 2007 (Zitiert: Bearbeiter, in: Immenga/Mestmäcker, Wettbewerbsrecht: GWB)

Inderst, Cornelia/Bannenberg, Britta/Poppe, Sina: Compliance, 2. Auflage, 2013 (Zitiert: Inderst/Bannenberg/Poppe, Compliance)

Jauernig, Othmar: Kommentar zum Bürgerlichen Gesetzbuch, 15. Auflage, 2014 (Zitiert: Bearbeiter, in: Jauernig, BGB)

Joecks, Wolfgang/Miebach, Klaus: Münchener Kommentar zum Strafgesetzbuch

- Band 1, 2. Auflage 2011
- Band 2, 2. Auflage 2012
- Band 3, 2. Auflage 2012
- Band 4, 2. Auflage 2012
- Band 5, 2. Auflage 2014
- Band 6, 2. Auflage 2013
- Band 6/1, 1. Auflage 2010
- Band 8, 2. Auflage 2013

(Zitiert: Bearbeiter, in: MüKo StGB)

Jos, Philip/Tompkins, Mark/Hays, Steven: In Praise of Difficult People: A Portrait of the Commited Whistleblower, PAR (Public Administration Review) 1989, Vol. 49, No. 6, S. 552–561 (Zitiert: Jos/Tompkins/Hays, PAR 1989, Vol. 49, No. 6)

Jüntgen, Alexander: Zur Verwertung von Kronzeugenerklärungen in Zivilprozessen, WuW 2007, 128–137 (Zitiert: Jüntgen, WuW 2007)

Jürgens, Robert/Seeliger, Daniela: Die geplante Neufassung der Kronzeugenregelung der Europäischen Kommission, EWS 2006, 337–342 (Zitiert: Jürgens/Seeliger, EWS 2006)

Kapp, Thomas: Grundsatz der Einzelabwägung sticht Gesetzgebungskompetenz aus – Anmerkung zu einer Entscheidung des EuGH, Urteil vom 6.6.2013 – Rs. C-536/11, BB 2013, 1556 (Zitiert: Kapp, BB 2013)

Kapp, Thomas: Das Akteneinsichtsrecht kartellgeschädigter Unternehmen: Bonn locuta, causa finita?, WuW 2012, 474–487 (Zitiert: Kapp, WuW 2012)

Kapp, Thomas/Hummel, Karin: Kartellrechts-Compliance in der Verbandsarbeit, CCZ 2013, 240–247 (Zitiert: Kapp/Hummel, CCZ 2013)

Kellerbauer, Manuel: Zugang zur Ermittlungsakte in Verfahren nach den Art. 101, 102 AEUV und der EU-Fusionskontrollverordnung – Wettbewerbsrecht vs. Transparenzverordnung, WuW 2011, 688–699 (Zitiert: Kellerbauer, WuW 2011)

Kersting, Christian: Anmerkung zur Entscheidung des EuGH – Urteil vom 14.06.2011 (C-360/09) – zur Frage des Akteneinsichtsrechts Dritter betreffend Kartelltätern, JZ 2012, 42–46 (Zitiert: Kersting, JZ 2012)

Kersting, Christian: Anmerkung zur Entscheidung des EuGH vom 6.06.2013 (C-536/11) – Zur Akteneinsicht Dritter in Kartellverfahren, JZ 2013, 737–740 (Zitiert: Kersting, JZ 2013)

Kersting, Christian: Auswirkungen des Sarbanes-Oxley-Gesetzes in Deutschland – Können deutsche Unternehmen das Gesetz befolgen?, ZIP 2003, 233–242 (Zitiert: Kersting, ZIP 2003)

Kersting, Christian: Perspektiven der privaten Rechtsdurchsetzung im Kartellrecht, ZWeR 2008, 252–271 (Zitiert: Kersting, ZWeR 2008)

Kießling, Axel/Kießling, Erik : Kontrolle durch interne Revision in Kreditinstituten – Aufsichtsrechtliche Anforderungen und deren gesellschaftliche Relevanz -, WM 2003, 513–523 (Zitiert: A. Kießling/E. Kießling, WM 2003)

Kiethe, Kurt: Vermeidung der Haftung von geschäftsführenden Organen durch Corporate Compliance, GmbHR 2007, 393–400 (Zitiert: Kiethe, GmbHR 2007)

Kiethe, Kurt: Zum Akteneinsichtsrecht des Verletzten (§ 406e StPO), wistra 2006, 50–54 (Zitiert: Kiethe, wistra 2006)

Kindhäuser, Urs/Neumann, Ulfrid/Paeffgen, Hans-Ullrich: Kommentar zum Strafgesetzbuch, 4. Auflage, 2013 (Zitiert: Bearbeiter, in: Kindhäuser/Neumann/Paeffgen, Kommentar StGB)

Király, Andrei: Der Beamte als Whistleblower, DÖV 2010, 894–897 (Zitiert: Király, DÖV 2010)

Király, Andrei: Der rechtliche Schutz von Whistleblowern – Lehren aus US-amerikanischen Regelungen, ZRP 2011, 146–149 (Zitiert: Király, ZRP 2011)

Kittner, Michael/Däubler, Wolfgang/Zwanziger, Bertram: Kommentar zum Kündigungsschutzrecht, 9. Auflage, 2014 (Zitiert: Bearbeiter, in: Kittner/Däubler/Zwanziger, Kündigungsschutzrecht)

Klein, Franz: Kommentar zur Abgabenordnung – einschließlich Steuerstrafrecht -, 12. Auflage, 2014 (Zitiert: Bearbeiter, in: Klein, AO)

Kleine, Maxim: Grenzenloser Zugang zu Dokumenten der Kartellbehörden?, ZWeR 2007, 303–318 (Zitiert: Kleine, ZWeR 2007)

Kleinlein, Kornelius/Schubert, Daniel: Die Auslobung als Möglichkeit zur Ausschaltung der Passing-on-Defence, WuW 2012, 345–354 (Zitiert: Kleinlein/Schubert, WuW 2012)

Kleinmann, Werner/Berg, Werner: Änderungen des Kartellrechts durch das „Gesetz zur Bekämpfung der Korruption" vom 13. 8. 1997, BB 1998, 277–284 (Zitiert: Kleinmann/Berg, BB 1998)

Knütel, Christian: Tätige Reue im Zivilrecht – Eine Untersuchung der Auswirkung nachträglichen Wohlverhaltens auf die Folgen der Verletzung zivilrechtlicher Pflichten und Obliegenheiten, 2000 (Zitiert: Knütel, Tätige Reue im Zivilrecht)

Koch, Arnd: Korruptionsbekämpfung durch Geheimnisverrat? Strafrechtliche Aspekte des Whistleblowing, ZIS 2008, 500–505 (Zitiert: Koch, ZIS 2008)

Koch, Jens: Compliance-Pflichten im Unternehmensverbund?, WM 2009, 1013–1020 (Zitiert: Koch, WM 2009)

Koch, Raphael: Rechtsdurchsetzung im Kartellrecht: Public vs. private enforcement, JZ 2013, 390–398 (Zitiert: Koch, JZ 2013)

Köhler, Helmut/Bornkamm, Joachim: Kommentar zum Gesetz gegen den unlauteren Wettbewerb (UWG), 32. Auflage, 2014 (Zitiert: Bearbeiter, in: Köhler/Bornkamm, UWG)

Kollmer, Norbert/Klindt, Thomas: Kommentar zum Arbeitsschutzgesetz, 2. Auflage, 2011 (Zitiert: Bearbeiter, in: Kollmer/Klindt, ArbSchG)

Kort, Michael: Individualarbeitsrechtliche Fragen des Whistleblowings, in: Hönn, Günther/Oetker, Hartmut/Raab, Thomas (Hrsg.), Festschrift zum 70. Geburtstag für Peter Kreutz, 2010, S. 247–262 (Zitiert: Kort, FS Kreutz)

Kort, Michael: Compliance-Pflichten und Haftung von GmbH-Geschäftsführern, GmbH 2013, 566–574 (Zitiert: Kort, GmbHR 2013)

Krause, Rüdiger: Maßregelungsverbot – Ausschluß von freiwilligen Leistungen, SAE 2003, 205–211 (Zitiert: Krause, SAE 2003)

Krebs, Peter/Eufinger, Alexander/Jung, Stefanie: Bußgeldminderung durch Compliance-Programme im deutschen Kartellbußgeldverfahren?, CCZ 2011, 213–217 (Zitiert: Krebs/Eufinger/Jung, CCZ 2011)

Kretschmer, Joachim: Der Begriff der Gefahr in § 34 StGB, Jura 2005, 662–666 (Zitiert: Kretschmer, Jura 2005)

Krüger, Carsten: Kartellregress: Der Gesamtschuldnerausgleich als Instrument der privaten Kartellrechtsdurchsetzung, Dissertation 2010 (Zitiert: Krüger, Kartellregress)

Krüger, Carsten: Überlegungen zur Gesamtschuld von Kartellmitgliedern im System der privaten Kartellrechtsdurchsetzung. Einer für alle, alle gegen einen?, in: Oberender, Peter (Hrsg.), Private und öffentliche Kartellrechtsdurchsetzung, 2012, S. 79–98 (Zitiert: Krüger, in: Oberender, Private und öffentliche Kartellrechtsdurchsetzung)

Krüger, Carsten: Die haftungsrechtliche Privilegierung des Kronzeugen im Außen- und Innenverhältnis gemäß dem Richtlinienvorschlag der Kommission, NZKart 2013, 483–488 (Zitiert: Krüger, NZKart 2013)

Kudlich, Hans: Grundfälle zum Rücktritt vom Versuch, JuS 1999, 449–452 (Zitiert: Kudlich, JuS 1999)

Küttner, Wolfdieter: Personalbuch, 21. Auflage, 2014 (Zitiert: Bearbeiter, in: Küttner, Personalbuch, Stichwort)

Lackhoff, Klaus/Schulz, Martin: Das Unternehmen als Gefahrenquelle? Compliance-Risiken für Unternehmensleiter und Mitarbeiter, CCZ 2010, 81–88 (Zitiert: Lackhoff/Schulz, CCZ 2010)

Lackner, Karl/Kühl, Kristian: Kommentar zum Strafgesetzbuch, 28. Auflage, 2014 (Zitiert: Bearbeiter, in: Lackner/Kühl, StGB-Kommentar)

Lanfermann, Georg/Maul, Silja: Auswirkungen des Sarbanes-Oxley Acts in Deutschland, DB 2002, 1725–1732 (Zitiert: Lanfermann/Maul, DB 2002)

Langen, Eugen/Bunte, Hermann-Josef: Kommentar zum deutschen und europäischen Kartellrecht

- Band 1: Deutsches Kartellrecht, 12. Auflage, 2014
- Band 2: Europäisches Kartellrecht, 12. Auflage, 2014

(Zitiert: Bearbeiter, in: Langen/Bunte, Deutsches KartellR)

Larenz, Karl: Methodenlehre, 6. Auflage, 1991 (Zitiert: Larenz, Methodenlehre)

Legner, Sarah: Gesamtschuldnerinnenausgleich zwischen Kartellanten, WRP 2014, 1163–1167 (Zitiert: Legner, WRP 2014)

Leipold, Klaus: Unternehmensstrafrecht – Eine rechtspolitische Notwendigkeit?, ZRP 2013, 34–37 (Zitiert: Leipold, ZRP 2013)

Leisinger, Klaus M.: Whistleblowing und Corporate Reputation Management, 2003 (Zitiert: Leisinger, Whistleblowing und Corporate Reputation Management)

Lelley, Jan Tibor: Compliance im Arbeitsrecht, 2010 (Zitiert: Lelley, Compliance im Arbeitsrecht)

Lelley, Jan Tibor: Auflösungsantrag bei Whistleblowing, BB 2012, 1866–1867 (Zitiert: Lelley, BB 2012)

Lensdorf, Lars: IT-Compliance – Maßnahmen zur Reduzierung von Haftungsrisiken von IT-Verantwortlichen, CR 2007, 413–418 (Zitiert: Lensdorf, CR 2007)

Leuchten, Alexius: Der gesetzliche Schutz für Whistleblower rückt näher, ZRP 2012, 142–145 (Zitiert: Leuchten, ZRP 2012)

Liese, Jens: Much Adoe About Nothing? oder: Ist der Vorstand einer Aktiengesellschaft verpflichtet, eine Compliance-Organisation zu implementieren?, BB Spezial 2008, Nr. 5, 17–22 (Zitiert: Liese, BB Spezial 2008, Nr. 5)

Loewenheim, Ulrich/Meessen, Karl/Riesenkampff, Alexander: Kommentar zum Kartellrecht, 2. Auflage, 2009 (Zitiert: Bearbeiter, in: Loewenheim/Meessen/Riesenkampff, Kartellrecht)

Logemann, Hans Philip: Der kartellrechtliche Schadensersatz – die zivilrechtliche Haftung bei Verstößen gegen das deutsche und europäische Kartellrecht nach Ergehen der VO Nr. 1/2003 und der 7. GWB-Novelle, Dissertation 2009 (Zitiert: Logemann, Der kartellrechtliche Schadensersatz)

Lutterbach, Carolin: Die strafrechtliche Würdigung des Whistleblowings, Dissertation 2010 (Zitiert: Lutterbach, Whistleblowing)

Lutz, Martin: Amnestie für aufklärungsbereite Kartellanten?, BB 2000, 677–684 (Zitiert: Lutz, BB 2000)

Mäger, Thorsten: Europäisches Kartellrecht, 2. Auflage, 2011 (Zitiert: Bearbeiter, in: Mäger, Europäisches Kartellrecht)

Mäger, Thorsten/Zimmer, Daniel/Milde, Sarah: Konflikt zwischen öffentlicher und privater Kartellrechtsdurchsetzung, WuW 2009, 885–899 (Zitiert: Mäger/Zimmer/Milde, WuW 2009)

Mahnhold, Thilo: „Global Whistle" oder „deutsche Pfeife" – Whistleblowing-Systeme im Jurisdiktionskonflikt, NZA 2008, 737–743 (Zitiert: Mahnhold, NZA 2008)

Maschmann, Frank: Corporate Compliance und Arbeitsrecht, 2009 (Zitiert: Bearbeiter, in: Maschmann, Corporate Compliance und Arbeitsrecht)

Maunz, Theodor/Dürig, Günter: Kommentar zum Grundgesetz, 70. Ergänzungslieferung, Stand: Dezember 2013 (Zitiert: Bearbeiter, in: Maunz/Düring, GG-Kommentar)

Mayer, Markus: Geschäfts- und Betriebsgeheimnis oder Geheimniskrämerei?, GRUR 2011, 884–888 (Zitiert: Mayer, GRUR 2011)

Meeßen, Gero: Der Anspruch auf Schadensersatz bei Verstößen gegen EU-Kartellrecht: Konturen eines Europäischen Kartelldeliktsrechts?, Dissertation 2011 (Zitiert: Meeßen, Schadensersatz)

Meier, Hans-Georg: Der Arbeitsvertrag des Compliance-Beauftragten – Rechtliche Notwendigkeiten und Möglichkeiten, NZA 2011, 779–782 (Zitiert: Meier, NZA 2011, 779)

Mengel, Anja: Der Gesetzesentwurf der SPD-Fraktion zum Whistleblowing, CCZ 2012, 146–155 (Zitiert: Mengel, CCZ 2012)

Mengel, Anja: Compliance und Arbeitsrecht, 2009 (Zitiert: Mengel, Compliance und Arbeitsrecht)

Mengel, Anja: Arbeitsrechtliche Besonderheiten der Implementierung von Compliance-Programmen in internationalen Konzernen, CCZ 2008, 85–91 (Zitiert: Mengel, CCZ 2008)

Mengel, Anja: EGMR: Kündigung wegen Whistleblowing, CCZ 2011, 229–231 (Zitiert: Mengel, CCZ 2011)

Menzies, Christof: Sarbanes-Oxley und Corporate Compliance, 2006 (Zitiert: Menzies, Sarbanes-Oxley und Corporate Compliance)

Meyer-Ladewig, Jens: Europäische Menschenrechtskonvention, Handkommentar, 3. Auflage, 2011 (Zitiert: Meyer-Ladewig, EMRK)

Meyer, Michael: Die Bindung der Zivilgerichte an Entscheidungen im Kartellverwaltungsrechtsweg – der neue § 33 IV GWB auf dem Prüfstand, GRUR 2006, 27–33 (Zitiert: Meyer, GRUR 2006)

Miceli, Marcia P./Near, Janet P.: Organizational dissidence: The case of whistleblowing, Journal of Business Ethics 1985, Vol. 4, Issue 1, P. 1–16. (Zitiert: Miceli/Near, Journal of Business Ethics 1985, Vol. 4, Issue 1)

Michalowski, Sabine: Schutz der Vertraulichkeit strafrechtlich relevanter Patienteninformationen, ZStW 109 (1997), 519–544 (Zitiert: Michalowski, ZStW 109 (1997))

Milde, Sarah: Schutz der Kronzeugen im Spannungsfeld von behördlicher Kartellrechtsdurchsetzung und privaten Schadensersatzklagen, Dissertation 2013 (Zitiert: Milde, Schutz der Kronzeugen)

Möhrenschlager, Manfred: Revision des Umweltstrafrechts – 2. Teil, NStZ 1994, 566–569 (Zitiert: Möhrenschlager, NStZ 1994)

Moll, Wilhelm: Münchener Anwaltshandbuch Arbeitsrecht, 3. Auflage, 2012 (Zitiert: Bearbeiter, in: Moll, Arbeitsrecht)

Moosmayer, Klaus: Compliance, Praxisleitfaden für Unternehmen, 2. Auflage, 2012 (Zitiert: Moosmayer, Compliance)

Müller-Glöge, Rudi/Preis, Ulrich/Schmidt, Ingrid: Erfurter Kommentar zum Arbeitsrecht, 14. Auflage, 2014 (Zitiert: Bearbeiter, in: Erfurter Kommentar)

Müller-Gugenberger, Christian/Bieneck, Klaus: Wirtschaftsstrafrecht – Handbuch des Wirtschaftsstraf- und -ordnungswidrigkeitenrechts, 5. Auflage, 2011 (Zitiert: Bearbeiter, in: Müller-Gugenberger/Bieneck, Wirtschaftsstrafrecht)

Müller, Michael: Whistleblowing – Ein Kündigungsgrund?, NZA 2002, 424–437 (Zitiert: Müller, NZA 2002)

Mundt, Andreas: Kartellverfolgung im 21. Jahrhundert: Erfolgreich durch wirksame Abschreckung, in: Bechtold, Stefan/Jickeli, Joachim/Rohe, Mathias (Hrsg), Recht, Ordnung und Wettbewerb, Festschrift zum 70. Geburtstag von Wernhard Möschel, 2011, S. 427–440 (Zitiert: Mundt, FS Möschel)

Mundt, Andreas/Meeßen, Gero: Private Kartellrechtsdurchsetzung – Deutsche und Europäische Initiativen aus Sicht des Bundeskartellamts, in: Oberender, Peter (Hrsg.) Private und öffentliche Kartellrechtsdurchsetzung, 2012, S. 39–54 (Zitiert: Mundt/Meeßen, in: Oberender, Private und öffentliche Kartellrechtsdurchsetzung)

Musielak, Hans-Joachim: Kommentar zur Zivilprozessordnung mit Gerichtsverfassungsgesetz, 11. Auflage, 2014 (Zitiert: Bearbeiter, in: Musielak, ZPO)

Nave, Jose A Campos/Bonenberger, Saskia: Korruptionsaffären, Corporate Compliance und Sofortmaßnahmen für den Krisenfall, BB 2008, 734–741 (Zitiert: Nave/Bonenberger, BB 2008)

Nestler, Claudia/Salvenmoser, Steffen/Bussmann, Kai-D.: Studie Wirtschaftskriminalität und Unternehmenskultur 2013, (Hrsg. PwC/Universität Halle-Wittenberg) (Zitiert: Nestler/Salvenmoser/Bussmann, Studie Wirtschaftskriminalität 2013)

Nicklisch, Annette Christina: Die Auswirkungen des Sarbanes-Oxley Act auf die deutsche Corporate Governance, Dissertation 2007 (Zitiert: Nicklisch, Auswirkungen des Sarbanes-Oxley Act)

Oetker, Hartmut: Ausgewählte Probleme zum Beschwerderecht des Beschäftigten nach § 13 AGG, NZA 2008, 264–270 (Zitiert: Oetker, NZA 2008)

Ohly, Ansgar/Sosnitza, Olaf: Kommentar zum Gesetz gegen den unlauteren Wettbewerb (UWG) mit Preisangabenverordnung, 6. Auflage, 2014 (Zitiert: Bearbeiter, in: Ohly/Sosnitza, UWG)

Otto, Harro: Die Reform des strafrechtlichen Schutzes gegen irreführende Werbung, GRUR 1982, 274–283 (Zitiert: Otto, GRUR 1982)

Otto, Harro: Verrat von Betriebs- und Geschäftsgeheimnissen, § 17 UWG, wistra 1988, 125–130 (Zitiert: Otto, wistra 1988)

Otto, Harro: Fehlgeschlagener Versuch und Rücktritt, Jura 1992, 423–431 (Zitiert: Otto, Jura 1992)

Pacholski, Robert: Kein Schadensersatzanspruch bei „whistleblowing“ gegen gutgläubigen Strafanzeigeerstatter (Urteilsanmerkung), öAT 2012, 19–20 (Zitiert: Pacholski, öAT 2012)

Palandt, Otto: Kommentar zum Bürgerlichen Gesetzbuch, 73. Auflage, 2014 (Zitiert: Bearbeiter, in: Palandt, BGB-Kommentar)

Palzer, Christoph: Stolperstein für die Kommission? Die Kronzeugenregelung auf dem Prüfstand der Transparenz-VO, EuR 2012, 583–607 (Zitiert: Palzer, EuR 2012)

Pampel, Gunnar: Die Bedeutung von Compliance-Programmen im Kartellordnungswidrigkeitenrecht, BB 2007, 1636–1640 (Zitiert: Pampel, BB 2007)

Pasewaldt, David: Zehn Jahre Strafbarkeit wettbewerbsbeschränkender Absprachen bei Ausschreibungen gemäß § 298 StGB, ZIS 2008, 84–91 (Zitiert: Pasewaldt, ZIS 2008)

Peglau, Jens: Neues zur „Kronzeugenregelung" – Beschränkung auf Zusammenhangstaten, NJW 2013, 1910–1913 (Zitiert: Peglau, NJW 2013)

Pfeiffer, Gerd: Kommentar zur Strafprozessordnung, 5. Auflage, 2005 (Zitiert: Pfeiffer, Kommentar StPO)

Rau, Charlotte: Compliance und Unternehmensverantwortlichkeit: materiellrechtliche Fragen der sanktionsrechtlichen Unternehmensverantwortlichkeit unter Berücksichtigung von Compliance-Maßnahmen, Dissertation 2010 (Zitiert: Rau, Compliance und Unternehmensverantwortlichkeit)

Reiter, Christian: Der Schutz der Whistleblowers nach dem Sarbanes-Oxley Act im Rechtsvergleich und im internationalen Arbeitsrecht, RIW 2005, 168–178 (Zitiert: Reiter, RIW 2005)

Rengier, Rudolf: Zur Reichweite von Sorgfaltspflichten und verwaltungsrechtlichen Pflichten im Umweltstrafrecht, in: Ebenroth, Carsten Thomas/Hesselberger, Dieter/Rinne, Manfred Eberhard (Hrsg.), Verantwortung und Gestaltung, Festschrift zum 65. Geburtstag für Karlheinz Boujong, 1996, S. 791–806 (Zitiert: Rengier, FS Boujong)

Rengier, Rudolf: Das moderne Umweltstrafrecht, 1992 (Zitiert: Rengier, Das moderne Umweltstrafrecht)

Rengier, Rudolf: Strafrecht, Besonderer Teil, Band II (Delikte gegen die Person und die Allgemeinheit), 15. Auflage, 2014 (Zitiert: Rengier, StrafR BT II)

Reufels, Martin/Deviard, Katja: Die Implementierung von Whistleblower-Hotlines aus US-amerikanischer, europäischer und deutscher Sicht, CCZ 2009, 201–209 (Zitiert: Reufels/Deviard, CCZ 2009)

Reufels, Martin/Molle, Karl: Neuere Rechtsprechung des EGMR im Arbeitsrecht, KSzW 2012, 3–9 (Zitiert: Reufels/Molle, KSzW 2012)

Reuter, Joachim: Verfall von Ethik und Moral und die Weisungsgebundenheit von Staatsanwälten, ZRP 2011, 104–108 (Zitiert: Reuter, ZRP 2011)

Richardi, Reinhard/Wißmann, Hellmut/Wlotzke, Otfried/Oetker, Hartmut: Münchener Handbuch zum Arbeitsrecht

- Band 1, 3. Auflage 2009
- Band 2, 3. Auflage 2009

(Zitiert: Bearbeiter, in: Münchener Handbuch zum Arbeitsrecht)

Richters, Swantje/Wodtke, Carolina: Schutz von Betriebsgeheimnissen aus Unternehmenssicht, NZA-RR 2003, 281–288 (Zitiert: Richters/Wodtke, NZA-RR 2003)

Ritter, Jan-Stephan/Käseberg, Thorsten: Der Regierungsentwurf zur 8. GWB-Novelle, WuW 2012, 661–670 (Zitiert: Ritter/Käseberg, WuW 2012)

Rohde-Liebenau, Björn/Peter, Gabriele: Whistleblowing – ein neues Thema für die Betriebsratsarbeit, AiB 2004, 615–621 (Zitiert: Rohde-Liebenau/Peter, AiB 2004)

Rolfs, Christian/Giesen, Richard/Kreikebohm, Ralf/Udsching, Peter: Beck'scher Online-Kommentar zum Arbeitsrecht, Edition 32, Stand: Juni 2014 (Zitiert: Bearbeiter, in: BeckOK Arbeitsrecht)

Röller, Jürgen: Küttner, Personalbuch 2014, 21. Auflage, 2014 (Zitiert: Küttner, Personalbuch, Stichwort)

Rosbach, Thorsten: Ethik in einem Wirtschaftsunternehmen – nützlich oder überflüssige Förmelei?, CCZ 2008, 101–104 (Zitiert: Rosbach, CCZ 2008)

Roxin, Claus: Strafrecht Allgemeiner Teil, Band I, 4. Auflage, 2006 (Zitiert: Roxin, AT I)

Rudkowski, Lena: Kernprobleme einer gesetzlichen Regelung zum Schutz von Whistleblowern, CCZ 2013, 204–209 (Zitiert: Rudkowski, CCZ 2013)

Rützel, Stefan: Illegale Unternehmensgeheimnisse?, GRUR 1995, 557–561 (Zitiert: Rützel, GRUR 1995)

Säcker, Franz Jürgen/Rixecker, Roland: Münchener Kommentar zum Bürgerlichen Gesetzbuch

- Band 1, 6. Auflage 2012
- Band 2, 6. Auflage 2012
- Band 3, 6. Auflage 2012
- Band 4, 6. Auflage 2012
- Band 5, 6. Auflage 2013
- Band 6, 6. Auflage 2013
- Band 7, 6. Auflage 2013
- Band 8, 6. Auflage 2012
- Band 9, 6. Auflage 2013

(Zitiert: Bearbeiter, in: MüKo BGB)

Sagan, Adam: Konvergenzen und Divergenzen zwischen EGMR, EuGH, BVerfG und BAG, DB 2012, Beilage zu Heft 6, 11–14 (Zitiert: Sagan, DB 2012, Beilage zu Heft 6)

Sänger, Marc: Whistleblowing in der börsennotierten Aktiengesellschaft, Dissertation 2011 (Zitiert: Sänger, Whistleblowing)

Schaefer, Hans/Baumann, Diethelm: Compliance-Organisation und Sanktionen bei Verstößen, NJW 2011, 3601–3605 (Zitiert: Schaefer/Baumann, NJW 2011)

Schaefer, Torsten: Akteneinsichtsrechte nach der StPO, NJW-Spezial 2007, 327–328 (Zitiert: Schaefer, NJW-Spezial 2007)

Schaub, Günter: Arbeitsrechtshandbuch, 15. Auflage, 2013 (Zitiert: Bearbeiter, in: Schaub, Arbeitsrechtshandbuch)

Schaub, Günter: Arbeitsrecht von A-Z, 19. Auflage, 2014 (Zitiert: Bearbeiter, in: Schaub, Arbeitsrecht von A-Z)

Schaupensteiner, Wolfgang: Rechtstreue im Unternehmen – Compliance und Krisenmanagement – Konzertiertes Vorgehen statt einzelbetriebliche Maßnahmen, NZA-Beilage 1/2011, 8–15 (Zitiert: Schaupensteiner, NZA-Beil. 1/2011)

Schemmel, Alexander/Ruhmannseder, Felix/Witzigmann, Tobias: Hinweisgebersysteme. Implementierung im Unternehmen, 2012 (Zitiert: Schemmel/Ruhmannseder/Witzigmann, Hinweisgebersysteme)

Schlachter, Monika: Kündigung wegen „Whistleblowing"? – Der Schutz der Meinungsfreiheit vor dem EGMR, RdA 2012, 108–112 (Zitiert: Schlachter, RdA 2012)

Schmitt-Schönenberg, Michael: Umweltschutz – Arbeitsvertragliche Rechte und Pflichten des Arbeitnehmers, AuR 1994, 281–286 (Zitiert: Schmitt-Schönenberg, AuR 1994)

Schmitt, Bettina: Whistleblowing – „Verpfeifen" des Arbeitgebers, Dissertation 2003 (Zitiert: Schmitt, Whistleblowing)

Schmolke, Klaus: Whistleblowing-Systeme als Corporate Governance-Instrument transnationaler Unternehmen, RIW 2012, 224–232 (Zitiert: Schmolke, RIW 2012)

Schneider, David: Die arbeitsrechtliche Implementierung von Compliance- und Ethikrichtlinien, Dissertation 2009 (Zitiert: Schneider, Implementierung)

Schneider, Moritz: Kronzeugenregelung im EG-Kartellrecht: die Praxis der Europäischen Kommission beim Erlaß und der Ermäßigung von Geldbußen in Kartellsachen unter Einbeziehung der „Leniency Notice" des US-Department of Justice und der „Bonusregelung" des Bundeskartellamtes, Dissertation 2004 (Zitiert: Schneider, Kronzeugenregelung)

Schneider, Uwe: Compliance als Aufgabe der Unternehmensleitung, ZIP 2003, 645–650 (Zitiert: Schneider, ZIP 2003)

Schneider, Uwe: Die 16 goldenen Regeln zur Haftung des Geschäftsführers einer Holding, GmbHR 2010, 1313–1318 (Zitiert: Schneider, GmbHR 2010)

Schneider, Uwe/Nowak, Claudia: Sind die Einrichtung einer Whistleblowing-Stelle und der Schutz des Whistleblowers Teil guter Corporate Compliance?, in: Hönn, Günther/Oetker, Hartmut/Raab, Thomas (Hrsg.), Festschrift zum 70. Geburtstag für Peter Kreutz, 2010, S. 855- 865 (Zitiert: Schneider/Nowak, FS Kreutz)

Schoene, Volker: Kartellschadensersatz wegen erhöhter Preise der Kartellaußenseiter, GRUR-Prax 2014, 293–294 (Zitiert: Schoene, GRUR-Prax 2014)

Schönke, Adolf/Schröder, Horst: Kommentar zum Strafgesetzbuch, 29. Auflage, 2014 (Zitiert: Bearbeiter, in: Schönke/Schröder, Kommentar StGB)

Schreiber, Alexander: Implementierung von Compliance-Richtlinien, NZA-RR 2010, 617–623 (Zitiert: Schreiber, NZA-RR 2010)

Schroeder, Dirk: Kronzeugenregelungen im Kartellrecht, in: Brinker, Ingo/Scheuing, Dieter/Stockmann, Kurt (Hrsg.) Recht und Wettbewerb, Festschrift zum 65. Geburtstag für Rainer Bechtold, 2006, S. 437–454 (Zitiert: Schroeder, FS Bechtold)

Schroll, Gregor: Der Einfluss interner und externer Faktoren auf die Effektivität der Kronzeugenprogramme der EU-Kommission und des Bundeskartellamtes, Dissertation 2012 (Zitiert: Schroll, Effektivität der Kronzeugenprogramme)

Schubert, Claudia: Whistle-Blowing after Heinisch v. Germany: Much Ado About Nothing?, German Yearbook of international Law 2011(54), S. 753–763 (Zitiert: Schubert, German Yearbook of international Law 2011 (54))

Schulz, Mike: Ethikrichtlinien und Whistleblowing – Arbeitsrechtliche Aspekte der Einführung eines Compliance-Systems, Dissertation 2010 (Zitiert: Schulz, Ethikrichtlinien)

Schulz, Mike: Compliance – Internes Whistleblowing, BB 2011, 629–634 (Zitiert: Schulz, BB 2011)

Schulze, Reiner: Kommentar zum Bürgerlichen Gesetzbuch, 8. Auflage, 2014 (Zitiert: Bearbeiter, in: Schulze, BGB-Kommentar)

Schürrle, Thomas/Fleck, Franziska: „Whistleblowing Unlimited" – Der U.S. Dodd-Frank Act und die neuen Regeln der SEC zum Whistleblowing, CCZ 2011, 218–221 (Zitiert: Schürrle/Fleck, CCZ 2011)

Schwalbe, Ulrich/Höft, Jan: Ausgestaltung von Kronzeugenprogrammen und private Kartellrechtsdurchsetzung, in: Bechtold, Stefan/Jickeli, Joachim/Rohe, Mathias (Hrsg), Recht, Ordnung und Wettbewerb, Festschrift zum 70. Geburtstag von Wernhard Möschel, 2011, S. 597–636 (Zitiert: Schwalbe/Höft, FS Möschel)

Seel, Henning-Alexander: Recht auf freie Meinungsäußerung – Wo sind die Grenzen des „Whistleblowing"?, MDR 2012, 9–12 (Zitiert: Seel, MDR 2012)

Seibt, Christoph: Deutscher Corporate Governance Kodex und Entsprechens-Erklärung (§ 161 AktG-E), AG 2002, 249–259 (Zitiert: Seibt, AG 2002)

Seitz, Claudia: Public over Private Enforcement of Competition Law? Zugleich Besprechung von AG Bonn, Beschluss v. 18.1.2012 – Gs 53/09 – Pfleiderer, GRUR-RR 2012, 137–142 (Zitiert: Seitz, GRUR-RR 2012)

Siekmann, Helmut: Verletzt die Weitergabe eines im Bundesanzeiger veröffentlichten Jahresabschlusses betreffend eine juristische Person ein dieser zustehendes, allgemeines Persönlichkeitsrecht, insbesondere das Recht auf informationelle Selbstbestimmung? ZIP 1994, 651–654 (Zitiert: Siekmann, ZIP 1994)

Simon, Oliver/Schilling, Jan Moritz: Kündigung wegen Whistleblowing?, BB 2011, 2421–2428 (Zitiert: Simon/Schilling, BB 2011)

Simonet, Michael: Notwendigkeit eines Gesetzes zum Schutz von Whistleblowern?, RdA 2013, 236–240 (Zitiert: Simonet, RdA 2013)

Soyez, Volker: Schadensersatzanspruch des indirekten Abnehmers von Kartellteilnehmern, EuZW 2012, 100–102 (Zitiert: Soyez, EuZW 2012)

Spindler, Gerald/Stilz, Eberhard: Kommentar zum Aktiengesetz, 3. Auflage, 2014 (Zitiert: Bearbeiter, in: Spindler/Stilz, AktG)

Stahlhacke, Eugen/Preis, Ulrich/Vossen, Reinhard: Kündigung und Kündigungsschutz im Arbeitsverhältnis, 10. Auflage, 2010 (Zitiert: Stahlhacke/Preis/Vossen, Kündigung und Kündigungsschutz)

Stelkens, Paul/Bonk, Heinz Joachim/ Sachs, Michael: Kommentar zum Verwaltungsverfahrensgesetz, 8. Auflage, 2014 (Zitiert: Bearbeiter, in: Stelkens/Bonk/Sachs, VwVfG)

Stern, Klaus/Becker, Florian: Kommentar zum Grundgesetz, 1. Auflage, 2010 (Zitiert: Bearbeiter, in: Stern/Becker, GG-Kommentar)

Stöber, Michael: Schadensersatzhaftung für Preisschirmeffekte bei Verstößen gegen deutsches und europäisches Kartellrecht, EuZW 2014, 257–261 (Zitiert: Stöber, EuZW 2014)

Strack,Guido: Unzureichender Schutz von Whistleblowern in Deutschland, Compliance-Berater 4/2014, 113–117 (Zitiert: Strack, Compliance-Berater 4/2014)

Stürner, Rolf: Der entgangene rechtswidrige oder sittenwidrige Gewinn, VersR 1976, 1012–1016 (Zitiert: Stürner, VersR 1976)

Sweeney, Brendan: The Role of Damages in Regulation Horizontal Price-Fixing: Comparing the Situation in the United States, Europe, and Australia, Melbourne University Law Review (2006), Vol. 30, Issue 3, S. 873–879 (Zitiert: Sweeney, Melbourne University Law Review (2006), Vol. 30, Issue 3)

Thomas, Heinz/Putzo, Hans: Kommentar zur Zivilprozessordnung, 35. Auflage, 2014 (Zitiert: Bearbeiter, in: Thomas/Putzo, ZPO)

Thomas, Stefan : Aufdeckung – Abschreckung – Kompensation. Wechselwirkungen und Zielkonflikte der privaten und öffentlichen Kartellrechtsdurchsetzung, in: Oberender, Peter (Hrsg.), Private und öffentliche Kartellrechtsdurchsetzung, 2012, S. 55–78 (Zitiert: Thomas, in: Oberender, Private und öffentliche Kartellrechtsdurchsetzung)

Thümmel, Roderich: Die Abweichung von der Entsprechenserklärung nach § 161 AktG – Neue Risiken für Vorstände und Aufsichtsräte?, CCZ 2008, 141–144 (Zitiert: Thümmel, CCZ 2008)

Thüsing, Gregor: Beschäftigtendatenschutz und Compliance, 2. Auflage, 2014 (Zitiert: Bearbeiter, in: Thüsing, Beschäftigtendatenschutz und Compliance)

Tietje, Christian/Nowrot, Karsten: Zugang zu Kartellrechtsakten nach der Transparenzverordnung als öffentliches Interesse, EWS 2006, 486–488 (Zitiert: Tietje/Nowrot, EWS 2006)

Többens, Hans: Wirtschaftsspionage und Konkurrenzausspähung in Deutschland, NStZ 2000, 505–512 (Zitiert: Többens, NStZ 2000)

Trips, Marco: Risikomanagement in der öffentlichen Verwaltung, NVwZ 2003, 804–811 (Zitiert: Trips, NVwZ 2003)

Umnuß, Karsten: Corporate Compliance Checklisten, 2. Auflage, 2012 (Zitiert: Bearbeiter, in: Umnuß, Corporate Compliance Checklisten)

Van Vormizeele, Philipp Voet: Kartellrechtliche Compliance-Programme im Rahmen der Bußgeldbemessung de lege lata und de lege ferenda, CCZ 2009, 41–49 (Zitiert: Van Vormizeele, CCZ 2009)

Vollrath, Christian: Das Maßnahmepaket der Kommission zum wettbewerbsrechtlichen Schadenersatzrecht, NZKart 2013, 434–446 (Zitiert: Vollrath, NZKart 2013)

Von Busekist, Konstantin/Fahrig, Stephan: Whistleblowing und der Schutz von Hinweisgebern, BB 2013, 119–124 (Zitiert: v. Busekist/Fahrig, BB 2013, 119)

Von Heintschel-Heinegg, Bernd: Beck'scher Online-Kommentar zum Strafgesetzbuch, Edition 24, Stand: Juli 2014 (Zitiert: Bearbeiter, in: BeckOK StGB)

Von Hoyningen-Huene, Gerrick/Linck, Rüdiger: Kommentar zum Kündigungsschutzgesetz, 15. Auflage, 2013 (Zitiert: Bearbeiter, in: v. Hoyningen-Huene/ Linck, KSchG)

Von Mangoldt, Hermann/Klein, Friedrich/Starck, Christian: Kommentar zum Grundgesetz, 6. Auflage, 2010 (Zitiert: Bearbeiter, in: v. Mangoldt/Klein/Starck, Kommentar zum GG)

Von Pelchrzim, Gero: Whistleblowing und der strafrechtliche Geheimnisschutz nach § 17 UWG, CCZ 2009, 25–29 (Zitiert: v. Pelchrzim, CZ 2009)

Von Staudinger, Julius: Kommentar zum Bürgerlichen Gesetzbuch, Buch 2 (Dienstvertragsrecht), 2011 (Zitiert: Bearbeiter, in: v. Staudinger, BGB-Kommentar)

Wabnitz, Heinz-Bernd/Janovsky, Thomas: Handbuch des Wirtschafts- und Steuerstrafrechts, 4. Auflage, 2014 (Zitiert: Bearbeiter, in: Wabnitz/Janovsky, Handbuch WstR)

Walter, Tonio: Submissionsbetrug, JZ 2002, 254–256 (Zitiert: Walter, JZ 2002)

Wedlich, Alexander Christian: Die strafrechtliche Würdigung von Submissionsabsprachen unter dem Gesichtspunkt des § 298 StGB, Dissertation 2004 (Zitiert: Wedlich, Submissionsabsprachen)

Wessels, Johannes/Beulke, Werner/Satzger, Helmut: Strafrecht Allgemeiner Teil, 43. Auflage, 2013 (Zitiert: Wessels/Beulke/Satzger, StrafR AT)

Westhoff, Jan Philipp: Der Zugang zu Beweismitteln bei Schadensersatzklagen im Kartellrecht: eine rechtsvergleichende Untersuchung, Dissertation 2010 (Zitiert: Westhoff, Zugang zu Beweismitteln bei Schadensersatzklagen im KartellR)

Wiedemann, Gerhard: Handbuch des Kartellrechts, 2. Auflage, 2008 (Zitiert: Bearbeiter, in: Wiedemann, Kartellrecht)

Wils, Wouter: Antitrust Compliance Programmes and Optimal Antitrust Enforcement, Journal of Antitrust Enforcement 2013, Vol. 1, No. 1, S. 52–81 (Zitiert: Wils, Journal of Antitrust Enforcement 2013, Vol. 1, No. 1)

Wisskirchen, Gerlin/Körber, Anke/Bissels, Alexander: „Whistleblowing" und „Ethikhotlines", BB 2006, 1567–1572 (Zitiert:Wisskirchen/Körber/Bissels, BB 2006)

Wittig, Petra: Wirtschaftsstrafrecht, 3. Auflage, 2014 (Zitiert: Wittig, Wirtschaftsstrafrecht)

Wolff, Heinrich Amadeus: Der verfassungsrechtliche Schutz der Betriebs- und Geschäftsgeheimnisse, NJW 1997, 98–101 (Zitiert: Wolff, NJW 1997)

Wolter, Jürgen: Systematischer Kommentar zum Strafgesetzbuch, 140. Ergänzungslieferung, Stand: Dezember 2013 (Zitiert: Bearbeiter, in: SK-StGB)

Wrase, Silvelyn/Fabritius, Christoph: Prämien für Hinweisgeber bei Kartellverstößen?, CCZ 2011, 69–73 (Zitiert: Wrase/Fabritius, CCZ 2011)

Wunderlich, Claudia: Die Akzessorietät des § 298 StGB zum Gesetz gegen Wettbewerbsbeschränkungen (GWB), Dissertation 2009 (Zitiert: Wunderlich, Die Akzessorietät des § 298 StGB zum GWB)

Wybitul, Tim: Interne Ermittlungen auf Aufforderung von US-Behörden – ein Erfahrungsbericht, BB 2009, 606–611 (Zitiert: Wybitul, BB 2009)

Zagrosek, Roman: Kronzeugenregelungen im US-amerikanischen, europäischen und deutschen Recht der Wettbewerbsbeschränkungen, Dissertation 2006 (Zitiert: Zagrosek, Kronzeugenregelungen)

Zimmer, Daniel/Logemann, Hans: Unterliegen „Altfälle" der verschärften Schadensersatzhaftung nach § 33 GWB?, WuW 2006, 982–990 (Zitiert: Zimmer/Logemann, WuW 2006)

Zimmer, Mark/Seebacher, Krikor R.: Whistleblowing – Wichtige Erkenntnisquelle oder gefährliches Pflaster?, CCZ 2013, 31–33 (Zitiert: Zimmer/Seebacher, CCZ 2013)

Zimmermann, Gernot: Strafrechtliche Risiken des „Whistleblowing", ArbRAktuell 2012, 58–61 (Zitiert: Zimmermann, ArbRAktuell 2012)

Zöller, Richard: Kommentar zur Zivilprozessordnung, 30. Auflage, 2014 (Zitiert: Bearbeiter, in: Zöller, ZPO)

Zöttl, Johannes/Schlepper, Lisa: Die private Durchsetzung von kartellrechtlichen Schadensersatzansprüchen – Status Quo in Deutschland, EuZW 2012, 573–576 (Zitiert: Zöttl/Schlepper, EuZW 2012)

Index

Veröffentlichungen des Instituts für deutsches und europäisches Wirtschafts-, Wettbewerbs- und Regulierungsrecht der Freien Universität Berlin

Herausgegeben von Franz Jürgen Säcker

Band 1 Franz Jürgen Säcker (Hrsg.): Deutsch-russisches Energie- und Bergrecht im Vergleich. Ergebnisse einer Arbeitstagung vom 31. März / 1. April 2006. 2007.

Band 2 Franz Jürgen Säcker / Walther Busse von Colbe (Hrsg.): Wettbewerbsfördernde Anreizregulierung. Zum Anreizregulierungsbericht der Bundesnetzagentur vom 30. Juni 2006. 2007.

Band 3 Dirk Zschenderlein: Die Gleichbehandlung der Aktionäre bei der Auskunftserteilung in der Aktiengesellschaft. Zum Problem der Zulässigkeit der Weitergabe von Informationen an einzelne Aktionäre und Dritte. 2007.

Band 4 Simone Kirchhain: Die Anwendung der Vertikal-GVO auf innerstaatliche Wettbewerbsbeschränkungen nach der 7. GWB-Novelle. 2007.

Band 5 Franz Jürgen Säcker: Der Independent System Operator. Ein neues institutionelles Design für Netzbetreiber? 2007.

Band 6 Stefanie Otto: Allgemeininteressen im neuen UWG. § 1 S. 2 UWG und die wettbewerbsfunktionale Auslegung. 2007.

Band 7 Jochen Eichler: Vertragliche Dritthaftung. Eine Auseinandersetzung mit der Frage der Dritthaftung von sogenannten Experten und anderen Auskunftspersonen im Rahmen des § 311 Abs. 3 BGB. 2007.

Band 8 Markela Stamati: Die Anforderungen der operationellen Entflechtung nach den Beschleunigungsrichtlinien der Europäischen Kommission. Umsetzung in Deutschland und Griechenland. 2008.

Band 9 Franz Jürgen Säcker: The Concept of the Relevant Product Market. Between Demand-Side Substitutability and Supply-Side -Substitutability in Competition Law. 2008.

Band 10 Renate Rabensdorf: Die Durchgriffshaftung im deutschen und russischen Recht der Kapitalgesellschaften. Eine rechtsvergleichende Untersuchung. 2009.

Band 11 Franz Jürgen Säcker: Der beschleunigte Ausbau der Höchstspannungsnetze als Rechtsproblem. Erläutert am Beispiel der 380-kV-Höchstspannungsleitung Lauchstädt – Redwitz – Grafenrheinfeld mit Querung des Rennsteigs im Naturpark Thüringer Wald. 2009.

Band 12 Helen Mahne: Eigentum an Versorgungsleitungen. 2009.

Band 13 Franz Jürgen Säcker (Hrsg.): Russisches Energierecht - Gesetzessammlung. 2009.

Band 14 Franz Jürgen Säcker / Maik Wolf: Integrierte Energieversorgung in geschlossenen Verteilernetzen. Zum Gestaltungsspielraum des Gesetzgebers zur Neuregelung des § 110 EnWG im Lichte des Dritten EG-Energiepakets. 2009.

Band 15 Franz Jürgen Säcker (Hrsg.): Das Dritte Energiepaket für den Gasbereich. Deutsch-Englische Textausgabe mit einer Einführung. 2009.

Band 16 Franz Jürgen Säcker (Hrsg.): Das Dritte Energiepaket für den Elektrizitätsbereich. Deutsch-Englische Textausgabe mit einer Einführung. 2009.

Band 17 Thomas Dörmer: Die Unternehmenspacht. Rechtsstellung der Vertragsparteien unter besonderer Berücksichtigung der Pflicht des Unternehmenspächters zur ordnungsgemäßen Unternehmensführung sowie der Rechtslage bei Vertragsbeendigung. 2010.

Band 18 Klaas Bosch: Die Kontrolldichte der gerichtlichen Überprüfung von Marktregulierungsentscheidungen der Bundesnetzagentur nach dem Telekommunikationsgesetz. 2010.

Band 19 Geng-Sook Leem: Einheitliche Corporate Governance-Grundsätze für die Europäische Aktiengesellschaft (SE). Eine rechtsvergleichende Untersuchung anhand der Ausgestaltung der SE im deutschen und britischen Recht. 2010.

Band 20 Wiebke Gebhardt: Gentechnik und Koexistenz nach der Gesetzesnovelle von 2008: Zivilrechtliche Haftung im Vergleich Deutschland und USA. 2010.

Band 21 Cathrin Isenberg: Die Geruchsmarke als Gemeinschaftsmarke. Schutzfähigkeit und Einsatzmöglichkeiten. 2010.

Band 22 Franz Jürgen Säcker / Jochen Mohr / Maik Wolf: Konzessionsverträge im System des europäischen und deutschen Wettbewerbsrechts. 2011.

Band 23 Judith Antonia Loeck: Die unzumutbare Belästigung nach der UWG Novelle 2008 und dem Gesetz zur Bekämpfung unerlaubter Telefonwerbung und zur Verbesserung des Verbraucherschutzes bei besonderen Betriebsformen. 2011.

Band 24 Jörg Jaecks: Konzernverrechnungsklauseln und verwandte einseitige Verrechnungsbefugnisse im Mehrpersonenverhältnis. 2011.

Band 25 Franz Jürgen Säcker: Marktabgrenzung, Marktbeherrschung, Markttransparenz und Machtmissbrauch auf den Großhandelsmärkten für Elektrizität. 2011.

Band 26 Susanne Wende: Die einheitliche Auslegung von Beihilfen- und Vergaberecht als Teilgebiete des europäischen Wettbewerbsrechts. 2011.

Band 27 Leonie Kempel: Die Anwendung von Art. 102 AEUV auf geistiges Eigentum und Sacheigentum. Die Voraussetzungen des Kontrahierungszwangs nach Art. 102 AEUV und der Essential-Facility-Doktrin unter besonderer Berücksichtigung der Unterschiede zwischen geistigem Eigentum und Sacheigentum. 2011.

Band 28 Christoph Schuldt: Werbejingles – schützenswerte Kompositionen!? Die urheberrechtliche und markenrechtliche Schutzfähigkeit von Werbejingles vor unbefugter Nachahmung. 2012.

Band 29 Lydia Scholz: Die Rechtfertigung von diskriminierenden umweltpolitischen Steuerungsinstrumenten. Eine Untersuchung der Reichweite der Warenverkehrsfreiheit und ihrer Begrenzung durch den Umweltschutz als Vertragsziel am Beispiel der deutschen Energiefördergesetze EEG und KWKModG. 2012.

Band 30 Franz Jürgen Säcker: Investitionen in Kraftwerke zur Sicherung einer zuverlässigen Elektrizitätsversorgung nach der Energiewende. Rechtliche und ökonomische Rahmenbedingungen. 2012.

Band 31 Florian Leib: Kartellrechtliche Durchsetzungsstrategien in der Europäischen Union, den USA und Deutschland. Eine rechtsvergleichende Darstellung. 2012.

Band 32 Holger Hoch: Marktverschlusseffekte und Effizienzen vertikaler Zusammenschlüsse. Kartellrechtliche Beurteilung nach europäischem und deutschem Recht. 2012.

Band 33 Sebastian Kemper: Gasnetzzugang in Deutschland und in Spanien. 2012.

Band 34 Elena Timofeeva: Unbundling in der russischen Elektrizitätswirtschaft im Vergleich zum deutschen und europäischen Energierecht. 2012.

Band 35 Gisela Drozella / Harald Krebs: Marktbeherrschung im Bereich Stromerzeugung / Stromgroßhandel. Eine kritische Analyse der neueren Sicht des Bundeskartellamts. 2013.

Band 36 Christian Rehm: Die Einzel- und Gesamtverantwortung der Vorstandsmitglieder der Aktiengesellschaft. Die Verantwortung für die Leitung und Geschäftsführung im mehrköpfigen Vorstand in der unabhängigen und der herrschenden AG. 2013.

Band 37 Ann-Christin Richter: Mengen- und umsatzbezogene Rabatte marktbeherrschender Unternehmen in den Grenzen des Art. 102 AEUV. 2013.

Band 38 Nikolaos Pitsos: Die erhebliche Behinderung des wirksamen Wettbewerbs (SIEC-Test) im Fusionskontrollrecht. Zugleich ein Beitrag zur Problematik der oligopolistischen Marktbeherrschung unter Vergleich mit dem amerikanischen SLC-Test. 2013.

Band 39 Mirko Heinemann: Das Markenzeichen als Anknüpfungspunkt des Markenimages. 2013.

Band 40 Sylvia Lorenz: Der Immaterialgüterschutz virtueller Ladengeschäfte. Ein Vorschlag für die rechtliche Erfassung virtueller Güter. 2014.

Band 41 Anke Reimers: Wettbewerbsrechtliche und regulatorische Analyse der Kapazitätenvergabe an Grenzkuppelstellen im Elektrizitätsbinnenmarkt. 2014.

Band 42 Franz Jürgen Säcker / Carsten König / Lydia Scholz: Der regulierungsrechtliche Rahmen für ein Offshore-Stromnetz in der Nordsee. 2014.

Band 43 David Preisendanz: Der Humor in der vergleichenden Werbung. 2014.

Band 44 Martin Gerig: Vollendung des EU-Energiebinnenmarktes vs. nationale Marktabschottungen. Europarechtskonformität mitgliedstaatlicher Fördermaßnahmen und Kapazitätsmärkte am Beispiel Deutschlands. 2014.

Band 45 Jan Patrick Ehinger: Dritthaftung für Kunstexpertisen und Aufnahmebestätigungen in den Catalogue raisonné. Ein Beitrag zur Expertenhaftung. 2014.

Band 46 Julia Gerzen: Das Recht der Gesellschafter-Fremdkapitalfinanzierung. Rechtsvergleichende Analyse der Behandlung von Gesellschafterdarlehen vor und in der Insolvenz der GmbH nach deutschem und russischem Recht. 2014.

Band 47 Laura Börger: Die gerichtliche Kontrolldichte bei der Überprüfung von Entscheidungen der Europäischen Kommission auf dem Gebiet der Fusionskontrolle. 2014.

Veröffentlichungen des Instituts für Energie- und Regulierungsrecht Berlin

Band 48 Daniel Simon Travers: Kopplungsangebote. Eine Studie zum Vertrags- und Lauterkeitsrecht. 2015.

Band 49 Anna Olbryś-Sobieszuk: EU-rechtliche Schranken der Glücksspielwerbung in Deutschland, Großbritannien und Polen. 2015.

Band 50 Franz Jürgen Säcker/Lydia Scholz/Thea Sveen: Renewable Energy Law in Europe. Challenges and Perspectives. 2015.

Band 51 Lina Berezgova: Festlegung von Stromnetzentgelten in Deutschland und Russland. Eine rechtsvergleichende Untersuchung der anreizbasierten Regulierungsvorgaben. 2015.

Band 52 Fengliang Jin: Aufgreifkriterien in der Fusionskontrolle. Eine rechtsvergleichende Untersuchung des europäischen (Art. 3 FKVO), deutschen (§ 37 GWB), taiwanesischen (§ 6 FHG) und festlandchinesischen (§ 20 AMG) Rechts. 2015.

Band 53 Vasileios Triantafyllidis: Die Einordnung der Passing-On-Defense in das allgemeine Schadensersatzrecht. Eine rechtsvergleichende, dogmatische Studie. 2015.

Band 54 Peipei Liu: Die Inhaltskontrolle von Allgemeinen Geschäftsbedingungen und formularmäßigen Vertragsklauseln im deutsch-chinesischen Vergleich. 2016.

Band 55 Evelyn Kozak: Zur Notwendigkeit eines arbeitsrechtlichen und haftungsrechtlichen Whistleblowerschutzes. 2016.

www.peterlang.com

Beiträge zum Controlling

Herausgegeben von Wolfgang Berens

Band 17

PETER LANG
Frankfurt am Main · Berlin · Bern · Bruxelles · New York · Oxford · Wien

Kundenwertorientierte Vertriebssteuerung im Firmenkundengeschäft der Kreditgenossenschaften

Florian Wüller

Kundenwertorientierte Vertriebssteuerung im Firmenkundengeschäft der Kreditgenossenschaften

Kundenwertmodellierung, -operationalisierung und konzeptionelle Gestaltungsempfehlungen

PETER LANG
Internationaler Verlag der Wissenschaften

Bibliografische Information der Deutschen Nationalbibliothek
Die Deutsche Nationalbibliothek verzeichnet diese Publikation in der Deutschen Nationalbibliografie; detaillierte bibliografische Daten sind im Internet über http://dnb.d-nb.de abrufbar.

Zugl.: Münster (Westfalen), Univ., Diss., 2012

Umschlaggestaltung:
© Olaf Gloeckler, Atelier Platen, Friedberg

Logo auf dem Umschlag:
Logo des Lehrstuhls für Betriebswirtschaftslehre, insb. Controlling der Westfälischen Wilhelms-Universität Münster

D 6
ISSN 1618-825X
ISBN 978-3-631-63724-1

© Peter Lang GmbH
Internationaler Verlag der Wissenschaften
Frankfurt am Main 2012

www.peterlang.de

Geleitwort

Der intensivierte Ertragsdruck drängt die Kreditinstitute im Firmenkundengeschäft dazu, ihre marktpolitischen Strategien weg von der Produktzentrierung hin zu einer beziehungsbezogenen Kundenorientierung anzupassen. Bei der Strategieumsetzung kommt dem Vertrieb als Schnittstelle zwischen Bank und Markt eine besondere Rolle zu. Die strategiekonsistente Ausrichtung des Vertriebs ist die zentrale Aufgabe der Vertriebssteuerung. Bestehende Vertriebssteuerungsansätze in den Banken sind jedoch ungeeignet, den Vertrieb im Sinne der Kundenorientierung zu steuern. In Wissenschaft und Praxis wird als konsistente Steuerungsgröße der Kundenwert genannt. Wie der Kundenwert in ein umfassendes Verständnis der Vertriebssteuerung in Banken als Ziel- und Steuerungsgröße integriert werden kann, ist bisher nicht wissenschaftlich untersucht worden. An dieses Defizit knüpft die Arbeit von Herrn Wüller an. Er fokussiert dabei auf die Kreditgenossenschaften, die etwa 60 Prozent der deutschen Banken ausmachen. Neben ihrer hohen Bedeutung weisen sie die Einzigartigkeit auf, dass sie ihre Mitglieder per Gesetz zu fördern haben und bei ihnen Kunden und Mitglieder in der Regel identisch sind. Daraus resultiert eine besondere Perspektive auf die Integration des Kundenwerts in die Vertriebssteuerung.

Aufbauend auf einer ausführlichen Darstellung des situativen Kontextes konzeptualisiert und operationalisiert Herr Wüller den Wert eines Firmenkunden aus Sicht der Kreditgenossenschaft und erarbeitet darauf aufbauend umfangreiche Empfehlungen für die Integration des Kundenwerts in die Vertriebssteuerung.

Eine besondere Würdigung verdient die fundierte Kundenwertmodellierung, die die Erkenntnisse der Kundenwertforschung auf den Kontext des Firmenkundengeschäfts überträgt. Hervorzuheben ist zudem die Praxisorientierung der Arbeit, die sich vor allem darin äußert, dass Kundenwertmodellierung und Gestaltungsempfehlungen sich nicht von etablierten Konzepten der Gesamtbanksteuerung in dieser Bankengruppe lösen, sondern diese konstruktiv und innovativ integrieren und ergänzen. Gleichzeitig berücksichtigt der Autor in bemerkenswerter Weise die genossenschaftlichen Besonderheiten bei der Herleitung seiner Ergebnisse.

Die Arbeit von Herrn Wüller liefert eine fundierte Grundlage für die Wissenschaft, um sich des Themas der kundenwertorientierten Vertriebssteuerung in Kreditgenossenschaften weiter anzunehmen. Für die Praxis liefert sie umfangreiche Impulse, eine kundenwertorientierte und damit strategiekonsistente Vertriebssteuerung zu implementieren. Ich wünsche ihr daher eine weite Verbreitung.

Münster, im Mai 2012 Prof. Dr. Wolfgang Berens

Vorwort

Die vorliegende Arbeit entstand während meiner Assistententätigkeit am Lehrstuhl für Betriebswirtschaftslehre, insb. Controlling sowie meiner Tätigkeit als Berater bei der BMS Consulting GmbH, Düsseldorf. Sie wurde im Wintersemester 2011/2012 als Dissertationsschrift an der Westfälischen Wilhelms-Universität in Münster angenommen. Besonderen Dank möchte ich meinem akademischen Lehrer und Erstgutachter Herrn Prof. Dr. Wolfgang Berens aussprechen. Die Möglichkeit zur Vereinbarkeit von Lehrstuhltätigkeiten, Beratungsaktivitäten und Promotion sowie die von ihm verfolgte Integration von Wissenschaft und Praxis haben die Promotionszeit zu einem lehrreichen Erlebnis werden lassen. Herrn Prof. Dr. Heinz-Lothar Grob danke ich recht herzlich für die Übernahme des Zweitgutachtens.

Großer Dank gilt auch Herrn Dr. Andreas Siemes und Herrn Dr. Klaus Segbers, die in unvergleichlicher Weise die notwendigen Rahmenbedingungen geschaffen haben, um die Promotion erfolgreich abzuschließen. Gleichzeitig bedanke ich mich bei den Lehrstuhlkollegen in Münster, die trotz räumlicher Distanz eine große Unterstützung bei den kleinen und großen Herausforderungen der Promotion waren, und den Arbeitskollegen in Düsseldorf, die ein gerüttelt Maß zu einer offenen und familiären Unternehmenskultur beitragen, die eine wichtige Stütze bei einem Promotionsvorhaben ist.

Sehr, sehr herzlich möchte ich mich bei meinen Freunden Thorsten Albers, Benjamin Becker, Frank Bellmann, Alexander Eiting, Martin Graumann, Karsten Kraume, Marcel Kolm, Stephan Meinecke, Jörg Micheel, Martin Schmidt und Florian Witt bedanken, die durch ihre Freundschaft bewusst und unbewusst Leuchttürme in dunklen Zeiten waren und so dazu beitrugen, dass ich den Kurs nicht aus den Augen verlor.

Meinen Eltern und meinen Geschwistern danke ich von Herzen dafür, dass sie stets an mich geglaubt, mich gefordert und gefördert haben. Ohne diesen Rückhalt wäre eine Promotion eine nicht zu meisternde Herausforderung gewesen.

Meinen tiefsten und herzlichsten Dank möchte ich an Lina Wenninger richten, die trotz zahlreicher Entbehrungen mit unendlichem Verständnis und Liebe in nicht ermesslicher Weise zu mir gestanden hat.

Ich widme diese Arbeit Anna Schwering, die an mich in unvergleichlicher Weise geglaubt hat, und Jacob, der unser nächstes großes Vorhaben ist.

Reken, im Mai 2012 Dipl.-Kfm. Florian Wüller, BSc

Inhaltsverzeichnis

Abbildungsverzeichnis

Tabellenverzeichnis

Abkürzungsverzeichnis

Abs.	Absatz
AG	Aktivgeschäfte
AGVR	Arbeitgeberverband der Deutschen Volksbanken und Raiffeisenbanken e.V.
Art.	Artikel
B2B	Business-to-Business
bspw.	beispielsweise
BVR	Bundesverband der Deutschen Volksbanken und Raiffeisenbanken e.V.
bzw.	beziehungsweise
c. p.	ceteris paribus
ca.	zirka
CLV	Customer Lifetime Value
d. h.	das heißt
DB	Deckungsbeitrag
DBW	Die Betriebswirtschaft
DEA	Data Envelopment Analysis
DGRV	Deutscher Genossenschafts- und Raiffeisenverband e. V.
DSGV	Deutscher Sparkassen- und Giroverband
dt.	deutsch
e. V.	eingetragener Verein
EDV	Elektronische Datenverarbeitung
eG	eingetragene Genossenschaft
Erg. d. Verf.	Ergänzung des Verfassers
et al.	et alii
EU	Europäische Union
EUR	Euro
f.	folgende Seite

ff.	fortfolgende Seiten
Fn.	Fußnote
GenG	Gesetz betreffend die Erwerbs- und Wirtschaftsgenossenschaften (Genossenschaftsgesetz)
ggf.	gegebenenfalls
HGB	Handelsgesetzbuch
hrsg.	herausgegeben
Hrsg.	Herausgeber
http	Hypertext Transfer Protocol
i. d. R.	in der Regel
i. H. v.	in Höhe von
i. V. m.	in Verbindung mit
IMF	International Monetary Fund (dt.: Internationaler Währungsfonds)
Inv./Invest.	Investierend
IRZ	Zeitschrift für internationale Rechnungslegung
IT	Informationstechnologie
Jg.	Jahrgang
KfW	Kreditanstalt für Wiederaufbau
KG aA	Kommanditgesellschaft auf Aktien
KStG	Körperschaftsteuergesetz
KWG	Gesetz über das Kreditwesen (Kreditwesengesetz)
MaRisk	Mindestanforderungen an das Risikomanagement
MbO	Management by Objectives
MbR	Management by Results
o. S.	ohne Seite
OECD	Organisation for Economic Co-operation and Development
Rn.	Randnummer
S.	Seite(n)

SCE	Societas Cooperativa Europaea (Europäische Genossenschaft)
SCE-VO	Verordnung über das Statut der Europäischen Genossenschaft (SCE)
SE-St	Statut der Sicherungseinrichtung (des BVR)
SG	Sonstige Geschäfte
SolVv	Solvabilitätsverordnung
Sp.	Spalte(n)
Spez.	Spezielle
URL	Uniform Resource Locator
z. B.	z. B.
ZfB	Zeitschrift für Betriebswirtschaft
ZfbF	Zeitschrift für betriebswirtschaftliche Forschung
ZfCM	Zeitschrift für Controlling und Management
ZfgG	Zeitschrift für das gesamte Genossenschaftswesen
ZfgK	Zeitschrift für das gesamte Kreditwesen
ZfO	Zeitschrift für Organisation

Symbolverzeichnis

α:	Anbieterspezifische Ausschöpfung des Kundenwerttreiberpotenzials (Loyalitätsgrad)
$\alpha_{c,p_c,t}$:	Loyalitätsgrad eines Kunden c für das Produkt p_c in der Periode t; mit $0 \leq \alpha_{c,p_c,t} \leq 1$
$\alpha_{c,q_c,t}$:	Loyalitätsgrad eines Kunden c für das Produkt q_c in der Periode t, mit $0 \leq \alpha_{c,q_c,t} \leq 1$ und $\alpha_{c,q_c,t} = \alpha_{c,p_c,t}$, wenn p_c und q_c in einer Substitutionsbeziehung im Sinne des Up-Selling stehen
$\alpha_{c,x_c^{BW},t}$:	Loyalitätsgrad eines Kunden c für das Produkt x_c^{BW} in der Periode t; $0 \leq \alpha_{c,x_c^{BW},t} \leq 1$
$\alpha_{c,x_c^{L},t}$:	Loyalitätsgrad eines Kunden c für das Produkt x_c^{L} in der Periode t; mit $0 \leq \alpha_{c,x_c^{L},t} \leq 1$
$\alpha_{c,x_c^{UE},t}$:	Loyalitätsgrad eines Kunden c für das Produkt x_c^{UE} in der Periode t; mit $0 \leq \alpha_{c,x_c^{UE},t} \leq 1$
$\alpha_{c,y_c^{BW},t}$:	Loyalitätsgrad eines Kunden c für das Produkt y_c^{BW} in der Periode t; mit $0 \leq \alpha_{c,y_c^{BW},t} \leq 1$
$\alpha_{c,y_c^{L},t}$:	Loyalitätsgrad eines Kunden c für das Produkt y_c^{L} in der Periode t; mit $0 \leq \alpha_{c,y_c^{L},t} \leq 1$
$\alpha_{c,y_c^{UE},t}$:	Loyalitätsgrad eines Kunden c für das Produkt y_c^{UE} in der Periode t; mit $0 \leq \alpha_{c,y_c^{UE},t} \leq 1$
$\beta_{c,p_c,t}^{\Delta BP_{c,p_c,t}}$:	Zusätzlicher Loyalitätsgrad des Kunden c für das Produkt p_c in der Periode t bezogen auf das zusätzliche Bedarfspotenzial $\Delta BP_{c,p_c,t}$; mit $0 \leq \beta_{c,p_c,t}^{\Delta BP_{c,p_c,t}} \leq 1$
$\beta_{c,q_c,t}^{\Delta BP_{c,q_c,t}}$:	Zusätzlicher Loyalitätsgrad des Kunden c für das Produkt q_c in der Periode t bezogen auf das zusätzliche Bedarfspotenzial $\Delta BP_{c,q_c,t}$; mit $0 \leq \beta_{c,q_c,t}^{\Delta BP_{c,q_c,t}} \leq 1$
$\delta_{c,t}^{kPK^{aP}}$:	Kundeninduzierte Quote des Kosteneinsparungspotenzials für kundenautonome Prozesse für einen Kunden c in der Periode t; mit $0 \leq \delta_{c,t}^{kPK^{aP}} \leq 1$

$\delta_{c,t}^{kPK^{liP}}$:	Kundeninduzierte Quote des Kosteneinsparungspotenzials für leistungsbezogen-integrative Prozesse für einen Kunden c in der Periode t; mit $0 \leq \delta_{c,t}^{kPK^{liP}} \leq 1$
$\delta_{c,t}^{kPK^{aaP}}$:	Kundeninduzierte Quote des Kosteneinsparungspotenzials für ablaufbezogen-integrative Prozesse für einen Kunden c in der Periode t; mit $0 \leq \delta_{c,t}^{kPK^{aaP}} \leq 1$
$\Delta\alpha_{c,p_c,t}$:	Zusätzlicher Loyalitätsgrads eines Kunden c für das Produkt p_c in der Periode t im Vergleich zur aktuellen Situation bezogen auf das Bedarfspotenzial $BP_{c,p_c,t}$; mit $\Delta\alpha_{c,p_c,t} + \alpha_{c,p_c,t} \leq 1$ und $\alpha_{c,p_c,t} \geq -1 \cdot \Delta\alpha_{c,p_c,t}$
$\Delta\alpha_{c,q_c,t}$:	Zusätzlicher Loyalitätsgrad eines Kunden c für das Produkt q_c in der Periode t im Vergleich zur aktuellen Situation bezogen auf das Bedarfspotenzial $BP_{c,q_c,t}$; mit $\Delta\alpha_{c,q_c,t} + \alpha_{c,q_c,t} \leq 1$ $\alpha_{c,q_c,t} \geq -1 \cdot \Delta\alpha_{c,q_c,t}$
$\Delta BP_{c,p_c,t}$:	Zusätzliches Bedarfspotenzial eines Kunden c für das Produkt p in der Periode t
$\Delta BP(\tau_{c,p_c,t}^{BGP})_{c,p_c,t}$:	Zusätzliches Bedarfspotenzial eines Kunden c für das Produkt p_c in der Periode t unter Berücksichtigung der Ausnutzung der für das Basisgeschäftspotenzial zugrunde gelegten Zahlungsbereitschaft $(= \Delta BP_{c,p_c,t})$
$\Delta BP(\tau_{c,p_c,t}^{max})_{c,p_c,t}$:	Zusätzliches Bedarfspotenzial eines Kunden c für das Produkt p_c in der Periode t unter Berücksichtigung der Ausnutzung der maximalen Zahlungsbereitschaft
$\Delta BP_{c,q_c,t}$:	Zusätzliches Bedarfspotenzial eines Kunden c für das Produkt q_c in der Periode t
$\varepsilon_{c,t}$:	Ausfallwahrscheinlichkeit des Kunden c in der Periode t
$\mu[KWT_t]$:	Erwartungswert der Verteilung der Ausprägungen der Kundenwerttreiber in der Periode t
$\rho_{c,pq}$:	Dummyvariable der Verbundbeziehung zwischen den Produkten p und q mit $\rho_{c,pq} = 1$, wenn eine Substitutionsbeziehung im Sinne des Up-Sellings für den Kunden c vorliegt und $\rho_{c,pq} = 0$, wenn keine Substitutionsbeziehung im Sinne des Up-Sellings für den Kunden c vorliegt

$\tau^{BGP}_{c,p_c,t}$:	Zahlungsbereitschaft in Form von Zins- oder Gebührensätzen des Kunden c für das Produkt p_c in der Periode t, die bei der Bewertung des Basisgeschäftspotenzials zugrunde gelegt wurde
$\tau^{max}_{c,p_c,t}$:	Maximale Zahlungsbereitschaft in Form von Zins- oder Gebührensätzen des Kunden c für das Produkt p_c in der Periode t
$\phi_{c,A}$:	Akquisitionswahrscheinlichkeit für den Kunden c in der Periode der erwarteten Akquisition A (t=A)
A:	Periode der erwarteten Akquisition (t=A)
AE_{ct}:	Autonome Erlöse, die durch den Kunden c in der Periode t verursacht werden
AK_c:	Akquisitionskosten zur Gewinnung des Kunden c (nur relevant bei Bewertung eines Neukunden)
$AK_{c,t}$:	Akquisitionskosten für den Kunden c in der Periode t
$AufK_c$:	Auflösungskosten der Kundenbeziehung mit dem Kunden c
B :	Umsatz-/Geschäftsvolumenabhängiger Bonussatz bzw. Rückvergütungssatz in Prozent
$BGP_{c,t}$:	Basisgeschäftspotenzial eines Kunden c in der Periode t
$BGP'_{c,t}$:	Unter Berücksichtigung von bankleistungsbezogenen Prozesskosten (Standardstückkosten) bewertetes Basisgeschäftspotenzial eines Kunden c in der Periode t
$BGP''_{c,t}$:	Unter Berücksichtigung von bankleistungsbezogenen Prozesskosten (Standardstückkosten) und Kreditrisikokosten bewertetes Basisgeschäftspotenzial eines Kunden c in der Periode t
BK :	Branchenüblicher absoluter Anteil an Bankkrediten
$BK_{zusätzlich}$:	zusätzlicher Bedarf an Bankkrediten
$BP_{c,p_c,t}$:	Bedarfspotenzial eines Kunden c für das Produkt p_c in der Periode t
$BP(\tau^{BGP}_{c,p_c,t})_{c,p_c,t}$:	Bedarfspotenzial eines Kunden c für das Produkt p_c in der Periode t unter Berücksichtigung der Ausnutzung der für das Basisgeschäftspotenzial zugrunde gelegten Zahlungsbereitschaft ($=BP_{c,p_c,t}$)

$BP(\tau^{max}_{c,p_c,t})_{c,p_c,t}$:	Bedarfspotenzial eines Kunden c für das Produkt p_c in der Periode t unter Berücksichtigung der Ausnutzung der leistungsbezogenen maximalen Zahlungsbereitschaft
$BP_{c,q_c,t}$:	Bedarfspotenzial eines Kunden c für das Produkt q_c in der Periode t
$BP_{c,x_c^{BW},t}$:	Bedarfspotenzial eines Kunden c für das Produkt x_c^{BW} in der Periode t
$BP_{c,x_c^{L},t}$:	Bedarfspotenzial eines Kunden c für das Produkt x_c^{L} in der Periode t
$BP_{c,x_c^{UE},t}$:	Bedarfspotenzial eines Kunden c für das Produkt x_c^{UE} in der Periode t
$BP_{c,y_c^{BW},t}$:	Bedarfspotenzial eines Kunden c für das Produkt y_c^{BW} in der Periode t
$BP_{c,y_c^{L},t}$:	Bedarfspotenzial eines Kunden c für das Produkt y_c^{L} in der Periode t
$BP_{c,y_c^{UE},t}$:	Bedarfspotenzial eines Kunden c für das Produkt y_c^{UE} in der Periode t
BS:	Bilanzsumme
C:	Kapitalwert
CE_{ct}:	Cross-Selling-Erlöse, die durch den Kunden c in der Periode t verursacht werden
CLV_c:	Customer Lifetime Value des Kunden c
$CSP_{c,t}$:	Cross-Selling-Potenzial eines Kunden c in der Periode t
$CSP'_{c,t}$:	Unter Berücksichtigung von bankleistungsbezogenen Prozesskosten (Standardstückkosten) bewertetes Cross-Selling-Potenzial eines Kunden c in der Periode t
$CSP''_{c,t}$:	Unter Berücksichtigung von bankleistungsbezogenen Prozesskosten (Standardstückkosten) und Kreditrisikokosten bewertetes Cross-Selling-Potenzial eines Kunden c in der Periode t
$CSP^{L}_{c,t}$:	Loyalitätsbezogenes Cross-Selling-Potenzial eines Kunden c in der Periode t

$CSP_{c,t}^{BW}$:	Bedürfnisweckendes Cross-Selling-Potenzial eines Kunden c in der Periode t
$CSP_{c,t}^{UE}$:	Unternehmensentwicklungsbezogenes Cross-Selling-Potenzial eines Kunden c in der Periode t
D :	Dividenden-/Thesaurierungsbetrag
$EP_{c,t}$:	Entwicklungspotenzial eines Kunden c in der Periode t
$EP'_{c,t}$:	Unter Berücksichtigung von bankleistungsbezogenen Prozesskosten (Standardstückkosten) bewertetes Entwicklungspotenzial eines Kunden c in der Periode t
$EP''_{c,t}$:	Unter Berücksichtigung von bankleistungsbezogenen Prozesskosten (Standardstückkosten) und Kreditrisikokosten bewertetes Entwicklungspotenzial eines Kunden c in der Periode t
FB :	jährlicher Finanzierungsbedarf
FI^{I} :	Kosten des Förderinstruments im Quadrant I
FP :	Potenzial für eine nachhaltige Förderung der aktuellen und zukünftigen Mitglieder
G :	Gewinn
GF_{m} :	Gewichtungsfaktor der Eigenschaft m
GF_{IP} :	Gewichtungsfaktor des Kundenteilscores für das Informationspotenzial
GF_{h}^{IP} :	Gewichtungsfaktor der Eigenschaft h in der Dimension Informationspotenzial
GF_{MVP} :	Gewichtungsfaktor des Kundenteilscores für das Member Value-Potenzial
GF_{l}^{MVP} :	Gewichtungsfaktor der Eigenschaft l in der Dimension Member Value Potenzial
GF_{RP} :	Gewichtungsfaktor des Kundenteilscores für das Referenzpotenzial
GF_{j}^{RP} :	Gewichtungsfaktor der Eigenschaft j in der Dimension Referenzpotenzial

$GGP''_{c,t}$::	Unter Berücksichtigung von bankleistungsbezogenen Prozesskosten (Standardstückkosten) und Risikokosten bewertetes Gesamtgeschäftspotenzial eines Kunden *c* in der Periode *t*
GS_c:	Gesamtscore des Kunden *c*
$HSL_{c,t}$:	Entwicklungspotenzial aus dem Verkauf höherwertiger substitutiver Leistungen eines Kunden *c* in der Periode *t*
$InfoP_{ct}$:	Informationspotenzial des Kunden *c* in der Periode *t* (sofern monetarisierbar)
$InnoP_{ct}$:	Innovationspotenzial des Kunden *c* in der Periode *t* (sofern monetarisierbar)
K:	Nicht in Standardstückkosten verrechnete Gemeinkosten
$KK_{c,t}^{KKP_{c,t}}$:	Ausgaben, die im Zusammenhang mit dem Kooperationspotenzial auf Seiten der Kreditgenossenschaft für den Kunden *c* in der Periode *t* anfallen (Kompetenzkosten)
$KoopP_{ct}$:	Kooperationspotenzial des Kunden *c* in der Periode *t* (sofern monetarisierbar)
$KP_{c,t}$:	Kooperationspotenzial für den Kunden *c* in der Periode *t*
KU:	branchenübliche Kapitalumschlagshäufigkeit
KV:	Konditionsvorteil für Mitglieder
KW_c:	Anbieterspezifischer Kundenwert des Kunden *c* aus Sicht der Gesamtbank
$KW_c^{Vertrieb}$:	Durch die Vertriebsbank beeinflussbarer Kundenwert des Kunden *c*
$KW_c^{Produktion}$:	Durch die Produktionsbank beeinflussbarer Kundenwert des Kunden *c*
$KW_c^{Steuerung}$:	Durch die Steuerungsbank beeinflussbarer Kundenwert des Kunden *c*
KW_c^{m}:	Monetärer periodenübergreifender Kundenwert unter Berücksichtigung des Kreditrisikos und des Kundenplanungsrisikos
KW_c^{mnKR}:	Monetärer periodenübergreifender Kundenwert unter Berücksichtigung des Kreditrisikos

$KW_{c}^{m,u}$:	Monetärer Kundenwert eines Kunden c aus für das Leistungsprogramms eines Unternehmens u der FinanzGruppe aus Sicht der Kreditgenossenschaft
$KW_{c,t}^{mvR}$:	Monetärer Kundenwert vor Integration des Risikos eines Kunden c in der Periode t
KW_{c}^{mvR}:	Periodenübergreifender monetärer Kundenwert vor Integration des Risikos eines Kunden c
KW_{c}^{nm}:	Nicht-monetärer Kundenwert des Kunden c
KWT:	(Anbieterspezifischer) Kundenwerttreiber
KTW_{c}^{IP}:	Kundenteilscore des Kunden c für das Informationspotenzial
KTW_{c}^{MVP}:	Kundenteilscore des Kunden c für das Member Value-Potenzial
KTW_{c}^{RP}:	Kundenteilscore des Kunden c für das Referenzpotenzial
$KWTP$:	Maximale Ausprägung des Kundenwerttreibers (Kundenwerttreiberpotenzial)
L:	durchschnittliche Kreditlaufzeit
$ME_{c,t}^{BGP}$:	Mengeneffekt mit Bezug zum Basisgeschäftspotenzial (*BGP*) eines Kunden c in der Periode t
$ME_{c,t}^{HSL}$:	Mengeneffekt mit Bezug zum Entwicklungspotenzials aus dem Verkauf höherwertiger substitutiver Leistungen (*HSL*) eines Kunden c in der Periode t
MF:	Mittelbare Förderung z. B. durch Dividenden und Verzinsung des Geschäftsguthabens, also Beiträgen aus der Einlegerbeziehung (auch: Mitgliedernutzen)
MK_{ct}:	Marketingkosten für den Kunden c in der Periode t (zur Erreichung der prognostizierten Erlöse und der Bindungsrate)
MV:	Member Value oder innerer Wert der Kreditgenossenschaft für das Mitglied
$MVP_{c,t}$:	Entwicklungspotenzial aus Mehrverkauf eines Kunden c in der Periode t
$MVP_{c,t}^{BGP}$:	Entwicklungspotenzial aus Mehrverkauf eines Kunden c in der Periode t bei Produkten des Basisgeschäftspotenzials (*BGP*)

$MVP_{c,t}^{HSL}$:	Entwicklungspotenzial aus Mehrverkauf eines Kunden c in der Periode t bei Produkten des Entwicklungspotenzials aus dem Verkauf höherwertiger substitutiver Leistungen (*HSL*)
$MX_{c,t}^{BGP}$:	Mixeffekt mit Bezug zum Basisgeschäftspotenzial (*BGP*) eines Kunden c in der Periode t
$MX_{c,t}^{HSL}$:	Mixeffekt mit Bezug zum Entwicklungspotenzials aus dem Verkauf höherwertiger substitutiver Leistungen (*HSL*) eines Kunden c in der Periode t
$OK_{c,t}^{KKP_{c,t}}$:	Kosten, die im Zusammenhang mit dem Kooperationspotenzial auf Seiten der Kreditgenossenschaft für den Kunden c in der Periode t anfallen (Opportunitätskosten)
$PP_{c,t}$:	Preispotenzial eines Kunden c in der Periode t (bezogen auf das Basisgeschäftspotenzial und das Mehrverkaufspotenzial)
$PP_{c,t}^{BGP}$:	Preispotenzial des Kunden c in der Periode t bezogen auf das Basisgeschäftspotenzial (*BGP*)
$PP_{c,t}^{MV^{BGP}}$:	Preispotenzial des Kunden c in der Periode t bezogen auf den Mehrverkauf der Produkte des Basisgeschäftspotenzial (MV^{BGP})
$PSLK_{ct}$:	Produkt-, Service- und Logistikkosten, die durch den Kunden c in der Periode t verursacht werden
PW_t:	Potenzialwert der Kreditgenossenschaft zum Zeitpunkt t
R_{ct}:	Bindungsrate des Kunden c in der Periode t
RA_t:	Subjektiver Risikoabschlag in der Periode t
RF:	refinanzierungsbedingter Kreditbedarf
RW_{ct}:	Referenzwert des Kunden c in der Periode t (sofern monetarisierbar)
S_{cm}:	Score des Kunden c bei der Eigenschaft m
$S_{c,h}^{IP}$:	Score des Kunden c bei der Eigenschaft h in der Dimension Informationspotenzial
$S_{c,j}^{RP}$:	Score des Kunden c bei der Eigenschaft j in der Dimension Referenzpotenzial
$S_{c,l}^{MVP}$:	Score des Kunden c bei der Eigenschaft l in der Dimension Member Value-Potenzial

$SÄ[KWT_t]$:	Sicherheitsäquivalent für einen Kundenwerttreiber in der Periode t
T:	Lebensdauer der Geschäftsbeziehung mit einem Kunden
U:	Jahresumsatz
UE_{ct}:	Up-Selling-Erlöse, die durch den Kunden c in der Periode t verursacht werden
UF:	Unmittelbare (reale bzw. naturale) Förderung des Mitglieds aus den Leistungsbeziehungen zur Kreditgenossenschaft, die zu vereinbarten Konditionen und Standards erfolgt (auch: Kundennutzen)
$USP_{c,t}$:	Up-Selling-Potenzial eines Kunden c in der Periode t
$VE_{c,t}^{BGP}$:	Verdrängungseffekt mit Bezug zum Basisgeschäftspotenzial (*BGP*) eines Kunden c in der Periode t
$VE_{c,t}^{HSL}$:	Verdrängungseffekt mit Bezug zum Entwicklungspotenzial aus dem Verkauf höherwertiger substitutiver Leistungen (*HSL*) eines Kunden c in der Periode t
VW_t:	Vermögenswert der Kreditgenossenschaft zum Zeitpunkt t
ZW_t:	Zerschlagungswert der Kreditgenossenschaft zum Zeitpunkt t
a_t:	(erwartete) Auszahlungen in der Periode t
$b_{c,t}$:	Kundenbindungsrate für den Kunden c in der Periode t
c:	Index für Kunden
d:	Dividenden-/Thesaurierungssatz in Prozent
e_t:	(erwartete) Einzahlungen in der Periode t
$f(\cdot)$:	Nutzenfunktion
$g(\cdot)$:	Kundenwertfunktion
h:	Index der Eigenschaften des Informationspotenzials
i:	Kalkulationszinssatz
i_t:	Periodisch differenzierter Diskontierungsfaktor
j:	Index der Eigenschaften des Referenzpotenzials

k:	Standardstückkosten
k^Q:	Qualitätsbedingter Stückkostenaufschlag
$kPK_{c,t}$:	Kundenbezogene Prozesskosten eines Kunden c in der Periode t
$kPK_{c,t}^{aP}$:	Kundenbezogene Prozesskosten eines Kunden c in der Periode t für autonome Prozesse
$kPK_{c,t}^{liP}$:	Kundenbezogene Prozesskosten eines Kunden c in der Periode t für leistungsbezogen-integrative Prozesse
$kPK_{c,t}^{aiP}$:	Kundenbezogene Prozesskosten eines Kunden c in der Periode t für ablaufbezogene-integrative Prozesse
kq:	branchenüblicher Anteil der Bankkredite an der Bilanzsumme
l:	Index der Eigenschaften des Member Value-Potenzials
m:	Index der Eigenschaften bzw. Bewertungskriterien im Punktbewertungsverfahren
n:	(erwartet) Restdauer der Geschäftsbeziehung
p:	Marktpreis je Stück
p_c:	Index der Produkte im Sinne des Basisgeschäftspotenzials aus Sicht des Kunden c
q_c:	Index der Produkte im Sinne des Up-Sellings aus Sicht des Kunden c
r_c:	Kundenspezifischer, risikoadäquater Zinssatz
s_i^P:	Strategiealternative i der Produktionsbank
s_i^S:	Strategiealternative i der Steuerungsbank
s_i^P:	Strategiealternative i der Produktionsbank
t:	Periodenindex
u:	Index der beteiligten Unternehmen der FinanzGruppe
w:	Wachstumsrate des Unternehmens bezogen auf die Bilanzsumme
x_c^{BW}:	Index der aktuell vorhandenen Produkte im Sinne des be-

dürfnisweckenden Cross-Selling-Potenzials aus Sicht des Kunden

x_c^L: Index der aktuell vorhandenen Produkte im Sinne des loyalitätsbezogenen Cross-Selling-Potenzials aus Sicht des Kunden

x_c^{UE}: Index der aktuell vorhandenen Produkte im Sinne des bedürfnisweckenden Cross-Selling-Potenzials aus Sicht des Kunden

y_c^{BW}: Index der aktuell noch nicht vorhandenen Produkte im Sinne des bedürfnisweckenden Cross-Selling-Potenzials aus Sicht des Kunden

y_c^L: Index der aktuell noch nicht vorhandenen Produkte im Sinne des loyalitätsbezogenen Cross-Selling-Potenzials aus Sicht des Kunden

y_c^{UE}: Index der aktuell noch nicht vorhandenen Produkte im Sinne des bedürfnisweckenden Cross-Selling-Potenzials aus Sicht des Kunden

z_c: Risikozuschlag für den Kunden c

A Einleitung

1 Problemstellung und Motivation der Arbeit

GEORG KOTTMANN, Bankier, berichtete in einem Vortrag über einen, in den Wirren der letzten Wochen des zweiten Weltkriegs vor den alliierten Truppen flüchtenden Vorstand eines thüringischen Kreditinstituts. Er fragte die Zuhörer, was sie dem unter Zeitdruck stehenden Bankvorstand geraten hätten, aus der Bank unter Berücksichtigung des Platzmangels beim Transport mitzunehmen. „Kasse“, „Tresor“, „Gold“, „Wertpapiere“ wurden genannt. KOTTMANNS Antwort war: „Die Kundenliste hat er mitgenommen! Und mit dieser Liste hat er das Geschäft in kurzer Zeit an anderer Stelle wieder aufgebaut.“[1] Seit Ende des zweiten Weltkriegs hat sich das Bankgeschäft in Deutschland stark gewandelt. Zentrales Charakteristikum dieses Wandels ist die Entwicklung des Bankenmarkts von einem Verkäufer- zu einem Käufermarkt. Diese Entwicklung, die aus dem Markteintritt so genannter Non- und Near-Banks,[2] dem Eintritt ausländischer Banken, den Fortschritten in der Informations- und Kommunikationstechnologie und der zunehmenden Autonomie sowie komplexeren Anforderungen der Kunden resultiert, wird als Begründung eines intensivierten Wettbewerb angesehen.[3]

Auch das Bankgeschäft mit gewerblichen Kunden[4] weist diese Intensivierungstendenzen auf.[5] Spezifische Einflussfaktoren sind die positive wirtschaftliche Entwicklung im Zeitraum 2003 bis zur Finanzmarktkrise in 2008, die neuen Herausforderungen an die Unternehmen in einem globalisierten Umfeld[6] sowie die strategischen Wechsel der Banken und Bankengruppen.[7] Bedingt durch die

---

1 Wiedergegeben nach *Plinke, W.* (1997), S. 115.

2 Zu den Near-Banks zählen z. B. Kreditkartengesellschaften, Anlageberater und Makler und den Non-Banks werden bspw. Versand- und Warenhäuser zugeordnet, die Bankleistungen im Kontext der eigenen Absatzförderung einsetzen. Vgl. dazu ausführlich *Wiedemann, A.* (1992), S. 72 ff.

3 Vgl. *BVR* (2007b), S. 4. Vgl. auch die Ausführungen in Kapitel C.1.5.

4 Für eine detaillierte Abgrenzung des gewerblichen und privaten Geschäfts vgl. Kapitel C.2.2.1.

5 So geben ein Drittel der im Rahmen einer Studie des FRAUNHOFER INSTITUTS befragten Banken an, einen intensivierten Wettbewerb im gewerblichen Kundensegment festzustellen. Vgl. *Engstler, M./Praeg, P./Voecke, C.* (2007), S. 7. Ähnlich ist die Einschätzung von *Rolfes, B.* (2002), S. 141.

6 Vgl. *Lippmann, I.* (2006), S. 396; *Schulz, T. C.* (2005), S. 71.

7 So nennt etwa DEHNE als einen Grund der Wettbewerbsintensivierung die Tatsache, dass die Großbanken die Attraktivität und Chancen des gewerblichen Kundensegments

Finanzmarktkrise wird erwartet, dass die Wettbewerbsintensität weiter steigen wird.[8] „Banks are renewing their focus on the core products and segments they neglected during the boom years. Corporate banks can expect to find more players trying harder to win the same clients.“[9] Vor dem Hintergrund der relativ zeitstabilen Marktanteilsstrukturen[10] in den letzten zehn Jahren auf dem deutschen Firmenkundenmarkt indiziert diese Skizze der aktuellen Ausgangslage zwei wesentliche Implikationen. Zum einen zeichnet sich ab, dass ob der verteilten Marktanteile bei gleichzeitig nur geringen Wachstumsperspektiven des Gesamtmarktes[11] die strukturellen Voraussetzungen für einen Preiswettbewerb gegeben sind und dieser von Theorie und Praxis konstatiert wird.[12] Folglich liegt ein zunehmender Ertragsdruck auf den Banken und Bankengruppen.[13] Zum anderen kann angenommen werden, dass Wachstumspotenziale für die Banken weniger in ressourcenintensiven Akquisitionsvorhaben als vielmehr in Strategien, den bestehenden Kundenbestand auszuschöpfen, liegen. Die wesentliche Konsequenz dieser Implikationen für die einzelnen Banken und Bankengruppen ist, dass ein strategisch-marktpolitisches Umdenken gefordert und umgesetzt wird.[14] Nachdem zuletzt vor allem Kostensenkungsprogramme realisiert wurden, rücken nun marktgerichtete Aktivitäten wieder in den Fokus, um die Zu-

---

neu entdeckt haben und zunehmend um diese Kunden werben. Vgl. *Dehne, T.* (2008), S. 53.

8 Vgl. *Osman, Y.* (2009), o. S.; *Hölzer, K./Schnarr, T.* (2010), S. 55.

9 *Boston Consulting Group* (2009a), S. 16.

10 Vgl. *Wiedemann, A.* (1992), S. 42. Gemessen an dem an Unternehmen und Selbständige ausgereichten Kreditvolumen hatte der Genossenschaftssektor einen Marktanteil von ca. 16 Prozent im Jahr 2000 und ca. 15 Prozent im Jahr 2010, der Sparkassensektor hatte 2000 einen Marktanteil von ca. 40 Prozent und in 2010 einen Anteil von ca. 43 Prozent und die Kreditbanken erzielten einen Marktanteil von ca. 31 Prozent in 2000 sowie einen Marktanteil von ca. 27 Prozent in 2010. Die übrigen kreditgebenden Banken hatten Marktanteile von ca. 13 Prozent in 2000 und ca. 15 Prozent in 2010. Vgl. *Deutsche Bundesbank* (2011a).

11 Die Boston Consulting Group prognostiziert als Auswirkung der Finanzmarktkrise geringe Wachstumsaussichten in den saturierten Industrieländern wie Deutschland. Vgl. *Boston Consulting Group* (2009b), S. 5.

12 Eine ausführliche Darstellung von Studien zur Profitabilität des Firmenkundengeschäfts erfolgt in Kapitel C.1.5.

13 Vgl. *Segbers, K.* (2007), S. 74.

14 So sahen nach einer Umfrage von Steria Mummert Consulting 54 Prozent der 100 befragten Entscheider aus Banken und Bankengruppen in Deutschland den Wettbewerb und 32 Prozent die Ertragseinbrüche als größte Herausforderungen für die Banksteuerung. Vgl. *Steria Mummert Consulting* (2007), S. 11 und S. 18; ähnlich auch *Schmoll, A.* (1997), S. 21 ff.

kunft der Banken zu gestalten.[15] Während es in ungesättigten Märkten sinnvoll ist, die Produktbekanntheit zu steigern und Kunden zu akquirieren, findet in den gegenwärtig wettbewerbsintensiven Märkten ein Wandel von der produktzentrierten zur kunden-orientierten Geschäftsausrichtung statt.[16] Somit ist es nur konsequent, wenn die Banken für sich in Anspruch nehmen, den „Wiedergewinn der Kundennähe“[17] zu forcieren. Doch die „jahrzehntelange Dominanz von Anbieter- und Produktzentrierung hat in vielen Banken die Kundenorientierung überlagert […].“[18] Auf Käufermärkten gelten der Kunde und dessen Bedarf als knappes Gut,[19] so dass das bisherige, von SCHÖNING als beinahe hoheitlich bezeichnete Verhalten der Banken nicht mehr adäquat ist.[20]

Werden die marktpolitischen Ansätze betrachtet, die gegenwärtig von den Banken umgesetzt werden, ist zu konstatieren, dass sich diese auf den auch in der Marketingwissenschaft vollzogenen Wechsel von der eher kurzfristigen Transaktionsorientierung, die kongruent zu einer produkt- und anbieterzentrierten Sicht ist, hin zur Beziehungsorientierung bewegen.[21] Nicht der schnelle und einmalige Produktverkauf wird als Ziel proklamiert, sondern die möglichst systematische Ausschöpfung des gesamten bankrelevanten Bedarfs einer bestehenden oder aufzubauenden Kundenbeziehung.[22] In der wissenschaftlichen Theorie und Praxis wurde und wird dieser Ansatz in Form des Beziehungsmarketings intensiv erforscht.[23] Nach DILLER wird beim Beziehungsmarketing, das hier nicht im generischen Sinn, sondern als strategische absatzpolitische Konzeption verstanden wird, der Markterfolg durch systematisches Management, d. h. Analyse, Planung, Organisation und Kontrolle individueller Kundenbeziehungen

15 Vgl. *Engstler, M./Praeg, P./Voecke, C.* (2007), S. 7.

16 Vgl. *Lippmann, I.* (2006), S. 392.

17 *Engstler, M./Praeg, P./Voecke, C.* (2007), S. 1.

18 *Benölken, H.* (2002), S. 24. Ähnlich auch *Pelzl, U.* (2006), S. 8.

19 Vgl. *Lütje, S.* (2009), S. 3.

20 Vgl. *Schöning, S.* (2003), S. 1043, der diese Attitüde seitens der Banken auch im allgemeinen Sprachgebrauch in Begriffen wie „Kreditantrag“ und „Bankbeamter“ wiederfindet.

21 Vgl. *Bruhn, M.* (2009), S. 1 ff.; *Gelbrich, K.* (2001), S. 10 ff.; *Plinke, W.* (1989), S. 305 ff. Siehe auch die Ausführungen zu den Vertriebsansätzen der Kreditgenossenschaften in Kapitel C.2.4.1.

22 Vgl. *Schmoll, A.* (1997), S. 23.

23 Vgl. für eine pointierte Darstellung des Beziehungsmarketings z. B. *Diller, H.* (1995b). In dieser Arbeit wird der Begriff des Beziehungsmarketing aufgrund der inhaltlichen Schnittmengen synonym zu den Begriffen Relationship Marketing und Customer Relationship Marketing verwendet. Analog *Diller, H.* (2006), S. 99.

gesucht.[24] Oberstes Ziel des Beziehungsmarketings ist, langfristige Kundenbeziehungen aufzubauen, mit denen ökonomische Vorteile für die Bank erreicht werden können. Als Grundlage langfristiger Beziehungen gilt die Kundenbindung, die vielschichtig begründet ist[25] und als Ergebnis von Kundenzufriedenheit aufgefasst wird. Diese wird wiederum beeinflusst durch kundenorientiertes bzw. kundennahes Verhalten eines Anbieters.[26] Dass „Nähe zum Kunden" einen wichtigen Erfolgsfaktor für Unternehmen darstellt, ist das Ergebnis der Untersuchungen von PETERS/WATERMAN.[27] Jedoch gilt es, das Ausmaß der Kundennähe als Optimierungsproblem zu verstehen.[28] Daher kann im Sinne des Beziehungsmarketing die Kundennähe und Kundenorientierung als effektivitätsorientiertes Vorgehen beschrieben werden, das lange Zeit im Fokus von Wissenschaft und Praxis stand.[29] Aufgrund der parallel – zunächst jedoch noch unverbunden mit diesen Entwicklungen im Marketing – zunehmenden Verbreitung der wertorientierten Unternehmensführung[30] wird in jüngerer Zeit auch für das Marketing und den Vertrieb die Notwendigkeit gesehen, markt- und kundengerichtete Maßnahmen im Rahmen von Unternehmenswertanalysen hinsichtlich ihrer Effizienz zu prüfen. Empirische Studien stellen dazu fest, dass Kunden in unterschiedlichem Ausmaß zum wirtschaftlichen Erfolg einer Bank beitragen (siehe Tabelle 1).

---

24 Vgl. *Diller, H.* (2006), S. 99; *Diller, H.* (1995b). Im generischen Sinn ist Marketing als bewusst marktorientierte Führung des gesamten Unternehmens oder marktorientiertes Entscheidungsverhalten zu verstehen. Vgl. *Meffert, H.* (1998), S. 7.

25 Vgl. dazu ausführlich *Eggert, A.* (1999).

26 Vgl. *Krafft, M.* (2007), S. 11 ff. Zu berücksichtigen ist, dass die Kundenorientierung u. a. die Kundennähe umfasst, die von KRAFFT als relevantes Konstrukt untersucht wird. Für eine ausführliche Erläuterung dieser beiden Konstrukte vgl. *Krafft, M.* (2007), S. 12 ff.

27 Vgl. Peters, T. J./Waterman, R. H. (1991), S. 35 f.

28 Vgl. z. B. die Studien von *Reinartz, W. J./Kumar, V.* (2000) und *Reinartz, W. J./Krafft, M.* (2001), die zeigen, dass Kundenbindung nicht zwingend mit Unternehmenserfolg oder Kundenerfolg verbunden ist.

29 Vgl. *Lissautzki, M.* (2007), S. 1 ff.

30 Als grundlegend gelten dazu die Arbeiten von RAPPAPORT. Vgl. z. B. *Rappaport, A.* (1999).

Tabelle 1: Rentabilitätsbezogene Kundenstruktur in ausgewählten Studien[31]

Quelle	Rentabilitätsbezogene Bankkundenstruktur
GANDY (2001) (mehrere Unternehmen)	52% der Kunden sind wertgenerierend, 23% sind wertvernichtend und 25% sind wertneutral
COSTANZO (1995) (mehrere Unternehmen)	20% der Kunden erwirtschaften 100% des Vorsteuerergebnisses, 30% vernichten 40% des möglichen Ergebnisses
SCHLENZKA (1987) (anonymes Beispiel)	4% der Kunden erwirtschaften 90% des Ergebnisses, 72% der Kunden vernichten 33% des möglichen Ergebnisses
HAGEMANN (1986) (Privatkundenbereich)	68% der Kunden vernichten fast zwei Drittel der durch die restlichen Kunden erwirtschafteten Gewinne

Als zentrale Steuerungsgröße richten Wissenschaft und Praxis vor diesem Hintergrund den Blick auf den Kundenwert, mithin auf die Frage, welche Erfolge mit dem Auf- und Ausbau von Kundenbeziehungen realisiert werden können. Es hat sich die Erkenntnis durchgesetzt, dass der Kundenwert als zentrale Größe für die Bewertung und Steuerung von Kundenbeziehungen zu nutzen sei.[32] Da nach herrschender Meinung der Beitrag eines Kunden zum Erfolg eines Unternehmens – entgegen den in Tabelle 1 genannten Studien – über seine rein monetäre Profitabilität hinausgeht, lag der Schwerpunkt der ersten wissenschaftlichen Beiträge zu diesem Thema in der Konzeptualisierung und Operationalisierung des Konstrukts Kundenwert.[33] Im weiteren Verlauf der Forschung wurden Fragen nach der Integration des Kundenwerts in das (Beziehungs-)marketing[34], das Kundenmanagement[35] sowie die Unternehmensführung[36] und -bewertung[37] erörtert. Hinsichtlich der vornehmlich betrachteten Entscheidungsfelder lässt sich konstatieren, dass vor allem Marketingentscheidungen der Segmentierung, der Selektion attraktiver Kunden oder Kundensegmente sowie der kundenwertsteigernden Auswahl kommunikationspolitischer Strategien analysiert wurden.[38] Es gibt zudem einige Arbeiten, die sich dem Kundenwert aus der Sicht des Control-

31 Eigene Darstellung.

32 Vgl. *Stüker, D.* (2008), S. 30 sowie die dort genannte Literatur.

33 Vgl. z. B. *Schemuth, J.* (1996), *Schleuning, C.* (1997), *Cornelsen, J.* (2000), *Rudolf-Sipötz, E.* (2001) und *Gelbrich, K.* (2001), wenngleich sich kein einheitliches Ergebnis feststellen lässt.

34 Vgl. z. B. *Tewes, M.* (2003).

35 Vgl. z. B. *Eberling, G.* (2002).

36 Vgl. z. B. *Lissautzki, M.* (2007) und *Stüker, D.* (2008).

37 Vgl. z. B. *Gupta, S./Lehmann, D. R./Stuart, J. A.* (2004); *Bauer, H.-H./Hammerschmidt, M.* (2005); *Wiesel, T./Skiera, B.* (2007).

38 Die einzige umfassende Analyse für das Marketing bietet *Tewes, M.* (2003).

lings nähern.[39] Eine in diesen Kontext aus Beziehungsmarketing und Kundenwert eingefasste explizite wissenschaftliche Analyse der Integration des Kundenwerts in andere Politiken des Marketings, wie z. B. die Produkt-, die Preis- oder die Vertriebspolitik, lassen sich jedoch kaum finden,[40] so dass auch die Integration des Kundenwerts in die Entscheidungsfelder und Steuerung dieser Politiken als noch weitgehend unerforscht bezeichnet werden kann.

Wird nun der Fokus auf den hier interessierenden Untersuchungsbereich des Firmenkundengeschäfts bei Banken gerichtet, ist festzustellen, dass der zentrale Träger der operativen und strategischen Aufgaben hinsichtlich des Beziehungsmarketings die Schnittstelle zwischen Bank und Firmenkunde, der Vertrieb ist.[41] Dabei muss dem Grundsatz „Banking ist und bleibt People Business"[42] folgend die besondere Stellung des persönlichen Vertriebs herausgehoben werden. Wie bei allen arbeitsteiligen Prozessen, in denen wirtschaftliche Akteure involviert sind, bedarf es Mechanismen zur zielgerichteten Entscheidungskoordination. Entscheidungen zur Ausgestaltung des Vertriebs leiten sich aus dem sachlich übergeordneten Marketing ab und werden in der Vertriebssteuerung zusammengefasst. Die Vertriebssteuerung dient dazu, die im Vertrieb gebündelten Ressourcen und Prozesse im Sinne der Gesamtbankzielsetzung abzustimmen und zu lenken. Es ist jedoch zu konstatieren, dass die in der Praxis gängigen Ansätze der Vertriebssteuerung nicht konsistent mit den gegenwärtigen marketingpolitischen Ansätzen sind. Grundsätzlich lassen sich drei typische Vertriebssteuerungsansätze unterscheiden, die in Abhängigkeit des Inhalts der Ziel- und Steuerungsgrößen als volumenorientiert, wertorientiert oder aktivitätenorientiert bezeichnet werden können.

In der volumenorientierten Vertriebsteuerung gilt die These, dass die Höhe kundenbezogener Bankbilanzaktiv- oder -passivpositionen mit dem Ertrag posi-

39 Vgl. z. B. *Schmöller, P.* (2001), *Weber, J./Lissautzki, M.* (2004), *Mödritscher, G. J.* (2008) und *Stüker, D.* (2008). Vgl. zum Controllingbegriff *Berens, W./Bertelsmann, R.* (2002), Sp. 280 ff.

40 Arbeiten zum Thema Kundenwert, die über die Segmentierungsentscheidungen hinausgehen, und den Kundenwert z. B. in Zielsysteme des Vertriebs integrieren, wurden vom Verfasser in den für die vorliegende Arbeit vorgenommenen Literaturrecherchen kaum gefunden. Lediglich die Arbeit von *Buhl, H. U./Dzienziol, J./Heidemann, J.* (2010) kann in diesem Zusammenhang genannt werden.

41 Vertriebsmitarbeiter werden in Banken typischerweise als Firmenkunden- oder Gewerbekundenberater oder -betreuer bezeichnet. Für eine theoretische Abgrenzung des Vertriebs vgl. Kapitel B.1 und für eine Darstellung des Vertriebs im Firmenkundengeschäft der Kreditgenossenschaften Kapitel C.2.4.

42 *Cromme, G.* (2008), S. 431.

tiv hoch korrelieren.[43] Folglich ist in der Vertriebssteuerung die zentrale Steuerungsgröße das Volumen, während Ertragspositionen weitestgehend ausgeblendet werden. Es ist evident, dass dieser Ansatz ungeeignet ist, bei erodierenden Margen zielgerichtete Steuerungsimpulse zu geben.

Als Ziel- und Steuerungsgröße im Vertrieb fokussiert die wertorientierte Steuerung auf den Deckungsbeitrag. Im Gegensatz zur volumenorientierten Steuerung wird nicht auf Bestands-, sondern auf Stromgrößen, die den betriebswirtschaftlichen Erfolg eines Geschäfts darstellen,[44] rekurriert. Da das Bezugsobjekt das einzelne Geschäft ist, gelingt es diesem eher kurzfristig orientierten Steuerungsansatz nicht, die Anforderungen des Beziehungsmarketing mit der über die einzelne Transaktion hinausgehenden Perspektive zu erfüllen.

Die aktivitätenorientierte Steuerung nutzt vertriebliche Aktivitäten als Bezugsobjekt. Es wird vereinfachend angenommen, je aktiver der Vertrieb ist, desto ertragreicher ist die Bank.[45] Dieser Steuerungsansatz wird i. d. R. als zusätzliche Komponente zur volumen- oder wertorientierten Steuerung implementiert. Aufgrund ihrer reinen Inputsicht ist sie in isolierter Form nicht geeignet, effizienzbezogene Aussagen im Kontext des Beziehungsmarketings zu machen. In kombinierter Form mit den outputorientierten Ansätzen der volumen- oder wertorientierten Steuerung müssen jedoch die bereits genannten Defizite dieser Ansätze berücksichtigt werden.

Mit diesen ungeeigneten Ansätzen geht einher, dass die Erwartungen an die Erfolgswirksamkeit des Beziehungsmarketings nicht erfüllt werden. So finden REINARTZ/ KRAFFT/HOYER in einer branchenübergreifenden Studie heraus, dass die „[.] implementation of CRM processes is more likely to improve performance when the company develops an incentive and organizational scheme to support CRM-compatible behaviour."[46] Basierend auf den neuen Rahmenbedingungen nennt WIMMER folglich als zentrales Problem der Vertriebssteuerung in praxi die Festlegung von Zielgrößen.[47] Zugleich stellen BURMANN und VON BOEHM-BEZING fest, dass die Anwendung einer kundenwertorientierten Steuerungsgröße vor allem bei Banken sinnvoll und wichtig sei.[48] Kreditgenossen-

---

43 Vgl. *Rehberg, M.* (2004), S. 10 f. Dies ist im industriellen Kontext gleichzusetzen mit der Annahme, Umsatz und Ertrag seien hoch positiv korreliert.

44 Vgl. *Schierenbeck, H.* (2003), S. 44 f.

45 Vgl. *Schmoll, A.* (2008), S. 52; *Rehberg, M.* (2004), S. 11.

46 *Reinartz, W. J./Krafft, M./Hoyer, W. D.* (2004), S. 302.

47 Vgl. *Wimmer, K.* (2009), S. 210.

48 Vgl. *Burmann, C.* (2003), S. 130; *von Boehm-Bezing, C.-L.* (2002), S. 90.

schaften[49], die im Fokus der vorliegenden Arbeit stehen, stellen eine besondere Form der Banken dar.

Aufgrund ihrer Rechtsform, der eingetragenen Genossenschaft (eG), wird den genossenschaftlichen Primärbanken in Deutschland eine besondere Ausprägung des Beziehungsmarketings bereits in die Wiege gelegt, die Mitgliederorientierung.[50] Wenngleich nicht jeder Kunde Mitglied der Genossenschaft ist, ist es ob dieser einmaligen Historie jedoch umso erstaunlicher, dass insbesondere im Genossenschaftssektor, der mit seinen fast 1.200 typischen und atypischen Kreditgenossenschaften[51] über 60 Prozent der rechtlich und wirtschaftlich selbständigen,[52] deutschen Kreditinstitute ausmacht, diese inadäquaten und inkonsistenten Vertriebssteuerungsansätze dominieren. Dies dürfte ein Grund dafür sein, warum sich der Wunsch der Genossenschaftsbanken nach einer besseren Kundenorientierung der Mitarbeiter nicht erfüllen lässt.[53] Der ARBEITGEBERVERBAND DER DEUTSCHEN VOLKSBANKEN UND RAIFFEISENBANKEN E.V. (AGVR) beurteilt es als Fehlentwicklung der kurzfristigen Deckungsbeitragssteuerung sowie der volumen- und mengenorientierten Produktsteuerung, wenn die Vertriebsmitarbeiter nach dem „Prinzip der verbrannten Erde" arbeiten und aus

49 Der Begriff der Kreditgenossenschaft wird in dieser Arbeit synonym mit den Begriffen Primärbank, VR-Bank und Genossenschaftsbank verwendet.

50 Vgl. dazu ausführlich Kapitel C.1.4. Kreditgenossenschaften entstanden in der zweiten Hälfte des 19. Jahrhunderts zur Bewältigung der Kreditprobleme kleiner und mittlerer Unternehmen bzw. Unternehmer. Im Laufe der Geschichte haben sich die daraus hervorgegangenen Volks- und Raiffeisenbanken bedingt durch die wirtschaftlichen Rahmenbedingungen von den ursprünglichen Selbsthilfeeinrichtungen zu Universalbanken entwickelt, die ihren Kunden und Mitgliedern eine breite Palette an Produkten und Dienstleistungen anbieten.

51 REITH unterscheidet bei den Kreditgenossenschaften „typische" und „atypische" Primärbanken. „Atypische" Primärbanken lassen sich dadurch abgrenzen, dass sie ihren Förderauftrag auf eine spezielle Zielgruppe ausrichten und/oder der Errichtungszweck der Förderung einer bestimmten Idee folgt und/oder ihr Geschäftsgebiet nicht geographisch begrenzt ist und/oder sie ein Verbundinstitut mit Spezialaufgaben sind und/oder nicht dem Genossenschaftssektor entstammen. Trifft mindestens eins der vorgenannten Kriterien zu, so gilt die betrachtete Bank als „atypische" Primärbank. Vgl. *Reith, D.* (1997), S. 87 ff. Dazu gehören bspw. die kirchlichen Kreditgenossenschaften oder die DEUTSCHE ÄRZTE- UND APOTHEKERBANK EG. Da diese Unterscheidung auf der Primärebene für die vorliegende Arbeit irrelevant ist, wird sie im weiteren Verlauf der Analyse nicht weiter berücksichtigt.

52 Aufgrund der engen Beziehungen zur FINANZGRUPPE gilt die wirtschaftliche Selbständigkeit nur de jure jedoch nicht de facto, da ohne die intensive Verknüpfung zur FINANZGRUPPE nur eine Minorität der Kreditgenossenschaften überlebensfähig wäre.

53 Vgl. *AGVR* (2001), S. 14.

kurzfristigen Anreizen die langfristige Kundenbindung vernachlässigen.[54] Und so sehen auch SCHIERENBECK/POHL die „Sicherstellung nachhaltiger Kundenbeziehungen als Herausforderung für bankbetriebliche Steuerungssysteme“[55].

Aufgrund ihrer großen Bedeutung für den Gesamtmarkt und ihren besonderen Eigenschaften durch die Rechtsform, d. h. insbesondere der Einheit von Kunde und Mitglied, richtet die vorliegende Arbeit den Blick auf die typischen Kreditgenossenschaften. Diese Auswahl wird auch damit begründet, dass es in Deutschland seit jeher eine intensive wissenschaftliche Auseinandersetzung mit dem Genossenschaftswesen gibt.[56] Deswegen und aufgrund des intensiven und vergleichsweise offenen Austauschs zwischen Wissenschaft und Praxis liegen umfangreiche Informationen zu den betriebswirtschaftlichen Konzepten in diesem Bankensektor vor.

Zu diesen betriebswirtschaftlichen Konzepten ist die Gesamtbanksteuerungskonzeption VR-CONTROL zu zählen. In Anlehnung an die wertorientierte Unternehmenssteuerung ist die Ausrichtung von VR-CONTROL die vermögensorientierte Gesamtbanksteuerung. Es heißt im WEIßBUCH DER GESAMTBANKSTEUERUNG, in dem diese Steuerungskonzeption beschrieben wird:

> „Der Gesamtwert einer Bank setzt sich aus zwei Komponenten zusammen:
> (1) Wert der Ist-Positionen, der als „Zerschlagungswert“ der Bank interpretiert werden kann.
> (2) Wert der erwarteten Geschäfte bzw. der Kundenbeziehungen („Goodwill“, „Potenzialwert“ der Bank).
> Diese Werte werden in VR-CONTROL nicht gemessen.“[57]

Aus dieser grundlegenden Ausrichtung heraus empfiehlt diese Gesamtbanksteuerungskonzeption eine wertorientierte Vertriebssteuerung auf Basis barwertiger Deckungsbeiträge. Die Einschränkung in Punkt (2) verdeutlicht jedoch, dass das bereits in den vorhergehenden Ausführungen identifizierte konzeptionelle Defizit zwischen der beziehungsorientierten Ausrichtung der Marketing- und Vertriebsaktivitäten der Kreditgenossenschaft auf der einen und der Steuerungskonzeption auf der anderen Seite auch mit Umsetzung von VR-CONTROL besteht.

Aufbauend auf der somit skizzierten Ausgangslage im Firmenkundengeschäft der Kreditgenossenschaften lässt sich die Problemstellung zusammenfassen als konzeptionelle Inkonsistenz zwischen der marketing- und vertriebspoliti-

---

54 Vgl. *AGVR* (2001), S. 36.

55 *Schierenbeck, H./Pohl, M.* (2010), S. 277.

56 Anzumerken ist, dass das Genossenschaftswesen neben den Kreditgenossenschaften auch z. B. Wohnungsbaugenossenschaften oder Handels- und Warengenossenschaften umfasst.

57 *BVR* (2002b), S. 11. In der ersten Auflage des WEIßBUCH DER BANKSTEUERUNG wird diese differenzierte Betrachtung des Vermögenswerts noch nicht vorgenommen.

schen Konzeption auf der einen und der sich daraus ableitenden Vertriebssteuerung auf der anderen Seite. Bisher liefert die wissenschaftliche Forschung trotz der unüberschaubaren Anzahl an Beiträgen zum Beziehungsmarketing und Kundenwert wenig Vorschläge, diese Inkongruenz zu beseitigen. Es besteht konkret für die Kreditgenossenschaften im Firmenkundengeschäft der Zweifel, mit den bestehenden Ansätzen und Instrumenten in der Vertriebssteuerung hinreichend ausgerüstet zu sein.[58] An diese Problemstellung und Forschungslücke knüpft diese Arbeit an.

2 Zielsetzungen der Arbeit

In den bisherigen Ausführungen wurde darauf hingewiesen, dass der Kundenwert zur ökonomischen Beurteilung und Steuerung von Geschäftsbeziehungen zu nutzen sei. Die Analyse von und die Ausrichtung der Vertriebsressourcen an Kundenpotenzialen ist nicht neu. Jedoch lassen sich bisher entwickelte Ansätze allenfalls als Partialkonzepte charakterisieren, die auf die Segmentierung oder andere Aspekte der Vertriebssteuerung fokussieren.[59] Ein konzeptionell umfassenderer Ansatz zur kundenwertorientierten Vertriebssteuerung im Firmenkundengeschäft liegt derzeit nicht vor. Auch beziehen sich die bisherigen Arbeiten vornehmlich auf das Privatkundengeschäft der Banken. So liefert SCHEER einen frühen Beitrag der Kundenpotenzialorientierung in der Steuerung im Privatkundengeschäft.[60] Und STORBACKA betrachtet das Privatkundengeschäft, fokussiert dabei aber explizit auf die Geschäftsbeziehung als Ganzes.[61] Aus diesem Grund betrachtet die vorliegende Arbeit das Geschäftsfeld der Firmenkunden, das einen maßgeblichen Anteil an der betriebswirtschaftlichen Situation einer Kreditgenossenschaft hat.

Nach REINARTZ/KRAFFT/HOYER lassen sich im Beziehungsmarketing zwei Fehlertypen unterscheiden.[62] Der erste Fehlertyp ist als Situation definiert, in der ein wertvoller Kunde nicht erkannt und daher nicht wertausschöpfend behandelt wird. Der zweite Fehler liegt darin, wertlose Kunden zu hoch zu bewerten und diesen folglich inadäquate Ressourceninvestitionen zukommen zu lassen. „The development of and reliance on a systematic approach that aids in the measurement of customer value and in the interaction with these heterogeneous customers decreases these errors."[63] Voraussetzung des effizienten Beziehungsmarke-

---

58 Vgl. ähnlich *Dauser, A.* (1995), S. 11 mit Bezug zur Versicherungsbranche.
59 Vgl. z. B. *Schmoll, A.* (2006), S. 62 ff.
60 Vgl. *Scheer, O.* (1989).
61 Vgl. *Storbacka, K.* (1993).
62 Vgl. *Reinartz, W. J./ Krafft, M./Hoyer, W. D.* (2004), S. 295.
63 *Reinartz, W. J./ Krafft, M./Hoyer, W. D.* (2004), S. 295.

tings ist es also, die Attraktivität von Kunden zu beurteilen. Daran knüpft das erste Forschungsziel dieser Arbeit an. Es sollen Modell und Metrik entwickelt werden, anhand derer die Attraktivität von Kunden ermittelt und beurteilt werden kann. Dabei wird darauf geachtet, dass die Ergebnisse steuerungsadäquat sind. Steuerungsadäquanz liegt vor, wenn der Kundenwert als Größe in der Planung, Durchführung und Kontrolle bankbetrieblicher Vertriebsmaßnahmen genutzt werden kann.

Das zweite Forschungsziel ist, auf Basis des hergeleiteten Kundenwertmodells Gestaltungsempfehlungen für eine Vertriebssteuerungskonzeption zu erläutern und zu erörtern, die es erlauben, die benannte Inkonsistenz zu beseitigen, so dass die Vorteile eines beziehungsorientierten Vorgehens in der kreditgenossenschaftlichen Praxis nicht an methodischen Steuerungsdefiziten scheitern.

Eine ausführliche Auseinandersetzung der ersten beiden Forschungsziele mit Blick auf die Besonderheiten der eingetragenen Genossenschaft stellt das dritte Forschungsziel dar. Damit will die Arbeit der wissenschaftlichen Tendenz entgegenwirken, Konzepte und Methoden widerspruchsfrei auf die besondere Rechtsform der eG zu übertragen.

Bei der Verfolgung dieser Ziele erhebt die vorliegende Arbeit einen praxeologischen Anspruch. Dieser konkretisiert sich jedoch nicht in einer Simplifizierung der instrumentellen und konzeptionellen Überlegungen, sondern wird dadurch fundiert, dass die gewonnenen Erkenntnisse soweit dies widerspruchsfrei möglich ist, in die bestehenden betriebswirtschaftlichen Steuerungsansätze der Kreditgenossenschaften integriert werden.

3 Gang der Analyse

Um die im vorangegangenen Kapitel aufgeworfenen Forschungsfragen zu beantworten, gliedert sich die Arbeit wie folgt. Aufgrund eines in der betriebswirtschaftlichen Literatur uneinheitlichen Verständnisses werden in Teil B zunächst die dieser Arbeit zugrunde liegenden begrifflichen Grundlagen der Vertriebssteuerung definiert. Dazu werden die Begriffe Vertrieb und Steuerung definiert und zur Vertriebssteuerung synthetisiert.

Aufbauend auf der Erkenntnis der Kontingenztheorie, dass die optimale Ausgestaltung einer Organisation von ihren situativen Umfeldbedingungen abhängt, werden in Teil C Implikationen für die Vertriebssteuerung aus dem Kontext des Geschäftsbereichs Firmenkunden der Kreditgenossenschaften abgeleitet.[64] Dazu werden die Kreditgenossenschaften zunächst in das deutsche Ban-

[64] Ein ähnliches Vorgehen wählt WIEDEMANN in seiner Dissertationsschrift aus. Auch er nutzt den konkreten situativen Kontext als Grundlage seiner weiteren konzeptionellen Überlegungen. Vgl. *Wiedemann, A.* (1992), S. 49 ff.

kensystem eingeordnet und detailliert beschrieben. Dabei wird insbesondere den kreditgenossenschaftlichen Besonderheiten aufgrund ihrer Rechtsform der eG Rechnung getragen. Weiterhin wird der Wettbewerb auf dem deutschen Bankenmarkt im Allgemeinen und im Firmenkundengeschäft im Speziellen betrachtet. Daran anknüpfend wird der Status quo der Geschäftseinheit „Firmenkundenbank" in den Kreditgenossenschaften mit Fokus auf die vertrieblichen Aspekte erläutert, bevor die strategische Ausrichtung skizziert sowie die ökonomische Bedeutung dieser Geschäftseinheit auf sekundärstatistischen Daten fundiert wird. Da die Vertriebssteuerung in die Gesamtbanksteuerung zu integrieren ist, werden die für die Arbeit wesentlichen Aspekte der kreditgenossenschaftlichen Gesamtbanksteuerungskonzeption VR-CONTROL vorgestellt. Abschließend werden die sich ergebenden Implikationen dargestellt und deren Bedeutung für die Ausgestaltung der Vertriebssteuerung deduziert.

In Teil A werden im ersten Schritt die Grundlagen einer kundenwertorientierten Vertriebssteuerung dargelegt. In concreto beinhaltet dies die Auseinandersetzung mit kundenbezogenen Wertbegriffen sowie die für die Arbeit notwendige Definition des Begriffs Kundenwert. Im zweiten Schritt werden die Erfolgsvoraussetzungen einer kundenwertorientierten Vertriebssteuerung erläutert und deren Vorhandensein für die Kreditgenossenschaften geprüft. Hieran schließt sich unmittelbar die Diskussion der Beziehung zwischen den Genossenschaftsprinzipien und der Kundenwertorientierung an. Anschließend wird ein geeignetes Kundenwertmodell für kreditgenossenschaftliche Firmenkunden hergeleitet, für das im Weiteren Messverfahren vorgestellt werden. Dieser Teil der Arbeit stellt die notwendige und hinreichende Bedingung der weiteren Analyse dar.

Im Teil E der Arbeit werden aufbauend auf den vorhergehenden Erkenntnissen Gestaltungsempfehlungen für die kundenwertorientierte Vertriebssteuerung im Firmenkundengeschäft der Kreditgenossenschaften vorgestellt. Dabei wird diskutiert, in welcher Form eine kundenwertorientierte Vertriebssteuerung ausgestaltet sein soll und wie sich diese in die bestehenden Prozesse und Strukturen der Gesamtbanksteuerung – mit besonderer Berücksichtigung der Steuerungskonzeption VR-CONTROL – einfügen lässt. Der Schwerpunkt der Betrachtung liegt weniger auf konkreten Instrumenten, sondern auf der Berücksichtigung der Kundenwertorientierung in Vertrieb und Vertriebssteuerung. Instrumentelle sowie institutionelle Aspekte werden, sofern und soweit dies notwendig ist, skizziert.

Die Arbeit schließt mit dem Teil F, in dem eine zusammenfassende Darstellung der Erkenntnisse erfolgt. Abschließend werden Ansätze für weitere Forschungsschwerpunkte thematisiert.

B Begriffliche und inhaltliche Grundlagen der Vertriebssteuerung

In der betriebswirtschaftlichen Forschung ist es zweckmäßig, schillernde Begriffe aus unterschiedlichen Perspektiven zu betrachten, um den Untersuchungsgegenstand zu konkretisieren.[65] Diese sind die funktionale, die prozessuale und die institutionelle Perspektive. Wenngleich diese Perspektiven nicht immer trennscharf abzugrenzen sind, dienen sie dazu, das Untersuchungsobjekt hinsichtlich unterschiedlicher Schwerpunkte zu analysieren. In der funktionalen Perspektive wird der Fokus auf die Ziele und Aufgaben gelegt, während in der prozessualen Perspektive der zeitliche Ablauf der Aufgaben im Mittelpunkt steht. Die institutionelle Sicht legt den Schwerpunkt auf die Akteure,[66] d. h. auf Aufgabenträger. Nachfolgend werden die Begriffe Vertrieb (Kapitel B.1) und Vertriebssteuerung (Kapitel B.2) aus diesen Perspektiven beschrieben, um sie zweckmäßig abzugrenzen. Anschließend werden die Instrumente der Vertriebssteuerung beleuchtet (Kapitel B.3) und abschließend eine zusammenfassende Darstellung der Vertriebssteuerung (Kapitel B.4) vorgenommen.

1 Begriffsabgrenzung Vertrieb

1.1 Funktionaler Begriffsinhalt

Der Begriff Vertrieb wird in der betriebswirtschaftlichen Literatur und Praxis uneinheitlich verwendet.[67] Insbesondere werden die Begriffe Marketing, Vertrieb, Absatz, Distribution und Verkauf mehrdeutig oder synonym verwendet. Daher werden diese Begriffe zunächst abgegrenzt und geordnet, bevor der Begriff Vertrieb funktional definiert wird.

Bis Mitte des 20. Jahrhunderts wurde der Begriff des Marketings primär distributionsorientiert verstanden.[68] Bedingt durch die beobachteten wirtschaftlichen Entwicklungen passte sich das Anspruchsspektrum des Marketings an und Marketing wird heute allgemein als die bewusst marktorientierte Führung des

65 Vgl. *Flacke, K.* (2006), S. 18 sowie die dort angegebene Literatur.

66 Die mangelnde Trennschärfe zeigt sich an dieser Stelle unter Rekurs auf HORVÁTH, der unter die institutionellen Aussagen sowohl Prozess- als auch Strukturaspekte subsumiert. Vgl. *Horváth, P.* (2003), S. 150.

67 Für eine ausführliche Auseinandersetzung mit den Inhalten des Begriffs Vertrieb und ähnlichen Begriffen vgl. *Winkelmann, P.* (2003), S. 1 ff. oder auch *Pepels, W.* (2002), S. 1 ff.

68 Vgl. dazu ausführlich *Meffert, H.* (1998), S. 4 ff.

Unternehmens definiert.[69] Mit dem gegenwärtigen Schwerpunkt auf Beziehungen zwischen Anbieterunternehmen und Unternehmensexternen definiert PEPELS Marketing als „Planung, Organisation, Umsetzung und Kontrolle aller Aktivitäten mit der Absicht der Erreichung psychografischer und/oder ökonomischer Ziele bei eigenen Kunden und deren Kunden durch Aufbau, Unterhalt, Ausbau oder Wiederherstellung von Geschäftsbeziehungen."[70] Die marktorientierte Führungskonzeption wird durch den Marketing-Mix in Richtung des Absatzmarktes konkretisiert. Dazu gehören die Produkt-, die Preis-, die Kommunikations- und die Distributionspolitik. Letztere ist somit nicht mehr zentraler Gegenstand des Marketings, sondern Teil des Marketing-Mix.

Die Distributionspolitik wird auch als Absatz-[71] oder Vertriebspolitik[72] bezeichnet und hat zwei abgrenzbare Aufgabenbereiche.[73] Erstens die akquisitorischen Aufgaben, die auf die Erzielung von Geschäftsabschlüssen gerichtet sind, und zweitens die physisch-distributiven Aufgaben, die sich mit der physischen Warenverteilung befassen.[74] Nur die akquisitorischen Aufgaben werden in dieser Arbeit unter Vertrieb subsumiert. Die physisch-distributiven Aufgaben werden unter den enger definierten Begriff der Distribution zusammengefasst und hier nicht näher betrachtet. Vertrieb ist somit derjenige Aufgabenbereich der Absatzpolitik, der sich mit konkreten Transaktionen zur Leistungsverwertung befasst und auf die Akquisition von, Interaktion mit und Nachbereitung bei Kunden ausgerichtet ist.[75] Ziel und Ergebnis des Vertriebs ist der Absatz, also die Übertragung der betrieblichen Leistungen an fremde Wirtschaftseinheiten im schuldrechtlichen Sinn.[76]

Nach diesem Verständnis ist der Vertrieb die Stufe der Wertschöpfung, in der die Aufgaben des Verkaufs und der Bereitstellung der produzierten Güter

---

69 Vgl. *Meffert, H.* (1998), S. 7. Hier ist insbesondere die Entwicklung der Verkäufer- zu Käufermärkten auf den meisten nationalen und internationalen Absatzmärkten zu nennen.

70 *Pepels, W.* (2002), S. 5. Diese Definition ist die Synthese der Begriffsverständnisse mehrerer Standardwerken zum Marketing.

71 Vgl. *Pepels, W.* (2002), S. 5.

72 Vgl. *Busch, R./Fuchs, W./Unger, F.* (2008), S. 12.

73 Vgl. *Olbrich, R.* (2006), S. 218 sowie die dort angegebene Literatur.

74 Vgl. *Busch, R./Fuchs, W./Unger, F.* (2008), S. 319; *Nieschlag, R./Dichtl, E./Hörschgen, H.* (2002), S. 1319; *Wöhe, G.* (1996), S. 595; *Pangratz, O.* (1979), S. 9.

75 Vgl. *Pepels, W.* (2002), S. 5.

76 Vgl. *Rinn, T.* (1991), S. 11. Der Begriff Absatz, der früher in Literatur und Praxis der herrschende Begriff für die betriebliche Funktion der Leistungsverwertung, mithin des Vertriebs war, wird heute nur noch für den mengenmäßigen Markterfolg des Unternehmens verwendet.

und Dienstleistungen zusammengefasst sind.[77] Der Verkauf ist Teil des funktionalen Vertriebs[78] und definiert als der „Vorgang des Kaufvertragsabschlusses einschließlich der zuvor erfolgten Anbahnung in Form der Güterdarbietung, der Kaufberatung und der Kaufverhandlung".[79] Er ist also der unmittelbare Vollzug der ökonomischen Interaktion im einzelnen Tauschakt, während dem Vertrieb eine die einzelne Transaktion übergreifende Sichtweise inhärent ist.[80]

Aufbauend auf der Rolle des Vertriebs, die Schnittstelle zwischen Anbieterunternehmen und Kunden zu sein,[81] können ihm die Aufgaben des Kundenmanagements zugeordnet werden. Das Kundenmanagement umfasst alle Maßnahmen der Analyse, Planung, Durchführung und Kontrolle, die der Initiierung, Stabilisierung, Intensivierung, Wiederaufnahme oder Beendigung von Geschäftsbeziehungen zu Kunden eines Unternehmens mit dem Ziel der gegenseitigen Nutzenstiftung dienen.[82] Zum Vertrieb gehört daher die Kundenidentifikation und -selektion, die Kundenakquisition und -bindung sowie die Kundentrennung.[83] Der Vertrieb wird als Träger marketingpolitischer Aufgaben verstanden. Er soll die Konzepte und Strategien des Marketings unmittelbar im Kundenkontakt umsetzen.[84]

Abhängig von den durch das Unternehmen angebotenen Leistungen ist der Vertrieb eine der Beschaffung und Produktion nachfolgende oder vorhergehende Wertschöpfungsstufe. Bei materiellen Produkten, die für einen anonymen Massenmarkt gefertigt werden, erfolgt die Beschaffung der Inputfaktoren und die Produktion i. d. R. vor dem Vertrieb. Bei immateriellen Gütern, die sich durch ihre Nichtlagerfähigkeit auszeichnen, und bei den meisten Investitionsgütern folgt die eigentliche Produktion während bzw. nach dem Vertrieb.[85] Es werden seitens des Anbieterunternehmens in diesem Fall Leistungs-potenziale, d. h. infrastrukturelle Kapazitäten zur Leistungserbringung, bereitgehalten.

---

77 Vgl. *Proeller, M.* (1996), S. 20.

78 KOTLER/BLIEMEL verwenden die Begriffe Verkauf und Vertrieb synonym. Vgl. *Kotler, P./Bliemel,F.* (1999).

79 *Ahlert, D.* (1996), S. 27.

80 Vgl. *Pepels, W.* (2002), S. 5 f.

81 Vgl. *Reichwald, R./Bastian, C./Lohse, C.* (2000), S. 6; *Dauser, A.* (1995), S. 13.

82 Vgl. *Bruhn, M.* (2009), S. 10; ähnlich *Diller, H.* (1995a), Sp. 1369.

83 Oft wird in der betriebswirtschaftlichen Literatur der Frage nach der Beendigung von Geschäftsbeziehungen und der Zuordnung dieser Aufgabe zum Vertrieb kaum Beachtung geschenkt. Vgl. z. B. *Dauser, A.* (1995), S. 13.

84 Vgl. *Bauer, R. A.* (2000), S. 43.

85 In diesem Zusammenhang wird von auftrags- oder kundenindividueller Fertigung gesprochen. Vgl. für eine ausführliche Vorstellung *Adam, D.* (1998), S. 10 f.

1.2 Institutioneller Begriffsinhalt

Institutionell bezeichnet der Begriff Vertrieb die aufbauorganisatorische Einheit eines Unternehmens, in der die personellen und organisationalen Aufgabenträger unternehmensinterner (z. B. Außendienstmitarbeiter) und unternehmensexterner Herkunft (z. B. Handelsvertreter), welche die Aufgaben im Sinne des funktionalen Begriffsverständnisses erfüllen, zusammengefasst sind.[86]

Aufbauorganisatorische Gestaltungsfragen des Vertriebs ergeben sich unter anderem in Hinblick auf die Art, in welcher wirtschaftlichen und rechtlichen Beziehung die Aufgabenträger zum Anbieterunternehmen stehen (Vertriebssystem oder Vertriebsorgane), in welcher Mittelbarkeit die Beziehungen oder Verbindungen zum Kunden erfolgen (Vertriebskanalstruktur) und wie das logistische System des Leistungstransfers ausgestaltet sein soll.[87] Letzteres wird hier, wie oben bereits ausgeführt, nicht weiter betrachtet.

Hinsichtlich des Vertriebssystems kann zwischen einem werkseigenen, einem werksgebundenen und einem offenen Vertriebssystem unterschieden werden.[88] Im werkseigenen Vertriebssystem können die Aufgabenträger entweder rechtlich und wirtschaftlich unselbständig (z. B. Reisende) oder rechtlich selbständig, aber wirtschaftlich abhängig sein (z. B. Handelsvertreter). Werksgebundene Aufgabenträger (z. B. Franchisenehmer) sind rechtlich eigenständig, jedoch vertikal eng mit dem Anbieter verbunden und wirtschaftlich abhängig. Im offenen Vertriebssystem sind die jeweiligen Akteure rechtlich und wirtschaftlich unabhängig (z. B. freie Vermögensverwalter).

In der Vertriebskanalsstruktur wird unterschieden, ob die Produkte den Kunden direkt vom Anbieterunternehmen bereitgestellt oder Absatzmittler eingeschaltet werden. Vertriebskanäle können ein- oder mehrstufig ausgestaltet sein.[89]

1.3 Prozessualer Begriffsinhalt

Fokus des prozessualen Begriffsinhalts sind die vertrieblichen Prozesse eines Unternehmens. Als Prozess gilt eine inhaltlich abgeschlossene, zeitliche und sachlogische Folge von einzelnen Aktivitäten, die zur Bearbeitung eines betriebswirtschaftlichen Objekts notwendig sind.[90] Vertriebsprozesse legen die Reihenfolge fest, in denen die Vertriebsaufgaben und -teilaufgaben im Sinne der

---

86 Vgl. *Nieschlag, R./Dichtl, E./Hörschgen, H.* (2002), S. 1320.

87 Vgl. *Schmalen, H.* (1999), S. 550 ff.; *Meffert, H.* (1998), S. 582.

88 Vgl. hier und nachfolgend *Schmalen, H.* (1999), S. 550 ff.

89 Vgl. für eine ausführliche Darstellung z. B. *Meffert, H.* (1998), S. 600 ff.

90 Vgl. *Becker, J./Kahn, D.* (2005), S. 6.

Vertriebs- und Unternehmensziele erfüllt werden sollen. Aufgrund der Heterogenität unternehmensspezifischer Kontextfaktoren gibt es keine allgemeingültigen Vertriebsprozesse.[91]

Allgemein gehören die Vertriebsprozesse mehrheitlich zu den Kernprozessen eines Unternehmens, da diese die wertschöpfenden Aktivitäten des Unternehmens aus Sicht des Kunden bündeln und eine direkte Leistungsbeziehung zum Kunden vorhanden ist. Jedoch sind nicht alle Prozesse im Rahmen des Vertriebs Kernprozesse. Basierend auf dem „Service-Blueprinting" können Vertriebsprozesse in fünf Prozessebenen unterteilt werden, die sich hinsichtlich des Kundenintegrationsgrads unterscheiden.[92] Diese sind Kunden-, Onstage- und Backstage-Prozesse sowie Support- und Facility-Prozesse.

Kundenprozesse sind vom Kunden autonom durchgeführte Prozesse, bei denen keine direkte Beteiligung des Anbieterunternehmens notwendig ist, so bspw. die Suche nach relevanten Produktanbietern für Finanzdienstleistungen. Onstage-Prozesse sind anbieterseitige Prozesse, die den Kunden integrieren und daher für diesen auch beobachtbar sind wie z. B. die Eröffnung eines Kontos. Backstage-Prozesse sind vergleichsweise autonome, vom Kunden weitgehend nicht wahrgenommene Anbieterprozesse, die der Gewährleistung der Onstage-Prozesse dienen. Dazu ist z. B. die Prüfung der Kreditantragsunterlagen durch Kreditsachbearbeiter zu zählen. Onstage- und Backstage-Prozesse werden durch Support-Prozesse unterstützt, die vorwiegend autonom sind und fast keinen direkten Bezug zur Absatzleistung haben. Prozesse im Rahmen der Weiterbildung von Mitarbeitern im Vertrieb oder in der Kreditsachbearbeitung gehören z. B. zu dieser Kategorie. Schließlich gibt es Facility-Prozesse. Diese schaffen die Voraussetzungen und Rahmenbedingungen für die vorgenannten Prozesse und schaffen selbst keinen Beitrag für die Absatzleistung (z. B. Prozesse bei der Bewirtschaftung der Bankimmobilie).

1.4 Zusammenfassende Begriffsabgrenzung Vertrieb

WINKELMANN definiert den Begriff Vertrieb zusammenfassend als „alle Tätigkeiten und Funktionen (Mitarbeiter und deren Aufgaben im Rahmen betrieblicher Stellen), Strukturen und Abläufe, Methoden und Systeme zur Leistungsverwertung".[93] Unter Berücksichtigung der vorhergehenden inhaltlichen Auseinandersetzung soll in Anlehnung an diese Definition der Vertriebsbegriff wie folgt abgegrenzt werden: Als Vertrieb gelten alle Aufgaben und Tätigkeiten,

---

91 Eine Übersicht verschiedener Vertriebsprozesse in verschiedenen Branchen findet sich bspw. bei *Reiners, J. O.* (2004), S. 120 ff.

92 Vgl. hier und nachfolgend *Kleinaltenkamp, M./Schweikart, J.* (1998), S. 114 f.

93 *Winkelmann, P.* (2003), S. 15.

Strukturen und Abläufe, Methoden und Systeme sowie alle darin involvierten Mitarbeiter des Unternehmens, die dem Kundenmanagement, dem Verkauf und der Bereitstellung von Gütern oder Dienstleistungen dienen.

Die Gesamtheit der Entscheidungen des Vertriebs wird als Vertriebspolitik bezeichnet.[94] Sie ist Bestandteil des Marketing-Mix mit den weiteren Bestandteilen der Produkt-, Preis-, Kommunikations- und Distributionspolitik[95] und beinhaltet sämtliche Funktionen, deren Gegenstand die Gestaltung der vielfältigen Beziehungen zwischen Unternehmen und Absatzmarkt im Allgemeinen und Kunden im Speziellen ist.[96] Die Vertriebspolitik gibt den operativen und strategischen Handlungsrahmen für den Vertrieb vor[97] und wird selbst wiederum von der Marketingstrategie determiniert.[98] Dabei gilt: „Erst der Vertrieb bringt die Marketingstrategien zum Laufen."[99]

2 Begriffsabgrenzung Vertriebssteuerung

Nachdem der Vertrieb begrifflich definiert ist, ist es für den weiteren Gang der Analyse notwendig, den Begriff der Vertriebssteuerung grundlegend für diese Arbeit zu konkretisieren. Da der Steuerungsbegriff in der betriebswirtschaftlichen Literatur häufig verwendet, jedoch nur selten eindeutig abgegrenzt wird,[100] wird zunächst das zugrunde zu legende Verständnis des Begriffs Steuerung herausgearbeitet, woran sich die Begriffssynthese aus den Untersuchungsgegenständen Vertrieb und Steuerung anschließt. Aufbauend auf diesem Verständnis erfolgt die Betrachtung der Vertriebssteuerung aus funktionaler, institutioneller und prozessualer Sicht. Bei der funktionalen Sicht wird zwischen den Zielen und Aufgaben differenziert.

2.1 Grundlegendes Steuerungsverständnis

Grundlage der begrifflichen Ableitung der Vertriebssteuerung ist die zweckorientierte Definition des Begriffs Steuerung. Dieser kann allgemein als zielorientierte Lenkung von Ressourcen und Prozessen interpretiert werden. Ressourcen sind alle Inputfaktoren, die mittelbar oder unmittelbar in die Leistungserstellung und -verwertung einfließen. Es ist evident, dass Steuerung originäre Aufgabe

---

94 Vgl. *Winkelmann, P.* (2003), S. 32.

95 Vgl. *Duderstadt, S.* (2006), S. 4.

96 Vgl. *Rinn, T.* (1991), S. 11.

97 Vgl. *Pangratz, O.* (1979), S. 8.

98 Vgl. *Hofbauer, G./Hellwig, C.* (2009), S. 62.

99 *Renker, C.* (2004), S.9

100 Vgl. *Dauser, A.* (1995), S. 40.

und gleichzeitig integraler Bestandteil der Führung ist. So definieren STEINMANN/SCHREYÖGG Führung als „ein Komplex von Steuerungsaufgaben"[101] und WILD beschreibt Führen als „eine Tätigkeit [.], die die Steuerung und Gestaltung des Handelns anderer Personen zum Gegenstand hat."[102] Eine isolierte Betrachtung der Steuerung ist somit nicht trennscharf von der Führung möglich. Daher soll zunächst der Zusammenhang zwischen Führung und Steuerung dargestellt werden.

Das Phänomen Führung tritt immer dort auf, wo mehrere Personen arbeitsteilig mit einem gemeinsamen extern vorgegebenen oder intern vereinbarten Ziel interagieren.[103] Dabei gilt es, das gemeinsame Handeln der beteiligten Personen in Hinblick auf das Ziel abzustimmen. Auch Führung kann institutionell, funktional oder prozessual betrachtet werden.[104] In der institutionellen Sicht wird gefragt, wer die Träger der Führung sind und wie diese im Unternehmen verankert werden. In der funktionalen Betrachtung gilt es, zu bestimmen, was die Aufgaben- und Tätigkeitsinhalte der Führung sind. Planung, Organisation und Kontrolle werden als die zentralen Führungsfunktionen verstanden.[105] Einigkeit besteht weiterhin darüber, dass sich die funktionale Führung über Entscheidungen konstituiert, wobei die hinreichenden Merkmale der Führungsentscheidungen selbst vielfältig diskutiert werden. Prozessual betrachtet, lässt sich Führung gedanklich in die beiden Phasen Willensbildung und Willensdurchsetzung gliedern.[106] Um Steuerung als Bestandteil der Führung herauszuarbeiten, ist der Führungsprozess näher zu betrachten.

Ausgangspunkt der Willensbildung sind Ziele, welche auf Basis der vorausschauenden Auseinandersetzung mit den künftigen Umfeldentwicklungen formuliert werden. Orientiert an einem oder mehreren Zielen werden die Maßnahmenentscheidungen getroffen, die sich auf die Gestaltung der künftigen Handlungen beziehen,[107] um diese Ziele zu erreichen. Die Gesamtheit der Maßnahmen wird als Strategie bezeichnet. Diese Tätigkeiten und Entscheidungen kon-

---

101 *Steinmann, H./Schreyögg, G.* (2005), S. 7.

102 *Wild, J.* (1974), S. 15.

103 Vgl. *Macharzina, K.* (2003), S. 35 f.

104 Vgl. *Flacke, K.* (2007), S. 18 ff.; *Macharzina, K.* (2003), S. 35 ff.

105 Als grundlegende Arbeit für die Definition von Führungsfunktionen wird die Arbeit von FAYOL aus dem Jahre 1916 angesehen, der die Funktionen Vorschau und Planung (prévoir), Organisation (organiser), Anweisung (commander), Koordination (coordonner) und Kontrolle (contrôller) unterscheidet. Vgl. *Steinmann, H./Schreyögg, G.* (2005), S. 8 f. sowie *Fayol, H.* (1929), S. 7 f.

106 Vgl. *Macharzina, K.* (2003), S. 38.

107 Vgl. *Horváth, P.* (2003), S. 168 sowie die dort genannte Literatur; *Weber, J./Schäffer, U.* (2008), S. 241.

stituieren zusammen die Planung. In der Phase der Willensdurchsetzung werden ausgehend von der Planung die zur Planerfüllung notwendigen Strukturen im Unternehmen geschaffen[108] und die zur Planrealisation notwendigen Maßnahmen initiiert. Dies beinhaltet zum einen die Organisation, welche die aufbau- und ablauforganisatorische Struktur sowie die dauerhaften und generellen Regelungen umfasst, und zum anderen das Einwirken durch die Führung auf die Geführten. Korrespondierend steht der Planung die Kontrolle gegenüber, die zusammen betrachtet von SCHREYÖGG/KOCH als „Zwillingsfunktionen“ [109] bezeichnet werden. Allgemein ist die Kontrolle „der Vergleich eines eingetretenen Ist mit einem vorgegebenen Soll“[110]. Ausgehend vom Vergleichsergebnis kann die Notwendigkeit resultieren, Maßnahmen einzuleiten, um die gewollte Zielerreichung zu gewährleisten. Dieser korrigierende Eingriff wird regelmäßig als Steuerung verstanden.[111] Durch die Steuerung soll das Verhalten einer abhängigen Größe, z. B. der Geführten, direkt beeinflusst werden.[112] In diesen Ausführungen zeigt sich, wie die Steuerung in die Führung integriert ist. Gleichzeitig wird deutlich, dass die Steuerung unmittelbar mit der Planung und der Kontrolle zusammenhängt bzw. auf diesen beiden Funktionen aufbaut und ohne sie gegenstandslos ist oder willkürlich wird.

Vor diesem Hintergrund stellt nach SCHWEITZER die Steuerung ein Führungsinstrument dar, das dann zum Einsatz kommt, wenn eine gestörte Realisation der Planung gegeben ist und dies zur Notwendigkeit von Anpassungsmaßahmen führt. Steuerung wird von ihm definiert als „ein geordneter, informationsverarbeitender Prozess zielführender Eingriffe (Anpassungsmaßnahmen) in die Planrealisation“[113]. Damit wird der Begriff der Steuerung begrenzt auf eine Einsatznotwendigkeit und -möglichkeit für den Fall, dass bei der Planrealisation in der Planung nicht berücksichtigte Entwicklungen, welche die Zielerreichung negativ beeinflussen, eingetreten sind oder sich abzeichnen und auf diese

---

108 Vgl. *Steinmann, H./Schreyögg, G.* (2005), S. 131 f. Dies entspricht der Strukturfolge-Hypothese, die aussagt, dass sich die Struktur der Strategie des Unternehmens anpasst. Vgl. dazu ausführlich *Schewe, G.* (1999), S. 61 ff.

109 *Schreyögg, G./Koch, J.* (2007), S. 11.

110 *Weber, J./Schäffer, U.* (2008), S. 242.

111 Vgl. *Macharzina, K.* (2003), S. 40.

112 Vgl. *Nieschlag, R./Dichtl, E./Hörschgen, H.* (2002), S. 949.

113 *Schweitzer, M.* (2001), S. 20. Nach DELLMANN liegt der Ursprung des Begriffs Steuerung in der Kybernetik, wo Steuerung als die Anweisung an ein Systemelement verstanden wird, Störungen entgegen zu wirken oder Störungen mit dem Ziel zu beseitigen, das System in den gewünschten Zustand zu bringen. Vgl. *Dellmann, K.* (1992), S. 117. Diese Ausführungen machen deutlich, dass es traditionell einen engen Bezug zwischen den Begriffen der Störung und der Steuerung gibt.

Kenntnisse reagiert wird.[114] Steuerung wird dadurch als reaktives Führungsinstrument definiert.

Eine breiteres Verständnis der Steuerung unterstellen SCHWEITZER/KÜPPER, die die Plandurchsetzung und die Vorgabe von Sollwerten ebenso der Steuerung zuordnen wie die Überwachung von Zielerreichungen und die darauf aufbauenden Sicherungsmaßnahmen im Einklang mit dem reaktiven Steuerungsverständnis (siehe Abbildung 1).[115]

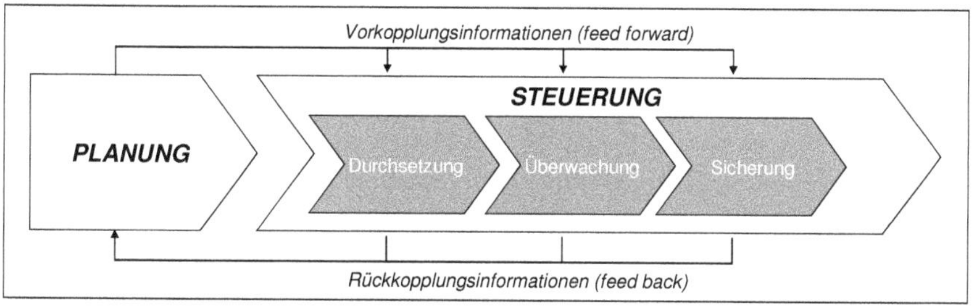

Abbildung 1: Struktur des Planungs- und Steuerungsprozesses[116]

In Anlehnung an dieses Begriffsverständnis kann Steuerung definiert werden als geordneter, informationsverarbeitender Prozess zur Durchsetzung, Überwachung und Sicherung der vorhergehenden Planung, der auf die zielorientierte Lenkung von Prozessen der Planrealisation und auf die Lenkung von Entscheidungs- und Verhaltensprozessen der Mitarbeiter ausgerichtet ist.[117] Im Rahmen der Lenkung der Prozesse der Planrealisation beinhaltet die Steuerung die Durchsetzung der Planvorgaben, die Feststellung von Abweichungen zwischen Soll- und Ist-Werten sowie die Auslösung von Maßnahmen zur Gegensteuerung auf allen Ebenen der Um- und Durchsetzung.[118]

Die Steuerung der Entscheidungs- und Verhaltensprozesse der Mitarbeiter basiert oft auf der Vorgabe oder Vereinbarung von Zielen, um den Handlungsrahmen oder das Handlungsergebnis der Mitarbeiteraktivitäten zu definieren. Damit wird deutlich, dass Steuerung nicht nur im Fall von Planungsabweichun-

---

114 Vgl. *Hoppen, D.* (1999), S. 105.

115 Vgl. *Schweitzer, M./Küpper, H.-U.* (2008), S. 4.

116 Vereinfachte Darstellung in Anlehnung an *Schweitzer, M./Küpper, H.-U.* (2008), S. 4. Ähnlich auch *Duderstadt, S.* (2006), S. 5.

117 Vgl. hier und nachfolgend *Schweitzer, M./Küpper, H.-U.* (2008), S. 3.

118 Eine ausführliche Betrachtung der Aufgaben der Vertriebssteuerung erfolgt in Kapitel B.2.3.2.

gen relevant ist, sondern auch schon bei der Planrealisation. Steuerung umfasst damit auch die zielgerichtete Gestaltung und den zielgerichteten Einsatz von Methoden, Maßnahmen und Funktionsträgern,[119] und damit Komponenten der Managementfunktion Organisation.

2.2 Ableitung des Begriffs der Vertriebssteuerung

Vertriebssteuerung kann auf Basis der vorhergehenden Ausführungen als informationsverarbeitender, geordneter Prozess, der die Vertriebsplanrealisation durchsetzen, überwachen und sichern soll und dabei antizipative und reaktive Steuerungsmaßnahmen im Rahmen der Planumsetzung zum Gegenstand hat, definiert werden. Objekte der Vertriebssteuerung sind in diesem Sinn alle Methoden, Informationen, Maßnahmen, Prozesse und Funktionsträger des Vertriebs.[120] Dies ist die Vertriebssteuerung im engeren Sinn.

Insgesamt machen die bisherigen Ausführungen deutlich, dass Planung und Steuerung untrennbare Prozesse respektive Aufgabenkomplexe sind, um Ziele zu erreichen. Eine von der Planung losgelöste Betrachtung der Steuerung kann also nicht erfolgen, da die Planung den grundsätzlichen Ankerpunkt und den, in Abhängigkeit der konkreten Planung, relevanten Bezugsrahmen der Steuerung vorgibt. In der betriebswirtschaftlichen Literatur und vor allem in praxisbezogenen Beiträgen finden sich daher Ansätze, die die Steuerung als einen Regelkreislauf aus Planung, Umsetzung und Kontrolle verstehen[121] und damit die Planung nicht wie bisher vor den zuvor erläuterten Steuerungsprozess setzen, sondern sie als einen Prozessschritt der Steuerung betrachten. So definieren KÖHLER/ROLVERING/GERMANN die Vertriebssteuerung als „Führungsinstrument, das sich mit der Planung, Steuerung und Kontrolle der Vertriebstätigkeit eines Unternehmens beschäftigt“ [122]. Grundsätzlich entspricht dies in weiten Teilen den Führungstätigkeiten so wie diese eingangs vorgestellt wurden, so dass Führung und Steuerung abzugrenzen sind.

Ein fundamentaler Differenzierungsaspekt dieser beiden Konstrukte findet sich in der Aufteilung der Führungsaufgaben in einen originären und derivativen Aufgabenbereich.[123] Danach lassen sich derivative Führungsfunktionen untertei-

119 Vgl. *Pangratz, O.* (1979), S. 10.

120 Ähnlich auch *Duderstadt, S.* (2006), S. 7.

121 Vgl. z. B. *Averbeck, C./Frère, E./Reuse, S.* (2008), S. 41; *Dauber, M.* (2008), S. 890; *Schmoll A.* (2008), S. 52; *Schmoll, A.* (2006), S. 147 oder *Recht, A./Holm, R.* (2005), S. 136 f.

122 *Köhler, A./Rolvering, H.-S./Germann, S.* (2005), S. 417. Ähnlich auch *Dihlmann, K.* (1991), S. 157.

123 Vgl. hier und nachfolgend *Bleicher, K./Meyer, E.* (1976), S. 37 ff.

len in sachbezogen-initiative und steuernde Aufgaben (Lokomotionsfunktion) und in motivierende und auf den Zusammenhalt ausgerichtete Aufgaben (Motivations- und Kohäsionsfunktion). Die Steuerung ist also der Lokomotionsfunktion zuzuordnen, so dass vor allem die personenbezogene Motivations- und Kohäsionsfunktion die Führung von der Steuerung abgrenzt.

Die Begriffsfestlegung der Vertriebssteuerung als Regelkreis trifft insbesondere auch auf praxisorientierte Definitionsansätze zu, so dass vor dem Hintergrund der praxeologischen Ausrichtung dieser Arbeit dieses Verständnis als Vertriebssteuerung im weiteren Sinn bezeichnet werden soll. Dieses Begriffsverständnis ist Grundlage der weiteren Ausführungen und soll nachfolgend im Vordergrund stehen, da sie eine ganzheitliche Sicht unterstützt, der praxisbezogenen Zielsetzungen der Arbeit entspricht und die oben definierte engere Auffassung der Vertriebssteuerung umfasst.

In der Literatur lässt sich häufig die synonyme Verwendung der Begriffe Vertriebssteuerung und Vertriebscontrolling beobachten.[124] Dies rührt daher, dass Planungs- und Kontrollaufgaben im Rahmen der Steuerung auch in der Controllingfunktion verankert werden. In der Literatur zum Controlling wird dieser Sachverhalt als Aufbau eines Planungs- und Kontrollsystems sowie der Verbindung desselben mit weiteren Führungsteilsystemen beschrieben.[125] Für die vorliegende Arbeit wird diese Identität nicht angenommen, sondern das Controlling unabhängig von der jeweiligen theoretischen Controllingkonzeption als Führungsunterstützung, aber steuerungsnahe Funktion verstanden.[126] „Die Vertriebssteuerung bedient sich der Dienstleistung des Controllingbereichs."[127] Es ist leicht nachvollziehbar, dass zwar eine isolierte Analyse der Funktion Controlling vorgenommen werden kann, eine isolierte Betrachtung der Steuerung ohne Controlling jedoch nicht möglich ist, denn das Vertriebscontrolling, dessen zentrale Aufgabe in der Koordination und der Informationsversorgung liegt, stellt einen wesentlichen Bestandteil des gesamthaften Vertriebsteuerungssystems dar.[128] Daher ist eine durchgängige Trennung von Vertriebssteuerung und -controlling nicht möglich. Abbildung 2 verdeutlicht diesen Gedanken.

---

124 Vgl. z. B. *Hummel, D.* (2007), S. 86 ff., *Schmoll, A.* (2006), S. 145 ff.; *Benölken, H.* (2002), S. 135 ff.

125 Vgl. z. B. *Horváth, P.* (2003), S. 165 ff.

126 Vgl. *Schildbach, T.* (1992), S. 23. Zu steuerungsnahen Funktionen vgl. *Rolfes, B.* (2008), S. 22.

127 *Dihlmann, C.* (1992), S. 157.

128 So auch LISSAUTZKI, der vor dem Hintergrund der engen inhaltlichen Verknüpfungen von Steuerung und Controlling ebenfalls Schwierigkeiten bei der isolierten Betrachtung der Steuerung sieht und in seiner Diskussion abschließend darauf verweist, „dass das

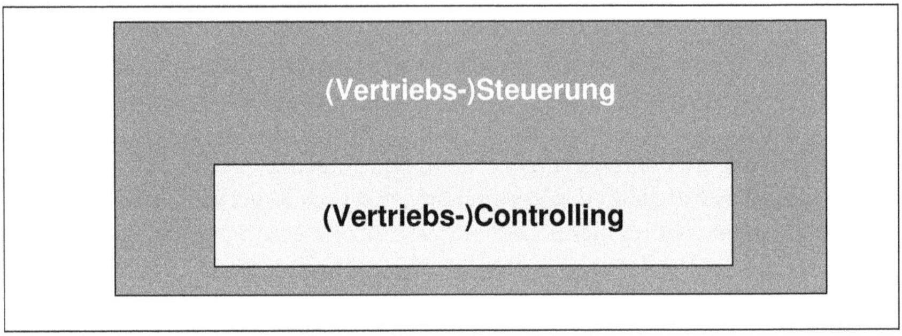

Abbildung 2: Vertriebssteuerung und -controlling als Elemente des Vertriebssteuerungssystems[129]

2.3 Funktionaler Begriffsinhalt

2.3.1 Ziele der Vertriebssteuerung

Vertriebssteuerung ist kein Selbstzweck, sondern richtet sich in ihrer Ausgestaltung an zu verfolgenden Zielen aus. Als Ziel sei in dieser Arbeit „ein bestimmter intendierter Sachverhalt [...] im Sinne von Lage oder auch Situation"[130] verstanden. Für die Vertriebsteuerung können direkte und indirekte Ziele unterschieden werden.[131] Direkte Ziele grenzen den Umfang der Aufgaben der Vertriebssteuerung ab. Indirekte Ziele beziehen sich auf die Unternehmensziele und präzisieren den Inhalt der von der Vertriebssteuerung wahrgenommenen Aufgaben. Art und Charakter der indirekten Ziele ergeben sich aus dem Verhältnis dieser Ziele zu den Zielen der übergeordneten Unternehmenssteuerung.[132]

Aus dem zugrunde liegenden Verständnis der Vertriebssteuerung lässt sich das direkte Ziel unmittelbar ableiten. Das zentrale und übergeordnete direkte Ziel der Vertriebssteuerung ist, ein wirkungsvolles Instrument zum Erreichen von Vertriebszielen zu sein.[133] Dazu gehört, die Effektivität und Effizienz des Vertriebs zu gewährleisten.

Hinsichtlich der indirekten Ziele der Vertriebssteuerung ist der Zusammenhang zwischen Unternehmenssteuerung und Vertriebssteuerung zu beachten.

Controlling sowohl inhaltlich als auch organisatorisch eine bedeutende Rolle in der Unternehmenssteuerung spielt." *Lissautzki, M.* (2007), S. 13.

129 Eigene Darstellung.

130 *Schmidt-Sudhoff, U.* (1967), S. 16.

131 Vgl. hier und nachfolgend *Schweitzer, M./Friedl, B.* (1992), S. 143.

132 Vgl. *Langguth, H.* (1994), S. 50.

133 Vgl. *Schweitzer, M.* (2001), S. 21.

Ausgangspunkt der Unternehmenssteuerung ist das Unternehmenszielsystem. Da die obersten Entscheidungsträger die einzelnen Ziele frei auswählen können, existiert kein allgemeingültiges Unternehmenszielsystem.[134] Ein elementares Unternehmensziel ist jedoch die nachhaltige Existenzsicherung. Um dieses Ziel zu erreichen, ist die conditio sine qua non die Verfolgung erfolgswirtschaftlicher Ziele „ohne die der Bestand selbständiger Unternehmungen schwer möglich ist."[135] Die Zielsetzung der Existenzsicherung lässt sich in eine operative und eine strategische Perspektive unterscheiden. Auf der operativen Ebene beinhaltet das Ziel der Existenzsicherung die Ausrichtung des Unternehmens und damit des Vertriebs an erfolgswirtschaftlichen Teilzielen. Klassischerweise sind diese liquiditäts- sowie umsatz- und kostenbezogene Zielkategorien. Für die Vertriebssteuerung resultiert auf der operativen Ebene das indirekte Ziel, die Vertriebsleitung bei der Erreichung der liquiditäts- sowie umsatz- und kostenbezogenen Ziele zu unterstützen.

Auf der strategischen Ebene gilt es, künftige Erfolge zu sichern. Grundlegend für künftige Erfolge sind Erfolgspotenziale.[136] Der Begriff Erfolgspotenzial wird dabei im Sinne GÄLWEILERS interpretiert, der damit „das gesamte Gefüge aller jeweils produkt- und marktspezifischen erfolgsrelevanten Voraussetzungen, die spätestens dann bestehen müssen, wenn es um die Erfolgsrealisierung geht"[137] bezeichnet und diese als Vorsteuergröße von Gewinn und Liquidität ansieht. Als indirektes Ziel der Vertriebssteuerung auf der strategischen Ebene wird daher die Unterstützung der Unternehmensführung bzw. Vertriebsleitung beim Schaffen und Erhalten von Erfolgspotenzialen verstanden.[138] Erfolgspotenziale werden in ihrer Gesamtheit durch strategische Erfolgsfaktoren determiniert.[139] Erfolgsfaktoren sind diejenigen unternehmensinternen und -externen Größen, welche die Erfolgsmöglichkeit über Ursache-Wirkungs-Beziehungen beeinflussen.

---

[134] Vgl. *Küpper, H.-U.* (2001), S. 18.

[135] *Küpper, H.-U.* (2001), S. 18 f.

[136] Vgl. *Schmid, S./Kutschker, M.* (2002), S. 1242; *Friedag, H. R./Schmidt, W.* (2001), S. 495.

[137] *Gälweiler, A.* (1987), S. 26.

[138] Vgl. *Baum, H.-G./Coenenberg, A. G./Günther, T.* (2007), S. 9; *Langguth, H.* (1994), S. 50 f.

[139] Vgl. *Baum, H.-G./Coenenberg, A. G./Günther, T.* (2007), S. 31; *Rehkugler, H.* (1989), S. 627.

2.3.2 Aufgaben der Vertriebssteuerung

Die aufgabenbezogene Betrachtung der Vertriebssteuerung beschäftigt sich mit den Inhalten und Obliegenheiten, die von der Vertriebssteuerung wahrzunehmen sind, um ihre direkten und indirekten Ziele zu erreichen. Basierend auf dem zugrunde gelegten Verständnis können die Aufgaben einzelnen Phasen der Vertriebssteuerung zugeordnet werden. So ist zwischen Aufgaben im Rahmen der Planung, der Durchsetzung, der Überwachung und der Sicherung zu unterscheiden. Erfüllungskritische Grundlage der einzelnen Aufgaben ist die Bereitstellung der jeweils phasenrelevanten Methoden und Informationen durch das Vertriebscontrolling, das zudem die Koordination der einzelnen Aufgaben übernimmt.

In der Planungsphase hat die Vertriebssteuerung drei zentrale Aufgaben zu erfüllen. Zum einen ist es ihre Aufgabe, die vereinbarten Unternehmensziele in operationale Vertriebsziele herunterzubrechen sowie die Koordination innerhalb der Vertriebsziele und mit anderen Funktionalbereichszielen vorzunehmen.[140] Diese impliziert die Prüfung der Beziehungen zwischen den verfolgten Vertriebszielen. Unterschieden werden komplementäre, konkurrierende und indifferente Zielbeziehungen.[141] Komplementäre Ziele liegen vor, wenn das Erreichen eines Ziels mit dem Erreichen eines anderen Ziels einhergeht. Konkurrierende Ziele liegen vor, wenn das Verfolgen eines Ziels nicht vereinbar ist mit dem Verfolgen eines anderen Ziels. Und indifferente Ziele liegen vor, wenn das Verfolgen des einen Ziels das Verfolgen eines anderen Ziels nicht beeinflusst. Zum zweiten muss die Vertriebssteuerung dafür Sorge tragen, dass die Vertriebsstrategie, mit der die Ziele erreicht werden sollen, definiert wird sowie deren Risiken geprüft und Alternativen vorgegeben werden.[142] Die Vertriebsstrategie fokussiert auf die Auswahl der Zielkunden und der relevanten Wettbewerbsvorteile, die Ausgestaltung des Vertriebssystems und der Vertriebswege sowie der Preispolitik.[143] Und zum dritten ist es Aufgabe der Vertriebssteuerung, die bereitgestellten Methoden der Planung zu verwenden sowie die Beurteilung der Planergebnisse vorzunehmen. Um die Planung durchzuführen, lassen sich drei Vorgehensweisen unterscheiden. Dies ist die Top-Down-Planung, bei der die Vertriebsleitung die Planung initiiert und Vorgaben setzt. Davon ausgehend werden die Teilpläne für untergeordnete Organisationseinheiten erarbeitet. In umgekehrter Richtung arbeitet die Bottom-Up-Planung. Bei diesem Vorgehen werden von der kleinsten Vertriebseinheit ausgehend die Teilpläne nach oben zu

140 Vgl. *Lissautzki, M.* (2007), S. 63.
141 Vgl. hier und nachfolgend *Wöhe, G.* (1996), S. 127 f.
142 Vgl. *Pufahl, M.* (2003), S. 339.
143 Vgl. *Homburg, C./Schäfer, H./Schneider, J.* (2010), S. 27 ff.

einem Gesamtplan aggregiert. Das Gegenstromverfahren kombiniert die beiden erstgenannten Vorgehensweisen und basiert auf einem Dialog- bzw. Aushandlungsprozess.[144]

Zur Gruppe der planrealisationsbezogenen Aufgaben gehört die Plandurchsetzung vertriebszielsetzungsorientiert auszugestalten sowie die Früherkennung und Beseitigung von Störungen bei der Planrealisation zu gewährleisten.[145] Störungen bei der Planrealisation resultieren bspw. aus der zunehmenden Diskontinuität und Dynamik umfeldbezogener Entwicklungen und der unternehmensinternen und unternehmensexternen Komplexität.[146] Die Aufgaben der Phase der Plandurchsetzung umfassen weiterhin alle Maßnahmen der Organisation und Information der von der Planrealisation betroffenen Mitarbeiter und der weiteren Personen innerhalb des Unternehmens.[147] In diesem Zusammenhang gilt es, bei den betroffenen Mitarbeitern Akzeptanz für die verfolgten Vertriebsziele und die Vertriebspläne zu schaffen, um mitarbeiterverhaltensbedingte Störungen der Planrealisation zu vermeiden oder zu reduzieren[148] und ein im Sinne der Vertriebsziele rollenkonformes Verhalten der Mitarbeiter zu erreichen.[149] Hierzu zählt der Aufbau eines zielkongruenten Anreizsystems.[150] Weiterhin ist es Aufgabe der Vertriebssteuerung, den Mitarbeitern die für die Umsetzung der Planung relevanten Informationen sowie Mittel und Instrumente zur Verfügung zu stellen. Dies betrifft die weisungsabhängigen Mitarbeiter ebenso wie die Führungskräfte.[151] Die Prüfung der aufbau- und ablauforganisatorischen Struktur des institutionellen Vertriebs vor dem Hintergrund der Plananforderungen ist ebenso Aufgabe der Vertriebssteuerung wie die Neugestaltung des Vertriebsbereichs, sofern sich aus einer unzulänglichen Struktur störende Einflüsse auf die Planrealisation ergeben können. Diese organisatorischen Aufgaben der Durch-

---

144 Vgl. *Horváth, P.* (2003), S. 196; *Küpper, H.-U.* (2001), S. 306; *Wimmer, K./Schilcher, N./Windbichler, C.* (2010), S. 93 ff.

145 Vgl. *Schweitzer, M.* (2001), S. 23.

146 Vgl. *Hopfenbeck, W.* (1997), S. 340.

147 Vgl. *Schweitzer, M.* (2001), S. 71.

148 Die dieser Beziehung impliziten Probleme der Zielidentifikation und der Anreizsysteme sowie deren Lösung werden unter anderem im Rahmen der Prinzipal-Agenten-Theorie und der verhaltenswissenschaftlichen Managementtheorie diskutiert. Eine Auseinandersetzung mit diesem Themenbereich wird an dieser Stelle nicht vorgenommen, sondern es sei auf die themenspezifische Literatur verwiesen. Vgl. auch die Ausführungen in Kapitel E.3.3.

149 Vgl. *Hopfenbeck, W.* (1997), S. 365.

150 Vgl. *Lissautzki, M.* (2007), S. 63.

151 Vgl. *Schmoll, A.* (2006), S. 160. An dieser Stelle zeigt sich die enge Verknüpfung zum Vertriebscontrolling, zu dessen direkten Aufgabenbereich die Informationsversorgung der Führung gehört.

setzungsphase betreffen auch die Sicherstellung der notwendigen Qualifikation der betroffenen Mitarbeiter. Schließlich ist der Aufbau eines geeigneten Frühwarnsystems im Rahmen der Vertriebssteuerung vorzunehmen, um rechtzeitig mögliche Störungen zu erkennen und Gegensteuerungsmaßnahmen ergreifen zu können.

In den Phasen der Überwachung und Sicherung sind durch das Vertriebssteuerungssystem Vertriebsplanabweichungen zu identifizieren, zu analysieren und für die Ableitung von Gegenmaßnahmen aufzunehmen.[152] Grundlage dieser Aufgaben sind verschiedene Kontrollen, die sich hinsichtlich der Art der Kontrollobjekte, der Komponenten des Planungsprozesses, des Informationscharakters der Vergleichsgrößen und der Informationsvermittlung sowie deren Verknüpfung unterscheiden lassen (siehe Abbildung 3).

Mit Blick auf die Kontrollobjekte lassen sich drei Kontrolltypen unterscheiden. Bei den Ergebniskontrollen wird die Zielerreichung, also der Vergleich zwischen einem realisierten und einen gewollten Zustand, geprüft. Bei den Verfahrenskontrollen wird geprüft, ob und wie der zur Zielerreichung definierte Prozess vollzogen wurde. Die Verhaltenskontrollen sollen informieren, ob die Menschen, die mit den Prozessen betraut sind, unter Berücksichtigung der beobachteten Rahmenbedingungen geeignete Maßnahmen getroffen haben.

<table>
<tr><th>Merkmale</th><th colspan="12">Ausprägungen</th></tr>
<tr><td>Art der Kontrollobjekte</td><td colspan="3">Verhaltens-kontrolle</td><td colspan="3">Verfahrens-kontrolle</td><td colspan="3">Teilergebnis-kontrolle</td><td colspan="3">Endergebnis-kontrolle</td></tr>
<tr><td>Komponenten des Planungsprozesses</td><td colspan="4">Prämissenkontrolle</td><td colspan="4">Planfortschritts-kontrolle</td><td colspan="4">Realisationskontrolle</td></tr>
<tr><td>Informationsarten</td><td colspan="2">Ist-Ist (Zeitvergleich, Sachvergleich)</td><td colspan="2">Soll-Soll (Zielkontrolle)</td><td colspan="2">Soll-Ist (Ergebnis-kontrolle)</td><td colspan="2">Soll-Wird (Planfort-schritts-kontrolle)</td><td colspan="2">Wird-Ist (Prämissen-kontrolle)</td><td colspan="2">Wird-Wird (Prognose-kontrolle)</td></tr>
<tr><td>Informationsvermittlung und -verarbeitung</td><td colspan="6">Persönlich</td><td colspan="6">Maschinell</td></tr>
</table>

Abbildung 3: Formen der Kontrolle[153]

[152] Vgl. hier und nachfolgend *Schweitzer, M.* (2001), S. 73 ff.; *Küpper, H.-U.* (2001), S. 170 ff.

[153] *Küpper, H.-U.* (2001), S. 174.

Anhand der Komponenten des Planungsprozesses können die Realisations-, die Planfortschritts- und die Prämissenkontrolle unterschieden werden. In der Realisationskontrolle werden Endergebnisse nach Ablauf der Planperiode überprüft. Aufgrund der nicht mehr bestehenden Beeinflussbarkeit der Planrealisation können die Ergebnisse nur als Information neuer Planungsprozesse dienen. Die Planfortschrittskontrolle prüft die Umsetzung einzelner Planbestandteile während der Planrealisation, so dass Prognosen über das Ergebnis am Ende des Planungshorizonts ermöglicht und ggf. Steuerungsmaßnahmen begründet werden können. Die Prämissenkontrolle untersucht, ob die in der Planung unterstellten Annahmen für die Planrealisation tatsächlich eintreten bzw. eingetreten sind.

Weiterhin lassen sich Kontrollen durch die Kombinationen von normativen, faktischen und prognostischen Größen (Informationsarten) unterscheiden. Die normativen Größen stellen Soll-Größen dar, die als Zielformulierung verstanden werden. Faktische Größen stellen Ergebnisse dar, die bei den betrachteten Größen tatsächlich realisiert wurden und prognostische Größen sind sich abzeichnende Ist-Größen zu einem bestimmten Zeitpunkt, deren tatsächliches Ergebnis jedoch noch nicht bestimmt ist. Werden diese Größen mit unterschiedlichem Zeitbezug kombiniert, lassen sich verschiedene Kontrollarten ableiten.[154]

Hinsichtlich der Informationsermittlung und der -verarbeitung wird die persönliche Kontrolle von der maschinellen Kontrolle unterschieden. Es ist anzumerken, dass nicht alle der erläuterten Kontrollformen stets maschinell möglich sind, so z. B. die Prämissenkontrolle.

Aufbauend auf den vorgenommen Kontrollen ist es die Aufgabe des Vertriebssteuerungssystems, die Analyse der Ursachen für etwaige Abweichungen vorzunehmen. Zweck ist es, die konkreten Störgrößen der Planrealisation zu identifizieren. Diese können unternehmensexogen oder unternehmensendogen verankert sein. Störgrößen können sowohl in der Planung selbst oder in der anschließenden Planrealisation begründet sein. Als Fehlerbereiche in der Vertriebsplanung lassen sich mit SCHWEITZER die Zielbildung, die Problemfeststellung, die Alternativensuche, die Prognose sowie die Bewertung und Entscheidung von Handlungsalternativen und in der Vertriebsplanrealisation die Vertriebsplandurchsetzung, die Ausführung, die Plankontrolle, die Plansicherung und die Änderungen der Planungsprämissen unterscheiden.[155]

Eine weitere Aufgabe zur Erreichung der Ziele der Vertriebssteuerung ist der Aufbau eines aussagekräftigen Berichtssystems, das dem Informationsbedarf des Vertriebsleiters und der Vertriebsmitarbeiter entspricht.[156] Dabei muss das

---

154 Vgl. dazu ausführlich *Küpper, H.-U.* (2001), S. 173 ff.

155 Vgl. *Schweitzer, M.* (2001), S. 77 ff.

156 Vgl. *Schmoll, A.* (2006), S. 162.

Postulat erfüllt werden, dass auf der operativen und der administrativen Ebene ein konsistentes Informations- und Berichtssystem als Grundlage der Vertriebssteuerung vorhanden ist.[157]

Die zentrale Aufgabenstellung der Sicherungsphase ist die Erarbeitung von Maßnahmen, die das Wirken von Störgrößen verhindern, deren negativen Einfluss reduzieren und die Störgrößen beheben bzw. vermeiden.[158] Weiterhin muss gewährleistet werden, dass die identifizierten Maßnahmen um- und durchgesetzt werden.

2.4 Institutioneller Begriffsinhalt

Der institutionelle Begriffsinhalt bezieht sich auf die Aufbauorganisation der Vertriebssteuerung und mithin auf die Frage, wer Träger der Vertriebssteuerungsfunktion ist. Ebenso wie die noch zu diskutierende Auswahl der Instrumente ist die konkrete Beantwortung dieser Frage kontextabhängig und vor dem Hintergrund des hier zugrunde gelegten praktisch orientierten Verständnisses der Vertriebssteuerung zu beantworten. I. d. R. werden die Aufgaben der Vertriebssteuerung von der Vertriebsleitung in Verbindung mit dem Controlling durchgeführt. Diese Aussage ordnet sich den bisherigen Ausführungen zum Vertriebscontrolling und der Steuerung als Bestandteil der Führung unter. Alternativ dazu gibt es Situationen, in denen eigenständige Abteilungen oder Stäbe mit Vertriebssteuerungsaufgaben betraut sind. So gibt es z. B. bei den hier im Fokus stehenden Kreditgenossenschaften häufig die Stabsabteilung Vertriebssteuerung, die jedoch vornehmlich für das Privatkundengeschäft zuständig ist und hinsichtlich der Aufgaben von den für diese Arbeit zweckmäßigerweise definierten Verständnissen der Vertriebssteuerung abweicht.

2.5 Prozessualer Begriffsinhalt

Der prozessuale Begriffsinhalt betrachtet den zeitlichen Ablauf der durchzuführenden Vertriebssteuerungsaufgaben, mithin also die Planungs- und Kontrollzyklen sowie die Anpassungsrhythmik, in der bereits vorhandene Planergebnisse überprüft und etwaige Steuerungsmaßnahmen entwickelt werden. Im Zentrum steht die Frage, wann welche Aufgaben durchgeführt werden. Grundsätzliche prozessuale Strukturen resultieren aus dem Verständnis der Vertriebssteuerung im weiteren Sinn, das zwischen der Planungsphase, der Phase der Umsetzung und Kontrollphase (Überwachung und Sicherung) unterscheidet. Jedoch laufen

---

157 Vgl. *Becker, J.* (1994), S. 4.

158 Vgl. *Schweitzer, M.* (2001), S. 80.

diese Phasen nicht zwingend chronologisch ab, sondern weisen zum Teil gewollte oder ungewollte Überschneidungen auf.

3 Instrumente der Vertriebssteuerung

Zu den Instrumenten der Vertriebssteuerung gehören alle Methoden und Modelle, die für die Erreichung der Vertriebssteuerungsziele oder die Erfüllung der Aufgaben der Vertriebssteuerung dienlich sind.[159] In diesem Zusammenhang kann konstatiert werden, dass die Vertriebssteuerung über keine originären Instrumente verfügt, sondern sich vielmehr auf einen breiten betriebswirtschaftlichen Instrumentenkanon stützt, der zweckorientiert eingesetzt wird.[160] Dieser Instrumentenkanon kann nach verschiedenen Kriterien strukturiert werden, so dass sich ebenso für verschiedene Situationen unterschiedliche Instrumente wie auch gleiche Instrumente identifizieren lassen. Im Folgenden soll Strukturierungsmöglichkeiten vorgestellt werden, wobei jedoch nicht der Anspruch auf Allgemeingültigkeit oder Vollständigkeit erhoben wird.

In Bezug auf den zugrunde gelegten Betrachtungszeitraum und den verfolgten Zielgrößen lassen sich strategische und operative Instrumente auf Basis der verwendeten Daten unterscheiden. Im Mittelpunkt der strategischen Instrumente stehen zukunftsorientierte, qualitative Daten, die mit dem Ziel des Aufbaus und des Erhalts von Erfolgspotenzialen analysiert werden. Dazu gehören Früherkennungs- und Analyseinstrumente wie bspw. das Environmental Scanning, (Cross-)Impact-Matrizen oder Checklisten.[161] Operative Instrumente dagegen nutzen vergangenheits- bzw. gegenwartsbezogene, quantitative Größen, um kurzfristige Vertriebssteuerungsentscheidungen zu fundieren. Diese Instrumente können weiter in Instrumente zur Problemidentifikation und Instrumente zur Entscheidungsvorbereitung unterschieden werden. Dazu gehören z. B. periodische Erfolgsrechnungen wie die Absatzsegmentrechnung oder die Einzelkundenkalkulation aber auch Kennzahlensysteme.

Weiterhin können die Instrumente der Vertriebssteuerung dahingehend unterschieden werden, welche Arten von Zielen verfolgt werden. So lassen sich bspw. wertmäßige Ziele, Mengenziele, Qualitätsziele, Zeitziele oder Verhaltensziele unterscheiden,[162] denen jeweils unterschiedliche Instrumente zugeordnet werden können. Instrumente zur Steuerung bei wertmäßigen Zielen sind etwa Kosten- und Erlösrechnungen, Budgets oder Wertzuwachskurven. Bei Mengen-

---

159 Vgl. *Schweitzer, M./Friedl, B.* (1992), S. 158.

160 Vgl. mit identischer Argumentation *Rinn, T.* (1991), S. 46 für das von ihm analysierte Vertriebs-Controlling.

161 Vgl. *Küpper, H.-U.* (2001), S. 416 ff.

162 Vgl. *Schweitzer, M.* (2001), S. 82.

zielen sind dies z. B. Fortschrittserhebungen und bei den Verhaltenszielen verschieden ausgeprägte Anreizsysteme.

Auch die Phasen des Planungs- und Steuerungsprozesses können als Ordnungsprinzip verwendet werden. So lassen sich planungsbezogene Instrumente wie z. B. Planrechnungen, Planungshandbücher oder Szenariotechniken, umsetzungsbezogene Instrumente wie z. B. Instrumente der organisatorischen Gestaltung wie Stellenbeschreibungen oder Anreizsysteme, überwachungsbezogene Instrumente wie bspw. Ergebniskontrollen oder Realisationskontrollen und sicherungsbezogene Instrumente wie z. B. Frühwarnsysteme oder Regeln und Normen unterscheiden. Phasenübergreifend können alle Instrumente der Informationskommunikation berücksichtigt werden. Dies ist z. B. ein an dem Informationsbedarf der Träger der Vertriebs- und Vertriebssteuerungsaufgaben orientiertes Berichtswesen.

Ohne dass dies eine abschließende Aufzählung ist, zeigt sich, dass die Auswahl und der Einsatz von Instrumenten kontextabhängig erfolgen müssen, um so die richtigen Instrumente zu nutzen. Folglich ist es nicht möglich, eine allgemein gültige Aussage über die einzusetzenden Instrumente in der Vertriebssteuerung zu treffen.

4 Zusammenfassende Darstellung der Vertriebssteuerung

In den vorangegangenen Abschnitten wurden Vertrieb und Vertriebssteuerung begrifflich abgegrenzt und die Vertriebssteuerung hinsichtlich der funktionalen, prozessualen und institutionellen Begriffsinhalte beschrieben. Es wurde dargestellt, dass der Vertrieb als Schnittstelle zum Kunden insbesondere mit den Aufgaben des Kundenmanagements betraut ist. Die Vertriebssteuerung, deren zentrales Ziel die Sicherstellung der Effektivität und Effizienz des Vertriebs ist, beinhaltet im hier zugrunde gelegten praxisorientierten Verständnis die Phasen der Planung, Plandurchsetzung, Kontrolle und Plansicherung auf strategischer und operativer Ebene und baut dabei unmittelbar auf dem Vertriebscontrolling auf. Um die direkten und indirekten Ziele der Vertriebssteuerung zu erfüllen, sind der Vertriebssteuerung konkrete Aufgaben zugeordnet, die sich phasenbezogen abgrenzen lassen und durch kontextabhängig ausgewählte Instrumente unterstützt werden.

Werden nun die Komponenten des Vertriebs und hier im Speziellen die Aufgaben des Kundenmanagements sowie die funktionellen Aspekte der Vertriebssteuerung kombiniert, ergibt sich der Bezugsrahmen der konkreten Ausgestaltung der Vertriebssteuerung im situativen Kontext des Betrachtungsgegenstands (siehe Abbildung 4). Dabei ist jedoch zu berücksichtigen, dass die inhaltlichen Schwerpunkte in der Differenzierung nach operativer und strategischer Ausgestaltung mal stärker und mal weniger stark zu Tragen kommen. So finden sich häufig große Schnittmengen zwischen der operativen und der strategischen Kontrolle, die eine klare Differenzierung nicht erlauben.

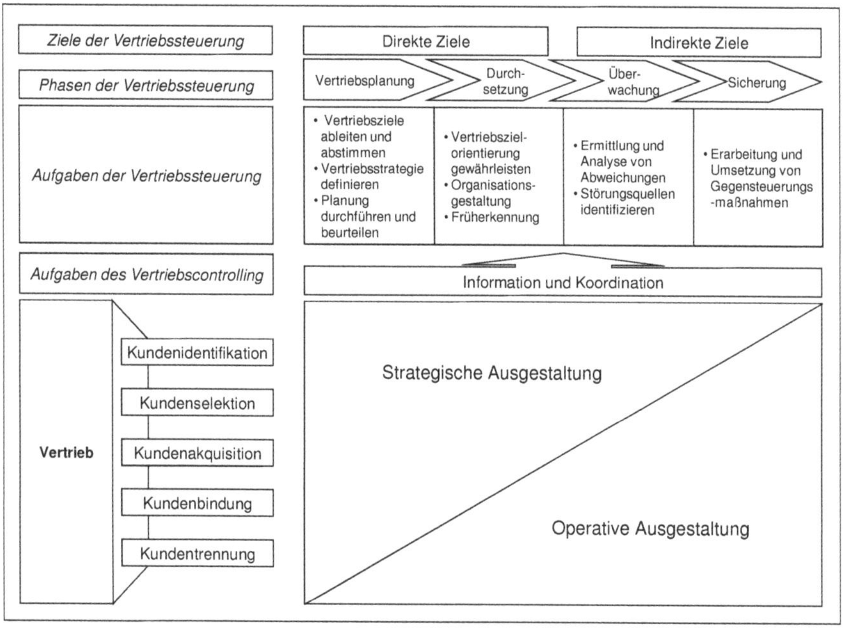

Abbildung 4: Zusammenfassende Darstellung der Vertriebssteuerung[163]

In den vorhergehenden Ausführungen ist mehrfach deutlich geworden, dass die konkrete Ausgestaltung der Vertriebssteuerung kontextspezifisch und -abhängig ist. Aus diesem Grund sollen im nachfolgenden Teil C die wesentlichen Implikationen herausgearbeitet werden, die aus dem situativen Kontext der etwa 1.200 Kreditgenossenschaften in Deutschland resultieren. Aufgrund der Vielzahl der eigenständigen Unternehmen und der damit verbundenen Heterogenität der Einzelfälle kann keine vollständige Darstellung erfolgen. Dennoch lassen sich für die Gruppe der Kreditgenossenschaften die zentralen, gemeinsamen Implikationen identifizieren, die im weiteren Gang der Analyse zu berücksichtigen sind. Dies begründet sich damit, dass die Kreditgenossenschaften zumindest bundesweit in einem einheitlichen ökonomischen und rechtlichen Marktumfeld agieren

163 Eigene Darstellung.

und der BVR zentrale strategische und operative Weichenstellungen für die Banksteuerung vorschlägt. Wenngleich die Kreditgenossenschaften nicht an diese Vorschläge gebunden sind, orientiert sich die Majorität der Vorstände daran. Damit erscheint es legitim und im Sinne der Arbeit zweckgemäß zu sein, die weiteren Betrachtungen und Analysen auf der Ebene der Gruppe der typischen Kreditgenossenschaften vorzunehmen.

C Ableitung kontextinduzierter Implikationen für die kundenwertorientierte Vertriebssteuerung im Firmenkundengeschäft der Kreditgenossenschaften

Im vorangegangenen Kapitel wurden allgemeine Grundlagen der Vertriebssteuerung erläutert. Um diese auf den konkreten Kontext der Kreditgenossenschaften zu übertragen, werden zunächst die Situationsspezifika und deren Implikationen für die Vertriebssteuerung vorgestellt. Begründet wird dieses Vorgehen mit der Aussage der Kontingenztheorie, dass es keine optimale, allgemeingültige Form einer effizienten Organisation gibt, sondern die Ausgestaltung von den vorliegenden Umfeldbedingungen abhängig sei.[164] Die ausschließliche Betrachtung der Umfeldbedingungen würde jedoch den Blickwinkel verengen und dazu führen, dass nicht alle relevanten Einflussfaktoren betrachtet werden. Aus diesem Grund wird der Blickwinkel um diejenigen internen Faktoren erweitert, die aus Sicht der Kreditgenossenschaft keine Umfeldbedingungen, sondern Resultate eigener Entscheidungen sind. Diese Entscheidungen stellen aus Sicht der Vertriebssteuerung ebenso Determinanten dar, die es zu berücksichtigen gilt. Statt von Umfeldbedingungen wird daher von vertriebssteuerungsrelevanten Rahmenbedingungen, die sowohl unternehmensextern als auch unternehmensintern bedingt sein können, gesprochen. Im Folgenden werden dazu zunächst die Kreditgenossenschaften in das deutsche Bankensystem eingeordnet (Kapitel C.1) und im Anschluss die Geschäftseinheit Firmenkunden (Kapitel C.2), dessen Strategie, Bedeutung und Marktanteile (Kapitel C.3) beschrieben. Darauf folgend werden die für die Vertriebssteuerung relevanten Bestandteile der Gesamtbanksteuerungskonzeption VR-Control vorgestellt (Kapitel C.4). Auf Basis dieser Ausführungen erfolgt die übergreifende und zusammenfassende Ableitung relevanter Implikationen (Kapitel C.5).

---

164 Vgl. *Flacke, K.* (2006), S. 95. So auch *Wiedemann, A.* (1992), S. 14, wenngleich sich dieser auf die Gesamtheit der Kreditgenossenschaften bezieht.

1 Einordnung der Kreditgenossenschaften in das deutsche Bankensystem und Wettbewerb auf dem deutschen Bankenmarkt

1.1 Begriffsabgrenzung Bank

Der Begriff Bank wird in der Literatur uneinheitlich abgegrenzt.[165] Im Folgenden soll der im Kreditwesengesetz (KWG) definierte Begriff „Kreditinstitut" vorgestellt werden. Da die Legaldefinition bezweckt, den Anwendungsbereich des Gesetzestextes festzulegen, ist sie für das Ziel dieser Arbeit unzureichend.[166] Daher wird zusätzlich eine zweckadäquate ökonomische Definition vorgestellt, die als Grundlage der Arbeit dient.

1.1.1 Legaldefinition von Banken

Aufgrund ihrer hohen Bedeutung für die Funktionsfähigkeit von Volkswirtschaften gehören Banken zu den stark regulierten Unternehmenstypen.[167] Dazu hat der deutsche Gesetzgeber die Legaldefinition für den Begriff Kreditinstitut (synonym: Banken[168]) erarbeitet, unter welchen die Kreditgenossenschaften zu subsumieren sind.

Gemäß § 1 Kreditwesengesetz (KWG) sind Kreditinstitute „Unternehmen, die Bankgeschäfte gewerbsmäßig oder in einem Umfang betreiben, der einen in kaufmännischer Weise eingerichteten Geschäftsbetrieb erfordert." Die dieser Definition zugrunde liegenden, erlaubnispflichtigen Bankgeschäfte werden in § 1 Abs. 1 KWG enumeriert.[169] Gemäß der 6. Novelle des KWG qualifiziert bereits das gewerbsmäßige Betreiben eines der genannten Bankgeschäfte ein Un-

---

165 Nach EILENBERGER lässt sich anhand der „Bemühungen einer Vielzahl von Autoren zur Bestimmung des Bankbegriffs [...] feststellen, dass kaum ein vergleichbares Erkenntnisobjekt sich ähnlich umfangreicher wissenschaftlicher Anstrengungen erfreuen kann." *Eilenberger, G.* (1996), S. 10.

166 Vgl. *Varmaz, A.* (2006), S. 16.

167 Vgl. *Hartmann-Wendels, T./Pfingsten, A./Weber, M.* (2010), S. 20.

168 Vgl. *Hartmann-Wendels, T./Pfingsten, A./Weber, M.* (2010), S. 20.

169 Siehe § 1 Abs. 1 Nr. 1 ff. KWG. Die genannten Geschäfte sind: das Einlagen-, das Pfandbrief-, das Kredit-, das Diskont-, das Finanzkommissions- sowie das Depotgeschäft, die Eingehung von Verpflichtungen, Darlehensforderungen vor Fälligkeit zu erwerben, das Garantie-, das Giro-, das Emissions- und das E-Geld-Geschäft sowie die Tätigkeit als zentraler Kontrahent im Sinne des § 1 Abs. 31 KWG. Eine Würdigung dieser Auflistung findet sich bei *Hartmann-Wendels, T./Pfingsten, A./Weber, M.* (2010), S. 21 f. Eine Erläuterung einzelner Bankgeschäfte findet sich in *Büschgen, H. E.* (1998), S. 12 ff. und *Hartmann-Wendels, T./Pfingsten, A./Weber, M.* (2010), S. 149 ff.

ternehmen als Kreditinstitut[170] und begründet damit die Anwendung der aufsichtsrechtlichen Vorgaben des KWG.[171] Derartige Vorgaben umfassen die Verpflichtung zur Eigenmittelunterlegung von Risikopositionen, Vorschriften über die Fristigkeiten der Aktiva und Passiva und Bestimmungen über organisatorische Mindeststandards. Weitere zentrale bankaufsichtsrechtliche Regelungen sind die Solvabilitätsverordnung (SolvV), die den § 10 KWG inhaltlich präzisiert, die Mindestanforderungen an das Risikomanagement (MaRisk) sowie die Einlagensicherungssysteme.[172] Eine ausführliche Darstellung dieser Vorschriften kann im Rahmen dieser Arbeit nicht erfolgen. Die Legaldefinition des KWG wird einer betriebswirtschaftlichen Betrachtungsweise nicht gerecht.[173] Weder kann dadurch die Dynamik des bankbetrieblichen Aktionsradius erfasst werden, noch ist die Definition geeignet, um eine Bank als einen im Wettbewerb befindlichen und verschiedenen Einflussfaktoren ausgesetzten Akteur darzustellen. Daher wird nachfolgend eine ökonomische Definition vorgenommen.

1.1.2 Ökonomische Definition von Banken

Eine ökonomische Definition des Bankenbegriffs ist aufgrund des verbundenen Zwecks globaler und abstrakter als die des Gesetzgebers.[174] Ökonomische Definitionen und konstituierende Beschreibungen von Banken erfolgen aus gesamt- oder einzelwirtschaftlicher Sicht.[175] Gesamtwirtschaftliche Definitionen beziehen sich auf das Wesen und die Bedeutung aller Banken für die Volkswirtschaft, betreffen also das Bankensystem. Schwerpunkte stellen hier die Transformations- und die Transaktionsfunktion dar, die sich aus der Eigenschaft der Banken als Finanzintermediäre ergeben.[176] Diese Arbeit betrachtet die Vertriebssteue-

---

170 Abweichend davon gibt es jedoch auch Unternehmen, die zwar Bankgeschäfte tätigen, auf die jedoch nur teilweise oder gar nicht die Normen des KWG Anwendung finden. Dies sind z. B. die DEUTSCHE BUNDESBANK oder die KREDITANSTALT FÜR WIEDERAUFBAU (KfW). Siehe § 2 KWG.

171 Diese aufsichtsrechtlichen Vorgaben dienen vor allem dazu, Ausfallrisiken für Gläubiger zu minimieren, indem die Risiken, die Banken eingehen können, begrenzt werden. Bankentypische Risiken sind Adress-, Sachwertausfall- und Preisrisiken sowie Liquiditätsrisiken und Großkredit-, Betriebs- und Informationsrisiken. Vgl. *Hartmann-Wendels, T./Pfingsten, A./Weber, M.* (2010), S. 412 ff. Vgl. auch Kapitel D.4.5.4.

172 Vgl. dazu ausführlich *Hartmann-Wendels, T./Pfingsten, A./Weber, M.* (2010), S. 412 ff.

173 Vgl. *Büschgen, H. E.* (1998), S. 33.

174 Vgl. *Büschgen, H. E.* (1998), S. 33.

175 Vgl. ausführlich z. B. *Büschgen, H. E.* (1998), S. 33 ff. oder *Hartmann-Wendels, T./ Pfingsten, A./ Weber, M.* (2010), S. 2 ff.

176 Aufgrund der für diese Arbeit nur nebensächlichen Bedeutung sei für eine theoretische Begründung der Existenz von Banken und der Erläuterung der einzelnen volkswirt-

rung in den einzelnen Kreditgenossenschaften, so dass die einzelwirtschaftliche Sicht relevant ist.

In der jüngeren Literatur vornehmlich zu findende Begriffsfestlegungen definieren Banken systemorientiert.[177] Dabei wird angenommen, dass Bankbetriebe zielgerichtete offene sozio-technische Systeme sind, die mit ihrer Umwelt durch ein Netz von Transaktionsbeziehungen bzw. Interaktionen verbunden sind.[178] Diese Begriffsauffassung basiert auf der Systemtheorie, nach der ein System als eine Gesamtheit von Elementen verstanden wird, zwischen denen (dauerhafte) Beziehungen bestehen. Die Elemente sind sowohl sozialer Natur (z. B. Personen) als auch technischer Natur (z. B. Bauten, EDV) und stehen als gemeinsam konstituiertes Subsystem in einem offenen Austausch mit subsystemexternen Elementen oder anderen Subsystemen. Die Zielgerichtetheit impliziert, dass zum einen das System Bank spezifische Institutsziele verfolgt und zum anderen die sozialen Systemelemente und Subsysteme individuelle Ziele verfolgen können. Der systemorientierte Bankbegriff berücksichtigt, dass sowohl der bankbetriebliche Leistungserstellungsprozess als auch dessen Ergebnisse von ebendiesen Zielen, den Organisationsmitgliedern und deren Verhalten, der Organisationsstruktur und den bankbetrieblichen Entscheidungen und Entscheidungsprozessen ebenso beeinflusst werden, wie auch von Umweltfaktoren (siehe Abbildung 5).[179]

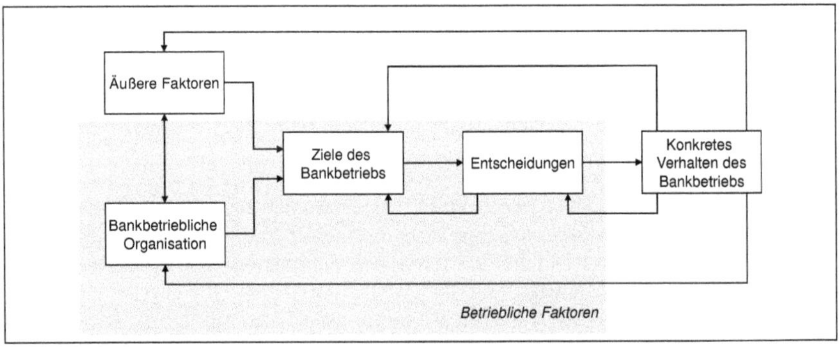

Abbildung 5: Haupteinflussgrößen bankbetrieblichen Verhaltens[180]

---

schaftlichen Funktionen auf die entsprechenden Kapitel in *Büschgen, H. E.* (1998) verwiesen.

177 Vgl. zur ausführlichen Diskussion verschiedener Begriffsinhalte *Priewasser, E.* (2001), S. 9 ff.

178 Vgl. hier und nachfolgend *Eilenberger, G.* (1996), S. 13.

179 Vgl. *Eilenberger, G.* (1996), S. 13.

180 Vgl. *Eilenberger, G.* (1996), S. 14.

Wenngleich der systemorientierte Bankbegriff für die Zielsetzung dieser Arbeit eine zweckgerechte Grundlage darstellt, ist er aufgrund der breiten Anwendbarkeit hinsichtlich der absatzmarktbezogenen Leistungen zu konkretisieren. Im Sinne einer aus der Kundenperspektive vorgenommenen Betrachtung können folgende spezifische Leistungsbereiche von Banken genannt werden:[181] Anbieten und Eröffnen von Zahlungsverkehrs-, Finanzierungs- und Kapitalanlagefazilitäten sowie zugehöriger Aktivitäten und Dienstleistungen sowie von sonstigen Dienstleistungen und Handel mit übrigen Finanzdienstleistungen. Der letztgenannte Punkt dieser Aufzählung orientiert sich an den tatsächlichen Tendenzen, dass Banken nicht nur als reine Dienstleistungsunternehmen zu verstehen sind. Denn obwohl Banken oft dem Dienstleistungssektor zugeordnet werden, muss abhängig von der jeweiligen Sortimentsbreite eine differenzierte Beurteilung erfolgen.[182] Unter Berücksichtigung des breiten und tiefen Leistungsangebots der deutschen Universalbanken, das nicht ausschließlich von den Banken selbst hergestellt wird, können diese folglich als Dienstleistungs- und als Handels- bzw. Vermittlungsunternehmen bezeichnet werden. Daraus resultiert, dass in der marktorientierten Institutsführung nicht nur die eigenen Interessen ausschlaggebend sind, sondern ggf. auch die Interessen der jeweiligen vertikalen Handels- und Vermittlungspartner. Hinsichtlich der Kreditgenossenschaften wird dieser Aspekt vor dem Hintergrund der genossenschaftlichen FINANZGRUPPE[183] in Kapitel C.1.4.4 aufgegriffen und weiter konkretisiert.

Die bankbetriebliche Leistungserstellung kann gedanklich in einen liquiditätsmässig-finanziellen (Wertbereich) und einen technisch-organisatorischen Bereich (Betriebsbereich) differenziert werden.[184] Der Wertbereich erwirtschaftet Erträge durch die Zinsmargen des eingezahlten und ausgeliehenen Kapitals und der Steuerung von Risiken. Der Betriebsbereich ist der Servicebereich, der den Kunden bei der Abwicklung der Bankgeschäfte unterstützt. Der Betriebsbereich ist Voraussetzung für Leistungen des Wertbereichs et vice versa, da nur in Kombination beider „Fertigungsbereiche" die bankbetriebliche Leistungserstellung möglich ist.[185]

---

181 *Büschgen, H. E.* (1998), S. 33.

182 Vgl. *Zollner, G.* (1995), S. 64 f.

183 Alternativ zum Begriff der FINANZGRUPPE wird auch der Begriff FINANZVERBUND verwendet.

184 Vgl. dazu *Deppe, H.-D.* (1969), S. 20 ff.

185 Vgl. *Knöbel, U.* (1997), S. 38 ff. sowie die dort angegebene Literatur.

1.2 Beschreibung des deutschen Bankensystems

Da grundsätzlich alle Kreditinstitute jede Art von Geschäft ausüben dürfen, wird das deutsche Bankensystem als Universalbankensystem bezeichnet.[186] Die Universalbank ist eng definiert ein Finanzintermediär, der Bankleistungen gemäß des im § 1 Abs. 1 KWG definierten Katalogs anbietet. Weiter definiert und dem deutschen Verständnis der Universalbank entsprechend, bieten Universalbanken auch Leistungen an, die traditionell nicht zu den Bankleistungen gezählt werden (z. B. Versicherungsleistungen, Reise- und Eigenkapitalvermittlung).[187] I. d. R. treten die Universalbanken für die Produkte, die nicht zu den legal definierten Bankleistungen zählen, als Händler oder Vermittler für ein drittes eigenständiges oder über einen Finanzverbund assoziiertes (z. B. bei Kreditgenossenschaften) oder in einen Bankkonzern (z. B. bei Großbanken) integriertes Finanzdienstleistungsunternehmen auf.

Universalbanken können anhand ihrer Rechtspersönlichkeit und der Art der Einbindung in einen größeren Verbund differenziert werden. Mit der Rechtspersönlichkeit ist eine für das geschäftspolitische Verhalten der jeweiligen Bank besonders relevante Determinante auf das Institutszielsystem zu berücksichtigen.[188] Bei den Universalbanken werden drei Sektoren unterschieden: die privatrechtlich organisierten Kreditbanken, die öffentlich-rechtlichen Sparkassen inkl. Girozentralen und die Kreditgenossenschaften inkl. genossenschaftlicher Zentralbanken.

Die erste Säule bilden die privatrechtlich organisierten Kreditbanken.[189] Diese firmieren überwiegend als Kapitalgesellschaften[190], und das Gewinnstreben wird als deren zentrales Unternehmensziel unterstellt.[191] Die Kreditbanken und hier vor allem die Großbanken, die auch als Prototypen der Universalbank bezeichnet werden,[192] zeichnen sich im Gegensatz zu den anderen beiden Säulen durch ihr vergleichweise großvolumiges Geschäft, ihre Überregionalität und Internationalität sowie der hohen Bedeutung des Effektengeschäfts und des In-

---

186 Vgl. *Varmaz, A.* (2006), S. 79; *Hartmann-Wendels, T./Pfingsten, A./Weber, M.* (2010), S. 30.

187 Vgl. *Hartmann-Wendels, T./Pfingsten, A./Weber, M.* (2010), S. 24.

188 Vgl. *Hartmann-Wendels, T./Pfingsten, A./Weber, M.* (2010), S. 30.

189 Vgl. z. B. *Deutsche Bundesbank* (2008), S. 15 ff. Zu den Kreditbanken werden in der Statistik der DEUTSCHEN BUNDESBANK die Großbanken, die Regionalbanken und die sonstigen Kreditbanken sowie die Zweigstellen ausländischer Banken gezählt.

190 Vgl. *Hartmann-Wendels, T./Pfingsten, A./Weber, M.* (2010), S. 33.

191 Vgl. *Hartmann-Wendels, T./Pfingsten, A./Weber, M.* (2010), S. 33; *Eim, A.* (2004), S. 16.

192 Vgl. *Priewasser, E.* (2001), S. 144.

vestmentbankings aus.[193] Weiteres typisches Kennzeichen der Großbanken ist die im Vergleich zu den anderen Bankensektoren geringe Bedeutung der Einlagenfinanzierung. In Tabelle 2 sind ausgewählte Kennzahlen der Kreditbanken dargestellt.

Tabelle 2: Struktur- und Ergebniskennzahlen der Säule der Kreditbanken[194]

	2001	2002	2003	2004[2]	2005	2006	2007	2008	2009
Anzahl Institute[1]	304	355	356	357	276	272	278	283	288
Anzahl Zweigstellen	5.576	5.122	5.105	14.750	14.012	11.548	11.286	11.277	11.496
Zahl der Beschäftigten	220.700	209.850	192.900	192.550	190.700	186.700	190.250	189.400	181.900
Ø-Bilanzsumme [in Mrd. €]	2.362	2.310	2.252	2.362	2.563	2.602	2.935	2.965	2.735
Zinsspanne[3]	1,15%	1,34%	1,17%	1,25%	1,27%	1,33%	1,30%	1,20%	1,20%
Provisionsüberschuss[4]	0,64%	0,60%	0,59%	0,57%	0,60%	0,63%	0,60%	0,54%	0,55%
Eigenkapitalrendite[5]	4,7%	1,0%	-6,2%	-0,4%	21,8%	11,2%	19,1%	-15,6%.	-5,83%

[1] Die Anzahl der Institute weicht ab von der Anzahl berichtender Institute, die als Grundlage der Ermittlung der jahresabschlussbezogenen Kennzahlen dient. Quelle: Bankstellenstatistik, in: Deutsche Bundesbank, Bankenstatistik, Statistisches Beiheft zum jeweiligen Monatsbericht. Kreditinstitutsbegriff auf § 1 KWG bezogen, insofern Abweichungen zu Angaben in der „Bilanzstatistik" und der „Statistik der Gewinn- und Verlustrechnungen".

[2] Ab 2004 wird die bis dahin nicht berücksichtigte Postbank AG der Gruppe der Kreditbanken zugeordnet.

[3] Die Zinsspanne entspricht dem Saldo der Zinserträge und Zinsaufwendungen jeweils in Relation zur durchschnittlichen Bilanzsumme.

[4] Der Provisionsüberschuss ist der Saldo der Provisionserträge und Provisionsaufwendungen jeweils in Relation zur durchschnittlichen Bilanzsumme.

[5] Die Eigenkapitalrendite ist definiert als Jahresüberschuss vor Steuern in % des durchschnittlichen Eigenkapitals.

Der Sparkassensektor liegt mehrheitlich in Trägerschaft der öffentlichen Hand und wird als öffentlich-rechtliche Säule bezeichnet.[195] Die Kreditinstitute der so genannten S-Finanzgruppe sind in einem dreistufigen Verbundsystem strukturiert, das hinsichtlich des Leistungsumfangs und des Marktauftritts der Stufen subsidiär am Markt agiert.[196] Dieser Verbund besteht aus den Sparkassen, den Landesbanken und der Deka-Bank sowie aus Landesbausparkassen, weiteren

---

193 Vgl. hier und nachfolgend *Büschgen, H. E.* (1998), S. 80; *Hartmann-Wendels, T./Pfingsten, A./Weber, M.* (2010), S. 33.

194 Eigene Darstellung auf Basis der Daten in *Deutsche Bundesbank* (2002), (2003), (2004), (2005), (2006a), (2007), (2008), (2009) und (2010).

195 Vgl. z. B. *Deutsche Bundesbank* (2008), S. 15 ff. Entstanden ist die heutige S-Finanzgruppe im letzten Viertel des 18. Jahrhunderts. Vgl. *Oellerking, C./Holzgrabe, M.* (1990), S. 9.

196 Vgl. *Büschgen, H. E.* (1996), S. 88.

Unternehmen (z. B. Factoring- und Leasinggesellschaften) und regionalen Sparkassen- und Giroverbänden.[197] Die Sparkassen sind dem in Sparkassengesetzen und -verordnungen formulierten öffentlichen Auftrag verpflichtet.[198] Die Sparkassengesetze enthalten zudem das Regionalprinzip und für einige Bundesländer den Grundsatz, dass Sparkassen zwar nach kaufmännischen Grundsätzen zu führen sind, jedoch die Erzielung von Gewinn nicht Hauptzweck des Geschäftsbetriebs sei.[199] Gleichwohl es nicht nach außen kommuniziert wird, steht das Gewinnstreben als praktisches Geschäftsziel im Vordergrund.[200] Tabelle 3 zeigt einige Kennzahlen der S-Finanzgruppe.

---

197 Vgl. *DSGV* (2009), o. S.

198 Dazu gehören, den Sparsinn und die Vermögensbildung der regionalen Bevölkerung zu fördern sowie die Kreditverfügbarkeit ebendieser Menschen zu gewährleisten. Vgl. *Hartmann-Wendels, T./Pfingsten, A./Weber, M.* (2010), S. 36.

199 Siehe z. B. § 3 Abs. des Sparkassengesetzes für das Land Nordrhein-Westfalen vom 10. September 2004 oder § 2 Abs. 3 des Sparkassengesetztes für das Land Brandenburg vom 29. Juni 1996, zuletzt geändert durch Artikel 17 des Gesetzes vom 23. September 2008.

200 Vgl. *Hartmann-Wendels, T./Pfingsten, A./Weber, M.* (2010), S. 36. So gab z. B. die Sachsen-Finanzgruppe den ihr angeschlossenen Sparkassen betriebswirtschaftliche Ziele vor. Danach war eine Eigenkapitalrendite vor Steuern von 15 Prozent und eine Aufwand-Ertrags-Quote von 60 Prozent zu erreichen. Vgl. *Padberg, T.* (2005), S. 6.

Tabelle 3: Struktur- und Ergebniskennzahlen des öffentlich-rechtlichen Bankensektors[201]

	2001	2002	2003	2004[5]	2005	2006	2007	2008	2009
Anzahl Institute[1]	534 (13)	519 (13)	489 (13)	477 (12)	463 (12)	457 (12)	446 (12)	438 (10)	432 (10)
Anzahl Zweigstellen	16.648 (604)	15.628 (553)	14.757 (571)	14.292 (549)	13.950 (580)	13.756 (496)	13.624 (485)	13.457 (482)	13.266 (482)
Zahl der Beschäftigten	282.850 (42.800)	278.800 (41.850)	271.900 (40.500)	265.400 (38.550)	260.800 (40.200)	257.000 (39.500)	253.700 (39.850)	251.400 (39.250)	249.600 (38.750)
Ø-Bilanzsumme [in Mrd. €]	948 (1.599)	975 (1.644)	981 (1.640)	986 (1.519)	995 (1.581)	1.007 (1.652)	1.019 (1.668)	1.042 (1.695)	1.061 (1.587)
Zinsspanne[2]	2,28% (0,60%)	2,38% (0,59%)	2,40% (0,63%)	2,35% (0,65%)	2,30% (0,63%)	2,23% (0,61%)	2,06% (0,65%)	2,00% (0,72%)	2,13% (0,72%)
Provisionsüberschuss[3]	0,50% (0,11%)	0,495 (0,11%)	0,53% (0,11%)	0,56% (0,11%)	0,56% (0,12%)	0,58% (0,13%)	0,60% (0,13%)	0,57% (0,13%)	0,55% (0,07%)
Eigenkapitalrendite[4]	9,2% (4,8%)	8,2% (2,8%)	10,9% (-4,3%)	9,7% (1,1%)	10,5% (6,4%)	8,9% (11,4%)	7,2% (1,5%)	4,0% 2,1%)	8,5% (-8,2%)

Die in Klammern eingetragenen Werte beziehen sich auf die Giro-Zentralen (Landesbanken) der öffentlich-rechtlichen Säule.

1 Die Anzahl der Institute weicht ab von der Anzahl berichtender Institute, die als Grundlage der Ermittlung der jahresabschlussbezogenen Kennzahlen dient. Quelle: Bankstellenstatistik, in: Deutsche Bundesbank, Bankenstatistik, Statistisches Beiheft zum jeweiligen Monatsbericht. Kreditinstitutsbegriff auf § 1 KWG bezogen, insofern Abweichungen zu Angaben in der „Bilanzstatistik" und der „Statistik der Gewinn- und Verlustrechnungen".

2 Die Zinsspanne entspricht dem Saldo der Zinserträge und Zinsaufwendungen jeweils in Relation zur durchschnittlichen Bilanzsumme.

3 Der Provisionsüberschuss ist der Saldo der Provisionserträge und Provisionsaufwendungen jeweils in Relation zur durchschnittlichen Bilanzsumme.

4 Die Eigenkapitalrendite ist definiert als Jahresüberschuss vor Steuern in % des durchschnittlichen Eigenkapitals.

5 Ab 2004 Umgliederung der NRW.BANK von der Bankengruppe „Landesbanken" zur Bankengruppe „Banken mit Sonderaufgaben".

Der Genossenschaftssektor stellt die dritte Säule des deutschen Bankenmarktes dar, welche die größte Anzahl rechtlich selbständiger Institute umfasst und flächendeckend in Deutschland präsent ist.[202] Zum Genossenschaftssektor zählen die 1.138 Kreditgenossenschaften[203] sowie die WGZ BANK AG als regionales Zentralinstitut und die DZ BANK AG als Spitzeninstitut. Während die Zentralinstitute als Aktiengesellschaft firmieren, ist die Rechtsform der meisten Kreditgenossenschaften die eingetragene Genossenschaft (eG).[204] Deren besonderes

201 Eigene Darstellung auf Basis der Daten in *Deutsche Bundesbank* (2002), (2003), (2004), (2005), (2006a), (2007), (2008), (2009) und (2010).

202 Vgl. z. B. *Deutsche Bundesbank* (2008), S. 15 ff.; *Reith, D.* (1997), S. 15.

203 Vgl. *BVR* (2011a), S. 41. Bezugsdatum ist der 31.12.2010.

204 Es gibt jedoch auch einige VR-Banken, die in der Rechtsform der AG firmieren, daher nicht als Kreditgenossenschaften zu bezeichnen sind. Einer der Einflussfaktoren auf die

Merkmal ist, dass die Eigentümer Mitglieder des Genossenschaftsbetriebs und originäre Quelle der externen Eigenfinanzierung mittels eingezahlter Geschäftsguthaben sind. Zusammen mit weiteren Kreditinstituten (z. B. DG HYP AG), Finanzdienstleistungsinstituten (z. B. Union Investment), Technik- und Dienstleistungsunternehmen (z. B. GAD eG) sowie den regionalen und bundesweiten Verbänden bilden die Kreditgenossenschaften und Zentralinstitute die genossenschaftliche FinanzGruppe.[205] Vergleichbar zu den Sparkassen haben die Kreditgenossenschaften aufgrund ihrer Rechtsform einen spezifischen Unternehmenszweck zu berücksichtigen, den so genannten genossenschaftlichen Förderauftrag, der in § 1 Abs. 1 GenG definiert ist.[206] Tabelle 4 zeigt ausgewählte Kennzahlen für die Banken des FinanzVerbunds.

Entscheidung zur Rechtsformwandlung liegt oftmals in der Prüfungsverbandspflicht, die im Falle der AG entfällt. In diesen Fällen kann die Bankleitung flexibel die für die Jahresabschlussprüfung zu beauftragende Prüfungsgesellschaft auswählen.

205 Strenggenommen wird zwischen dem FINANZVERBUND bzw. der FINANZGRUPPE als Genossenschaftsverbund im engeren Sinn, der aus den Zentralbanken, den Kreditgenossenschaften und den Verbundunternehmen besteht, und dem Genossenschaftsverband im weiteren Sinn, der aus der FINANZGRUPPE und den Verbänden besteht, unterschieden. Vgl. *Wiedemann, A.* (1992), S. 93 f. Auf diese Differenzierung wird in dieser Arbeit aufgrund der umgangssprachlichen Identität, der Irrelevanz für die Analyse und aus Vereinfachungsgründen verzichtet.

206 Der Förderauftrag wird detailliert in Kapitel C.1.4.2 behandelt.

Tabelle 4: Struktur- und Ergebniskennzahlen des genossenschaftlichen Bankensektors[207]

	2001	2002	2003	2004	2005	2006	2007	2008	2009
Anzahl Institute[1]	1.621 (2)	1.490 (2)	1.394 (2)	1.338 (2)	1.293 (2)	1.259 (2)	1.234 (2)	1.199 (2)	1.160 (2)
Anzahl Zweigstellen	14.584 (18)	13.889 (12)	13.201 (12)	12.967 (11)	12.722 (11)	12.583 (11)	12.477 (11)	12.344 (12)	12.144 (11)
Zahl der Beschäftig-ten	169.900 (6.950)	168.950 (6.050)	168.250 (5.400)	164.200 (5.050)	162.550 (4.950)	161.200 (4.900)	160.750 (4.900)	159.250 (5.100)	158.300 (5.00)
Ø-Bilanzsumme [in Mrd. €]	534 (239)	548 (214)	557 (204)	567 (194)	579 (220)	596 (234)	614 (254)	641 (273)	677 (262)
Zinsspanne[2]	2,41% (0,62%)	2,49% (0,66%)	2,51% (0,46%)	2,51% (0,49%)	2,47% (0,47%)	2,30% (0,43%)	2,15% (0,50%)	2,06% (0,58%)	2,23% (0,45%)
Provisionsüber-schuss[3]	0,58% (0,15%)	0,57% (0,14%)	0,61% (0,17%)	0,65% (0,16%)	0,67% (0,16%)	0,66% (0,14%)	0,67% (0,12%)	0,63% (0,11%)	0,58% (0,14%)
Eigenkapitalrendite[4]	7,5% (4,4%)	9,7% (4,6%)	10,6% (0,7%)	10,3% (2,91%)	13,8% (5,3%)	11,0% (4,5%)	8,1% (-4,0%)	5,5% (-4,4%)	9,0% (7,2%)

Die in Klammern eingetragenen Werte beziehen sich auf die genossenschaftlichen Zentralbanken.

1 Die Anzahl der Institute weicht ab von der Anzahl berichtender Institute, die als Grundlage der Ermittlung der jahresabschlussbezogenen Kennzahlen dient. Quelle: Bankstellenstatistik, in: Deutsche Bundesbank, Bankenstatistik, Statistisches Beiheft zum jeweiligen Monatsbericht. Kreditinstitutsbegriff auf § 1 KWG bezogen, insofern Abweichungen zu Angaben in der „Bilanzstatistik" und der „Statistik der Gewinn- und Verlustrechnungen".

2 Die Zinsspanne entspricht dem Saldo der Zinserträge und Zinsaufwendungen jeweils in Relation zur durchschnittlichen Bilanzsumme.

3 Der Provisionsüberschuss ist der Saldo der Provisionserträge und Provisionsaufwendungen jeweils in Relation zur durchschnittlichen Bilanzsumme.

4 Die Eigenkapitalrendite ist definiert als Jahresüberschuss vor Steuern in % des durchschnittlichen Eigenkapitals.

Aufgrund der Unterscheidung in drei Sektoren wird das deutsche Bankensystem als Drei-Säulen-Modell oder Drei-Säulen-System bezeichnet[208], welches in der Form als weltweit einzigartig gilt.[209] Diese Einzigartigkeit und insbesondere die mangelnde Möglichkeit einer intersektoralen Konsolidierung des deutschen Bankenmarktes wurde nach Veröffentlichung der Einschätzung der Finanzsystemstabilität für Deutschland durch den INTERNATIONAL MONETARY FUND (IMF)[210] intensiv diskutiert. In dieser Konsolidierungsrestriktion wird ein wesentlicher Grund für die im internationalen Vergleich geringe Profitabilität des deutschen Bankensektors vermutet, die im Zusammenhang mit dem hohen

---

207 Eigene Darstellung auf Basis der Daten in *Deutsche Bundesbank* (2002), (2003), (2004), (2005), (2006a), (2007), (2008), (2009) und (2010).

208 Vgl. z. B. *Pieper, B.* (2006), S. 452.

209 Vgl. *Eim, A.* (2004), S. 11.

210 Vgl. *IMF* (2003), S. 7 ff.

Wettbewerb und den vielen rechtlich selbständigen, aber kleinen Banken in Deutschland gesehen wird.[211]

Da es das Ziel der vorliegenden Arbeit ist, den Besonderheiten der Kreditgenossenschaften bei der Konzeption der kundenwertorientierten Vertriebssteuerung Rechnung zu tragen, erfolgt zunächst eine detaillierte Vorstellung dieser Besonderheiten und den damit verbundenen Implikationen. Anschließend werden Untersuchungen vorgestellt, um dem deutschen Bankenmarkt und insbesondere den Markt für das Firmenkundengeschäft vor dem Hintergrund der Aussagen des IWF zu evaluieren.

1.3 Entstehungsgeschichte und Entwicklung der Kreditgenossenschaften

Die Kenntnis der Entstehungsgeschichte[212] der heutigen Kreditgenossenschaften hilft, die spezifischen Anforderungen, Strukturen und Prinzipien zu verstehen, die das Selbstverständnis dieser Banken wesentlich prägen. Die von SCHULZE-DELITZSCH und RAIFFEISEN begründeten Besonderheiten sind als fundamentale und konstituierende Unternehmensmerkmale zu verstehen, die das Genossenschaftsgesetz (GenG) ergänzen.[213]

Erste Kreditgenossenschaften wurden Mitte des 19. Jahrhunderts in der Zeit des industriellen Wandels gegründet.[214] Die mit diesem Wandel verbundenen tiefgreifenden Strukturänderungen betrafen vor allem Handwerk, Gewerbe und Landwirtschaft, denen es an Kreditverfügbarkeit mangelte. Die Kreditgenossenschaften wurden zu dem Zweck gegründet, um als von Mitgliedern getragene Selbsthilfeeinrichtungen diese Kreditprobleme zu beheben. So konnten die Mit-

---

211 Vgl. z. B. die Ausführungen bei *Duhnkrack, T.* (2002), S. 154 f. So umfasste der deutsche Bankenmarkt im Jahr 2009 etwa 2.000 Banken. Im Vergleich zum Jahr 2000 mit etwa 2.700 Banken entspricht dies einer Verringerung um ca. 26 Prozent, die vor allem aus innersektoralen Zusammenschlüssen resultiert.

212 Für eine ausführliche Darstellung vgl. z. B. *Kluge, A. H.* (1991).

213 Vgl. *Oellerking, C./Holzgrabe, M.* (1990), S. 75; *Wiedemann, A.* (1992), S. 3 f. In der Historie der Kreditgenossenschaften gab es zwei Hauptströmungen: Die gewerblich orientierten Volksbanken, die auf SCHULZE-DELITZSCH zurückgehen und die ländlichen Raiffeisenbanken, die aus der Aufbauarbeit RAIFFEISENS hervorgingen. Erst im Jahre 1971, etwa einhundert Jahre nach dem sog. Systemstreit zwischen SCHULZE-DELITZSCH und RAIFFEISEN, begannen diese Hauptströmungen, die sich in Gestalt von getrennten Organisationen offen als innergenossenschaftliche Konkurrenten gegenüberstanden, zu kooperieren. Seit 1972 sind sie als genossenschaftliche Banken im BUNDESVERBAND DER DEUTSCHEN VOLKSBANKEN UND RAIFFEISENBANKEN E.V. (BVR) vereinigt. Vgl. *Wiedemann, A.* (1992), S. 5; *Reith, D.* (1997), S. 20.

214 Vgl. hier und nachfolgend *Schramm, B.* (1982), S. 11 ff.

glieder der Kreditgenossenschaften erstmals notwendige Bankleistungen in Anspruch nehmen, und das mit deutlichen Konditionsvorteilen. Zudem ersetzte aufgrund der Regionalität die persönliche Bonität die dinglichen Sicherheiten.[215] Dieser Zweck wird als klassischer oder traditioneller Förderauftrag der Kreditgenossenschaften bezeichnet.[216]

Wie sich in der Entstehungsgeschichte der Kreditgenossenschaften zeigt, kompensierten diese insbesondere in ländlichen Regionen Angebotsdefizite bei der Kreditversorgung. Aufgrund dessen und wegen des Erfolgs der Genossenschaftsidee hatte sich die Anzahl der Kreditgenossenschaften seit den Gründertagen stark entwickelt. Während 1948 etwa 12.000 Kreditgenossenschaften in Deutschland gezählt wurden, hat sich deren Anzahl heute deutlich verringert.[217] Gegenwärtig gibt es etwa 1.100 typische Kreditgenossenschaften. Ursache dafür ist der innersektorale Konzentrationsprozess, der teilweise extern bedingt ist, wenn Kreditgenossenschaften nicht mehr wettbewerbsfähig sind, und teilweise strategisch forciert wird.[218]

Die Kreditgenossenschaften wiesen Ende 2010 im Mittel eine Bilanzsumme von etwa 621 Mio. EUR auf. Im Vergleich zu 1970, als diese bei etwa 5,6 Mio. EUR lag, entspricht dies einer Steigerung um etwa das hundertzehnfache.[219] Wird die Verteilungsfunktion der Kreditgenossenschaften hinsichtlich des Merkmals Bilanzsumme betrachtet, kann konstatiert werden, dass fast 70 Prozent der Banken eine Bilanzsumme bis 500 Mio. EUR haben.[220] Im Vergleich zu den anderen beiden Säulen des deutschen Bankenmarktes sind die Kreditgenossenschaften nicht nur die zahlenmäßig meisten selbständigen Institute, sondern zugleich auch die anhand der Bilanzsumme gemessenen kleinsten Banken. Die mittlere Bilanzsumme der Sparkassen lag 2010 etwa vier Mal so hoch bei ca. 2,5 Mrd. EUR.[221]

---

215 Vgl. *Hahn, O.* (1980), S. 20; *Wiedemann, A.* (1992), S. 5 und S. 81. Die damals bereits existenten Sparkassen schieden als mögliche Kreditgeber aus, da sie die Zwecksetzung verfolgten, die Ersparnisse der Bevölkerung *sicher* auszuleihen. Aufgrund der fehlenden dinglichen Sicherheiten der potenziellen Kreditnehmer aus Handwerk, Gewerbe und Landwirtschaft war dies nicht möglich. Vgl. *Hahn, O.* (1980), S. 19 f.

216 Vgl. *Grosskopf, W.* (1990), S. 21; *Bonus, H.* (1998), S. 324.

217 Vgl. Oellerking, C./Holzgrabe, M. (1990), S. 88.

218 So wird in dem Strategiepapier BÜNDELUNG DER KRÄFTE: DIE GEMEINSAME STRATEGIE vom BVR ein Zielwert von 800 Kreditinstituten angegeben. Vgl. *BVR* (2001b), S. 6.

219 Vgl. *BVR* (2011b). Hier handelt es sich um nicht-inflationsbereinigte Zahlen.

220 Vgl. *BVR* (2001a), S. 117.

221 Eigene Berechnungen auf Basis *DSGV* (2011), o. S.

1.4 Besondere Charakteristika der Kreditgenossenschaften

Wie ausgeführt, umfasst der entwicklungsgeschichtliche Genossenschaftsbegriff mehr Merkmale als die, die im GenG rechtlich definiert sind. Aus dem allgemeinen übergesetzlichen Genossenschaftsbegriff werden die genossenschaftlichen Grundsätze abgeleitet.[222] Diese sind auch heute noch Grundlage des kreditgenossenschaftlichen Selbstverständnisses und müssen als zentrale Einflussfaktoren der Unternehmensstruktur berücksichtigt werden.

1.4.1 Grundsätze und Prinzipien der Genossenschaft

Das absolute Wesensprinzip der Genossenschaften ist das Förderprinzip.[223] Dieses ist in § 1 Abs. 1 GenG festgelegt und stellt den alleinigen Zweck sowie uno actu die hinreichende Bedingung für die Existenzberechtigung der Genossenschaft als Organisations- und Rechtsform dar. Aussage des Förderprinzips ist, dass der genossenschaftliche Betrieb dazu beitragen soll, die wirtschaftlichen, kulturellen oder sozialen Belange der Mitglieder zu unterstützen. Dabei ist weniger einzigartig, dass der Betrieb die Mitglieder fördert als vielmehr die Vorgabe der Art und Weise der Förderung. Diese soll durch gemeinsamen Geschäftsbetrieb erfolgen. Dem absoluten Wesensprinzip, das in Kapitel C.1.4.2 näher betrachtet wird, sind weitere Grundsätze subsumiert. Diese Grundsätze haben organisationstrukturdeterminierenden Charakter (Strukturprinzipien).[224]

Die drei wesentlichen Prinzipien im Genossenschaftsgesetz sind der Selbsthilfegrundsatz, der Selbstverwaltungsgrundsatz und der Selbstverantwortungsgrundsatz.[225] Gegenstand des Selbsthilfegrundsatzes ist, dass sich die Mitglieder zusammenschließen, um durch den gemeinschaftlichen Geschäftsbetrieb Synergieeffekte zu heben, die für den Einzelnen im Rahmen seiner wirtschaftlichen Tätigkeiten nicht realisierbar sind. BONUS ET AL. bezeichnen diese als Kooperationserlöse.[226] Der Selbstverwaltungsgrundsatz konkretisiert sich dadurch, dass die Mitglieder mittels des höchsten Organs, der General- oder Vertreterversammlung, direkt oder indirekt an der Entscheidungsfindung für den Genossenschaftsbetrieb und der Kontrolle desselbigen beteiligt sind (§§ 43, 43a GenG). Aus der Selbstorganschaft nach § 9 GenG resultiert, dass die Mitglieder der Pflichtorgane des Vorstands (§ 24 GenG) und des Aufsichtsrats (§ 36 GenG)

---

222 Vgl. *Beuthien, V./Dierkes, S./Wehrheim, M.* (2008), S. 9 f.
223 Vgl. *Beuthien, V./Dierkes, S./Wehrheim, M.* (2008), S. 156.
224 Vgl. *Beuthien, V./Dierkes, S./Wehrheim, M.* (2008), S. 156.
225 Vgl. *Engelhard, H. A.* (1989), S. 32.
226 Vgl. *Bonus, H. et al.* (1999), S. 8.

ebenfalls aus dem Kreis der Mitglieder der Genossenschaft stammen müssen und von den Mitgliedern gewählt und abberufen werden (Personalhoheit gem. §§ 24, 36 GenG). Der Grundsatz der Selbstverantwortung legt fest, dass die Mitglieder für die Verbindlichkeiten der Genossenschaft einstehen (§ 23 GenG). Sie haften mindestens mit dem von ihnen eingezahlten Geschäftsguthaben und können darüber hinaus zu einem beschränkten oder unbeschränkten Nachschuss verpflichtet sein. Diese Grundsätze werden ergänzt um das Demokratie- und Identitätsprinzip.[227] Im Rahmen der Selbstverwaltung üben die Mitglieder der Kreditgenossenschaft ihre Entscheidungs- und Kontrollrechte durch einfache Mehrheitsentscheidungen aus. In diesem Zusammenhang gilt unabhängig vom Betrag des eingezahlten Geschäftsguthabens grundsätzlich „ein Mitglied – eine Stimme" (§ 43 GenG). Dieser Punkt unterscheidet die eingetragenen Genossenschaften von den Aktiengesellschaften. Direkt aus dem Förderprinzip und dem Selbsthilfegrundsatz leitet sich das Identitätsprinzip ab, das den Rollenpluralismus der Mitglieder einer Kreditgenossenschaft umfasst. Als (Mit-)Eigentümer sind die Eigenkapitalgeber, also die Mitglieder, Träger der Entscheidungs- und Kontrollrechte. Gleichzeitig treten die nutzenden Mitglieder als Kunden der eG auf. Durch das Identitätsprinzip soll sichergestellt werden, dass nutzerfremde (Kapital-)Interessen keinen bestimmenden Einfluss gewinnen.

Die beschriebenen Strukturprinzipien bilden ergänzend zum übergeordneten Förderprinzip den Kern des genossenschaftlichen Wesens. Aus diesem Grund sind sie auch nur in geringem Umfang veränderlich. Nur wenn es zu einer besseren Erfüllung des Förderprinzips führt, kann per Satzung von diesen tragenden Strukturprinzipien abgewichen werden.[228]

Konkretisierende Verhaltensnormen werden für die Kreditgenossenschaften durch die variablen Verfahrensprinzipien definiert. Ziel dieser Normen ist, eine mitgliederorientierte Geschäftspolitik zu gestalten.[229] Dem Selbstverständnis der Kreditgenossenschaften folgend, sollen sich die Mitglieder vornehmlich aus denjenigen Personen und Unternehmen rekrutieren, die in leistungswirtschaftlichen Beziehungen zur Genossenschaft stehen (sog. Fördergeschäftsverkehr) und

---

227 Vgl. hier und nachfolgend *Beuthien, V./Dierkes, S./Wehrheim, M.* (2008), S. 157 f.

228 So kann bspw. durch satzungsmäßige Bestimmung die Delegation der Berufung des Vorstands auf den Aufsichtsrat gem. § 24 Abs. 2 Satz 2 GenG die strikte Selbstverwaltung abschwächen. Die Selbstverantwortung wird z. B. durch die Begrenzung oder den Ausschluss der Nachschusspflicht beschränkt (§ 105 Abs. 1 und § 6 Nr. 3 GenG). Die Möglichkeit das Mehrstimmrecht (maximal drei Stimmen je Mitglied) einzuführen, geht zu Lasten des Demokratieprinzips. Und durch die Aufnahme des so genannten Nichtmitgliedergeschäfts (§ 8 Abs. 1 Nr. 5 GenG) oder die Aufnahme so genannter investierender Mitglieder (§ 8 Abs. 2 GenG) wird stark vom Identitätsprinzip abgewichen.

229 Vgl. hier und nachfolgend *Beuthien, V./Dierkes, S./Wehrheim, M.* (2008), S. 158 f.

keine reinen Anleger sind.[230] In diesem Sinne gilt das Prinzip der begrenzten Kapitalbeteiligung. Damit wird vermieden, dass Genossenschaften zu „entarteten“ Dividendengenossenschaften degenerieren, welche genossenschaftsrechtlich unzulässig sind. Erwirtschaftet eine Genossenschaft Überschüsse, die über das zur Erhaltung des künftigen Förderpotenzials notwendige Maß hinausgehen und somit nicht in der Genossenschaft verbleiben, sollen diese gemäß dem Prinzip der leistungsorientierten Überschussverwendung an die Mitglieder ausgeschüttet werden. Dabei sollen nicht die Höhe der Kapitalbeteiligung, sondern der Umfang und die Intensität der leistungswirtschaftlichen Beziehungen, mithin also das Volumen des Fördergeschäftsverkehrs maßgebend sein.

Über diese Grundsätze und Prinzipien hinaus gibt es das in Kapitel C.1.4.3 zu erläuternde Regionalitätsprinzip sowie die in Kapitel C.1.4.4 beschriebenen Prinzipien der Kooperation und Subsidiarität. Diese Prinzipien werden separat betrachtet, da sie einen nennenswerten Einfluss auf die Ausgestaltung der kundenwertorientierten Steuerung im Firmenkundengeschäft der Kreditgenossenschaften haben. Ebenso sind der Förderauftrag und die bereits mehrfach angesprochene Mitgliederorientierung von hoher Bedeutung für die kundenwertorientierte Vertriebssteuerung.

1.4.2 Förderauftrag und Mitgliederorientierung

Das zentrale Alleinstellungsmerkmal der Kreditgenossenschaften ist die Förderung ihrer Mitglieder durch den gemeinschaftlichen Geschäftsbetrieb.[231] Häufig wird dies verkürzt mit Mitgliedsorientierung oder Förderauftrag[232] bezeichnet.[233] Zugleich ist es diese Eigenschaft, die der Rechtsform der eingetragenen Genossenschaft ihre Existenzberechtigung gibt. In § 1 Abs. 1 GenG ist die Rechtsform der eingetragenen Genossenschaft legal definiert:

> „Gesellschaften von nicht geschlossener Mitgliederzahl, deren Zweck darauf gerichtet ist, den Erwerb oder die Wirtschaft ihrer Mitglieder oder deren soziale oder kulturelle Belange durch gemeinschaftlichen Geschäftsbetrieb zu fördern (Genossenschaften) erwerben die Rechte einer eingetragenen Genossenschaft nach Maßgabe dieses Gesetzes.“[234]

---

230 Diesem Grundsatz steht die Aufnahme investierender Mitglieder nach § 8 Abs. 1 GenG entgegen. Diese sind jedoch in ihren Möglichkeiten der Mitbestimmung begrenzt. Siehe § 8 Abs. 1 GenG.

231 Vgl. *Beuthien, V./Hanrath, S./Weber, H.-O.* (2008), S. 1; *Hammerschmidt, M.* (2000).

232 Die Begriffe Förderauftrag, Förderungsauftrag und Förderzweck werden synonym verwendet.

233 Vgl. z. B. *Geschwandtner, M./Kuttler, H.-J.* (2009), S. 264 oder *Pfeifer, H.* (2009), S. 269; *Bonus, H. et al.* (1999), S. 27 f.

234 Siehe § 1 Abs. 1. GenG.

Während mit Erwerb die Förderung der beruflichen Lebenssphäre des Mitglieds gemeint ist,[235] umfasst der Begriff Wirtschaft den privaten Bereich desselben[236]. Die eG ist nicht die einzige gesellschaftsrechtliche Vereinigungsform, deren Zweck in der Mitgliederförderung liegt.[237] Ihre Besonderheit liegt im Prinzip der kollektiven Selbstförderung, die durch die gemeinsame bzw. genossenschaftliche Errichtung eines förderwirtschaftlichen Geschäftsbetriebs gelingt. Als förderwirtschaftlich gelten dabei alle Leistungen, die sachlich mit den Mitgliederwirtschaften zusammenhängen und dazu geeignet sind, diesen zu dienen. Dazu bedient sich die eG des Fördergeschäftsverkehrs mit den Mitgliedern, die mithin gleichzeitig Kunden der eG sind (förderwirtschaftliches Identitätsprinzip). Daraus resultieren zwei gesellschaftsrechtliche Strukturgrenzen: Erstens das Gebot der Mitgliedernützlichkeit, das beinhaltet, dass der Vereinigungszweck weder Dritte anstelle der Mitglieder fördern noch ausschließlich oder überwiegend gemeinnützig sein darf. Die zweite Strukturgrenze ist das Verbot der Dividendenwirtschaft, das besagt, dass unabhängig vom konkreten Unternehmenszweck dieser nicht verfolgt werden darf, um eine resultierenden Gewinn aus Geschäften mit beliebigen Dritten an die Mitglieder zu verteilen.

Die dem Fördergeschäftsverkehr zugrundeliegende Fördergeschäftsbeziehung kann gesell-/ genossenschaftsrechtlicher oder schuldrechtlicher Art sein. Bei Mitgliedern werden die schuldrechtlichen Verpflichtungen gesellschaftsrechtlich überlagert.[238] So hat die eG den Gleichbehandlungsgrundsatz zu beachten und unterliegt einem etwaigen grundsätzlichen Kontrahierungszwang, der jedoch aus triftigen Gründen, bspw. bei unzureichender Kreditwürdigkeit, negiert werden kann. Dieser Grundsatz betrifft jedoch nicht das Nichtmitgliedergeschäft, das seit 1973 für die Genossenschaften möglich ist.

Zur Gründungszeit der Kreditgenossenschaften war der Förderauftrag klar umrissen, da der gemeinschaftliche Betrieb, den im Wesentlichen gewerblichen Mitgliedern die Möglichkeit eröffnen sollte, Probleme bei der Kreditverfügbarkeit zu beheben (traditioneller Förderauftrag).[239] Wenngleich das Gesetz zwar den Förderauftrag zur conditio sine qua non der eG macht, ist heute die inhaltliche Ausgestaltung des Förderauftrags insbesondere bei den Kreditgenossenschaften offen. Konkrete Vorgaben sind im GenG nicht definiert.[240] Die Satzung der jeweiligen Genossenschaft regelt, welche Förderleistungen sie im Einzelnen

---

235 Vgl. *Hettrich, E./Pöhlmann, P.* (2001), § 1, Rn. 5; *Lang, J./Weidmüller, L.* (2005), § 1, Rn. 27.

236 Vgl. *Paulick, H.* (1956), S. 52; *Hettrich, E./Pöhlmann, P.* (2001), § 1, Rn. 7.

237 Vgl. hier und nachfolgend *Beuthien, V./Dierkes, S./Wehrheim, M.* (2008), S. 1 ff.

238 Vgl. Beuthien, V./Dierkes, S./Wehrheim, M. (2008), S. 65 f.

239 Vgl. *Grosskopf, W.* (1990), S. 21; *Bonus, H.* (1998), S. 324.

240 Vgl. *Lamprecht, D./Donschen, A.* (2006), S. 8.

erbringen soll, denn die Förderung wird von den Mitgliedern bezweckt und damit dem subjektiven Urteil der Mitglieder unterstellt.[241] Mangels rechtlicher Konkretisierung ist der Förderauftrag ein unbestimmter Rechtsbegriff.[242] Daher wird in Theorie und Praxis intensiv über die inhaltliche Konkretisierung des Förderauftrags diskutiert.[243]

In der Forschung zum Genossenschaftswesen haben sich unterschiedliche Ansätze der Konzeptualisierung des Förderauftrags herausgebildet.[244] Der Ansatz der Förderkreisläufe stellt einen vergleichsweise umfassenden Ansatz dar und soll stellvertretend vorgestellt werden.[245] Er basiert auf der Einteilung der Kunden in Mitglieder und Nicht-Mitglieder (siehe Abbildung 6), wobei nur die nutzenden Mitglieder (aktive Kunden) betrachtet werden, da ausschließlich mit diesen die für den Fördergeschäftsverkehr notwendige und relevante Leistungsbeziehung vorliegt. Bei den anderen Mitgliedern handelt es sich ausschließlich um eine Einlegerbeziehung.

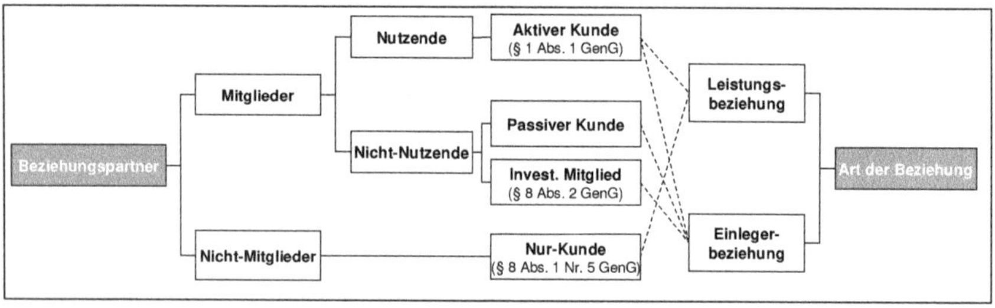

Abbildung 6: Beziehungspartner und -arten bei Kreditgenossenschaften[246]

Die Kreditgenossenschaft fördert ihre Mitglieder dem Ansatz der Förderkreisläufe entsprechend auf drei Wegen (siehe Abbildung 7).[247] Erstens über die un-

[241] Vgl. *Dülfer, E.* (1980), S. 47.

[242] Vgl. *Monßen, H.-G.* (1998), S. 293.

[243] Vgl. *Pfeifer, H.* (2009), S. 269; *Beuthien, V./Hanrath, S./Weber, H.-O.* (2008), S. 27 ff.; *Hanrath, S./Weber, H.-O.* (2008), S. 254 ff.; *Algner, M. R./Bolsinger, H. J.* (2006), S. 180 ff.; *Grosskopf, W.* (1990); *Hahn, O.* (1980), S. 19 ff. und S. 45 ff. GROSSKOPF weist darauf hin, dass „es im Grunde unmöglich ist, das 'wahre' Interesse der Genossenschaftsmitglieder und damit den genossenschaftlichen Förderauftrag verallgemeinert zu bestimmen." *Grosskopf, W.* (1990), S. 25.

[244] Vgl. dazu auch *Harbrecht, W.* (2005), S. 99.

[245] Vgl. hier und nachfolgend *Beuthien, V./Hanrath, S./Weber, H.-O.* (2008), S. 7 ff.

[246] In Anlehnung an *Beuthien, V./Hanrath, S./Weber, H.-O.* (2008), S. 8 f.

[247] Vgl. *Beuthien, V./Hanrath, S./Weber, H.-O.* (2008), S. 8 f.

mittelbare materielle (ökonomische) Förderung im Rahmen des banktypischen Leistungsangebots, indem dem Mitglied entweder wettbewerbsgleiche Produkte zu günstigeren Konditionen oder bessere Produkte zu wettbewerbsgleichen Konditionen angeboten werden.[248] Diese Art der Förderung, die im Sinne des § 1 Abs. 1 GenG erfolgt und sich auf die Funktion des Mitglieds als Kunde bezieht, wird dem inneren Förderkreislauf der eG zugeordnet und entspricht im Wesentlichen dem traditionellen Förderauftrag, der auf die Leistungsbeziehung zwischen eG und Mitglied fokussiert. Darüber hinaus kann dem inneren Förderkreislauf auch die Honorierung der Kapitalbeteiligung zugeordnet werden, die dann materiell und mittelbar ist und sich auf die Einlegerbeziehung bezieht. Dazu zählt, wenn Dividenden ausgeschüttet oder Geschäftsguthaben verzinst werden.[249] „Ebenso zählen hierzu – wenn auch zeitversetzt und nachträglich gewährt – die klassische genossenschaftliche Rück- bzw. Nachvergütung (§ 22 KStG) oder – in Anlehnung hieran – rückvergütungsähnlich ausgestaltete Mitglieder-Bonussysteme, die in der Praxis im Vordringen sind."[250] Eine den inneren Förderkreislauf ergänzende Art der Förderung erfolgt innerhalb des erweiterten Förderkreislaufs. In diesem Fall bietet die Kreditgenossenschaft dem Mitglied das Leistungsangebot Dritter günstiger oder mit besserer Qualität im Vergleich zu den Angeboten für Nichtmitglieder an. Der erweiterte Förderkreislauf bezieht sich nicht mehr direkt auf die Leistungsbeziehung, da die Kreditgenossenschaft hier als Initiatorin oder Maklerin auftritt.[251] Die Allgemeinförderung ist die dritte Form der Förderung, wobei diese direkt oder indirekt erfolgt und ideeller Art ist. Hierbei handelt es sich z. B. um die Förderung regionaler, kultureller Einrichtungen. Dies sind keine Förderungsleistungen im Sinne des § 1 Abs. 1 GenG, werden aber auch nicht untersagt, sofern auch der innere Förderungskreislauf substanziell ist.[252]

---

[248] Vgl. *Beuthien, V./Hanrath, S./Weber, H.-O.* (2008), S. 9.

[249] Vgl. *Beuthien, V./Hanrath, S./Weber, H.-O.* (2008), S. 10.

[250] *Hanrath, S./Weber, H.-O.* (2008), S. 250.

[251] Differenziert zu betrachten ist in diesem Zusammenhang die Einordnung der Leistungen der Verbundpartner. Nutzt die Kreditgenossenschaft diese als Erfüllungsgehilfen bzw. handelt sie mit den Produkten, sind sie dem inneren Förderkreislauf zuzuordnen. Tritt dagegen die Genossenschaft nur als Vermittlerin auf, so sind sie dem erweiterten Förderkreislauf zuzuordnen.

[252] Vgl. *Beuthien, V./Hanrath, S./Weber, H.-O.* (2008), S. 11.

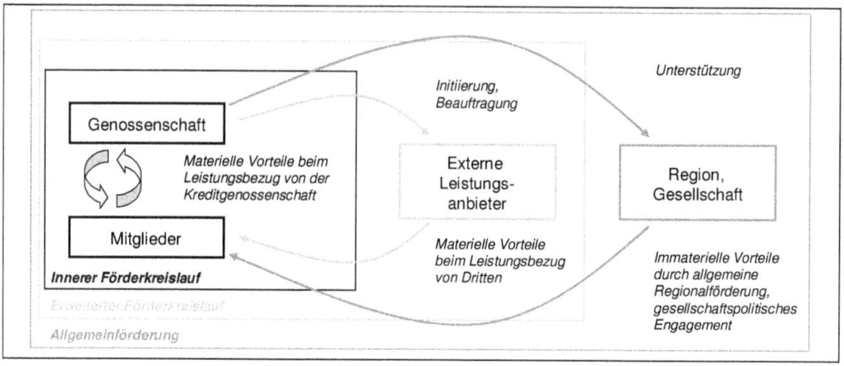

Abbildung 7: Förderkreisläufe zur Konzeptualisierung des Förderauftrags[253]

Die Frage nach der inhaltlichen Festlegung innerhalb dieses Ordnungsrahmens, also nach der messbaren Ausgestaltung des Förderauftrags, bleibt der Kreditgenossenschaft überlassen. In diesem Zusammenhang gibt es eine intensive Auseinandersetzung über die richtigen Förderinstrumente.[254] Unabhängig davon, wie der Förderauftrag konkretisiert ist, lässt sich erkennen, dass den Mitgliedern der Kreditgenossenschaft ein Nutzen aus dem Identitätsprinzip entsteht bzw. entstehen soll. Dieser Nutzen wird als Förderplus[255], Netto-Rendite[256] oder Member Value[257] bezeichnet. Wird angenommen, der Grundnutzen einer Leistungsbeziehung zwischen einem Kunden und einer Kreditgenossenschaft ist für Mitglieder-Kunden und Nur-Kunden identisch und der Zusatznutzen aus der Einlegerbeziehung für Mitglieder-Kunden und investierende Mitglieder ebenfalls, dann lassen sich die Nutzenbestandteile für die Mitglieder in Abhängigkeit der Beziehungsebene zur Kreditgenossenschaft den Nur-Kunden vergleichend gegenüberstellen (siehe Abbildung 8).[258] Demnach resultiert der Nutzen des Mitglieder-Kunden aus drei Quellen: dem Grundnutzen der Leistungsbeziehung, dem Zusatznutzen aus der Leistungsbeziehung sowie dem Zusatznutzen aus der Einlegerbeziehung. Investierende Mitglieder erzielen einen Nutzen nur aus der

253 *Beuthien, V./Hanrath, S./Weber, H.-O.* (2008), S. 11.
254 Vgl. dazu z. B. *Grosskopf, W.* (1990).
255 Vgl. *Dülfer, E.* (1995), S. 58.
256 Vgl. *Grosskopf, W.* (2001), S. 241.
257 Vgl. *Lamprecht, D./Donschen, A.* (2006), S. 15; *Theurl, T.* (2002), S. 84 f.
258 Dabei wird vereinfachend unterstellt, dass der Zusatznutzen aus der Leistungsbeziehung dem Zusatznutzen der Einlegerbeziehung entspricht. Diese Annahme ist vor dem Hintergrund der Zieldivergenzen der einzelnen Mitglieder kritisch zu hinterfragen, schränkt die Aussage der Darstellung und die argumentative Verwendung dieser Überlegung jedoch nicht ein.

Einlegerbeziehung und Nur-Kunden nur aus der Leistungsbeziehung. Unter der genannten Prämisse der Nutzenidentitäten kann geschlossen werden, dass der Mitglieder-Kunde den höchsten Nutzen realisiert bzw. realisieren sollte. Eine wertende Aussage für die Nur-Kunden und Investierenden Mitglieder ist nicht allgemeingültig möglich.

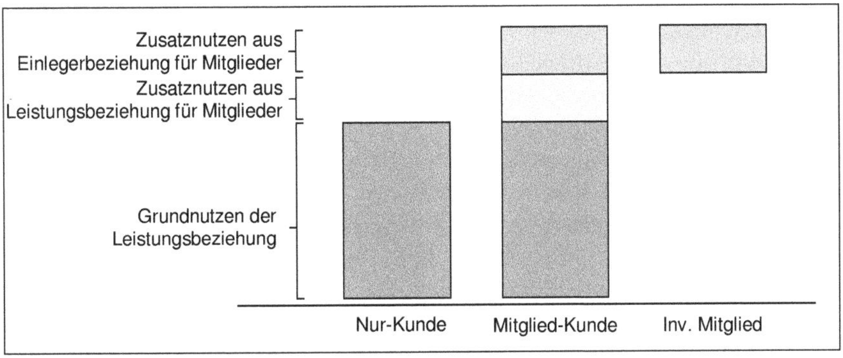

Abbildung 8: Nutzenkomponenten der genossenschaftlichen Beziehungsebenen[259]

1.4.3 Regionalprinzip

Der Absatzmarkt der Kreditgenossenschaften ist regional begrenzt.[260] Das Regionalprinzip begründet sich im historischen Ursprung der Kreditgenossenschaften, da das relevante Marktgebiet die jeweilige Stadt, das Dorf oder der Vorort waren.[261] Ökonomisch wurde das Regionalprinzip vor allem damit begründet, dass den Kreditgenossenschaften die jeweiligen regionalen Strukturen und wirtschaftlichen Entwicklungen bekannt und diese daher eher in der Lage waren, gute und schlechte Bonitäten zu unterscheiden als dies bei einem großen anonymen Markt möglich wäre. Dass die regionale Begrenzung des Geschäftsgebiets auch heute noch forciert wird, wird im Strategiepapier des BVR verdeutlicht, nach dem sich die Kreditgenossenschaften als regionale Allfinanzdienstleister positionieren.[262] Ziel ist es, neben der informationsbezogenen Vor-

259 Eigene Darstellung.

260 Bereits die Firma der Kreditgenossenschaft, die regelmäßig den lokalen oder regionalen Bezug herstellt, weist auf dieses Prinzip hin, so bspw. bei der SPAR- UND DARLEHNSKASSE REKEN EG.

261 Vgl. *Hahn, O.* (1980), S. 30.

262 Vgl. *Kring, T.* (2002), S. 8.

teilhaftigkeit im Kreditgeschäft, den intergenossenschaftlichen Wettbewerb[263] zu reduzieren und die Marktnähe als Wettbewerbsvorteil zu nutzen.

Das heutige Regionalprinzip leitet sich aus § 6 Abs. 2 des STATUTS DER SICHERUNGSEINRICHTUNG (SE-St) und Ziffer 5 der VERFAHRENSREGELN ZUM STATUT DER SICHERUNGSEINRICHTUNG ab. Dort heißt es, dass mit dem Zweck der Sicherungseinrichtung vereinbaren Bankgeschäfte „grundsätzlich vorrangig konzentriert auf den angestammten, d. h. durch den Sitz und die Zweigstellen einer Primärbank bestimmten Geschäftsbereich“ sind.[264] Stellen die jeweiligen Verbandsprüfer eine Abweichung von diesem Grundsatz fest, kann der Ausschluss aus dem genossenschaftlichen Garantiefonds erfolgen.[265]

Aus dem Regionalprinzip resultieren nennenswerte Einflüsse auf die Geschäftspolitik der Kreditgenossenschaft. Zum einen kann es gerade im Firmenkundenkundengeschäft Kreditkonzentrationen und insbesondere regionale Sektorkonzentrationen begünstigen.[266] Mit diesen sog. „Klumpenrisiken“ kann das vergleichsweise höhere Engagement dieser Institute bei gewerblichen Kreditkunden in Krisenphasen erklärt werden. Da Arbeitnehmer des Firmenkunden ggf. (Kredit-)Kunden der Kreditgenossenschaft sind, ergeben sich dadurch nicht nur kreditrisikobezogene, sondern auch ertragsbezogene Risiken.[267] Zum zweiten sind die Wachstumsmöglichkeiten einer Kreditgenossenschaft prinzipienbedingt begrenzt und können nur über Zusammenschlüsse mit Nachbarinstituten im signifikanten Ausmaß erreicht werden.

Auf ihrem regionalen Markt vertreten die Kreditgenossenschaften auch die Interessen der Verbundunternehmen, was sich in der Verbundorientierung, die das Kooperations- und das Subsidiaritätsprinzip umfasst, konkretisiert.

1.4.4 Verbundorientierung

Die Kreditgenossenschaften sind Elemente der genossenschaftlichen FinanzGruppe,[268] mit der sich das Kooperations- und das Subsidiaritätsprinzip

---

263 Vgl. zum intergenossenschaftlichen Wettbewerb *Reith, D.* (1997), S. 31 ff.

264 Für die überregional tätigen bzw. atypischen Kreditgenossenschaften wie z. B. der DEUTSCHE APOTHEKER- UND ÄRZTEBANK EG gelten diese Vorgaben nicht, weshalb sie als grundsätzliche Bestimmung formuliert sind und durch die strikte Auslegung der Prüfungsverbände konkretisiert werden. Vgl. *Reith, D.* (1997), S. 26 f.

265 Durchbrochen wird das Regionalprinzip durch die Intensivierung des Onlinebankings. Vgl. *BVR* (2001b), S. 22.

266 Vgl. *Deutsche Bundesbank* (2006b), S. 38.

267 Vgl. *Reith, D.* (1997), S. 47 sowie die dort genannte Literatur.

268 Die Schreibweise mit dem Großbuchstaben beim zweiten Substantiv erfolgt grundsätzlich im Rahmen der Begriffs- und Namensgebung in der genossenschaftlichen Kommu-

verbinden. Die weiteren Elemente der FinanzGruppe und deren Beziehungen zueinander werden in Abbildung 9 dargestellt. In regionaler Hinsicht kann die FinanzGruppe in eine lokale Ebene, in eine regionale Ebene und in die Bundesebene aufgeteilt werden.[269] Die FinanzGruppe wird aufgrund der gesellschaftsrechtlichen Zusammenhänge, nach denen die Mitglieder die Eigentümer der Kreditgenossenschaft sind und diese wiederum Träger der weiteren Verbundunternehmen sind, auch als „inverser Konzern"[270] bezeichnet.

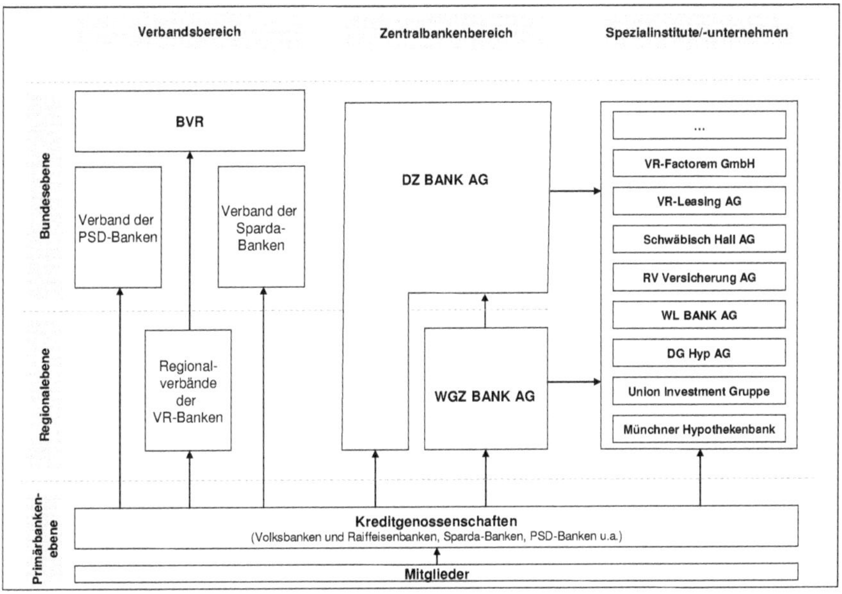

Abbildung 9: Vereinfachte Darstellung der Struktur der FinanzGruppe[271]

Das Kooperationsprinzip sagt aus, dass die einzelne Kreditgenossenschaft zur Stärkung ihrer Wettbewerbsfähigkeit und zur Stärkung der Fähigkeit, die

nikationspolitik und ist zentraler Bestandteil des Corporate Design. Aus diesem Grund soll dies in der vorliegenden Arbeit berücksichtigt werden.

269 Die regionale Ebene ergibt sich aus der WGZ BANK AG und den fünf Genossenschaftsverbänden.

270 *Brixner, U.* (2006), S. 62.

271 In Anlehnung an *Polster, D.* (2001), S. 295. Der Fokus dieser Darstellung liegt auf den für das Firmenkundengeschäft besonders relevanten Komponenten der FinanzGruppe. Die Pfeilrichtungen geben die Richtung der Willensbildung bzw. der Kapitalbeteiligung wieder.

Grundsätze der Selbsthilfe, Selbstverwaltung und Selbstverantwortung umzusetzen, auf lokaler, regionaler, nationaler und internationaler Ebene kooperieren soll.[272] Dabei ist das Subsidiaritätsprinzip einzuhalten, das postuliert, dass die übergeordneten Einheiten, nur die Aufgaben übernehmen, die auf der untergeordneten Ebene nicht oder nicht effizient erfüllt werden können. Diese Prinzipien greifen z. B. bei Krediten für Unternehmenskunden, welche die Kreditbereitstellungsfähigkeit der Kreditgenossenschaft übersteigen.[273] In diesem Fall treten die Zentralbanken als subsidiärer Kooperationspartner für die Bereitstellung der Leistung ein. Aus diesen beiden Grundsätzen hat sich ein (strategisches) Netzwerk entwickelt, in welchem die einzelne Kreditgenossenschaft auf ein breites Angebotsspektrum zurückgreifen kann.[274]

Ziel dieses Netzwerkes ist es, Kostenvorteile und Netzwerksynergien zu realisieren, so dass die Vorteile der Größe mit den Vorteilen der Lokalität, Dezentralität und Kundennähe kombiniert werden.[275] So konstatiert der BVR, dass nicht nur die Größe der einzelnen Kreditgenossenschaft entscheidend sei, sondern die Fähigkeit mittels des Verbunds die Wertkette der Banken so zu dekonstruieren, dass Effizienzvorteile in einzelnen Wertschöpfungsstufen erreicht werden.[276] Verbunden damit sieht KRING spezielle Anforderungen an das Management der Kreditgenossenschaften, die sich speziell aus den resultierenden Leistungsverflechtungen und der begrenzten Autonomie der einzelnen Kreditgenossenschaft ergeben.[277] Die Leistungsverflechtungen unterscheidet er dabei in eine Produkt- und eine Dienstleistungsebene.[278] Während letztere sich vor allem auf das Auslagern bestimmter Unternehmensaktivitäten bezieht (z. B. IT-Services auf die Rechenzentralen) und für die vorliegende Arbeit weniger von Relevanz ist, ist die Produktebene bedeutend.

---

272 Vgl. *Beuthien, V./Dierkes, S./Wehrheim, M.* (2008), S. 158.

273 Dies gilt z. B. aus risikopolitischen Erwägungen oder den Großkreditbestimmungen des §§ 13 ff. KWG. Ein Großkredit liegt vor, wenn das Gesamtkreditvolumen an einen Kreditnehmer 10 Prozent des haftenden Eigenkapitals einer Bank überschreitet. Die Großkrediteinzelobergrenze liegt bei 25 Prozent des haftenden Eigenkapitals. Vgl. *Hartmann-Wendels, T./Pfingsten, A./Weber, M.* (2010), S. 632. Die Großkreditgesamtobergrenze, die bei 800 Prozent des haftenden Eigenkapitals lag, soll im Zuge der Änderungen der EU-Bankenrichtlinie abgeschafft werden. Vgl. *Deutsche Bundesbank* (2009), S. 69.

274 Vgl. *Kring, T.* (2005), S. 268; *Bonus, H. et al.* (1999), S. 35 ff.

275 Vgl. *Pleister, C.* (2006), S. 116.

276 Vgl. *BVR* (2001b), S. 3. Dies gilt vor dem Hintergrund der Bestrebungen nach effizienten Betriebsgrößen erst ab einer gewissen Größenordnung uneingeschränkt.

277 Vgl. *Kring, T.* (2005), S. 264 f.

278 Vgl. *Kring, T.* (2005), S. 268 f.

Im Fokus der Produktebene steht zwar ebenfalls die Auslagerung von Aktivitäten. Jedoch betrifft dies die Auslagerung absatzmarktrelevanter Produktionsdienstleistungen. Die Rolle der Kreditgenossenschaften im Rahmen dieser Kooperation ist dergestalt, dass sie sich in ihrem Allfinanzangebot auf die Verbundprodukte konzentriert,[279] den Vertrieb für die Produkte übernimmt oder insoweit unterstützt, als sie Kontakte zwischen Kunden und Verbundspezialisten herstellt und koordiniert. „Die lokal agierenden Volksbanken sind uneingeschränkt für den Vertriebserfolg auf ihrem Marktgebiet verantwortlich"[280] und sind somit mit ihrer Reputation für die Geschäfte der Verbundpartner verantwortlich.[281] Die Kreditgenossenschaften stellen für die meisten Spezialunternehmen und Verbundpartner einen zentralen oder teilweise ausschließlichen Vertriebskanal dar. Sind die Erfolge der Kreditgenossenschaft jedoch nicht ausreichend, haben die Verbundunternehmen die Möglichkeit, direkt die jeweilige Marktregion bearbeiten zu können.[282] Auch im Zusammenhang mit den begrenzten Möglichkeiten der Kreditgenossenschaften, die aus ihrer (relativen) Kleinheit resultieren, wird das Subsidiaritätsprinzip beschränkt. So können und sollen die Zentralbanken vor allem im Firmenkundengeschäft unmittelbar vor Ort tätig werden, wenn die Kundengröße oder die spezifischen Kundenbedarfe dies erfordern.[283] Auf Seiten der Kreditgenossenschaft liegt dagegen die Möglichkeit der Beschränkung des Kooperations- und Subsidiaritätsprinzips vor, wenn entweder die Wettbewerbsfähigkeit des Verbundangebots nicht gegeben ist oder wenn der Kunde explizit ein verbundfremdes Produkt nachfragt.[284]

Nachdem nun die Kreditgenossenschaften in das deutsche Bankensystem eingeordnet und die besonderen Eigenschaften der Kreditgenossenschaften erörtert wurden, soll der Blick auf den Absatzmarkt der Kreditgenossenschaften gerichtet werden. Wie benannt, wird regelmäßig die geringe Profitabilität des deutschen Bankensektors im Allgemeinen und des Firmenkundensektors im Speziellen konstatiert und auf den hohen Wettbewerb und die fehlenden Konsolidierungsmöglichkeiten zurückgeführt. Nachfolgend sollen empirische Erkenntnisse vorgestellt werden, die helfen, die Konzentration, den Wettbewerb sowie die Rentabilität auf dem deutschen Bankenmarkt zu evaluieren.

---

279 Vgl. *Harbrecht, W.* (2005), S. 101.
280 *Pleister, C.* (2006), S. 116.
281 Vgl. *Bonus, H. et al.* (1999), S. 13.
282 Vgl. *BVR* (2001b), S. 8 f.
283 Vgl. *BVR* (2001b), S. 7.
284 Vgl. *BVR* (2001b), S. 28.

1.5 Wettbewerb und Rentabilität auf dem deutschen Bankenmarkt

Beim Structure-Conduct-Performance-Paradigm (SCP) werden statistische Konzentrationsmaße[285] (z. B. die Konzentrationsrate und der Hirschmann-Herfindahl-Index) genutzt, um die Marktkonzentration zu messen und so den Wettbewerb zu beurteilen.[286] Das SCP unterstellt, dass die Marktstruktur (Konzentration) Einfluss auf das Marktverhalten und so auf das Marktergebnis hat. Dabei steht eine hohe Konzentration für die Dominanz weniger großer Unternehmen (Structure), deren Marktmacht sich in kollusivem Verhalten niederschlägt (Conduct), welches den wettbewerbsinduzierten Druck auf die Gewinnmargen und Kosten verhindert (Performance).[287] Neben der Bankenanzahl wird bei Banken auch die Bilanzsumme als Größenindikator und Messvariable genutzt. VARMAZ stellt unter Anwendung verschiedener Konzentrationsmaße für den deutschen Bankenmarkt eine zunehmende Konzentration im Analysezeitraum von 1993 bis 2005 fest, bezeichnet den Konzentrationsgrad insgesamt aber als „mäßig“.[288] Die DEUTSCHE BUNDESBANK konstatiert auch, dass ein „im Ganzen geringer Konzentrationsgrad“ [289] vorliegt. Der Logik des SCP folgend ist der Wettbewerb auf dem deutschen Bankenmarkt als eher hoch zu bezeichnen.

Das SCP wird in der Literatur kritisiert.[290] Vor allem die Widersprüchlichkeit der unterstellten Wirkungsweise mit den realiter beobachtbaren Entwicklungen schränke die Aussagekraft ein.[291] Zudem seien mittels der nationalen Konzentrationsmaße keine Aussagen zu lokalen Strukturen möglich.[292] Die Neue Empirische Industrieökonomik bietet, aufbauend auf der Kritik am SCP, Verfahren an, um die Wettbewerbssituation auf Märkten zu analysieren. Sie ver-

---

285 Vgl. für einen ausführlichen Überblick *Varmaz, A.* (2006), S. 124 ff. Grundsätzlich bauen diese Indizes auf den Begriff der Konzentration aus der deskriptiven Statistik auf und messen den prozentualen Marktanteil einer bestimmten Anzahl von Unternehmen, z. B. der 3, 5 oder 10 größten Banken. Vgl. auch *Deutsche Bundesbank* (2001), S. 61.

286 Vgl. hier und nachfolgend *Varmaz, A.* (2006), S. 121 ff.; *Deutsche Bundesbank* (2001), S. 58 ff.

287 Vgl. *Schäfer, S./Weistroffer, C.* (2007), S. 603.

288 *Varmaz, A.* (2006), S. 234. Zu ähnlichen Ergebnissen vgl. *Deutsche Bundesbank* (2001), S. 60.

289 *Deutsche Bundesbank* (2001), S. 59.

290 Der Argumentation des SCP steht z. B. die Efficient-Structure-Hypothesis gegenüber, wonach Konzentration das Ergebnis des Erfolgs relativ effizienter Unternehmen ist, die die Konkurrenten vom Markt verdrängt haben. Vgl. *Schäfer, S./Weistroffer, C.* (2007), S. 603. Zur ausführlichen methodischen und inhaltlichen Kritik am SCP vgl. *Varmaz, A.* (2006), S. 142 ff.

291 Vgl. *Deutsche Bundesbank* (2001), S. 64 f.

292 Vgl. *Deutsche Bundesbank* (2001), S. 60.

sucht indirekt über die Schätzung von Verhaltensgleichungen der Banken, Rückschlüsse auf ihr Wettbewerbsverhalten zu ziehen.[293] Dazu nutzt sie Daten, die auf Ebene der einzelnen Bank vorhanden sind. PANZAR/ROSSE modellieren eine (vereinfachte) Ertragsfunktion, anhand derer Faktorpreiselastizitäten Rückschlüsse auf den vorherrschenden Wettbewerb möglich sind.[294] Liegt die Summe der Elastizitäten bei 1, liegt ein vollständiger Wettbewerb vor. Bei einem Wert kleiner 1 wird ein Monopol mit Randwettbewerb angenommen und bei einer Elastizität von kleiner oder gleich 0 ein Monopol oder perfekte Kollusion.[295] Angewendet auf den deutschen Bankenmarkt kann HEMPELL trotz der kontinuierlich rückläufigen Anzahl an Banken kein Wettbewerbsverhalten feststellen, das eine perfekte Kollusion oder einen vollständigen Wettbewerb belegen kann.[296] Ihre Ergebnisse zeigen weiter, dass die Sparkassen und Genossenschaftsbanken zunehmende Marktmacht gewinnen und weniger kompetitiv auftreten als die Kreditbanken.[297] Eine Begründung für dieses Ergebnis sieht sie im Regionalprinzip, da hier – zumindest theoretisch – der innersektorale Wettbewerb nicht relevant ist. Dieses Ergebnis indiziert, dass sich kleine Banken eher monopolistisch verhalten als große. Dies "contradicts the conventional wisdom that larger market shares correspond with the wielding of greater market power."[298] In diesem Kontext ergänzen FEHL/KUHN, dass nicht nur regionale Marktmacht, sondern auch der Wissensvorteil dieser Banken erklärend für höhere, durchgesetzte Zinsmargen ist.[299] Im Einklang mit den Ergebnissen wird der deutsche Bankenmarkt unter Berücksichtigung der Verbundbeziehungen als oligopolistischer Markt bezeichnet.[300] Nachfolgend soll daran anknüpfend ein detaillierter Blick auf empirische Untersuchungen zur Rentabilität der deutschen Banken geworfen werden.

---

293 Vgl. hier und nachfolgend *Deutsche Bundsbank* (2001), S. 65.

294 Vgl. dazu ausführlich *Panzar, J./Rosse, J.* (1987), S. 443 ff.; *Varmaz, A.* (2006), S. 150 ff; *Hempell, H. S.* (2002), S. 6 ff.

295 Vgl. *Reichel, R.* (2005), S. 287. Die Summe der Faktorpreiselastizitäten ergibt die H-Statistik im Modell von PANZAR/ROSSE.

296 Vgl. *Hempell, H. S.* (2002), S. 32 f.

297 Ähnlich auch *Deutsche Bundesbank* (2001), S. 65. Zu einem gegensätzlichen Ergebnis kommt REICHEL, der die relative Marktmacht der einzelnen Bankengruppen anhand eines vorwärts-rekursiven Almon-lag-Modells auf Basis der Korrelation von Zinsspanne und Umlaufsrendite festverzinslicher Wertpapiere schätzt. Danach ist die Marktmacht der Kreditgenossenschaften im Zeitraum von 1990 bis 2002 gesunken. Vgl. *Reichel, R.* (2005), S. 283 ff.

298 *Hempell, H. S.* (2002), S. 32.

299 Vgl. *Fehl, U./Kuhn, N.* (2005), S. 204.

300 Vgl. *Reith, D.* (1997), S. 22. Dies begründet sich auch damit, dass ein dominanter Marktführer fehlt.

Eine oft verwendete Kennzahl zur Darstellung der Profitabilität ist die Eigenkapitalrentabilität.[301] Werden die Eigenkapitalrentabilitäten der Jahre 2001-2009 betrachtet,[302] ist auffällig, dass die Kreditgenossenschaften und Sparkassen relativ zu den Großbanken deutlich stabilere Renditen aufweisen und in den Jahren von 2001-2004 sowie 2008-2009 rentabler waren als die Kreditbanken. Seit 2005 ist ein rückläufiger Trend bezüglich der Höhe der Eigenkapitalrendite dieser Bankengruppen zu erkennen, der sich in 2009 jedoch nicht fortsetzte.[303]

Die insgesamt als gering beurteilte Marktrentabilität wird u. a. damit begründet, dass Sparkassen und Kreditgenossenschaften aufgrund ihres spezifischen Zwecks nur ein gering ausgeprägtes Rentabilitätsstreben aufweisen.[304] In einem ausführlichen Vergleich verschiedener Länder für die Jahre 1996-2003 stellen SCHÜLLER/SCHNEIDER für deutsche Banken fest, dass diese eine strukturelle Ertragsschwäche aufweisen, da deren Ergebnisse für die Eigenkapitalrentabilität und die so genannte Aufwands-Ertrags-Quote (Cost-Income-Ratio) nachhaltig hinter den Werten von Ländern wie Frankreich und Großbritannien liegen.[305] PIEPER analysiert auf Basis der statistischen Daten der Organisation for Economic Co-operation and Development (OECD) für den gleichen Zeitraum verschiedene Werttreiber der Profitabilität von Banken, um die Ursachen für die strukturelle Ertragsschwäche der deutschen Institute im internationalen Vergleich zu identifizieren.[306] Anhand der Bruttoerträge[307] je Mitarbeiter, bei denen deutsche Banken vergleichsweise gute Werte erzielen, sieht er die These, dass Deutschland „overbanked" sei, nicht bestätigt. Auch in Bezug auf die Produktivität, die er mittels der Cost-Income-Ratio brutto[308] ermittelt, befindet sich

301 In der deutschsprachigen Literatur wird gelegentlich zwischen den Inhalten der Begriffe Profitabilität und Rentabilität unterschieden. Während ersterer absolute Saldogrößen umfasst, werden mit letzterer relative Kennzahlen, meist ergebnisbezogene Beziehungszahlen, beschrieben. Die Eigenkapitalrentabilität ist hier definiert als der Jahreserfolg vor Steuern in Prozent des jahresdurchschnittlichen bilanziellen Eigenkapitals.

302 Vgl. Tabelle 2; Tabelle 3 und Tabelle 4 in Kapitel C.1.2.

303 Dies erklärt sich partiell durch die steiler gewordene Zinsstrukturkurve, die sich positiv auf den aus der Fristentransformation resultierenden Strukturbeitrag auswirkt.

304 Vgl. *Schüller, S./Schneider, M.* (2006), S. 7; *Duhnkrack, T.* (2002), S. 154.

305 Vgl. *Schüller, S./Schneider, M.* (2006), S. 4 ff.

306 Vgl. *Pieper, B.* (2006), S. 452. Als Vergleichsländer nutzt er Frankreich, Spanien, Italien, Großbritannien, die Schweiz und die USA.

307 Als Bruttoerträge ist die Summe aus Zins- und Provisionserträgen, Handelsergebnis und Saldo der sonstigen betrieblichen Erträge und Aufwendungen definiert. Vgl. *Pieper, B.* (2006), S. 454, Fn. 14.

308 Als Cost-Income-Ratio brutto ist der Quotient aus der Summe von Personal und sonstigem Verwaltungsaufwand zu Zins- und Bruttoerträgen, Handelsergebnis sowie Saldo

Deutschland auf einem Spitzenplatz.[309] Beim Vergleich der realisierten Margen[310] zeigt sich jedoch, dass die deutschen Banken vergleichsweise schlecht abschneiden, da sie nur geringe Margen durchsetzen können, was er auf einen effizienten Wettbewerb auf dem deutschen Bankenmarkt zurückführt.[311] Interessanterweise sind die deutschen Kreditzinsen jedoch höher als im Durchschnitt der Währungsunion.[312] Ausgehend von der Struktur des Zinssatzes untersuchen SCHÄFER/WEISTROFFER die Determinanten des Kreditzinsniveaus. Diese sind die Wettbewerbsintensität, die Effizienz des Bankensystems, die Risiko- und die Refinanzierungsstruktur.[313] Im Ergebnis vermuten sie, dass der Wettbewerb um Einlagen als Refinanzierungskomponente und die im internationalen Vergleich höheren Risikoprämien bei Unternehmenskrediten den größten Erklärungsgehalt für das Zinsniveau haben.[314] Auch SCHÜLLER/SCHNEIDER kommen zu dem Ergebnis, dass die Verzehrquote, die das Verhältnis von Risikovorsorge zum Teilbetriebsergebnis bezeichnet, in Deutschland im internationalen Vergleich zu hoch sei.[315] Aufgrund der strukturellen Unterschiede auf den Bankenmärkten kommen die Autoren jedoch insgesamt zu dem Ergebnis, „die Gründe für den internationalen Rückstand sind vielmehr in den Ertragspotenzialen und der mangelnden Fähigkeit der Banken, diese in ausreichendem Maße zu heben, zu suchen“[316]. Diese Erkenntnisse bestätigen FEHL/KUHN in ihrer produktmarktspezifischen Sicht.[317] Aufgrund der institutionell gewachsenen Strukturen fokussiere sich der Wettbewerb für bankenorientierte Finanzsysteme wie das deutsche auf das eher bedeutende, aber margenschwache Einlagen- und Kreditgeschäft. Ausweichmöglichkeiten auf kapitalmarktorientierte Produkte fehlten (noch).

---

der sonstigen betrieblichen Erträge und Aufwendungen definiert. Vgl. *Pieper, B.* (2006), S. 453, Fn. 8 und Fn. 10.

309 Vgl. *Pieper, B.* (2006), S. 454.

310 Als Marge definiert er den Quotient aus Bruttoerlösen und Nettoerlösen. Vgl. *Pieper, B.* (2006), S. 455.

311 Vgl. *Pieper, B.* (2006), S. 455 und S. 459.

312 Vgl. *Schäfer, S./Weistroffer, C.* (2007), S. 600; *Rehbock, T.* (2007), S. 34 f.

313 Vgl. *Schäfer, S./Weistroffer, C.* (2007), S. 601 f. Der Zinssatz umfasst die Gewinnmarge sowie die Bearbeitungs-, Standardrisiko-, Eigenkapital- und die Refinanzierungskosten. Vgl. auch Kapitel C.4.3.

314 Vgl. *Schäfer, S./Weistroffer, C.* (2007), S. 606.

315 Vgl. *Schüller, S./Schneider, M.* (2006), S. 15 f.

316 *Schüller, S./Schneider, M.* (2006), S. 17. Ähnlich argumentieren *Nirschl, M./Paul, V.* (2007), S. 180 und *Duhnkrack, T.* (2002), S. 157.

317 Vgl. hier und nachfolgend *Fehl, U./Kuhn, N.* (2005), S. 198.

Diese Diagnose gilt vor allem für das Firmenkundengeschäft, das regelmäßig als „Wertvernichter" bezeichnet wird.[318] Nach einer Studie von MERCER OLIVER WYMAN lag der Rohertrag, d. h. die Summe aus Zins- und Provisionsüberschuss, für das Jahr 2003 in Deutschland bei EUR 7,2 Mrd.[319] Aufgrund relativ geringer Margen ist dazu ein deutlich höheres Finanzierungsvolumen als im internationalen Vergleich notwendig. Werden vom Rohertrag die operativen Aufwendungen, die Standardrisiko- sowie die Eigenkapitalkosten subtrahiert, bleibt ein negativer Beitrag von etwa EUR -1,0 Mrd. Ursachen für dieses Ergebnis liegen neben der Marktstruktur auch im Produktmix der Banken im Firmenkundengeschäft, das in Deutschland vor allem seine Schwerpunkte im klassischen, aber als margenschwach geltenden Kreditgeschäft hat. Dies belegen auch die jüngsten Schätzungen von OLIVER WYMAN, wonach im Jahr 2009 65 Prozent des Rohertrags im Firmenkundengeschäft aus dem Finanzierungsgeschäften, 23 Prozent aus dem Transaction Banking und 12 Prozent aus sonstigen Bankleistungen resultierten.[320] Risiken aus derartigen Ertragskonzentrationen sind in den in 2009 aktualisierten Mindestanforderungen an das Risikomanagement (MaRisk) mit dem Hinweis, dass diese in der Risikostrategie zu berücksichtigen sind, aufgenommen worden.[321]

Insgesamt zeigen die Ergebnisse der genannten Studien, dass die als vergleichsweise gering beurteilte Profitabilität nicht nur das Ergebnis der Wettbewerbsintensität ist, sondern auch in den institutsspezifischen Zielsetzungen sowie dem grundsätzlichen marktpolitischen Verhalten der Banken begründet ist. Daher besteht die nicht nur aus den MaRisk resultierende Notwendigkeit im Privat- wie auch im hier betrachteten Firmenkundengeschäft, diversifizierte Ertragspotenziale identifizieren und realisieren zu müssen. Wie die vorhergehenden Ausführungen zeigen, gilt diese Anforderung für alle Bankengruppen, die im Firmenkundengeschäft tätig sind. Nachfolgend soll die Geschäftseinheit Firmenkunden aus der Perspektive der Kreditgenossenschaften beschrieben werden.

---

318 Vgl. *Krauß, C.* (2005), o. S. [Vorwort]; *Hampl, H.* (1999), S. 17. Der Begriff „Wertvernichter" ist dabei im Kontext der von den meisten (kapitalmarktorientierten) Großbanken verfolgten Shareholder Value-Orientierung zu sehen, bei dem der Unternehmenswert aus Sicht der Anleger (Eigentümer) die zentrale Zielgröße ist. Wertzuwachs wird in diesem Verständnis z. B. in Form des Economic Value Added operationalisiert. Vgl. dazu auch *Rolfes, B.* (2002), S. 141 ff. Motiviert wird dies durch den Wettbewerb um knappes Kapital. Vgl. *Duhnkrack, T.* (2002), S. 155 ff.

319 Vgl. hier und nachfolgend *Hölzer, K.* (2006), S. 49 ff.

320 Vgl. *Hölzer, K./Schnarr, T.* (2010), S. 54. Zum Transaction Banking gehören Produkte des Leistungen des Zahlungsverkehrs, des Cash-Managements und des Handels.

321 Vgl. *Bundesanstalt für Finanzdienstleistungsaufsicht* (2010), AT 4.2, Textziffer 2.

2 Deskription der Geschäftseinheit Firmenkunden der Kreditgenossenschaften

2.1 Theoretischer Bezugsrahmen

Nach dem Drei-Ebenen-Konzept, das zwischen der Unternehmens-, der Geschäftseinheiten- und der Funktionalbereichsebene unterscheidet, können strategische Entscheidungen, deren Ziel die Schaffung komparativer Konkurrenzvorteile sind,[322] nur auf Ebene der (strategischen) Geschäftseinheiten erfolgen.[323] Strategische Geschäftseinheiten stellen die gedankliche Zusammenfassung eines strategischen Geschäftsfelds und deren korrespondierende Verankerung innerhalb der Organisationsstruktur des Unternehmens dar.[324] Regelmäßig lassen sich bei den Kreditgenossenschaften die strategischen Geschäftseinheiten der Privatkunden- und der Firmenkundenbank finden.[325]

Strategische Geschäftseinheiten können auf vielfache Weise abgegrenzt werden und sind generell dadurch gekennzeichnet, dass sie eine von anderen Geschäftseinheiten unabhängige Marktaufgabe haben, dass sich für sie eine eigenständige Strategie formulieren und umsetzen lässt und dass sie einen weitestgehend unabhängigen Beitrag zur Sicherung und Steigerung des Erfolgspotenzials des Unternehmens leisten.[326]

Vereinfachend werden die den Geschäftseinheiten zugrundeliegenden Geschäftsfelder oft zweidimensional als Produkt-Markt-Kombinationen aufgefasst.[327] ABELL nutzt einen dreidimensionalen Bezugsrahmen aus den Merkmalen „customer groups" (Abnehmergruppe), „customer function" (Funktionserfüllung, Kundenbedürfnis) und „alternative technology" (Technologieangebot oder allgemeiner: Lösungsangebot) um Geschäftsfelder abzugrenzen.[328] Die Produktdimension der einfachen, zweidimensionalen Abgrenzung wird bei ihm in einen

322 Vgl. zum Begriff der komparativen Konkurrenzvorteile *Backhaus, K.* (1999), S. 34 ff.

323 Vgl. *Hax, A. C./Majluf, N. S.* (1996), S. 5 ff.

324 Die externe und interne Abgrenzung muss jedoch nicht unbedingt deckungsgleich sein, denn es kann ein Geschäftsfeld von mehreren Geschäftseinheiten betreut werden et vice versa. Vgl. *Meffert, H.* (1998), S. 228. So trifft es bspw. in der kreditgenossenschaftlichen Praxis regelmäßig zu, dass die Privatperson des Unternehmers in der Firmenkundenbank betreut wird oder Firmenkunden aufgrund der bankinternen Zuordnung über Mitarbeiter der Privatkundenbank betreut werden.

325 Vgl. *Börner, C. J.* (2005), S. 34.

326 Vgl. *Backhaus, K.* (1999), S. 228; *Meffert, H.* (1998), S. 225 f.

327 Vgl. z. B. *von der Oelsnitz, D./Nirsberger, I.* (2007), S. 1288.

328 Vgl. *Abell, D. F.* (1980).

Problembezug und einen Problemlösungsansatz aufgespaltet.[329] Mittels der Dimension Abnehmergruppe wird festgelegt, wessen Bedürfnisse befriedigt werden sollen. Die Dimension der „customer function" basiert auf der Kombination von Bedürfnis und Produkt. Sie thematisiert, welche Bedürfnisse des Kunden durch die angebotenen Leistungen befriedigt werden sollen. Die dritte Dimension beschreibt Möglichkeiten, wie die Bedürfnisse befriedigt werden können. Vorteilhaft an diesem Ansatz ist, dass Zielgruppen, Kundenbedürfnisse und Lösungsansätze integriert betrachtet werden und somit eine Integration der Nachfrager- und Anbieterperspektive erfolgt.[330] Es wird jedoch kritisiert, dass die Dimensionen nicht unabhängig voneinander seien[331] und die regionale Abgrenzung fehle[332]. Trotz der Kritik ist das gering zu modifizierende Konzept von ABELL ein zweckmäßiger Bezugsrahmen zur Strukturierung der nachfolgenden Ausführungen.

Da das Regionalprinzip die regionale Abgrenzung beinhaltet, ist dieser Kritikpunkt irrelevant. Jedoch soll der Kritik der fehlenden Unabhängigkeit der Dimensionen insofern Rechnung getragen werden, als eine Umwidmung der Dimensionen stattfindet, womit gleichzeitig der Anwendungskontext von Geschäftsfeldern auf Geschäftseinheiten erweitert wird. So soll zunächst konform zum Ansatz von ABELL die Abnehmergruppe bzw. Zielgruppe als ein Kriterium verwendet werden. Entsprechend der Empfehlung von MEFFERT soll dabei von einem zunächst hohen Abstraktionsniveau zur grundsätzlichen Abgrenzung der Firmenkunden von den Privatkunden ausgegangen werden (Makro-Abgrenzung), das durch die Segmentierung (Mikro-Abgrenzung) ergänzt und konkretisiert wird.[333] Die Betrachtung der Bedürfnisse bzw. Bedarfe wird ebenfalls diesem Ansatz sowie der derzeitigen Ausgestaltung in der Firmenkundenpraxis folgend zunächst auf Bedarfsfeldebene und dann detailliert – wenngleich jedoch nicht abschließend – auf Produktebene konkretisiert. Statt der Dimension des Lösungsangebots, die im originären Ansatz eng verbunden mit der Dimension der Bedürfnisse ist und sich eher für Industriegütermärkte eignet, wie die Beispiele bei ABELL zeigen,[334] soll dem inhaltlichen Schwerpunkt dieser Arbeit folgend der Vertrieb betrachtet werden. Somit bilden die beiden letztgenannten Dimensionen die interne Perspektive der Bildung strategischer Geschäftseinheiten in zweifacher Weise ab: Erstens auf Basis des Leistungsprogramms und

329 Vgl. *Birkelbach, R.* (1988), S. 232.
330 Vgl. *Backhaus, K.* (1999), S. 226.
331 Vgl. *Backhaus, K.* (1999), S. 228 f.
332 Vgl. *Meffert, H.* (1998), S. 228.
333 Vgl. *Meffert, H.* (1998), S. 227.
334 Vgl. *Abell, D. F.* (1980), S. 112 oder *Engelhardt, W. H./Kleinaltenkamp, M.* (1989), S. 11.

zweitens auf Basis des Vertriebs.[335] Zusammengefasst ergibt sich der Bezugsrahmen zur Deskription des Firmenkundengeschäfts bei den Kreditgenossenschaften aus den Dimensionen Zielgruppen, Leistungsprogramm und Vertrieb.

Wie erinnerlich sind Kreditgenossenschaften rechtlich und wirtschaftlich selbständige Unternehmen. Daher ist es nicht möglich, alle institutsindividuellen Realisationen der ausgewählten Dimensionen aufzugreifen. Vielmehr soll die folgende Darstellung auf den Empfehlungen des BVR, der regelmäßig die Situation der Kreditgenossenschaften mit dem Ziel der Optimierung der betriebswirtschaftlichen Leistungsfähigkeit analysiert, und den Praktikerbeiträgen aus dem genossenschaftlichen Umfeld basieren.

2.2 Zielgruppen

Als Zielgruppen gelten allgemein die aktuellen und potenziellen Adressaten, die mit einer oder mehreren bestimmten marktpolitischen Aktivität(en) angesprochen werden sollen. Nachfolgend wird die Zielgruppenabgrenzung auf Ebene der Makro- und der Mikroabgrenzung vorgenommen. Anschließend werden sekundärstatistische Daten zur empirischen Struktur der Zielgruppe im Firmenkundengeschäft der Kreditgenossenschaften vorgestellt. Abschließend werden Kundenverbünde im Firmenkundengeschäft betrachtet.

2.2.1 Makro-Abgrenzung

Ausgangspunkt der Makro-Abgrenzung ist die Frage nach der inhaltlichen Konkretisierung der Firmenkunden auf abstrakter Ebene. FRETER bezeichnet dies auch als den relevanten Markt.[336] In der bankbetriebswirtschaftlichen Literatur gibt es keine einheitliche Definition des Begriffs Firmenkunde.[337] Eine Vielzahl von Beiträgen zum Firmenkundengeschäft von Banken verzichtet vollständig auf eine inhaltliche Abgrenzung des Begriffs oder grenzt diesen implizit ab.[338]

---

335 Ähnlich *Meffert, H.* (1998), S. 227.

336 Vgl. *Freter, H.* (1998), S. 233. Andere Autoren betrachten den relevanten Markt als Zusammenfassung von Kunden, Leistungen und Wettbewerbern. Vgl. z. B. *Backhaus, K.* (1999), S. 204 ff. Diese Vorgehensweise würde sich jedoch in weiten Teilen mit der Abgrenzung des strategischen Geschäftsfelds überschneiden, weshalb hier dem Verständnis von FRETER gefolgt werden soll.

337 Eine Auseinandersetzung mit dem Begriff Kunde folgt detailliert in Kapitel D.1.2.2 Die Begriffe Firmenkunde und Unternehmenskunde werden in dieser Arbeit synonym verwendet.

338 Vergleiche z. B. *Raddao, N.* (2008), *Brost, H./Dahmen, A./Lippmann, I.* (2006) und *Schmoll, A.* (2006) als Beispiele für Beiträge ohne Abgrenzung des Begriffs Firmenkunde und *Püthe, T.* (2009), *Ahrendt, B.* (2007) sowie *Segbers, K.* (2007) für Beiträge

Der BVR unterscheidet in seinem Strategiepapier BÜNDELUNG DER KRÄFTE: DIE GEMEINSAME STRATEGIE zwischen Privat- und Firmenkunden,[339] wobei Firmenkunden knapp beschrieben sind als „alle wirtschaftlich Selbständigen, die ein Gewerbe betreiben, mit wesentlicher Erwerbsquelle ihrer Tätigkeit, ohne Mindestgröße“[340]. Dabei ist es irrelevant, ob es sich um juristische oder natürliche Personen handelt, jedoch von Bedeutung, dass die Kundenbeziehung erwerbsbedingt ist.[341]

Um das Segment der Firmenkunden detaillierter zu beschreiben, nutzt HUTH einen umfangreichen Merkmalskatalog (siehe Tabelle 5).[342] Er nennt bspw. Unterschiede hinsichtlich der Anzahl potenzieller Merkmalsträger oder spezifische Strukturunterschiede mit Blick auf die Geschäftsbeziehung zwischen dem Kunden und der Bank. Es sei darauf hingewiesen, dass nicht in jedem Fall diese Unterscheidung trennscharf ist und zutrifft. Vielmehr sollte ein solcher Katalog als tendenzielle Aussage begriffen werden. Zusammenfassend wird konstatiert, dass sich das Firmenkundensegment stark vom Privatkundengeschäft unterscheidet, da es insgesamt individueller und komplexer in der Leistungs- und der Beziehungsdimension ist.[343]

---

mit impliziter Definition, die sich aus der Abgrenzung des Begriffs mittelständische Unternehmen ergibt.

339 Vgl. *BVR* (2001b), S. 14.

340 *BVR* (2001c), S. 29 f.

341 Vgl. *Scheer, O.* (1989), S. 5. Damit wird ausgeschlossen, dass Selbständige, die ihr Privatkonto bei einer Bank führen, als gewerbliche Kunden bzw. Firmenkunden definiert werden.

342 Vgl. *Huth, H. O.* (1985), S. 37 f. sowie die dort angegebene Literatur. Einen vergleichbaren Merkmalskatalog zur Unterscheidung von Firmen- und Privatkunden stellt KRAUß auf. Vgl. *Krauß, C.* (2005), S. XVI [Anhang].

343 Vgl. auch *BVR* (1999b), S. 13.

Tabelle 5: Qualitative Kriterien zur Unterscheidung von Privat- und Unternehmenskunden[344]

Privatkunden	Unternehmenskunden
- große Zahl - Kunde ist anonym oder der Bank nur wenig bekannt - bankungewohnt - keine eindeutige Orientierung am Gewinn - Interesse an Bequemlichkeit und Service dominiert - eng begrenzter Bereich der nachgefragten Leistungen - kleinere Umsätze - unregelmäßige Kontakte von relativ geringer Häufigkeit - mehr persönlich orientierte Geschäftsbeziehung (Person-Person) - individuelle Ansprüche leichter standardisierbar - legt Wert auf lokale Präsenz und persönlichen Kontakt	- überschaubare, begrenzte Zahl - Kunde ist i. d. R. der Bank bekannt - bankgewohnt - auf Gewinnerzielung bedachte Denk- und Verhaltensweise - Interesse an reibungsloser, schneller Abwicklung der Geschäfte - breite Streuung der nachgefragten Leistungen - größere Umsätze - häufige, regelmäßige, oft tägliche Kontakte - sachlich orientierte Geschäftsbeziehung (Unternehmung-Unternehmung) - individuelle Ansprüche schwer standardisierbar - legt Wert auf gute Kommunikationsmöglichkeiten

In ähnlicher Weise geht BRAUN vor, der die Firmenkunden anhand der Faktoren der Zugehörigkeit zu einer bestimmten Personengruppe oder Gesellschaftsform, ihrer Stellung im Erwerbsleben sowie den nachgefragten Bankgeschäften charakterisiert.[345] Konsequent zählt er zu den Unternehmenskunden alle wirtschaftlich selbständigen natürlichen Personen sowie Personen- und Kapitalgesellschaften, die sich dadurch auszeichnen, dass sie weitgehend wirtschaftlich unabhängig sind, die Arbeitgeberfunktion tragen und Risiken übernehmen. Daraus lässt sich ein zentrales Differenzierungsmerkmal ableiten. Als Firmenkunden können alle Wirtschaftsubjekte begriffen werden, die unabhängig von Rechtsform und Größe, Sach- und/oder Dienstleistungen auf Märkten für Dritte anbieten.[346] Als abgrenzendes Merkmal ist also die Tätigkeit des Firmenkunden als Anbieter auf (anderen) Märkten zu betrachten. Damit geht die Feststellung einher, dass die Leistungen der Bank als Inputfaktor im weitesten Sinne für die Wertschöpfung des Firmenkunden gelten bzw. die Nachfrage nach Bankleistungen derivativ ist.

344 In enger Anlehnung an *Huth, H. O.* (1985), S. 38.

345 Vgl. hier und nachfolgend *Braun, P. A.* (1981), S. 25.

346 Vgl. *Picot, A./Neuburger, R.* (2000), S. 16. Auch BENÖLKEN definiert Firmenkunden als „alle Selbständigen inklusive der Freiberufler sowie der Personen- und Kapitalgesellschaften, sofern die Haupteinkommensart aus gewerblicher/freiberuflicher Tätigkeit fließt.“ *Benölken, H.* (2002), S. 30.

“Der Tante Emma-Laden und Siemens haben aus Banksicht eines gemeinsam: Sie sind Firmenkunden”[347] Kreditgenossenschaften und Sparkassen gelten als klassische Finanzierungspartner mittelständischer Unternehmen.[348] Der Begriff der mittelständischen Unternehmen muss in diesem Kontext erläutert werden, denn die Heterogenität dieser Gruppe von Firmenkunden führt zum Problem der inhaltlichen Abgrenzung.[349] In der Literatur werden als begriffskonstituierende Kriterien sowohl qualitative als auch quantitative Merkmale verwendet, um die mittelständischen Unternehmen von Großunternehmen abzugrenzen.[350] Qualitative Kriterien werden i. d. R. im Bündel als Merkmalskataloge[351] verwendet, da eine eindimensionale Abgrenzung unzureichend ist.[352] Dabei wird aber nicht die kumulative Erfüllung aller Kriterien postuliert, sondern eine Tendenz in Richtung mittelständischer Unternehmen als ausreichend angesehen.[353] Die qualitative Abgrenzung fokussiert auf die Betriebstypenbestimmung, d. h. sie stellt auf das Wesen und somit u. a. auf die Wirtschaftsweise, die Struktur der Unternehmensführung und die Finanzierungssituation ab.[354] Vor allem die Einheit von Eigentum und Leitung gilt als herausragendes Merkmal.[355] HAMER spricht in diesem Zusammenhang von der Schicksalsgemeinschaft zwischen Unternehmen und Unternehmer.[356] Obwohl die qualitativen Merkmale mittelständische Unternehmen akkurater beschreiben als quantitative, werden deren schwere Erfassbarkeit und die daraus resultierende geringe Eignung für empirische Arbeiten bemängelt.[357] Daher werden, um mittelständische Unternehmen in Bezug auf die Betriebsgröße von Großunternehmen abzugrenzen, zusätzliche quantitative (Hilfs-)Kriterien angewendet.[358] Typischerweise werden Größen wie Umsatz, Bilanzsumme oder Beschäftigtenzahl genutzt. So werden Unternehmen mit einem Umsatz von bis zu 50 Mio. EUR und einer Beschäftigtenzahl von bis zu

---

347 *Juncker, K.* (1979), S. 7.

348 Vgl. *Lubitz, K.-J.* (1984), S. 10; *Juncker, K.* (1979), S. 11.

349 Vgl. *Krämer, W.* (2003), S. 5; *Bussiek, J.* (1996), S. 16.

350 Vgl. *Berens, W./Högemann, B./Segbers, K.* (2005), S. 8.

351 Vgl. *Mugler, J.* (1995), S. 18.

352 Vgl. *Weber, W.* (1981), S. 18.

353 Vgl. *Kunesch, H.* (1996), S. 630.

354 Vgl. die Merkmalsnennungen bei *Steiner, J.* (1980), S. 5; *Middermann, F.* (1987), S. 16; *Schuster, P.* (1991), S. 9; *Bussiek, J.* (1996), S. 18; *Bramsemann, R.* (1997), S. 29.

355 Vgl. *Wossidlo, P. R.* (1993), Sp. 2890.

356 Vgl. *Hamer, E.* (1987), S. 62 f.

357 Vgl. *Weber, W.* (1981), S. 20.

358 Vgl. *Daschmann, H.-A.* (1993), S. 50.

499 Personen als mittelständisch bezeichnet.[359] Diese Grenzwerte gelten allgemein jedoch nicht als unumstößlich, sondern können auch durch die qualitativen Merkmale dominiert werden, so dass die quantitative Abgrenzung obsolet ist.[360] Diese Begriffsabgrenzung der Literatur hat prägenden Charakter für die Praxis der Kreditgenossenschaften.[361] So wurden in einer empirischen Untersuchung des FRAUNHOFER INSTITUTS 365 Kreditgenossenschaften nach den Zielgruppenprioritäten im Firmenkundengeschäft gefragt.[362] Während 60 Prozent die Zielgruppe der Mittelstandsunternehmen priorisieren, werden die Freiberufler von 30 Prozent und die Kleinunternehmen von 20 Prozent als Zielgruppe im Vertrieb bevorzugt.[363] Lediglich 5 Prozent der Genossenschaftsbanken fokussieren auf die Gruppe der Großunternehmen. Da die Autoren keine Angaben über die Definition der jeweiligen Zielgruppen machen, kann aus diesen Ergebnissen nur die Tendenzaussage deduziert werden, dass sich die Kreditgenossenschaften vornehmlich auf mittelständische Unternehmen ausrichten.[364] Dies entspricht der subsidiären Aufgabenteilung in der FinanzGruppe zwischen den Primärbanken und den Zentralbanken, bei der die letztgenannten das Großkundengeschäft betreiben und ansonsten den Kreditgenossenschaften als Kooperationspartner zur Verfügung stehen.[365]

2.2.2 Mikro-Abgrenzung

Die Ausgestaltung der Marketing-Mix-Instrumente, zu denen die Vertriebspolitik gehört, bedarf im Kontext des Geschäftsfelds Firmenkunden einer weiteren

---

359 Vgl. dazu die Diskussionen zur Begriffsabgrenzung bei *Püthe, T.* (2009), S. 9 ff., *Flacke, K.* (2006), S. 11 ff. und *Segbers, K.* (2007), S. 60 ff.

360 So bezeichnet SIMON z. B. Unternehmen mit 2.904 Mitarbeitern wegen ihrer qualitativen Eigenschaften noch als mittelständisch. Vgl. *Simon, H.* (1992), S. 116.

361 Vgl. z. B. *Duhnkrack, T.* (2006), S. 38. Zu berücksichtigen ist, dass die einzelnen Banken und Bankengruppen unterschiedliche Definitionen für das Segment der Mittelstandskunden verwenden. So beginnt der gewerbliche Mittelstand für die DEUTSCHE BANK AG bei 15 Mio. EUR Jahresumsatz .Vgl. für eine Übersicht *Heckl, D./Moormann, J.* (2005), S. 13.

362 Vgl. hier und nachfolgend *Engstler, M./Praeg, P./Vocke, C.* (2007), S. 22 f. Da die Studienergebnisse lediglich in Schaubildern ohne Wertangaben präsentiert werden, handelt es sich um gerundete Werte.

363 In diesem Fall ist damit ein spezifisches Kundensegment bezeichnet. Die begriffliche Verwendung weicht von der erläuterten Begriffsabgrenzung ab. Nach dieser werden die Segmente der Mittelstands- und Kleinunternehmen sowie der Freiberufler als mittelständischen Unternehmen subsumiert.

364 Vgl. *Lubitz, K.-J.* (1984), S. 10; *Juncker, K.* (1979), S. 11.

365 Vgl. die Ausführungen in Kapitel C.1.4.4.

Differenzierung des relevanten Marktes (Mikro-Abgrenzung).[366] Denn zum einen ist der heterogenen Struktur der mittelständischen Firmenkunden Rechnung zu tragen, da diese Vielfalt mit variierenden Bedürfnissen und Anforderungen an eine Bank einhergeht.[367] Zum anderen ist die Leistungsfähigkeit der Bank zu berücksichtigen, die einerseits eine Fokussierung auf bestimmte Segmente des Firmenkundenmarkts aufgrund knapper Vertriebsressourcen bedingt und andererseits eine segmentspezifische Spezialisierung interner Ressourcen begründet. So ist bspw. aufgrund aufsichtsrechtlicher Vorgaben das Kreditvergabevolumen einer Kreditgenossenschaft restringiert, so dass sie spezifische Finanzierungsbedarfe größerer Firmenkunden nicht befriedigen kann. Die grundsätzliche Vorteilhaftigkeit spezialisierter Vertriebsmitarbeiter auf Branchen oder Unternehmensgrößen zur Nutzung von Synergiepotenzialen ist evident. Die notwendige Mikro-Abgrenzung wird mittels der Markt- respektive Kundensegmentierung vorgenommen und kann sachlich oder räumlich erfolgen.[368] Eine räumliche Abgrenzung ist aufgrund des Regionalprinzips nicht weiter nötig. Es kann angenommen werden, dass das jeweilige Marktgebiet der Primärbank fokussiert wird. Die sachliche Segmentierung wird im Weiteren vorgestellt.

Bei der Marktsegmentierung ist es wichtig, die Kundenseite als Adressaten und die marketingpolitische Instrumentenseite als Aktionsparameter für die Adressaten integrativ zu betrachten. Eine isolierte Betrachtung nur der Kunden- oder nur der Marktbearbeitungsseite birgt die Gefahr zweckinadäquater Ergebnisse.[369] Daher beinhaltet die Marksegmentierung neben der Markterfassung auch die Marktbearbeitung.[370] Dies wird als managementorientierter Ansatz aufgefasst (siehe Abbildung 10). Ziel der Marktsegmentierung ist die Aufteilung eines heterogenen Gesamtmarktes in homogene Teilmärkte, um diese zielsetzungsgerecht bearbeiten zu können. Die Homogenität der Teilmärkte ergibt sich aus der Ähnlichkeit der Kunden bei den zur Segmentierung herangezogenen Kriterien.[371] An die zu verwendenden Kriterien ist eine Reihe von Anforderungen zu stellen, um zum einen die Zweckmäßigkeit der Segmentierung zu ge-

---

366 Vgl. *Backhaus, K.* (1999), S. 210; *Meffert, H.* (1998), S. 225.

367 Vgl. *BVR* (1999b), S. 5; *Freter, H.* (1998), S. 231.

368 Vgl. *Backhaus, K.* (1999), S. 210. Die von KRAFFT/ALBERS vorgenommene semantische Unterscheidung zwischen der Markt- und der Kundensegmentierung wird aufgrund der methodischen und zielbezogenen Ähnlichkeiten nicht berücksichtigt. Vgl. *Krafft, M./Albers, S.* (2000), S. 516.

369 Vgl. *Freter, H.* (1998), S. 231 f.

370 Vgl. *Meffert, H.* (1998), S. 174 sowie die dort angegebene Literatur.

371 Vgl. *Freter, H.* (1998), S. 232.

währleisten und zum anderen die Vielzahl möglicher Kriterien einzugrenzen.[372] Dabei gilt, dass die Anforderungen der Messbarkeit und zeitlichen Stabilität von jedem einzelnen Kriterium erfüllt sein müssen, während die übrigen Anforderungen durch den insgesamt verwendeten Kriterienkatalog weitgehend erfüllt sein sollten.[373]

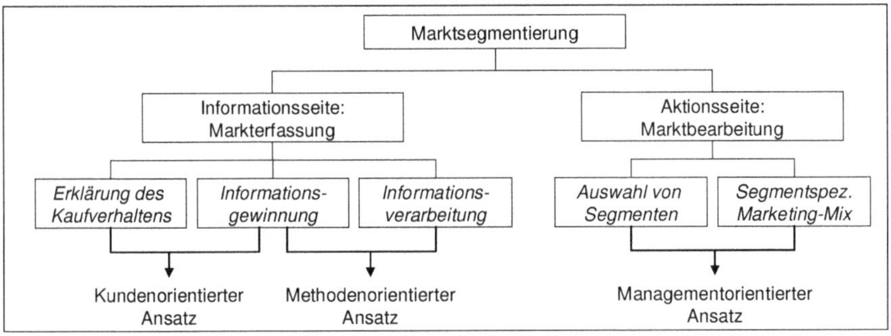

Abbildung 10: Komponenten der Marktsegmentierung[374]

In der Marketingtheorie wird aufgrund der immer stärkeren Individualisierung und Fragmentierung der Nachfrageseite in Verbindung mit dem Trend zum Beziehungsmarketing von einer Notwendigkeit zum „segment of one" gesprochen.[375] Es ist jedoch zu berücksichtigen, dass die Marktsegmentierung ein Optimierungsproblem darstellt.[376] Denn bei nur rudimentärer Segmentierung leistet sich ein Unternehmen Streuverluste bei der Marktbearbeitung, so dass bei adäquater Segmentierung der Erfolg pro Einheit eingesetzter Ressourcen steigt. Bei einer zu großen Anzahl von Segmenten dagegen, steigen die Kosten der segmentspezifischen Marketing-Mix-Gestaltung bei abnehmenden Grenzerlösen. Vor diesem Hintergrund gilt es, eine optimale Anzahl an Segmenten zu definieren.

Hinsichtlich der Marktbearbeitung ist zudem die zentrale Forderung aufzustellen, dass die Segmentierung nicht nur eine Entscheidungsgrundlage für die effektive Ausgestaltung des Marketing-Mix, sondern auch für dessen effizienten

372 Vgl. ausführlich *Meffert, H.* (1998), S. 178 f.; *Freter, H.* (1998), S. 238 f.; *Freter, H.* (1983), S. 43 f.

373 Vgl. *Meffert, H.* (1998), S. 179.

374 *Freter, H.* (1983), S. 14.

375 Vgl. *Vogel, V.* (2006), S. 44; *Krafft, M.* (2007), S. 51.

376 Vgl. hier und nachfolgend *Meffert, H.* (1998), S. 210 ff.

Einsatz sein sollte. Damit geht u. a. die Frage nach der Beurteilung der Segmente oder der Kunden in den Segmenten einher.

Im Strategiepapier BÜNDELUNG DER KRÄFTE: DIE GEMEINSAME STRATEGIE empfiehlt der BVR eine Segmentierung insoweit, als im Geschäftsfeld der Firmenkunden zwischen den gewerblichen Kunden und mittelständischen Firmenkunden differenziert werden sollte.[377] Eine Begründung der Auswahl oder die Erläuterung der zugrundeliegenden Kriterien erfolgt nicht. Weiterhin gibt es einen konkreteren Segmentierungsansatz für die gesamte FinanzGruppe von Seiten des BVR, der in der Situationsanalyse BÜNDELUNG DER KRÄFTE: EIN VERBUND - EINE STRATEGIE[378] beschrieben ist.[379] Als Segmentierungskriterien werden die Unternehmensgröße und das Banknutzungsverhalten gewählt. Die Unternehmensgröße wird anhand des Umsatzes bzw. des Vermögens operationalisiert und dient auch als Indikator für das potenzielle Geschäftsvolumen der Firmenkunden. Das Banknutzungsverhalten wird anhand der Spezifität des Beratungsbedarfs gemessen. Als Grundbedarf gelten bspw. Produkte wie Zahlungsverkehrsleistungen oder Betriebsmittelkredite. Spezifische Bedarfe entsprechen Produkten wie Cash-Management-Systemen oder Beratungsleistungen. Im Ergebnis resultiert der in Abbildung 11 dargestellte Segmentierungsansatz.

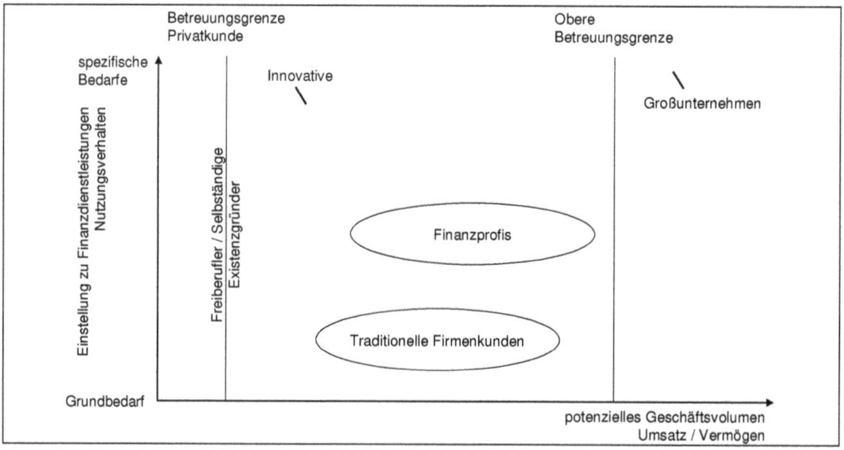

Abbildung 11: Segmentierungsansatz im Firmenkundengeschäft nach BVR[380]

[377] Vgl. *BVR* (2001b), S. 14.

[378] Vgl. dazu ausführlich *BVR* (1999a) und *BVR* (1999b).

[379] Vgl. hier und nachfolgend *BVR* (1999b), S. 5 ff.

[380] Vgl. *BVR* (1999b), S. 5.

Die beiden eingezeichneten Grenzlinien stellen im linken Bereich eine Empfehlung hinsichtlich der Zuordnung dieses Segments zu internen Vertriebseinheiten und im rechten Bereich einen Hinweis auf die „bestehenden Probleme bei der Betreuung großer Unternehmen durch Primärbanken"[381] dar. Für die einzelnen Segmente werden knappe Handlungsempfehlungen gegeben sowie jeweils eine Bewertung vorgenommen. So wird etwa für die Großunternehmen empfohlen, Verbundlösungen zu entwickeln, die in einem gemeinschaftlichen Antritt den Unternehmen kommuniziert werden können.[382] Problematisch an diesem Ansatz ist die alleinige Erklärung des Geschäftspotenzials auf Basis des Umsatzes des Firmenkunden, da auch dessen Branche, Internationalität und andere Kontextfaktoren des Firmenkunden, aber auch die Charakteristika der Entscheidungsträger in den Unternehmen erheblichen Erklärungsgehalt für das Absatzpotenzial haben dürften. Implizit setzt diese Segmentierung den Anreiz, große Unternehmen zu gewinnen, was jedoch vor dem Hintergrund der Positionierung und der Restriktionen der Kreditgenossenschaften schwierig sein dürfte.[383]

Den vorgestellten Ansatz ergänzt der BVR im Zusammenhang mit der Vertriebskonzeption VR-FINANZPLAN MITTELSTAND hinsichtlich der Verbindung von Potenzial und Bedarfsstruktur eines Firmenkunden dahingehend, dass die Betreuungsart und die Betreuungsintensität in Abhängigkeit des Geschäftspotenzials festgelegt werden, wobei den Kleinstgewerbetreibenden/Geschäftskunden und Freiberuflern eine inhaltlich enge und standardisierte sowie anlassbezogene Beratung und den Unternehmen des oberen Mittelstands eine vollumfängliche, situationsspezifische und regelmäßige Betreuung zugeordnet wird.[384]

Aufbauend auf der Empfehlung des BVR konkretisieren sich die Segmentierungsansätze der Kreditgenossenschaften für das Firmenkundengeschäft. Wobei die Transmission der Empfehlungen in die Praxis oftmals durch verbundeigene Beratungsgesellschaften begleitet wird. Dabei wird zuweilen argumentiert, der „Segmentierungsprozess darf nicht zur ‚Wissenschaft' werden, sondern muss nach klaren, einfachen Kriterien erfolgen".[385] Folglich werden im Rahmen der praktischen Segmentierungsansätze vor allem Kriterien verwendet, die zwar den Anforderungen der Wirtschaftlichkeit und der Operationaliät genügen, die aber keine adäquate Segmentierung erlauben. Denn klassischerweise erfolgt die Marktsegmentierung im Firmenkundengeschäft in praxi auf Basis zweier Kriterien.[386] In Betracht kommen dabei das Kredit- oder Einlagenvolumen des Fir-

381 *BVR* (1999b), S. 7.
382 Vgl. *BVR* (1999b), S. 7.
383 Vgl. dazu die Ausführungen in Kapitel C.1.2 und Kapitel C.2.2.3.
384 Vgl. *BVR et al.* (2006), S. 17.
385 *Schmoll, A.* (2004), S. 28.
386 Vgl. *Oelrich, J.-F.* (2008), S. 25 ff.

menkunden oder dessen Umsatz.[387] Dies belegen die Ergebnisse der Studien von Strauß für Kreditgenossenschaften und Sparkassen sowie von Krauß für ostdeutsche Kreditgenossenschaften. So findet Strauß in seiner Studie heraus, dass das Kreditvolumen sowie der Firmenumsatz zu den priorisierten Segmentierungskriterien gehören.[388] In den Ergebnissen der Befragung von Krauß zeigt sich ebenfalls, dass das Kreditvolumen (63,5 Prozent), die Umsatzgröße und das Einlagevolumen (jeweils 32,4 Prozent) die am häufigsten genutzten Kriterien sind.[389] Die Orientierung an Volumengrößen weist vor dem Hintergrund der Zielsetzungen der Marktsegmentierung jedoch vielfache Defizite auf, deren größte die fehlende Kaufverhaltensrelevanz und Zukunftsorientierung sind.[390]

Die Ausführungen zur Mikro-Abgrenzung machen deutlich, dass ausgehend von den theorieorientierten Empfehlungen des BVR über die Transmission eine zunehmende Simplifizierung der Kundensegmentierungsansätze im Firmenkundengeschäft der Kreditgenossenschaften zu beobachten ist.[391] Insbesondere die Frage nach der Bewertung von Firmenkunden aus Sicht der Kreditgenossenschaft erfolgt dabei sehr wenig fundiert. Dabei wird jedoch von Expertenseite konstatiert, dass „Unterschiede in der Qualität der Kundensegmentierung [...] ein signifikantes Merkmal für Rentabilitätsdifferenzen im Vergleich zwischen Banken sind“.[392]

2.2.3 Empirische Kundenstruktur

Die aktuellste vorliegende Studie zur Kundenstruktur im Firmenkundengeschäft der Kreditgenossenschaften wurde von TNS INFRATEST in 2010 vorgestellt.[393] Aufbauend auf einer Befragung von 3.105 Unternehmen mit einer Beschäftig-

387 Vgl. *Heine, J. M.* (2007), S. 15.

388 Vgl. *Strauß, M. R.* (2006), S. 338. In seiner Studie zu den Erfolgsfaktoren im Firmenkundengeschäft prüft er den Einfluss verschiedener Segmentierungskriterien auf die Ziele Gewinn, Wachstum, Sicherheit und außerökonomischen Erfolg. Seiner Studie liegen die Daten von 62 Instituten zugrunde.

389 Vgl. *Krauß, C.* (2005), S. 147. Diesen Ergebnissen liegt eine Befragung von 74 VR-Banken aus Ostdeutschland zugrunde. Vgl. auch *Dreyer, K.-J.* (2000) als Beispiel für den Sparkassensektor.

390 Vgl. *Eichelmann, T./Bufka, J./Lehmann, M.* (2001), S. 778.

391 An dieser Stelle sei darauf hingewiesen, dass in den Primärbanken die Segmente mit relativ geringer Merkmalsausprägung als Geschäfts- oder Gewerbekunden und mit relativ hoher Merkmalsausprägung als Mittelsstands- oder Firmenkunden bezeichnet werden. Im weiteren Verlauf der Arbeit wird diese Differenzierung nicht übernommen, sondern allgemein von Firmenkunden gesprochen.

392 *von Boehm-Bezing, C. L.* (2002), S. 89.

393 Vgl. *tns infratest* (2010).

tenanzahl zwischen 2 und 499 Personen ergibt sich eine Kundenstruktur im Firmenkundengeschäft der Kreditgenossenschaften mit deutlichem Schwerpunkt bei den Unternehmen bis 6 Mio. EUR Jahresumsatz (siehe Abbildung 12).

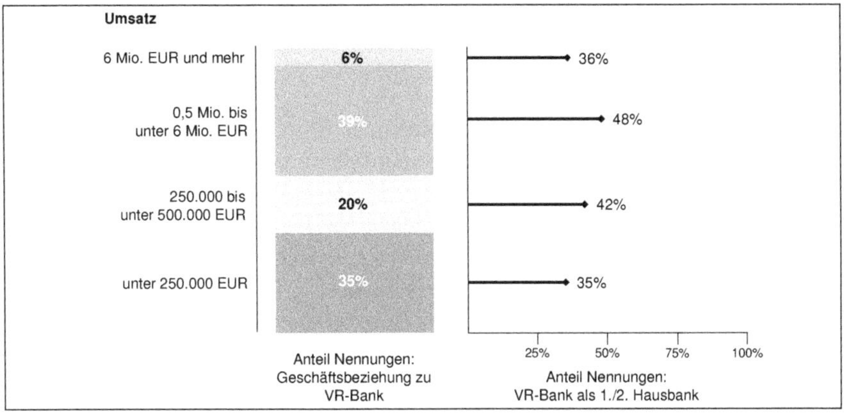

Abbildung 12: Kundenstruktur und Anteile Hausbankstatus nach Umsatzgrößenklassen[394]

Weiterhin zeigen die Ergebnisse der Studie, dass die Kreditgenossenschaften bei zunehmendem Umsatz zunehmend seltener als relevante Hausbank betrachtet werden und dass die Anzahl der Bankverbindungen steigt.[395] So hat etwa die Hälfte der Firmenkunden mit einem Umsatz von bis 250 TEUR zwei bis drei Bankverbindungen, während dies auf etwa 90 Prozent der Unternehmen zutrifft, die einen Umsatz von über 6 Mio. EUR erwirtschaften.[396] Etwa ein Drittel weisen sogar fünf oder mehr Bankverbindungen auf. Verbunden mit den Aussagen zum Hausbankbeziehungsstatus kann konstatiert werden, dass der potenzielle Wettbewerb bei den größeren Firmenkunden intensiver ist, als bei den kleineren Unternehmen. Gleichwohl ist es der Vorschlag des BVR u. a. aufgrund der unzureichenden Volumina im firmenkundenspezifischen Auslandsgeschäft, verstärkt größere Firmenkunden zu akquirieren und die bestehenden Kundenbezie-

394 Eigene Darstellung auf Basis der Daten von *tns infratest* (2010), S. 16 und S. 28 (n=3.105). In dieser Untersuchung wurden die Zentralbanken den VR-Banken zugeordnet, was eine Verzerrung der Ergebnisse in der Art beinhalten kann, dass entsprechend der subsidiären Verbundkooperation mehr größere Unternehmen als Kunden der VR-Banken aufgeführt werden.

395 Vgl. *tns infratest* (2010), S. 28 und S. 23.

396 Vgl. *tns infratest* (2010), S. 23.

hungen zu größeren Unternehmen professioneller zu betreuen.[397] Dabei wird vor allem auf die Kooperation mit den Zentralbanken gesetzt, die hinsichtlich der Kundenreichweite eine gegenläufige Tendenz zu den Kreditgenossenschaften aufweisen und mit zunehmender Unternehmensgröße höhere Kundenreichweiten realisieren,[398] wobei diese jedoch mit Werten unter 10 Prozent noch als gering zu beurteilen sind.

2.2.4 Kundenverbünde

Wie festgestellt wurde, ist ein konstituierendes Merkmal mittelständischer Unternehmen die so genannte Schicksalsgemeinschaft zwischen Unternehmen und Unternehmer.[399] Daher ist es aus Sicht der Kreditgenossenschaften naheliegend, als Firmenkunden nicht nur den gewerblichen Nukleus, der aus einem oder mehreren Unternehmen oder Betrieben besteht, zu verstehen, sondern auch den Unternehmer als Privatperson sowie dessen Familie.

Wie die Darstellung der empirischen Struktur der Firmenkunden der Kreditgenossenschaften zeigt, ist die Mehrheit der Bestandskunden den kleineren Unternehmen und Selbständigen (insb. Freiberufler und Kleingewerbekunden) zuzuordnen.[400] Nach Einschätzung des BVR ist daran vorteilhaft, dass hier eine enge Verzahnung von privater und geschäftlicher Bedarfsstruktur vorliegt (gemeinsame Haushaltsebene), die gemeinsam mittels des Allfinanzangebots der Kreditgenossenschaften bedient werden kann. Das „Firmenkundengeschäft [ist; Erg. d. Verf.] deshalb von so großer Bedeutung [.], weil es eine entscheidende Quelle der Akquisition für das Privatkundengeschäft darstellt.“[401] Denn wesentliche Ertragspotenzialanteile schätzt die Bankenberatung ZEB/, liegen bei diesen Kunden vor allem in der privaten Ebene (siehe Tabelle 6).

---

397 Vgl. *BVR* (2001b), S. 33.
398 Vgl. *Käser, B./Ritzenhofen, A.* (2009), S. 5.
399 Vgl. *Hamer, E.* (1987), S. 62 f.
400 Vgl. die Abbildung 12 in Kapitel C.2.2.3 sowie *BVR et al.* (2006), S. 24.
401 *Rometsch, S.* (1999), S. 170.

Tabelle 6: Segmentspezifische Aufteilung von Ertragspotenzialen im Firmenkundengeschäft[402]

Segmentbezeichnung	Segmentdefinition (Umsatz)	Ertragspotenzial-anteil betrieblich	Ertragspotenzial-anteil privat
Geschäftskunden	< 0,25 Mio. EUR	31 %	69 %
Gewerbekunden	0,25 - 1,0 Mio. EUR	63 %	37 %
Kleine Firmenkunden	1,0 - 5,0 Mio. EUR	81 %	19 %
Große Firmenkunden	> 5,0 Mio. EUR	89 %	11 %

Um dieser Situation Rechnung zu tragen, werden in den Kreditgenossenschaften Kundenverbünde gebildet. Neben den aus risikopolitischen Vorgaben (§ 19 Abs. 2 KWG) heraus definierten Kundenverbünden werden auch marketing- und vertriebspolitisch motivierte Kundenverbünde, sog. Betreuungseinheiten, festgelegt, die z. B. ein Unternehmen und dessen Gesellschafterfamilien zusammenfassen.[403] Dies erfolgt mit dem Ziel, alle vertrieblich relevanten Informationen in einem Kundenverbund zu bündeln und somit eine auf das ganzheitliche Optimum ausgerichtete Beratung und Betreuung anbieten zu können.[404]

Problematisch daran ist, dass die risiko- und vertriebspolitisch definierten Kundenverbünde sich nicht entsprechen müssen, so dass keine einheitlichen Bezugsgrößen für die integrierte Risiko-Ertrags-Steuerung vorliegen. In diesen Fällen ist von Seiten der Kreditgenossenschaft abzuwägen, welche kundengerichtete Bezugsgröße Gegenstand von risikopolitischen und vertriebspolitischen Maßnahmen ist.

2.3 Bankleistungsprogramm

Im Folgenden soll als Bankleistung eine bankbetriebliche Marktleistung verstanden werden, die eine Kombination aus einem Bankprodukt und einer Bankdienstleistung ist.[405] Das Bankprodukt stellt eine auf Geld lautende Information dar und ist das materielle oder immaterielle Gut, das eine Bank anbietet und zu dessen Verkauf sie ihre Tätigkeiten ausübt und ausrichtet (z. B. Kredit). Die

---

402 Eigene Darstellung basierend auf den Daten in *Heine, J. M.* (2007), S. 15. Es werden in der Quelle keine konkreten Angaben zu den absoluten Werten sowie zur Methodik der Ergebnisermittlung gemacht.

403 Nach einer Studie von PAUL/PAUL/STEIN haben etwa 66 Prozent der befragten Mittelständler auch privat mit der Bank ihres Unternehmens zu tun. Vgl. *Paul, M./Paul, S./Stein, S.* (2003), S. 2 und S. 6.

404 Dies betrifft z. B. die steuerliche Sachverhaltsgestaltung oder die Vermögensübertragung.

405 Vgl. hier und nachfolgend *BVR* (2003c), S. 8 ff.; ähnlich auch *Galasso, G.* (1999), S. 100.

Bankdienstleistung ist das aus einem bankbetrieblichen Arbeitsprozess resultierende Ergebnis, das i. d. R. mit dem Ziel des Verkaufes eines Bankprodukts erstellt wird (z. B. Aktivitäten im Zusammenhang mit dem Kredit). GALASSO ergänzt dieses Verständnis dahingehend, dass Bankleistungen als das Zurverfügungstellen von Möglichkeiten und Potenzialen, die vom Nachfrager genutzt werden, zu charakterisieren sind.[406]

Das Bankleistungsprogramm, also die Gesamtheit der angebotenen Bankleistungen der Geschäftseinheit Firmenkunden der Kreditgenossenschaften umfasst Bankleistungen, die entweder von den Kreditgenossenschaften selbst erstellt, von den Verbundunternehmen bezogen und gehandelt oder für die Verbundpartner vermittelt werden.[407] Bevor das konkrete Leistungsangebot der Kreditgenossenschaften im Firmenkundengeschäft erläutert wird, erfolgt eine theoretische Auseinandersetzung mit den Bankleistungen.

Ausgangspunkt der folgenden Ausführungen ist die Frage nach den Besonderheiten der Bankleistungen, die als spezielle Form von Dienstleistungen zu verstehen sind. Nach Stand der Forschung gibt es keine eindeutige und allgemeingültige Abgrenzung zwischen Dienst- und Sachleistungen.[408] Die ursprünglich als konstitutiv definierten Merkmale, die jedoch heute als betriebswirtschaftlich bedeutend, aber nicht als abschließend trennscharfe Abgrenzungskriterien gelten, sind die Existenz des Leistungspotenzials, die Immaterialität der Leistungsergebnisse, die Integration externer Faktoren sowie die Individualität und die anbieter- und nachfragerseitige Unsicherheit hinsichtlich des Verhaltens des jeweiligen Gegenüber und die Unsicherheit über den Transaktionsgegenstand.

Das Leistungspotenzial repräsentiert die Leistungsfähigkeit und -bereitschaft eines Anbieterunternehmens. Damit verbunden ist, dass die für die Erstellung der (Dienst-)Leistung notwendigen Kapazitäten durch den Anbieter kontinuierlich vorgehalten werden. Erst nachdem das Unternehmen mit einem Nachfrager kontrahiert hat, wird auf diese Kapazitäten zwecks Leistungserstellung zurückgegriffen.[409] Für die Dienstleistungen gilt weiter, dass diese nicht stofflich, sondern immateriell sind und somit nicht gelagert werden können, sondern abge-

---

406 Vgl. *Galasso, G.* (1999), S. 100; auch *Nader, G.* (1995), S. 7 ff.

407 Vgl. *Wiedemann, A.* (1992), S. 162 ff. Grundsätzlich muss unterschieden werden zwischen Bankleistungen auf der einen und Finanzdienstleistungen auf der anderen Seite. Aus Vereinfachungsgründen werden diese Begriffe in der vorliegenden Arbeit jedoch synonym verwendet.

408 Vgl. hier und nachfolgend *Reckenfelderbäumer, M.* (2005), S. 34.

409 In der industriellen Fertigung gibt es mit Bezug zu den Sachleistungen das Pendant der Auftragsfertigung. Vgl. *Adam, D.* (1998), S. 8.

setzt werden müssen.[410] Für das anbietende Unternehmen resultiert aus diesem Zusammenfallen von Herstellung und Nutzung die Notwendigkeit, die zur Leistungserstellung benötigten Kapazitäten in Quantität und Qualität so zu dimensionieren, dass „die negativen Wirkungen einer unzureichenden Kapazität bei Nachfragespitzen niedriger bleiben, als diejenigen der Unterbeschäftigung bei geringerer"[411] Nachfrage. Im bankbetrieblichen Kontext bedeutet dies bspw. das Vorhalten von Gebäuden und qualifizierten Mitarbeiterkapazitäten. Auf der einzelnen Leistungsebene impliziert dies auch, dass z. B. an einen Kunden ein Produkt (z. B. Kontokorrentvertrag) verkauft wurde, jedoch erst eine gewisse Zeit später mittels Kundenaktivitäten (laufende Transaktionen) die eigentliche Dienstleistung erfolgt. Gleichwohl muss für den verhandelten Kontokorrentrahmen bankintern die für die Leistungsbereitschaft notwendige Faktorreserve im Sinne von Zentralbankgeld, personeller und edv-technischer Ausstattung ab dem Zeitpunkt des Vertragsschlusses zweck- bzw. kundengebunden vorgehalten werden. Die Planung dieser Kapazitätsdimensionen ist abhängig von den absatzmarktbezogenen Zielen, mithin also den Ergebnissen der Vertriebsplanung.

Die Integrativität des externen Faktors, d. h. des Kunden selbst, ihm zugehöriger Verfügungsrechte oder Informationen in den Produktionsprozess (Bankdienstleistung) gilt als besonders bedeutsam. Die Kunden, die die Bankleistung nutzen wollen, werden zum Bestandteil der Bankleistung und des bankbetrieblichen Arbeitsprozesses.[412] Die Problematik, dass die Erstellung der Leistung ohne die Bereitschaft des Kunden, seinen Input in die Prozesse des Anbieters einzubringen, nicht möglich ist,[413] soll an einem typischen Beispiel aus dem Firmenkundengeschäft verdeutlicht werden. Im Rahmen des Kreditgeschäfts nimmt die Bank – entweder geschäftspolitisch begründet oder aufgrund der rechtlichen Vorgaben – vor Abschluss eines Kreditvertrags, eine Kreditwürdigkeitsprüfung vor. Liefert der Kunde für diesen Vorgang nicht die benötigten Unternehmensinformationen, die als Grundlage der Bonitätsprüfung dienen, so kann die Kreditwürdigkeit nicht überprüft und ein Kreditvertrag nicht abgeschlossen werden.[414] Vor diesem Hintergrund zeigt sich auch die Bedeutung der Eigenschaft der kundenbezogenen Individualität der Dienstleistungen, die sich vor allem auf den Leistungserstellungsprozess bezieht und aus der nicht oder nur gering planbaren Integration des externen Faktors herrührt. Für Bankleistungen gilt gar, dass diese per se kaum definitive Bestandteile aufweisen.[415]

---

410 Vgl. *Büschgen, H. E.* (1998), S. 311.

411 *Büschgen, H. E.* (1998), S. 311.

412 Vgl. *Galasso, G.* (1999), S. 99 f.

413 Vgl. *Reckenfelderbäumer, M.* (2005), S. 35.

414 Vgl. *Segbers, K.* (2007), S. 82.

415 Vgl. *Schöning, S.* (2003), S. 1043.

Wie bereits ausgeführt, sind die genannten und bisher erläuterten Merkmale nicht als trennscharfe Abgrenzungskriterien zu verstehen, sondern spannen gemeinsam einen mehrdimensionalen Raum auf, in dem Dienst- und Sachleistungen entsprechend ihrer Merkmalsausprägungen positioniert werden können. Es ist jedoch keine eindeutige Aussage darüber möglich, wo Dienstleistungen anfangen und Sachleistungen aufhören.[416] Aufbauend auf der angebotsorientierten,[417] zweidimensionalen Typologie von ENGELHARDT/KLEINALTENKAMP/ RECKENFELDERBÄUMER[418] mit den Dimensionen der Stofflichkeit des Leistungsergebnisses (Ergebnisdimension) und der Integrationsnotwendigkeit des externen Faktors (Prozessdimension) positioniert SEGBERS beispielhaft ausgewählte Bankprodukte (siehe Abbildung 13). SEGBERS weist darauf hin, dass die Einordnung von Bankleistungen mit einer gewissen Willkür verbunden ist und daher lediglich Tendenzaussagen darstellen kann.[419]

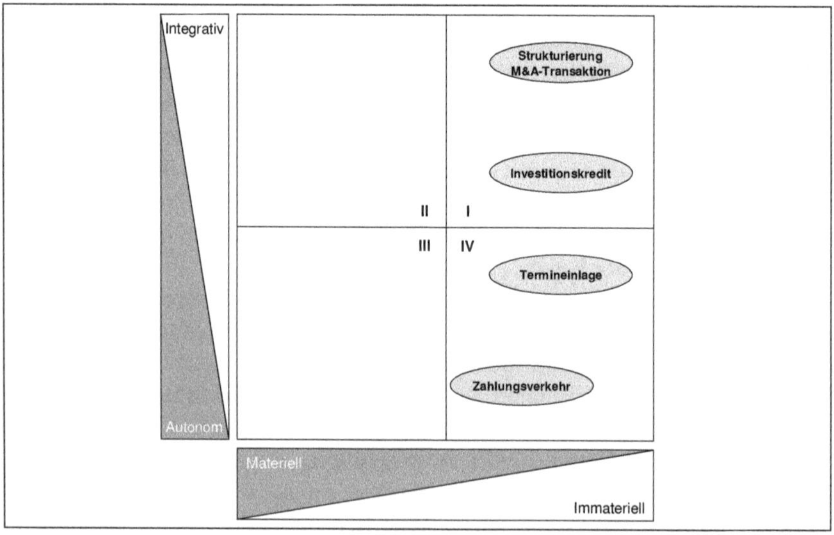

Abbildung 13: Exemplarische Leistungstypologie von Bankdienstleistungen[420]

416 Vgl. *Reckenfelderbäumer, M.* (2005), S. 34.

417 Als angebotsorientierte Typologien werden diejenigen Gütertypologien verstanden, die die Anbieterseite in den Fokus der Typologie stellen. Für eine Übersicht von angebotsorientierten Gütertypologien im Industriegütermarketing vgl. *Backhaus, K.* (1999), S. 284 ff.

418 Vgl. *Engelhardt, W. H./Kleinaltenkamp, M./Reckenfelderbäumer, M.* (1993).

419 Vgl. *Segbers, K.* (2007), S. 84.

420 *Segbers, K.* (2007), S. 84 sowie die dort genannte Literatur.

Grundsätzlich kann für die Bankleistungen ausgeführt werden, dass diese tendenziell einen größeren Anteil immaterieller Eigenschaften aufweisen. Jedoch wäre es unzulässig, alle Bankleistungen als intagibel zu verallgemeinern.[421] Hinsichtlich der Integrationsnotwendigkeit des externen Faktors ist dagegen eine tendenzielle Aussage unmöglich.

Die Verhaltensunsicherheit im Kontext der Dienstleistungen begründet sich insbesondere in ihrer Immaterialität und kann aus Perspektive des Anbieters und aus Perspektive des Nachfragers betrachtet werden. Aus Anbietersicht betrifft dies vor allem die Unsicherheit darüber, wie der Leistungserstellungsprozess ablaufen wird und zu welchen konkreten Ergebnissen er führt.[422] Es resultieren im Firmenkundenkreditgeschäft bspw. die intensiv untersuchten Probleme aus Informationsasymmetrien, die sich im theoretischen Kontext der Prinzipal-Agenten-Theorie in „adverse selection“, „moral hazard“ oder „hold up“ konkretisieren.[423]

Aus Sicht des Nachfragers sind es vor allem die Beschaffungsunsicherheiten, die zu thematisieren sind. Es wird unterstellt, dass mit Zunahme der Immaterialität und Integrationsnotwendigkeit das Kaufrisiko des Nachfragers steigt.[424] Als Möglichkeit diesen Sachverhalt für die Bankprodukte anschaulich zu strukturieren, bieten sich die informationsökonomischen Ansätze der Marketingtheorie an. Diese knüpfen an den Unsicherheiten im Transaktionsprozess an und leiten in Abhängigkeit der konkreten Leistung bzw. deren informationsökonomischer Eigenschaften sowie dem Beurteilungsverhalten des Käufers und den Beurteilungsmöglichkeiten des Leistungsangebots variable Informationsdefizite auf Seiten des Nachfragers ab.[425] Diese Informationsdefizite bedingen Informationsbeschaffungskosten, welche sich erhöhend auf die Transaktionskosten auswirken und damit die Transaktionsprozesse selbst in unterschiedlicher Weise beeinflussen. Ein auf diesen Erkenntnissen zurückgehender, nachfrageorientierter Typologisierungsansatz wurde von WEIBER/ADLER entwickelt.[426] Danach können Leistungen in Abhängigkeit ihres subjektiv wahrgenommenen Gehalts hinsichtlich ihrer Such-, Erfahrungs- und Vertrauenseigenschaften unterschiedlich typologisiert und der jeweilige Typ als Entscheidungsgrundlage für marke-

---

421 Vgl. *Nader, G.* (1995), S. 10.

422 Vgl. *Nader, G.* (1995), S. 11.

423 Eine pointierte Erörterung dieser, aus Informationsasymmetrien resultierenden Problemfelder im Kontext des Firmenkundengeschäfts von Banken findet sich bei *Püthe, T.* (2009), S. 52 ff.

424 Vgl. *Segbers, K.* (2007), S. 85.

425 Vgl. *Meffert, H.* (1998), S. 23.

426 Vgl. hier und nachfolgend *Weiber, R./Adler, J.* (1995a), S. 54 ff.; *Weiber, R./Adler, J.* (1995b), S. 61 ff.

tingpolitische Maßnahmen genutzt werden. Sucheigenschaften sind dadurch gekennzeichnet, dass sie vom Nachfrager vor dem Kauf vollständig durch Inspektion des Leistungsangebots oder entsprechende Informationssuche beurteilt werden können. Die ex ante stattfindende Suche wird abgebrochen, wenn der Informationsstand durch den Nachfrager als ausreichend oder die weitere Informationssuche als zu kostspielig angenommen wird. Erfahrungseigenschaften sind dagegen dadurch charakterisiert, dass die Beurteilung erst nach dem Kauf erfolgt. Ursächlich dafür ist, dass die Beurteilung entweder erst nach dem Kauf möglich ist oder aufgrund der subjektiven Wahrnehmung des Käufers bewusst auf Erfahrung bei Ge- bzw. Verbrauch eines Produkts verlagert wird. Abweichend von den erstgenannten Eigenschaften zeichnen sich die Vertrauenseigenschaften dadurch aus, dass sie durch den Nachfrager weder vor noch nach dem Kauf vollständig beurteilt werden können. Dies begründet sich in der Annahme, dass es dem Nachfrager am entsprechenden Know-how zur sachgerechten Beurteilung mangelt und dieser es nicht in vertretbarer Zeit aufbauen kann bzw. will oder die Kosten der Beurteilung subjektiv als prohibitiv einschätzt. Auf Basis dieser Eigenschaftsdimensionen können Bankdienstleistungen im so genannten informationsökonomischen Dreieck positioniert werden (siehe Abbildung 14).

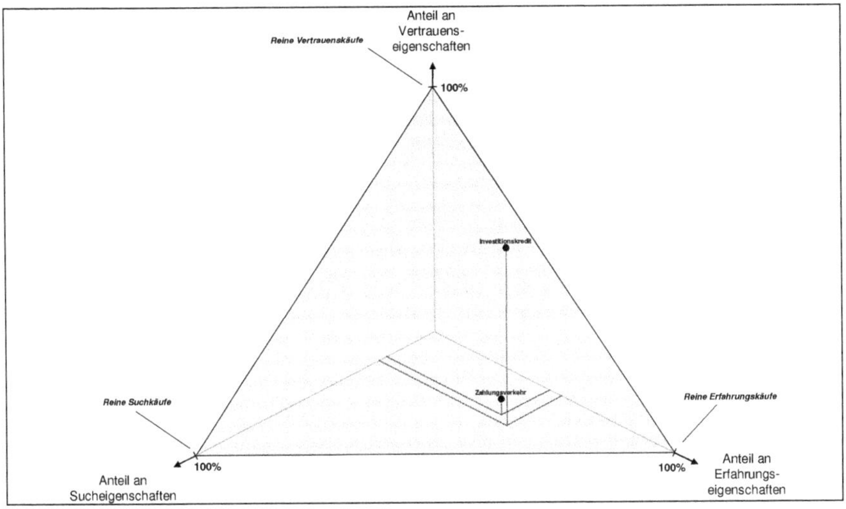

Abbildung 14: Einordnung ausgewählter Bankleistungen im informationsökonomischen Dreieck[427]

Grundsätzlich gilt, dass Bankleistungen einen geringen Anteil an Sucheigenschaften und hohe Anteile an Vertrauens- und Erfahrungseigenschaften aufweisen.[428] Bei der Einordnung der Leistungen sind jedoch zwei wesentliche Aspekte zu beachten. Erstens ist bei konsequenter Anwendung des Ansatzes das Leistungsprogramm aufgrund der hohen Bedeutung der subjektiven Wahrnehmung kundenindividuell zu positionieren. Diese kann sich hinsichtlich der einzelnen Leistungsbestandteile deutlich unterscheiden, wobei auch der zweite Punkt angesprochen ist. Da Dienstleistungen aus verschiedenen Leistungskomponenten bestehen (z. B. Preis, Bearbeitungsdauer, Umsetzungsqualität und ggf. Leistungsstörungsbearbeitung im Sinne des Beschwerdemanagements) orientiert sich eine Positionierung an einer dekompositionellen Analyse der betrachteten Leistung. So lässt sich regelmäßig für einen Investitionskredit hinsichtlich der einzelnen Leistungsbestandteile aus Sicht des Unternehmens feststellen, dass z. B. der Zinssatz einen hohen Anteil Sucheigenschaften hat,[429] während die ver-

[427] Darstellung auf Basis WEIBER/ADLER und in Anlehnung an SEGBERS in Bezug auf die exemplarische Einordnung der Bankprodukte. Vgl. *Weiber, R./Adler, J.* (1995a), S. 63 und *Segbers, K.* (2007), S. 85.

[428] Vgl. *Nader, G.* (1995), S. 11.

[429] Dies gilt insbesondere seitdem die Transparenz erhöhenden Vergleichsmöglichkeiten durch die Fortschritte in der Informations- und Kommunikationstechnologie die Vergleichbarkeit schnell und kostengünstig ermöglichen. So ist es grundsätzlich leicht für

tragsgemäße Abwicklung des Kreditvertrags hohe Anteile an Erfahrungseigenschaften und die zieladäquate Strukturierung der Finanzierung selbst hohe Anteile an Vertrauenseigenschaften aufweisen.[430] Aufgrund der kundenindividuell subjektiven Wahrnehmung ist diese Einschätzung jedoch nicht objektiv bestimmt. Tendenziell gilt für Bankleistungen, dass es sich bei ihnen vor allem aufgrund der Immaterialität, Integrativität und der Komplexität häufig um Güter handelt, deren Kaufprozesse als Erfahrungs- oder Vertrauenskäufe zu bezeichnen sind.[431] Daraus resultiert unmittelbar die hohe Bedeutung der so genannten Mund-zu-Mund-Propaganda und persönlicher Informationsquellen bei den vor Vertragsabschluss stattfindenden Informationsbeschaffungen.[432]

Aufbauend auf den genannten Merkmalen von Dienstleistungen im Allgemeinen und Bankleistungen im Besonderen lassen sich mit BRUHN Besonderheiten des Marketing und Vertriebs von Bankleistungen ableiten.[433] Diese lassen sich in dienstleistungsorientierte, extern orientierte und intern orientierte Besonderheiten unterscheiden. Zu den dienstleistungsorientierten Besonderheiten zählt, dass es nur ein geringes Standardisierungspotenzial des Vertriebs gibt und die zentrale Herausforderung in der Sicherstellung einer konstanten Qualität liegt. Extern orientierte Besonderheiten zeigen sich in der Bedeutung dauerhafter Kundenbeziehungen und der hohen Bedeutung des Bankimages als kaufentscheidendem Faktor. Die Relevanz einer hohen Qualifikation und Motivation der Mitarbeiter im Vertrieb ist die zentrale intern orientierte Besonderheit.

Die Betrachtung des im Firmenkundengeschäft der Kreditgenossenschaften relevanten Bankleistungsprogramms basiert auf der gegenwärtig in Umsetzung befindlichen Vertriebskonzeption VR-FINANZPLAN MITTELSTAND.[434] Im Rahmen dieser Konzeption wird unter dem Ansatz der ganzheitlichen Betreuung die aktuelle und potenzielle Bedarfsstruktur des Firmenkunden in einzelne Bedarfsfelder zusammengefasst und diesen Bedarfsfeldern Leistungen der FinanzGruppe zugeordnet. Die Ganzheitlichkeit bezieht sich dabei auf die vollständige angebotsseitige Abdeckung sowohl der gewerblichen als auch der privaten Sphäre des Firmenkunden. Abbildung 15 stellt ausgewählte Bedarfsfelder dar.[435]

Unternehmen, für Kreditanfragen mehrere Kreditinstitute anzusprechen und deren Angebote hinsichtlich des Zinssatzes zu vergleichen.

430 Vgl. *Segbers, K.* (2007), S. 86.

431 Vgl. *Segbers, K.* (2007), S. 87 sowie die dort angegebene Literatur.

432 Vgl. *Galasso, G.* (1999), S. 99; *Nader, G.* (1995), S. 12.

433 Vgl. hier und nachfolgend *Bruhn, M.* (1997), S. 14.

434 Vgl. hier und nachfolgend *BVR et al.* (2006), S. 13.

435 Vgl. für eine Auflistung relevanter Bankleistungen im Firmenkundengeschäft auch *Kasten, L./Reents, B.* (o. J.), S. 28.

Bedarfsfeld	Leistungen	Involvierte Verbundmitglieder
Investition & Finanzierung	Betriebsmittel- und Investitionskredite	Kreditgenossenschaft, Zentralbank
	Mobilien- und Immobilienleasing	VR Leasing Gruppe
	Factoring, Forderungsverkauf	VR Factorem Gruppe
	Bürgschaften	Kreditgenossenschaft, R+V Versicherung
Liquidität & Zahlungsverkehr	Kontokorrentkonto	Kreditgenossenschaft
	Elektronischer Zahlungsverkehr	Rechenzentrale
	Kurzfristige Geldanlagen	Kreditgenossenschaft, Union-Investment
	Dokumentärer Zahlungsverkehr	Zentralbank, R+V Versicherung
Risiko & Absicherung	Sachversicherungen	R+V Versicherung
	Ausfall- und Kautionsversicherungen	VR-Factorem, R+V Versicherung
	Zinssicherungsgeschäfte	Zentralbank
	Währungsabsicherungen	Zentralbank
Betriebliche Altersvorsorge	Direktversicherungen	R+V Versicherung, Compertis
	Pensionsfonds	
	Unterstützungskasse	
	Pensionskasse	
Vermögensanlage (des Unternehmens)	Tages- und Termingelder	Kreditgenossenschaft
	Geldmarktfonds, Fonds	Union Investment
	Rentenpapiere	Zentralbank, Kreditgenossenschaft
	Zertifikate und Aktien	Zentralbank
Unternehmens-management	Brancheninformationen	DG Verlag
	Länderinfos und Research	DG Verlag
	MinD.unternehmer	Zentralbank, Kreditgenossenschaft, Rechenzentrale
Existenzgründung	Gründungsplaner	DG Verlag, Zentralbank
	Geno-Star	Kreditgenossenschaft, Zentralbank
Unternehmensnachfolge	Plan B (Notfallordner)	DG Verlag
	M & A	Zentralbank-Gruppe
Private Finanzplanung (Finanzplanung des Unternehmers)	Private Altersvorsorge	R+V Versicherung, Union-Investment
	Real Estate Management	Kreditgenossenschaft

Abbildung 15: Bankleistungsstruktur des Firmenkundengeschäfts[436]

In diesem Ansatz zeigt sich deutlich sowohl die Ausrichtung als Allfinanzanbieter als auch die bereits diskutierte Verbundorientierung.[437] Die Kreditgenossenschaften beanspruchen damit für sich, alles aus einer Hand anzubieten. Empiri-

436 Eigene Darstellung auf Basis der Angaben in *BVR et al.* (2006), o. S.

437 Vgl. dazu die Ausführungen in Kapitel C.1.4.4.

sche Ergebnisse zeigen jedoch, dass die tatsächliche Nutzungsbreite bei den Produkten, vor allem aber die tatsächlich beobachtbare Bedarfsabdeckung beim Firmenkunden durch die Kreditgenossenschaften, diesen Anspruch (noch) nicht bestätigen kann.[438]

2.4 Vertrieb

Analog zum Vorgehen in Teil B wird bei der Betrachtung des Vertriebs eine funktionale, eine prozessuale und eine institutionelle Perspektive eingenommen. Zunächst wird dazu aus funktionaler Sicht die Darstellung des Drei-Banken-Modells vorgenommen und anschließend werden die transaktionalen und relationalen Vertriebsansätze vorgestellt. Darauf folgend wird der institutionelle Vertrieb im Firmenkundengeschäft erläutert. Abschließend werden ausgewählte Vertriebsprozesse dargestellt.

2.4.1 Funktionaler Vertrieb

2.4.1.1 Vertriebsbank im Drei-Banken-Modell

Der funktionale Vertriebsbegriff im Firmenkundengeschäft ist gekennzeichnet durch eine dem einzelnen Geschäftsfeld übergeordnete Aufteilung der bankbetrieblichen Wertschöpfungsstufen. Unterstützt durch informationstechnologische Innovationen und vor dem Hintergrund des Wettbewerbdrucks setzt sich in Banken der Trend durch, die vollständige Wertschöpfungskette nicht mehr integriert vorzuhalten, sondern sie aufzubrechen und partiell auszulagern.[439] Grundlage derartiger Maßnahmen ist die gedankliche oder tatsächliche Aufteilung der Gesamtbank in drei Teilbanken: die Vertriebs-, die Produktions- und die Steuerungsbank (Drei-Banken-Modell).[440] Ziel der damit verbundenen Trennung von Vertrieb, Produktion und Steuerung ist die Konzentration der Vertriebsmitarbeiter auf das Kundengeschäft, indem „Aufgaben jenseits des direkten Kundenprozesses in den Bereich der Produktionsbank [verlagert werden, Erg. d. Verf.]"[441]. Die Vertriebsbank ist Träger der Vertriebspolitik. Die Abwicklung eines abgeschlossenen Geschäfts ist in der Produktionsbank[442] gebündelt, die entweder gleichzeitig auch Produktanbieter ist oder lediglich die Pro-

---

[438] Vgl. dazu die Ausführungen in Kapitel C.3.3 und dort insbesondere Abbildung 22.

[439] Vgl. *Moormann, J./Schmidt, G.* (2007), S. 21; *Segbers, K.* (2007), S. 113.

[440] Vgl. *BVR et al.* (2006), S. 23.

[441] *BVR et al.* (2006), S. 23.

[442] Synonyme Begriffe, die in der Literatur verwendet werden, sind Transaktionsbank, Back-Office-Fabrik oder Abwicklungsbank. Vgl. *Moormann, J./Schmidt, G.* (2007), S. 21.

duktion im Sinne der weiteren Bearbeitung übernimmt.[443] Die Steuerungsbank übernimmt die Aufgaben der Portfoliosteuerung für die einzelnen Anlageklassen und Geschäftsarten, das Risikomanagement sowie weitere zentrale Aufgaben.[444] Theoretisch stellen diese Teilbanken eigenständige Organisationen dar. Praktisch wird diese Trennung bei den Kreditgenossenschaften innerhalb einer Organisation vorgefunden.[445] Gemeint ist mit der Vertriebsbank dann die kundenorientierte Unternehmensführung in Verbindung mit einem aktiven Vertrieb im Privat- und Firmenkundengeschäft.[446]

Die Vertriebsbank hat die Funktion, umfassend die Kundenbetreuung im Sinne des Kundenmanagement vorzunehmen. Das bedeutet, sie ist für die Anbahnung, Entwicklung sowie für Qualität und Quantität der Geschäftsverbindung mit den Firmenkunden verantwortlich.[447] Dies beinhaltet neben den Aufgaben, die sich auf Transaktionen beziehen, auch Aufgaben im Rahmen der Kreditüberwachung, so z. B. beim Rating. Bei risikorelevanten Kreditgeschäften ist es Aufgabe der Vertriebsmitarbeiter (Markt), eine partielle Entscheidung (Erstvotum) zu verfassen.[448] Diese wird ergänzt durch das Votum der Marktfolge. Abhängig von innerorganisatorischen Strukturen darf ein Kredit nur bei konsensualer Votierung durch die am Vergabeprozess beteiligten Mitarbeiter aus Markt und Marktfolge erfolgen. Daraus ergeben sich Auswirkungen auf den Vertrieb von Kreditprodukten, auf die im Rahmen der prozessualen Betrachtung eingegangen wird. Für Kreditentscheidungen, die unter Risikogesichtspunkten

---

443 Vgl. *Segbers, K.* (2007), S. 114. Zu den Aufgaben der Produktionsbank zählen z. B. die Bereiche Marktfolge Aktiv und Marktfolge Passiv und Organisation. Vgl. *Schax, E.* (2007), Fn. 155 auf S. 39.

444 Dazu zählen z. B. die Bereiche Personal, Bankmanagement, Controlling und Marketing. Vgl. *Schax, E.* (2007), Fn. 155 auf S. 39. Zum Marketing gehören auch die Aufgaben der Preis- und Produktpolitik. Vgl. *BVR et al.* (2006), S. 23.

445 So z. B. bei *Armbruster, H./Hallberg, I.* (2006), S. 59; *Kramer, F.* (2006), S. 58; *Vomhoff, S./Richter, K.* (2006), S. 22; *Richter, K./Sterk, T.* (2004), S. 36; *Wings, H./Benölken, H.* (2004), S. 37.

446 Vgl. *Berhorst, U.* (2008), S. 36. ROLFES spricht in diesem Zusammenhang vom Wandel von der ‚Bestands- zur Vertriebsbank', der sich jedoch aufgrund der genossenschaftlichen Unternehmenskultur nur langsam vollzieht. Diese beurteilt er als Entwicklungshemmnis. Vgl. *Rolfes, B.* (2006), S. 121.

447 Vgl. *Schmoll, A.* (2006), S. 123.

448 Die getrennte Votierung erfolgt nach MaRisk in Abhängigkeit von Art, Umfang, Komplexität und Risikogehalt eines Kreditengagements. Vgl. *Bundesanstalt für Finanzdienstleistungsaufsicht* (2010), BTO 1.1, Ziffer 2. Die Konkretisierung der Risikorelevanzgrenzen ist von den Kreditgenossenschaften im Rahmen der aufsichtsrechtlichen Vorgaben individuell festzulegen.

als nicht wesentlich zu beurteilen sind, kann institutsintern festgelegt werden, dass nur ein Votum ausreichend ist.[449]

Weiterhin ist der Funktionsumfang der Vertriebsbank abhängig von der zugrundeliegenden Vertriebsorientierung, die sich zwischen den Kategorien des transaktionalen und des relationalen Vertriebs bewegt. Diese beiden Kategorien sollen nachfolgend erläutert werden. Ausgangspunkt der Überlegungen ist der Ansatz, dass Marketing – und damit auch der Vertrieb als integraler Bestandteil – die marktorientierte Ausgestaltung von Austauschbeziehungen ist.[450] Sämtliche Austauschbeziehungen zwischen der Bank und dem Firmenkunden bewegen sich im Spannungsfeld zwischen diskreten Einzeltransaktionen und transaktionsübergreifenden Geschäftsbeziehungen.

2.4.1.2 Transaktionale Vertriebsansätze

Die transaktionalen Vertriebsansätze leiten sich aus dem „transaction marketing" ab und werden mit Blick auf den bankbetrieblichen Kontext auch als „transaction banking"[451] oder „transactional banking" bezeichnet.[452] Sie sind darauf ausgerichtet, mit bestimmten Kunden einmalige, zeitpunktbezogene und im Wesentlichen auf den ökonomischen Inhalt beschränkte Leistungsaustausche (Transaktion) vorzunehmen.[453] Folglich liegt der Erfolgsmaßstab bei den transaktionalen Ansätzen im Verkauf und dem damit unmittelbar verbundenen ökonomischen Transaktionserfolg. Das Kundenverständnis ist dadurch geprägt, dass dieser weitgehend anonym ist und Unabhängigkeit zwischen Firmenkunden und Bank besteht.[454] Faktoren der inneren Bindung zwischen Firmenkunde und Bank wie z. B. die persönliche Sympathie zwischen den beteiligten Personen sind von inferiorer Bedeutung. Ziel dieses Ansatzes ist die kurzfristige und dependenzfreie Erzielung von Umsatzerfolgen bzw. der Kundenakquisition.[455] Zwar sind

---

449 Vgl. *Bundesanstalt für Finanzdienstleistungsaufsicht* (2010), BTO 1.1, Ziffer 4.

450 Vgl. *Cornelsen, J.* (2000), S. 14.

451 Darüber hinaus wird der Begriff des "transaction banking" im Bankenkontext auch verwendet, um transaktionsbezogene Aktivitäten in der Produktionsbank (synonym Transaktionsbank) zu bezeichnen. Diese liegen dabei jedoch außerhalb des Kundenkontakts. Dazu gehören etwa die Back-Office-Tätigkeiten in der Zahlungsverkehrs- und Wertpapierabwicklung. Vgl. *Hepp, C.* (2008), S. 50. Da diese Doppeldeutigkeit vorliegt, wird im Weiteren von transaktionalen Vertriebsansätzen gesprochen.

452 Vgl. *Segbers, K.* (2007), S. 138; *Flacke, K.* (2006), S. 66; *Degryse, H./Ongena, S.* (2004), S. 32; *Töfferl, S.* (1997), S. 123.

453 Vgl. *Lütje, S.* (2009), S. 79; *Cornelsen, J.* (2000), S. 16.

454 Vgl. *Galasso, G.* (1999), S. 107.

455 Vgl. *Diller, H.* (2006), S. 99; *Lütje, S.* (2009), S. 79.

derartige Austauschbeziehungen in ihrer reinen Form selten anzutreffen,[456] dennoch sind die tendenzielle Ausrichtung und die zugrundeliegende Intention von Relevanz für die Marktbearbeitung. Der transaktionale Ansatz fokussiert auf den konkreten aktuellen Geschäftsabschluss und betrachtet folglich nicht mögliche künftige Transaktionen zwischen den beiden Akteuren.[457] Der transaktionale Ansatz kann gleichgesetzt werden mit einem produktorientierten Vertrieb. „Der Kunde interessiert in erster Linie im Zusammenhang mit dem Verkauf oder Abschluß eines Geschäfts."[458]

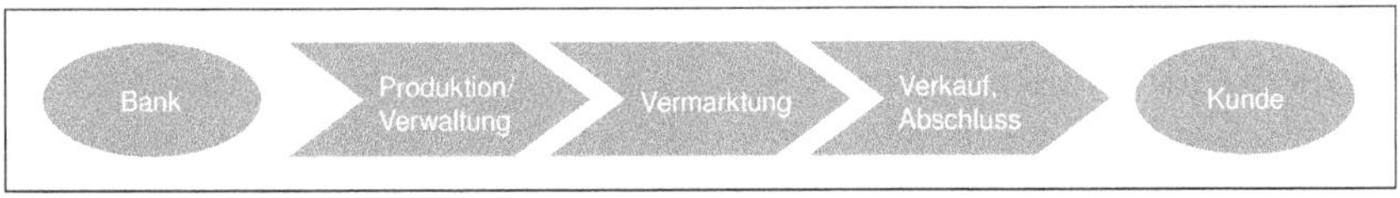

Abbildung 16: Produktorientierung im Sinne der transaktionalen Vertriebsansätze[459]

Die transaktionalen Ansätze haben im Firmenkundengeschäft die Tendenz, dass sie am Bringgeschäft des Kunden oder an Kampagnen ausgerichtet sind. Im ersten Fall liegen die Aufgabenschwerpunkte in der reaktiven Bearbeitung von Kundenanfragen. Der Aufgabeninhalt beschränkt sich i. d. R. auf das für die konkrete Produktanfrage notwendige Informationsgespräch mit dem Firmenkunden, der folgenden Abstimmung im Hinblick auf die Produkteigenschaften und der Koordination der Integration des externen Faktors in den Dienstleistungserstellungsprozess. Alle weiteren Aufgaben der Dienstleistungsproduktion sind in der Produktionsbank angesiedelt. Bei den Kampagnen findet im Vorfeld eine Selektion für den Produktabsatz affiner Kunden statt, die darauf hin von den Firmenkundenberatern gezielt auf diese Produkte angesprochen werden. In diesem Fall übernimmt der Vertrieb neben den bereits genannten Aufgaben im Kontext des Bringgeschäfts auch akquisitorische Aufgaben.

Nach SEGBERS können die transaktionalen Ansätze aus Perspektive des strategischen Managements mit der Preisstrategie im PORTER'schen Sinn gleichgesetzt werden. In diesem Fall ist konsequenterweise der Preis respektive das vom Firmenkunden wahrgenommene Preis-Leistungs-Verhältnis das zentrale Kaufkriterium.[460] Die Entscheidung zugunsten der transaktionalen Ansätze sieht er

456 Vgl. *Backhaus, K./Aufderheide, D./Späth, G.-M.* (1994), S. 28 ff.; *MacNeil, I. R.* (1979), S. 60.

457 Vgl. *Vogel, V.* (2006), S. 9.

458 *Wiedemann, A.* (1992), S. 116.

459 In Anlehnung an *Wiedemann, A.* (1992), S. 116.

460 Vgl. *Segbers, K.* (2007), S. 138.

unter Rekurs auf die theoretische Arbeit von BOOT/THAKOR und den empirischen Beiträgen von ELSAS und DEGRYSE/OGENA in Abhängigkeit des wettbewerblichen Umfelds.[461] Aufbauend auf den Ergebnissen dieser Arbeiten kommt SEGBERS zu dem Fazit, dass auf Kreditmärkten, auf denen Banken als relevante Wettbewerbsakteure agieren, mit steigender Wettbewerbsintensität die Wahrscheinlichkeit der Auswahl transaktionaler Ansätze sinkt. Begründet wird dies mit der Möglichkeit der Banken, sich über beziehungsorientierte Ansätze vom Wettbewerb zu differenzieren. Allerdings läuft dieser Ansatz nicht zwingend streng monoton ab, so dass ab einer bestimmten Wettbewerbsintensität die Bereitschaft zu transaktionalen Ansätzen wieder zunimmt.[462] RECHT/HOLM argumentieren, dass die transaktionalen Vertriebsansätze, bei denen „Banken nur einzelgeschäftsbezogen mit ihren Kunden in Geschäftsbeziehung treten [.] für Universalbanken weitgehend zugunsten der auf Dauer angelegten Geschäftsbeziehung"[463] verdrängt wurden.

Die in Kapitel C.1.5 beschriebene Wettbewerbssituation auf dem deutschen Bankenmarkt im Firmenkundengeschäft und die Tatsache, dass es sich bei den Kreditgenossenschaften um Universalbanken handelt, fundieren, dass seitens des BVR dem Beziehungsmanagement, mithin also dem relationalen Vertrieb eine größere Bedeutung beigemessen wird.[464]

2.4.1.3 Relationale Vertriebsansätze

Den relationalen (beziehungsorientierten) Vertriebsansätzen liegt das Vorhandensein oder Erreichen wollen einer langfristigen Austauschbeziehung, mithin also einer Geschäftsbeziehung zwischen Firmenkunde und Bank als zentrale Zielsetzung zugrunde.[465] Die Definitionen einer Geschäftsbeziehung von DILLER und PLINKE integrierend leitet SEGBERS eine Begriffsdefinition zum Zweck der Analyse der Geschäftsbeziehung zwischen einer Bank als Anbieter und einem Firmenkunden als Nachfrager von Bankleistungen her.[466] Danach ist eine Ge-

---

461 Vgl. *Boot, A. W. A./Thakor, A. V.* (2000); *Elsas, R.* (2005); *Degryse, H./Ongena, S.* (2004).

462 Vgl. *Segbers, K.* (2007), S. 138 ff.

463 *Recht, A./Holm, R.* (2005), S. 136.

464 Vgl. *BVR* (1999b), S. 28; *BVR et al.* (2006), S. 30.

465 Zu den damit verbundenen ökonomischen Vorteilen vgl. Kapitel D.4.5.2.

466 Vgl. *Segbers, K.* (2007), S. 10 ff. Die explizite Nennung der Rollen in den betrachteten Geschäftsbeziehungen geschieht vor dem Hintergrund, dass es realiter vorkommen kann, dass ein Unternehmen sowohl Kunde der Bank als auch Anbieter für die Bank ist. So ist es z. B. bei Cateringfirmen möglich, dass diese bei einer Volksbank Zahlungsverkehrsleistungen nachfragen und gleichzeitig die Versorgung von Cateringleistungen bei Generalversammlungen vornehmen.

schäftsbeziehung eine an ökonomischen Zielen ausgerichtete Folge von Interaktionsprozessen zwischen Unternehmen mit personalen Kontakten, zwischen denen eine innere Verbindung besteht, die auf wechselseitigen, aber nicht unbedingt symmetrischen Bindungen zwischen den Interaktionspartnern basiert.[467] Die Folge von Interaktionsprozessen bezieht sich unter anderem auf die bewusst intendierte Wiederholung von Einzeltransaktionen. Kennzeichnend für Geschäftsbeziehungen ist, dass sie dynamische Gebilde sind, deren Strukturen, Inhalte und Intensität im Zeitablauf variieren können.

Der Blick auf langfristige Geschäftsbeziehungen wird in der wissenschaftlichen Literatur im Anwendungskontext des Firmenkundengeschäfts der Banken vor allem mit Blick auf die aus Informationsasymmetrien resultierenden Probleme in Form von „adverse selection", „moral hazard„ und „hold up" vorgenommen.[468] In diesem Zusammenhang gilt der Aufbau einer Geschäftsbeziehung als Möglichkeit, die genannten Risiken zu reduzieren oder gar zu vermeiden, da in Abhängigkeit der Dauer und Intensität der Geschäftsbeziehung proprietäre Informationen des Firmenkunden (Agenten) gewonnen werden können, die helfen, die Informationssymmetrien zugunsten der Bank (Prinzipal) abzubauen.[469] Dies gilt als wichtiger ökonomischer Vorteil einer Geschäftsbeziehung, da insbesondere für kleine und mittlere Unternehmen – also den typischen Firmenkunden der Kreditgenossenschaften – gilt, dass es für eine Bank aufgrund der nur in einem geringem Ausmaß vorliegenden Publizitätspflicht dieser Unternehmen aufwendig ist, relevante Informationen über sie zu beschaffen.[470] Die hier betrachteten beziehungsbezogenen Ansätze werden mit verschiedenen Begriffen in der bankbetrieblichen Literatur behandelt, so z. B. als „relationship lending", „relationship banking" oder „Hausbankprinzip".[471]

Das „relationship lending" fokussiert sehr eng auf die Kreditbeziehung zwischen Unternehmen und Kreditgeber.[472] Es weist aufgrund des fehlenden Bankbezugs einen höheren Neutralitätsgrad auf als das „relationship banking", welches hinsichtlich der Leistungsdimension, d. h. der Frage welche Leistungen in

---

467 Vgl. *Segbers, K.* (2007), S. 12.

468 Vgl. dazu ausführlich *Püthe, T.* (2009), S. 52 ff.

469 Vor diesem Hintergrund lassen sich diese beziehungsbezogenen Ansätze als konkrete Ausgestaltungen der übergeordneten Theorie der Finanzintermediation verstehen, in der Banken effiziente Institutionen zur Verringerung von Informationsasymmetrien zwischen dem Kollektiv der Einlegenden und Ausleihenden sind.

470 Vgl. *Baas, T./Schrooten, M.* (2006), S. 127.

471 Vgl. für eine ausführliche Auseinandersetzung mit den verschiedenen Begriffen *Püthe, T.* (2009), S. 40 ff. und *Elsas, R.* (2001), S. 12 ff.

472 Vgl. *Berlin, M. S.* (1996), S. 1; *Petersen, M. A./Rajan, R. G.* (1994), S. 5. Ähnlich wird der Begriff des „relationship financing" definiert. Vgl. *Aoki, M./Dinç, S.* (1997), S. 5.

der Analyse des Austausches zwischen Bank und Firmenkunde betrachtet werden, weiter gefasst ist.[473] MORIARTY/KIMBALL/GAY definieren relationship banking als "recognition that a bank can increase its earning by maximizing the profitability of the total customer relationship over time, rather than seeking to extract the most profit from any one individual or transaction."[474] Grundlage der Ertragsmaximierung ist, dass die Bank in die Beschaffung der Informationen vor und während der Geschäftsbeziehung investieren muss und die Profitabilität dieser Investitionen anhand von mehrfachen Interaktionen über die Zeit und über das angebotene Produktprogramm bewertet.[475] Im Fokus steht dabei die intertemporale Wiederverwendung von Informationen.

Der Begriff der Hausbankbeziehung ist eng verbunden mit den bereits erläuterten Begriffen und resultiert aus den engen Beziehungen zwischen Unternehmen und Banken auf dem deutschen Kapital- bzw. Kreditmarkt.[476] Als konstitutiv für eine Hausbankbeziehung gilt, dass die Hausbank den dominanten Anteil an der Kreditfinanzierung des Firmenkunden hat und auch die meisten sonstigen Finanzdienstleistungen durchführt, die Beziehung auf Langfristigkeit ausgelegt ist, in der sich ein „besonderes Vertrauensverhältnis" entwickelt, die Bank einen Informationsvorteil gegenüber Wettbewerbern hat und in Unternehmenskrisen eine „besondere Verantwortung" trägt.[477] Die Existenz der Hausbankbeziehung wird wie beim „relationship banking" mit dem besonderen Informationsvorteil der Hausbank begründet.[478] Dieser Informationsvorteil wird auch als schwer imitierbares und in weiten Teilen implizites Kundenwissen bezeichnet.[479]

Die den Ansätzen gemeinsame Dominanz des Faktors Information begründet sich darin, dass überwiegend das Kreditgeschäft analysiert wird, insofern also aus Bankensicht eine risikoorientierte Beurteilung und Verwendung von Informationen stattfindet.[480] Weniger im Vordergrund stehen die Betrachtung vertriebsrelevanter Informationen und die Zielsetzung der vertriebsmotivierten (selektiven) Kundenbindung, die den Ansätzen des Beziehungsmarketing inhärent ist.[481] Zwar wird mit den intertemporalen Dependenzen von Erlösströmen

---

473 Vgl. *Boot, A. W. A.* (2000), S. 9.

474 *Moriarty, R. T./ Kimball, R. C./Gay, J. H.* (1983), S. 4.

475 Vgl. *Boot, A. W. A.* (2000), S. 10.

476 Aufgrund der hohen inhaltlichen Kongruenz werden die Begriffe des relationship banking und der Hausbank auch häufig synonym verwendet. Vgl. *Püthe, T.* (2009), S. 48; *Segbers, K.* (2007), S. 122; *Flacke, K.* (2006), S. 66.

477 Vgl. *Fischer, K.* (1990), S. 3.

478 Vgl. *Segbers, K.* (2007), S. 120.

479 Vgl. *Fehl, U./Kuhn, N.* (2005), S. 196.

480 Vgl. *Boot, A. W. A.* (2000), S. 8.

481 Vgl. *Diller, H.* (2006), S. 100.

argumentiert, was insbesondere die „besondere Verantwortung“ der Hausbank begründet, die in wirtschaftlich schwierigen Zeiten des Firmenkunden auf Gewinne verzichtet, um in wirtschaftlich besseren Zeiten an Mehrgewinnen zu partizipieren.[482] Welche Konsequenzen diese Annahme jedoch für den Vertrieb hat und mit welchen Instrumentarien eine derartige Entscheidung fundiert wird, bleibt weitgehend offen.[483] Weiterhin nicht betrachtet wird die Situation eines wirtschaftlich gut situierten Unternehmens, dass aufgrund dessen eher Akquisitionsbemühungen von Wettbewerbern attrahiert. Damit verschärft sich die Wettbewerbssituation für die Hausbank und kann aufgrund der allgemeinen Erwartungshaltung hinsichtlich der langfristigen Profitabilität der Geschäftsbeziehung in einem Preiswettbewerb münden.[484] Aufbauend auf den bestehenden beziehungsorientierten Analyseansätzen wird im Weiteren der Blickwinkel auf das Beziehungsmarketing als Grundlage der relationalen Vertriebsansätze ausgeweitet. Von zentraler Bedeutung ist auch bei diesen Ansätzen das implizite Kundenwissen, das aus kontinuierlichen Interaktionen resultiert.[485] Im Gegensatz zu den transaktionalen Ansätzen ist der Erfolgsmaßstab bei den relationalen Vertriebsansätzen definiert als der ökonomische Vorteil einer gesamten Geschäftsbeziehung,[486] der unter anderem durch die Tiefe und Breite der Kundendurchdringung determiniert wird.[487] Insbesondere dieser Aspekt begründet auch die hohe Bedeutung der relationalen Ansätze für die Kreditgenossenschaften. Dabei ist zu konstatieren, dass den relationalen Vertriebsansätzen aufgrund ihrer grundsätzlichen Ausrichtung eine höhere Kundenorientierung als grundlegende Philosophie innewohnt. Im direkten Kontrast zum produktorientierten Ansatz lässt sich dies wie folgt darstellen (siehe Abbildung 17).

---

482 Vgl. *Fischer, K.* (1990), S. 46 f.

483 Zudem ist die Bestimmung eines erfolgsorientierten Zeitwerts als Saldo barwertiger, erwarteter Zahlungsgrößen im Falle des Problemkreditmanagements in praxi nicht vorkommend. Vgl. *Rübenacker, C.* (2009), S. 135 und S. 250.

484 Diese Entwicklung wird durch die Zunahme öffentlich zugänglicher Informationen wie z. B. über das elektronische Handelsregister im Internet oder spezifische Unternehmensdatenbanken verstärkt.

485 Vgl. *Lütje, S.* (2009), S. 79.

486 Vgl. *Diller, H.* (2006), S. 99; *Wiedemann, A.* (1992), S. 119.

487 Vgl. *Galasso, G.* (1999), S. 107.

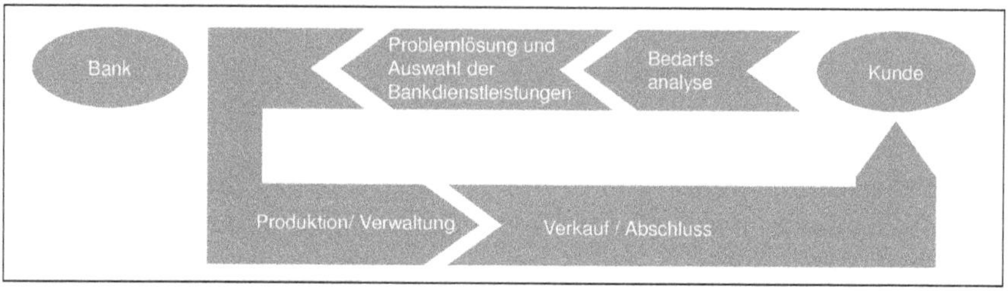

Abbildung 17: Kundenorientierung im Sinne der relationalen Vertriebsansätze[488]

Ausgangspunkt der folgenden Ausführungen ist, dass zum einen aus der intensiven Geschäftsbeziehung zum Firmenkunden dieser eher bereit ist, umfangreiche Informationen in Qualität und Quantität bereitzustellen[489] und zum anderen eine Geschäftsbeziehung, die auf mehreren verschiedenen Produktbeziehungen basiert, als stabiler gilt, da wesentliche Bindungsfaktoren wie Wechselkosten stärker zum Tragen kommen. In einer schematischen Darstellung illustriert DUHNKRACK die Ergebnisstruktur im Firmenkundengeschäft (siehe Abbildung 18) und verdeutlicht dabei, dass die Profitabilität des Firmenkundengeschäfts nicht aus dem originären Bankgeschäft, nämlich dem Firmenkreditgeschäft resultiert, sondern aus dem Provisions-, Dienstleistungs- und Passivgeschäft.[490]

---

488 In Anlehnung an *Wiedemann, A.* (1992), S. 116.

489 Vgl. *Boot, A. W. A.* (2000), S. 12.

490 Die in den letzten Jahren zunehmende Bedeutung des Provisionsgeschäfts für die Kreditgenossenschaften insgesamt ist die grundlegende Ausgangshypothese von WIEDEMANN. Vgl. *Wiedemann, A.* (1992).

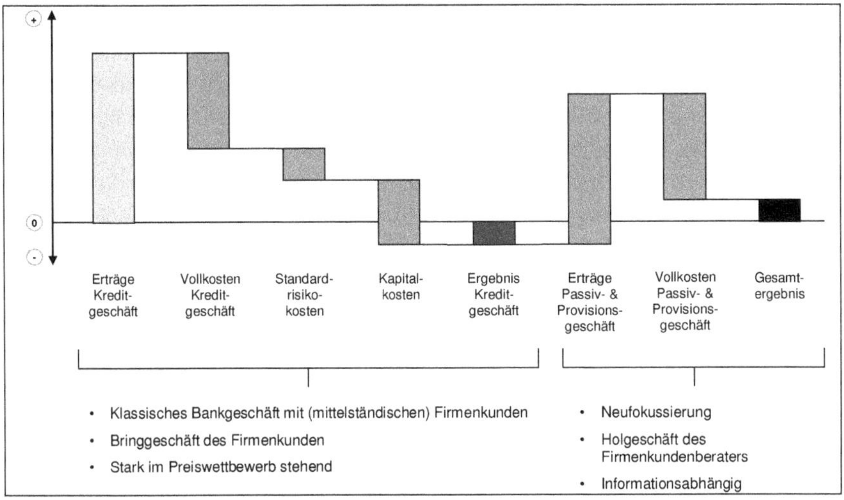

Abbildung 18: Illustrative Ergebnisstruktur im Firmenkundengeschäft[491]

Für die Ableitung der Vertriebsaufgaben in den relationalen Ansätzen ist nun zu berücksichtigen, dass die Firmenkunden das Leistungsprogramm der Kreditgenossenschaften gegenwärtig noch nicht vollumfänglich in Anspruch nehmen.[492] Um dieser Situation entgegen zu wirken und die Breite der Kundendurchdringung zu erhöhen, ist es für den Vertrieb notwendig, die relevanten Bedarfe des Firmenkunden zu identifizieren und institutseigene Leistungen proaktiv anzubieten. Dazu müssen bedarfs- und vertriebsrelevante Informationen vom Kunden beschafft werden. Die Bedeutung dieses Vorgehens wird unterstrichen durch die, aufgrund des intensiven Wettbewerbs geringen Möglichkeiten der Neukundenakquisition. Damit geht einher, dass organische Wachstumsquellen im nennenswerten Umfang nur in der Kundenpenetration, also der Ausschöpfung der Tiefe und Breite der Kundendurchdringung im Kundenstamm liegen. Vor diesem Hintergrund lassen sich zentrale Aufgaben des Vertriebs im Firmenkundengeschäft ableiten, die zusätzlich zu den beschriebenen Aufgaben, die auch bei den transaktionalen Ansätzen anfallen, zu erfüllen sind. Diese konkretisieren sich darin, dass erstens die langfristige Bindung des Kunden an die Bank forciert wird und zweitens die Suche nach Zusatzgeschäften zur Steigerung der Profitabilität der Geschäftsbeziehung zentrale Aufgabe der Firmenkundenberater ist. Implizit beinhaltet dies also eine Abkehr vom kreditfokussierten Bringgeschäft

491 In Anlehnung an *Duhnkrack, T.* (2002), S. 154.

492 Vgl. z. B. die empirische Untersuchung der *WGZ BANK* (2008).

der Kunden, das initiiert wird durch Finanzierungsanfragen des Unternehmens, hin zu einem Holgeschäft durch den Vertrieb der Kreditgenossenschaften, das mit stärker durch die Bank initiierten Interaktionen zwischen Bank und Firmenkunden einhergeht. Die bedeutendsten Aufgaben dabei sind die Koordination der Geschäftsbeziehung mit dem Firmenkunden sowie die Beschaffung relevanter Informationen zur Identifikation zusätzlicher Geschäfte. Eingebettet ist dies in einen inhaltlich weiter gefassten Ansatz des Beziehungsmanagements,[493] der die bankspezifischen Ansätze und das systematische Management individueller Kundenbeziehungen beinhaltet.[494]

Zur Bewältigung dieser Herausforderung hat der BVR dem Vertriebskonzept VR-FINANZ-PLAN MITTELSTAND ein Instrument zur strukturierten Informationsbeschaffung beigefügt, das den Firmenkundenberater in die Lage versetzen soll, die gesamte Bedarfssituation des Kunden systematisch zu analysieren.[495] Konsistent zum Allfinanzangebot in der Produktpolitik wird somit die Vertriebspolitik auf eine „ganzheitliche Finanzberatung"[496] ausgerichtet. Der Vertriebsmitarbeiter ist also zentraler Ansprechpartner in der Bank für die Firmenkunden hinsichtlich aller finanzierungs- und dienstleistungsrelevanten Bedarfe.

2.4.2 Institutioneller Vertrieb

Aufbauend auf dem Grundsatz „Banking ist und bleibt People Business"[497] kann im Firmenkundengeschäft der persönliche Vertrieb in Form des einstufigen Vertriebskanals als die dominante Ausprägung des institutionellen Vertriebs betrachtet werden. Vor diesem Hintergrund gilt es vor allem, der Organisationseinheit bzw. Stelle des Firmenkundenbetreuers oder auch Firmenkundenberaters Beachtung zu schenken. Diese sind unternehmensinterne Vertriebsmitarbeiter. In größeren Kreditgenossenschaften gibt es zudem produktspezifische Experten, da die Komplexität einiger Bankprodukte nicht mehr von einem einzelnen Mitarbeiter beherrscht werden kann.[498]

---

493 Vgl. dazu *Rieker, S.* (1995), S. 42.

494 Vgl. *Diller, H.* (2006), S. 99.

495 Bei diesem Instrument handelt es sich um einen strukturierten Gesprächsleitfaden, der in Anlehnung an die Bedarfsfelder aus Abbildung 15 zum Teil geschlossene und zum Teil offene Fragen beinhaltet. Auf Basis dieser systematisch thematisierten vertriebsrelevanten Gesprächsinhalte soll der Firmenkundenberater in die Lage versetzt werden, die Bedarfssituation des Firmenkunden vollumfänglich einschätzen zu können. Neben dem VR-FINANZPLAN MITTELSTAND werden durch die Kreditgenossenschaften zum Teil standardisierte und zum Teil individuelle Gesprächsleitfäden verwendet.

496 *Dehne, T.* (2008), S. 54.

497 *Cromme, G.* (2008), S. 431.

498 Vgl. *Recht, A./Holm, R.* (2005), S. 133.

Die Aufbauorganisation im Firmenkundengeschäft erfolgt häufig in einer Mischform aus Funktional- und Divisionalorganisation.[499] Funktional werden die Tätigkeiten des Vertriebs aufgeteilt in die direkte Kundenbetreuung und unterstützende Tätigkeiten. Letztere Aufgaben werden oftmals von einer (Vertriebs-)Assistenz übernommen,[500] die häufig mehreren Firmenkundenbetreuern zugeordnet ist, die wiederum divisional nach Zielgruppensegmenten aufgeteilt sind. Häufig erfolgt auch eine innerorganisatorische Trennung nach Leistungsbereichen, bspw. zwischen Produkten für die private und die gewerbliche Sphäre. Die Rolle des Firmenkundenbetreuers können sowohl weisungsgebundene (Rolle = Stelle) als auch weisungsbefugte Mitarbeiter (Rolle als Bestandteil der Stelle Vorstand) innehaben. Die konkrete Zuordnung hängt vom situativen Kontext der Bank ab, und in diesem Zusammenhang vor allem von der Struktur und Menge der Firmenkunden sowie der Größe und geschäftspolitischen Ausrichtung. In Abhängigkeit der Segmentierung werden die Vertriebsmitarbeiter als Firmenkunden- oder Gewerbekundenbetreuer oder -berater bezeichnet. Diesen Stellen bzw. Vertriebsmitarbeitern sind i. d. R. Kunden fest zugeordnet. Firmenkunden werden auch durch Privatkundenbetreuer vertrieblich betreut.[501] Dies erfolgt entweder losgelöst von der Firmenkundenbetreuung, in Form einer Tandembetreuung oder nach dem Verständnis des Firmenkundenberaters als Kundenbeziehungsmanager, der fakultativ Spezialisten in die Betreuungsprozesse integriert. Eine weitere Aufteilung in der Organisation der Firmenkundenbetreuung erfolgt nach Risikoaspekten.[502] So werden risikoreiche Firmenkundenkreditengagements in Spezialabteilungen wie der Intensivbetreuung, der Sanierung oder Abwicklung betreut. Diese Abteilungen sind nicht zwingend dem institutionelle Vertrieb zuzuordnen. Eine Übergabe der Firmenkunden aus der normalen in die Intensivbetreuung erfolgt auf Basis von Kreditausfallindikatoren und wird durch die MaRisk vorgegeben.[503]

Mit Blick auf den Verbund werden weiterhin unternehmensexterne Vertriebsmitarbeiter der jeweiligen Verbundunternehmen als Produktspezialisten in die Vertriebsorganisation integriert. Vor diesem Hintergrund verstärkt sich das Verständnis des Firmenkundenbetreuers als Kundenmanager, der in Abhängig-

---

499 Vgl. *Krauß, C.* (2005), S. 220.

500 Vgl. *Kasselmann, J./Melter, O.* (2006), S. 26; *Krebber, M./Melter, O./Sträter, J.* (2006), S. 42.

501 In diesem Zusammenhang gibt es vom BVR eine Empfehlung, dass erst eine Kreditgenossenschaft mit einer Bilanzsumme über 300 Mio. DM (ca. 153 Mio. EUR) eine hinreichende Betriebsgröße für das qualifizierte Firmenkundengeschäft und den Aufbau einer eigenen Firmenkundenabteilung aufweist. Vgl. *BVR* (1999b), S. 65 ff.

502 Vgl. *Krauß, C.* (2005), S. 230.

503 Vgl. *Bundesanstalt für Finanzdienstleistungsaufsicht* (2010), BTO 1.2.

keit der kommunizierten Bedarfssituation des Firmenkunden auf die spezialisierten Vertriebsressourcen zurückgreift.[504]

Für das kreditbezogene Firmenkundengeschäft ist bedingt durch die Mindestanforderungen an das Risikomanagement (MaRisk), ein weiterer Organisationsaspekt zu beachten. Bei der Aufbauorganisation wird eine bis zur Geschäftsleitung gehende klare Trennung zwischen den so genannten Bereichen Markt und Marktfolge gefordert.[505] Der Bereich Markt ist die Organisationseinheit, die für die Geschäftsinitiierung verantwortlich ist, mithin also im Kundenkontakt steht.[506] Damit ist der Terminus Markt grundsätzlich begrifflich identisch zum institutionellen Vertrieb.[507] Gleichwohl kann der Markt im Kreditgeschäft nicht immer unabhängig von der Marktfolge agieren, so dass diese den Vertrieb bedingt.[508]

2.4.3 Prozessualer Vertrieb

Für die Betrachtung des prozessualen Vertriebs ist voranzustellen, dass die in Kapitel B.1.3 beschriebenen allgemeinen Vertriebsprozesse auch auf den Vertrieb im Firmenkundengeschäft der Kreditgenossenschaften übertragen werden können. Im Folgenden sollen zwei Besonderheiten vorgestellt werden, die sich aus regulatorischen Anforderungen und dem VR-FINANZPLAN MITTELSTAND bzw. artverwandter aktiver Vertriebskonzepte ergeben. Diese sind der für alle Banken geltende, durch aufsichtsrechtliche Vorgaben beeinflusste Kreditprozess und der kreditgenossenschaftlich spezifische Kundenbetreuungsprozess.

Bei den Kreditprozessen sind die Vorgaben der MaRisk bezüglich der ablauforganisatorischen Gestaltung zu berücksichtigen.[509] Unterschieden werden die Teilprozesse der Kreditbearbeitung, welcher die Kreditgewährung und Kreditweiterbearbeitung umfasst, der Kreditbearbeitungskontrolle, der Intensivbetreuung, der Problemkreditbetreuung und der Risikovorsorge. Hinsichtlich der direkten Relevanz für den Vertrieb im Firmenkundengeschäft der Banken ist vor

---

504 Vgl. *Segbers, K.* (2007), S. 77.

505 RUSS unterscheidet darüber hinaus zwischen Markt, Marktfolge und Kreditrisikocontrolling. Das Kreditrisikocontrolling soll als marktunabhängige Einheit eine prozessintegrierte Überwachung der betriebenen Kreditgeschäfte sicherstellen. Vgl. *Russ, A.* (2003), S. 351.

506 Vgl. *Bundesanstalt für Finanzdienstleistungsaufsicht* (2010), BTO Ziffer 2.

507 Vgl. *Strauß, M. R.* (2006), S. 123. Da der Kredit im Firmenkundengeschäft breiten Raum einnimmt, wird in der Praxis häufig der organisatorisch definierte Vertriebsbereich synonym als Markt bezeichnet.

508 Vgl. dazu die Ausführungen in Kapitel C.2.4.1.1 sowie Kapitel C.2.4.3.

509 Vgl. hier und nachfolgend *Bundesanstalt für Finanzdienstleistungsaufsicht* (2010), BTO 1.2.

allem die Kreditbearbeitung zu nennen, da die übrigen Teilprozesse entweder erst zeitlich nachgelagert und nur bei so genannten Leistungsstörungen erfolgen (Intensivbetreuung, Problemkreditbetreuung, Risikovorsorge) oder nur mittelbar auf den Vertrieb von Kreditprodukten Einfluss nehmen (Kreditbearbeitungskontrolle). Gleichwohl sind diese Prozesse mittelbar für den Vertrieb relevant, da Kundenbeziehungen, die in die Intensiv- oder Problemkreditbetreuung überführt werden, außerhalb des in Kapitel C.2.4.2 beschriebenen institutionellen Vertriebs, d. h. dem eigentlichen Marktbereich betreut werden. Wie erinnerlich, ist eine zentrale Aufgabe des Vertriebs, die Produkte der Bank abzusetzen. Vor dem Hintergrund der in den MaRisk enthaltenen Anforderungen muss jedoch für das risikorelevante Kreditgeschäft zwischen einem bedingten Vertrieb und dem eigentlichen Vertrieb unterschieden werden (siehe Abbildung 19).[510] So erfolgt die Geschäftsanbahnung und -vorbereitung durch den Bereich Markt inklusive des so genannten Erstvotums. Diese Aktivitäten können zusammengefasst als bedingter Vertrieb betrachtet werden. Erst nachdem ein gleichlautendes (positives) Zweitvotum durch die Marktfolge vorliegt, wird ein Kredit gewährt und kann somit verkauft werden. Dies kann als eigentlicher Vertrieb verstanden werden. Zu beachten ist jedoch, dass die Marktfolge institutionell und funktionell der Produktions- und nicht der Vertriebsbank zugeordnet ist.

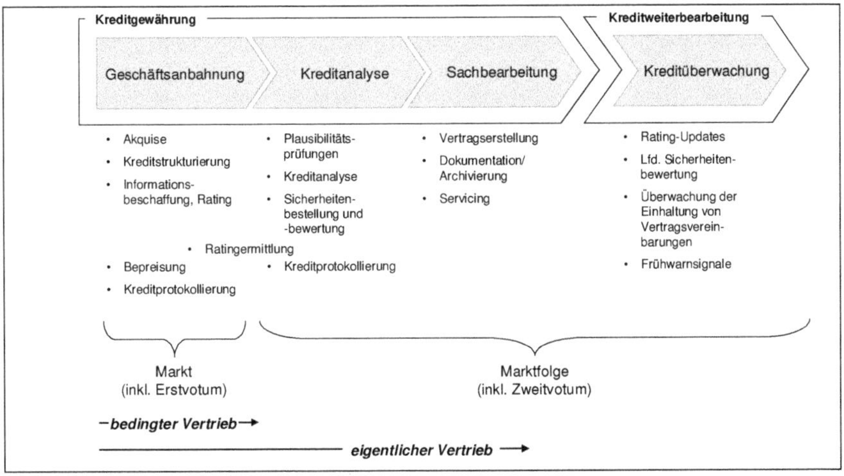

Abbildung 19: Beispielhafter MaRisk-konformer Kreditprozess aus Vertriebssicht[511]

[510] Zum Begriff des risikorelevanten Kreditgeschäfts vgl. Kapitel C.2.4.2.

[511] In Anlehnung an *Berens, W. et al.* (2005), S. 785.

Als besonders relevanter aktiver Betreuungsprozess soll die prozessuale Dimension des VR-FINANZPLAN MITTELSTAND vorgestellt werden.[512] Deren Kern ist ein systematischer Betreuungsprozess, der einen nach (Produkt-)Bedarfsfeldern strukturierten Gesprächsleitfaden für die Kundengespräche beinhaltet (siehe Abbildung 20). Entsprechend den konzeptionellen Vorgaben des Vertriebsansatzes wird mit ausgewählten Kunden ein mindestens zweistufiger Betreuungsansatz umgesetzt.[513] Im ersten zukunftsgerichteten Dialog wird die Ausgangssituation des Firmenkunden auf betrieblicher und privater Ebene besprochen. Darauf aufbauend erstellt der Firmenkundenberater ein Stärken-Schwächen-Profil, das dem Kunden betriebswirtschaftliche Handlungsbedarfe aufzeigt. Diejenigen Handlungsbedarfe, für die das Allfinanzangebot geeignete Lösungen beinhaltet, werden ggf. unter Integration eines Produktspezialisten mit dem Kunden intensiver besprochen und im Idealfall der Produktverkauf realisiert.[514] Um der zu forcierenden Verbundorientierung gerecht zu werden, sieht der Prozess in einzelnen Schritten die fakultative Integration der Verbundpartner als Produktspezialisten vor. Deutlich zu erkennen ist, dass dieser Betreuungsprozess dem grundsätzlichen Schema der relationalen Vertriebsansätze entspricht (siehe Abbildung 17).

---

512 Neben diesen aktiven Vertriebsprozessen sind auch reaktive Vertriebsprozesse zu nennen. Bei diesen kommt der Firmenkunde initiativ auf die Kreditgenossenschaft mit einem manifesten Bedarf zu. Die aktiven Vertriebsprozesse entsprechen im größeren Umfang den akquisitorischen Aufgaben des Vertriebs. Vgl. dazu Kapitel B.1.1.

513 Vgl. hier und nachfolgend *Dehne, T.* (2008), S. 53 ff.

514 Vgl. für eine praxisfallbezogene Beschreibung des Ansatzes *Segbers, K./Wirtz, H.* (2008), S. 47 f. Den unterschiedlichen Lebenszyklusphasen des Unternehmens werden dabei unterschiedliche Bedarfsschwerpunkte zugeordnet, die unmittelbar mit dem in Kapitel C.2.3 beschriebenen Leistungsprogramm verbunden sind. Vgl. dazu z. B. für das Bedarfsfeld „Investition & Finanzierung“ *WGZ Initiativkapital* (2005), S. 11.

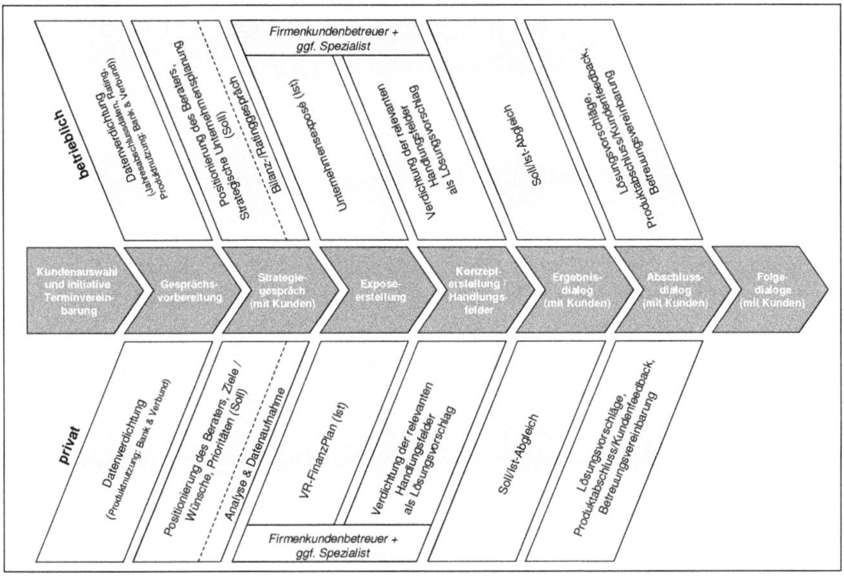

Abbildung 20: Ganzheitlicher Beratungs- und Verkaufsprozess[515]

Dieser Betreuungsprozess wird durch eine Vielzahl von Subprozessen konstituiert (z. B. Gesprächsvorbereitung und -nachbereitung), deren abschließende Darstellung jedoch nicht zielführend für diese Arbeit ist und daher nicht vorgenommen werden soll.

[515] *Dehne, T.* (2008), S. 56. Der dargestellte Prozess bezieht sich auf das Segment Mittelstand.

3 Strategische Ausrichtung, wirtschaftliche Bedeutung und Marktanteile der Geschäftseinheit Firmenkunden

3.1 Strategische Ausrichtung

Eines der konstituierenden Merkmale strategischer Geschäftseinheiten ist die Formulierbarkeit einer weitgehend eigenständigen Strategie. Aufgrund der aufgezeigten charakteristischen Merkmale der Kreditgenossenschaften und hier insbesondere ihrer (relativen) Kleinheit ist dieses Merkmal nur mit Einschränkungen für die Geschäftseinheit Firmenkundenbank gegeben. So begrenzen die Strukturparameter wie etwa die limitierte Ressourcenverfügbarkeit beim Humankapital die Freiheitsgrade der Strategieauswahl auf Gesamtbank- und Geschäftseinheitsebene. Als grundsätzliche strategische Ausrichtung gelten nach PORTER die Kosten-Preisführerschaft oder die Differenzierungsstrategie.[516] Unter Annahme der Gültigkeit des strategischen Imperativs in seiner Konkretisierung als Fähigkeitsstruktur-Hypothese[517] muss eine bankweite, die einzelnen strategischen Geschäftseinheiten übergreifende strategische Ausrichtung als Kosten-Preisführer für die Kreditgenossenschaften negiert werden.[518] Für die Kreditgenossenschaften gilt als strategische Ausrichtung gemeinhin die Qualitätsführerschaft[519] auf Basis einer partnerschaftlichen Beziehung zum Kunden und einer ganzheitlichen Begleitung.[520] Dieser strategische Ansatz, der auf Ebene der Geschäftseinheiten als inhaltlich zu konkretisierender Rahmen zu verstehen ist, gilt sowohl für das Firmenkunden- als auch das Privatkundengeschäft.

3.2 Wirtschaftliche Bedeutung

Nach Aussage des BVR gehört das Firmenkundengeschäft zu den Kerngeschäftsfeldern der FinanzGruppe.[521] Diese qualitative Aussage soll anhand ausgesuchter Daten fundiert werden. Da dass Kreditgeschäft zu den fokussierten

---

516 Vgl. *Porter, M. E.* (1997).

517 Vgl. zur Fähigkeitsstruktur-Hypothese *Schewe, G.* (1999), S. 66.

518 Vgl. *BVR* (2007b), S. 4. Eine ausführliche Begründung für die nicht vorhandene Umsetzbarkeit der Kosten-Preisführerschaft bei den Kreditgenossenschaften findet sich bei *Wiedemann, A.* (1992), S. 112 ff.

519 Bei der Qualitätsführerschaft handelt es sich um eine spezielle Art der Differenzierungsstrategie nach PORTER. Vgl. dazu auch *Börner, C. J.* (2005), S. 40.

520 Vgl. *Jacobs, B./Krauß, C.* (2005), S. 25.

521 Vgl. *BVR* (2009b), S. 14 f.; *Fehl, U./Kuhn, N.* (2005), S. 204; *Krauß, C.* (2005), S. 62.

Produktfeldern der Kreditgenossenschaften gehört[522] und der Kredit auch das bedeutendste Produkt im Firmenkundengeschäft für alle Universalbanken ist,[523] soll das Kreditgeschäft als ein Indikator der Bedeutung des Firmenkundengeschäfts für die Kreditgenossenschaften herangezogen werden.

Am Ende des Jahres 2010 betrug der Wert der Kredite, die von den Kreditgenossenschaften an Unternehmen und wirtschaftlich selbständige Privatpersonen ausgegeben wurden,[524] über 134 Mrd. EUR.[525] Dies entspricht einem Anteil von ca. 19 Prozent an der aggregierten Bilanzsumme der berichtenden Kreditgenossenschaften. Bezogen auf das gesamte Kreditgeschäft mit inländischen Nichtbanken liegt der Anteil bei ca. 30 Prozent. Die Betrachtung im Zeitverlauf zeigt jedoch, dass die Bedeutung des gewerblichen Kreditgeschäfts am gesamten Kreditgeschäft zurückgegangen ist.[526] Trotz der rückläufigen Bedeutung des gewerblichen Kreditgeschäfts in der Vergangenheit stellt das Firmenkundengeschäft mit etwa einem Drittel an den gesamten Forderungen eine wichtige Säule der Geschäftspolitik der Kreditgenossenschaften dar. Diese Beurteilung wird unterstützt, wenn die Diversifikationseffekte einer heterogenen Geschäftsfeldstruktur betrachtet werden. Denn während die Kreditgenossenschaften im Privatkundengeschäft einen zunehmenden Wettbewerb durch andere Banken und Non- und Near-Banks erleben, stellten in einer Befragung von 33 Kreditgenossenschaften aus dem Jahre 2007 diejenigen Institute mit ausgeprägten Firmenkundengeschäft fest, dass bedingt durch die positive Entwicklung der Firmenkunden eine Ausweitung des Geschäfts beobachtbar gewesen sei.[527] Daraus kann geschlossen werden, dass das Firmenkundengeschäft stabilisierend für die Kreditgenossenschaften sein kann.[528]

In Bezug auf den Beitrag des Firmenkundengeschäfts zur Profitabilität der Kreditgenossenschaften lässt sich ein ähnliches Ergebnis ableiten. DUHNKRACK

---

522 Vgl. *Varmaz, A.* (2006), S. 96; *Wittmann, G.* (2006), S. 13; *Juncker, K.* (1997), S. 16.

523 Vgl. *Hölzer, K./Schnarr, T.* (2010), S. 54; *Ahrendt, B.* (2007), S. 41; *Duhnkrack, T.* (2007), S. 1222.

524 Aufgrund der Kundensystematik der DEUTSCHEN BUNDESBANK sind hier Privatiers und Rentiers enthalten. Per definitionem sollten die an diese Teilgruppe ausgegebenen Kredite jedoch vernachlässigbar sein.

525 Vgl. hier und nachfolgend *Deutsche Bundesbank* (2011b). Die Begrenzung auf das inländische Kreditgeschäft erfolgt, da die Kreditgenossenschaften aufgrund des Regionalprinzips regelmäßig nur national tätig sind. Vgl. dazu die Ausführungen in Kapitel C.1.4.3.

526 Vgl. *Deutsche Bundesbank* (2011a). Die rückläufige Bedeutung erklären KASTEN/REENTS für kleine Universalbanken aufgrund des Wegfalls kleiner Firmenkunden. Vgl. *Kasten, L./Reents, B.* (o. J.), S. 82.

527 Vgl. *Steria Mummert Consulting* (2007), S. 4.

528 Vgl. *Hölzer, K.* (2006), S. 47.

beziffert den Anteil der Zinserträge aus dem gewerblichen Geschäft an den Gesamtzinserträgen auf etwa 40 bis 60 Prozent.[529] Einen aggregierten Überblick über die relative Ertragsstruktur differenziert nach Privat- und Firmenkundengeschäft zeigt die Geschäftsfeldrechnung zum 31.12.2009 für die Kreditgenossenschaften (siehe Tabelle 7).

Tabelle 7: Geschäftsfeldrechnung der Kreditgenossenschaften zum 31.12.2009[530]

	DB I	Risikokosten	DB II	Aufwand	DB III
Privatkunden	62%	25%	66%	75%	51%
- Kredite	18%	25%	17%		
- Einlagen	25%	-	28%		
- Provisionen	20%	-	22%		
Firmenkunden	38%	75%	34%	25%	49%
- Kredite	21%	75%	14%		
- Einlagen	11%	-	12%		
- Provisionen	6%	-	7%		
Ergebnis Kundengeschäft	100%	100%	100%	100%	100%

Es ist zu erkennen, dass der Anteil des Firmenkundengeschäfts am Deckungsbeitrag III der Kreditgenossenschaften bei etwa 49 Prozent liegt.[531] Auffallend ist, dass sich dieses Ergebnis vor allem aufgrund der zum Privatkundengeschäft vergleichsweise günstigen Situation bei den Sach- und Personalaufwendungen ergibt, die für das Firmenkundengeschäft bei etwa einem Viertel der gesamten betrachteten Aufwendungen liegen. Diametral umgekehrt verhält sich dies bei den Risikokosten, die zu drei Vierteln dem Firmenkundengeschäft zugerechnet werden. Es ist davon auszugehen, dass diese Werte für die einzelnen Kreditgenossenschaften unter anderem in Abhängigkeit des Tätigkeitsschwerpunktes und der Größe variieren werden. So wird etwa für eine typisierte Volksbank mit 700 Mio. EUR Bilanzsumme ein Ergebnisbeitrag des Firmenkundengeschäfts – gemessen am Deckungsbeitrag II[532] – von etwa 52 Prozent angegeben.[533] Basie-

529 Vgl. *Duhnkrack, T.* (2007), S. 1220.

530 Eigene Darstellung auf Basis der Daten in *BVR* (2010a), S. 19, inkl. „atypischer“ Institute.

531 Der Deckungsbeitrag III ist die Summe der Zins- und Provisionsergebnisse abzüglich der Risikokosten und den zugerechneten Verwaltungskosten (Sach- und Personalkosten). Vgl. dazu Kapitel C.4.3. Im Jahr 2008 lag dieser Wert bei nur 39 Prozent.

532 Der Deckungsbeitrag II ist allgemein definiert als Konditionsbeitrag zzgl. Provisionsergebnis abzgl. Risikoprämie.

533 Vgl. *BVR et al.* (2006), S. 22.

rend auf einer kundenstrukturbezogenen Betrachtung werden 72 Prozent der Erträge im Firmenkundengeschäft mit Unternehmen mit einem Jahresumsatz bis EUR 5 Mio. erwirtschaftet.[534]

Insgesamt zeigen diese Daten, dass das Firmenkundengeschäft für die Kreditgenossenschaften eine hohe wirtschaftliche Bedeutung hat und damit gleichzeitig eine relative große Abhängigkeit von diesem Geschäftsfeld zu konstatieren ist. Dieser Eindruck wird verstärkt durch die Aussage, dass das Firmenkundengeschäft in der FinanzGruppe ca. 48 Prozent zum Deckungsbeitrag III beiträgt, wobei die Kreditgenossenschaften zu ca. 78 Prozent beitragen.[535] Neben diesen wirtschaftlichen Beiträgen wird auch die Rolle des Firmenkundengeschäfts als „Imagebringer" für die Kreditgenossenschaften als Begründung der Bedeutung angeführt.[536]

3.3 Marktanteile

Die Ermittlung des Marktanteils der Kreditgenossenschaften im Firmenkundengeschäft ist nur mit Restriktionen möglich. Diese Restriktionen beziehen sich vor allem auf den relevanten Absatzmarkt. Bezogen auf den relevanten Markt wird die Betrachtung auf den gesamten Firmenkundenmarkt vorgenommen, was aus folgenden Gründen problematisch ist. Zum einen liegt, wie in Kapitel C.2.2 gezeigt wurde, der Schwerpunkt auf dem Geschäft mit mittelständischen Unternehmen. Zum anderen ist die aggregierte Berücksichtigung der Kreditgenossenschaften, die gemäß dem Regionalprinzip lediglich regionale Teilmärkte bedienen, zu berücksichtigen. Daher ist zu erwarten, dass die regionalen Marktanteile deutlich von den hier betrachteten Werten abweichen. Diese Restriktionen sind bei den abzuleitenden Aussagen auf Basis der nachfolgenden Ausführungen zu berücksichtigen.

Die Indikation von Marktanteilen im Firmenkundengeschäft erfolgt regelmäßig auf Basis der Bankenstatistik der DEUTSCHEN BUNDESBANK, in welcher die Aktivseite der einzelnen Banken und Bankengruppen nach Kunden differenziert ausgewiesen wird.[537] Anhand dieser Daten zeigt sich eine relativ hohe zeit-

---

534 Vgl. *Heine, J. M.* (2007), S. 7.

535 Vgl. *BVR* (2010a), S. 15 ff.

536 Vgl. *Krauß, C.* (2005), o. S. [Vorwort].

537 Dieses Vorgehen wählen z. B. auch *Segbers, K.* (2007), S. 71 f. oder *Juncker, K.* (1979), S. 16 ff. Aufgrund der hohen Bedeutung des Aktivgeschäfts im Firmenkundengeschäft der Kreditgenossenschaften erscheint eine ausschließliche Betrachtung des bilanziellen Kreditgeschäfts legitim. Für die Gesamtbankbetrachtung könnten auch die bilanziellen Größen Geschäftsvolumen sowie Kundeneinlagen (Passivgeschäft) betrachtet werden. So z. B. *Wiedemann, A.* (1992), S. 42 ff.

liche Invarianz der Marktanteile im Firmenkundengeschäft (siehe Abbildung 21). Eine Ausnahme stellt der Sparkassensektor dar. Dieser konnte seit 1990 deutliche Marktanteile gewinnen. Dies begründet sich vor allem in den überdurchschnittlichen Zuwächsen der Landesbanken, die in dem betrachteten Zeitraum das Kreditvolumen um ca. 266 Prozent steigerten, während die Steigerung des gesamten betrachteten Marktvolumens nur ca. 80 Prozent betrug.[538]

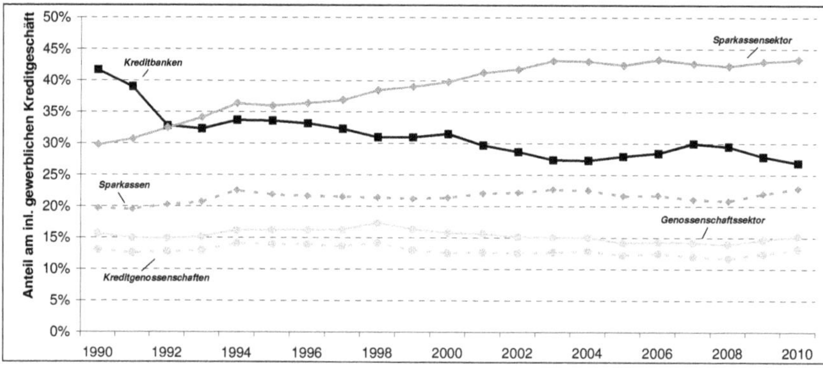

Abbildung 21: Marktanteile im inländischen Firmenkundengeschäft nach Bankenstatistik[539]

Der relativ geringe Marktanteil der Kreditgenossenschaften erklärt sich, wenn die Struktur der Firmenkunden berücksichtigt wird. Denn kleinere Unternehmen fragen vergleichsweise geringe Kreditvolumina nach, was auch durch den Mengeneffekt, d. h. die Anzahl der Firmenkunden, nicht kompensiert wird. Wird der Genossenschaftssektor, also die Kreditgenossenschaften und die Zentralbanken, betrachtet, ist der Marktanteil nur geringfügig höher. Eine Erklärung dafür kann der begrenzte relevante Absatzmarkt der regionalen WGZ BANK auf der einen und die strenge Subsidiarität der DZ BANK, die kaum direktes Firmenkundengeschäft betreibt, auf der anderen Seite sein.

Werden ergänzend einzelne Produktmärkte betrachtet, ist erkennbar, dass die Kreditgenossenschaften recht heterogene Marktpositionierungen aufweisen (siehe Abbildung 22). Die Sparkassen weisen grundsätzlich bei den ausgewählten Produktfeldern die höchsten Produktreichweiten aus, i. d. R. dicht gefolgt

538 Vgl. *Deutsche Bundesbank* (2011a). Das Marktvolumen ist definiert als das Volumen der Kredite an Unternehmen und wirtschaftlich selbständige Unternehmen aller Bankengruppen.

539 Eigene Berechnung und Darstellung auf Basis der Daten in *Deutsche Bundesbank* (2011a).

von den Kreditgenossenschaften. Die nur geringe Produktreichweite der Genossenschaftsbanken bei den Derivaten erklärt sich zum einen durch die Firmenkundenstruktur, da die bedeutenden Segmente tendenziell keinen Bedarf an diesen Produkten haben. Und zum anderen an der ggf. fehlenden Kompetenzvermutung für die Genossenschaften seitens der Unternehmen.[540]

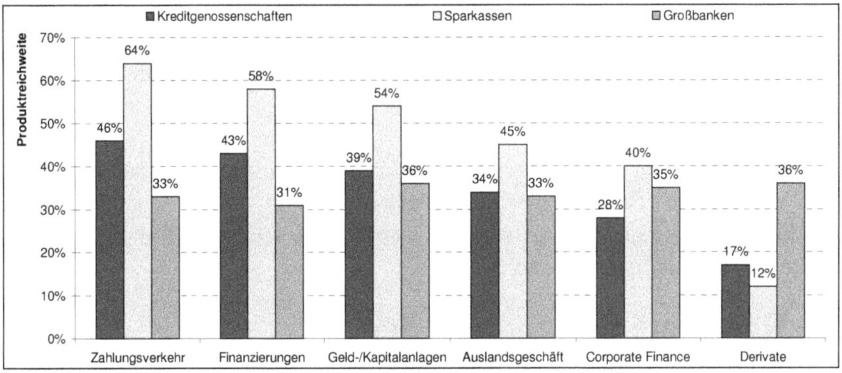

Abbildung 22: Marktanteile in ausgewählten Produktbereichen[541]

Andere Studien zeigen, dass mit zunehmender Unternehmensgröße die Bedeutung der Kreditgenossenschaften als relevanter Anbieter von Bankleistungen abnimmt. Ohne auf die Ergebnisse im Detail einzugehen, zeigt sich, dass bei den nach Umsatz größeren Firmenkunden die Großbanken die einzelnen Produktmärkte dominieren.[542] Diese Tendenz der abnehmenden Bedeutung der Kreditgenossenschaften bei steigender Umsatzgröße der Unternehmen spiegelt sich auch in der Intensität der Geschäftsbeziehung wider. In einer auf diesen Untersuchungsgegenstand ausgerichteten Studie von SCHMIDT zeigt sich, dass die Kreditgenossenschaften mit zunehmender Umsatzgröße der Unternehmen seltener als Hausbank beurteilt werden (siehe Tabelle 8).

---

540 Vgl. *Duhnkrack, T.* (2007), S. 1221; *BVR et al.* (2006), S. 25.

541 Eigene Darstellung auf Basis der Daten in *tns infratest* (2008), S. 32 f.

542 Vgl. ausführlich *Käser, B./Ritzenhofen, A.* (2009), S. 6.

Tabelle 8: Marktanteile der Kreditgenossenschaften im Firmenkundengeschäft[543]

Umsatz des Unternehmens	< 0,5 Mio. €	0,5-2,5 Mio. €	2,5-12,5 Mio. €	12,5-50 Mio. €	> 50 Mio. €
Marktanteil (Hauptbank-verbindung)	26%	27%	19%	13%	5%

Diese älteren Daten werden in ihrer Tendenz bestätigt durch jüngere Veröffentlichungen, die ebenfalls darauf hindeuten, dass die Kreditgenossenschaften im Segment der umsatzstärkeren Unternehmen seltener als Haus- oder Hauptbank präsent sind (siehe auch Abbildung 12).[544]

Nachdem damit die Geschäftseinheit der Firmenkunden im Wesentlichen für die vorliegende Forschungsarbeit beschrieben und deren wirtschaftlichen Merkmale dargelegt wurden, werden nachfolgend relevante Bestandteile der Gesamtbanksteuerung vorgestellt.

4 Relevante Bestandteile der Gesamtbanksteuerungskonzeption VR-CONTROL

Im Sinne der Zielsetzung dieser Arbeit in diesem Kapitel für die Vertriebssteuerung relevante Bestandteile der Gesamtbanksteuerungskonzeption VR-CONTROL vorgestellt. Begründet wird dies damit, dass sich die im weiteren Verlauf der Arbeit konzeptionell zu entwickelnde kundenwertorientierte Vertriebssteuerung in die Gesamtbanksteuerung einfügen muss, um für die Praxis umsetzbar zu sein. Die Vorstellung der relevanten Bestandteile von VR-CONTROL basiert vornehmlich auf den Empfehlungen des BVR. Denn aufgrund der hohen Komplexität des allgemeinen Steuerungsthemas wird in der FinanzGruppe vermutet, dass „eine im Verbund koordinierte Vorgehensweise gegenüber individuellen Lösungen effektiver"[545] sei. Insofern wurden mit dem WEIßBUCH DER UNTERNEHMENSSTEUERUNG und VR-CONTROL WISSEN Grundlagenwerke als „Basis für eine verbundeinheitliche Steuerung des Bankgeschäftes einer Kreditgenossenschaft"[546] geschaffen. Die folgende Vorstellung stellt eine allgemeine Beschreibung dar, die im Einzelfall variieren wird.

543 In Anlehnung an *Schmidt, T.* (2001), S. 45 f.

544 Vgl. auch *tns infratest* (2010), S. 28.

545 *Darfener, S./Lüders, U.* (2002), S. 38. Vergleiche für eine ausführliche Begründung der zentralen Entwicklung aus Sicht des BVR *Pleister, C.* (2002), S. 4 f. und aus Sicht einer Kreditgenossenschaft *Kalefeld, K.* (2002), S. 44 f.

546 *BVR et al.* (o. J.b), S. 1.

4.1 Grundlagen der Gesamtbanksteuerung

Allgemein kann unter der Gesamtbanksteuerung der Prozess des Managements eines Kreditinstituts als Ganzes verstanden werden.[547] Insoweit entspricht sie inhaltlich dem Begriff der Unternehmenssteuerung. Aufgrund der Transformationsfunktion von Banken, der exponierten Bedeutung von Risiken und den damit einhergehenden aufsichtsrechtlichen Regelungen wird jedoch häufig die Risikodimension expliziert,[548] so dass als Gesamtbanksteuerung „ein System zur Messung, Überwachung sowie zur aktiven Einflussnahme auf die Ertrags-, Vermögens- und Risikolage der Bank“[549] verstanden wird. Diese integrierte Sicht von Risiko- und Rentabilität gilt heute als „Inbegriff moderner Gesamtbanksteuerung.“[550]

Die Gesamtbanksteuerung in den Kreditgenossenschaften ist als duales Steuerungsmodell konzipiert, das aus einer zentralen Struktur-/Porfoliosteuerung und einer dezentralen Geschäftssteuerung besteht.[551] Diese Aufteilung lässt sich damit begründen, dass diejenigen Steuerungsbereiche, die nur aus einer übergeordneten Warte der Gesamtbank vollumfänglich überblickt und zielsetzungsgerecht beurteilt werden können, nur in einer mit entsprechenden Kompetenzen ausgestatteten Zentralinstanz zielorientiert ausgeführt werden können.[552] In der Struktur-/Portfoliosteuerung wird die Bank in verschiedene Subportfolios disaggregiert, denen spezifische Risiken und Erträge zugeordnet werden können.[553] In VR-CONTROL wird auf der obersten Portfolioebene das Nicht-

---

547 Vgl. *Tolkmitt, V.* (2007), S. 379.

548 Vgl. *Schierenbeck, H./Lister, M./Kirmße, S.* (2008), S. 1.

549 *Küpker, H./Pudig, A.* (2001), S. 575.

550 *Schierenbeck, H./Lister, M./Kirmße, S.* (2008), S. 1.

551 Vgl. *BVR et al.* (o. J.b), S. 5 ff. Vgl. zu den theoretischen Grundlagen der dualen Steuerung auch *Schierenbeck, H./Lister, M./Kirmße, S.* (2008), S. 14 sowie *Schierenbeck, H.* (2003), S. 293 ff. Die duale Steuerung ist konzeptionell eingebettet in den Steuerungsansatz VR-CONTROL, der aufbauend auf den Ergebnissen der strategischen Analyse BÜNDELUNG DER KRÄFTE: EIN VERBUND – EINE STRATEGIE als zentrales Thema zur Professionalisierung und Modernisierung der Gesamtbanksteuerung sowie zur Steigerung der nachhaltigen Profitabilität der Kreditgenossenschaften seit 2002/2003 in den Instituten implementiert wird.

552 Vgl. *Schierenbeck, H.* (2003), S. 297 Kalkulatorische Grundlage des dualen Steuerungsmodells in der bankinternen Kosten- und Leistungsrechnung ist die Marktzinsmethode, die eine Erfolgsaufspaltung in Strukturbeitrag und Konditionsbeitrag erlaubt. Erst auf Basis dieser Aufteilung ist eine differenzierte Steuerung nach dem dualen Ansatz möglich, denn so ist der Strukturbeitrag zentral zu verantworten und der Konditionsbeitrag dezentral zu steuern. Vgl. *Schierenbeck, H.* (2003), S. 293.

553 Vgl. hier und nachfolgend *BVR et al.* (o. J.b), S. 6 ff.

kundengeschäft und das Kundengeschäft unterschieden. Letzteres wird weiter aufgeteilt in das Privat- und Firmenkundengeschäft. Das Nichtkundengeschäft umfasst das Management von Risiken mittels Handel und Disposition von risikobehafteten Kontrakten auf dem Geld- und Kapitalmarkt (Disposition oder Treasury).

Zentrale Steuerungsgröße ist der Teil des Vermögenswerts der Bank, der sich aus der Differenz der zu Marktwerten bewerteten Aktiva, Passiva und den außerbilanziellen Geschäften ergibt („Zerschlagungswert“[554]).[555] Die Maximierung eines so definierten Unternehmenswertes unter Berücksichtigung festgelegter Risikolimite wird als oberstes Ziel der Bankführung in VR-CONTROL angesehen.[556] Die Veränderung des Vermögens wird in das Kundengeschäftsergebnis, das Risikoergebnis, das Produktivitätsergebnis und Overheadergebnis aufgeteilt.[557] Das Kundengeschäftsergebnis spiegelt die Veränderungen des Vermögenswertes aufgrund von Neugeschäften wieder, während das Risikoergebnis den (Miss-)Erfolg aus der Übernahme von Marktpreis- und Adressrisiken bezeichnet. Das Produktivitätsergebnis reflektiert die Differenz zwischen den kalkulatorischen Bearbeitungsprämien aus dem Kundengeschäft sowie den realiter angefallenen Sach- und Personalkosten. Der Ergebnisbeitrag nicht direkt zugeschlüsselter Kostenkomponenten ist das Overheadergebnis.

Der Ermittlung der Vermögenswertveränderung liegt eine barwertige Betrachtung der Erträge, Kosten und Risiken zugrunde.[558] Dies bedingt jedoch zum Zwecke der externen Rechnungslegung eine Umrechnung der barwertigen Werte in periodische Werte. Aufgrund der historisch bedingten hohen Bedeutung der handelsrechtlichen und somit periodischen Größen für die Gesamtbanksteuerung gibt es einen bisher nicht abgeschlossenen wissenschaftlichen und praktischen Diskurs darüber, welche methodeninhärenten Anreize eine reine Barwertsteuerung oder eine reine Periodenwertsteuerung als inadäquat erscheinen lassen.[559] Aus diesem Grund wird empfohlen, einen kombinierten Ansatz umzusetzen. Auf Gesamtbankebene ist dies so konzipiert, dass der Vermögenswert als Ziel- und Steuerungsgröße dient und um periodische Zielwerte als strenge Restriktion

---

[554] Der Vermögenswert der Bank setzt sich zusammen aus dem „Zerschlagungswert“ und dem „Good-Will-Wert“ oder „Potenzialwert“ der Bank. Vgl. *BVR* (2002b), S. 11.

[555] Zur Marktwertbestimmung wird die Veräußerung am Markt als wertbestimmende Grundlage verwendet. Vgl. *BVR et al.* (o. J.b), S. 8.

[556] Vgl. *BVR et al.* (o. J.b), S. 8.

[557] Vgl. hier und nachfolgend *BVR* (2002b), S. 6 ff.

[558] Vgl. hier und nachfolgend *BVR* (2002b), S. 4 ff.; *BVR et al.* (o. J.b), S. 4 ff.

[559] Vgl. z. B. *Wimmer, K.* (2010), S. 15 ff.; *Averbeck, C./Frère, E./Reuse, S.* (2008), S. 41 ff.; *Averbeck, C.* et al. (2008), S. 406 ff.

ergänzt wird.[560] Grundsätzlich wird dieser Ansatz auch auf Ebene der Einzelgeschäftssteuerung vorgenommen.[561]

4.2 Vertriebssteuerung als Baustein der Gesamtbanksteuerung

Da zentrale Gesamtbanksteuerungsmodelle aufgrund ihrer Komplexität allgemein in der Praxis nicht umsetzbar sind,[562] erfolgt regelmäßig im Rahmen des dualen Steuerungsmodells eine Modelldisaggregation, was zur Implementierung und Abstimmung entsprechender Partialsteuerungsmodelle führt. Bei den Kreditgenossenschaften wird eine konzeptionelle Aufteilung der Gesamtbanksteuerung in die Säulen Rating, Vertriebssteuerung (auch: Kundengeschäftssteuerung), Marktpreisrisikosteuerung, Adressrisikosteuerung und Produktivitätssteuerung vorgenommen (siehe Abbildung 23).

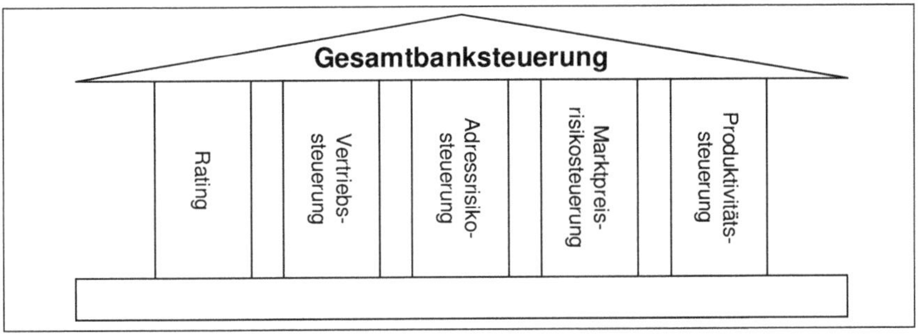

Abbildung 23: Partialsteuerungsbereiche in Kreditgenossenschaften[563]

Die Säule des Ratings beinhaltet die verbundeinheitliche Messung von Kreditnehmerrisiken anhand eines standardisierten Ratingverfahrens (VR-RATING) und entspricht damit den Anforderungen der ersten Säule aus Basel II sowie den MaRisk.[564] Das Ratingergebnis ist eine der zentralen Einflussfaktoren auf die Kalkulation von Kreditkonditionen. Die Produktivitätssteuerung fokussiert auf das Produktivitätsergebnis und damit auf die Ressourceneffizienz des kreditge-

560 Vgl. *BVR et al.* (o. J.b), S. 6.
561 Vgl. dazu die Ausführungen in Kapitel C.4.3.
562 Vgl. *Wimmer, K.* (2010), S. 18.
563 *Hensel, V./Wild, C.* (2003), S. 35.
564 Vgl. *BVR* (2003b), Kapitel 4.4.

nossenschaftlichen Geschäftsbetriebs.[565] Aufgrund der Ausgestaltung der im folgenden Abschnitt zu erläuternden Kundengeschäftskalkulation stellt das Produktivitätsergebnis ein „Korrektiv" dar, dessen Berücksichtigung die Überführung der Teilergebnisse in das Gesamtbankergebnis sichert. Gegenstand der Adressrisikosteuerung ist das Management derjenigen Risiken, die sich aus den bonitätsmäßig bedingten Verlusten aus Geschäften mit Dritten (Kreditnehmer) ergeben, die über dem erwarteten Wert liegen (unerwartete Verluste).[566] Die Marktpreisrisikosteuerung fokussiert dagegen auf diejenigen Risiken, die sich aus Wertänderungen bei offenen Zins-, Währungs-, Aktien- oder Volatilitätspositionen ergeben können.[567] Während sowohl die Produktivitäts-, die Adressrisiko- und die Marktpreisrisikosteuerung im Rahmen des dualen Steuerungsmodells zentral und auf Portfolioebene erfolgen, wird die Kundengeschäfts- bzw. Vertriebssteuerung dezentral vorgenommen.

Aufgrund der Disaggregation und der bestehenden Wechselbeziehungen zwischen den Steuerungsbereichen, erfolgt die Koordination über Verrechnungspreismodelle. Dabei beziehen sich die Wechselbeziehungen insbesondere auf die Risiken, die im Verbund gemäß dem WEIßBUCH DER UNTERNEHMENSSTEUERUNG wie folgt systematisiert werden (siehe Abbildung 24).

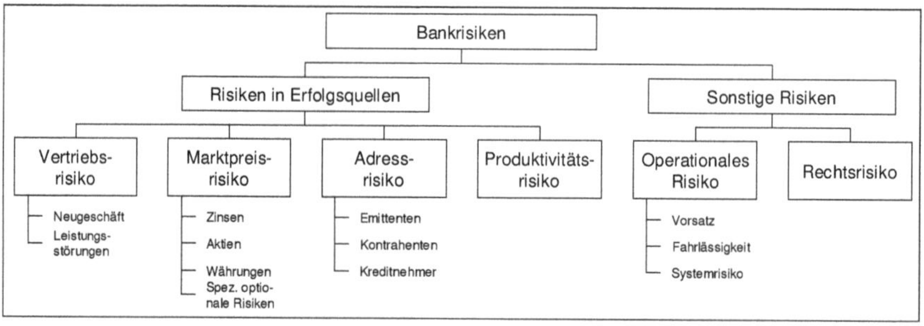

Abbildung 24: Systematisierung von Bankrisiken bei Kreditgenossenschaften[568]

---

565 Vgl. hier und nachfolgend *BVR et al.* (o. J.b), S. 52; *BVR* (2003b), Kapitel 5.1.

566 Vgl. *BVR et al* (o. J.b), S. 38 ff.; *BVR* (2003b), Kapitel 4.2, Kapitel 4.3, Kapitel 4.5 und Kapitel 4.6.

567 Vgl. *BVR et al.* (o. J.b), S. 19 ff. sowie *BVR* (2003b), Kapitel 3.1.

568 *BVR et al.* (o. J.b), S. 19. Das Liquiditätsrisiko wird in diesem Systematisierungsansatz nicht explizit genannt, da es als „Meta-Risiko" und ständige strenge Nebenbedingung betrachtet wird. Zudem wird angenommen, dass durch „einen verantwortungsvollen Umgang mit den anderen Risiken [.] gleichzeitig dem Liquiditätsrisiko [vorgebeugt wird; Erg. d. Verf.]". *BVR et al.* (o. J.b), S. 20.

Zu den Vertriebsrisiken zählen nach diesem Ansatz die Gefahr, dass das erwartete Kundengeschäftsergebnis nicht erreicht wird, da der kontrahierte Konditionsbeitrag im Neugeschäft nicht dem geplanten Betrag entspricht, dass negative Leistungsstörungen in bestehenden Kundengeschäften über den Erwartungen liegen und dass das geplante Kundengeschäftsvolumen nicht erreicht wird.[569] Dahinter steht die Idee, dass Adressrisiken und Marktpreisrisiken nicht direkt zu den vom Vertrieb verantwortbaren Risiken gehören.[570] Aus diesem Grund werden im Rahmen des Verrechnungspreismodells die betroffenen Geschäfte bzw. Zahlungsflüsse aus den dezentralen Marktbereichen an die zentrale Disposition (Treasury) veräußert. So werden bspw. aus dem Marktbereich Firmenkunden die Zinsgeschäfte zum Marktwert auf der Basis der abschlussstichtagsbezogenen Zinsstrukturkurve an die Zentraldisposition weitergereicht.[571] Dem veräußernden Marktbereich werden die jeweiligen ökonomischen Wertbeiträge in Form und Höhe des Margenbarwerts ((Zins-)Konditionsbeitragbarwert) gutgeschrieben. Die ökonomischen Konsequenzen aus der offenen Zinsposition betreffen den empfangenden Bereich, in dessen Aufgabenbereich die Steuerung der Risiken aus den offenen Positionen fällt (Transformations- oder Strukturbeitrag). Diese Unterteilung von Konditionsbeitrag und Transformationsbeitrag aus zinsabhängigen Geschäften ist Ergebnis der Marktzinsmethode,[572] welche die Basis der in den Kreditgenossenschaften implementierten Deckungsbeitrag-III-Rechnung ist. Die Grundlagen dieses Kalkulationsschemas werden im Folgenden vorgestellt

4.3 Einzelgeschäftskalkulation im Deckungsbeitrag-III-Schema

Die klassische Produktkalkulation des industriellen Kontextes wird in den Kreditgenossenschaften als Einzelgeschäftskalkulation bezeichnet. Die Grundlage der Einzelgeschäftskalkulation ist das Deckungsbeitrag-III-Schema (siehe Abbildung 25) mit dem die Vorteilhaftigkeit eines Kundengeschäfts bestimmt

---

569 Vgl. *BVR et al.* (o. J.b), S. 20. Für eine abweichende Risikosystematik bei Banken vgl. z. B. *Krammer, C. K. et al.* (2001), S. 1453.

570 Vgl. *Wimmer, K.* (2010), S. 19.

571 Vgl. *Wimmer, K.* (2010), S. 19; *BVR et al.* (o. J.b), S. 27 ff. Analog ist dieses Vorgehen für die Adressrisiken, die der Markt gegen Zahlung der barwertigen Risikoprämie an die zentrale Risikosteuerung verkauft und sich damit der Adressrisiken entledigt. Vgl. *Wimmer, K.* (2010), S. 21.

572 Vgl. *Rolfes, B./Dartsch, A.* (1997), S. 926 ff.

werden soll.[573] Die DB-III-Rechnung ist als stufenweise Deckungsbeitragsrechnung ausgestaltet, die sich an der Trennung des Leistungsprozesses der Bank in einen Wert- und einen Betriebsbereich orientiert.

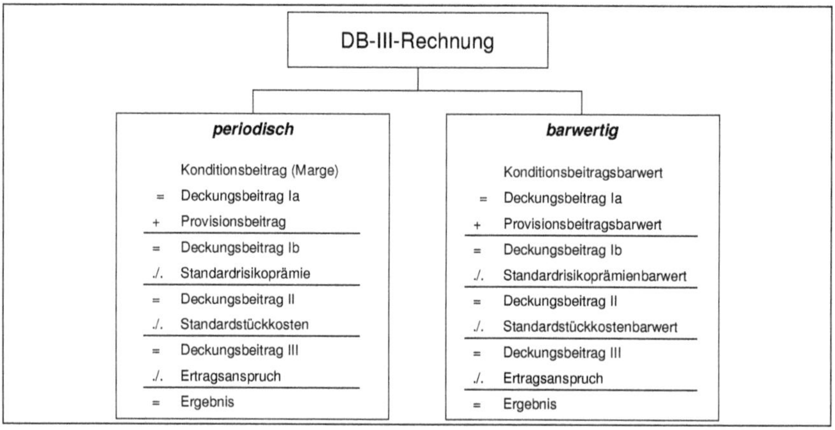

Abbildung 25: Schema der DB-III-Rechnung in VR-CONTROL[574]

Der Konditionsbeitrag(sbarwert) ist der Ertragsbeitrag(sbarwert), der sich aus der Differenz des Kundengeschäfts und einer strukturkongruenten Alternativanlage oder Refinanzierung am Geld- und Kapitalmarkt ergibt.[575] Er resultiert aus zinsabhängigen Geschäften, die von der Bank selbst erstellt werden und wird mit der Marktzinsmethode berechnet.[576]

Für Bank- oder Finanzdienstleistungen, die für den Verbund vermittelt werden, erhalten die Banken eine einmalige oder laufende Einzahlung als Vermittlungsprovision. Diese Beiträge, aber auch die erhaltenen Gebühren aus Dienstleistungen wie dem Zahlungsverkehr, werden dem Provisionsbeitrag zugeord-

---

573 Vgl. hier und nachfolgend *Batz, C./Hensel, V.* (2007), S. 102 ff. und *Kroon, G./Pool, F.* (2002), S. 27 ff.

574 Eigene Darstellung.

575 Vgl. *Batz, C./Hensel, V.* (2010), S. 110. Aufgrund ihrer unsicheren Zahlungsströme lassen sich für variable Produkte, bei denen die Dauer der Kapitalbindung und die Volumenshöhe zeitlich variant sind, aus praktischen Erwägungen heraus keine Refinanzierungs- oder Wiederanlagegeschäfte zuordnen. In diesen Fällen wird die Kalkulation des Konditionsbeitrags mit alternativen Verfahren vorgenommen. Vgl. *Herrmann, J./Lüders, U.* (2004), S. 48.

576 Vgl. zur Marktzinsmethode z. B. *Rolfes, B./Dartsch, A.* (1997), S. 926 ff.

net, der auch die mit dem Provisionsgeschäft direkt verbundenen Kosten umfasst.[577]

Für die risikobehafteten Aktivgeschäfte der Bank besteht die Notwendigkeit, das Risiko eines Ausfalls der ausstehenden Zins- und Tilgungsleistungen (Adressrisiko) kalkulatorisch zu berücksichtigen und damit in die Preisfindung zu integrieren. Die Standardrisikoprämie, die den erwarteten Verlust aus dem Geschäft aus Sicht der Bank darstellt, wird auf der Basis der konzeptionellen Grundlage des sog. Ausfallmodells berechnet.[578] Die Berechnung der Prämie erfolgt in diesem Modell verbundeinheitlich auf Basis des VR-RATING, das die Ausfallwahrscheinlichkeit für den Betrachtungszeitraum von einem Jahr gemäß den aufsichtsrechtlichen Vorgaben berechnet.[579] Die Gesamtheit der Risikoprämien bildet einen Fonds, der die tatsächlichen Verluste aus dem adressrisikotragenden Geschäft übernimmt.

Die Kosten des Betriebsbereichs werden in dieser Rechnung anhand einer prozessorientierten Standardeinzelkostenrechnung ermittelt und als Standardstückkosten integriert.[580] Dabei werden neben den direkt zurechenbaren variablen Einzelkosten auch auf Basis der Prozesskostenrechnung verrechnete Gemeinkosten berücksichtigt. Da die Standardstückkosten produktindividuell ermittelt werden, liegen der Prozesskostenrechnung annahmegemäß standardisier-

---

577 Vgl. *BVR* (2002b), S. 19; *BVR* (2003b), Kapitel 2.1, S. 47.

578 Das Ausfallmodell ist eine von zwei grundsätzlichen Methoden zur Risikoquantifizierung. Dabei wird der erwartete Verlust, der die Grundlage der Risikoprämienkalkulation ist, als Produkt aus den Faktoren Kreditvolumen bzw. Kreditäquivalent, Ausfallwahrscheinlichkeit und Verlustquote berechnet. Vgl. *BVR* et al. (o. J.b), S. 42 ff. Das Kreditäquivalent stellt die auf den Zeitpunkt der Berechnung diskontierten ausstehenden Zins- und Tilgungsleistungen dar und die Verlustquote den Anteil des Kreditäquivalents, der durch die Sicherheitenverwertung nicht abgedeckt ist (Abschreibungsbedarf). Vgl. dazu auch *Batz, C./Hensel, V.* (2007), S. 114 f. und *BVR et al.* (o. J.b), S. 42 ff.

579 Das VR-RATING entspricht dem so genannten IRB-Basisansatz, wobei IRB für Internal Ratings Based steht. Vgl. allgemein zum IRB-Basisansatz *Basler Ausschuss für Bankenaufsicht* (2004), S. 14 ff. und zum VR-RATING bzw. dessen Vorläufer dem BVR-II-RATING z. B. *Nowak, H.* (2002), S. 22 ff. und *Altenbäumker, D./Andritzky, U.* (2007), S. 26 ff. Ziel dieser mit dem Begriff Basel II verbundenen Vorgaben ist eine risikoorientierte Eigenkapitalhinterlegung. Nach Studien des BVR nutzen etwa 40 Prozent der Kreditgenossenschaften diesen IRB-Basisansatz und 60 Prozent den Standardansatz, der ein einheitliches Risikogewicht nutzt, und vor allem für das kleinvolumige Mengengeschäft der Institute als Erleichterung gilt. Vgl. *BVR* (2005b), S. 1. Bei den i. d. R. häufiger vorkommenden mehrjährigen Krediten werden die jährlichen Ausfallwahrscheinlichkeiten mit Hilfe von im Verbund erhobenen Migrationsmatrizen ermittelt. Vgl. *Batz, C./Hensel, V.* (2010), S. 112 f.

580 Vgl. hier und nachfolgend *Batz, C./Hensel, V.* (2010), S. 115 ff.; *Hensel, V./Wild, C.* (2003), S. 36 ff.

te Vertriebs- und Produktionsprozesse für die einzelnen Produkte zugrunde. Dabei werden die Tätigkeiten im Rahmen der Eröffnungs-, Auflösungs- und laufenden Betreuungsprozesse betrachtet, so dass erfahrungsgemäß etwa 20 Prozent der gesamten Bankkosten in der Kundengeschäftskalkulation enthalten sind. Für einen Großteil der Produkte werden diese Standardstückkosten aus den Ergebnissen einer verbundweiten Erhebung übernommen, so dass nur für wenige Produkte institutsindividuelle Werte genutzt werden. Grundlage der Personalkostenzurechnung in diesem Prozesskostenrechnungsansatz sind Mitarbeiterzeiten, die im klassischen personenbezogenen Filialvertrieb anfallen.[581]

Da nicht alle anfallenden Kosten der Bank berücksichtigt werden, ist ein Ertragsanspruch zu definieren.[582] Dieser soll die Kosten der Sicherungseinrichtung, die sonstigen Overheadkosten und die Eigenkapitalkosten umfassen. Die Beiträge zur Sicherungseinrichtung dienen der Aufrechterhaltung der verbundinternen Stabilität einzelner Mitglieder auf Basis einer Solidarversicherung und sind grundsätzlich bei allen Produkten zu berücksichtigen. Die Eigenkapitalkosten betreffen dagegen nur die Aktivgeschäfte und sind als Verzinsung des ökonomischen Risikokapitals, definiert als Höhe des unerwarteten Verlusts, zu verstehen.[583] WIMMER bezeichnet dies als Risikoübernahmeprämie.[584] Die Eigenkapitalkosten sollte der Ergebnisanspruch mindestens umfassen.[585]

Bei der Verwendung der richtigen Steuerungsanreize wird diskutiert, ob die Ausgestaltung der DB-III-Rechnung barwertig oder periodisch erfolgen soll. Als Vorteile der barwertigen Steuerung werden die stringente Ausrichtung am Neugeschäft, die Vermeidung von Problemen der Zurechnung von Erfolgen bei Beraterwechseln bei gleichzeitiger Universalität des Konzepts im Sinne von Ergebnisaggregation und Vergleichbarkeit, die Steigerung der Ertragsorientierung bei den Vertriebsmitarbeitern sowie die Sicherung künftiger Periodenerfolge gesehen.[586] Die barwertige Steuerung vereinfacht zudem Planungs- und Control-

---

581 Vgl. *BVR* (2003c), S. 25.

582 Vgl. hier und nachfolgend *Batz, C./Hensel, V.* (2010), S. 116 f.; *Hensel, V./Wild, C.* (2003), S. 36.

583 Die Eigenkapitalkosten werden auf Basis des so genannten Eigenkapitalmultiplikatoransatzes mit Bezug zum Risikogehalt des Portfolios berechnet. Formal entspricht dies dem Produkt aus dem Quotienten des unerwarteten und erwarteten Verlusts des Portfolios, dem erwarteten Verlust des betrachteten Neugeschäfts und dem Verzinsungsanspruch für das ökonomische Risikokapital. Vgl. *BVR* (2002b), S. 28; *BVR* (2003b), Kapitel 2.4; *Batz, C./Hensel, V.* (2010), S. 116 ff.

584 Vgl. *Wimmer, K.* (2010), S. 28.

585 Vgl. *BVR* (2002b), S. 26.

586 Vgl. *Averbeck, C./Frere, E./Reuse, S.* (2008), S. 40 ff.; *Wimmer, K.* (2010), S. 15 ff.; *Batz, C.* (2004), S. 32; *Dokter, M./Lüders, U.* (2003), S. 28.

lingprozesse, da Geschäfte nur einmal geplant, gemessen und berichtet werden, und leistet eine Beitrag zur Kapazitätsteuerung.[587] Dem stringenten Einsatz stehen jedoch wesentliche Probleme entgegen. So ist festzustellen, dass die Firmenkundenberater aufgrund der höheren Komplexität nicht hinreichend mit der barwertigen Steuerung vertraut sind[588] und damit die Akzeptanz barwertiger Ergebnisse bei diesen eingeschränkt sein soll.[589] Zudem wird kritisiert, dass wichtige Barwertdeterminanten gerade bei zins- und laufzeitvariablen Produkten auf Basis von Annahmen prognostiziert und damit bei abweichenden Ist-Situationen (Leistungsstörungen) die kalkulierten Ergebnisse korrigiert werden müssen.[590] Dies führt zu Problemen bei der verursachungsgerechten Zuordnung von Vertriebserfolgen.[591] Die implizite Neugeschäftsfokussierung kann zudem eine Vernachlässigung der Bestandskunden ohne aktuellen Produktbedarf sowie Margendumping begründen.[592] Dies wirkt vor allem im Kontext der langfristigen Geschäftsbeziehungen zu Firmenkunden kontraproduktiv. Schließlich unterliegen Kreditgenossenschaften perioden- und volumenbezogenen Eigenkapitalrestriktionen und damit periodischen Ertragsanforderungen, die im Barwertkonzept nicht erfasst werden.[593] Aus der Abwägung der Vor- und Nachteile resultiert analog zur Gesamtbanksteuerung die Forderung, die barwertige und periodische Steuerung integrierend umzusetzen.[594] In der Praxis wird jedoch vornehmlich noch die periodische Steuerung eingesetzt.

---

587 Vgl. *Rolfes, B./Dartsch, A.* (1997), S. 938.

588 Vgl. *Batz, C.* (2004), S. 32; *Rehberg, M.* (2004), S. 12.

589 Vgl. *Dokter, M./Lüders, U.* (2003), S. 30.

590 Vgl. *Batz, C./Hensel, V.* (2010), S. 118 f.

591 WIMMER schlägt vor, Ereignisse wie Prolongationen von Geschäften oder Leistungsstörungen als Neugeschäfte zu definieren und dem jeweiligen Kundenberater als (Miss-) Erfolg zuzuordnen. Vgl. *Wimmer, K.* (2010), S. 28.

592 Vgl. *Dokter, M./Lüders, U.* (2003), S. 30. Beim Margendumping wird einer langen Laufzeit mit geringen Margen einer kurzen Laufzeit mit hohen Margen der Vorzug gegeben.

593 Vgl. *Rolfes, B./Dartsch, A.* (1997), S. 940.

594 Vgl. z. B. *Klemm, V./Mönke, R.* (2009), S. 48 ff.; *Krob, B.* (2007), S. 57.

5 Zusammenfassende Erörterung der Implikationen

Nachdem die wesentlichen Kontextspezifika erläutert wurden, soll diesen Teil C abschließend eine zusammenfassende Erörterung der resultierenden Implikationen erfolgen. Für die weitere Arbeit ergibt sich daraus ein Anforderungskatalog an eine zielsetzungsgerechte und zweckorientierte kundenwertorientierte Vertriebssteuerung. Zudem lassen sich erste Erkenntnisse über die Möglichkeiten und Grenzen der Ausgestaltung ebendieser Steuerungsphilosophie erkennen.

Aus der rechtlichen Definition der Kreditgenossenschaften resultiert,[595] dass diese besonderen aufsichtsrechtlichen Vorgaben unterworfen sind. Die kundenwertorientierte Vertriebssteuerung darf in ihrer formellen und materiellen Ausgestaltung zu diesen nicht im Widerspruch stehen. Aus der ökonomischen Definition und der hohen Bedeutung des personalisierten Vertriebs lässt sich schlussfolgern, dass Institutsziele und Individualziele der Vertriebsmitarbeiter divergieren können.[596] Die kundenwertorientierte Vertriebssteuerung sollte daher in der Lage sein, das Verhalten der Vertriebsmitarbeiter konsistent zu den Institutszielen zu steuern.

Hinsichtlich der Ausgestaltung der Vertriebssteuerung ist auch die Größe der hier betrachteten Kreditgenossenschaften zu berücksichtigen.[597] Die relative Kleinheit dieser Kreditinstitute ist als limitierender Faktor für deren Freiheitsgrade in der Ausgestaltung und Implementierung betriebswirtschaftlicher Konzepte zu verstehen. Daher ist eine strenge Pragmatik der Vertriebssteuerung zu fordern, was jedoch keinen Praktizismus in der Entwicklung bedeutet.[598] Dies ist als besondere Anforderung an die Adaption bestehender und Entwicklung neuer konzeptionell-theoretischer Ansätze zu begreifen. In diesem Zusammenhang kann jedoch auf die in der kreditgenossenschaftlichen Praxis bewährte Vorgehensweise der zentralen Entwicklung und dezentralen Nutzung rekurriert werden,[599] die dieses Postulat relativiert. Darüber hinaus gilt es, auf die bestehenden

---

595 Vgl. dazu die Ausführungen in Kapitel C.1.1.1.

596 Vgl. dazu die Ausführungen in Kapitel C.1.1.2 und Kapitel C.2.4.2.

597 Vgl. dazu die Ausführungen in Kapitel C.1.2.

598 Als Praktizismus gilt die Neigung, bei der praktischen Arbeit die theoretisch-ideologischen Grundlagen zu vernachlässigen.

599 Neben der bereits mehrfach angesprochenen Gesamtbanksteuerungskonzeption VR-CONTROL gibt es weitere methodische Entwicklungen, die federführend durch den BVR vorgenommen werden. Dazu zählt z. B. auch das in der FinanzGruppe eingesetzte Ratingverfahren VR-RATING. Der BVR hat zu diesem Zweck seit 2005 Fachräte installiert, die sich mit entsprechenden Schwerpunktthemen aus der Regionalbankenpraxis auseinandersetzen. Vgl. *BVR* (2004b).

Ansätze der Gesamtbanksteuerungskonzeption VR-CONTROL aufzusetzen.[600] Das bedeutet, dass die konzeptionellen Ergebnisse der Arbeit widerspruchsfrei zu den Merkmalen und Inhalten von VR-CONTROL in den Kreditgenossenschaften eingesetzt werden können. Dies betrifft konkret die Ausrichtung an der Zielsetzung der Vermögenswertorientierung sowie die Verwendung bestehender kalkulatorischer Grundlagen.

Aus den besonderen Charakteristika der Kreditgenossenschaften ergeben sich weitere Implikationen.[601] Die Besonderheit der mitgliederorientierten Unternehmensführung und Corporate Governance impliziert, dass Kunden, an denen sich die kundenwertorientierte Vertriebssteuerung als zentrale Bezugsebene orientiert, gleichzeitig Eigentümer (Mitglieder) der Genossenschaft sein können. Damit sind unmittelbar der mitgliederorientierte Förderauftrag sowie die damit verbundenen Strukturgrenzen verbunden. Sobald sich die kundenwertorientierte Vertriebssteuerung und die sich daraus gezielt ergebenden Anreize an die Vertriebseinheiten konträr zu diesen wesentlichen Merkmalen der Genossenschaft verhalten, könnte den Kreditgenossenschaften eine Wesensverfremdung vorgeworfen werden. Träfe dies zu, würde der Genossenschaftsbank de jure die Existenzberechtigung entzogen werden können. Denn wenn diese nicht-genossenschaftliche Förderzwecke im Sinne des § 1 Abs. 1 GenG verfolgt, kann sie gemäß § 81 GenG aufgelöst werden.[602] Eine mögliche Verquerung des Förderauftrags könnte darüber hinaus insofern negative Konsequenzen für die Kreditgenossenschaft haben, als dadurch das Alleinstellungsmerkmal unterminiert und so auch aus Sicht der Kunden letztlich irrelevant wird. Insbesondere gilt es, dass durch die Implementierung einer kundenwertorientierten Vertriebssteuerung keine unzulässige Einschränkung der Mitgliederorientierung und des Förderzwecks aus Sicht der Mitglieder erfolgt.

Das Regionalprinzip begrenzt den absatzmarktbezogenen Handlungsraum der Kreditgenossenschaften.[603] In Abhängigkeit der jeweils vorliegenden Marktanteilsstruktur zwischen den relevanten Wettbewerbern impliziert dies eine starke Akquisitions- oder Bindungs- und Ausschöpfungsausrichtung als vertrieblichen Schwerpunkt. Die kundenwertorientierte Vertriebssteuerung muss die jeweils richtigen Ansätze und Anreize liefern können, um den jeweiligen Maßnahmenschwerpunkt zielgerecht zu unterstützen. Konkret bedeutet dies, dass

---

600 Vgl. dazu die Ausführungen in Kapitel C.4

601 Vgl. dazu die Ausführungen in Kapitel C.1.4 und Kapitel C.1.4.2.

602 In diesem Fall haften die Mitglieder des Vorstands und Aufsichtsrats aufgrund der Verletzung der Sorgfaltspflicht durch mangelnde Erfüllung des Förderzwecks für den daraus entstandenen Schaden gem. § 34 Abs. 2 GenG.

603 Vgl. dazu die Ausführungen in Kapitel C.1.4.3.

Bestands- und Nichtkunden Bezugspunkte der Kundenwertorientierung sein müssen.

Auch aus der Verbundorientierung resultieren spezifische Anforderungen.[604] Transaktionskostentheoretisch ist die FinanzGruppe eine hybride Koordinationsform[605] aus „selbständigen Unternehmen mit eigenen Zielen“[606]. Dass dabei Zielinkongruenz zwischen den einzelnen Unternehmen vorliegen kann,[607] lässt sich bereits anhand der unterschiedlichen Rechtsformen vermuten. So stehen in diesem Netzwerk die mit dem Förderzweck beauftragten Kreditgenossenschaften neben den primär erwerbswirtschaftlich ausgerichteten Aktiengesellschaften, der Rechtsform der meisten Verbundunternehmen.[608] Darüber hinaus können sich auch auf der funktionalen Ebene des Vertriebs Unterschiede im Zielsystem ergeben und damit Einfluss auf die netzwerkinterne Koordination der arbeitsteiligen Aufgabenerfüllung haben (siehe Abbildung 26). Denn die meisten Spezialinstitute sind fokussiert auf einen bestimmten Produktbereich, was eine gegenüber den Kreditgenossenschaften produktbezogene Koordinationsausrichtung begünstigt (z. B. durch Produktabsatzziele in der Vertriebsabstimmung).[609] Die Kreditgenossenschaften dagegen stehen als Allfinanzanbieter dem Bankkunden gegenüber,[610] was eine vom Produktziel abweichende Zielformulierung im Vertrieb begünstigt.[611] Dies kann zu Konflikten in der Koordination und Steuerung führen.

---

604 Vgl. dazu die Ausführungen in Kapitel C.1.4.4.

605 Vgl. *Brixner, U.* (2006), S. 66.

606 *Pleister, C.* (2006), S. 114.

607 Vgl. *Pleister, C.* (2006), S. 115.

608 Zwar gilt, dass der Förderauftrag auch für die Verbundunternehmen gilt, da diese von den Kreditgenossenschaften gegründet wurden, um diesen zu dienen. Dieser Förderauftrag gilt aber nur im Verhältnis von Verbundunternehmen zu Kreditgenossenschaft. Vgl. *Wiedemann, A.* (1992), S. 97 ff.

609 Vgl. dazu auch die Aussage des BVR, Produkte und Produktvolumina in die Zielvereinbarungen zwischen Kreditgenossenschaft und Verbundunternehmen aufzunehmen. Vgl. *BVR* (2001b), S. 26.

610 Vgl. dazu die Ausführungen in Kapitel C.2.3.

611 Vgl. *Kring, T.* (2002), S. 33. Vgl. auch die Empfehlung einer Verbundmatrixorganisation bei *Wiedemann, A.* (1992), S. 104, die den genannten steuerungsrelevanten Zielkonflikt bereits implizit beinhaltet.

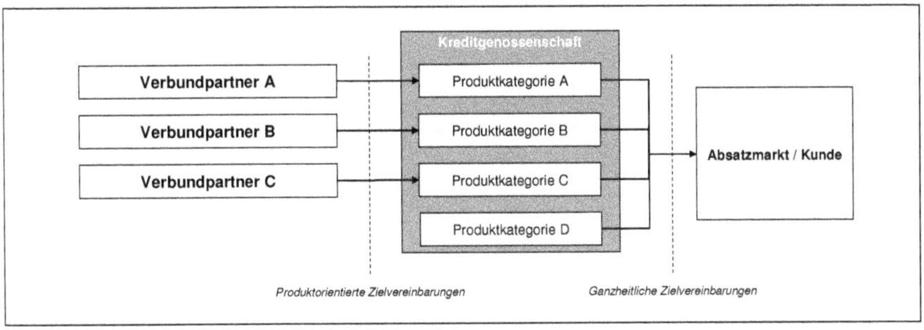

Abbildung 26: Vertriebskoordination über Zielvereinbarungen in der FinanzGruppe[612]

So stellt etwa das Vorstandsmitglied der VOLKSBANK TRIER EG, HORST SCHREIBER, fest, dass über die Genossenschaftsbanken kein reiner Produktverkauf stattfinden könne, sondern Platz für individuelle und ganzheitliche Lösungen für den Kunden sein müsse.[613] Gleichwohl gaben in einer empirischen Untersuchung von KRING von 653 teilnehmenden Kreditgenossenschaften 47,9 Prozent an, dass die produktbereichsbezogenen Zielvereinbarungen mit den Verbundunternehmen einen starken bis sehr starken Einfluss auf die eigene Strategie ausüben.[614] Und so verbindet auch der BVR in dem zentralen Strategiepapier BÜNDELUNG DER KRÄFTE: DIE GEMEINSAME STRATEGIE die Verbundorientierung der Kreditgenossenschaften mit der Notwendigkeit zum „Abschluss von schriftlichen Vertriebsvereinbarungen über Vertriebsziele, Produkte, Provisionen, Form der Zusammenarbeit und Vertriebscontrolling“[615]. Dabei ist aber zu beachten, dass die Verbundpartner ein zum Eigenangebot der Kreditgenossenschaften teilweise substitutives Leistungsprogramm bereitstellen (z. B. Factoring (Verbund) und Betriebsmittelkredit (Kreditgenossenschaft)). Diese Substitutionsbeziehungen müssen ebenfalls zielsetzungsgerecht in der Vertriebssteuerung berücksichtigt werden können. Für die weitere Untersuchung sind also die vertriebsbezogene Verbundstruktur, die Absatzverbundbeziehungen der gesamten Verbundproduktbreite sowie der Koordinationsbedarf zwischen Verbundunternehmen und Kreditgenossenschaft von wesentlicher Bedeutung. Damit ist die Anforderung verbunden, die Produktvielfalt in Verbindung mit der Verbundorientierung zweckmäßig und zielsetzungsgerecht in die Vertriebssteuerung zu integrieren.

612 Eigene Darstellung.
613 Zitiert nach *Theurl, T.* (2004), S. 2.
614 Vgl. *Kring, T. I.* (2002), S. 33.
615 *BVR* (2001b), S. 20.

Vor dem Hintergrund der hohen Wettbewerbsintensität und der Erwartungshaltung einer weiteren Zunahme dieses Wettbewerbs aufgrund der Finanzmarktkrise ist es die vornehmliche Aufgabe der Vertriebssteuerung,[616] die Effektivität und Effizienz des Vertriebs sicherzustellen. Während die Effektivität das Ausmaß der Zielerreichung beschreibt, bezieht sich die Effizienz auf die Wirtschaftlichkeit der Zielerreichung. In diesem Zusammenhang deuten die Ergebnisse der zitierten empirischen Studien daraufhin, dass manifeste und latente Ertragspotenziale entwickelt und ausgeschöpft werden müssen. Für die kundenwertorientierte Vertriebssteuerung bedeutet dies, dass ihr ein entsprechender Analyse- und Steuerungsansatz immanent sein muss. In diesem Kontext gilt es auch, die bestehenden Segmentierungsansätze unter Rekurs auf die Kundenwertorientierung anzupassen.

Aus den Ausführungen zu bestehenden Zielgruppenabgrenzungen wird deutlich,[617] dass diese Verfahren nicht hinreichend sind, um einen effektiven und effizienten Vertrieb zu unterstützten. Die kundenwertorientierte Vertriebssteuerung sollte daher Instrumente beinhalten, die dieses Defizit kompensieren können. Wie anhand empirischer Daten gezeigt wurde, stellen kleine und mittlere sowie mittelständische Unternehmen über 90 Prozent der Firmenkunden der Kreditgenossenschaften dar.[618] Für diese ist es typisch, dass nur in geringem Umfang Unternehmensdaten öffentlich zugänglich sind. Wie in den weiteren Ausführungen noch deutlich werden wird, ist es jedoch der Kundenwertorientierung inhärent, dass diese den Blick aus dem Unternehmen heraus richtet und somit auf externe Daten und Informationen angewiesen ist. Diesem Spannungsverhältnis aus Informationsangebot und Informationsbedarf muss entsprechend Rechnung getragen werden. Weiterhin fordern die Verbindung von privater und unternehmerischer Sphäre sowie die Bildung von vertriebsrelevanten Betreuungseinheiten eine gewisse Relativität der Vertriebssteuerung.

Weitere wichtige Implikationen für die Vertriebssteuerung ergeben sich aus den Besonderheiten des Leistungsprogramms, konkret aus den Eigenschaften von Bankdienstleistungen.[619] Denn in Unternehmen, in denen der Vertrieb der Produktion zeitlich vorgelagert ist, ist der Vertriebsplan die zentrale Determinante der anderen Teilpläne.[620] So richtet sich entsprechend der Kapazitätsplan an den mittel- und langfristigen Vertriebsplänen aus. Damit ist gleichzeitig die Gestaltung der bedeutenden Gemeinkostenbereiche der Dienstleistungsunter-

---

616 Vgl. dazu die Ausführungen in Kapitel C.1.5.
617 Vgl. dazu die Ausführungen in Kapitel C.2.2.2.
618 Vgl. dazu die Ausführungen in Kapitel C.2.2.3.
619 Vgl. dazu die Ausführungen in Kapitel C.2.3.
620 Vgl. *Homburg, C. et al.* (2008), S. 635.

nehmen betroffen. In der kurzfristigen Perspektive jedoch gelten die Gemeinkostenbereiche als weitgehend remanent. Folglich muss die Vertriebsplanung hinreichend konkrete Informationen in Bezug auf die wahrscheinliche Kapazitätsauslastung geben, so dass die Bankleitung in der Lage ist, frühzeitig Maßnahmen zu ergreifen. Aufgrund der Dienstleistungseigenschaften ergibt sich jedoch ein Steuerungsproblem, das aus der Integration des externen Faktors resultiert und die Planbarkeit der Produktion beschränkt. Weiterhin resultieren Probleme im Rahmen kostenrechnerischer Bewertungen, durch die hohe Bedeutung der Gemeinkosten, die eine produktbezogene Kostenzurechnung erschweren.[621]

Es ist evident, dass die vertriebliche Vorgehensweise im Kontext der beziehungsorientierten Ansätze ebenfalls besondere Implikationen für die Vertriebssteuerung beinhaltet. Zum einen besteht aufgrund der ressourcenintensiven Prozessstruktur die Notwendigkeit den Ressourceneinsatz effizient vorzunehmen. Dies bedeutet, dass die knappen Personalressourcen der Vertriebsmitarbeiter, d. h. der Firmenkundenbetreuer in die ertragreichste Verwendungsrichtung alloziiert werden. Gleichzeitig ist aufgrund des relationalen Ansatzes eine ausschließliche Fokussierung auf die Profitabilität des einzelnen Geschäfts nicht zwingend maßgeblich, wird sogar im Ansatz der Hausbankbeziehung als kontraproduktiv angesehen.[622] Vielmehr muss über die gesamte restliche Dauer der Geschäftsbeziehung ein positives Ergebnis für die Kreditgenossenschaft resultieren. Damit muss die Vertriebssteuerung die unterschiedlichen Ansätze im Vertrieb beachten und gleichzeitig bei den relationalen Ansätzen die Beziehung in den Vordergrund der Bewertung vertrieblicher Maßnahmen rücken.

Schließlich sind auch die institutionellen Besonderheiten zu beachten, was vor allem dann für die Vertriebssteuerung wesentliche Implikationen beinhaltet, wenn verschiedene Organisationseinheiten der Kreditgenossenschaft in die Betreuung involviert sind oder gar Bankmitarbeiter und Verbundmitarbeiter gleichermaßen in die Vertriebsarbeit investieren.[623] Für die Vertriebssteuerung bedeutet dies, dass Entscheidungsgrundlagen hinreichend transparent sind und Anreize aus Sicht von allen involvierten Akteuren institutszielsetzungsgerecht ausgestaltet sind.

Die strategische Ausrichtung ist ebenfalls als Orientierungspunkt zu berücksichtigen. Für die Kreditgenossenschaften wird eine qualitätsorientierte Differenzierungsstrategie konstatiert. Dieser strategische Ansatz muss bezüglich seiner konkreten Maßnahmen in die Planungs-, Steuerungs- und Kontrollaktivitäten integriert werden. Damit ist verbunden, dass die Vertriebssteuerung dazu

---

621 Vgl. *Reckenfelderbäumer, M.* (2005), S. 37.

622 Vgl. *Doberanzke, V.* (1993), S. 23.

623 Vgl. dazu die Ausführungen in Kapitel C.2.4.3.

beiträgt, strategiekonsistente Anreize zu geben, um eine möglichst hohe Kongruenz zwischen Strategie und Vertriebsmaßnahmen herzustellen.

Die in Tabelle 9 zusammenfassend dargestellten Implikationen konstituieren einen Anforderungskatalog, der für die Entwicklung einer kundenwertorientierten Vertriebssteuerung zu beachten ist. Dabei gilt, dass nicht alle der genannten Anforderungen unmittelbare Auswirkungen auf die konzeptionellen Ausführungen in den Teilen A und E haben und sich daher dort expressis verbis wiederfinden. Einige dieser Anforderungen sind vielmehr so zu verstehen, dass eine kundenwertorientierte Vertriebssteuerung diesen Anforderungen bzw. deren Umsetzung in der Praxis nicht entgegenstehen darf.

Tabelle 9: Zusammenfassende Darstellung der Implikationen für die Vertriebssteuerung[624]

Kontextfaktor	Implikationen für Vertriebssteuerung
Definition von Banken	• (Aufsichts-)rechtliche Anforderungen berücksichtigen • Zielunterschiede berücksichtigen
Kleinheit der Kreditgenossenschaften	• Pragmatismus, aber kein Praktizismus • Orientierung an Gesamtbanksteuerungskonzeption VR-CONTROL insbesondere im Hinblick auf Zielgrößen und Kalkulationsgrundlagen
Wesenszweck (Mitgliederorientierung und Förderauftrag)	• Integration des genossenschaftlichen Förderauftrags als Strukturdeterminante • Beachtung des Identitätsprinzips
Regionalprinzip	• Sicherstellung der Ausschöpfung von Ertragspotenzialen im Firmenkundenbestand • Ausrichtung der Akquisitionstätigkeiten auf potenzialstarke Nicht-Kunden
Verbundorientierung und Bankleistungsprogramm	• Abbildung des Allfinanzangebots im Sinne der Verbundorientierung • Harmonisierende Integration von verschiedenen Unternehmenszielsetzungen und Koordinationsansätzen • Allfinanzangebot als Ganzes berücksichtigen
Wettbewerb (im Firmenkundengeschäft) und Ertragsschwäche	• Sicherstellung der nachhaltigen Effektivität und Effizienz des Vertriebs • Fokussierung auf manifeste und latente Ertragspotenziale
Zielgruppenabgrenzung und Kundenstruktur	• Identifikation attraktiver bzw. werthaltiger Firmenkunden und Integration dieses Wertes in Segmentierungsverfahren • Knappheit der Informationen bei mittelständischen Firmenkunden berücksichtigen • Berücksichtigung von Kundenverbünden als Bezugsgröße vertrieblicher Maßnahmen
Eigenschaften von Dienstleistungen (Leistungstypologisierung)	• Vertriebsplanung muss als „Engpasssektor" die zweckmäßige Verbindung von Unternehmen und Markt sein (Grundlage anderer Teilpläne) • Integration des externen Faktors als Unsicherheitsdeterminante der Leistungserstellung • Berücksichtigung des hohen Anteils an Gemeinkosten
Vertriebsansätze	• Differenzierte Steuerung für transaktionale und relationale Ansätze • Bei relationalen Ansätzen Fokussierung auf Geschäftsbeziehung als relevanten Betrachtungsgegenstand
Vertriebsorganisation	• Berücksichtigung der Arbeitsteilung im Vertrieb bei Erfolgszurechnung
Strategische Ausrichtung	• Sicherung der Umsetzung der strategischen Maßnahmen und einer strategiekonsistenten Ausgestaltung des Vertriebs

624 Eigene Darstellung.

D Grundlagen einer kundenwertorientierten Vertriebssteuerung für das Firmenkundengeschäft der Kreditgenossenschaften

Nachdem im vorhergehenden Teil C die kontextspezifischen Implikationen einer kundenwertorientierten Vertriebssteuerung bei den Kreditgenossenschaften herausgearbeitet wurden, sollen in diesem Teil A die Grundlagen einer kundenwertorientierten Vertriebssteuerung erörtert und definiert werden. Dies betrifft zunächst die Abgrenzung des Begriffs Kundenwert (Kapitel D.1). Anschließend wird geprüft, ob die in der wissenschaftlichen Literatur definierten Erfolgsvoraussetzungen einer kundenwertorientierten Steuerung bei den Kreditgenossenschaften vorliegen (Kapitel D.2). Im Weiteren werden das Zielsystem der Kreditgenossenschaften und insbesondere die Vereinbarkeit von genossenschaftlichen Prinzipien und Kundenwertorientierung analysiert (Kapitel D.3). Im Anschluss daran wird die Prüfung bestehender und die Konzeptualisierung eines geeigneten Kundenwertmodells vorgenommen (Kapitel A.1) woran sich die für die Vertriebssteuerung notwendige Operationalisierung anschließt (Kapitel D.5). Der Teil A schließt mit kritischen Würdigung der Erkenntnisse (Kapitel D.6).

1 Begriffsabgrenzung Kundenwert

1.1 Abgrenzung ausgewählter kundenbezogener Wertbegriffe

In der Literatur werden verschiedene kundenbezogene Wertbegriffe mehrdeutig verwendet. Nachfolgend sollen in der Literatur häufig vorkommende Begriffe abgegrenzt werden.[625] Gemeinsame Basis dieser Begriffe ist der Zweck jedweder tauschbasierten Transaktion, wonach die beteiligten Akteure von einem tauschinduzierten Nutzen- oder Wohlfahrtsgewinn ausgehen.[626]

Aus Nachfragersicht beschreibt der Begriff Customer Value den Nutzen, den ein Kunde aus dem Konsum eines Produkts bzw. einer Dienstleistung oder

---

[625] Siehe dazu auch die Ausführungen bei *Wünschmann, S.* (2007), S. 27 ff. Für eine ausführliche Diskussion verwendeter Begriffe vgl. z. B. *Rudolf-Sipötz, E.* (2001), S. 14 ff., *Eberling, G.* (2002), S. 35 ff. oder *Wille, K.* (2005), S. 34 ff.

[626] Vgl. *Eggert, A.* (2006), S. 48, der dies als Wertgewinn bezeichnet.

der Interaktion mit einem Anbieter erzielen kann.[627] Der Customer Value ist das Ergebnis eines kognitiven Abgleichs zwischen der jeweils subjektiven Nutzen- und Aufwandssumme, die der Kunde im Rahmen einer Transaktion erwartet.[628] Im deutschsprachigen Raum wir dies auch als Kundennettonutzen bezeichnet.[629] In jüngeren Publikationen wird auch der Begriff des Value Equity verwendet, der „represents a customer's balancing of what is given up (price) and what is received (value)."[630] Werden diese Komponenten nur monetär verstanden, wird er dem Begriff Value for Money gleichgesetzt.[631]

Der Begriff Customer Value wird selten auch aus Anbieterperspektive verwendet und ist definiert als die Bedeutung des Kunden für ein Anbieterunternehmen.[632] Verbunden mit dem Ziel langfristiger Geschäftsbeziehungen wird der eher statische Begriffsinhalt auf den Begriff Customer Lifetime Value übertragen und somit dynamisiert.[633] Der Customer Lifetime Value oder auch Kundenlebenswert[634] wird als investitionstheoretischer und somit rein monetärer Kundenwert operationalisiert.[635] Den Begriff Customer Value anbieterseitig zu verwenden schließen BARTH/WILLE aus. „Ein solches Begriffsverständnis ist [.] hinsichtlich seiner Blickrichtung inkonsistent zu den übrigen Wertbegriffen [...]".[636] Wenn der Shareholder Value den Wert eines Unternehmens(anteils) für den Anteilseigner bedeutet, dann muss der Customer Value der Wert für den Kunden sein.[637]

---

627 Vgl. *Woodruff, R. B.* (1997), S. 140. Der Begriff Konsum steht hierbei stellvertretend für den Ge- oder Verbrauch einer Leistung.

628 Vgl. *Lam, S. Y. et al.* (2004), S. 295 sowie die dort angegebene Literatur.

629 Vgl. z. B. *Vogel, V.* (2006), S. 13 ff. Nutzen- und Aufwandssumme setzen sich aus verschiedenartigen Teilnutzen- (z. B. Produktnutzen) und Aufwandskomponenten (z. B. Kosten) zusammen.

630 *Vogel, V./Evanschitzky, H./Ramaseshan, B.* (2008), S. 104.

631 Vgl. *Wille, K.* (2005), S. 34 sowie die dort angegebene Literatur.

632 Vgl. *Krafft, M.* (2007), S. 44.

633 Vgl. *Dwyer, F. R.* (1997), S. 6 ff.

634 Vgl. *Wiesel, T./Skiera, B.* (2007), S. 706 ff.

635 Vgl. *Lissautzki, M.* (2007), S. 21 ff.; *Weber, J./Lissautzki, M.* (2004), S. 10; *Hempelmann, B./Lürwer, M.* (2003), S. 337. Alle dem Kunden zurechenbaren Einzahlungsüberschüsse über die erwartete Dauer der Geschäftsbeziehung werden auf den Bewertungszeitpunkt mit einem risikoadjustierten Diskontfaktor abgezinst. Unterschieden wird dabei, ob der für die Ermittlung der Einzahlungsüberschüsse relevante Zeitraum nur prospektiv, vgl. z. B. *Weber, J./Lissautzki, M.* (2004), S. 10, oder die gesamte Dauer der Geschäftsbeziehung umfasst. Vgl. z. B. *Kumar, V./George, M.* (2007), S. 157.

636 *Barth, K./Wille, K.* (2000), S. 6, Fn. 46.

637 Vgl. *Wille, K.* (2005), S. 36.

Als weiterer, etablierter kundenbezogener Wertbegriff ist Customer Equity zu nennen.[638] Auch dieser Begriff wird anbieter- und nachfragerseitig verwendet. Ausgehend von der Überlegung, dass der Wert eines Unternehmens vor allem durch den Wert der Kundenbeziehungen determiniert wird, wird der Kundenstamm als intangibler Vermögensgegenstand des Unternehmens verstanden.[639] Nach RUST/ZEITHAML/LEMON ist „A firm's Customer Equity [.] the total of the discounted lifetime values of all of its customers“[640]. WAYLAND/COLE dagegen nutzen den Begriff Customer Equity, um den Wert einer einzelnen Kundenbeziehung zu benennen und bezeichnen den aggregierten Wert als Customer Portfolio Value.[641] DORSCH/CARLSON und DORSCH ET AL. sehen die Kunden als Investoren in eine Geschäftsbeziehung,[642] wenn diese durch loyales Verhalten Autonomie und Handlungsspielräume einschränken, um künftig diese Nachteile durch die Vorteile aus der Beziehung zum betrachteten Unternehmen überzukompensieren.[643] Customer Equity wird zum Wert der Investitionen des Kunden in die Unternehmung und damit ein Perspektivwechsel von Beurteilungsobjekt und -subjekt vorgenommen.[644]

Den Begriff des (monetären) Kundenwerts nutzen GIERL/KURBEL.[645] Ihr Begriffverständnis beinhaltet neben den Bestandteilen des Customer Lifetime Value auch das informatorische Nutzenpotenzial von Kunden, dass sich für das Anbieterunternehmen z. B. darin zeigt, dass Kundenbeschwerden Produktverbesserungen initiieren können. Da dieser Ansatz durch eine dynamische und zahlungsorientierte Berechnungsvorschrift operationalisiert wird, müssen alle Größen monetarisiert werden.

---

638 Vgl. z. B. *Blattberg, R. C./Deighton, J.* (1996), S. 137 f.; *Kumar, V./George, M.* (2007), S. 157 ff.

639 So argumentieren SRIVASTAVA/SHERVANI/FAHEY in diesem Zusammenhang von Kundenbeziehungen als „market-based assets“. Vgl. *Srivastava, R. K./Shervani, T. A./Fahey, L.* (1998), S. 2 ff.; siehe auch *Wiesel, T./Skiera, B.* (2007), S. 709; *Kumar, V./George, M.* (2007), S. 157.

640 *Rust, R. T./ Zeithaml, V. A./Lemon, K. N.* (2000), S. 4. Dabei werden sowohl die aktuellen als auch die potenziellen Kunden berücksichtigt. Vgl. *Rust, R. T./Lemon, K. N./Zeithaml, V. A.* (2004), S. 110.

641 Vgl. *Wayland R. E./Cole, P. M.* (1997), S. 8. So auch KRAFFT, der den Begriff Customer Equity aus der Analogie zum Markenwert, der synonym zum Brand Equity verwendet wird, ableitet, damit aber sowohl den Wert eines Kunden als auch einer Kundengruppe bezeichnet. Vgl. *Krafft, M.* (2007), S. 44.

642 Vgl. *Dorsch, M. J./Carlson, L.* (1996), S. 253; *Dorsch, M. J. et al.* (2001), S. 158.

643 Vgl. *Bell, D. et al.* (2002), S. 80.

644 Vgl. *Wille, K.* (2005), S. 33 f.

645 Vgl. *Gierl, H./Kurbel, T.* (1997), S. 176 ff.; so auch *Stanger, M.* (2007), S. 7.

Gemeinsamkeit der bisher betrachteten kundenbezogenen Wertbegriffe aus Anbietersicht ist, dass diese monetär ausgeprägt sind.[646] In der Literatur herrscht jedoch Uneinigkeit darüber, ob der Kundenwert konzeptionell rein monetär dimensioniert sein sollte. Insbesondere jüngere Beiträge aus dem deutschsprachigen Raum verwenden den Begriff Kundenwert, um mehrdimensionale, monetäre und nicht-monetäre Bestandteile des Werts eines Kunden für ein Unternehmen zu modellieren.[647] Diese Begriffe umfassen die vorgestellten kundenorientierten Wertbegriffe aus Anbietersicht und ergänzen sie um qualitative, nicht-monetäre Faktoren.

Die Heterogenität der kundenbezogenen Wertbegriffe bezeichnet WILLE als das normative Begriffsproblem der Kundenbewertung,[648] denn für „den Begriff Kundenwert gibt es [.] keine eindeutige, allgemeingültige Definition."[649] Auffällig ist, dass ein Teil der kundenorientierten Wertbegriffe aus Anbietersicht rein monetäre Grundlagen hat, während andere, und hier insbesondere die in die Forschung eingeführten Kundenwertbegriffe jüngerer deutschsprachiger Publikationen, auch nicht-monetäre Aspekte berücksichtigen. Aus diesem Grund ist es für den weiteren Verlauf der Arbeit geboten, eine zweckmäßige Begriffsauffassung als Grundlage der weiteren Analyse abzugrenzen.

1.2 Abgrenzung eines zweckmäßigen Begriffsverständnisses des Kundenwerts

Der abzugrenzende Begriff Kundenwert ist ein Kompositum der Begriffe Kunde und Wert. Zunächst wird der Wertbegriff abgegrenzt, bevor der Kunde als das zu bewertende Objekt näher betrachtet wird.

1.2.1 Grundlegendes Verständnis des Begriffs Wert

Der Begriff Kundenwert konkretisiert inhaltlich einen abstrakten Wertbegriff, der im Allgemeinen die zwischen einem Gegenstand (Objekt) und einem Maßstab durch den wertenden Menschen (Subjekt) hergestellte Beziehung be-

---

646 So sehen auch HENSELER/HOFFMANN das Ziel der Kundenbewertung in der Bestimmung eines monetären Werts der Kundenbeziehung. Dieser monetäre Wert wird als Beitrag des Kunden zur Erreichung des Unternehmensziels angesehen. Vgl. *Henseler, J./Hoffmann, T.* (2003), S. 17.

647 Vgl. z. B. *Cornelsen, J.* (2000), S. 37 ff.; *Gelbrich, K.* (2001), S. 102 ff.; *Rudolf-Sipötz, E.* (2001), S. 90 ff.; *Eberling, G.* (2002), S. 36 ff.; *Böhrs, S.* (2004), S. 54 ff.; *Foehn, P.* (2006), S. 47 ff.; *Howald, B.* (2007), S. 32.

648 Vgl. *Wille, K.* (2005), S. 31 ff.

649 *Mengen, A./Mettler, A.* (2008), S. 30.

nennt.[650] In der wirtschaftswissenschaftlichen Literatur existieren zahlreiche Wertbegriffe. So unterscheidet die Volkswirtschaftslehre u. a. zwischen einem (subjektiven) Gebrauchswert (value of use) und einem (objektiven) Tauschwert (value of exchange).[651] Der Tauschwert eines Gutes ist im Sinne der damaligen Vertreter bestimmt durch objektiv messbare Inputfaktoren, wie Arbeit, Boden und Kapital, die den Wert des Outputs direkt bedingen. In der Neoklassik wurde der Wertbegriff durch den subjektiven Nutzen eines Gutes geprägt. Letztgenannte Auffassung entspricht dem in der Betriebswirtschaftslehre verbreiteten Wertbegriff.[652] Die Betriebswirtschaftslehre hat sich vor allem intensiv in der Unternehmensbewertungs- und der Entscheidungstheorie mit dem Wertbegriff auseinandergesetzt.[653] Die Ergebnisse der Theorie der Unternehmensbewertung[654] sind jedoch weniger Begriffsbestimmungen als vielmehr Wertermittlungsverfahren,[655] die zweckorientiert zur Bewertung von Vermögensgegenständen (Unternehmen) herangezogen werden.

In der entscheidungstheoretischen Wertlehre wird die situative Bedingtheit zum Fundament eines eigenen Wertkonzepts gemacht. Dabei orientiert sie sich an dem Nutzen, den das Bewertungsobjekt dem Bewertenden stiftet.[656] Als Basis des Wertbegriffs für den Kundenwert bezieht sich die Literatur in diesem

---

650 Vgl. *Baran, P.* (1990), S. 816.

651 In historischer Chronologie kann hier zunächst im Merkantilismus auf den Franzosen PETTY (1662) verwiesen werden, der eine duale Werttheorie mit Boden und Arbeit als Wertdeterminanten zugrunde legt. CANTILLON (1755) dagegen definiert den Wert der nützlichen Güter nur als den Anteil des zur Herstellung verwendeten Bodens. SMITH unterscheidet später zwischen dem natürlichen Preis und dem Marktpreis, wobei er ersteren im Sinne der Arbeitswertlehre definiert. „Labour […] is the only universal, as well as the only accurate measure of value, or the only standard by which we can compare the values of different commodities at all times and at all places". *Smith, A.* (1776) zitiert nach *Söllner, F.* (2001), S. 29. Dabei unterstellt SMITH, dass der aus Angebot und Nachfrage resultierende Marktpreis sich dem natürlichen Preis angleicht. MARX (1818-1883) folgt im Wesentlichen der Werttheorie von RICARDO (1772-1823), welche wiederum auf dem SMITH'schen „natürlichen Preis" basiert und eine geschlossenere Darstellung der Arbeitswertlehre darstellt. Vgl. *Söllner, F.* (2001), S. 15 ff.

652 Vgl. dazu z. B. ausführlich *Stützel, W.* (1976), Sp. 4407 ff.

653 Ein weiteres in der jüngeren Vergangenheit intensiv in der wissenschaftlichen Diskussion stehendes Wertkonzept ist der Markenwert (Brand Equity). Dieses kann als eine alternative, eher produktorientierte Sichtweise auf den Wert eines Unternehmens aufgefasst werden. Vgl. z. B. *Aaker, D. A.* (1991).

654 Vgl. für eine kurze Darstellung und Erläuterung der Begriffe *Eberling, G.* (2002), S. 31 f. sowie die dort angegebene Literatur.

655 Vgl. *Wöhe, G.* (1996), S. 794.

656 Nutzenvorstellungen, die mit einem bestimmten Sachverhalt verbunden sind, gelten als Agens wirtschaftlichen Handelns. Vgl. *Meyer, P. W.* (1990), S. 18.

Kontext regelmäßig auf den von ENGELS erläuterten gerundiven Wert.[657] Nach diesem Begriffsverständnis ist der Wert ein allgemeiner Maßstab, mit dem der Nutzen eines zu bewertenden Objekts, Subjekts oder einer Aktion quantifiziert werden kann.[658] So ist der „Wert [.] ein Maßstab der Vorziehenswürdigkeit von Aktionsmöglichkeiten oder Gegenständen“[659]. In dieser entscheidungstheoretischen Betrachtung handelt es sich um einen subjektiven Wertbegriff, der in einem konkreten und situationsspezifischen Verwendungszusammenhang steht und in so genannte sekundäre Werturteile eingeht, die als Finalrelationen bezeichnete Zweck-Mittelbeziehungen feststellen.[660] Der gerundive Wert misst den Beitrag, den ein Objekt zur Erreichung der Ziele eines Subjekts leistet.[661] Die Vorziehenswürdigkeit bemisst sich anhand der Zielsetzungen des Bewertenden, der möglichen Alternativen und der Umweltbedingungen, unter denen das Bewertungsobjekt betrachtet wird.[662] Während Alternativen und Umweltvariablen als exogene Daten angenommen werden,[663] sind die Zielsetzungen durch

---

657 So z. B. *Mödritscher, G. J.* (2008), S. 9, *Wiedmann, K.-P./Hennigs, N./In der Stroth, A.-B.* (2006), S. 11, *Wille, K.* (2005), S. 32 f., *Böhrs, S.* (2004), S. 54, *Eberling, G.* (2002), S. 33, *Cornelsen, J.* (2000), S. 28 und *Krüger, S. M.* (1997), S. 105. ENGELS selbst, der diesen Begriff vorstellt, bezieht sich dabei auf einen Beitrag von TOULMIN, in welchem dieser Überlegungen über die Zusammenhänge von Werttheorie und Entscheidungstheorie anstellt. Dessen Theorie wird von ENGELS eingeschränkt übernommen, denn während es TOULMIN um den tatsächlich richtigen Wert geht, diskutiert ENGELS den richtigen Wert bei vorgegebener Zielsetzung. ENGELS stellt dazu der gerundiven Werttheorie die objektive und die subjektive Werttheorie gegenüber. In der objektiven Werttheorie ist der Wert eine Eigenschaft eines Gutes, die von allen Bewertenden identisch wahrgenommen wird. Entsprechend haben die Aussagen „Blut ist rot“ und „das Beil ist 5 DM wert“ die gleiche logische Struktur. Unter Annahme der Gültigkeit dieser Werttheorie ist es ökonomisch schwierig, Austauschaktionen zu erklären. In der subjektiven Werttheorie drückt der Wert dagegen die Vorliebe des Bewertenden aus. In diesem Fall sind die Aussagen „Schokolade schmeckt gut“ und „das Beil ist 5 DM wert“ logisch äquivalent. Wird dieser Ansicht gefolgt, stellt sich die Frage nach der Existenzberechtigung der Betriebswirtschaftslehre, da in diesem Fall die Aussagen zweier Personen „das Beil ist 5 DM wert“ und „das Beil ist 10 DM wert“ keine Gegensätze sind. In diesem Fall fehlt jedoch die Willkür vermeidende Regelung von durchzuführenden Maßnahmen. Vgl. *Engels, W.* (1962), S. 5 ff. sowie die dort angegebene Literatur.

658 Vgl. *Engels, W.* (1962), S. 11 ff.

659 *Engels, W.* (1962), S. 1.

660 Vgl. zur ausführlichen Darstellung *Wöhe, G.* (1996), S. 53 ff.

661 Vgl. *Diller, H.* (2001), S. 6, der den gerundiven Wert auch als teleologischen Wert bezeichnet.

662 Vgl. *Cornelsen, J.* (2000), S. 28.

663 Vgl. *Meyer, A./Kantsperger, R./Wilkoszewski, A.* (2005), S. 300; *Cornelsen, J.* (2000), S. 28 und implizit auch *Diller, H.* (2001), S. 6. Weder MEYER/KANTSPERGER/ WILKOS-

den Bewertenden subjektiv bestimmt. Konsequenterweise stellen die jeweiligen Ziele die zentralen Wertdeterminanten dar und erlauben es, den subjektiven Wert aufgrund der Prüfung des Zielerreichungsgrads als Indikator zu objektivieren.[664] Implizit wird dabei davon ausgegangen, mit operationalen, d. h. in sachlicher, zeitlicher und räumlicher Hinsicht hinreichend definierten Zielen zu argumentieren. Denn nur in diesem Fall kann der Zielerreichungsbeitrag des zu bewertenden Gegenstands oder Sachverhalts bestimmt werden.

Der gerundive Wertbegriff soll auch für die vorliegende Arbeit als Verständnisgrundlage des Begriffs Kundenwert dienen, denn er integriert in die Bewertung die für die Steuerung vorauszusetzenden Ziele des bewertenden Subjekts.[665] Aufgrund seiner inhaltlichen Breite erlaubt der gerundive Wertbegriff zugleich verschiedene Zielinhalte zu berücksichtigen. Dies ist vor allem deshalb für diese Arbeit von Bedeutung, da eine rein erwerbswirtschaftliche Zielformulierung bei den Kreditgenossenschaften aufgrund ihres kraft Gesetzes vorgegeben Unternehmenszwecks der Mitgliederförderung de jure unmöglich und de facto nicht sinnvoll ist.[666] Dieser Wertbegriff determiniert gleichzeitig auch den Begriff Erfolg. Als Erfolg gilt konsequenterweise das erreichte Ausmaß der Zielerreichung unbegrenzter und begrenzter Ziele im Verhältnis zu einem durch Entscheidungsträger im Prozess der Planung und Zielformulierung definierten Mindestniveau.[667]

Nachdem damit der zugrunde zulegende Wertbegriff allgemein abgegrenzt ist, wird im Folgenden das Bezugsobjekt der Bewertung, also der Kunde betrachtet.

1.2.2 Grundlegendes Verständnis des Begriffs Kunde

Ein Ansatz, der eine zweckadäquate strukturelle Basis für den zu verwendenden Kundenbegriff legt,[668] ist das numerische Kundenpotenzial nach LIPPMANN.[669]

ZEWSKI noch CORNELSEN geben für diese Annahme jedoch eine Begründung an und negieren damit die Tatsache, dass auch die Umfelddeterminanten durch ein Unternehmen beeinflusst werden können.

664 Vgl. *Mödritscher, G. J.* (2008), S. 9 f.; *Meyer, A./Kantsperger, R./Wilkoszewski, A.* (2005), S. 300.

665 Vgl. dazu die Ausführungen in Kapitel B.2.1.

666 Vgl. dazu die Ausführungen in Kapitel C.1.4.2 sowie die noch folgende Diskussion des Zielsystems in Kreditgenossenschaften in Kapitel D.3.1.

667 Vgl. *Schmidt-Sudhoff, U.* (1967), S. 141. Bei unbegrenzten Zielen wird ein maximaler Zielerreichungsgrad und bei begrenzten Zielen das Erreichen eines bestimmten Anspruchsniveaus angestrebt. Vgl. *Wöhe, G.* (1996), S. 124 ff.

668 Die Notwendigkeit, den Begriff des Kunden zu definieren, wird nicht in allen relevanten Beiträgen gesehen. Vgl. z. B. *Wille, K.* (2005) oder *Gelbrich, K.* (2001). Explizit auf

Das numerische Kundenpotenzial umfasst die relevante Menge möglicher Kunden eines Anbieterunternehmens (siehe Abbildung 27).

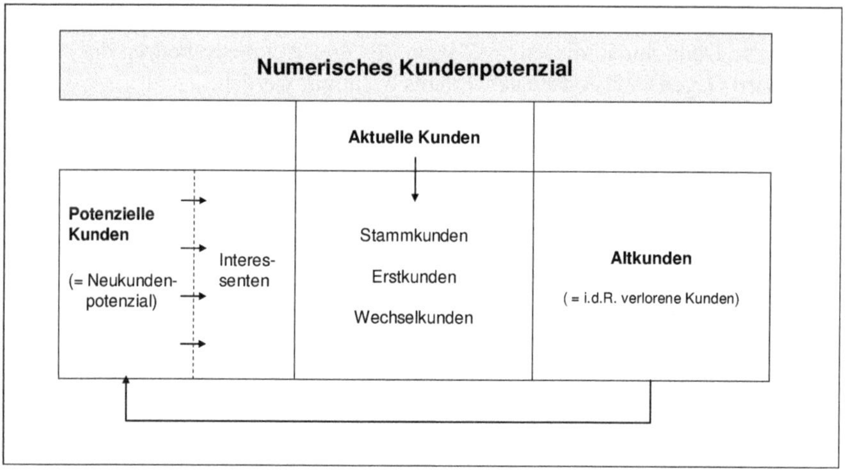

Abbildung 27: Numerisches Kundenpotenzial[670]

Neben den aktuellen Kunden, die in Abhängigkeit ihres Beziehungsstatus in Stammkunden, Erstkunden und Wechselkunden eingeteilt werden, berücksichtigt das numerische Kundenpotenzial auch die potenziellen Kunden, die in Abhängigkeit ihrer Affinität zum Unternehmen oder dessen Leistungsangebot auch als Interessenten bezeichnet werden. Die Gruppe der Altkunden umfasst jene Kunden, mit denen das Anbieterunternehmen in der Vergangenheit mindestens einen Leistungsaustausch vorgenommen hat. In diesem Kontext ist zu diskutieren, ob und ab wann diese (wieder) als potenzielle Kunden betrachtet werden können oder sollten. Kriterien, die in diesem Zusammenhang verwendet werden, sind etwaige Vertragsdauern[671], die Regelmäßigkeit der Käufe[672], die Länge der Zeitintervalle zwischen den Käufen[673] oder die Gründe einer abgebrochenen Be-

den Kunden als zu definierenden Begriff gehen dagegen z. B. *Friederichs-Schmidt, S.* (2006) oder *Foehn, P.* (2006) ein.

669 Vgl. hier und nachfolgend *Lippmann, H.* (1992), S. 4.

670 *Lippmann, H.* (1992), S. 4.

671 Vgl. *Fischer, T. M./Schmöller, P.* (2006), S. 485 f.

672 Vgl. z. B. *Nötzel, R.* (1982), S. 453.

673 Vgl. z. B. *Krüger, S. M.* (1997), S. 15; *Dwyer, F. R.* (1997), S. 6 ff.

ziehung[674]. Nahezu alle Bankleistungen sind vertragsgebunden, wobei jedoch die noch nicht beendete Vertragsdauer nicht in jedem Fall einen verlässlichen Indikator für einen aktiven Kunden darstellt. Einige Bankleistungen werden z. B. über Rahmenverträge kontrahiert, die ggf. ungekündigt sind, der Kunde dennoch keine geschäftliche Interaktion mit der Bank mehr vornimmt.[675] Daher soll mit GADDE/MATTSON von der beobachtbaren Beendigung einer Geschäftsbeziehung ausgegangen werden, wenn „no commercial exchange has taken place during a specific time period"[676]. Problematisch ist, diesen spezifischen Zeitraum festzulegen. Für die vorliegende Arbeit sollen dazu typisierte Wiederkaufintervalle der in der Geschäftsbeziehung kontrahierten Bankleistungen genutzt werden.[677] In Abhängigkeit der konkreten Bankleistung kann dabei zwischen einem dauerhaften und einem situationsabhängigen Bedarf unterschieden werden.[678] Eine Geschäftsbeziehung gilt dann als beendet, wenn kumulativ für alle relevanten Bankleistungen typische Wiederkaufintervalle ohne Wiederkauf oder Interaktion (z. B. in Form von Angebotsanfragen) verstrichen sind. Zwar kann dies aufgrund zufälliger Abweichungen der spezifischen Dauer zu Fehlentscheidungen führen. Dennoch gilt, dass aus „Sicht eines Lieferanten [.] aber davon auszugehen [ist; Erg. d. Verf.], dass die Wahrscheinlichkeit, einen Kunden verloren zu haben, mit der über das mittlere Wiederkaufintervall hinausgehenden Länge der Inaktivität zunimmt."[679] Die durch Verträge bestimmte Dauer von Bankleistungen (z. B. Darlehensverträge mit definierter Laufzeit sowie Tilgungs- und Zinsmodalitäten) dient bei diesen Produkten als relevante Determinante des Wiederkaufintervalls. Hinsichtlich der Gründe abgebrochener Beziehungen unterscheiden STAUSS/FRIEGE sechs Gruppen:[680] Absichtlich und unabsichtlich vertriebene Kunden, abgeworbene Kunden, weggekaufte Kunden, ungewollt ausscheidende Kunden und sich entfernte Kunden. Grundsätzlich sollen Altkunden, bei denen ein Ende der Geschäftsbeziehung festgestellt wurde, als potenzielle Kunden verstanden werden. Ausnahmen bilden Situationen, in denen die

674 Vgl. *Stauss, B./Friege, C.* (2006), S. 517.

675 Der Kundenwille ist hierbei aus Anbietersicht nicht prägnant. Vgl. *Günter, B./Helm, S.* (2003), S. 52.

676 *Gadde, L.-E./Mattson, L.-G.* (1987), S. 30.

677 Vgl. dazu *Günter, B./Helm, S.* (2003), S. 52.

678 Vgl. *Schulz, T. C.* (2005), S. 81. Wiederkaufintervalle können daher zwischen den einzelnen Bankleistungen stark variieren. Während bei Finanzierungsleistungen das vertragliche Ende den Wiederkaufintervall determiniert, ist das Wiederkaufintervall bei Produkten wie dem Zahlungsverkehr oder Geldanlagen institutsindividuell zu bestimmen.

679 *Reinartz, W. J./ Krafft, M.* (2001), S. 1268.

680 Vgl. hier und nachfolgend *Stauss, B./Friege, C.* (2006), S. 517.

Existenz des Kunden nicht mehr gegeben ist (sich entfernte Kunden), so etwa bei Beendigung der unternehmerischen oder gewerblichen Tätigkeit, oder bei Beziehungsbeendigungen, denen ein fundamentaler Dissens zwischen den Geschäftspartnern zugrunde liegt (absichtlich vertriebene Kunden). Zu diesen zählen auch Geschäftsbeziehungen, die aus Sicht der Kreditgenossenschaft auf absehbare Zeit nicht fortgeführt werden sollen, so z. B. aufgrund der Risikosituation des Kunden.

Vereinfachend soll aufbauend auf dem numerischen Kundenpotenzial im weiteren Verlauf der Arbeit die Gruppe der aktuellen Kunden als Kunden im engeren Sinn von der Gruppe der aktuellen und potenziellen Kunden, die als Kunden im weiteren Sinn bezeichnet wird, abgegrenzt werden.[681] Das numerische Potenzial stellt die Anzahl und Art der Alternativen im Sinne des gerundiven Werts für das Unternehmen dar. Damit wird zudem der Anforderung Rechnung getragen, dass die kundenwertorientierte Vertriebssteuerung sowohl Bestands- als auch Nichtkunden berücksichtigen soll.

Eine weitere inhaltliche Abgrenzung resultiert aus der Betrachtung der Definition des Begriffs Kunde nach DILLER, der auf den Entscheidungseinfluss als begriffskonstitutives Merkmal abstellt.[682] Das Problem findet sich z. B. bei multipersonellen Kaufentscheidungen oder als mögliches Ergebnis asymmetrischer Lieferbeziehungen.[683] Die Multipersonalität von Kaufentscheidungen ist im Kontext des für diese Arbeit relevanten Firmenkundengeschäfts, das auch als Business-to-Business-Geschäft bezeichnet werden kann, grundsätzlich denkbar, erhöht jedoch die Abgrenzungsprobleme deutlich. Zwar ist die Kenntnis über Eigenschaften von so genannten Buying-Centern bei organisationalen Beschaffungen wichtig, um den Marketing-Mix zielsetzungsgerecht auszugestalten.[684] Aus zwei Gründen soll diese Besonderheit jedoch nicht berücksichtigt werden. Erstens kann empirisch begründet angenommen werden, dass die entscheidende Persönlichkeit auf Seiten der Firmenkunden der Unternehmer oder ein angestellter Geschäftsführer selbst ist. Denn aufgrund der Größe der Firmenkunden findet bei diesen Stellen häufig eine entsprechende Aufgabenkonzentration statt.[685] Zweitens ist die Vertriebssteuerung eine dem Marketing sachlich nachgelagerte Funktion, für die die Struktur des Kunden von untergeordneter Bedeutung ist. Daher soll im Folgenden grundsätzlich vereinfachend von individuellen Kunden gesprochen werden.[686]

---

681 Vgl. *Tomczak, T./Rudolf-Sipötz, E.* (2006), S. 129.

682 Vgl. *Diller, H.* (1992), S. 583.

683 Vgl. dazu z. B. *Rieker, S.* (1995), S. 47 f.

684 Vgl. dazu ausführlich *Backhaus, K.* (1999), S. 60 ff.

685 Vgl. *Segbers, K.* (2007), S. 67 f. und S. 220.

686 Vgl. ähnlich *Eberling, G.* (2002), S. 26.

Im Folgenden ist es in Anlehnung an die empirische Struktur und die Definition des Segmentes Firmenkunden irrelevant,[687] ob es sich bei den Nachfragern um natürliche (z. B. im Falle eines Einzelunternehmers) oder juristische Personen (z. B. im Fall einer Kapitalgesellschaft) handelt und ob es einzelne Personen, Gruppen, Institutionen oder Organisationen sind.[688] Ausschlaggebend für die weitere Berücksichtigung eines Kunden im Rahmen der kundenwertorientierten Vertriebssteuerung ist, dass diese der Geschäftseinheit der Firmenkunden in der Kreditgenossenschaft zugeordnet werden.

1.2.3 Ableitung des Verständnisses des Begriffs Kundenwert

Wie in Kapitel D.1.2.1 dargelegt wurde, ist der gerundive Wert die Grundlage des für diese Arbeit abzuleitenden Begriffs des Kundenwerts. In diesem Zusammenhang wurde festgestellt, dass die unternehmensindividuellen Ziele zentrale Wertdeterminanten sind. Soll nun der Kundenwert begrifflich abgegrenzt werden, nimmt er also direkt Bezug zum Zielsystem der Unternehmung. Für den Unternehmenskontext ist jedoch zu konstatieren, dass es kein allgemeingültiges Zielsystem gibt.[689] In der Literatur werden mit Rekurs auf marktwirtschaftliche Wirtschaftssysteme dem Ziel der (langfristigen) Gewinnmaximierung und der Unternehmenswertmaximierung die größte Bedeutung und zugleich ein Absolutheitsanspruch zuerkannt.[690] Gleichwohl zeigen empirische Ergebnisse, dass dies nicht widerspruchsfrei aufrechterhalten werden kann.[691] So hat das Gewinnziel oft eine relative Bedeutung im Sinne eines Mindestgewinns wobei gleichzeitig weitere Ziele verfolgt werden. Unterschieden werden können in diesem Zusammenhang monetäre und nicht-monetäre Ziele.[692] Nicht-monetäre Ziele können sowohl ökonomischer, außerökonomischer oder nicht-ökonomischer Natur sein (siehe Abbildung 28).

---

687 Vgl. dazu die Ausführungen in Kapitel C.2.2.

688 Vgl. *Diller, H.* (1992), S. 583. Unter die Kundengruppen fallen auch die Kundenverbünde. Vgl. dazu die Ausführungen in Kapitel C.2.2.4.

689 Vgl. *Küpper, H.-U.* (2001), S. 18.

690 Vgl. z. B. *Wöhe, G.* (1996), S. 124; *Eberling, G.* (2002), S. 43 ff.; *Henseler, J./Hoffmann, T.* (2003), S. 17; *Tewes, M.* (2003), S. 38 f.; *Wille, K.* (2005), S. 45 ff.; *Schroeder, N.* (2006), S. 11 ff.

691 So stellt KROL in einer Befragung von 214 kleineren und mittleren Unternehmen fest, dass finanzielle Zielgrößen wie die Eigenkapitalrendite aus Sicht der Befragten den nicht-finanziellen Zielen wie z. B. der Kunden- oder der Mitarbeiterzufriedenheit untergeordnet sind. Vgl. *Krol, F.* (2009), S. 5.

692 Vgl. *Wöhe, G.* (1996), S. 125. Monetäre Ziele sind in Geldeinheiten bewertet.

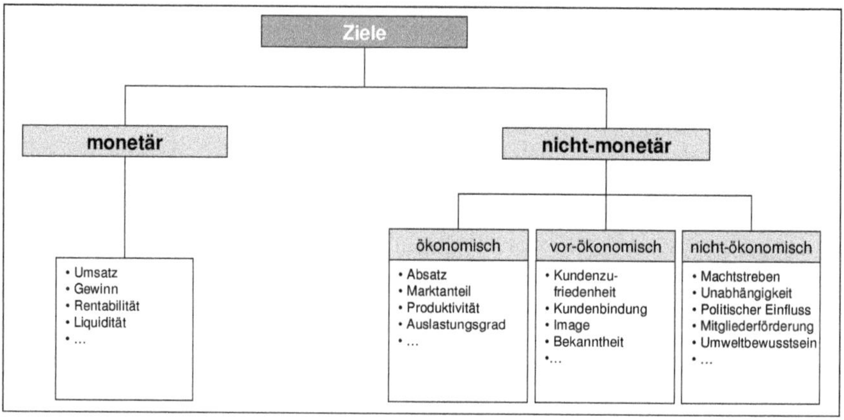

Abbildung 28: Zielsystematik[693]

Mit Bezug zu nicht näher konkretisierten Unternehmenszielen definiert CORNELSEN den Kundenwert als „ein Maß für die ökonomische Bedeutung eines Kunden, d. h. dessen direkten und/oder indirekten Beitrag zur Zielerreichung eines Anbieters."[694] Im Rahmen der theoretisch-konzeptionellen Vorüberlegungen seiner Arbeit kritisiert EBERLING vor dem Hintergrund der praktischen Existenz nicht-ökonomischer Ziele diese auf die ökonomische Bedeutung reduzierte Betrachtung CORNELSENs und anderer Autoren,[695] und legt seiner Arbeit die Definition zugrunde, dass der Kundenwert „eine Größe zur Beurteilung der eingesetzten Mittel für eine und des resultierenden Nutzens aus einer Geschäftsbeziehung hinsichtlich ihres Beitrags zur Erfüllung des Zielsystems des bewertenden Unternehmens"[696] ist. Neben der weiter gefassten Bedeutung sieht er die Kundenbeziehung statt des Kunden als relevanten Bewertungsgegenstand an.[697] Diese Sicht basiert auf der Feststellung, dass „Wert […] durch die vielfältigen dyadischen oder multipersonalen Austauschbeziehungen unter der Annahme, dass diese nicht nur rein deterministischer Natur, sondern *gestaltbar* und zumindest von *Dauer* sind"[698], entsteht. An der Definition von EBERLING kritisiert WILLE,

[693] In Anlehnung an *Cornelsen, J.* (2000), S. 30.

[694] *Cornelsen, J.* (2000), S. 43. Diese Definition legt auch VOGEL ihrer Arbeit zur Beziehung von Kundenbindung und Kundenwert zugrunde. Vgl. *Vogel, V.* (2006), S. 45.

[695] Vgl. *Eberling, G.* (2002), S. 36.

[696] *Eberling, G.* (2002), S. 36.

[697] So auch WIEDMANN/HENNIGS/IN DER STROTH, die jedoch ihre Begriffsauffassung auch auf die rein ökonomische Bedeutung begrenzen. Vgl. *Wiedmann, K.-P./Hennigs, N./In der Stroth, A.-B.* (2006), S. 12.

[698] *Stahl, H. K.* (1996), S. 387.

dass nur monetäre Ziele zu berücksichtigen seien.[699] Denn nach seiner Auffassung können auch bei einer weiter gefassten Zielbetrachtung nur die monetären Ziele dominant sein. Zwar erkennt er an, das dass Zielsystem der Unternehmung die Grundlage im Kundenwertkonzept ist, aber er verweigert mit dem Hinweis auf den Anspruch der Betriebswirtschaftslehre, rationale Entscheidungen zu begründen, die Möglichkeit nicht-monetäre Ziele zu berücksichtigen. Weiterhin argumentiert er, dass das erwerbswirtschaftliche Prinzip GUTENBERGS der wissenschaftlichen Forschung helfe, dieses normative Problem der Zieldefinition zu lösen. Nach seiner Interpretation unterstellt dieses Prinzip, dass alleinig die Gewinnmaximierung in Bezug auf das eingesetzte Kapital rational ist. Daraus folgert er als allgemeines Ziel „die Sicherstellung und Maximierung eines profitablen Wachstums durch eine simultane Optimierung von Customer Value und Shareholder Value“[700]. Kritisch an dieser Vorgehensweise ist die nicht zu beantwortende aber fundamentale Frage, ob die implizit enthaltene Shareholder Value-Orientierung allgemeingültig und ist und z. B. auf die zahlenmäßig stark vertretenen mittelständischen Unternehmen im Allgemeinen und auch auf Kreditgenossenschaften im Speziellen übertragen werden kann. Auf diese Frage wird später bei der Betrachtung des kreditgenossenschaftlichen Zielsystems noch detailliert eingegangen werden.[701]

Für diese Arbeit soll, abgeleitet aus den Begriffsdefinitionen für Kunde und Wert, unter Kundenwert der vom Anbieter subjektiv wahrgenommene Beitrag zu dessen Zielerreichung verstanden werden, der sich aus der Geschäftsbeziehung zu einem aktuellen oder potenziellen Kunden ergibt oder zu erwarten ist.[702] Dieses Verständnis kann sowohl auf Firmen- als auch auf Privatkunden übertragen werden, daher ist die Definition mit Blick auf den Kunden allgemeingültig. Der Kundenwert kann als Indikator über das Ausmaß, in dem ein Kunde bzw. die Geschäftsbeziehung zu diesem Kunden dazu beiträgt, die monetären und nicht-monetären Ziele eines Anbieters zu erfüllen, interpretiert werden.[703] Im Gegensatz zu den Definitionen von WILLE und SCHROEDER wird der Begriff Kundenwert bewusst nicht auf eine ökonomische Saldogröße,[704] die nur Erlöse

[699] Vgl. hier und nachfolgend *Wille, K.* (2005), S. 43 ff.

[700] *Wille, K.* (2005), S. 46.

[701] Vgl. dazu die Ausführungen in Kapitel D.3.1.

[702] Vgl. mit ähnlicher Definition des Begriffs Kundenwert *Helm, S./Günter, B.* (2006), S. 7; *Tomczak, T./Rudolf-Sipötz, E.* (2006), S. 129. Begrifflich synonym wird auch der Begriff Customer-Lifetime-Value genutzt. Vgl. *Günter, B.* (2006), S. 244. Da jedoch dieser Begriff für eines der im Folgenden zu erläuternden Verfahren der Messung des Kundenwerts genutzt wird, soll auf eine synonyme Verwendung verzichtet werden.

[703] Vgl. *Cornelsen, J.* (2000), S. 38; *Schemuth, J.* (1996), S. 21.

[704] Vgl. *Wille, K.* (2005), S. 44; *Schroeder, N.* (2006), S. 31.

und Kosten berücksichtigt, beschränkt, da dies letztendlich vom unternehmensindividuellen respektive kreditgenossenschaftlichen Zielsystem abhängt. Indem diese Beschränkung nicht vorliegt, kann der möglichen genossenschaftlichen Vielfalt besser Rechnung getragen werden. Mit dieser Definition wird weiterhin impliziert, dass nicht der Kunde selbst das zentrale Untersuchungs-, Beurteilungs- und Bewertungsobjekt ist, sondern die mit dem Kunden bestehende Geschäftsbeziehung, die als gestaltbar und andauernd angenommen wird.[705] In Anlehnung an SCHEER und in Verbindung mit obigen Aussagen zum numerischen Kundenpotenzial erfolgt eine begriffliche Differenzierung. Für aktuelle Kunden wird der interne Kundenwert und für potenzielle Kunden der externe Kundenwert bestimmt.[706]

Weiterhin ist es geboten, eine zusätzliche begriffliche Differenzierung vorzunehmen. WÜNSCHMANN empfiehlt, „den Kundenwert (im Sinne einer Potentialgröße) nicht an der aktuellen Qualität der Kundenbeziehung abzulesen. [...] Vielmehr interessiert, welchen Wert der Kunde hätte, wenn er seinen gesamten Bedarf bei dem Anbieter nachfragen würde“[707]. Das Potenzial selbst sieht er als anbieterexogene Größe an.[708] Anders ist dies bei TEWES, der den maximal möglichen Wertbeitrag eines Kunden als Kundenwertpotenzial bezeichnet, das den anbieterspezifischen Kundenwert umfasst.[709] Der Kundenwert, definiert als Beitrag zur Zielerreichung, ergibt sich folglich als Ausschöpfungsgrad dieses Potenzials. Dieses Verständnis entspricht der hier zugrunde gelegten Definition des Kundenwerts und führt daher zur folgenden weiteren Begriffsdifferenzierung. Zusätzlich zum Begriff des Kundenwerts ist der Begriff des Kundenwertpotenzials einzuführen. Dieses entspricht dem zu einem bestimmten Bewertungszeitpunkt maximal möglichen Kundenwert.[710] Mit dieser inhaltlichen Differenzierung des Begriffs Kundenwert wird der Annahme Rechnung getragen, dass der Kundenwert als vom Anbieterunternehmen beeinflussbare Größe zu verstehen ist.[711] Anbieterseitige Möglichkeiten, die auch als Kundenwerthebel bezeichnet werden, sind bspw. Marketing-Mix-Entscheidungen oder Preise, Volumina oder

---

705 Vgl. *Helm, S.* (2006), S. 105; *Stahl, H. K.* (1996), S. 387.

706 Vgl. *Scheer, O.* (1989), S. 36.

707 *Wünschmann, S.* (2007), S. 32.

708 Ähnlich auch *Gelbrich, K.* (2001), S. 62 und *Gelbrich, K./Müller, S.* (2006), S. 460.

709 Vgl. *Tewes, M.* (2003), S. 80.

710 STÜKER spricht in diesem Zusammenhang auch vom Kundenpotenzial. Er bezeichnet aber den Kundenwert als kundenbezogenes Erfolgspotenzial, wobei der Begriff des Erfolgspotenzials im Sinne GÄLWEILERS verwendet wird. Vgl. *Stüker, D.* (2008), S. 32. Ähnlich auch *Wille, K.* (2005), S. 52 f.

711 Vgl. *Stahl, H. K. et al.* (2009), S. 261; *Howald, B.* (2007), S. 51.

Stückkosten.[712] Auch RUDOLF-SIPÖTZ argumentiert mit Potenzialen, die aktuell bestehende und künftige bzw. gegenwärtig noch nicht bestehende Möglichkeiten der Realisierung von Kundenwertbeiträgen umfassen.[713] Nach WILLE handelt es sich dabei um einen „Prozess, der (objektiv) vorhandene Kundenpotenziale in subjektiven Nutzen überführt."[714] Abbildung 29 fasst diese Überlegungen zusammen.

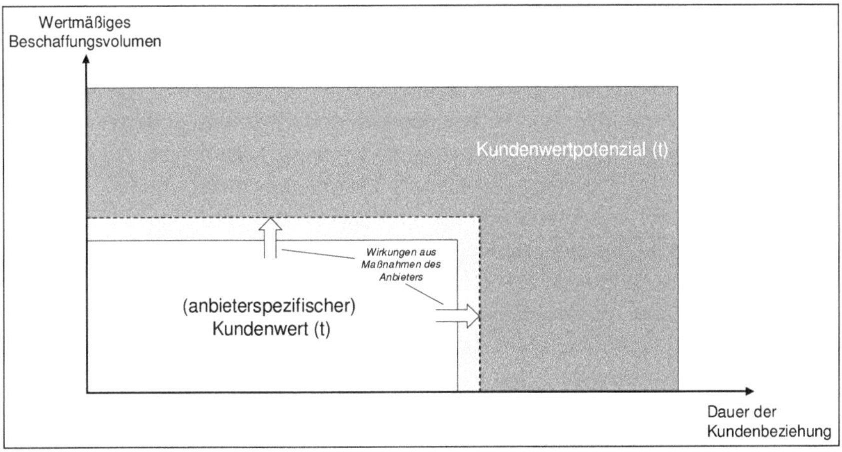

Abbildung 29: Zusammenhang von Kundenwert und Kundenwertpotenzial[715]

Nachdem das grundlegende Verständnis des Begriffs Kundenwert abgeleitet ist, soll dieser hinsichtlich seiner möglichen Ausprägungsdimensionen betrachtet werden.

1.2.4 Dimensionen des Kundenwertbegriffs

Um den Begriff des Kundenwerts sowie des Kundenpotenzials näher zu konkretisieren, lassen sich in Anlehnung an EBERLING drei Dimensionen unterscheiden: Die personell/institutionelle oder auch objektbezogene Dimension, die zeitliche Dimension und die sachliche Dimension.[716]

---

712 Vgl. *Juncker, K.* (2006), S. 346.
713 Vgl. *Rudolf-Sipötz, E.* (2001), S. 22.
714 *Wille, K.* (2005), S. 52.
715 In Anlehnung an *Kiwitz, K./Scheffler, C.* (2009), S. 36.
716 Vgl. *Eberling, G.* (2002), S. 34 ff.

1.2.4.1 Objektbezogene Dimension

Bei der objektbezogenen Dimension des Kundenwerts steht die Frage im Fokus, was als Bewertungsobjekt festgelegt wird. Wenngleich die in dieser Arbeit zugrunde gelegte Definition des Begriffs Kundenwert auf einen einzelnen Kunden bzw. die Geschäftsbeziehung mit diesem Kunden fokussiert, stellt sich auf der Ebene der Wertermittlung die Frage, für welchen Differenzierungsgrad diese stattfinden soll.[717] So ist es offensichtlich, dass für einen auf einem Massenmarkt tätigen Produzenten von Konsumgütern die Kundenwertermittlung auf Ebene des einzelnen Kunden aufwendiger ist als für einen spezialisierten Auftragsfertiger, der nur eine geringe Anzahl Kunden beliefert. Während ersterer aus wirtschaftlichen Erwägungen eher auf einem höheren Aggregationsniveau Kundenwerte ermittelt, wird letzterer jeden Kunden einzeln bewerten (können).

In der Literatur werden drei Bezugsgrößenebenen unterschieden.[718] Auf der untersten Ebene ist der individuelle Kundenwert angesiedelt. Vorteilhaft an der individuellen Kundenbewertung ist neben der Kongruenz zum Begriff des Kundenwerts vor allem die Möglichkeit des Rückschließens auf die einzelne Geschäftsbeziehung im Sinne eines „segment-of-one".[719] Damit ist der größte Informationsgehalt verbunden. Bei großen Kundenanzahlen übersteigt dieses Vorgehen jedoch den ökonomisch vertretbaren Ermittlungsaufwand.[720] DILLER postuliert dennoch, dass eine dynamische Konzeption des Kundenwerts aufgrund der Notwendigkeit individueller Annahmen und Daten einer Individualkundenbetrachtung bedarf.[721]

Werden die Kunden segmentiert, so ist eine die Komplexität der Aufgabenstellung reduzierende segmentbezogene oder kundengruppenbezogene Beurteilung als Aggregationsebene möglich. Angenommen wird dabei, dass die individuellen Kundenwerte in den Segmenten homogen sind.[722] Zwar reduziert dieses Vorgehen die Kosten der Kundenwertermittlung, hat jedoch den gravierenden Nachteil, dass es mit großen Informationsverlusten verbunden ist. Denn durch die notwendige Verwendung von statistischen Durchschnittswerten werden in Wahrheit unterschiedlich wertvolle Kunden egalisiert, wodurch es zu Fehlinter-

---

717 Vgl. hier und nachfolgend *Eberling, G.* (2002), S. 37; *Vogel, V.* (2006), S. 44.

718 Vgl. *Diller, H.* (2001), S. 25.

719 Vgl. *Vogel, V.* (2006), S. 44; *Krafft, M.* (2007), S. 51.

720 Vgl. *Diller, H.* (2001), S. 25. Es kann jedoch keine allgemeingültige Aussage darüber getroffen werden, ab welcher Anzahl an Kunden dieses Vorgehen ökonomisch nicht mehr vertretbar wird.

721 Vgl. *Diller, H.* (2001), S. 25.

722 Vgl. *Eberling, G.* (2002), S. 37.

pretationen und Fehlanreizen kommen kann.[723] Bei statischen Konzeptionen des Kundenwerts gilt dies als akzeptabel.[724]

Auf dem maximalen Aggregationsniveau wird der Kundenstamm bewertet. Einige Autoren unterstellen in diesem Kontext, dass die Einzelkundenwerte zu segmentbezogenen Werten summiert werden können und/oder die Summe der segmentbezogenen Werte dann dem Kundenbestandswert des Unternehmens entspricht.[725] Dies basiert auf der Annahme der Wertadditivität des Kundenwerts, die im Rahmen der Kundenwertmodellierung und Kundenwertermittlung zu berücksichtigen ist.[726] Diese Vorgehensweise wird vor allem dazu genutzt, den Wert junger Unternehmen zu bestimmen.[727] Hinsichtlich der Vor- und Nachteile dieser Bezugsebene sei auf die segmentbezogene Kundenbewertung verwiesen. Aufgrund der Trennung von Privat- und Firmenkundengeschäft ist diese Bezugsebene für die Vertriebssteuerung im Firmenkundengeschäft nur von untergeordneter Bedeutung, für die vor allem strategische Gesamtbanksteuerung jedoch sehr interessant.

Grundsätzlich gilt für das Firmenkundengeschäft der Banken, wie die Ausführungen zur Zielgruppe[728] zeigen, dass die für die Kundenbewertung relevanten Rahmenbedingungen dergestalt vorliegen, dass tendenziell eine einzelkundenbezogene Bewertung vorgenommen werden kann und sollte. Das bedeutet nicht zwingend, dass für jeden Kunden der Kundenwert spezifisch ermittelt werden muss, sondern dass auf Ebene einzelner Firmenkunden ein Kundenwert

---

723 Vgl. *Beeser, A. J.* (2003), S. 56.

724 Vgl. *Diller, H.* (2001), S. 25.

725 Vgl. z. B. *Eberling, G.* (2002), S. 37; *Schroeder, N.* (2006), S. 140 f.; *Lissautzki, M.* (2007), S. 34. Der Kundenbestandswert wird in diesem Zusammenhang auch regelmäßig als rechnerische Wertdeterminante des Unternehmenswerts definiert. Vgl. auch *Wiesel, T./Skiera, B.* (2007), S. 709. Für eine vergleichende Darstellung verschiedener Ansätze der Berechnung des Kundenbestandswerts (Customer Equity) sei auf den Beitrag von KUMAR/GEORGE verwiesen. Vgl. *Kumar, V./George, M.* (2007).

726 Aus steuerungstheoretischer Sicht muss diese Annahme nicht getroffen werden. Diese Annahme wird getroffen, um aus den Kundenwerten den Unternehmenswert ermitteln zu können. Damit wird auch die Fokussierung auf monetäre Kundenwertbeiträge begründet, da dies modelltheoretisch konsistent ist und so eine Vielzahl von Problemen des Kundenwerts umgangen werden kann. Denn diese Annahme impliziert, dass Kundenwerte eindeutig sind. Die Frage der Eindeutigkeit tritt z. B. bei Kundenteilwerten auf, die beziehungsübergreifend sind, wie der Referenzwert. Vgl. z. B. *Helm, S.* (2003).

727 Vgl. *Stüker, D.* (2008), S. 2; *Gupta, S./Lehmann, D. R./Stuart, J. A.* (2004), S. 7 ff.; *Rehbach, S.* (2003), S 4 ff; *Eberling, G.* (2002), S. 48 f.

728 Vgl. die Ausführungen in Kapitel C.2.2.

ermittelt wird.[729] Ergänzend sei darauf hingewiesen, dass auch bei einer individuellen Kundenwertermittlung im kreditgenossenschaftlichen Kontext das Bewertungsobjekt der relevante, vertrieblich motiviert Kundenverbund sein kann.[730]

1.2.4.2 Zeitliche Dimension

Bezüglich der zeitlichen Dimension des Kundenwerts ist zu entscheiden, welcher Betrachtungszeitraum und welche Periodizität der Kundenwertermittlung zugrunde gelegt werden soll (siehe Abbildung 30). Ausgehend vom Bewertungszeitpunkt kann die Bewertung des Kunden retrospektiv (vergangenheitsorientiert), prospektiv (zukunftsorientiert), beziehungsbezogen oder beziehungsübergreifend erfolgen. Zudem können die Bewertungen in periodischer oder periodenübergreifender Form vorgenommen werden. Bei der prospektiven Bewertung werden die Beiträge der Kunden bzw. der Geschäftsbeziehung ab dem Bewertungszeitpunkt und in der retrospektiven Bewertung bis zum Bewertungszeitpunkt berücksichtigt. Soll die Kundenbewertung dahingehend erfolgen, den Totalwert zu bestimmen, so ist ein beziehungsdauerbezogener bzw. ein beziehungsdauerübergreifender Zeitraum zugrunde zu legen. Im Unterschied zur streng prospektiven Betrachtung werden dabei auch vergangene Beiträge des Kunden zur Zielerreichung berücksichtigt. Bei der Periodizität wird unterschieden, ob Beiträge einzelnen Perioden zugeordnet werden oder ob eine Periodisierung nicht stattfindet. Im erstgenannten Fall sind zur Kundenwertermittlung periodische oder periodisierte Größen zu nutzen. Bei einigen qualitativen Bestandteilen des Kundenwerts lässt sich der Bezug zu einzelnen Perioden nicht eindeutig vornehmen.

---

729 Das bedeutet, dass die individuellen Kundenwerte auch mittels Analogieschlüssen oder Durchschnittswerten ermittelt werden können. Diese Vorgehensweise ist jedoch nicht zu präferieren.

730 Vgl. dazu die Ausführungen in Kapitel C.2.2.4. Ähnlich geht STRÄTER im Rahmen der Kundenbewertung im Privatkundengeschäft vor. Statt einer Einzelkundenbewertung misst er die Kundenertragswerte auf Basis von Haushaltsgrößen. Begründet werden kann dies damit, dass Haushalte die aus Sicht der Bank relevante Betreuungseinheit des Vertriebs darstellen. Zudem liegen auf Haushaltsebene die für das Modell notwendigen statistischen Daten vor. Vgl. *Sträter, J.* (2005), S. 95.

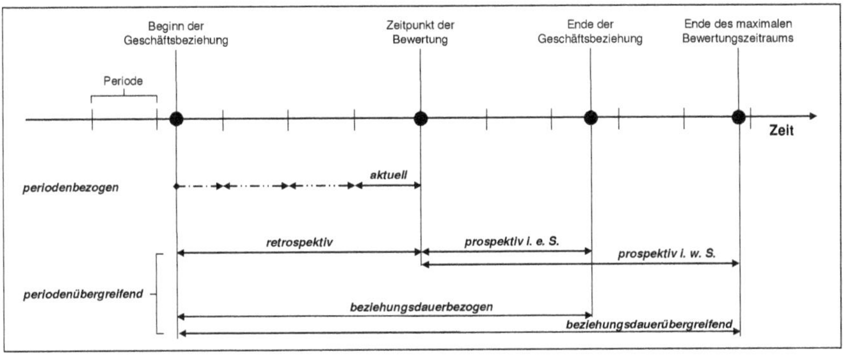

Abbildung 30: Zeitliche Differenzierung bei der Kundenwertbestimmung[731]

Die Auswahl des relevanten Bezugszeitraumes für die Kundenbewertung orientiert sich an der Datenverfügbarkeit sowie dem Zweck der Kundenbewertung. Während die Planung kundenwertorientierter Maßnahmen auf Grundlage prospektiver Betrachtungen erfolgt, basiert die Kontrolle der Erfolgswirksamkeit auf der prospektiven sowie retrospektiven und meist periodischen Betrachtung des Kundenwerts.[732] Um vertriebliche oder kundengerichtete Marketing-Maßnahmen entscheidungs- und steuerungstheoretisch zu fundieren, wird der Kundenwert daher prospektiv betrachtet. In diesem Kontext weicht die in Abbildung 30 dargestellte Differenzierung von den in der Literatur vorgestellten Ansätzen ab.[733] Während in der Literatur der prospektive Zeitabschnitt lediglich bis zum Ende der Geschäftsbeziehung reicht, wird dieser Zeitraum als prospektiv im engeren Sinne bezeichnet. Denn für die Bewertung des Kunden zum Bewertungszeitpunkt sind – in Abhängigkeit der berücksichtigten Wertdeterminanten – alle die Unternehmenszielerreichung unterstützenden Maßnahmen des Kunden zu berücksichtigen. Werden in der Kundenbewertung lediglich transaktionsbezogene monetäre Größen wie etwa der Deckungsbeitrag betrachtet, so ist die Begrenzung auf den prospektiven Zeitraum im engeren Sinn, der mit der Geschäftsbeziehung endet, richtig. Werden jedoch auch bspw. Referenztätigkeiten als Beiträge des Kundenwerts berücksichtigt, so kann ein aktueller Kunde im Sinne des Unternehmens agieren, auch wenn keine aktive Geschäftsbeziehung zum Unternehmen mehr besteht. Zu denken ist hier etwa im Firmenkundengeschäft der Banken an Unternehmer, die bei der Nachfolgeregelung ihre Geschäftsanteile verkaufen. Sofern keine private Geschäftsbeziehung besteht, ist

731 In Anlehnung an *Wille, K.* (2005), S. 56.
732 Vgl. *Eberling, G.* (2002), S. 38 f.
733 Vgl. *Eberling, G.* (2002), S. 39; *Wille, K.* (2005), S. 56.

per definitionem die Geschäftsbeziehung zu ihm beendet. Gleichwohl kann er positive Referenzen für die Bank aussprechen. In einer stringenten Auslegung des Referenzpotenzials wäre dies ebenfalls zu berücksichtigen. In diesem Fall ist es zweckmäßig, eine prospektive Betrachtung bis zum Ende des maximalen Bewertungszeitraums vorzunehmen. Diese wird als prospektiv im weiteren Sinne bezeichnet.[734]

An dieser Stelle sei darauf hingewiesen, dass auch der Zeitbezug der relevanten Ziele relevant ist. Bisherige Arbeiten gehen i. d. R. implizit davon aus, dass die Zeitbezüge der Ziele und des Kundenwerts identisch sind. Aber ist das Ziel eines Unternehmens die Gewinnmaximierung in der Periode t so ist grundsätzlich zu hinterfragen, ob der Kundenwert in der Periode $t+1$ einen tatsächlichen Zielerreichungsbeitrag leistet und somit zu berücksichtigen ist. Grundsätzlich gilt, dass der Kundenwert eine prospektive und beziehungsdauerübergreifende Größe ist, da vergangenheitsbezogene Ziele per definitionem nicht sinnvoll sind.

1.2.4.3 Sachliche Dimension

Die sachliche Dimension des Kundenwerts bezieht sich zum einen auf die Differenzierungsstufe des Kunden und zum anderen auf die Art der Beiträge zur Zielerreichung.[735] Die zu bewertenden Kunden können horizontal und vertikal differenziert betrachtet werden.[736] In der horizontalen Differenzierung ist zu entscheiden, welche Stufe des Absatzkanals[737] als Bewertungsobjekt betrachtet wird. Beim einstufigen, direkten Vertrieb ist dies eindeutig. Im Fall des indirekten, mehrstufigen Vertriebs ist zu entscheiden, welche Absatzstufe(n) (Absatzintermediäre und/oder Endkunden) bewertet werden soll(en).[738] Hier zeigt sich die

---

734 Diese Überlegung steht nicht, wie ggf. auf den ersten Blick zu vermuten ist, in einer konträren Position zur in dieser Arbeit verwendeten Begriffsdefinition des Kundenwerts. Denn die Grundlage der auch nach der Geschäftsbeziehung positiven Beiträge des Kunden basiert auf einer gemeinsamen Geschäftsbeziehung und ist somit auch beziehungsinduziert.

735 Vgl. *Eberling, G.* (2002), S. 40; *Wille, K.* (2005), S. 57, der diese Fragestellung jedoch der objektbezogenen Definition zuordnet. In diesem Fall liegt eine begrifflich abweichende aber inhaltlich identische Betrachtung vor.

736 Vgl. *Cornelsen, J.* (2000), S. 66.

737 Vgl. zum Begriff des Absatzkanals z. B. *Meffert, H.* (1998), S. 596 ff.

738 Bezogen auf den genossenschaftlichen Bankensektor sei an dieser Stelle beispielhaft auf die unterschiedlichen Vorgehensweisen des Produktabsatzes der Zentralbanken, WGZ BANK und DZ BANK, im mittelständischen Firmenkundengeschäft verwiesen. Während erstere sowohl direkt mit mittelständischen Unternehmen in Geschäftsbeziehungen tritt als auch indirekt, setzt die DZ BANK vornehmlich die Kreditgenossenschaften als Absatzmittler ein.

Verbindung zur objektbezogenen Dimension. Hat ein Anbieterunternehmen eine handvoll Absatzintermediäre, aber eine große Anzahl an Konsumenten am Ende des Absatzkanals, so stellt sich erstens die Frage nach der relevanten Absatzkanalstufe und zweitens nach dem für die Kundenwertermittlung relevanten Aggregationsniveau. Für die Kreditgenossenschaften erfolgt die Kundenwertermittlung auf der Stufe des Firmenkunden. Zwar können auch diese als Absatzmittler von Leistungen bspw. im Fall von Absatzfinanzierungen für Kunden des Automobilhandels sein, jedoch ist die Bedeutung dieser Rolle zu gering, als dass sie zu berücksichtigen ist.

Die vertikale Differenzierung ist immer dann zu diskutieren, wenn Kunden in Verbünden auftreten. So ist es bspw. regelmäßig im mittelständischen Firmenkundengeschäft der Banken der Fall, dass Kundenverbünde sich durch Kunden aus der gewerblichen (Unternehmen) und der privaten Sphäre (Unternehmer als Privatperson) zusammensetzen.[739] In der gewerblichen Sphäre ist es zudem möglich, dass Konzernstrukturen vorliegen, so dass die Bank mit einer Vielzahl einzelner Vertragspartner kontrahiert. Schwierigkeiten ergeben sich in diesem Fall aufgrund etwaiger unterschiedlicher Ergebnisse der Kundenbewertung und der darauf aufbauenden Behandlung des Kunden durch das Anbieterunternehmen. Auch hier gibt es keine generell richtige Vorgehensweise, sondern nur eine fallweise Entscheidung.

Die Auswahl der relevanten Zielerreichungsbeiträge erfolgt in direkter Abhängigkeit von den verfolgten Zielen. Wie in Abbildung 28 dargestellt, können Unternehmen eine Vielzahl unterschiedlicher Ziele mit unterschiedlichen Dimensionen verfolgen. Für die inhaltliche Ausgestaltung des Kundenwerts bedeutet dies eine zielkonsistente Vorgehensweise bei der Modellierung und Messung des Kundenwerts. So können bei monetären Zielen wie z. B. Umsatzzielen, Kundenwerte bspw. als Kundenumsatz relativ zum Gesamtunternehmensumsatz gemessen werden. „Die Konzeptualisierung des Kundenwerts hat in den letzten Jahren einen erheblichen Perspektivenwechsel erlebt. Während lange Zeit ausschließlich monetäre Größen [...] im Vordergrund standen, hat in jüngerer Zeit eine Öffnung des Begriffs stattgefunden, um der Realität besser zu entsprechen."[740] Wird dieser Entwicklung gefolgt und werden auch nicht-monetäre und nicht-ökonomische Ziele verfolgt, so sind vielfältige Komponenten in den Kundenwert zu integrieren (z. B. das Image des Kunden). In diesen Fällen muss ggf. auch von einer verhältnisskalierten Messung abgewichen werden, was zu etwaigen Informationsverlusten führen kann.[741] Dies ist vor allem dann der Fall, wenn

---

739 Vgl. dazu die Ausführungen in Kapitel C.2.2.4.

740 *Meyer, A./Kantsperger, R./Wilkoszewski, A.* (2005), S. 300.

741 Vgl. *Eberling, G.* (2002), S. 14.

monetäre und nichtmonetäre Kundenwertbeiträge in einer Kundenwertkennzahl zusammengefasst werden sollen (z. B. mittels Scoring-Modellen). Auf die Frage der relevanten Wertbeiträge wird in Kapitel A.1 der Arbeit zurückgekommen.

2 Erfolgsvoraussetzungen einer kundenwertorientierten Vertriebssteuerung

Die Mehrzahl der einschlägigen Beiträge zur Kundenwertorientierung fokussiert einseitig auf die Beziehungsorientierung zwischen Kunde und Unternehmen als Begründung für ein kundenwertorientiertes Management.[742] LISSAUTZKI dagegen formuliert darüber hinaus gehende Erfolgsvoraussetzungen, die die Zweckmäßigkeit der Kundenwertorientierung in der Unternehmenssteuerung begründen.[743]

Als notwendige Bedingungen fasst er die Relevanz der Dimension Kunde sowie die heterogenen Kundencharakteristika als Voraussetzungen des Marktumfelds zusammen. Im Sinne der engpassorientierten Führungsphilosophie nach GUTENBERG ist eine kundenwertorientierte Steuerung vorteilhaft, wenn Kunden ein knappes Gut und damit einen Engpassfaktor für die Erreichung der Unternehmensziele darstellen. Dies ist bspw. bei monopolistischen Marktstrukturen oder effektiven Kollusionen nicht der Fall. Wie in Kapitel C.1.5 gezeigt wurde, entspricht dies nicht der Marktstruktur auf dem deutschen Bankenmarkt, der eher als oligopolistisch etikettiert wurde. Praxisbeiträge argumentieren zudem, dass sich der Firmenkundenmarkt vom Verkäufer- zum Käufermarkt entwickelt und das anbieterseitige Bemühen um Firmenkunden in den vergangenen Jahren zugenommen hat, was sich insbesondere im preisbasierten Wettbewerb artikuliert.[744] Es ist somit zu konstatieren, dass der Firmenkunde als knappes Gut zu bezeichnen ist und somit die Relevanz der Dimension Kunde gegeben ist.

Weiterhin fordert LISSAUTZKI, dass sich die Heterogenität der Kunden auch in den Ausprägungen der Kundenwerte widerspiegelt.[745] Problematisch daran ist jedoch, dass die Prüfung dieser Forderung nur möglich ist, wenn alle Kunden hinsichtlich ihres Kundenwerts bewertet werden und folglich eine entsprechende Kundenwertorientierung vorliegt. Daher formuliert er Indikatoren, die sich auf die Ausprägung des Kundenwerts auswirken. Gilt für diese Größen eine ausreichend plausibel vermutete Verschiedenheit zwischen Kunden oder Kundenseg-

---

742 Vgl. z. B. *Eberling, G.* (2002), S. 50 ff.

743 Vgl. hier und nachfolgend *Lissautzki, M.* (2007), S. 46 ff.

744 Vgl. dazu ausführlich z. B. *Schmoll, A.* (2006), S. 11 ff. sowie die Ausführungen in Kapitel A.1.

745 Vgl. hier und nachfolgend *Lissautzki, M.* (2007), S. 51 ff. Ähnlich auch *Eberling, G.* (2002), S. 41.

menten, kann die Forderung als erfüllt angesehen werden. Die Indikatoren beziehen sich auf den nachgefragten Leistungsmix und die Leistungsmenge, auf preisliche und vertriebliche Differenzierungen sowie auf Dauern und Risiken der Geschäftsbeziehungen. Mit Blick auf die Tabelle 1 und vor dem Hintergrund der grundsätzlichen Heterogenität der Firmenkunden der Kreditgenossenschaften und der differenzierten vertrieblichen Beratung und Betreuung in der Praxis der Kreditgenossenschaften,[746] kann diese Heterogenität begründet unterstellt werden.

Als hinreichende Bedingungen einer kundenwertorientierten Steuerung müssen nach LISSAUTZKI die Wertorientierung als Hauptzielsetzung im Zielsystem des Unternehmens verankert sein, müssen Kundeninformationen vorliegen oder ermittelbar sein, soll das Unternehmen keine nischenfokussierte Positionierung am Markt haben, muss die kundenwertorientierte Steuerung dem Postulat der Wirtschaftlichkeit entsprechen und muss eine nach kundenwertorientierten Aspekten differenzierte Steuerung de jure und de facto umsetzbar sein.[747] Die strikte Ausrichtung des Zielsystems auf wertorientierte Ziele muss zunächst relativiert werden, da der Anforderungskatalog von LISSAUTZKI im Kontext einer kundenwertorientierten Steuerung in Verbindung mit einer kapitalmarktorientierten Unternehmenssteuerung zu sehen ist. Da dies für die Kreditgenossenschaften nicht gilt, soll zunächst die Wertorientierung weniger eng kapitalmarktorientiert erfolgen. Nach EBERLING ist das Ziel der wertorientierten Unternehmensführung, „die Steigerung eines (wie auch immer gearteten) Unternehmenswertes.“[748] Diesen setzt er mit einem an den Anspruchsgruppen orientierten Nutzenpotenzial gleich. Unter Berücksichtigung dieser weiten Auslegung der wertorientierten Unternehmensführung kann mit Verweis auf die Vermögenswertorientierung in VR-CONTROL begründet werden, dass dies grundsätzlich auch für Kreditgenossenschaften gilt. Wenngleich hier aufgrund des Förderauftrags weitere Überlegungen anzustrengen sind.

Aufgrund der den bereits heute eingesetzten relationalen Vertriebsansätzen inhärenten großen Bedeutung von Informationen und den relativ umfangreichen Möglichkeiten den Kundeninformationsstand in Banken aufzubauen,[749] kann die Forderung nach hinreichend Kundeninformationen tendenziell bejaht werden.[750] Aufgrund ihres Allfinanzangebots kann zudem plausibel angenommen werden, dass die Kreditgenossenschaften keine Nischenstrategie im Sinne der fokussier-

---

[746] Vgl. dazu die Ausführungen in den Kapiteln C.2.2 und C.2.4.

[747] Vgl. hier und nachfolgend *Lissautzki, M.* (2007), S. 54 ff.

[748] *Eberling, G.* (2002), S. 45.

[749] Vgl. dazu z. B. *Segbers, K.* (2007), S. 91 ff. und S. 165 ff.

[750] Vgl. die Ausführungen in Kapitel C.2.4.1.3

ten Zielgruppen oder Bankleistungen verfolgen.[751] Das Wirtschaftlichkeitspostulat ist in diesem Zusammenhang als Anforderung an die Konzeption einer kundenwertorientierten Vertriebssteuerung zu verstehen und insofern erst am Ende dieser Arbeit final zu beantworten.

Neben der noch nicht konkretisierten Definition des Zielsystems ist weiter auch die de jure und de facto Umsetzbarkeit einer kundenwertorientierten Vertriebssteuerung vor dem Hintergrund des spezifischen Zwecks der Kreditgenossenschaften intensiver zu diskutieren. Denn der Förderauftrag fokussiert auf die Beziehung zwischen Mitglieds-Firmenkunde und Kreditgenossenschaft. Und auf eben diese ist auch die kundenwertbedingte Differenzierung der anbieterseitigen Maßnahmen im Sinne des Marketing-Mixes im Allgemeinen und des Vertriebs im Speziellen gerichtet. Insofern ist zu prüfen, ob hier widerstreitende Einflüsse aus den beiden Konzepten vorliegen und in wie weit diese harmoniert werden müssen und können.

3 Genossenschaftsprinzipien und Kundenwertorientierung

Wie in Kapitel C.1.4.2 dargestellt, lässt sich die Beziehung zwischen einem Mitglied und der Kreditgenossenschaft basierend auf dem Identitätsprinzip in zwei Beziehungsebenen differenzieren.[752] Dies sind die Leistungsbeziehungsebene, bei der das Mitglied Abnehmer der genossenschaftlichen Leistungen ist, und die Einlegerbeziehungsebene, bei der die Gesamtheit der Mitglieder als Eigentümer und Träger der Verfügungsrechte in Form von Koordinations- (Planung, Organisation, Kontrolle), Aneignungs- (Gewinn und Verlust) und Veräußerungsrechten betrachtet wird.[753] Auf diese beiden Beziehungsebenen kann der Förderauftrag nach § 1 GenG wirken. Die Art und Weise der Förderung ist jedoch unbestimmt und muss institutsindividuell festgelegt werden.[754] Diese Festlegung ist ein zentraler Einflussfaktor auf das Zielsystem der Genossenschaftsbank. Die Ziele der Mitglieder als Eigentümer der Genossenschaft sind im Ziel-

---

751 Vgl. die Ausführungen in Kapitel C.2.3 und C.2.2. Die Beschränkung auf mittelständische Unternehmen bzw. die empirisch belegte Kundenstruktur mit Fokus auf kleine und mittlere Unternehmen schränken diese Aussage nicht ein, da über den subsidiären Verbundansatz auch größere Unternehmen als potenzielle Kunden verstanden werden können.

752 Das Identitätsprinzip umfasst den Rollenpluralismus der Mitglieder einer Kreditgenossenschaft, die i. d. R. gleichzeitig Kunden der Kreditgenossenschaft sind bzw. sein sollen. Vgl. Kapitel C.1.4.1.

753 Vgl. *Theurl, T.* (2002), S. 59.

754 Vgl. dazu die Ausführungen in Kapitel C.1.4.2.

system und folglich in der Steuerungsphilosophie der Kreditgenossenschaft zu verankern.[755] Die Steuerungsphilosophie umfasst die Vertriebssteuerung, die sich, wie im Weiteren erörtert wird, auf den Fördergeschäftsverkehr auswirken kann.

Um die Mitglieder als wichtigste Eigenkapitalgeber zu gewinnen und zu binden, ist es wirksam, die Unternehmensziele an den Mitgliederzielen auszurichten. Damit wird ferner die Bindungswirkung der Mitgliedschaft an die Kreditgenossenschaft intensiviert. Der Umfang der Förderung gilt als zentrale Determinante des Werts der Genossenschaft für das Mitglied und dessen Beitrittsentscheidung.[756]

Vor dem Hintergrund dieses so skizzierten Wirk- und Beziehungsgefüges sind für die kundenwertorientierte Vertriebsteuerung die bereits mit der Diskussion der Erfolgsvoraussetzungen angerissenen folgenden zwei Aspekte näher zu analysieren: Dies ist zum einen die Frage nach dem relevanten Zielsystem der Kreditgenossenschaft und zum zweiten die Frage nach der Vereinbarkeit von kundenwertorientierter Vertriebssteuerung und genossenschaftlichen Prinzipien. Bei der letzten Frage werden die Beziehungen zu den Mitglieder-Kunden fokussiert, da nur hier genossenschaftsspezifische Wirkbeziehungen möglich sind. Da keine Lösungsalternative darin zu sehen ist, ausschließlich das Nicht-Mitglieder-Kundengeschäft zu forcieren, um damit die Frage nach der Vereinbarkeit zu umgehen, ist die Notwendigkeit einer solchen Diskussion evident.

Die diesen Einzelaspekten übergeordnete Fragestellung nach dem Einfluss der genossenschaftlichen Prinzipien auf die kundenwertorientierte Steuerung ist pointiert mit dem Disput der normativen Kraft des Faktischen und der faktischen Kraft des Normativen beschrieben.[757] Daher ist eine Betrachtung und Gegenüberstellung theoretischer Ausführungen und tatsächlicher Realitäten notwendig.[758]

---

755 Nach LICHTSTEINER/SCHWARZ lassen sich in Abhängigkeit der verfolgten Ziele spezifische Logiken von Strategien und Maßnahmen differenzieren. Sie sehen auf der einen Seite die Marktlogik, die sich in einer strengen Gewinnorientierung manifestiert, und auf der anderen Seite die Sachzwecklogik, die sich am Angebot einer spezifischen Leistung zur Deckung eines bestimmten Bedarfs orientiert. Vgl. *Lichtsteiner, H./Schwarz, P.* (2008), S. 197 f. In Abhängigkeit der jeweils dominanten Logik sind alle nachgelagerten Funktionen auszugestalten, so auch Steuerungs- und Controllingfunktionen.

756 Vgl. *Schöffling, B.* (1991), S. 5 ff.

757 Vgl. im ähnlichen genossenschaftlichen Kontext *Sperling, D.* (1984), S. 33.

758 Vgl. auch *Theurl, T.* (2002), S. 51.

3.1 Zielsystem der Kreditgenossenschaften

3.1.1 Identifikation relevanter Ziele

Aufgrund des verwendeten Wertverständnisses sind Ziele die zentralen Determinanten des Kundenwerts.[759] Daher ist es von außerordentlicher Bedeutung für den weiteren Verlauf der Arbeit, die relevanten Ziele der Kreditgenossenschaften zu identifizieren.[760] Abweichend von den bisherigen wissenschaftlichen Beiträgen zum kundenwertorientierten Management müssen neben den organisationstyp-neutralen Zielen (z. B. Unternehmenswert-, Volumen-, Rentabilitäts- sowie Risiko-, Mitarbeiter- und kundenbezogene Ziele) auch organisationsspezifische Ziele explizit berücksichtigt werden.[761] Organisationsspezifische Ziele ergeben sich aus den faktischen oder juristischen Besonderheiten eines Unternehmens. Bei den Kreditgenossenschaften ergeben sich diese Ziele vor allem aus dem Identitätsprinzip und dem Förderzweck.

Eine typischerweise vorgenommene Zielverengung auf erwerbswirtschaftliche Ziele begründet tendenziell die Betrachtung einer Dividendengenossenschaft, was jedoch die gesetzlichen Strukturgrenzen der Kreditgenossenschaft verletzen würde.[762] Gleichfalls ist es eine realitätsfremde Annahme, das Zielsystem nur auf die organisations-spezifischen Ziele reduzieren zu können.[763] Um ein weitgehend allgemeingültiges Zielsystem für die Kreditgenossenschaften zu formulieren, muss das spezifische Spannungsfeld aus ihrer Eigenschaft als Kreditinstitut und als Genossenschaft berücksichtigt werden.

WIEDEMANN unterscheidet vor diesem Hintergrund zwischen materiellen Zielen des Genossenschaftsbetriebs als Kreditinstitut und den immateriellen Zielen der Genossenschaft, die sich aus dem Förderzweck für die Mitglieder als Träger des Genossenschaftsbetriebs ergeben.[764] Für die weitere Diskussion wird auf das vom konkreten Förderinhalt zunächst abstrahierende Verständnis des zusätzlichen Nutzens aus der Mitgliedschaft als Förderzweck abgestellt.[765] Das Konstrukt des Member Value stellt in diesem Kontext einen sinnvollen Ansatz dar, den Förderauftrag in einer modernen Interpretation als immaterielle Ziel-

---

759 Vgl. dazu die Ausführungen in Kapitel D.1.2.1 und Kapitel D.1.2.3.

760 Dabei kann es jedoch nicht gelingen, alle von den etwa 1.100 typischen Kreditgenossenschaften verfolgten Ziele zu berücksichtigen.

761 Vgl. hier und nachfolgend *Ringle, G.* (2006), S. 213 f.

762 Vgl. dazu die Ausführungen in Kapitel C. 1.4.1.

763 Vgl. *Blome-Drees, J.* (2008), S. 14.

764 Vgl. hier und nachfolgend *Wiedemann, A.* (1992), S. 78 f.

765 Dieser wird z. B. als Förderplus oder Member Value bezeichnet. Vgl. Kapitel C.1.4.2.

größe heranzuziehen.[766] Er wird als Ergebnis einer mitgliederbezogenen Nutzenfunktion mit drei Komponenten modelliert (siehe Formel [D.1]):

$$MV = f(UF, MF, FP) \qquad [D.1]$$

mit:

MV: Member Value oder innerer Wert der Kreditgenossenschaft für das Mitglied

f(·): Nutzenfunktion

UF: Unmittelbare (reale bzw. naturale) Förderung des Mitglieds aus den Leistungsbeziehungen zur Kreditgenossenschaft, die zu vereinbarten Konditionen und Standards erfolgt (auch: Kundennutzen)

MF: Mittelbare Förderung z. B. durch Dividenden und Verzinsung des Geschäftsguthabens, also Beiträgen aus der Einlegerbeziehung (auch: Mitgliedernutzen)

FP: Potenzial für eine nachhaltige Förderung der aktuellen und zukünftigen Mitglieder

Im Konsens mit der Literatur[767] sieht WIEDEMANN Rentabilität, Wachstum und Risiko als zentrale materielle Zielgrößen des Kreditinstituts. Die Rentabilität gilt dabei als dominante Größe und strenge Nebenbedingung für die Wachstumsziele. Die Risikoziele sind begrenzt durch die Risikotragfähigkeit eines Instituts und stellen vor dem Hintergrund der Verbindung von Rendite und Risiko eine Nebenbedingung des Renditeziels dar. Neben dieser theoretischen Betrachtung des Zielsystems muss auch das in der Praxis konzeptionell ausgerichtete Zielsystem berücksichtigt werden. Wie in den Ausführungen zur Gesamtbanksteuerung in Kapitel C.4.1 deutlich wurde, stellt der Vermögenswert aus Sicht des BVR ET AL. die zentrale Ziel- und Steuerungsgröße für die Kreditgenossenschaften nach Maßgabe der Gesamtbanksteuerungskonzeption VR-CONTROL dar.[768] In kombinierter Betrachtung stellen die materiellen Ziele nach WIEDEMANN die Instrumentalziele des Fundamentalziels Vermögenswertsteigerung in der Steuerungskonzeption VR-CONTROL dar. In beiden Sichtweisen gilt implizit, dass durch Erfolg bei der Erreichung der genannten Ziele, Sicherheits- und Li-

---

766 Vgl. hier und nachfolgend *Tschöpel, M.* (2010), S. 5 f.; *Lamprecht, D./Donschen, A.* (2006), S. 14 ff.; *Theurl, T.* (2005), S. 139.

767 Vgl. z. B. *Schierenbeck, H.* (2003), S. 1; *Tebroke, H.-J.* (1998), S. 333; *Grosskopf, W.* (1990), S. 13.

768 Vgl. die Ausführungen in Kapitel C.4.1.

quiditätsziele, die aufsichtsrechtlich von den Banken gefordert werden, ebenfalls erreicht werden.[769] Daher werden diese Ziele nicht explizit aufgeführt.

In der wissenschaftlichen Diskussion, und vor allem im Diskurs zwischen der genossenschaftlichen Wissenschaft und Praxis ist umstritten, wie die einzelnen Zielgrößen in Beziehung stehen oder stehen sollten. Daher wird nachfolgend analysiert, von welche Beziehungen realistischerweise auszugehen ist und wodurch diese Beziehungen beeinflusst werden können.

3.1.2 Erörterung ausgewählter Zielsysteme

3.1.2.1 Auswahl relevanter Zielsysteme

Der gesetzliche Förderauftrag gilt theoretisch als das Hauptziel des Genossenschaftsbetriebs.[770] Es wird jedoch vielfach angemerkt, dass dieses organisationsspezifische Ziel weitgehend verdrängt ist, da sich die Genossenschaftsbanken stark den Wettbewerbern angenähert hätten.[771] Dem entgegenhaltend wird jedoch – vor allem von theoretischer Seite – argumentiert, zweckfreies Gewinnstreben oder Gewinnmaximierung seien im genossenschaftlichen Kontext inadäquate Zielvorstellungen.[772] In der genossenschaftlichen Tradition der Uneigennützigkeit formuliert BONUS, es sei nicht Zweck der Genossenschaft, Gewinn zu machen.[773] VIEHOFF argumentiert, dass der Gewinn einer Genossenschaft ein spezifisches Gewinnminimum darstelle, welcher neben der wachstumsbedingten und risikoabdeckenden Rücklagenbildung zur Zahlung einer marktgerechten Dividende auf das Eigenkapital der Mitglieder benötigt werde.[774] Fokussiert wird dabei vor allem auf ein ausgeprägtes Kosten- und Risiko-

769 „Durch einen verantwortungsvollen Umgang mit den anderen Risiken wird gleichzeitig dem Liquiditätsrisiko vorgebeugt." *BVR et al.* (o. J.b), S. 20.

770 Vgl. *Beuthien, V./Dierkes, S./Wehrheim, M.* (2008), S. 177; *Vogel, W.* (2002), S. 69 ff. sowie die dort zitierte Literatur; *Hammerschmidt, M.* (2000), S. 21.

771 Vgl. z. B. *Steiner, J.* (1995), S. 36; *Lürig, R.* (1985b), S. 24; *Hahn, O.* (1980), S. 21 ff.

772 Vgl. z. B. *Ringle, G.* (2006), S. 217; *Algner, M. R.* (2006), S. 187; *Hildner, R.* (2003), S. 6; *Grosskopf, W.* (1990), S. 13; *Großfeld, B./Jäger, W./Lenfers, G.* (1989), S. 90; *Engelhard, H. A.* (1989), S. 26. Die Betrachtung der Instrumentalziele resultiert daher, dass die Vermögenswertorientierung erst in der jüngeren Vergangenheit Gegenstand der Gesamtbanksteuerung in den Kreditgenossenschaften wurde. Aus Vereinfachungsgründen wird die theoretisch richtige Unterscheidung von Gewinn und Vermögenswert im Weiteren nicht weiter stringent eingehalten. Mit diesem Vorgehen geht keine Einschränkung der weiteren Analyse einher.

773 Vgl. *Bonus, H.* (1998), S. 324.

774 Vgl. *Viehoff, F.* (1979), S. 272 ff. So auch *Aubin, P.* (2011), S. 44.

management und weniger auf eine ausgeprägte Erlösorientierung.[775] Da der Genossenschaftsbetrieb jedoch auf Dauer angelegt ist und die Existenz- und Förderfähigkeit des Unternehmens langfristig sicherzustellen sind, führt dies nach THEURL zwingend zur Ertragsorientierung,[776] die der „Erwirtschaftung nachhaltiger und risikoangepasster Renditen für Kapitalgeber und Eigentümer“[777] dient. WIEDEMANN begründet das ertragsorientierte Zielsystem mit dem „Primat der Existenzsicherung“[778], das auch für die Kreditgenossenschaften gilt.[779] Nach BROSCHART ET AL. gilt dies nur für die Zielgrößen Liquidität und Sicherheit, während das Rentabilitätsziel durch den Zweck selbst also den Förderauftrag respektive den Member Value tangiert wird und daher nicht isoliert betrachtet werden kann.[780] Sie sehen folglich im Nebeneinander von Rentabilitätsstreben und Umsetzung des Förderauftrags das Besondere der genossenschaftlichen Bankpolitik. Insofern gelten die materiellen Ziele als Nebenbedingung und Grundlage des immateriellen Oberziels. Ein Konflikt zwischen den Zielen ist zu konstatieren, wenn das „Primat der Existenzsicherung“ als einzuhaltende Restriktion das Erreichen des immateriellen Oberziels einschränkt. Sobald die Existenzsicherung gewährleistet ist, gilt es nach diesem Verständnis, ausschließlich das Oberziel zu verfolgen, d. h. den Member Value zu erhöhen.

Ohne konkreten Bezug zu etwaigen Förderzielen begründen der BVR ET AL. den Vermögenswert als zentrale Zielgröße der Steuerungskonzeption VR-CONTROL für die Kreditgenossenschaften und dessen Maximierung als konkretes Ziel,[781] da „die Maximierung des Vermögenszuwachses unter Einhaltung der definierten Risikolimite [.] ein wesentliches Ziel sämtlicher Bankaktivitäten“[782] ist. Einen Konflikt zum immateriellen Ziel, d. h. dem Förderauftrag, sehen sie darin nicht, da mit der „Vermögensmehrung eine wesentliche Voraussetzung für die nachhaltige Bestandssicherung der Bank geschaffen wird“[783] und nur durch den Bestand des Genossenschaftsbetriebs der Förderauftrag dauerhaft wahrge-

---

775 Dies zeigt sich hinsichtlich der Erlösquellen z. B. in der Aussage von AUBIN, dem Vorstandssprecher der VOLKSBANK GÖPPINGEN EG, der argumentiert: „Möglichst weitgehender Verzicht auf Gewinn aus dem Kundengeschäft ist das wesentliche Merkmal der Volksbanken Raiffeisenbanken und damit entscheidende Grundlage für ihren Erfolg.“ *Aubin, P.* (2011), S. 42.

776 Vgl. *Theurl, T.* (2002), S. 79.

777 *Theurl, T./Böttiger, J.-M.* (2007), S. 9.

778 *Wiedemann, A.* (1992), S. 79.

779 So auch *Theurl, T./Böttiger, J.-M.* (2007), S. 7.

780 Vgl. *Broschart, J. et al.* (2002b), S. 12.

781 Vgl. *BVR et al.* (o. J.b), S. 6. Zur Definition des Begriffs Vermögenswert vgl. Kapitel C.4.1.

782 *BVR et al.* (o. J.b), S. 8.

783 *BVR et al.* (o. J.b), S. 6.

nommen werden kann. Damit ist der Förderzweck nicht mehr zwingend Bestandteil des kreditgenossenschaftlichen Zielsystems.

Vor dem Hintergrund dieser unterschiedlichen Auffassungen lassen sich zwei grundsätzliche Arten von Zielsystemverständnissen skizzieren. Aus Gründen der einfacheren Darstellung werden in Abbildung 31 die materiellen Ziele als Gewinn-/ Vermögenswert bezeichnet. Die immateriellen Ziele werden über den Member Value inhaltlich konkretisiert. Dabei wird berücksichtigt, dass die Existenz der Kreditgenossenschaft einem definierten Zweck dienen muss,[784] womit eine zwecklose Existenz nicht als anzustrebendes Ziel gilt. Daher schließt sich eine ausschließliche Betrachtung des Gewinn-/ Vermögenswertziels aus.

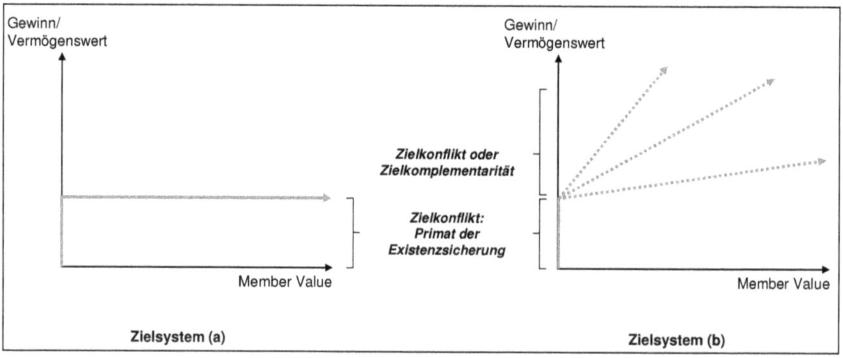

Abbildung 31: Skizze grundsätzlicher Zielsystemverständnisse für Kreditgenossenschaften[785]

Während beiden Zielsystemen gemein ist, dass zur Existenzsicherung ein Mindestgewinn erwirtschaftet werden muss, wird im Zielsystem (a) davon ausgegangen, dass alle weiteren Anstrengungen der Kreditgenossenschaft sich auf den Member Value als Oberziel beschränken. Im Zielsystem (b) dagegen wird das Gewinn-/ Vermögenswertziel angestrebt, was auch dem Member Value zugute kommen kann (Zielkomplementarität), jedoch nicht muss (Zielkonflikt).[786] Die Art der Zielbeziehung ist abhängig davon, welche Komponente des Member Value am höchsten gewichtet wird und mit welchen Maßnahmen der Member Va-

784 Vgl. *Blome-Drees, J.* (2008), S. 17.

785 Eigene Darstellung. Beide Zielsysteme bestehen aus zwei Zielen. Zwar stellt das Zielsystem (a) einen konkreten Fall des Zielsystems (b) dar. Es soll jedoch aufgrund seiner extremen Ausprägung und den damit verbundenen Implikationen als eigenständiges Zielsystem verstanden werden.

786 Die konkrete Zielbeziehung hängt letztlich von der konkreten Ausgestaltung des Förderauftrags ab. Vgl. dazu die Ausführungen in Kapitel D.3.1.2.4.

lue erreicht und/oder gesteigert werden soll. Es gilt in diesem Zusammenhang zu erörtern, welches Gewicht im Zielsystem (b) die beiden Zielgrößen haben, da davon wiederum die Konzeptualisierung des Kundenwerts als Beitrag zur Zielerreichung abhängt.[787]

TEWES argumentiert, dass Unternehmen selbst keine Ziele haben, sondern die Ziele für das Unternehmen sich aus den Individualzielen der an ihm beteiligten Personen ableiten.[788] Basierend auf der Anreiz-Beitrags-Theorie und dem methodologischem Individualismus geht er davon aus, dass das Ziel der betrachteten Personen die Maximierung ihres Nutzens aus der Beziehung zum Unternehmen ist. In wie weit nun die jeweiligen Individualziele berücksichtigt werden, hängt davon ab, wie groß das Einflusspotenzial eines Individuums oder einer Gruppe zielkongruenter Individuen auf die Zielbildung ist. Das resultierende Zielsystem wird als eben noch hinzunehmender Kompromiss verstanden. Im Folgenden sollen vor dem Hintergrund dieser Argumentation kreditgenossenschaftsspezifische Einflussfaktoren dahingehend erörtert werden, wie diese auf die notwendige Konsensfindung einwirken und welche Bedeutung diese auf das relevante Zielsystem und die Zielbeziehungen haben.

3.1.2.2 Mitgliederstruktur als Einflussfaktor auf das Zielsystem

In der traditionellen Bewertungslehre wird als „innerer Wert“ eines Unternehmens dessen finanzieller Nutzen für die Anteilseigner verstanden, wobei die Gewinnmaximierung als einzige Zielformulierung verwendet wird.[789] Es herrscht Einigkeit in der Literatur, dass die Kapitalanlage unter Renditegesichtspunkten für den Beitritt als Mitglied nicht ausschließlich relevant ist, sondern wesentliche Beitrittsanreize aus der wirtschaftlichen Förderung im Rahmen der

---

787 In diesem Sinne würde ein Zielgewicht von 100 Prozent für den Förderauftrag (Member Value) dazu führen, dass die Zielsystem (a) und (b) identisch sind. Im umgekehrten Fall, bei einem Zielgewicht von 100 Prozent für das Gewinn-/Vermögenswertziel, würde die Zielfunktion eine Parallele zur Ordinate sein. Abweichend von den dargstellten Verläufen ist es in der Praxis ebenso denkbar, dass nicht stetige Verläufe vorkommen. Z. B., wenn unterstellt wird, es gibt ein operationales Ziel für den Förderauftrag. Ist dies erreicht, dann würde die Kurve wieder in Richtung des Gewinnziels streben. Anders argumentieren GESCHWANDTNER/KUTTLER, die aufbauend auf den gesetzlichen Vorgaben argumentieren, eine Zieldiskussion erübrige sich. Vgl. *Geschwandtner, M./Kuttler, H.-J.* (2009), S. 266. Jedoch zeigen sich in ihrem Beitrag vielfältige Inkonsistenzen, wie etwa die unbedingte Mitgliederförderung bei gleichzeitiger Widmung der Mitgliedschaft als Kundenbindungsinstrument und der Unterordnung der Mitgliedschaft unter dem Wirtschaftlichkeitsprinzip.

788 Vgl. hier und nachfolgend *Tewes, M.* (2003), S. 16 ff. so wie die dort angegebene Literatur.

789 Vgl. *Schöffling, B.* (1991), S. 5.

Leistungsbeziehung resultieren.[790] Folglich ist es unzureichend, ausschließlich den Gewinn als Determinante des inneren Werts für das Mitglied der Kreditgenossenschaft zu nutzen. Der innere Wert – im Sinne des Gesamtgenossenschaftswertes – muss weitere Determinanten umfassen. Das Konstrukt des Member Value, das als relevantes immaterielles Ziel verwendet wird, stellt in diesem Kontext einen geeigneten Ansatz dar, den Wert oder Nutzen der wirtschaftlichen Aktivitäten des genossenschaftlichen Gewerbebetriebs für die Mitglieder darzustellen (siehe Formel [D.1]).[791]

Die Komponente *FP* kann auch als Substanzwert oder zukunftsorientiert als Ertragswert verstanden werden. In Anlehnung an MONßEN ist ein auf Basis des Discounted Cash-flow-Verfahrens[792] bestimmter Unternehmenswert als Fördersubstanz interpretierbar, die die Sicherung der Genossenschaftsbank als förderndes Unternehmen in der Zukunft gewährleistet.[793]

Die drei Komponenten stehen in interdependenter Beziehung zueinander. Wird eine Komponente festgelegt, schränkt dies die Freiheitsgrade bei den anderen beiden ein.[794] Problematisch ist nun, die Komponenten zu konkretisieren, zu gewichten und zu bewerten. Dabei ist zu berücksichtigen, dass dies auf die Zielbeziehung zwischen materiellen und immateriellen Zielen Einfluss hat. Wird etwa das zukünftige Förderpotenzial *FP* stark gewichtet, wird zu Lasten des Ziels Förderauftrag, der der Komponente *UF*, mithin der unmittelbaren naturalen Förderung entspricht,[795] dem Gewinn-/Vermögenswertziel entsprochen.[796] Steht dagegen die reale oder die mittelbare Förderung im Mittelpunkt, geht dies zu Lasten des zukünftigen Förderpotenzials *FP* und ggf. der gesamten Zukunftsfähigkeit der Genossenschaft.

Die Entscheidung, wie die Komponenten zu gewichten sind, wird von der Kreditgenossenschaft individuell getroffen und basiert auf den Präferenzen der

---

790 Vgl. z. B. *Bornemann, S./Schaaff, C./Pfingsten, A.* (2009), S. 28 f.; *Baumgärtler, T.* (2000), S. 99; *Grauer, F.* (1998), S. 206 f.

791 Vgl. hier und nachfolgend *Lamprecht, D./Donschen, A.* (2006), S. 14 ff.; *Theurl, T.* (2005), S. 139. Eine ähnliche Konzeptualisierung nimmt SCHÖFFLING vor, der ausgehend vom Identitätsprinzip die finanziellen und nicht-finanziellen Nutzenkomponenten aus der Leistungs- und der Einlegerbeziehung zum Gesamtwert einer Genossenschaft zusammenfasst. Vgl. *Schöffling, B.* (1991), S. 7.

792 Das Discounted Cash-flow-Verfahren bezeichnet eine Methode zur Ermittlung eines Unternehmenswerts. Vgl. dazu ausführlich *Stüker, D.* (2008), S. 72 ff. sowie die dort angegebene Literatur.

793 Vgl. Monßen, H.-G. (1998), S. 295; Algner, M. R./Bolsinger, H. J. (2006), S. 189.

794 Vgl. *Theurl, T.* (2002), S. 85.

795 Vgl. dazu die Ausführungen in Kapitel C.1.4.2.

796 Vgl. *Theurl, T.* (2002), S. 86.

jeweiligen Mitglieder.[797] Diese Entscheidungen können grundsätzlich auf zwei Arten erfolgen. Entweder so, wie WIEDEMANN es in einem Phasenschema darstellt (siehe Abbildung 32). Danach formuliert der Vorstand die Ziele sowohl materieller als auch immaterieller Art und steuert deren Umsetzung mittels des Genossenschaftsbetriebs. Ihre Zustimmung oder Ablehnung können die Mitglieder auf grundsätzlich drei Arten artikulieren. Erstens über den Grad und Umfang der Inanspruchnahme der genossenschaftsbetrieblichen Leistungen, zweitens über die Mitbestimmung in der Generalversammlung und drittens über die Kündigung der Mitgliedschaft. Vorteilhaft an diesem mitgliederpassiven Vorgehen ist, dass der Vorstand aufgrund seiner vergleichsweise höheren Informationen über die Ressourcen der Kreditgenossenschaft die materiellen und immateriellen Ziele in Bezug auf ihre Umsetzbarkeit beurteilen und entsprechend berücksichtigen kann. Dies spricht tendenziell für eine ausgewogene komplementäre Zielbeziehung zwischen den genannten Zielen, die sich an den Möglichkeiten und Grenzen ausrichtet. Nachteilig daran ist jedoch die fehlende Mitgliederintegration ex ante.

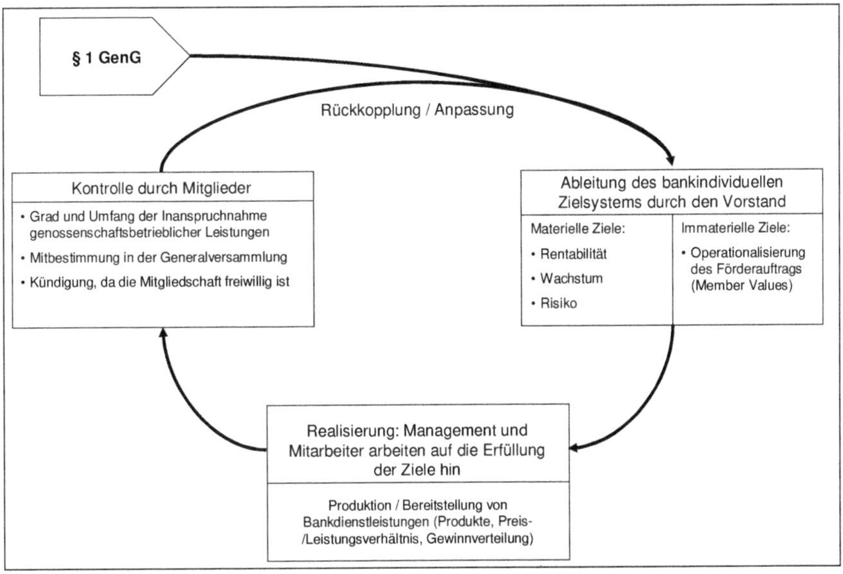

Abbildung 32: Phasenschema zur Operationalisierung des Förderauftrags[798]

[797] Vgl. hier und nachfolgend *Wiedemann, A.* (1992), S. 78 sowie die Ausführungen in Kapitel C.1.4.2.

[798] In Anlehnung an *Wiedemann, A.* (1992), S. 78.

Abweichend von diesem Ablaufschema sieht THEURL die Aufgabe, die drei genannten Einflussfaktoren des Member Value zu operationalisieren, zu bewerten und zu gewichten aktiv beim Mitgliederkollektiv.[799] Die Entscheidung des Mitgliederkollektivs dient dann der Genossenschaftsleistung als Handlungsleitlinie, um den Genossenschaftsbetrieb zu führen. Dies kann als mitgliederaktives Vorgehen bezeichnet werden.

Unabhängig davon ob die Mitglieder ex post über den Vorschlag des Vorstands entscheiden oder ex ante die Ziele festlegen, sind viele Akteure in diesen Prozess involviert. Daher kann der Nutzen des einzelnen Mitglieds und damit der Wert der Genossenschaft aus individueller Mitgliedersicht durch die Befriedigung der Ziele anderer Parteien gemindert werden.[800] Wesentlichen Einfluss auf die Kollektivfestlegung des Förderauftrags hat also der Grad der Ziel- oder Interessenhomogenität der Mitgliedergruppe. DÜLFER unterscheidet anhand dieses Kriteriums drei Gruppen.[801] In einer homogenen Gruppen stimmen die Interessen der einzelnen Mitglieder weitgehend überein, während in der heterogenen Gruppe für Teile der Gesamtheit nicht alle Interessen von gleicher Bedeutung sind und bei einer gespaltenen Gruppe von Teilen der Gesamtheit völlig verschiedene Ziele verfolgt werden.

Für die Kreditgenossenschaften soll nun die Art der Gruppe auf aggregierter Ebene festgestellt werden, da keine aktuellen Erkenntnisse zu den Interessen oder Zielen Einzelner vorliegen.[802] Eine approximative Bestimmung erfolgt auf Basis der Mitgliederstruktur aus erwerbswirtschaftlicher Perspektive. Diese hat sich seit der Gründungszeit stark verändert (siehe Abbildung 33). So ist der Anteil der gewerblichen Mitglieder, also der Unternehmen und der Selbständigen in freien Berufen von 77 Prozent im Jahr 1895 auf unter ca. 6 Prozent im Jahr 2010 zurückgegangen. Heute dominieren Privatpersonen die Mitgliedstruktur der Kreditgenossenschaften. Aufgrund der insgesamt heterogenen Mitgliederstruktur kann begründet davon ausgegangen, dass auch die förderauftragsbezogenen Anforderungen der Mitglieder an die Genossenschaft differierend ausgestaltet sein dürften.[803]

799 Vgl. *Theurl, T.* (2002), S. 85.

800 Vgl. *Schöffling, B.* (1991), S. 17.

801 Vgl. hier und nachfolgend *Dülfer, E.* (1995), S. 195.

802 Verwiesen sei hier auf die etwas ältere Studie von BAKONYI aus dem Jahr 1983, der heterogene und situationsspezifische Erwartungen der Mitglieder an die Bankgenossenschaft identifiziert. Vgl. *Bakonyi, E.* (1983). Es ist begründet davon auszugehen, dass sich dieser Trend bis zum gegenwärtigen Zeitpunkt verstärkt hat. Vgl. *Vogel, W.* (2002), S. 74.

803 Vgl. *Broschart, J. et al.* (2002b), S. 13.

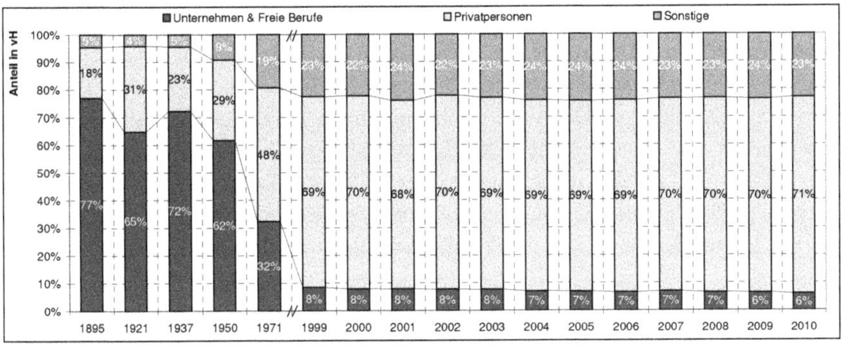

Abbildung 33: Entwicklung der Mitgliederstruktur der Kreditgenossenschaften[804]

Wird zu dieser strukturellen Entwicklung zusätzlich die Entwicklung der Mitgliederanzahl betrachtet, wird deutlich, dass diese Faktoren zusammengenommen die Annahme einer homogenen Gruppe nicht unterstützen. So hat sich die Mitgliederanzahl von ca. 6 Mio. in 1970 auf ca. 16,7 Mio. in 2010 um etwa 10 Mio. erhöht. Gleichzeitig hat sich durch innergenossenschaftliche Fusionsaktivitäten die Anzahl der Mitglieder je Kreditgenossenschaft im gleichen Zeitraum von 872 auf 14.665 deutlich erhöht.[805]

Insgesamt betrachtet, kann von einer inhomogenen Interessenslage der Mitglieder ausgegangen werden, so dass auch nicht von einer homogenen Gruppe auszugehen ist.[806] Vor dem Hintergrund der aufgezeigten Struktur der Mitglieder ist vielmehr eine heterogene Gruppe anzunehmen. Folglich ist es nötig, eine in der Mitgliederschaft mehrheitlich akzeptierte und allgemeingültige Lösung hinsichtlich der Förderziele zu finden,[807] wenngleich diese nicht pareto-optimal ist. Theoretisch erklärt werden kann dies durch das ARROW'sche Unmöglich-

---

804 Eigene Berechnung und Darstellung auf Basis der Daten in *BVR* (1999c), (2000), (2001a), (2002a), (2003a), (2004a), (2005a), (2006), (2007a), (2008), (2009a), (2010b), (2011a) sowie *Lürig, R.* (1985b), S. 34 f.

805 Eigene Berechnung auf Basis der Daten des *BVR* (2011b). Enthalten sind in dieser Berechnung auch die so genannten „atypischen" Kreditgenossenschaften, die aufgrund mangelnder Datenverfügbarkeit nicht für den gesamten Betrachtungszeitraum identifiziert werden können. Da die Anzahl der „atypischen" Kreditgenossenschaften doch insgesamt relativ gering ist, dürfte dieser Sachverhalt die Aussagekraft nicht stark eingrenzen.

806 Etwas anders CARIO, die in den nutzenden Mitgliedern eine homogene Gruppe sieht. Vgl. *Cario, D.* (2005), S. 150. Dabei geht sie jedoch nicht auf die Struktur der Mitgliederschaft ein, sondern trennt nur die Interessen von Mitglieder-Kunden und Investorenmitgliedern.

807 Vgl. *Grauer, F.* (1998), S. 196.

keitstheorem, wonach es unmöglich ist, die Präferenzen aller relevanten Entscheidungsträger zu aggregieren.[808] Naheliegend ist in diesem Zusammenhang eine finanziell orientierte Operationalisierung des Förderauftrags in der Praxis in Form von Dividenden. Denn die in Geldeinheiten lautenden Dividendenzahlungen können für die Mitglieder als allgemeingültige Gutscheine aufgefasst werden. Mit Blick auf die identifizierten Ziele der Kreditgenossenschaft bedeutet dies, dass Dividenden die Anreize erhöhen, den Gewinn/Vermögenswert als Ziel stärker zu gewichten als ein Ziel auf Basis eines alternativ ausgestalteten Förderauftrag.

Für eine tendenziell stärkere Dividendenorientierung spricht auch die seit der Novelle des GenG in 2006 bestehende Möglichkeit zur Aufnahme von investierenden Mitgliedern nach § 8 Abs. 2 GenG, welchen eine höhere Renditeorientierung unterstellt wird,[809] da diesen kein gesellschaftsrechtlicher Anspruch auf die naturale Förderung aus der Leistungsbeziehung zusteht und somit kompensierende Nutzenkomponenten im Fokus dieser Mitglieder stehen.[810]

Daher lässt sich die Erkenntnis festhalten, dass die heterogene Mitgliederstruktur sowie die Anreizdifferenzen zwischen nutzenden und investierenden Mitgliedern dazu führen, dass eher allgemeingültige und kurzfristige Nutzenkomponenten im Sinne des Member Value von den Kreditgenossenschaften zu forcieren sind. Diese für beide Mitgliedergruppen vorliegende Allgemeingültigkeit kann vor allem in Dividenden, mithin also der Nutzenkomponente *MF* in Formel [D.1] gesehen werden, wodurch das Gewinnziel tendenziell stärker gewichtet wird, jedoch in diesem Kontext einen engen Bezug zum Förderauftrag aufweist und eine komplementäre Zielbeziehung begründen würde. Dies kann erklären, warum in der Studie von BROSCHART ET AL. etwa 90 Prozent der befragten Kreditinstitute explizit keine speziellen Ziele zum Förderauftrag haben.[811]

In diesen Überlegungen wird implizit den investierenden Mitgliedern zuerkannt, den gleichen Einfluss auf die Führung des Kreditgenossenschaftsbetriebs zu haben wie die nutzenden Mitglieder. Deren Einfluss auf die Kreditgenossenschaft wird jedoch genossenschaftsrechtlich begrenzt, da der Mitgliedsaufnahme die Generalversammlung zuzustimmen und die Satzung sicherzustellen hat, dass die zu fördernden Mitglieder nicht von investierenden Mitgliedern überstimmt werden sowie Beschlüsse der Generalversammlung, die einer qualifizierten Mehrheit von drei Vierteln bedürfen, nicht durch die investierenden Mitglieder

808 Vgl. *Arrow, K. J.* (1970).
809 Vgl. ausführlich *Cario, D.* (2005), S. 150 f.
810 Vgl. *Beuthien, V.* (2006), S. 57.
811 Vgl. *Broschart, J. et al.* (2002a), S. 22.

verhindert werden können.[812] Anknüpfend an diesen Punkt stellt sich insgesamt die Frage, in welcher Form die Mitglieder Einfluss auf die Zielbildung in den Kreditgenossenschaften haben.

3.1.2.3 Mitgliederpartizipation als Einflussfaktor auf das Zielsystem

Wie in Kapitel C.1.4.1 dargelegt, ist die Selbstverwaltung einer der zentralen Grundsätze der Kreditgenossenschaften. Da die Mitglieder auch Träger der Verfügungsrechte sind,[813] ist deren Einfluss auf das Zielsystem der Kreditgenossenschaft zu diskutieren.

Für Genossenschaftsbanken lassen sich vielfache Prinzipal-Agent-Beziehungen identifizieren,[814] wobei die zentrale Prinzipal-Agent-Beziehung die zwischen Mitgliedern und dem Vorstand der Kreditgenossenschaft ist.[815] Damit verbunden ist, dass die Mitglieder nicht sicher darüber sein können, dass die Vorstände die Kreditgenossenschaft im Sinne der Mitglieder führen.[816] Dies indizieren empirische Ergebnisse zur Integration der genossenschaftlichen Prinzipien in die Geschäftspolitik der Kreditgenossenschaften. In einer kleinzahligen Studie von BROSCHART ET AL. antworten alle befragten Banken, dass das Mitgliederkollektiv nicht in die Strategiefindung involviert ist und lediglich drei von zehn Banken orientieren sich dabei an den vermuteten Mitgliederinteressen.[817] Weiterhin geben neun von zehn Banken an, dass das Ziel der Existenzsicherung Oberziel der Strategieformulierung ist, während eine Bank explizit die Mitgliederförderung und keine Bank die Gewinnmaximierung nennt. Auch nur eine Bank hat förderzweckspezifische Ziele. In der Studie von KRING spiegelt sich ein ähnliches Bild. Danach beziehen 24,7 Prozent der befragten Kreditgenossenschaften ihre Mitglieder nur in geringfügigem Ausmaß direkt oder indirekt und 58,6 Prozent ihre Mitglieder gar nicht in den Prozess der Strategieformulierung ein. Weiterhin geben nur 10 Prozent an, dass der Förderauftrag als Schwerpunkt der Strategie gesehen wird.[818] Es ist unstrittig, dass eine vollumfängliche Integ-

---

812 Siehe § 8 Abs. 2 GenG; auch *Zieger, T.* (2007), S. 316.

813 Zu den Verfügungsrechten zählen das Recht auf Gebrauch, das Recht auf Erträge aus der Nutzung, das Recht auf Veränderung, das Recht auf vollständige oder teilweise Übertragung sowie das Recht andere von der Nutzung auszuschließen.

814 So z. B. zwischen Mitgliedern und Aufsichtsrat, Aufsichtsrat und Vorstand, Vorstand und Mitarbeiter, Mitglieder und Vorstand. Damit noch nicht betrachtet sind die Beziehungen zwischen Kreditgenossenschaft und Kunden im Kreditgeschäft.

815 Vgl. *Lamprecht, D./Donschen, A.* (2006), S. 19 ff.

816 Vgl. *Lamprecht, D./Donschen, A.* (2006), S. 22; *Vogel, W.* (2002), S. 76; *Tebroke, H. J.* (1998), S. 334.

817 Vgl. *Broschart, J. et al.* (2002a), S. 40 ff. (Anhang).

818 Vgl. *Kring, T.* (2002), S. 30 und S. 39.

ration der Mitglieder in Strategieprozesse nicht realisierbar ist. Gleichwohl könnte eine Koordination mit den Mitgliedern über die Zustimmung zu den Förderzielen erfolgen. Dazu stellt MÜNKER jedoch fest, dass 92 Prozent der Kreditgenossenschaften, ihre Mitglieder ex post nicht nach der Beurteilung der Förderleistung fragen.[819] Daraus lässt sich ableiten, dass das immaterielle Ziel nur von untergeordneter Bedeutung für das Zielsystem der heutigen Kreditgenossenschaften ist und dies in einer Ziel- und Interessendistanz zwischen Mitglied und Vorstand begründet ist.

Wie weit die Distanz zwischen den Interessen des Vorstands und den Interessen der Mitglieder ist, resultiert in weiten Teilen daraus, wie groß die Ähnlichkeit in den Idealen, Werten, Normen und Zielen ist und in wie weit die Mitglieder als Prinzipale die Möglichkeit haben, den Vorstand zu kontrollieren, mithin also im Sinne des Selbstverwaltungsgrundsatzes an der Unternehmenslenkung der Kreditgenossenschaft zu partizipieren.

Nach vielfacher Meinung steigt mit zunehmender Größe der Kreditgenossenschaft, die ein Ergebnis der innersektoralen Konzentrationsprozesse ist,[820] die Distanz zwischen Bankleitung und Mitgliedern.[821] LÜRIG bezeichnet dies als Ausfluss eines Ökonomisierungsprozesses der Genossenschaftsbanken.[822] Eine mögliche Distanz soll nach § 9 Abs. GenG, wonach Vorstände Mitglieder der Kreditgenossenschaft sein müssen, reduziert werden. Verbunden mit den gesetzlichen Vorgaben an die Leitung eines Kreditinstituts bezüglich der fachlichen Eignung und dem Grundsatz, dass der Vorstand die Genossenschaft in eigener Verantwortung leitet,[823] resultieren Tendenzen dahingehend, dass der Vorstand einer Kreditgenossenschaft grundsätzlich nicht anders profiliert ist als ein Leitungsmitglied einer Kapitalgesellschaft. Aufgrund dessen und wegen der der Prinzipal-Agent-Beziehung inhärenten Probleme werden vor Vertragsabschluss mit einem potenziellen Vorstandsmitglied Informationen über diesen und dessen Eigenschaften gesucht.[824] Da die Möglichkeiten der ex-ante Informationsbeschaffung jedoch beschränkt sein dürften und die Genossenschaftsanteile mit nur geringen Kosten und Risiken verbunden sind, ist davon auszugehen, dass mit diesen Maßnahmen die Distanz zwischen Bankleitung und Mitgliedern nicht hinreichend sicher reduziert werden kann. Somit steht die ex post Kontrolle des Managements im Fokus der weiteren Betrachtung.

---

[819] Vgl. *Münkner, H.-H.* (1990), S. 69. Grundlage ist eine Befragung von 725 Kreditgenossenschaften.

[820] Vgl. dazu die Ausführungen in Kapitel C.1.3 und Kapitel C.1.5.

[821] Vgl. z. B. *Pfaffenberger, K.* (2009), S. 279.

[822] Vgl. *Lürig, R.* (1985b), S. 24 f.

[823] Siehe § 27 Abs. 1 GenG sowie § 14 Abs. 1 der Mustersatzung des BVR für VR-Banken.

[824] Vgl. dazu z. B. *Tebroke, H.-J.* (1998), S. 334 f.

Die verfügbaren Instrumente der ex post Managementkontrolle lassen sich unterscheiden in marktliche Kontrollmechanismen und die Kontrolle durch Mitgliederpartizipation.[825] Zur ersten Gruppe zählen der Wettbewerb auf dem Finanzdienstleistungsmarkt, der Stellenmarkt für Manager und die Kontrolle über den Kapitalmarkt. Diese Kontrollmechanismen sind jedoch für die vorliegende Fragestellung weitgehend irrelevant. Während die Kontrolle über den Kapitalmarkt aufgrund der eingeschränkten Fungibiliät der Genossenschaftsanteile und des fehlenden Sekundärmarktes substanzlos ist, bleibt die Kündigung der Mitgliedschaft nach § 65 GenG als Alternative. Die Effektivität dieses Instruments wird jedoch durch lange Kündigungsfristen und die relative Bedeutung der Rücklagen am Eigenkapital begrenzt. Denn je höher der Anteil der Rücklagen am Eigenkapital ist, desto geringer ist der Einfluss der Kündigung, da damit nur geringfügige Kapitalverluste einhergehen. Nach TEBROKE gibt es nicht wenige Kreditgenossenschaften, die einen Anteil von über 80 Prozent der unkündbaren Rücklagen am Eigenkapital aufweisen.[826] Und auch der Stellenmarkt für Manager ist aufgrund der Intransparenz über vakante Stellen und der relativen Unattraktivität als Kontrollinstrument schwach.[827] Alleinig dem Wettbewerb auf dem Finanzdienstleistungsmarkt ist ein Kontrollbeitrag zu attestieren. Aussagekräftig ist diesbezüglich eine differenzierte Analyse des Teils der Marktperformance, die sich auf das Mitgliedergeschäft beschränkt, da damit die Kongruenz der leistungsbeziehungsbezogenen Ziele der Mitglieder mit denen des Vorstands als Entscheidungsgremium über das Angebot beurteilt werden kann. Jedoch gilt zu berücksichtigen, welche Bedeutung das Mitgliedergeschäft für die Kreditgenossenschaft hat und in welcher Form hier tatsächlich förderspezifische Komponenten integriert sind. Gegenwärtig ist die Anzahl der Mitglieder mit ca. 16,7 Mio. nur etwa halb so groß wie die Anzahl der Kunden mit über 30 Mio., was nicht für einen dominanten Stellenwert des Mitgliedergeschäfts spricht[828] und so die Bedeutung der Mitglieder für die Kreditgenossenschaften zumindest aus Sicht des operativen Geschäfts relativiert.[829]

---

825 Vgl. hier und nachfolgend *Tebroke, H.-J.* (1998), S. 335 ff.

826 Vgl. *Tebroke, H.-J.* (1998), S. 336. So z. B. die VR-BANK WESTMÜNSTERLAND EG, bei der das gezeichnete Kapital – Geschäftsguthaben der Mitglieder – ca. 18 Prozent und die Ergebnisrücklagen ca. 79 Prozent des bilanziellen Eigenkapitals ausmachen. Vgl. *VR-Bank Westmünsterland eG* (2010).

827 Vgl. hier und nachfolgend *Tebroke, H.-J.* (1998), S. 335 ff.

828 Weitere mögliche Ursachen sind nach BEUTHIEN/HANRATH/WEBER die Institutsgröße, die Ausrichtung des Instituts als reine Vertriebsbank oder integrierte Bank sowie die Struktur der Leitungsorgane. Vgl. *Beuthien, V./Hanrath, S./Weber, H.-O.* (2008), S. 3.

829 Vgl. *Grauer, F.* (1998), S. 193.

Zur Sicherstellung der Integration der Mitgliederziele in der Geschäftspolitik der Kreditgenossenschaft sind also aufgrund der bisherigen Ausführungen vor allem die Kontrollen durch Mitgliederpartizipation weiter zu analysieren. Kontrollinstrumente in diesem Zusammenhang sind die Generalversammlung und der Aufsichtsrat. Oberstes Willensbildungsorgan ist die Generalversammlung, in der die Mitglieder ihre Rechte in den Angelegenheiten der Kreditgenossenschaft ausüben.[830] Dazu gehören Beschlüsse über die Gewinnverwendung sowie über die Besetzung von Vorstand und Aufsichtsrat. Damit sind die wichtigsten Entscheidungsbefugnisse theoretisch bei den Mitgliedern angesiedelt, die damit die geschäftspolitische Ausrichtung der Kreditgenossenschaft wesentlich beeinflussen können.[831] Praktisch lassen sich jedoch vielfältige Probleme nennen, die diese Möglichkeiten der Mitglieder einschränken. Denn Grundlage der Entscheidungen der Generalversammlung sind die Informationen des Vorstands. Nach TEBROKE divergieren jedoch der entscheidungsrelevante Informationsbedarf seitens der Generalversammlung und das Informationsangebot des Vorstands.[832] Problematisch ist dabei, dass der objektive Informationsbedarf mangels Qualifikation und zeitlicher Verfügbarkeit der Mitglieder i. d. R. von diesen nicht hinreichend bestimmt werden kann. Weiterhin steuert der Vorstand die zeitliche Struktur der Informationsweitergabe, da er bei ordentlichen Generalversammlungen die Tagesordnung vorgibt.[833] Darüber hinaus können die Generalversammlungen nach § 43a Abs. 1 GenG ab einer Mitgliederzahl von 1.500 Personen in Form von Vertreterversammlungen durchgeführt werden. Dadurch wird ein Teil der Mitglieder – abgesehen von der Vertreterwahl – aus dem Willensbildungsprozess weitgehend ausgeschlossen. Im Durchschnitt gehören aktuell 14.665 Mitglieder zu einer Bank, so dass davon auszugehen ist, dass bei vielen Banken Vertreterversammlungen durchgeführt werden. Zudem ist aufgrund eines unausgewogenen mitglieder-/vertreterindividuellen Kosten-Nutzen-Verhältnisses und der Evidenz eines Free-Rider-Problems nicht mit einer aktiven Vorbereitung und Teilnahme zu rechnen.[834] Folglich ist das Interesse der

830 Siehe § 43 Abs. 1 GenG.

831 Vgl. *Pfaffenberger, K.* (2009), S. 279; *Tebroke, H.-J.* (1998), S. 336.

832 Vgl. *Tebroke, H.-J.* (1998), S. 336.

833 Siehe § 28 Abs. 1 der Mustersatzung des BVR für VR-Banken

834 GROSSEKETTLER wertet die Kontrolle der Genossenschaftsleitung als öffentliches Gut, da die Kontrollbemühungen eines Genossen (Mitglied oder Vertreter) allen übrigen ebenfalls zugute kommt. Daher nehmen nach seiner Ansicht nur wenige die Kosten und Mühen intensiver Kontrolle auf sich. Vgl. *Wiemeyer, J./Carstensen, C.* (1984), S. 61. Für eine theoretische Erklärung dieses Sachverhalts vgl. *Engels, M.* (1997), S. 678 ff. oder *Kammlott, C./Schiereck, D.* (2000), S. 267. Weiterhin konstatiert TEBROKE, dass aufbauend auf den Ergebnissen von DÜLFER/KRAMER, in denen die Versammlung keine

Mitglieder am Mitspracherecht von „untergeordneter Bedeutung [.], wobei die Manager der Genossenschaftsbanken sich dieser Tatsache bewusst sind.“[835] Dies bestätigen auch empirische Ergebnisse von HAMMERSCHMIDT, wonach für nur 7 Prozent der Befragten das Mitspracherecht ein Grund für die Mitgliedschaft ist.[836] Dieses Disengagement wird auch als Ursache für die Notwendigkeit gesehen, dass der Vorstand den Förderauftrag inhaltlich festlegt.[837] Dies entspricht also der mitgliederpassiven Zielformulierung.[838]

Neben der direkten Mitgliederpartizipation ist der Aufsichtsrat als Vertretungsorgan der Mitglieder institutionalisiert. Im Sinne der Prinzipal-Agent-Struktur wird der Aufsichtsrat installiert, um den Vorstand in der Geschäftsführung in allen Bereichen zu überwachen.[839] Dazu hat er sich vom Vorstand unterrichten zu lassen. Im Sinne der vorliegenden Fragestellung ist zu untersuchen, welchen Einfluss der Aufsichtsrat auf die Zielformulierung in der Kreditgenossenschaft hat. Der Aufsichtsrat hat in der Praxis regelmäßig die von der Generalversammlung delegierte Aufgabe, den hauptamtlichen Vorstand zu bestellen.[840] Zentrale Aufgabe des Aufsichtsrats ist auch dessen Kontrolle, wobei er dem Vorstand nicht übergeordnet ist und ihm keine Weisungen erteilen darf und selbst weisungsfrei von der Generalversammlung agiert.[841] Es gehört zu den Kontrollinhalten des Aufsichtsrats, die Mitgliedernützlichkeit der Geschäfte zu überwachen. Mitgliedernützlichkeit bedeutet, dass der Vereinigungszweck der eG nicht darauf gerichtet ist, Dritte statt der Mitglieder zu fördern oder überwiegend gemeinnützig zu sein.[842] An der Zielformulierung ist der Aufsichtsrat direkt nicht beteiligt. Kritisiert wird an der Bestellung des Aufsichtsrats in Kreditgenossenschaften – insbesondere bei Flächenbanken – der Regionalproporz, die aufgrund des geringen Entgelts unzureichende Incentivierung von qualifizierten

---

Mitverantwortung für wirtschaftliche Schieflagen ausgesprochen wird, der Generalversammlung keine Relevanz bei der Kontrolle zuerkannt wird. Zudem mutmaßt er, dass die Mitglieder oder Vertreter nicht im vollen Umfang das Ausmaß der Jahresabschlussfeststellung kennen, da damit auch eine ordnungsgemäße Geschäftsführung im Sinne des Förderzwecks bescheinigt wird. Vgl. *Tebroke, H.-J.* (1998), S. 337.

835 *Gerstenmaier, V.* (1992), S. 73. Vgl. auch die Ausführungen in *Bonus, H.* (1994), S. 89 ff.

836 Vgl. *Hammerschmidt, M.* (2000), S. 14. Der Befragung liegen Antworten von 666 Mitgliedern zugrunde, wobei der Autor keine Aussage darüber macht, zu welchen Anteilen es sich um private oder gewerbliche Mitgliederkunden handelt.

837 Vgl. *Grauer, F.* (1998), S. 194.

838 Vgl. dazu die Ausführungen in Kapitel D.3.1.2.2.

839 Siehe § 38 Abs. 1 GenG sowie § 22 Abs. 1 Mustersatzung des BVR für VR-Banken.

840 Vgl. *Tebroke, H.-J.* (1998), S. 337.

841 Vgl. *Beuthien, V./Dierkes, S./Wehrheim, M.* (2008), S. 45.

842 Vgl. *Beuthien, V./Dierkes, S./Wehrheim, M.* (2008), S. 3.

Mitgliedern sowie die vornehmliche Besetzung mit Selbständigen, die häufig auch große Kreditkunden der Genossenschaftsbank sind.[843] Neben der mangelnden Bankleiterqualifikation impliziert dies den Konflikt zwischen Kontrollaufgabe und Abhängigkeit. So könnte die Vermutung, nicht dem Willen des Vorstands entsprechendes Verhalten könne zu einem erschwerten Zugang zu Krediten führen, angepasstes Abstimmungsverhalten begünstigen.[844] Unter anderem aus diesen Gründen ist die Kontrolleffizienz des Aufsichtsrats in Kreditgenossenschaften insgesamt als gering zu beurteilen und ein Einfluss auf die Zielformulierung im Sinne der Mitglieder zu negieren.

Vor dem Hintergrund dieser theoretischen und praktischen Defizite in der Corporate Governance von Kreditgenossenschaften hat die Wissenschaft Vorschläge zur Kompensation vorgetragen. So hat BOETTCHER etwa den Förderplan als organisationszweckspezifische Ergänzung der Unternehmensplanung angeregt, der am Ende des Geschäftsjahres durch den Förderbericht ergänzt wird.[845] Auch THEURL/BÖTTIGER und LAMPRECHT/DONSCHEN empfehlen mit dem Member Value-Reporting ein Instrument, das im Sinne der Mitglieder deren spezifische Zielvorstellungen in der Geschäftspolitik sicherstellen soll.[846] Jedoch haben diese Instrumente nach Kenntnisstand des Verfassers bisher keine weite Verbreitung in der kreditgenossenschaftlichen Praxis gefunden.

Kreditgenossenschaften lassen sich vor dem Hintergrund der gemachten Ausführungen zusammenfassend als managementkontrollierte Unternehmen auffassen, in denen die Vorstände ihre Interessen weitgehend autonom umsetzen können.[847] PFAFFENBERGER argumentiert, dass „die heutigen genossenschaftlichen Organe im Wesentlichen durch die jeweiligen Vorstände oktroyiert werden."[848] So konstatiert auch ENGELS eine im Vergleich zu Kapitalgesellschaften stärkere Verwässerung von Verfügungsrechten, wobei er das Koordinationsrecht unmittelbar dem Vorstand zugesteht und das Aneignungsrecht sowie das Veräußerungsrecht zwar bei den Mitgliedern sieht, diese jedoch als stark eingeschränkt bewertet.[849] Vor allem durch die Unbestimmtheit des Förderauftrags sieht er das Aneinungsrecht verwässert und faktisch auf die Manager übertragen,

---

843 Vgl. *Tebroke, H.-J.* (1998), S. 337.

844 Vgl. *Pfaffenberger, K.* (2009), S. 283.

845 Vgl. *Boettcher, E.* (1979), S. 198 ff.

846 Vgl. *Theurl, T./Böttiger, J.-M.* (2007), S. 4 ff.; *Lamprecht, D./Donschen, A.* (2006), S. 18 ff. Weitere Instrumente in diesem Zusammenhang sind die Sozialbilanz und die Förderbilanz. Vgl. *Lamprecht, D./Donschen, A.* (2006), S. 39 ff.

847 Eine empirische Bestätigung dieser Aussage liefert *Topp, S.* (1999), S. 93; siehe auch *Vogel, W.* (2002), S. 78 sowie die dort zitierte Literatur.

848 *Pfaffenberger, K.* (2009), S. 283.

849 Vgl. hier und nachfolgend *Engels, M.* (1997), S. 674 ff.

wodurch deren Handlungsspielraum erhöht wird.[850] Dies wird noch durch die Imparität der Teilhabe der Genossen, die Gewinne tendenziell eher zu thesaurieren und für Verluste theoretisch unbeschränkt zu haften, verstärkt.[851]

Insgesamt lässt sich festhalten, dass der Vorstand das zentrale und einflussreichste Gremium im Willensbildungsprozess der Kreditgenossenschaften ist.[852] Dessen Ziele und Vorstellungen bestimmen das Zielsystem und so die Zielbeziehung zwischen immateriellen und materiellen Zielen. Aufbauend auf der Erkenntnis, dass in der Praxis das mitgliederpassive Vorgehen zur Zieldefinition zu erwarten ist, stellt sich nun zunächst die Frage, welche Implikationen auf die Zielbeziehung aus der vorstandsseitigen Operationalisierung des Förderauftrags resultiert.

3.1.2.4 Förderinstrumente als Einflussfaktoren auf das Zielsystem

Der Förderauftrag wird in der kreditgenossenschaftlichen Praxis und in der Theorie intensiv mit dem Ergebnis diskutiert, dass es eine Vielfalt an Konkretisierungsansätzen gibt. Einige beurteilen die Existenz der Kreditgenossenschaft und die damit verbundenen Wettbewerbsergebnisse in Form von günstigen Konditionen und breiten Leistungsprogrammen als erfüllten Förderauftrag.[853] Andere wiederum sehen den Förderauftrag erst dann erfüllt, wenn preis- und konditionenpolitisch zugunsten der Mitglieder differenziert wird.[854] Dritte dagegen definieren den Förderauftrag als Verlässlichkeit in prekären Bindungen.[855] Es wurde vor dem Hintergrund dieser Vielfalt bereits verallgemeinernd auf einen Zusatznutzen aus der Mitgliedschaft rekurriert, der mit THEURL als Member Value

---

850 „In der Terminologie der Property-Rights-Theorie wird das sich auf die Aneignung des Unternehmenserfolges erstreckende Rechtsbündel durch den Übergang von der rein monetären Gewinnbetrachtung auf die Leerformel des Förderungsauftrages für die Genossen dermaßen verwässert, daß auch dieses Recht faktisch auf die Manager übergeht und sie durch dessen Disposition ihren eigenen diskretionären Handlungsspielraum erweitern.“ *Engels, M.* (1997), S. 675.

851 Vgl. *Engels, M.* (1997), S. 676.

852 Vgl. *Pfaffenberger, K.* (2009), S. 282; *Licht, W.* (1980), S. 13; *Heckt, J. G.* (1977), S. 34.

853 Vgl. *Beuthien, V.* (2006), S. 53; *Hahn, O.* (1980), S. 18. Dies würde jedoch implizieren, dass diese Wettbewerbsergebnisse, die sich vor allem im Preiswettbewerb konkretisieren, und aufgrund dessen Bedeutung für die Ertragssituation der Genossenschaftsbanken die Existenz der Kreditgenossenschaften bedrohen, gewollt sind. Dies erscheint jedoch nur schwer nachvollziehbar, da sich hierin ein nicht lösbarer Widerspruch zeigt.

854 Vgl. *Hahn, O.* (1980), S. 19.

855 Vgl. *Bonus, H.* (1994), S. 83 ff. In diesem Fall würde mit Bezug zum Firmenkundengeschäft der Förderauftrag eine derartig weite inhaltliche Kongruenz zum Hausbankprinzip haben, dass es keine kreditgenossenschaftliche Eigenheit sein kann.

konzeptualisiert wird.[856] Für die vorliegende Fragestellung ist relevant, ob und wenn ja, in wie weit die inhaltliche Ausgestaltung der einzelnen Komponenten des Member Value im Sinne einer Konkretisierung des Förderauftrags Einfluss auf das Zielsystem hat. Da die Vielfalt der einzelnen Förderinstrumente groß ist,[857] soll auf ein adäquates übergeordnetes Ordnungsprinzip abgestellt werden, aus dem sich die Einflüsse auf die Zielbeziehung ableiten lassen.

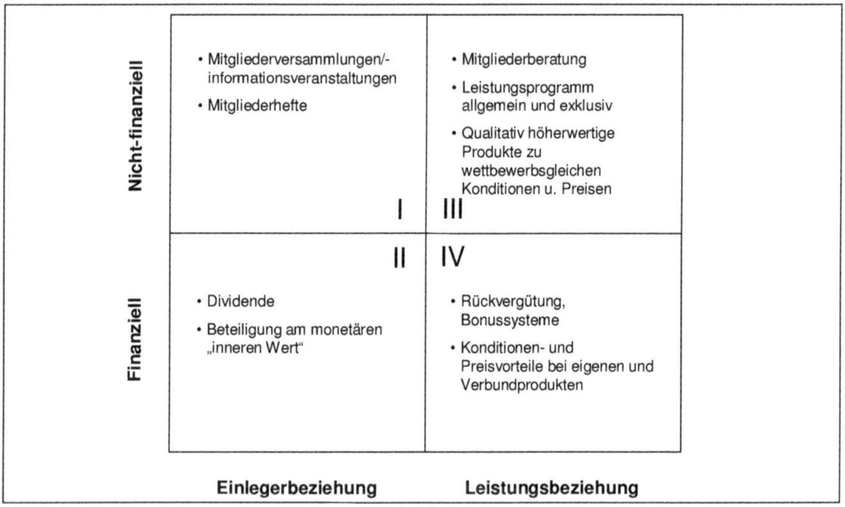

Abbildung 34: Mögliche Ordnungsstruktur für Förderinstrumente[858]

Als Ordnungskriterien soll zum einen die Beziehungsebene herangezogen werden. Danach wird zwischen Förderinstrumenten, die sich auf die Leistungsbeziehung und Förderinstrumenten die sich die Einlegerbeziehung beziehen, unterschieden. Zum anderen soll zwischen finanziellen und nicht-finanziellen Förderinstrumenten unterschieden werden. Finanzielle Förderinstrumente führen zu einer Erhöhung der Einzahlungen oder Verringerung von Auszahlungen eines Mitglieds, während die nicht-finanziellen Förderinstrumente Leistungen der

---

856 Vgl. die Ausführungen in Kapitel C.1.4.2 und Kapitel D.3.1.2.2.

857 Vgl. dazu z. B. die Untersuchungen von *Beuthien, V./Hanrath, S./Weber, H.-O.* (2008), S. 27 ff.

858 Eigene Darstellung. Dabei ist zu berücksichtigen, dass die Förderinstrumente nicht immer trennscharf den einzelnen Quadranten zugeordnet werden können. Aus diesem Grund wurden die aufgeführten Förderinstrumente schwerpunktgemäß zugeordnet.

Kreditgenossenschaft darstellen, die nicht direkt zu Vermögensänderungen des Mitglieds führen. Durch Gegenüberstellung dieser Merkmalsausprägungen lassen sich nun die Förderinstrumente des inneren und erweiterten Förderkreislaufs ordnen (siehe Abbildung 34).[859]

Die betrachteten Instrumente lassen sich weiterhin den einzelnen Komponenten des Member Value zuordnen. Dies erfolgt im Folgenden im Rahmen der Diskussion einzelner Instrumente. Um unter Berücksichtigung dieser Zuordnung die Analyse auch auf Ebene der einzelnen Förderinstrumente vorzunehmen, ist es sinnvoll, die Komponenten in eine nutzenbasierte Rangfolge zu bringen. Da bisher keine empirischen Ergebnisse zu nutzentheoretischen Bewertungen der Komponenten für die Kreditgenossenschaften vorliegen, erfolgt dies über plausible Annahmen. Es soll angenommen werden, dass bei gleichem Input die unmittelbare Förderung *UF* einen größeren Nutzen stiftet als die mittelbare Förderung *MF*.[860] Dies begründet sich damit, dass die Leistungsbeziehung genuine Quelle der Förderung ist und den eigentlichen Anreiz für einen Mitglieds-Kunden ausmacht, mit der Kreditgenossenschaft eine Geschäftsbeziehung aufzubauen. Weiterhin soll angenommen werden, dass das Förderpotenzial *FP* einen inferioren Nutzenbeitrag gegenüber den anderen beiden Komponenten hat. Diese Annahme wird damit begründet, dass eine intertemporale Präferenzrate zugunsten der Gegenwart vorliegt und, dass das Förderpotenzial auch künftigen Mitgliedern zugute kommt, wobei diese jedoch gegenwärtig noch keinen Beitrag zu diesem Potenzial z. B. in Form von thesaurierten Gewinnen leisten. Zudem gilt aus Sicht des einzelnen Mitglieds, dass Beiträge zum Förderpotenzial in der Zukunft nicht zwingend zum eigenen Nutzen abgerufen werden können. Es besteht nämlich kein Anspruch auf das Förderpotenzial.[861] Insgesamt ergibt sich auf Basis dieser Annahmen die folgende Rangordnung zwischen den Nutzenbeiträgen der Komponenten:

$$UF \succ MF \succ FP \qquad [D.2]$$

[859] Dabei ist zu berücksichtigen, dass die beispielhaft und nicht abschließend angeführten Förderinstrumente nicht hinsichtlich ihrer Förderfähigkeit oder -zweckmäßigkeit beurteilt werden. Es handelt sich um in der theoretischen und praktischen Literatur genannte Förderinstrumente. Vgl. z. B. *Grosskopf, W.* (1990), S. 41 ff. Die Förderinstrumente der Allgemeinförderung werden hier nicht betrachtet.

[860] Plakativ an einem Beispiel verdeutlicht heißt dies etwa, dass EUR 100 direkter Konditionsvorteil dem Mitglied einen höheren Nutzen stiften als EUR 100 Dividende oder EUR 100 Kosten je Mitglied, die für eine Mitgliederzeitschrift aufgewendet werden.

[861] Dies zeigt sich z. B. darin, dass Mitglieder keinen Anspruch auf die das Förderpotenzial u. a. konstituierenden Ergebnisrücklagen haben. Vgl. *Cario, D.* (2005), S. 149.

Um Ergebnisse begründen zu können, ist es auch nötig, die betrachteten Förderinstrumente selbst nutzentheoretisch zu beurteilen. Da keine empirischen Ergebnisse zu nutzentheoretischen Urteilen vorliegen, sind jeweils weitere Annahmen zu treffen.

Um den Einfluss der Konkretisierung des Förderauftrags auf das Zielsystem und die Zielbeziehung zu identifizieren, ist nun zu prüfen, ob das jeweilige konkrete Förderinstrument gleichgerichtet oder konfliktär zum Gewinnziel zu beurteilen ist. Ausgangspunkt der weiteren Analyse ist eine vereinfachte Gewinnfunktion (siehe Formel [D.3]):

$$G = (p - k) \cdot x - K \qquad [D.3]$$

mit:

G: Gewinn

p: Marktpreis je Stück

k: Standardstückkosten

K: Nicht in Standardstückkosten verrechnete Gemeinkosten

Förderinstrumente im Quadrant I, die nicht-finanziell sind und auf die Einlegerbeziehung abstellen, können als fixe Kostenkomponente in die Formel [D.3] integriert werden und können der Komponente *MF* des Member Value, die als mittelbare Förderung oder Mitgliedernutzen definiert ist, zugeordnet werden.

$$G = (p - k) \cdot x - K - FI^{I} \qquad [D.4]$$

mit:

FI^{I}: Kosten des Förderinstruments im Quadrant I

Beispiele für *FI* sind etwa die Kosten, die auf Ebene der Kreditgenossenschaft anfallen, um eine Mitgliederzeitschrift herzustellen und an die Mitglieder zu verteilen. Diese Instrumente sind formal-analytisch zunächst konfliktär zum Gewinn-/Vermögenswertziel, da sie den Gewinn reduzieren. Es ist jedoch davon auszugehen, dass Mitglieder mit Blick auf diese Förderinstrumente saturierungsfähig sind und der Nutzen aus diesem Förderinstrument ein Maximum hat. Vor dem Hintergrund der Saturierungsfähigkeit kann eine nach oben begrenzte Kostenkomponente angenommen werden, weil bspw. eine über dem Nutzenmaximum liegende Anzahl an Mitgliederzeitschriften keinen positiven Einfluss auf den Fördernutzen des einzelnen Mitglieds hat. Es liegt also nach dem Maximum keine gegenläufige Wirkrichtung auf die Gewinnfunktion vor und ein derartiges Förderinstrument widerspricht dem Gewinnziel nicht weiter.

Instrumente des Quadranten II sind finanziell und beziehen sich auf die Einlegerbeziehung. Diese Förderinstrumente lassen sich wie folgt mit der Formel [D.3] verbinden:[862]

$$D = G \cdot d = ((p-k) \cdot x - K) \cdot d \qquad [D.5]$$

mit:

D: Dividenden- / Thesaurierungsbetrag

d: Dividenden- / Thesaurierungssatz in Prozent

Die Ableitung der Ergebnisse für die Förderinstrumente, die auf der Einlegerbeziehung basieren und finanzieller Art sind, muss differenziert vorgenommen werden. Zum einen hinsichtlich der Zuordnung zu den Komponenten des Member Value, denn während die Dividende der mittelbaren Förderung bzw. dem Mitgliedernutzen *MF* zugeordnet wird, ist die Gewinnthesaurierung als Bestandteil des „inneren Werts" dem Förderpotenzial *FP* zuzurechnen. Zum anderen muss hinsichtlich der damit verbundenen Mittelbarkeit des Nutzens für das Mitglied differenziert werden. Denn während die Dividende direkt zugunsten des Mitglieds fließt und damit direkt und effektiv Nutzen stiftet, ist dies für die Thesaurierung als Ansatz der Operationalisierung des „inneren Werts" nicht der Fall. Denn dabei ist zu berücksichtigen, dass Mitglieder beim Ausscheiden aus der Genossenschaft de jure grundsätzlich keinen Anspruch auf Eigenkapitalbestandteile in Form von Rücklagen etc. haben.[863] Und während der laufenden Mitgliedschaft besteht kein mitgliedsindividueller Anspruch auf dieses Förderpotenzial. Unter Berücksichtigung der Rangfolge der Komponenten des Member Value aus nutzentheoretischer Sicht (siehe Formel [D.2]) würde ein Mitglied die Dividende dem „inneren Wert" vorziehen. Um den Nutzen aus Sicht der Mitglieder zu erhöhen, kann aufgrund des funktionalen Zusammenhangs entweder der Dividendensatz oder der Gewinn gesteigert werden.[864] Da das Mitglied letztlich das Produkt dieser beiden Größen als relevante Komponente in sein Nutzenfunktion integriert, dürfe es indifferent bezüglich der Frage sein, welche Komponente erhöht wird. Unter Berücksichtigung der begrenzten Möglichkeiten des einzelnen Mitglieds auf die Höhe des Dividendensatzes einzuwirken, kann zu-

[862] Zu beachten ist hier, dass die Integration dieses Förderinstruments in die Gewinngleichung abweichend von den sonstigen Vorgehensweisen ist. Denn die Dividende oder die Thesaurierung stellen keine Kosten- oder Aufwandsgrößen im Sinne der Rechnungslegung dar, sondern sind eine Größe der Gewinnverwendung.

[863] Auf diese kreditgenossenschaftlichen Besonderheiten im Zusammenhang mit dem „inneren Wert" wird in Kapitel D.3.1.2.5 noch näher eingegangen.

[864] Es handelt sich bei der funktionalen Darstellung zwar um eine vereinfachte Vorgehensweise, dennoch ist auch bei einer komplexeren Darstellung der Aussagenkern identisch.

dem diese Größe als Datum angenommen werden. Insgesamt ist daher zu konstatieren, dass Förderinstrumente in Quadrant II komplementär zum Gewinn-/Vermögenswertziel sind.

Hinsichtlich des Quadranten III muss eine weiter differenzierte Betrachtung erfolgen. Während etwa die auf das Leistungsprogramm ausgerichteten Förderinstrumente bedingt durch eine zunehmende Komplexität der Vertriebs- und Produktionstätigkeiten als fixe Gemeinkosten zum gleichen Ergebnis führen dürften wie Instrumente aus Quadrant I, sollen die Mitgliederberatung und die Qualitätsvorteile näher betrachtet werden. Diese Förderinstrumente sind im Sinne des Member Value der unmittelbaren Förderung *UF* zuzuordnen und können wie folgt in die Formel [D.3] integriert werden:

$$G = (p - (k + k^Q)) \cdot x - K \qquad [D.6]$$

mit:

k^Q: Qualitätsbedingter Stückkostenaufschlag

Aufgrund der Eigenschaften von Bankleistungen können fast alle die Bankleistungsqualität steigernden Maßnahmen auf der Leistungserstellungsebene als Aufschlag auf die Stückkosten operationalisiert werden.[865] Es ist leicht erkennbar, dass dieses Förderverständnis tendenziell zu Zielkonflikten führt. Denn eine mit einem steigenden Nutzen für die Mitglieder verbundene Ausweitung dieses Förderinstruments reduziert c. p. die Gewinnsituation der Kreditgenossenschaft. Somit sind Gewinn-/Vermögenswertziel und Förderauftrag nicht gleichzeitig zu verfolgende Zielgrößen. Insgesamt zeigt sich, dass Instrumente im Quadrant III eine konfliktäre Zielbeziehung zwischen Förderauftrags- und Gewinn-/Vermögenswertziel begründen.

Im abschließend zu betrachtenden Quadranten IV sind die Förderinstrumente zusammengefasst, die sich auf die Leistungsbeziehungsebene beziehen, daher der unmittelbaren Förderung *UF* zugeordnet werden können und finanzieller Art aus Sicht des Mitglieds sind. Aus diesen Instrumenten werden hier die Konditionsvorteile sowie die Boni oder Rückvergütungen betrachtet. Diese können wie folgt in die grundlegende Gewinnfunktion aus Formel [D.3] integriert werden:

---

865 Vgl. für die Beschreibung der Bankleistungen Kapitel C.2.3. So können etwa eine intensivere Beratung als Gemeinkosten in Verbindung mit der Deckungsbeitragsrechnung über die Standardstückkosten ebenso als Kostenaufschlag verrechnet werden wie z. B. hochwertige Trägermedien (z. B. Kreditkarten).

$$G = (p \cdot (1 - KV) - k) \cdot x - K \qquad [D.7]$$

$$G = (p - k) \cdot x - (p \cdot x) \cdot B - K \qquad [D.8]$$

mit:

KV: Konditionsvorteil für Mitglieder in Prozent

B: Umsatz-/Geschäftsvolumenabhängiger Bonussatz bzw. Rückvergütungssatz in Prozent

Während der Konditionsvorteil auf den Marktpreis der Bankleistung zielt und diese für das Mitglied günstiger macht, basiert der Bonus bzw. die Rückvergütung auf der Idee, dass nach Abschluss einer Abrechnungsperiode auf das insgesamt in dieser Periode von einem Mitglied nachgefragte Volumen ein Bonus an das Mitglied gezahlt wird. Zwar wurden aus Gründen der Nachvollziehbarkeit diese Instrumente jeweils in eigener Struktur in den Formeln [D.7] und [D.8] dargestellt, jedoch lassen sich diese in eine gleiche Struktur bringen und somit gemeinsam diskutieren. Es ist leicht ersichtlich, dass die Konditionenvorteile und Boni bzw. Rückvergütungen im Konflikt mit dem Gewinnziel stehen, da eine Ausweitung dieser Instrumente c. p. mit einer Verringerung des Gewinns einhergehen. Zusammengefasst begründen die Instrumente im Quadranten IV konfliktäre Beziehungen zwischen Förderauftrags- und Gewinn-/ Vermögenswertziel.

Zwar können die genannten Instrumente detaillierter im Hinblick auf die Nutzenfunktionen diskutiert werden und es können auch weitere Bedingungen eingefügt werden, die die Bedeutung auf die Zielbeziehung teilweise kompensieren. Aber insgesamt zeigen die Beispiele, dass die leistungsbeziehungsbezogenen Förderinstrumente eher dem Gewinnziel entgegen stehen und die einlegerbezogenen finanziellen Instrumente zu diesem komplementär sind. Bevor vor dem Hintergrund dieser Erkenntnisse die in der Praxis vorfindbaren Konkretisierungen des Förderauftrags betrachtet werden, soll noch ein wesentlicher Aspekt thematisiert werden, der die Mitglieder-Kunden betrifft. Der Rollenpluralismus führt in Verbindung mit dem Zielsystem und der Konkretisierung des Förderauftrags tendenziell zu einem intrapersonellen Zielkonflikt, der letztlich nur über die nutzenbasierten Rangfolge der Komponenten des Member Value in Gleichung [D.2] aufgehoben werden kann. Auf der einen Seite zahlt das Mitglied die Preise für die erhaltenen Leistungen und auf der anderen Seite erhält es eine gewinnabhängige Dividende auf das Geschäftsguthaben. Aufgrund des unmittelbaren Zusammenhangs von Preis und Gewinn wird der Zielkonflikt deutlich. Bevorzugt das Mitglied günstige Konditionen (*UF*) so sinkt c. p. der Preis, damit der Gewinn und damit wiederum die Dividende (*MF*). Bevorzugt das Mitglied jedoch die Dividende, steigt der Preis relativ, damit der Gewinn und auch die Dividende.

Empirische Ergebnisse zum Einsatz von Förderinstrumenten zeigen, dass vornehmlich Instrumente eingesetzt werden, die auf die Einlegerbeziehungsebene abstellen und finanzieller und nicht-finanzieller Art sind. So finden GROSSKOPF und BONUS, dass Konditionenvorteile aus Sicht der Kreditgenossenschaften kaum noch als Förderinstrumente realisierbar sind, während überwiegend diejenigen Vorteilsgewährungen anerkannt und umgesetzt werden, die mit der Trägerfunktion der Mitglieder verknüpft sind (z. B. Dividenden, gezielte Informationen oder Verzehrgutscheine für Mitgliederversammlungen).[866] Selten werden für das Bankgeschäft genossenschaftliche Rückvergütungen oder Beteiligungen am inneren Wert als Förderinstrumente praktiziert. Diese Ergebnisse verwundern nicht, wenn die Argumente betrachtet werden, die von den Mitarbeitern im Kundengespräch für eine Mitgliedschaft vorgebracht werden.[867] So ergaben Studien, dass für 85 Prozent der Mitarbeiter die vergleichsweise hohe Dividende aus Sicht der Kunden ein wichtiges Argument ist.[868] Unter Berücksichtigung der Beobachtung, dass jeder zweite aufgrund der Bankempfehlung Mitglied einer Kreditgenossenschaft wird,[869] führt die so praktizierte Förderpolitik zu einer erwartungsgemäßen Operationalisierung des Förderauftrags. GRAUER sieht in dieser Entwicklung einen Strukturwandel der Genossen, der aufgrund der nahezu ausschließlich dividendenbasierten Incentivierung von Mitgliedern den Transformationsprozess zu einer erwerbswirtschaftlichen Unternehmung beschleunigt.[870]

Während die meisten der genannten Instrumente undifferenziert allen Mitgliedergruppen angeboten werden, sollen mitgliedergruppenspezifische Operationalisierungen für das Firmenkundengeschäft gesondert betrachtet werden. Häufig wird in diesem Zusammenhang vor allem auf die Verlässlichkeit der Kreditgenossenschaft in prekären Beziehungen rekurriert.[871] Dies konkretisiert sich bspw. in geschäftspolitischen Erwägungen in Krisenphasen, wo nach diesem Verständnis die Mitgliedsförderung im Vordergrund stehen solle. Jedoch kann der Förderauftrag dem Firmenkunden-Mitglied gegenüber nicht als der alleinige

---

866 Vgl. hier und nachfolgend *Grosskopf, W.* (1990), S. 41 ff. Vgl. auch die Ausführungen der Bankenvorstände von Kreditgenossenschaften in *Bonus, H.* (1994), S. 94 ff.

867 Vgl. z. B. *Eckstein, B.* (1998), S. 5 ff.; *Erben, E.* (1970), S. 12.

868 Vgl. *Hammerschmidt, M.* (2000), S. 15. So auch die Ergebnisse bei MÜNKNER, wonach für 80,3 Prozent der Befragten die Dividende eine höhere Entscheidungsbedeutung hat als 10 Jahre vor dem Befragungszeitpunkt. Vgl. *Münkner, H.-H.* (1990), S. 68.

869 Vgl. *Hammerschmidt, M.* (2000), S. 14.

870 Vgl. *Grauer, F.* (1998), S. 202.

871 Vgl. *Broschart, J.* et al. (2002b), S. 31; *Bonus, H.* (1994), S. 77.

Erklärungsfaktor dafür angesehenen werden, das Mitgliedsunternehmen in der Krise zu begleiten. Eine Geschäftsbeziehung mit einem Firmenkunden zu beenden, der in einer Krisensituation ist, führt zwangsläufig zu durch das Regionalprinzip beeinflusste Imponderabilien. Zu denken ist dabei an die Angestellten des Unternehmens, die gegebenenfalls (Kredit-)Kunden der Kreditgenossenschaft sind und der Außenwirkung auf Dritte, die in einem regionalen Markt bedeutend höher sein dürfte als bei einem überregionalen Markt.[872] Zudem ist der so operationalisierte Förderauftrag der Kreditgenossenschaft den konstituierenden Merkmalen der Hausbankbeziehung identisch, was seine anzustrebende Besonderheit ausschließt.

Als weitere Möglichkeit den Förderauftrag für gewerbliche Mitglieder zu konkretisieren, wird in der betriebswirtschaftlichen Beratung des Unternehmers gesehen.[873] Zweifelhaft ist jedoch, ob dies vor dem Hintergrund knapper Ressourcen und des mit der Beratung verbundenen erheblichen Zeitaufwands möglich ist. Fraglich ist zudem, in wie weit der Firmenkundenberater in der Lage ist, den Mitglieder-Firmenkunden betriebswirtschaftlich zu beraten, da dazu neben der klassischen betriebswirtschaftlichen Kompetenz auch Branchen- und Marktwissen gehören, was eine zeitintensive Auseinandersetzung mit diesen Rahmenbedingungen voraussetzt. Für das Firmenkundengeschäft ist insgesamt von einer relativ größeren Unbestimmtheit des Förderauftrags auszugehen.

Zusammenfassend lässt sich konstatieren, dass die in der Praxis am häufigsten eingesetzten Instrumente den Förderauftrag zu operationalisieren, komplementär zum Gewinnziel stehen. Darüber hinaus gilt, dass im Mitgliedergeschäft mit gewerblichen Kunden eine relativ hohe Unbestimmt hinsichtlich der Operationalisierung des Förderauftrags vorliegt.

3.1.2.5 Eigenfinanzierungsoptionen als Einflussfaktor auf das Zielsystem

Die Kreditgenossenschaften haben aufgrund ihrer Größe, Struktur und Rechtsform nur begrenzte Möglichkeiten der Eigenfinanzierung, insbesondere ist der Kapitalmarktzugang rechtsformbedingt limitiert. Daher wird den Genossenschaftsbanken auch „eine rechtsformimmanente Eigenkapitalbildungsproblematik“[874] nachgesagt. Die Bildung von haftendem Eigenkapital ist jedoch für das Wachstum insbesondere im Kreditgeschäft erforderlich. Gleichzeitig stellt die notwendige Ausweitung des haftenden Eigenkapitals bzw. der Eigenmittelba-

---

872 Vgl. dazu die Ausführungen in Kapitel D.4.5.3.2.1.

873 Vgl. *Lürig, R.* (1985c), S. 46.

874 *Algner, M. R.* (2006), S. 186.

sis[875] im Sinne der regulatorischen Vorgaben auch bei den Kreditgenossenschaften eine nennenswerte Herausforderung dar.[876] Mögliche Optionen zum Erhalt und Ausbau der Eigenkapitalbasis und deren Würdigung durch STRIEDER, HOFINGER/VAN HUSEN und BORNEMANN/SCHAAF/PFINGSTEN sind in der folgenden Tabelle 10 aufgeführt:[877]

---

875 Die für alle Banken gültige Definition der Eigenmittel ist in § 10 Abs.2-7 KWG enthalten. Zu unterscheiden ist zwischen dem bilanziellen Eigenkapital und dem haftenden Eigenkapital, also der Eigenmittel. Eigenmittel bestehen aus dem Kernkapital, dem Ergänzungskapital 1. und 2. Klasse sowie Drittrangmitteln. Für eine ausführliche Erläuterung der Eigenmittelbestandteile vgl. *Hartmann-Wendels, T./Pfingsten, A./Weber, M.* (2010), S. 429 ff.

876 Zu den relevanten Veränderungen, die der BASLER AUSSCHUSS FÜR BANKENAUFSICHT im Dezember 2009 fordert, gehören die Steigerung des Eigenkapitals bei gleichzeitig höherer Qualitätsanforderung, die Verschärfung von Abzügen vom Kernkapital sowie quantitative Liquiditätsvorschriften. Vgl. *Basler Ausschuss für Bankenaufsicht* (2009). Obwohl sich diese, zusammen als Basel III bezeichneten, postulierten Änderungen noch in der Konsultationsphase befinden, gehen einigen Autoren davon aus, diese nicht mehr wesentlich geändert werden. Vgl. z. B. *Hofmann, G.* (2010), S. 666.

877 Für eine ausführliche Würdigung der einzelnen Eigenfinanzierungsoptionen vgl. *Strieder, T.* (2000); *Hofinger, H./Van Husen, R.* (2004); *Bornemann, S./Schaaff, C./Pfingsten, A.* (2009). Vorzugsaktien werden nicht betrachtet, da Sie für die Kreditgenossenschaften nicht einsetzbar sind. Aufgrund ihrer begrenzten Ausbaufähigkeit werden auch die Vorsorgereserven nach § 340f HGB nicht betrachtet. Auch die Möglichkeit zur Bildung stiller Reserven wird aufgrund ihrer begrenzten Möglichkeiten und der gegenwärtigen Entwicklungen in Bezug auf die Harmonisierung der Rechnungslegungsvorschriften auf Ebene der EU nicht betrachtet.

Tabelle 10: Maßnahmen zum Erhalt und Aufbau des Eigenkapitals[878]

Maßnahme	**Beurteilung**
Geschäftsguthabenbezogene Maßnahmen	
1. Akquisition neuer Mitglieder-Kunden	begrenzt, weitestgehend ausgereizt
2. Erhöhung der verpflichtenden finanziellen Beteiligung durch die Mitglieder a. Erhöhung der Pflichtbeteiligung b. Erhöhung des Geschäftsanteils	Mitgliederakzeptanz fraglich, da Kündigung von Mitgliedern erwartet wird
3. Mindestkapital gem.§ 8a GenG	Mitgliederakzeptanz fraglich, Rechtsformvorteil (Haftungsbegrenzung bei nicht festem Kapitalstock) wird nivelliert
4. Aufnahme investierender Mitglieder	Potenzieller Konflikt zu genossenschaftlichen Prinzipien (Förder-, Identitäts- und Demokratieprinzip); Attraktivität aus Sicht investierender Mitglieder gering
Maßnahmen der Innenfinanzierung	
5. Thesaurierung erzielter Gewinne/Bildung von Rücklagen	Mitgliederakzeptanz fraglich, da Forderung zur Weitergabe an Mitglieder erwartet wird; beeinflusst durch Höhe erzielter Gewinne
Mitgliederferne Maßnahmen der externen Eigenfinanzierung	
6. Hereinnahme nachrangiger Haftungsmittel a. Stille Beteiligungen b. Ausgabe von Genussrechten	 Erfolgsanspruch der stillen Gesellschafter widerspricht § 48 Abs. 1. S. 2 GenG zulässig, aber abhängig von Ausgestaltung, ggf. günstiger als Geschäftsguthaben
7. Hereinnahme nachrangiger Darlehen	strittig, ob Eigenkapital, da abhängig von Ausgestaltung
Strukturverändernde Maßnahmen	
8. Umwandlung in genossenschaftliche Aktiengesellschaft	Vereinbarkeit mit genossenschaftlichen Prinzipien fraglich, verbesserter Zugang zu potenziellen Investoren, höhere Attraktivität für Investoren; Mitgliederakzeptanz fraglich
9. Generierung von Eigenkapital am Kapitalmarkt durch Konzerngesellschaften	
a. Genossenschaftliche Holding	zulässig, aber strittig, ob vollumfänglich vereinbar mit genossenschaftlichen Prinzipien; Zugang zum Kapitalmarkt; Mitgliederakzeptanz fraglich
b. Genossenschaftliche KG aA	zulässig, Zugang zum Kapitalmarkt; Mitgliederakzeptanz fraglich

[878] Eigene Darstellung.

Da die genannten strukturverändernden Maßnahmen eine im Vergleich zum bisherigen, an den Besonderheiten der Kreditgenossenschaften orientierten Gang der Analyse vollständig andere Vorgehensweise bedingen würden,[879] sollen diese nicht weiter betrachtet werden. Der Fokus richtet sich daher auf die geschäftsguthabenbezogenen Maßnahmen, die Maßnahmen der Innenfinanzierung sowie die als mitgliederfern bezeichneten Maßnahmen der externen Eigenfinanzierung.[880]

Die eingezahlten Geschäftsguthaben der Mitglieder sind dem gezeichneten Kapital bei Kapitalgesellschaften gleichzustellen.[881] Da jedoch die Geschäftsguthaben nicht in Abhängigkeit der geschäftspolitischen Erfordernisse wie bei Kapitalgesellschaften gesteuert werden können und gleichzeitig aufsichtsrechtliches Kernkapital darstellen, welches seinerseits wiederum als wesentliches Regulativ für die Geschäftstätigkeit mit Risikoaktiva anzusehen ist, kann es zum Engpassfaktor werden.[882] Die beschränkte Steuerbarkeit ergibt sich aus dem begrenzten Mitgliederpotenzial, den aus Anreizgründen meist recht gering gehaltenen Beträgen für Geschäftsanteile sowie dem Kündigungsrecht der Mitglieder gem. § 65 Abs. 1 GenG womit regelmäßig ein Kapitalabfluss in Höhe des Geschäftsguthabens verbunden ist. Daher werden die Geschäftsguthaben auch als nicht festes Eigenkapital bezeichnet.[883] Die Stabilität des gezeichneten Eigenkapitals kann de facto durch entsprechende Anreize für die aktuellen und potenziellen Mitglieder-Kunden erreicht werden. Hinsichtlich der Anreize, die durch den Förderauftrag determiniert werden, und ihren Auswirkungen auf das Zielsystem sei auf die Ausführungen in Kapitel D.3.1.2.4 verwiesen. Doch auch bei richtig gesetzten Anreizen stellt das Regionalprinzip eine „natürliche" Grenze

---

879 Die strukturverändernden Maßnahmen hätten wesentlichen Einfluss auf die Frage danach, ob die in Kapitel C.1.4 beschriebenen Besonderheiten der Kreditgenossenschaften noch gültig sind.

880 Durch entsprechende Einschränkungen des relevanten Adressatenkreises auf die Gruppe der Mitglieder würde die Mitgliederferne als Charakteristikum entfallen. In diesem Fall wäre die weitere Diskussion hochgradig kongruent zu den geschäftsguthabenbezogenen Maßnahmen. Um eine differenzierte Analyse zu ermöglichen, wird diese Begrenzung des Adressatenkreises nicht vorgenommen.

881 Siehe § 337 Abs. 1 Satz 1 HGB i. V. m. § 25 Abs. 1 Satz 1 RechKredV.

882 Vgl. hier und nachfolgend *Bornemann, S./Schaaff, C./Pfingsten, A.* (2009), S. 19.

883 Vgl. *Strieder, T.* (2000), S. 214. Da die Geschäftsguthaben als geringwertige Nominalpapiere ausgestaltet sind, ist jedoch die Gefährdung der Kapitalbasis der Kreditgenossenschaft aus Mitgliedskündigungen begrenzt. Vgl. *Hofinger, H./van Husen, R.* (2004), S. 90.

dar. Zudem wird unterstellt, dass das Mitgliederpotenzial tendenziell ausgeschöpft ist.[884]

Bei konstantem Mitgliederbestand bestünde weiterhin die Möglichkeit, die Pflichtbeteiligung oder die Geschäftsanteile zu erhöhen, um die Eigenmittel der Kreditgenossenschaft zu steigern.[885] Für beide Maßnahmen wäre eine Satzungsänderung mit einer Mehrheit von drei Viertel gem. § 16 Abs. 2 Nr. 2 bzw. 3 GenG erforderlich, was die Akzeptanz der Mitglieder bzw. Vertreter voraussetzt. Diese dürfte dadurch geschmälert werden, dass neben der zusätzlichen finanziellen Belastung, die Haftungssumme je Mitglied steigt und gleichzeitig bei konstanter Dividendenhöhe die Rendite der genossenschaftlichen Beteiligung sinkt. Vor diesem Hintergrund ist es fraglich, ob die notwendige mehrheitliche Mitgliederakzeptanz erreicht werden kann.

Die Stabilität des gezeichneten Kapitals kann de jure auch über das Mindestkapital gem. § 8a GenG erreicht werden.[886] In diesem Fall wird die Entscheidung über die Auszahlung gekündigter Geschäftsguthaben auf die Gesellschaftsorgane der Kreditgenossenschaft verlagert. Dies dürfte jedoch die Attraktivität der Mitgliedschaft aus Sicht aktueller und potenzieller Mitglieder einschränken. Aus Sicht dieser ist auch die Haftungsbegrenzung als wesentliches Merkmal der eG als Anreiz zu sehen. Da aktuell jedoch auch die Beseitigung eines Mindestkapitals als Haftsubstanz bei Kapitalgesellschaften diskutiert wird, was die Einführung der haftungsbeschränkten Unternehmergesellschaft belegt, würde die „Beseitigung dieses Rechtsformvorteils [...] bei Schaffung einer solchen in anderen Rechtsformen [.] die Stellung von Genossenschaften im Wettbewerb um Beteiligungskapital drastisch [verschlechtern; Erg. d. Verf.]“[887]. Die für diese Maßnahme notwendige Mitgliederakzeptanz dürfte nur schwierig zu erreichen sein.

Dem Ausbau der Mitgliederschaft und damit die Erhöhung des gezeichneten Kapitals dienlich, wäre auch die Aufnahme investierender Mitglieder nach § 8 Abs. 2 GenG, denen eine stärkere Renditeorientierung zugesprochen wird.[888] Aufgrund einer ungleichen Verteilung zwischen Anreizen und Beiträgen kommen BORNEMANN/SCHAAFF/ PFINGSTEN hinsichtlich dieser Mitgliedergruppe

---

[884] Vgl. *Strieder, T.* (2000), S. 214; *Hofinger, H./van Husen, R.* (2004), S. 90; *Bornemann, S./Schaaff, C./Pfingsten, A.* (2009), S. 27.

[885] Vgl. hier und nachfolgend *Strieder, T.* (2000), S. 214 f.

[886] Vgl. hier und nachfolgend *Bornemann, S./Schaaff, C./Pfingsten, A.* (2009), S. 26.

[887] *Bornemann, S./Schaaff, C./Pfingsten, A.* (2009), S. 28. Die gleiche Argumentation gilt für die denkbare Ausweitung der Haftsumme der Mitglieder, die als Haftsummenzuschlag im Ergänzungskapital zu berücksichtigen wäre.

[888] Vgl. *Bornemann, S./Schaaff, C./Pfingsten, A.* (2009), S. 27; *Cario, D.* (2005), S. 150; *Baumgärtler, T.* (2000), S. 99.

jedoch zu dem Ergebnis, dass die „Attraktivität einer genossenschaftlichen Beteiligung [...] aus der Perspektive investierender Mitglieder nicht sehr hoch [ist; Erg. d. Verf.]"[889] Zudem sind die in den vorhergehenden Kapiteln bereits erläuterten möglichen Zielkonflikte zu berücksichtigen, die als konfliktär zu den genossenschaftlichen Prinzipien interpretiert werden könnten.

Da die geschäftsguthabenbezogenen Maßnahmen in der Gesamtschau nur ein geringes Potenzial zum Ausbau des Eigenkapitals beinhalten und daraus resultierende Zielkonflikte vor allem auf Ebene der Anreize und damit durch Mitgliederstruktur und Konkretisierung des Förderauftrags beeinflusst werden, sollen diese Maßnahmen hinsichtlich ihrer Wirkung auf die Zielbeziehung nicht weiter betrachtet werden.

Aus den genannten Gründen dürften die mitgliederfernen Formen der externen Eigenfinanzierung an Bedeutung gewinnen. Vermögenseinlagen stiller Gesellschafter gehören unter bestimmten Bedingungen gem. § 10 Abs. 2 KWG zum Kernkapital.[890] Aus kreditgenossenschaftlicher Sicht ist jedoch zu berücksichtigen, dass es nach § 48 Abs. 1 S. 2 GenG grundsätzlich keine Beteiligung Dritter am Jahresüberschuss der Kreditgenossenschaft geben darf. Da das Gewinnbezugsrecht des stillen Gesellschafters zwar eingeschränkt, aber nicht vollständig ausgeschlossen werden kann, bleiben Bedenken bestehen. Für die Genossenschaftsbanken gilt gegenwärtig darüber hinaus, dass der Betrag der stillen Einlagen in Relation zu den Geschäftsguthaben und Rücklagen vernachlässigbar ist.[891] Gleiches gilt für die nachrangigen Darlehen, die unter bestimmten Bedingungen zum Ergänzungskapital der Klasse 2 zu zählen sind.[892] Nachteilig an dieser Finanzierungsform ist jedoch, dass die Anrechnung als haftendes Eigenkapital begrenzt und die Verzinsung vergleichsweise erhöht ist.[893] Genussrechtskapital kann nach der Verordnung über die Rechnungslegung der Kreditinstitute und Finanzdienstleistungsinstitute (RechKredV) nicht dem bilanziellen Eigenkapital zugerechnet, wohl aber im Sinne der aufsichtsrechtlichen Vorgaben als Ergänzungskapital der Klasse 1 betrachtet werden. Diese Finanzierungs-

---

889 *Bornemann, S./Schaaff, C./Pfingsten, A.* (2009), S. 28.

890 Vgl. hier und nachfolgend *Strieder, T.* (2000), S. 217 ff.

891 Vgl. *Padberg, T.* (2005), S. 76. Diese Erkenntnis basiert auf einer Vollerhebung der Jahresabschlüsse des Geschäftsjahres 2003 der „typischen" und „atypischen" Genossenschaftsbanken. Die Hälfte der stillen Einlagen wurde dabei bei der DEUTSCHE APOTHEKER- UND ÄRZTEBANK EG festgestellt, einer atypischen Kreditgenossenschaft. Bei anderen Instituten dagegen können die stillen Einlagen von besonderer Bedeutung sein. Vgl. z. B. *Hartmann-Wendels, T./Pfingsten, A./Weber, M.* (2010), S. 434.

892 Vgl. *Padberg, T.* (2005), S. 85. Nach dessen externer Jahresabschlussanalyse machen die nachrangigen Verbindlichkeiten etwa 5 Prozent des haftenden Eigenkapitals aus.

893 Vgl. *Strieder, T.* (2000), S. 220.

form wird relativ häufig eingesetzt, wenngleich nicht sehr umfangreich.[894] Zudem ist aufgrund der begrenzten Anrechenbarkeit als Eigenmittel die aufsichtsrechtliche positive Wirkung begrenzt. Kritisiert wird aus Sicht potenzieller Kapitalgeber jedoch, dass der fehlenden mitgliedschaftlichen Beteiligung die Übernahme unternehmerischen Risikos gegenübersteht. Um dieses Ungleichgewicht zu kompensieren, sind Vergütungen auf die Genussrechte vorzunehmen, die oberhalb der Dividendenzahlung an die Mitglieder liegen. Da diese, wie in Kapitel D.3.1.2.4 gezeigt wurde, häufig der einzige materielle Vorteil aus der Mitgliedschaft sind, werden dadurch Anreize gesetzt, aus der Kreditgenossenschaft auszutreten und gleichzeitig genossenschaftliche Genussscheine zu erwerben. Damit würde das Identitätsprinzip unterminiert, was in Verbindung mit der unterstellten bindenden Wirkung der Mitgliedschaft für Mitglieder-Kunden und der positiven Wirkung auf die Kundendurchdringungsraten tendenziell negative Auswirkungen auf das Geschäftsvolumen der genossenschaftlichen Primärbank haben kann. Auch für diese Maßnahmen kann vor dem Hintergrund der angestellten Überlegungen die weitere Diskussion aufgrund der Unerheblichkeit entfallen.

Abschließend sind die genannten Maßnahmen der Innenfinanzierung zu betrachten. Dazu zählt vor allem die Bildung von Rücklagen.[895] Aufgrund des Nominalwertprinzips stellen Kapitalrücklagen für Genossenschaftsbanken keine relevante Kapitalposition dar.[896] Die Rücklagen zählen zum Kernkapital, was deren unbegrenzte Anrechenbarkeit begründet. Zudem gilt, dass die Bildung von Rücklagen zur bedeutsamsten Finanzierungsform der genossenschaftlichen Eigenfinanzierung avanciert ist.[897] Problematisch dürfte jedoch sein, die Bildung von Rücklagen mit den Interessen der Mitglieder zu vereinbaren, da diesen grundsätzlich kein Anspruch auf diese Beträge zusteht.[898] Bezüglich der Rücklagenbildung verweisen THEURL/BÖTTIGER und BEUTHIEN darauf, dass förderwirtschaftlich nicht benötigte positive Jahreserfolge an die Mitglieder auszuschütten sind.[899] VOLK/VOLK werten eine übermäßige Rücklagenbildung bei

---

894 Vgl. hier und nachfolgend *Bornemann, S./Schaaff, C./Pfingsten, A.* (2009), S. 31.

895 Da die offenen Reserven in Form des Sonderpostens für allgemeine Bankrisiken einen wirtschaftlich ähnlichen Charakter haben wie die Rücklagen, werden diese zusammen betrachtet. Vgl. *Hartmann-Wendels, T./Pfingsten, A./Weber, M.* (2010), S. 434.

896 Aufgrund des im § 73 Abs. 2 S. 2 GenG niedergelegten Nominalwertprinzips kommt bei der Abgabe von Genossenschaftsanteilen kein Agio in Betracht. Vgl. *Strieder, T.* (2000), S. 214.

897 Vgl. *Algner, M. R.* (2006), S. 192 sowie die dort genannte Literatur.

898 Siehe § 73 Abs. 2 GenG.

899 Vgl. *Theurl, T./Böttiger, J.-M.* (2007), S. 7; *Beuthien, V.* (2006), S. 56 f.

Genossenschaften sogar als Verstoß gegen den Förderauftrag.[900] Damit soll vermieden werden, den Mitgliedern Vermögen zu entziehen und in eine Form vorstandsverwalteten Stiftungsvermögens zu transformieren.[901] Unter die förderwirtschaftlich benötigten Jahreserfolge sind dabei auch die Einstellungen in die gesetzlichen und statuarischen Rücklagen zu verstehen, die dazu dienen, die Eigenkapitalvorschriften zu erfüllen. Erst darüber hinausgehende Bestandsteile sollen an die Mitglieder ausgeschüttet werden. Neben dieser periodischen Beteiligung der Mitglieder an den Erfolgen, die als Dividende ausgezahlt wird, besteht nach § 73 Abs. 3 GenG die Möglichkeit in der Satzung festzulegen, eine Ergebnisrücklage zu bilden, die den Anspruch der ausscheidenden Mitglieder an der Erfolgsentwicklung der Kreditgenossenschaft beinhaltet. Dies wird in der Praxis jedoch kaum umgesetzt.[902] Der Zugriff auf die Rücklagen ist den Mitgliedern also weitgehend verwehrt. Aus Sicht der Vorstände besteht dagegen ein Interesse an vergleichsweise hohen und für die Kreditgenossenschaft gesicherten Rücklagen, da damit zum einen die Eigenkapitalsituation von der Mitgliederentwicklung teilweise emanzipiert werden kann und zum anderen das Drohpotenzial ausscheidender Mitglieder reduziert wird.[903] Wie im Phasenschema von WIEDEMANN hinterlegt, stellt der Austritt aus der Kreditgenossenschaft die Ultima Ratio eines Mitglieds bei Unzufriedenheit mit der förderwirtschaftlichen Leistung der Kreditgenossenschaft dar (siehe Abbildung 32). Um den mit dem Mitgliederaustritt verbundenen Einfluss auf die Kreditgenossenschaft zu reduzieren, und damit das Drohpotenzial abzuschwächen, bietet sich folglich die Rücklagenbildung an. Zudem ist es aufgrund des Informationsvorteils des Vorstands und der wenig effizienten Kontrollmöglichkeiten der Mitglieder und des Aufsichtsrates schwierig, den förderwirtschaftlich benötigten Anteil und damit zu thesaurierenden Jahreserfolg vollständig objektiv zu bestimmen.[904] Denn neben den gesetzlichen Eigenkapitalvorschriften sind Planungen für das Aktivgeschäft in Bezug auf Volumen und Risiko wichtige Einflussgrößen auf die Höhe des vorzuhaltenden Eigenkapitals. Ein weiterer wichtiger Anreiz für die Bildung

---

900 Vgl. *Volk, K. O./Volk, G.* (1989), S. 151.

901 GRAUER bezeichnet diesen Sachverhalt als „dem Zugriff der Genossen entzogenes ‚Sozialkapital'". *Grauer, F.* (1998), S. 204.

902 Vgl. *Schöffling, B.* (1991), S. 25 sowie die dort angegebene Literatur.

903 Diese Aussage wird unterstützt durch die Pecking-Order-Theorie, wonach Vorstände oder Geschäftsführer vor allem die Finanzierungsformen auswählen, die ihnen die höchstmögliche Autonomie gewähren. Vgl. zu ähnlichen Aussagen im genossenschaftlichen Kontext *Algner, M. R.* (2006), S. 192. Nach TEBROKE gibt es nicht wenige Kreditgenossenschaften, die einen Anteil von über 80 Prozent der unkündbaren Rücklagen am Eigenkapital aufweisen. Vgl. *Tebroke, H.-J.* (1998), S. 336.

904 Vgl. ähnlich *Kammlott, C./Schiereck, D.* (2000), S. 266 f.

von Rücklagen ist in der von vielen Kreditgenossenschaften beanspruchten Dividendenkontinuität zu sehen. Dies vor allem in der Form, dass eine im Marktvergleich überdurchschnittliche Verzinsung des Betrags, der in das Geschäftsguthaben durch das Mitglied investiert wurde, erreicht wird. Damit diese auch bei nicht hinreichenden Jahreserfolgen ausgezahlt werden kann, müssen Rücklagen als Manövriermasse gebildet werden. Zusammenfassend lässt sich feststellen, dass Rücklagen eine wesentliche Rolle für den Erhalt und Aufbau der Eigenmittelposition der Kreditgenossenschaften spielen. Damit geht notwendigerweise eine hohe Bedeutung des Gewinn-/Vermögenswertziels einher.[905]

3.1.2.6 Weitere Einflussfaktoren auf das Zielsystem

Neben den bereits erörterten Einflussfaktoren auf das Zielsystem lassen sich im kreditgenossenschaftlichen Kontext weitere finden. Dies sind die Prüfungsinhalte der Regionalverbände, das STATUT DER SICHERUNGSEINRICHTUNG (SE-St), die Intensität des Nicht-Mitgliedergeschäfts, die Verbundorientierung sowie die gegenwärtige Interpretation des Mitgliederstatus als Marketing- und Kundenbindungsinstrument. Diese sollen im Folgenden kursorisch vorgestellt werden.

Gemäß § 54 GenG muss sich eine Kreditgenossenschaft einem genossenschaftlichen Prüfungsverband anschließen, der bei seiner Vermögens- und Geschäftsführungsprüfung nach § 53 GenG darauf zu achten hat, ob und inwieweit der Bankenvorstand den mitgliederbezogenen Förderauftrag einhält.[906] Prüfungsziel ist dabei die Kreditgenossenschaft in ihrem Unternehmensziel, ihrem betrieblichen Aufbau und Ablauf eigens „als genossenschaftliche Personenvereinigung zu durchleuchten."[907] Prüfbausteine sind die Förderzweckprüfung, die Förderprogrammprüfung und die Fördererfolgsprüfung.[908] Die Prüfungsempfehlung des DEUTSCHEN GENOSSENSCHAFTS- UND RAIFFEISENVERBANDS (DGRV) beschränkt sich demgegenüber darauf, die wirtschaftlichen Verhältnisse der Kreditgenossenschaft als Grundlage für ihre Leistungen an die Mitglieder zu beurteilen und dazu im Einzelnen deren Vermögens-, Finanz-, Ertrags- und Risikolage zu bewerten.[909] Anhand dieser Daten wird die Förderfähigkeit bewertet,[910] wobei dies unabhängig von der tatsächlichen Förderung geschieht. Aufgrund der empfohlenen Prüfinhalte ergibt sich ein Anreiz für die relativ stärkere

905 Dies gilt auch, wenn der Förderauftrag als Dividende operationalisiert wird. Siehe Kapitel D.3.1.2.4.

906 Vgl. dazu ausführlich *Beuthien, V./Hanrath, S.* (2008), S. 85 ff.

907 *Müller, K.* (1998), § 1 Rn. 28 zitiert nach *Beuthien, V./Hanrath, S.* (2008), S. 86.

908 Vgl. *Beuthien, V./Hanrath, S.* (2008), S. 86.

909 Vgl. *DGRV* (2005), S. 81.

910 Vgl. *Beuthien, V./Hanrath, S.* (2008), S. 87.

Gewichtung des Gewinn-/Vermögenswertziels, da bei daraus resultierender positiver wirtschaftlicher Lage die Förderfähigkeit attestiert wird.

Ein weiterer Einflussfaktor zugunsten der Dominanz der materiellen Ziele resultiert aus den formalen Vorgaben aus dem STATUT DER SICHERUNGSEINRICHTUNG (SE-St) und den damit verbundenen Verfahrensregeln. Aufgabe der Sicherungseinrichtung ist nach § 1 SE-St „drohende oder bestehende wirtschaftliche Schwierigkeiten bei den dem Garantiefonds nach § 3 angeschlossenen Instituten abzuwenden oder zu beheben (Institutsschutz) und Beeinträchtigungen des Vertrauens in die genossenschaftlichen Institute zu verhüten.“[911] Die genossenschaftliche Sicherungseinrichtung funktioniert als solidarische Versicherung der angeschlossenen Banken, deren jeweilige Höhe der „Versicherungsprämie“ sich auf Basis von Kennzahlen zur Vermögens-, Ertrags- und Risikolage ergibt.[912] In Abhängigkeit der Klassifizierung sind die Beitragszu- oder -abschläge auf den Ausgangsbeitrag für die Banken gestaffelt. Der Beitrag ergibt sich als Produkt des Beitragsfaktors mit dem berechneten Ausgangsbeitrag.[913]

Tabelle 11: Beitragsfaktoren in Abhängigkeit der Klassifizierung[914]

Klasse	A++	A+	A	A-	B+	B	B-	C	D
Beitragsfaktor (in %)	80	90		100			110	120	140

Mit dieser Methodik werden die zu fokussierenden Ziele zunächst auf die materiellen Zielbereiche gelegt, damit die wirtschaftliche Situation der Bank so ausgestaltet wird, dass die Beitragszahlung gering ausfällt. Damit wird die Substanz der Bank geringer belastet, womit c. p. die Förderfähigkeit gesichert wird.

Wie bereits genannt wurde, ist die Anzahl der Mitglieder etwa halb so groß wie die Anzahl der Kunden von Kreditgenossenschaften. Seit der Gesetzesnovelle von 1973 ist den Genossenschaften möglich, das Nicht-Mitglieder-Geschäft zu betreiben, was vor allem mit der notwendigen Auslastung von Kapazitäten begründet wurde.[915] Mit dem Wachstum des Geschäftsvolumens ist nach BONUS eine Fixierung auf Managementinstrumente verbunden, die die Genossenschaftsidee aus dem Blickfeld und folglich aus dem Zielsystem ver-

911 § 1 Abs. 1 SE-St.

912 Siehe § 3a SE-St i.V.m. §4 Abs. 6 SE-St und Ziffer 1b der Verfahrensregeln zum SE-St.

913 Der Ausgangsbeitrag ist das Produkt der Summe bestimmter Kreditpositionen der Bilanz und dem Grunderhebungssatz, der für die Kreditgenossenschaften bei 0,05 ‰ liegt. Siehe § 4 Abs. 1 und 4 SE-St.

914 Eigene Darstellung auf Basis der Daten in § 4 Abs. 6 SE-St.

915 Vgl. *Lipfert, H.* (1988), S. 199.

drängt.[916] BEUTHIEN argumentiert, dass ein zu weitgehendes Nichtmitgliedergeschäft den Verlust des maßgeblichen Bezugs des Geschäftsbetriebs zum Mitglied bedingt und zum Status einer Dividendengenossenschaft führt.[917] Da dies mit einer Wesensverfremdung einhergeht, die außerhalb der genossenschaftsrechtlichen Strukturgrenzen liegt, müsste sich dies in praxi in einer Vielzahl von juristisch begründeten Rechtsformänderungen niederschlagen, was jedoch nicht der Fall ist. Dass dies nicht so ist, kann damit begründet werden, dass entweder die Entwicklung von den betroffenen Adressaten nicht angezweifelt wird oder es sich aufgrund der weiten Interpretationsspielräume nicht zweifelsfrei bestimmen lässt, ob reine Dividendengenossenschaften vorliegen. In jedem Fall jedoch gilt, dass die Zunahme des Nichtmitgliedergeschäfts die Mitgliederorientierung zulasten der allgemeinen Kundenorientierung verdrängt hat.[918] So argumentiert auch der BUNDESFINANZHOF, dass es sich u. a. aus diesem Grund bei den Genossenschaftsbanken umsatzsteuerrechtlich um Erwerbsgesellschaften handele.[919] Dies kann eine geringere Bedeutung des Förderauftragsziels begründen.

In den Kapiteln C.1.4.4 und C.5 wurden bereits die Verbundorientierung sowie deren Einflüsse auf die einzelne Kreditgenossenschaft thematisiert. Auch wurde gezeigt, welchen Einfluss diese auf den Vertrieb und die Vertriebssteuerung haben. Bezüglich des Zielsystems der Kreditgenossenschaften ist dieser Aspekt ebenfalls zu berücksichtigen. Denn zum einen ist aufgrund des hohen Einflusses der Verbundunternehmen auf die Kreditgenossenschaften begründet davon auszugehen, dass organisationstyp-spezifische Ziele geringer gewichtet werden. Zwar wird argumentiert, der Förderauftrag gelte auch für die Verbundunternehmen, da diese von den Kreditgenossenschaften zur Erfüllung ihres Förderauftrags gegründet wurden.[920] Es ist jedoch zweifelhaft, dass ein wenig konturiertes Konstrukt wie der Förderauftrag über Organisationsgrenzen hinweg Bestand haben kann. So postuliert etwa BRIXNER, dass es eines gemeinsamen Wertefundaments bedarf, „das nicht die Maximierung des individuellen Eigennutzens des Unternehmens, sondern die Interessenten des gesamten Verbunds in den Vordergrund stellt.“[921] Der individuelle Eigennutzen der Kreditgenossenschaften entspricht dem Wert der Kreditgenossenschaft aus Sicht des Mitglieds, der sich aus den verschiedenen Komponenten zusammensetzt. Durch ein verbundbezogenes Wertverständnis löst sich der genossenschaftliche Verbund aber gerade davon.

916 Vgl. *Bonus, H.* (1998), S. 319 f.
917 Vgl. *Beuthien, V.* (2006), S. 58.
918 Vgl. *Wiedemann, A.* (1992), S. 81.
919 Vgl. *Bundesfinanzhof* (2009).
920 Vgl. *Wiedemann, A.* (1992), S. 97 f.
921 *Brixner, U.* (2006), S. 85.

Die Möglichkeit zur Mitgliedschaft wird in praktischen und theoretischen Beiträgen als Alleinstellungsmerkmal der Kreditgenossenschaften gewertet.[922] Insbesondere vor dem Hintergrund des für Dienstleistungen typischen hohen Grads an Austauschbarkeit der Leistungen und des intensiven Wettbewerbs gewinnen anbieterbezogene Alleinstellungsmerkmale an Bedeutung. Aus diesem Grund wird die Mitgliederorientierung intensiv aus Sicht marketingstrategischer Überlegungen betrachtet.[923] Dies betrifft zwar vornehmlich das Privatkundengeschäft, ist aber dennoch von Relevanz, da es den Status des Mitglieds und den Status der Mitgliederorientierung aus Sicht der Kreditgenossenschaft kennzeichnet. Im Fokus steht die Zweck-Mittel-Beziehung zwischen Mitglied und Kreditgenossenschaft, wobei das Mitglied bzw. die Mitgliederorientierung als Mittel zum Zweck dient, wobei der Zweck vor allem über die materiellen Ziele definiert ist.[924] So sieht etwa WIEDEMANN in der Intensivierung der Kundenbeziehung durch die Mitgliedschaft die Vorteile, akquisitorisches Potenzial aufzubauen, die Geschäftsbeziehung auf weitere Produkte auszuweiten und dabei Kostenvorteile für die Bank im Vergleich zu Neukundenakquisitionen zu realisieren.[925] GESCHWANDTNER/KUTTLER sehen die „Mitgliedschaft als gute Möglichkeit [.], Gelder wie Kunden zu binden und langfristig erfolgreich zu wirtschaften."[926] Konsequent argumentieren HAMMERSCHMIDT/HELLINGER, die „Mitgliedschaft sollte nicht allen Kunden offeriert werden. Es erscheint sinnvoll, nur jene Kunden als Mitglieder zu werben, die der Genossenschaft langfristig von Nutzen sind. Diese könnten Kunden sein, […] die ein umfassendes Geschäftsvolumen mit der Kreditgenossenschaft abwickeln."[927] Insbesondere den Faktoren der Bindung und des Cross-Sellings werden auch von der Praxis hohe Bedeutungen beigemessen. Damit ließe sich erklären, dass in vielen Kreditge-

---

922 Vgl. z. B. *Geschwandtner, M./Kuttler, H.-J.* (2009), S. 264; *Weegen, M.* (2006), S. 26; *Hammerschmidt, M.* (2000), S. 1 ff.; *Hammerschmidt, M./Hellinger, C.* (1998), S. 5 f.

923 Für die Integration der Mitgliedschaft in kommunikationspolitische Ansätze vgl. z. B. *Bookhagen, B.* (2006), S. 38 ff.

924 Theoretisch möglich ist auch, die Mitgliederorientierung als Mittel für die Erfüllung des Förderauftragziels zu sehen, wobei diese Zweck-Mittel-Relation genuiner im Sinne der Entstehungsgeschichte und Besonderheiten der Kreditgenossenschaften wäre. In diesem Fall würde jedoch die Integration der Mitglieder in die Corporate Governance anders ausgestaltet sein, als es die Erörterungen in den vorhergehenden Kapiteln erwarten lassen.

925 Vgl. *Wiedemann, A.* (1992), S. 83.

926 *Geschwandtner, M./Kuttler, H.-J.* (2009), S. 265. Ähnlich auch THEURL/BÖTTIGER, die jedoch die aus der Mitgliedschaft resultierende Stabilisierung der Geschäftsbeziehung Vorteile für Kunden und Genossenschaft sehen. Vgl. *Theurl, T./Böttiger, J.-M.* (2007), S. 7 f.

927 *Hammerschmidt, M./Hellinger, C.* (1998), S. 20.

nossenschaften das „Produkt" Mitgliedschaft in den volumenbezogenen Zielvorgaben in der Vertriebssteuerung enthalten ist und aktuell postuliert wird, die Kennzahl share-of-member, d. h. den Anteil des Geschäftsvolumens, den das Mitglied mit der Kreditgenossenschaft macht, als Steuerungsziel zu nutzen.[928] Es liegen jedoch bisher noch keine empirischen Erkenntnisse darüber vor, dass die Mitgliedschaft ein wesentlicher Erklärungsfaktor für die Loyalität das Produktnutzungsverhalten oder die Bindung eines Kunden ist. Unter Würdigung der kreditgenossenschaftlichen Praxis, guten Kunden die Mitgliedschaft anzubieten, kann vielmehr auch von einer umgekehrten Kausalität ausgegangen werden.

3.1.3 Abgeleitetes Zielsystem der Kreditgenossenschaft

Werden die in den vorhergehenden Abschnitten gewonnen Erkenntnisse zum Zielsystem der Kreditgenossenschaften und deren Einflussfaktoren zusammengefasst, so ist in der Gesamtschau zu konstatieren, dass das Zielsystemverständnis (a) in Abbildung 31 von geringerer Relevanz für die weitere Untersuchung ist. Es kann begründet angenommen werden, dass die beschriebenen Einflussfaktoren einen Anreiz aufweisen, auch bei Erreichen eines für die Existenz ausreichenden Gewinn-/Vermögenswertminimums nicht ausschließlich den Förderauftrag als oberstes Ziel zu verfolgen.

Für das Zielsystemverständnis (b) in Abbildung 31 wurde die Frage aufgeworfen, welche Gewichte dem Förderziel und dem Gewinn-/Vermögenswertziel beizumessen sind. Zwar ist eine eindeutige Antwort nicht möglich, doch lassen sich die Ergebnisse dahingehend zusammenfassen, dass das Gewinn-/Vermögenswertziel deutlich stärker gewichtet wird als das Förderziel. Auf Ebene der Gesamtbank sind die materiellen Ziele priorisiert. Vornehmliches Ziel der Kreditgenossenschaft ist die Steigerung des Gewinns bzw. des Vermögenswertes, wobei hier keine Begrenzung zugunsten des Förderziels vorliegt. Das Förderziel ist aber dennoch im Zielsystem enthalten. Aufgrund der beschriebenen Strukturen und Anreize gilt, dass dieses in Abhängigkeit des Gewinn-/Vermögenswertziels steht, dieses aber nicht beeinflusst. Bezüglich der Instrumentalziele (Rentabilität, Wachstum, Risiko, Sicherheit, Liquidität) gelten die in Kapitel D.3.1.1 gemachten Ausführungen.

Wie in Kapitel D.1.2.3 erläutert, sind die Ziele zentrale Determinante des Kundenwerts. Die einzelnen Ziele und Zielbeziehungen sowie etwaige zeitliche Differenzierungen wirken sich folglich direkt in die noch vorzunehmende Konzeptualisierung des Kundenwerts aus. Bevor jedoch diese Analyseschritte vorgenommen werden können, gilt es zu klären, ob die kundenwertorientierte Dis-

---

[928] Vgl. *Geschwandtner, M./Kuttler, H.-J.* (2009), S. 266.

kriminierung im Vertrieb mit dem Mitgliedergleichbehandlungsgebot[929] der Kreditgenossenschaften konform geht.

3.2 Vereinbarkeit von Gleichbehandlungsgebot und Kundenwert

Die kundenwertorientierte Vertriebssteuerung hat unter Annahme divergierender Kundenwerte für die einzelnen Mitglieder-Kunden einer Kreditgenossenschaft zum Ergebnis, dass Mitglieder vertrieblich ungleich behandelt werden. Diese in der Literatur auch als Diskriminierungsproblematik[930] bezeichnete Konsequenz ist nicht neu, wird jedoch durch die explizite Ausrichtung marketing- und vertriebspolitischer Entscheidungen i. V. m. dem Förderauftrag und dem damit korrespondierenden Förderanspruch und dem Identitätsprinzip für die vorliegende Untersuchung relevant.

Nach VON USSLAR konkretisiert sich in § 11 der Mustersatzung des BVR, in dem die Inanspruchnahme der kreditgenossenschaftlichen Leistungen als Mitgliederrecht definiert ist,[931] das Gleichbehandlungsgebot,[932] das sich allgemein aus dem Förderzweck und der Treuepflicht der eG gegenüber ihren Mitgliedern ergibt.[933] Im Gegensatz dazu wird auch die Meinung einer sachlich gerechtfertigten Ungleichbehandlung der Mitglieder vertreten. Dabei differenziert BONUS hinsichtlich der Gleichbehandlung zwischen der Leistungs- und der Einlegerbeziehungsebene.[934] Auf der Einlegerbeziehungsebene sieht er die Gleichbehandlung als unverzichtbaren Bestandteil des kulturellen Kerns einer modernen Genossenschaft an. Auf der Leistungsbeziehungsebene kann dagegen eine Ungleichbehandlung sachlich gerechtfertigt werden, wenn dadurch alle besser bedient werden können.[935] Zugleich warnt er aber davor, dass die Ungleichbehandlung zu Vertrauensverlust führen kann und daher „in großer Offenheit mit den Mitgliedern besprochen werden und für diese *einsehbar* sein [muss; Erg. d. Verf.]“[936]. Auch BEUTHIEN/DIERKES/WEHRHEIM halten die Ungleichbehandlung

---

929 Vgl. dazu die Ausführungen in Kapitel C. 1.4.1 und Kapitel C. 1.4.2.

930 Vgl. z. B. *Hansen, U.* (2006), S. 160; *Homburg, C./Droll, M./Totzek, D.* (2008), S. 110 ff.; *Droll, M.* (2008), S. 2 ff.

931 „Jedes Mitglied hat das Recht, nach Maßgabe des Genossenschaftsgesetzes und der Satzung die Leistungen der Genossenschaft in Anspruch zu nehmen“. § 11 Mustersatzung des BVR für VR-Banken.

932 Vgl. *von Usslar, L.* (2007), S. 44.

933 Vgl. *Beuthien, V./Dierkes, S./Wehrheim, M.* (2008), S. 73.

934 Vgl. *Bonus, H.* (1994), S. 75.

935 Vgl. *Bonus, H.* (1994), S. 66.

936 *Bonus, H.* (1994), S. 75.

von Mitglieder-Kunden in Abhängigkeit des Geschäftsumfangs als sachlicher Rechtfertigung für möglich, da bei geringem Geschäftsumfang die relativen Kosten der Leistungsbeziehung höher sind.[937] Sie unterscheiden dabei zwischen dem absoluten Gleichbehandlungsgebot hinsichtlich der satzungsmäßigen Regelungen, die bei allen Mitgliedern gleich sein müssen (z. B. Höhe und maximale Anzahl der Geschäftsanteile) und dem aus dem Verfassungsrecht bekannten Prinzip der relativen Gleichbehandlung, welches auf die Kundenbeziehung mit Mitgliedern anzuwenden ist. Danach wird Gleiches gleich und Ungleiches ungleich behandelt, wobei die Unterscheidung nicht willkürlich sein darf. Möglich ist bspw. die unterschiedliche Preisgestaltung in Abhängigkeit der nachgefragten Menge.[938] Möglich ist auch die differenzierte Verzinsung bzw. Dividende auf Geschäftsguthaben für investierende Mitglieder und Mitglieder-Kunden.[939] Vor diesem Hintergrund sind drei Fragen näher zu betrachten: Erstens die Frage danach, ob der Vertrieb Gegenstand der vom grundsätzlichen Gleichbehandlungsgebot betroffenen Leistungen ist. Ist dies der Fall, stellt sich zweitens die Frage, ob trotz des Gleichbehandlungsgebots der Kundenwert als sachliche Rechtfertigung für die Anwendung des relativen Gleichbehandlungsgebots herangezogen werden kann. Drittens ist zu fragen, unter welchen Voraussetzungen eine kundenwertorientierte Vertriebssteuerung im Fördergeschäftsverkehr zwischen Mitglied und Genossenschaftsbank auch aus Sicht des Mitglieds akzeptabel ist.

Originär sind die in § 11 Mustersatzung des BVR für Genossenschaftsbanken genannten Leistungen der Produktpolitik einer Kreditgenossenschaft zuzuordnen.[940] Die Produktpolitik beinhaltet alle Entscheidungstatbestände, die sich auf die marktgerechte Gestaltung aller vom Unternehmen im Absatzmarkt angebotenen Leistungen beziehen.[941] Innerhalb der Koordinationsaufgabe des Marketings werden Produkte bestimmten Segmenten zugeordnet, wobei dem Vertrieb die Aufgabe innewohnt, gemäß dieser Vorgabe die Leistungen des Unternehmens zu verwerten. In dieser instrumentellen Argumentation tangiert die Vertriebspolitik nur mittelbar den Gleichheitsgrundsatz, da sie lediglich ausführend tätig ist. Wird im übergeordneten Marketing das vollständige Leistungsprogramm den Mitgliedern zugeordnet, liegt hier kein Gegensatz zum Gleich-

937 Vgl. hier und nachfolgend *Beuthien, V./Dierkes, S./Wehrheim, M.* (2008), S. 74. So auch *Wiedemann, A.* (1992), S. 98 f. Anderer Meinung ist GRAUER, der in diesem Verhalten eine fundamentale Abkehr von der genossenschaftlichen Grundidee sieht. Vgl. *Grauer, F.* (1998), S. 200 sowie die dort zitierte Literatur.

938 Vgl. *Beuthien, V./Dierkes, S./Wehrheim, M.* (2008), S. 74.

939 Vgl. *Schulte, G.* (2006), § 8 Rn. 16, § 19 Rn. 16.

940 Siehe § 11 der Mustersatzung des BVR für Genossenschaftsbanken.

941 Vgl. *Meffert, H.* (1998), S. 317.

heitsgrundsatz vor,[942] da der Vertrieb nur dem Zweck dient, Leistungen der Kreditgenossenschaft an die Mitglieder zu verkaufen. Da jedoch dem Vertrieb Entscheidungen im Rahmen des Kundenmanagement zugewiesen sind, kanalisiert er die Verwertung von Leistungen auf ausgewählte Kunden in den jeweiligen Segmenten und trifft Entscheidungen über die Akquisition, Intensität und Beendigung von Kundenbeziehungen bzw. Fördergeschäftsbeziehungen.[943] Im Kontext dieser Kanalisierung kann es folglich dazu kommen, dass einem Mitglied bestimmte Leistungen nicht angeboten werden oder gar eine Mitglieder-Kundenbeziehung beendet wird. Damit wird allerdings dem Gleichbehandlungsgebot nur mittelbar widersprochen. Fragt das Mitglied eine bestimmte Leistung nach und wird ihm diese seitens der Kreditgenossenschaft nicht verwehrt, so wird das Gleichheitsbehandlungsgebot nicht nachhaltig beschädigt.[944] Anders ist dies, wenn vertriebliche Aufgaben wie etwa die Beratung zum Gegenstand des Förderauftrags deklariert werden und die differenzierte Beratung durch den Vertrieb nicht zu einer besseren Förderung aller führt oder dem Förderanspruch jedes einzelnen Mitglieds entspricht. In der Untersuchung von MÜNKNER bezeichnen 91,3 Prozent die intensive Beratung der Mitglieder als geeignetes Förderinstrument und 85,9 Prozent praktizieren dies entsprechend.[945] Es kann bezweifelt werden, dass in diesen Fällen die von BONUS formulierte Bedingung, einer aus der differenzierten Behandlung resultierende Besserstellung aller Mitglieder, erfüllt ist.

Analog zur Logik der genossenschaftlichen Rückvergütung kann nun im Sinne der relativen Gleichbehandlung argumentiert werden, dass auch bei Förderinstrumenten differenziert vorgegangen wird und der Kundenwert eine sachliche Begründung für diese Ungleichbehandlung darstellt. Dabei muss aber beachtet werden, dass die genossenschaftliche Rückvergütung auf Basis realisierter Leistungsvolumina angewendet wird, während der Kundenwert als gegenwarts- und zukunftsorientiertes, mehrdimensionales Konstrukt sowohl außerhalb des eigentlichen Leistungsvolumens stehende Wertdeterminanten als auch künftig mögliche, aber noch nicht realisierte Erträge berücksichtigt. Ein Ansatz, der den Kundenwert zu einer Größe macht, die vergleichbar zum Ansatz der genossenschaftlichen Rückvergütung ist, ist sinnlos, da er gerade die Vorteile dieser Steuerungsgröße konterkariert. Daher ist zu konstatieren, dass der Kundenwert keine sachliche Rechtfertigung für eine Ungleichbehandlung darstellt, wenn er

---

942 Vgl. *Lürig, R.* (1985a), S. 16.

943 Vgl. dazu die Ausführungen in Kapitel B.1.1.

944 Auch die bankseitig initiierte Beendigung einer Mitglieder-Kundenbeziehung ist gerechtfertigt, sofern damit Schaden von der Kreditgenossenschaft und somit eine Schlechterstellung aller Mitglieder abgewendet werden kann.

945 Vgl. *Münkner, H.-H.* (1990), S. 63.

seine Vorteile als Steuerungsgröße behalten soll. Auch das relative Gleichheitsgebot steht somit einer kundenwertdifferenzierten Mitgliederbehandlung im Vertrieb eher konträr gegenüber.

Unter Annahme dieser Ergebnisse muss daher nach einem Ansatz gesucht werden, der die kundenwertorientierte Lenkung von Vertriebsressourcen ermöglicht. Zunächst ist daran zu denken, die kundenwertorientierte Vertriebssteuerung lediglich auf das Nichtmitgliedergeschäft zu beschränken. Unter Gültigkeit der ökonomischen Interessen der Kreditgenossenschaft in Bezug auf die Mitgliedschaft,[946] streben diese ein Wachstum der Mitgliederzahlen an. Gleichzeitig werden dem Kundenwert superiore ökonomische Steuerungseigenschaften gegenüber den traditionell verwendeten Größen im beziehungsorientierten Marketing zugesprochen. Konsequenterweise stehen sich diese Aspekte widerstreitend gegenüber. Zudem impliziert eine derartig differenzierte Vorgehensweise wesentliche Probleme bei der praktischen Umsetzung. So müssten bspw. zwei Vertriebssteuerungsansätze in einer vergleichsweise kleinen Unternehmung nebeneinander implementiert und gepflegt werden, was nicht zuletzt dem formulierten Wirtschaftlichkeitspostulat widerspricht.[947] Vielmehr soll die kundenwertorientierte Vertriebssteuerung für den gesamten Kundenbestand gelten. Folglich handelt es sich um eine kunden- und mitgliederwertorientierte Vertriebssteuerung. Grundlage dafür könnte die Einführung eines mitgliederseitig akzeptierten und unterstützten Duldungsprinzips, das die genossenschaftlichen Prinzipien dahingehend ergänzt, dass mitgliederorientierte Bausteine der Unternehmenskultur und -steuerung gegen kundenorientierte Bausteine substituiert werden, sein. Damit dies jedoch nicht die Rechte der Mitglieder einschränkt oder ihren Interessen widerspricht, sind strikte Restriktionen zu formulieren.

Bezogen auf den konkreten Fall der kundenwertorientierten Vertriebssteuerung sind zur Sicherung der Mitgliederinteressen zwei zentrale Schranken einzubauen. Die aus dem Kundenwert abgeleitete Vorziehenswürdigkeit eines Nur-Kunden oder eines Mitglieds darf in keinem Fall zu einer Einschränkung des vom nicht-bevorzugten Mitglied wahrgenommen Fördernutzens führen, wenn dies mit dem Kundenwert begründet wird.[948] Weiterhin darf der Kundenwert nicht als Argument dazu führen, dass ein Mitglied ausgeschlossen wird.[949] Im

---

946 Vgl. dazu die Ausführungen in Kapitel D.3.1.2.6.

947 Vgl. dazu die Ausführungen in Kapitel D.2.

948 Unberührt sind davon also die genossenschaftliche Rückvergütung oder ähnliche Förderinstrumente, die das Förderausmaß an die Intensität der Beziehung zwischen Mitglied und Kreditgenossenschaft koppeln.

949 Die Ausschlussmöglichkeiten aus § 9 der Mustersatzung des BVR für VR-Banken bleiben davon unberührt, da diese sich auf die tatsächliche Nutzung des Genossenschaftsbetriebs bezieht. Ebenso bleiben davon diejenigen Entscheidungen zulasten einer Mitglie-

Sinne des Member Value bedeutet dies, dass die kundenwertorientierte Vertriebssteuerung in der tatsächlichen Umsetzung keinen negativen Einfluss auf dieses Nutzenkonstrukt haben darf. Damit unterscheidet sich die Kundenwertorientierung fundamental von der Mitgliederwertorientierung. Während in der Gruppe der Nur-Kunden eine Trennung im Sinne der erwerbswirtschaftlichen Ziele liegt und sogar geboten ist,[950] gilt dies für Mitglieder-Kunden eingeschränkt. Es gilt nur insofern, als die Förderung eines einzelnen Mitglieds nicht die Förderung der anderen Mitglieder einschränkt oder unmöglich macht.[951] Dies gilt nach VON USSLARS Argumentation auch, wenn aktuelle Förderleistungen für einzelne Mitglieder dazu führen, dass künftig die Förderfähigkeit der Kreditgenossenschaft für alle Mitglieder eingeschränkt wird.

3.3 Zwischenfazit

In den vorhergehenden Kapiteln wurde dargelegt, dass die materiellen Ziele die immateriellen Ziele einer Kreditgenossenschaft dominieren und dass Nur-Kunden und Mitglieder-Kunden in Abhängigkeit ihres erwarteten Beitrags zur Zielerreichung vertrieblich differenziert behandelt werden können. Bedingung für Letzteres ist, dass dies von den Mitgliedern geduldet wird und dadurch der Mitgliedernutzen nicht unbegründet negativ beeinflusst wird. Diese Ergebnisse können im Widerspruch zu weiten Teilen der Meinung in der Genossenschaftsforschung gesehen werden. VOGEL bemängelt etwa „den die Förderbeziehung auf den Kopf stellenden Kunstbegriff des fördernden Mitglieds“[952], welcher der Idee des Kundenwerts mit Blick auf die Mitglieder-Kunden sehr ähnlich ist und den RINGLE als unheilbaren Eingriff in das Identitätsprinzip wertet.[953] Jedoch ist in diesem Zusammenhang zu konstatieren, dass der allgemein fehlende theoretische Konsens sowie die Distanz zwischen Wissenschaft und Praxis zwar nachvollziehbar sind, aber keinen Lösungsbeitrag zu den Problemen der (kredit-)genossenschaftlichen Praxis liefern. Zudem wird argumentiert: „Der Fördergedanke ist in (vielen) Kreditgenossenschaften ohnehin nur noch eine leere Worthülle. Die rechtsformspezifische Ideologie ist nicht mehr vorhanden und die Erwartungshaltung der Mitglieder schon jetzt eher auf Renditeaspekte als auf Förderung ausgerichtet. Die traditionelle Identität von Mitgliedern und Kunden ist nur noch in sehr geringem Maße ausgebildet und damit nicht mehr prägend für Ge-

der-Kundenbeziehung unberührt, die dazu dienen, einen Schaden von der Kreditgenossenschaft abzuwenden.

950 Vgl. *Scheer, O.* (1989), S. 27.

951 Vgl. *von Usslar, L.* (2007), S. 45 f.

952 *Vogel, W.* (2002), S. 78.

953 Vgl. *Ringle, G.* (1996), S. 254.

nossenschaftsbanken."[954] Weit früher hat LÜRIG aus seiner praktischen Erfahrung festgestellt, dass an „die Stelle der Ideologie [.] auf beiden Seiten die ökonomische Rationalität getreten [ist; Erg. d. Verf.]"[955], was vor allem für die gewerblichen Mitglieder gelte.

Der Genossenschaftsbetrieb war Mittel zum Zweck und auf die Förderung der Mitglieder ausgerichtet.[956] Bei den modernen Großgenossenschaften mit einem hohen Anteil an Nichtmitgliedergeschäften, stark heterogener Mitgliederstruktur sowie eingeschränkter Mitgliederdemokratie, verschärft sich die Problematik der Distanz zwischen ideellen und ökonomisch rationalen Merkmalen. Z. B. kann eine primär absatzmarktorientierte Unternehmensführung, wie sie die Entwicklung zur Vertriebsbank bei den Kreditgenossenschaften letztlich darstellt, zu völlig von den Mitgliederinteressen abweichenden Marketingstrategien führen, welchen von den Mitgliedern kaum gegensteuert werden kann. Obwohl sich das Management aus Eigeninteresse weiterhin verbal zu den genossenschaftlichen Zielen bekennen dürfte, wird es sich bei der Formulierung der in seinem eigenverantwortlichen Entscheidungsbereich liegenden marketingbezogenen Unternehmensstrategien doch konsequenterweise an erfolgreichen Vorbildern nichtgenossenschaftlicher Unternehmen orientieren.[957] Dies ist auch aus Sicht der Mitglieder nicht zu beanstanden, wenn dies eine Orientierung an vergleichsweise besseren Praktiken führt und diese, wie hier im Ansatz der kundenwertorientierten Vertriebssteuerung, mit einem speziellen „Schutzschirm" für die Interessen der Mitglieder ausgestattet ist.

Für den weiteren Gang der Analyse ist das in Kapitel D.3.1.3 beschriebene Zielsystem Grundlage der weiteren Überlegungen, den Kundenwert zu konzeptualisieren und zu operationalisieren. Weiterhin gilt, dass die kundenwertorientierte Vertriebssteuerung mit ihren Implikationen für Mitglieder-Kunden unter den genannten Bedingungen grundsätzlich mit der Mitgliederorientierung und dem Förderauftrag auf der Leistungsbeziehungsebene vereinbar ist, d. h. dem relativen Gleichbehandlungsgrundsatz nicht widerspricht.

---

954 *Bornemann, S./Schaaff, C./Pfingsten, A.* (2009), S. 27 f.

955 *Lürig, R.* (1985b), S. 26.

956 Vgl. hier und nachfolgend *Grauer, F.* (1998), S. 191 ff.

957 Vgl. dazu z. B. *Gloger, K./Lehmann, S.* (2008), S. 64 ff., *Lehmann, S.* (2008), S. 64 f. und *Krob, B./Zimmermann, Y.* (2007), S. 22 f.

4 Konzeptualisierung eines zweckspezifischen Kundenwertmodells

Anknüpfend an die semantische Festlegung des Kundenwerts in Kapitel D.1.2 und den Erkenntnissen zum Zielsystem der Kreditgenossenschaften bedarf es nun der am Anwendungskontext orientierten, zweckmäßigen Konzeptualisierung bzw. Modellierung des Kundenwerts, um ihn für den weiteren Verlauf der Arbeit nutzbar zu machen.[958] In Anlehnung an HOMBURG/PFLESSER soll unter der Konzeptualisierung die „Erarbeitung der relevanten Konstruktdimensionen"[959] verstanden werden.

GALASSO fordert, dass die Analyse aktueller und potenzieller Kundenbeziehungen hinsichtlich ihres Beitrags zur Zielerreichung einer Kreditgenossenschaft insbesondere die folgenden zwei Schritte umfassen muss:[960]

1. Schritt: Identifikation der wertbestimmenden Faktoren bzw. Kriterien einer Kundenbeziehung und
2. Schritt: Definition der Methode zur Ermittlung des Kundenwerts

Mit konkretem Blick auf das Firmenkundengeschäft ist Schritt 1 Gegenstand dieses Kapitels D.4 und in Kapitel D.5 wird der Schritt 2 thematisiert.

4.1 Allgemeines Strukturmodell

Der Kundenwert ist ein komplexes und in Abhängigkeit der zugrundeliegenden Vielfalt der Unternehmensziele sowie seiner betrachteten Komponenten nicht direkt messbares theoretisches Konstrukt.[961] Daher ist es notwendig, den Kundenwert mittels eines Kundenwertmodells zu konzeptualisieren. Als Kundenwertmodell sind alle modelltheoretischen Konzeptionen zu verstehen, deren Ziel in der Bestimmung des Wertes einer einzelnen Kundenbeziehung oder von Kundengruppen und der Einflussfaktoren des Kundenwertes liegt.[962] Um ein Kundenwertmodell aufzubauen, müssen zunächst diejenigen Faktoren identifiziert werden, die ihn beeinflussen. Dazu ist es hilfreich, auf ein grundlegendes Ordnungsprinzip in Form eines allgemeinen Strukturmodells zu rekurrieren. In

958 Vgl. zum Zweckbezug von Kundenwertmodellen *Cornelsen, J.* (2000), S. 55 und *Beeser, A. J.* (2003), S. 26.

959 *Homburg, C./Pflesser, C.* (2000), S. 415.

960 Vgl. *Galasso, G.* (1999), S. 205.

961 Vgl. *Wünschmann, S.* (2007), S. 30; *Rudolf-Sipötz, E.* (2001), S. 22. Vgl. auch die Ausführungen zum Kundenwert in Kapitel D.1.2

962 Vgl. *Dittmar, M.* (2000), S. 13.

der Literatur ist kein allgemeingültiges Modell vorhanden. Ein vergleichsweise allgemeines Strukturmodell entwickeln WEBER/LISSAUTZKI und LISSAUTZKI, die dieses jedoch zweckmäßigerweise aus Sicht der Unternehmenssteuerung in Form einer Werttreiberhierarchie strukturieren (siehe Abbildung 35).[963]

Abbildung 35: Allgemeines Strukturmodell des Kundenwerts in Form einer Werttreiberhierarchie[964]

Auf der obersten Ebene steht der Kundenwert als zu bestimmende, zentrale Größe. Dieser lässt sich separieren in einzelne Kundenwertdimensionen die auch als Determinanten, Facetten, Bestimmungsfaktoren[965] oder Teilwerte bezeichnet werden und den (zielbezogenen) Wertbeiträgen aus der Kundenbeziehung entsprechen.[966] Die Autoren unterteilen in ihrem konkreten Modell die Kunden-

[963] Vgl. hier und nachfolgend *Weber, J./Lissautzki, M.* (2004), S. 18 ff.; *Weber, J./Lissautzki, M.* (2005), S. 309 ff., *Lissautzki, M.* (2007), S. 118 ff. und *Stüker, D.* (2008), S. 169 ff. In der wissenschaftlichen Literatur zum Kundenwert gibt es nach Kenntnisstand des Verfassers nur diese genannten Beiträge, die eine allgemeingültige Systematik ihrer Konzeptualisierung von Kundenwerttreibern voranstellen. Aus diesem Grund lässt sich eine Vielzahl von konkreten Kundenwertmodellen finden, aus denen sich jedoch unterschiedlich strukturierte allgemeine Kundenwertmodelle induktiv ableiten ließen.

[964] *Weber, J./Lissautzki, M.* (2004), S. 19.

[965] Da im Sinne der hier vorgestellten und im Weiteren genutzten Kundenwerttreiberhierarchie der Begriff Bestimmungsfaktor anderweitig bestimmt ist, wird eine mögliche synonyme Verwendung nicht vorgenommen.

[966] Vgl. *Eberling, G.* (2002), S. 127. Anders LISSAUTZKI, der die hier definierten Bestimmungsfaktoren als Werttreiber bezeichnet und als Bestimmungsfaktoren „die eigentli-

wertdimensionen in Einzahlungen, Auszahlungen und Risiken. Dies setzt eine entsprechende rein monetäre Zieldefinition auf Unternehmensebene voraus.

Diese Dimensionen werden wiederum durch eine Vielzahl von (mehrstufigen) Kundenwerttreibern im engeren Sinn beeinflusst. Als solche gelten alle, die jeweilige Dimension positiv oder negativ beeinflussenden Größen,[967] wobei einzelne Kundenwerttreiber auf mehrere Dimensionen und auch aufeinander wirken können. Der Zusammenhang zwischen Kundenwerttreiber(n) und Kundenwertdimensionen ist sowohl kunden- als auch werttreiberindividuell festzustellen.[968] Beispiele für Kundenwerttreiber sind die Bestandswahrscheinlichkeit einer Kundenbeziehung, das Referenzpotenzial, Akquisitions- und Bindungskosten, Cross- und Up-Selling-Optionen oder der Anteil des Anbieters am Beschaffungsvolumen des Kunden.[969]

Die Kundenwerttreiber im engeren Sinn basieren wiederum auf so genannten Bestimmungsfaktoren. Diese stellen z. B. psychografische Dispositionen eines Kunden dar. Beispiele sind Kundenzufriedenheit, Lebenszyklusstand oder Innovationstätigkeit eines Kunden.[970]

In der Kundenwertforschung lassen sich, wie angedeutet, viele verschiedene Kundenwertmodelle finden (siehe für eine Auswahl Tabelle 12). Diese können bezüglich der Anzahl der betrachteten Dimensionen in ein- oder mehrdimensionale Modelle eingeteilt werden. Auch ist die Anzahl der berücksichtigten Stufen der Wertreiberebenen als Einflussfaktoren auf die Kundendimensionen unterschiedlich. Einige Modelle ermitteln den Kundenwert ausschließlich anhand der Ebene der Dimensionen bzw. der Kundenwerttreiber im engeren Sinn, während andere Modelle im Sinne von Marktreaktionsmodellen darüber hinausgehende mehrstufige Wirkbeziehungen berücksichtigen.[971] Eine weitere Unterscheidung der Modelle lässt sich auf Basis der Verknüpfung der Dimensionen zur Bestimmung des Kundenwerts als relevante Kennzahl, Ziel- oder Steuerungsgröße vornehmen. Die Mehrzahl der englischsprachigen Literatur verknüpft die Teilwerte und Einflussfaktoren in Form eines Kapitalwerts mit dem Zweck der Berech-

---

chen Ursachen der Kundenwertdifferenzen“ definiert. *Lissautzki, M.* (2007), S. 119. Damit meint er etwa qualitative Größen wie die Kundenzufriedenheit.

967 Einige Autoren nutzen statt der zielbezogenen Teilwerte die beeinflussenden Faktoren als Grundlage der Bestimmung und Messung des Kundenwerts. Vgl. z. B. *Wünschmann, S.* (2007), S. 51 ff.

968 An dieser Stelle sei darauf hingewiesen, dass Kundenwertmodelle kontextabhängig zu definieren sind, da weder eine unternehmensübergreifende noch eine branchenübergreifende Allgemeingültigkeit postuliert werden kann. Vgl. *Lissautzki, M.* (2007), S. 120.

969 Vgl. *Stüker, D.* (2008), S. 169; *Lissautzki, M.* (2007), S. 119.

970 Vgl. *Stüker, D.* (2008), S. 169; *Lissautzki, M.* (2007), S. 119.

971 Vgl. *Burmann, C.* (2003), S. 113 ff.

nung des Customer Lifetime Values oder des Customer Equity.[972] Relevante Determinanten sind dabei i. d. R., die kundenbezogenen Ein- und Auszahlungen, die Bindungsrate sowie ein adäquater Diskontierungsfaktor. Dieses Verfahren wird vor allem dann angewendet, wenn der Kundenwert mit der wertorientierten Unternehmensführung verknüpft wird.[973] Andere Beiträge operationalisieren den Kundenwert oft mittels Scoring-Modellen[974] oder anhand von Portfolio-Ansätzen[975].

Aufgrund der Vielfalt an Modellen bedarf es sinnvoller Kriterien, um bestehende Kundenwertmodelle hinsichtlich ihrer Eignung für die vorliegende Arbeit würdigen zu können oder, falls es keine hinreichenden Modelle im gegenwärtigen Stand der Forschung und Praxis gibt, Leitlinien für die Entwicklung eines Kundenwertmodells zu haben. Dabei ist es evident, dass die Entwicklung und Messung von Kundenwerten zweckgebunden sind. Daher werden nachfolgend neben den allgemeinen Anforderungen auch untersuchungszweckspezifische Anforderungen thematisiert. Diese ergänzen die für die gesamte Vertriebssteuerung geltenden Implikationen insofern, als sie konkrete Anforderungen an die Ziel- und Steuerungsgröße Kundenwert in einer kundenwertorientierten Vertriebssteuerung sind. Die Anforderungen gelten für die Konzeptualisierung und Operationalisierung des Kundenwerts.

4.2 Anforderungen an Kundenwertmodelle

4.2.1 Allgemeine Anforderungen an Kundenwertmodelle

Die allgemeinen Anforderungen zeichnen sich dadurch aus, dass sie losgelöst vom konkreten Anwendungskontext zu erfüllen sind.

Die Anforderung der Ganzheitlichkeit postuliert, dass alle kundenwertbestimmenden Komponenten zu berücksichtigen sind, da sonst erhebliche Wertbeiträge missachtet werden und Fehlentscheidungen resultieren können.[976] Aufbauend auf den Erkenntnissen von RUDOLF-SIPÖTZ und CORNELSEN bedeutet dies, dass nicht nur die direkten Beiträge eines Kunden den Kundenwert aus Anbietersicht determinieren, sondern auch darüber hinausgehende indirekte Fak-

---

972 Vgl. dazu die Ausführung in Kapitel D.1.1

973 Vgl. dazu z. B. *Lissautzki, M.* (2007); *Schroeder, N.* (2006); *Wille, K.* (2005); *Tewes, M.* (2003).

974 Vgl. z. B. *Foehn, P.* (2006).

975 Vgl. z. B. *Götz, P./Diller, H.* (1991).

976 Vgl. *Schneider, N. C.* (2007), S. 86; *Foehn, P.* (2006), S. 72; *Friederichs-Schmidt, S.* (2006), S. 176.

toren wie etwa Referenztätigkeiten oder Informationsaktivitäten.[977] Die Ganzheitlichkeit betrachtet neben der sachlichen auch die zeitliche Perspektive. Ebenfalls zur Ganzheitlichkeit gehört, dass der Kundenwert nicht nur die aus Sicht des Unternehmens wahrscheinlichen Zielbeiträge, sondern auch die ggf. darüber hinausgehenden potenziell mit Kunden realisierbaren Wertbeiträge erfassen kann, da auch diese Potenziale steuerungsrelevant sind. Ein weiterer wichtiger Aspekt ist, dass sich die Ganzheitlichkeit am jeweiligen Zielsystem orientiert.

Die Auswahl der relevanten Bestimmungsfaktoren soll sich an der interindividuellen Variabilität orientieren.[978] Das bedeutet, dass die Determinanten des Kundenwerts für die zu bewertenden aktuellen und potenziellen Kunden hinsichtlich ihrer Höhe verschieden sind, so dass unterschiedliche Zielbeiträge identifiziert werden können, woraus sich dann differenzierte Anbietermaßnahmen ableiten lassen.

Das Kundenwertmodell bzw. dessen einzelne Bausteine sollen dynamisch prognostizierbar sein, da sonst eine zukunftsgerichtete Kundenbewertung nicht möglich ist.[979] Dynamik bezieht sich in diesem Fall sowohl auf die im Zeitablauf prognostizierte Kundenentwicklung aufgrund des sich im Zeitablauf variierenden Verhaltens als auch auf die Art der Bewertungsmethode.[980] Dies impliziert die Möglichkeit unterschiedliche Zeiträume zu betrachten.[981] Diese Anforderung ist komplementär zur zeitlichen Ganzheitlichkeit.

Weiterhin ist der Kundenwert bezugsgrößenflexibel zu modellieren, so dass mit dem Kundenwertmodell grundsätzlich sowohl aktuelle als auch potenzielle Kunden bewertet werden können.[982]

---

977 Vgl. *Cornelsen, J.* (2000), S. 170 ff.; *Rudolf-Sipötz, E.* (2001), S. 108 ff.

978 Vgl. *Wünschmann, S.* (2007), S. 31 f.

979 Vgl. *Beeser, A. J.* (2003), S. 8; *Wünschmann, S.* (2007), S. 32; *Mödritscher, G. J.* (2008), S. 285.

980 Vgl. Kiwitz, K./Scheffler, C. (2009), S. 10.

981 So zeigen HOMBURG/STEINER/TOTZEK, dass es nennenswerte Unterschiede zwischen einer statischen, also zeitpunktbezogenen, und dynamischen, also zeitraumbezogenen, Segmentierung geben kann. Diese Unterschiede können dazu führen, dass Ressourcenzuordnungen auf Basis statischer Segmentierungen suboptimal im Vergleich zu den Anforderungen bei dynamischer Segmentierung sind. Einfach lässt sich dies etwa anhand von Deckungsbeiträgen erklären. Kunden mit relativ hohen Deckungsbeiträgen (und am voraussichtlichen Ende des Beziehungslebenszyklus) werden nach statischer Segmentierung höher bewertet als Kunden mit geringen Deckungsbeiträgen (und in der Anfangsphase des Beziehungslebenszyklus). Vgl. *Homburg, C./Steiner, V. V./Totzek, D.* (2009).

982 Vgl. *Kiwitz, K./Scheffler, C.* (2009), S. 9.

Insgesamt ist das Kundenwertmodell so auszugestalten, dass die Kosten zur Messung des Kundenwerts geringer sind als der Nutzen.[983] Die Forderung der Wirtschaftlichkeit ist eine gravierende Restriktion für die Modellierung des Kundenwerts in der Praxis.[984]

4.2.2 Anforderungen aus Sicht der Vertriebssteuerung

Ziel der Arbeit ist es, den Kundenwert als Ziel- und Steuerungsgröße in der Vertriebssteuerung zu nutzen. Wie die Ausführungen zum Wettbewerb auf dem deutschen Bankenmarkt zeigen, schöpfen hiesige Kreditinstitute Ertragspotenziale nicht ausreichend aus. Der Fokus liegt folglich auf der Ableitung eines handlungsorientierten Kundenwertmodells, das neben der Messung des Kundenwerts auch Ansatzpunkte für die Entwicklung und Ausschöpfung von kundenwertbezogenen Potenzialen liefert. Das bedeutet, dass die relevanten Bausteine des Kundenwerts steuerungs- und entscheidungsrelevant sein müssen, um so als Entscheidungsgrundlage für Vertriebsmaßnahmen dienen können.[985] Dies begründet eine hinreichend differenzierte Betrachtung der Kundenwertbestimmungsfaktoren sowie eine prospektive Betrachtung des Kundenwerts.[986]

Ein vertriebssteuerungszweckbezogenes Kundenwertmodell muss weiterhin der Anforderung genügen, operational zu sein.[987] Dies gilt sowohl auf Ebene des Einzelkunden als auch auf Ebene von Kundengruppen. Dabei muss die Messgröße nicht zwangsläufig monetär sein.[988] Die Operationalisierbarkeit ist vor allem für den Kundenwert als Planungs-, Steuerungs- und Kontrollgröße in der Vertriebssteuerung von herausragender Bedeutung.[989]

4.2.3 Anforderungen aus Sicht der Kreditgenossenschaften

Neben den allgemeinen und vertriebssteuerungsspezifischen Anforderungen sind Anforderungen aus dem Kontext der Kreditgenossenschaften zu berücksichtigen. Diese ergeben sich aus ihren Unternehmensmerkmalen sowie in Anlehnung an die Grundprinzipien der Banksteuerungskonzeption VR-CONTROL.

Zunächst ist festzustellen, dass die Kreditgenossenschaften als Gruppe heterogen sind. Diese Heterogenität konkretisiert sich u. a. in der Unternehmensgrö-

983 „Die (sicheren) Kosten der Informationsbeschaffung dürfen den (unsicheren) Informationsnutzen nicht übersteigen." *Mödritscher, G. J.* (2008), S. 286.

984 Vgl. *Lissautzki, M.* (2007), S. 84; *Friederichs-Schmidt, S.* (2006), S. 176.

985 Vgl. *Lissautzki, M.* (2007), S. 81 f.

986 Vgl. *Bruhn, M./Georgi, D./Wöhle, C. B.* (2005), S. 385.

987 Vgl. *Lissautzki, M.* (2007), S. 82 f.; *Foehn, P.* (2006), S. 72.

988 Vgl. *Schneider, N. C.* (2007), S. 86; *Foehn, P.* (2006), S. 72.

989 Vgl. *Friederichs-Schmidt, S.* (2006), S. 176.

ße sowie in unterschiedlichen Wettbewerbs- und Kundenstrukturen. Daraus ergibt sich die zentrale Anforderung, das Kundenwertmodell skalierbar auszugestalten. Das betrifft sowohl die Konzeptualisierung der Kundenwerttreiber und deren Messung sowie die Operationalisierung des Kundenwertmodells als Gesamtmodell. Es widerspräche dem Ziel der Arbeit, wenn das resultierende Modell und die darauf aufbauende Vertriebssteuerung nicht grundsätzlich für alle Kunden und bei allen Kreditgenossenschaften eingesetzt werden könnte.

Des Weiteren ist zu berücksichtigen, dass die Kreditgenossenschaften in der Majorität als kleine bzw. mittlere Unternehmen zu bezeichnen sind. Daraus resultieren limitierte Ressourcen auf personaler, sachlicher und finanzieller Ebene.[990] Diese Limitationen führen häufig zur Knappheit der Ressource „Information“, welche von besonderer Bedeutung bei der Kundenbewertung ist. Das impliziert, dass die unreflektierte Übernahme bestehender Ergebnisse abzulehnen ist. Es gilt „a small business is not a little big business“[991]. Daraus leitet sich die Anforderung der Adäquanz ab.

In Anlehnung zur vermögensorientierten Banksteuerungskonzeption VR-CONTROL lässt sich ableiten, dass Kundenwerte vermögensorientiert zu messen und zu steuern sind. Daran bauen die folgenden konzeptionellen Anforderungen auf:[992]

- Grenznutzenorientierte Einzelbewertung
- Zutreffende Ergebnisinformation
- Integrierte Ergebnisrechnung
- Akzeptanz der Ergebnisinformation

Der auf den Kundenwert adaptierte Inhalt der grenznutzenorientierten Einzelbewertung besagt, dass die Kundenwertinformation eines jeden einzelnen Kunden den mit ihm verbundenen zusätzlichen Grenznutzen widerspiegeln soll. Der Kundenwert sollte Aussagen darüber zulassen, wie sich die Zielerreichung der Bank verändert und verändern wird, wenn dieser Kunde akquiriert bzw. gebunden oder nicht akquiriert bzw. nicht gebunden werden kann.

Das Postulat der zutreffenden Ergebnisinformation besagt, dass die Erfolgswirksamkeit von Entscheidungen der Realität entsprechend wiedergegeben wird. Im Kundenwertkontext ist vor allem der darin impliziten Forderung nach Objektivität zu entsprechen. Dabei gilt es zu berücksichtigen, dass der Kundenwert als Konstrukt und zukunftsbezogene Größe und vor dem Hintergrund der

---

990 Vgl. *Flacke, K.* (2006), S. 16 f. sowie die dort angegebene Literatur.

991 *Welsh, J. A./White, J. F.* (1981), S. 18.

992 Vgl. hier und nachfolgend *Schierenbeck, H.* (2003), S. 45 ff. und *Kroon, G.* (2009), S. 25 ff.

genannten Limitationen auf Seiten der Kreditgenossenschaften nicht mit vollständiger Objektivität ermittelt werden kann. Daher ist die Objektivität als „intersubjektive Vergleichbarkeit“ bei der Ermittlung der Kundenwerte zu konkretisieren.

Hinsichtlich der integrierten Ergebnisrechnung geht es um die Einbindung des Kundenwerts in das gesamte Steuerungssystem der Bank. Wenngleich in dieser Arbeit nur die Vertriebssteuerung als relevanter Teilbereich der Gesamtbanksteuerung betrachtet wird, ist die Offenheit des Kundenwerts zu berücksichtigen. Dieses Postulat beinhaltet auch, dass der Kundenwert und dessen Bestimmungsfaktoren für die Planung und Kontrolle auf gleichen Entscheidungsgrundlagen beruhen.

Die Akzeptanz der Kundenwertinformation besagt, dass diese von den involvierten Akteuren nachvollzogen und verstanden werden können. Diese Anforderung steht im potenziellen Konflikt zu den anderen Anforderungen, da Kundenwertermittlungen so komplex ausgestaltet sein können, dass sie die Akzeptanz negativ beeinflussen.

Nachdem die zentralen Anforderungen ohne Anspruch auf Vollständigkeit benannt sind, sollen bestehende Kundenwertmodelle hinsichtlich ihrer Eignung und ihres Erkenntnisbeitrags für den weiteren Verlauf der Arbeit untersucht werden.

4.3 Stand der Forschung und Praxis zu Kundenwertmodellen

4.3.1 Stand der Forschung zu Kundenwertmodellen

4.3.1.1 Überblick und Auswahl bisheriger Forschungsbeiträge

Die Forschungsbeiträge zum Kundenwert sind so mannigfaltig, dass eine vollumfängliche Diskussion einzelner Modelle den Rahmen dieser Arbeit überschreitet. Einen mit dem Schwerpunkt auf deutschsprachiger Literatur liegenden, nicht abschließenden Überblick über Kundenwertmodelle gibt Tabelle 12. Deutlich ist zu erkennen, dass es bisher keine einheitliche Gliederung der Kundenwerttreiber gibt. So liegen unterschiedliche Vorgehensweisen vor, diese Einflussgrößen zu systematisieren. Eine Möglichkeit bezieht sich auf die Art der berücksichtigten Erfolgsbeiträge. Hier kann zunächst zwischen monetären und nicht-monetären Erfolgsbeiträgen unterschieden werden.[993] Können die Kundenwerttreiber direkt in Geldeinheiten bewertet werden, so handelt es sich um monetäre Determinanten (z. B. Umsatzerlöse). Nicht-monetäre Größen dagegen

[993] Vgl. z. B. bei *Krüger, S. M.* (1997), S. 114 ff.

sind originär nicht in Geldeinheiten zu bewerten (z. B. die Mund-zu-Mund-Propaganda eines Kunden). Aufgrund der Problematik, nicht-monetäre Faktoren konsistent in ein Messinstrumentarium für den Kundenwert zu integrieren, werden diese Faktoren zwar regelmäßig theoretisch postuliert, in praktischen Ansätzen aber nur selten berücksichtigt. Insbesondere dann werden sie jedoch berücksichtigt, wenn sich die nicht-monetären Kundenwertdeterminanten monetarisieren lassen. CORNELSEN bspw. stellt einen Ansatz vor, den Referenzwert eines Kunden zu monetarisieren.[994] VON WANGENHEIM trennt daher zwischen monetären und monetarisierbaren Einflussgrößen.[995]

HOMBURG/SCHNURR unterscheiden quantitative und qualitative Facetten[996] des Kundenwerts und EBERLING sowie HERRMANN/FÜDERER ökonomische und vor-ökonomische Größen.[997] Die beiden erstgenannten Kriteriengruppen unterscheiden sich von der letztgenannten dadurch, dass sich hier die Notwendigkeit der Mess- und Operationalisierbarkeit der Kundenwerttreiber in der Konzeptualisierung niederschlägt.[998] Während die Messbarkeit bei den quantitativen Kriterien, die metrisch skaliert sind, unproblematisch ist, ist dies bei den qualitativen Kriterien, die nominal oder ordinal skaliert sind, schwierig (so z. B. beim Lead-User-Konzept[999]). Qualitative Kriterien werden berücksichtigt, wenn quantitative Informationen nicht oder nur vergleichsweise schwer ermittelt werden können oder um diese zu ergänzen.[1000] Bei der Einteilung der Kundenwerttreiber in ökonomische und vorökonomische Größen werden die im jeweiligen Kundenwertmodell betrachteten Faktoren in Bezug auf ihren Einflusses auf die ökonomischen Ziele des Unternehmens untersucht. Während die ökonomischen Größen unmittelbar auf die Erreichung dieser Ziele (Erfolg) einwirken, wirken die vorökonomischen Größen nur mittelbar.

Eine weitere Differenzierung der Kundenwerttreiber stellt SCHEMUTH vor, der nach dem Funktionsbereich (z. B. Produktion, Vertrieb) differenziert, in dem anbieterseitig die Wertbeiträge konkretisiert werden.[1001] Dies ähnelt der Systematisierung von TOMCZAK/RUDOLF-SIPÖTZ, bei denen die Kundenwertfaktoren in marktbezogene und ressourcenbezogene Gruppen eingeteilt werden.[1002] Wäh-

---

994 Vgl. *Cornelsen, J.* (2000), S. 194 ff.

995 Vgl. *von Wangenheim, F.* (2003), S. 34.

996 Vgl. *Homburg, C./Schnurr, P.* (1998), S. 169 ff.

997 Vgl. *Eberling, G.* (2002), S. 127 ff.; *Hermann, A./Füderer, R.* (1997), S. 352 ff.

998 Vgl. *Eberling, G.* (2002), S. 127.

999 Vgl. zum Lead-User-Konzept den grundlegenden Beitrag von *von Hippel, E.* (1986), S. 791 ff.

1000 Vgl. *Rieker, S.* (1995), S. 57.

1001 Vgl. *Schemuth, J.* (1996), S. 24.

1002 Vgl. *Tomczak, T./Rudolf-Sipötz, E.* (2006), S. 130 ff.

rend marktbezogene Kundenwerttreiber die Wertbeiträge umfassen, die sich aus der Leistungsabnahme des Kunden ergeben, sind ressourcenbezogene Einflussgrößen Inputgrößen, die helfen, die Wertschöpfung des Anbieterunternehmens zu verbessern.[1003]

SCHMÖLLER unterscheidet direkt und indirekt wirkende Einflussgrößen auf den Kundenwert. Ein Einflussfaktor gilt als direkt, wenn er die ökonomischen Charakteristika der betrachteten Kundenbeziehung beeinflusst. Eine indirekte Wirkung liegt dagegen vor, wenn der Faktor auf andere Kundenbeziehungen wirkt (z. B. Referenzpotenzial).[1004]

Tabelle 12: Ausgewählte Kundenwertmodelle[1005]

Autor(en) (Jahr)	Kundenwertmodell	Anmerkungen
CANNING (1982)	*Customer Profit*; *Value of the Customer beyond Profit*: Growth, Technological (lead user), Industry leadership (referral value), Share of customer volume;	entwickelt im industriellen B2B-Kontext
KRAPFEL/ SALMOND/ SPEKMAN (1991)	Criticality of the good purchased by the buyer; Quantity of the seller's output consumed by the buyer; Replaceability of the buyer; Cost savings resulting from the buyer's practices and procedure	entwickelt im B2B-Kontext
STORBACKA (1993)	Relationship Volume, Relationship Revenue, Relationship Cost	entwickelt im Kontext des Privatkundengeschäfts von Banken
SCHLEUNING (1994)	Monetärer Kundenrentabilitätswert; kommunikativ/akquisitorischer Wert; informatorischer Wert	entwickelt im Dialogmarketing
RIEKER (1995)	*Höhe der relativen Ressourcenabhängigkeit: Quantitative Kriterien*: Umsatz, Deckungsbeitrag; *Qualitative Kriterien*: Entwicklungspotenzial, Ausstrahlungspotenzial, Innovationspotenzial, Einflusspotenzial, Kooperationspotenzial, Koordinationspotenzial	entwickelt im industriellen B2B-Kontext
SCHULZ (1995)	*Ökonomische Größen*: Umsatz, Kundendeckungsbeitrag, Kundenbonität; *Vor-ökonomische Größen*: Referenzpotenzial, Innovationspotenzial, Loyalitätspotenzial	entwickelt als allgemeines Modell

(Fortsetzung auf der nächsten Seite)

1003 Vgl. *Eberling, G.* (2002), S. 127.
1004 Vgl. *Schmöller, P.* (2001), S. 120 ff.
1005 Eigene Darstellung.

Autor(en) (Jahr)	Kundenwertmodell	Anmerkungen
WALTER (1996)	*Ökonomische Bezugsgrößen:* Umsatz, Kreditwürdigkeit, Kundendeckungsbeitrag, Prozesskostenbeitrag; *Außerökonomische Größen*: Multiplikator	entwickelt im Kontext des Privatkundengeschäfts von Kreditinstituten
DWYER (1997)	Discounted Benefits; Discounted Burdens	entwickelt im Kontext des Direct Marketing
KNÖBEL (1997)	Wertpotenzial; Servicepotenzial; Risikopotenzial; Zukunftspotenzial	entwickelt im Kontext des Retailbanking von Banken
KRÜGER (1997)	*Kundenbeziehungsdauer*; *Monetäre Determinanten:* Umsatzerlöse, Gewinne aus Weiterempfehlungen, Kosten der Kundengewinnung, Kosten der Leistungserstellung; *Nicht-monetäre Determinanten:* z. B. Meinungsführerschaft, Preissensibilität, Qualitätsbewusstsein, Risikobereitschaft, Abwanderungsgefährdung, Beziehungsintensität	entwickelt als allgemeines Modell
PLINKE (1997)	Menge der vom Kunden bereitgestellten Ressourcen; Art/Nutzen der Ressourcen; Beherrschbarkeit der Ressourcen	entwickelt im B2B-Kontext
HOMBURG/ SCHNURR (1998)	*Quantitative Facetten*: Umsatzbezogene Größen, Deckungsbeitragsgrößen; *Qualitative Facetten*: Innovationspotenzial, Kooperationspotenzial, Informationspotenzial, Referenzpotenzial	entwickelt als allgemeines Modell (ohne Anspruch auf Integrativität)
CORNELSEN (2000)	Umsatz; Kundenerfolg; Referenzwert; Informationswert; Cross-Selling-Wert	entwickelt als allgemeines Modell
RUST/ZEITHAML/ LEMON (2000)	Value Equity; Brand Equity; Retention Equity	entwickelt als allgemeines Modell
DILLER (2001)	*Sicherheitszielbezogene Kundenwertkomponenten*: Basiswert (Retention), Informationswert (Feedback); *Wachstumszielbezogene Kundenwertkomponenten*: Penetrationswert, Cross-Selling-Wert, Referenzwert; *Profitabilitätszielbezogene Kundenwertkomponenten*: Preispremium, Kundengewinnungs- und Bindungskosten, Transaktionskosten	entwickelt als allgemeines Modell

(Fortsetzung auf der nächsten Seite)

Autor(en) (Jahr)	Kundenwertmodell	Anmerkungen
GELBRICH (2001)	*Monetäre Nutzenbestandteile*: Künftiges Umsatzpotenzial; *Nicht-monetäre Nutzenbestandteile*: Cross-Selling-Potenzial, Generische Markentreue, Wechselbereitschaft, Referenzpotenzial Verschuldungsbereitschaft, Kostenbestandteile, (Kredit-) Ausfallwahrscheinlichkeit	entwickelt im Kontext der Automobilfinanzierung von Autobanken
RUDOLF-SIPÖTZ (2001)	*Marktpotenzial*: Ertragspotenzial, Entwicklungspotenzial, Cross-Buying-Potenzial, Loyalitätspotenzial; *Ressourcenpotenzial*: Referenzpotenzial, Informationspotenzial, Kooperationspotenzial, Synergiepotenzial	entwickelt als allgemeines Modell
SCHMÖLLER (2001)	*Marktpotenziale*: Ertragspotenzial, Cross-Buying-Potenzial, Kooperationspotenzial; *Ressourcenpotenziale*: Referenzpotenzial, Strategische Relevanz, Informationspotenzial, Innovationspotenzial, Kooperationspotenzial	entwickelt als allgemeines Modell
EBERLING (2002)	*Ökonomische Bestimmungsfaktoren:* Erlöse, Kosten; *Vorökonomische Bestimmungsfaktoren:* Informationswert, Referenzwert, Sicherheitswert	entwickelt als Synthese aus bisherigen allgemeinen Kundenwertmodellen
VON WANGENHEIM (2003)	*Monetäre Größen:* Transaktionswert; *Monetarisierbare Größen*: Weiterempfehlungswert, Informationswert, Referenzwert, Kooperationswert	entwickelt als allgemeines Modell
BÖHRS (2004)	*Direkte Einflussgrößen*: Umsatzpotenzial, Umsätze aus Add-on-Leistungen, Cross-Selling-Potenzial, Kostenpotenzial; *Indirekte Einflussgrößen*: Referenzpotenzial, Informationspotenzial	entwickelt im Kontext der Verkehrsdienstleistungsbranche
RENKER (2004)	Umsatzvolumen/Wachstumspotenzial; Gewinn-/Deckungsbeitrag; Risikoentlastungspotenzial; Personal- und Sachaufwandsreduktionspotenzial; Informationspotenzial; Loyalitätspotenzial; Cross-Selling-Potenzial; Referenzpotenzial	entwickelt als allgemeines Modell
BAUER/ HAMMERSCHMIDT (2005)	Acquisition Costs, Rentention Rate, Autonmous Revenue, Up-Selling-Revenue, Cross-Selling-Revenue, Reference Value, Marketing Costs, Service Costs, Termination Costs	entwickelt als allgemeines Modell
WILLE (2005)	*Transaktionswert*: Künftiger Kundendeckungsbeitrag; *Ressourcenpotenzial*: Referenzwert. Informationswert, Kooperationswert	entwickelt als allgemeines Modell

(Fortsetzung auf der nächsten Seite)

Autor(en) (Jahr)	Kundenwertmodell	Anmerkungen
FOEHN (2006)	*Monetäre Kriterien:* Gegenwärtige finanzielle Beiträge, Cross-Selling-Potenzial *Vor-Monetäre Kriterien*: Referenzpotenzial, Informationspotenzial, Garantiepotenzial	entwickelt im Kontext des Private Banking von Kreditinstituten
FREILING (2006)	Kunde als Ressourcenlieferant; Absatzbezogenes Potenzial des Kunden; Kundenbezogenes Kostensenkungspotenzial; Kunde als Impulsgeber der Ressourcen- und Kompetenzentwicklung; Kundenbeziehungen als generative Faktoren	entwickelt als allgemeines Modell auf Basis des Resource-Based View
FRIEDERICHS - SCHMIDT (2006)	*Prospektiver Deckungsbeitrag*: Lebens-/Krankenversicherung, Schadensversicherung; Referenz-/Informationswert	Entwickelt im Kontext der Versicherungsbranche
SCHMOLL (2006)	*Internes Kundenpotenzial* (vergangenheitsorientiert): Kundendeckungsbeitrag; *Externes Kundenpotenzial* (zukunftsorientiert): Entwicklungspotenzial, Cross-Selling-Potenzial	entwickelt im Kontext des Firmenkundengeschäfts bei Kreditinstituten
SCHROEDER (2006)	Einzahlungsüberschüsse; Wiederkaufverhalten; Referenzwert	entwickelt als allgemeines Modell
LISSAUTZKI (2007)	*Einzahlungen:* Umsatz der Basisperiode, Loyalitätswert; Entwicklungswert; Bestandswahrscheinlichkeit, Cross-Selling-Wert; *Auszahlungen:* Kundenlebenskosten; *Risiken:* Kundenverlustrisiko, Zahlungsausfallrisiko, Kundenplanungs-/-informationsrisiko	entwickelt als allgemeines Modell
WÜNSCHMANN (2007)	Finanzpotenzial; Referenzpotenzial; Informationspotenzial; Loyalitätspotenzial; Effektivitätspotenzial; Effizienzpotenzial	entwickelt als allgemeines Modell
FRESE (2009)	*Marktpotenzial*: Ertragspotenzial, Entwicklungspotenzial, Cross-Buying-Potenzial, Loyalitätspotenzial; *Ressourcenpotenzial*: Referenzpotenzial, Informationspotenzial, Kooperationspotenzial, Synergiepotenzial	übertragen auf den Kontext des Firmenkundengeschäfts bei Kreditinstituten
RADDAO (2009)	*Ertragsdimension:* Status, Potenzial; *Risikodimension:* Bonitätsnote, Sicherheitennote	entwickelt im Kontext des Firmenkundengeschäfts bei Kreditinstituten

Mit Blick auf den hier interessierenden Bankenkontext ist festzustellen, dass sich nur vergleichsweise wenige Arbeiten mit dem Kundenwert in diesem konkreten Untersuchungsfeld auseinandersetzen. Weiterhin fällt auf, dass die Majorität dieser Beiträge sich auf das Retailbanking oder das Private Banking beziehen.[1006]

FOEHN entwickelt ein Kundenwertmodell für das Private Banking mit den Kundenwerttreibern gegenwärtige finanzielle Beiträge, Cross-Selling-Potenzial, Referenzpotenzial, Informationspotenzial und Garantiepotenzial.[1007] Die ersten zwei genannten Wertreiber bezeichnet er als monetäre und die letzten drei Teilwerte als vormonetäre Wertbeiträge. Die gegenwärtigen finanziellen Beiträge umfassen die bisher mit dem Kunden realisierten Erlöse. Das Cross-Selling-Potenzial beinhaltet diejenigen Produktbedarfe des Kunden, die durch die Bank noch nicht gedeckt sind. Alle monetären oder monetarisierbaren Effekte aus Referenzaktivitäten bilden das Referenzpotenzial und zweckgebundenes Wissen vom oder über den Kunden das Informationspotenzial. Das Garantiepotenzial betrifft den Kundenbindungsgrad bzw. die Abwanderungswahrscheinlichkeit eines Kunden. Die einzelnen Potenziale werden über Indikatoren gemessen und in einem Scoringmodell zu einer Kennzahl Kundenwert zusammengeführt.

Im Beitrag von KNÖBEL für das Retailbanking wird der Wert einer Geschäftsbeziehung bestimmt durch die Kundenwerttreiber Wertpotenzial, Servicepotenzial, Zukunftspotenzial und Sicherheitspotenzial.[1008] Das Wertpotenzial bezeichnet die Zinserträge aus den Kundengeschäften und das Servicepotenzial die Erträge aus entgeltlichen Dienstleistungen (z. B. Zahlungsverkehrsleistungen). Die Möglichkeit eines teilweisen Kreditausfalls wird im Risikopotenzial abgebildet und die zu erwartenden, zukünftigen Cash-flows sind dem Zukunftspotenzial zugeordnet.

Ebenfalls für das Retailbanking entwickelt STORBACKA ein Kundenwertmodell, das die Determinanten „relationship volume", „relationship revenue" und „relationship cost" umfasst.[1009] Das „relationship volume" ist der anbieterbezogene Anteil der Gesamtausgaben eines Kunden in einer bestimmten Branche und die „relationship revenue" umfassen die Zins- und Dienstleistungserträge. Letztere ergeben sich funktional aus den Zinsmargen und dem „relationship volume". Die „relationship costs" umfassen die Kosten, die anfallen um die Kundenbeziehung aufrechtzuerhalten und zu entwickeln. STORBACKA saldiert Erlöse und Kosten, um den Wert der Geschäftsbeziehung zu berechnen.

Speziell für Firmenkunden sollen nachfolgend die Kundenwertmodelle von SCHMOLL, RADDAO und FRESE detailliert vorgestellt und auf Basis der definier-

---

1006 Während das Private Banking sich auf vermögende Privatkunden bezieht, wird als Retailbanking das klassische Mengengeschäft im Privatkundensegment verstanden.

1007 Vgl. hier und nachfolgend *Foehn, P.* (2006), S. 115 ff. Er lehnt sich dabei stark an *Eberling, G.* (2002), S. 127 ff.

1008 Vgl. hier und nachfolgend *Knöbel, U.* (1997), S. 152 ff.

1009 Vgl. hier und nachfolgend *Storbacka, K.* (1993), S. 49 ff.

ten Anforderungen gewürdigt werden. Das Modell von SCHMOLL wird ausgewählt, da es in seiner Struktur, Basis vieler in der kreditgenossenschaftlichen Praxis umgesetzter Kundenwertmodelle ist. Das Kundenwertmodell von RADDAO wird berücksichtigt, da es im Rahmen einer praxisorientierten wissenschaftlichen Arbeit entwickelt und wissenschaftlich honoriert wurde. Das Kundenwertmodell von FRESE basiert auf dem vielzitierten theoretischen Modell von RUDOLF-SIPÖTZ, das mittels einer branchenübergreifenden Studie empirisch validiert wurde.

4.3.1.2 Kundenwertmodell von SCHMOLL

Das Kundenwertmodell von SCHMOLL unterscheidet zwischen dem internen und dem externen Kundenpotenzial.[1010] Das interne Kundenpotenzial umfasst einen Kundenwerttreiber, der dem bisher mit dem Firmenkunden realisierten Deckungsbeitrag pro Jahr entspricht. Diese Dimension sagt „etwas über die Leistungsfähigkeit des Firmenkunden in der Vergangenheit aus."[1011] Das externe Kundenpotenzial fokussiert unter der Prämisse, dass die bisher abgesetzten Produkte des Allfinanzangebots weiterhin bzw. wieder vom Firmenkunden nachgefragt werden, auf die zusätzlich möglichen Geschäfte bzw. das zusätzliche Ergebnisbeitragspotenzial.[1012] Kundenwerttreiber des externen Potenzials sind die Entwicklungschancen des Firmenkunden sowie die daraus resultierenden Möglichkeiten des Cross-Selling. Als Einflussfaktoren des Entwicklungspotenzials nennt er das Branchenpotenzial, die Marktposition innerhalb der Branche, das Expansionspotenzial des Unternehmens, das Umsatzpotenzial, die Auslandsorientierung, die Innovationskraft und die Investitionsneigung. Werden die Aussagen SCHMOLLS zusammengefasst, kann aufbauend auf dem allgemeinen Strukturmodell das folgende konkrete Kundenwertmodell visualisiert werden (siehe Abbildung 36).

[1010] Vgl. *Schmoll, A.* (2006), S. 66 ff. Diesem Schritt geht die Aufteilung der Firmenkunden in Umsatzgrößenklassen anhand des Firmenumsatzes der Unternehmen voran. Da dieser Schritt jedoch nur für die Segmentierung von Bedeutung ist, um segmentspezifische Sortimentsbreiten zu definieren, und keinen Einfluss auf die Kundenbewertung hat, wird er an dieser Stelle nicht weiter berücksichtigt.

[1011] *Schmoll, A.* (2006), S. 69.

[1012] Vgl. hier und nachfolgend *Schmoll, A.* (2006), S. 66 ff.

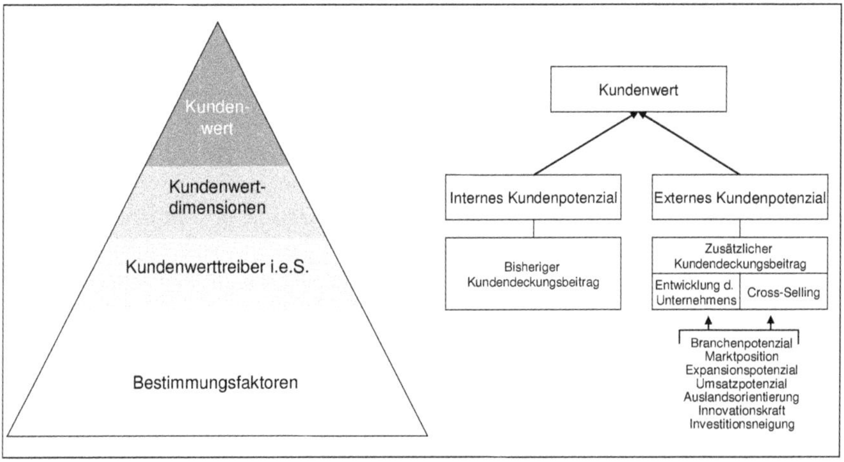

Abbildung 36: Kundenwertmodell nach SCHMOLL[1013]

Die Zusammenfassung des internen und des externen Kundenpotenzials zum Kundenwert erfolgt über einen Portfolioansatz, bei dem die Ordinate dem externen Kundenpotenzial und die Abszisse dem internen Kundenpotenzial entspricht.[1014] Beide Achsen werden in zwei Bereiche eingeteilt.[1015] Den resultierenden vier Portfoliofeldern werden Normstrategien zugeordnet.[1016] Der Kundenwert selbst ergibt sich aus der Zuordnung eines Kunden zu einem Quadranten im Portfolio, so dass vier Kundenwertklassen zu unterscheiden sind.

Auf Basis der genannten Anforderungen ist dieses Kundenwertmodell nun zu würdigen. Dabei ist zu berücksichtigen, dass es lediglich als handlungsleitende Struktur für Banken intendiert ist und die konkrete Operationalisierung diesen überlassen bleibt. Gleichwohl können die strukturellen Bestandteile der Konzeptualisierung des Kundenwerts einer kritischen Analyse unterzogen werden.

1013 Eigene Darstellung.

1014 Vgl. *Schmoll, A.* (2006), S. 71.

1015 Es wird keine Vorgabe gemacht, wie diese Achsenaufteilung erfolgen soll.

1016 SCHMOLL unterscheidet zwischen Standardkunden, die nur wenig internes und externes Potenzial aufweisen und standardisiert betreut werden sollen, Beobachtungskunden, die viel internes, aber nur wenig externes Potenzial aufweisen und hinsichtlich Konditionen und Risiko regelmäßig geprüft werde sollen, Potenzialkunden, die wenig internes, aber viel externes Potenzial aufweisen und daher ausgebaut werden sollen und Idealkunden, die viel internes und externes Potenzial aufweisen und intensiv betreut werden sollen. Vgl. *Schmoll, A.* (2006), S. 70 f.

Mit Blick auf die allgemeinen Anforderungen ist zunächst festzustellen, dass das Kundenwertmodell von SCHMOLL dem Postulat der Ganzheitlichkeit nicht genügt. So werden weder die Risiken noch die Kundenbindung als Kundenwerttreiber betrachtet. Beide Größen stellen jedoch relevante Einflussfaktoren auf die Werthaltigkeit eines Kunden dar. Auch fehlen Aussagen zu Wertbeiträgen, die nicht unmittelbar aus der Geschäftsbeziehung resultieren, wie etwa das Referenzpotenzial. Vor dem Hintergrund des für die Kreditgenossenschaften hergeleiteten Zielsystems fehlt zudem die Möglichkeit, den Förderauftrag als Zielgröße im Kundenwertmodell zu berücksichtigen, da sich das Modell ausschließlich auf die erwerbswirtschaftlichen Ziele beschränkt. Aufgrund des Zukunftsbezugs der Potenzialdimension kann tendenziell von einer zeitlichen Ganzheitlichkeit ausgegangen werden. Fraglich ist jedoch, inwieweit der Portfolioansatz als Bewertungsmethode zweckgerecht ist, da die Dynamik des Kundenverhaltens nur unzureichend berücksichtigt wird.[1017] Die betrachteten Größen sind interindividuell variabel und grundsätzlich im Sinne der Bezugsgrößenflexibilität auf potenzielle Kunden anwendbar. Zudem ist davon auszugehen, dass das Modell dem Postulat der Wirtschaftlichkeit genügen kann.

Aus Sicht der vertriebssteuerungsbezogenen Anforderungen kann das Kundenwertmodell zwar als tendenziell handlungsorientiert beurteilt werden, da es explizit auf zusätzliche kundenbezogene Geschäftspotenziale abzielt und diese in Bestandsgeschäft, Entwicklungspotenzial und Cross-Selling differenziert. Anzumerken ist aber, dass eine derartig differenzierte Betrachtung der Kundenwerttreiber für die Anwendung des Messmodells nicht zwingend ist. Sie ist jedoch nötig, um den Kundenwert als steuerungsrelevante Größe einsetzen zu können. Problematisch ist zudem, den anhand dieses Modells bestimmten Kundenwert als Planungs- und Kontrollgröße zu nutzen, da die ordinalskalierten Portfoliodimensionen für diese Zwecke nicht hinreichend operational sind und der Kundenwert nur vier Merkmalsausprägungen hat.

Aufgrund der von SCHMOLL beschriebenen ordinalen Darstellung der Dimensionen, die zwar der Skalierbarkeit und Adäquanz des Kundenwertmodells zugute kommt, ist die grenznutzenorientierte Einzelbewertung vor dem Hintergrund der vermögensorientierten Gesamtbanksteuerungskonzeption zu hinterfragen. Denn eine direkte Übertragung des Kundenwerts, definiert über die Quadrantenzugehörigkeit im Portfolioansatz, auf den Einfluss auf die Zielerreichung aus Gesamtbanksicht ist aufgrund der unterschiedlichen Dimensionalität (ohne Dimension versus Euro) nicht möglich. Aus dem genannten Grund ist auch die Integration des so ermittelten Kundenwerts in eine monetäre Gesamt-

---

1017 Vgl. *Kiwitz, K./Scheffler, C.* (2009), S. 10; *Homburg, C./Steiner, V. V./Totzek, D.* (2009), S. 70.

banksteuerungskonzeption nach VR-CONTROL fraglich. Die dem Kundenwertmodell zugrundeliegenden Informationsbeschaffungs- und –verarbeitungsmethoden können intersubjektiv nachprüfbar ausgestaltet werden. Zudem dürften die Modellergebnisse aufgrund ihrer verständlichen Visualisierung und der möglichen Einfachheit der Ermittlung akzeptiert werden. Tabelle 13 fasst die vorstehende Würdigung zusammen.

Tabelle 13: Würdigung des Kundenwertmodells von SCHMOLL[1018]

Anforderungen	**Würdigung**	**Anmerkung**
I. Allgemeine Anforderungen		
Ganzheitlichkeit	-	Risiko und Kundenbindung fehlt, Kundenwerttreiber außerhalb der Kundenbeziehung fehlen, Förderauftrag als mögliche Zielgröße nicht integrierbar
Interindividuelle Variabilität	+	
Dynamik	o	Dynamik wird nicht/kaum berücksichtigt; Zukunft wird pauschal betrachtet
Bezugsgrößenflexibilität	+	
Wirtschaftlichkeit	+	
II. Vertriebssteuerungsbezogene Anforderungen		
Handlungsorientierung	o	stärkere Differenzierung des Kundenwerttreiber im engeren Sinn nötig
Operationalität	-	problematisch als kundenbezogene Planungs-, Steuerungs- und Kontrollgröße
III. Kreditgenossenschaftliche Anforderungen		
Skalierbarkeit	+	
Adäquanz	+	
Grenznutzenorientierte Einzelbewertung	o	Übertragung des Kundenwerts auf Einfluss auf Zielerreichung kaum möglich
Intersubjektive Vergleichbarkeit	+	
Integrierte Ergebnisrechnung	-	unmöglich den Kundenwert in das Gesamtbanksteuerungskonzept nach VR-CONTROL zu integrieren
Akzeptanz	+	

Zusammenfassend kann für das Modell von SCHMOLL konstatiert werden, dass es aufgrund seiner Struktur und methodischen Grundlagen zunächst vor allem als strategische Entscheidungsgrundlage in die Vertriebssteuerung integriert

[1018] Eigene Darstellung.

werden kann. Eine geschlossene kundenwertorientierte Vertriebssteuerung nach dem hier vorliegenden Verständnis ist darauf aufbauend jedoch nicht möglich.

4.3.1.3 Kundenwertmodell von Raddao

Das Kundenwertmodell von Raddao wurde im Rahmen der Tätigkeit des Autors in einer Kreditgenossenschaft als Thema einer wissenschaftlichen Arbeit entwickelt und weist damit einen hohen praktischen Bezug auf. Es orientiert sich grundsätzlich an der Struktur des Kundenwertmodells von Schmoll, konkretisiert und erweitert dieses jedoch in wesentlichen Punkten.[1019] Kern des Kundenbewertungsmodells ist ein zweidimensionaler Ansatz, der zwischen dem Status als retrospektiven bzw. gegenwärtigen und dem Potenzial als prospektiven Kundenwerttreiber unterscheidet.[1020] Für beide Dimensionen werden dieselben Einflussgrößen berücksichtigt, die aufgeteilt werden in eine betriebliche und eine private Sphäre und anhand folgender Größen gemessen werden (vgl. Tabelle 14).

1019 Vgl. hier und nachfolgend *Raddao, N.* (2009), S. 40 ff.

1020 Ergänzend betrachtet Raddao auch eine Risiko- und Ressourcendimension. Diese sind jedoch nicht Bestandteil des Kundenwerts im hier verstandenen Sinne, sondern werden im Rahmen eines managementorientierten Vorgehens der Ertragsdimension gegenübergestellt, um zu prüfen, ob Risikosituation oder Ressourcenzuordnung in Abhängigkeit des Kundenwerts erfolgt oder hier inkongruente Situationen vorliegen, die entsprechende Korrekturmaßnahmen begründen. Vgl. *Raddao, N.* (2009), S. 55 ff.

Tabelle 14: Einflussgrößen der Status- und Potenzialbeurteilung[1021]

Einflussgröße	Messung Status	Messung Potenzial
Ergebnisbeitrag	Intensität der Bankverbindung	Bankverbindung – Entwicklungspotenzial
	Deckungsbeitrag II der Betreuungseinheit p. a.	Einschätzung der Entwicklung des Deckungsbeitrags II
	Deckungsbeitrag II p. a. relativ zu Umsatz des Firmenkunden	
Leistungsanalyse der Beziehung zum Unternehmen	Produktabdeckung bei Produkten aus dem Angebot für Firmenkunden	
Leistungsanalyse der Beziehung zum Unternehmer	Produktabdeckung bei Produkten aus dem Angebot für Privatkunden	
Branchensituation	Umsatzentwicklung	Aktuelle Umsatzprognose
	Umsatzentwicklung im Vergleich zur Branchenentwicklung	Kundenprognose im Vergleich zur Branche
Volkswirtschaftliche Rahmenparameter – Unternehmen	KfW-IFO-Mittelstandsbarometer Lagebeurteilung	KfW-IFO-Mittelstandsbarometer Erwartungen
Volkswirtschaftliche Rahmenparameter – Unternehmer	Regionale Kaufkraftkennziffer	Prognose – Kaufkraftkennziffer (regional)
	Entwicklung der Sparquote	Prognose – Entwicklung der Sparquote

In Verbindung mit dem allgemeinen Strukturmodell lässt sich damit das folgende konkrete Kundenwertmodell skizzieren (siehe Abbildung 37).

1021 Eigene Darstellung auf Basis der Angaben in *Raddao, N.* (2009), S. 49 und S. 93.

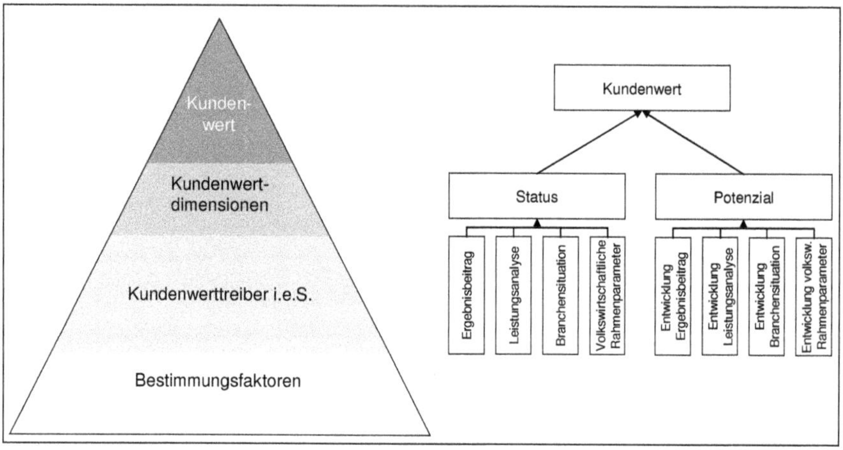

Abbildung 37: Kundenwertmodell nach RADDAO[1022]

In Anlehnung an die Operationalisierung bei SCHMOLL wird der aggregierte Kundenwert über ein Portfoliomodell ermittelt. Dabei werden die in Tabelle 14 genannten Messgrößen mittels eines Scoringmodells je Dimension aggregiert und in einen Scoringwert transformiert.[1023] Der Scorewert wird dazu verwendet, den Kunden in dem Portfolio zu positionieren. Durch die Aufteilung des gesamten Wertebereichs in vier Quadranten gibt es, wie bei SCHMOLL, vier verschiedene Kundenwerte, denen Normstrategien zugeordnet sind.[1024]

Aufgrund der Orientierung an den Vorarbeiten von SCHMOLL gelten mit Blick auf die definierten Anforderungen weitgehend die gleichen Würdigungsergebnisse für dieses Kundenwertmodell. Nachfolgend sollen daher nur die abweichenden Anforderungserfüllungen thematisiert werden. Mit Blick auf die allgemeinen Anforderungen kann konstatiert werden, dass die Berücksichtigung der Intensität der Bankbeziehung das Spektrum der relevanten Kundenwerttrei-

1022 Eigene Darstellung.

1023 Vgl. dazu ausführlich *Raddao, N.* (2009), S. 51 ff.

1024 In Anlehnung an das Marktanteils-Marktwachstums-Portfolio benennt RADDAO die Quadranten als Standardkunden, wenn die Bewertung von Potenzial und Status gering sind, als Potenzialkunden, wenn die Bewertung des Potenzials hoch und des Status niedrig ist, als Starkunden, wenn sowohl Potenzial als auch Status hoch bewertet werden und Cash-Cow-Kunden, wenn die Potenzialbewertung gering und die Statusbewertung hoch ist. Vgl. *Raddao, N.* (2009), S. 55. Die Normstrategien unterscheidet er in eine Leistungsstrategie, die das Produktangebot der Bank umfasst, und eine Betreuungsstrategie, die die vertriebliche Betreuung durch die Bank umfasst. Vgl. *Raddao, N.* (2009), S. 76 ff.

ber des Kundenwerts erweitert. Auch die Steuerungsrelevanz bzw. Handlungsorientierung wird durch die differenzierte Leistungsanalyse im Scoringmodell erhöht, wenngleich dies jedoch nicht mehr in der Ergebnisdarstellung explizit zur Geltung kommt. Abweichend von der bei SCHMOLL allgemein gehaltenen Operationalisierung wird bei diesem Modell das Vorgehen der Operationalisierung konkretisiert und erläutert. Es ist somit das Verdienst von RADDAO, das Modell von SCHMOLL zu substantiieren. Daraus resultiert jedoch, dass die Anforderung einer grenznutzenorientierten Einzelbewertung durch das vorliegende Kundenwertmodell nicht gewährleistet ist. Zusammenfassend ergibt sich die in Tabelle 15 dargestellte Würdigung des Kundenwertmodells von RADDAO.

Tabelle 15: Würdigung des Kundenwertmodells nach RADDAO[1025]

Anforderungen	**Würdigung**	**Anmerkung**
I. Allgemeine Anforderungen		
Ganzheitlichkeit	-	Risiko nur mittelbar betrachtet, Kundenwerttreiber außerhalb der Kundenbeziehung werden nicht beachtet, Förderauftrag als mögliche Zielgröße nicht integrierbar
Interindividuelle Variabilität	+	
Dynamik	o	Dynamik wird nicht/kaum berücksichtigt; Zukunft wird pauschal betrachtet
Bezugsgrößenflexibilität	o	vergleichsweise schwierig auf potenzielle Kunden anwendbar
Wirtschaftlichkeit	+	
II. Vertriebssteuerungsbezogene Anforderungen		
Handlungsorientierung	o	stärkere Differenzierung des Kundenwerttreiber im engeren Sinn nötig
Operationalität	-	problematisch als kundenbezogene Planungs-, Steuerungs- und Kontrollgröße
III. Kreditgenossenschaftliche Anforderungen		
Skalierbarkeit	+	
Adäquanz	+	
Grenznutzenorientierte Einzelbewertung	-	Übertragung des Kundenwerts auf Einfluss auf Zielerreichung nicht möglich
Intersubjektive Vergleichbarkeit	+	
Integrierte Ergebnisrechnung	-	unmöglich den Kundenwert in das Gesamtbanksteuerungskonzept nach VR-CONTROL zu integrieren
Akzeptanz	+	

1025 Eigene Darstellung.

Auch das Kundenwertmodell von RADDAO dient aufgrund seiner Struktur und Operationalisierung eher dazu, strategische Kundenselektionsentscheidungen zu unterstützen. Dabei ist jedoch zu berücksichtigen, dass es nicht alle Kundenwerttreiber berücksichtigt. Zudem ist eine geschlossene kundenwertorientierte Vertriebssteuerung nach dem hier vorliegenden Verständnis darauf aufbauend unmöglich.

4.3.1.4 Kundenwertmodell von FRESE/RUDOLF-SIPÖTZ

Abschließend soll das Kundenwertmodell, das FRESE als Grundlage für die Kundenwertorientierung in Universalbanken verwendet, vorgestellt und diskutiert werden.[1026] Es weicht insofern von den bisherigen Modellen ab, als es explizit Teilwerte betrachtet, die nicht im direkten Zusammenhang mit der dyadischen Beziehung zwischen dem Firmenkunden und der Bank stehen. Es sei darauf hingewiesen, dass dieses Modell nicht firmenkundenspezifisch ist und sich stark am Kundenwertmodell von RUDOLF-SIPÖTZ orientiert.[1027] Da dieses Kundenwertmodell in einer branchenübergreifenden empirischen Studie validiert wurde, soll es trotz des nur mittelbaren Bezugs zum Firmenkundengeschäft als Beispiel des aktuellen Forschungsstands vorgestellt werden.

Ausgangspunkt des Modells ist der kundenbezogene Rollenpluralismus im Hinblick auf die Wertbeiträge aus Sicht der Bank.[1028] Der Kunde hat neben der Rolle des Nachfragers und damit als Ertrags- und Kostenfaktor auch die Funktionen als Informationslieferant bei der Produktentwicklung und Bedürfniserfassung, als Partner und Co-Produzent im Leistungserstellungsprozess, als Referenzträger zur Akquisition von Neukunden und als Träger von Erlös- und Kostenwirkungen, die sich erst im Verbund mit den anderen Kunden zeigen. Dies sind z. B. economies of scale und economies of scope im Kundenstamm. Diese Rollen werden zwei Dimensionen, dem Markt- und das Ressourcenpotenzial, zugeordnet.

Das Marktpotenzial ist der monetäre Beitrag eines Kunden zum Unternehmenserfolg.[1029] Das Marktpotenzial umfasst im Einzelnen die Teilwerte Ertragspotenzial, Entwicklungspotenzial, Cross-Buying-Potenzial sowie das Loyalitätspotenzial. Das Ertragspotenzial bezieht sich auf den gegenwärtigen monetären Beitrag des Kunden, während das Entwicklungspotenzial die künftig zu erwartenden monetären Beiträge über die gesamte Geschäftsbeziehung beinhaltet. Das Cross-Buying-Potenzial bezeichnet die mögliche Nachfrage eines Kunden nach

[1026] Vgl. *Frese, B.* (2009), S. 10 ff.
[1027] Vgl. *Rudolf-Sipötz, E.* (2001), S. 95 ff.
[1028] Vgl. hier und nachfolgend *Rudolf-Sipötz, E.* (2001), S. 90 ff.
[1029] Vgl. hier und nachfolgend *Rudolf-Sipötz, E.* (2001), S. 95 ff.

Produkten, die bisher nicht Gegenstand von Transaktionen waren, und das Loyalitätspotenzial beschreibt die Affinität eines Kunden zur Kontinuität einer Beziehung mit einem bestimmten Anbieter. Das Cross-Buying-Potenzial und das Loyalitätspotenzial werden durch eine Vielzahl kundenspezifischer Einflussfaktoren bestimmt.

Das Ressourcenpotenzial repräsentiert die indirekten Beiträge eines Kunden zum Unternehmenserfolg, indem er aktiv oder passiv als Unternehmensressource dient.[1030] Dem Ressourcenpotenzial werden das Referenz-, das Informations-, das Kooperations- und das Synergiepotenzial zugeordnet. Das Referenzpotenzial wird determiniert durch die Anzahl potenzieller Kunden, die der zu bewertende Kunde, innerhalb eines bestimmten Zeitraums aufgrund seines Weiterempfehlungsverhaltens und Einflussvermögens sowie der Größe, Art, Kontakthäufigkeit und -intensität seiner sozialen Beziehungsnetzwerke erreichen kann. Es kann positiv, negativ oder neutral sein. Das Informationspotenzial umfasst alle Informationen, die der Kunde einem Anbieter liefert und die vom Anbieter in weiterer Folge zur erfolgreichen Unternehmensführung genutzt werden können. Das Kooperationspotenzial eines Kunden ist seine Bereitschaft und Fähigkeit, auf begrenzte Zeit Produktionsfaktoren in den Dispositionsbereich der Bank einzubringen. Löst der Kunde aktiv oder passiv Wechselwirkungen im Kundenstamm aus, wird dies dem Synergiepotenzial zugerechnet. Als Beispiel seien hier economies of scale durch die kundeninduzierte Kapazitätsauslastung genannt. Wie erkenntlich, hängen einige der Kundenwerttreiber von weiteren Bestimmungsfaktoren ab. Ohne diese ausführlich zu diskutieren,[1031] ergibt sich das Kundenwertmodell in Abbildung 38.

1030 Vgl. hier und nachfolgend *Rudolf-Sipötz, E.* (2001), S. 108 ff.

1031 Dazu sei auf die Ausführungen bei *Rudolf-Sipötz, E.* (2001), S. 95 ff. verwiesen.

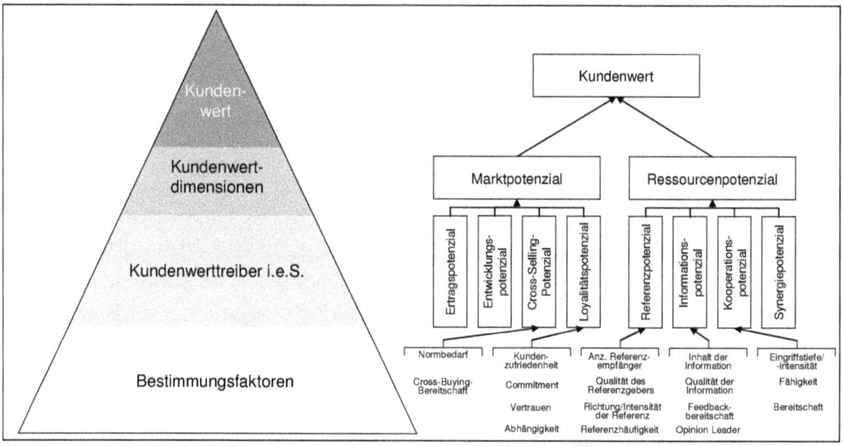

Abbildung 38: Kundenwertmodell nach FRESE/ RUDOLF-SIPÖTZ[1032]

FRESE macht keine Ausführungen dazu, wie aufbauend auf diesem Modell, der Kundenwert konkret gemessen werden soll. Da das Modell weitgehend auf der Arbeit von RUDOLF-SIPÖTZ aufbaut, soll dieser Beitrag als Grundlage für die Erläuterung einer Operationalisierung dienen. RUDOLF-SIPÖTZ entwickelt einen dreidimensionalen Portfolioansatz, den Kundenkubus, um den Kundenwert aufbauend auf ihrer Konzeptualisierung zu messen.[1033] Die Kunden werden anhand der Dimensionen/Kriterien gegenwärtiges Erfolgspotenzial (niedrig bis hoch), künftiges Erfolgspotenzial (niedrig bis hoch) und komplementärer Wertbeitrag (tief bis hoch) positioniert. Die Skalierung ist im konkreten Anwendungsfall festzulegen, so dass monofaktorielle Dimensionen oder multifaktorielle Dimensionen unterstützt werden.[1034] Die Kombinationen aus den drei Dimensionen können durch zweckgerechte Aufteilungen der Kubusachsen ebenfalls situativ bestimmt werden, so dass der Wertebereich relevanter Kundenwerte, definiert als Zuordnung eines Kunden zu den anhand der Aufteilung resultierenden Subkuben, nicht allgemein definiert ist.

Hinsichtlich der Anforderung der Ganzheitlichkeit kann ein im Vergleich zu beiden bisher vorgestellten Kundenwertmodellen ambivalentes Ergebnis festge-

---

[1032] Eigene Darstellung.

[1033] Vgl. *Rudolf-Sipötz, E.* (2001), S. 192 ff.

[1034] So kann z. B. das gegenwärtige Erfolgspotenzial mittels des kumulierten Kundendeckungsbeitrags monokriteriell und metrisch skaliert sein, während das künftige Erfolgspotenzial mittels eines Scoringmodells multifaktoriell bestimmt wird. Darüber hinaus erläutert RUDOLF-SIPÖTZ, dass der komplementäre Wertbeitrag auch nominalskaliert sein könne . Vgl. *Rudolf-Sipötz, E.* (2001), S. 193.

stellt werden. Auf der einen Seite zeigen dieses Kundenwertmodell und dessen empirische Prüfung, dass zwar das Marktpotenzial, dem in etwa die benannten Kundenwerttreiber der beiden anderen Modelle zugeordnet werden können, wesentlichen Einfluss auf den Kundenwert hat. Jedoch zeigen die empirischen Ergebnisse erstens auch, dass die gegenwartsgerichteten Kundenwerttreiber nur unwesentlich das Marktpotenzial beeinflussen.[1035] Und zweitens ergänzt RUDOLF-SIPÖTZ die relevanten Kundenwerttreiber um mittelbare Einflussgrößen wie das Referenz-, das Kooperations-, das Informations- und das Synergiepotenzial. Auf der anderen Seite ist jedoch zu kritisieren, dass die wichtige Risikodimension nicht berücksichtigt ist. Weiterhin muss auch hier konstatiert werden, dass keine Berücksichtigung des Förderauftrags als Zielsetzung erfolgt. Aufgrund der vorgeschlagenen Operationalisierung des konzeptionellen Ansatzes ist zu bemerken, dass auch die dynamische Prognostizierbarkeit nur gering ausgeprägt ist. Zwar kann für die ausgewählten Kundenwerttreiber begründet angenommen werden, dass diese interindividuell variabel sind, jedoch gilt die Kritik, dass das Modell schwierig aufgrund der notwendigen Informationsbasis auf potenzielle Kunden übertragen werden kann. Aufgrund des in der Arbeit dargestellten praktischen Einsatzes ist davon auszugehen, dass das Wirtschaftlichkeitspostulat eingehalten werden kann.

Aus Sicht der Vertriebssteuerung kann zunächst beschieden werden, dass das Modell einen handlungsorientierten differenzierteren Ausweis vornimmt. Aufgrund des eher ordinalen Messansatzes ist der so bestimmte Kundenwert nur schwer als Planungs-, Steuerungs- und Kontrollgröße einzusetzen.

Obwohl das Modell skalierbar ist und als adäquat bezeichnet werden kann, sind die kreditgenossenschaftlichen Anforderungen nicht vollumfänglich erfüllt. Es ist keine direkte grenznutzenorientierte Einzelbewertung möglich und zudem gelingt es nicht, den so modellierten und gemessenen Kundenwert in das Gesamtbanksteuerungskonzept VR-CONTROL zu integrieren. Aufgrund der klar definierten Struktur ist der gemessene Kundenwert intersubjektiv nachprüfbar und dürfte Akzeptanz bei den betroffenen Mitarbeitern finden. Problematisch könnte sein, dass die Auswahl der Kundenwerttreiber nicht für alle Mitarbeiter nachvollziehbar ist. Dies dürfte sich jedoch nach einer Eingewöhnungsphase relativieren.

Zusammenfassend ergibt sich die in Tabelle 16 dargestellte Würdigung des Kundenwertmodells von FRESE bzw. RUDOLF-SIPÖTZ.

[1035] Vgl. *Rudolf-Sipötz, E.* (2001), S. 175. In den Modellen von RADDAO und SCHMOLL ist aufgrund der Zusammenführung der Dimensionen über das Portfoliomodell davon auszugehen, dass den gegenwartsgerichteten Dimensionen (Internes Kundenpotenzial bzw. Status) messtheoretisch eine relativ hohe Bedeutung beigemessen wird.

Tabelle 16: Würdigung des Kundenwertmodells nach FRESE/RUDOLF-SIPÖTZ[1036]

Anforderungen	Würdigung	Anmerkung
I. Allgemeine Anforderungen		
Ganzheitlichkeit	o	Risiko fehlt, Förderauftrag als mögliche Zielgröße nicht integriert
Interindividuelle Variabilität	+	
Dynamik	o	Dynamik wird nicht/kaum berücksichtigt; Zukunft wird pauschal betrachtet
Bezugsgrößenflexibilität	o	vergleichweise schwierig auf potenzielle Kunden anwendbar
Wirtschaftlichkeit	+	
II. Vertriebssteuerungsbezogene Anforderungen		
Handlungsorientierung	+	
Operationalität	-	problematisch als kundenbezogene Planungs-, Steuerungs- und Kontrollgröße
III. Kreditgenossenschaftliche Anforderungen		
Skalierbarkeit	+	
Adäquanz	+	
Grenznutzenorientierte Einzelbewertung	-	Übertragung des Kundenwerts auf Einfluss auf Zielerreichung nicht möglich
Intersubjektive Vergleichbarkeit	+	
Integrierte Ergebnisrechnung	-	unmöglich den Kundenwert in das Gesamtbanksteuerungskonzept nach VR-CONTROL zu integrieren
Akzeptanz	+	

Auch hier kann zusammenfassend konstatiert werden, dass aufbauend auf diesem Modell keine geschlossene kundenwertorientierte Vertriebssteuerung nach dem hier vorliegenden Verständnis entwickelt werden kann.

Zusätzlich zu den diskutierten theoretischen Modellen werden nun ergänzend Kundenwertmodelle vorgestellt, die in der kreditgenossenschaftlichen Praxis eingesetzt werden. Auch diese werden anhand der definierten Anforderungen gewürdigt.

4.3.2 Stand der Praxis zu Kundenwertmodellen

Mit Blick auf den Stand der Praxis sei darauf hingewiesen, dass der Kundenwert in dem dieser Arbeit zugrundeliegenden Verständnis selten definiert und als Steuerungsgröße implementiert ist. Um dennoch Aussagen darüber zu treffen, wie die kreditgenossenschaftliche Praxis den Wert eines Kunden zur Grundlage

1036 Eigene Darstellung.

ihrer Entscheidungen macht, wird ein etwas breiteres Verständnis des Kundenwerts toleriert. Und wenngleich das praktische Vorgehen der betrachteten Kundenwertorientierung im Allgemeinen und der kundenwertorientierten Vertriebssteuerung im Speziellen weitgehend entspricht, muss konstatiert werden, dass grundlegende Überlegungen, Maßnahmen und Messgrößen stark abweichen. Im Ergebnis zeigt dies auf, dass eine kundenwertorientierte Vertriebssteuerung im Sinne der vorliegenden Arbeit gegenwärtig nicht in den Kreditgenossenschaften umgesetzt ist und sein kann, da vor allem das Verständnis des Kundenwerts als fundamentale Bezugsgröße und dessen konsequente Implementierung fehlt. Häufigstes Einsatzfeld der praktischen Kundenwertorientierung ist die Kundenselektion.

Unter dem Begriff der potenzial- oder (kunden-)wertorientierten Segmentierung werden in den Kreditgenossenschaften Kundengruppen gebildet, um „die Intensität der Betreuung ertrags- und kostenorientiert zu steuern und den Arbeits- und Personaleinsatz zu optimieren.“[1037] Typisch für das Gros der Kreditgenossenschaften ist der nachfolgend skizzierte Ansatz aus der VOLKSBANK TRIER EG (siehe Tabelle 17).

Tabelle 17: Kundensegmentierung der VOLKSBANK TRIER EG[1038]

Zielgruppe / Kriterium	Betreuungs-Firmenkunden (inkl. Freiberufler)			Beratungs-Firmenkunden (inkl. Freiberufler)	Gewerbekunden
1. Aktivvolumen ODER	> 0,25 T€			0,05 - 0,25 T€	< 0,05 T€ (Haupteinnahmequelle ist Gewerbebetrieb)
2. G+V-Umsatz (alt.: kumulierter Habenumsatz)	A	B	C	0,15 - 1,0 T€	< 0,15 T€
Fertigungsbetriebe und Dienstleister	> 2,5 T€	> 1,25 T€	> 1,0 T€		
Handelsbetriebe	> 5,0 T€	> 2,5 T€	> 1,0 T€		

Als Kundenwerttreiber i. e. S. gelten das Aktivvolumen, d. h. die Summe der an den Firmenkunden ausgegebenen Kredite[1039] oder alternativ dazu der Umsatz

1037 *Schreiber, H.* (2007), S. 43.

1038 *Schreiber, H.* (2007), S. 44.

1039 Dabei wird in den Instituten unterschieden, ob das Volumen der Inanspruchnahme oder der (historischen) Zusage verwendet wird.

des Firmenkunden bzw. dessen kumulierter Habenumsatz auf dem Geschäftskonto bei der Bank sowie die Branchenzugehörigkeit des Kunden.[1040] Im Ergebnis resultieren aus dem genannten Praxisansatz fünf Kundenwertklassen: Betreuungs-Firmenkunden in den Klassen A, B und C, Beratungs-Firmenkunden sowie Gewerbekunden. Weitere Kundenwerttreiber werden nicht oder nicht systematisch berücksichtigt.

Hinsichtlich der allgemeinen Anforderungen, ist zu urteilen, dass die Anforderung der Ganzheitlichkeit nicht erfüllt ist. Es fehlt in diesem Modell die explizite Berücksichtigung der Ertragsdimension. Es gilt die implizite Annahme, je höher das Aktivvolumen bzw. je höher das Umsatzvolumen des Firmenkunden, desto höher ist sein Zielerreichungsbeitrag für die Bank. Darüber hinaus fehlt das Risiko aus und in der Kundenverbindung, das bspw. aus der Kundenbindung oder der Bonität des Kunden resultiert.[1041] Weiterhin kann kritisiert werden, dass beziehungsübergreifende Erfolgsbeiträge bspw. in Form von Referenztätigkeiten nicht berücksichtigt werden. Weiterhin ist zu kritisieren, dass die bereits erwähnte implizit unterstellte Kausalität von Aktivvolumen und Ertrag nicht gilt. Denn dass dies hinsichtlich des Gewinnziels nicht zwingend gilt, zeigen die in Tabelle 1 aufgeführten Ergebnisse empirischer Untersuchungen.[1042] Darüber hinaus fehlt auch hier – und dass ist umso überraschender, da es sich um kreditgenossenschaftliches Modell handelt – die Berücksichtigung des Förderauftrags als Ziel der Volksbank. Aufbauend auf der Tatsache, dass die betrachteten Größen einen ausschließlichen Vergangenheits- bzw. Gegenwartsbezug haben, ist deren Verwendung zur zukunftsgerichteten Kundenbewertung in diesem Ansatz nicht intendiert und daher zu kritisieren. Auch besteht nicht die Möglichkeit, das Potenzial eines Kunden zu betrachten, da ausschließlich anbieterbezogene Größen betrachtet werden. Aufgrund der einfachen Ermittelbarkeit der relevanten Kundenwerttreiber kann das Modell jedoch vergleichsweise einfach auf potenzielle Kunden übertragen werden, so dass aktuelle und potenzielle Kunden be-

---

[1040] Im Sinne eines zweistufigen Vorgehens dient die Kundensegmentierung aus Tabelle 17 in einigen wenigen, größeren Instituten der FinanzGruppe als erste Stufe bzw. als so genannte Grobsegmentierung, die um eine Feinsegmentierung als zweite Stufe ergänzt wird. In der Feinsegmentierung werden grobsegmentindividuell weitere Kundenwerttreiber definiert und für die Bewertung des Kunden eingesetzt. I. d. R. werden auf dieser Stufe Scoringmodelle eingesetzt, um die einzelnen unterschiedlich skalierten Kundenwerttreiber zusammenzufassen.

[1041] Das verwendete Aktivvolumen stellt nur einen schwachen Indikator des Risikos dar, da es nicht unterscheidet zwischen einer hohen und einer geringen Bonität.

[1042] Aufgrund der Aussage die Betreuungsressourcen ertrags- und kostenorientiert zu steuern, kann begründet davon ausgegangen werden, dass die VOLKSBANK TRIER EG das Gewinnstreben als dominante Grundlage ihres Zielsystems definiert hat.

wertet werden können. Aus dem gleichen Grund kann unterstellt werden, dass das Modell wirtschaftlich eingesetzt werden kann.

Tabelle 18: Zusammenfassende Würdigung des Kundenwertmodells aus der Praxis[1043]

Anforderungen	**Würdigung**	**Anmerkung**
I. Allgemeine Anforderungen		
Ganzheitlichkeit	-	Ertrags- und Risikodimensionen fehlen, Kundenwerttreiber außerhalb der Kundenbeziehung werden nicht beachtet, Förderauftrag als mögliche Zielgröße nicht integriert
Interindividuelle Variabilität	+	
Dynamik	-	Vergangenheits- und Gegenwartsbezug der betrachteten Kundenwerttreiber erlauben keine dynamische Prognose
Bezugsgrößenflexibilität	+	
Wirtschaftlichkeit	+	
II. Vertriebssteuerungsbezogene Anforderungen		
Handlungsorientierung	-	stärkere Differenzierung des Kundenwerttreiber im engeren Sinn nötig
Operationalität	o	möglich als Planungs- und Kontrollgröße, aber Inkongruent zu den Zielen auf Gesamtbankebene
III. Kreditgenossenschaftliche Anforderungen		
Skalierbarkeit	+	
Adäquanz	+	
Grenznutzenorientierte Einzelbewertung	-	Übertragung des Kundenwerts auf Einfluss auf Zielerreichung nicht möglich
Intersubjektive Vergleichbarkeit	+	
Integrierte Ergebnisrechnung	-	unmöglich den Kundenwert in das Gesamtbanksteuerungskonzept nach VR-Control zu integrieren
Akzeptanz	+	

Aus der Auswahl der relevanten Kundenwerttreiber folgt die Kritik eines kaum bis gar nicht handlungsorientierten Kundenwertmodells. Ausreichend wäre diese Modellierung nur, wenn die dominante Zielgröße das Aktivgeschäft mit den (Firmen-)Kunden wäre und auch die Steuerungssysteme volumenorientiert ausgestaltet wären.[1044] Für ein vertriebssteuerungsrelevantes Kundenwertmodell ist jedoch eine größere Breite und stärkere Differenzierung der Kundenwerttrei-

1043 Eigene Darstellung.

1044 Vgl. für die volumenorientierte Steuerung die Ausführungen in Kapitel A.1.

ber notwendig. Aufgrund seiner ausschließlich verhältnisskalierten Kriterien könnte – trotz der konzeptionellen Kritik – die Operationalisierbarkeit im Sinne der Planungs- und Kontrollverfahren festgestellt werden. Dies wäre jedoch nur gültig, wenn das Zielsystem ausschließlich bzw. vornehmlich volumenbezogen wäre. Da dies aus begründeten Annahmen jedoch nicht der Fall ist, wird dieses Urteil durch die fehlende Kongruenz dieser Größe zu den Zielgrößen der Volksbank widerlegt.

Werden abschließend die kreditgenossenschaftlichen Anforderungen betrachtet, so kann festgestellt werden, dass diese weitgehend erfüllt werden. Unzureichend ist jedoch, dass durch das Kundenwertmodell weder die grenznutzenorientierte Einzelbewertung unterstützt, noch in die Ergebnisrechnung integriert werden kann. Zusammenfassen lässt sich in Tabelle 18 dargestellt Würdigung ableiten. Das Fazit dieser Würdigung ist, dass auch die in der kreditgenossenschaftlichen Praxis typischerweise eingesetzten Kundenwertmodelle nicht dazu geeignet sind, eine kundenwertorientierte Vertriebssteuerung zu unterstützen.

4.3.3 Zwischenfazit

Wird der Stand der allgemeinen Kundenwertforschung betrachtet, lässt sich feststellen, dass die bisher definierten Kundenwertmodelle, wenngleich sie teilweise Allgemeingültigkeit beanspruchen,[1045] nicht unmittelbar auf das Firmenkundengeschäft übertragbar sind. Spezifische Forschungsbeiträge und Praxismodelle dagegen widersprechen in wesentlichen Teilen den Anforderungen, die eine kundenwertorientierte Vertriebssteuerung an ein solches Modell stellt.

Aus diesen Gründen wird nachfolgend ein Kundenwertmodell für das Firmenkundengeschäft der Kreditgenossenschaften entwickelt. Dazu wird zunächst untersucht, ob es theoretisch fundierte Leitlinien aus bisherigen Forschungsbeiträgen gibt, welche die Ableitung, Präzisierung und Differenzierung relevanter Kundenwerttreiber fundieren. Dazu sollen die in der Tabelle 12 dargestellten ausgewählten Beiträge herangezogen werden. Darüber hinaus werden Beiträge betrachtet, die sich mit der theoretischen Fundierung des Kundenwerts auseinandersetzen.

4.4 Stand der Forschung zur Herleitung von Kundenwerttreibern

In der betriebswirtschaftlichen Forschung gab es schon früh erste kritische Überlegungen zu denjenigen Modellen des Kundenwerts, die ausschließlich auf mo-

1045 So z. B. *Rudolf-Sipötz, E.* (2001), S. 222.

netären Größen basieren.[1046] Spätere Beiträge führten daher konsequent an, dass qualitative, nicht-monetäre Faktoren zur ganzheitlichen Darstellung des Kundenwerts notwendig seien.[1047] Wobei auch die Betrachtung von der dyadischen Kundenbeziehung auf Interdependenzen zwischen Kunden im weiteren Sinn ausgeweitet wird.[1048] „Erst die Kenntnis aller Dimensionen und Aspekte des Kundenwertes erlaubt eine Priorisierung der „wertvollsten" Kunden sowie eine konsequent auf die Erhöhung des Kundenbestandswerts ausgerichtetes Beziehungsmanagement"[1049] und somit eine effektive kunden(wert)orientierte Vertriebssteuerung.[1050] Die sich aus diesen Überlegungen und Erkenntnissen ergebende Frage, welche Kundenwerttreiber in einem konkreten Kundenwertmodell zu berücksichtigen sind, wird in Forschung und Praxis unterschiedlich und zweckbezogen beantwortet. Auch die Art, wie die Werttreiber hergeleitet werden, ist unterschiedlich. Für die weitere Analyse nützlicher Leitlinien für die Ableitung von Kundenwerttreibern ist zu berücksichtigen, dass alle bisher in der Literatur vorgestellten und diskutierten Modelle von dem erwerbswirtschaftlichen Ziel der Gewinnmaximierung ausgehen. Alternative Ziele werden nicht berücksichtigt oder implizit als integrierbar in einzelne ökonomische Kundenwerttreiber angesehen.[1051]

Eine Vielzahl der betrachteten Beiträge, die sich mit der Definition von Kundenwerttreibern im Rahmen der Modellierung des Kundenwerts auseinandersetzen, leitet die betrachteten Kundenwerttreiber auf Basis von Plausibilitätsüberlegungen her.[1052] Dabei werden in einigen Fällen unterschiedliche Systema-

---

1046 Vgl. z. B. *Plinke, W.* (1989), S. 306 f.

1047 Vgl. z. B. *Wünschmann, S.* (2007), S. 14 ff. sowie *Günter, B.* (2006), S. 242; *Tomczak, T./Rudolf-Sipötz, E.* (2006), S. 130; *von Wangenheim, F.* (2003), S. 31; *Bayón, T./Gutsche, J./Bauer, H.-H.* (2002), S. 216 ff.; *Rudolf-Sipötz, E.* (2001), S. 93 f.; *Cornelsen, J.* (2000), S. 30; *Homburg, C./Schnurr, P.* (1998), S. 171; *Plinke, W.* (1997), S. 126 f.; *Krüger, S. M.* (1997), S. 109 f.; *Schleuning, C.* (1994), S. 161 ff.; *Canning, G.* (1982), S. 91.

1048 Vgl. *Cornelsen, J.* (2006), S. 191.

1049 *von Wangenheim, F.* (2003), S. 31 f.

1050 Diese Frage stellen auch RUST/ZEITHAML/LEMON: „Of all of the potential levers that a company might pull [...] which will yield the best return on investment?" *Rust, R. T./Zeithaml, V. A./Lemon, K. N.* (2000), S. 7.

1051 So etwa WILLE, der ausgehend vom erwerbswirtschaftlichen Prinzip die nicht-ökonomischen Bestimmungsfaktoren vernachlässigt. Vgl. *Wille, K.* (2005), S. 42.

1052 Vgl. dazu die Kundenwertmodelle und deren Herleitungen von *Canning, G.* (1982); *Storbacka, K.* (1993); *Schleuning, C.* (1994); *Schulz, B.* (1995); *Walter, G.* (1996); *Dwyer, F. R.* (1997), wobei dessen Überlegungen aufgrund der Definition des Kundenwerts als investitionsrechnerischer Größe methodisch inhärent sind; *Knöbel, U.* (1997); *Krüger, S. M.* (1997); *Homburg, C./Schnurr, P.* (1998); *Rust, R. T./Zeithaml, V.*

tisierungsansätze genutzt, um die Überlegungen zu strukturieren. WILLE gliedert die betrachteten Kundenwerttreiber anhand eines dreiteiligen Prozessschemas, das zwischen dem Transaktionsprozess, dem Informationsprozess und dem Kooperationsprozess unterscheidet.[1053] Die Herleitung des Schemas selbst wird jedoch nicht begründet. Wie bereits in Kapitel D.4.1 beschrieben, resultieren die Kundenwerttreiber bei LISSAUTZKI aus der zugrunde gelegten Wertreibersystematik, die eine formallogische Differenzierung der Werttreiber bedingt.[1054] Auf Basis interaktionstheoretischer Überlegungen leitet CORNELSEN drei Kundenwertquellen ab.[1055] Dies sind Transaktionen, Outgoing- und Ingoing-Informationen. Die Auswahl der Werttreiber erfolgt nach Meinung DILLERS „allerdings nur sehr vage“[1056]. GELBRICH begründet die Auswahl und Differenzierung der Kundenwerttreiber dagegen kaum, fundiert jedoch anhand verschiedener theoretischer Ansätze, auf welche Weise diese ausgewählten Kundenwerttreiber den Kundenwert determinieren.[1057]

Eine große Anzahl an Beiträgen orientiert sich an einem rein investitionsrechnerischen Kundenwert, dem Customer Lifetime Value, und reduziert methodisch inhärent konsequent die betrachteten Kundenwerttreiber auf Deckungsbeiträge, Kosten und Risiken bzw. Wahrscheinlichkeitsaussagen zur Dauer der Geschäftsbeziehung.[1058] Praxisorientierte Arbeiten, die einen direkten Bezug zu bestimmten Branchen oder Unternehmen aufweisen, begründen die Auswahl entweder kaum oder übernehmen allgemein formulierten Modelle vorheriger Arbeiten.[1059] Jüngere theoretische Arbeiten, die auf spezifische Aspekte fokussieren (z. B. das Weiterempfehlungsverhalten und dessen Einfluss auf den Kundenwert) oder das kundenwertorientierte Management thematisieren, nutzen die Erkenntnisse der vorhergehenden Kundenwertforschung. Diese Arbeiten modellieren den Kundenwert auf Basis literaturanalysengestützter Synthesen einzelner Kundenwertmodelle.[1060]

A./Lemon, K. N. (2000); *Böhrs, S.* (2004); *Renker, C.* (2004); *Günter, B.* (2006); *Schmoll, A.* (2006); *Schroeder, N.* (2006); *Raddao, N.* (2009).

1053 Vgl. *Wille, K.* (2005), S. 151 ff.

1054 Vgl. *Lissautzki, M.* (2007), S. 118 ff.

1055 Vgl. *Cornelsen, J.* (2000), S. 170 ff.

1056 *Diller, H.* (2001), S. 7.

1057 Vgl. *Gelbrich, K.* (2001), S. 52 ff.

1058 Vgl. z. B. *Stüker, D.* (2008), S. 167 ff.

1059 So z. B. FRIEDERICHS-SCHMIDT, die den Kundenwert in der Versicherungsbranche betrachtet und auf dem Modell von RUDOLF-SIPÖTZ aufbaut. Vgl. *Friederichs-Schmidt, S.* (2006), S. 27 ff.

1060 Vgl. z. B. *Schmöller, P.* (2001), S. 110 ff.; *Eberling, G.* (2002), S. 127 ff.; *Tewes, M.* (2003), S. 80 ff.; *von Wangenheim, F.* (2003), S. 31 ff.; *Bauer, H.-H./Hammerschmidt, M.* (2005), S. 331 ff.; *Schneider, Y.* (2007), S. 42 ff.; *Wünschmann, S.* (2007), S. 22 ff.

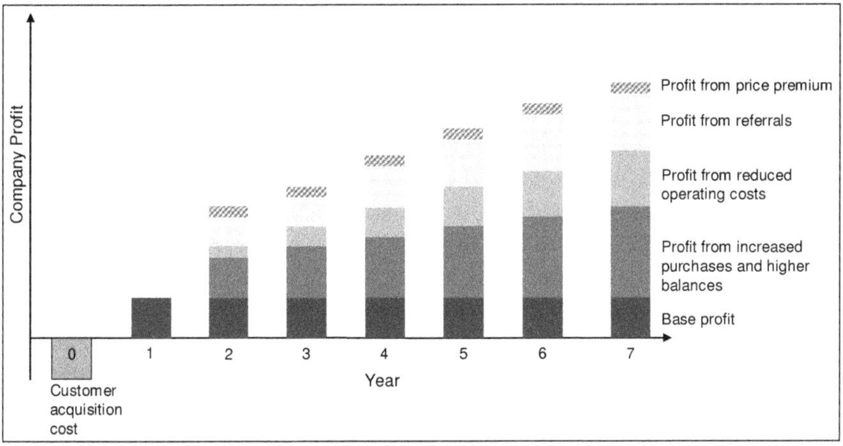

Abbildung 39: Kundenprofitabilität in Abhängigkeit der Zeit (Kundenbeziehungsdauer)[1079]

FORNELL ET AL. stellen auf Basis empirischer Daten fest, dass die langfristige Profitabilität eines Unternehmens aus Effizienzvorteilen langfristiger Kundenbeziehungen bestimmt wird.[1080] KRÜGER prüft den Einfluss der Kundenbeziehungsdauer auf den Umsatz, das Weiterempfehlungsverhalten in der Vergangenheit, die Weiterempfehlungsabsichten und die erforderliche Betreuungsintensität (Beratungsdauer) bei Kunden eines Reisebüros anhand von Mittelwertvergleichen. Sie kommt zu dem tendenziell die Aussagen von REICHHELD/SASSER bestätigenden Ergebnis, dass eine längere Kundenbeziehung mit höheren Umsätzen, häufigeren Empfehlungen bzw. Empfehlungsabsichten und kürzeren Beratungszeiten verbunden ist. Zwar sind die Ergebnisse signifikant, jedoch die festgestellten Unterschiede zwischen lang- und kurzfristigen Kundenbeziehungsdauern nur schwach ausgeprägt.[1081]

Abweichend von diesen Ergebnissen, die unterstellen, dass langfristige Geschäftsbeziehungen profitabler als vergleichsweise kurzfristige (oder transaktionale) sind, finden KRAFFT und REINARTZ/KRAFFT im Rahmen einer Studie zur Kundenprofitabilität in Versandhandelsunternehmen heraus, dass die These eines positiven Zusammenhangs zwischen Kundenlebenszeit und Ertragswert einer Kundenbeziehung und eines steigenden Ertragswerts von Kundenbeziehungen mit zunehmender Kundenlebenszeit nicht verallgemeinerbar ist. So zeigt sich in ihren Ergebnissen, dass sowohl langfristig, aber auch kurzfristig tätige

---

1079 *Reichheld, F. F./Sasser, W. E.* (1990), S. 108.

1080 Vgl. *Fornell, C. et al.* (1996), S. 11.

1081 Vgl. *Krüger, S. M.* (1997), S. 233 ff.

Kunden wesentlich zum Unternehmenserfolg beitragen.[1082] Zu einem ähnlichen Ergebnis kommen REINARTZ/KUMAR, die den Einfluss der Dauer der Kundenbeziehung auf den Customer Lifetime Value in einem so genannten „nocontractual setting" prüfen.[1083]

Für das Firmenkundengeschäft im Bankenkontext argumentieren dagegen WALTER, STECK/KUNDISCH und STRÄTER, dass aufgrund der hohen anfänglichen Investitionen in Kundenbeziehungen z. B. in Form von Informationsbeschaffungen und Bonitätsanalysen, diese sich erst im Laufe der Zeit amortisieren.[1084] Auch SCHLEUNING vermutet einen negativen Zusammenhang zwischen Zeitpunkt des kundenseitigen Beziehungsabbruchs und der Kundenbeziehungsrentabilität.[1085] Nach BOOT gilt zudem zu berücksichtigen: „Shorter Relationships inhibit the reusability of information and thus diminish the value of information."[1086] Uneinheitliche Ergebnisse liefern auch die empirischen Studien im Bankenkontext, die den Einfluss der Dauer einer Geschäftsbeziehung auf den Zinssatz als zentrale ökonomische Größe in der Geschäftsbeziehung zwischen Bank und Firmenkunde prüfen.[1087] So kommen die Studien von PETERSEN/RAJAN, HARHOFF/KÖRTING und ELSAS/KRAHNEN zu dem Ergebnis, dass die Dauer der Bankbeziehung keinen Einfluss auf den Zinssatz hat.[1088] Einen negativen Einfluss der Dauer der Bankbeziehung auf den Zinssatz finden dagegen

---

1082 Vgl. *Reinartz, W. J./Krafft, M.* (2001), S. 1271 ff.; *Krafft, M.* (2007), S. 192 ff.

1083 Vgl. *Reinartz, W. J./Kumar, V.* (2000), S. 17 ff. „Noncontractual settings" sind dadurch gekennzeichnet, dass die Geschäftsbeziehung zwischen einem Kunden und einem Anbieter nicht vertraglich determiniert ist, wie z. B. bei Versandhändlern.

1084 Vgl. *Walter, G.* (1996), S. 34; *Steck, W./Kundisch, D.* (2005), S. 444 f.; *Sträter, J.* (2005), S. 46.

1085 Vgl. *Schleuning, C.* (1997), S. 148.

1086 *Boot, W. A.* (2000), S. 18.

1087 Der verengte Blickwinkel auf den Zinssatz (den Preis für Kredite) resultiert aus zwei Gründen. Zum einen ist nicht der mehrdimensionale Kundenwert im Sinne der vorliegenden Arbeit Gegenstand der Untersuchung, sondern es handelt sich vornehmlich um Studien, die sich mit den Informationsasymmetrien und den Konsequenzen in den Geschäftsbeziehungen zwischen mittelständischen Firmenkunden und Hausbanken auseinandersetzen. Zum zweiten ist der Zinssatz die dafür zweckmäßige und leicht beobachtbare Größe.

1088 Vgl. *Petersen, M. A. /Rajan, R. G.* (1994), S. 12; *Harhoff, D./Körting, T.* (1998), S. 1342; *Elsas, R./Krahnen, J. P.* (1998), S. 1302. Die Studien untersuchen dabei den Darlehenszinssatz.

BERGER/UDELL und BLACKWELL/WINTERS.[1089] Und einen positiven Einfluss messen schließlich DEGRYSE/VAN CAYSEELE und ANGELINI/DI SALVO/FERRI.[1090]

Vor dem Hintergrund der uneinheitlichen empirischen Ergebnisse, lässt sich resümieren, dass die zusätzlichen Erfolge langfristiger Kundenbeziehungen aus Unternehmenssicht kein zeitabhängiger Determinismus sind, sondern kundenindividuell variante Potenziale darstellen, deren Abschöpfung von den Fähig- und Fertigkeiten und dem Willen des Anbieterunternehmens in Verbindung mit der Bereitschaft des Kunden und der Art des Geschäftsmodells abhängen. Gleichwohl kann konstatiert werden, dass die grundlegende Bedingung dafür, diese Potenziale zu entwickeln und zu heben, eine lange Dauer der Beziehung zwischen Anbieterunternehmen und Kunden ist.[1091]

Die Beziehungsdauer wird in der marketingwissenschaftlichen Forschung auf die Konstrukte der Kundenbindung und der Kundenloyalität bzw. Kundentreue zurückgeführt.[1092] Da vor allem die Konsequenzen der Kundenbindung aus Anbietersicht relevant sind, sei an dieser Stelle auf eine umfassende Erläuterung der Einflussfaktoren des Konstrukts Kundenbindung verzichtet und auf die umfangreiche spezifische Literatur verwiesen.[1093] Stellvertretend sei der Konzeptualisierungsansatz von PETER genutzt, um die Determinanten der Kundenbindung überblicksartig aufzuzeigen (siehe Abbildung 40). Auf einzelne Aspekte dieses Konzeptualisierungsansatzes wird im weiteren Verlauf knapp eingegangen.

1089 Vgl. *Berger, A. N./Udell, G. F.* (1995), S. 353; *Blackwell, D. W./Winters, D. B.*(1997), S. 283. Diese Studien nutzen den Kontokorrentkreditzinssatz, der i. d. R. kurzfristig und variabel ist.

1090 Vgl. *Degryse, H./Van Cayseele, P.* (2000), S. 99; *Angelini, P./Di Salvo, R./Feri, G.* (1998), S. 941. Während erste den Zinssatz für Darlehen verwenden, beziehen sich letztgenannte auf den Kontokorrentkreditzinssatz.

1091 Vgl. *Tewes, M.* (2003), S. 219; *Rudolf-Sipötz, E.* (2001), S. 104; *Schemuth, J.* (1996), S. 35; *Dorsch, M. J./Carlson, L.* (1996), S. 254.

1092 Vgl. *Tewes, M.* (2003), S. 110 sowie die dort angegebene Literatur.

1093 Vgl. dazu ausführlich *Krafft, M.* (2007), S. 29 ff. sowie die dort genannte Literatur.

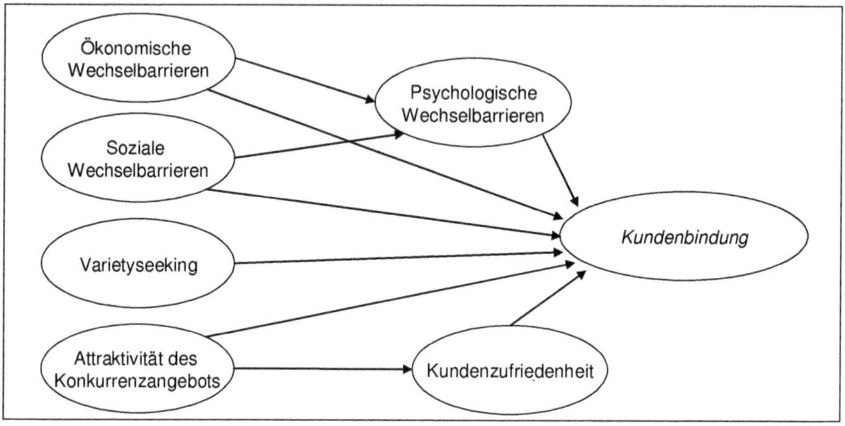

Abbildung 40: Determinanten der Kundenbindung[1094]

Als Kundenbindung oder „customer retention" werden der Aufbau und die Aufrechterhaltung einer Geschäftsbeziehung als Folge intendierter und damit nicht zufälliger Markttransaktionen zwischen einem Anbieter und einem Nachfrager bezeichnet.[1095] Intendiert bedeutet im Kontext der vorliegenden Untersuchung, dass sowohl Firmenkunde als auch Kreditgenossenschaft Gründe haben, um eine Geschäftsbeziehung aufzubauen, zu erhalten und auszubauen. Es kann in diesem Zusammenhang folglich zwischen einer anbieter- und einer nachfragerbezogenen Perspektive unterschieden werden. Bei ersterer Sichtweise hat die Kundenbindung vornehmlich instrumentellen Charakter in Form von systematischen Bindungsaktivitäten, während letztere Perspektive die Kundenbindung dagegen eher verhaltens- und ergebnisorientiert interpretiert. Anbieterbezogene Maßnahme lassen sich aus dem instrumentellen Verständnis ableiten, sind für die weitere Erläuterung dieses Kundenwerttreibers jedoch von nachgelagerter Bedeutung. Denn hinsichtlich der Wahrscheinlichkeit der Fortführung der Geschäftsbeziehung zur Bank als wesentlicher Einflussfaktor auf die Realisierbarkeit von Kundenwerten ist insbesondere die nachfragerbezogene Perspektive zu betrachten, auf die letztlich die anbieterbezogenen Maßnahmen gerichtet sind.

MEYER/OEVERMANN definieren Kundenbindung nachfragerbezogen als „das bisherige Kauf- und Weiterempfehlungsverhalten und [.] die zukünftigen Wiederkauf-, Zusatzkauf- (Cross-Selling) und Weiterempfehlungs-Absichten

---

1094 In Anlehnung an *Peter, S.* (1999), S. 126.

1095 Vgl. *Vogel, V.* (2006), S. 33; *Krüger, S. M.* (1997), S. 22; *Rieker, S.* (1995), S. 11. Vgl. in diesem Zusammenhang auch die Definition der Geschäftsbeziehung in Kapitel C.2.4.1.3

(Goodwill) eines Kunden gegenüber einem Anbieter und dessen Leistungen, die aus psychologischen, situativen, rechtlichen, ökonomischen oder technologischen Bindungsursachen resultieren."[1096]

In jüngeren Forschungsbeiträgen, die sich mit der Beziehung zwischen mittelständischen Unternehmen und Banken auseinandersetzen,[1097] wird Kundenbindung in Anlehnung an EGGERT über zwei Bindungszustände aus Kundensicht konkretisiert, den Zustand der Verbundenheit und den Zustand der Gebundenheit.[1098] Die Verbundenheit des Kunden resultiert aus der Situation eines „Nicht-Wechseln-Wollens" eines Kunden, während sich in der Gebundenheit die beschränkte bis nicht gegebene Möglichkeit des Wechseln-Könnens niederschlägt.[1099] SEGBERS bezeichnet diese Bindungszustände als ökonomische Bindung (Gebundenheit) und als psychologische Bindung (Verbundenheit).[1100] Wenngleich SEGBERS die Argumentation dyadisch aus Anbieter- und Nachfragersicht vornimmt, soll nachfolgend der Fokus auf Seiten des Kunden liegen, da dessen Entscheidung über die Fortführung der Beziehung darüber entscheidet, ob der mit ihr verbundene prospektive Kundenwert, der letztlich die ökonomische Motivation des Aufbaus oder der Fortführung einer Geschäftsbeziehung zu einem Firmenkunden aus Sicht der Kreditgenossenschaft darstellt, realisiert werden kann.[1101]

Verbundenheit fasst den affektiven und normativen Bindungszustand eines Kunden zusammen.[1102] Affektive Bindungszustände liegen vor, wenn der Kunde die Aufgabe oder den Verlust einer Geschäftsbeziehung persönlich bedauert.[1103] Derartige Bindungszustände münden darin, dass Geschäftsbeziehungen unabhängig von ihrem ökonomischen Stellenwert an Bedeutung gewinnen. Normati-

---

1096 *Meyer, A./Oevermann, D.* (1995), Sp. 1341. Siehe dazu auch die Determinanten der Kundenbindung in Abbildung 40.

1097 Vgl. *Püthe, T.* (2009), S. 162 ff.; *Segbers, K.* (2007), S. 334 ff. In beiden Arbeiten wird dabei der Begriff der Kundenbindung synonym zum Begriff Commitment verwendet. Zu berücksichtigen ist diesbezüglich, dass Commitment eine anbieter- und nachfragerseitige Bereitschaft zur Fortführung der Kundenbeziehung impliziert, während Kundenbindung eher aus Kundensicht verstanden wird.

1098 Vgl. *Eggert, A.* (1999), S. 99. Diese Unterscheidung wird z. B. auch von STANLEY/MARKMAN vorgenommen, die zwischen einer „dedication-based relationship maintenance" und einer „constraint-based relationship maintenance" in persönlichen Beziehungen unterscheiden. Vgl. *Stanley, S. M./Markman, H. J.* (1992), S. 595 ff.

1099 Vgl. *Bliemel, F./Eggert, A.* (1998), S. 39.

1100 Vgl. *Segbers, K.* (2007), S. 335. TOMCZAK/DITTRICH unterscheiden dagegen zwischen der psychologischen und der faktischen Bindung. Vgl. *Tomczak, T./Dittrich, S.* (1997).

1101 Vgl. *Tewes, M.* (2003), S. 113.

1102 Vgl. *Eggert, A.* (1999), S. 130.

1103 Vgl. hier und nachfolgend *Püthe, T.* (2009), S. 164 ff. sowie die dort genannte Literatur.

ve Bindung ist Ausdruck dessen, dass der Kunde im Einklang mit persönlich und gesellschaftlich vorgegebenen Werten lebt. PÜTHE vermutet, dass in Kunde-Bank-Beziehungen die normative Bindung primär bankseitig relevant sei, denn so würde eine Bank im Falle einer wirtschaftlichen Krise des Firmenkunden nicht nur eigene Interessen verfolgen, sondern in ihren Entscheidungen auch Rücksicht auf potenziell gefährdete Arbeitsplätze nehmen.[1104] Im Hinblick auf die Kreditgenossenschaften, deren Geschäftsbetrieb einem spezifischen und von Kreditbanken abgrenzbaren Zweck folgt, kann jedoch auch ein Einfluss auf die normative Bindung des Kunden identifiziert werden. Nämlich dann, wenn ein Firmenkunde sich stark mit den „altruistischen" Komponenten des Förderverständnisses identifiziert.[1105] Vor diesem Hintergrund kann die Mitgliedschaft eines Firmenkunden bei der Kreditgenossenschaft als beobachtbarer Ausdruck der normativen Bindung verstanden werden.[1106] Im Wirkmodell der Bankbeziehung von SEGBERS wird die Verbundenheit weiterhin beeinflusst durch die Konstrukte Vertrautheit, Vertrauen und durch positive oder negative Emotionen aufgrund der subjektiv wahrgenommenen Qualität von Bankleistungen.[1107] Die Qualität, die von SEGBERS differenziert im Sinne des Kano-Modells analysiert wird,[1108] beeinflusst die Kundenzufriedenheit, die wiederum positiv auf die

[1104] Vgl. *Püthe, T.* (2009), S. 165. Dies muss für Kreditgenossenschaften mit Blick auf die Motivation derartigen Handelns sicherlich insoweit relativiert werden, als das Regionalprinzip ebenfalls großen Einfluss auf derartige Entscheidungen haben dürfte. Denn jeder Arbeitnehmer eines ortsansässigen Unternehmens könnte auch (Kredit-)Kunde der Genossenschaftsbank sein, so dass kumulative Effekte bei der Beendigung von Geschäftsbeziehungen aufgrund finanzieller Schieflagen des Firmenkunden zu erwarten sind und somit ins Kalkül einbezogen werden müssen. Siehe zum Regionalprinzip Kapitel C.1.4.3.

[1105] Im Sinne der Förderkreisläufe aus Kapitel C.1.4.2 ist dies z. B. eine ausgeprägte Förderung im Sinne der Allgemeinförderung, die seitens der Kreditgenossenschaften oft mit regionaler Förderung verbunden wird.

[1106] Vgl. auch die ähnliche Argumentation bei *Theurl, T.* (2003), S. 13. Die Mitgliedschaft in einem Kundenbindungsprogramm, zu dem die genossenschaftliche Mitgliedschaft aufgrund ihrer praktischen Ausgestaltung zu zählen ist, gehört in der Praxis zu den allgemein akzeptierten Indikatoren von Kundenbindung. Vgl. *Rudolf-Sipötz, E.* (2001), S. 107.

[1107] Vgl. zu den Konstrukten Vertrauen und Vertrautheit ausführlich *Segbers, K.* (2007), S 331 ff. sowie die dort angegebene Literatur.

[1108] Das Kano-Modell unterscheidet Basisanforderungen, Leistungsanforderungen und Begeisterungsanforderungen. Während Basisanforderungen asymmetrisch wirken, d. h. bei Nichterfüllung Unzufriedenheit begründen, aber nicht umgekehrt wirken, liegt zwischen der Erfüllung von Leistungsanforderungen und der Kunden(un-)zufriedenheit ein linearfunktionaler Zusammenhang vor. Begeisterungsanforderungen dagegen wirken ebenfalls wieder asymmetrisch und führen bei Nichterfüllung nicht zu Unzufriedenheit stei-

Kundenbindung wirkt[1109] und als Vergleich der präkonsumtiven Erwartung hinsichtlich der Leistungen eines Unternehmens mit den tatsächlichen Erfahrungen in der theoretischen Forschung konzeptualisiert wird.[1110] Dabei wird mit Blick auf die Geschäftsbeziehungen die Kundenzufriedenheit nicht auf eine einzelne Transaktion bezogen, sondern auf die Geschäftsbeziehung als übergeordnetes Bezugsobjekt.[1111] Nach BLUT „vermag die Kundenzufriedenheit durch ihren Einfluss auf das Wiederkaufverhalten langfristig die zukünftigen Zahlungsströme eines Unternehmens sicherzustellen“[1112]. Und nach HERRMANN/JOHNSON vor allem „je intensiver der Wettbewerb in einem Sektor ist“[1113].Der Zusammenhang zwischen Kundenzufriedenheit und Kundenbindung ist dabei nicht zwingend linear, sondern eher nichtlinear und asymmetrisch.[1114]

Die Gebundenheit zwischen Bank und Firmenkunde wird als Ergebnis eines rationalen Kosten-Nutzen-Kalküls verstanden. Gebundenheit resultiert demnach daraus, dass eine Vorteilhaftigkeitsanalyse der bestehenden Geschäftsbeziehung zu dem Ergebnis führt, einen Bankwechsel nicht als nutzensteigernd aus Perspektive des Firmenkunden zu erachten.[1115] Im Rahmen des Vergleichs werden alle Nutzen- und Kostenkomponenten der Beziehung ins Kalkül gezogen. Aus Sicht des Firmenkunden sind dies neben den Kosten, die in Form von Zinsen und Gebühren anfallen, auch Kosten der Kommunikation und Koordination sowie die Kosten der Suche nach alternativen Bankbeziehungen und die Kosten eines partiellen oder vollständigen Bankwechsels. Wenngleich SEGBERS dies nicht explizit aufführt, ist die Vorteilhaftigkeitsanalyse erst vollständig, wenn alternative Kosten-Nutzen-Kalküle, d. h. die Bewertung alternativer Bankbeziehungen in die Bewertung der aktuellen Bankbeziehung involviert sind. Im Sinne

---

gern aber die Zufriedenheit überproportional bei Erfüllung. Vgl. dazu ausführlich *Kano, N.* (1984), S. 39 ff.; *Segbers, K.* (2007), S. 310 ff.

1109 Vgl. *Herrmann, A./Johnson, M. D.* (1999), S. 579 ff. sowie die dort genannten Studien.

1110 Das in der Literatur am weitesteten verbreitete Konzept zur Erklärung und Entstehung von (Un-)Zufriedenheit findet sich unter der Bezeichnung Confirmation/Disconfirmation (C/D)-Paradigma, wonach die Kundenzufriedenheit eine Funktion der kundenseitigen Erwartungen ist. Vgl. *Blut, M.* (2008), S. 23 ff. sowie die dort angegebene Literatur.

1111 Vgl. *Backhaus, K./Bauer, M.* (2003), S. 108 ff. Aus diesem Grund sprechen die Autoren auch von Beziehungszufriedenheit, wodurch der bezugslose Begriff Kundenzufriedenheit inhaltlich spezifiziert wird. Diese Argumentation führen auch STECK/KUNDISCH, die Beziehungszufriedenheit als „relational benefit“ bezeichnen. Vgl. *Steck, W./Kundisch, D.* (2005), S. 453 f.

1112 *Blut, M.* (2008), S. 23.

1113 *Herrmann, A./Johnson, M. D.* (1999), S. 580.

1114 Vgl. *Herrmann, A./Johnson, M. D.* (1999), S. 584 ff.; *Rudolf-Sipötz, E.* (2001), S. 106.

1115 Vgl. *Segbers, K.* (2007), S. 335.

der Gebundenheit wird die Beziehung als instrumentell beschrieben.[1116] Dieses auch als kognitiver Bindungszustand bezeichnete Ergebnis resultiert zum Teil aus Sachzwängen und dem Bewusstsein der Abhängigkeit. Diese Form der Bindung liegt bspw. auch dann vor, wenn der Kunde mangels Alternativen nicht wechseln kann oder vertraglich an eine Bank gebunden ist.[1117] Aus diesem Grund muss das beobachtbare Wiederkaufverhalten des Kunden nicht unbedingt auf einer positiven Einstellung diesem Anbieter gegenüber beruhen.[1118]

Es ist zu beachten, dass der Bindungszustand mit unterschiedlichen Auswirkungen auf das Verhalten und die Verhaltensabsichten eines Kunden verbunden ist und Einfluss auf die Stabilität einer Geschäftsbeziehung hat. So zeigen empirische Studien, dass im Zustand der Verbundenheit Kunden eher bereit sind, ihren Anbieter weiterzuempfehlen und die Geschäftsbeziehung zu intensivieren. Auch suchen die verbundenen Kunden weniger intensiv nach Alternativen und haben eine geringere Wechselabsicht.[1119] Folglich bedeutet dies, dass in Abhängigkeit des vorliegenden Bindungszustands differenziert zu analysieren ist, ob durch die Kundenbindung die eingangs erwähnten Vorteile tatsächlich realisiert werden können. Insbesondere ist also zu Fragen, ob die Gebundenheit freiwillig oder unfreiwillig erfolgt. Denn bei unfreiwilliger Gebundenheit wird die Wiederabschlusswahrscheinlichkeit negativ beeinflusst, nimmt die Suche nach Alternativen zu und die Bereitschaft zur Intensivierung der Geschäftsbeziehung z. B. über Cross-Selling ab. Zudem steigt die (negative) Referenzaktivität des Kunden insbesondere bei Unzufriedenheit mit der Bindungssituation.[1120]

Bezüglich des prospektiven Kundenwerts haben die jeweiligen Bindungszustände somit unterschiedliche Konsequenzen. Während die Gebundenheit zumindest für den Zeitraum der Bindung, z. B. aufgrund eines bestehenden Kreditvertrags ohne Sondertilgungsoptionen, die Kreditgenossenschaft in die Lage versetzt, vergleichsweise sicher, die daraus resultierenden Erträge zu schätzen, hängt die Wiederabschlusswahrscheinlichkeit im Falle eines erneuten Kreditfinanzierungsbedarfs wohl eher davon ab, inwieweit der Firmenkunde sich der Kreditgenossenschaft verbunden fühlt.[1121] Diese freiwillige Kundentreue konzeptualisieren HOMBURG/GIERING/HENTSCHEL über die Dimensionen bisheriges Verhalten und Verhaltensabsicht und ordnen die von MEYER/OEVERMANN[1122]

1116 Vgl. *Segbers, K.* (2007), S. 335 f.

1117 Vgl. dazu auch die Ausführungen bei *Tewes, M.* (2003), S. 112 f.

1118 Vgl. *Blut, M.* (2008), S. 19 sowie die dort angegebene Literatur.

1119 Vgl. *Eggert, A.* (2006), S. 44 sowie die dort angegebene Literatur.

1120 Vgl. *Tewes, M.* (2003), S. 113.

1121 Vgl. *Tewes, M.* (2003), S. 113.

1122 *Meyer, A./Oevermann, D.* (1995), Sp. 1341.

genannten Wirkungen der Kundenbindung diesen Dimensionen zu (siehe Abbildung 41).

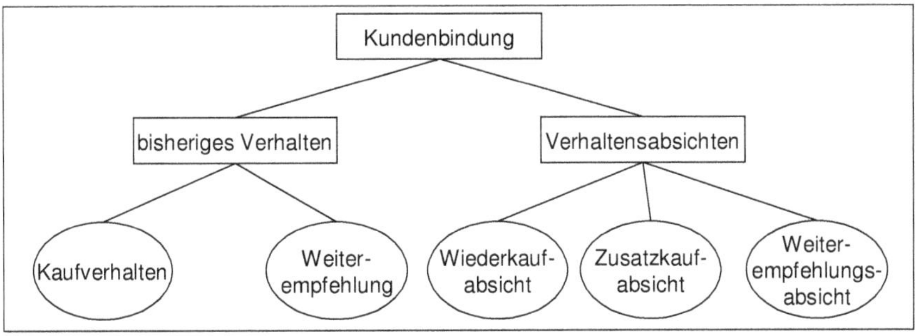

Abbildung 41: Konzeptualisierung des Konstrukts Kundenbindung[1123]

Werden vor dem Hintergrund der bisherigen Ausführungen empirische Studien aus dem Bankenkontext herangezogen, so zeigt sich ein heterogenes Bild. Den Ergebnissen von PAUL/PAUL/STEIN zufolge handelt es sich bei den Beziehungen zwischen mittelständischen Unternehmen und Hausbanken allgemein um vergleichsweise stabile Gebilde, „die durch Loyalität und Langfristigkeit gekennzeichnet“[1124] sind. Indiziert wird dieses Ergebnis aufgrund der Stichprobenstruktur hinsichtlich der Dauer der Geschäftsbeziehung, wonach 47 Prozent der befragten Unternehmen seit über 25 Jahren mit der Hausbank eine Geschäftsbeziehung unterhalten, 59 Prozent der Unternehmen nicht mehr als zwei Bankbeziehungen haben und 80 Prozent der Befragten künftig keine weiteren Beziehungen zu zusätzlichen Kreditinstituten aufbauen wollen. Die aktuelle Studie von TNS INFRATEST weist für die Volksbanken Raiffeisenbanken 35 Prozent stark gebundene und 23 Prozent gebundene Kunden auf.[1125] Zu anderen Ergebnissen kommt

1123 *Homburg, C./Giering, A./Hentschel, F.* (1999), S. 89. Die Autoren sprechen von Kundenbindung, die in dieser Arbeit inhaltlich mit der Kundentreue bzw. der Verbundenheit gleichgesetzt wird. TEWES dagegen spricht von Kundenloyalität. Vgl. *Tewes, M.* (2003), S. 113 ff. Da in dieser Arbeit der Begriff der Loyalität in einem anderen Kontext verwendet wird, soll der Begriff der Kundentreue verwendet werden.

1124 *Paul, M./Paul, S./Stein, S.* (2003), S. 5. Im Rahmen dieser Befragung wurden 490 mittelständische Unternehmen zu Aspekten der Geschäftsbeziehung zu ihrer Hausbank befragt.

1125 Vgl. *tns infratest* (2010), S. 58.

die Unternehmensberatung ZEB/.[1126] Danach wählen Unternehmen im Schnitt 3 bis 5 Bankverbindungen, was von 91,4 Prozent der Teilnehmer damit begründet wird, die eigene Unabhängigkeit zu erhöhen. Weiterhin geben etwa ein Drittel an, mehrere Hausbanken zu haben. Und gut jedes zweite Unternehmen gibt an, die Hausbankverbindung zu wechseln oder schon mal gewechselt zu haben. Und auch STECK/KUNDISCH kommen auf Basis einer Studie aus dem Jahr 2001 zu dem Ergebnis, dass die Kundenbindung bei deutschen Banken eher schwach ist.[1127] Wobei jedoch auch bei dieser Studie auffällt, dass die Kreditgenossenschaften über 50 Prozent „verbundener" bis „verwurzelter" Kundenverbindungen haben. Gleichwohl darf auf Seiten der Kreditgenossenschaften nicht davon ausgegangen werden, diese hohen Kundenbindungsraten seien zeitlich invariant. So weisen etwa die Ergebnisse der vorhergehenden Studie der TNS INFRATEST 39 Prozent stark gebundene und 29 Prozent gebundene Kunden auf,[1128] so dass im Zeitvergleich eine tendenziell sinkende Kundenbindung zu konstatieren ist. Auch die Gründe der relativ langen Kundenbeziehungen, die in habitualisierten Strukturen, gegenwärtig hohen Wechselkosten und potenziellen Außenwirkungen eines Bankwechsel begründet sind, zeigen, dass eher Gebundenheit, denn Verbundenheit als Bindungszustand anzunehmen ist.[1129] Dass dies den Banken bewusst ist, darauf weisen die Ergebnisse von ENGSTLER/PRAEG/VOCKE hin, wonach 64,4 Prozent der befragten Kreditgenossenschaften gegenwärtig Schwerpunkte der vertrieblichen Investitionen in Kundenbindungsprogramme planen.[1130]

Vor dem Hintergrund der getätigten Ausführungen soll das Bindungspotenzial als die kundenseitige Bereitschaft zur Fortführung einer Geschäftsbeziehung mit einem Anbieter verstanden werden. Dabei kann die Bereitschaft sowohl in der ökonomischen Bindung (Gebundenheit) als auch in der psychologischen Bindung (Verbundenheit) begründet sein.

Obwohl inhaltlich ähnlich bis identisch zum hier erläuterten Bindungspotenzial wird in bisher entwickelten Kundenwertmodellen vornehmlich vom Loyalitätspotenzial gesprochen. So definiert RUDOLF-SIPÖTZ das Loyalitätspotenzial in ihrem Kundenwertmodell als Affinität eines Kunden zur Kontinuität der Beziehung mit einem bestimmten Anbieter.[1131] Nach WÜNSCHMANN ist das Loyali-

1126 Vgl. hier und nachfolgend *Jansen, S.* (2009), S. 14, *Käser, B.* (2009), S. 11 und *Käser, B./Ritzenhofen, A.* (2009), S. 15. Grundlage ist eine Stichprobe von 674 in Deutschland tätigen Unternehmen.

1127 Vgl. *Steck, W./Kundisch, D.* (2005), S. 452.

1128 Vgl. *tns infratest* (2008), S. 48.

1129 Vgl. *KPMG* (2009), S. 9 ff. sowie die dort genannte Literatur.

1130 Vgl. *Engstler, M./Praeg, P./Vocke, C.* (2007), S. 17.

1131 Vgl. *Rudolf-Sipötz, E.* (2001), S. 104.

tätspotenzial eine zeitbezogene Größe, die die Konsumlebensdauer eines Kunden reflektiert[1132] und bei LISSAUTZKI ist die Bestandswahrscheinlichkeit formalanalytisch der Grad der Kundenbindung, den er als Gegenwahrscheinlichkeit einer Kundenabwanderung definiert.[1133] In dieser Arbeit wird zwischen Kundenbindung und Kundenloyalität unterschieden. Während Kundenbindung als Bereitschaft zur Fortführung verstanden wird, ist die Kundenloyalität ein Maß für die Intensität der Kundenbeziehung. So kann bspw. ein gebundener Kunde über Jahre einen geringen Anteil seines Zahlungsverkehrs mit einer Bank abwickeln. Die Loyalität beziffert in diesem Kontext den Anteil des Zahlungsverkehrs. Die Loyalität kann also produktspezifisch betrachtet werden.[1134] Diese Differenzierung kann anhand empirischer Ergebnisse für die Kreditgenossenschaft begründet werden. Obwohl in einer repräsentativen Studie die Kreditgenossenschaften sehr hohe Kundenbindungswerte erreichen, erzielen sie nur einen Anteil von 30 Prozent beim Zahlungsverkehrsvolumen und nur knapp ein Viertel des Volumens an Geld- und Kapitalanlagen von Firmenkunden.[1135]

4.5.2.2 Akquisitionspotenzial

Für die Bewertung von potenziellen Kunden ist neben dem Bindungspotenzial relevant, mit welcher Wahrscheinlichkeit diese akquiriert oder wiedergewonnen werden können. Dieser Aspekt ist Gegenstand des Akquisitionspotenzials.

Bei der inhaltlichen Bestimmung des Akquisitionspotenzials ist zunächst ein differenzierter Blick auf die relevanten Kunden zu richten. Wie in der Abgrenzung des Kundenbegriffs in Kapitel D.1.2.2 erläutert wurde, ergibt sich das numerische Kundenpotenzial einer Kreditgenossenschaft aus den aktuellen und den potenziellen Kunden, die dem Segment der Firmenkunden zuzuordnen sind. In der Gruppe der potenziellen Kunden können Altkunden ebenso enthalten sein wie Kunden, mit denen die Kreditgenossenschaft bisher keine geschäftliche Beziehung hatte. Unter Berücksichtigung dieser Tatsache lässt sich das Akquisitionspotenzial differenziert betrachten für die Akquisition von Neukunden und die Rückgewinnung von Altkunden.

1132 Vgl. *Wünschmann, S.* (2007), S. 39.

1133 Vgl. *Lissautzki, M.* (2007), S. 133.

1134 Ähnlich differenziert LISSAUTZKI die Begriffe Kundenbindung und Kundenloyalität. Vgl. *Lissautzki, M.* (2007), S. 145.

1135 Vgl. *AGVR et al.* (2008), S. 14 ff. Gemessen wird die Kundenbindung mittels des sog. TRI* M-Index. Dieser von der TNS INFRATEST ermittelte Index bildet in einer Kennzahl die Zufriedenheit, die Weiterempfehlungs- sowie die Fortführungsabsicht und den unter Berücksichtigung von Alternativen subjektiv wahrgenommenen Nettonutzen ab. Nach Aussage dieses Indexes weisen die Genossenschaftsbanken mit 68 Prozent den höchsten Anteil „gebundener" Kunden auf.

Die Akquisition von Neukunden ist insbesondere für die Kreditgenossenschaften vor dem Hintergrund des bisher traditionell etablierten Bringgeschäfts der Kunden eine vergleichsweise neue Herausforderung.[1136] Dass die Akquisitionsbemühungen auch im Vergleich zu den Kundenbindungsbemühungen vergleichsweise stiefmütterlich durchgeführt werden, resultierte nicht zuletzt aus der oftmals geäußerten Einschätzung, es sei relativer günstiger bestehende Geschäftsbeziehungen auszuschöpfen als neue Kunden zu akquirieren.[1137] Gleichwohl darf aus Sicht der Kreditgenossenschaften nicht vergessen werden, dass die Akquisition wertvoller Firmenkunden zentraler Bestandteil der langfristigen Existenzsicherung des Genossenschaftsbetriebs ist. Denn es kann begründet angenommen werden, dass im Zeitverlauf bei der Ausschöpfung von Kundenwertpotenzialen „natürliche“ Grenzen erreicht werden. Sofern diese Ausschöpfung nicht effizient im Sinne der geforderten Rentabilität aus Mitgliedersicht und nachhaltig, also andauernd ist, müssen neue Kundenwertquellen identifiziert werden. Dass zudem die Kundenbeziehungsdauern nicht unendlich sind, liegt im Bereich der statistischen Erwartung. So ist davon auszugehen, dass Unternehmen ihren Lebenszyklussen entsprechend in Degenerationsphasen eintreten und so ihre Existenz verlieren. Auch kann es sein, dass sich Firmenkunden aufgrund dynamischen Wachstums aus der Positionierung und Kompetenz der Kreditgenossenschaften heraus entwickeln und Geschäftsbeziehungen zu den vergleichsweise größeren Kreditbanken aufnehmen.[1138] Es liegen bisher noch keine fundierten Erkenntnisse über erfolgreich Akquisitionsprozesse im Firmenkundengeschäft der Banken vor. Es dürfte jedoch allgemeingültig gelten, dass die Einschätzung über die Erfolgswirksamkeit einer Akquisitionsbemühung in umgekehrter Richtung davon abhängt, wie hoch das Bindungspotenzial eines anvisierten Akquisekunden gegenüber seinen bisherigen wettbewerblichen Bankverbindungen ist.

1136 So wurden im Jahr 2010 von 3.105 mittelständischen Unternehmen mit einer Beschäftigtenzahl von 2-499 Mitarbeitern 20 Prozent von einem Mitarbeiter eines Geldinstituts, mit dem der Kunde zu der Zeit keine geschäftliche Zusammenarbeit hatte, angesprochen. Davon waren in nur 10 Prozent der Fälle die Mitarbeiter bei einer VR-Bank beschäftigt. Vgl. *tns infratest* (2010), S. 87.

1137 Vgl. *Sheth,J. N./Partvatiyar, A.* (1995), S. 265; *Vogel, V.* (2006), S. 1 sowie die dort genannte Literatur. „Für das Firmenkundengeschäft der Banken fehlen genaue Untersuchungen.“ *Renker, C.* (2005), S. 249.

1138 Dies indizieren auch die Ergebnisse der TNS INFRATEST-Untersuchung zum Firmenkundengeschäft in Deutschland aus dem Jahr 2008. Demnach gibt es einen negativen Zusammenhang zwischen der Größe eines Firmenkunden gemessen am Umsatz und dem Marktanteil einer Kreditgenossenschaft im jeweiligen Segment, Vgl. *tns infratest* (2008), S. 6 und S. 22.

Neben der Akquisition von potenziellen Kunden sind auch Maßnahmen des sog. „Corporate Customer Recovery Management“, d. h. der Kundenrückgewinnung,[1139] zu berücksichtigen. Für die Rückgewinnung ist vor allem von Relevanz, aus welchen Gründen Firmenkunden die Kreditgenossenschaft vollständig oder partiell bspw. in Form der Nachfrage nach Produkten, die bisher bei der Kreditgenossenschaft kontrahiert wurden, und nun bei anderen Anbietern bezogen werden, wechseln. RENKER teilt die Abwanderungsursachen in drei Gruppen ein: bankinduzierte, firmenkundeninduzierte und bankbetreuerinduzierte Ursachen. An einer Studie aus dem Sparkassensektor zeigt er auf, dass vor allem Firmenschließungen, das aktivere Akquisitionsverhalten von Wettbewerbern und ein gestörtes Verhältnis zum Firmenkundenbetreuer als relativ wichtige Abwanderungsgründe genannt werden.[1140] Die Ergebnisse von TNS INFRATEST zeigen zudem, dass vor allem die wahrgenommene Unflexibilität und Schwerfälligkeit der Kreditgenossenschaften einen Wechsel durch den Firmenkunden begründen.[1141] Dass die Rückgewinnung von Altkunden zwar relativ teuerer als die Bindung bestehender Kunden und relativ günstiger als die Akquisition von Nichtkunden, aber dennoch erfolgversprechend ist, wird damit begründet, dass diese profitabler und treuer sein können als Kunden mit kontinuierlicher Geschäftsbeziehung (siehe Abbildung 42).[1142] STAUSS/FRIEGE zeigen in einer Fallstudie, dass der Return on Investment eines Neukunden 23 Prozent im Vergleich zu 214 Prozent bei einem zurückgewonnenen Altkunden beträgt.[1143] Die vergleichsweise höhere Kundentreue und -profitabilität begründet RENKER damit, dass zurück gewonnene Firmenkunden zur Rechtfertigung der revidierten Entscheidung eine Bestätigung benötigen, die sich in intensiveren Geschäftsvolumina konkretisiert.[1144]

1139 Vgl. hier und nachfolgend *Renker, C.* (2005), S. 244 ff.

1140 Zudem argumentiert er, die Notwendigkeit und die Chancen eines erfolgreichen Rückgewinnungsprozesses würden von vielen Firmenkundenbetreuern nicht erkannt. So gaben von 620 befragten Kreditinstituten im Jahr 2003 nur 2 von 61 der antwortenden Banken an, die Rückgewinnung von abgebrochenen Geschäftsbeziehungen systematisch anzustreben. Vgl. *Renker, C.* (2005), S. 247.

1141 Vgl. *tns infratest* (2010), S. 85.

1142 Vgl. *Renker, C.* (2005), S. 248 sowie die dort genannte Literatur; *Sauerbrey, C./Henning, R.* (2000), S. 17. SAUERBREY/HENNING gehen in ihren Untersuchungen davon aus, dass die Kosten der Rückgewinnung etwa halb so hoch sind wie die Kosten der Kundenakquisition und dreimal so hoch wie die Kundenbindungskosten. Für das Firmenkundengeschäft fehlen genaue Untersuchungen dazu jedoch noch.

1143 Vgl. *Stauss, B./Friege, C.* (1999), S. 359. Es handelt sich hierbei jedoch um eine zum Anwendungskontext dieser Arbeit branchenfremde Fallstudie.

1144 Vgl. *Renker, C.* (2005), S. 248.

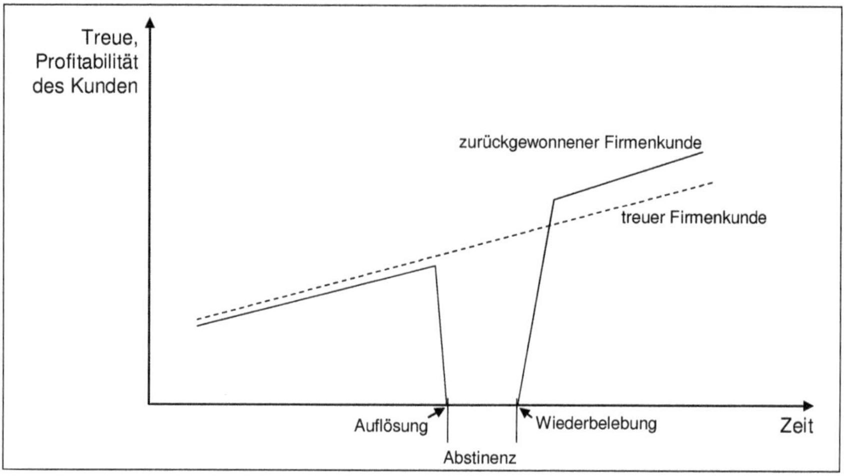

Abbildung 42: Profitabilität und Treue zurückgewonnener und treuer Firmenkunden[1145]

Werden die bisherigen Überlegungen zur Bindungsdimension zusammengefügt, wird deutlich, dass für aktuelle Kunden das Bindungspotenzial und für potenzielle Kunden die gleichzeitige Betrachtung des Akquisitions- und des Bindungspotenzials relevant ist. Die bisher betrachteten Kundenwerttreiber sind noch nicht ausreichend, um die Kundenwertermittlung zu ermöglichen. Da sie noch keine Aussage über die tatsächlichen Zielbeiträge treffen, sind entsprechende Kundenwerttreiber der Ertragsdimension zu betrachten.

4.5.3 Kundenwerttreiber der Ertragsdimension

Während die Beziehungsdimension, die im vorhergehenden Kapitel festgelegt und diskutiert wurde, als Kundenwerttreiber umstritten ist, hat die Ertragsdimension in allen Kundenwertmodellen eine herausragende Bedeutung.[1146] Basierend auf dem Gewicht der materiellen Ziele der Kreditgenossenschaften[1147] ist

1145 In Anlehnung an *Renker, C.* (2005), S. 248.

1146 Dabei ist zu berücksichtigen, dass die Spezifizierung der Ertragsdimension unterschiedlich ausfällt. Während in einer Vielzahl von Beiträgen die strenge und theoretisch postulierte liquiditätsorientierte Sicht vertreten wird, gibt es ebenso viele Beiträge, die aus pragmatischen Gründen auf Größen des internen Rechnungswesens wie z. B. den Deckungsbeitrag abstellen. Unterschiedlich ist in diesen Arbeiten jedoch die Definition des Deckungsbeitrags, da dieser unterschiedliche Kostengrößen umfasst.

1147 Vgl. die Ausführungen in Kapitel D.3.1.

diese Dimension auch für das Firmenkundenwertmodell der Kreditgenossenschaften zwingend zu berücksichtigen.

Aus Steuerungsperspektive ist es zielführend, die Ertragsdimension in einzelne Kundenwerttreiber, die verschiedene Maßnahmen zur Kundenwertrealisierung fundieren und Analysen in Bezug auf relevante Potenziale unterstützen, zu differenzieren. Vor dem Hintergrund des jedoch realiter vorliegenden Zusammenwirkens dieser Komponenten lehnt sich die folgende Betrachtung in dieser Arbeit an die Aussage von KAMINSKY an: „Vom analytischen Standpunkt [...] aus erlaubt das Zusammenwirken [...] keinen Verzicht auf die getrennte Betrachtung der Sphären."[1148] Daher wird zum einen zwischen bestehenden und zusätzlichen Produktbeziehungen unterschieden. Zum zweiten wird eine differenzierte Betrachtung von positiven (Erträge) und negativen Wertbeiträgen (Kosten) vorgenommen. Wie sich aus der Übersicht ausgewählter Kundenwertmodelle in Tabelle 12 zeigt, betrachten die meisten Autoren nicht nur diejenigen ökonomischen Erfolgsgrößen, die unmittelbar und direkt aus der Kundenbeziehung resultieren. Vielmehr werden auch Faktoren betrachtet, die einem Kunden zugerechnet werden können, sich jedoch nicht auf die betrachtete Kundenbeziehung auswirken. Mit Blick auf die besondere Bedeutung der Geschäftsbeziehung sollen die Kundenwerttreiber der Ertragsdimension in beziehungsbezogene und beziehungsübergreifende Größen unterteilt werden. Während erstere ihren erfolgsrelevanten Beitrag in der dyadischen Beziehung zwischen Firmenkunde und Kreditgenossenschaft leisten, berücksichtigen letztere auch das Zusammenspiel mit Dritten, insbesondere auf der Kundenseite.

4.5.3.1 Beziehungsbezogene Kundenwerttreiber der Ertragsdimension

4.5.3.1.1 Basisgeschäftspotenzial

Unter dem Basisgeschäftspotenzial soll im Folgenden das aus Sicht der Kreditgenossenschaft mit einem Firmenkunden über einen zu bestimmenden in der Zukunft liegenden Zeitraum kontrahierbare Leistungsvolumen in zum Bewertungszeitpunkt gegebenen Produktbeziehungen verstanden werden. Dieser Kundenwerttreiber umfasst die künftigen Erlöse bzw. Erträge, die aus einer etablierten Kundenbeziehung zu erwarten sind. Der Ertragsbegriff wird durch die Provi-

1148 *Kaminsky, S.* (1955), S. 30. Dabei bezieht sich KAMINSKY in seiner Argumentation ursprünglich auf die differenzierte Betrachtung des Werte- und des Betriebsbereichs in der Kosten- und Erfolgsrechnung der Kreditinstitute, obwohl diese nur gemeinsam eine mit einem Preis dotierte Marktleistung erzeugen. Siehe dazu auch die Ausführungen in Kapitel C.4.3.

sionsbeiträge und Konditionsbeiträge konkretisiert.[1149] DILLER bezeichnet diesen Bestimmungsfaktor in seinem Kundenwertmodell als Penetrationswert[1150] und WEBER/LISSAUTZKI als Loyalitätswert.[1151] Das Basisgeschäftspotenzial stellt den gedanklichen Sockelwert des ökonomischen Kundenwerts dar.[1152] BAUER/HAMMERSCHMIDT und BAUER/HAMMERSCHMIDT/BRÄHLER subsumieren unter dem Basisgeschäftspotenzial den autonomen Umsatz, welcher sich durch Kundenumsätze aufgrund von Standardmarketingmaßnahmen ergibt.[1153] ANDON/BAXTER/ BRADLEY definieren noch enger: „Base Profits are profits that will be earned from customers purchases, unaffected by factors such as time, loyalty and efficiency, over the period of a customer relationship.“[1154] Dieser Kundenwerttreiber steht in direkter Beziehung zu den Wachstums- und Profitabilitätszielen der Kreditgenossenschaft.

Die produktbezogene Abgrenzung des Basisgeschäftspotenzials resultiert aus der notwendigen Trennung von den später noch zu erläuternden Cross-Selling- und Up-Selling-Potenzialen. Während letztgenannte Trennung regelmäßig auch in der Literatur vorgenommen wird,[1155] setzen sich nur wenige Beiträge damit auseinander, wie der Umfang der durch diesen Kundenwerttreiber berücksichtigten Produkte bzw. Bankleistungen abgegrenzt werden soll. Daher soll auf diesen Punkt kurz eingegangen werden.

1149 Während im industriellen Kontext eine explizite Differenzierung zwischen Erlösen (ohne Berücksichtigung von Kosten) und Erträgen (mit Berücksichtigung von Kosten) sinnvoll ist, soll nachfolgend bei zinstragenden Geschäften ausschließlich von Erträgen gesprochen werden. Dies begründet sich damit, dass der Begriff des Ertrags in der bankbetrieblichen Betriebswirtschaftslehre weiter verbreitet ist. Er impliziert zudem, dass aufbauend auf der Marktzinsmethode Finanzierungs- bzw. Opportunitätskosten bereits in Abzug gebracht sind. Da diese i. d. R. nicht vom Kunden beeinflusst werden können, sondern durch das bankinterne Treasury, und auch dort i. d. R. verantwortet werden, sind die Auswirkungen auf die nachfolgenden Ausführungen gering bis nicht vorhanden. Analog wird der Saldo aus Provisionserträgen und -aufwendungen als Provisionsbeitrag bezeichnet und verwendet.

1150 Vgl. *Diller, H.* (2001), S. 10.

1151 Vgl. *Weber, J./Lissautzki, M.* (2005), S. 313.

1152 Vgl. *Stahl, H. K./Matzler, K./Hinterhuber, H. H.* (2006), S. 428 f. Die Autoren bezeichnen dies auch als Sockelgeschäft der Kundenbeziehung und gehen von einer bestimmten Bedarfsstruktur des Kunden über die gesamte Beziehungsdauer aus. Ähnlich auch *Hippner, H.* (2004), S. 25, der diesen Teilwert als Basisvolumen bezeichnet.

1153 Vgl. *Bauer, H.-H./Hammerschmidt, M.* (2005), S. 335; *Bauer, H.-H./Hammerschmidt, M./Brähler, M.* (2002), S. 328 ff.; so auch *Schneider, N. C.* (2007), S. 67 und *Homburg, C.* (2007), S. 409.

1154 *Andon, P./Baxter, J./Bradley, G.* (2003), S. 303.

1155 Vgl. z. B. *Rudolf-Sipötz, E.* (2001), S. 95 ff.; *Cornelsen, J.* (2000), S. 169 ff.

Ausgangspunkt der folgenden Überlegungen ist die Feststellung, dass der Absatz einer bestimmten Bankleistung in der Vergangenheit der Kundenbeziehung stattgefunden hat und nun festzulegen ist, ob dieses Produkt bzw. diese Produktbeziehung zugrunde gelegt werden soll, um das Basisgeschäftspotenzial zu bestimmen. Grundsätzlich lässt sich der betrachtete Zeitraum als Abgrenzungskriterium verwenden, womit drei prinzipielle Arten der Produktabgrenzung verbunden sind (siehe Abbildung 43). Diese Möglichkeiten sind:

- Möglichkeit 1: Die Betrachtung des Basisgeschäftspotenzials bezieht sich auf alle Produkte, die in einem bestimmten Zeitraum, z. B. der letzten (Abrechnungs-)Periode, mit dem Kunden kontrahiert wurden (Produkte B, C und D werden berücksichtigt).
- Möglichkeit 2: Der relevante Zeitraum wird über die gesamte bisherige Kundenbeziehungsdauer definiert, so dass alle mindestens einmal seit Beginn der Kundenbeziehung kontrahierten Produkte betrachtet werden (alle Produkte werden berücksichtigt).
- Möglichkeit 3: Es werden der Zeitraum der bisherigen Kundenbeziehung betrachtet und nur die Produkte berücksichtigt, für die aufbauend auf dem letzten vorliegenden Produktabschluss und einem typisierten oder bedarfsspezifischen Wiederkaufszyklus[1156] ein Wiederkauf für den prospektiven Bewertungszeitraum zu erwarten ist (Produkte A und D werden berücksichtigt. Produkt B scheidet aus, da ein Wiederkauf außerhalb des Betrachtungszeitraumes liegt. Zudem scheidet C aus, da das Produkt beim letzten erwarteten Kauf nicht beim Anbieterunternehmen bezogen wurde).

1156 Wiederkaufzyklen ergeben sich aus der Annahme eines wiederkehrenden produktspezifischen Bedarfs eines Kunden. Sie sind nicht gleichzusetzen mit der kundenseitigen Entscheidung den Bedarf beim betrachteten Anbieterunternehmen zu decken.

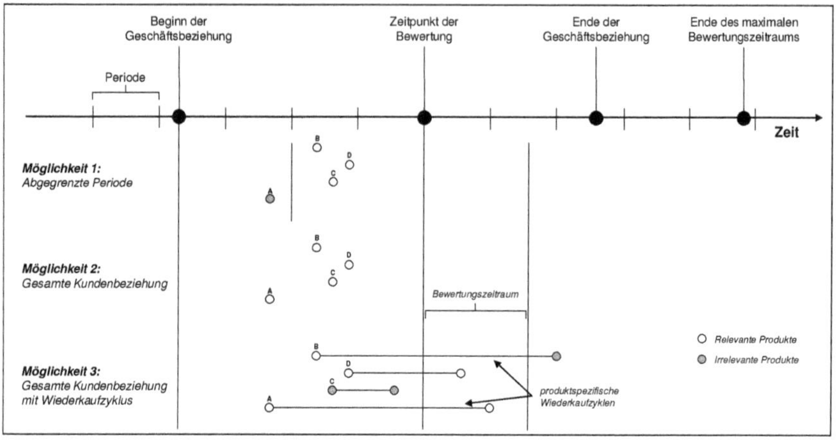

Abbildung 43: Möglichkeiten zur Produktabgrenzung für das Basisgeschäftspotenzial[1157]

Wird realistischerweise eine kunden- oder anbieterseitige Dynamik im Zeitablauf unterstellt (z. B. weil das Nachfrageverhalten des Firmenkunden sich verändert oder die Kreditgenossenschaft im Rahmen von Sortimentsentscheidungen Leistungen künftig nicht mehr anbietet), führt dies zwingend zum Ausschluss der Möglichkeiten 1 und 2, welche in der Literatur oft implizit angenommen werden. Gelingt es für die in der Vergangenheit kontrahierten Produkte Wiederkaufzyklen zu bestimmen, so sollte prinzipiell die Möglichkeit 3 angewendet werden. Problematisch ist dabei, dass kundenspezifisch bestimmte Bankprodukte kontinuierlich (z. B. die Dienstleistungen des Zahlungsverkehrs) und andere Produkte diskontinuierlich (z. B. Investitionskredite) nachgefragt werden.[1158] Typische Wiederkaufzyklen wie sie bei den meisten Gebrauchsgütern wie z. B. bei Automobilen vorliegen, gibt es für Bankleistungen nicht.[1159] In diesem Fällen sollte auf bedarfsspezifische Wiederkaufzyklen rekurriert werden, die sich aus den bestehenden Verträgen und der künftigen, erwarteten Bedarfssituation eines Kunden ergeben. Implizit bedeutet dies ein wertloses Basisgeschäftspotenzial, wenn keine Wiederkaufrelevanz in der relevanten künftigen Periode vorliegt.[1160]

1157 Eigene Darstellung. Die jeweils grau kolorierten Produkte werden nicht berücksichtigt.

1158 Vgl. *Lippmann, I.* (2006), S. 397; *Eichelmann, T./Bufka, J./Lehmann, M.* (2001), S. 779.

1159 So hängt der künftige Bedarf eines Kredits nicht nur von Fälligkeit eines bestehenden Kredits ab, sondern auch davon, welches Auszahlungsvolumen das Unternehmen in der Zukunft plant und welche Finanzierungsmöglichkeiten vorliegen. Bei ausreichender Innenfinanzierungskraft ist die Fremdfinanzierung über Kredite gegebenenfalls obsolet.

1160 Vgl. *Diller, H.* (2001), S. 11.

Korrigiert wird das so abgegrenzte Produktprogramm durch zwei Faktoren. Dies ist erstens die geschäftspolitische Ausrichtung der Bank, mithin also die im Planungs- oder Bewertungszeitraum liegenden bzw. hier umzusetzenden Sortimentsentscheidungen. Ggf. ist zu berücksichtigen, dass Bankleistungen nicht mehr angeboten werden oder dass Leistungsvarianten mit gleichem Grundnutzen angeboten werden sollen. Liegt letzterer Fall vor, sollten die entsprechenden Leistungsvarianten betrachtet werden. Zweitens ist der erwartete Bedarf des Kunden bezüglich dieser Produkte zu berücksichtigen. Spricht mehr gegen den konkreten Produktbedarf als dafür, so sind diese Produkte auszuschließen. Dies betrifft z. B. Bankleistungen für das Auslandsgeschäft, wenn das Unternehmen seine Auslandsaktivitäten bekanntermaßen einstellt. Diese Aspekte sind in der Prognose des Basisgeschäftspotenzials zu berücksichtigen.

Zentrale Einflussfaktoren auf das Basisgeschäftspotenzial sind die Entwicklung des Kunden über den Planungs- oder Bewertungszeitraum, das daraus abgeleitete Bedarfspotenzial[1161] sowie der kundenseitige Loyalitätsgrad. Das Bedarfspotenzial des Firmenkunden nach Finanzdienstleistungsprodukten und Bankleistungen kann im Sinne des industriellen Marketings als derivativer Bedarf oder abgeleitete Nachfrage verstanden werden.[1162] In lexikalischer Definition ist die abgeleitete Nachfrage dadurch charakterisiert, dass sie die Nachfrage nach Produktionsfaktoren ist, die sich aus der Nachfrage nach Konsumgütern ergibt.[1163] Der Definitionsaspekt der Konsumgüter ist zu relativieren, da sonst nur Firmenkunden betrachtet werden können, die auf Konsumgütermärkten tätig sind. Der Charakter der abgeleiteten Nachfrage resultiert vielmehr daher, dass in diskreter Form die Nachfrage nach Produkten des Firmenkunden dessen Existenz und damit auch den Bedarf nach Bankleistungen begründet. In stetiger Form hat die Entwicklung der Absatz- und Beschaffungsmärkte des Kunden Einfluss darauf, in welchem Umfang er etwa Zahlungsverkehrsdienstleistungen oder alternative Produkte in Anspruch nimmt. Das Ausmaß des Einflusses der Marktentwicklung auf die Unternehmensentwicklung bzw. die Stärke und Richtung der Korrelation zwischen diesen Variablen wird moderiert durch die Managementkompetenz. Die Managementkompetenz drückt die Fähig- und Fertigkeiten der Geschäftsleitung eines Unternehmens aus, die Marktentwicklung im Sinne der Unternehmenszielsetzung zu nutzen. Eine weitere moderierende Variable im Kontext der Ableitung des konkreten Bedarfspotenzials ist die Qualität

1161 In Anlehnung an die im Marketing etablierten Begriffsverwendungen wird an dieser Stelle zwischen dem kundenspezifischen Bedarfspotenzial und dem anbieterspezifischem Anteil am Bedarfspotenzial, dem Bedarfsvolumen unterschieden.

1162 Vgl. *Barth, K./Wille, K.* (2000), S. 10.

1163 Vgl. *Gabler Wirtschaftslexikon* (2000), Band A- Be, S. 10.

des Informationsprozesses, mittels dessen die Informationen über die Markt- und Unternehmensentwicklung an die Kreditgenossenschaft fließen. Liegen hinreichende Informationen über die bisher genannten Variablen vor, bedarf es eines geeigneten Transformationsmodells, dass die konkrete Unternehmensentwicklung in das Bedarfspotenzial nach Bankdienstleistungen übersetzt.[1164]

Der kreditgenossenschaftliche Anteil des Bedarfspotenzials eines Kunden wird als Bedarfsvolumen bezeichnet. Dieses ergibt sich aus der funktionalen Verknüpfung des Bedarfspotenzials mit dem Loyalitätsgrad. Der Loyalitätsgrad ist in Anlehnung an WEBER und LISSAUTZKI als absatzmengenbezogener Anteil definiert, der mathematisch Werte zwischen 0 und 1 (100%) annehmen kann.[1165] Um das Basisgeschäftspotenzial zu bestimmen, wird der Loyalitätsgrad in Abgrenzung zum Entwicklungspotenzial als kundenseitig gegeben (konstant oder dynamisch veränderlich) für den Planungs- oder Bewertungszeitraum angenommen.

Werden diese Überlegungen auf das industrielle Geschäft übertragen, determinieren Bedarfsvolumen und Bedarfsstruktur das „Mengengerüst" des zu bestimmenden Basisgeschäftspotenzials. Dieses ist mit einem geeigneten Bewertungsfaktor (Preis oder Zinssatz) zu gewichten, um das „Wertgerüst" zu erhalten. In diesem Zusammenhang kann unterschieden werden, ob der Bewertungsfaktor kundenbezogen ausgestaltet sein soll. Im Falle eines kundenbezogenen Bewertungsfaktors lassen sich aus der Historie der Geschäftsbeziehung approximativ kundenindividuelle Zahlungsbereitschaften im Vergleich zur Zahlungsbereitschaft des Gesamtkundenbestands anhand des kundendurchschnittlichen Zinssatz- und/oder Konditionsniveaus ermitteln,[1166] so dass die geschäftspolitisch motivierten Standardkonditionen kundenindividuell relativiert und die Bewertung des „Mengengerüsts" vorgenommen werden kann. Alternativ können auch die festgelegten Standardkonditionen verwendet werden, um das Basisgeschäftspotenzial zu bewerten. Vor dem Hintergrund der expliziten Berücksichtigung einer noch nicht vollständig abgeschöpften Zahlungsbereitschaft eines

---

1164 Ein beispielhaftes Modell wird in Kapitel E.1.2.4 vorgestellt.

1165 Vgl. *Lissautzki, M.* (2007), S. 145 ff.; *Weber, J./Lissautzki, M.* (2005), S. 313 ff. Damit grenzen die Autoren die Kundenbindung von der Loyalität ab. Während die Kundenbindung eine Aussage über das Ob trifft, bezieht sich die Loyalitätsrate auf die Intensität im Fall eines Ob. Vgl. für ein plakatives Beispiel *Lissautzki, M.* (2007), S. 145 ff.

1166 Dabei ist zu berücksichtigen, dass sowohl Produktspezifika wie etwa Laufzeiten, Volumina wie auch Kundenspezifika wie unterschiedliche Ausfallrisikoklassen und Besicherungen Einfluss auf die Konditionsgestaltung haben. Daher ist die Auswahl einer geeigneten Vergleichsgruppe aus dem Gesamtkundenbestand dahingehend auszuwählen, dass eine hinreichend aussagekräftige Vergleichbarkeit gegeben ist.

Firmenkunden, sollte ein kundenbezogener Bewertungsfaktor auf Basis bisheriger Transaktionen genutzt werden.

4.5.3.1.2 Entwicklungspotenzial

Das Entwicklungspotenzial fasst zwei Möglichkeiten zusammen, um über das Basisgeschäftspotenzial hinaus Erträge in bestehenden Kundenbeziehungen zu realisieren. Es umfasst zum einen das Up-Selling-Potenzial und zum anderen das Preispotenzial. Wie auch das Basisgeschäftspotenzial bezieht sich das Entwicklungspotenzial auf die bestehenden Produktbeziehungen, die somit auch kongruent zum Basisgeschäftspotenzial abzugrenzen sind.

Das Up-Selling-Potenzial basiert auf dem Marketing- und Vertriebskonzept des Up-Selling, wobei der Begriff Up-Selling in der betriebswirtschaftlichen Forschung nicht eindeutig definiert ist. Aus einer Gegenüberstellung ausgewählter Begriffsdefinitionen leitet POHLKAMP drei wesentliche Merkmale ab, die aus seiner Sicht in den Vordergrund treten:[1167]

- Up-Selling umfasst Transaktionen innerhalb einer bestehenden Geschäftsbeziehung. Dies setzt mindestens eine Transaktion vor dem als Up-Selling zu bezeichnenden Verkauf voraus (Einstiegsprodukt).
- Up-Selling zielt auf den Verkauf von höherwertigen Produkten und Dienstleistungen im Sinne von qualitativ besser und teurer. Teilweise ist jedoch unklar, welcher der beiden Bezugspunkte einer höheren Wertigkeit heranzuziehen ist.
- Up-Selling ist der Verkauf von Produkten und Dienstleistungen, die in einer substitutiven Beziehung zueinander stehen. Das Up-Selling-Produkt entstammt damit der gleichen Leistungskategorie und stiftet dem Kunden den gleichen Grundnutzen. Es resultieren aus dem Up-Selling folglich Kannibalisierungseffekte.

Weiterhin wird in der Literatur auch der Mehrverkauf eines bisher schon verkauften Produkts[1168] und die Ausschöpfung zunehmender Zahlungsbereitschaften im Zeitablauf der Kundenbeziehung als Up-Selling definiert.[1169] Ersteres bezieht sich auf das Ziel, eine Ausweitung der Kundendurchdringungsrate bzw. des Loyalitätsgrads zu realisieren.[1170] Letzteres lässt sich damit begründen,

---

[1167] Vgl. *Pohlkamp, A.* (2009), S. 14 ff.

[1168] Vgl. *Stüker, D.* (2008), S. 171; *Homburg, C.* (2007), S. 409; *Schneider, N. C.* (2007), S. 117; *Vogel, V.* (2006), S. 47; *Bauer, H.-H./Hammerschmidt, M.* (2005), S. 335; *Kamakura, W. et al.* (2005), S. 283.

[1169] Vgl. *Pohlkamp, A.* (2008), S. 27; *Hempelmann, B./Lürwer, M.* (2003), S. 338; *Reichheld, F. F./Sasser, W. E.* (1990), S. 105 f.

[1170] Vgl. *Eisenächer, H. W./Backofen, O. J./Hilverkus, S.* (2006), S. 792 und S. 795.

dass Up-Selling unter anderem auf die kundenindividuelle Zahlungsbereitschaften abzielt, um kundenseitige Preisobergrenzen auszunutzen.[1171] Im Rahmen des Entwicklungspotenzials ist dieses das Preispotenzial, das ein enges Verständnis der Abschöpfung von Zahlungsbereitschaften nutzt. Zusammenfassend lassen sich für das Entwicklungspotenzial die folgenden Facetten ableiten (siehe Abbildung 44).

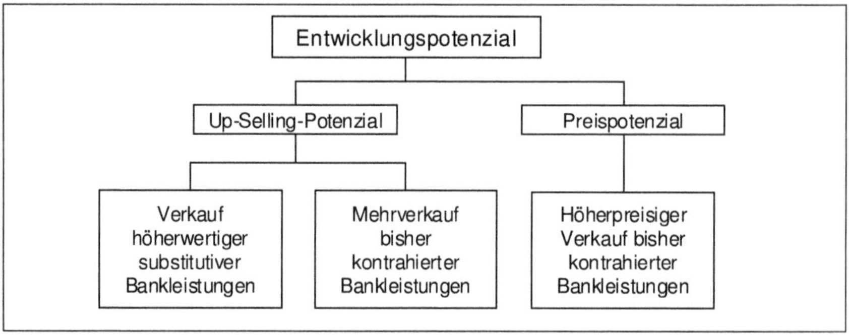

Abbildung 44: Facetten des Entwicklungspotenzials[1172]

Diese einzelnen Facetten sind das Gerüst der nachfolgenden Erläuterung des Entwicklungspotenzials als Werttreiber des Kundenwerts. Dabei ist zu berücksichtigen, dass das Up-Selling-Potenzial bisher nur rudimentär in der Literatur behandelt wurde,[1173] so dass eine konzeptionelle Auseinandersetzung vor dem Hintergrund eines zweckadäquaten Verständnisses für das hier zu entwickelnde Kundenwertmodell vorgenommen werden muss. Denn so ist insbesondere in wettbewerbsintensiven Märkten dem Up-Selling als Möglichkeit der Bestandsausschöpfung eine besondere Bedeutung einzuräumen[1174] und zu konstatieren, dass im Vergleich zu ausländischen Banken dieses Potenzial in deutschen Kreditinstituten nahezu ungenutzt ist.[1175]

Um den Verkauf höherwertiger, substitutiver Leistungen im Kundenwertmodell berücksichtigen zu können, also bei gleichem Grundnutzen ein höherwertiges Produkt anzubieten, ist zunächst eine bedarfs- bzw. nutzenbasierte Produktgruppenbildung vorzunehmen. Es sind alle Produkte, die einen ähnlichen

1171 Vgl. *Pohlkamp, A.* (2009), S. 21.

1172 Eigene Darstellung.

1173 Vgl. *Pohlkamp, A.* (2009), S. 13; *Vogel, V.* (2006), S. 47; *von Wangenheim, F.* (2003), S. 44.

1174 Vgl. *Duttenhöffer, S.* (2004), S. 514.

1175 Vgl. *Heppelmann, S./Middelmann, M. K.* (2006), S. 5.

Grundnutzen wie das in der Vergangenheit kontrahierte Produkt aufweisen, einer Produktgruppe zuzuordnen.[1176] Grundlegende empirische Erkenntnisse dazu liegen für das Firmenkundengeschäft der Banken nicht vor. Diese würden sich aus entsprechenden empirischen Kundenstudien ableiten lassen, wonach die befragten Kunden das Leistungsangebot der Banken hinsichtlich ihres Grundnutzens strukturieren. Trotz dieses Defizits ist als Grundlage des Up-Sellings im Sinne des Verkaufs höherwertiger Leistungen davon auszugehen, dass eine entsprechend ausgerichtete Leistungsstruktur in den Kreditgenossenschaften vorliegt bzw. durch institutsinterne Plausibilitätsüberlegungen auch ohne Integration der Kundensicht ermittelt werden kann. Die in Abbildung 15 dargestellte Systematik des BVR kann einen ersten groben Anhaltspunkt für eine derartige Produktgruppenbildung sein. Jedoch muss im Sinne des Up-Selling eine Leistungsstruktur definiert werden, die auch den Aspekt der Höherwertigkeit, d. h. der relativen Wertunterschiede aus Sicht der Kreditgenossenschaft berücksichtigt.[1177] Dieses ist in einem zweiten Schritt vorzunehmen. Hier empfiehlt es sich, vergleichbare Ausprägungsmerkmale zugrunde zu legen (so z. B. Volumina, Laufzeiten und Bonitätsklassen für Aktivgeschäfte). Als Beispiel nennt DUDERSTADT unterschiedliche Ausprägungen von Kreditkarten.[1178]

Der Mehrverkauf bisher kontrahierter Leistungen kann kunden- oder anbieterinitiiert sein, wenngleich dies im Ergebnis mit Blick auf die Wirkung auf den Kundewert weitgehend identisch ist. DILLER bezeichnet diese Facette als Penetrationswert.[1179] Der kundeninitiierte Einfluss wird konzeptionell im Basisgeschäftspotenzial berücksichtigt. Liegt die Initiative auf Anbieterseite, ist hierunter der vertriebliche Ansatz zu verstehen, das kundenindividuelle Bedarfsvolu-

---

1176 Dies gilt jedoch nicht für gebündelte Produkte, die den gleichen Nutzen stiften wie ein vorhergehendes Basisprodukt. Dies betrifft z. B. im Firmenkundengeschäft den Fall eines festverzinslichen Kredits, der prinzipiell nutzenäquivalent zu einem synthetischen Festzinskredit ist, der sich aus einem variablen Kredit und einem Zinsderivat (Zins-Cap oder Zins-Swap) zusammensetzt.

1177 Offen ist dabei vor allem die Beurteilung der Wertigkeit einzelner Produkte anhand derer diese in Beziehung gesetzt werden können. In diesem Punkt zeigt sich auch die fundamentale Kritik an bisherigen Arbeiten, die Up-Selling auf der reinen Erlösebene im Kundenwert berücksichtigen, aber keine Aussage darüber treffen, wie die Höherwertigkeit bestimmt wird. Z. B. *Bauer, H.-H./Hammerschmidt, M.* (2005), S. 335. Das Problem, die Höherwertigkeit zu bestimmen, gilt insbesondere für Bankleistungen, die hochgradig von den kundenindividuell ausgestalteten Produktparametern und bei risikotragenden Geschäften von der aktuellen und zu erwartenden Risikosituation des Kunden abhängen. Eine abschließende Klärung dieser Frage kann im Rahmen dieser Arbeit jedoch nicht erfolgen.

1178 Vgl. *Duderstadt, S.* (2006), S. 54, Fn. 243.

1179 Vgl. *Diller, H.* (2001), S. 10.

men zu steigern. Dieser Effekt resultiert daraus, andere Banken zu verdrängen (Verdrängungseffekt), indem bei gegebenem Bedarfspotenzial der Loyalitätsgrad erhöht wird, und/oder daraus, die Nachfrage des Kunden zu stimulieren, mithin das Bedarfspotenzial zu erhöhen und proportional, also mit konstantem Loyalitätsgrad, daran zu partizipieren (Mengeneffekt). Aus Kombination beider Alternativen resultieren Mixeffekte, die nicht eindeutig zugeordnet werden können. Im Zusammenhang mit dem Verdrängungseffekt ist die durch Anbietermaßnahmen zu entwickelnde und für den Planungs- oder Bewertungszeitpunkt zu prognostizierende Größe der Loyalitätsgrad in Verbindung mit dem gegebenen leistungsbezogenen Basisgeschäftspotenzial. Der Mengeneffekt resultiert bspw. aus normativen Bedarfsschätzungen, die sich im Firmenkundengeschäft für einzelne Produkte anhand betriebswirtschaftlicher Überlegungen begründen lassen. So z. B. die Substitution von für den Firmenkunden relativ teureren Lieferantenkrediten durch Kontokorrentkredite.[1180] Die relevante Größe, die in diesem Kontext zusätzlich zum Loyalitätsgrad zu bestimmen ist, ist das perioden- und produktspezifische zusätzliche Bedarfspotenzial des Firmenkunden. Hier ist zu berücksichtigen, dass der zusätzliche Bedarf vollständig oder nur quotal von der Kreditgenossenschaft gedeckt wird (letzteres z. B. bei Poolverträgen). Abbildung 45 visualisiert die vorhergehenden Überlegungen an einem einfachen Beispiel. Dabei wird unterstellt, dass die Ausweitung des Bedarfspotenzials von TEUR 100 auf TEUR 125 vollständig durch die Kreditgenossenschaft gedeckt wird. Es liegen drei Effekte vor: die Ausweitung des Loyalitätsgrads von 40 Prozent auf 60 Prozent für das bestehende Bedarfspotenzial führt zum Verdrängungseffekt. Die Steigerung des Bedarfspotenzials von TEUR 100 auf TEUR 125 führt bei konstantem gegebenen Loyalitätsgrad von 40 Prozent zum Mengeneffekt. Der Mixeffekt ist analytisch zweigeteilt und basiert auf der Ausweitung des Bedarfspotenzials bei im Beispiel vollständiger Deckung durch die Genossenschaftsbank.[1181] Diese Effekte sind im Sinne des Kundenwerts auch durch das Preispotenzial beeinflussbar.

[1180] Vgl. *Perridon, L./Steiner, M.* (1999), S. 416 ff.

[1181] Dies kann etwa dann der Fall sein, wenn ein Kunde bestehende Betriebsmittelkredite hat und einen neuen Betriebsmittelkredit bei der Kreditgenossenschaft aufnimmt.

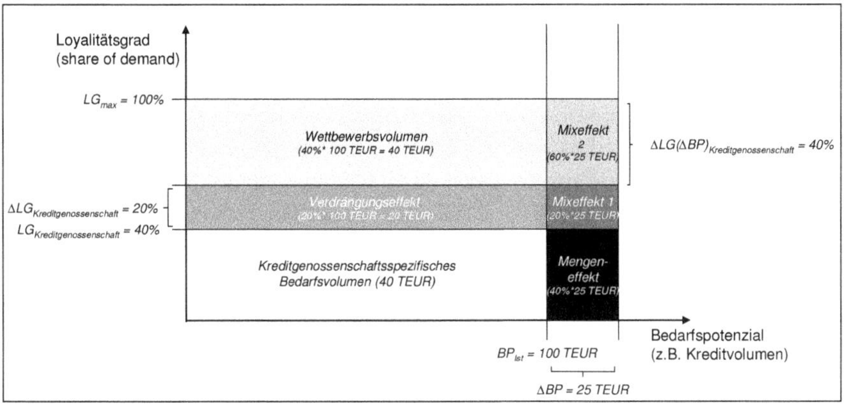

Abbildung 45: Verdrängungs-, Mengen- und Mixeffekt im Kontext des Up-Selling-Potenzials[1182]

Die Ergebnisse der Studie der TNS INFRATEST indizieren erhebliche Beiträge aus dieser Facette des Entwicklungspotenzials. Denn so erreichen die Kreditgenossenschaften bei einer Kundenreichweite von 52 Prozent einen produktübergreifenden Loyalitätsgrad von 26 Prozent, woraus eine Ausschöpfungsquote von 50 Prozent resultiert.[1183]

Das Preispotenzial konkretisiert sich in der Ausschöpfung einer im Zeitablauf gestiegenen,[1184] aber auch einer gegenwärtig nicht abgeschöpften Zahlungsbereitschaft des Kunden. Es knüpft an die Feststellung an, dass aufgrund des zunehmenden Wettbewerbs ein Preisverfall mit sinkenden Margen zu beobachten ist.[1185] Denn dass dies nicht ausschließlich und zwingend wettbewerbsbedingt ist, begründen SCHIERENBECK/POHL, die argumentieren, dass die bisherige Preispolitik „unprofitables Verhalten der Kunden erst ermöglicht."[1186] Zwar sieht SCHEER bei Firmenkunden eine höhere Preissensibilität und Verhand-

---

1182 Eigene Darstellung. Hier handelt es sich um eine einperiodige Betrachtung.

1183 Vgl. *tns infratest* (2010), S. 8 und S. 24. Die Ausschöpfungsquote berechnet sich als Quotient aus Loyalitätsgrad und Kundenreichweite. Eine Ausschöpfungsquote von 100 Prozent entspräche einer Situation, in der der Anteil der Kundenreichweite und der Loyalitätsgrad gleich sind.

1184 Häufig wird unterstellt, die Zahlungsbereitschaft eines Kunden steige mit zunehmender Dauer der Geschäftsbeziehung. Vgl. die vielzitierten Ausführungen von *Reichheld, F. F./Sasser, W. E.* (1990), S. 108 sowie *Homburg, C./Hoyer, W. D./Koschate, N.* (2005), S. 46 f. und *Choi, D. H. et al.* (2006), S. 938.

1185 Vgl. *Pohle, A./Schlütken, R.* (2009), S. 40. Siehe auch die Ausführungen in Kapitel C.1.5.

1186 *Schierenbeck, H./Pohl, M.* (2010), S. 290.

lungsmacht als bei den Privatkunden.[1187] Jedoch zeigen unterschiedliche Studien, dass oft kleine Preisänderungen zu einer nachhaltigen Ergebnisveränderung führen können.[1188] So zeigt PAUL für die Kreditgenossenschaften auf Basis der konsolidierten Jahresabschlüsse des Jahres 2006 welche Bedeutung der Absatzpreis auf das Ergebnis haben kann.[1189] Isolierte preisbedingte Änderungen im Zinsgeschäft um 10 Prozent hätten zu Veränderungen der Reingewinnspanne von über einem Drittel geführt.[1190] Veränderungen bei Personal- oder Sachkosten oder der Risikoaufwendungen im gleichen Ausmaß hätten dagegen einen deutlich geringeren Einfluss von lediglich 19 Prozent, 13 Prozent bzw. 7 Prozent gehabt. Erhebliche Preisgestaltungsspielräume werden bei entsprechender Preispolitik gerade im Geschäft mit Firmenkunden gesehen, „da der Preis für jeden Kunden einzeln bestimmt werden kann“.[1191] Auf Basis branchenübergreifender Daten findet STOCK heraus, dass zudem die Komplexität und Spezifität der Leistungen Einfluss auf die Preissensitivität haben.[1192] Sie begründet bei gegebenen positiv konnotierten Geschäftsbeziehungen die geringe Preissensitivität mit Transaktionskostenvorteilen auf Seiten des Kunden.[1193] Aufbauend auf den Aussagen aus den Kapiteln C.2.4.1 und C.3.1 lässt sich feststellen, dass die kreditgenossenschaftliche Qualitätsführerschaftsstrategie in Verbindung mit den relationalen Vertriebsansätzen mit der Identifikation und Ausnutzung von Zahlungsbereitschaften auf Seiten der Firmenkunden durch preispolitische Maßnahmen im Sinne des Kundenwerts korrespondiert.

1187 Vgl. *Scheer, O.* (1989), S. 46.

1188 Vgl. *Völckner, F.* (2006), S. 34.

1189 Vgl. hier und nachfolgend *Paul, S.* (2008), S. 403.

1190 Die Reingewinnspanne ist eine bilanzanalytische Kennzahl zur Analyse der Erfolgslage eines Finanzdienstleistungsunternehmens. Sie findet auch Anwendung in der ROI-Analyse und ist definiert als der Jahresüberschuss vor Steuern in Relation zur Bilanzsumme. Als Bezugsgröße kommen jedoch auch andere bilanzielle Größen in Frage. Vgl. *Schierenbeck, H.* (2003), S. 422 ff.

1191 *Wübker, G./Baumgarten, J.* (2004), S. 578.

1192 Vgl. *Stock, R.* (2003), S. 340 ff. Die Preissensitivität darf jedoch nicht mit der Preisbereitschaft verwechselt werden, obwohl diese inhaltlich sehr ähnlich sind. Erstere ist definiert als Grad, in dem ein Kunde seine Kaufentscheidung von der Höhe des Preises einer Leistung abhängig macht. Die Preisbereitschaft dagegen ist die grundsätzliche Absicht in einer künftigen Kaufsituation höchstens einen bestimmten Preis für eine bestimmte Leistung zu akzeptieren. Vgl. *Stock, R.* (2003), S. 335.

1193 Basis dieser Aussage ist die Annahme eines maximalen Kostenbudgets für eine Transaktion, das für den Preis und die Transaktionskosten kalkuliert wird. Sinken die Transaktionskosten, kann dies zugunsten des Preises genutzt werden. Vgl. *Stock, R.* (2003), S. 337.

Von großer Bedeutung für das Preispotenzial ist die Preisdifferenzierung ersten Grades nach PIGOU,[1194] die sich auf die kundenindividuelle Abschöpfung der Preisbereitschaft bezieht. Ziel ist, die Preisbereitschaft, verstanden als maximale Preisobergrenze,[1195] eines Kunden für die Kreditgenossenschaft zu nutzen. Aufgrund der Vielzahl an Kunden und der zusätzlichen internen Koordinationsaufwendungen, die aus einem kundenindividuellen Konditionssystem resultieren, dürfte die Preisdifferenzierung zweiten Grades leichter umzusetzen sein.[1196] Dabei wird eine bestimmte Anzahl von Segmenten mit Maximalpreisen festgelegt und der Firmenkunde ordnet sich mittels Selbstselektion dem relevanten Segment zu. Für die Kundenwertermittlung ist zu berücksichtigen, dass sich diese Preisbereitschaft auf das insgesamt prognostizierte Bedarfsvolumen im Planungs- oder Bewertungszeitraum bezieht und angenommen wird, dass die Substitutionsprodukte zu maximalen kundenindividuellen Preisen kalkuliert werden. Folglich wird angenommen, dass die maximale Preisobergrenze abgeschöpft werden kann, ohne dass Einflüsse auf die Absatzmenge resultieren. Problematisch an der praktischen Berücksichtigung dieses Kundenwerttreibers ist nach JUNCKER, dass es im Alltag des Firmenkundengeschäfts oft an den Voraussetzungen einer differenzierten Preispolitik fehlt. „So sind Preissensibilitäten, Bedürfnisse und Präferenzen der Kunden und die resultierenden Preissteigerungsspielräume durchweg nicht bekannt.“[1197] Gelingt es jedoch die Preisbereitschaften verlässlich zu schätzen, lassen sich durch Abschöpfung der Konsumentenrente zusätzliche Erträge erzielen, die positiv auf den Kundenwert wirken.[1198]

Es ist evident, dass das Basisgeschäftspotenzial und das Entwicklungspotenzial bei dynamischer Betrachtung interagierende Potenziale des Kundenwerts sind, die im Zeitablauf negativ korrelieren.[1199] Denn Bestandteile des Entwicklungspotenzials einer Periode stellen Bestandteile des Basisgeschäftspotenzials der folgenden Periode dar, sofern in beiden Perioden eine Kundenwertbestimmung vorgenommen wird.

Während sich das Basisgeschäftspotenzial und das Entwicklungspotenzial auf die bestehenden Produktbeziehungen beziehen, sollen im folgenden Kapitel die Potenziale aus dem Verkauf zusätzlicher Leistungen erörtert werden.

1194 Vgl. hier und nachfolgend *Pigou, A. C.* (1960), S. 279.
1195 Vgl. *Woratschek, H./Roth, S.* (2006), S. 372.
1196 Vgl. hier und nachfolgend *Pigou, A. C.* (1960), S. 279.
1197 *Juncker, K.* (2006), S. 350.
1198 Vgl. *Woratschek, H./Roth, S.* (2006), S. 373; *Diller, H.* (2001), S. 16.
1199 Vgl. *Diller, H.* (2001), S. 11.

4.5.3.1.3 Cross-Selling-Potenzial

Das Cross-Selling-Potenzial gehört zu den intensiv diskutierten Wertreibern des Kundenwerts und wird in der Literatur als der realisierbare zusätzliche Erlös aus dem als Cross-Selling bezeichneten Überkreuz-Verkauf verstanden.[1200] Cross-Selling bezeichnet die Aktivierung von vorhandenen Kundenkontakten für weitere Produktverkäufe oder Nutzung von weiteren Dienstleistungen eines Unternehmens.[1201] Dieser Vertriebsansatz zielt darauf ab, produktübergreifende Verkaufschancen zu realisieren.[1202] Während sich die Literatur in diesem Punkt weitgehend einig ist, liegt keine einheitliche inhaltliche Konkretisierung des Begriffs Cross-Selling vor. Es besteht Uneinigkeit darüber, ob sich Cross-Selling nur auf bestehende Kundenbeziehungen bezieht, ob es durch eine zeitliche Distanz zwischen dem initialen und dem nachfolgenden Verkauf gekennzeichnet ist und ob ein Verbund zwischen den Produktkategorien bestehen muss. Weiterhin unterscheiden einige Autoren dahingehend, ob das Cross-Selling nur das vom Anbieterunternehmen selbst hergestellte Produktprogramm oder aber auch von Dritten bezogene und den Kunden angebotene Sortimentsbestandteile umfasst. SCHÄFER setzt die inhaltliche Konkretisierung hinsichtlich des Vorhandenseins einer bestehenden Geschäftsbeziehung mit der Frage des Zeitbezugs gleich,[1203] so dass sich die Begriffsdefinition auf drei Aspekte konkretisiert: die zeitliche Abgrenzung, die sachliche Abgrenzung und die Abgrenzung im Hinblick auf das Kriterium der Eigenerstellung oder Fremdfertigung. Diese Aspekte sollen mit Blick auf die bisherige Diskussion des Kundenwertmodells zweckadäquat bestimmt werden.

1200 Vgl. z. B. *Lissautzki, M.* (2007), S. 153 ff.; *Foehn, P.* (2006), S. 128 ff.; *Friederichs-Schmidt, S.* (2006), S. 105 ff.; *Böhrs, S.* (2004), S. 58 f.; *Tewes, M.* (2003), S. 87 ff.; *Diller, H.* (2001), S. 14 f.; *Gelbrich, K.* (2001), S. 57 ff.; *Rudolf-Sipötz,* E. (2001), S. 100 ff. Siehe auch die genannten Kundenwertmodelle in Tabelle 12. Während das Cross-Selling die Aktivitäten aus Anbietersicht umfasst, wird für die Nachfragerperspektive der Begriff Cross-Buying verwendet. Vgl. *Cornelsen, J.* (2000), S. 175. RUDOLF-SIPÖTZ legt ihrem Kundenwertmodell der Outside-in Perspektive des Marketing entsprechend das Cross-Buying-Potenzial zugrunde. Vgl. *Rudolf-Sipötz, E.* (2001), S. 100. Da sich der Begriff Cross-Selling etabliert hat und die Cross-Buying-Bereitschaft als ein Bestimmungsfaktor des Cross-Selling-Potenzials verstanden wird, wird diesem Vorgehen hier nicht gefolgt.

1201 Vgl. *Wittmann, G.* (2006), S. 10.

1202 Vgl. hier und nachfolgend *Homburg, C./Schäfer, H.* (2006), S. 159 ff.; *Schäfer, H.* (2002), S. 52 ff.

1203 Vgl. *Schäfer, H.* (2002), S. 55.

Mit Bezug auf die Kriterien Verbund und Zeitbezug unterscheidet CORNELSEN die folgenden vier Arten des Verbundverkaufs, von denen nach seinen Ausführungen das Cross-Selling eine Ausprägung ist (vgl. Tabelle 19).

Tabelle 19: Einordnung des Cross-Selling in verschiedene Formen des Verbundverkaufs[1204]

Zeitbezug / Produkt-/ Nutzenverbund	**Zeitpunkt (statisch)**	**Zeitraum (dynamisch)**
ja	substitutional / complementary bundling	Anschlussverkauf
nein	independent bundling	*Cross-Selling*

Nach dieser Betrachtung läge Cross-Selling nur dann vor, wenn zwischen den Verkäufen eine zeitliche Distanz liegen würde. Dies entspricht jedoch nicht dem praktischen Verständnis in Kreditinstituten, die auch den Verkauf einer Versicherung im Rahmen eines Finanzierungsgeschäfts als Cross-Selling betrachten. Dies gleicht eher dem von CORNELSEN als bundling, d. h. der Bündelung verbundener, aber auch unverbundener Leistungen, bezeichneten Vorgehen. Nach HOMBURG/SCHNURR kann Cross-Selling dagegen sowohl zeitlich versetzt als auch zeitgleich stattfinden.[1205] Dieser Aussage wird hier gefolgt. Denn insbesondere der (zeitgleiche) Verkauf von Produktbündeln wird als erfolgversprechende Möglichkeit angesehen, Cross-Selling im Firmenkundengeschäft erfolgreich zu implementieren.[1206] Daher wird hier im Gegensatz zu CORNELSEN bundling als spezielle Form des Cross-Selling verstanden.[1207]

Im Hinblick auf die sachliche Abgrenzung, also der Frage nach der Beziehung zwischen den vom Cross-Selling erfassten Produkten wird unterschieden, ob diese komplementär[1208] zueinander oder unabhängig[1209] voneinander sind. BEUTIN/KLENK argumentieren beim Cross-Selling im Bankenkontext für einen komplementären Verwendungsverbund zwischen Einstiegs- oder Kernleistungen und Zusatzleistungen.[1210] Sie führen als Beispiel für das Firmenkundenge-

---

1204 *Cornelsen, J.* (2000), S. 173.

1205 Vgl. *Homburg, C./Schnurr, P.* (1998), S. 172. Diese Festlegung übernimmt auch *Schäfer, H.* (2002), S. 55.

1206 Vgl. *Wittmann, G.* (2006), S. 9 ff.

1207 So auch *Heppelmann, S./Middelmann, M.* (2006), S. 6.

1208 Vgl. z. B. *Diller, H.* (2001), S. 14.

1209 Vgl. *Cornelsen, J.* (2000), S. 173 sowie die Tabelle 19.

1210 Vgl. *Beutin, N./Klenk, P.* (2005), S. 51; so auch *Wittmann, G. et al.* (2007), S. 16.

schäft den Kontokorrentkredit als Einstiegsprodukt und Beratungsgespräche zum Rating als Zusatzleistungen an.[1211] PRAXMARER dagegen sieht in der komplementären Verbundbeziehung keine zwingende Voraussetzung für Cross-Selling, sondern lediglich den Vorteil eines leichter realisierbaren Cross-Selling-Erfolgs im Bankenkontext.[1212] Insofern können auch bezüglich des Verwendungsverbunds vollständig unverbundene Produkte Gegenstand des Cross-Selling sein. In der bankbetrieblichen Praxis herrscht zudem das Verständnis vor, dass Cross-Selling-Erfolge auch bei unverbundenen Produkten realisiert werden können.[1213] Dies zeigt sich in der Kennzahl der Cross-Selling-Rate, die eine der zentralen kundengerichteten Erfolgskennzahlen ist.[1214] Sie ist zeitlich statisch definiert, bezieht sich auf das gesamte Produktprogramm und berücksichtigt den Absatzverbund von Produkten nicht.[1215] Die Cross-Selling-Rate ist „the number of different product categories a customer has purchased."[1216] Aus diesen Gründen liegt im Weiteren dem Cross-Selling der Verkauf sowohl komplementär verbundener als auch unverbundener Produkte zugrunde. Nachfolgend gelten als Einstiegsprodukte die Produkte, die zum Bewertungszeitpunkt Transaktionsgegenstand der Geschäftsbeziehung sind. Zusatzprodukte sind dann diejenigen Produkte, die im Zuge des Cross-Selling zusätzlich zu den Einstiegsprodukten verkauft werden und diese nicht substituieren.[1217] In diesem Punkt zeigt sich eine Interdependenz zur Produktabgrenzung für das Basisgeschäftspotenzial und das Entwicklungspotenzial. Alle Bankleistungen, die dort in der Betrachtung ausgegrenzt werden, werden im Cross-Selling-Potenzial unter Berücksichtigung der genannten Korrekturfaktoren (Geschäftspolitik und manifester Bedarf des Kunden sowie keine Berücksichtigung im Entwicklungspotenzial als höherwertiges Substitutionsgut) in die Betrachtung integriert. Dies begründet gleichzeitig, dass beim Cross-Selling-Potenzial keine Kannibalisierungseffekte zu berücksichtigen sind.[1218] Die Bankprodukte, die bisher nicht an den Kunden

---

1211 Der Nutzenverbund ergibt sich in diesem Beispiel daraus, dass bei Kreditgeschäften die Risikosituation des Kunden eine Einflussgröße auf den zu entrichtenden Zinssatz ist. Dies wird als risikoadjustierte Bepreisung bezeichnet. Die Beratung zum Rating beinhaltet die Analyse der Risikosituation sowie ggf. Empfehlungen zu positiven Veränderungen.

1212 Vgl. *Praxmarer, M. A.* (1993), S. 260.

1213 Hier könnte der Verbund bei den selbsterstellten Dienstleistungen aus Sicht des Kunden in der Identität des Anbieters der einzelnen Leistungen gesehen werden.

1214 Vgl. hier und nachfolgend *Ritter, D.* (1993), S. 140.

1215 Die Cross-Selling-Quote ermittelt sich als Mittelwert der Anzahl der verschiedenen mit einem Kunden kontrahierten Produkte über alle relevanten Kunden.

1216 *Venkatesan, R./Kumar, V.* (2004), S .111.

1217 Vgl. *Schäfer, H.* (2002), S. 15 und S. 56 sowie die dort jeweils angegebene Literatur.

1218 Vgl. *Tewes, M.* (2003), S. 90; *Rudolf-Sipötz, E.* (2001), S. 132.

abgesetzt wurden und in einer Substitutionsbeziehung stehen, werden bei einem Preisklassenwechsel dem Entwicklungspotenzial und ohne vorliegendem Preisklassenwechsel dem Basisgeschäftspotenzial zugeordnet.

Vor dem Hintergrund der ausgeprägten Kooperationsstruktur der Finanz-Gruppe und mit Blick auf die große Anzahl externer Leistungen, die die Kreditgenossenschaften den Firmenkunden anbieten,[1219] wird nachfolgend das gesamte Sortiment unabhängig davon, ob selbst erstellt oder extern bezogen, als relevantes Sortiment für das Cross-Selling angesehen.

Auf Basis der bisherigen Ausführungen wird das Cross-Selling-Potenzial allgemein als der bewertete, mögliche Verkauf zusätzlicher, bisher nicht verkaufter und nicht substitutiver Leistungen verstanden.[1220] Es ist ein Maßstab dafür, in wie weit eine bestehende Geschäftsbeziehung auf andere Bankleistungen der Kreditgenossenschaft ausgeweitet werden kann.[1221] Das Cross-Selling-Potenzial als Einflussgröße des Kundenwerts zu berücksichtigen, wird vor allem in stagnierenden Märkten empfohlen, denn es wird als viel versprechende Wachstumsoption betrachtet.[1222] Aber auch die Positionierung der Kreditgenossenschaften als Allfinanzanbieter[1223] drängt die Berücksichtigung eines speziell auf die Breite des Leistungsprogramms ausgerichteten Kundenwerttreibers auf. Gleichzeitig kann durch die veränderte Produktnachfrage der Firmenkunden eine marktseitige Möglichkeit zum Cross-Selling indiziert werden.[1224] Aus Steuerungssicht hat das Cross-Selling-Potenzial die Funktion, die enge Kreditfokussierung der Firmenkundenberater[1225] zugunsten des Allfinanzangebots zu überwinden. Denn obwohl in der gesamten Bankenbranche Cross-Selling im Firmenkundengeschäft intensiv betrieben wird,[1226] zeigen empirische Ergebnisse,

---

1219 Vgl. dazu die Ausführungen in den Kapiteln C.1.4.4 und C.2.3. So auch *Duhnkrack, T.* (2007), S. 1221.

1220 So auch *Schneider, N. C.* (2007), S. 118; *Schäfer, H.* (2002), S. 56; *Eberling, G.* (2002), S. 134.

1221 Vgl. *Bauer, H.-H./Hammerschmidt, M.* (2005), S 335; *Cornelsen, J.* (2000), S. 172.

1222 Vgl. *Homburg, C./Schäfer, H.* (2006), S. 159; *Kamakura, W. A./Ramaswami, S. N./Srivastava, R. K.* (1991), S. 329.

1223 Vgl. *Duhnkrack, T.* (2007), S. 1221.

1224 Vgl. *Engstler, M./Praeg, P./Vocke, C.* (2007), S. 8; *Heine, J. M.* (2007), S. 5.

1225 Vgl. *Groß, T./Michaelis, H.* (2002), S. 164 ff.

1226 So gaben in einer Studie der Unternehmensberatung IBM 95 Prozent der befragten Kreditinstitute an, Cross-Selling aktiv zu betreiben. Vgl. *Pfeiffer, M./Pirngruber, N./Mährle, C.* (2003), S. 5. Dabei scheint das Cross-Selling eine im Zeitablauf zunehmend größere Bedeutung zu bekommen. Eine Studie des FRAUNHOFER INSTITUTS zeigt, dass im Jahr 2006 60,5 Prozent und im Jahr 2007 87,2 Prozent der Befragten, die Vertriebsintensivierung mittels Cross- und Up-Selling-Ansätzen konkretisieren wollten. Vgl. *Engstler, M./Praeg, P./Vocke, C.* (2007), S. 10 f.

dass die realisierten Erfolge noch „deutlich hinter den Möglichkeiten“[1227] bleiben.[1228] Speziell für die Kreditgenossenschaften zeigen Studien, dass die Cross-Selling-Raten nur bei etwa 3,3 liegen.[1229]

Der positive Einfluss des Cross-Selling-Potenzials auf den Kundenwert wird in der Literatur vielfältig begründet.[1230] Unterschieden wird dabei zwischen beziehungsbezogenen und beziehungsübergreifenden Effekten. Zu den beziehungsbezogenen Effekten zählen vor allem die Auswirkungen auf die Erlöse und Kosten der Geschäftsbeziehung, wobei von einem positiven Einfluss ausgegangen wird,[1231] die aus transaktionsbezogenen economies of scope[1232] resultieren.[1233] Zudem wird vermutet und ist in einzelnen Branchen empirisch belegt, dass das Ausmaß, zu dem ein Kunde über verschiedene Produktkategorien hinweg bei einem Anbieter kauft, positiven Einfluss auf die Stabilität und Dauer der Geschäftsbeziehung hat.[1234] Dies wird begründet mit dem Aufbau von ökonomischen Wechselbarrieren sowie mit der Qualität und Intensität der persönlichen Beziehungen zwischen den interagierenden Personen auf Anbieter- und Kundenseite.[1235] Auch der Diversifikationseffekt hinsichtlich der produktbasierten Erlösströme auf Einzelkundenebene wird als stabilisierender Einfluss aus dem Cross-Selling angesehen.[1236] Mit Blick auf die beziehungsübergreifenden Effekte werden vor allem economies of scale (Mengeneffekte) und Diversifikationseffekte als Vorteil des Cross-Selling angeführt.[1237] Konträre Effekte resultieren vor allem daraus, dass Kunden aufgrund der Anbieterkonzentration eine

---

[1227] *Gawlik, H./Rathmann, B.* (2008), S. 18 f.

[1228] Vgl. *Paul, M./Paul, S./Stein, S.* (2003), S. 6; *Beutin, N./Klenk, P.* (2005), S. 52.

[1229] Vgl. *Wittmann, G. et al.* (2007), S. 43. Als Vergleichsmaßstab sei auf das firmenkundenspezifische Produktprogramm in Kapitel C.2.3 verwiesen.

[1230] Eine ausführliche Übersicht über Studien zu Auswirkungen des Cross-Sellings liefert *Schäfer, H.* (2002), S. 20 ff.

[1231] Vgl. *Schmöller, P.* (2001), S. 120; *Stanger, M.* (2007), S. 20.

[1232] Die economies of scope sind Kostensynergieeffekte, die sich im Kontext des Cross-Selling aus im Vergleich zu Neukundenakquisitionen geringeren Anbahnungskosten sowie Vereinbarungs-, Anpassungs- oder Kontrollkosten ergeben. Vgl. *Schäfer, H.* (2002), S. 156.

[1233] Vgl. *Wünschmann, S.* (2007), S. 43; *Schneider, N. C.* (2007), S. 118; *Tewes, M.* (2003), S. 90; *Schäfer, H.* (2002), S. 154 ff.; *Cornelsen, J.* (2000), S. 180.

[1234] Vgl. *Reinartz, W. J./Kumar, V.* (2000), S. 24; *Venkatesan, R./Kumar, V.* (2004), S. 118; *Kamakura, W. A./Ramaswami, S. N./Srivastava, R. K.* (1991), S. 329.

[1235] Vgl. *Schäfer, H.* (2002), S. 155.

[1236] Vgl. *Cornelsen, J.* (2000), S. 181.

[1237] Vgl. *Schäfer, H.* (2002), S. 155; *Matzler, K./Stahl, H. K./Hinterhuber, H. H.* (2009), S. 11.

höhere Verhandlungsmacht aufbauen können, die zu Preiszugeständnissen auf Seiten des Anbieters führen können.[1238]

Das Cross-Selling-Potenzial einer Geschäftsbeziehung wird bestimmt durch den gegebenen und künftigen Bedarf des Firmenkunden, der bisherigen Ausschöpfung und der Bereitschaft des Kunden weitere Produkte von dem Anbieter zu beziehen.[1239] Die in der Literatur aufgeführten Einflussgrößen dieser Bestimmungsfaktoren ordnet SCHÄFER folgenden fünf Bereichen zu: Merkmale des Kunden (z. B. Komplexität des Kunden), Merkmale der Umwelt des Kunden (z. B. Marktdynamik), Merkmale der Produkte (z. B. Produktverbundbeziehungen), Merkmale der Geschäftsbeziehung (z. B. Vertrauen) und Merkmale des Anbieters (z. B. Breite des Produktprogramms).[1240] „Innerhalb dieser fünf Bereiche kann prinzipiell eine Vielzahl möglicher Einflussfaktoren des Cross-Selling-Potenzials identifiziert werden."[1241] Daher sollen nur die am häufigsten diskutierten näher betrachtet werden.

Die Cross-Buying-Bereitschaft wird vor allem dadurch beeinflusst, wie zufrieden der Kunde mit den bisherigen Leistungen des Anbieters ist und wie stark seine Diversifikations- und Variety-Seeking-Bedürfnisse ausgeprägt sind.[1242] Das Diversifikationsbedürfnis eines Kunden wird durch dessen Risikotoleranz beeinflusst. Je höher die Risikotoleranz ist, desto geringer ist dessen Diversifikationsbedürfnis, da die Abhängigkeit von einem oder wenigen Anbietern für ihn unproblematisch ist. Als Variety-Seeking wird der Wunsch nach Abwechslung bezeichnet, der dazu führen kann, dass auch ein zufriedener Kunde den Anbieter wechselt, um etwas Neues auszuprobieren.[1243] Dieses Persönlichkeitsmerkmal ist vor allem in Konsumgütermärkten relevant. Inwieweit es vor dem Hintergrund der Besonderheiten des Firmenkundengeschäfts für diese von Relevanz ist, ist jedoch fraglich. Die Cross-Buying-Bereitschaft eines Firmenkunden ist als limitierender Faktor bei der Ausschöpfung des Cross-Selling-Potenzials zu verstehen, denn häufig sinkt diese mit zunehmender Anzahl der vom selben Lieferanten gekauften Produkte.[1244] Zudem fragen nach Erkenntnissen von MEYER/DORNACH nur wirklich überzeugte Kunden zusätzliche Leistungen bei einem Anbieter nach.[1245]

1238 Vgl. *Schäfer, H.* (2002), S. 155.
1239 Vgl. *Homburg, C./Schäfer, H.* (2006), S. 162; *Schäfer, H.* (2002), S. 108.
1240 Vgl. *Schäfer, H.* (2002), S. 108 f.
1241 *Schäfer, H.* (2002), S. 109.
1242 Vgl. *Tewes, M.* (2003), S. 89 f.; *Rudolf-Sipötz, E.* (2001), S. 100 ff.
1243 Vgl. *Tewes, M.* (2003), S. 90, Fn. 289.
1244 Vgl. *Stahl, H. K. et al.* (2009), S. 259.
1245 Vgl. *Meyer, A./Dornach, F.* (1997), S. 66.

Während die Cross-Buying-Bereitschaft durch geeignete Marketing-Mix-Entscheidungen insgesamt zu fördern ist,[1246] ist die Frage nach dem relevanten Bedarf für die steuerungsbezogen zweckmäßige Betrachtung des Cross-Selling-Potenzials zu stellen. Um den relevanten Bedarf abzugrenzen, soll auf das steuerungsorientierte Konzept von LISSAUTZKI rekurriert werden.[1247] Er unterscheidet anhand der Kriterien a) Art des Kundenbedarfes (aktuell vorhanden, latent vorhanden, später vorhanden), b) bisheriges Kundenverhalten (bisher nicht nachgefragt, Nachfrage bei Wettbewerbern gedeckt) und c) Bestandteil des Leistungsprogramms (im Sortiment des Anbieters vorhanden, zukünftig in Sortiment aufzunehmen) verschiedene Cross-Selling-Ansätze (vgl. Abbildung 46), die als steuerungsorientierte Dimensionen des Cross-Selling-Potenzials betrachtet werden sollen.

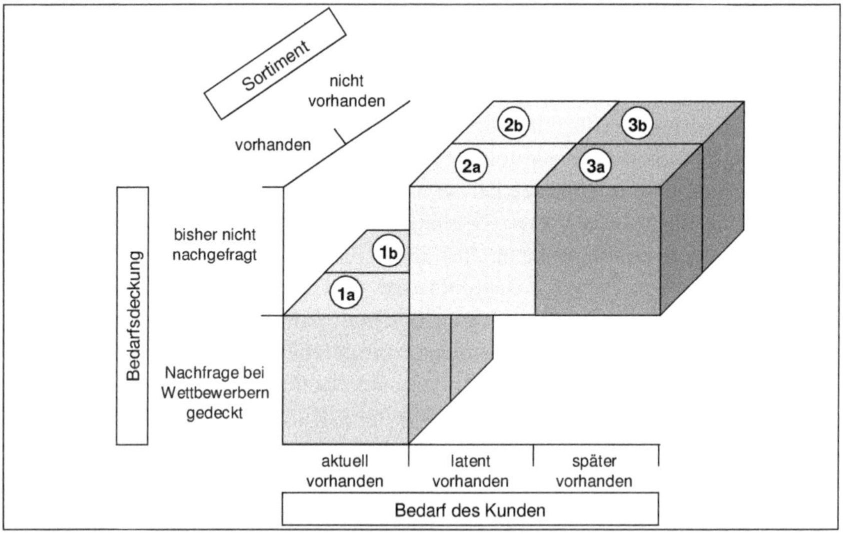

Abbildung 46: Mögliche Unterscheidungen im Cross-Selling[1248]

[1246] In seiner branchenübergreifenden Studie stellt SCHÄFER fest, dass die Bereitschaft zum Cross-Buying direkt von der Beschaffungsverbundenheit der angebotenen Leistungen sowie dem Vertrauen des Kunden zum Anbieter positiv beeinflusst wird. Vgl. *Schäfer, H.* (2002), S. 128 f. Damit sind Ansatzpunkte definiert, um das Cross-Selling-Potenzial auszuschöpfen, dass Cross-Selling-Potenzial als solches wird dadurch jedoch nicht gemessen.

[1247] Vgl. hier und nachfolgend *Lissautzki, M.* (2007), S. 156 ff.

[1248] *Lissautzki, M.* (2007), S. 157.

Das loyalitätsbezogene Cross-Selling resultiert aus dem Verkauf von Produkten, für die der Kunde bereits einen vorhandenen Bedarf hat, der jedoch bei einem Wettbewerber gedeckt wird und sich entweder auf Bankleistungen des aktuellen Sortiments (1a) oder auf neue Bankleistungen (1b) der Kreditgenossenschaft bezieht. Typisch im Firmenkundengeschäft sind etwa Versicherungsprodukte oder Leasingdienstleistungen. Aus diesem Ansatz leitet sich das loyalitätsbezogene Cross-Selling-Potenzial ab.

Handelt es sich um Produkte, für die der Kunde lediglich einen latenten Bedarf hat und daher dieser noch nicht gedeckt ist, so wird dies als bedürfnisweckendes Cross-Selling bezeichnet. Auch hier wird unterschieden, ob die Kreditgenossenschaft bedarfsdeckende Leistungen im Produktprogramm hat (2a) oder nicht (2b). Dies können bspw. Produkte des elektronischen Zahlungsverkehrs sein, die ab einer bestimmten Größe das Cash-Management unterstützen und den beleghaften Zahlungsverkehr ersetzen. Damit wird das bedürfnisweckende Cross-Selling-Potenzial konstituiert.

Als dritten Ansatz definiert LISSAUTZKI das lebenszyklusbezogene Cross-Selling. Da der Begriff des Lebenszyklus im Zusammenhang mit Privatkunden verwendet, jedoch nicht idiomatisch für das Firmenkundengeschäft ist,[1249] soll dieser Ansatz als unternehmensentwicklungsbezogenes Cross-Selling bezeichnet werden. Im Gegensatz zum latenten Cross-Selling, bei dem der Bedarf durch die Kreditgenossenschaft identifiziert und an den Firmenkunden kommuniziert wird, resultiert das unternehmensentwicklungsbezogene Cross-Selling-Potenzial aus einer auch für den Firmenkunden absehbaren Bedarfsentwicklung (3a, 3b). Dies liegt z. B. vor, wenn der Firmenkunde Aktivitäten im Ausland vornimmt und damit der Bedarf nach Produkten des Auslandszahlungsverkehrs verbunden ist.

Das loyalitätsbezogene Cross-Selling-Potenzial kann durch die Analyse der produktspezifischen Wettbewerbssituation eines Kunden ermittelt werden kann. Dazu ist eine Analyse der zum Bewertungszeitpunkt bestehenden Drittbeziehungen geeignet. Wie bereits im Zusammenhang mit dem Basisgeschäftspotenzial ausgeführt, ist die Grundlage der Bestimmung der anderen beiden Potenzialdimensionen die Analyse der Unternehmensentwicklung. Aus den Analyseergebnissen sind dann die Bedarfsstruktur und die Bedarfsvolumina abzuleiten. Hinsichtlich der einzelnen Dimensionen ist dabei der Loyalitätsgrad der Ver-

---

1249 Im Vergleich zu den Privatkunden und ihrem Lebenszyklus sind im Finanzdienstleistungsbereich die Firmenkunden und der ihnen zuortenbare Unternehmenslebenszyklus vergleichsweise weniger deterministisch, so dass keine leicht abgrenz- und erkennbaren Lebenszyklusphasen vorhanden sind.

gangenheit per definitionem Null bzw. mathematisch unbestimmt.[1250] Aus diesem Grund ist für die Ermittlung des Cross-Selling-Potenzials die Höhe des Loyalitätsgrads zu bestimmen, mit dem das prognostizierte Bedarfsvolumen zugunsten der Kreditgenossenschaft ermittelt wird. Im Sinne des Potenzialgedankens wird angenommen, dass die maximale Zahlungsbereitschaft des Kunden ausgeschöpft werden kann. Hinsichtlich der Prognosemöglichkeiten stellen DILLER jedoch fest, dass das Cross-Selling-Potenzial kundenindividuell nur schwer zu bestimmen sei.[1251]

Die bisher diskutierten Kundenwerttreiber berücksichtigen außerhalb der Refinanzierungs- bzw. Opportunitätskosten bei zinstragenden Geschäften und direkten Provisionsaufwendungen keine zu investierenden Ressourcen in die Kundenbeziehung bzw. in die einzelnen Transaktionen. Dies begründet sich damit, dass die Leistungserstellung und ihre Kosten im Firmenkundengeschäft von der Integration des externen Faktors abhängen. Damit besteht die Möglichkeit die Kundenintegration als kostenverursachenden Einflussfaktor explizit zu berücksichtigen. Diesem Gedanken tragen die nächsten Kundenwerttreiber Rechnung.

4.5.3.1.4 Basiskostenpotenzial

Die Deckungsbeitragskalkulation der Kreditgenossenschaften gemäß VR-CONTROL differenziert zwischen den Standardstückkosten und der Standardrisikoprämie des jeweiligen Einzelgeschäfts.[1252] Bei den nachfolgenden Ausführungen zum Basiskostenpotenzial wird die Standardrisikoprämie nicht betrachtet, sondern nur die leistungs- und kundenbezogenen Kosten des Betriebsbereichs, die aus der Leistungserstellung und -verwertung resultieren. Die explizite Betrachtung des bewerteten Ressourcenverbrauchs des Betriebsbereichs in diesem Kundenwertmodell beruht auf der Intention, die kundeninduzierten Kosten als eigenständigen und gestaltbaren Kundenwerttreiber des Kundenwerts zu betrachten. Dabei soll analog zur differenzierten Betrachtung bei den ertragsbezogenen Kundenwerttreibern zwischen zwei Komponenten unterschieden werden. Zum einen werden die erwarteten bzw. geplanten leistungs- und kundenbezogenen Kosten, die sich aus dem Basisgeschäfts-, Entwicklung- und Cross-Selling-Potenzial ergeben im Kontext des Basiskostenpotenzials betrachtet. Zum ande-

---

1250 Vgl. *Lissautzki, M.* (2007), S. 157 f.

1251 Vgl. *Diller, H.* (2001), S. 15. So auch *Stahl, H. K. et al.* (2009), S. 259.

1252 Vgl. die Ausführungen in Kapitel C.4.3. Die Standardrisikoprämie hängt ab von der Kundenbonität, dem Umfang der kreditspezifischen Sicherheiten sowie dem Kreditvolumen. Das bedeutet, sie ist nur zum Teil produktspezifisch, da sich die Kundenbonität auch auf andere Kreditverträge auswirkt. Die Standardrisikoprämie als negative Einflussgröße wird im Rahmen der Risikodimension näher betrachtet.

ren werden die Möglichkeiten, die Kosten durch eine bessere Integration des externen Faktors in die Bankleistungserstellung zu reduzieren im Rahmen des Kooperationspotenzials betrachtet. Denn die Integration des Kunden wird als das betriebswirtschaftliche Kernproblem der kundenintegrierenden Leistungserstellung angesehen, da dieser den Leistungserstellungsprozess mit steuert und so Einfluss auf die Effizienz und Effektivität des Unternehmens hat.[1253]

Ausgangspunkt der folgenden Überlegungen ist, zunächst den Umfang der zu berücksichtigenden Kosten festzulegen. Die Literatur ist sich einig, dass mindestens die Einzelkosten der Kundenbeziehung betrachtet werden müssen.[1254] Grundlage ist die relative Einzelkostenrechnung nach RIEBEL.[1255] Diese basiert auf dem Einzelkosten-, dem damit zwingend verbundenen Deckungs- sowie dem Grundrechnungsprinzip. Das Einzelkostenprinzip besagt, dass alle im Unternehmen anfallenden Kosten als Einzelkosten eines speziellen Bezugsobjekts – einer einzelnen Entscheidung – ausgewiesen werden, auf dessen Disposition sie zurückzuführen und dem sie daher logisch zwingend zuzurechnen sind.[1256] Auf-

---

1253 Vgl. *Kleinaltenkamp, M./Schweikart, J.* (1999), S. 111; *Gelbrich, K.* (2001), S. 72. Nach SCHIERENBECK sind produktbezogene Kosten des Betriebsbereichs über die Vertragslaufzeit einer Bankleistung derart zu verteilen, dass sie zu Beginn sehr hoch, im weiteren Verlauf relativ niedrig und gegen Vertragsende wieder relativ hoch sind. Vgl. *Schierenbeck, H.* (2003), S. 309. Dies ist unter anderem darauf zurückzuführen, dass zu Beginn intensive Abstimmungen und Vorbereitungen mit dem Kunden vorgenommen werden müssen. Dazu auch *BVR* (2003b), Kapitel 2.1, S. 65.

1254 Vgl. z. B. *Stüker, D.* (2008), S. 49 ff.; *Tewes, M.* (2003), S. 129; *Cornelsen, J.* (2000), S. 98.

1255 Vgl. hier und nachfolgend *Riebel, P.* (1992), S. 247 ff. Im Gegensatz zum Direct Costing basiert die Einzelkostenrechnung auf der Kostenspaltung in Abhängigkeit der Zurechenbarkeit. Beim Direct Costing und dessen Aufbaustufe der stufenweisen Fixkostendeckungsrechnung ist die konzeptionelle Grundlage die Kostenspaltung nach dem Verhalten bei Variation eines Kosteneinflussfaktors. I. d. R. wird als Kosteneinflussfaktor die Beschäftigung des Unternehmens betrachtet. Damit liegt der Fokus der stufenweisen Fixkostendeckungsrechnung tendenziell auf Produkten, während die Einzelkostenrechnung sich auf „beliebige" Bezugsgrößen beziehen kann. Vgl. zur Kostenspaltung z. B. *Coenenberg, A. G.* (1999), S. 49 ff. und zur stufenweisen Fixkostendeckungsrechnung z. B. *Agthe, K.* (1959).

1256 Um die Kosten einem Bezugsobjekt zuzurechnen nennt RIEBEL das Relevanz- und das Identitätsprinzip als zwingende Anforderungskriterien. Nach dem Relevanzprinzip sind nur die durch eine Entscheidung zusätzlich ausgelösten positiven und negativen Wirkungen zu berücksichtigen, wenn diese Entscheidung beurteilt werden soll. Das Identitätsprinzip wechselt die Blickrichtung im Vergleich zum Relevanzprinzip und fragt nach, welcher identische Entstehungszusammenhang zu negativen und positiven Wirkungen geführt hat. Nach dem Identitätsprinzip „lassen sich zwei Größen untereinander oder einem Bezugsobjekt dann und nur dann eindeutig zurechnen, wenn diese auf den-

bauend auf einer Bezugsobjekthierarchie ist die Unterscheidung zwischen Einzel- und Gemeinkosten relativ. Die strikte Anwendung des Einzelkostenprinzips führt dazu, dass Gemeinkosten nicht auf nur gemeinsam zurechenbare Bezugsgrößen geschlüsselt werden und folglich eine Netto-Ergebnisrechnung i. d. R. nicht möglich ist.[1257] Ersetzt wird die Netto-Ergebnisrechnung durch die Deckungsbeitragsrechnung, die sachlich und zeitlich abschließend auszuführen ist. Das bedeutet, dass z. B. ein über Leistungen und Perioden gemeinsamer Deckungsbeitrag nicht aufgeteilt werden darf. Der Deckungsbeitrag ist als eine durch eine bestimmte Maßnahme ausgelöste Erfolgsänderung zu verstehen.[1258] Die vorgenannten RIEBEL'schen Prinzipien bedürfen einer, keinen einzelnen Zweck bevorzugenden Datenbasis, die als Grundrechnung zu konzipieren ist. Die Grundrechnung stellt sachlich begründete Zuordnungsbeziehungen für die Kostenrechnung her und gibt an, welchen Bezugsobjekten welche Kosten gemäß dem Identitätsprinzip unmittelbar zugerechnet werden können.[1259] Bezugsobjekte können alle interessierenden Sachverhalte sein, die hierarchisch so geordnet werden, dass Gemeinkosten einer Ebene auf einer hierarchisch höheren Ebene als Einzelkosten zugerechnet werden können. Für eine kundenorientierte, sachliche Bezugsgrößenhierarchie gibt es in der Literatur viele Beispiele, die sich vornehmlich auf den industriellen Kontext beziehen.[1260] Für das Firmenkundengeschäft der Banken sind die vorgestellten Bezugsgrößenhierarchien jedoch ungeeignet. Eine sachliche Bezugsgrößenhierarchie für das Firmenkundengeschäft könnte dem in Abbildung 47 dargestellten Aufbau entsprechen. Dabei wird besonders auf die Verbundeffekte auf Kundenseite eingegangen, die darin bestehen, dass Einzelkunden zu Engagements zusammengeschlossen werden und diese Adressaten vertriebspolitischer Maßnahmen sind. Diese Bezugsgrößenhierarchie ist sowohl für die Kostenzuordnung als auch für die Zuordnung der bereits

---

selben dispositiven Ursprung, dieselbe Entscheidung zurückgeführt werden können." *Riebel, P.* (1992), S. 259.

1257 Theoretisch ist dies denkbar bei einem Unternehmen, dass für einen Kunden innerhalb einer typischen Abrechnungsperiode ein Produkt herstellt, dafür gegründet wurde und nach Fertigstellung des Produkts liquidiert wird und sich daher alle Kosten des Unternehmens für diese Bezugsgrößen Einzelkosten sind. Da dies in der Realität nur selten anzutreffen ist (bestimmte Projektgesellschaften außen vor gelassen), kann die Aussage RIEBELS zur „Abkehr von der Netto-Erfolgsrechnung" als Konsequenz der Einzelkostenrechnung bestätigt werden.

1258 Dies entspricht dem Prinzip der Einzelgeschäftskalkulation in VR-CONTROL wobei jedoch durch die Barwertrechnung statt der Erfolgsänderung eine Vermögenswertänderung fokussiert wird.

1259 Vgl. *Tewes, M.* (2003), S. 256.

1260 Vgl. z. B. *Lissautzki, M.* (2007), S. 171; *Wille, K.* (2005), S. 98; *Tewes, M.* (2003), S. 257.

zogen-integrativen Prozesse können, müssen aber nicht, kundenindividuell ablaufen. Die ablaufbezogen-integrativen Prozesse sind dadurch gekennzeichnet, dass der Kunde ganz oder teilweise während des Prozessablaufs in diesen eingreift und dabei einen oder mehrere Prozessparameter verändert. Auf Produktebene liegt dies z. B. vor, wenn mit dem Firmenkunden spezielle Finanzierungen oder sonstige Bankdienstleistungen erarbeitet werden. Auf Kundenebene gehört dazu etwa die kundenindividuelle Ausgestaltung der Beratungsgespräche, die in der Bank oder beim Kunden stattfinden.

In der vorliegenden Arbeit soll kein vollständiger konzeptioneller Rahmen für eine derartige kundenbezogene Prozesskostenrechnung erarbeitet werden, für die weitere Untersuchung sind die skizzierten Basiselemente jedoch ausreichend. Denn unter Berücksichtigung dieser Ausführungen lässt sich die in den Kreditgenossenschaften implementierte Systematik der Prozesskostenrechnung im Sinne der kundenwertorientierten Steuerung erweitern (siehe Abbildung 49).

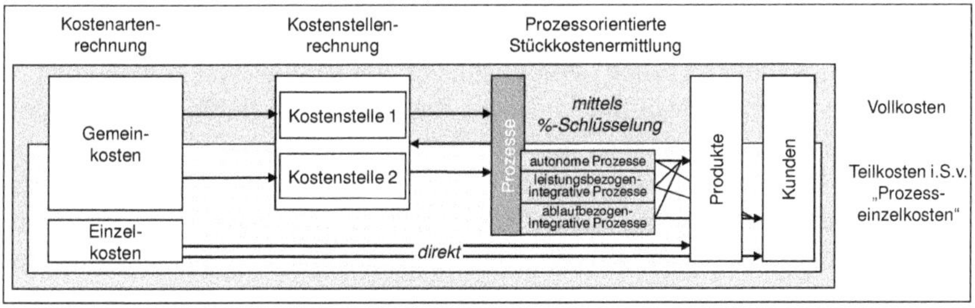

Abbildung 49: Integration der Prozessdifferenzierung in die Systematik der prozessorientierten Kostenrechnung der Kreditgenossenschaften[1280]

Aufbauend auf diesem Ansatz der Prozessanalyse und den bereits bestimmten Kundenwerttreibern lässt sich das Basiskostenpotenzial nun näher konkretisieren.[1281] Das Basiskostenpotenzial stellt eine im Zusammenhang mit dem Basisgeschäfts-, dem Entwicklungs- und dem Cross-Selling-Potenzial stehende Maßgröße für die in die Geschäftsbeziehung zu investierenden Ressourcen dar. Es setzt sich somit zusammen aus den jeweiligen Einzelkosten und den produktbe-

1280 Eigene Darstellung.

1281 Statt von Potenzial spricht DILLER vom Transaktionskostenwert. Vgl. *Diller, H.* (2001), S. 16. Aufgrund der Verbindungen zu den schon definierten Kundenwerttreibern, die als Potenzialgrößen definiert sind, soll jedoch vom Basiskostenpotenzial gesprochen werden. Zudem wäre wie die Ausführungen in diesem Kapitel zeigen eine enge Fokussierung auf die einzelne Transaktion nicht zielgerecht.

zogenen Prozesskosten der einzelnen vorgenannten Kundenwerttreiber und den kundenbezogenen Prozesskosten, die produktübergreifend sind und somit auch nicht diesen Kundenwerttreibern zugeordnet werden können. Konsequent zur produktbezogenen Bestimmung des Basis-, Entwicklungs- und Cross-Selling-Potenzials lassen sich die kundenbezogenen Prozesskosten also übergreifend verstehen.

Ein Dilemma bei der Betrachtung der Berücksichtigung der Kosten liegt darin, dass der Kundenwert Entscheidungsgrundlage dafür ist, ob in eine Geschäftsbeziehung Ressourcen zur Ausschöpfung der Potenziale investiert werden sollen und gleichzeitig diese zu investierenden Ressourcen den prospektiven Kundenwert beeinflussen. Bezogen auf das hier definierte Basiskostenpotenzial entspricht dies den kundenbezogenen Einzel- und Prozesskosten z. B. in Form der Betreuungskosten. Auf diesen Aspekt ist im Zusammenhang mit den kundenwertorientierten Vertriebsplanungsaktivitäten weiter einzugehen.

Das Basiskostenpotenzial umfasst diejenigen Kosten, die produktspezifisch auf Basis der erwarteten Erlösträger und dem Kundenverhalten als beeinflussende Variable resultieren. Dabei wird davon ausgegangen, für die zukünftigen Perioden das Kundenverhalten bezüglich der Integrationsbereitschaft und -fähigkeit als „konstant" anzunehmen. Im Rahmen des nachfolgend erläuterten Kooperationspotenzials wird ein Kundenwerttreiber fokussiert, der sich dadurch ergibt, dass der Kunde im Sinne der Kreditgenossenschaft bezüglich seines Verhaltens „entwickelt" werden kann. Damit stellt das Kooperationspotenzial ein Korrektiv des Basiskostenpotenzials dar, beinhaltet darüber hinaus jedoch weitere Vorteile für die Primärbank.

4.5.3.1.5 Kooperationspotenzial

Die Literatur zum Kundenwert kennt das Kooperationspotenzial als Bereitschaft und Fähigkeit eines Kunden, auf begrenzte Zeit Produktionsfaktoren in den Dispositionsbereich des Anbieters einzubringen.[1282] Die Berücksichtigung eines derartigen Kundenwerttreibers ist insbesondere bei Geschäftsbeziehungen relevant, die durch ein hohes Maß an Integrativität gekennzeichnet sind, wie es bei Dienst- und Bankleistungen der Fall ist.[1283] Denn diese Integrativität bestimmt u. a. die Kosten der von der Kreditgenossenschaft erbrachten Bankleistungen. Et-

---

1282 Vgl. *Rudolf-Sipötz, E.* (2001), S. 121. Etwas anders RIEKER, der das Kooperationspotenzial im Kontext von Anbieter-Kunden-Beziehungen als Möglichkeit, langfristig zusammenzuarbeiten, definiert. Vgl. *Rieker, S.* (1995), S. 60 f. Als Synergieeffekte der vertikalen Kooperation sieht auch GELBRICH das Kooperationspotenzial. Vgl. *Gelbrich, K.* (2001), S. 72.

1283 Vgl. *Rudolf-Sipötz, E.* (2001), S. 126. Vgl. auch Kapitel C.2.3.

wa 55 Prozent der Verwaltungskosten der Kreditgenossenschaften resultieren aus kundenbezogenen Prozessen.[1284]

Basis der weiteren Überlegungen zum Kooperationspotenzial sind die Besonderheiten der Bankleistungen, die in Kapitel C.2.3 herausgearbeitet wurden. Zentrales Konstitutionsmerkmal des Kooperationspotenzials ist die Kooperation zwischen zwei oder mehr Akteuren, die als Zusammenarbeit mit dem Ziel des Erreichens individueller Vorteile definiert wird.[1285] Als Vorteile werden Kooperationsgewinne unterstellt, die sich bezüglich der Kundenbindung und -zufriedenheit, im Rahmen der Produktentwicklung und des Produktionsprozesses und gemeinsamer Marktaktivitäten von Kunden und Herstellern ergeben.[1286] Ob diese Vorteile realisiert werden können, hängt davon ab, ob der Kunde in der Lage ist, sich wertschaffend in den Dienstleistungserstellungsprozess einzubringen und der Anbieter die dafür erforderlichen Fähigkeiten aufbringt. Weiterhin müssen Kunden und Anbieter bereit sein, miteinander zu kooperieren. Ein Ansatz, Kooperationsoptionen zwischen Anbieter und Kunde für den Dienstleistungsbereich darzustellen, ist die Isoleistungslinie von CORSTEN (siehe Abbildung 50). Die Isoleistungslinie stellt dar, bei welchen Aktivitätsgraden von Anbieter und Nachfrager ein gleichartiger Leistungsoutput erstellt werden kann.

Die Isoleistungslinie visualisiert, dass sowohl die Integration des Kunden (Externalisierung) als auch die Übernahme von Aufgaben des Kunden durch den Anbieter (Internalisierung) Ansatzpunkte liefern kann, um kundenwertschaffende Maßnahmen zu identifizieren. Dies sei an drei Beispielen aus dem Bankenbereich verdeutlicht, wobei nicht beachtet wird, ob dadurch die Isoleistungslinie parallel verschoben wird.

Durch die Nutzung des Online-Bankings oder der Selbstbedienungseinrichtungen, wie Geldausgabeautomaten und Kontoauszugsdrucker, übernimmt der Bankkunde traditionelle Aufgabenfelder der Bank (Externalisierung). In Abhängigkeit bestimmter Einflussvariablen des Kundenverhaltens kann dies auf Seiten der Bank zu höheren Kundenprofitabilitäten führen.[1287]

1284 Vgl. *BVR* (2007b), S. 20.

1285 Vgl. *Schneider, Y.* (2007), S. 120.

1286 Vgl. dazu ausführlich *Schneider, Y.* (2007), S. 122 ff.

1287 Vgl. *Gensler, S./Skiera, B./Böhm, M.* (2007). Einflussfaktoren sind der Kanaleffekt, der auftritt, wenn ein kausaler Effekt der Nutzung eines Vertriebskanals auf das Produktnutzungsverhalten besteht, und der Selbstselektionseffekt, der die Unterschiede der Kundencharakteristika beschreibt.

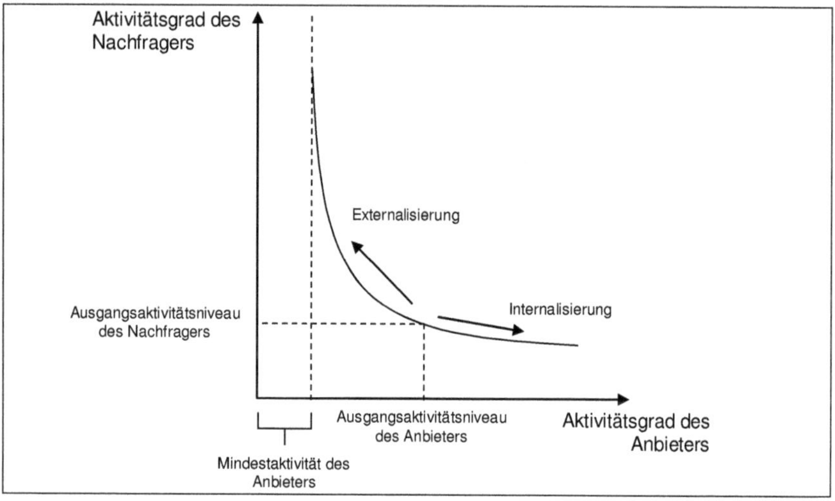

Abbildung 50: Isoleistungslinie[1288]

Das zweite Beispiel der bankenseitigen Externalisierung knüpft an die Darstellung des Kreditprozesses in Abbildung 19 an. Aus den Erkenntnissen der Prinzipal-Agenten- i. V. m. der Transaktionskostentheorie und dem Relationship Banking folgt,[1289] dass für den Bankenkontext solche (Kredit-)Vertragsbezie-hungen in besonderer Weise relevant sind, die sich durch Unsicherheit und Informationsdefizite des Prinzipals (Bank) dem Agenten (Firmenkunde) gegenüber auszeichnen. In Abhängigkeit des Ausmaßes und der Art der Informationsasymmetrie sowie den Maßnahmen zu dessen Abbau (z. B. Screening, Monitoring) fallen Transaktionskosten auf Seiten des Prinzipals an. Unberührt von der informationsökonomischen Begründung der Existenz von Banken[1290] stellt das Kooperationspotenzial eines Firmenkunden in diesem Zusammenhang einen Einflussfaktor auf die Höhe der Transaktionskosten und die Effizienz der Informationsproduktion auf Seiten der Bank dar. Dies soll an einem konkreten Beispiel verdeutlicht werden. Im Rahmen der Kreditwürdigkeitsprüfung ist die Beschaffung von

---

1288 *Corsten, H.* (2000), S. 151.

1289 Vgl. dazu z. B. *Püthe, T.* (2009), S. 40 ff. sowie *Segbers, K.* (2007), S. 27 ff. und S. 120 ff.

1290 Informationsökonomische Modelle kommen zu dem Ergebnis, dass die Existenz von Banken sich darin begründet, dass diese komparative Kostenvorteile bei der Kompensation von Informationsasymmetrien gegenüber anderen Finanzintermediären haben. Vgl. *Püthe, T.* (2009), S. 61 sowie die dort angegebene Literatur.

qualitativen Ratinginformationen des Kreditkunden mit vielfältigen Problemfeldern versehen (z. B. Medienbrüche, Notwendigkeit zur häufigen Kundenansprache).[1291] Einen Lösungsansatz beschreiben BERENS ET AL. damit, dass auf Seiten des Firmenkunden eine Software implementiert wird, mittels der die relevanten Informationen erfasst und digital an die Bank übermittelt werden können. Auch hier übernimmt der Kunde in kooperativer Weise Aufgaben der Bank, um Schnittstellenprobleme und Probleme der Informationsbeschaffung zu lösen und uno actu für eine höhere Prozessqualität und bessere Ratinggrundlage zu sorgen.

Ein Beispiel für die bankenseitige Internalisierung von Aufgaben besteht in der sog. Wertschöpfungskettenkreuzung, die die selektive Auslagerung von (Teilen von) Finanzprozessen an Banken umfasst.[1292] Der von KÖNIG ET AL. beschriebene Ansatz der Wertschöpfungskettenkreuzung besteht darin, dass die Bank mittels eines eingebetteten Informationssystems Zugriff auf Rechnungsdaten des Firmenkunden hat und Aufgaben des Forderungs- und Liquiditätsmanagement übernimmt. Die bankseitige Vorteilhaftigkeit wird von den Autoren nicht dargelegt. Es kann jedoch vermutet werden, dass diese sich in einer stärkeren Gebundenheit des Kunden aufgrund zusätzlicher (System-)Wechselkosten, einer höheren risikorelevanten Transparenz über die Liquiditätslage des Unternehmens sowie dem Angebot zusätzlicher Finanzdienstleistungen zeigen. Dagegen müssen jedoch Kosten für die Integration- und Interaktion ebenso berücksichtigt werden, wie Kosten, die im Rahmen der Akquisition sowie der Entwicklung der Kooperationsbereitschaft und -fähigkeit (Qualifikation) des Firmenkunden entstehen.

Die Qualifikation der Kunden sieht GOUTHIER als zwingend notwendig an, um die Kunden optimal in den Dienstleistungserstellungsprozess integrieren zu können.[1293] Diese Aufgabe obliegt dem Anbieterunternehmen, das damit die Vorteile eines integrationsbereiten und -fähigen Kunden nutzen kann. GOUTHIER unterscheidet aus Sicht des Unternehmens zwei Extremformen von Kundenbildern.[1294] Dies ist der Kunde als Störfaktor und der Kunde als Ressource. An diesen zwei Leitbildern richtet er Integrationsziele und -strategien aus. Während

---

1291 Vgl. hier und nachfolgend *Berens, W. et al.* (2005), S. 786 f.

1292 Vgl. *König, W. et al.* (2007), S. 133 ff. Zu beachten ist, dass bisher keine derartigen Angebote zur Wertschöpfungskettenkreuzung im (mittelständischen) Firmenkundengeschäft vorliegen. Bei großen Unternehmen ist zudem eher der Trend zur Disintermediation zu beobachten, der sich darin zeigt, dass Finanzprozesse in die Unternehmen integriert werden, so dass Kreditinstitute als Finanzintermediäre obsolet werden.

1293 Vgl. hier und nachfolgend *Gouthier, M. H. J.* (2003), S. 11.

1294 Vgl. hier und nachfolgend *Gouthier, M. H. J.* (2003), S. 179 ff. Ausprägungen zwischen diesen beiden Extremen werden von GOUTHIER anerkannt jedoch nicht weiter thematisiert

beim Kunden als Störfaktor das Ziel des Abbaus des Integrationsgrades über eine Strategie der Internalisierung seiner Aufgaben zu erreichen ist, verhält es sich beim Kunden als Ressource umgekehrt. Hier soll der Integrationsgrad über eine Externalisierungsstrategie ausgebaut werden. Dem Nutzen der Kooperationsfähigkeit und -bereitschaft stehen somit i. d. R. Kosten der Kundenentwicklung gegenüber.[1295] So berichtet RUDOLF-SIPÖTZ von einem Fall im Bankenbereich, bei dem „unprofessionelles" Agieren des Kunden zu Haftungsansprüchen gegenüber der Bank führte.[1296] Anschließend versuchte die Bank mittels geeigneter Maßnahmen die Kunden zu schulen und damit das Kooperationspotenzial zu steigern. Die Entscheidung, ob eine Maßnahme zur Kundenentwicklung durchgeführt werden soll, knüpft GOUTHIER an die Frage, ob die Kosten der Inkompetenz größer sind als die Kosten der Kompetenzbeschaffung zzgl. etwaiger kompetenzbedingter Opportunitätskosten, die z. B. aus einem intensiveren Preiskampf aufgrund besser informierter Kunden resultieren.[1297] Diese Kostenarten werden zusammen als Kompetenzkosten bezeichnet.

Im Gegensatz zu den bisherigen Kundenwerttreibern, die sich durch konkrete Bankleistungen als Fixpunkte der Zuordnung von Erlösen und Kosten messen lassen können, besteht beim Kooperationspotenzial eine größere Unbestimmtheit der resultierenden Nutzen- und Kostengrößen. Nach SCHNEIDER lassen sich für die Größen Kooperationsfähigkeit und -bereitschaft des Kunden und des Anbieters und dem produktbedingten Bedarf an Zusammenarbeit Kennzahlen ableiten, die geplant werden können.[1298] VON WANGENHEIM sieht das Wertpotenzial der Kooperation „in den realisierbaren Synergien und Wertsteigerungen, die durch die Zusammenlegung von Ressourcen erreicht werden können."[1299] Gleichwohl dürfte die Messung dieses Kundenwerttreibers problematisch bis nicht hinreichend valide realisierbar sein.

Zusammenfassend lässt sich festhalten, dass das Kooperationspotenzial die Möglichkeiten beinhaltet, wie durch die Gestaltung der Integration des externen Faktors, also des Firmenkunden, die produkt- und kundenbezogenen Kosten reduziert werden können. Daraus resultieren unmittelbare Auswirkungen auf die ökonomischen Charakteristika der zu bewertenden Kundenbeziehung. Synergien aus der vertikalen Zusammenarbeit von Kreditgenossenschaft und Bank schaffen somit den Wert des Kooperationspotenzials.[1300] Die Höhe wird grds. durch die Eingriffstiefe, d. h. die Stelle im Wertschöpfungsprozess, an der der Eingriff

---

1295 Vgl. *Gouthier, M. H. J.* (2003), S. 456 ff.
1296 Vgl. hier und nachfolgend *Rudolf-Sipötz, E.* (2001), S. 126.
1297 Vgl. hier und nachfolgend *Gouthier, M. H. J.* (2003), S. 461 ff.
1298 Vgl. *Schneider, Y.* (2007), S. 158.
1299 *von Wangenheim, F.* (2003), S. 54.
1300 Vgl. *Schmöller, P.* (2001), S. 120 ff.; *Meffert, H./Bruhn, M.* (2009), S. 424.

stattfindet, sowie die Eingriffsintensität, die durch die Zahl der betroffenen Prozesse sowie der erforderlichen Intensität beeinflusst wird, bestimmt.[1301]

Mit dem Kooperationspotenzial sind die sich auf den Güter- und Warenstrom beziehenden Kundenwerttreiber der Ertragsdimension abschließend behandelt. Bereits im Rahmen der Diskussion zum Kooperationspotenzial wurde deutlich, dass Informationsströme ebenfalls inhärenter Bestandteil der Interaktionsprozesse zwischen Firmenkunden und Kreditgenossenschaft sind. Neben den unmittelbar mit der Leistungserstellung verbundenen Informationen können diese Informationsströme auch Informationen beinhalten, die sich nur mittelbar oder auch gar nicht auf die betrachteten Interaktionsprozesse auswirken, gleichwohl nutzenstiftend für die Kreditgenossenschaft sein können.

4.5.3.1.6 Informationspotenzial

Der begriffliche Inhalt des Informationspotenzials ist in der Literatur uneinheitlich.[1302] In dieser Arbeit umfasst das Informationspotenzial diejenigen Informationen, die von Seiten des Kunden an das Anbieterunternehmen fließen und von diesem zur erfolgreichen Unternehmensführung (z. B. Prozessoptimierung, Produktentwicklung, Kundenorientierung) eingesetzt werden können.[1303] Nach BAUER ist der Kunde „die wichtigste Quelle eines Unternehmens für strategisch relevante Informationen.“[1304] „Das Informationspotential beschreibt ganz allgemein das Potential eines Kunden durch sein Informationsverhalten positive Kosten- und Nutzen-Wirkungen herbeizuführen und damit zur Zielerreichung eines Anbieters beizutragen.“[1305] Abzugrenzen ist das Informationspotenzial vom Kooperationspotenzial insofern, als das Kooperationspotenzial die Informationen als Inputfaktor in einem konkreten kundenintegrativen Leistungserstellungsprozess umfasst. Im Informationspotenzial dagegen sollen nun diejenigen Informationen von einem Firmenkunden erfasst werden, die über den konkreten Leistungserstellungskontext hinaus nutzenstiftend für die Kreditgenossenschaft sind. In diesem Kontext sind Informationen nicht zweckorientiertes Wissen[1306], son-

---

1301 Vgl. *Schmöller, P.* (2001), S. 124 sowie die dort angegebene Literatur.

1302 Vgl. z. B. *Wünschmann, S.* (2007), S. 37 ff., *Schneider, N. C.* (2007), S. 159 ff., *Howald, B.* (2007), S. 52, *Friederichs-Schmidt, S.* (2006), S. 139 ff., *Foehn, P.* (2006), S. 136 ff., *Böhrs, S.* (2004), S. 62 ff., *von Wangenheim, F.* (2003), S. 51 ff., *Tewes, M.* (2003), S. 94 ff., *Eberling, G.* (2002), S. 154 ff., *Rudolf-Sipötz, E.* (2001), S. 113 ff. und *Cornelsen, J.* (2000), S. 224 ff.

1303 Vgl. ähnlich *Rudolf-Sipötz, E.* (2001), S. 113.

1304 *Bauer, R. A.* (2000), S. 41.

1305 *Tewes, M.* (2003), S. 94.

1306 Vgl. dazu die Ausführungen in Kapitel E.1.2.1.

dern „wertschaffender Inputfaktor“[1307]. GELBRICH verwendet dazu den Begriff des Know-how- oder Innovationspotenzials.[1308] Da jedoch auch Beschwerden des Kunden Gegenstand des Informationspotenzials sein können,[1309] soll dieser Begriffsverwendung nicht gefolgt werden.[1310] Vielmehr wird das Innovationspotenzial in Anlehnung an RUDOLF-SIPÖTZ unter das hier weiter gefasste Informationspotenzial subsumiert.[1311]

RUDOLF-SIPÖTZ/TOMCZAK sehen in dem Inhalt und der Qualität der Informationen sowie der Bereitschaft zur Informationsabgabe zentrale Bestimmungsfaktoren des Informationspotenzials.[1312] Informationen, die Gegenstand des Informationspotenzials sind, beziehen sich auf unterschiedliche Bereiche und Themen.[1313] Diese sind z. B. Kundenbedürfnisse- und -situationen, Prozessoptimierungen oder Produktentwicklungen und -verbesserungen, Entwicklung neuer/Überarbeitung bestehender Konzepte in Beratung oder Service, Service- und Produktqualität, Ansprüche der Kunden an die Mitarbeiter, Konkurrenz- und Marktanalysen oder Determinanten der Loyalität oder Abwanderungsgründe. Wertvoll aus Sicht der Kreditgenossenschaft sind diejenigen Informationen, die sich nutzenstiftend einsetzen lassen, z. B. wenn sie zu Verbesserungen auf der Prozess- oder Leistungsebene führen.[1314] Das bedeutet zugleich, dass sich auf Basis der erhaltenen Informationen für die Kreditgenossenschaft realisierbare Veränderungen initiieren lassen.

Hinsichtlich der Qualität der Informationen lassen sich faktische und normative Informationen unterscheiden. Faktische Informationen beinhalten Aussagen über objektiv prüfbare Sachverhalte, während normative Informationen subjektive Bewertungen und Meinungen von Kunden sind.[1315] Faktische Informationen gelten als wertvoller, da die Inhalte aufgrund ihrer Objektivität leicht auf andere Kunden übertragbar bzw. für die Bank nutzbar sind.[1316] Normative Informationen führen erst dann zu Handlungen, wenn Anregungen gleicher Art mehrfach

---

1307 *Schneider, Y.* (2007), S. 92.

1308 Vgl. *Gelbrich, K.* (2001), S. 72. Ähnlich auch *Rieker, S.* (1995), S. 59 und *Schulz, B.* (1995), S. 115.

1309 Vgl. *Meffert, H./Bruhn, M.* (2009), S. 424.

1310 Für eine ausführliche Auseinandersetzung mit dem Beschwerdeverhalten im Kontext des Kundenwerts vgl. *Wünschmann, S.* (2007), S. 93 ff.

1311 Vgl. *Rudolf-Sipötz, E.* (2001), S. 114. So auch *Schneider, Y.* (2007), S. 91 f.

1312 Vgl. *Rudolf-Sipötz, E./Tomczak, T.* (2001), S. 39 f.

1313 Vgl. *Rudolf-Sipötz, E.* (2001), S. 114 sowie die dort angegebene Literatur.

1314 Vgl. *Tewes, M.* (2003), S. 95.

1315 Vgl. *Schneider, N. C.* (2007), S. 92.

1316 Vgl. *Tewes, M.* (2003), S. 94.

an das Unternehmen herangetragen werden.[1317] Auf Ebene des Einzelkunden erhöhen diese die Transparenz über und das Verständnis für den Kunden. Während normative Informationen häufig in Form von Beschwerden artikuliert werden, werden faktische Informationen durch initiative Maßnahmen auf Seiten des Anbieters generiert wie z. B. Kundenbefragungen oder Kundenworkshops. CORNELSEN kombiniert Kundeninformationen in Abhängigkeit des Inhalts (Leistungen und Prozesse) und der Qualität. Übertragen auf den hier fokussierten Bankenkontext lassen sich darauf aufbauend die folgenden Informationsarten ableiten.

Tabelle 20: Kundeninformationen in Abhängigkeit von Qualität und Inhalt[1318]

Inhalt / Art	Bankleistung	Prozess
Normativ (subjektiv)	„Produktive" Informationen: • Konditionshöhe • Leistungsindividualität • Bedarfssituation/ Künftige Planung • …	„Produktive" Informationen: • Qualifikation der Berater • Art und Inhalt der Betreuung • …
Faktisch (objektiv)	Anregungen/Verbesserungsvorschläge: • Beratungsinhalte • Leistungsmerkmale • …	Anregungen/Verbesserungsvorschläge: • Kreditprozessstruktur • Bearbeitung von Anfragen • …

Die Bereitschaft des Firmenkunden, Informationen über den konkreten Leistungserstellungsprozess hinaus an die Kreditgenossenschaft abzugeben, stellt die conditio sine qua non des Informationspotenzials dar. Als Einflussfaktoren dieser Bereitschaft werden in der Literatur insbesondere das Commitment und das Involvement des Kunden genannt.[1319] Commitment wird als die innerlich empfundene Verpflichtung, sich in die Beziehung einzubringen, verstanden.[1320] Nach MORGAN/HUNT führt ein erhöhtes Commitment zu einer höheren Kooperationsbereitschaft,[1321] die sich auch auf die Weitergabe von Informationen bezieht.[1322] Das Involvement beschreibt das innere Engagement eines Kunden, sich

1317 Vgl. hier und nachfolgend *Cornelsen, J.* (2000), S. 225 f.

1318 In Anlehnung an *Cornelsen, J.* (2000), S. 226.

1319 Vgl. *Tewes, M.* (2003), S. 96.

1320 Vgl. *Tewes, M.* (2003), S. 96. Eine ausführliche Diskussion verschiedener Definitionen des Begriffs Commitment findet sich bei *Püthe, T.* (2009), S. 162 ff.

1321 Vgl. *Morgan, R. M./Hunt, S. D.* (1994), S. 23.

1322 Vgl. *Tewes, M.* (2003), S. 96. DWYER/SCHURR/OH messen Commitment unter anderem anhand des Inputs, worunter sie den signifikanten Austausch von ökonomischen, kom-

einer subjektiv als wichtig empfundenen Sache zu widmen. Unterschieden werden kann zwischen dem product-involvement, bei dem die Leistungen des Anbieters Objekt des Engagements sind, dem self-involvement, unter dem ein persönliches Bedürfnis nach Anerkennung verstanden wird, und dem other-involvement, das auf die Unterstützung Dritter ausgerichtet ist. Unabhängig davon, welches Involvement vorliegt, wird davon ausgegangen, dass ein hohes Involvement mit einer höheren Bereitschaft zur Abgabe von Informationen einhergeht.[1323] Darüber hinaus werden auch auskunftsfördernde Persönlichkeitsmerkmale des Kunden, situative Einflussfaktoren, Unzufriedenheit, Erfolgsaussichten auf Veränderungen aus Sicht des Kunden sowie produkt- und dienstleistungsspezifische Faktoren diskutiert.[1324]

Informationen zu generieren und zu nutzen, kann eine Quelle von Wettbewerbsvorteilen sein, da es Außenstehenden nur schwer gelingen dürfte, die Kundeninformationen und das im Unternehmen bereits enthaltene Wissen in ähnlicher Weise zu verknüpfen.[1325] Herausragende Bedeutung wird daher den Kundeninformationen und demzufolge der Berücksichtigung des Informationspotenzials in einem Kundenwertmodell in Märkten mit hohem Konkurrenzdruck und intensivem Wettbewerb beigemessen.[1326] Wie die Ausführungen in Kapitel C.1.5 zum Wettbewerb und der Konzentration zeigen, kann davon für das Firmenkundengeschäft der Kreditgenossenschaften ausgegangen werden. Je besser ein Unternehmen seine Kunden versteht, desto besser kann es auf deren Bedürfnisse eingehen und den Leistungserstellungsprozess auf die Herstellung kundenorientierter Leistungen ausrichten.[1327] Ein weiterer mit dem Kundenverständnis verbundener Vorteil ist, das Kundenplanungs- und -informationsrisiko zu reduzieren.[1328] Dieses Risiko besteht aus dem Verlustpotenzial, dass sich daraus ergibt, dass auf Basis der informationsschwachen, aber annahmenintensiven Prognose von Kundenwerten Entscheidungen getroffen werden, die bei Abweichungen von den erwarteten Werten zu negativen Konsequenzen für die Kreditgenossenschaft führen. Dies ist klassischerweise der Fall, wenn spezifisch in Kundenbeziehungen investiert wird. Aus diesen Gründen soll das Informationspo-

---

munikativen und/oder emotionalen Ressourcen verstehen. Vgl. *Dwyer, F. R/Schurr, P. H./Oh, S.* (1987), S. 19.

1323 Vgl. *Schneider, Y.* (2007), S. 102 f.; *Tewes, M.* (2003), S. 96.

1324 Vgl. ausführlich dazu *Schneider, Y.* (2007), S. 101 ff.

1325 Vgl. *Schneider, N. C.* (2007), S. 95.

1326 Vgl. *Kleinaltenkamp, M./Dahlke, B.* (2006), S. 218 f.

1327 Vgl. *Schneider, Y.* (2007), S. 94 sowie die dort angegebene Literatur.

1328 Vgl. dazu die Ausführungen in Kapitel D.4.5.4.1.2.

tenzial eines Kunden in einem Kundenwertmodell für das Firmenkundengeschäft der Kreditgenossenschaften explizit berücksichtigt werden.[1329]

Die Höhe und Wirkrichtung des Informationspotenzials hängt davon ab, welchen Nutzen die Informationen für die Kreditgenossenschaft haben und welche Kosten mit deren Beschaffung, Aufbereitung und Verwendung verbunden sind. Der Nutzen kann aus Sicht der Kreditgenossenschaft bspw. in Ertragssteigerungen, Kostenreduzierungen oder Kundenbindungsintensivierungen gesehen werden. Der Nutzen des Informationspotenzials wird dadurch beeinflusst, in wie weit die Kreditgenossenschaft fähig ist, die Informationen zu sammeln, zu systematisieren und auszuwerten.[1330] Relevante Kosten entstehen durch die Schaffung und Aufrechterhaltung von Infrastrukturen, die als Kundenfeedbacksysteme bezeichnet werden können,[1331] sowie den anfallenden Kosten im Rahmen der Umsetzung von induzierten Veränderungen. VON WANGENHEIM schlägt daher vor, die Bewertung des Informationspotenzials auf Basis dessen künftiger Verwendungsmöglichkeiten vorzunehmen.[1332] Dies können z. B. Erlöszuwächse aus auf Kundenanregung entwickelten Produktvarianten oder -innovationen oder Kosteneinsparungen aus Prozessoptimierung, Zeitreduzierung und Produktivitätssteigerung sein.[1333] Er stimmt jedoch KLEINALTENKAMP/DAHLKE zu, die argumentieren, dass „dieser Wert nicht per se existiert“[1334]. Denn die Verwendung des durch die Kundeninformation veränderten Wissens im Unternehmen hängt von einer Vielzahl weiterer Faktoren ab. „Eine Zurechnung der Kosten-Nutzen-Effekte von Kundeninformationen auf einzelne Kunden ist [.] sehr problematisch [...] [, da sie; Erg. d. Verf.] multiple Wirkungs- und Verbundeffekte entfalten [können; Erg. d. Verf.]“[1335].

Im Gegensatz zu den bisherigen Bestimmungsfaktoren, sind die Messung des Informationspotenzials und dessen etwaige monetäre Abbildung schwierig.[1336] SCHNEIDER entwickelt zwar ein theoretisches Modell eines monetären

---

1329 Anders *Krafft, M.* (2007), S. 203, der die Notwendigkeit qualitative Aspekte im Kundenwert zu berücksichtigen nicht sieht, da diese, obgleich sie eine wesentliche Rolle bei Entscheidungen spielen, auch ohne explizite Integration intuitiv berücksichtigt werden. Ähnlich auch *Lissautzki, M.* (2007), S. 249 f., der qualitative Aspekte nicht berücksichtigt, da sie nur schwer zu monetarisieren sind.

1330 Vgl. *Gelbrich, K.* (2001), S. 58.

1331 Vgl. *Tewes, M.* (2003), S. 98.

1332 Vgl. *von Wangenheim, F.* (2003), S. 52.

1333 Vgl. *Cornelsen, J.* (2000), S. 230.

1334 *Kleinaltenkamp, M./Dahlke, B.* (2006), S. 233.

1335 *Tewes, M.* (2003), S. 99.

1336 Vgl. *Diller, H.* (2001), S. 10.

Informationswerts.[1337] Der monetäre Informationswert wird jedoch in der praktischen Anwendung in ihrer empirischen Studie durch einen ordinalen Wert ersetzt.[1338] Daher wird bei der Messung des Informationspotenzials weniger auf den tatsächlichen monetären Nutzen der Information, sondern auf die Kunden- und Informationsmerkmale, die das Informationspotenzial beeinflussen, eingegangen. Dieses sind die bereits genannten Grundvoraussetzungen Commitment und Involvement des Kunden, der Aktivitätsgrad des Kunden (reaktive oder selbstinitiierte Informationsübermittlung), die Intensität und Häufigkeit mit der Kunden Informationen übermitteln, die Qualität der Information (normative oder faktische Informationen) sowie die Relevanz der Inhalte der Information.[1339] Die Informationsinhalte sind im Rahmen einer Messung kreditgenossenschaftsindividuell auszugestalten. Relevante Inhaltsdimensionen können Markt- und Konkurrenzinformationen auf Seiten der Kreditgenossenschaft, Informationen zur gegenwärtigen und geplanten, zukünftigen Unternehmensentwicklung oder Informationen zu risikorelevanten Bereichen, die sich nicht in direkt in den kommunizierten Unternehmensplanungen des Firmenkunden niederschlagen, sein.

Es ist die Aufgabe des Kundenmanagements in der Kreditgenossenschaft und folglich auch des Vertriebs bzw. der Firmenkundenbetreuung, die unterschiedlichen Informationsinhalte zu benennen, zu strukturieren und den Informationsfluss aktiv zu fördern. Erst dadurch kann eine zweckgerechte und verstärkte Ausschöpfung des Informationspotenzials des Firmenkunden erreicht werden.[1340]

Mit dem Informationspotenzial ist die Ableitung der beziehungsbezogenen Kundenwerttreiber der Ertragsdimension abgeschlossen. Nachfolgend werden die Kundenwerttreiber erörtert, die über die dyadische Beziehung hinaus einen Beitrag zur Zielerreichung des Anbieterunternehmens leisten können.

4.5.3.2 Beziehungsübergreifende Kundenwerttreiber der Ertragsdimension

Die beziehungsübergreifenden Kundenwerttreiber zeichnen sich dadurch aus, dass sie sich nicht auf die direkten ökonomischen Charakteristika der Beziehung zwischen Firmenkunde und Kreditgenossenschaft auswirken. Ihren wertschaffenden Beitrag erzielen sie dadurch, dass die Beziehung zu einem Firmenkunden

---

1337 Vgl. *Schneider, N. C.* (2007), S. 181.

1338 Vgl. *Schneider, N. C.* (2007), S. 210 ff.

1339 Vgl. *Tewes, M.* (2003), S. 94 ff. Ähnlich auch die Einflussfaktoren des Informationspotenzials bei *Rudolf-Sipötz, E.* (2001), S. 120.

1340 Vgl. *Rudolf-Sipötz, E.* (2001), S. 121.

Einfluss nimmt auf aktuelle oder potenzielle Geschäftsbeziehungen zu anderen Firmenkunden.

4.5.3.2.1 Referenzpotenzial

Auch das Referenzpotenzial ist ein uneinheitlich definierter Kundenwerttreiber.[1341] Die Begriffsvielfalt lässt sich sinnvoll anhand eines einfachen Beziehungsgefüges erläutern. Das relevante Beziehungsgefüge besteht aus dem Anbieterunternehmen sowie einer Menge von aktuellen und/oder potenziellen Kunden. Es wird davon ausgegangen, dass alle drei genannten Akteure miteinander kommunizieren können. Für die weiteren Ausführungen soll vereinfachend von zwei Kunden ausgegangen werden. Anhand dieses gedanklich skizzierten Beziehungsdreiecks können die verschiedenen Ansätze des Referenzpotenzials als Kundenwerttreiber diskutiert werden. Gemeinsamkeit aller Ansätze ist, dass es eine verbale oder non-verbale Kommunikation auf Ebene der Kunden gibt, deren Gegenstand das Anbieterunternehmen oder Leistungen des Anbieterunternehmens sind. Die Informationen, die über die Kommunikation ausgetauscht werden, können positiv, negativ oder neutral sein.[1342] Unterschiede der Ansätze ergeben sich aus dem Verständnis von Referenzen und der Integration des Anbieters in den Referenzprozess. DILLER unterscheidet dazu zwischen aktiven und passiven Referenzen.[1343]

Im Industriegüterbereich handelt es sich bei einer Referenz typischerweise „um einen anbieterseitigen Verweis auf bereits bestehende Anlagen, Systeme oder Projekte."[1344] Das bedeutet, das Anbieterunternehmen kommuniziert mit einem potenziellen oder aktuellen Kunden derart, dass es auf einen anderen Kunden als Referenz verweist, um damit ein Signal der eigenen Leistungsfähigkeit zu geben. „Man leiht sich [.] den guten Ruf des Referenzgebers und signalisiert auf diese Weise Selbstbindung, was Anbahnungskosten spart."[1345] Dabei kann es sich bei dem Referenznehmer sowohl um einen potenziellen Kunden handeln mit dem das Anbieterunternehmen eine Geschäftsbeziehung aufbauen will als auch um einen bestehenden Kunden, dem bisher nicht von ihm nachgefragte oder neu entwickelte Leistungen angeboten werden. Die beiden Kunden weisen dabei nicht zwingend eine eigenständige Beziehung auf. In diesem Kontext sind die Referenz akquisitorisch in ihrer Wirkung und der Referenzgeber

1341 Vgl. z. B. *Cornelsen, J.* (2000), S. 186 ff. sowie *Rudolf-Sipötz, E.* (2001), S. 108 ff.

1342 Vgl. *Cornelsen, J.* (2000), S. 190.

1343 Vgl. *Diller, H.* (2001), S. 13.

1344 *von Wangenheim, F.* (2003), S. 54.

1345 *Stahl, H. K./Matzler, K./Hinterhuber, H. H.* (2006), S. 429.

passiv. In der Terminologie DILLERS wären dies passive Referenzen.[1346] RUDOLF-SIPÖTZ bezeichnet diesen Wertbeitrag als Ausstrahlungseffekt eines Schlüsselkunden.[1347]

Im Rahmen der Kundenwertforschung wird vor allem die aktive Referenz im Sinne der intendierten oder unintendierten Kommunikation zwischen zwei Kunden über ein Anbieterunternehmen oder dessen Leistungen betrachtet. So definiert CORNELSEN Referenzen als „eine Form der direkten, verbalen Kommunikation zwischen zwei Personen [.], bei der positive, negative oder neutrale Informationen über einen Anbieter bzw. dessen Angebotsleistung [...] ausgetauscht werden, ohne dass dabei kommerzielle Interessen im Vordergrund stehen."[1348] Derartige Referenzen werden entweder aktiv von Referenzgeber bzw. -empfänger initiiert oder passiv erfragt. Nach SCHMÖLLER und STRÄTER sind Referenzempfänger stets Nichtkunden.[1349] VON WANGENHEIM rekurriert nicht auf den Begriff der Referenz, da er diesen als – wie oben beschrieben – etabliert im Industriegütermarketing ansieht. Er nutzt den Begriff der Weiterempfehlung, den er als „Abgabe positiver und negativer Informationen über wahrgenommene Merkmale eines Anbieters oder eine Anbieterleistung durch einen nicht kommerziell interessierten Kommunikator an mehrere Rezipienten"[1350] definiert.[1351] In Abgrenzung zu CORNELSENs Definition beschränkt VON WANGENHEIM die Informationsübergabe nicht auf die verbale Kommunikation.[1352] In beiden Definitionen ist das Anbieterunternehmen nicht direkt involviert, sondern lediglich Kommunikationsinhalt, wobei jedoch die Referenzübergabe aktiv durch die Kunden verläuft und keine kommerzielle Motivation aufweist. Häufig liegen daher Beziehungen zwischen den Kunden vor. Durch die Referenz oder Weiterempfehlung erfolgt eine Beeinflussung des faktischen oder potenziellen Kaufverhaltens anderer Akteure. VON WANGENHEIM fokussiert zudem im Gegensatz zu SCHMÖLLER und STRÄTER sowohl auf Nichtkunden als auch auf Bestandskunden. Aufbauend auf diese weite Definition können die Einflusseffekte von aktiven Referenzen differenziert werden. Die Wirkung der Referenzen sind bei potenziellen Kunden so zu verstehen, dass diese hinsichtlich der Entscheidung

---

1346 Vgl. *Diller, H.* (2001), S. 13.

1347 Vgl. *Rudolf-Sipötz, E.* (2001), S. 111; auch *Rieker, S.* (1995), S. 58 f. und *Schmöller, P.* (2001), S. 126, die in diesem Zusammenhang den Begriff der Strategischen Relevanz eines Kunden nutzt.

1348 *Cornelsen, J.* (2000), S. 190.

1349 Vgl. *Schmöller, P.* (2001), S. 125; *Sträter, J.* (2005), S. 76.

1350 *von Wangenheim, F.* (2003), S. 55.

1351 Eine ähnliche Definition nutzen *von Wangenheim, F./Bayón, T./Herrmann, A.* (2006), S. 306.

1352 Vgl. *von Wangenheim, F.* (2003), S. 55.

eine Geschäftsbeziehung mit einem Anbieterunternehmen aufzubauen, beeinflusst werden (Akquisitionseffekt). Bei aktuellen Kunden dagegen wirken die Referenzen auf die Dauer der verbleibenden Kundenbeziehung[1353] (Bindungseffekt) und/oder auf das Kaufvolumen und die Kauffrequenz[1354] (Intensitätseffekt).

Die unbestrittene Bedeutung von Referenzen liegt in ihrer Eigenschaft, als Informationsquelle aus Konsumentenperspektive und als werbliche Kommunikationsform aus Anbieterperspektive zu fungieren und der damit verbundenen nicht unbedingt intendierten Beeinflussung des Verhaltens oder der Einstellungen des Referenznehmers.[1355] Die Bedeutung von Referenzen nimmt auch deshalb zu, da klassische Kommunikationsstrategien als relativ ineffektiv gelten,[1356] so dass referenzaktive Kunden bei positiven Referenzen helfen, Akquisitionskosten zu sparen.[1357] Insbesondere für die Finanzdienstleistungsbranche gelten aktive Referenzen in Form der Weiterempfehlungen als wesentlicher Erfolgsfaktor.[1358] Aber auch die passiven Referenzen entwickeln signifikante Multiplikatoreffekte. So nennt RUDOLF-SIPÖTZ das Beispiel einer Universalbank um die Ausstrahlungseffekte einer Kündigung der Bankbeziehung zu verdeutlichen.[1359] Danach haben 29,1 Prozent der befragten Kontoabbrecher angegeben, dass weitere Personen aus dem Familien- oder Bekanntenkreis mit ihnen die Bank verlassen hatten. Zwar beziehen sich diese Belege für die Bedeutung von Referenzen auf Geschäftsbeziehungen im Privatkundengeschäft und können damit nur bedingt auf das Firmenkundengeschäft übertragen werden. Dennoch kann plausibel davon ausgegangen werden, dass aktive und passive Referenzen auch für das Firmenkundengeschäft der Kreditgenossenschaften von gehobener Relevanz sind. Dabei sind vor allem das Regionalprinzip und die Eigenschaften von Bankleistungen als relevante Kontextfaktoren zu beachten.[1360] Aufgrund des Regio-

---

1353 Vgl. *von Wangenheim, F.* (2003), S. 73 ff.

1354 Vgl. *Tewes, M.* (2003), S. 105.

1355 Vgl. *Cornelsen, J.* (2000), S. 190 ff.; *von Wangenheim, F./Bayón, T./Herrmann, A.* (2006), S. 305. Ein Ergebnis dieser Verhaltensbeeinflussung ist z. B. in der Geschwindigkeit von Diffusionsprozessen zu sehen, die durch Referenzen bzw. Weiterempfehlungen beschleunigt oder gebremst werden können. Vgl. *Hogan, J. E./Lemon, K. N./Libai, B.* (2003), S. 196 ff.

1356 Vgl. *Rudolf-Sipötz, E.* (2001), S. 109 f.

1357 Vgl. *Gelbrich, K.* (2001), S. 59; *Stahl, H. K./Matzler, K./Hinterhuber, H. H.* (2006), S. 429.

1358 Vgl. *Lissautzki, M.* (2007), S. 109; *Walter, G.* (1996), S. 17; *Schulz, B.* (1995), S. 114; *Nader, G.* (1995), S. 2 f.

1359 Vgl. hier und nachfolgend *Rudolf-Sipötz, E.* (2001), S. 110 f.

1360 Vgl. zum Regionalprinzip Kapitel C.1.4.3 und zum Bankleistungsprogramm Kapitel C.2.3.

nalprinzips sind – wenngleich durch Fusionen größer werdend – die Absatzmärkte der Kreditgenossenschaften von begrenzter und überschaubarer Ausdehnung. Damit geht einher, dass die physische Distanz zwischen den Kunden ebenfalls gering ist. Dadurch ist die Wahrscheinlichkeit relativ hoch, dass sich Firmenkunden in Netzwerken wie z. B. regionalen Unternehmerverbänden begegnen und austauschen. Es kann begründet angenommen werden, dass auch die örtlichen Banken bzw. deren Leistungen Gegenstand der Kommunikation sind. Somit können die notwendigen Voraussetzungen aktiven Referenzgebens als gegeben angenommen werden. Kommunikationsinhalt dürfte dabei auch das Verhalten von Hausbanken in Krisensituationen sein, da hier deren geschäftspolitische Grundausrichtung am Einzelfall beurteilt werden kann. Gerade dieses Verhalten stellt für mittelständische Unternehmen ein wesentliches Motiv für den Auf- oder Ausbau einer Bankbeziehung dar, ist jedoch ex ante unbekannt. Hinzu kommen die informationsökonomischen Eigenschaften von Bankleistungen, die tendenziell einen hohen Grad an Vertrauens- und Erfahrungseigenschaften aufweisen.[1361] Nach BLACK neigen Menschen in unklaren und durch Unsicherheit geprägten Situationen dazu, den Rat von anderen zu suchen.[1362] Und CUNNIGHAM zeigt, dass Kunden mit hohem Risiko öfter das Gespräch mit anderen Kunden suchen.[1363]

Eine weitere Überlegung ist aus Sicht der Kreditgenossenschaft mit Blick auf die strategische Positionierung anzustellen. Wie gezeigt wurde,[1364] hat die empirische Kundenstruktur ihren Schwerpunkt bei Kleinst- und kleinen mittelständischen Unternehmen. Sollen nun auch größere Unternehmen als Firmenkunden akquiriert werden, so muss sich die Kreditgenossenschaft als leistungsfähiger Partner in der Wahrnehmung dieser Kunden positionieren. Dabei helfen Referenzkunden, die sowohl passiv – sofern Informationen über bestehende Geschäftsbeziehungen publizitätsfähig sind – als auch aktiv als Referenzgeber im Sinne der Kreditgenossenschaft auftreten. Aber auch auf Ebene der einzelnen Bankleistungen können referenzbezogene Überlegungen ansetzen. Aufgrund der Immaterialität der Leistung und den damit verbundenen hohen Anteilen an Erfahrungs- und Vertrauenseigenschaften, dienen Referenzen, als interpersonelle Informationen, als glaubwürdigere Quelle befürwortender Aussagen.[1365]

Mit Blick auf die Gesamtbank ist jedoch zu berücksichtigen, dass eine rege, positive Referenztätigkeit auch Risiken in sich birgt. Etwa dann, wenn aus den

---

1361 Vgl. dazu die Ausführungen in Kapitel C.2.3.

1362 Vgl. *Black, J. S.* (1982), S. 169 f.

1363 Vgl. *Cunningham, S. M.* (1967), S. 279 f. So auch *Helm, S.* (2003), S. 125.

1364 Vgl. dazu die Ausführungen in Kapitel C.2.2.3.

1365 Vgl. *Rudolf-Sipötz, E.* (2001), S. 109.

Referenzaktivitäten eine einseitige Kundenstruktur resultiert, die zu Klumpenrisiken führt.[1366]

Aufbauend auf den Erläuterungen zum Referenzbegriff und deren Bedeutung kann das Referenzpotenzial definiert werden als der aktuelle oder potenzielle Beitrag zur Zielerreichung der Kreditgenossenschaft, der sich dadurch ergibt, dass aktuelle oder potenzielle Firmenkunden aktiv oder passiv als Referenzgeber auftreten und dadurch das Verhalten anderer aktueller oder potenzieller Firmenkunden positiv oder negativ beeinflussen. Im Gegensatz zur Begriffsdefinition von GELBRICH, die ausschließlich einen positiven Wertebereich für das Referenzpotenzial sieht,[1367] wird mit diesem Verständnis auch der Möglichkeit Raum gegeben, dass Referenzen der Kreditgenossenschaft schaden. Es gilt, dass das Referenzpotenzial die ökonomischen Merkmale anderer Kundenbeziehungen beeinflusst. Betriebswirtschaftlich betrachtet, können Referenzen zusammenfassend insofern auf die Zielerreichung einer Kreditgenossenschaft wirken, als sie das Ertragsniveau heben oder senken, die Dauerhaftigkeit von Kundenbeziehungen verlängern oder verkürzen und Akquisitionskosten sowie Kundenbindungskosten erhöhen oder reduzieren können. Aufgrund dessen wird auf das Risiko hingewiesen, das Referenzpotenzial eines Kunden nicht im Kundenwert zu berücksichtigen, da dadurch fehlerhafte Priorisierungen resultieren können.[1368] Die Integration des Referenzpotenzials in den Kundenwert gilt jedoch als schwierig, da er mangels objektiver Messbarkeit leicht manipulierbar ist[1369] und die Messung hohe Anforderungen stellt. „If a „true" referral value were to be measured, models would need to be even more complicated."[1370]

In wie weit das Referenzpotenzial seine Wirkung entfaltet, ist schwierig zu prognostizieren und für das Firmenkundengeschäft von Banken bisher kaum wissenschaftlich untersucht. Nach RUDOLF-SIPÖTZ sind relevante Einflussfaktoren, die Anzahl der potenziellen Referenzempfänger (soziales Netz), die Qualität des Referenzträgers, die sich konkretisiert in Fachwissen, Persönlichkeit und Kundenzufriedenheit, die Richtung und Intensität der Referenz, die abhängt von der Kundenzufriedenheit sowie die Referenzhäufigkeit, die durch Persönlichkeit, Kundenzufriedenheit und Verhalten in der Vergangenheit bestimmt

---

1366 Vgl. *Diller, H.* (2006), S. 110.

1367 Vgl. *Gelbrich, K.* (2001), S. 59.

1368 Vgl. *von Wangenheim, F.* (2003), S. 2.

1369 „In fact, the lifetime value of a loyal customer can be astronomical, especially when referalls are added to the economies of customer retention and repeat purchases of related products." *Heskett, J. L et al.* (1994), S. 164.

1370 *Helm, S.* (2003), S. 130.

wird.[1371] GELBRICH fügt produktbezogene Einflussgrößen hinzu, die sich im Fachwissen über das konkrete Produkt und hohes Involvement in Bezug auf das Produkt manifestieren.[1372] Im Einklang mit der überwiegenden Mehrzahl der empirischen Studien[1373] belegt VON WANGENHEIM für den B2B-Kontext, dass die Kundenzufriedenheit, das situative Involvement und der eigene Empfang einer positiven Weiterempfehlung (sog. Schneeballeffekt) signifikanten Einfluss auf die Weitergabe positiver Weiterempfehlungen haben. Negative Weiterempfehlungen werden dagegen durch Unzufriedenheit, Produktbedeutung, situatives Involvement, Entscheidungsunsicherheit und dem eigenen Empfang negativer Weiterempfehlungen signifikant beeinflusst.[1374]

Neben dem Referenzpotenzial werden in der Literatur weitere beziehungsübergreifende Kundenwerttreiber diskutiert. Diese lassen sich aber grundsätzlich mit den bereits für das hier zu entwickelnde Kundenwertmodell ausgeschlossenen Verbundeffekten gleichsetzen.[1375]

4.5.3.2.2 Member Value-Potenzial

Wie die Ausführungen in den Kapiteln C.1.4.2 und D.3.1.2.4 darlegen, ist der Förderauftrag ein unbestimmter Rechtsbegriff, der in der Praxis vielfältig materiell ausgestaltet ist. Im Gegensatz zu den bisherigen Werttreibern des Kundenwerts ist ein auf den Förderauftrag ausgerichteter Werttreiber zunächst kontraintuitiv, da er die definitorische Wirkrichtung des Kundenwerts umkehrt. Während der Förderauftrag in die Richtung wirkt, welchen Beitrag die Kreditgenossenschaft zum wirtschaftlichen Erfolg des Mitglieds leisten kann, fragt das auf einen Mitgliedskunden ausgerichtete Maß des Kundenwerts, welchen Beitrag dieses Mitglied der Kreditgenossenschaft zur Zielerreichung leisten kann. Beim Versuch diese Überlegungen zusammenzuführen, wird schnell evident, dass dies nicht bei Betrachtung der dyadischen Beziehung zwischen Unternehmen in seiner als Rolle Mitgliedskunde, und Genossenschaftsbank, sondern nur unter Ein-

---

1371 Vgl. *Rudolf-Sipötz, E.* (2001), S. 112 in enger Anlehnung an *Cornelsen, J.* (2000), S. 199 ff.

1372 Vgl. *Gelbrich, K.* (2001), S. 63.

1373 Vgl. *von Wangenheim, F./Bayón, T./Herrmann, A.* (2006), S. 304 ff.

1374 Vgl. *von Wangenheim, F.* (2003), S. 201 ff. Interessant ist zudem, dass die Aussage von REICHHELD/SASSER damit widerlegt wird, da das Weiterempfehlungsverhalten am höchsten bei Kunden ist, die erst vor kurzem zu einem Anbieter gewechselt sind, was in diesem Studiendesign einem hohen situativen Involvement entspricht. REICHHELD/SASSER dagegen sagen aus, dass Kunden, die seit längerer Zeit in einer Geschäftsbeziehung mit dem Anbieterunternehmen stehen, einen höheren Erfolgsbeitrag aus Referenzen leisten. Vgl. *Reichheld, F. F./Sasser, W. E.* (1990).

1375 Vgl. dazu die Ausführungen in Kapitel D.4.5.1.

bezug dritter Geschäftsbeziehungen erfolgen kann. Denn nur so lässt sich der scheinbare Widerspruch auflösen und ein Kundenwerttreiber mit Bezug zum Förderauftrag ableiten. Diese Facette des Kundenwerts zu berücksichtigen, leitet sich unmittelbar aus dem immateriellen Ziel der Kreditgenossenschaften ab.[1376] Das immaterielle Ziel wurde mittels des Member Values operationalisiert, so dass im Weiteren vom Member Value-Potenzial gesprochen wird.

Das Member Value-Potenzial stellt die Frage, in wie weit ein Firmenkunde einen Beitrag dazu leistet oder leisten kann, das Member Value-Ziel der Kreditgenossenschaft für Dritte zu erreichen. Es bezieht sich also nicht primär auf den Member Value, den der konkret betrachtete Mitglieds-Firmenkunde selbst wahrnimmt. Es wird als beziehungsübergreifender Kundenwerttreiber der Ertragsdimension verstanden, da der Member Value als Mittel gesehen wird, um die Mitgliederbindungen und die damit verbundenen wirtschaftlichen Vorteile aus einer Mitgliederbeziehung zu erhöhen. Die Mitgliedschaft wird als explizites Zeichen der Verbundenheit zwischen Kreditgenossenschaft und Kunden angesehen,[1377] woraus ein positiver Einfluss auf das Bindungspotenzial und die ökonomischen Beiträge der jeweiligen Kundenbeziehung resultieren dürfte.

Die einzelnen Wertbeiträge des Member Value-Potenzials sind in Abhängigkeit davon, ob der zu bewertende Firmenkunde Mitglieds- oder Nur-Kunde, zu betrachten. Der Wertbeitrag einer Mitgliedschaft aus Sicht der Kreditgenossenschaft resultiert aus verschiedenen Quellen. Erstens stellen Mitglieder der Kreditgenossenschaft Eigenkapital zur Verfügung, so dass dadurch die Förderfähigkeit bzw. das Förderpotenzial *FP* der Kreditgenossenschaft gestützt wird. Weiterhin wird angenommen, die Mitgliedschaft intensiviere die Geschäftsbeziehung zwischen Firmenkunden und Kreditgenossenschaft, woraus positive Einflüsse auf die Kundenwerttreiber der Ertragsdimension resultieren. Schließlich kann argumentiert werden, dass in Unternehmen weitere potenzielle Mitglieder arbeiten, die aus den genannten Gründen als Mitglieder rekrutiert werden können, denen wiederum die bereits genannten positiven Wertbeiträge zugeordnet werden können.

Ein Beitrag des Mitglieds- oder Nur-Firmenkunden zur Zielerreichung der Kreditgenossenschaft liegt mit Blick auf § 1 GenG dann vor, wenn dieser einen Beitrag dazu leistet, dass die Kreditgenossenschaft ihren Förderauftrag relativ besser nachkommen kann und somit in der Lage ist, den Member Value für andere Mitglieder zu erhöhen. Wird der in Abbildung 7 dargelegte Ansatz zur Konzeptualisierung des Förderauftrags zugrunde gelegt, so kann dieser Zielerreichungsbeitrag mit Blick auf den inneren und den erweiterten Förderkreislauf

---

[1376] Vgl. dazu die Ausführungen in Kapitel D.3.1.3.

[1377] Vgl. *Theurl, T.* (2003), S. 13.

oder auf die Allgemeinförderung diskutiert werden.[1378] Die jeweiligen Maßnahmen innerhalb dieser Förderkreisläufe erzeugen dabei einen Nutzen im Sinne des Member Value, der aus der unmittelbaren Förderung *UF*, der mittelbaren Förderung *MF* oder dem Förderpotenzial *FP* resultieren kann.

Erfolgsbeiträge bei der Förderung im Sinne des inneren Förderkreislaufs resultieren daraus, dass der betrachtete Firmenkunde positiven Einfluss auf die Leistungsbeziehungsebene zwischen der Kreditgenossenschaft und einem dritten Kunden nimmt. Es wurde oben beschrieben, dass ein unmittelbarer Fördernutzen sich in mitgliedergerichteten Leistungs- oder Preisvorteilen niederschlägt. Es ist leicht ersichtlich, dass dies vor allem dann gelingt, wenn der zu bewertende Firmenkunde in den bereits diskutierten Kundenwerttreibern positive Bewertungen erfährt.[1379] In diesem Fall würde der Firmenkunde auch einen Beitrag zu den Möglichkeiten der mittelbaren Förderung und des Förderpotenzials leisten. Würde dies nun dem Member Value-Potenzial zugrunde gelegt, würde jedoch eine Doppelzählung dieser Kundenwerttreiber resultieren, so dass eine verzerrte Kundenbewertung folgen würde.

Gegenstand des erweiterten Förderkreislaufs ist z. B. das Angebot Dritter zu günstigeren Konditionen oder mit besserer Qualität. Letztgenanntes lässt sich gut als Teilwert des Member Value-Potenzials nutzen. Wird der kreditgenossenschaftliche Förderauftrag darüber festgelegt, den Mitgliedern aus einem Netzwerk von Kooperationspartnern Leistungen mit entsprechenden Vorteilen anbieten zu können, so ist den Firmenkunden, die diesem Netzwerk angehören und aus Sicht der anderen Mitglieder attraktive Leistungen anbieten, ein positives Förderpotenzial zuzuweisen. Dies vor allem dann, wenn dadurch auch die Attraktivität einer kreditgenossenschaftlichen Mitgliedschaft gesteigert werden kann, woraus selbst wieder Vorteile für die Kreditgenossenschaft resultieren können, wenn es sich um wertvolle Mitglieder-Kundenbeziehungen handelt. Diese Facette des Member Value-Potenzials ist abhängig von der Bereitschaft des Firmenkunden, einem derartigen Kooperationsnetzwerk beizutreten. Ein konkretes Beispiel liefert die BERLINER VOLKSBANK EG.[1380] Im Rahmen eines

---

[1378] Vgl. dazu die Ausführungen in Kapitel C.1.4.2 und Kapitel D.3.1.2.4.

[1379] So ist z. B. davon auszugehen, dass bei langfristig ertragsmaximaler und kostenminimaler Betreuung von Firmenkunden, was einem hohen Kundenwert gemäß der bisher diskutierten Kundenwerttreiber im Kern entspricht, die Gesamtertragslage der Kreditgenossenschaft positiv beeinflusst wird. Daraus wiederum lassen sich positive Wirkungen auf den Förderauftrag ableiten, wenn z. B. für Mitglieder-Kunden Preisvorteile durch Quersubventionierung ermöglicht werden. Dieser Gedanke gilt analog für die an dieser Stelle der Analyse noch nicht betrachtete Risikodimension.

[1380] Vgl. hier und nachfolgend *Algner, M. R./Bolsinger, H. J.* (2006), S. 12 sowie die dort angegebene Literatur.

Bonusprogramms können Mitglieder bei über 700 Firmenkunden vergünstigt einkaufen.

Die dritte Form der Förderung ist die Allgemeinförderung, die unter anderem die Förderung regionaler, kommunaler Einrichtungen umfasst. In diesem Zusammenhang wird seitens der Kreditgenossenschaften oftmals argumentiert, die Bereitstellung von Krediten an mittelständische Unternehmen der Region sei dem Förderauftrag zuzurechnen, der auch mit dem Schlagwort „regionale Verantwortung" überschrieben wird. Darüber hinaus liege eine besondere Förderleistung vor, wenn Unternehmen auch in wirtschaftlich angespannten Situationen die Finanzierung bereitgestellt wird. So würden Arbeitsplätze in der Region gesichert und somit die Region gefördert. Es ist offensichtlich, dass dies Allgemeinförderungskonzept keine Bestandteil des Member Value-Potenzials eines Firmenkunden sein kann. Wenngleich es auch denkbare Ansätze gibt, wie ein Firmenkunde auch die Ansätze in der Allgemeinförderung im Sinne der Kreditgenossenschaft unterstützen kann, so z. B. wenn sich Firmenkunden auf Initiative der Kreditgenossenschaft an Maßnahmen zur Allgemeinförderung beteiligen.

Wie die Ausführungen zeigen, ist es für die Bestimmung des Member Value-Potenzials eines Firmenkunden von grundlegender Bedeutung eine hinreichend materielle Konkretisierung des Förderauftrags vorzulegen. Doch auch bei Vorhandensein einer derartigen Grundlage, wie sie etwa in einem Förderplan vorgenommen wird,[1381] dürfte die Messung des Förderpotenzials schwierig sein.

4.5.4 Kundenwerttreiber der Risikodimension

Obwohl ohne die Einschätzung von Risiken die Kundenbewertung unvollständig wäre,[1382] erfolgt die Betrachtung von Risiken in der Kundenwertforschung bisher nur rudimentär.[1383] Dadurch resultiert die Gefahr, dass risikoreiche Kunden mehr Ressourcen zugeordnet bekommen als es ökonomisch vertretbar wäre.[1384] Um ein vollständiges Kundenwertmodell zu entwickeln, soll der Fokus der weiteren Betrachtung auf die Risikodimension gerichtet werden. Als Risiko wird im Folgenden die Gefahr einer negativen Abweichung der Ausprägung einer Größe von einer ex ante geplanten Referenzausprägung der Größe verstanden.[1385]

1381 Vgl. *Boettcher, E.* (1979), S. 33 ff.

1382 Vgl. *Mödritscher, G. J.* (2008), S. 213.

1383 Vgl. *Stüker, D.* (2008), S. 177; *Weber, J./Lissautzki, M.* (2005), S. 311; *Eberling, G.* (2002), S. 157.

1384 Vgl. *Eberling, G.* (2002), S. 157.

1385 Vgl. *Siemes, A.* (2002), S. 8. Korrespondierend dazu ist die positive Abweichung der Istausprägung von der Planausprägung als Chance zu verstehen. Nicht weiter betrachtet

Nur wenige Beiträge zum Kundenwert nennen risikogerichtete Kundenwerttreiber. WEBER/LISSAUTZKI, LISSAUTZKI und MÖDRITSCHER unterscheiden zwischen dem Kundenverlustrisiko, dem Zahlungsausfalls- oder Bonitätsrisiko[1386] sowie dem Kundenplanungs- und -informationsrisiko als kundenwertrelevante Risiken auf Ebene des Einzelkunden.[1387] Aus Sicht der Kreditgenossenschaft kommt dem Kundenverlustrisiko die größte Bedeutung zu, da eine Abwanderung zu einem kompletten Verlust der kundenbezogenen Zielbeiträge führt.[1388] Das Kundenverlustrisiko ist Bestandteil des Bindungspotenzials, das im Kontext der Beziehungsdimension diskutiert wurde.[1389] Aus diesem Grund soll diese Risikoart hier nicht weiter betrachtet werden. Das Bonitätsrisiko umfasst die Fähigkeit eines Kunden, seine Verpflichtungen gegenüber dem Anbieterunternehmen nachzukommen, was insbesondere im Aktivgeschäft der Kreditgenossenschaften von herausragender Bedeutung ist. Es ist aufgrund seines strengen Kundenbezugs (Adressrisiko) zwingend von Relevanz und soll daher im Weiteren detailliert betrachtet werden. Das Kundenplanungs- und -informationsrisiko steht im Zusammenhang mit der Prognose der zukunftsgerichteten Kundenwerttreiber. Es steht für Risiken, die in Fehleinschätzungen bezüglich dieser Größen liegen.[1390]

Mit Blick auf das hier zu entwickelnde Kundenwertmodell sollen das Kundenplanungs- und -informationsrisiko sowie das Bonitätsrisiko näher betrachtet werden. Da sich das Bonitätsrisiko sowohl beziehungsbezogen als auch bezie-

---

werden die unterschiedlichen Risikoeinstellungen. Vgl. ebenfalls *Siemes, A.* (2002), S. 8 ff. sowie die dort genannte Literatur.

1386 Es sei darauf hingewiesen, dass die synonyme Verwendung der Begriffe (Zahlungs-) Ausfallrisiko und Bonitätsrisiko nicht der bankbetriebswirtschaftlichen Terminologie entspricht. Während das Zahlungsausfallsrisiko den insolvenzinduzierten Verlust umfasst, bezieht sich das Bonitätsrisiko auf den bonitätsmigrationsinduzierten Verlust. Vgl. *Schierenbeck, H./Lister, M./Kirmße, S.* (2008), S. 6.

1387 Vgl. *Weber, J./Lissautzki, M.* (2005), S. 311; *Lissautzki, M.* (2007), S. 204 ff. *Mödritscher, G. J.* (2008), S. 213.

1388 Vgl. *Ryals, L.* (2003), S. 170.

1389 LISSAUTZKI weist darauf hin, dass das Kundenverlustrisiko sowie der Kundenbindungsgrad unterschiedliche Aspekte des Kundenwerts sind. Vgl. *Lissautzki, M.* (2007), S. 207. Den Unterschied konkretisiert er darin, dass der Kundenbindungsgrad Einfluss auf die Höhe des erwarteten Kundenwerts hat. Das Kundenverlustrisiko dagegen beschreibt, die Höhe des Verlusts, der sich aufgrund einer unerwarteten Beendigung der Geschäftsbeziehung ergibt. Dieser Verlust entspricht in seinem weiten Verständnis dem entgehenden Kundenwert, stellt somit lediglich die negativ formulierte Aussage des Kundenwerts dar.

1390 Vgl. *Lissautzki, M.* (2007), S. 205.

hungsübergreifend analysieren lässt,[1391] soll diese Differenzierung hier ebenfalls vorgenommen werden. In der Bankbetriebslehre ist das Bonitätsrisiko eine Ausprägung des Kreditrisikos. Daher wird im Folgenden der umfassendere Begriff des Kreditrisikos verwendet.

4.5.4.1 Beziehungsbezogene Kundenwerttreiber der Risikodimension

4.5.4.1.1 Kreditrisiko

Kreditrisiken treten in drei Ausprägungen auf, als Ausfallrisiko, als Bonitätsrisiko und als Spread-Risiko (siehe Tabelle 21). Ein Ausfallrisiko liegt vor, wenn ein Kreditschuldner seinen Zins- und/oder Tilgungs- und/oder Liefer-/Erfüllungsverpflichtungen gar nicht oder nicht vollständig nachkommt.[1392] Das Bonitätsrisiko bezeichnet die Gefahr, dass ein Kreditnehmer aufgrund einer veränderten Bonität in eine schlechtere Ratingklasse migriert.[1393] Durch diese Bonitätsveränderung und der damit verbundenen höheren Risikoaufschläge auf den Diskontierungsfaktor resultieren ein geringerer (barwertiger) Marktwert der ausstehenden Position und damit ein geringerer Vermögenswert für die Kreditgenossenschaft et vice versa.[1394] Das Spread-Risiko resultiert aus einem erhöhten Risikoaufschlag auf den risikolosen Zins bei unveränderter Bonitätseinstufung (Rating). Ursachen eines abweichenden Spreads sind bspw. Erwartungen oder das Verhalten von Marktteilnehmern. Sowohl das Bonitäts- als auch das Spread-Risiko wirken sich negativ auf den Marktwert eines Kredites aus und können unter dem Begriff Marktwertrisiken subsumiert werden.

1391 „For example, in the financial services industry in the management of credit relationships that […] customers have with banks, where the measurement and management of the risk as well as return of the overall portfolio is taken very seriously." *Ryals, L.* (2003), S. 167.

1392 Vgl. *Kroon, G.* (2009), S. 11 sowie die dort angegebene Literatur. Unterschieden werden kann zwischen dem Vorleistungsrisiko, dass sich auf die Zins- und Tilgungsleistungen bezieht und dem Eindeckungsrisiko, das sich bei Swap-Transaktionen, Termingeschäften und Optionen auf die Erfüllungs- und Lieferverpflichtung bezieht.

1393 Das kreditnehmerindividuelle Risiko eines Kreditausfalls wird mittels einer ex ante Schätzung auf Basis eines Ratingverfahrens ermittelt. Dabei ist zu berücksichtigen, dass Veränderungen der Architektur des Ratingverfahrens bei gleichen Inputdaten zu unterschiedlichen Ergebnissen führen. Vgl. *Bernet, B./Westerfeld, S.* (2008), S. 1013. Als Ratingklasse wird das kategoriale Ergebnis eines Ratingverfahrens bezeichnet. Dazu wird die intervallskaliert ermittelte Ausfallwahrscheinlichkeit unter Berücksichtigung weiterer Aspekte wie z. B. der bereitstehenden Sicherheiten des Kreditnehmers in eine kategoriale Messgröße transformiert.

1394 Vgl. hier und nachfolgend *Kroon, G.* (2009), S. 11 ff.

Tabelle 21: Ausprägungen des Kreditrisikos[1395]

			Marktwertrisiken	
		Ausfallrisiko	**Bonitätsrisiko**	**Spread-Risiko**
Ex ante	**Erwartete Verluste**	Erwarteter Kreditausfall	Erwartete Bonitätsverschlechterung	Erwartete Spread-Ausweitung
	Unerwartete Verluste	Über den erwarteten Verlust hinausgehender möglicher…		
		…insolvenzbedingter Kreditausfall	…bonitätsinduzierter Wertverlust	…spread-induzierter Wertverlust
Ex post		Ist-(Ausfall-) Risikokosten	Ist-(Bonitäts-) Risikokosten	Ist-(Spread) Risikokosten
Aus bilanziellen und außerbilanziellen (Geld- und Kreditleihe-)Geschäften mit in- und ausländischen Privatkunden, Firmenkunden (Kreditschuldner) und aus Eigengeschäften (Anleiheemittenten oder Kontrahenten)				

Das Spread-Risiko und Bonitätsrisiko werden nicht explizit betrachtet, da sie vor allem bei Forderungen vorliegen, die am Markt gehandelt werden, was für die Firmenkundenkredite der Kreditgenossenschaften i. d. R. (noch) nicht der Fall ist.

Das ex ante Kreditrisiko eines Kreditnehmers lässt sich, wie in Tabelle 21 dargestellt, differenzieren in den erwarteten Verlust und den unerwarteten Verlust der ausstehenden Forderungen. Aus diesem Grund ist es sinnvoll, die im statistischen Erwartungsbereich liegenden Kreditverluste nicht mit „echten" Verlustüberraschungen zu vermischen.[1396] Denn der erwartete Verlust eines Kreditnehmers stellt strenggenommen kein Risiko im zugrunde gelegten Verständnis dar, wenn als Risiko die Abweichung eines tatsächlichen Ergebnisses von einem erwarteten Ergebnis definiert wird.[1397] AVERBECK ET AL. argumentieren jedoch, dass erst durch die Kalkulation der Ausfallrisiken die Bewertung der Vertriebsleistung und die Steuerung des Kreditgeschäfts ermöglicht werden.[1398] Aus diesem Grund und aufgrund der kalkulatorischen Verknüpfung von Konditionsmarge und erwartetem Verlust wird der erwartete Verlust betrachtet, um alle relevanten Kostenarten des Kundenwerts im weiteren Verlauf der Analyse zu erfassen. Erwartete Verluste sollen durch vom Kunden geschäftsbezogen zu entrichtende Risikoprämien abgedeckt werden. Die so von der Kreditgenossenschaft eingenommenen Risikoprämien fließen nach dem versicherungsmathematischen Prinzip in einen Topf und sollen in ihrer Gesamtheit die erwarteten Verluste der Kreditgenossenschaft decken. Dabei wird kein periodischer Ausgleich,

---

1395 *Kroon, G.* (2009), S. 14.

1396 Vgl. *Schierenbeck, H./Lister, M./Kirmße, S.* (2008), S. 154.

1397 Vgl. *Wald, M.* (2002), S. 79.

1398 Vgl. *Averbeck, C. et al.* (2008), S. 407.

sondern ein bei sachgerechter Kalkulation und Berücksichtigung in den Konditionen langfristig kumulierter Ausgleich angenommen.[1399] Im Falle eines tatsächlichen Ausfalls eines Kreditnehmers dürfte jedoch die vereinnahmte Risikoprämie nicht ausreichen, den effektiven Verlust zu decken. Bei vertragsgemäßer Rückführung des Kredits wird die Risikoprämie als Ertrag vereinnahmt.[1400] Die Höhe der Risikoprämie ist abhängig von der Ausfallwahrscheinlichkeit eines Kreditnehmers, dem Kreditvolumen bzw. der Inanspruchnahme, der Laufzeit des abgeschlossenen Geschäfts und der Höhe und Art der Besicherung sowie den diese Größen beeinflussenden Faktoren (siehe Abbildung 51).[1401]

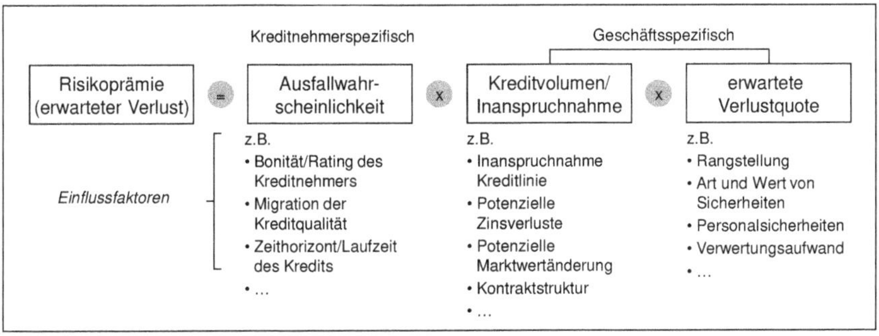

Abbildung 51: Einflussfaktoren der Risikoprämie[1402]

Das eigentliche Kreditrisiko einer Bank ist der unerwartete Verlust, der nicht über die Summe der vereinnahmten Risikoprämien, sondern mittels des sog. ökonomischen Eigenkapitals bzw. Risikokapitals getragen wird.[1403] Der unerwartete Verlust kann nur auf Ebene des Kreditportfolios gemessen werden.[1404] In Bezug auf den Kundenwert hat der unerwartete Verlust zwei relevante Implikationen. Das ist zum einen die Berücksichtigung der Kosten, die anfallen, um risikoabhängige Geschäfte mit ökonomischem Eigenkapital zu hinterlegen. Und zum zweiten der Beitrag des Firmenkunden zur Kreditportfoliostruktur bzw. zum Credit-Value-at-Risk. Die Betrachtung der Kosten des ökonomischen Eigenkapitals ergänzt den Umfang relevanter kundenbezogener Kosten ebenso wie die im vorhergehenden Abschnitt erläuterte Standardrisikoprämie für erwartete

1399 Vgl. *Döhring, J./Hromadka, J.* (2002), S. 37 f.

1400 Vgl. *BVR* (2003b), Kapitel 4.2, S. 4.

1401 Vgl. *BVR* (2003b), Kapitel 2.1, S. 48.

1402 *BVR* (2003b), Kapitel 2.1, S. 49.

1403 Vgl. *Kroon, G.* (2009), S. 15; *BVR* (2003b), Kapitel 2.4, S. 3.

1404 Vgl. *Kroon, G.* (2009), S. 15.

Verluste. Ausgangspunkt der Kalkulation der Kosten für das ökonomische Kapital ist der Verzinsungsanspruch auf das bereitgestellte Eigenkapital sowie korrespondierend dazu die Berechnung der Kapitalkosten auf Ebene des einzelnen risikoabhängigen Geschäfts. In der Steuerungskonzeption VR-CONTROL werden die Kapitalkosten mittels des sog. Multiplikatoransatzes ermittelt.[1405] In diesem Ansatz wird der erwartete Verlust eines Einzelgeschäfts mit einem Eigenkapitalmultiplikator multipliziert und somit der unerwartete Verlust berechnet. Der Eigenkapitalmultiplikator ist der Quotient aus unerwartetem zu erwartetem Verlust eines repräsentativen Referenzkreditportfolios (für ein bestimmtes Konfidenzniveau und eine bestimmte Haltedauer).[1406] Er soll bankindividuell festgelegt und nach Privat- und Firmenkunden differenziert werden. Abbildung 52 visualisiert die Zusammenhänge. Die Zielrendite zur Ermittlung der Eigenkapitalkosten ist institutsindividuell vom Vorstand festzulegen.[1407]

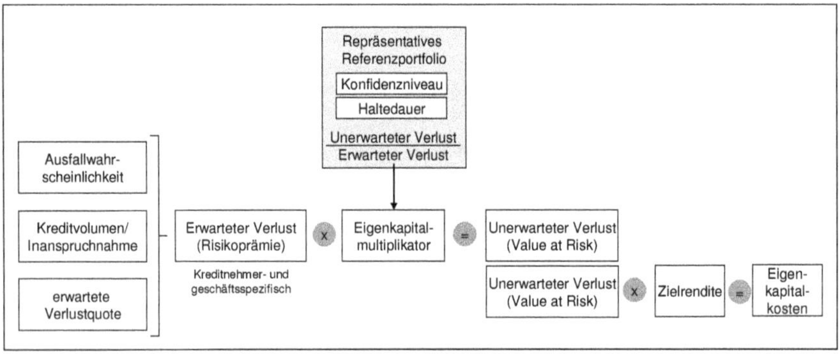

Abbildung 52: Multiplikatoransatz (erwarteter Verlust)[1408]

Aufgrund der funktionalen Verknüpfung mit dem erwarteten Verlust sind teilweise identische Einflussfaktoren des unerwarteten Verlusts zu berücksichtigen.

1405 Vgl. hier und nachfolgend *BVR* (2003b), Kapitel 2.4, S. 10. Der Multiplikatoransatz orientiert sich am marginalen Value-at-Risk-Ansatz. Danach ergeben sich die Eigenkapitalkosten aus einer Differenzbetrachtung des ökonomisch notwendigen Kapitals mit dem zusätzlichen Geschäft und ohne das zusätzliche Geschäft. Vgl. dazu *BVR* (2003b), Kapitel 2.4, S. 9.

1406 Vgl. *Kroon, G.* (2009), S. 51; *BVR* (2003b), Kapitel 2.4, S. 10. Mit diesem Ansatz wird unterstellt, dass ein linearer Zusammenhang zwischen der Variation der Standardabweichung des Referenzportfolios und der Standardabweichung des Value-at-Risk bei Aufnahme eines weiteren Kreditnehmers vorliegt. Vereinfachend wird darauf verzichtet, die Veränderung der Standardabweichung des Referenzportfolios zu berechnen.

1407 Vgl. *BVR* (2002b), S. 29.

1408 In Anlehnung an *BVR* (2003b), Kapitel 2.4, S. 10.

Dies sind die Ausfallwahrscheinlichkeit, das Kreditvolumen, die Inanspruchnahme sowie die erwartete Verlustquote.

Neben den Kosten, die für ex ante erwarteten und unerwarteten Verluste in der Kundenkalkulation zu berücksichtigen sind, sind weitere Einflüsse des kundenspezifischen Ausfallrisikos zu nennen. Zum einen resultiert aus dem Umstand, dass einige produkt- bzw. bedarfsspezifische Potenziale nur in Abhängigkeit bestimmter Voraussetzungen hinsichtlich der Bonität des Kreditnehmers realisierbar sind, ein weiterer Einfluss auf den Kundenwert. Dies ist z. B. der Fall bei der Ausweitung von Kreditlinien, die aufgrund des Bedarfs des Kunden zwar möglich, aber vor dem Hintergrund der Bonität des Kunden nicht realisiert werden sollen. Folglich resultiert hier ein mittelbarer Einfluss des Adressrisikos auf den Kundenwert. Dies muss im Rahmen der konkreten kundenwertorientierten Vertriebsplanung berücksichtigt werden. Zum anderen besteht das Risiko eines insolvenzbedingten Kundenausfalls auch darin, dass künftige, nach dem Kreditausfallereignis liegende Kundenwerte nicht mehr realisierbar sind.[1409]

Der Beitrag des Firmenkunden zur Kreditportfoliostruktur bzw. zum Credit-Value-at-Risk ist aufgrund der notwendigen Betrachtung des gesamten Kreditrisikoportfolios und damit weiterer Kundenbeziehungen als beziehungsübergreifender Kundenwerttreiber zu verstehen. Dieser ist Gegenstand des Kapitels D.4.5.4.2.

So wie hier geschehen, betrachten einige Autoren das Kreditrisiko als eigenständiges Risiko in einem Kundenwertmodell.[1410] Bei näherer Betrachtung ist jedoch zu konstatieren, dass das Kreditrisiko eine besondere Ausprägung des Kundenplanungsrisikos ist.[1411] Denn die Wahrscheinlichkeit einer Zahlungsunfähigkeit des Kunden und dessen Einfluss auf die Kundenwerttreiber ließen sich vollständig in das nachfolgend beschriebene Kundenplanungsrisiko überführen. Aufgrund seiner herauszuhebenden Bedeutung für das Bankgeschäft soll das Kreditrisiko im Einklang mit der Literatur dennoch separat betrachtet werden.[1412]

1409 Aufgrund der Insolvenz des Firmenkunden ist die Struktur des Konstrukts Firmenkunde aus Sicht der Kreditgenossenschaft zu überdenken. In der Regel sind dann Kundenverbünde neu zu strukturieren, so dass das Konstrukt Kundenverbund keine das Insolvenzereignis überdauernde Qualität hat.

1410 Vgl. z B. *Stüker, D.* (2008), S. 178; *Lissautzki, M.* (2007), S. 204 ff.

1411 Zu dieser Erkenntnis kommt auch *Eberling, G.* (2002), S. 160.

1412 STEIN/KALTOFEN/PAUL bezeichnen das Kreditrisiko als das bedeutendste Einzelrisiko von Banken. Vgl. *Stein, S./Kaltofen, D./Paul, S.* (2009), S. 352.

4.5.4.1.2 Kundenplanungsrisiko

Wie in den vorhergehenden Kapiteln beschrieben, liegt dem hier zu entwickelnden Kundenwertmodell eine prospektive Betrachtung zugrunde, da nur dadurch die aus Steuerungssicht geforderte Entscheidungsrelevanz erreicht wird. Damit geht zwingend einher, dass die Kundenwerttreiber für einen in der Zukunft liegenden Zeitraum prognostiziert werden müssen, womit wiederum das Risiko von Schätzfehlern verbunden ist.[1413] Das Kundenplanungsrisiko bezieht sich in diesem Kontext auf die Unwägbarkeiten, die aus einer notwendigen Prognose der bisher diskutierten Kundenwerttreiber resultieren.

SRIVASTAVA/SHERVANI/FAHEY unterscheiden als Ausprägungen dieser Unwägbarkeiten die „vulnerability" und die „volatility". Die „[...] vulnerability is defined simply as any occurence that negatively affects cash flow, while volatitity is any occurence that creates fluctuations."[1414] STAHL/MATZLER/ HINTERHUBER übersetzen den Begriff der "vulnerability" mit Reagibilität, worunter sie das Ausmaß der Reaktion der Kundenwerttreiber auf externe Einflüsse verstehen.[1415] Beide Autorenschaften unterscheiden die makroökonomische Umwelt, die Branche, das Anbieterunternehmen und den Kunden als Bezugsebenen, auf denen die Volatilität und die Reagibilität beeinflusst werden. Letztgenannte Autorenschaft differenziert auf der Kundenebene zwischen zwei weiteren untergeordneten Bezugsebenen. Dies sind zum einen die soziodemografischen und zum zweiten die psychografischen und verhaltensbezogenen Merkmale des Kunden.[1416] Erstere sind für Firmenkunden einer Kreditgenossenschaft z. B. die Ertragslage sowie die Finanz- und Vermögenslage des Unternehmens, aber auch die Marktposition. Letztere sind z. B. Verhaltensabsichten hinsichtlich der Beziehungsgestaltung mit der Kreditgenossenschaft. Für das Kundenplanungsrisiko sind vor allem diejenigen Faktoren relevant, die sich unterschiedlich auf die Ausprägungen der Kundenwerttreiber auswirken. So betreffen z. B. alle Firmenkunden einer Branche politische oder rechtliche Änderungen auf der makroökonomischen Ebene gleichermaßen. Unterschiede könnten sich dennoch hier ergeben, wenn die Reagibilität unterschiedlich hoch ausfällt. Dies z. B. dann, wenn

---

1413 Vgl. *Buhl, H. U./Heinrich, B.* (2008), S. 1.

1414 *Srivastava, R. K./Shervani, T. A./Fahey, L.* (1997), S. 52.

1415 Vgl. hier und nachfolgend *Stahl, H. K./Matzler, K./Hinterhuber, H. H.* (2006), S. 433; *Srivastava, R. K./Shervani, T. A./Fahey, L.* (1997), S. 52.

1416 Tatsächlich unterscheiden STAHL/MATZLER/HINTERHUBER die vier Ebenen Makro-Umwelt, Branche, Kunde und Share-Of-Wallet. Damit ersetzen sie im Vergleich zu SRIVASTAVA/SHERVANI/FAHEY die Ebene Anbieterunternehmen durch die Kundenebene und ergänzen die Share-of-Wallet-Ebene.

ein Firmenkunde die Veränderungen schnell und für sich erfolgreich adaptieren kann, während ein anderes Unternehmen hier Probleme hat.[1417]

Es ist begründet anzunehmen, dass die Einschätzung des Kundenplanungsrisikos bei Bestandskunden aufgrund der Geschäftsbeziehungshistorie im Vergleich zu Nichtkunden mit weniger Schwierigkeiten verbunden und die Risikohöhe geringer ist.[1418]

4.5.4.2 Beziehungsübergreifende Kundenwerttreiber der Risikodimension

Die kreditnehmerindividuellen Risiken der Firmenkunden aggregieren sich auf Ebene der Bank zum Firmenkreditportfoliorisiko und zusammen mit den Krediten und Risiken der anderen Bankportfolien zum Gesamtbankportfoliorisiko. Durch den Einzug der Portfoliotheorie in das Kreditgeschäft der Banken findet ein Wandel im Umgang mit Krediten statt. Nicht mehr der Ertrag als einzig bestimmende Größe eines Kreditgeschäfts ist entscheidend, sondern die Relation des mit dem Ertrag einhergehenden Risikos des Grundgeschäfts und die Einbettung des Geschäftes in das gesamte Kreditportfolio.[1419] Das Kreditportfolio wird in den Kreditgenossenschaften mittels eines speziellen Kreditportfoliomodells abgebildet.[1420] Das Ist-Kreditprotfolio dient als Grundlage der Festlegung von Adressrisikolimiten und der Steuerung des Kreditportfolios.

Im Rahmen der Adressrisikoplanung wird ein Soll-Portfolio definiert, das durch Eigenschaften für das zu erzielende Neugeschäftsvolumen (z. B. Kreditvolumen Anzahl Engagements, Ratingklassenstruktur etc.) und der Risikostatistik (z. B. erwarteter Verlust, unerwarteter Verlust bzw. Value-at-Risk etc.) charakterisiert wird.[1421] Um das Soll-Portfolio zu erreichen, wird unter Berücksichtigung des Ist-Portfolios ein Plan-Portfolio für den Planungszeitraum definiert. Für die Erreichung dieses Plan-Portfolios wird in der operativen Geschäftspolitik eine darauf aufbauende Limitsystematik auf Portfolioebene, Strukturebene und Einzelengagementebene definiert. Auf Portfolioebene wird der Value-at-Risk, d. h. der mit einer bestimmten Wahrscheinlichkeit innerhalb des Planungshorizonts nicht zu überschreitende Verlustbetrag als Limit bestimmt. Dieser orientiert sich am Gesamtbanklimit und dem damit verbundenen absoluten Anteil des Risikokapitals für das Adressrisiko. Die Höhe des vorzuhaltenden Risikokapitals wird von der Anzahl Kreditnehmer, der Volumenstruktur sowie

---

1417 Auswirkungen auf den Kundenwert der Unternehmen ergeben sich z. B. aus Erträgen aus Finanzierungsprodukten oder aus dem Zahlungsverkehr, da sich diese Bedarfe derivativ aus dem Erfolg des Firmenkunden ableiten.

1418 Vgl. *Lissautzki, M.* (2007), S. 218.

1419 Vgl. *Stricker, P./Leßmann, J.* (2008), S. 1.

1420 Zu den methodischen Grundlagen vgl. *BVR* (2003b), Kapitel 4.5.

1421 Vgl. hier und nachfolgend *BVR* (2003b), Kapitel 4.3, S. 9 ff.

den Korrelationsbeziehungen zwischen den Krediten bestimmt.[1422] Daraus resultieren Vorgaben im Hinblick auf das Kreditrisikoportfolio beeinflussende Strukturgrößen, die sich in Strukturlimiten in Form von Blankovolumenlimiten je Branchen und Ratingklassen konkretisieren. Weiterhin wird auf Einzelengagementebene ein Blankovolumen definiert, um Risiken aus dem gleichzeitigen Ausfall mehrerer großer Kreditengagements zu begrenzen.

Der Beitrag eines Kunden im Sinne des Kundenwerts liegt nun darin, welche Wirkung seine Existenz im Ist-Portfolio auf die Risiko-Struktur des gesamten Portfolios hat. In Kreditgenossenschaften sind oftmals die 10 gemessen am Aktivvolumen größten Kunden für 60 Prozent des Value-at-Risk verantwortlich.[1423] Bezüglich des prospektiven Kundenwerts können sich vor dem Hintergrund der Limitsystematik Restriktionen bei denjenigen Teilwerten des Kundenwerts ergeben, die auf Basis eines über das Einzelengagementlimit hinaus gehenden Kreditengagements realisiert werden könnten. Insofern kann die Kreditrisikostrategie der Kreditgenossenschaft restriktiven Einfluss auf den Kundenwert haben. Dies ist im Kontext der Prognose und Integration des Kundenwerts in die Vertriebsplanung zu berücksichtigen.

Ein weiterer Einfluss der Portfoliobetrachtung resultiert aus dem Beitrag des einzelnen Kreditengagements zu einem nicht adäquat diversifizierten Portfolio, mit der Konsequenz von Konzentrationsrisiken.[1424] Zu den Konzentrationsrisiken gehören Großkreditrisiken, deren Ausfälle starke negative Ertragsauswirkungen induzieren können. Bei Kreditgenossenschaften sind dies vor allem jedoch Branchenrisiken und geographisch bedingte Risiken aufgrund unzureichender Diversifikation bzw. aufgrund des Regionalprinzips fehlender Möglichkeiten zur Streuung.[1425] Um das Kreditportfolio zu steuern, stehen den Kreditgenossenschaften grds. verschiedene Instrumente zur Verfügung. Dies sind neben den bereits genannten ex ante definierten adressrisikospezifischen Vorgaben für Neugeschäfte auch ex post Möglichkeiten des Kreditrisikotransfers sowie der Einsatz von Kreditderivaten.[1426] Es kann auch direkt in die Geschäftsbeziehung eingegriffen werden, so z. B. bei der Nutzung anstehender Fälligkeiten. Nachteilig am letztgenannten Vorgehen ist, dass die Handlungsspielräume aufgrund der kreditvertraglichen Bindungen bei problemlosen Kreditbeziehungen nur gering sind und einen massiven Eingriff in eine Kundenbeziehung darstellen.[1427] Denn wenn aufgrund von Steuerungsmaßnahmen in der Portfoliostruktur ansonsten

---

1422 Vgl. *Rolfes, B.* (1999), S. 403.

1423 Vgl. *Geilmann-Ebbert, A./Heine, S.* (2006), S. 1147.

1424 Vgl. *Koneberg, M.* (2006), S. 7.

1425 Vgl. *Geilmann-Ebbert, A./Heine, S.* (2006), S. 1147.

1426 Vgl. *Koneberg, M.* (2006), S. 11.

1427 Vgl. *Ridder, T.* (2002), S. 22.

bezüglich Bonität und bisheriger oder zu erwartender Profitabilität intakte Kundenbeziehungen beendet werden, führt dies ggf. zu Unverständnis auf Seiten des Firmenkunden mit potenziell negativen Auswirkungen auf seine Referenztätigkeit. Daher wird empfohlen auf kapitalmarktorientierte oder verbundinterne Maßnahmen der Adressrisikosteuerung zurückzugreifen, durch die die Geschäftsbeziehung nicht direkt tangiert wird. Voraussetzung für diese Möglichkeiten ist, dass ein hinreichendes Maß an Transparenz über die Qualität des zu handelnden Kreditrisikos vorhanden ist. Im Gegensatz zu den börsennotierten Unternehmen basiert diese Transparenz im Mittelstandssegment auf den Ergebnissen eines standardisierten Ratings. Problematisch ist dabei für die Kreditgenossenschaften jedoch, dass der Handel von einzelnen Krediten stark eingeschränkt ist und ein liquider Markt bisher nicht existiert.[1428] Aus diesem Grund sind vor allem verbundinterne Lösungen wie z. B. WGZ Loop von der WGZ BANK AG und VR-Circle von der DZ BANK AG als Möglichkeiten des Kreditrisikotransfers von gehobener Relevanz für die Kreditgenossenschaften.[1429]

Werden Kreditrisiken nicht gehalten, sondern weitergegeben, ergeben sich keine Eigenkapitalkosten für unerwartete Verluste. In diesem Fall sind Kosten für das Hedging des unerwarteten Verlusts, d. h. der Teil der Kosten aus der Sekundärmarkttransaktion, der nicht dem erwarteten Verlust zuzurechnen ist, als Kosten für unerwartete Verluste zu berücksichtigen.

4.6 Zusammenfassung der Erkenntnisse

In Kapitel D.4.5 wurden die Dimensionen und Kundenwerttreiber eines für das Firmenkundengeschäft der Kreditgenossenschaften adäquaten Kundenwertmodells begründet hergeleitet. Dabei wurden drei Dimensionen unterschieden: die Beziehungsdimension, die Ertragsdimension und die Risikodimension. Diesen Dimensionen wurden wiederum verschiedene Kundenwerttreiber zugeordnet. Dabei wurde deutlich, dass die Quellen des Kundenwerts eines Firmenkunden für die Kreditgenossenschaften differieren. Die Zusammenfassung des gesamten Kundenwertmodells wird aufbauend auf dem allgemeinen Strukturmodell in Abbildung 53 dargestellt.

[1428] Vgl. *Geilmann-Ebbert, A./Heine, S.* (2006), S. 13.

[1429] Vgl. dazu ausführlich *Koneberg, M.* (2006), S. 205 ff.

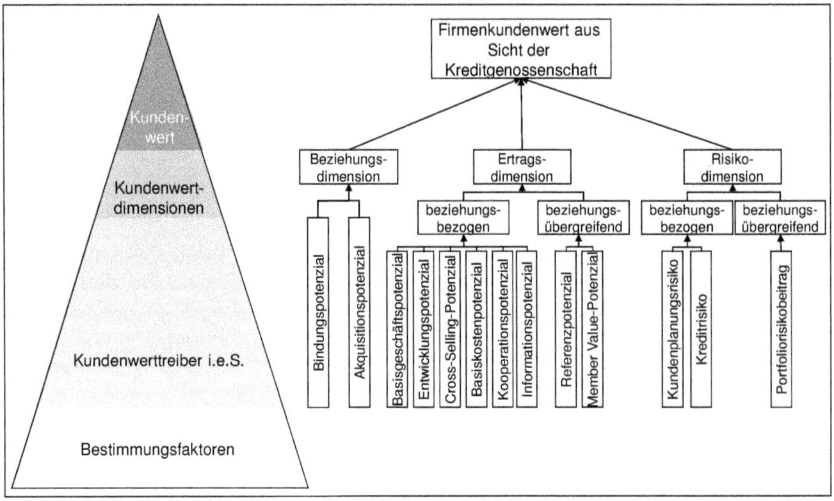

Abbildung 53: Kundenwertmodell für das Firmenkundengeschäft der Kreditgenossenschaften[1430]

Das Kundenwertmodell weist hinsichtlich der Kundenwerttreiber einen hohen Differenzierungsgrad aus, der wie in Kapitel D.4.2.2 bereits dargelegt aus, vertriebssteuerungsbezogenen Anforderungen resultiert. Ein Blick in das Schrifttum zeigt, dass grundsätzlich noch weitere Determinanten konzeptualisierbar sind.[1431] Jedoch ist zu berücksichtigen, dass eine noch stärker differenzierte Konzeptualisierung des Kundenwerts den Aufwand erhöht, die einzelnen Kundenwerttreiber zu messen und in der Unternehmens- und Funktionalsteuerung zu berücksichtigen. Es ist davon auszugehen, dass die Konzeptualisierung des Kundenwerts einem Optimierungsproblem entspricht, dass es also einen optimalen Grad an differenzierter Erfassung (Kosten) und Verwendung (Nutzen) der Kundenteilwerte gibt. Es kann begründet angenommen werden, dass mit zunehmender Differenziertheit des Kundenwertmodells die zusätzlichen Kosten gegenüber dem zusätzlichen Nutzen überproportional ansteigen, so dass aus

1430 Eigene Darstellung. Um die Darstellung übersichtlich zu halten, wurde die Ebene der Bestimmungsfaktoren nicht berücksichtigt. Es sei dazu auf die entsprechenden Unterkapitel von Kapitel D.4.5 verwiesen.

1431 So z. B. weitere Verbundeffekte im Kundenstamm, die sich z. B. in steigender oder geringerer Komplexität eines Unternehmens durch das Vorhandensein eines Kunden zeigen. Jedoch sind diese gemeinschaftlich resultierenden Verbundeffekte nur schwierig einzelnen Kunden zuzuordnen, so dass diese nicht weiter betrachtet werden. Vgl. zu einer ausführlichen Darstellung der Verbundeffekte *Lissautzki, M.* (2007), S. 255 ff.

ökonomischer Sicht kein Nutzen resultiert. Zu diskutieren ist, ob der Optimalpunkt, an dem Grenzkosten und Grenznutzen identisch sind, bei dem entwickelten Firmenkundenmodell noch nicht erreicht oder bereits überschritten wurde. Diese Frage vermag entweder eine empirische Überprüfung des Kundenwertmodells oder der praktische Einsatz desselben zu beantworten. Für den weiteren Verlauf der Untersuchung soll – begründet durch die jeweiligen Ausführungen zu den Teilwerten – angenommen werden, dass der Kundenwert in diesem Kontext optimal modelliert ist.

Weiterhin ist es notwendig, eine Differenzierung des Kundenwertmodells im Firmenkundengeschäft für aktuelle und potenzielle Kunden vorzunehmen. Denn so sind die einzelnen Determinanten des Kundenwerts primär für Bestandskunden bestimmt. Um dennoch die Möglichkeit zu bieten, auch potenzielle Kunden zu berücksichtigen, muss die differenzierte Betrachtung zwischen Basis-, Entwicklungs- und Cross-Selling-Potenzial aufgegeben werden.[1432]

Abschließend sei darauf hingewiesen, dass Marktpreisrisiken etwa in Form von Zinsrisiken, die im Zusammenhang mit Kundengeschäften entstehen, nicht dem Kundenwert zugeordnet werden, da die am Markt verfügbaren Interbankengeschäfte es grundsätzlich erlauben, diese Risiken zu eliminieren. Die Entscheidung, dies nicht zu tun, unterliegt dem Risiko-Chancen-Kalkül der Kreditgenossenschaft.[1433] Die Frage nach dem Aufwand der entsteht, wenn der „vielschichtige" Kundenwert als Ziel- und Steuerungsgröße eingesetzt wird, hängt zudem davon ab, wie er gemessen werden soll. Mit dem Kundenwert als Messproblem setzt sich daher das folgende Kapitel auseinander, in dem ein instrumenteller Messansatz entwickelt wird.

[1432] Vgl. dazu die Ausführungen in Kapitel D.5.5.1.

[1433] Vgl. *BVR* (2002b), S. 9 f.

5 Operationalisierung des zweckspezifischen Kundenwertmodells

Nachdem das Kundenwertmodell im vorhergehenden Kapitel hergeleitet wurde, soll nun die Metrik zur Messung des Kundenwerts erläutert werden. Die weitere Vorgehensweise für die Messung der Kundenwerttreiber und des Kundenwerts lehnt sich konsistent an die begrifflichen Grundlagen aus Kapitel D.1.2.3 an, wo die Begriffe (anbieterspezifischer) Kundenwert und Kundenwertpotenzial differenziert wurden. Übertragen auf die Kundenwerttreiber kann somit unterschieden werden, ob diese im Sinne des Potenzials oder des anbieterspezifischen Ergebnisses, also des Werts, beschrieben werden. Vereinfacht lässt sich dieser Gedanke funktional wie folgt darstellen (Formel [D.9]).[1434]

$$KWT = g(KWTP, a) \qquad [D.9]$$

mit:

KWT: (Anbieterspezifischer) Kundenwerttreiber

$g(\cdot)$: Kundenwertfunktion

KWTP: Maximale Ausprägung des Kundenwerttreibers (Kundenwerttreiberpotenzial)

α: Anbieterspezifische Ausschöpfung des Kundenwerttreiberpotenzials (Loyalitätsgrad)

Daraus folgt, dass die Messung des Kundenwerts und der Kundenwerttreiber derart vorgenommen werden soll, dass mit gleichen Messinstrumentarien sowohl das Kundenwertpotenzial als auch der anbieterspezifische Kundenwert gemessen werden können. Es sei bereits an dieser Stelle darauf hingewiesen, dass dies nicht bei allen Kundenwerttreibern gelingt.

5.1 Überblick und Auswahl der Methodik zur Messung des Kundenwerts

Ebenso vielfältig wie die in der Forschung und Praxis diskutierten Kundenwertmodelle sind die Metriken, die zur Messung des Kundenwerts eingesetzt werden. Hierzu moniert WÜNSCHMANN, dass die „Flut von operational orientierten Publikationen [.] eine Vielfalt von Formeln und Modellen [hinterlässt; Erg. d. Verf.]“[1435], mit denen Kunden bewertet und beurteilt werden können. Aus die-

---

1434 Aus Vereinfachungsgründen wird nur ein einziger Kundenwerttreiber berücksichtigt.

1435 *Wünschmann, S.* (2007), S. 46.

sem Grund sollen relevante Verfahren der Messung des Kundenwerts zunächst begründet ausgewählt werden.[1436]

Aufgrund der Vielzahl an Messansätzen werden in der Literatur verschiedene Systematisierungsansätze vorgestellt, um Metriken zur Kundenwertmessung zu strukturieren.[1437] Diese Ansätze bedienen sich i. d. R. der Dimensionen des Kundenwerts als Systematisierungskriterien.[1438] Dabei ist zu beobachten, dass diese Ansätze nicht zwingend zu trennscharfen Einordnungen der Verfahren führen.[1439] Einen Überblick häufig angewendeter Verfahren zeigt Abbildung 54.

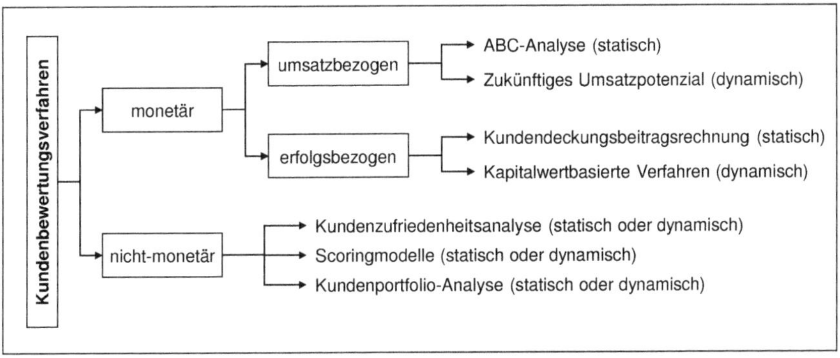

Abbildung 54: Mögliche Systematisierung ausgewählter Kundenbewertungsverfahren[1440]

Während die Majorität der Systematisierungsansätze die Messmethoden anhand der Anzahl und der Dimension der Inputvariablen sowie der zeitlichen und der objektbezogenen Dimension strukturiert, klassifiziert BURMANN die Metriken danach, welches Steuerungspotenzial die Verfahren innehaben.[1441] Steuerungsaufgaben sieht er in der Selektion von Neukunden, der Allokation von Ressour-

---

[1436] Vgl. *Mödritscher, G. J.* (2008), S. 155.

[1437] Vgl. z. B. *Krafft, M.* (2007), S. 75; *Homburg, C.* (2007), S. 400; *Schroeder, N.* (2006), S. 38.

[1438] Vgl. dazu die Ausführungen in Kapitel D.1.2.4.

[1439] Vgl. *Schemuth, J.* (1996), S. 43.

[1440] *Homburg, C.* (2007), S. 400. Eine alternative Systematisierung findet sich bei z. B. *Mödritscher, G. J.* (2008), S. 157 oder bei *Wünschmann, S.* (2007), S. 47.

[1441] Vgl. hier und nachfolgend *Burmann, C.* (2003), S. 113 ff. Obwohl BURMANN in seinem Beitrag den Schwerpunkt auf investitionstheoretische Verfahren legt, ist die der Systematisierung zugrunde liegende Frage verallgemeinerbar. Sie Stellt mit Blick auf die Systematisierung in Abbildung 54 auch insofern eine Ergänzung dar, als sie nach der Tiefe der Verfahren fragt.

cen auf Bestandskunden, der Ausgestaltung von Marktbearbeitungsstrategien (Instrumenteneinsatz) und der Unterstützung zur Unternehmensbewertung. Er unterscheidet im Ergebnis drei Gruppen von Operationalisierungen des Kundenwerts. Als erstes nennt er die Black-Box-Modelle, bei denen die rein monetären Inputdaten modellexogen bestimmt werden und teilweise auf Durchschnittswerten basieren. Diese Modelle haben ein begrenztes Steuerungspotenzial für die Neukundenselektion und Unternehmensbewertung, da deren Fokus auf den Bestandskunden liegt und zudem keine zusätzlichen Impulse für die unternehmenswertorientierte Steuerung resultieren. Potenzial haben diese Modelle bei der Ressourcenallokation für Bestandskunden und kein Potenzial hinsichtlich des Instrumenteneinsatzes, da sie nicht ermitteln können, wie der Kundenwert auf deren Einsatz reagiert. Als zweite Gruppe werden verhaltenstheoretisch orientierte Modelle geprüft, deren Inputdaten mehrdimensional sind. Bei diesen werden vereinzelte Kundenwerttreiber modellendogen ermittelt.[1442] Diese Modelle haben zwar Steuerungspotenzial bezüglich der Ressourcenallokation bei Bestandskunden, da sie in der Kundensegmentierung angewendet werden können. Auch haben sie bedingt gegebenes Steuerungspotenzial bei der Akquisitionsentscheidung mit Blick auf potenzielle Kunden und dem Instrumenteneinsatz des Marketings. Hinsichtlich der Unternehmensbewertung attestiert ihnen BURMANN jedoch kein Steuerungspotenzial, da Kundenrisiken nicht berücksichtigt werden. Lediglich den hybriden Modellen, die investitionstheoretische Verfahren mit dem Einsatz von Marktbearbeitungsinstrumenten verknüpfen und somit die relevanten Bestimmungsfaktoren und deren Auswirkungen auf die Kundenwerttreiber modellendogen umfassen, bescheinigt BURMANN ein umfängliches Steuerungspotenzial für die vier genannten Steuerungsaufgaben.

Werden BURMANNS Ergebnisse zugrunde gelegt, empfiehlt sich die Messung des Kundenwerts auf Basis eines hybriden Modells. Damit ließe sich der Einsatz vertriebspolitischer Maßnahmen und Instrumente in einem Totalmodell abbilden, was durchaus im Sinne der kundenwertorientierten Vertriebssteuerung ist (z. B. für die Berücksichtigung des Einflusses der Besuchsaktivitäten des Firmenkundenberaters auf den anbieterspezifischen Kundenwert). Aus folgenden Gründen wird dies hier jedoch nicht gemacht. Der Forschungsstand zu den hybriden Modellen muss gegenwärtig noch als vergleichsweise niedrig beschrieben werden, zumal in vielen Teilbereichen eines solchen Totalmodells bank- und kreditgenossenschaftsspezifische Erkenntnisse fehlen. Die Entwicklung eines umfassenden Messmodells ist zudem nicht der zentrale Untersuchungsgegenstand dieser Arbeit. Vielmehr steht die auf Basis des Kundenwerts

---

[1442] Als Beispiel nennt BURMANN das Referenzwertmodell von CORNELSEN. Vgl. *Cornelsen, J.* (2000), S. 169 ff.

ausgestaltete Vertriebssteuerung im Fokus. Dazu ist die Messung des Kundenwerts zwar notwendig, jedoch lassen sich die wesentlichen Gestaltungsvorschläge und Implikationen auch auf Basis der Black-Box-Modelle ableiten und diskutieren, ohne den Erkenntnisgewinn der Arbeit zu beeinträchtigen. Zwar wird damit in Kauf genommen, die Steuerungsaufgaben bezüglich der Unternehmensbewertung zu vernachlässigen. Dies ist jedoch irrelevant, da diese Steuerungsaufgaben nicht zum Aufgabenkatalog der Vertriebssteuerung gehören[1443] und die Bewertung genossenschaftlicher Unternehmen nur bedingt vergleichbar mit Unternehmen anderer Rechtsformen ist.[1444] Des Weiteren müssen für die praktische Anwendung der hybriden sowie auch der verhaltenstheoretisch orientierten Modelle umfangreiche kundenbezogene Informationen erhoben werden. Dabei ist zum einen offen, ob es den relativ und absolut eher kleinen Kreditgenossenschaften gelingt, diese Informationen valide zu erheben. Und zum zweiten ist fraglich, ob die stärker detaillierte Analyse zwingend einen positiven Beitrag zum Kosten-Nutzen-Verhältnis der kundenwertorientierten Vertriebssteuerung leistet.[1445] Nach MEFFERT/BRUHN nimmt mit zunehmender Berücksichtigung vorökonomischer Faktoren die Reliabilität und Validität der Messergebnisse ab und die Komplexität der Definition des Kundenwerts zu.[1446] Und VOGEL/EVANSCHITZKY/RAMASESHAN stellen in diesem Kontext fest, dass trotz der wissenschaftlichen Entwicklungen „the results are still inconclusive as to the relationship among marketing inputs, customer attitudes, and customer behavior."[1447] Mit Blick auf die Werttreiberhierarchie, die hier als strukturelle Basis für den Kundenwert dient (siehe Abbildung 35), erläutert auch LISSAUTZKI, dass auf Ebene der Bestimmungsfaktoren große Schwierigkeiten vorlägen, die Zusammenhänge zwischen den einzelnen Faktoren zu identifizieren.[1448] Für den Gang seiner Analyse, die den Kundenwert in die gesamthafte wertorientierte Unternehmenssteuerung integriert und dazu eines logisch-konsistenten Berechnungssystems bedarf, sei eine vollständige und eindeutige Darstellung der Zusam-

---

1443 Vgl. dazu die Ausführungen in Kapitel B.2.3.

1444 Vgl. *Schöffling, B.* (1991), S. 3 ff.

1445 So zeigen z. B. WÜBBEN/VON WANGENHEIM, dass bei wesentlichen Analysen im Rahmen der Prognose und Messung kundenwertrelevanter Informationen wie dem Aktivitätszustand eines Kunden (Bindungspotenzial), der Vorhersage künftigen Kaufverhaltens und der Identifikation der Top-10- und Top-20-Kunden Managementheuristiken grundsätzlich gleichwertig zu komplexeren statistischen Methoden sind. Lediglich bei der Vorhersage des künftigen Kaufverhaltens erwiesen sich die statistischen Methoden als geringfügig leistungsfähiger. Vgl. *Wübben, M./von Wangenheim, F.* (2008), S. 90 ff.

1446 Vgl. *Meffert, H./Bruhn, M.* (2009), S. 425.

1447 *Vogel, V./Evanschitzky, H./Ramaseshan, B.* (2008), S. 98.

1448 Vgl. hier und nachfolgend *Lissautzki, M.* (2007), S. 120.

menhänge notwendig. Aus diesem Grund berücksichtigt er in seinem Messmodell nur die Ebenen, für die er den Zusammenhang und die Stärke quantifizieren kann und schließt somit die Ebene der Bestimmungsfaktoren weitgehend aus. Diese Argumentation lässt sich auch auf die vorliegende Arbeit übertragen. Aus den genannten Gründen sollen Messansätze diskutiert werden, die tendenziell der Gruppe der Black-Box-Modelle zuzuordnen sind.

Aufbauend auf den bisherigen Erkenntnissen dieser Arbeit ist ein dynamisches Verfahren zu präferieren, da die statischen Verfahren nicht geeignet sind, das Bewertungsobjekt Kundenbeziehung adäquat zukunftsorientiert zu messen.[1449] Dass rein umsatzbezogene Verfahren mit Blick auf das Kundenwertmodell nicht hinreichend leistungsfähig sind, ist leicht ersichtlich.[1450] Hinsichtlich des Aggregationsniveaus können Messmodelle auch dahingehend unterschieden werden, ob sie den Kundenwert auf Einzelkundenebene oder auf Gruppen- bzw. Gesamtkundenstammebene messen.[1451] Der Fokus liegt nachfolgend auf den individuellen Messansätzen, da dies dem Postulat der grenznutzenorientierten Einzelbewertung am ehesten entspricht und in der Literatur gefordert wird.[1452] Gleichwohl sollen auch die Möglichkeiten einer gruppenbasierten Messung skizziert werden. Dies erfolgt vor dem Hintergrund, dass es in der Vertriebsplanung sinnvoll sein kann, die Messung von Kundenwerten anhand differierender Vorgehensweisen vorzunehmen.[1453] Aus diesen Überlegungen resultiert, dass die investitionstheoretischen im Allgemeinen und die kapitalwertbasierten Verfahren im Speziellen Gegenstand der weiteren Analyse sind.[1454]

Jedoch zeigen bereits die Ausführungen zu einigen der Kundenwerttreiber eines kreditgenossenschaftlichen Kundenwertmodells für Firmenkunden, dass eine in einem einzigen Messansatz vorzunehmende und ausschließlich monetäre

---

1449 Vgl. *Homburg, C.* (2007), S. 406; *Homburg, C./Steiner, V. V./Totzek, D.* (2009), S. 70 ff.

1450 So berücksichtigen diese Verfahren z. B. die Ressourcenverbräuche nicht, die sich aus dem Basiskostenpotenzial i. V. m. dem Kooperationspotenzial und der Risikodimension ergeben.

1451 An dieser Stelle sei darauf hingewiesen, dass mit Einzelkunde sowohl ein Einzelstamm als auch ein Kundenverbund im Sinne einer Betreuungseinheit gemeint sein kann. Vgl. dazu die Ausführungen in Kapitel C.2.2.4.

1452 HOEKSTRA/HUIZINGH argumentieren, „in customer-based marketing the focus is on individual customers […]. LTV should be computed for each individual customer." *Hoekstra, J. C./Huizingh, E.* (1999), S. 263.

1453 Vgl. dazu die Ausführungen in Kapitel E.2.

1454 Vgl. für die Anwendung der kapitalwertbasierten Verfahren bei der Bewertung von Existenzgründungen als Investitionsprojekt aus Sicht einer Bank auch *Böhm, A.* (1999), S. 90 ff. Vor dem Hintergrund der kundenindividuellen Bewertung kann auch von einer partiellen Investitionsrechnung gesprochen werden.

Messung des Kundenwerts gegenwärtig praktisch unmöglich erscheint.[1455] Aus diesem Grund soll auf das Ergebnis von SCHMÖLLER und TEWES aufgebaut werden.[1456] Diese differenzieren den Kundenwert in einen monetären und einen nicht-monetären Kundenteilwert.[1457] Während der erstgenannte anhand eines kapitalwertbasierten Verfahrens gemessen wird, wird für die Messung des nichtmonetären Kundenwerts ein Scoringmodell genutzt.[1458] Wird eine solche Trennung vorgenommen, ist im Weiteren zu prüfen, ob es Möglichkeiten gibt, die Kundenteilwerte zielsetzungsgerecht zusammenzufassen.

5.2 Messung des monetären Kundenteilwerts

5.2.1 Kapitalwertbasierte Verfahren als konzeptionelle Grundlage

Gegenwärtig erfahren die kapitalwertbasierten Verfahren in der Literatur besondere Aufmerksamkeit.[1459] In einer Reihe von Studien wurden diese Verfahren auf ihre theoretische Fundierung, praktische Tauglichkeit und empirische Relevanz hin untersucht.[1460] VENKATESAN/KUMAR stellen fest, dass derart gemessene Kundenwerte gegenüber anderen Metriken überlegen sind.[1461] RYALS zeigt in zwei Fallstudien der Finanzdienstleistungsbranche die durch diese Messung erzeugten Veränderungen in der Marktbearbeitung, die durch die kundenwertbezogenen Informationen eingeleitet werden, sowie deren positive Wirkungen auf den Unternehmenserfolg.[1462] Die empirischen Ergebnisse von KRAFFT, HOM-

---

[1455] Vgl. *Homburg, C.* (2007), S. 410. Dies gilt für das Informationspotenzial, das Referenzpotenzial, das Member Value-Potenzial sowie die beziehungsübergreifenden Kundenwerttreiber der Risikodimension. Zwar zeigen einige Autoren wie z. B. SCHNEIDER auf, wie die Messung und Integration des Informationswerts in den CLV theoretisch vorgenommen werden kann. In der empirischen Anwendung jedoch wird dieser Ansatz nicht umgesetzt. Vgl. *Schneider, N.* (2007), S. 159 ff. und S. 210 ff. Für eine Übersicht und Würdigung von Metriken für die Messung des Referenzwerts vgl. *Schneider, N.* (2007), S. 143 ff. und *Helm, S.* (2001), S. 124 ff.

[1456] Vgl. *Tewes, M.* (2003), S. 159; *Schmöller, P.* (2001), S. 165.

[1457] GELBRICH/MÜLLER sprechen sich dagegen bei der Ermittlung eines prospektiven Kundenwerts für einen rein monetären Kundenwert aus. Vgl. *Gelbrich, K./Müller, S.* (2006), S. 471.

[1458] Die Bewertung nicht-monetärer oder schwer bis nicht monetarisierbarer Kundenwerttreiber mittels Punktbewertungsmodellen wird auch von *Hoberg, P.* (2007), S. 1248 als gängige Vorgehensweise beurteilt.

[1459] Vgl. *Rust, R. T./Lemon, K. N./Zeithaml, V. A.* (2004), S. 110.

[1460] Vgl. *Kiwitz, K./Scheffler, C.* (2009), S. 2 ff.; *Mödritscher, G. J.* (2008), S. 157; *Homburg, C.* (2007), S. 413; *Sträter, J.* (2005), S. 182 ff.

[1461] Vgl. *Venkatesan, R./Kumar, V.* (2004),S. 118 f.

[1462] Vgl. *Ryals, L.* (2005), S. 254 ff.

BURG ET AL., REINECKE/TOMCZAK und SCHÄFFER/STEINERS zeigen jedoch, dass diese Verfahren in der Praxis bisher nur in geringem Umfang angewendet werden.[1463] Methodische Grundlage dieser Verfahren ist die Kapitalwertformel [D.10]:[1464]

$$C = \sum_{t=0}^{t=n} \frac{e_t - a_t}{(1+i)^t} \qquad \text{[D.10]}$$

mit:

C: Kapitalwert

e_t: (erwartete) Einzahlungen in der Periode t

a_t: (erwartete) Auszahlungen in der Periode t

i: Kalkulationszinssatz

t: Periodenindex

n: (erwartete) Restdauer der Geschäftsbeziehung

Der Wert eines Kunden entspricht demnach dem Saldo der auf den Bewertungszeitpunkt diskontierten Ein- und Auszahlungen, die dem betrachteten Kunden direkt zurechenbar sind und im zugrundeliegenden Planungshorizont anfallen. Inhaltlich entspricht der Kundenkapitalwert also dem anbieterspezifischen Kundenwert.[1465] Er wird auch als Vermögenswert eines Kunden bezeichnet[1466] und ist kongruent zur vermögensorientierten Steuerung gemäß der Gesamtbanksteuerungskonzeption VR-CONTROL. Es ist erkennbar, dass der Einsatz dieses Verfahrens dem Verständnis der Kundenbeziehung als Investitionsobjekt entspricht.[1467] Die Höhe des Kundenkapitalwerts wird in diesem Modell durch die (erwarteten) periodischen Cash-flows, die (restliche) Kundenbindungsdauer sowie den Diskontierungsfaktor bestimmt.[1468] Wie in Kapitel D.1.1 erläutert, hat sich als allgemeine Kategoriebezeichnung dieser Verfahren der Begriff des Customer Lifetime Value (CLV) etabliert. „Although a considerable number of CLV models have been developed so far, no generally accepted superior approach exists."[1469] JAIN/SINGH unterteilen die bestehenden Ansätze in Berech-

[1463] Vgl. *Krafft, M.* (1997), S. 9; *Homburg, C. et al.* (2000), S. 247; *Reinecke, S./Tomczak, T.* (2001) S. 81; *Schäffer, U./Steiners, D.* (2005), S. 322.

[1464] Vgl. *Mödritscher, G. J.* (2008), S. 188; *Homburg, C.* (2007), S. 407.

[1465] Dies gilt nur, wenn alle Dimensionen und Kundenwerttreiber monetär abgebildet werden.

[1466] Vgl. *Link, J./Hildebrand, V. G.* (1997), S. 164.

[1467] Vgl. *Gupta, S./Lehmann, D. R.* (2003), S. 9.

[1468] Vgl. *Kiwitz, K./Scheffler, C.* (2009), S. 34; *Rosset, S. et al.* (2003), S. 322.

[1469] *Bauer, H.-H./Hammerschmidt, M.* (2005), S. 333.

nungsmodelle des CLV, Analysemodelle des Kundenstamms sowie in Modelle, die sich mit der Analyse des CLV beschäftigen und dessen Einfluss auf Entscheidungsprozesse analysieren.[1470] KUMAR/GEORGE hingegen unterscheiden zwischen Modellen, die auf Ebene des Einzelkunden (bottom-up approach) oder auf Ebene des Kundenstamms (top-down approach) ansetzen.[1471] Zu den letzteren Ansätzen sind auch die Methoden zu zählen, die den CLV auf Kundensegmentebene berechnen. Eine ähnliche Aufteilung nehmen KUMAR/RAMANI/ BOHLING vor, die zwischen einem „Average CLV Approach" und einem „Individual Level CLV Approach" unterscheiden.[1472] Während beim ersten Ansatz Durchschnittswerte berechnet werden, wird beim zweiten Ansatz ein kundenindividueller CLV ermittelt. Diese Vielfalt kann als Indiz gewertet werden, dass die Modellierung des CLV „eine der zentralen Herausforderungen der Kundenwertforschung"[1473] ist.

Nach der klassischen Formel wird der Kundenwert als barwertiger Saldo aus undifferenzierten Ein- und Auszahlungen ermittelt.[1474] Möglich ist auch eine differenzierte Berechnung, die nützlicher als Kontroll- und Steuerungsinstrument ist.[1475] Um die für die Vertriebssteuerung und unter Berücksichtigung des hergeleiteten Kundenwertmodells notwendige differenzierte Messung des Kundenwerts vornehmen zu können, haben einige Autoren differenzierte Formeln des CLV entwickelt,[1476] die zugleich versuchen, vollumfänglich die einzelnen Bestandteile des Kundenwerts zu integrieren. Ein umfassendes Rahmenkonzept stellen z. B. BAUER/HAMMERSCHMIDT/BRÄHLER vor (siehe Formel [D.11]):[1477]

1470 Vgl. *Jain, D./Singh, S. S.* (2002), S. 34 ff.

1471 Vgl. *Kumar, V./George, M.* (2007), S. 157 f.

1472 Vgl. *Kumar, V./Ramani, G./Bohling, T.* (2004), S. 62 ff.

1473 *Fischer, M./Albers, S.* (2007), S. 170.

1474 Vgl. z. B. *Homburg, C./Beutin, N.* (2001), S. 229.

1475 Vgl. *Henseler, J./Hoffmann, T.* (2003), S. 39. HENSELER/HOFFMANN schlagen zur Differenzierung vor, Erlöse und Kosten separat zu betrachten.

1476 So z. B. GIERL/KURBEL, die zwischen dem Wert der eigenen Nachfrage, dem Wert aus der Unterstützung bei der Neukundenakquisition und dem Wert aus der Unterstützung bei der Verhinderung von Abwanderung unterscheiden. Vgl. *Gierl, H./Kurbel, T.* (1997), S. 177.

1477 Vgl. *Bauer, H.-H./Hammerschmidt, M./Brähler, M.* (2002), S. 330. Zur einheitlichen Verwendung von Indizes wird der von HAMMERSCHMIDT/BRÄHLER verwendete Index *i* für Kunden in das in dieser Arbeit verwendete *c* geändert.

$$CLV_c = -AK_{cc} + R_{ct} \cdot \sum_{t=0}^{T} \frac{(AE_{ct} + UE_{ct} + CE_{ct} + RW_{ct})}{(1+r_c)^t} - R_{ct} \cdot \sum_{t=0}^{T} \frac{(MK_{ct} + PSLK_{ct})}{(1+r_c)^t} + R_{ct} \cdot \sum_{t=0}^{T} \frac{(InfoP_{ct} + KoopP_{ct} + InnoP_{ct})}{(1+r_c)^t} - \frac{AufK_c}{(1+r_c)^T} \quad \text{[D.11]}$$

mit:

CLV_c:	Customer Lifetime Value des Kunden *c*
AK_c:	Akquisitionskosten zur Gewinnung des Kunden *c* (nur relevant bei Bewertung eines Neukunden)
T:	Lebensdauer der Geschäftsbeziehung mit einem Kunden *c*
R_{ct}:	Bindungsrate des Kunden *c* in der Periode *t*
AE_{ct}:	Autonome Erlöse, die durch den Kunden *c* in der Periode *t* verursacht werden
UE_{ct}:	Up-Selling-Erlöse, die durch den Kunden *c* in der Periode *t* verursacht werden
CE_{ct}:	Cross-Selling-Erlöse, die durch den Kunden *c* in der Periode *t* verursacht werden
RW_{ct}:	Referenzwert des Kunden *c* in der Periode *t* (sofern monetarisierbar)
$InfoP_{ct}$:	Informationspotenzial des Kunden *c* in der Periode *t* (sofern monetarisierbar)
$KoopP_{ct}$:	Kooperationspotenzial des Kunden *c* in der Periode *t* (sofern monetarisierbar)
$InnoP_{ct}$:	Innovationspotenzial des Kunden *c* in der Periode *t* (sofern monetarisierbar)
MK_{ct}:	Marketingkosten für den Kunden *c* in der Periode *t* (zur Erreichung der prognostizierten Erlöse und der Bindungsrate)
$PSLK_{ct}$:	Produkt-, Service- und Logistikkosten, die durch den Kunden *c* in der Periode *t* verursacht werden
r_c:	Kundenspezifischer, risikoadäquater Zinssatz
$AufK_c$:	Auflösungskosten der Kundenbeziehung mit dem Kunden *c*

Dieser differenzierte Ansatz sensibilisiert den Anwender stärker als die verdichteten Berechnungsformeln, die Vielzahl an möglichen Einflussfaktoren, die den

Wert des Kunden beeinflussen, zu berücksichtigen.[1478] Somit könnte, sofern die einzelnen Kundenwerttreiber monetär gemessen werden können, der Customer Lifetime Value als geeignetes Messmodell angewendet werden. Nachfolgend soll daher eine für die kundenwertorientierte Vertriebssteuerung im Firmenkundengeschäft der Kreditgenossenschaften adäquate kapitalwertbasierte Metrik entwickelt werden.

5.2.2 Entwicklung eines kapitalwertbasierten Kundenwertmessmodells für Kreditgenossenschaften

5.2.2.1 Integration der Kundenwerttreiber der Ertragsdimension

Um die Zielsetzung dieser Arbeit zu erreichen, die kundenwertorientierte Vertriebssteuerung als Erweiterung der Steuerungskonzeption VR-CONTROL zu entwickeln, werden für die Messung des Kundenwerts die folgenden Festlegungen getroffen. Erstens sollen alle bankleistungsbezogenen Erfolgsbeiträge mittels des Schemas der Einzelgeschäftskalkulation aus VR-CONTROL bewertet werden.[1479] Konkret bedeutet dies, dass die Bewertung der Kundenwerttreiber auf Basis der Einzelgeschäftskalkulation erfolgt. Dies ist methodisch problematisch,[1480] jedoch gängig in praktischen und wissenschaftlichen Beiträgen.[1481] Mit eben dieser methodischen Unschärfe stellt z. B. BARTH für den Bankenkontext eine einfache formale Definition des CLV auf.[1482] Damit wird eine unmittelbare Verknüpfung zum internen Rechnungswesen der Bank erreicht.[1483] In Konsequenz bedeutet dieses Vorgehen eine entsprechend leichtere Einbindung in die Gesamtbanksteuerung. In diesem Zusammenhang wird zudem von einer barwertigen Deckungsbeitragsrechnung ausgegangen.[1484] Nachfolgend werden zunächst die einzelnen Kundenwerttreiber hinsichtlich ihrer Messung vorgestellt.

Für das Basisgeschäfts-, das Entwicklungs- und das Cross-Selling-Potenzial gilt, dass sich der anbieterspezifische Wert als Ergebnis aus jeweils bankleistungsbezogenem Bedarf des Kunden (Bedarfspotenzial) und Loyalitätsgrad ermitteln lässt. Wird der Loyalitätsgrad nicht berücksichtigt bzw. auf den Faktor 1 bestimmt, so kann das gesamte Kundenwerttreiberpotenzial gemessen werden. In letztgenannten Fall sind Bedarfspotenzial und Bedarfsvolumen eines Kunden

---

1478 Vgl. *Homburg, C.* (2007), S. 411.
1479 Vgl. dazu die Ausführungen in Kapitel C.4.3.
1480 Vgl. *Mödritscher, G. J.* (2008), S. 188.
1481 Vgl. *Barth, K./Wille, K.* (2000), S. 33.
1482 Vgl. *Barth, W.* (2009), o. S.
1483 Vgl. dazu die Ausführungen in Kapitel C.4.3. So auch *Beeser, A. J.* (2003), S. 44 ff.
1484 Vgl. *Bruhn, M./Georgi, D./Wöhle, C. B.* (2005), S. 386 ff.

identisch. Das Bedarfspotenzial und das Bedarfsvolumen sind Ertragsgrößen, d. h. mittels ausgewählter (Teil-)Ergebnisse der Einzelgeschäftskalkulation bewertet.[1485] Anders als im industriellen Kontext gilt bei Bankleistungen nicht immer der einfache funktionale Zusammenhang, dass der Umsatz das Produkt aus Preis und Menge ist. Dies begründet sich mit der etablierten Anwendung der Marktzinsmethode für zinstragende Geschäfte.[1486] Ohne auf die Kalkulation einzelner Bankleistungen detailliert einzugehen, soll dies beispielhaft erläutert werden. Angenommen wird, dass in einem in der Zukunft liegenden Jahr der betrachtete Firmenkunde eine Kreditfinanzierung vornehmen möchte. Der gesamte Kreditbedarf beläuft sich auf TEUR 1.000 (Bedarfspotenzial). Für die Kalkulation wird ein Ratenkredit mit jährlicher Tilgung, einem Nominalzinssatz von 10 Prozent und einer Laufzeit von 3 Jahren unterstellt. Als Refinanzierungsgeschäfte werden struktur- und fristenkongruente Geldmarktgeschäfte abgeschlossen, die je Laufzeitband einen individuellen Zinssatz haben. Basierend auf diesen Annahmen wird der Zinskonditionsbeitrag ermittelt (siehe Tabelle 22), der als Bewertungsmaßstab herangezogen wird.

Tabelle 22: Beispielhafte Ermittlung des Zinskonditionsbeitragsbarwerts[1487]

		t=0	t=1	t=2	t=3
Kredit	Auszahlung	-900			
	Tilgung		300	300	300
	Zinsen (10%)		90	60	30
	Kapitaldienst		390	360	330
	Cash-flow	-900	390	360	330
Refinanzierung	3-jährige Refinanzierung (8%)	306	-24	-24	-330
	2-jährige Refinanzierung (6%)	317	-19	-336	
	1-jährige Refinanzierung (5%)	330	-347		
	Summe	52	0	0	0
	Zinskonditionsbeitragsbarwert	**52**			

Das bewertete Bedarfspotenzial beträgt somit 52 EUR. Würde nun der Loyalitätsgrad auf 50 Prozent festgelegt, so dass davon ausgegangen wird, dass die Kreditgenossenschaft nicht das gesamte Bedarfspotenzial decken kann, liegt das

1485 Das bedeutet konkret, dass nicht das Nachfragevolumen z. B. in Form von Kreditfinanzierungsbedarf in Höhe von 250.000 EUR für 10 Jahre, sondern die aus diesen und weiteren zu prognostizierenden/schätzenden Größen resultierende Größe in Form des Deckungsbeitrags I, Deckungsbeitrags II oder Deckungsbeitrags III zugrunde gelegt wird.

1486 Vgl. *Rolfes, B./Dartsch, A.* (1997), S. 926 ff.; *Schierenbeck, H.* (2003), S. 70 ff.

1487 Eigene Darstellung.

bewertete Bedarfsvolumen bei 26 EUR.[1488] Diese Verfahrensweise gilt in analoger Weise auch für Provisionsbeiträge und soll daher als grundlegendes Vorgehen für die Ermittlung des Bewertungsmaßstabs verwendet werden. Diese Festlegung gilt für alle Bedarfspotenziale und Bedarfsvolumina in der entwickelten Metrik, daher wird in den Formellegenden selbst auf die Festlegung nicht mehr hingewiesen. Vor dem Hintergrund dieser Überlegungen gilt für das Basisgeschäftspotenzial Formel [D.12]:

$$BGP_{c,t} = \sum_{p_c=1}^{P_c} BP_{c,p_c,t} \cdot \alpha_{c,p_c,t} \cdot (1-\rho_{c,pq}) \qquad \text{[D.12]}$$

mit:

$BGP_{c,t}$: Basisgeschäftspotenzial eines Kunden c in der Periode t

$BP_{c,p_c,t}$: Bedarfspotenzial eines Kunden c für das Produkt p_c in der Periode t

$\alpha_{c,p_c,t}$: Loyalitätsgrad eines Kunden c für das Produkt p_c in der Periode t; mit $0 \leq \alpha_{c,p_c,t} \leq 1$

$\rho_{c,pq}$: Dummyvariable der Verbundbeziehung zwischen den Produkten p und q mit

$\rho_{c,pq} = 1$, wenn eine Substitutionsbeziehung im Sinne des Up-Sellings für den Kunden c vorliegt und

$\rho_{c,pq} = 0$, wenn keine Substitutionsbeziehung im Sinne des Up-Sellings für den Kunden c vorliegt

c: Index für Kunden

p_c: Index der Produkte im Sinne des Basisgeschäftspotenzials aus Sicht des Kunden c

q_c: Index der Produkte im Sinne des Up-Sellings aus Sicht des Kunden c

t: Periodenindex

Aufgrund der Interdependenzen zwischen dem Basisgeschäftspotenzial und dem Up-Selling-Potenzial ist in Formel [D.12] eine Dummyvariable eingefügt, so dass es im Falle einer Substitutionsbeziehung zwischen den relevanten Bankleistungen zu keiner Verzerrung im Sinne einer Doppelzählung kommt.

Die Messung des Entwicklungspotenzials erfolgt gemäß der Modellkonzeption mehrstufig. Dazu gehören der Verkauf höherwertiger substitutiver Leistun-

[1488] Aufgrund der mathematischen Struktur der Ermittlung des Zinskonditionsbeitrags lässt sich das bewertete Bedarfsvolumen in Verbindung mit dem Loyalitätsgrad als Linearkombination des bewerteten Bedarfspotenzials darstellen.

gen, der Mehrverkauf (zusammen Up-Selling-Potenzial) und der höherpreisige Verkauf bisher kontrahierter Leistungen (Preispotenzial) (siehe Formel [D.13]).

$$\begin{aligned} EP_{c,t} &= USP_{c,t} + PP_{c,t} \\ &= HSL_{c,t} + MVP_{c,t} + PP_{c,t} \end{aligned} \qquad [D.13]$$

mit:

$EP_{c,t}$: Entwicklungspotenzial eines Kunden c in der Periode t

$USP_{c,t}$ Up-Selling-Potenzial eines Kunden c in der Periode t

$PP_{c,t}$: Preispotenzial eines Kunden c in der Periode t (bezogen auf das Basisgeschäftspotenzial und das Mehrverkaufspotenzial)

$HSL_{c,t}$: Entwicklungspotenzial aus dem Verkauf höherwertiger substitutiver Leistungen eines Kunden c in der Periode t

$MVP_{c,t}$: Entwicklungspotenzial aus Mehrverkauf eines Kunden c in der Periode t

Das Entwicklungspotenzial aus dem Verkauf höherwertiger substitutiver Leistungen basiert auf einer Substitutionsbeziehung zwischen einer Bankleistung, die dem Basisgeschäftspotenzial zugeordnet ist und einer nutzenidentischen, jedoch aus Banksicht höherwertigen Bankleistung, die nicht dem Basisgeschäfts- sondern dem Entwicklungspotenzial zugeordnet ist. Diese Interdependenz wird in Formel [D.14] berücksichtigt, so dass im Falle einer Substitutionsbeziehung zwischen zwei Bankleistungen das leistungsspezifische Basisgeschäftspotenzials durch die Dummyvariable korrigiert wird. Für die Berechnung des Entwicklungspotenzials aus dem Verkauf höherwertiger substitutiver Leistungen gilt, dass zunächst ein zur Substitutionsleistung gleichwertiger leistungsspezifischer Loyalitätsgrad angesetzt wird:

$$HSL_{c,t} = \sum_{q_c=1}^{Q_c} BP_{c,q_c,t} \cdot \alpha_{c,q_c,t} \cdot \rho_{c,pq} \qquad [D.14]$$

mit:

$BP_{c,q_c,t}$: Bedarfspotenzial eines Kunden c für das Produkt q_c in der Periode t

$\alpha_{c,q_c,t}$: Loyalitätsgrad eines Kunden c für das Produkt q_c in der Periode t; mit $0 \leq \alpha_{c,q_c,t} \leq 1$ und $\alpha_{c,q_c,t} = \alpha_{c,p_c,t}$, wenn p_c und q_c in einer Substitutionsbeziehung im Sinne des Up-Selling stehen

Um das Entwicklungspotenzial aus dem Mehrverkauf zu messen, ist zu berücksichtigen, dass dies über das Bedarfspotenzial und/oder den Loyalitätsgrad beeinflusst wird. Die jeweiligen Veränderungen können isoliert (Verdrängungs- und Mengeneffekt) oder zusammen auftreten (Mixeffekt). Weiterhin ist zu beachten, dass der Mehrverkauf sich entweder auf das Basisgeschäftspotenzial oder das bereits dargestellte Entwicklungspotenzial aus dem Verkauf höherwertiger substitutiver Leistungen (siehe Formel [D.14]) bezieht:

$$\begin{aligned} MVP_{c,t} &= MVP_{c,t}^{BGP} + MVP_{c,t}^{HSL} \\ &= VE_{c,t}^{BGP} + ME_{c,t}^{BGP} + MX_{c,t}^{BGP} + VE_{c,t}^{HSL} + ME_{c,t}^{HSL} + MX_{c,t}^{HSL} \end{aligned} \qquad \text{[D.15]}$$

mit:

$MVP_{c,t}^{BGP}$:	Entwicklungspotenzial aus Mehrverkauf eines Kunden c in der Periode t bei Produkten des Basisgeschäftspotenzials (*BGP*)
$MVP_{c,t}^{HSL}$:	Entwicklungspotenzial aus Mehrverkauf eines Kunden c in der Periode t bei Produkten des Entwicklungspotenzials aus dem Verkauf höherwertiger substitutiver Leistungen (*HSL*)
$VE_{c,t}^{BGP}$:	Verdrängungseffekt mit Bezug zum Basisgeschäftspotenzial (*BGP*) eines Kunden c in der Periode t
$ME_{c,t}^{BGP}$:	Mengeneffekt mit Bezug zum Basisgeschäftspotenzial (*BGP*) eines Kunden c in der Periode t
$MX_{c,t}^{BGP}$:	Mixeffekt mit Bezug zum Basisgeschäftspotenzial (*BGP*) eines Kunden c in der Periode t
$VE_{c,t}^{HSL}$:	Verdrängungseffekt mit Bezug zum Entwicklungspotenzial aus dem Verkauf höherwertiger substitutiver Leistungen (*HSL*) eines Kunden c in der Periode t
$ME_{c,t}^{HSL}$:	Mengeneffekt mit Bezug zum Entwicklungspotenzials aus dem Verkauf höherwertiger substitutiver Leistungen (*HSL*) eines Kunden c in der Periode t
$MX_{c,t}^{HSL}$:	Mixeffekt mit Bezug zum Entwicklungspotenzials aus dem Verkauf höherwertiger substitutiver Leistungen (*HSL*) eines Kunden c in der Periode t

Der Kundenwertbeitrag aus dem Mehrverkauf von Bankleistungen aus dem Basisgeschäftspotenzial lässt sich anhand der folgenden Formel [D.16] in den Messansatz integrieren:

$$MV_{c,t}^{BGP} = \sum_{p_c=1}^{P_c} BP_{c,p_c,t} \cdot \Delta\alpha_{c,p_c,t} \cdot (1-\rho_{c,pq}) + \sum_{p_c=1}^{P_c} \Delta BP_{c,p_c,t} \cdot \alpha_{c,p_c,t} \cdot (1-\rho_{c,pq}) + \sum_{p_c=1}^{P_c} \Delta BP_{c,p_c,t} \cdot \beta_{c,p_c,t}^{\Delta BP_{c,p,t}} \cdot (1-\rho_{c,pq}) + \sum_{p_c=1}^{P_c} \Delta BP_{c,p_c,t} \cdot \Delta\alpha_{c,p_c,t} \cdot (1-\rho_{c,pq}) \qquad [D.16]$$

mit:

$\Delta BP_{c,p_c,t}$: Zusätzliches Bedarfspotenzial eines Kunden c für das Produkt p in der Periode t

$\Delta\alpha_{c,p_c,t}$ Zusätzlicher Loyalitätsgrads eines Kunden c für das Produkt p_c in der Periode t im Vergleich zur aktuellen Situation bezogen auf das Bedarfspotenzial $BP_{c,p_c,t}$;

mit $\Delta\alpha_{c,p_c,t} + \alpha_{c,p_c,t} \leq 1$ und $\alpha_{c,p_c,t} \geq -1 \cdot \Delta\alpha_{c,p_c,t}$

$\beta_{c,p_c,t}^{\Delta BP_{c,p_c,t}}$: Zusätzlicher Loyalitätsgrad des Kunden c für das Produkt p_c in der Periode t bezogen auf das zusätzliche Bedarfspotenzial $\Delta BP_{c,p_c,t}$;

mit $0 \leq \beta_{c,p_c,t}^{\Delta BP_{c,p_c,t}} \leq 1$

und $\alpha_{c,p_c,t} + \Delta\alpha_{c,p_c,t} + \beta_{c,p_c,t}^{\Delta BP_{c,p_c,t}} \leq 1$

Der Mehrverkauf ist auch für Produkte möglich, die Gegenstand des Entwicklungspotenzials aus dem Verkauf höherwertiger substitutiver Leistungen sind. Für dieses Potenzial kann daher in analoger Weise der Kundenwertbeitrag aus dem Mehrverkauf von Leistungen wie folgt in die Messung integriert werden:

$$MV_{c,t}^{HSL} = \sum_{q_c=1}^{Q_c} BP_{c,q_c,t} \cdot \Delta\alpha_{c,q_c,t} \cdot \rho_{c,pq} + \sum_{q_c=1}^{Q_c} \Delta BP_{c,q_c,t} \cdot \alpha_{c,q_c,t} \cdot \rho_{c,pq} + \sum_{q_c=1}^{Q_c} \Delta BP_{c,q_c,t} \cdot \beta_{c,q_c,t}^{\Delta BP_{c,q,t}} \cdot \rho_{c,pq} + \sum_{q_c=1}^{Q_c} \Delta BP_{c,q_c,t} \cdot \Delta\alpha_{c,q_c,t} \cdot \rho_{c,pq} \qquad [D.17]$$

mit:

$\Delta BP_{c,q_c,t}$: Zusätzliches Bedarfspotenzial eines Kunden c für das Produkt q_c in der Periode t

$\Delta\alpha_{c,q_c,t}$ Zusätzlicher Loyalitätsgrad eines Kunden c für das Produkt q_c in der Periode t im Vergleich zur aktuellen Situation bezogen auf das Bedarfspotenzial $BP_{c,q_c,t}$; mit $\Delta\alpha_{c,q_c,t} + \alpha_{c,q_c,t} \leq 1 \quad \alpha_{c,q_c,t} \geq -1 \cdot \Delta\alpha_{c,q_c,t}$

$\beta_{c,q_c,t}^{\Delta BP_{c,q_c,t}}$: Zusätzlicher Loyalitätsgrad des Kunden c für das Produkt q_c in der Periode t bezogen auf das zusätzliche Bedarfspotenzial $\Delta BP_{c,q_c,t}$;

mit $0 \leq \beta_{c,q_c,t}^{\Delta BP_{c,q_c,t}} \leq 1$

und $\alpha_{c,q_c,t} + \Delta\alpha_{c,q_c,t} + \beta_{c,q_c,t}^{\Delta BP_{c,q_c,t}} \leq 1$

Abschließend ist das Preispotenzial in die Messung zu integrieren. Aus dynamischer Sicht kann argumentiert werden, dass sich das Preispotenzial sowohl auf das Basisgeschäftspotenzial als auch auf die anderen Facetten des Entwicklungspotenzials bezieht. Da das Entwicklungspotenzial aus dem Verkauf höherwertiger substitutiver Leistungen jedoch schon eine preispolitische Komponente im Sinne der Abschöpfung der Zahlungsbereitschaft des Kunden beinhaltet, soll nachfolgend lediglich das Basisgeschäftspotenzial und der Mehrverkauf bisher kontrahierter Leistungen betrachtet werden. Hinsichtlich des Preispotenzials ist zu berücksichtigen, dass dies bereits in der Einzelgeschäftskalkulation für eine Bankleistung enthalten ist. Anders als bei gängigen Vorgehensweisen, bei denen sich der Umsatz als Produkt aus Preis und Menge ergibt, erlaubt die Einzelgeschäftskalkulation eine einfache formelmäßige Abbildung nicht. Um die Darstellung nachvollziehbar zu halten, soll daher an dieser Stelle darauf verzichtet werden, die vollständige Kalkulation darzustellen.[1489] Die Messung des Preispotenzials wird unter Berücksichtigung dieses Aspekts wie folgt in den Messansatz integriert (siehe Formel [D.18]):[1490]

1489 Für eine ausführliche konzeptionelle Darstellung der Kalkulationen des Deckungsbeitrags III in VR-Control sei auf BVR (2003b), Kapitel 2.1 verwiesen.

1490 Dabei gilt die Annahme, dass es keine zu berücksichtigende Wechselwirkung zwischen Höhe der Kondition bzw. Gebühr einer Bankleistung und der nachgefragten Menge gibt. Es wird also angenommen, dass die Ausnutzung der maximalen Zahlungsbereitschaft auf Einzelkundenebene bedarfsvolumenneutral erfolgt.

$$
\begin{aligned}
PP_{c,t} &= PP_{c,t}^{BGP} + PP_{c,t}^{MV^{BGP}} \qquad \text{[D.18]}\\
&= \sum_{p_c=1}^{P_c}\Big(\big(BP(\tau_{c,p_c,t}^{max})_{c,p_c,t} - BP(\tau_{c,p_c,t}^{BGP})_{c,p_c,t}\big)\cdot \alpha_{c,p_c,t}\cdot\big(1-\rho_{c,pq}\big)\Big)\\
&+ \sum_{p_c=1}^{P_c}\Big(\big(BP(\tau_{c,p_c,t}^{max})_{c,p_c,t} - BP(\tau_{c,p_c,t}^{BGP})_{c,p_c,t}\big)\cdot \Delta\alpha_{c,p_c,t}\cdot\big(1-\rho_{c,pq}\big)\Big)\\
&+ \sum_{p_c=1}^{P_c}\Big(\big(\Delta BP(\tau_{c,p_c,t}^{max})_{c,p_c,t} - \Delta BP(\tau_{c,p_c,t}^{BGP})_{c,p_c,t}\big)\cdot \alpha_{c,p_c,t}\cdot\big(1-\rho_{c,pq}\big)\Big)\\
&+ \sum_{p_c=1}^{P_c}\Big(\big(\Delta BP(\tau_{c,p_c,t}^{max})_{c,p_c,t} - \Delta BP(\tau_{c,p_c,t}^{BGP})_{c,p_c,t}\big)\cdot \beta_{c,p_c,t}^{\Delta BP_{c,p_c,t}}\cdot\big(1-\rho_{c,pq}\big)\Big)\\
&+ \sum_{p_c=1}^{P_c}\Big(\big(\Delta BP(\tau_{c,p_c,t}^{max})_{c,p_c,t} - \Delta BP(\tau_{c,p_c,t}^{BGP})_{c,p_c,t}\big)\cdot \Delta\alpha_{c,p_c,t}\cdot\big(1-\rho_{c,pq}\big)\Big)
\end{aligned}
$$

mit:

$PP_{c,t}^{BGP}$:	Preispotenzial des Kunden c in der Periode t bezogen auf das Basisgeschäftspotenzial (*BGP*)
$PP_{c,t}^{MV^{BGP}}$:	Preispotenzial des Kunden c in der Periode t bezogen auf den Mehrverkauf der Produkte des Basisgeschäftspotenzial (MV^{BGP})
$BP(\tau_{c,p_c,t}^{max})_{c,p_c,t}$:	Bedarfspotenzial eines Kunden c für das Produkt p_c in der Periode t unter Berücksichtigung der Ausnutzung der leistungsbezogenen maximalen Zahlungsbereitschaft
$BP(\tau_{c,p_c,t}^{BGP})_{c,p_c,t}$:	Bedarfspotenzial eines Kunden c für das Produkt p_c in der Periode t unter Berücksichtigung der Ausnutzung der für das Basisgeschäftspotenzial zugrunde gelegten Zahlungsbereitschaft ($= BP_{c,p_c,t}$)
$\Delta BP(\tau_{c,p_c,t}^{max})_{c,p_c,t}$:	Zusätzliches Bedarfspotenzial eines Kunden c für das Produkt p_c in der Periode t unter Berücksichtigung der Ausnutzung der maximalen Zahlungsbereitschaft
$\Delta BP(\tau_{c,p_c,t}^{BGP})_{c,p_c,t}$:	Zusätzliches Bedarfspotenzial eines Kunden c für das Produkt p_c in der Periode t unter Berücksichtigung der Ausnutzung der für das Basisgeschäftspotenzial zugrunde gelegten Zahlungsbereitschaft ($= \Delta BP_{c,p_c,t}$)
$\tau_{c,p_c,t}^{BGP}$:	Zahlungsbereitschaft in Form von Zins- oder Gebührensätzen des Kunden c für das Produkt p_c in der Periode t, die bei der Bewertung des Basisgeschäftspotenzials zugrunde gelegt wurde

$\tau^{max}_{c,p_c,t}$: Maximale Zahlungsbereitschaft in Form von Zins- oder Gebührensätzen des Kunden c für das Produkt p_c in der Periode t

und $\alpha_{c,p_c,t} + \Delta\alpha_{c,p_c,t} + \beta_{c,p_c,t}^{\Delta BP_{c,p_c,t}} \leq 1$

Als weiterer Kundenwerttreiber ist das Cross-Selling-Potenzial in die Metrik zu integrieren. In dynamischer Sicht kann gefordert werden, dass ein Cross-Selling-Produkt gemäß den Ausführungen in Kapitel D.4.5.3.1.3 in einer Periode t bei erfolgreichem Abschluss in der nachfolgenden Periode $t+1$ nicht mehr dem Cross-Selling-Potenzial zugeordnet werden darf. Durch dieses Vorgehen würde jedoch die Komplexität der Metrik weiter gesteigert werden, ohne dass ein signifikanter Nutzen resultieren würde. Denn für die Vertriebssteuerung sollen zum Planungszeitpunkt, der uno actu der Bewertungszeitpunkt ist, Maßnahmenpläne erarbeitet werden, die das Cross-Selling umfassen. Aktualisierte Maßnahmenpläne bedingen jedoch aktualisierte Entscheidungsgrundlagen, die unter anderem aus ermittelten Kundenwerten bestehen. Stattdessen wird daher vom Zeitpunkt der Messung ausgegangen und festgelegt, ob es sich mit Bezug zur gegenwärtigen Situation um ein Cross-Selling-Produkt handelt. Aus dynamischer Sicht besteht zudem die Fragestellung, ob die Ausweitung des Geschäftsvolumens aus Cross-Selling-Produkten durch im Entwicklungspotenzial zusammengefasste Effekte sinnvoller Weise in der Metrik zu berücksichtigen ist. Welche Vorgehensweise zu präferieren ist, kann nicht eindeutig beantwortet werden.[1491] Da zum Zeitpunkt der Bewertung – anders als beim Basisgeschäftspotenzial – keine aktuellen, proprietären Informationen über das Produktnutzungsverhalten eines Kunden bei der Kreditgenossenschaft vorliegen und somit eine Messung desselben relativ schwieriger ist, soll das Cross-Selling-Potenzial ohne diese Effekte gemessen werden. Das bedeutet auch, dass das Cross-Selling-Potenzial annahmegemäß unter Berücksichtigung der maximalen Zahlungsbereitschaft des Kunden für die relevanten Bankleistungen erfolgt. Das Cross-Selling-Potenzial wird damit wie folgt in die Metrik des Kundenwerts integriert, wobei der in Abbildung 46 vorgenommenen Differenzierung Rechnung getragen wird (siehe Formel [D.19]:

$$CSP_{c,t} = CSP^{L}_{c,t} + CSP^{BW}_{c,t} + CSP^{UE}_{c,t} \qquad [D.19]$$

mit:

$CSP_{c,t}$: Cross-Selling-Potenzial eines Kunden c in der Periode t

---

1491 Vgl. *Lissautzki, M.* (2007), S. 154 ff.

$CSP^{L}_{c,t}$: Loyalitätsbezogenes Cross-Selling-Potenzial eines Kunden c in der Periode t

$CSP^{BW}_{c,t}$: Bedürfnisweckendes Cross-Selling-Potenzial eines Kunden c in der Periode t

$CSP^{UE}_{c,t}$: Unternehmensentwicklungsbezogenes Cross-Selling-Potenzial eines Kunden c in der Periode t

Für die formale Darstellung der einzelnen Summanden des Cross-Selling-Potenzials ist zu berücksichtigen, dass sich diese sowohl auf aktuell angebotene als auch auf aktuell noch nicht angebotene Bankleistungen beziehen. Das loyalitätsbezogene Cross-Selling-Potenzial wird gemäß folgender Formel [D.20] berechnet:

$$CSP^{L}_{c,t} = \sum_{x_c^L=1}^{X_c^L} BP_{c,x_c^L,t} \cdot \alpha_{c,x_c^L,t} + \sum_{y_c^L=1}^{Y_c^L} BP_{c,y_c^L,t} \cdot \alpha_{c,y_c^L,t} \qquad [D.20]$$

mit:

$BP_{c,x_c^L,t}$: Bedarfspotenzial eines Kunden c für das Produkt x_c^L in der Periode t

$BP_{c,y_c^L,t}$ Bedarfspotenzial eines Kunden c für das Produkt y_c^L in der Periode t

$\alpha_{c,x_c^L,t}$: Loyalitätsgrad eines Kunden c für das Produkt x_c^L in der Periode t; mit $0 \leq \alpha_{c,x_c^L,t} \leq 1$

$\alpha_{c,y_c^L,t}$: Loyalitätsgrad eines Kunden c für das Produkt y_c^L in der Periode t; mit $0 \leq \alpha_{c,y_c^L,t} \leq 1$

x_c^L: Index der aktuell vorhandenen Produkte im Sinne des loyalitätsbezogenen Cross-Selling-Potenzials aus Sicht des Kunden

y_c^L: Index der aktuell noch nicht vorhandenen Produkte im Sinne des loyalitätsbezogenen Cross-Selling-Potenzials aus Sicht des Kunden

Das bedürfnisweckende Cross-Selling-Potenzial wird wie folgt berechnet:

$$CSP^{BW}_{c,t} = \sum_{x_c^{BW}=1}^{X_c^{BW}} BP_{c,x_c^{BW},t} \cdot \alpha_{c,x_c^{BW},t} + \sum_{y_c^{BW}=1}^{Y_c^{BW}} BP_{c,y_c^{BW},t} \cdot \alpha_{c,y_c^{BW},t} \qquad [D.21]$$

mit:

$BP_{c,x_c^{BW},t}$: Bedarfspotenzial eines Kunden c für das Produkt x_c^{BW} in der Periode t

$BP_{c,y_c^{BW},t}$: Bedarfspotenzial eines Kunden c für das Produkt y_c^{BW} in der Periode t

$\alpha_{c,x_c^{BW},t}$: Loyalitätsgrad eines Kunden c für das Produkt x_c^{BW} in der Periode t; mit $0 \leq \alpha_{c,x_c^{BW},t} \leq 1$

$\alpha_{c,y_c^{BW},t}$: Loyalitätsgrad eines Kunden c für das Produkt y_c^{BW} in der Periode t; mit $0 \leq \alpha_{c,y_c^{BW},t} \leq 1$

x_c^{BW}: Index der aktuell vorhandenen Produkte im Sinne des bedürfnisweckenden Cross-Selling-Potenzials aus Sicht des Kunden

y_c^{BW}: Index der aktuell noch nicht vorhandenen Produkte im Sinne des bedürfnisweckenden Cross-Selling-Potenzials aus Sicht des Kunden

Das unternehmensentwicklungsbezogene Cross-Selling-Potenzial wird wie folgt berechnet:

$$CSP_{c,t}^{UE} = \sum_{x_c^{UE}=1}^{X_c^{UE}} BP_{c,x_c^{UE},t} \cdot \alpha_{c,x_c^{UE},t} + \sum_{y_c^{UE}=1}^{Y_c^{UE}} BP_{c,y_c^{UE},t} \cdot \alpha_{c,y_c^{UE},t} \qquad [D.22]$$

mit:

$BP_{c,x_c^{UE},t}$: Bedarfspotenzial eines Kunden c für das Produkt x_c^{UE} in der Periode t

$BP_{c,y_c^{UE},t}$ Bedarfspotenzial eines Kunden c für das Produkt y_c^{UE} in der Periode t

$\alpha_{c,x_c^{UE},t}$: Loyalitätsgrad eines Kunden c für das Produkt x_c^{UE} in der Periode t; mit $0 \leq \alpha_{c,x_c^{UE},t} \leq 1$

$\alpha_{c,y_c^{UE},t}$: Loyalitätsgrad eines Kunden c für das Produkt y_c^{UE} in der Periode t; mit $0 \leq \alpha_{c,y_c^{UE},t} \leq 1$

x_c^{UE}: Index der aktuell vorhandenen Produkte im Sinne des bedürfnisweckenden Cross-Selling-Potenzials aus Sicht des Kunden

y_c^{UE}: Index der aktuell noch nicht vorhandenen Produkte im Sinne des bedürfnisweckenden Cross-Selling-Potenzials aus Sicht des Kunden

Die bisher für die Messung des Kundenwerts betrachteten Kundenwerttreiber bezogen sich hinsichtlich des in Kapitel 4.5 hergeleiteten Kundenwertmodells auf ertragsbezogene Komponenten, die außer den Refinanzierungs- bzw. Opportunitätskosten und direkten Provisionsaufwendungen keine Kosten berücksichtigen. Aufgrund der oben getroffenen Festlegung, dass im Rahmen der Messung auf Ergebnisgrößen der Deckungsbeitragskalkulation nach VR-Control rekurriert wird, könnte auch eine andere Ergebnisgröße als Bewertungsmaßstab für das Bedarfspotenzial und das Bedarfsvolumen herangezogen werden. In Abhängigkeit der ausgewählten Ergebnisgröße sind unterschiedliche Volumina an Kosten zu betrachten. Dabei ist zu berücksichtigen, dass die Kalkulation nach VR-Control gegenwärtig nicht einer notwendigen kundenbezogenen Prozesskostenrechnung entspricht.[1492]

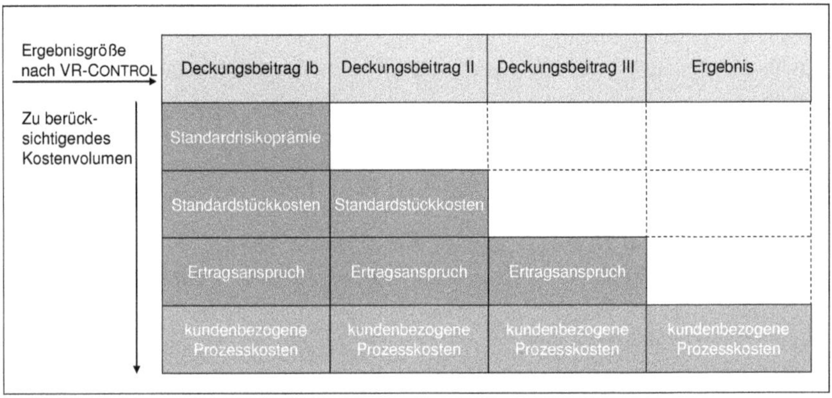

Abbildung 55: Kostenvolumen in Abhängigkeit der verwendeten Ergebnisgröße[1493]

Wie in Kapitel D.4.5.4.1.1 geschildert, sind die Standardrisikoprämie und die Eigenkapitalkosten – als Komponente des Ertragsanspruchs – adressrisikobezogene Größen, die folglich nur bei risikotragenden Aktivgeschäften von Bedeutung sind. Auf die Integration der adressrisikobezogenen Kosten wird in Kapitel D.5.2.2.4 eingegangen werden. Im Ertragsanspruch sind auch Overheadkosten und weitere Kosten enthalten. Diese werden für die Kundenbewertung nicht weiter berücksichtigt, da sie unabhängig von der Entscheidung sind, ob und wie mit einem Kunden eine Geschäftsbeziehung ausgestaltet sein soll. Sie gelten im

[1492] Zur Erläuterung der einzelnen Kostengrößen vgl. die Ausführungen in Kapitel C.4.3.
[1493] Eigene Darstellung.

vorliegenden Kontext als entscheidungsirrelevant. Der Fokus der weiteren Betrachtung liegt nunmehr auf den Standardstückkosten und den kundenbezogenen Prozesskosten.

Gemäß der Einzelgeschäftskalkulation in VR-CONTROL sind die Standardstückkosten fester Bestandteil der bankleistungsbasierten Deckungsbeitragsrechnung, also der Einzelgeschäftskalkulation. Dies gilt auch für die gemäß den Ausführungen zur kundenbezogenen Prozesskostenrechnung ermittelten bankleistungsbezogenen Prozesskosten, die auch als Standardstückkosten betrachtet werden können. Die Integration dieser Kosten in die Messmethode kann auf folgende Arten erfolgen. Erstens kann der Zinskonditions- bzw. Provisionsbeitrag um die Standardstückkosten reduziert werden, um dann die Saldogröße als Bewertungsmaßstab des Bedarfsvolumens zu nutzen. In diesem Fall sind die Formeln [D.12] bis [D.22] zu korrigieren, da der funktionale Zusammenhang zwischen bewerteten Bedarfsvolumen und bewerteten Bedarfspotenzial, der über den Loyalitätsgrad hergestellt wird, nicht mehr gilt. Stattdessen müsste zunächst das nicht bewertete Bedarfsvolumen bestimmt werden und dann mittels des um die Standardstückkosten reduzierten Konditionsbeitrags bzw. Provisionsbeitrags bewertet werden. Zweitens kann alternativ dazu je Kundenwerttreiber, der mit dem Konditions- bzw. Provisionsbeitrag bewertet wurde, die Summe der Standardstückkosten, die den Produkten zugerechnet werden und die dem jeweiligen Kundenwerttreiber zugeordnet sind, in Abzug gebracht werden. Dabei sind die Wechselwirkungen zwischen Basisgeschäftspotenzial und Entwicklungspotenzial zu beachten. Drittens besteht die Möglichkeit die Standardstückkosten en bloc in die Metrik zu integrieren. Den differenziertesten Ausweis stellt die erste Vorgehensweise dar. Konform zu den bisherigen formalen Ausführungen wäre jedoch eine Integration gemäß der zweitgenannten Alternative. Im Weiteren soll davon ausgegangen werden, dass die Bewertung des Bedarfsvolumens auf Basis eines um die bankleistungsbezogenen Standardstückkosten reduzierten Konditions- bzw. Provisionsbeitrags erfolgt. Die formelmäßige Abbildung entspricht somit so weitgehend den bisher gemachten Ausführungen in der Formeln [D.12] bis [D.22], dass eine wiederholte Darstellung nicht vorgenommen werden soll.

Weiterhin zu betrachten sind die kundenbezogenen Prozesskosten. Die kundenbezogenen Prozesskosten stellen per definitionem für die einzelne Bankleistung Gemeinkosten dar, so dass abweichend von den bisherigen Ausführungen die formelmäßige Abbildung nicht an bankleistungsbezogenen Kundenwerttreibern orientiert. Unter der Annahme, dass sich die kundenbezogenen Prozesskosten auf Basis der in Kapitel D.4.5.3.1.4 genannten Prozessstruktur ermitteln lassen, können diese Kosten wie folgt formal beschrieben werden:

$$kPK_{c,t} = kPK_{c,t}^{aP} + kPK_{c,t}^{liP} + kPK_{c,t}^{aiP} \qquad [D.23]$$

mit:

$kPK_{c,t}$: Kundenbezogene Prozesskosten eines Kunden c in der Periode t

$kPK_{c,t}^{aP}$: Kundenbezogene Prozesskosten eines Kunden c in der Periode t für autonome Prozesse

$kPK_{c,t}^{liP}$: Kundenbezogene Prozesskosten eines Kunden c in der Periode t für leistungsbezogen-integrative Prozesse

$kPK_{c,t}^{aiP}$: Kundenbezogene Prozesskosten eines Kunden c in der Periode t für ablaufbezogene-integrative Prozesse

Unmittelbar an das Basiskostenpotenzial knüpft das Kooperationspotenzial an. Da sich das Kooperationspotenzial auf die Integration des externen Faktors, mithin des Kunden ausrichtet, wird die Wirkung auf den Kundenwert über die kundenbezogenen Prozesskosten integriert. Bezogen auf diese Kosten stellt das Kooperationspotenzial eine Korrekturgröße dar, die absolut oder relativ in die Metrik eingefügt werden kann. Für die Berücksichtigung im vorliegenden Messansatz soll eine prozentuale Größe verwendet werden, so dass unter Berücksichtigung der Ausführungen in Kapitel D.4.5.3.1.5 dieser Kundenwerttreiber wie folgt in die Metrik für die Messung des Kundenwerts integriert wird. Die Kompetenzkosten, die sich aus den Kosten für die Kompetenzbeschaffung und den Opportunitätskosten der Kompetenz des Kunden zusammensetzen, sind als Einflussfaktoren ebenfalls zu berücksichtigen. Die mathematische Formulierung erfolgt dabei so, dass das Ergebnis der Berechnung als Korrekturgröße verstanden wird:

$$\begin{aligned} KP_{c,t} = {} & KK_{c,t}^{KKP_{c,t}} + OK_{c,t}^{KKP_{c,t}} \\ & - \delta_{c,t}^{kPK^{aP}} \cdot kPK_{c,t}^{aP} \\ & - \delta_{c,t}^{kPK^{liP}} \cdot kPK_{c,t}^{liP} \\ & - \delta_{c,t}^{kPK^{aiP}} \cdot kPK_{c,t}^{aiP} \end{aligned} \qquad [D.24]$$

mit:

$KP_{c,t}$: Kooperationspotenzial für den Kunden c in der Periode t

$KK_{c,t}^{KKP_{c,t}}$: Ausgaben, die im Zusammenhang mit dem Kooperationspotenzial auf Seiten der Kreditgenossenschaft für den Kunden c in der Periode t anfallen (Kompetenzkosten)

$OK_{c,t}^{KKP_{c,t}}$: Kosten, die im Zusammenhang mit dem Kooperationspotenzial auf Seiten der Kreditgenossenschaft für den Kunden c in der Periode t

anfallen (Opportunitätskosten)

$\delta_{c,t}^{kPK^{aP}}$: Kundeninduzierte Quote des Kosteneinsparungspotenzials für kundenautonome Prozesse für einen Kunden *c* in der Periode *t*; mit $0 \leq \delta_{c,t}^{kPK^{aP}} \leq 1$

$\delta_{c,t}^{kPK^{liP}}$: Kundeninduzierte Quote des Kosteneinsparungspotenzials für leistungsbezogen-integrative Prozesse für einen Kunden *c* in der Periode *t*; mit $0 \leq \delta_{c,t}^{kPK^{liP}} \leq 1$

$\delta_{c,t}^{kPK^{aaP}}$: Kundeninduzierte Quote des Kosteneinsparungspotenzials für ablaufbezogen-integrative Prozesse für einen Kunden *c* in der Periode *t*; mit $0 \leq \delta_{c,t}^{kPK^{aaP}} \leq 1$

Bei den bisher für die Messung betrachteten Kundenwerttreibern handelt es sich dem Wesen nach um monetäre Größen, die zusammengefasst den periodischen monetären Wertbeitrag des anbieterspezifischen Kundenwerts vor der Integration des Risikos ergeben:

$$KW_{c,t}^{mvR} = BGP'_{c,t} + EP'_{c,t} + CSP'_{c,t} - kPK_{c,t} - KP_{c,t} \qquad [D.25]$$

mit:

$KW_{c,t}^{mvR}$ Monetärer Kundenwert vor Integration des Risikos eines Kunden *c* in der Periode *t*

$BGP'_{c,t}$: Unter Berücksichtigung von bankleistungsbezogenen Prozesskosten (Standardstückkosten) bewertetes Basisgeschäftspotenzial eines Kunden *c* in der Periode *t*

$EP'_{c,t}$: Unter Berücksichtigung von bankleistungsbezogenen Prozesskosten (Standardstückkosten) bewertetes Entwicklungspotenzial eines Kunden *c* in der Periode *t*

$CSP'_{c,t}$: Unter Berücksichtigung von bankleistungsbezogenen Prozesskosten (Standardstückkosten) bewertetes Cross-Selling-Potenzial eines Kunden *c* in der Periode *t*

Der in Formel [D.25] berechnete monetäre Kundenwert stellt eine periodische Größe dar. Bevor die periodenübergreifende Messung des monetären Kundenwerts dargestellt wird, werden die Kundenwerttreiber der Bindungsdimension integriert.

5.2.2.2 *Integration der Kundenwerttreiber der Beziehungsdimension*

Zur Beziehungsdimension zählen das Bindungs- und das Akquisitionspotenzial. Letzteres ist nicht relevant für aktuelle Kunden, d. h. Bestandskunden. Für Nichtkunden wird das bisher und nachfolgend für Bestandskunden entwickelte Messverfahren im Weiteren anzupassen sein.[1494]

Hinsichtlich der Integration des Bindungspotenzials finden sich in der Literatur verschiedene Verfahren, die eine metrisch-skalierte Bindungsrate z. B. auf Basis von Diffusionsmodellen, Markov-Ketten oder Kundenloyalitätsmodellen schätzen.[1495] Ergänzend zur klassischen Kapitalwertformel (siehe Formel [D.10]) wird bei kapitalwertbasierten Verfahren eine Kundenbindungsrate eingefügt, um das Bindungspotenzial zu integrieren. Dies entspricht dem von DWYER in die wissenschaftliche Diskussion eingebrachten „customer retention"-Modellen (Kundenloyalitätsmodelle).[1496] Fundamentale Annahme dieser Modelle ist, dass „customers have made long-term commitments to a vendor because switching vendors is costly and assets dedicated to the transaction cannot be redeployed easily […]. If a customer decides to leave a supplier, the account is lost forever."[1497] Derartige Situationen werden als "Lost-for-Good"-Situationen bezeichnet und i. d. R. bei Kundenbeziehungen angewandt, die auf Verträgen basieren (contractual settings). Demgegenüber stehen die „customer migration"-Modelle (Kundenmigrationsmodelle), die bei so genannten „Always-a-Share"-Situationen angewendet werden. Diese Situationen zeichnen sich dadurch aus, dass „customers may rely on several vendors and can adjust their share of business done with each."[1498] Kundenbeziehungen ohne vertragliche Basis werden diesen Situationen zugeordnet. Aufgrund ihrer strikten Dichotomie sollen beide Modelle als theoretisches Grundschema einer Bewertung angesehen werden.[1499]

Nach STRÄTER und FISCHER/ALBERS lassen sich Kunde-Bank-Beziehungen eher durch „Lost-for-Good"-Situationen beschreiben, so dass Bewertungsmodelle in Form der „customer retention"-Modelle angewendet werden können.[1500] Problematisch ist, dass die Situationstypisierung auf Ebene der Bezugsgröße Anbieterunternehmen stattfindet, was implizit der Annahme von Einproduktun-

---

1494 Vgl. dazu die Ausführungen in Kapitel D.5.5.1.

1495 Vgl. *Homburg, C.* (2007), S. 407 so wie die dort angegebene Literatur.

1496 Vgl. *Dwyer, F. R.* (1997), S. 6 ff.

1497 *Dwyer, F. R.* (1997), S. 8. "If a customer is won back after termination, the company treats the customer as a new customer and ignores its history with the customer." *Venkatesan, R./Kumar, V.* (2004), S. 108.

1498 *Dwyer, F. R.* (1997), S. 8.

1499 Vgl. *Sträter, J.* (2005), S. 57.

1500 Vgl. *Sträter, J.* (2005), S. 57; *Fischer, M./Albers, S.* (2007), S. 170 und S. 173. So auch *Böhm, A.* (1999), S. 227 ff.

ternehmen entspricht. Die hier betrachteten Kreditgenossenschaften dagegen bieten einerseits Bankleistungen an, die wie etwa Investitionskredite hochgradig durch Verträge determiniert sind, und andererseits Produkte wie Geschäftskonten, deren erlösrelevante Nutzungsintensität i. d. R. nicht durch Verträge bestimmt ist.[1501] So folgert STRÄTER, dass in einem praxistauglichen Modell die Komplexität deutlich erhöht werden muss.[1502] BEESER argumentiert, dass Geschäftsbeziehungen zu Firmenkunden eher als „Always-a-Share"-Situationen zu interpretieren sind, denn diese nutzen i. d. R. mehr als eine Bankverbindung.[1503] Nach BRUHN ET AL. ist für die Auswahl einer relevanten Verfahrensklasse zugrunde zu legen, welche Marktform und welche Marketingziele verfolgt werden. So sind „customer-retention"-Modelle vor allem in Märkten sinnvoll anzuwenden, bei denen das Relationship Marketing im Mittelpunkt steht.[1504] Und BAUER/HAMMERSCHMIDT argumentieren, dass dieses Vorgehen in den Fällen sinnvoll ist, wo Wechselkosten hoch und das Commitment der Beziehungspartner langfristig ist.[1505] Diese Situationscharakteristika gelten für die Geschäftsbeziehungen zwischen Firmenkunden und Kreditgenossenschaften.[1506] Daher wird nachfolgend von einer „Lost-for-Good"-Situation und damit einem „customer retention"-Modell ausgegangen.

Für die Integration des Bindungspotenzials wird in diesen Verfahrensklassen eine periodische Bindungsrate („retention rate") geschätzt und in die Berechnung des periodenübergreifenden monetären Kundenertragswerts integriert. Die Kundenbindungsrate wird als die Wahrscheinlichkeit des Bestands einer Kundenbeziehung in einer bestimmten Periode verstanden. „Die Kundenbindungswahrscheinlichkeit ist ein grundlegender Faktor [...], denn nur diese kann die Beziehung zwischen Kunde und Unternehmen sowie die darin enthaltenen Chancen und Risiken im erweiterten investitionstheoretischen Ansatz bewerten."[1507] Die „retention rate" bildet die Unsicherheit des Bestands der Geschäfts-

---

1501 Vgl. dazu auch die Ausführungen in Kapitel D.1.2.2. Bei den Geschäftskonten kann es sein, dass die vertragliche Basis noch existiert, jedoch eine andauernde Inaktivität seitens des Kunden festzustellen ist.

1502 Vgl. *Sträter, J.* (2005), S. 57.

1503 Vgl. *Beeser, A. J.* (2003), S. 50. Diese Möglichkeit sehen auch *Fischer, M./Albers, S.* (2007), S. 171.

1504 Vgl. *Bruhn, M. et al.* (2000), S. 168.

1505 Vgl. *Bauer, H.-H./Hammerschmidt, M.* (2005), S. 334.

1506 Vgl. z. B. die Ergebnisse von PAUL/PAUL/STEIN, die zeigen, dass die Geschäftsbeziehungsdauer zwischen Banken und Firmenkunden in 75 Prozent der Fälle mehr als 10 Jahre beträgt. Vgl. *Paul M./Paul, S./Stein, S.* (2003), S. 5.

1507 *Meffert, H./Bruhn, M.* (2009), S. 420.

beziehung ab.[1508] In vielen „customer retention“-Modellen wird eine konstante Bindungsrate unterstellt. Empirische Untersuchungen stützen in vielen Fällen diese Annahme, da die beobachteten Bindungsraten um einen bestimmten Wert schwanken.[1509] Einige Autoren nutzen dagegen eine im Zeitablauf sinkende kumulative Kundenbindungsrate, so dass ferner in der Zukunft liegende Kundenwertbeiträge mit einem geringeren Beitrag in den monetären Kundenwert eingehen als näher in der Zukunft liegende.[1510] DILLER kritisiert diese Vorgehensweisen mit dem Hinweis, dass dadurch Präferenzänderungen ebenso ausgeblendet würden, wie Konkurrenzmaßnahmen.[1511] Dieses Defizit ließe sich kompensieren, in dem eine periodenindividuelle Bindungsrate geschätzt wird, zu deren Einflussgrößen diese Aspekte zählen.[1512] Für Banken argumentieren z. B. BLATTBERG/DEIGHTON könne aufgrund der durch zusätzliche Produktbeziehungen im Zeitablauf steigenden Komplexität auch von einer steigenden Kundenbindungsrate ausgegangen werden.[1513] Mit der periodenindividuellen Schätzung der Bindungsrate unter Berücksichtigung vielfältiger Einflussfaktoren wird jedoch der Prognoseaufwand weiter erhöht. Die Verwendung der Kundenbindungsrate unterstellt nach MÖDRITSCHER implizit, „dass über die Dauer hinweg die Entscheidungen des Management zur Abwendung des Abbruchs beschränkt sind.“[1514]

Ob eine periodenindividuelle Kundenbindungsrate in die Berechnung des monetären Kundenwerts eingefügt wird, ist kreditgenossenschaftlich festzulegen und abhängig vom jeweiligen Institutskontext. Unabhängig davon resultiert unter Berücksichtigung der Kundenbindungsrate die folgende Formel:

$$KW_{c,t}^{mvR} = \left(BGP'_{c,t} + EP'_{c,t} + CSP'_{c,t} - kPK_{c,t} - KP_{c,t}\right) \cdot b_{c,t} \qquad [D.26]$$

mit:

1508 Vgl. *Meffert, H./Bruhn, M.* (2009), S. 418; *Stüker, D.* (2008), S. 211.

1509 Vgl. *Fischer, M./Albers, S.* (2007), S. 174 sowie die dort genannte Literatur.

1510 Vgl. *Bruhn, M. et al.* (2000), S. 172; *Stüker, D.* (2008), S. 211 ff. *Bauer, H.-H./Hammerschmidt, M.* (2005), S. 334 ff.; *Lissautzki, M.* (2007), S. 134. Ein einfaches Modell, um eine mittlere Kundenbindungsrate abzuleiten, nennen ROSSET ET AL. in Form einer Survivalfunktion der Form $S(t)=r^t$. r wird dabei als mittlere Kundenverlustquote definiert, die sich als Quotient von Kundenabwanderungen eines Jahres zum Gesamtkundenbestand am Anfang des Jahres berechnet. Vgl. *Rosset, S. et al.* (2003), S. 323.

1511 Vgl. *Diller, H.* (2001), S. 20 ff.

1512 MEFFERT/BRUHN schlagen dazu die Verwendung eines Scoringmodells vor, da damit alle relevanten Determinanten der Kundenbindung berücksichtigt werden können. Vgl. *Meffert, H./Bruhn, M.* (2009), S. 421.

1513 Vgl. *Blattberg, R. C./Deighton, J.* (1996), S. 138 f.

1514 *Mödritscher, G. J.* (2008), S. 205.

$b_{c,t}$: Kundenbindungsrate für den Kunden c in der Periode t; mit $0 \leq b_{c,t} \leq 1$

Wird Formel [D.26] zugrunde gelegt, um den periodenübergreifenden monetären Kundenwert zu ermitteln, ist darauf hinzuweisen, dass der Einfluss der Kundenbindungsrate auf den Kundenwert erheblich sein kann.[1515] Somit wird zwar durch die Kundenbindungsrate das Abwanderungsrisiko eines Kunden explizit in die Metrik integriert. Die Kundenbindungsrate selbst stellt jedoch einen unsicheren Parameter dar, was es zu berücksichtigen gilt.[1516] Basierend auf Formel [D.26] ergibt sich für die Berechnung des periodenübergreifenden monetären Kundenwerts die folgende Berechnung:

$$KW_c^{mvR} = \sum_{t=1}^{T} \frac{\left(BGP'_{c,t} + EP'_{c,t} + CSP'_{c,t} - kPK_{c,t} - KP_{c,t}\right) \cdot b_{c,t}}{(1+i)^t} \qquad [D.27]$$

mit:

KW_c^{mvR}: Periodenübergreifender monetärer Kundenwert vor Integration des Risikos eines Kunden c

i: Kalkulationszinssatz

Neben der Kundenbindungsrate wird in Formel [D.27] für die mehrperiodige Betrachtung auch der Diskontierungsfaktor i eingefügt. Dieser bildet die Zeitpräferenzrate des Bewertenden ab[1517] Die Auswahl des Diskontierungsfaktors ist Gegenstand des nächsten Kapitels.

5.2.2.3 Bestimmung des Diskontierungsfaktors

Die Konkretisierung der Zeitpräferenzrate in Form des Diskontierungsfaktors, der seit jeher als eine der kritischen Größen investitionstheoretischer Modelle

---

[1515] So zeigen beispielhafte Berechnung von STÜKER, dass die Veränderung der Kundenbindungsrate um 10 Prozentpunkte zu Veränderungen des berechneten Kapitalwerts c. p. von über 200 Prozent bzw. zum Vorzeichenwechsel führen kann. Vgl. *Stüker, D.* (2008), S. 214.

[1516] Vgl. *Stüker, D.* (2008), S. 215.

[1517] Die Zeitpräferenz ist zu berücksichtigen, wenn Ereignisse zu unterschiedlichen Zeitpunkten anfallen und aufgrund dessen für den Entscheider von unterschiedlicher Bedeutung sind. Die Zeitpräferenzrate kann auch als Grenzrate der Substitution von Zukunfts- und Gegenwartskonsum verstanden werden. Vgl. zur Zeitpräferenz *Berens, W./Delfmann, W.* (2002), S. 58. Regelmäßig wird angenommen, dass gegenwärtiger Konsum dem zukünftigen Konsum vorgezogen wird.

gilt,[1518] wird mal mehr mal weniger fundiert vorgenommen. Dies gilt auch für die investitionstheoretische Bewertung von Kunden.[1519] Die theoretisch richtige Konkretisierung und damit Interpretation der Zeitpräferenzrate stellt eine große Herausforderung für den Bewertenden dar.[1520] Mögliche Interpretationen klassifiziert ROLFES wie folgt:

Tabelle 23: Interpretationsansätze des Kalkulationszinsfußes[1521]

Finanzierungsorientiert			**Opportunitätsorientiert**			
				Engpassorientiert		
Eigenkapitalkosten	Fremdkapitalkosten	Mischkosten	Rendite einer alternativen Anlage	Rendite der nächstgünstigsten Investition	Grenzrendite aus Investitions- und Finanzierungsmöglichkeiten	Dualvariable aus einem Totalmodell

Bei den finanzierungsorientierten Interpretationen wird der Kalkulationszins aus den für die Investition bereitstehenden Finanzierungsmitteln abgeleitet.[1522] In der Regel wird dazu ein Mischzins herangezogen, der als gewichtetes arithmetisches Mittel berechnet wird („Weighted Average Cost of Capital"). Während für das Fremdkapital die Zinsen als exogene und objektive Größe bestimmt werden können, ist die Ermittlung der Eigenkapitalkosten diffiziler, da diese nicht vertraglich fixiert sind. Eigenkapitalkosten setzen sich aus Sicht der Finanzierungstheorie aus einem risikolosen Zins und einer investitionsobjektspezifischen Risikoprämie zusammen.[1523] Probleme resultieren bei diesem Vorgehen daraus, die Finanzierungsstruktur für das einzelne Investitionsobjekt festzulegen und die Höhe der Risikoprämie zu bestimmen. Empfohlen und angewandt wird das

---

1518 Vgl. *Grob, H. L.* (1989), S. 187 f. sowie die dort angegebene Literatur.

1519 Zu diesem Ergebnis kommt auch *Homburg, C.* (2007), S. 410. Bei BAUER/HAMMERSCHMIDT gilt z. B.: „the discount rate appropriate for marketing investments". *Bauer, H.-H./Hammerschmidt, M.* (2005), S. 337. REINARTZ/KRAFFT argumentieren bei der Auswahl des Diskontierungsfaktors mit dem Hinweis auf vergleichbare Studien. Vgl. *Reinartz, W. J./Krafft, M.* (2001), S. 1269. VENKATESAN/KUMAR gehen von einem Diskontierungsfaktor von 15 Prozent aus, ohne dies weiter zu begründen. Vgl. *Venkatesan, R./Kumar, V.* (2004), S. 108.

1520 Vgl. *Grob, H. L.* (1989), S. 188.

1521 *Rolfes, B.* (1993), S. 693.

1522 Vgl. hier und nachfolgend *Wille, K.* (2005), S. 73 ff. sowie die dort angegebene Literatur.

1523 Vgl. z. B. *Perridon, L./Steiner, M.* (1999), S. 497 ff.

„Capital Asset Pricing Model" (CAPM), um den Eigenkapitalkostensatz zu bestimmen.[1524] Zentrale Annahme dieses Modells ist, dass sich das Risiko einer Anlage in ein unsystematisches Risiko, das durch Diversifikationsmaßnahmen verringert werden kann, und ein systematisches Risiko, das nicht vermieden werden kann, aufteilen lässt. Für das systematische Risiko erhält der Investor eine marktübliche Risikoprämie, die als Produkt aus Risikomenge und Risikopreis berechnet wird. Der Risikopreis wird als Differenz aus durchschnittlicher Marktrendite und risikolosem Zins ermittelt. Die Risikomenge, die vereinfachend als Gewichtungsfaktor verstanden werden kann, ist ein Ausdruck dafür wie stark die Investitionsrendite um die Marktrendite schwankt und wird als Betafaktor bezeichnet.[1525] Problematisch bei der Anwendung dieses Verfahrens in der Kundenbewertung ist, dass die für den Einsatz des CAPM notwendigen Betafaktoren auf Ebene der Kundenbeziehung schwer bis gar nicht ermittelt werden können.[1526] GLEIẞNER kommt zu dem Schluss, dass dafür „keine sinnvollen risikoadäquaten Diskontierungszinsen aus Kapitalmarktdaten ableitbar sind."[1527] Die logischen Widersprüche bei der Anwendung dieses Modells für die Bewertung von Investitionen nennt WILLE als Grund dafür, dass es „nicht für die partielle Investitionsrechnung und somit auch nicht für die Bestimmung des [.] [monetären Kundenwert; Erg. d. Verf.] geeignet"[1528] ist. Darüber hinaus ist von Zurechnungsproblemen der Finanzierung auszugehen,[1529] so dass die finanzierungsorientierten Diskontierungsfaktoren insgesamt als inadäquat zu beurteilen sind.

Die opportunitätsorientierte Interpretation leitet den Diskontierungsfaktor aus einem Vergleich der Investition mit alternativen und somit verdrängten In-

---

1524 Vgl. *Barth, K./Wille, K.* (2000), S. 39; *Sträter, J.* (2005), S. 111.

1525 Schwankt die Investitionsrendite stärker um die Marktrendite führt dies zum einem Betafaktor von größer 1, so dass die Risikoprämie über dem marktüblichen Abstand zum risikolosen Zins liegt. Im umgekehrten Fall ist der Betafaktor kleiner 1, so dass die Risikoprämie kleiner ist als die Risikoprämie des Marktes.

1526 Mögliche Vorgehen finden sich bei *Schmöller, P.* (2001), S. 160 ff., *Dhar, R./Glazer, R.* (2003), S. 88 f. und *Barth, K./Wille, K.* (2000), S. 39 ff. Zur kritischen Auseinandersetzung vgl. *Stüker, D.* (2008), S. 225 f., *Gleißner, W.* (2005), S. 412 ff. und *Eberling, G.* (2002), S. 189 ff.

1527 *Gleißner, W.* (2005), S. 413. So auch *Tewes, M.* (2003), S. 280.

1528 *Wille, K.* (2005), S. 77. Im Original spricht WILLE vom CE („Customer Equity"). Aufgrund abweichender Begriffsverwendungen wurde dieser Begriff an die verwendete Terminologie angepasst. Vgl. zum Begriff des „Customer Equity" Kapitel D.1.1.

1529 Vgl. *Schmöller, P.* (2001), S. 157. TEWES abstrahiert bei der Kundenwertmessung von Fremdfinanzierungen und legt einen reinen Eigenkapitalkostensatz zugrunde. Vgl. *Tewes, M.* (2003), S. 277 f.

vestitionsprojekten ab.[1530] Die engpassbezogenen Ansätze nutzen dazu ein Totalmodell.[1531] Das führt zu dem Zirkelproblem, dass für das Totalmodell ein Kalkulationszinsfuß feststehen muss, um die Investitionen in eine Reihenfolge zu bringen.[1532] Dies ist jedoch nicht zwingend für die Ansätze notwendig, die nicht auf einen Kapitalengpass aufbauen und bei denen die Rendite einer alternativen Finanzanlage herangezogen wird (Opportunitätsansatz). Eine zum Opportunitätsansatz alternative Argumentation bietet das Gegenseitenkonzept, bei dem mit korrespondierenden und strukturgleichen Gegengeschäften argumentiert wird.[1533] In der Marktzinsmethode, die der Deckungsbeitragsberechnung zugrunde liegt, wird so vorgegangen, wobei zusätzlich von Risikofreiheit und Laufzeitkongruenz ausgegangen wird. Im Sinne der partiellen Investitionsrechnung, der auch die hier vorgenommene Einzelkundenbewertung auf Basis kapitalwertbasierter Verfahren zuzuordnen ist, gilt, dass sich aus der Summe der Erfolge der Einzelentscheidungen der Gesamterfolg konstituiert. Daraus leitet sich ab, dass eine partielle Entscheidung unabhängig von (früheren) Entscheidungen ist. Diesen Gedanken, der auch der Marktzinsmethode unterliegt, kann auf den Kalkulationszinsfuß übertragen werden.[1534] Nach ROLFES stellt die Rendite einer alternativen Finanzanlage den einzig zulässigen Kalkulationszins dar, da diese sicher und im Entscheidungszeitpunkt aktuell ist, darüber hinaus extern gegeben und somit objektiv ist und einen echten Grenzkostensatz darstellt.[1535] Darüber hinaus berücksichtigt dieses Vorgehen die Tatsache, dass Kreditinstitute unabhängig vom Kundengeschäft am Geld- und Kapitalmarkt agieren können. Die Verwendung derartiger laufzeitkongruenter Zinssätze bei der kapitalwertbasierten Kundenbewertung wird z. B. von BÖHM und STRÄTER vorgenommen.[1536] Dann gilt, dass der Diskontierungsfaktor i periodisch differenziert und in der Regel mittels eine Zerobondabzinsungsfaktors dargestellt wird.[1537]

1530 Vgl. *Wille, K.* (2005), S. 77 f. sowie die dort angegebene Literatur.

1531 Vgl. *Grob, H. L.* (1989), S. 191.

1532 Vgl. *Grob, H. L.* (1989), S. 188; *Wille, K.* (2005), S. 77.

1533 Im Falle nicht gespaltener Zinssätze am Geld- und Kapitalmarkt führen beide Argumentationen zu gleichen Ergebnissen. Vgl. hier und nachfolgend *Schierenbeck, H.* (2003), S. 84 ff.

1534 Tatsächlich wird dies bei der barwertigen Deckungsbeitragsrechnung vorgenommen. Dabei werden die künftigen Erfolge mittels laufzeitspezifischer Zinssätze diskontiert.

1535 Vgl. dazu ausführlich *Rolfes, B.* (1993), S. 696.

1536 Vgl. *Böhm, A.* (1999), S. 95 ff.; *Sträter, J.* (2005), S. 16 f.

1537 Vgl. *Kroon, G.* (2009), S. 30 f. Zerobandabzinsungsfaktoren stellen den in Prozent des Rückzahlungsbetrages ausgedrückten Anlagebetrag dar (et vice versa).

Mit Auswahl dieser Größe kann der Zeitpräferenz des Entscheiders Rechnung getragen werden. Jedoch muss auch die Risiko-Präferenz in die Messung integriert werden.

5.2.2.4 Integration der Kundenwerttreiber der Risikodimension

Als relevante Risikodimensionen wurden das Kreditrisiko sowie das Kundenplanungsrisiko definiert. Wie im Vorhergehenden erläutert,[1538] wird das Kreditrisiko in einen erwarteten und einen unerwarteten Verlust unterschieden. In der Deckungsbeitragskalkulation nach VR-CONTROL, die dem Kundenertragswert zugrunde liegt, werden diese Verlustarten entsprechend differenziert berücksichtigt, haben jedoch gemein, dass sie als Kostenkomponente kalkuliert werden.

Für den erwarteten Verlust wird auf Basis des in Abbildung 51 dargestellten Modells eine Standardrisikoprämie berechnet, die im Sinne einer Versicherungsprämie vom Kunden für das erwartete Ausfallrisiko zu entrichten ist. Sie wird für alle Aktivgeschäfte berechnet und wird als produktbezogene Kostenart berücksichtigt. Basis der Kalkulation ist die Verknüpfung der erwarteten Ausfallwahrscheinlichkeit mit der erwarteten Rückflussquote. Die periodische Ausfallwahrscheinlichkeit (marginale Ausfallrate) beschreibt die Wahrscheinlichkeit, dass der Kunde frühestens am ersten und spätestens am letzten Tag einer betrachteten Periode ausfällt. Die periodischen Ausfallwahrscheinlichkeiten werden mittels eines Ratings zum Bewertungszeitpunkt in Verbindung mit Migrationswahrscheinlichkeiten zwischen den Ratingklassen ermittelt.[1539] Letztere werden regelmäßig für einen Zeitraum von einem Jahr ermittelt. Über Migrationsmatrizen werden darauf aufbauend mehrjährige Ausfallwahrscheinlichkeiten berechnet.[1540] Unter Annahme der Gültigkeit der Markov-Eigenschaft der 1-Jahres-Matrix wird diese n-fach potenziert.[1541] Aus der kumulativen Ausfallwahrscheinlichkeit bis zum 2. Jahr, kann durch Subtraktion der Ausfallwahrscheinlichkeit des 1. Jahres die marginale Ausfallrate des 2. Jahres berechnet

---

1538 Vgl. dazu die Ausführungen in Kapitel D.4.5.4.1.1.

1539 Bei der Ermittlung der Ausfallraten auf Basis historischer Daten wird das Ausfallereignis als erstmalige Bildung einer Einzelwertberichtigung im Jahresabschluss der Kreditgenossenschaft definiert.

1540 Vgl. zur Methodik auch *BVR* (2003b), Kapitel 2.1, S. 51 f.

1541 Die Markov-Eigenschaft besagt, dass die Wanderungswahrscheinlichkeit eines Jahres unabhängig von der der jeweiligen Vorjahre ist. Vgl. *Kroon, G.* (2009), S. 40 sowie die dort genannte Literatur. Markov-Ketten werden auch in Kundenmigrationsmodellen verwendet. Vgl. z. B. *Beeser, A. J.* (2003) oder *Homburg,C./Steiner, V. V/Totzek, D.* (2009), S. 70 ff.

werden. Unterjährige Ausfallraten lassen sich mittels linearer Interpolation ermitteln.[1542]

Die kumulierten ausstehenden Forderungen der Kreditgenossenschaft gegenüber dem Firmenkunden werden als Kreditäquivalent bezeichnet.[1543] Bedingt durch die Sicherheitenstellung des Firmenkunden ist jedoch nicht das Kreditäquivalent selbst maßgeblich zur Berechnung der Standardrisikoprämie, sondern es ist um den Wert der Sicherheiten zu korrigieren. Der Quotient aus Sicherheitenwert und Kreditäquivalent ist die Besicherungsquote[1544] oder Rückflussquote[1545]. Wird diese Quote von 1 subtrahiert, resultiert die erwartete Verlustquote. Wird die Verlustquote mit dem Kreditäquivalent mit der zum Zeitpunkt des Ausfalls erwarteten Ausfallwahrscheinlichkeit multipliziert, ergibt sich der Ausfallschaden für die Periode t. Die barwertige Betrachtung über den gesamten relevanten Zeitraum ist die Basis zur Ermittlung der Standardrisikoprämie.

Unerwartete Verluste werden nicht über die vereinnahmten Risikoprämien getragen, sondern über das sog. Risikokapital ausgeglichen. Die Höhe des zu hinterlegenden Eigenkapitals wird mittels des in Abbildung 52 dargestellten formalen Zusammenhangs berechnet. Die auf das zu hinterlegende Eigenkapital zu entrichtende Zielrendite wird institutsindividuell festgelegt[1546] und in der Deckungsbeitragskalkulation als Kostengröße berücksichtigt. Wie die Ausführungen verdeutlichen, kann die konkrete Integration dieser Risikodimension in die Metrik als bankleistungsbezogene Kostengröße erfolgen.[1547] Analog zu den bankleistungsbezogenen Prozesskosten (Standardstückkosten) können die Standardrisikoprämie und die Eigenkapitalkosten über verschiedene Alternativen in die Metrik integriert werden.[1548] Auch für diese Größen soll für den weiteren Verlauf angenommen werden, dass sie Gegenstand des Bewertungsmaßstabs für das Bedarfsvolumen sind.[1549]

1542 Vgl. *BVR* (2003b), Kapitel 2.1, S. 52.

1543 Zu den Kreditäquivalenten zählen die Nominalvolumina der Aktivgeschäfte sowie die dazu gehörigen Zinserträge. Vgl. *BVR* (2003b), Kapitel 2.1. Das Kreditäquivalent ist in der barwertigen Betrachtung der Barwert der noch ausstehenden periodischen Tilgungen und Zinszahlungen in der restlichen vertraglich definierten oder fiktiv angenommenen Laufzeit der aktivischen Bankleistung.

1544 Vgl. *Böhm, A.* (1999), S. 223.

1545 Vgl. *BVR* (2003b), Kapitel 2.1, S. 49 sowie die Abbildung 51.

1546 Vgl. *BVR* (2002b), S. 29. STRÄTER legt seinen Berechnungen z. B. eine Zielrendite von 15 Prozent zugrunde. Vgl. *Sträter, J.* (2005), S. 123.

1547 Sie können auch als kalkulatorische Wagniskosten bezeichnet werden. Vgl. *Baumeister, A.* (2003), S. 223.

1548 Vgl. dazu die Ausführungen in Kapitel D.5.2.2.1.

1549 Es sei darauf hingewiesen, dass sich die formale Darstellung um die risikobezogenen Komponenten reduziert, wenn die Bewertung des Bedarfspotenzials respektive des Be-

Neben der kalkulatorischen Wirkung hat das mit dem Kreditrisiko bezeichnete Ausfallrisiko eine weitere Implikation für die kapitalwertbasierte Kundenwertmessung. Im Falle eines Ausfalls eines Kunden in einer Periode t können die in den Perioden $t+1$ bis T prognostizierten Erfolgsbeiträge des Kunden aufgrund des Ausfallereignisses des Kunden nicht mehr realisiert werden, da damit die Kundenbeziehung wie sie der Kundenbewertung zugrunde lag, in der Periode t beendet wird.[1550] Das bedeutet, dass es neben der Kundenbindung noch eine weitere Einflussgröße gibt, die ursächlich für die Beendigung einer der Bewertung zugrunde gelegten Geschäftsbeziehung sein kann.[1551] Wird begründet angenommen, dass die beiden Ereignisse „Abwanderung" und „Kreditausfall" als Beendigungsgrund nicht gemeinsam auftreten können, so ist die Wahrscheinlichkeit dieses kombinierten Ereignisses gleich Null.[1552] Für die übrigen drei Ereignisse bzw. Ereigniskombinationen lassen sich die Wahrscheinlichkeiten anhand einer Vierfeldertafel ableiten.[1553]

Tabelle 24: Vierfeldertafel mit Wahrscheinlichkeiten der Ereignisse 1 bis 4[1554]

	Ausfall	**Kein Ausfall**	**Σ**
Abwanderung	*Ereignis 1* W(Abw.∩Ausf.) = 0	*Ereignis 2* W(Abw.∩K. Ausf.) = (1-b)	W(Abw.) = (1-b)
Keine Abwanderung	*Ereignis 3* W(K. Abw.∩Ausf.) = ε	*Ereignis 4* W(K. Abw.∩K. Ausf.) = b-ε	W(Keine Abw.) = b
Σ	W(Ausf.) = ε	W(Kein Ausf.) = (1-ε)	**100 %**

darfsvolumens mittels einer entsprechenden Deckungsbeitragsgröße vorgenommen wird.

1550 Vgl. *Böhm, A.* (1999), S. 224.

1551 Es ist zwar zu berücksichtigen, dass mit dieser Situation nicht zwingend die Folge verbunden sein muss, dass alle mit dem Firmenkunden verbundenen Kundenbeziehungen zur Kreditgenossenschaft beendet sind. So ist es durchaus wahrscheinlich, dass Privatkunden noch Kunden der Bank bleiben oder auch das das Unternehmen selbst noch Kunde der Bank bleibt. In diesen Fällen ist jedoch eine so gravierende Veränderung der Geschäftsbeziehungscharakteristika (z. B. in Form geänderter Inhaber/Gesellschafter oder eine Überführung in die Problemkreditbetreuung) zu beobachten, dass eine Neubetrachtung der einzelnen Kundenbeziehungen geboten ist.

1552 Vgl. auch *Böhm, A.* (1999), S. 274. Diese Annahme entspricht der grundsätzlichen Annahme der "Lost-for-Good"-Modelle auf der einen Seite und der Annahme des Kreditrisikos als die Struktur einer Geschäftsbeziehung veränderndes Ereignis auf der anderen Seite.

1553 Vgl. *Böhm, A.* (1999), S. 274 f.; *Bleymüller, J./Gehlert, G./Gülicher, H.* (1998), S. 30.

1554 Eigene Darstellung.

Auf Basis der vorgehenden Ausführungen berechnet sich der periodenübergreifende monetäre Kundenwert unter Berücksichtigung des Kreditrisikos wie folgt:

$$KW_c^{mnKR} = \sum_{t=1}^{T}\left(\frac{\left(BGP''_{c,t}+EP''_{c,t}+CSP''_{c,t}-kPK_{c,t}-KP_{c,t}\right)\cdot\left(b_{c,t}-\varepsilon_{c,t}\right)}{\left(1+i_t\right)^t}\right) \quad [D.28]$$

mit:

KW_c^{mnKR}: Monetärer periodenübergreifender Kundenwert unter Berücksichtigung des Kreditrisikos

$BGP''_{c,t}$: Unter Berücksichtigung von bankleistungsbezogenen Prozesskosten (Standardstückkosten) und Kreditrisikokosten bewertetes Basisgeschäftspotenzial eines Kunden c in der Periode t

$EP''_{c,t}$: Unter Berücksichtigung von bankleistungsbezogenen Prozesskosten (Standardstückkosten) und Kreditrisikokosten bewertetes Entwicklungspotenzial eines Kunden c in der Periode t

$CSP''_{c,t}$: Unter Berücksichtigung von bankleistungsbezogenen Prozesskosten (Standardstückkosten) und Kreditrisikokosten bewertetes Cross-Selling-Potenzial eines Kunden c in der Periode t

$\varepsilon_{c,t}$: Ausfallwahrscheinlichkeit des Kunden c in der Periode t

i_t: Periodisch differenzierter Diskontierungsfaktor

In Abgrenzung zum Ausfallrisiko, das analog wie das Risiko der Kundenabwanderung auf den Kundenwert wirkt, stellt das noch zu integrierende Planungsrisiko auf die Unsicherheit hinsichtlich der Höhe der Kundenwerttreiber ab. Gleichwohl sind diese Risiken abhängig voneinander. Die Frage nach der Höhe der Kundenwerttreiber erfolgt unter der Bedingung, dass der Kunde nicht abwandert oder ausfällt. Im Falle der Abwanderung oder des Ausfalls sind die Kundenwerttreiber der nachfolgenden Perioden annahmegemäß gleich Null.

Wie die Ausführungen in Kapitel D.4.5.4.1.1 deutlich machen, ist die Bewertung einer Kundenbeziehung auf Basis zukunftsgerichteter Größen mit Unwägbarkeiten verbunden. Läge Sicherheit[1555] über die künftige Höhe der relevanten Kundenwerttreiber vor, kann eine deterministische, d. h. einwertige Messung vorgenommen werden. Es ist jedoch begründet davon auszugehen, dass bei der Messung von Kundenwerten die Ausprägungen der Kundenwerttreiber nur

1555 (Entscheidungs-)Situationen bei Sicherheit sind dadurch gekennzeichnet, dass für jede Aktion der Realisationsgrad der verfolgten Zielgrößen eindeutig feststeht. Vgl. zu Entscheidungen bei Sicherheit z. B. *Berens, W./Delfmann, W.* (2002), S. 59 ff.

unter Unsicherheit[1556] geschätzt werden können. In diesen Fällen ist eine stochastische, d. h. mehrwertige Betrachtung unter Berücksichtigung von Eintrittswahrscheinlichkeiten sinnvoll.[1557] Dies wird in der Metrik darüber berücksichtigt, dass für alle prognostizierten Größen Erwartungswerte angesetzt werden.[1558] Um Redundanzen in der Darstellung zu vermeiden, soll eine explizite formale Abbildung nicht vorgenommen werden. Wird der so ermittelte Erwartungswert des monetären Kundenwerts als Entscheidungsgrundlage verwendet, impliziert dies, dass von einem risikoneutralen Entscheider ausgegangen wird, der die Streuung der mehrwertigen Ergebnisse nicht berücksichtigt.[1559] Kreditgenossenschaften gelten jedoch als (leicht) risikoavers.[1560] D. h., aus Sicht der Kreditgenossenschaft dürfte bei gleichem Erwartungswert ein Firmenkunde mit geringem Planungsrisiko einen höheren Kundenwert haben als ein Kunde mit einem hohen Planungsrisiko.[1561]

Die Abbildung dieser Risikopräferenz kann über zwei Vorgehensweisen erfolgen:[1562] Mittels der Sicherheitsäquivalentmethode oder mittels eines Risikozuschlags auf den Diskontierungsfaktor. Bei der Sicherheitsäquivalentmethode werden die Kundenwerttreiber adjustiert, um der mit ihnen verbundenen Unsicherheit Rechnung zu tragen. Das Sicherheitsäquivalent ist der sichere Betrag, welcher dem unsicheren Betrag, d. h. der Wahrscheinlichkeitsverteilung der Ausprägungen der Kundenwerttreiber, aus Sicht eines risikoaversen Entscheiders gleichgesetzt wird.[1563] Mittels der Lage- und Streuungsparameter der Wahrscheinlichkeitsverteilungen lässt sich die Risikostruktur der betrachteten Größe transparent machen. Das Sicherheitsäquivalent wird wie folgt bestimmt:[1564]

1556 Unsicherheit beschreibt die Tatsache, dass die künftige Entwicklung nicht gewiss ist und somit verschiedene Umweltentwicklungen bestehen, deren Eintrittswahrscheinlichkeit unbekannt ist. Können Wahrscheinlichkeiten für die einzelnen Umweltentwicklungen angegeben werden, so dies als Risiko beschrieben. Vgl. *Berens, W./Delfmann, W.* (2002), S. 66 und S. 72.

1557 Vgl. *Stüker, D.* (2008), S. 216.

1558 Vgl. *Schroeder, N.* (2006), S. 57 f. Vgl. für die Prognose der kundenwertrelevanten Einflussfaktoren als Gegenstand des kundenwertorientierten Vertriebssystems die Ausführungen in Kapitel E.1.

1559 Vgl. *Schroeder, N.* (2006), S. 57; *Berens, W./Delfmann, W.* (2002), S. 74.

1560 Vgl. *Buzziol, S.* (2004), S 17.

1561 Vgl. *Stahl, H. K./Matzler, K./Hinterhuber, H. H.* (2006), S. 230; *Srivastava, R. K./Shervani, T. A./Fahey, L.* (1997), S. 51.

1562 Vgl. hier und nachfolgend *Schmöller, P.* (2001), S. 159 sowie die dort angegebene Literatur.

1563 Vgl. *Stüker, D.* (2008), S. 185; *Perridon, L./Steiner, M.* (1999), S. 15.

1564 GLEIßNER unterstellt für das Umsatzrisiko, das dem Basisgeschäftspotenzial gleichzusetzen möglich ist, eine Normalverteilung, während er für das Cross-Selling-Potenzial

$$SÄ[KWT_t] = \mu[KWT_t] - RA_t \qquad [D.29]$$

mit:

$SÄ[KWT_t]$: Sicherheitsäquivalent für einen Kundenwerttreiber in der Periode t

$\mu[KWT_t]$: Erwartungswert der Verteilung der Ausprägungen der Kundenwerttreiber in der Periode t

RA_t: Subjektiver Risikoabschlag in der Periode t

Das Sicherheitsäquivalent ist die durch Risikoeinstellung des Entscheiders geprägt. Um die Sicherheitsäquivalente zu bestimmen, bedarf es eines Entscheidungskriteriums, dass den Anforderungen der Eindeutigkeit, Konsistenz sowie der Umsetzbarkeit und Wirtschaftlichkeit gerecht wird.[1565] STÜKER schlägt dazu das μ-σ-Prinzip für den Kontext der Kundenbewertung vor. Für den Einsatz im Rahmen der vorliegenden Metrik sind zunächst die Ausprägungen der Kundenwerttreiber unter Berücksichtigung der Kundenbindungsrate zu ermitteln, um erst Erwartungswert und Standardabweichung zu ermitteln und dann das Sicherheitsäquivalent. Bei der Nutzung der Sicherheitsäquivalentmethode erfolgt die Diskontierung mittels des risikolosen Zinssatzes.

Bei der Risikozuschlagsmethode wird der Diskontierungsfaktor zur Bezugsgröße der Risikopräferenz. Um einen risikoadjustierten Diskontierungsfaktor abzuleiten, wird im Schrifttum häufig auf das hier bereits ausgeschlossene CAPM verwiesen. Alternativ dazu werden qualitative Risikobewertungsmodelle diskutiert, die die kundenindividuellen Planungsrisikofaktoren berücksichtigen können.[1566] In Abhängigkeit der kundenindividuellen Ausprägungen der Risikofaktoren werden Risikoauf- oder -abschläge auf den Diskontierungsfaktor berechnet. Da als Diskontierungsfaktor hier der risikolose Zins ausgewählt wurde, sind nur Risikoaufschläge zu berücksichtigen.[1567] Der Risikoaufschlag wird häufig mittels eines Scoringmodells ermittelt.[1568] Als Risikoeinflussfaktoren lassen sich z. B. die Entwicklung der Vorsteuerungsgrößen der bankleistungsbezogenen Nachfrage des Firmenkunden, dessen Fähigkeit diese für das eigene Unternehmen umzusetzen sowie das Ausmaß an Sicherheit, mit dem diese Entwick-

---

eine digitale Verteilung annimmt. Vgl. *Gleißner, W.* (2005), S. 429. Neben stetigen Wahrscheinlichkeiten sind auch diskrete Wahrscheinlichkeitsverteilungen hinreichend leistungsfähig, um die Risikostruktur transparent zu machen. Vgl. *Stüker, D.* (2008), S. 76. Vgl. dazu auch das Beispiel in *Grob, H. L.* (1999), S. 428 ff.

1565 Vgl. hier und nachfolgend *Stüker, D.* (2008), S. 77 ff. und 219 ff.

1566 Vgl. z. B. *Tewes, M.* (2003), S. 281 ff.; *Ryals, L.* (2003), S. 169 f.

1567 In der Argumentation von TEWES wird auf die Marktrendite als Ausgangsgröße rekurriert. Vgl. *Tewes, M.* (2003), S. 282.

1568 Vgl. *Fischer, T./von der Decken, T.* (2001), S. 319.

lungen aus Sicht der Bank prognostiziert werden können, nutzen. Die Ermittlung von planungsrisikobezogenen Faktoren nach dieser Vorgehensweise ist schnell und einfach durchführbar und fällt als Quasi-Kuppelprodukt der Kundenwertprognosetätigkeiten der Bank an. Aufgrund der Kritik der methodeninhärenten Subjektivität und der für die Kundenwertmessung noch weitgehend ungeklärten methodisch richtigen Herangehensweise kommen GELBRICH/ MÜLLER jedoch zu dem Ergebnis, dass in diesem Kontext „die Frage nach dem [risikoadjustierten; Erg. d. Verf.] Diskontierungszinssatz großen Spielraum zur Manipulation lässt.“[1569]

HOGAN ET AL. weisen darauf hin, dass bei Nutzung der Kundenbindungsrate das Risiko einer Kundenabwanderung berücksichtigt wird und daher nicht zusätzlich im risikoadjustierten Diskontierungsfaktor erfasst werden darf.[1570] STÜKER führt dies zu der Aussage, dass die Sicherheitsäquivalentmethode im Fall der im Zähler der Metrik berücksichtigten Kundenbindungsrate der Risikozuschlagsmethode überlegen ist, da eine separate Berücksichtigung des Abwanderungsrisikos und des Planungsrisikos erfolgen kann.[1571] Weitere Vorteile der Sicherheitsäquivalentmethode sieht er darin, dass mit diesem Verfahren auf das für die Prognose der Kundenwerte vorhandene Wissen aufgebaut wird und keine zusätzlichen Daten beschafft werden müssen. Er trifft dabei die Annahme, dass die Risikozuschlagsmethode auf Basis des CAPM erfolgt.[1572] Bei einem alternativen Vorgehen, wie es das qualitative Risikobewertungsmodell darstellt, ist ebenfalls eine getrennte Berücksichtigung möglich. Zudem kann mit diesem Vorgehen der Diskontierungsfaktor einfach an die kundenindividuellen Planungsrisiken angepasst werden. In diesen Fällen ist jedoch auf eine strikte Trennung der Risikoarten zu achten. Zudem sollten nach EBERLING nur diejenigen Risiken über den Diskontierungsfaktor abgebildet werden, die einen symmetrischen Einfluss auf die Ausprägungen der Kundenwerttreiber haben.[1573] Da die Risikozuschlagsmethode über den Diskontierungsfaktor zwar subjektiv, jedoch im praktischen Einsatz leichter vermittel- und durchführbar sein dürfte, ist dieses Verfahren bei der Kundenbewertung der Firmenkunden in Kreditgenossenschaften zu empfehlen. Die Verfahrensschwächen sollte sich der Bewerter je-

---

[1569] *Gelbrich, K./Müller, S.* (2006), S. 472.

[1570] Vgl. *Hogan, J. E. et al.* (2002), S. 31.

[1571] Vgl. hier und nachfolgend *Stüker, D.* (2008), S. 223.

[1572] In diesem Fall lassen sich die kundenindividuellen Beta-Faktoren hinsichtlich des berücksichtigten Risikos nicht differenzieren.

[1573] Vgl. *Eberling, G.* (2002), S. 190. Ein asymmetrisches Risiko ist z. B. das Ausfallrisiko, da nur eine Verschlechterung der Situation möglich ist.

doch stets vergegenwärtigen.[1574] Für die Metrik wird vereinfachend kein periodenspezifischer Risikozuschlag unterstellt:

$$KW_c^m = \sum_{t=1}^{T}\left(\frac{\left(BGP''_{c,t}+EP''_{c,t}+CSP''_{c,t}-kPK_{c,t}-KP_{c,t}\right)\cdot\left(b_{c,t}-\varepsilon_{c,t}\right)}{\left(1+i_t+z_c\right)^t}\right) \qquad [D.30]$$

mit:

KW_c^m: Monetärer periodenübergreifender Kundenwert unter Berücksichtigung der Risikodimension

z_c: Risikozuschlag für den Kunden c

Zur Plausibilitätskontrolle lässt sich der ermittelte Risikozuschlag in ein entsprechendes Sicherheitsäquivalent umwandeln.[1575] Es ist zu fordern, dass das mit dem risikolosen Zins diskontierte Sicherheitsäquivalent dem Erwartungswert der betrachteten Größe, die mit dem risikoadjustierten Zinssatz diskontiert wird, betragsmäßig entspricht.

Der Beitrag einer Kundenbeziehung zur Kreditrisikostruktur des Kreditinstituts, der als beziehungsübergreifender Kundenwerttreiber der Risikodimension in Kapitel D.4.5.4.2 beschrieben wurde, wird nicht in die Messung des monetären Kundenwerts integriert. Denn die entscheidungsbezogene Würdigung dieses Kundenwerttreibers kann nur auf Basis der gesamthaften Betrachtung des Portfolios erfolgen. Dieser übergeordnete Kundenwerttreiber, der konträre Entscheidungen zu Ergebnissen der Messung des Kundenwerts bedingen kann, soll im Kontext der Vertriebsplanung berücksichtigt werden.[1576]

5.2.3 Festlegung des Messzeitraums

Werden Kunden mit dem oben hergeleiteten kapitalwertbasierten Modell bewertet, so zwingt dies dazu, den Messzeitraum *T* festzulegen. MÖDRITSCHER fordert zwar, dass die dynamische Berechnung anhand eines abgrenzbaren, plausiblen Handlungszeitraums erfolgt.[1577] Die Auseinandersetzung mit dem für die Messung des monetären Kundenwerts relevanten Zeitraum erfolgt im Schrifttum jedoch nur rudimentär und die Festlegung desselbigen sehr unterschiedlich. Werden die einzelnen Alternativen klassifiziert, kann zwischen steuerungstheoretisch ausgewählten Zeiträumen und Zeiträumen mit Bezug zur Kundenbezie-

---

[1574] Vgl. *Stüker, D.* (2008), S. 226.

[1575] Vgl. dazu ausführlich *Tewes, M.* (2003), S. 282 ff. sowie *Stüker, D.* (2008), S. 226 sowie die jeweils angegebene Literatur.

[1576] Vgl. dazu die Ausführungen in Kapitel E.2.

[1577] *Mödritscher, G. J.* (2007), S. 284.

hungsdauer unterschieden werden. Beiden liegt eine prospektive Betrachtung zugrunde.[1578]

Bei steuerungstheoretischen Zeiträumen wird die erwartete Beziehungsdauer zum Kunden nicht berücksichtigt. Es kann argumentiert werden, dass der Messzeitraum dem Zeitraum des Planungshorizonts entsprechen sollte.[1579] Dies begründet sich damit, dass der Kundenwert als Entscheidungsgrundlage für den Planungshorizont dient und somit dazu zeitlich kongruent auszugestalten ist. Nach BERGER/NASR ist die Auswahl des Messzeitraums branchenabhängig.[1580] Für das Marketing zeigen die Ergebnisse von REINECKE/REIBSTEIN, dass der durchschnittliche Planungshorizont der befragten Unternehmen bei 3,6 Jahren liegt.[1581] In der kreditgenossenschaftlichen Praxis beträgt der Planungshorizont im Vertrieb der Firmenkundenbank häufig nur ein Jahr. Die mittlere Kundenbeziehungsdauer im Firmenkundengeschäft liegt jedoch deutlich darüber.[1582] Wird der Messzeitraum für den Kundenwert so definiert, kann die Kundenbeziehungsdauer zu einem Kunden länger oder kürzer sein als dieser Messzeitraum. Im Falle einer längeren Kundenbeziehungsdauer wird von dem abgeschnittenen, hinter dem Messzeitraum liegenden Zeitraum abstrahiert. Dessen ggf. gewichtiger Einfluss auf den Kundenwert bleibt unberücksichtigt.[1583] Damit würden Maßnahmen fundiert, die erst nach Ende des Planungshorizonts wertvolle Kunden nicht hinreichend würdigen und somit Fehlentscheidungen implizieren. Gleichwohl werden in vielen Anwendungsfällen in der Literatur diese Messzeiträume genutzt.[1584]

Die Festlegung des Messzeitraums mit Bezug zur Kundenbeziehungsdauer erfolgt in der Literatur ebenfalls uneinheitlich. HELM/GÜNTER nutzen „die durchschnittliche Dauer einer Geschäftsbeziehung“[1585]. Andere Autoren gehen

---

[1578] Vgl. dazu auch die Ausführungen in Kapitel D.1.2.4.2.

[1579] Vgl. z. B. *Venkatesan, R./Kumar, V.* (2004), S. 108.

[1580] Vgl. *Berger, P. D./Nasr, N. I.* (1998), S. 21.

[1581] Vgl. *Reinecke, S./Reibstein, D. J.* (2002), S. 20.

[1582] Etwa drei Viertel der befragten Unternehmen in der Studie von PAUL/PAUL/STEIN gaben an, seit mehr als 10 Jahren mit der Hausbank eine Geschäftsbeziehung zu haben. Und fast jedes zweite Unternehmen hat seit mehr als 25 Jahren eine Bankbeziehung zur Hausbank. Vgl. *Paul, M./Paul, S./Stein, S.* (2003), S. 5. Die Studie der KPMG bestätigt dies auch für Unternehmen bis 5 Mio. EUR Jahresumsatz. Danach haben fast zwei Drittel der befragten Unternehmen eine seit mehr als 10 Jahren andauernde Geschäftsbeziehung. Vgl. *KMPG* (2009), S. 10 f.

[1583] Vgl. *Adam, D.* (2000), S. 40.

[1584] Vgl. *Reinartz, W. J./Krafft, M.* (2001), S. 1269 berechnen Kundenwerte für einen Zeitraum von 24 Monaten.

[1585] *Helm, S./Günter, B.* (2006), S. 22. Ähnlich auch *Meyer, M./Shaffu, N.* (2007), S. 57.

von der kundenindividuellen Dauer der Geschäftsbeziehung aus.[1586] Vor dem Hintergrund der Beziehungsorientierung sollte der Kundenwert über die gesamte erwartete Kundenbeziehungsdauer betrachtet werden.[1587] Dabei besteht das Problem, dass mit zunehmendem Messzeitraum, das Prognoserisiko steigt.[1588] Daher wird im Schrifttum für die Bewertung des Kundenstamms eine aus der Unternehmensbewertung zweigeteilte zeitliche Messung vorgeschlagen.[1589] Neben einer detaillierten Planungsphase, die der Dauer der Vertriebs- oder Marketingstrategie entspricht, wird ein Restwert für einen unendlichen Zeitraum (ewige Rente) angenommen.[1590] Während bei Privatkunden nicht von einer unendlichen Lebensdauer als Voraussetzung dieses Vorgehens ausgegangen werden kann, kann für das Business-to-Business-Geschäft die Möglichkeit eines solchen Vorgehens als theoretisch möglich betrachtet werden.[1591] Zentrale Annahme ist, das Unternehmen nicht zwingend aufhören müssen zu existieren.[1592] Die Definition von Firmenkunden als Kundenverbünde aus privaten und juristischen Personen und die Einzelkundenbetrachtung[1593] stehen diesem Vorgehen jedoch konträr gegenüber.[1594] Dennoch scheint dieses Vorgehen ein akzeptabler Kompromiss zu sein, um zeitlich vertikale Abhängigkeiten in der Messung des Kundenwerts zu berücksichtigen. Besondere Aufmerksamkeit ist dann dem Restwert zu widmen.[1595] In der Unternehmensbewertung kann der Restwert ein Veräußerungswert, ein Liquidationswert oder eine unendliche Rente mit oder ohne Ver-

---

1586 Vgl. z. B. *Schröder, N.* (2006), S. 48; *Hempelmann, B./Lürwer, M.* (2003), S. 357; *Pelzl, U.* (2006), S. 11; *Barth, W.* (2009), o. S.

1587 Vgl. *Bruhn, M. et al.* (2000), S. 169 sowie die dort angegebene Literatur.

1588 Vgl. *Mulhern, F. J.* (1999), S. 29; *Rehbach, S.* (2003), S. 41. Das Prognoserisiko sinkt jedoch, je höher der Diskontierungsfaktor ist, da dadurch der Einfluss der später anfallenden Erfolgsbeiträge sinkt.

1589 Vgl. *Barth, K./Wille, K.* (2000), S. 36; *Reiners, J. O.* (2004), S. 43 f.; *Wiesel, T./Skiera, B.* (2007), S. 712.

1590 So z. B. *Homburg, C./Steiner, V. V./Totzek, D.* (2009), S. 79 ff., die einen für einen Zeitraum von 4 Jahren für jedes Jahr detailliert planen und für den danach liegenden, unendlichen Zeitraum eine ewige Rente unterstellen. Vgl. auch *Schmöller, P.* (2001), S. 157 ff.

1591 Vgl. *Schmöller, P.* (2001), S. 158.

1592 Vgl. *Reiners, J. O.* (2004), S. 43.

1593 Im Gegensatz Einzelkundenbetrachtung können bei der Betrachtung des gesamten Kundenstamms kompensatorische Effekte in der Art angenommen werden, dass im Prognosezeitraum beendete Kundenbeziehungen durch neue Kundenbeziehungen kompensiert werden.

1594 Vgl. dazu die Ausführungen in Kapitel C.2.2.4 sowie Kapitel D.1.2.4.3.

1595 Vgl. *Barth, K./Wille, K.* (2000), S. 36.

änderung sein.[1596] Da der Veräußerungs- und der Liquidationswert als Optionen für die Kundenbewertung irrelevant sind, soll von einer ewigen Rente ausgegangen werden. In wieweit diese Veränderungen unterworfen ist, ist kundenindividuell zu entscheiden. So ist z. B. bei jungen Unternehmen eher mit einem Wachstum der ewigen Rente als bei etablierten Unternehmen zu rechnen. Weiterhin ist festzulegen, wie mit der Kundenbindungsrate und der Ausfallwahrscheinlichkeit bei diesem Vorgehen umgegangen werden soll (siehe Formel [D.30]). Zunächst ist festzustellen, dass die Verwendung einer ewigen Rente von diesen Größen konzeptionell abstrahiert, wenn der Zähler der letzten Periode für die Berechnung der ewigen Rente herangezogen wird.[1597] Wird bei der Herleitung der Kundenbindungsrate unterstellt, dass sich diese modellendogen als exponentielles Produkt über die Zeit ergibt,[1598] so ist zu beachten, dass die Kundenbindungsrate mit zunehmender Dauer in Richtung des Wertes Null konvergiert. Für die Ausfallwahrscheinlichkeit hingegen liegt eine Abhängigkeit von den zugrundeliegenden Migrationsmatrizen vor, so dass die Entwicklung mit zunehmender Dauer unbestimmt ist. Es erscheint vor diesem Hintergrund plausibel, diese beiden Größen modellexogen vorzugeben. Problematisch ist, einen konkreten Wert festzulegen, der das Ergebnis der Berechnung nicht (zu stark) verzerrt.

Vor dem Hintergrund der genannten Argumente für und wider die einzelnen Vorgehensweisen muss konstatiert werden, dass eine allgemein- und endgültige Empfehlung nicht vorgenommen werden kann. „Typically the length-of-time-period decision should be made based on the time-related aspects of customer lifetime, interpurchase times, and an organization's planning cycle."[1599] Für die Vertriebssteuerung ist es jedoch sinnvoll, bei der Kundenwertermittlung von einem zweigeteilten Messzeitraum auszugehen, der eine detaillierte Detailplanung und eine ewige Rente aufweist. Hinsichtlich der Verzerrungsproblematik ist mittels Sensitivitätsanalysen zu ermitteln, wie stark der Einfluss variierender Größen der Kundenbindungsrate und der Ausfallwahrscheinlichkeit auf die Entscheidungen der Vertriebssteuerung sind, so dass mindestens Transparenz über deren Einflüsse besteht, wenn sie schon nicht ausgeschlossen werden können.

5.3 Messung des nicht-monetären Kundenteilwerts

Im vorhergehenden Kapitel D.5.2 wurde ein Messinstrument entwickelt, in dem die ihrer Natur nach monetären Kundenwerttreiber berücksichtigt werden. Wie

---

1596 Vgl. *Stüker, D.* (2008), S. 84.

1597 Vgl. *Reiners, J. O.* (2004), S. 43.

1598 Mathematisch formuliert bedeutet dies $r_1 = 0{,}8$ und $r_2 = r_1 \cdot r_1 = 0{,}8 \cdot 0{,}8 = 0{,}64$.

1599 *Mulhern, F. J.* (1999), S. 29.

bereits die Ausführungen zu den ihrem Wesen nach nicht-monetären Kundenwerttreiber – also dem Informations-, dem Referenz- und dem Member Value-Potenzial – verdeutlichen, ist eine monetäre Messung dieser Größen nur schwierig bis gar nicht möglich. Es entspräche zwar teilweise den formulierten Anforderungen an die Kundenwertmodellierung,[1600] widerspräche aber dem überzuordnenden Ziel der Arbeit, wenn nun Ansätze beschrieben würden, die eine monetäre Messung dieser Kundenwerttreiber unterstützen. Denn der Forschungsstand im Allgemeinen und im Bankenkontext im Speziellen ist noch zu gering, als dass hier auf hinreichend ausgearbeitete und praktikable Verfahren rekurriert werden könnte. Für diesen als Bewertungsdefekt zu charakterisierenden Strukturdefekt des Kundenwertmodells schlägt ADAM vor, eine sachgerechte Transformation des defekten Ausgangsproblems in nicht defekte Teilprobleme zu überführen.[1601] Aus diesem Grund wird eine monetäre Messung dieser Kundenwerttreiber nicht vorgenommen. Als geeignete Alternative im Kontext der Kundenbewertung wurden bereits Punktbewertungsverfahren genannt und in Wissenschaft und Praxis verwendet. Dieses Instrument ist die Grundlage der folgenden Ausführungen zur Messung des nicht-monetären Kundenwerts.

5.3.1 Grundlegendes Punktbewertungsverfahren

Punktbewertungsverfahren, auch als Scoringmodelle bezeichnet, werden in der Kundenbewertung aufgrund ihrer breiten Anwendbarkeit auch bei differierenden Messskalen und Heterogenität der Inputgrößen, d. h. der Kundenwerttreiber oft eingesetzt. FOEHN hat ein Scoringmodell entwickelt, mit dem der Wert eines Kunden im Private Banking gemessen werden kann.[1602] RADDAO hat ein Scoringmodell entworfen, um Firmenkunden zu beurteilen.[1603] Scoringmodelle dienen dazu, „die Wertschätzung einer Alternative im Hinblick auf alle relevanten Bewertungskriterien durch inhaltlich nicht interpretierbare und damit gegenseitig aufrechenbare Punktwerte zum Ausdruck zu bringen."[1604] Das grundlegende eindimensionale Scoringmodell lässt sich formal wie folgt darstellen:[1605]

1600 So wurde in Kapitel D.4.2.3 die Anforderung benannt, dass die Kundenwertmodellierung eine grenznutzenorientierte Einzelbewertung unterstützt. Dieser Anforderung wird die nicht-monetäre Messung einzelner Kundenwerttreiber nicht gerecht. Dieselbe Aussage gilt für die Anforderungen der zutreffenden Ergebnisinformation und der integrierten Ergebnisrechnung.

1601 Vgl. *Adam, D.* (1996), S. 15.

1602 Vgl. *Foehn, P.* (2006), S. 115 ff.

1603 Vgl. *Raddao, N.* (2009), S. 29 ff.

1604 *Diller, H.* (1998), S. 253.

1605 Vgl. *Adam, D.* (1996), S. 413.

$$GS_c = \sum_{m=1}^{M} S_{cm} \cdot GF_m \qquad [D.31]$$

mit:

GS_c: Gesamtscore des Kunden c

S_{cm}: Score des Kunden c bei der Eigenschaft m

GF_m: Gewichtungsfaktor der Eigenschaft m

m: Index der Eigenschaften bzw. Bewertungskriterien im Punktbewertungsverfahren

Wie Formel [D.31] verdeutlicht, sind Scoringmodelle grundsätzlich einfach strukturierte, jedoch differenzierbare Bewertungsverfahren.[1606] Beim Aufbau eines Punktbewertungsverfahrens werden zunächst die relevanten Bewertungskriterien ausgewählt und die Messskala festgelegt. Anschließend werden den Bewertungskriterien Bedeutungsgewichte zugewiesen, die der Bedeutung des jeweiligen Kriteriums im Bewertungszusammenhang entsprechen. In einem weiteren Schritt wird das Verfahren für die Wertsynthese der Produkte aus Messwert und Gewicht je Bewertungskriterium festgelegt. Auf Basis dieser Struktur werden dann für die Bewertungskriterien die Messwerte für die Bewertungsobjekte erfasst und Gesamtpunktwerte je Bewertungsobjekt ermittelt.

Kritisiert wird an den Scoringmodellen, dass sie zahlreichen subjektiven Einflüssen unterliegen und eine für die Abbildung der Realität inadäquate lineare Modellstruktur aufweisen.[1607] Diese Schwäche kann partiell kompensiert werden, wenn bei Aufbau und Anwendung eines Scoringmodells darauf geachtet wird, sich am Rationalitätsbegriff nach WEBER zu orientieren. Dieser definiert „Rationalität als herrschende Meinung von Fachleuten hinsichtlich einer bestimmten Zweck-Mittel-Situation“[1608]. Damit verbundene Bedingungen sind, dass die Fachleute als solche anerkannt werden (das dürfte z. B. bei erfahrenen Firmenkundenberatern der Fall sein) und dass es die Möglichkeit und Fähigkeit eines offenen Diskurses bei unterschiedlichen Meinungen gibt, der ein konsen-

---

[1606] Vgl. *Helm, S./Günter, B.* (2006), S. 17; *Adam, D.* (1996), S. 412 f.

[1607] Insbesondere die Auswahl der Bewertungskriterien, deren Gewichtung sowie die konkrete Bewertung eines Merkmalsträgers unterliegen häufig der Subjektivität. Vgl. *Gelbrich, K./Müller, S.* (2006), S. 479 f.; *Stüker, D.* (2008), S. 61 f.; *Barth, W./Wille, K.* (2000), S. 25 und S. 29. Darüber hinaus kritisiert ADAM, dass statt der Realität die empfundene Differenz zwischen einer subjektiv wahrgenommenen Realität und einer subjektiven, nicht explizierten Idealvorstellung des Bewertenden gemessen wird. Vgl. *Adam, D.* (1996), S. 83.

[1608] *Weber, J.* (2004), S. 51.

suales Ergebnis liefern kann.[1609] Die enge Orientierung an diesem Rationalitätsbegriff würde nur zu einem wirtschaftlichen Messmodell führen. Gemeint ist mit Orientierung, dass der Definitionskern, der sich darin zeigt, dass die herrschende und konsensuale Meinung von einer Gruppe von Fachleuten als Fixpunkt sinnvollen Verhaltens verstanden werden kann, einer stark subjektiven Modellbildung entgegen wirken kann.

Weiterhin wird eine Genauigkeit der metrischen Ergebnisse, die der Operationalisierung nicht entspricht, kritisiert.[1610] Es resultiert keine absolute Vorziehenswürdigkeit, sondern nur eine relative in Form eines Ranking.[1611] Dies unterscheidet die Kundenbewertung mit Punktbewertungsverfahren von der Bewertung mit monetären Verfahren. Wird im vorliegenden Anwendungskontext ein Scoringmodell verwendet, impliziert dies, dass von den Zielen der Gesamtbanksteuerung hinsichtlich ihrer Dimensionalität abgewichen wird und die formalanalytische Verknüpfung von Gesamtbank- und Vertriebszielen nicht erreicht werden kann.[1612] Gleichwohl ist zu konstatieren, dass sich Punktbewertungsverfahren in ihrer Ausbaubarkeit inhaltlich zu den Gesamtbankzielen konsistent ausgestalten lassen. Aus diesem Grund sind sie grundsätzlich auch für die Messung von Kunden(teil)werten geeignet.

Weiterhin werden ein etwaiger Skalenwechsel nach unten, der zu einem Informationsverlust führt, sowie die additive Verknüpfung, die aufgrund ihrer methodeinhärenten Kompensationsmöglichkeit von Merkmalsausprägungen zu unsinnigen Ergebnissen führen kann, kritisiert.[1613] Die Kompensationsgefahr kann reduziert werden, wenn z. B. Mindestwerte für zentrale Kriterien definiert werden. Werden diese nicht erreicht, gilt ein Kunde als unattraktiv bzw. das Bewertungskriterium wird negativ beurteilt.[1614] Dies gilt z. B. dann, wenn der Kunde negativ über die Kreditgenossenschaft spricht, woraus ein negatives Referenzpotenzial resultiert.

1609 Vgl. *Weber, J.* (2004), S. 51. WEBER weist in seiner weiteren Argumentation darauf hin, dass Rationalität verstanden als herrschende Meinung in Wissenschaft und Praxis weit verbreitet ist. Dennoch sind auch hier Irrtümer vorbehalten, und zwar immer dann, wenn Paradigmenwechsel aus individuellen Leistungen resultieren. Schwierig ist jedoch, dass diese nicht ex ante erkannt werden können.

1610 Vgl. *Gelbrich, K./Müller, S.* (2006), S. 480.

1611 Vgl. *Stüker, D.* (2008), S. 61; *Stanger, M.* (2007), S. 28; *Adam, D.* (1996), S. 82.

1612 So ist z. B. das Ziel der Vermögenswertsteigerung für eine Kreditgenossenschaft, das anhand eines in Euro gemessenen Zielwerts definiert ist, nicht mit einem Kundenscore zu verknüpfen, der per definitionem dimensionslos ist.

1613 Vgl. *Gleißner, W.* (2005), S. 410.

1614 Vgl. *Tewes, M.* (2003), S. 147.

Diesen Nachteilen stehen einige gewichtige Vorteile gegenüber, die dazu führen, dass Kundenscoringmodelle in der Praxis eine weite Verbreitung haben.[1615] Dazu zählen ihre Einfachheit sowie die frei bestimmbare Anzahl von Bewertungskriterien und Dimensionen.[1616] Weiterhin sind Scoringmodelle aufgrund ihrer leichten Erfassbarkeit aussagekräftige Diskussionsgrundlagen, die zu „einer kritischen Auseinandersetzung mit unternehmerischen Erfolgsfaktoren“[1617] führen. Im Gegensatz zu der impliziten Berücksichtigung dieser Faktoren[1618] in der Kundenbewertung haben Punktbewertungsverfahren den Vorteil, dass die Faktoren expliziert[1619] und deren Einbeziehung intersubjektiv nachvollziehbar wird.[1620]

5.3.2 Punktwertbasiertes Kundenwertmessmodell für das Firmenkundengeschäft der Kreditgenossenschaften

Der erste Schritt beim Aufbau eines punktwertbasierten Messmodells ist die Auswahl der Bewertungskriterien. Diese resultieren aus dem hergeleiteten Kundenwertmodell und sind das Informations-, das Referenz- sowie das Member Value-Potenzial. Diese können als drei Dimensionen eines mehrdimensionalen Punktbewertungsverfahrens genutzt werden, denen jeweils ein Bedeutungsgewicht zugeordnet werden kann. Wird keine dimensionsspezifische Gewichtung vorgenommen, so gelten die Dimensionen als gleichgewichtig aus Sicht des Kreditinstituts. Für die genannten drei Dimensionen sind anschließend die konkreten Bewertungskriterien festzulegen, die unabhängig voneinander sein müssen. Weitere Anforderungen sind, dass die Bewertungskriterien mindestens ordinal skaliert sind und deren Messung intersubjektiv nachprüfbar ist.[1621] Für die einzelnen Bewertungskriterien können die Bedeutungsgewichte festgelegt wer-

---

1615 In einer älteren, branchenübergreifenden Studie von HARNISCHFEGER gaben etwa 36 Prozent der Befragten an, ein Kundenscoringmodell zu nutzen; und etwa 64 Prozent planten ein derartiges Modell einzusetzen. Vgl. *Harnischfeger, U.* (1996), S. 20.

1616 Vgl. *Helm, S./Günter, B.* (2006), S. 17; *Gelbrich, K./Müller, S.* (2006), S. 479.

1617 *Barth, W./Wille, K.* (2000), S. 29.

1618 So formuliert KRAFFT: „Qualitative Aspekte wie bspw. das Referenzpotenzial [...] werden bewusst vernachlässigt [...]. Diese nicht unmittelbar monetären Erwägungen fließen nach Kenntnis des Verfassers bei konkreten Fragestellungen aber ohnehin in die Entscheidung von Managern und Kundenbetreuern ein und wirken somit als Korrektiv der [...] ökonomischen Entscheidungshilfen.“ *Krafft, M.* (2007), S. 201.

1619 Postuliert wird dieses Vorgehen auch von *Schmöller, P.* (2001), S. 176 f., *Hoberg, P.* (2007), S. 1248.

1620 Vgl. *Rieker, S.* (1995), S. 69 sowie die dort angegebene Literatur.

1621 Vgl. *Adam, D.* (1996), S. 416.

den.[1622] Basierend auf den Ausführungen zum Informationspotenzial in Kapitel D.4.5.3.1.6 lässt sich ein Scoringmodell anhand der in Abbildung 56 gezeigte Struktur aufbauen.

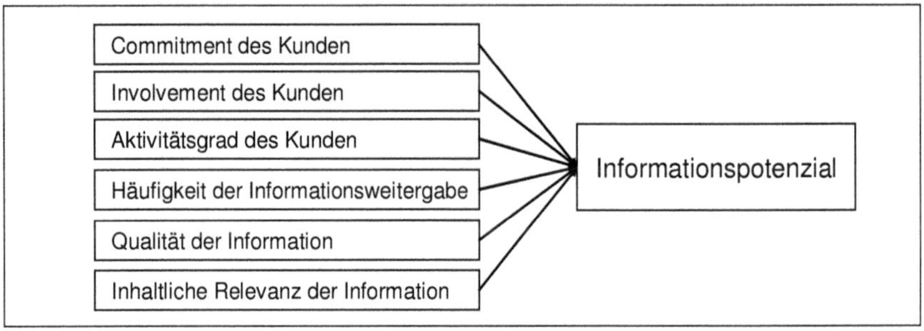

Abbildung 56: Beispielhafte Struktur eines Scoringmodells für das Informationspotenzial[1623]

Analog ist für das Referenzpotenzial vorzugehen, für das sich basierend auf den Ausführungen in Kapitel D.4.5.3.2.1 ein Scoringmodell ableiten lässt, das der in Abbildung 57 dargestellten Struktur entspricht.[1624] Für das Referenzpotenzial ist zu berücksichtigen, dass die Richtung der Referenz, also die Frage ob es sich um eine positive oder negative Äußerung bezüglich der Kreditgenossenschaft handelt ggf. als KO-Kriterium oder als negativwertige Einflussgröße zu berücksichtigen ist. Damit kann beachtet werden, dass ein Kunde, der sich negativ über die Kreditgenossenschaft oder deren Leistungsangebot äußert, gerade keinen Beitrag zur Zielerreichung leistet, sondern sich konträr dazu verhält.

1622 Ein auf Expertenwissen basierendes Verfahren, das konsistente Ergebnisse für die Bedeutungsgewichte liefert beschreibt *Adam, D.* (1996), S. 417 f. Alternativ lassen sich auch multivariate Analyseverfahren wie z. B. das Conjoint Measurement oder die Delphi-Methode nutzen. Letztere wird von SCHMÖLLER in diesem Anwendungskontext empfohlen, „da es sich bei der Bewertung von Kunden zur besseren Objektivierung immer um eine Gruppenentscheidung handeln sollte." *Schmöller, P.* (2001), S. 175.

1623 Eigene Darstellung.

1624 In der Literatur finden sich Modelle, mit denen das Referenzpotenzial auch monetär ermittelt werden kann. Vgl. für eine vergleichende Übersicht und Kritik *Helm, S.* (2003), S. 124 ff. Diese Modelle gehen jedoch von einem im Vergleich zur vorliegenden Arbeit abweichenden Begriff des Referenzpotenzials aus.

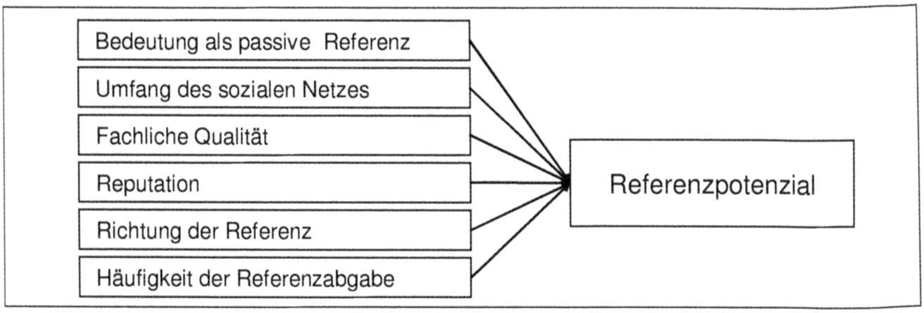

Abbildung 57: Beispielhafte Struktur eines Scoringmodells für das Referenzpotenzial[1625]

Für das Member Value-Potenzial ließe sich vor dem Hintergrund der Ausführungen in KapitelD.4.5.3.2.2 ein beispielhaftes Scoringmodell nutzen, das in Abbildung 58 gezeigt wird.

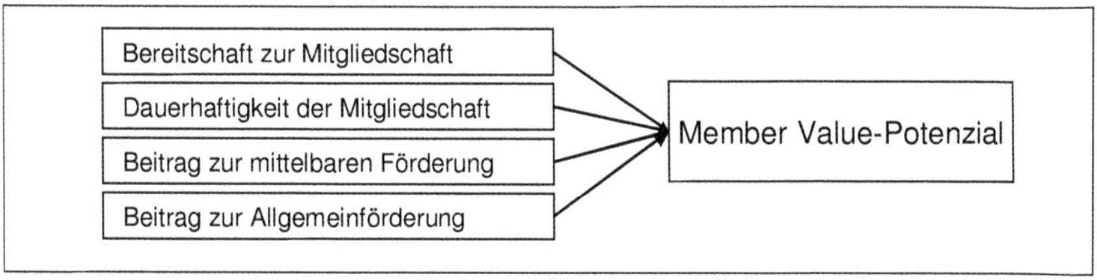

Abbildung 58: Beispielhafte Struktur eines Scoringmodells für das Member Value-Potenzial[1626]

Wichtig ist, gleiche Punktskalen für die Bewertung zu verwenden. Sofern sich die Anzahl der Einflussfaktoren unterscheidet, sind die sich ergebenden Punktwerte zu relativieren.[1627] Liegen für alle drei Kundenwerttreiber Punktwerte vor, ist eine Wertsynthese zu einem Gesamtpunktwert vorzunehmen. Neben der in Formel [D.31] dargestellten additiven gibt es auch die multiplikative sowie die multilineare Aggregation.[1628] Das additive Aggregationsmodell ist das in praxi

[1625] Eigene Darstellung.

[1626] Eigene Darstellung.

[1627] Punktwerte werden relativiert, indem der erreichte Punktwert durch die Anzahl der Bewertungskriterien dividiert wird. Vgl. *Schmöller, P.* (2001), S. 175.

[1628] Bei Vorliegen einer messbaren Einzelpräferenzfunktion stellt die additive Aggregation einen Spezialfall der multiplikativen Aggregation und diese wiederum einen Spezialfall der multilinearen Aggregation dar. Alle Syntheseverfahren basieren auf dem additiven

am weitesten verbreitete Syntheseverfahren, das als ausreichend gewertet wird, da es meistens gelingt, die Artenpräferenz durch die intuitive additive Verknüpfung hinreichend genau abzubilden.[1629] RIEKER empfiehlt eine multiplikative Verknüpfung, wenn es essenzielle Bewertungskriterien gibt. Wird die Null als Wert der Messskala berücksichtigt, resultiert über diese Kombination das „Knock-Out-Kriterium".[1630] Dies wäre z. B. Beispiel mit Blick auf das Referenzpotenzial bei einer bekanntermaßen negativen Kommunikation des Kunden sinnvoll. In wie weit dies notwendig ist, ist institutsindividuell zu entscheiden. Für den einfachen Fall, lässt sich der Gesamtpunktwert für den nicht-monetären Kundenwert wie folgt ermitteln:

$$\begin{aligned} KW_c^{nm} &= KTW_c^{IP} \cdot GF_{IP} + KTW_c^{RP} \cdot GF_{RP} + KTW_c^{MVP} \cdot GF_{MVP} \\ &= \left(\sum_{h=1}^{h} GF_h^{IP} \cdot S_{c,h}^{IP} \right) \cdot GF_{IP} \\ &+ \left(\sum_{j=1}^{J} GF_j^{RP} \cdot S_{c,j}^{RP} \right) \cdot GF_{RP} \\ &+ \left(\sum_{l=1}^{L} GF_l^{MVP} \cdot S_{c,l}^{MVP} \right) \cdot GF_{MVP} \end{aligned} \qquad [D.32]$$

mit:

$$GF_{IP} + GF_{RP} + GF_{MVP} = 1$$

KW_c^{nm}: Nicht-monetärer Kundenteilwert des Kunden *c*

KTW_c^{IP}: Kundenteilscore des Kunden *c* für das Informationspotenzials

GF_{IP}: Gewichtungsfaktor des Kundenteilscores für das Informationspotenzial

KTW_c^{RP}: Kundenteilscore des Kunden *c* für das Referenzpotenzial

GF_{RP}: Gewichtungsfaktor des Kundenteilscores für das Referenzpotenzial

KTW_c^{MVP}: Kundenteilscore des Kunden *c* für das Member Value-Potenzial

GF_{MVP}: Gewichtungsfaktor des Kundenteilscores für das Member Value-Potenzial

GF_h^{IP}: Gewichtungsfaktor der Eigenschaft *h* in der Dimension Informationspotenzial

---

Grundansatz der Präferenzaggregation. Vgl. dazu ausführlich *Siebert, J.* (2009), S. 38 ff.

1629 Vgl. *Siebert, J.* (2009), S. 30 und S. 38 sowie die dort angegebene Literatur.

1630 Vgl. *Rieker, S.* (1995), S. 69.

$S_{c,h}^{IP}$: Score des Kunden *c* bei der Eigenschaft *h* in der Dimension Informationspotenzial

GF_{j}^{RP}: Gewichtungsfaktor der Eigenschaft *j* in der Dimension Referenzpotenzial

$S_{c,j}^{RP}$: Score des Kunden *c* bei der Eigenschaft *j* in der Dimension Referenzpotenzial

GF_{l}^{MVP}: Gewichtungsfaktor der Eigenschaft *l* in der Dimension Member Value Potenzial

$S_{c,l}^{MVP}$: Score des Kunden *c* bei der Eigenschaft *l* in der Dimension Member Value-Potenzial

h: Index der Eigenschaften des Informationspotenzials

j: Index der Eigenschaften des Referenzpotenzials

l: Index der Eigenschaften des Member Value-Potenzials

Beispielhaft lässt sich der nicht-monetäre Kundenteilwert damit wie folgt bestimmen:

Tabelle 25: Beispielhafte Ermittlung des Gesamtpunktwerts für den nicht-monetären Kundenwert[1631]

Bewertungsdimension	**Gewicht**	**Punktwert**	**Teilscore**
Informationspotenzial	0,3	2,00	$0{,}3 \cdot 2{,}00 = 0{,}60$
Referenzpotenzial	0,4	1,75	$0{,}4 \cdot 1{,}75 = 0{,}70$
Member Value-Potenzial	0,3	1,50	$0{,}3 \cdot 1{,}50 = 0{,}45$
Nicht-monetärer Kundenwert (Gesamtpunktwert):			**1,75**
Maximal möglicher Wert:			**3,00**
Prozentuale Bewertung:			**58,3 %**

Um das Schätzrisiko im Scoringmodell zu berücksichtigen, können statt einwertiger Merkmalsausprägungen auch mehrere Ausprägungen zugelassen werden, so dass in Verbindung mit jeweiligen Eintrittswahrscheinlichkeiten ein Erwartungswert ermittelt werden kann.[1632]

Werden der monetäre und der nicht-monetäre Kundenteilwert auf Basis der hier vorgestellten Metriken ermittelt, so ist darauf hinzuweisen, dass die Berücksichtigung der Dynamik in den beiden Metriken unterschiedlich ist. Während bei dem kapitalwertbasierten Verfahren eine periodenspezifische Betrachtung

---

1631 Eigene Darstellung.

1632 Vgl. *Adam, D.* (1996), S. 413.

für die Zukunft erfolgt, ist das Punktbewertungsverfahren statisch bzw. nicht periodisch differenziert.[1633] Es bildet nicht die zu berücksichtigende Veränderung der Bewertungskriterien im weiteren Verlauf der Kundenbeziehung ab. So wäre es etwa vorstellbar, dass ein Firmenkunde künftig eine exponierte Stellung in einem regionalen Verein oder Verband einnehmen wird. Daraus ließe sich ein positiver Einfluss auf sein Referenzpotenzial ableiten. Grundsätzlich besteht die Möglichkeit, ein mehrperiodisches Scoringmodell zu nutzen. In diesem Fall müsste dann auch der Zeitpräferenz der Entscheider Rechnung getragen werden. In Analogie zur Diskontierung im kapitalwertbasierten Verfahren, wäre auch hier ein Diskontierungsfaktor plausibel herzuleiten. Für das weitere Vorgehen wird von einem periodenunabhängigen nicht-monetären Kundenteilwert ausgegangen.

5.4 Zusammenführung des monetären und des nicht-monetären Kundenteilwerts

Im Ergebnis dieses Abschnitts zur Messung des Kundenwerts liegen zwei auf unterschiedlichen Skalen gemessene Kundenteilwerte vor. Fraglich ist nun, ob und wie diese im Sinne der Steuerungsadäquanz zusammengefasst werden sollten und können.

Die Zusammenführung des monetären und nicht-monetären Kundenwerts in einer Größe dient im Schrifttum vor allem dem Zweck, einen einzelnen übergeordneten kundenbezogenen Entscheidungswert zu erhalten.[1634] Die mit dieser Größe in Zusammenhang stehenden Entscheidungssituationen sind die Selektion werthaltiger Kunden aus Sicht des Anbieterunternehmens sowie darauf aufbauend die Zuordnung von Ressourcen. Darüber hinaus ist es für den vorliegenden Kontext zwingend erforderlich, einen steuerungsadäquaten Entscheidungswert für die Vertriebsplanung, -steuerung und -kontrolle kundenwertgerichteter Maßnahmen zu haben. Denn nur dann ist der gemessene Kundenwert hinreichend, um als Vertriebssteuerungsgröße genutzt zu werden.

Problematisch ist nun vor allem die sinnvolle Verknüpfung des monetären mit dem nicht-monetären Kundenwerts, da „für letzteren Näherungswerte eines in Geldeinheiten ausgedrückten Äquivalents gefunden werden müssen, um eine Gewichtung zu ermitteln."[1635] SCHMÖLLER empfiehlt für die Aggregation der beiden Werte die Verwendung eines Scoringmodells, das um Kundenprofile ergänzt werden soll.[1636] Die Berücksichtigung der Kundenprofile erfolgt mit dem

---

[1633] Vgl. *Tewes, M.* (2003), S. 158.

[1634] Vgl. z. B. *Mengen, A./Metler, A.* (2008), S. 30 ff.

[1635] *Schmöller, P.* (2001), S. 181.

[1636] Vgl. *Schmöller, P.* (2001), S. 181 ff.

Ziel, eine informatorische Grundlage zu haben, um die Kundenwerte transparent darzustellen. Dieses Vorgehen eignet sich zwar gut im Rahmen der Planung, um werttreiberbasierte Maßnahmen zu planen. In Verbindung mit der vermögenswertorientierten Zielsetzung auf Gesamtbankebene führt die dann fehlende Verknüpfbarkeit von Gesamtbank- und Vertriebssteuerung jedoch zu Problemen, die zu fehlender praktischer Akzeptanz dieses Vorgehens führen dürften.

Als Aggregationsverfahren schlägt NAGEL vor, den qualitativen, nicht-monetären Kundenwert als Gewichtungsfaktor für den monetären Kundenwert zu verwenden.[1637] In diesem Fall besteht nicht die Möglichkeit einer kompensatorischen Beziehung zwischen den Kundenteilwerten. Zwar kann einem so gemessenen Kundenwert die Steuerungsadäquanz analog zu anderen monetären Größen zuerkannt werden. Ein etwaiger negativer monetärer Kundenwert eines Kunden, der aufgrund des hohen nicht-monetären Kundenwerts dennoch wertvoll ist, kann bei diesem Vorgehen nicht mehr kompensiert werden.[1638] Damit schränkt dieses Vorgehen den Wertebereich der Entscheidungsgrundlagen unbegründet ein. Aus diesem Grund ist dieses Verfahren abzulehnen.

GELBRICH und GELBRICH/MÜLLER nutzen die Data Envelopement Analysis (DEA), um den aggregierten Kundenwert zu akquirierender Kunden zu messen.[1639] Der aggregierte Kundenwert wird als Effizienzgröße definiert. Die DEA ist ein Verfahren, das auf Basis verschiedener inhaltlicher Spezifikationen der eingehenden Größen (Inputs und Outputs) angewendet werden kann. Sie hat jedoch den Vorteil, dass die Bedeutungsgewichte nicht subjektiv bestimmt, sondern „als Ergebnis eines integrierten mathematischen Verfahrens“[1640] ermittelt werden. Das Ergebnis der DEA ist eine dimensionslose Kennzahl. Damit ist das Problem verbunden, dass eine Verknüpfung mit der vermögenswertorientierten Gesamtbanksteuerung nicht möglich ist. Zudem liegt der vornehmliche Zweck der DEA darin, Handlungsalternativen hinsichtlich ihrer Effizienz zu messen.[1641] Dieser Zweck ist für die Auswahl und Kontrolle von Vertriebsmaßnahmen im Zusammenhang mit der Kundenbewertung interessant, stellt jedoch nur einen Ausschnitt der relevanten Aufgaben der Vertriebssteuerung dar.[1642]

1637 Vgl. *Nagel, D.* (2008), S. 190.

1638 Durch die multiplikative Verknüpfung dieser beiden Größen ist kein kompensierender Vorzeichenwechsel möglich. Dafür kann es passieren, dass ein nicht-intendierter Vorzeichenwechsel resultiert, wenn beide Kundenteilwerte negativ sind.

1639 Vgl. *Gelbrich, K./Müller, S.* (2006), S. 481 f.; *Gelbrich, K.* (2001), S. 95 ff.

1640 *Hagenloch, T.* (2008), S. 1374.

1641 Vgl. *Schneider, N. C.* (2007), S. 101.

1642 Vgl. dazu die Ausführungen in Kapitel B.2.3.2.

RUDOLF-SIPÖTZ nutzt einen dreidimensionalen Vektor, um die von ihr berücksichtigten Kundenteilwerte zu einer Entscheidungsgröße zu aggregieren.[1643] Diesen überführt sie zur besseren Veranschaulichung in ein dreidimensionales Portfoliomodell, einen Kundenkubus. Ihrem Vorgehen liegt die Aussage zugrunde, dass es aus Managementperspektive nicht das Ziel der Kundenbewertung sein kann, „den Kundenwert in einer einzigen Zahl auszudrücken."[1644] Im Ergebnis resultiert eine über die Ziehung von Grenzwerten im Kundenkubus frei definierbare Vielzahl an ordinalen Kundenrängen. Dieses Vorgehen lässt keine eindeutige Rangreihung der einzelnen Firmenkunden zu. Das Problem der analytischen Verbindung der beiden Kundenteilwerte wird durch grafische Mittel umgangen.[1645] Darin liegt auch der wesentliche Nutzen dieses Vorgehens.[1646] Für die Einzelkundenbetrachtung ist es wenig geeignet, da es eher für die Analyse der Gesamtheit der Kunden ausgerichtet ist,[1647] wodurch der Nutzen dieses Vorgehens eingeschränkt ist.

Es ist zu konstatieren, dass alle genannten Verfahren Schwächen haben, die eine ausschließliche Anwendung wenig zweckmäßig werden lassen. Aus diesem Grund wird für die Konzeption einer kundenwertorientierten Vertriebssteuerung anwendungs- und entscheidungszweckbezogen mit den Kundenteilwerten umgegangen. Statt der Auswahl eines einzelnen Verfahrens bietet dieses Vorgehen die Möglichkeit, die Verfahren entsprechend ihrer Zweckmäßigkeit im Sinne der Aufgaben der Vertriebssteuerung einzusetzen.

5.5 Besonderheiten der Messung des Kundenwerts

In den vorhergehenden Kapiteln wurden für die Messung des Kundenwerts Metriken beschrieben, die aufgrund ihrer konzeptionellen Struktur nur für Bestandskunden auf Einzelkundenebene sinnvoll eingesetzt werden können. Nachfolgend wird ergänzend erläutert, wie die Messung des monetären Kundenwerts für potenzielle Kunden und wie eine kundengruppenbasierte Messung erfolgen kann. Weiterhin wird erläutert, wie die Aggregation der auf Basis unterschiedlicher

---

1643 Während das Kundenwertmodell zwischen dem Markt- und dem Ressourcenpotenzial unterscheidet, differenziert RUDOLF-SIPÖTZ bei der Messung des Kundenwerts zwischen dem gegenwärtigen Marktpotenzial, dem künftigen Marktpotenzial und dem Ressourcenpotenzial. Vgl. hier und nachfolgend *Rudolf-Sipötz, E.* (2001), S. 178 ff.

1644 *Rudolf-Sipötz, E.* (2001), S. 178.

1645 Vgl. *Gelbrich, K./Müller, S.* (2006), S. 476.

1646 Vgl. *Stüker, D.* (2008), S. 66.

1647 Dies bedeutet auch, dass die Anwendung eines Kundenkubus auf einen einzelnen Kunden ohne Aussage ist.

objektbezogener Dimensionen ermittelten Werte auf Gesamtbankebene erfolgen kann.

5.5.1 Messung der Kundenwerte für Nichtkunden

Eine der allgemeinen Anforderungen an Kundenwertmodelle ist, dass diese bezugsgrößenflexibel sein sollen. Das bedeutet, dass sie sich auf aktuelle und potenzielle Kunden gleichermaßen anwenden lassen. Grundsätzlich ist dies mit den vorgestellten Messverfahren möglich, wenngleich Anpassungen im Kundenwertmodell und in der Metrik notwendig sind. Denn aufgrund der fehlenden Historie ist die vorgenommene Differenzierung der Kundenwerttreiber nicht vollumfänglich möglich. Hinsichtlich der Kundenwerttreiber der Ertragsdimension ist festzuhalten, dass die grundlegende bankleistungsbezogene Aufteilung in das Basisgeschäfts-, das Entwicklungs- und das Cross-Selling-Potenzial nicht möglich ist. Statt dieser Differenzierung soll von einem gesamthaften Geschäftspotenzial gesprochen werden. Dabei ist zu beachten, dass im Entwicklungspotenzial das Preispotenzial gegenstandslos ist, wenn davon ausgegangen wird, zu Beginn der Geschäftsbeziehung die Zahlungsbereitschaft des Kunden abzuschöpfen. Wird im Rahmen der Akquisition von einer Penetrationsstrategie ausgegangen, die mit geringen Preisen respektive Gebühren und Zinsen ausgeht, so ist in der dynamischen Betrachtung das Preispotenzial zu berücksichtigen. Ergänzend zu den bereits definierten Kosten sind darüber hinaus auch Kosten für die Kundenbeziehungsanbahnung zu betrachten. Diese sollen für die Bewertung von Nichtkunden als Akquisitionskosten in die Metrik eingehen.[1648]

Bei der Risikodimension bleiben die genannten Kundenwerttreiber erhalten. Es sei jedoch darauf hingewiesen, dass mangels der Anwendbarkeit interner Ratingverfahren wie dem VR-RATING auf alternative Verfahren zur Bestimmung der Ausfallwahrscheinlichkeit sowie der weiteren relevanten Einflussfaktoren für die Standardrisikoprämie und die Eigenkapitalkosten zurückgegriffen werden muss. Hier sind strukturell identische Verfahren zu nutzen, die annähernd identische und verlässliche Ergebnisse liefern wie die Instrumente bei den Bestandskunden, da ansonsten die Vergleichbarkeit der Ergebnisse sowie die darauf aufbauende Rangreihung der Kundenbeziehungen nach ihren Kundenwerten nicht sinnvoll möglich ist.

Mit Blick auf die Beziehungsdimension wurde das Akquisitionspotenzial thematisiert, das sowohl für die Nichtkunden als auch die wiederzugewinnenden

---

1648 Die Kosten im Zusammenhang mit der Akquisition von Kunden können auch stärker differenziert berücksichtigt werden. So unterscheidet LISSAUTZKI zwischen den Werbekosten, den Kosten des Direktmarketings sowie den Verhandlungs-, Entscheidungs- und Vereinbarungskosten. Vgl. *Lissautzki, M.* (2007), S. 231.

Kunden gilt. In Analogie zur Kundenbindungsrate soll das Akquisitionspotenzial über eine Erfolgswahrscheinlichkeit der vorzunehmenden Akquisitionsbemühungen berücksichtigt werden.[1649]

Werden diese Überlegungen auf die Metrik übertragen, kann der monetäre externe Kundenwert wie folgt gemessen werden:

$$KW_c^m = \sum_{t=1}^{A} \frac{-AK_{c,t}}{(1+i_t+z_c)^t} + \left(\frac{\phi_{c,A} \cdot \left(\sum_{t=A}^{T} \left(\frac{(GGP''_{c,t} - kPK_{c,t} - KP_{c,t}) \cdot (b_{c,t} - \varepsilon_{c,t})}{(1+i_t+z_c)^{t-A}} \right) \right)}{(1+i_t+z_c)^A} \right) \quad [D.33]$$

mit:

$AK_{c,t}$: Akquisitionskosten für den Kunden *c* in der Periode *t*

$GGP''_{c,t}$ Unter Berücksichtigung von bankleistungsbezogenen Prozesskosten (Standardstückkosten) und Risikokosten bewertetes Gesamtgeschäftspotenzial eines Kunden *c* in der Periode *t*

$\phi_{c,A}$: Akquisitionswahrscheinlichkeit für den Kunden *c* in der Periode der erwarteten Akquisition *A (t=A)*

A: Periode der erwarteten Akquisition (t=*A)*

Mit diesem Messmodell können die monetären Kundenwerte der Nichtkunden des numerischen Kundenpotenzials ermittelt werden. Für die Anwendung des Scoringmodells, um den nicht-monetären Kundenteilwert zu messen, sind keine Anpassungen nötig, sofern beim Aufbau des Punktbewertungsverfahrens darauf geachtet wird, dass keine Bewertungskriterien verwendet werden, für die die Daten nur für Bestandskunden vorliegen können.

Eine weitere Besonderheit der Kundenwertmessung ist die kundengruppenbasierte Messung, die Gegenstand des folgenden Kapitels ist.

5.5.2 Kundengruppenbasierte Messung des Kundenwerts

Während die bisherigen Ausführungen und Überlegungen ihr Fundament in der Einzelkundenbetrachtung bzw. der Betrachtung vertriebspolitisch motivierter

---

[1649] Hinsichtlich der Erfolgswahrscheinlichkeit der Akquisitionsbemühungen ließe sich ebenfalls eine in Abhängigkeit des Fortschritts des Anbahnungsprozesses differenzierte Betrachtung vornehmen.

Kundenverbünde haben und als direkte Messung des Kundenwerts bezeichnet werden, wird für die kundengruppenbasierte Messung eine andere Ebene der objektbezogenen Dimension des Kundenwerts betrachtet.[1650]

Die indirekte, kundengruppenbasierte Messung des Kundenwerts ist dann relevant, wenn informationsökonomische Gründe eine Bewertung auf Basis kundenindividueller Informationen ausschließen.[1651] Für SACHENBACHER verdeutlicht bereits der Versuch, die Frage nach der individuellen Kundenentwicklung zu beantworten, dass zukunftsorientierte Kalkulationsmethoden nicht auf kundenindividueller Basis erfolgen können.[1652] Die Unwägbarkeiten je Einzelkundenverbindung seien dafür zu groß. Bei der kundengruppenbasierten Messung von Kundenwerten wird demnach nicht für jeden einzelnen Kunden der Kundenwert gemessen, sondern für eine zweckmäßig zu definierende Gruppe oder ein Segment von Kunden. Dadurch lassen sich einige Vorteile in der Kundenwertorientierung erreichen. Zunächst ist von einem geringeren Prognose- und Schätzaufwand auszugehen. Zudem werden Prognoserisiken auf Ebene des Einzelkunden durch die Bewertung eines größeren Portfolios teilweise oder ganz diversifiziert. Die Bewertung von Kundengruppen und des gesamten Kundenstamms ist daher bspw. bei strategischen Entscheidungen auf Unternehmens- oder Ebene der strategischen Geschäftseinheiten und für Zwecke der kundenwertbasierten Unternehmensbewertung relevant.

Für die kundengruppenbasierte Kundenwertermittlung ist im ersten Schritt der Kundengesamtbestand in homogene Segmente aufzuteilen. Für die jeweiligen Segmente werden im zweiten Schritt die relevanten Kundenwerttreiber ermittelt und darauf aufbauend der Kundenwert berechnet. Für den Kundenwert auf Ebene des einzelnen Kunden wird der Kundenwert der betrachteten Gruppe durch die Anzahl der Kunden in der Gruppe dividiert, so dass eine Durchschnittsgröße resultiert.[1653] Es ist offensichtlich, dass dieses Vorgehen sich nur für den monetären Kundenteilwert eignet.

Den genannten Vorteilen stehen jedoch auch einige Nachteile gegenüber. Werden gemittelte Kundenwerte auf Einzelkundenebene betrachtet, so geht damit ein großer Informationsverlust einher, der insbesondere für die Vertriebssteuerung im Firmenkundenkontext nachteilig sein kann. Denn für die Ableitung von mitarbeiterindividuellen Vertriebszielen auf Basis des monetären Kundenteilwerst dürften diese gemittelten Größen wenig Akzeptanz finden. Zudem

---

1650 Vgl. dazu die Ausführungen in Kapitel D.1.2.4.1.

1651 Vgl. *Hempelmann, B./Lürwer, M.* (2003), S. 339.

1652 Vgl. *Sachenbacher, H. U.* (1990), S. 10.

1653 Ein alternatives Vorgehen liegt darin, direkt mittlere Werte für die einzelnen Kundenwerttreiber zu ermitteln und diese dann mit der Anzahl der Kunden in einer Kundengruppe zu multiplizieren. Vgl. *Hempelmann, B./Lürwer, M.* (2003), S. 340.

führt die Betrachtung von Durchschnittswerten zu Ineffizienzen in der kundenwertbasierten Ressourcenzuordnung, so dass die Gefahr besteht, relativ wertvollen Kunden relativ zu wenig Ressourcen zuzuweisen und relativ wertgeringen Kunden relativ zu viele Ressourcen zuzuweisen. Gerade dies sollte jedoch vermieden werden.

Tendenziell ist aufgrund der genannten Nachteile die kundengruppenbasierte Messung des Kundenwerts für die Vertriebssteuerung im Verständnis dieser Arbeit von nachgelagerter Bedeutung, da strategische Entscheidungen über die Auswahl von strategischen Geschäftsfeldern auf Ebene des Gesamtunternehmens und auf Ebene des der Vertriebspolitik übergeordneten Marketings sind. Dennoch sind kundengruppenbasierte Messungen des monetären Kundenwerts nicht per se abzulehnen. Ihr Einsatz ist zweckmäßig vorzunehmen und kann in begründeten Einzelfällen die Ermittlung eines individuellen monetären Kundenteilwerts ersetzen. Einen dieser Ausnahmefälle stellt etwa die Messung von monetären Kundenteilwerten für Nichtkunden dar. So ließen sich die in Formel [D.33] verwendeten Kundenwerttreiber mittels multivariater Analyseverfahren fundieren und logische Analogieschlüsse für die Bewertung ableiten.

5.5.3 Aggregation der Kundenwerte

Die bisherigen Ausführungen bezogen sich darauf, wie der Wert einer einzelnen Kundenbeziehung oder einer Gruppe von Kunden gemessen werden kann. Die Ausführungen im vorherigen Kapitel machen bereits deutlich, dass die objektbezogene Dimension des Kundenwerts in Abhängigkeit der konkreten Aufgabenstellung der Vertriebssteuerung variiert.[1654] Vor allem für strategische Aufgaben ist die Kenntnis aggregierter Kundenwerte für den gesamten Kundenbestand von zentraler Bedeutung, während für die operative Planung und Vertriebssteuerung Kundenwerte auf Einzelkundenebene relevant sind.

Um aggregierte Kundenwerte zu ermitteln, wird in der Literatur für den monetären Kundenteilwert die Wertadditivität angenommen.[1655] Die Prämisse der Wertadditivität unterstellt, dass sich der aggregierte monetäre Kundenteilwert des gesamten Kundenbestands aus den monetären Kundenteilwerten der einzel-

---

1654 So auch *Schroeder, N.* (2006), S. 44 f.

1655 Vgl. z. B. *Blattberg, R. C./Deighton, J.* (1996), S. 137 f.; *Hogan, J. E. et al.* (2002), S. 6 ff.; *Jain, D./Singh, S. S.* (2002), S. 40 f.; *Gupta, S./Lehmann, D. R.* (2003), S. 19 ff.; *Burmann, C.* (2003), S. 122 ff.; *Bauer, H.-H./Hammerschmidt, M.* (2005), S. 338 ff. Die letztgenannten Autoren addieren mit dem Ziel der kundenwertbasierten Unternehmensbewertung sowohl die Bestandskunden als auch die potenziellen Neukunden, um den monetären Kundenteilwert der Kundenbasis zu erhalten.

nen Kundenbeziehungen durch Summation ermitteln lässt.[1656] Ihr liegt die Annahme zugrunde, dass die einzelnen bewerteten Kundenbeziehungen unabhängig voneinander sind, so dass durch die Mischung der Kundenbeziehungen kein zusätzlicher Wert entsteht. Zwar ist diese Annahme kritisch zu sehen, da in Analogie zur Kreditportfoliosteuerung unterschiedliche Mischungen von Kundenbeziehungen mit unterschiedlichen Kundenplanungsrisiken und damit stabilen monetären Kundenteilwerten aus Sicht der Kreditgenossenschaft verbunden sind. Jedoch liegt zum einen in diesem Zusammenhang noch ein erheblicher Forschungsbedarf vor[1657] und zum zweiten sind diese Aspekte von nachgelagerter Bedeutung, da deren Auswirkungen vor allem bei am Markt gehandelten Vermögensgegenständen relevant sind, dies jedoch weder für Geschäftsanteile an Kreditgenossenschaften noch für Kundenbeziehungen gilt. Aus diesen Gründen soll vereinfachend von Wertadditivität ausgegangen werden, so dass sich der Wert des gesamten Kundenbestands aus der Summe der direkt ermittelten monetären Kundenteilwerte der einzelnen Kundenbeziehungen ermitteln lässt.

6 Zusammenfassung und kritische Würdigung der Erkenntnisse

Wie die vorhergehenden Ausführungen zeigen, liegen für die Kundenwertorientierung mannigfaltige Konzeptualisierung- und Operationalisierungsansätze vor. Aufbauend auf einem grundlegenden Verständnis des Begriffs Kundenwert wurde geprüft, ob die Erfolgsvoraussetzungen der Kundenwertorientierung für das Firmenkundengeschäft der Kreditgenossenschaften vorliegen. Dies wurde mit Rekurs auf die Implikationen aus Teil C der Arbeit bestätigt. Besonders das Zusammenspiel von Kundenwertorientierung und genossenschaftlichen Prinzipien musste einer detaillierten Analyse unterworfen werden, um zu prüfen, ob daraus Besonderheiten erwachsen, die zwingend zu berücksichtigen sind und wie ein allgemein gültiges Zielsystem für Kreditgenossenschaften formuliert werden kann. Hier zeigte sich, dass im Mitgliedergeschäft eine unbedingte Übertragung der Kundenwertorientierung wie sie in bisherigen Beiträgen zur Kundenwertforschung vorgenommen wird, nicht möglich ist. Der genossenschaftliche Wesenszweck, die genossenschaftliche Selbstverwaltung und der Rollenpluralismus der Mitgliederkunden führen zu Einflüssen, die zwingend in der Umsetzung der Kundenwertorientierung der Kreditgenossenschaften zu berücksichtigen sind. Anderenfalls würde eine weitere Distanzierung vom kulturellen Kern der Kreditgenossenschaften unterstützt werden, die in einer weiteren

---

[1656] Vgl. *Breuer, W.* (1997), S. 1148.

[1657] Vgl. *Schroeder, N.* (2006), S. 64 sowie die dort genannte Literatur.

Verkehrung des genossenschaftlichen Förderauftrags, die sich darin zeigt, das der Mitgliedskunde der Kreditgenossenschaft zu nutzen habe und nicht umgekehrt, mündet. Dieses wichtige Ergebnis ist auch für die in Teil E zu entwickelnde Vertriebssteuerungskonzeption zu berücksichtigen.

Weiterhin wurde die Konzeptualisierung und Operationalisierung eines Kundenwertmodells vorgenommen, das sich zweckmäßig an der Situation der Kreditgenossenschaften ausrichtet. Dies war nötig, da bisherige Beiträge, die Kundenwertmodelle entwickelt haben, entweder branchenfremd waren oder den Anforderungen, die an Kundenwertmodelle zu stellen sind, nicht vollumfänglich standhalten konnten. Das in den Kapiteln D.4 und D.5 konzeptualisierte und operationalisierte Kundenwertmodell soll abschließend mittels dieser Anforderungen gewürdigt werden.

Aus Sicht des Ganzheitlichkeitspostulats kann festgestellt werden, dass durch die Betrachtung der drei Dimensionen Beziehung, Ertrag und Risiko sowie der jeweiligen Differenzierung zwischen beziehungsbezogenen und beziehungsübergreifenden Kundenwerttreibern ein ganzheitliches Kundenwertmodell konzeptualisiert wurde. Diejenigen beziehungsübergreifenden Kundenwerttreiber, die fallweise in der Literatur genannt werden, wurden mangels Eindeutigkeit in der kundenbezogen Zuordnung nicht berücksichtigt. Aufgrund des breiten und differenzierten Zielsystems, das der Herleitung des Kundenwertmodells zugrunde liegt, kann konstatiert werden, dass alle relevanten Ziele berücksichtigt wurden. Zudem bieten die verwendeten Metriken aufgrund ihrer Struktur die Möglichkeit, sowohl den anbieterspezifischen Kundenwert als auch den das Kundenwertpotenzial zu messen. Bei letzterem ist jedoch zu berücksichtigen, dass semantisch richtig von einem anbieterspezifischen Kundenwertpotenzial zu sprechen ist. Dies ergibt sich daraus, dass die institutsspezifischen Kostenstrukturen gegenständlich für die Kundenbewertung sind. Weichen die Kostenstrukturen eines dritten Anbieters davon ab, muss er zwingend zu einer abweichenden Einschätzung des Kundenwerts kommen. Das Kundenwertmodell begrenzt die Betrachtung des relevanten Zeitraums auf den prospektiven Kundenwert, was jedoch als zweckadäquat zu beurteilen ist. Aufgrund der Auswahl der Kundenwerttreiber und des einzelkundenbezogenen Vorgehens kann begründet davon ausgegangen werden, dass die Voraussetzungen der interindividuellen Variabilität gegeben sind. Hinreichend ist diese Anforderung jedoch erst erfüllt, wenn die tatsächliche Prognose der Kundenwerttreiberausprägungen von gemittelten oder zu stark standardisierten Werten abstrahiert. Da das Messmodell des monetären Kundenwerts auf einem dynamischen Investitionsrechenverfahren aufbaut, ist der Dynamik bei diesem Kundenteilwert hinreichend Rechnung getragen. Dies gilt jedoch nicht für den mittels eines Punktbewertungsverfahrens gemessenen nicht-monetären Kundenteilwert, da dieser keine dynamische Berechnung ermöglicht. Mit den genannten Anpassungen für die Messung des Kundenwerts auf Ebene der potenziellen Kunden kann die Anforderung der Bezugsgrößenflexibilität als erfüllt bewertet werden. Hinsichtlich des wirtschaftlichen Einsatzes des Kundenwertmodells kann richtigerweise vermutet werden, dass ein größerer Erfassungs- und Prognoseaufwand als im Vergleich zu den diskutierten Modellen und bisherigen Vertriebssteuerungsansätzen zu unterstellen ist. Dieser zusätzliche Aufwand dürfte jedoch durch die höhere Steuerungsadäquanz des operationalisierten Kundenwerts im Kontext der vorhandenen marktstrategischen Ansätze kompensiert werden. Zudem ist die Auseinandersetzung mit dem Kundenwert und dessen Kundenwertreibern kein Selbstzweck,[1658] sondern Teil einer systematischen Auseinandersetzung mit kundenbezogenen Erfolgspotenzialen im Sinne des strategischen Managements.

Aufgrund des sehr differenzierten Ausweises der Erfolgsquellen des Kundenwerts ist aus Sicht der vertriebssteuerungsbezogenen Anforderungen festzustellen, dass das Postulat der Handlungsorientierung erfüllt ist. Für die Anforderung der Operationalität kann dies uneingeschränkt für den monetären Kundenteilwert konstatiert werden. Als problematisch ist dies jedoch für den nicht-monetären Kundenteilwert zu sehen, da dieser nur schwierig als Planungs-, Steuerungs- und Kontrollgröße genutzt werden kann.

Wie schon die Ausführungen zu den Möglichkeiten des Bewertungsmaßstabs zeigen, kann für das Kundenwertmodell eine hinreichende Skalierbarkeit in Abhängigkeit der Möglichkeiten der Kreditgenossenschaft unterstellt werden. Im einfachsten Fall erfolgt die Bewertung auf Ebene der Konditionsbeiträge. Diese Vereinfachung kann ergänzt werden, indem die Kostengrößen nicht betrachtet werden. Dies ist zwar problematisch, da es den Kundenwert Ziel- und Steuerungsgröße verzerrt, aber es ist möglich. Da die gesamte Herleitung des Kundenwertmodells und dessen Messung das Ziel verfolgt, von den Kreditgenossenschaften eingesetzt werden zu können, ist begründet davon auszugehen, dass das Kundenwertmodell adäquat ist. Durch den Rekurs auf die Einzelgeschäftskalkulation werden auch deren Eigenschaften auf den monetären Kundenwert übertragen. Das betrifft die Erfüllung des Postulats der grenznutzenorientierten Einzelbewertung ebenso wie die Einbindung in die Gesamtbanksteuerung. Nicht oder nur zu Teilen erfüllt sind diese beiden Anforderungen durch den nicht-monetären Kundenteilwert. Hier zeigt sich deutlich die Begründetheit der Forderung von STAHL ET AL., die es als vordringlich notwendig absehen Ansätze zur adäquaten Bewertung von nicht-monetären Größen zu entwickeln.[1659]

[1658] Vgl. *Reiners, J. O.* (2004), S. 68.

[1659] Vgl. *Stahl, H. K.* et al. (2009), S. 260.

Aufgrund des inhaltlichen Zielbezugs kann eine tendenzielle Integrationsfähigkeit des nicht-monetären Kundenteilwerts in die Gesamtbanksteuerung vorgenommen werden. In wie weit das entwickelte Kundenwertmodell in den Kreditgenossenschaften akzeptiert wird, kann nicht abschließend beantwortet werden, da eine entsprechende empirische Überprüfung im Rahmen dieser Arbeit nicht vorgenommen werden soll.

Wie in Kapitel A.1 erläutert, setzt sich der Wert einer Kreditgenossenschaft aus dem „Zerschlagungswert“ und dem „Potenzialwert“ zusammen.[1660] Der „Potenzialwert“ wird unter anderem substantiiert durch das hergeleitete Kundenwertmodell. Dieser Gedanke lässt sich mit dem Förderpotenzial *FP* des Member Value zusammenbringen. Denn der „Potenzialwert“ ist eine der zentralen Größe, die den Umfang des Förderpotenzials determinieren. Daher kann abschließend ausgeführt werden, dass der firmenkundenspezifische Kundenwert in toto wesentlicher Bestandteil des „Förderpotenzials“ im Sinne des Member Value ist.

[1660] Vgl. Kapitel A.1.

E Gestaltungsempfehlungen einer kundenwertorientierten Vertriebssteuerung im Firmenkundengeschäft der Kreditgenossenschaften

Aufbauend auf dem konzeptualisierten und operationalisierten Kundenwertmodell und unter Berücksichtigung der herausgearbeiteten Implikationen aus dem konkreten Anwendungskontext werden in Teil E Gestaltungsempfehlungen einer kundenwertorientierten Vertriebssteuerung erläutert. Dazu wird zunächst der Blick auf das Vertriebsinformationssystem gelegt (Kapitel E.1), das die informatorische Grundlage für den Vertrieb und die Vertriebssteuerung ist. Die explizite Betrachtung des Vertriebsinformationssystems, das originär dem Vertriebscontrolling zuzuordnen ist,[1661] erfolgt, da Kunden- und Kundenwertdaten von essentieller Bedeutung für die kundenwertorientierte Vertriebssteuerung sind, und im Gegensatz zu etablierten Vertriebssteuerungsansätzen nicht ausschließlich aus institutsinternen Quellen generiert werden können. Empirische Studien zeigen, dass die Qualität kundenbezogener Daten in einer Vielzahl von Kreditinstituten mangelhaft ist.[1662] Die Beschaffung der relevanten Informationen wird daher gleichzeitig zum inhaltlichen Gegenstand der Vertriebssteuerung. Anknüpfend an die informatorischen Grundlagen wird die Vertriebssteuerung bestehend aus Vertriebsplanung (Kapitel E.2), Vertriebsplanrealisation (Kapitel E.3) und Überwachung und Sicherung (Kapitel E.4) konzeptionell dargestellt und hinsichtlich ihrer Ausgestaltungsmöglichkeiten diskutiert. Abschließend werden die Ergebnisse zusammengefasst und gewürdigt (Kapitel E.5).

1661 Vgl. zum Zusammenhang von Vertriebssteuerung und -controlling die Ausführungen in Kapitel B.2.2.

1662 Vgl. *Pfeiffer, M./Pirngruber, N./Mährle, C.* (2003), S. 6. So gaben 72 Prozent der in der Studie befragten Institute an, dass ihre Firmenkundendaten fehlerbehaftet sind. Grundlage der Studie sind 44 der 400 größten deutschen Banken. Wenn diese Ergebnisse für die großen Banken gelten, dürften sie auch – ggf. tendenziell verstärkt – für die kleinen und kleineren Kreditgenossenschaften Gültigkeit haben.

1 Kundenwertorientiertes Vertriebsinformationssystem

1.1 Definition des Vertriebsinformationssystems

Die Voraussetzung einer erfolgreichen Vertriebssteuerung ist ein zielgerichtetes Vertriebsinformationssystem, dessen zentrale Aufgabe in der Bereitstellung entscheidungsrelevanter Informationen für das Vertriebsmanagement und die Vertriebsmitarbeiter liegt.[1663] Informationen sind zweckbezogenes, entscheidungsrelevantes Wissen, das dazu dient, zielsetzungsorientiertes Handeln vorzubereiten.[1664] Wissen wird differenziert in explizites und implizites Wissen.[1665] Explizites Wissen ist konkret und auf geeigneten Medien übermittelbar. Implizites Wissen lässt sich nur unvollständig in Worte fassen. Es umfasst kognitive (mentale Modelle von Individuen wie Schemata, Paradigmen und Überzeugungen) und technische Elemente (individuelle Fähig- und Fertigkeiten, die kontextspezifisch zu erfolgreichen Problemlösungen führen). Implizites Wissen ist nur schwer in explizites Wissen wandelbar.

Wie in Kapitel B.2.1 erläutert wurde, ist die Steuerung integraler Bestandteil der Führung, die in die Phasen der Willensbildung und Willensdurchsetzung aufgeteilt wird. Willensbildungsprozesse können intuitiv oder reflexiv erfolgen.[1666] Intuitive Willensbildungsprozesse basieren auf begrenztem und oft implizitem Wissen, während reflexive Willensbildungsprozesse mit explizitem Wissen fundiert werden. Das explizite Wissen wird für spezifische Entscheidungssituationen durch eine gezielte Informationssuche generiert. Entscheidungen, die intuitiv erfolgen, können nur schwer intersubjektiv nachvollzogen und somit kaum bis gar nicht in einen konzeptionellen Steuerungsansatz integriert werden. Vertriebssteuerungshandlungen basieren daher auf explizitem Wissen, dessen Beschaffung, Speicherung, Aufbereitung und Bereitstellung Aufgaben der Vertriebsinformationssysteme sind.[1667] Unter einem Vertriebsinformationssystem wird die planvoll entwickelte und geordnete Gesamtheit von Informationen und Regelungen hinsichtlich der Träger der informatorischen Aufgaben, die Informationswege zwischen ihnen sowie die Methoden und Verfahren der In-

---

[1663] Vgl. *Duderstadt, S.* (2006), S. 9 ff. und S. 39.

[1664] Vgl. *Wittmann, W.* (1959), S. 8 und S. 14.

[1665] Vgl. hier und nachfolgend *Schreyögg, G./Geiger, D.* (2005), S. 434 f. sowie die dort genannte Literatur.

[1666] Vgl. hier und nachfolgend *Tewes, M.* (2003), S. 236 sowie die dort angegebene Literatur.

[1667] Vgl. *Weber, J.* (2004), S. 105.

formationsbeschaffung und -verarbeitung verstanden.[1668] Das Vertriebsinformationssystem ist primär auf den Informationsbedarf der Führungs- und Steuerungsfunktion ausgerichtet, unterstützt jedoch auch die Funktionen des ausführenden Vertriebs.[1669] Das Vertriebsinformationssystem liefert sowohl deskriptive Aussagen (z. B. Anzahl Unternehmen resp. Firmenkunden zzgl. Kundenwert im regionalen Marktgebiet der Kreditgenossenschaft) als auch normative und präskriptive Aussagen (z. B. kundenwertbezogene Bereichs- und Individualziele oder Maßnahmenpläne).[1670]

Aufbauend auf diesem Verständnis des Vertriebsinformationssystems kann zwischen Instrumenten der Informationsbeschaffung, -aufbereitung und -bereitstellung sowie den funktionalen und institutionellen Aspekten des Vertriebsinformationssystems unterschieden werden. Nachfolgend sollen funktionale und instrumentelle Aspekte knapp skizziert werden. Da sich die instrumentellen den funktionellen Aspekten zu- und unterordnen lassen, sollen letztere strukturgebend für die folgenden Ausführungen sein. Um die funktionalen Aspekte zu strukturieren, bietet sich die Gliederung der Aufgaben nach den Stufen des Informationsprozesses an. Dieser gliedert sich in der Literatur in die Stufen der Informationsbeschaffung, Informationsaufbereitung bzw. -verarbeitung, Informationsspeicherung und Informationsübermittlung.[1671]

1.2 Kundenwertorientierte Informationsbeschaffung

1.2.1 Informationsbedarf als Determinante der Informationsbeschaffung

Ausgangspunkt der Informationsbeschaffung ist die Analyse des Informationsbedarfs. Aufgabenorientiert ist der Informationsbedarf definiert als die „Art, Menge und Qualität der Informationsgüter, die ein Informationssubjekt im gegebenen Informationskontext zur Erfüllung einer Aufgabe in einer bestimmten Zeit und innerhalb eines gegebenen Raumgebietes benötigt bzw. braucht.“[1672] Spezifiziert auf den konkreten Untersuchungskontext ist der Informationsbedarf definiert als Art, Menge und Qualität der Kundenwertinformationen, die im

---

[1668] Vgl. *Diller, H.* (1975), S. 7.

[1669] Vgl. *Dauser, A.* (1995), S. 130, der sich in seiner Argumentation auf die institutionellen Aspekte und nicht so wie hier allgemeiner geschehen auf die funktionellen Aspekte bezieht.

[1670] Vgl. zur Unterscheidung deskriptiver, präskriptiver und normativer Aussagen *Dauser, A.* (1995), S. 130.

[1671] Vgl. *Eberling, G.* (2002), S. 84 sowie die dort angegebene Literatur.

[1672] *Szyperski, N.* (1980), Sp. 904.

Rahmen des Vertriebs und der Vertriebssteuerung von den Aufgabenträgern zur Erfüllung ihrer jeweiligen Aufgaben benötigt werden. Der Informationsbedarf wird als objektivierte, intersubjektiv bestimmbare Größe verstanden, die von der Informationsnachfrage des einzelnen Akteurs abweichen kann.[1673] Die Informationsnachfrage ist geprägt durch das Verhalten und die Einstellungen des Aufgabenträgers und kann aufgrund dessen stark vom für die Aufgabenerfüllung notwendigen Informationsbedarf abweichen (Felder 7 und 5 in Abbildung 59). In Abweichungen zwischen Informationsnachfrage und -bedarf sowie dem instrumentell determinierten Informationsangebot und -bedarf liegen Ineffizienzen des Informationssystems begründet. Die Informationsbereitstellung wird ineffizient, wenn Informationen angeboten werden, die für die Erfüllung der Aufgabe nicht von Bedeutung sind (Felder 5 und 6 in Abbildung 59).[1674]

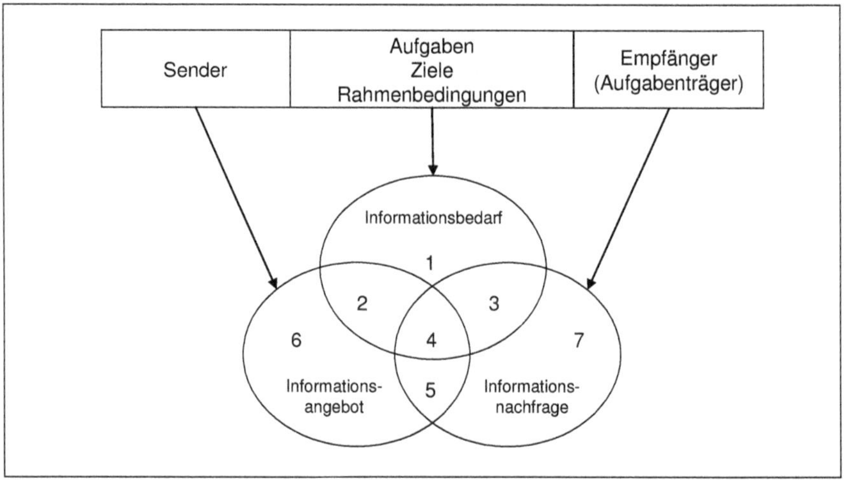

Abbildung 59: Informationsbedarf, -angebot und -nachfrage[1675]

Die Qualität des kundenwertorientierten Vertriebsinformationssystems ist dadurch gekennzeichnet, in wie weit es gelingt, Informationsbedarf, -nachfrage und -angebot des Vertriebs und der Vertriebssteuerung zur Deckung zu bringen (Feld 4 in Abbildung 59). Um zu vermeiden, dass das Informationssystem als

1673 Vgl. *Szyperski, N.* (1980), Sp. 905.

1674 Für eine ausführliche Diskussion der Ineffizienzen hinsichtlich des kundenwertorientierten Informationsbedarfs vgl. *Eberling, G.* (2002), S. 86 ff. Für eine allgemeine Diskussion vgl. *Horváth, P.* (2003), S. 362 f.

1675 *Berthel, J.* (1992), Sp. 875.

Ganzes oder in Teilen ineffizient ist, weil irrelevante Informationen angeboten oder nachgefragt werden, ist der Informationsbedarf hinsichtlich Inhalt und Eigenschaften zu konkretisieren. Die unabhängige Ermittlung des objektivierten Informationsbedarfs soll gewährleisten, dass Informationsineffizienzen nicht vorkommen. Zudem ist sie nötig, um Aufgabenträger auf relevante Informationen und die damit verbundenen Chancen und Risiken hinzuweisen.[1676] Inhaltlich ergibt sich der Informationsbedarf aus den weiter unten diskutierten Determinanten sowie den Aufgaben der Vertriebssteuerung. Um den Informationsbedarf zu konkretisieren reicht es nicht, die Informationsinhalte zu bestimmen, sondern es müssen auch deren relevante Merkmale beschrieben werden. Dazu können folgende Eigenschaftsmerkmale herangezogen werden:[1677]

- Verdichtungsgrad,
- Darstellungsform,
- Aktualitätsgrad,
- Art,
- Bedeutung,
- Menge,
- Häufigkeit,
- Termindringlichkeit,
- Genauigkeit,
- Verwendungszweck,
- Qualität,
- Zuverlässigkeit,
- Sicherheit,
- Kosten und
- Messbarkeit.

Die je Merkmal notwendigen oder angestrebten Ausprägungen der kundenwertorientierten Informationen sollten zwar im Vorfeld des Einsatzes des Kundenwerts als Ziel- und Steuerungsgröße in der Vertriebsplanung formuliert werden, jedoch ist davon auszugehen, dass nur Mindestanforderungen verlässlich formuliert werden können. Nicht zuletzt aufgrund gegebener Budgetbeschränkungen ist zudem anzunehmen, dass die Informationseigenschaften nur in submaximaler Intensität angestrebt werden können.[1678]

1.2.2 Ermittlung des kundenwertorientierten Informationsbedarfs

Die Ermittlung des für ein effektives Vertriebsinformationssystem detaillierten Informationsbedarfs ist kein triviales Problem, da genau spezifiziert werden muss, welche Informationen zu welchen Zwecken benötigt werden. Zudem gilt: „Der Bedarf an Kundenwertinformationen kann in der Realität [.] a priori nicht präzise ermittelt werden."[1679] Denn es ist davon auszugehen, dass der Informationsbedarf zeitlich variant ist. In Abhängigkeit des Fortschreitens der jeweiligen Vertriebs- und insbesondere Vertriebssteuerungsaufgaben, für welche der Informationsbedarf zu ermitteln ist, wird sich dieser erst hinreichend konkretisie-

---

1676 Vgl. *Eberling, G.* (2002), S. 89.

1677 Vgl. *Horváth, P.* (2003), S. 362 f.

1678 Vgl. *Berthel, J.* (1992), Sp. 874.

1679 *Eberling, G.* (2002), S. 95.

ren.[1680] Aus diesem Grund wird an dieser Stelle keine abschließende Definition des kundenwertorientierten Informationsbedarfs vorgenommen, sondern lediglich eine grobe Struktur vorgestellt.

Grob strukturiert sind drei Schwerpunkte des kundenwertorientierten Informationsbedarfs zu nennen. Erstens die kundenbezogenen, vertrieblich relevanten Rohdaten, die zur Kundenbewertung im Vertriebscontrolling genutzt werden (z. B. Kundenstammdaten, Kundenplanungen, Produktnutzungsdaten).[1681] Ein zweiter Schwerpunkt sind der Kundenwert und die Kundenwerttreiber. Den relevanten Bezugsrahmen stellen die Metriken für den monetären und nichtmonetären Kundenteilwert dar. Die zu den Kundenteilwerten aggregierten und verarbeiteten Rohdaten sind die informatorische Basis der kundenwertorientierten Vertriebssteuerung. Der dritte Schwerpunkt liegt in den aus dem Kundenwert abgeleiteten präskriptiven und normativen Aussagen im Kontext der Vertriebssteuerung. Diese dreiteilige Grobstruktur ist als Grundlage einer systematischen und detaillierten Ermittlung des kundenwertbezogenen Informationsbedarfs zugrunde zu legen. Der konkrete Informationsbedarf ist institutsindividuell und unter Berücksichtigung nachfolgend erläuterter Determinanten zu ermitteln.

Die relevanten Determinanten des Informationsbedarfs sind aufgaben- und subjektbezogen oder ergeben sich aus betrieblichen Rahmenbedingungen.[1682] Aufgabenbezogene Determinanten sind der Aufgabeninhalt und die -ziele, die Handlungsrhythmik und die -ebene sowie die verwendeten Methoden. Der Aufgabeninhalt bestimmt in wesentlicher Weise den kundenwertorientierten Informationsbedarf. Für Planungs- und Steuerungsaufgaben sind bspw. prospektive und für Kontrollaufgaben auch retrospektive Kundenwertinformationen relevant.[1683] Weiterhin hat das Zielsystem einer Aufgabe Einfluss auf den Informationsbedarf. Werden alle Kundenwerttreiber aktiv in Zielbildungsprozessen berücksichtigt, induziert dies einen umfassenderen Informationsbedarf als im Fall von selektiven Zielbildungsprozessen (z. B. nur das Cross-Selling-Potenzial).

[1680] Vgl. *Horváth, P.* (2003), S. 364.

[1681] Vgl. dazu die Ausführungen in Kapitel D.4 und Kapitel D.5.

[1682] Vgl. hier und nachfolgend *Eberling, G.* (2002), S. 91 ff. sowie die dort angegebene Literatur.

[1683] Grundsätzlich können auch für Kontrollaufgaben zum Bewertungsstichtag prospektive Kundenwertinformationen relevant sein. So z. B. in denjenigen Fällen, in denen Maßnahmen zur Entwicklung des prospektiven Kundenwerts initiiert werden. Als Beispiel sei die Steigerung des Loyalitätsgrads des Kunden genannt. Eine Steigerung dieser Größe kann c. p. bei zwei aufeinander folgenden Bewertungszeitpunkten zu steigenden prospektiven Kundenwerten führen.

Abhängig ist der Informationsbedarf auch davon, ob die Kundenwerte in die operative oder die strategische Handlungsebene eingebunden werden sollen.[1684] Während operative Informationen hinreichend exakt und konkretisiert sein müssen, sind strategische Informationen vergleichsweise unbestimmt. Die Handlungsrhythmik korrespondiert mit den Informationseigenschaften der Häufigkeit und Termindringlichkeit und ist eng verbunden mit der Handlungsebene. Werden Aufgaben häufiger durchgeführt, sinkt der Netto-Informationsbedarf (Differenz zwischen Wissensbestand und Informationsbedarf) auf Seiten des Aufgabenträgers. Liegen termindringliche Aufgaben vor, wird der Informationsbedarf mangels zeitlicher Kapazitäten reduziert. Die zur Aufgabenerfüllung verwendeten Methoden führen zu konkreten Anforderungen an den Informationsbedarf. So benötigt etwa die Firmenkundenbewertung anhand eines Scoringmodells andere Informationen als die Berechnung eines kapitalwertbasierten Kundenwerts. In diesem Kontext sei darauf hingewiesen, dass hier die mehrfache Nutzung von Informationen in der Geschäftsbeziehung zum Firmenkunden evident wird.[1685] Neben den Aufgaben des Vertriebs- und der Vertriebssteuerung können in einer erweiterten Betrachtung auch Aufgaben des Risikomanagements betrachtet werden. Denn z. B. für die Nutzung aktiver Instrumente der Portfoliosteuerung der Kreditrisiken, die insbesondere bei Bewertung des beziehungsübergreifenden Kundenwerttreibers der Risikodimension relevant werden können, ist es zwingend notwendig, die für den Sekundärmarkt erforderliche Transparenz über den Kreditnehmer herzustellen. Mangels externen Ratings bei mittelständischen Unternehmen, erfordert dies zwingend den Einsatz der internen Ratingverfahren,[1686] die ebenfalls auf externen Kundendaten aufbauen. Dieser Sachverhalt ist vor allem im Kontext der Informationsbeschaffung relevant, um diese effektiv und effizient auszugestalten.

Hinsichtlich der subjektbezogenen Determinanten des Informationsbedarfs lassen sich der Wissensstand, das Risikoverhalten und das Informationsverhalten des Aufgabenträgers als relevante Einflussfaktoren unterscheiden. Der Wissensstand des Aufgabenträgers bestimmt den (Netto-)Informationsbedarf in wesentlicher Weise. Studien zeigen, dass Experten weniger Informationen zur Lösung eines Problems benötigen als Laien.[1687] So kann ein erfahrener Firmenkundenberater das Wertpotenzial eines Kunden und die Möglichkeiten dieses auszuschöpfen wahrscheinlich besser und mit weniger zusätzlichen Informationen

---

1684 Vgl. hier und nachfolgend *Eberling, G.* (2002), S. 91 ff. sowie die dort angegebene Literatur.

1685 Vgl. dazu die Ausführungen in Kapitel C.2.4.1.3.

1686 Vgl. *Ridder, T.* (2002), S. 22.

1687 Vgl. *Gemünden, H. G.* (1993), S. 1733 f.

einschätzen als ein junger Firmenkundenberater oder vertriebsferner Mitarbeiter der Kreditgenossenschaft. Das Informationsverhalten äußert sich in der Aufnahmefähigkeit und den informationskommunikativen Präferenzen (z. B. bildhafte, textliche oder verbale Informationen) eines Informationsadressaten. Dies muss bei der Informationsbedarfsermittlung ebenfalls berücksichtigt werden. Das Risikoverhalten eines Aufgabenträgers beeinflusst den Informationsbedarf derart, dass risikoaverse Aufgabenträger tendenziell mehr Informationen benötigen als risikofreudige Aufgabenträger.

Die Determinanten, die den betrieblichen Rahmenbedingungen zugeordnet werden, sind die betriebsinternen und -externen Ressourcen sowie etwaige Restriktionen. Insgesamt ermöglichen und beschränken die betrieblichen Rahmenbedingungen die kundenwertorientierten betrieblichen Aktivitäten und beeinflussen dadurch den Informationsbedarf. In Teil C dieser Arbeit wurden die kreditgenossenschaftsspezifischen Kontextfaktoren beschrieben, die in diesem Zusammenhang zu berücksichtigen sind. Zu den betriebsinternen Ressourcen gehören z. B. IT-Systeme (Data-Warehouses, CRM-Systeme) oder Organisationsstrukturen wie Vertriebskonzeptionen, die eine differenzierte und individualisierte Bearbeitung von Kunden ermöglichen und bezwecken und somit den Informationsbedarf an kundenwertorientierten Informationen begründen. Betriebsexterne Ressourcen wie bspw. die weiteren Unternehmen der FinanzGruppe erweitern den Pool an beschaffbaren Kundeninformationen, die in die Bewertung einfließen können, und bedingen zugleich die Informationsbereitstellung (z. B. für die Verbundpartner relevante Komponenten des Basisgeschäfts-, Entwicklungs- oder Cross-Sellings-Potenzials). Schließlich sind auch Restriktionen zu berücksichtigen, die den kundenwertorientierten Informationsbedarf limitieren und ökonomischer (z. B. Budgetbeschränkungen), juristischer (z. B. Datenschutzgesetze)[1688], technischer (z. B. inkompatible Informationssysteme), politischer (z. B. Einflussnahme von Mitgliedern) oder sozialer Art (z. B. Fairness-Prinzipien i. V. m. dem Förderauftrag) sein können.

1.2.3 Quellen kundenwertorientierter Informationen

Ist der relevante Bedarf an Kundenwertinformationen definiert, besteht die zentrale Aufgabe darin, die Informationsbeschaffung zu strukturieren. Ziel der Informationsbeschaffung ist es, effektiv und effizient, die für die Kundenbewertung und kundenwertorientierte Steuerung relevanten Daten aus internen und externen Quellen zu beschaffen.[1689] Ausgangspunkt der Informationsbeschaffung ist die Frage nach dem aktuellen Informationsstand im Unternehmen, an

---

1688 Vgl. dazu ausführlich *Wölfel, J./Denzler, M.* (2008), S. 1 ff.

1689 Vgl. *Stüker, D.* (2008), S. 300.

den sich die zweite Frage nach den Alternativen, die fehlenden Informationen zu beschaffen, anschließt. Diese betrifft sowohl die Quellen der Informationen als auch die Träger der Informationsbeschaffung.[1690] Alternative Informationsquellen lassen sich danach unterscheiden, ob sie institutsintern oder institutsextern sind.[1691] Mit Bezug zum hergeleiteten Kundenwertmodell listet Tabelle 26 exemplarisch und nicht abschließend einige Quellen auf.[1692]

Tabelle 26: Überblick über ausgewählte Informationsquellen[1693]

Interne Quellen	• Interne Datenbanken mit Kundenstammdaten (Kernbankensoftware) • Kosten-Leistungs-Rechnung (Prozesskostenrechnung) • Kontodaten / Produktdaten • Zahlungsverkehrsdaten • Befragung von Mitarbeitern der Kreditgenossenschaft aus Markt und Marktfolge • Daten des gesamten Kundenbestands (Vergleichsgruppenanalysen)
Externe Quellen	• Jahresabschlussdaten (e-bundesanzeiger) • Planungsrechnungen • Vermögensübersichten • Branchenstudien des BVR • Kundenbezogene Daten der Verbundpartner • Firmendatenbanken (HOPPENSTEDT, DAFNE, MARKUS) • Befragung und Selbstauskünfte der Firmenkunden • Kundenbezogene Beobachtungen

Probleme können bei der Informationsbeschaffung auftreten, wenn die aussichtsreichen Beschaffungsquellen nicht bekannt sind, die Beschaffung einer Information nicht möglich ist oder die Beschaffung bestimmter Informationen von den Beteiligten nicht gewollt wird.[1694] Traditionell stehen aus vertriebspolitischer Sicht[1695] vor allem interne Quellen im Fokus. So finden WITTMANN ET AL. in einer empirischen Studie heraus, dass gegenwärtig vor allem interne und

1690 Vgl. *Berthel, J.* (1975), S. 59.

1691 Vgl. *Walter, G.* (1996), S. 47 ff.; *Dauser, A.* (1995), S. 140.

1692 Vgl. für eine alternative Übersicht *Segbers, K.* (2007), S. 92 und S. 95. Der Schwerpunkt dieser Übersichten liegt jedoch in der Finanzkommunikation zum Abbau der Informationsasymmetrien, die in Hausbankbeziehungen mit Schwerpunkt im risikotragenden Kreditgeschäft vorliegen.

1693 Eigene Darstellung.

1694 Vgl. *Horváth, P.* (2003), S. 353.

1695 Diese Aussage gilt nicht aus risikopolitischer Sicht. Vgl. dazu *Segbers, K.* (2007), S. 92 und S. 95.

produktbezogene Daten im Vertrieb erfasst und gespeichert werden.[1696] Gleichzeitig moniert BERGHAUS für die Kreditgenossenschaften, dass die Kenntnis über die tatsächliche Produktnutzung zu gering sei, um systematisch Kundenpotenziale auszuschöpfen.[1697] Dazu sind kundenseitige Informationen, z. B. in Form von Aussagen über kontrahierte Finanzdienstleistungen bei anderen Anbietern, zur Vervollständigung des vertrieblich relevanten Informationsstands notwendig, zu berücksichtigen und zu beschaffen.[1698]

Für die Kundenbewertung ist der Kunde als Informationsquelle von herausragender Bedeutung.[1699] Dies begründet sich mit den Ausführungen zum Kundenwertmodell und dessen Ganzheitlichkeit und Zukunftsorientierung. Neben der Notwendigkeit den gesamten latenten und manifesten Finanzdienstleistungsbedarf eines Firmenkunden zu erfassen, besteht die Notwendigkeit, die Entwicklung des bankleistungsbezogenen Bedarfspotenzials einzuschätzen. Es wurde bereits darauf hingewiesen, dass das künftige Bedarfspotenzial eines Firmenkunden wesentlich von dessen Unternehmensentwicklung abhängt.[1700] Diese hängt von den Umfeldentwicklungen sowie der Antizipations- und Adaptionsfähigkeit des Unternehmens mithin der Managementkompetenz im Unternehmen ab. Das Ergebnis aus erwarteter Umfeldentwicklung und Managementkompetenz mündet in der Strategie und Planung des Firmenkunden. Bei kapitalmarktorientierten Unternehmen stellen Informationen zur Strategie und Planung in Teilen öffentliche Informationen dar, die aufgrund von Informationsprozessen des Unternehmens allgemein erhältlich. Bei allen anderen Unternehmen sind diese Informationen proprietär, also private Informationen, die nicht öffentlich verfügbar sind.[1701] Die Majorität der Firmenkunden der Kreditgenossenschaften ist aufgrund der Größe, Rechtsform und den rechtlichen Vorgaben nur in einem vergleichsweise geringem Maße dazu verpflichtet, umfangreiche Informationen

---

1696 Vgl. *Wittmann, G. et al.* (2007), S. 32 und S. 46. In dieser Studie gaben 63 Prozent der Befragten an, (produkt-) nutzungsbezogene Daten zu erheben. Und 44 Prozent gaben an, dass die vorhandenen Kundendaten umfassend, aktuell und verlässlich sind.

1697 Vgl. *Berghaus, U.* (2009), S. 36.; ähnlich auch *Berghaus, U.* (2008), S. 40 ff.

1698 Dies wird auch von HOEKSTRA/HUIZINGH gefordert. Vgl. *Hoekstra, J. C./Huizingh, E.* (1999), S. 264.

1699 Vgl. *Bauer, R. A.* (2000), S. 41.

1700 Vgl. dazu die Ausführungen in D.4.5.3.1.1. Während dies für das Unternehmen als Nukleus des betrachteten Kundenverbunds unstrittig ist, kann dies aufgrund der „Schicksalsgemeinschaft“ zwischen Unternehmen und Unternehmer sowie dessen Familie nur angenommen werden. Diese Annahme soll für diese Argumentation zugrunde gelegt werden. Vgl. zum Begriff der Schicksalsgemeinschaft von Unternehmen und Unternehmer *Hamer, E.* (1987), S. 62 f.

1701 Vgl. *Segbers, K.* (2007), S. 116.

öffentlich zu machen.[1702] Daher müssen die relevanten Informationen durch die Kreditgenossenschaft direkt beim Kunden beschafft werden. Zwar ist der Vertrieb nicht der alleinige Träger der Aufgaben im Sinne der Informationsbeschaffung, jedoch kommt ihm dabei als Schnittstelle zum Kunden eine zentrale Bedeutung, wenn nicht sogar die wichtigste Rolle zu.[1703] Die Forderung WALTERS, der es für Banken als besonders wichtig erachtet, den eigenen Datenbestand zu qualifizieren, d. h. ihn mit weiteren, kundenwertrelevanten Merkmalen zu versehen, stellt daher vor allem eine Anforderung an den institutionellen Vertrieb der Kreditgenossenschaften dar.[1704]

Für die Informationsbeschaffung ist es bei bankexternen und subjektiven Informationen notwendig, die erhaltenen Informationen im Rahmen der bankinternen Informationsprozesse zu evaluieren und zu plausibilisieren.[1705] Daraus resultiert z. B. die Frage, in wie weit die vom Kunden übermittelten Informationen zur Unternehmensplanung aus Sicht der Kreditgenossenschaft als realistisch eingeschätzt werden. Wird als Firmenkunde ein Verbund aus betrieblichen und privaten Kunden betrachtet, so ist neben der zusätzlichen isolierten Betrachtung der privaten Kundenebene auch die Transmission von kundenwertrelevanten Einflussgrößen zwischen der betrieblichen und der privaten Ebene zu berücksichtigen.[1706]

Diesen Überlegungen tragen die Kreditgenossenschaften bereits grundsätzlich Rechnung, da sie mittels zukunftsorientierter Dialoge versuchen, betriebli-

---

1702 Gem. § 325 HGB sind Kapitalgesellschaften und Personenhandelsgesellschaften ohne natürliche Person als persönlich haftendem Gesellschafter dazu verpflichtet ihren Jahresabschluss im elektronischen Bundesanzeiger zu veröffentlichen. Für kleine und mittelgroße Kapitalgesellschaften gem. § 267 HGB gibt es nach §§ 326, 327 HGB reduzierte Offenlegungspflichten. Darüber hinaus sind nach § 241a HGB Einzelkaufleute, die in zwei aufeinanderfolgenden Jahren einen Umsatz von weniger als TEUR 500 haben von den Buchführungspflichten aus §§ 238-241 HGB befreit. Vor dem Hintergrund der in Kapitel C.2.2.3 dargelegten empirischen Kundenstruktur der Kreditgenossenschaften ist begründet davon auszugehen, dass die Möglichkeiten der externen Informationsbeschaffung begrenzt sind.

1703 Vgl. *Bauer, R. A.* (2000), S. 41; *Segbers, K.* (2007), S. 95 f., der den Blick in diesem Zusammenhang auf den Firmenkundenberater als Ausdruck des institutionellen Vertriebs richtet, dessen zentrale Aufgabe in der Plausibilisierung der erfassten Informationen liegt.

1704 Vgl. *Walter, G.* (1996), S. 49.

1705 Vgl. *Segbers, K.* (2007), S. 93.

1706 So sind z. B. positive Cash-flows, die im Unternehmen bleiben, in der Kundenwertermittlung anders zu handhaben als die an die Unternehmenseigner ausgeschütteten Jahresüberschüsse.

che und private bedarfsrelevante Informationen von den Kunden zu erhalten.[1707] Eine institutionalisierte Plausibilisierung der erhaltenen Informationen liegt jedoch, anders als beim risikopolitisch motivierten Informationsbedarf im Kontext der internen Ratingverfahren, nicht vor. Es ist daher zu empfehlen, ein solches, institutionalisiertes Verfahren einzurichten und anzuwenden. Dabei sollte dringend auf den Datenfundus und die Erfahrung Bezug genommen werden, die in der genossenschaftlichen FinanzGruppe an unterschiedlichen Stellen vorliegen. Im gegenwärtigen Vorgehen der Kreditgenossenschaften liegt zudem die Ursache eines möglichen Zirkularitätsproblems, denn die Auswahl der mittels eines zukunftsorientierten Dialogs wie dem VR-FINANZPLAN MITTELSTAND und somit einer ressourcenintensiven Maßnahme zu betreuenden Kunden erfolgt auf Basis von angenommenen Kundenwerten und gleichzeitig zum Zweck der Ermittlung von Kundenwerten. Dieses Problem könnte vermieden werden, indem institutsinterne, vergangenheitsorientierte Inputdaten verwendet werden, um den Kundenwert zu ermitteln. Wie jedoch später noch begründet wird, ist dies im Sinne eines prospektiven Kundenwerts problematisch. Auf das Zirkularitätsproblem wird in Kapitel E.1.2.4 weiter eingegangen.

Eine wichtige Quelle kundenwertrelevanter Informationen sind auch die intern vorliegenden Zahlungsverkehrsdaten, die als Kuppelprodukt der Zahlungsverkehrsabwicklung für den Firmenkunden anfallen. Mittels geeigneter Analysen können hier wichtige Informationen zum Kunden und dessen aktuelle und künftige Bedarfe erhoben werden. Diese Informationen können z. B. in die Ermittlung des Cross-Selling-Potenzials fließen. Zu nennen sind hier bspw. Überweisungen an Drittbanken oder verbundfremde Versicherungs- oder Leasinggesellschaften.[1708]

Die Ausführungen zur Struktur und zum Umfang der FinanzGruppe[1709] machen deutlich, dass aus dem Verbund heraus umfangreiche Möglichkeiten vorliegen, fehlende Kundeninformationen z. B. für die Bewertung von Nichtkunden, zu generieren. Als weitere wichtige Quellen sind daher bank- und verbundinterne Kundenbestandsdaten, die für Längs- und Querschnittsanalysen genutzt werden können, zu nennen.[1710]

1707 Vgl. dazu die Ausführungen in Kapitel C.2.4.1.3 und Kapitel C.2.4.3 zum Thema VR-FINANZPLAN MITTELSTAND.

1708 Beispiele für weitere Potenzialanalysen auf Basis der bankinternen Produktnutzungsdaten finden sich bei *Scheer, O.* (1989), S. 111 ff. und S. 210 ff. Zwar beziehen sich die Beispiele auf das Privatkundengeschäft, sie lassen sich aber mit geringem Aufwand auch auf das Firmenkundengeschäft anpassen.

1709 Vgl. dazu die Ausführungen in Kapitel C.1.2 und Kapitel C.1.4.4.

1710 Vgl. zu den damit verbundenen Möglichkeiten zur kundenbezogenen Analyse *Schulz, B,* (1995), S. 71 ff.

Insgesamt ist davon auszugehen, dass unterschiedliche Informationsquellen den Kenntnisstand der Kreditgenossenschaft zu steigern vermögen. Besondere Bedeutung hat jedoch der Kunde selbst als Quelle zukunftsorientierter kundenwertrelevanter Informationen. In diesem Zusammenhang sei auf das Informationspotenzial des Kunden verwiesen, das direkt mit dieser Aufgabe korrespondiert. Je höher die Bereitschaft eines Kunden ist, Informationen an die Kreditgenossenschaft zu liefern, desto besser ist die Informationslage zu diesem Kunden.

1.2.4 Methoden der Informationsbeschaffung

In der Literatur findet sich eine Vielzahl von Methoden, die der Informationsbeschaffung dienen. Teilweise werden diese Verfahren nicht der Informationsbeschaffung, sondern der Informationsaufbereitung zugeordnet.[1711] Dies begründet sich mit der Entscheidung, an welcher Stelle des Informationsprozesses von Informationen im Sinne der aufgabenbezogenen Definition nach WITTMANN gesprochen wird. Wird in der Informationsbeschaffung nur die Sammlung von Daten verstanden,[1712] sind alle datenverarbeitenden Aktivitäten sowie Prognosen der Informationsaufbereitung zuzuordnen. Werden jedoch auch Prognoseergebnisse als deskriptive Informationen verstanden, so können die Prognoseaktivitäten, d. h. die Verwendung von Prognoseinstrumenten selbst der Informationsbeschaffung zugeordnet werden. Dieses breitere Verständnis wird hier zugrunde gelegt. Folglich sind unter Methoden der Informationsbeschaffung alle Verfahren zu subsumieren, die dazu dienen, kundenwertorientierte Informationen extern und/oder intern zu sammeln und ggf. auf Basis bestehender Daten und/oder Informationen neue zu erzeugen. Dieses breite Verständnis umfasst demnach ebenso die Instrumente, die eingesetzt werden, um Informationen vom Kunden zu erfassen,[1713] sowie Instrumente, um diese Informationen zu verarbeiten und in neue Informationen zu transformieren. Im Fokus der weiteren Ausführungen stehen Methoden, die sich auf den Informationsbedarf des konkreten Kundenwerts und dessen Messansätze beziehen. Der darüber hinausgehende Informationsbedarf aus Sicht der kundenwertorientierten Vertriebssteuerung wird im weiteren Verlauf der Arbeit situativ aufgegriffen.

Vor dem Hintergrund des hergeleiteten Kundenwertmodells und dessen Messansätzen für den monetären und nicht-monetären Kundenteilwert lassen sich eine Vielzahl von Instrumenten finden, Kundenwertinformationen zu be-

---

1711 So z. B. *Stüker, D.* (2008), S. 303 ff.

1712 Daten können in diesem Zusammenhang im Gegensatz zu Informationen als zweckneutrales Wissen definiert werden.

1713 Dazu gehören etwa der Fragenkatalog des Ratings oder die strukturierten Gesprächsleitfäden des VR-FINANZPLAN MITTELSTAND. Vgl. *BVR et al.* (2006).

schaffen.[1714] Ohne auf den in der kundenwertorientierten Literatur hinreichend diskutierten und gewürdigten Katalog en detail einzugehen,[1715] werden besondere Methoden und Aspekte besprochen. Zur Strukturierung bietet es sich an, zwischen dem nicht-monetären und monetären Kundenteilwert zu unterscheiden, sowie bei letzterem eine prozessuale Betrachtung vorzunehmen.

Für den nicht-monetären Kundenteilwert, der mittels eines Punktbewertungsverfahrens gemessen wird, sind die Merkmalsausprägungen der relevanten Bewertungskriterien kundenindividuell zu erheben (z. B. Häufigkeit, Qualität und Inhalt von Informationen). In Abhängigkeit der Art und Messkala der Bewertungskriterien kommen unterschiedliche Verfahren zur Anwendung. Bei einigen der Verfahren ist dazu die Interaktion mit dem Kunden notwendig (z. B. Fragen nach der Größe des sozialen Netzes bei der Beurteilung des Referenzpotenzials) bei anderen Kriterien wiederum kann auf interne Daten zurückgegriffen werden (z. B. dokumentierte Kundenanregungen im Sinne des Informationspotenzials). Ein ausschließlich auf leicht beobachtbaren Merkmalen des Firmenkunden ausgerichtetes Punktbewertungsmodell der so gemessenen Kundenwerttreiber reduziert den Bewertungsaufwand. Weiterhin liegen für einige Kriterien objektive Daten vor, während für andere Bewertungskriterien eine subjektive Einschätzung vorgenommen werden muss. Kritisch ist die Informationsbeschaffung, wenn sie nicht auf objektiver Beobachtbarkeit resultiert, sondern Meinungen und Urteile von Personen zugrunde gelegt werden. In der Auswahl der Methoden der Informationsbeschaffung ist es daher wichtig, Subjektivität weitgehend zu vermeiden. Dazu erscheint es sinnvoll, den Begriff der Subjektivität näher zu betrachten. Subjektivität zeichnet sich durch einen situativen Subjektbezug aus. Das bedeutet, dass die Aussage über einen zu beschreibenden oder zu bewertenden Gegenstand durch die Persönlichkeitsstruktur des Bewertenden geprägt ist. Daraus folgt, dass zwei Personen in äußerlich gleichen Entscheidungs- oder Bewertungssituationen zu unterschiedlichen Ergebnissen kommen können. Das damit verbundene zentrale Problem liegt in der mangelnden Nachvollziehbarkeit der Ergebnisfindung durch Dritte. Eine Verbindung zum Begriff des impliziten Wissens wird an dieser Stelle evident. Im Gegensatz dazu ist Objektivität dadurch gekennzeichnet, dass statt des Subjektbezugs ein ausschließlicher Objektbezug vorliegt und folglich zwei Personen zu gleichen Ergebnissen kommen. Werden die Begriffe Objektivität und Subjektivität als Extrema eines Kontinuums aufgefasst, lässt sich als Punkt zwischen den Extrema die intersub-

---

[1714] Für eine allgemeine Übersicht von Verfahren und Instrumenten, die leicht auf die Kundenwertorientierung übertragen werden können vgl. *Horváth, P.* (2003), S. 380 ff.

[1715] Vgl. dazu ausführlich z. B. *Schulz, B.* (1995), S. 71 ff. und *Eberling, G.* (2002), S. 110 ff.

jektive Nachprüfbarkeit finden. Ein Ergebnis ist intersubjektiv nachprüfbar, wenn zwei Personen unabhängig voneinander basierend auf für die Ergebnisfindung geltenden Vorgaben, Inputs und Methoden zu einem gleichen oder sehr ähnlichen Ergebnis kommen, auch wenn keine objektive Bestimmung möglich ist. Methoden der Informationsbeschaffung für die Messung des nicht-monetären Kundenteilwerts sollen insbesondere dieser Anforderung genügen.

Für den monetären Kundenteilwert lässt sich in allgemeiner Struktur ein mehrstufiges Verfahrensmodell definieren, das als Grundlage der Methodenauswahl genutzt werden kann. Im ersten Schritt ist festzustellen, welche bankleistungsbezogenen Bedarfe auf Seiten des Kunden zu welchen Zeitpunkten vorliegen und welche Volumina (z. B. Kreditvolumen, Zahlungsverkehrsvolumen, Anzahl Firmenkreditkarten) damit verbunden sind. Im zweiten Schritt ist das erkannte Bedarfspotenzial zu bewerten.[1716]

Wie oben erwähnt, ist die originäre Quelle der dafür benötigten Informationen der Kunde selbst. Es wurde bereits im Rahmen der Herleitung der Kundenwertreiber und der relevanten Informationsquellen auf den Umstand hingewiesen, dass die Nachfrage eines Firmenkunden nach Bankleistungen als derivative Nachfrage verstanden werden kann, die von der Unternehmensentwicklung abhängt. Demnach gilt es zunächst über geeignete Kommunikationsprozesse festzustellen, welche Unternehmensentwicklung die Geschäftsleitung des Unternehmens anstrebt. Die Kommunikationsprozesse müssen dabei nicht zwingend persönlicher Art sein, sondern können vor dem Hintergrund der Ausführungen zum Kooperationspotenzial auch auf Basis technologischer Verfahren erfolgen. Die generierten Informationen, die im Wesentlichen Plandaten und strategische Überlegungen des Unternehmens sind, sind mittels geeigneter Verfahren in bankleistungsbezogene Bedarfe zu übersetzen sowie die damit verbundenen Bedarfspotenziale zu schätzen.

Aufgrund der Metrik für den monetären Kundenteilwert ist festzustellen, dass bei der Informationsbeschaffung Instrumente in den Vordergrund rücken, die die Ausprägungen der Kundenwertreiber für die Zukunft dynamisch prognostizieren. Es gibt zwei grundsätzliche Vorgehensweisen, die dazu angewendet werden können. Erstens der Rückgriff auf Daten der Vergangenheit und deren Extrapolation in die Zukunft und zweitens die Prognose künftiger Entwicklungen auf Basis von Indikatoren.[1717] Für die Verwendung von Prognosen, die auf Vergangenheitsdaten basieren, spricht das zentrale Problem der Prognose, das in der richtigen Schätzung der Dauer und des Verlaufes der Kundenbeziehung,

---

1716 Vgl. zur Bewertung und möglicher Bewertungsmaßstäbe auch die Ausführungen in Kapitel D.5.2.2.1.

1717 Vgl. *Tewes, M.* (2003), S. 270.

mithin also der gemessenen Ausprägungen der Kundenwerttreiber in der Zukunft liegt.[1718] Jedoch sind sie nicht in der Lage, Diskontinuitäten hinreichend zu verarbeiten.[1719] Wie beispielhafte Anwendungsfälle zeigen, führen schon kleine Variationen der Daten zu massiven Änderungen der Kundenbewertung mit bemerkenswerten Konsequenzen.[1720] MALTHOUSE/BLATTBERG stellen in ihrer Untersuchung zwei Regelmäßigkeiten vor, die dabei bedeutend sind. Erstens die 20/55-Regel, wonach 55 Prozent der tatsächlich 20 wertvollsten Kunden nicht als solche erkannt werden. Und zweitens die 80/15-Regel, die besagt, dass 15 Prozent der tatsächlich durchschnittlich wertvollen Kunden nicht als solche identifiziert werden. Werden auf Basis der fehlerhaften Kundenwerte den Kundenbeziehungen Ressourcen zugeordnet, resultieren folglich erhebliche Ineffizienzen.[1721] BEESER führt für den Bankenkontext aus, dass bei der Bewertung von Bankkunden keine reine Fortschreibung der Vergangenheit erfolgen dürfe, da die Dynamik der Kundenbeziehung im Zeitablauf vom Lebenszyklus abhängt.[1722] Nach ihrer Ansicht bedarf es vielmehr einer differenzierten, auf ökonomischen Überlegungen beruhenden Prognose zukünftiger Kundenwerttreiber.[1723] Zwar bezieht sich diese Aussage auf das Privatkundengeschäft, jedoch dürfte der Lebenszyklus eines Privatkunden vergleichsweise einfacher zu schätzen sein als die Dynamik der Unternehmensentwicklung.[1724] Für den Business-to-Business-Bereich, dem das Firmenkundengeschäft der Banken zuzuordnen ist, bieten sich nach STÜKER als Basis Entwicklungspotenzialanalysen an, die Informationen über die Stellung des Firmenkunden auf seinen Märkten relativ zum Wettbewerb sowie zu den Zukunftschancen dieser Märkte liefern können.[1725] „Calculations should [.] also be based on data concerning future situations, such as purchase intentions or sales potential."[1726]

In Abhängigkeit der berücksichtigten Einflussfaktoren auf die zu prognostizierende Größe wird zwischen Entwicklungs- und Wirkungsprognosen unter-

---

1718 Vgl. *Wimmer, K.* (2009), S. 210; *Mödritscher, G. J.* (2008), S. 205; *Henseler, J./ Hoffmann, T.* (2003), S. 39; *Burmann, C.* (2003), S. 115.

1719 Vgl. *Stüker, D.* (2008), S. 305.

1720 Vgl. *Mödritscher, G. J.* (2008), S. 206 sowie die dort genannte Literatur.

1721 Vgl. *Malthouse, E. C./Blattberg, R. C.* (2005), S. 11.

1722 Vgl. *Beeser, A. J.* (2003), S. 14.

1723 Vgl. *Beeser, A. J.* (2003), S. 47.

1724 So lassen sich für einen Privatkunden Eckpunkte des Lebenszyklus wie z. B. Ausbildung, Berufseinstieg, Heirat, Renteneintritt vergleichweise einfacher definieren und abschätzen als dies bei der Unternehmensentwicklung der Fall ist.

1725 Vgl. *Stüker, D.* (2008), S. 170.

1726 *Hoekstra, J. C./Huizingh, E.* (1999), S. 262.

schieden.[1727] Entwicklungsprognosen zeigen die zu prognostizierenden Größen in Abhängigkeit von Variablen, die das Unternehmen nicht beeinflussen kann. Bezugsgröße dieses Vorgehens sind z. B. das Bedarfspotenzial eines Firmenkunden für das Basisgeschäfts-, das Entwicklungs- und das Cross-Selling-Potenzial (bzw. das Gesamtgeschäftspotenzial bei Nichtkunden), dessen Preiselastizität sowie dessen Prädispositionen hinsichtlich des Basiskosten- und des Kooperationspotenzials. Bei den Wirkungsprognosen wird die abhängige Größe in Abhängigkeit von Variablen, die das Unternehmen direkt beeinflussen kann, gezeigt. Bezugsgröße dieser Prognosen ist z. B. das bewertete Bedarfsvolumen im Sinne des Kundenwertmodells oder eine Einflussgröße des Referenzpotenzials, die durch institutsseitige Maßnahmen beeinflusst wird. Zu den gestaltbaren Variablen gehören unter anderem die vertriebspolitischen Maßnahmen, die in der Vertriebsplanung festgelegt werden. In diesem Punkt zeigt sich eine enge Verzahnung des kundenwertorientierten Informations- und des kundenwertorientierten Planungssystems, die auf das bereits angesprochene Zirkularitätsproblem rekurriert. Dieses begründet sich damit, dass Maßnahmen, die den Einsatz von Ressourcen bedingen und den Wert einer Kundenbeziehung beeinflussen, auf Basis des Kundenwerts fundiert werden (siehe Abbildung 60).

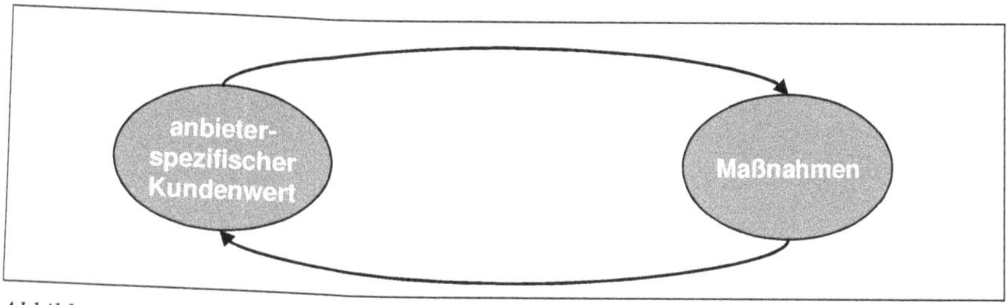

Abbildung 60: Zirkularitätsproblem des Kundenwerts[1728]

Dieses Problem kann auf unterschiedliche Arten gelöst werden. Es ist offensichtlich, dass das Zirkularitätsproblem einen geeigneten Ansatzpunkt für ein mathematisches Lösungsverfahren darstellt. Dazu müsste der anbieterspezifische Kundenwert jedoch mit entsprechenden Kundenreaktionsmodellen verknüpft werden, die die Reagibilität der Kundenwerttreiber auf anbieterseitige Maßnahmen funktional abbilden. Bereits bei der Operationalisierung des Kunden[illegible] in Kapitel D.5.1 wurde darauf hingewiesen, dass für deren Einsatz ent[illegible] Erkenntnisse in den Kreditgenossenschaften noch fehlen. Zudem

[illegible]d nachfolgend *Meffert, H.* (1998), S. 166.

[illegible]lung.

wirkt der deutlich höhere Komplexitätsgrad einer Anwendung entgegen.[1729] Daher ist zunächst zu empfehlen, praktikable Alternativen zu nutzen. Eine Alternative ist, statt des anbieterspezifischen Kundenwerts das Kundenwertpotenzial als Bezugsgröße in der Vertriebssteuerung zu nutzen. In diesem Fall werden jedoch die Charakteristika der Kundenbeziehung nicht beachtet, so dass ein hohes Risiko ineffizienter Entscheidungen vorliegt.[1730] TEWES schlägt als Alternative vor, den anbieterspezifischen Kundenwert auf Basis gegebener Strategien und Planungen zu prognostizieren und dann diesen nach der Planung auf Basis der geänderten Strategie und Planung erneut zu ermitteln.[1731] Zu dieser von ihm benannten direkten Vorgehensweise schlägt er auch eine indirekte Vorgehensweise vor.[1732] Danach wird zunächst das Kundenwertpotenzial ermittelt. Anschließend wird im Rahmen der Planung die Strategie zur Realisierung des anbieterspezifischen Kundenwerts festgelegt. Zentrale Bezugsgröße der Strategie ist die Ausschöpfungsquote des Kundenwertpotenzials, die sich unter anderem in erwarteten Loyalitätsgrad, realisierbarer Preisbereitschaft sowie die Integrationsfähigkeit und -bereitschaft konkretisiert.[1733] Durch Verknüpfung dieser Größen kann das bewertete Bedarfsvolumen und der anbieterspezifische Kundenwert ermittelt werden, der dann Grundlage der Vertriebsplanung ist. Abbildung 61 stellt die relevante Vorgehensweise auf Basis der bisherigen Ausführungen grafisch dar.

[1729] Vgl. *Woratschek, H./Roth, S.* (2006), S. 370.

[1730] Denkbar ist der praktische Fall, einen hinsichtlich des Bedarfspotenzials attraktiven Firmenkunden intensiv zu betreuen. Da die Kreditgenossenschaft aus Sicht des Firmenkunden jedoch nur eine Nebenbankbeziehung darstellt, kann diese das Potenzial des Kunden nicht abschöpfen. Aufgrund des Bezugs auf das Bedarfspotenzial würde dieser Aspekt jedoch nicht berücksichtigt werden.

[1731] Vgl. *Tewes, M.* (2003), S. 269.

[1732] Vgl. *Tewes, M.* (2003), S. 270.

[1733] Der Loyalitätsgrad entspricht dem Anteil des Bedarfs, der auf die Kreditgenossenschaft entfällt. Er ist daher für fast alle beziehungsbezogenen Kundenwerttreiber der Ertragsdimension gleichermaßen direkt (Basisgeschäfts-, Entwicklungs- und Cross-Selling-Potenzial) oder indirekt (Basiskostenpotenzial) relevant. Die Preiselastizität dagegen betrifft nur das Entwicklungspotenzial bzw. dies Facette des Preispotenzials und die Integrationsfähigkeit und -bereitschaft das Kooperationspotenzial.

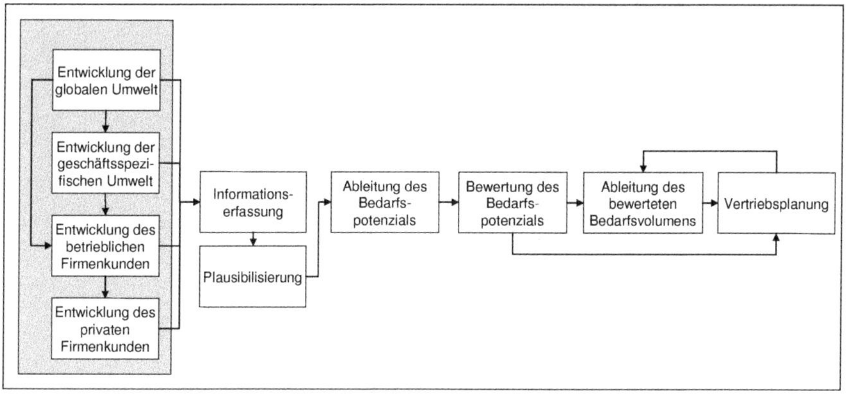

Abbildung 61: Vorgehensweise zur Ermittlung des monetären Kundenteilwerts[1734]

Für die Prognose des Bedarfspotenzials stehen unterschiedliche Methoden zur Verfügung, die sich in quantitative und qualitative Verfahren unterscheiden lassen.[1735] Ohne auf diese Verfahren dezidiert einzugehen, seien beispielhaft einige Verfahren skizziert, die im bankbetrieblichen Kontext eingesetzt werden können. Wichtig sind dabei insbesondere drei Fragestellungen. Erstens die Frage nach dem zeitlichen Anfall des Kundenbedarfs. Zweitens die Frage nach der Höhe des Bedarfspotenzials bzw. des Bedarfsvolumens. Und drittens die Frage nach der konkreten Ausprägung der Bedürfnisbefriedigung mithin also der konkreten Bankleistung.

Um zu erkennen, welche bankleistungsbezogenen Bedarfe auf Seiten des Firmenkunden in dynamischer Sicht künftig vorliegen können, kann z. B. ein idealisierter Unternehmenslebenszyklus eingesetzt werden. Dabei werden den einzelnen Phasen des Unternehmens Bankleistungen zugeordnet. Beispielhaft sei ein auf Finanzierungsprodukte ausgerichteter Ansatz der WGZ BANK vorgestellt (siehe Abbildung 62). Danach wird der Frühphase des Unternehmens ein anderer Produktmix zugeordnet als der Expansionsphase und der Phase der strategischen Weiterentwicklung.

1734 Eigene Darstellung.

1735 Vgl. *Meffert, H.* (1998), S. 166 ff.; *Tewes, M.* (2003), S. 271 f. Einen ausführlichen Überblick über derartige Methoden findet sich bei *Schulz, B.* (1995), S. 71 ff.

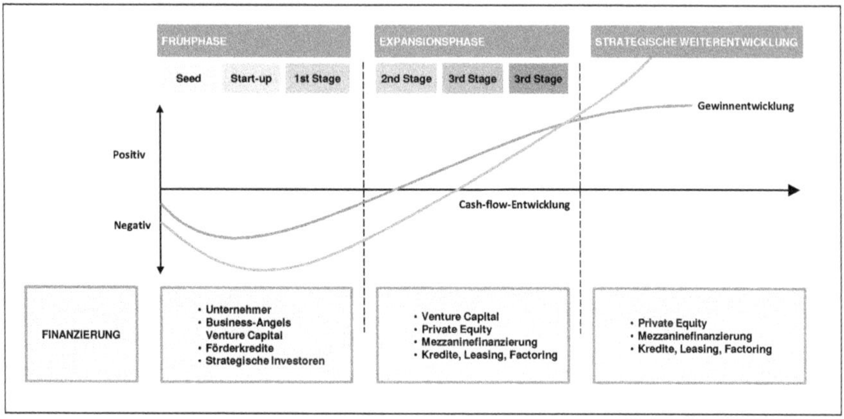

Abbildung 62: Finanzierungsmittel für mittelständische Unternehmen im Lebenszyklus[1736]

Es sei darauf hingewiesen, dass es problematisch ist, zu erkennen, in welcher Phase des Lebenszyklus' sich ein Unternehmen befindet. Aus diesem Grund könnte auch ein Unternehmenslebenszyklus etwa mit den Phasen Existenzgründung, Etablierung, Internationalisierung und Nachfolge genutzt werden. Dabei müssen die einzelnen Phasen nicht zwingend in der genannten Chronologie ablaufen. Wichtig ist, dass Merkmale benannt sind, die es erlauben, einen Firmenkunden frühzeitig einer Phase zuzuordnen und dass den einzelnen Phasen spezifische Produktbedarfe verlässlich zugewiesen werden können. Ein ähnliches Vorgehen lässt sich auf Basis von qualitativen Expertensystemen vornehmen, die aufbauend auf normativen Vorgaben in Verbindung mit konkreten Kundendaten ausgeben, ob für bestimmte Bankleistungen ein Bedarf bei einem Firmenkunden vorliegt. In der Literatur wird zudem vorgeschlagen, Sollproduktnutzungs- oder Musterkundenprofile zu verwenden, um Cross-Selling-Potenziale zu identifizieren.[1737] Diese können auf Basis von Längs- oder Querschnittsanalysen des Kundenbestands ermittelt werden. Ebenfalls zweckmäßig sind in diesem Kontext Analysen der leistungsspezifischen Affinitäten in Form von (bedingten) Wahrscheinlichkeitsmodellen.[1738] Jedoch können hieraus keine unmittelbaren Bedarfspotenziale abgelesen werden, da diese Modelle nur indizieren, dass für

1736 *WGZ Initiativkapital* (2005), S. 11.

1737 Vgl. *Frese, B.* (2009), S. 45; *Duderstadt, S.* (2006), S. 114; *Freiberger, T./Kornatz, C.* (2007), S. 55.

1738 Vgl. *Hagemann, H.* (1986), S. 20. Ergebnis dieser Analysen ist die Aussage, dass Kunden, die ein Produkt A nutzen mit einer bestimmten Wahrscheinlichkeit auch das Produkt B nutzen.

eine bestimmte Bankleistung ein Bedarf vorliegen könnte, jedoch nicht in welchem Umfang oder Volumen. Gleichwohl können derartige Analysen ergänzende Hilfestellungen geben, um die bankleistungsbezogenen Bedarfe zu erkennen, die dann jedoch weiter zu quantifizieren sind.

Mittels des relativ einfachen Modells von GLEIßNER/JAHN kann z. B. das Bedarfpotenzial künftiger Kreditfinanzierungen prognostiziert werden.[1739] Annahmegemäß resultiert das gesamte Kreditfinanzierungspotenzial aus drei Quellen: der Anschlussfinanzierung bestehender Kredite, die im Prognosezeitraum getilgt werden, der ggf. möglichen Steigerung der Finanzierungsquote auf ein branchenübliches Niveau und künftigen fremdfinanzierten Erweiterungsinvestitionen des Kunden. Für die Anwendung benötigt das Modell keine weiteren kundenindividuellen Informationen als die Branchenzugehörigkeit und einen ggf. geschätzten Jahresumsatz U. Unter Rückgriff auf eine branchenübliche Kapitalumschlagshäufigkeit KU lässt sich auf die Bilanzsumme BS schließen, indem der Umsatz durch die Kapitalumschlagshäufigkeit dividiert wird.[1740] Auf Basis der Analyse vorhandener Jahresabschlussdaten der Branche wird der Anteil kq an der Bilanzsumme bestimmt, den ein Unternehmen der betrachteten Branche typischerweise durch Bankkredite BK finanziert. Für den nächsten Schritt wird eine durchschnittliche Kreditlaufzeit L geschätzt, mit der sich der jährliche Refinanzierungsbedarf RF ermitteln lässt. Dabei wird eine konstante Bankkreditquote unterstellt. Zusätzlich zum jährlichen Refinanzierungsbedarf ergibt sich bei wachsenden Unternehmen ein Bedarf an zusätzlichen Bankkrediten, der von der Wachstumsrate w des Unternehmens abhängig ist. Für die Berechnung werden eine konstante Finanzierungsstruktur und ein konstanter Kapitalumschlag angenommen sowie eine rentabilitätsabhängige Gewinnthesaurierung nicht berücksichtigt. Insgesamt errechnet sich auf Basis dieser Annahmen und Daten der gesamte jeweilige (jährliche) Bedarf an Bankkrediten eines betrachteten Kunden gemäß Formel [E.1]:

[1739] Vgl. hier und nachfolgend *Gleißner, W./Jahn, C.* (2005), S. 154 ff. Bei der Modellskizze handelt es sich um einen Ausschnitt eines weitergehenden Modells, das nach Angaben der Verfasser in der IKB DEUTSCHE INDUSTRIEBANK AG eingesetzt wird.

[1740] Da der Umsatz im Gegensatz zur Bilanzsumme kein Bestandteil der Publizitätspflicht kleiner und mittelgroßer Kapitalgesellschaften gem. § 267 HGB ist, kann auch direkt von der aus öffentlich zugänglichen Daten bekannten Bilanzsumme ausgegangen werden. Vgl. Fn. 1702.

$$FB = RF + BK_{zusätzlich} = \frac{BK}{L} + BS \cdot kq \cdot w = \frac{BS \cdot k}{L} + BS \cdot kq \cdot w \qquad [E.1]$$

$$= \frac{\frac{U}{KU} \cdot kq}{L} + \frac{U}{KU} \cdot kq \cdot w = \frac{U}{KU} \cdot kq \cdot \left(w + \frac{1}{L} \right)$$

mit:

FB:	jährlicher Finanzierungsbedarf
RF:	refinanzierungsbedingter Kreditbedarf
$BK_{zusätzlich}$:	zusätzlicher Bedarf an Bankkrediten
BK:	branchenüblicher absoluter Anteil an Bankkrediten
L:	durchschnittliche Kreditlaufzeit
BS:	Bilanzsumme
kq:	branchenüblicher Anteil der Bankkredite an der Bilanzsumme
w:	Wachstumsrate des Unternehmens bezogen auf die Bilanzsumme
U:	Jahresumsatz
KU:	branchenübliche Kapitalumschlagshäufigkeit

Dieses einfache Modell stellt die Grundstruktur eines weiter differenzierbaren Modells dar, bei dem die genannten Annahmen (kundenindividuell) variiert werden können. Die empirische Überprüfung der Prognosefähigkeit dieses Modells ergab, dass branchendurchschnittliche Werte für den Anteil der Kreditfinanzierung unzureichend sind und zusätzlich nach der Unternehmensgröße differenziert werden sollte.[1741] Prognosemodelle wie dieses sollten in analoger Weise auch für die anderen relevanten Bankleistungen entwickelt werden. Dabei ist vor dem Hintergrund der Verbundorientierung eine nach Bankleistungen differenzierte Vorgehensweise einzufordern, da ansonsten die kundenwertbasierte Verbundkoordination erschwert wird. Vergleichsweise einfach dürften sich Modelle für den Bedarf an Zahlungsverkehrsdienstleistungen ableiten lassen. Bis auf das dargestellt Modell von GLEIßNER/JAHN gibt es nach Kenntnis des Verfassers keine weiteren im Schrifttum beschriebenen Verfahren, so dass hier ein breites Forschungsfeld zu identifizieren ist. Mit Blick auf die etwa 1.100 eigenständigen Kreditgenossenschaften, die fast alle im Firmenkundengeschäft tätig sind, kann angenommen werden, dass für die Entwicklung derartiger Modelle eine breite Datenbasis vorhanden ist, die es zweckgemäß zu nutzen gilt.

Sowohl das Bedarfspotenzial als auch das Bedarfsvolumen stellen die bankleistungsbezogenen „Mengengerüste" der Kundenwertermittlung dar und müs-

---

[1741] Vgl. *Gleißner, W./Jahn, C.* (2005), S. 157 ff.

sen für die Messung des monetären Kundenwerts bewertet werden. Wie bei der Herleitung des Kundenwertmodells ausgeführt, sind dazu weitere Informationen zu beschaffen und zu nutzen. Dazu zählen z. B. Informationen zur Zinsentwicklung, zur Preiselastizität des Firmenkunden, für deren Ermittlung eine Vielzahl von Methoden zur Verfügung steht,[1742] oder Informationen zu bankleistungsbezogenen und kundenbezogenen Prozesskosten, die eventuell unter Berücksichtigung weiterer Einflussfaktoren wie etwa Kostensenkungsprogrammen zu erheben sind.[1743]

Die Anzahl relevanter Methoden hängt davon ab, wie differenziert und breit der Bewertungsmaßstab (DB Ib bis DB III)[1744] für das Bedarfspotenzial und das Bedarfsvolumen sein soll. Eine besondere Problematik stellt in diesem Zusammenhang die Parametrisierung der Bankleistungen dar. So stellt sich z. B. beim Modell von GLEIßNER/JAHN die Frage, mit welchen konkreten Bankleistungen der Kreditbedarf gedeckt werden soll. Dazu gehören etwa Fragen zu Zins- und Tilgungsmodalitäten, Sicherheitenstellungen etc. Für die Ermittlung des monetären Kundenteilwerts sind diese Aspekte, wie das Beispiel in Tabelle 22 verdeutlicht, notwendig. Es ist evident, dass Abweichungen zwischen den der Kundenwertprognose zugrunde liegenden Parametern und den tatsächlichen Parametern einer abgesetzten Bankleistung ggf. signifikante Veränderungen des ex-ante und ex-post gemessenen Kundenwerts begründen. Eine Möglichkeit die Prognosetätigkeiten für den Kundenwert zu vereinfachen und zu unterstützen, ist die Nutzung von "Defaults" in der Prognose und in der späteren, darauf aufbauenden Kundenbearbeitung sowie die Straffung des Bankleistungsprogramms in der Tiefe.[1745] "Defaults" stellen bei grundsätzlich konfigurier- oder parametrisierbaren Leistungen wie z. B. Investitionsdarlehen hinsichtlich Laufzeit, Verzinsungsart und etwaigen Optionen (Sondertilgungen), Vorgaben (harte Defaults) oder Empfehlungen (weiche Defaults) des Herstellers dar. Dass "Defaults" grundsätzlich einen nennenswerten Einfluss auf das Entscheidungsverhalten von Kunden haben zeigen POLAK ET AL. anhand eines Produktkonfigurators im Internet und führen dabei aus, dass ihre Erkenntnisse auch auf andere Produktbereiche übertragen werden können.[1746] Zwar liegen noch keine Erkenntnisse dazu für das Firmenkundengeschäft der Banken im Allgemeinen oder der Kreditge-

---

1742 Vgl. für einen Überblick z. B. *Völckner, F.* (2006), S. 34 ff.

1743 Vgl. dazu z. B. *Lissautzki, M.* (2007), S. 197 ff.

1744 Vgl. dazu die Ausführungen in Kapitel C.4.3.

1745 Während die Breite eines Bankleistungssortiments angibt, wie viele verschiedenartige Produkte und Dienstleistungen angeboten werden, beschreibt die Tiefe des Bankleistungssortiments die Anzahl an Differenzierungen und Varianten der einzelnen Bankleistungen.

1746 Vgl. *Polak B. et al.* (2008), S. 1033 ff.

nossenschaften im Speziellen vor. Aufgrund der hohen Bedeutung des persönlichen Vertriebs und der tendenziellen Informationsasymmetrie hinsichtlich der Bankleistungen zwischen Firmenkundenberater und Firmenkunde kann jedoch begründet angenommen werden, dass Vorgaben und Empfehlungen der Bank in diesem Kontext ebenfalls in der Art wirken, dass die prognostizierten und realisierten Parameter abgesetzter Bankleistungen in relativ geringerem Ausmaß divergieren. Unberührt davon resultiert eine vereinfachte Prognose dadurch, dass das Bankleistungsprogramm in der Tiefe reduziert wird. Entsprechende Ansätze sind gegenwärtig in der FinanzGruppe im Rahmen der Strategie BÜNDELUNG DER KRÄFTE: EIN VERBUND – EINE STRATEGIE zu beobachten.[1747]

Im Kapitel D.5.2.2.4 wurde bereits auf die Notwendigkeit einer mehrwertigen Prognose der Kundenwerttreiber hingewiesen. In diesem Zusammenhang wird in der Literatur regelmäßig eine dreistufige Szenarioanalyse verwendet,[1748] mit der ein optimistisches, ein neutrales und ein pessimistisches Szenario Eingang in die Bewertung finden. Zu beachten ist, dass nicht der gesamte monetäre Kundenteilwert mit einer Szenariowahrscheinlichkeit multipliziert wird, sondern die Ausprägungen der einzelnen Kundenwerttreiber, die in diese Größe Eingang finden.

Wird die Vielfalt der Informationen und der Informationsquellen betrachtet, die für die Kundenbewertung benötigt werden, ist zu konstatieren, dass nur ein ausgewählter Methodenmix in Frage kommt, den Informationsbedarf zu decken.[1749] Dabei ist in Abhängigkeit der verfügbaren Informationen zwischen Bestands- und Nichtkunden zu unterscheiden. Es ist unstrittig, dass dieses Vorgehen besondere Anforderungen an das Vertriebscontrolling in den Kreditgenossenschaften stellt, so dass es sinnvoll erscheint, nachfolgend wesentliche Probleme der Informationsbeschaffung zu thematisieren.

1.2.5 Probleme der Informationsbeschaffung

BRUHN ET AL. weisen darauf hin, dass mit dem auch hier verwendeten kapitalwertbasierten Verfahren zur Berechnung des Kundenwerts auf Basis der Investitionstheorie vor allem Zeit-, Qualitäts- und Kostenprobleme verbunden sind.[1750]

1747 Hinsichtlich der Breite des Bankleistungssortiments gibt es im Sinne der Positionierung der Allfinanzanbieter keine Alternative. Vgl. *BVR* (1999b), S. 13. Hinsichtlich der Tiefe empfiehlt der BVR Produkte zur Abdeckung des Grundbedarfs zu standardisieren. Selten nachgefragte Bankleistungen des Spezialbedarfs dagegen sollten maßgeschneidert angeboten werden.

1748 Vgl. z. B. *Tewes, M.* (2003), S. 273; *Stüker, D.* (2008), S. 333 ff.

1749 Vgl. *Gleißner, W.* (2005), S. 420.

1750 Vgl. dazu ausführlich *Bruhn, M. et al.* (2000), S. 184 ff.

Das Zeitproblem hat zwei Facetten. Erstens liegen für die Berechnung hinreichend aussagekräftige Daten nicht immer zum Zeitpunkt der Messung vor (z. B. für den Loyalitätsgrad eines Kunden). Zweitens ist es schwierig, die Kosten und Nutzengrößen den richtigen Zeiträumen in der Zukunft zuzuordnen. Das Qualitätsproblem fokussiert auf die Datenqualität und die Abhängigkeit der Kundenwertmessung von den richtigen Daten. Nach SACHENBACHER ist die Beschaffung der relevanten Informationen für Banken aufgrund deren Besonderheiten bedeutend schwieriger als bei industriellen Vorhaben.[1751] Gleichzeitig führt er mit Blick auf das Firmenkundengeschäft aus, dass "die Firmenkundschaft gegenüber der Privatkundschaft zahlenmäßig wesentlich niedriger ist"[1752], wodurch sich die Bedeutung dieses Problems reduziert. Jedoch ist die geringere Anzahl Firmenkunden nicht allein ausschlaggebend. Im Rahmen der Informationsbeschaffung ist der Firmenkundenberater von besonderer Bedeutung. Daher kommt es auch darauf an, welche Zeiträume diesem für derartige Tätigkeiten bleiben. Zentrale Einflussfaktoren dafür sind die Betreuungsspanne sowie die so genannte Vertriebszeit. Die maximale Anzahl betreuter Unternehmen je Firmenkundenbetreuer ist insbesondere bei Sparkassen und Genossenschaftsbanken mit Werten bis hin zu 180 Unternehmen sehr hoch, was eine individuelle Betreuung und Erfassung kundenwertrelevanter Informationen sehr schwierig macht.[1753] Der Mittelwert liegt einer umfangreicheren Studie von EULER HERMES zufolge bei 105 Unternehmen je Firmenkundenberater,[1754] worin BENÖLKEN die maximale Betreuungsspanne sieht, die in seiner Studie von 54 Prozent der befragten Kreditinstitute nicht überschritten wird.[1755] Die Vertriebszeit lässt eine Aussage darüber zu, zu welchem Anteil der effektiven Arbeitszeit ein Firmenkundenberater mit oder mit Blick auf einen Firmenkunden vertrieblich tätig ist. Nur die kombinierte Betrachtung von Betreuungsspanne und Vertriebszeit erlaubt ein aussagefähiges Urteil über die Probleme der Beschaffung kundenwertrelevanter Informationen mittels des persönlichen Vertriebs. Gegenwärtig liegen die durchschnittlichen Zeitanteile für die Beschäftigung mit dem Kunden bei etwa 25 bis 30 Prozent, was nach Expertenmeinung als zu gering angesehen wird.[1756] KÄSER/RITZENHOFEN nennen als Benchmarkwert für die anteilige Ver-

[1751] Vgl. *Sachenbacher, H.-U.* (1990), S. 16.

[1752] *Sachenbacher, H.-U.* (1990), S. 224.

[1753] Vgl. *Segbers, K.* (2007), S. 227.

[1754] Vgl. *Euler Hermes* (2005), S. 19.

[1755] Vgl. *Benölken, H.* (2002), S. 122. Grundlage der Ergebnisse ist die Befragung von 137 Regionalbanken, also Sparkassen und Kreditgenossenschaften.

[1756] Vgl. *Schmoll, A.* (2006), S. 133 f.

triebszeit 50 Prozent, da sie einen tendenziell positiven Zusammenhang zwischen Vertriebszeit und Erfolg im Firmenkundengeschäft feststellen.[1757]

Vor diesem Hintergrund ist es bedeutend, das Stellen- und Aufgabenprofil des Firmenkundenberaters kritisch zu hinterfragen.[1758] DAHMEN nennt zwei Gründe, warum die intensive Auseinandersetzung mit den Firmenkunden notwendiger Bestandteil des Aufgabengebiets der Firmenkundenberater ist.[1759] Zum einen kann der Firmenkundenberater nur über diesen Weg mit dem Firmenkunden auf Augenhöhe und in dessen Sprache kommunizieren sowie bedarfsadäquate Lösungen anbieten. Zum zweiten bedingt die Nutzung interner Ratingverfahren, die für die Bonitätsbeurteilung und somit Hinterlegung von Eigenkapital bedeutend ist, sich mit dem Geschäftsmodell seiner aktuellen und potenziellen Kunden auseinanderzusetzen. Dazu ist es nicht nur notwendig, sich mit den Kundenmärkten der Firmenkunden zu beschäftigen, sondern auch mit deren Wettbewerbern und den Geschäftsprozessen, die dem Geschäftsmodell zugrunde liegen. Es sei an dieser Stelle darauf hingewiesen, dass die Erfassung von Rohdaten im Vordergrund steht. Die Integration dieser Daten in die Prognose- oder andere Informationsaufbereitungsmodelle für die Kundenwerttreiber des Kundenwerts kann institutionell an anderer Stelle platziert sein.

Zwar kann allgemein konstatiert werden, dass im Business-to-Business-Bereich die Einschätzung des Bedarfspotenzials leichter als im Privatkundengeschäft möglich ist.[1760] Eines der zentralen Probleme der Informationsbeschaffung liegt darin, dass sich die künftigen Transaktionen hinsichtlich Zeitpunkt, Volumen und konkretisierter Bankleistung nur schwer abschätzen lassen, und so die Ermittlung des monetären Kundenteilwerts den objektiv messbaren Bereich verlässt und subjektive Einschätzungen in die Beurteilung der Kundenattraktivität einfließen.[1761] MULHERN stellt dazu fest„models incorporating predicted future purchases are subject to a great deal of forecasting error"[1762].

[1757] Vgl. *Käser, B./Ritzenhofen, A.* (2009), S. 32. Diese definieren die Vertriebszeit als Zeitanteil für persönliche und telefonische Gespräche mit Kunden inkl. Zeit für Gesprächsvor- und -nachbereitung (ohne administrative Aufgaben aus dem Kreditprozess). Grundlage der Vergleichswertherleitung ist die Gegenüberstellung von Nettovertriebszeit und Deckungsbeitrag von 168 Kreditinstituten.

[1758] Vgl. *Schmoll, A.* (2006), S. 134 ff. Relevante Einflussgrößen auf die Vertriebszeit sind z. B. aufbauorganisatorische Änderungen (z. B. Vertriebsassistenzen), ablauforganisatorische Änderungen (z. B. Service Level Agreements oder automatisierte Früherkennungsverfahren) sowie ein laufendes Prozesscontrolling. Vgl. *Käser, B./Ritzenhofen, A.* (2009), S. 32.

[1759] Vgl. hier und nachfolgend *Dahmen, A.* (2006), S. 20 f.

[1760] Vgl. *Barth, W./Wille, K.* (2000), S. 17.

[1761] Vgl. *Sträter, J.* (2005), S. 46; *Beeser, A. J.* (2003), S. 47.

[1762] *Mulhern, F.* (1999), S. 28.

Ein Kostenproblem resultiert vor allem daraus, dass Zeit für die Informationsbeschaffung aufgewandt werden muss und dass interne Informationssysteme für die Berechnung und Verwendung des Kundenwerts neu beschafft oder angepasst werden müssen.

Anknüpfend an den letzten Problempunkt sei noch einmal auf die Integration der FinanzGruppe in die Kundenwertorientierung aufmerksam gemacht. Es ist im Lichte des Kooperationsprinzips gegenwärtige Praxis, dass die Verbundunternehmen oder die Kreditgenossenschaften in konzertierter Aktion nach Lösungen streben, um kosteneffiziente Verfahrensweisen und -methoden im Verbund zu implementieren. Gegenstand solcher Maßnahmen könnten daher sowohl die Ermittlung geeigneter Prognoseverfahren als auch die Übernahme konkreter Aufgaben der Kundenwertbeschaffung sein, sofern diese nicht effizienter durch die Kreditgenossenschaften selbst erledigt werden können.[1763]

Von Seiten der Praxis wird als Problem angeführt, dass eine kundenindividuelle und differenzierte Potenzial- bzw. in diesem Fall Kundenwertanalyse unwirtschaftlich sei, da mit dem zunehmenden Detaillierungsgrad die Kosten überproportional zunähmen und in der Regel prohibitive Ausmaße erreichen. In diesen Fällen kann natürlich eine Vereinfachung der Informationsbeschaffung vorgenommen werden, die konsistent zur Nutzung kundengruppenbasierter Messungen des Kundenwerts ist. Es besteht in diesen Fällen jedoch die Gefahr, werthaltige Kunden nicht zu erkennen und somit Kundenwertpotenziale nicht realisieren zu können. Weitere Möglichkeiten der Vereinfachung basieren auf der einfacheren Bewertung von Bedarfspotenzialen, die statt mit barwertigen Deckungsbeiträgen mit periodischen Margen, Volumen und Laufzeiten bewertet werden. In diesen Fällen ist aber auf die richtige Berücksichtigung der Kostengrößen zu achten.

Abschießend sei auf das Problem hingewiesen, dass der Firmenkunde eventuell nicht bereit ist, Informationen preiszugeben.[1764] Gelingt es nicht, den Firmenkunden dazu zu motivieren, die relevanten Informationen mitzuteilen, sind ebenfalls alternative Verfahren der Informationsbeschaffung z. B. in Form von Analogieschlüssen vorzunehmen.

Trotz der genannten Probleme ist BLATTBERG/GETZ/THOMAS zuzustimmen, die argumentieren, dass es besser ist, „to be roughly right and to utilize […] customer equity than not to measure it and to operate an enterprise using insuffi-

---

[1763] Dabei geht es weniger um die Aufgaben der Informationsbeschaffung des persönlichen Vertriebs als vielmehr um die Informationsbedarfe bei der Prognose und Ermittlung des Kundenwerts und der Kundenwerttreiber.

[1764] Dieses Problemfeld thematisieren z. B. *Treiblmaier, H.* (2007), S. 45 ff. und *Hansen, U.* (2006), S. 146 ff. nach deren Aussage, im Beziehungsmarketing zwingend auch die Verbraucherperspektive eingenommen werden müsse.

cient indicators, such as current sales and profits. Such indicators may be more accurately measured, but they are less relevant managerially and strategically."[1765]

1.3 Kundenwertorientierte Informationsaufbereitung

Dieser Prozessschritt des Vertriebsinformationssystem begründet sich damit, dass die zur Lösung von Entscheidungsproblemen benötigten Informationen i. d. R. nicht so vorliegen, dass sie unmittelbar im Entscheidungsprozess eingesetzt werden können und somit die Aufbereitung der Informationen nötig ist.[1766] In der Literatur finden sich einige Beispiele wie die kundenwertbezogenen Informationen aufbereitet werden können. Bereits im Zusammenhang mit der Operationalisierung des Kundenwerts wurden Portfoliomodelle als eine Möglichkeit genannt. Eine weitere Möglichkeit stellen Kennzahlen oder Kennzahlensysteme dar. Zu nennen ist hier z. B. die so genannte Customer Scorecard,[1767] die die Idee der Balanced Scorecard auf die Kundenwertorientierung überträgt.

Ohne auf die einzelnen Möglichkeiten der Informationsaufbereitung im Detail einzugehen, seien die wesentlichen Anforderungen an diesen Aspekt des Vertriebsinformationssystems genannt.[1768] Um Missverständnissen vorzubeugen, muss einheitlich und konsistent vorgegangen werden, d. h. konkret, Gleiches muss gleich benannt werden. Weiterhin muss dafür Sorge getragen werden, dass die Informationen richtig und verlässlich sind, sowie zeitnah aufbereitet und bereitgestellt werden können. Dabei ist darauf zu achten, dass die Nachvollziehbarkeit der aufbereiteten Informationen gewährleistet ist und problem- und benutzeradäquat erfolgt. Von zentraler Bedeutung ist, dass der Informationsadressat die Form der Aufbereitung mitbestimmen kann. Nur in diesem Fall wird den bereits für den Informationsbedarf relevanten subjektbezogenen Determinanten hinreichend stringent Rechnung getragen. Des Weiteren ist zu fordern, dass schon bei der Informationsversorgung die Bedeutung der Informationen beurteilt wird, um adressatengerechte Impulse geben zu können. Dabei ist jedoch darauf zu achten, keine verzerrende Darstellung vorzunehmen. Zudem sollte die Informationsversorgung den Informationsstand der Adressaten berücksichtigen. Dies betrifft im vorliegenden Fall nicht nur die Ergebnisse der Kundenwertermittlung, sondern auch die Kenntnisse über die Verfahrensweisen, die

[1765] *Blattberg, R. C./Getz, G./Thomas, J. S.* (2001), S. 211.

[1766] Vgl. *Erichson, B./Hammann, P.* (1985), S. 165.

[1767] Vgl. für eine an der Balanced Scorecard orientierte kundenwertbezogene Scorecard *Friederichs-Schmidt, S.* (2006), S. 199 ff.

[1768] Vgl. hier und nachfolgend *Horváth, P.* (2003), S. 379 f. und *Weber, J.* (2004), S. 130 ff. sowie die dort angegebene Literatur.

eingesetzt werden, um den Kundenwert zu ermitteln. Die Informationsaufbereitung ist zudem mit der Speicherung und Weitergabe der Informationen abzustimmen, so dass keine Inkongruenzen auftreten.

1.4 Speicherung der Kundenwertinformationen

Kundenwertrelevante Informationen und Kundenwertinformationen bedürfen dann der Speicherung, wenn die Informationsverwendung und -verfügbarkeit zeitlich auseinander fallen. Insbesondere die Instrumente zur Prognose von einzelnen Größen des Kundenwerts basieren auf umfangreichen Daten, die zu unterschiedlichen Zeiten erfasst, aufbereitet und verwendet werden, so dass das zeitliche Auseinanderfallen begründet angenommen werden kann. Im Rahmen der Informationsspeicherung ist daher zunächst auszuwählen, welche Informationen in welcher Aufbereitung gespeichert werden sollen. Diese Auswahl ist davon abhängig, welche Informationen künftig verwendet werden sollen und ist somit institutsspezifisch. Neben der Auswahl der relevanten Informationen sind auch die Dauer der Speicherung sowie die Zugriffsnotwendigkeiten beeinflussende Größen. Aufgrund der weiten und zunehmenden Verbreitung von EDV-Systemen für den Austausch von Informationen bieten sich vor allem elektronische Speichermedien an.[1769]

Bei den Kreditgenossenschaften erfolgt die Speicherung von kundenbezogenen Informationen entweder zentral bei den Rechenzentralen oder dezentral und lokal in den Bankhäusern.[1770] Vor dem Hintergrund der zunehmenden Bemühungen der Kreditgenossenschaften, Möglichkeiten der Synergiehebung durch die gemeinschaftliche Ausnutzung von Ressourcen zu realisieren, ist davon auszugehen, dass die systematische Speicherung von Kundenwertinformationen auf Ebene der Rechenzentralen erfolgt. Die damit verbundenen Koordinationsaufgaben und -probleme können an dieser Stelle nicht diskutiert werden, vielmehr sollen allgemeine Anforderungen genannt werden. Für die Belange der kundenwertorientierten Steuerung ist es erforderlich, dass die Systeme:

- die integrierte und zentrale Sicht aller relevanten Kundendaten als Voraussetzung ganzheitlicher Bewertungen und Vergleiche,

---

1769 So müssen z. B. nach § 5b EStG i. V. m. §§ 4 Abs. 1, 5 und 5 a EStG steuerpflichtige, bilanzierende Unternehmen ihre Jahresabschlüsse für die Wirtschaftsjahre, die nach dem 31.12.2011 beginnen, elektronisch an die Finanzverwaltung übermitteln. Somit firmenkundenseitig bestehende Schnittstellen könnten auch von den Kreditgenossenschaften verwendet werden.

1770 Von einer weiteren dezentralen Speicherung auf Ebene der Filialen wird hier abgesehen.

- die Speicherung unterschiedlicher Daten über einen längeren Zeitraum als Voraussetzung für Prognosen und periodenübergreifende Bewertungen und
- die automatisierte, EDV-gestützte Datenanalyse (Data-Mining) als Voraussetzung der Verarbeitung komplexer und umfangreicher Datenmengen

erlauben.[1771]

Tabelle 27: Exemplarische Datenarten einer Kundendatenbank[1772]

Datenart	**gibt Informationen über...**	**Datenbeispiele**
Grund-/ Stammdaten	längerfristig gleichbleibende und weitgehend produktunabhängige Kundendaten	Namen, Firma, Gründungsjahr, Geburtstag, Geschlecht, Beruf, Ausbildung, Branche etc.
Wirtschaftliche Verhältnisse	periodisch variante weitgehend produktunabhängige Kundendaten	Jahresabschluss und -bericht, Vermögens- und Schuldenübersichten, Bankenspiegel, Einkommensteuererklärungen und -bescheide etc.
Transaktionsdaten (eigen und Wettbewerb)	bankleistungs-, verbundleistungs- und fremdleistungsbezogene Nachfragedeckung	Volumina, Tilgungs- und Zinsmodalitäten, Fälligkeiten, Transaktionspartner, Zahlungsverkehrsdaten etc.
Risiko-, Rating- und Bonitätsdaten	risikobezogene Aspekte im Sinne der Risikodimension des Kundenwerts	Value-at-Risk, Besicherungsumfang und -struktur, quantitative und qualitative Ratingdaten, Ratingklasse etc.
Controllingdaten	bisherige Erträge und Kosten in Verbindung mit einer Kundenbeziehung	Deckungsbeiträge, Standardstück- und Risikokosten, kundenbezogene Prozesskosten etc.
Institutsseitige Aktionsdaten	kundenbezogene (vertriebspolitische) Maßnahmen in Bezug auf Art, Intensität, Häufigkeit, Zeitpunkt und Kosten	Kontaktarten und -dauern (wann durch wen, in welcher Form?), geplante Kontakte, Anschreiben und Einladungen etc.
Kundenseitige Aktions- und Reaktionsdaten	kundenbezogene Aktionen und Kontaktaufnahmen sowie Reaktionen auf (vertriebspolitische) Maßnahmen und Verhaltensweisen des Kunden	Teilnahmen, Ablehnungsgründe, Kundenanfragen, Kundenanregungen, Kundenkorrespondenz etc.
Kundenwertbezogene Daten	Attraktivität eines Kunden aus Sicht der Kreditgenossenschaft	prognostizierter monetärer Kundenteilwert, prognostizierte Ausprägungen der einzelnen Kundenwerttreiber, nicht-monetärer Kundenteilwert, Inputdaten des nicht-monetären Kundenteilwerts etc.

[1771] Vgl. *Eberling, G.* (2002), S. 232.

[1772] Eigene Darstellung.

Aufgrund der vielfältigen Informationsquellen und der mehrfachen Nutzung des kundenbezogenen Wissens im Kontext der Geschäftsbeziehung zu einem Firmenkunden ist es zweckmäßig, den Einsatz von Datenbanken zu forcieren, die rohe und aufbereitete Kunden- und Kundenwertdaten zweck- und prozessentkoppelt speichern und in der Lage sind, diese zweck- und prozessbezogen zusammenzuführen. In der Literatur gibt es eine Vielzahl von möglichen Datenbankstrukturen für Data-Warehouses, die dieser Anforderung Rechnung tragen. Relevante Datenfelder eines solchen Data-Warehouses stellt die Tabelle 27 exemplarisch dar.[1773] Eine derartige Datenbank gilt als Haupttressource des Marketings und Vertriebs, wobei sie ohne Wert ist, wenn die Mitarbeiter keinen Zugriff auf diese haben.[1774]

1.5 Übermittlung von Kundenwertinformationen

Die Informationsentstehung und -verwendung fallen je nach Grad der Zentralisation des Vertriebs und der Vertriebssteuerung in der Kreditgenossenschaft räumlich auseinander. Daher müssen zwischen den Stellen der Kundenwertermittlung und der Kundenwertverwendung Informationsübermittlungsvorgänge stattfinden.[1775] Diese Vorgänge fallen im Allgemeinen in den Bereich der betrieblichen bzw. in concreto der vertrieblichen Berichtssysteme, deren Gegenstand Berichte in physischer, in mündlicher oder elektronischer Form sind. Das Berichtswesen dient den Zwecken der Dokumentation, der Initiierung, der Kontrolle und der Entscheidungsvorbereitung.[1776]

Die zentralen Berichtseigenschaften (Zweck, Inhalt, Verdichtungsgrad, Genauigkeit, Adressaten und Absender, Termine und Bearbeitungszeiten) ergeben sich zum Teil bereits direkt aus der Informationsbedarfsermittlung. Folglich entspricht das vertriebliche Berichtswesen einer dem vertrieblichen Informationsbedarf angepassten, geordneten Struktur aller Vertriebsberichte.[1777] Der Aufbau des Vertriebsberichtssystems orientiert sich an der zeitlichen Frequenz in Abhängigkeit der Fristigkeiten für Vertriebsplanung und -kontrolle, an den Phasen des Vertriebssteuerungsprozesses, welcher in die Phasen der Planung und der Steuerung bestehend aus Durchsetzung, Überwachung und Sicherung aufgeteilt

[1773] Vgl. *Galasso, G.* (1999), S. 236 sowie die dort genannte Literatur; *Preißner, A.* (2003), S. 21 ff.

[1774] Vgl. *Homburg, C./Schäfer, H.* (2000), S. 43.

[1775] Vgl. *Horváth, P.* (2003), S. 607.

[1776] Vgl. *Dauser, A.* (1995), S. 143.

[1777] Vgl. *Eberling, G.* (2003), S. 236; *Küpper, H.-U.* (2001), S. 152.

wird,[1778] und an der jeweiligen Adressatengruppe (Vorstand, Vertriebsleitung und -steuerung, Firmenkundenberater).

Mit Blick auf die hier betrachtete Vertriebssteuerung sind drei Berichtsformen zu unterscheiden.[1779] Dies sind Standardberichte, Abweichungsberichte und Bedarfsberichte. Standardberichte bilden den Kern des vertrieblichen Berichtswesens. Sie zeichnen sich dadurch aus, dass sie in regelmäßigen Abständen erstellt werden und eine fest definierte Menge an Informationen enthalten. Im Rahmen der kundenwertorientierten Vertriebssteuerung sind dies etwa die kundenindividuellen Berichte, die die Prognosewerte für Determinanten des Kundenwerts enthalten und in Verbindung mit den mitarbeiterindividuellen Zielen als Basis der dezentralen Vertriebssteuerung dienen. Aufgrund ihres Standardisierungsgrads können diese Berichte mehr Informationen enthalten, als für den konkreten Empfänger relevant sind.

Liegen Situationen vor, in denen definierte Meldegrenzen überschritten werden, so werden Abweichungsberichte eingesetzt. Sie werden bspw. erstellt, wenn von kundenwertorientierten Zielen abgewichen wird, und dadurch ggf. Einzelfallentscheidungen initiiert werden müssen. Regelmäßig kann dies etwa auftreten, wenn sich im unterjährigen Verlauf die Bonität eines Firmenkunden deutlich verschlechtert und somit z. B. geplante Kundenwerte nicht mehr realisierbar werden.

Da Informationsbedürfnisse im Verlauf der Planrealisation variieren können, treten Situationen auf, in denen die Standardberichte keine ausreichende Informationsbasis begründen. In diesen Fällen sind Bedarfsberichte notwendig. Sie decken ein momentanes Informationsbedürfnis und können Abweichungsberichte ergänzen. Dies kann bspw. dann der Fall sein, wenn im Laufe einer Realisationsphase neue Verbundleistungen zu berücksichtigen sind. In diesen Fällen könnten Bedarfsberichte darüber Auskunft geben, welche verbundleistungsbezogenen Kundenwerte im Absatzmarkt der jeweiligen Kreditgenossenschaft unter Berücksichtigung des strategischen Status quo zu erwarten sind.

Das kundenwertorientierte Vertriebsinformationssystem stellt in Summe die Basis einer kundenwertorientierten Vertriebssteuerung dar. Gestaltungsempfehlungen für die Phasen der Vertriebsplanung, -steuerung i. e. S. und Überwachung werden in den nachfolgenden Kapiteln erörtert. Die folgenden Ausführungen orientieren sich an der prozessualen Sicht auf die Vertriebssteuerung, wobei die einzelnen Phasen nicht immer trennscharf abzugrenzen sind und Abhängigkeiten oder Rückkopplungen möglich und wahrscheinlich sind.

[1778] Vgl. dazu die Ausführungen in Kapitel B.2.

[1779] Vgl. hier und nachfolgend *Horváth, P.* (2003), S. 609 ff.; *Küpper, H.-U.* (2001), S. 153 f.

2 Kundenwertorientierte Vertriebsplanung

Der erste Schritt der kundenwertorientierten Vertriebssteuerung ist die kundenwertorientierte Vertriebsplanung, die in eine strategische, d. h. langfristige und eine operative, d. h. kurzfristige Planungsebene differenziert werden kann.

2.1 Strategische Vertriebsplanung

In der strategischen Vertriebsplanung werden langfristige und aggregiert formulierte Vorgaben für die Entwicklung von Vertriebseinheiten und -leistungen erarbeitet. Der in der strategischen Planung verwendete abstrakte Begriff des Erfolgspotenzials[1780] wird durch die Aspekte Kundenwert und Vertrieb inhaltlich konkretisiert und beschreibt das gesamte Gefüge vertrieblicher und kundenwertspezifischer Voraussetzungen, die vorliegen müssen, um die prognostizierten und zum Zielinhalt gemachten Kundenwerte zu realisieren. Die strategische Vertriebsplanung zeichnet sich durch eine große Irreversibilität der Entscheidungen und durch ihren strukturbildenden Charakter aus.[1781] Sie ist inhaltlich mit der strategischen Gesamtbankplanung verbunden und wird durch diese determiniert. Die möglichen Anknüpfungspunkte der kundenwertorientierten strategischen Vertriebsplanung an die strategische Gesamtbankplanung werden zunächst benannt und im Weiteren ausführlich erörtert, um darauf aufbauend Gestaltungsempfehlungen ableiten zu können.

2.1.1 Anknüpfungspunkte an die strategische Gesamtbankplanung

Die strategische Vertriebsplanung ist Bestandteil des strategischen Managements, dessen zentraler Gegenstand die Formulierung, Implementierung und Kontrolle von Strategien durch Gestaltung und Steuerung der Unternehmensstrukturen und -prozesse ist.[1782] Eine Strategie ist ein bedingter, langfristiger und globaler Verhaltensplan, mit dem festgelegte Ziele erreicht werden sollen.[1783] Strategien beziehen sich in Anlehnung an das Drei-Ebenen-Konzept von HAX/MAJLUF auf die Ebenen der Gesamtbank, der strategischen Geschäftseinheiten und der Funktionalbereiche.[1784] Dabei gilt, dass die Gesamtbankstrategie den Rahmen für die Geschäftseinheiten und die Strategie der Geschäftseinheiten den Rahmen für die Funktionalbereiche vorgibt.[1785] Strategische Entscheidungen

---

1780 Vgl. dazu die Ausführungen in Kapitel B.2.3.1.

1781 Vgl. *Tewes, M.* (2003), S. 187 sowie die dort angegebene Literatur.

1782 Vgl. *Börner, C. J.* (2005), S. 33.

1783 Vgl. *Meffert, H* (1998), S. 60.

1784 Vgl. *Hax, A. C./Majluf, N. S.* (1996), S. 5 ff.

1785 Vgl. hier und nachfolgend *Duderstadt, S.* (2006), S. 64 f.

auf Ebene der Gesamtbank sind u. a. die Auswahl und Definition der strategischen Geschäftseinheiten.[1786] Auf Ebene der Geschäftseinheiten wird über die strategische Positionierung ebendieser entschieden. Und auf Funktionalbereichsebene wird darüber entschieden, wie die konkrete Ausgestaltung der bankbetrieblichen Funktion sein muss, um die gewünschte Positionierung zu erreichen. Es ist somit erforderlich, die jeweiligen Strategien vertikal zwischen den Ebenen und horizontal untereinander auf einer Ebene und mit den Zielen auf Gesamtbankebene abzustimmen.[1787] Unabhängig davon auf welcher Ebene Strategien formuliert werden, richten sie sich an der Erschließung und dauerhaften Sicherung von Erfolgspotenzialen aus.[1788]

Erfolgspotenziale zu schaffen und zu erhalten, ist der zentrale Zweck des strategischen Planungsprozesses, der darauf ausgerichtet ist, die Existenz der Kreditgenossenschaft langfristig zu sichern.[1789] Er lässt sich grob in die Analyse- und die Planungsphase, in der strategische Handlungsalternativen formuliert, bewertet und ausgewählt werden, differenzieren.[1790] In der Analysephase wird die Ausgangssituation der Kreditgenossenschaft hinsichtlich der externen (Umweltanalyse) und internen Rahmenbedingungen (Unternehmensanalyse) systematisch analysiert.[1791] In der Umweltanalyse werden die Chancen und Risiken der Kreditgenossenschaft, die sich aus der Entwicklung der externen Rahmenbedingungen ergeben, prognostiziert und beurteilt. Um die Vielfalt möglicher Einflussfaktoren zu strukturieren, empfiehlt der BVR den Primärbanken die Umwelt in die Wirkungsfelder Megatrends, Region, Wettbewerb, Konjunktur, Geld- und Kapitalmarkt sowie Recht und Politik aufzuteilen.[1792] Die Einflüsse von beobachteten oder erwarteten Entwicklungen dieser Wirkungsfelder sind differenziert für das Privat- und Firmenkundengeschäft zu analysieren.

[1786] Im Weiteren wird davon ausgegangen, dass die Entscheidung über die strategischen Geschäftseinheiten zugunsten des Firmenkundengeschäfts vorliegt.

[1787] Vgl. *Schierenbeck, H.* (2003), S. 535.

[1788] Vgl. *Börner, C. J.* (2005), S. 33.

[1789] Vgl. *Schierenbeck, H.* (2003), S. 531.

[1790] Vgl. *Hoffmann, W./Klien, W./Unger, M.* (1996), S. 216; *Hinterhuber, H. H.* (1997), S. 203; *Baum, H.-G./Coenenberg, A. G./Günther, T.* (2004), S. 23; *Schierenbeck, H.* (2003), S. 531 f.

[1791] Dies ist auch Gegenstand der MaRisk. Vgl. *Bundesanstalt für Finanzdienstleistungsaufsicht* (2010), AT 4.2, Textziffer 1.

[1792] Vgl. dazu und für Beispiele zu den einzelnen Wirkungsfeldern *BVR* (2007b), S. 9 ff. Megatrends zeichnen sich dadurch aus, dass sie langfristig wirken und einen nachhaltigen und durchgreifenden Wandel begründen. Beispiel sind die demographische Entwicklung, die Globalisierung sowie die zunehmende Individualisierung und Digitalisierung. Vgl. *BVR* (2007b), S. 10 f.

Die Unternehmensanalyse hat zum Ziel, die Stärken und Schwächen des Kreditinstituts zu erkennen. Um Wettbewerbsvor- und -nachteile ableiten zu können, erfolgt diese Analyse häufig relativ zum Wettbewerb, indem die Ausprägungen strategischer Erfolgsfaktoren aus Sicht der Kreditgenossenschaft und des Wettbewerbs beurteilt werden. Die Ausprägungen der strategischen Erfolgsfaktoren beeinflussen stark die Zielerreichungen hinsichtlich der Rentabilität und des Wachstums der Kreditgenossenschaft.[1793] Zu diesen zählt SCHIERENBECK unter anderem das Marktpotenzial, den Marktanteil, die Konkurrenzsituation, die Kundengewohnheiten und -bedürfnisse, die Preiselastizität der Kunden, das Image der Bank, das Vertriebsnetz oder die Akquisitionsperformance der Primärbank. Darüber hinaus wird vom BVR empfohlen, die einzelnen Funktionsbereiche der Bank systematisch zu analysieren. Zu diesen Funktionsbereichen werden das Management der Bank, der Vertrieb, die Produktion, das Portfoliomanagement und der Service gezählt.[1794] Zur Analyse des Funktionsbereichs Vertrieb wird vom BVR vorgeschlagen, Deckungsbeiträge, Vertriebskennzahlen und Marktpotenzialdaten zu nutzen.[1795]

Die Ergebnisse der Umwelt- und der Umfeldanalyse werden in der SWOT-Matrix kombiniert,[1796] um Erfolgspotenziale zu identifizieren[1797]. Aufbauend auf den Analyseergebnissen, werden in der Planungsphase die strategischen Ziele formuliert sowie die zur Zielerreichung vorzunehmenden Handlungsalternativen (Strategien) entwickelt und hinsichtlich ihrer Erfolgswirksamkeit bewertet. Gemäß VR-CONTROL ist die zentrale Ziel- und Steuerungsgröße der Gesamtbanksteuerung der Vermögenswert der Bank, der sich aus dem „Zerschlagungswert" und dem „Potenzialwert" konstituiert, wobei jedoch im Status quo nur der „Zerschlagungswert" adressiert wird.[1798] Diese beiden Wertgrößen werden beeinflusst durch die Ertrags-, Sicherheits- und Wachstums- sowie Förderauftragsziele der Kreditgenossenschaft, die als strategische Ziele in dieser Phase für den strategischen Planungszeitraum zu konkretisieren sind.

Anknüpfend an die Ergebnisse der SWOT-Analyse wird für die strategische Planung der Kreditgenossenschaften empfohlen, die Planungsphase in die Teil-

---

1793 Vgl. *Schierenbeck, H.* (2003), S. 533.

1794 Vgl. *BVR* (2007b), S. 15 f. Konkret wird die Gegenüberstellung der genannten Funktionsbereiche mit den Ebenen der Balanced Scorecard von KAPLAN/NORTON vorgeschlagen. Zu den Ebenen zählen Finanz-, Kunden- und Mitarbeiterebene sowie die Ebene der internen Prozesse. Vgl. *Kaplan, R. S./Norton, D. P.* (1997), S. 46 ff.

1795 Vgl. *BVR* (2007b), S. 17.

1796 SWOT steht für Strengths (dt.: Stärken), Weaknesses (dt.: Schwächen), Opportunities (dt.: Chancen) und Threats (dt.: Risiken).

1797 Vgl. *Börner, C. J.* (2005), S. 33 und S. 35.

1798 *BVR* (2002b), S. 11. Vgl. dazu die Ausführungen in Kapitel D.3.1.

prozesse strategische Vision, strategische Positionierung, strategische Planung, strategische Handlungsfelder und strategisches Controlling aufzuteilen.[1799] Die strategische Vision umfasst die Formulierung eines qualitativen Sollwerts für das Wesen der jeweiligen Kreditgenossenschaft, an der sich alle Strategien auszurichten haben.[1800] Im Rahmen der strategischen Positionierung werden für die Gesamtbank, die strategischen Geschäftseinheiten und die Funktionalbereiche strategische Ziele und Strategien definiert,[1801] wobei die Ziele erst in der Phase der strategischen Planung in konkret messbare Zielvorgaben transformiert werden.[1802] Anschließend werden strategische Handlungsfelder identifiziert.[1803] Strategische Handlungsfelder zeichnen sich dadurch aus, dass für bestimmte Aspekte der Kreditgenossenschaft die gegenwärtige Situation nicht hinreichend gestaltet ist, um die strategischen Ziele zu erreichen. Für die strategischen Handlungsfelder werden Maßnahmenpläne erarbeitet und umgesetzt. Sie stellen also letztlich, den Implementierungsaspekt des strategischen Managements dar. Den Abschluss des strategischen Planungsprozesses stellt die Auswahl der strategischen Handlungsalternativen zur Erreichung der strategischen Ziele dar, die im Weiteren in enger Verzahnung zur operativen Planung und Steuerung implementiert und kontrolliert wird. Nur durch die strategiekonsistente Ausgestaltung der operativen Planung, Steuerung und Kontrolle kann eine erfolgreiche Strategieumsetzung gewährleistet werden.[1804] Das strategische Controlling umfasst die Koordination des strategischen Planungsprozesses, die Identifikation und Analyse von Zielabweichungen sowie die Einleitung von Gegenmaßnahmen.[1805]

Basierend auf dieser Skizze des strategischen Planungsprozesses auf Gesamtbankebene lassen sich nun die Anknüpfungspunkte zwischen kundenwertorientierter Vertriebssteuerung und Gesamtbanksteuerung benennen, für die Gestaltungsempfehlungen zu erörtern sind. Es wird bewusst von Anknüpfungspunkten gesprochen, da damit ausgesagt wird, dass die Kundenwertorientierung ein Aspekt in diesem Bezugsrahmen auf Gesamtbankebene darstellt. Die folgenden Ausführungen stellen somit Überlegungen dar, wie die Gesamtbanksteuerung durch die Kundenwertorientierung zielführend ergänzt werden kann. Der Fokus liegt dabei auf dem Funktionalbereich des Vertriebs.

1799 Vgl. *BVR* (2007b), S. 3.

1800 Vgl. *BVR* (2007b), S. 26 ff. Dies betrifft sowohl die Strategie auf Gesamtbankebene als auch die Strategien auf Geschäftseinheiten- oder Funktionalbereichsebene.

1801 Vgl. *BVR* (2007b), S. 32 ff.

1802 Vgl. *BVR* (2007b), S. 43 ff.

1803 Vgl. hier und nachfolgend *BVR* (2007b), S. 50 ff.

1804 Vgl. *Börner, C. J.* (2005), S. 36.

1805 Vgl. *BVR* (2007b), S. 54 ff. Diese Phase wird im Weiteren nicht explizit als strategisches Controlling untersucht, sondern der strategischen Vertriebskontrolle zugeordnet.

Im Rahmen der strategischen Analyse ist das Kundenwertmodell als zusätzliches Analyseraster der strategischen Umweltanalyse mit Fokus auf das Firmenkundengeschäft zu nutzen. Darüber hinaus ist der Vertrieb als Funktionalbereich Gegenstand der Unternehmensanalyse. Mit dem Kundenwert kann für die strategische Geschäftseinheit Firmenkunden ein Beurteilungsmaßstab der Stärken und Schwächen des Vertriebs genutzt werden.

Die strategische Vision bietet keinen besonderen Anknüpfungspunkt für die Kundenwertorientierung im Allgemeinen oder die kundenwertorientierte Vertriebssteuerung und wird daher nicht näher betrachtet.

In den Phasen der strategischen Positionierung und der strategische Planung ist der Kundenwert die zentrale Bezugsgröße für Ziele auf Ebene des Funktionalbereichs Vertrieb und neben dem Risiko ergänzende Zielbezugsgröße auf der Ebene der strategischen Geschäftseinheit Firmenkunden. Weiterhin sind der Kundenwert und dessen Werttreiber Bezugsgrößen für die Ableitung und Formulierung von strategischen Handlungsalternativen auf Ebene der strategischen Geschäftseinheit und des Funktionalbereichs Vertrieb. Im Weiteren werden diese Anknüpfungspunkte detailliert erörtert.

2.1.2 Analyse der strategischen Ausgangssituation

2.1.2.1 Chancen-Risiken-Analyse

Im Rahmen der Umfeldanalyse werden die Entwicklungen der globalen und geschäftsspezifischen Umwelt prognostiziert. Wie oben beschrieben, wird die Umwelt der Kreditgenossenschaft in einzelne Wirkungsfelder unterteilt. Die Diagnose der Umfeldentwicklungen soll Chancen und Risiken in der Schaffung und Sicherung von Erfolgspotenzialen benennen.[1806] Wichtig für die strategische Planung ist es daher, die Umfeldentwicklungen aus Sicht der Kreditgenossenschaft zu beurteilen. Da der Kundenwert die externe Bezugsgröße der Erfolgspotenziale darstellt, bietet es sich für die Firmenkundenbank an, relevante Entwicklungen der einzelnen Wirkungsfelder auf Basis ihrer Einflüsse auf diese Größe zu beurteilen, indem diese dem Kundenwert als Ganzem und den einzelnen Kundenwerttreibern gegenübergestellt werden. Im Zentrum der Gegenüberstellung stehen die Fragen nach der Wirkrichtung und Wirkstärke der Entwicklungen auf den Kundenwert. Für die Wirkrichtung lassen sich die Kategorien Chance (positive Wirkrichtung) und Risiko (negative Wirkrichtung) nutzen. Die Bedeutung von Chancen und Risiken kann mittels der Wirkstärke (hoch, mittel, gering) beurteilt werden. Werden Entwicklungen in den einzelnen Wirkungsfeldern auf diese Weise dahingehend untersucht, welchen Einfluss sie auf die Kun-

---

[1806] Vgl. *Eberling, G.* (2002), S. 275.

denwert oder einzelne Kundenwerttreiber haben, können darauf aufbauend Chancen und Risiken begründet werden (siehe Abbildung 63). Diese Analyse kann sowohl auf Basis des gesamten numerischen Kundenpotenzials, des Bestands an aktuellen Kunden, einzelner Kundengruppen als auch auf Ebene der einzelnen Kundenbeziehung der Firmenkundenbank erfolgen. Letzteres ist zu empfehlen, da die Formulierung strategischer Ziele und vertrieblicher Strategien auf Basis einzelner Kundenbeziehungen erfolgen und die Grundlage dieser Schritte unter anderem die Chancen-Risiken-Analyse ist.

Wirkungsfelder → / Kundenwert	Bindungspotenzial	Basisgeschäftspotenzial	Entwicklungspotenzial	Cross-Selling-Potenzial	Basiskostenpotenzial	Kooperationspotenzial	Informationspotenzial	Referenzpotenzial	Förderpotenzial	Kreditrisiko	Kundenplanungsrisiko	Portfoliorisikobeitrag
Megatrends	Richtung	Richtung	Richtung	Richtung	Richtung	Richtung	Richtung	Richtung	Richtung	Richtung	Richtung	Richtung
	Stärke	Stärke	Stärke	Stärke	Stärke	Stärke	Stärke	Stärke	Stärke	Stärke	Stärke	Stärke
Region	Richtung	Richtung	Richtung	Richtung	Richtung	Richtung	Richtung	Richtung	Richtung	Richtung	Richtung	Richtung
	Stärke	Stärke	Stärke	Stärke	Stärke	Stärke	Stärke	Stärke	Stärke	Stärke	Stärke	Stärke
Wettbewerb	Richtung	Richtung	Richtung	Richtung	Richtung	Richtung	Richtung	Richtung	Richtung	Richtung	Richtung	Richtung
	Stärke	Stärke	Stärke	Stärke	Stärke	Stärke	Stärke	Stärke	Stärke	Stärke	Stärke	Stärke
Konjunktur	Richtung	Richtung	Richtung	Richtung	Richtung	Richtung	Richtung	Richtung	Richtung	Richtung	Richtung	Richtung
	Stärke	Stärke	Stärke	Stärke	Stärke	Stärke	Stärke	Stärke	Stärke	Stärke	Stärke	Stärke
Geld- und Kapitalmarkt	Richtung	Richtung	Richtung	Richtung	Richtung	Richtung	Richtung	Richtung	Richtung	Richtung	Richtung	Richtung
	Stärke	Stärke	Stärke	Stärke	Stärke	Stärke	Stärke	Stärke	Stärke	Stärke	Stärke	Stärke
Recht und Politik	Richtung	Richtung	Richtung	Richtung	Richtung	Richtung	Richtung	Richtung	Richtung	Richtung	Richtung	Richtung
	Stärke	Stärke	Stärke	Stärke	Stärke	Stärke	Stärke	Stärke	Stärke	Stärke	Stärke	Stärke

Abbildung 63: Beurteilungsraster zur Ableitung kundenwertbezogener Chancen und Risiken[1807]

Diese allgemeinen Überlegungen sollen an einigen Beispielen konkretisiert werden. In Teilen der Literatur und Praxis wird argumentiert, dass es eine Bedeutungsabnahme des Hausbankenprinzips von Seiten der Firmenkunden gäbe.[1808] Dies hat sichtbar unmittelbaren Einfluss auf das Bindungspotenzial, mithin also auf die Bereitschaft der Firmenkunden, langfristig und intensiv eine Geschäftsbeziehung mit einer Bank einzugehen. Darüber hinaus impliziert dies, dass sich

1807 Eigene Darstellung.

1808 Vgl. z. B. *Börner, C. J.* (2005), S. 58.

die damit anzunehmende reduzierte Cross-Selling-Bereitschaft der Kunden negativ auf das Cross-Selling-Potenzial auswirkt.

Veränderungen auf dem Geld- und Kapitalmarkt in Form von Leitzinsänderungen durch die Europäische Zentralbank können Einfluss auf das Basisgeschäfts- und das Entwicklungspotenzial haben. Reduziert die Zentralbank die Leitzinsfazilitäten dürfte einerseits eine entsprechende Erwartungshaltung zur Konditionsanpassung auch bei den Firmenkunden gegenüber der Kreditgenossenschaft beobachtbar sein.[1809] Darüber hinaus dürfte es schwieriger sein, Zinserhöhungen durchzuführen. Andererseits bietet eine derartige Situation die Möglichkeit, Bankdienstleistungen des Zinsmanagement zur Absicherung des niedrigen Zinssatzes anzubieten, was einen positiven Einfluss auf das Cross-Selling-Potenzial haben kann.

Ebenfalls Einfluss auf das Entwicklungs- und Cross-Selling-Potenzial können die Markteintritte oder Expansionsstrategien von Wettbewerbern haben. Verfolgen diese bspw. das Ziel, schnell Marktanteile zu gewinnen, so kann dies mit einer Penetrationsstrategie begleitet werden, wodurch tendenziell die Marktpreise sinken.[1810] Auch die veränderten Angebote von Automobilherstellerbanken, die Veränderungen der produktbegleitenden Finanzdienstleistungen vornehmen, und damit Einfluss auf den Kundenwert haben können, sind als Beispiel anzuführen.[1811]

Dieses kundenwertbezogene Analyse- und Beurteilungsraster bietet Gewähr, dass die strategische Umfeldanalyse sich auf den Kundenwert als zentrale Ziel- und Steuerungsgröße ausrichtet. Gleichsam sind die Ergebnisse dieser Analyse in Verbindung mit der Ableitung von Vertriebsstrategien Voraussetzung für die Prognose von Kundenwerten und Kundenwerttreibern.[1812]

2.1.2.2 Stärken-Schwächen-Analyse

Die kundenwertbezogene Stärken-Schwächen-Analyse betrifft im Wesentlichen zwei Aufgabenkomplexe. Zum einen die Analyse der strategischen Ausgangssituation mit Blick auf gegenwärtige Kundenwerte und zum zweiten die Schwach-

---

1809 Dies gilt vornehmlich für die variabel verzinslichen Produkte wie z. B. Kontokorrentkredite. Bei Zinssteigerungen betrifft dies vor allem passivlastige Kunden z. B. aufgrund von Tagesgeldkonditionen.

1810 Bei der Penetrationsstrategie sollen mit relativ niedrigen Konditionen und Gebühren schnell Märkte erschlossen werden. Vgl. dazu ausführlich *Meffert, H.* (1998), S. 548 ff.

1811 So bietet der Hersteller MERCEDES BENZ neben der Anschaffungsfinanzierung auch Kfz-Versicherungen für gewerbliche Kunden an. Vgl. *Mercedes Benz* (2011), o. S.

1812 Vgl. dazu auch die Ausführungen in Kapitel E.1.2.4.

stellen- inkl. Ursachenanalyse.[1813] Bezugsobjekte der Schwachstellenanalyse sind sowohl der Kundenwert als auch der Vertrieb.

Die direkte Diagnose der anbieterspezifischen Kundenwerte und Kundenwertpotenziale steht naturgemäß im Zentrum der kundenwertbezogenen Stärken-Schwächen-Analyse.[1814] Daher wird im Rahmen des ersten Aufgabenkomplexes eine Bewertung der gegenwärtigen Kundenbeziehungen im Status quo vorgenommen.[1815] Dies ist im Sinne der integrierten Rendite-Risiko-Steuerung zweistufig vorzunehmen. Im ersten Schritt ist die Kreditrisikostrategie der Gesamtbank festzulegen und hinsichtlich ihrer Auswirkungen auf die Firmenkundenbank zu eruieren. Die Notwendigkeit dazu resultiert daraus, dass Firmenkunden, die gegenwärtig oder aufgrund der zu erwartenden Entwicklung auch unter Berücksichtigung der Möglichkeiten der Portfoliorisikosteuerung nicht mehr zum risikopolitischen Profil passen, schon zu diesem Zeitpunkt der Planung selektiert werden. Für diese ergibt sich konsequenterweise eine Strategie der Reduzierung oder Beendigung der Kundenbeziehung.[1816]

Der um die selektierten Adressrisiken im Sinne der Kreditrisikostrategie bereinigte Bestand der Firmenkunden ist im zweiten Schritt näher zu analysieren. Dazu werden sowohl die Kundenwertpotenziale als auch die anbieterspezifischen Kundenwerte der einzelnen Kundenbeziehungen unter der Prämisse zum Bewertungszeitpunkt vorliegender Pläne und Strategien ermittelt. Die Analysen können differenziert nach den Kundenteilwerten ein- oder mehrdimensional vorgenommen werden.

Für die eindimensionale Analyse kann für den monetären Kundenteilwert eine Rangreihung der Bestandskunden vorgenommen und mittels der Lorenzkurve visualisiert werden. Aufgrund der Möglichkeit negativer Kundenwerte kann die Form der Lorenzkurve den charakteristischen Verlauf einer „whalecurve" annehmen,[1817] die sich dadurch auszeichnet, dass nach Erreichen des Maximums von 100 Prozent die Kurve eine negative Steigung aufweist. Typischerweise wird diese Darstellung mit der normativen Ableitung von Wertgruppen für die Kunden vorgenommen. So sind z. B. in der exemplarischen Abbildung 64 31 Prozent A-Kunden, 23 Prozent B-Kunden, 20 Prozent C-Kunden und 26 Prozent aufgrund ihres negativen monetären Kundenteilwerts D-

---

[1813] Vgl. hier und nachfolgend insbesondere *Stüker, D.* (2008), S. 317 ff.; *Tewes, M.* (2003), S. 188 ff.; *Cornelsen, J.* (2000), S. 269 ff.; *Eberling, G.* (2002), S. 264 f.

[1814] Vgl. *Eberling, G.* (2002), S. 266.

[1815] Vgl. *Meyer, A./Kantsperger, R./Wilkoszewski, A.* (2005), S. 309.

[1816] Vgl. dazu auch die Ausführungen in Kapitel E.2.1.3.2.

[1817] Vgl. *Kaplan, R. S./Narayan, V. G.* (2001), S. 7.

Kunden.[1818] Durch diese Analyse wird deutlich, dass knapp ein Drittel der Firmenkunden für 80 Prozent des „Potenzialwerts" und des „Förderpotenzials" im Sinne des Member Value der Bank aus dem Firmenkundengeschäft verantwortlich sind. Abweichend von der normativen Festlegung der Wertgruppen kann auch eine mathematische Herleitung erfolgen, indem z. B. parallel zur Diagonalen an die Lorenzkurve eine Tangente angelegt wird. Das Lot vom Tangentialpunkt teilt den Kundenbestand in unterdurchschnittliche Kundenwerte (rechts vom Lot) und überdurchschnittliche Kundenwerte (links vom Lot).[1819] Unabhängig davon, wie kundenwertbasierte Gruppen gebildet werden, liegt die zentrale Aussage dieser Analyse darin, zu erkennen, wie sich monetäre Kundenwerte unterschiedlicher Höhe im Kundenstamm verteilen. Daraus lassen sich wiederum Aussagen zu Abhängigkeiten oder strukturellen Schwächen begründen. Nach den MaRisk ist diese Information wesentlich für die Risikostrategie, die mit der Gesamtbank- und der Geschäftsfeldstrategie abzustimmen ist.[1820]

In der Regel wird argumentiert, die D-Kunden seien als Wertvernichter abzustoßen, C-Kunden standardisiert und B- und A-Kunden intensiv und individuell zu betreuen. Da es sich hier jedoch um den anbieterspezifischen Kundenwert unter Annahme bestehender Pläne und Strategien handelt, ist noch keine Aussage darüber möglich, ob eine zum Status quo alternative Bearbeitung der jeweiligen Kunden kundenwertsteigernde Ergebnisse hätte. Der Fokus richtet sich folglich auf die Frage, ob „noch bestehende Wertsteigerungspotentiale durch strategische und operative Veränderungen der Kundenbearbeitung ausgeschöpft werden können."[1821] Folglich ist im Weiteren die Analyse der Potenzialausschöpfung von Interesse. Mögliche Bezugsgrößen dieser Analyse sind:

- die Kundenwerttreiber der gleichen Kundenbeziehungen zu einem vergangenen Zeitpunkt,
- die Kundenwerttreiber anderer Kundenbeziehungen zum gleichen Zeitpunkt und
- das Kundenwertpotenzial der Kundenbeziehung.[1822]

1818 Bei der vorgenommenen Darstellung stellen 100 Prozent des Kundenwerts nicht die Summe der einzelnen Kundenwerte dar, sondern das Maximum der kumulierten Kundenwerte.

1819 Vgl. *Rieker, S.* (1995), S. 53.

1820 Vgl. *Bundesanstalt für Finanzdienstleistungsaufsicht* (2010), AT 4.2, Textziffer 2.

1821 *Tewes, M.* (2003), S. 190.

1822 Vgl. hier und nachfolgend *Eberling, G.* (2002), S. 268 ff.

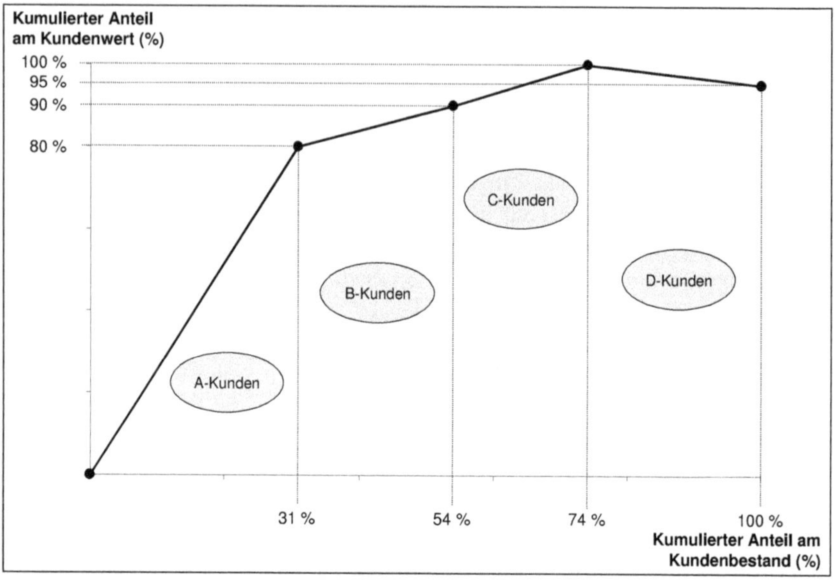

Abbildung 64: Beispiel einer kundenwertbasierten Lorenzkurve[1823]

Mittels des Zeitvergleichs lässt sich die bisherige Entwicklung der Kundenbeziehung im Lichte des Kundenwerts feststellen. Diese Analyse ist insbesondere für langfristige Kundenbeziehungen unter Anwendung der relationalen Vertriebsansätze von Interesse. Der Vergleich mit den Kundenwerttreibern anderer Geschäftsbeziehungen des Kundenstamms ermöglicht es, Ansatzpunkte der bankseitig induzierten Entwicklung des anbieterspezifischen Kundenwerts zu finden. CORNELSEN bezeichnet dies als internen Kundenwert-Wettbewerb. Alternativ dazu nennt er den externen Kundenwert-Wettbewerb, wenn die vergleichende Analyse mit bankexternen Kundenbeziehungen vorgenommen wird.[1824]

Der Vergleich des anbieterspezifischen Kundenwerts mit dem Kundenwertpotenzial stellt den herausragenden Analyseschritt in der kundenwertorientierten Stärken-Schwächen-Analyse dar. Das Kundenwertpotenzial stellt die Obergrenze realisierbarer Wertbeiträge auf Basis der gegenwärtigen Strategien und Planungen dar. Im Sinne einer GAP-Analyse[1825] ist daher zu empfehlen, die Kun-

---

1823 *Stüker, D.* (2008), S. 319.

1824 Vgl. *Cornelsen, J.* (2000), S. 272 f.

1825 Die GAP-Analyse gehört zu den klassischen Instrumenten der strategischen Planung. Sie basiert auf der Idee, dass zwischen den geplanten Zielen und der Prognose der Zielerreichung unter Beibehaltung der bisherigen Aktivitäten aufgrund der Veränderungen

denpotenzialwerte den anbieterspezifischen Kundenwerten gegenüber zu stellen. Dadurch wird eine Aussage darüber möglich, welche anbieterspezifischen Kundenwerte möglich wären. Dies führt zu Überlegungen, warum die anbieterspezifischen Kundenwerte in ihrer beobachteten Höhe und Struktur vorliegen und warum diese von den Kundenpotenzialwerten abweichen. Möglich ist eine derartige Gegenüberstellung in Form einer Portfoliodarstellung. Auf der Abszisse wird dazu das Kundenwertpotenzial und auf der Ordinate der Ausschöpfungsgrad, der sich als Quotient aus anbieterspezifischem Kundenwert und Kundenwertpotenzial ergibt, dargestellt (siehe Abbildung 65).[1826] Jede Kundenbeziehung kann mittels ihrer konkreten Werte in dieser Darstellung positioniert werden. So lässt sich bspw. herausstellen, bei welchen Kunden gegenwärtig die Ausschöpfungsgrade unterdurchschnittlich sind (Quadrant 3 und 4). Liegt bei diesen ein hohes Kundenwertpotenzial vor, ist zu fragen, warum der Ausschöpfungsgrad gering ist und wie dieser ggf. erhöht werden kann. Bei geringen Kundenwertpotenzialen dagegen ist zu erörtern, ob sich ein Bemühen darum, den Ausschöpfungsgrad zu steigern, bei knappen Vertriebsressourcen lohnt. Bei Kunden mit überdurchschnittlichen Kundenwertpotenzialen und überdurchschnittlichen Ausschöpfungsgraden (Quadrant 2) ist zu diskutieren, wie eine solche Situation genutzt und abgesichert werden kann, während im Fall von unterdurchschnittlichen Kundenwertpotenzialen und überdurchschnittlichen Ausschöpfungsgraden (Quadrant 1) auch zu fragen ist, ob vertriebliche Ressourcen hier entzogen werden können, um an anderer Stelle eingesetzt zu werden.

Die Gegenüberstellung von Kundenwertpotenzial und anbieterspezifischen Kundenwert kann ebenso auf Ebene der einzelnen Kundenwerttreiber der Ertragsdimension erfolgen,[1827] um so detaillierte Auskünfte über mögliche Erfolgsquellen zu erhalten. Dieses Vorgehen knüpft unmittelbar an die Diagnose an, dass im Firmenkundengeschäft die Ertragspotenziale nicht differenziert und vollumfänglich analysiert und genutzt werden.[1828]

---

von Stärken und Schwächen eine im Zeitablauf immer größer werdende Ziellücke (engl.: Gap) entstehen wird. Vgl. *Horváth, P.* (2003), S. 386.

1826 Vgl. *Cornelsen, J.* (2000), S. 272. CORNELSEN bezeichnet das Kundenwertpotenzial als externe Dimension und den Ausschöpfungsgrad als interne Dimension der Portfoliodarstellung.

1827 Vgl. *Eberling, G.* (2002), S. 269.

1828 Vgl. dazu die Ausführungen in Kapitel C.1.5.

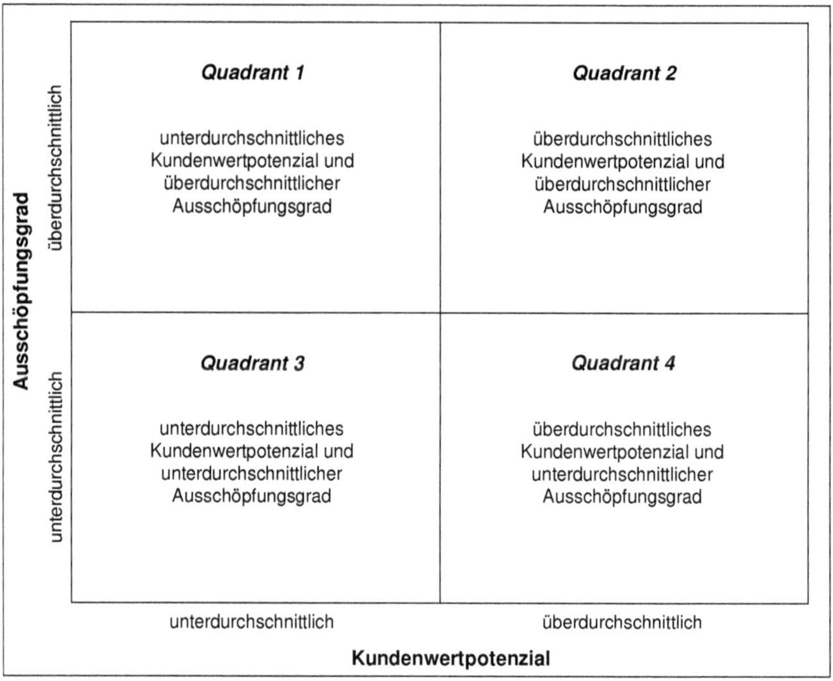

Abbildung 65: Kundenwertpotenzial-Ausschöpfungsgrad-Portfolio[1829]

Mit Blick auf das hergeleitete Kundenwertmodell ist es für die Stärken-Schwächen-Analyse empfehlenswert, eine dreidimensionale Analyse mittels eines Kundenkubus, der zwischen den Dimensionen Ertragspotenzial, Bindungspotenzial und Risiko unterscheidet, vorzunehmen. In der Dimension Ertragspotenzial wird das mit dem risikolosen Zins diskontierte Ergebnis aus Basisgeschäfts-, Entwicklungs-, Cross-Selling-, Basiskosten- und Kooperationspotenzial für den ausgewählten Bewertungszeitraum abgetragen. Für die Dimension des Bindungspotenzials kann die über den Bewertungszeitrum additiv kumulierte Bindungsrate zur Operationalisierung herangezogen werden. Für die Risikodimension kann das Scoringmodell zur Herleitung des Risikoaufschlags auf den Zins um die Kreditrisikogrößen erweitert werden, so dass es eine einzige die Risikodimension darstellende Größe gibt.[1830] Werden nun die einzelnen Kunden-

1829 Eigene Darstellung.

1830 Vgl. dazu auch die Ausführungen in Kapitel D.5.2.2.4. Hinsichtlich der Ertrags- und der Risikodimension kann theoretisch zwar keine Unabhängigkeit angenommen werden. Jedoch zeigen empirische Ergebnisse, dass das im vorliegenden Kontext diese Abhän-

beziehungen in dem so aufgespannten Kundenkubus (siehe Abbildung 66) positioniert, lassen sich in einer einfachen Interpretation relative Stärken und Schwächen der Kundenbeziehungen mit Blick auf den monetären Kundenwert ableiten.

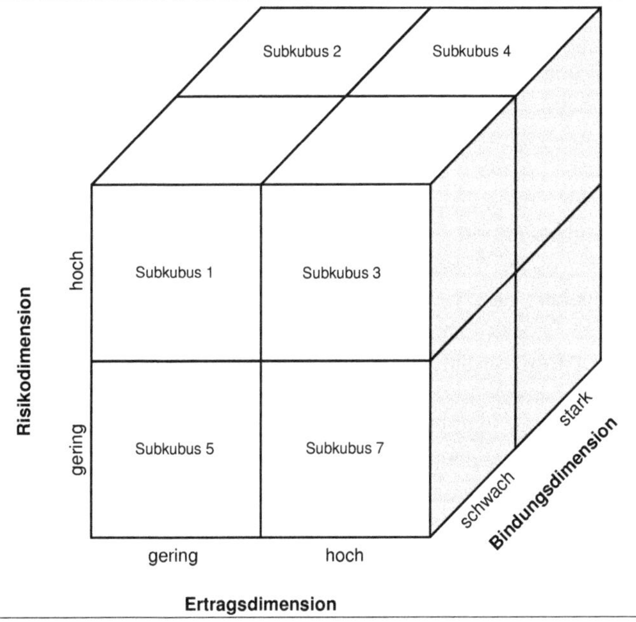

Abbildung 66: Dreidimensionaler Kundenkubus auf Basis des hergeleiteten Kundenwertmodells[1831]

In den Subkuben 1 und 2 sind die Kundenbeziehungen positioniert, deren Schwächen in einem hohen Risiko und einem geringen Ertragspotenzial liegen. Kundenbeziehungen in den Subkuben 3 und 4 zeichnen sich durch die Stärke eines hohen Ertragspotenzials aus, weisen aber auch hohe Risiken auf. In wie weit diese Situation im Sinne der Kreditgenossenschaft geändert werden kann, ist Gegenstand der individuellen Analyse der Kundenbeziehungen. Wird die kompensatorische Interdependenz zwischen Risiko und Ertrag akzeptiert, sind

gigkeit vor allem beeinflussende risikoadjustierte Pricing in den Kreditgenossenschaften nicht im ökonomisch geforderten und regulatorisch empfohlenen Sinne umgesetzt wird. Vgl. dazu *Püthe, T.* (2009), S. 310 f.

1831 Eigene Darstellung.

die Beziehungen im Subkubus 3 durch ein schwaches Bindungspotenzial geprägt. Zu prüfen ist, ob durch geeignete Vertriebsmaßnahmen, die Bindung des Kunden erhöht werden kann. Kundenbeziehungen in den Subkuben 5 und 6 weisen als Stärke ein geringes Risiko auf, so dass die zwar nur geringen Erträge vergleichsweise sicher beurteilt werden können. Derartige Kunden können als Auslastungskunden betrachtet werden. Wesentlich ist, die Bindungsschwächen zu beheben. Zu thematisieren ist zudem, ob derartige Kunden durch geeignete Maßnahmen in die Subkuben 7 oder 8 überführt werden können. Die Kundenbeziehungen in den Subkuben 7 und 8 zeichnen sich durch ihre Ertragspotenzialstärke und geringen Risiken aus. Gelingt es, alle Kunden aus dem Subkubus 7 um ihre Bindungsschwäche zu kurieren, so liegen Kundenbeziehungen mit dem größten Ausmaß an relativen Stärken vor.

Es sei jedoch darauf hingewiesen, dass diese Darstellungen auf dem gegenwärtigen Kundenwertpotenzial und anbieterspezifischem Kundenwert basieren. Sie stellen den ersten Schritt der strategischen Vertriebsplanung dar.[1832] Es ist davon auszugehen, dass veränderte Strategien und operative Aktionspläne Einfluss auf diese Größen haben. Folglich dienen diese Analysen im ersten Schritt lediglich einer Bestandsaufnahme und kritischen Diskussion möglicher Ansatzpunkte. Weitergehende Erkenntnisse liefert die Analyse von Schwachstellen auf Ebene der einzelnen aktuellen Kundenbeziehungen. Dadurch wird eine detaillierte Analyse des Beziehungsstands erreicht, die eine fundierte Grundlage der Ableitung von Strategien und Zielen ist, da auf die einzelnen Kundenwerttreiber und deren Bestimmungsfaktoren eingegangen wird.[1833]

Anleihen nimmt die Schwachstellenanalyse an den aus der wertorientierten Unternehmensführung bekannten Werttreiberanalysen. In der Literatur wird in diesem Zusammenhang vor allem die Analyse der Schwächen forciert.[1834] Gleichwohl sollte sich die Analyse nicht darauf beschränken, sondern auch die Stärken berücksichtigen. Dies begründet sich damit, dass stärken- und schwächenorientierte Strategieansätze unterschieden werden können.[1835] Zu den stärkenorientierten Strategieansätzen sind alle Strategien zu zählen, die im Kontext der SWOT-Analyse versuchen, durch Stärken Chancen zu nutzen oder Risiken zu vermeiden. Schwächenorientierte Ansätze fokussieren, ob und wie Schwächen vor dem Hintergrund von Chancen und Risiken wirken und kompensiert werden müssen und können. Entscheidend dafür, welcher Ansatz zu verfolgen

---

1832 Vgl. *Cornelsen, J.* (2000), S. 275.

1833 Vgl. *Tewes, M.* (2003), S. 190; *Eberling, G.* (2002), S. 270; *Cornelsen, J.* (2000), S. 275.

1834 Vgl. *Tewes, M.* (2003), S. 190; *Cornelsen, J.* (2000), S. 275.

1835 Vgl. hier und nachfolgend *BVR* (2007b), S. 22.

ist, ist die Frage nach der Erfolgswirksamkeit. Ist es betriebswirtschaftlich sinnvoller, Schwachstellen zu beheben, da die damit verbundenen Wirkungen auf den Kundenwert die Kosten weit stärker überwiegen als bei der Nutzung und dem Ausbau der Stärken, sind schwächenorientierte Strategien sinnvoll et vice versa. Kundenwertbezogene Schwachstellen liegen vor, wenn Ausprägungen der Bestimmungsfaktoren des Kundenwerts vorliegen, die zu einem im Verhältnis zum Kundenwertpotenzial geringen anbieterspezifischen Kundenwert führen. Dies kann z. B. die Situation sein, dass der betrachtete Kunde nur eine geringe Bereitschaft dazu hat, mehr als Zahlungsverkehrs- und Finanzierungsleistungen bei der Kreditgenossenschaft zu beziehen. Dadurch wird das Cross-Selling-Potenzial des Kunden negativ beeinflusst. Bei der Betrachtung der Schwachstellen und Stärken ist zu differenzieren, ob es sich um kundenbezogene oder bankbezogene Bestimmungsfaktoren handelt. So lassen sich aus der Analyse des Cross-Selling-Potenzials bspw. Hinweise ableiten, durch die Erweiterung des Bankleistungssortiments das Kundenwertpotenzial und den anbieterspezifischen Kundenwert zu erhöhen.[1836] Diese Analyse erfolgt individuell für die einzelnen Kundenwerttreiber, wird an dieser Stelle jedoch nicht weiter im Detail vertieft. Es sei daher auf die entsprechende Literatur verwiesen.[1837]

Werden vor allem die bankbezogenen Bestimmungsfaktoren betrachtet, lässt sich die kundenwertbezogene Stärken-Schwächenanalyse für den Funktionsbereich des Vertriebs vornehmen. Ziel ist es, zu bestimmen, wie die Fähigkeiten des Vertriebs zur aktuellen und zukünftigen Realisierung von Wertbeiträgen aus Kundenbeziehungen zu beurteilen sind.[1838] Wie bereits mehrfach erwähnt, ist der Vertrieb die Schnittstelle zum Kunden, die eine zentrale Rolle bei der Einschätzung, Schaffung und Nutzung von Kundenwerten im Firmenkundengeschäft spielt. Daher ist es für die strategische Vertriebsplanung von besonderer Bedeutung, die Stärken und Schwächen des Vertriebs zu kennen. Die Struktur eines für die Wissensgenerierung einzusetzenden Analyserasters basiert zum einen auf dem Kundenwert, den Kundenwerttreibern und deren Bestimmungsfaktoren sowie zum anderen auf der differenzierten Betrachtung des Vertriebs. Letzteres kann bspw. erreicht werden, in dem der Vertrieb in seinen funktionalen, prozessualen und institutionellen Dimensionen betrachtet wird. In der funktionalen Sicht ist z. B. von Interesse, in wie weit die akquisitorischen Aufgaben tatsächlich effektiv erfüllt werden. Dieser Aspekt lässt sich weiter verfeinern, so

---

1836 Vgl. dazu die Ausführungen in Kapitel D.4.5.3.1.3. So gibt es z. B. eine Reihe von Kreditgenossenschaften, die Tochterunternehmen haben, die als Reisebüros tätig sind. So z. B. die VR-BANK UCKERMARK-RANDOW EG oder die VR-BANK AALEN EG.

1837 Vgl. *Tewes, M.* (2003), S. 190 ff.; *Cornelsen, J.* (2000), S. 275 ff.

1838 Vgl. *Eberling, G.* (2002), S. 271.

dass zu z. B. zu fragen ist, wie intensiv das Allfinanzangebot im Vertrieb kommuniziert wird, ob die ganzheitliche Beratung im Sinne der relationalen Vertriebsansätze erfolgt oder ob die Schnittstelle zwischen Markt und Marktfolge reibungslos funktioniert. Auch lässt sich fragen, in wie weit der Vertrieb die Informationen eines Kunden aufnimmt und intern weiterleitet oder ob der Vertrieb hinreichend ausgerüstet ist, die verschlechterte Risikosituation eines Firmenkunden frühzeitig zu erkennen und geeignete Maßnahmen einzuleiten. Hinsichtlich der prozessualen Dimension ist bspw. zu fragen, in wie weit es gelingt, die Vertriebsprozesskosten zu reduzieren oder die Firmenkunden in die Vertriebsprozesse zu integrieren. In der institutionellen Sicht ist etwa eine Frage, in wie weit Firmenkundenberater und Firmenkunden zueinander passen oder ob den einzelnen Firmenkundenberatern eine adäquate Quantität und Qualität an Firmenkunden zugeordnet ist. Ziel dieser Analyse ist es also, herauszufinden, welchen Einfluss die vertriebliche Disposition in funktioneller, prozessualer und institutioneller Hinsicht auf den Kundenwert als Ganzes, auf einzelnen Kundenwerttreiber und/oder deren Bestimmungsfaktoren hat. Vor allem das Verfahren des Benchmarking ist ein aussagekräftiger Ansatz bei der Analyse der Stärken und Schwächen des Vertriebs, da es nicht nur die Ergebnisse vergleicht, sondern darüber hinaus danach fragt, wie die Ergebnisse zustande gekommen sind und worin sich die Unterschiede bei den Benchmarkobjekten begründen (siehe Tabelle 28).[1839]

Tabelle 28: Mögliche Formen des kundenwertbezogenen Benchmarkings des Vertriebs[1840]

Parameter	**Ausprägungen des Parameters**		
Objekt	funktionaler Vertrieb	prozessualer Vertrieb	institutioneller Vertrieb
Zielgrößen	Kundenwert		
	Kundenwerttreiber		
	Bestimmungsfaktoren		
Vergleichs-partner	verschiedene Vertriebseinheiten	andere Primärbanken	

Die vorgestellten Analysen dürften in der Regel auf Basis des jeweiligen kreditgenossenschaftlichen Kundenstamms der Firmenkundenbank erfolgen. Insofern

1839 Vgl. *Horváth, P.* (2003), S. 413 ff. Zentrale Prämisse eines solchen Vorgehens ist eine die Benchmarkobjekte übergreifende einheitliche Konzeptualisierung und Operationalisierung des Kundenwerts z. B. anhand des in dieser Arbeit vorgestellten Modells.

1840 In Anlehnung an *Horváth, P.* (2003), S. 413.

kann die relative Analyse der Stärken und Schwächen durch das bisherige marktpolitische Verhalten der Primärbank verzerrt sein. Wichtige Impulse lassen sich zusätzlich gewinnen, wenn kundenwertbezogene Bezugsgrößen verbundweit erhoben und genutzt werden können. Verbundinterne Benchmarkings, die Voraussetzung eines Benchmarkings zwischen den einzelnen Primärbanken sind, sind dazu zentral zu entwickeln. Da es solche Benchmarkings zu anderen Themenfeldern schon gibt, bietet sich die Möglichkeit, auf bestehenden Infrastrukturen aufzusetzen.

Im Anschluss an die Analysephase werden die Vertriebsziele geplant und strategische Handlungsalternativen formuliert. In wie weit dies im Kontext des Kundenwerts erfolgen kann, wird nachfolgend beschrieben.

2.1.3 Vertriebszielplanung und -strategieformulierung

Es wurde in Kapitel C.3.1 dargestellt, dass die Kreditgenossenschaften grundsätzlich die generische Differenzierungsstrategie in Form der Qualitätsführerschaft verfolgen. Innerhalb dieser strategischen Vorgabe bewegen sich die nachfolgenden Ausführungen.

2.1.3.1 Vertriebszielplanung

Für die Herleitung von Strategien ist die Definition von strategischen Zielen als Orientierungspunkt notwendig. Das Fundamentalziel ist die Steigerung des Vermögenswerts. Die damit verbundenen Instrumentalziele wurden in Kapitel D.3.1.3 als Grundlage der Kundenwertmodellierung hergeleitet. In Kapitel E.2.1.1 wurde der Vermögenswert als die zentrale Ziel- und Steuerungsgröße hinsichtlich seiner Relevanz für die strategische Planung auf Gesamtbankebene konkretisiert. Wie bereits erläutert, konstituiert sich der Vermögenswert einer Bank aus dem „Zerschlagungswert“ und dem „Potenzialwert“.[1841] Vereinfachend gilt für den gesamtbanksteuerungsbezogenen Wert der Kreditgenossenschaft der in Formel [E.2] dargestellte Zusammenhang:

$$VW_t = ZW_t + PW_t \qquad \text{[E.2]}$$

mit:

VW_t : Vermögenswert der Kreditgenossenschaft zum Zeitpunkt t

ZW_t : Zerschlagungswert der Kreditgenossenschaft zum Zeitpunkt t

PW_t : Potenzialwert der Kreditgenossenschaft zum Zeitpunkt t

[1841] Vgl. *BVR* (2002b), S. 11.

Die Kundenwerte, die durchgängig aus in der Zukunft liegenden Transaktionen mit Firmenkunden oder auf nicht in monetären Größen messbaren Erfolgsbeiträgen des Kunden resultieren, sind sachlogisch dem „Potenzialwert" zuzuordnen. Diese Größe ist mangels konzeptioneller Grundlagen bisher selten oder gar nicht Gegenstand der strategischen Planung in der Kreditgenossenschaft.[1842] Wird der Auffassung gefolgt, dass sich die bankbetriebswirtschaftlichen Steuerungskonzepte der Kreditgenossenschaften weiter an die des Kreditbankensektors, für den die strenge Shareholder Value Orientierung angenommen wird, angleichen, so dürfte es langfristig so sein, dass der „Potenzialwert" als Bezugsgröße der Gesamtbanksteuerung im Allgemeinen und der strategischen Gesamtbanksteuerung im Besonderen aufgrund der Verknüpfung von Kundenwert und Unternehmenswert an Bedeutung gewinnen wird.[1843] Im Folgenden wird daher davon ausgegangen, dass die strategische Gesamtbankplanung um diese Komponente des Vermögenswerts erweitert wird. Damit wird die größtmögliche Kongruenz zwischen strategischer Planung auf Ebene der Gesamtbank, der Ebene der strategischen Geschäftseinheiten und den Funktionalbereichen erreicht, wenn für die Firmenkundenbank und den Vertrieb die Kundenwerte als strategische Zielgröße betrachtet werden. Gleichwohl gelingt es auch bei einer geringeren Kongruenz, die strategischen Ziele der Gesamtbank hinsichtlich des „Zerschlagungswerts" mit der kundenwertorientierten Zielformulierung für den Vertrieb der Firmenkundenbank zu harmonisieren. So lässt sich ein strategisches Ziel auf Basis des barwertigen Deckungsbeitrags III, der dem Vermögenszuwachs aus Kundengeschäften entspricht,[1844] mit zu realisierenden anbieterspezifischen Kundenwerten gleichsetzen. Insgesamt ist in diesen Fällen jedoch ein höherer Abstimmungsaufwand zwischen Gesamtbank- und Vertriebszielen für das Bank- und das Vertriebscontrolling zu konstatieren.

1842 Genau in diese Lücke ordnet sich diese Arbeit ein. Wenngleich an dieser Stelle keine gesamtbankbezogene Sicht eingenommen wird. Weiterhin ist in der Praxis festzustellen, dass in der Planung der Kreditgenossenschaften vor allem handelsrechtliche, periodische Größen der Gewinn- und Verlustrechnung und der Bilanz betrachtet werden. Von diesem Aspekt wird im Weiteren jedoch abstrahiert, da von der langfristigen Dominanz der betriebswirtschaftlichen Konzeption von VR-CONTROL ausgegangen wird.

1843 So sieht EBERLING die Kundenwertorientierung als zwingende Konsequenz einer irgendwie ausgeprägten Wertorientierung der Unternehmensführung. Vgl. *Eberling, G.* (2002), S. 43 ff. Ähnlich auch *von Wangenheim, F.* (2003), S. 11 ff. Und für SCHROEDER ist der Kundenwert die zentrale Steuerungsgröße der wertorientierten Unternehmenssteuerung. Vgl. *Schroeder, N.* (2006). Vgl. für die Verknüpfung von Kundenwert und Unternehmenswert z. B. *Schmöller, P.* (2001), S. 151 ff.; *Tewes, M.* (2003), S. 162 ff.; *Wille, K.* (2005), S. 182 ff.; *Stüker, D.* (2008), S. 31 ff.

1844 Vgl. *BVR* (2002b), S. 16.

Um den Vermögenswert der Kreditgenossenschaft zu steigern, kann diese anstreben, den „Zerschlagungswert“ oder den „Potenzialwert“ oder beide Werte zu erhöhen. In der strategischen Planung sind Erfolgspotenziale Gegenstand der Zielplanungsaktivitäten. Erfolgspotenziale sind im Kontext der Kundenwertorientierung determiniert durch die anbieterspezifischen Kundenwerte und Kundenwertpotenziale. Aus diesem Grund sind die strategischen Zielplanungen auf Ebene der „Potenzialwerte“, mithin unter anderem auf Ebene der firmenkundenspezifischen Kundenwerte zu konkretisieren. Der kreditgenossenschaftliche Vorstand definiert mit Blick auf den Vermögenswert der Bank den Entwicklungspfad sowie das zu erreichende Zielniveau des „Potenzialwerts“ am Ende des strategischen Planungszeitraums. Daraus lassen sich die „Zerschlagungswerte“ ableiten, die aus den Kundengeschäften resultieren.

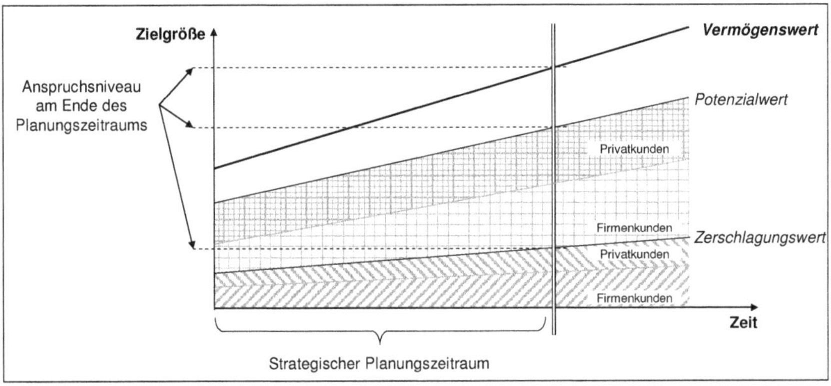

Abbildung 67: Zielgrößen der strategischen Gesamtbankplanung[1845]

Ausgehend von den Gesamtbankzielen leiten sich die strategischen Ziele der Firmenkundenbank und des Vertriebs ab. Für die Ableitung von kundenwertorientierten Vertriebszielen ist es sinnvoll, die Ziele näher zu konkretisieren, d. h. den Geltungsbereich, den Zielinhalt, das Zielausmaß sowie den Zeitbezug zu präzisieren.[1846]

1845 Eigene Darstellung. Nicht betrachtet wird das Treasury. Die Integrale unterhalb der Kurven für den Zerschlagungs- und Potenzialwert stellen die Ziele für die Firmen- bzw. Privatkundenbank dar. Die Addition der Kurven von Zerschlagungs- und Potenzialwert führt zur aggregierten Vermögenswertkurve. Die Abbildung erhebt keinen Anspruch auf eine realistische relative Struktur der Zielgrößen.

1846 Vgl. *Eberling, G.* (2002), S. 249 ff.

Der Geltungsbereich eines Ziels bestimmt, für welchen Gestaltungsraum ein Ziel Steuerungsfunktion besitzt.[1847] Als Gliederungskriterien der Präzisierung bieten sich neben der objektbezogenen Dimension des Kundenwerts auch die Aufbauorganisation der Kreditgenossenschaft und die Fristigkeit der Planung an. Bei der objektbezogenen Dimension wird zwischen Zielen auf Ebene des Kundenstamms und der Menge der Nichtkunden, Zielen auf Ebene von Kundengruppen und Zielen auf Ebene einzelner Kundenbeziehungen unterschieden. Es ist konsistent zum Verständnis der Kundenwertorientierung in dieser Arbeit, wenn strategische Ziele auf Ebene der einzelnen Kundenbeziehung formuliert werden. In vertikaler Sicht kann bei der Aufbauorganisation zwischen der Firmenkundenbank und Privatkundenbank sowie den Teilbanken der Produktions-, Steuerungs- und Vertriebsbank unterschieden werden. Der jeweilige Geltungsbereich determiniert auf Funktionalbereichsebene die Zielinhalte. Bei der Ableitung von Zielen für die einzelnen Geltungsbereiche ist darauf zu achten, dass diese widerspruchsfrei sind. So ist z. B. die Summe der Ziele der einzelnen Kundenbeziehungen kompatibel zu den Vertriebszielen zu formulieren, wobei die Vertriebsziele im Top-Down-Ansatz den Anspruch definieren, der wiederum durch die Gesamtbankplanung determiniert ist.

Der Zielinhalt gibt darüber Auskunft, welche Soll-Größe im Planungsprozess angestrebt wird. Um den auf den gesamten Kundenwert bezogenen Zielinhalt zu präzisieren, bieten sich zwei grundlegende Vorgehensweisen an:

- Zielableitung anhand definitionslogischer Beziehungen oder
- Zielableitung anhand sachlogischer Zweck-Mittel-Beziehungen.[1848]

Bei den definitionslogischen Zielbeziehungen stellen die untergeordneten Ziele Partialziele dar, aus denen sich mittels mathematischer Verknüpfung das Oberziel ableiten lässt. Dies wäre z. B. für die einzelnen Größen in Formel [D.30] für den monetären oder in Formel [D.32] für den nicht-monetären Kundenwert möglich. Sachlogisch abgeleitete Zielsysteme stützen sich auf vermutete oder empirisch ermittelte Zweck-Mittel-Relationen. So ließen sich z. B. für das Bindungspotenzial Ziele für das Ausmaß der Gebundenheit und/oder der Verbundenheit definieren, die positiv auf die Bindung des Kunden wirken. Im Rahmen der strategischen Planung sollten Ziele aufbauend auf den ermittelten Kundenwertpotenzialen auf Ebene des anbieterspezifischen Kundenwerts formuliert werden. In der operativen Planung, die Gegenstand des Kapitels E.2.2 ist, werden konkrete kundenwerttreiberbasierte Zielformulierungen erörtert. Diese Trennung begründet sich damit, dass die operative Vertriebsplanung konkreter und detaillierter als die strategische Vertriebsplanung ist und es daher als ausrei-

---

[1847] Vgl. *Eberling, G.* (2002), S. 249.

[1848] Vgl. hier und nachfolgend *Eberling, G.* (2002), S. 252.

chend angesehen werden kann, strategische Vertriebsziele auf Basis des anbieterspezifischen Kundenwerts unter Rekurs auf das Kundenwertpotenzial zu formulieren.

Hinsichtlich des Zielausmaßes werden unbegrenzte und begrenzte Ziele unterschieden. Bei unbegrenzten Zielen wird ein maximaler Zielerreichungsgrad und bei begrenzten Zielen das Erreichen eines bestimmten Anspruchsniveaus angestrebt.[1849] Als eine besondere Ausprägung eines unbegrenztes Ziels kann z. B. die Erhöhung des monetären Kundenwerts in fünf Jahren um mindestens 10 Prozent verstanden werden. Dies wird auch als satisfizierendes Zielausmaß bezeichnet. Ein begrenztes Ziel wäre die Steigerung des Kundenwerts um 10 Prozent in fünf Jahren und wird auch als fixiertes Zielausmaß bezeichnet. Für kundenwertorientierte Ziele werden satisfizierende Ziele empfohlen, da die mit der Kundenbewertung verbundenen Unsicherheiten fixierte Ziele konzeptionell inadäquat sind.[1850]

Allgemeine Ziele wie z. B. den Kundenwert steigern, sind zeitraumübergreifend und zeitraumunabhängig. Ziele sind erst dann als Steuerungsgrößen leistungsfähig, wenn sie zeitlich operationalisiert sind. Korrespondierend zur Trennung zwischen strategischer und operativer Planung, sind auch die Ziele strategisch oder operativ, wobei die strategischen Ziele meist über einen mehrjährigen Zeitraum zu erreichen sind und die operativen Ziele für ein Jahr gelten. Die konkreten Zeiträume sind institutsindividuell festzulegen. Für die strategische Planung gilt, dass sich die Zielvorgaben auf einen in der weiteren Zukunft liegenden Zeitraum beziehen.

Da der Kundenwert aus Sicht der Gesamtbank definiert ist, ist für die untergeordnete Ebene der Funktionalbereiche resp. Teilbanken zu konstatieren, dass strategische Funktionalbereichsziele auf Basis des gesamthaften Kundenwerts konzeptionell inadäquat sind. Dem Vertrieb wird i. d. R. keine Verantwortlichkeit über die produktionsbezogenen Standardstückkosten einzelner Bankleistungen, die Einfluss auf den monetären Kundenwert haben, zuzuweisen sein. Ebenso gilt nach dem Prinzip der dualen Steuerung, dass dem Vertrieb keine Verantwortung für die Risikostruktur der Firmenkundenbank zugewiesen werden kann.[1851] Daher sollte der Kundenwert in teilbankenbezogene Kundenwerte differenziert werden (siehe Formel [E.3]).[1852]

1849 Vgl. *Wöhe, G.* (1996), S. 126.

1850 Vgl. *Eberling, G.* (2002), S. 260.

1851 Vgl. dazu die Ausführungen Kapitel C.4.

1852 Die dargestellte additive Verknüpfung der teilbankenbezogenen Kundenwerte entspricht nicht den tatsächlichen vorliegenden funktionalen Verknüpfungen, die auch multiplika-

$$KW_c = KW_c^{Vertrieb} + KW_c^{Produktion} + KW_c^{Steuerung} \quad [E.3]$$

mit:

KW_c:	Anbieterspezifischer Kundenwert des Kunden *c* aus Sicht der Gesamtbank
$KW_c^{Vertrieb}$:	Durch die Vertriebsbank beeinflussbarer Kundenwert des Kunden *c*
$KW_c^{Produktion}$:	Durch die Produktionsbank beeinflussbarer Kundenwert des Kunden *c*
$KW_c^{Steuerung}$:	Durch die Steuerungsbank beeinflussbarer Kundenwert des Kunden *c*

Strategische Kundenwertziele des Vertriebs der Firmenkundenbank beziehen sich somit auf den durch die Vertriebsbank beeinflussbaren Kundenwert, der auch als eigenständiger monetärer Kundenteilwert gemessen werden kann, und diejenigen Einflussgrößen des monetären und nicht-monetären Kundenwerts, die in dessen unmittelbaren oder mittelbaren Einflussbereich liegen. Dazu gehören mit Blick auf die Beziehungsdimension vor allem Ziele über die Bindung und Akquisition von Firmenkunden. Auf Ebene der Kundenwerttreiber der beziehungsbezogenen Ertragsdimensionen gehören dazu kundenwertbezogene Ziele, die sich auf den Absatz von Bankleistungen, die damit verbundenen Vertriebskosten und vertriebliche Ansätze zur Hebung des Kooperationspotenzials (z. B. über die Gestaltung von Vertriebskanälen bei der Beratung und Betreuung von Firmenkunden) beziehen. Hinsichtlich der beziehungsübergreifenden Kundenwerttreiber beziehen sich die Ziele auf die Möglichkeiten des Vertriebs das Informations- (z. B. Beschaffung und Aufnahme von Informationen), das Referenz- (z. B. Akquisition und Bindung referenzpotenzieller Unternehmen) und das Member Value-Potenzial (z. B. Akquisition von gewerblichen Mitgliedern) zu erhöhen. Hinsichtlich der Risikodimension können strategische Ziele für das Kundenplanungsrisiko und die Adressrisiken definiert werden, sofern diese durch den Vertrieb beeinflusst werden können.[1853] Die Präzisierung der Zielinhalte umfasst auch die Betrachtung der Bestimmungsfaktoren, die sich als Zielgrößen strategischer Ziele eignen, sofern diese durch den Vertrieb beeinflusst werden können. Dazu gehören vor allem vorökonomische Ziele wie z. B. Kundenzufriedenheit oder Dienstleistungsqualität in der Firmenkundenbetreuung. Insgesamt wird bei der Auswahl kundenwertbezogener Ziele deutlich: „Short-

tiv sein können. Sie ist eine vereinfachende Darstellung zur Verdeutlichung des grundlegenden Gedankens.

1853 So sieht BATZ die Möglichkeit eines entscheidenden Einflusses auf die Bonitätsentwicklung beim Markt aufgrund dessen relativ höchster Kundenkenntnis. Vgl. *Batz, C.* (2004), S. 35.

term objectives such as sales volume are [.] replaced by long-term objectives such as customer satisfaction; sales and profits are considered as being the rewards for delivering customer satisfaction and realizing customer value [...]. Also, transaction-related objectives such as market share [...] are replaced by relational objectives such as share of customer [...]. In replacing transaction-oriented objectives by relational objectives, an organization focuses on the quality rather than the quantity of the customer base."[1854]

Nachdem die Vertriebsziele festgelegt sind, sind die Strategien und Aktionspläne zu formulieren, mit denen diese Ziele erreicht werden sollen. Dies ist Gegenstand des nächsten Kapitels.

2.1.3.2 Vertriebsstrategieformulierung

Die Vertriebsstrategie beinhaltet die langfristig wirkende funktionelle, prozessuale und institutionelle Ausrichtung des Vertriebs und fokussiert auf die Auswahl der Zielkunden und der relevanten Wettbewerbsvorteile, die Ausgestaltung des Vertriebssystems und der Vertriebswege sowie der Preispolitik.[1855] Vor dem Hintergrund der Kundenwertorientierung bietet es sich an, für die Formulierung von Vertriebsstrategien zwei Ebenen zu unterscheiden, denn in der Literatur wird empfohlen, kundenwertbezogene Strategien im Allgemeinen und damit auch Vertriebsstrategien auf Ebene der einzelnen Kundenbeziehung und nicht auf Ebene von Kundengruppen oder des gesamten Kundenstamms zu entwickeln.[1856] In Verbindung mit dem in den Kapiteln C.2.1 und E.2.1.1 genannten Drei-Ebenen-Konzept resultiert somit ein Vier-Ebenen-Konzept, das dazu zwingt, die strategischen Entscheidungen der Funktionalbereichsebene des Vertriebs und der Ebene der Kundenbeziehungen abzugrenzen.[1857]

Auf der Ebene des gesamten Vertriebs sind langfristig bindende Entscheidungen hinsichtlich der institutionellen, funktionellen und prozessualen Merkmale des Vertriebs zu treffen. Dazu zählen z. B. Entscheidungen hinsichtlich der aufbauorganisatorischen Gestaltung des Vertriebs und der Art und Anzahl relevanter Vertriebskanäle, aber auch Entscheidungen über das grundsätzliche Vorgehen in der Firmenkundenbank. Entscheidungen dieser Ebene sind kundenübergreifend gültig, formell und haben eine besonders hohe Bindungswirkung

---

1854 *Hoekstra, J. C./Huizingh, E. K. R.* (1999), S. 262.

1855 Vgl. *Homburg, C./Schäfer, H./Schneider, J.* (2010), S. 27 ff.

1856 Vgl. *Stüker, D.* (2008), S. 327.

1857 Vgl. *Payne, A./Frow, P.* (2005), S. 170.

für künftige Entscheidungen, da nachträgliche Änderungen mit ggf. hohen Anpassungskosten verbunden sein können.[1858]

Auf der Ebene der einzelnen Kundenbeziehungen werden dagegen die materiellen Inhalte der Vertriebsstrategie festgelegt. Zu diesen zählen vor allem die Selektion und Segmentierung der Firmenkunden, die Handlungsalternativen der Kundenbearbeitung sowie die inhaltliche Ausgestaltung von Instrumenten der Vertriebsunterstützung (z. B. der VR-FINANZPLAN MITTELSTAND). Entscheidungen der materiellen Vertriebsstrategie haben stärker konzeptionellen Charakter und können daher vergleichsweise einfacher geändert werden. Gleichwohl sind die formelle und die materielle Ebene abhängig voneinander, wodurch die Anpassungsmöglichkeiten begrenzt sind. So z. B. wenn aufbauend auf der Kundensegmentierung und der Kundenselektion institutionelle Vertriebsstrukturen definiert wurden.

Hinsichtlich der formellen Aspekte sind die strategischen Entscheidungsalternativen institutsindividuell zu entwickeln und zu bewerten. Allgemeingültige Aussagen hierzu lassen sich nicht oder nur schwer ableiten. In der Literatur werden daher vornehmlich die Aspekte der materiellen Strategieebene im Lichte des Kundenwerts diskutiert, da hier eher allgemeingültige Aussagen zur Strategieentwicklung getroffen werden können. Ausgangspunkt der Formulierung von Vertriebsstrategien ist die Frage, ob die strategischen Kundenwertziele des Vertriebs mit der gegenwärtigen Strategie und den damit verbundenen Maßnahmenprogrammen und Aktionsplänen erreicht werden können. Im Sinne der GAP-Analyse kann z. B. dem Ziel, den Gesamtfirmenkundenwert im Planungszeitraum um 20 Prozent auf 120 zu erhöhen, mit dem prognostizierten Ergebnissen auf Basis der gegenwärtigen Strategie und Planung gegenübergestellt werden. Zeigt sich, dass das formulierte Ziel nicht erreicht werden kann, ist zu fragen, welche Strategieänderungen oder -wechsel notwendig sind, um das Ziel zu erreichen. Der erste Blick richtet sich dabei auf die Bestandskunden und deren Kundenwertpotenzial sowie den Möglichkeiten, dies auszuschöpfen. Zeigt sich, dass eine Intensivierungsstrategie nicht ausreichend ist, fällt der Blick auf die Nichtkunden, so dass Überlegungen über eine Extensivierungsstrategie anzustrengen sind.[1859] Diese beiden strategischen Stoßrichtungen sind die grundsätzlichen Leitlinien der weiteren Überlegungen für die Vertriebsstrategie.

[1858] So z. B. wenn das Vertriebssystem von einem werkseigenen zu einem offenen Vertriebssystem geändert werden soll und damit die Notwendigkeit einhergeht, Personalstrukturen zu ändern.

[1859] Die Begriffe der Intensivierungs- und Extensivierungssstrategie leiten sich aus einer von SCHMOLL für das Firmenkundengeschäft adaptierten Ansoff-Matrix ab. Vgl. *Schmoll, A.* (2006), S. 78 f.

Es wird zwischen zwei grundsätzlichen strategischen Handlungsalternativen mit Blick auf die Bestandskunden unterschieden, die sich auf die Quantität des Ressourceneinsatzes beziehen. Zum einen der Erhalt und die Entwicklung von auch ggf. unter gegenwärtigen Planungen und Maßnahmen wertlosen Kundenbeziehungen. Zum zweiten die konsequente Beendigung von Kundenbeziehungen, die einen negativen Kundenwert aufweisen.[1860] Die Auswahl von Strategien erfolgt auf Basis wertbezogener Kriterien, d. h. dem anbieterspezifischen Kundenwert aus gegenwärtiger Perspektive, dem gesamten Kundenwertpotenzial sowie der Wahrscheinlichkeit, dieses Potenzial durch eine veränderte Strategie zu erschließen.[1861] Der kundenwertbezogene Blick auf die Kunden stellt eine an der Effizienz ausgerichtete Determinante der Strategieformulierung dar, die die an der Effektivität orientierte Kundensegmentierung nach (kauf-) verhaltensorientierten Merkmalen ergänzt.[1862]

Zur Ableitung von Strategien haben sich Portfolioanalysen als das klassische Instrument etabliert[1863] und stellen auch zur Steuerung der vielfältigen Aktivitäten gegenüber Kunden ein gängiges methodisches Hilfsmittel dar.[1864] Mit Blick auf die Portfoliodarstellung in Abbildung 65 lassen sich bspw. kundenwertbasierte Normstrategien formulieren. Kunden im Quadranten 3, die sich durch einen unterdurchschnittlichen Ausschöpfungsgrad und ein unterdurchschnittliches Kundenwertpotenzial auszeichnen, stellen Kandidaten für eine potenzielle Beendigung der Kundenbeziehung dar, wenn es unwahrscheinlich ist, das Kundenwertpotenzial oder den anbieterspezifischen Kundenwert durch bankbezogene Einflussfaktoren zu erhöhen.[1865] TOMCZAK/REINECKE/ FINSTERWALDER unterscheiden zwei Basisstrategien, die bei diesen Geschäftsbeziehungen angewendet werden können.[1866] Bei einer Teilausgrenzungsstrategie wird

---

1860 Vgl. *Tewes, M.* (2003), S. 196. Ähnlich auch HILGENSTOCK, der noch die Beobachtungsstrategie als Zwischenstufe eingefügt. Vgl. *Hilgenstock, G. P.* (1995), S. 811.

1861 Vgl. *Tewes, M.* (2003), S. 197.

1862 Vgl. ähnlich *Blattberg, R. C./Deighton, J.* (1996), S. 136 ff. In Praktikerbeiträgen wird regelmäßig postuliert, eine Segmentierung auf Basis von Kundenwerten oder Potenzialen vorzunehmen. Eine Berücksichtigung sonstiger Kriterien, um Kundengruppen zu definieren, die die Grundlage der Auswahl vertrieblicher Strategien und Maßnahmen sind, wird dabei nicht vorgenommen. Es ist dagegen jedoch zu empfehlen, die klassische Kundensegmentierung anhand kaufverhaltens- oder bedarfsrelevanter Kriterien, um die ökonomische Segmentierung mit Rekurs zum Kundenwert zu ergänzen. Vgl. dazu *Homburg, C./Schäfer, H./Schneider, J.* (2010), S. 34 ff. Die Autoren sprechen von Kundensegmentierung (Effektivität) und Kundenpriorisierung (Effizienz).

1863 Vgl. *Horváth, P.* (2003), S. 386; *Götz, P./Diller, H.* (1991), S. 25.

1864 Vgl. *Götz, P./Diller, H.* (1991), S. 1.

1865 Vgl. *Weber, J./Lissautzki, M.* (2004), S. 40 f.; *Günther, B./Helm, S.* (2003), S. 61.

1866 Vgl. hier und nachfolgend *Tomczak, T./Reinecke, S./Finsterwalder, J.* (2000), S. 413 ff.

die Geschäftsbeziehung zwar nicht beendet, dem Kunden wird jedoch nur ein begrenztes Leistungsspektrum offeriert. Dies impliziert also nicht, dass keine Transaktionen mit dem Kunden mehr vorgenommen werden. Es beinhaltet vielmehr, dass aus Sicht der Kreditgenossenschaft die Bereitschaft fehlt, eine Geschäftsbeziehung mit dem Kunden aktiv aufrechtzuerhalten.[1867] Statt eines relationalen Vertriebsansatzes wird für diese Kunden ein transaktionaler Vertriebsansatz verfolgt, so dass der Umfang zu investierender Ressourcen in die Kundenbeziehung auf Null gesenkt wird. Die Totalausgrenzung, bei der die Kundenbeziehung zum Kunden insgesamt aktiv beendet wird, ist die zweite Basisstrategie.[1868] Zwar wird ein derartiges Vorgehen als „ultima ratio“ beurteilt,[1869] dennoch bedingen begrenzte Ressourcen der Kreditgenossenschaft diese Kundenselektions-[1870] und Kundenpriorisierungsentscheidungen[1871]. Insbesondere aufgrund des Regionalprinzips und der Kommunikationsprozesse zwischen Firmenkunden sollten Kreditgenossenschaften jedoch hinreichend vorsichtige kundenwertbezogene Maßstäbe anlegen, um Kundenselektionsentscheidungen zu begründen.[1872] Dazu gehört auch, eine dezidierte Analyse der Auswirkungen einer Kundenbeendigung auf den gesamten Kundenstamm vorzunehmen.[1873] Weiterhin ist zu berücksichtigen, dass die in Kapitel D.3.2 diskutierten Besonderheiten bei Mitglieder-Kunden zu beachten sind, so dass ggf. aufgrund des Förderauftrags die Kundenbeziehung zu einem Mitglieder-Kunden trotz eines geringen Kundenwerts nicht beendet werden sollte.

Bei Kunden im Quadranten 4, die gekennzeichnet sind durch ein hohes Kundenwertpotenzial und einen unterdurchschnittlichen Ausschöpfungsgrad, sind strategische Handlungsalternativen zu eruieren, wie der anbieterspezifische Kundenwert gesteigert werden kann. Ansatzpunkte dazu lassen sich z. B. mittels des Kundenkubus in Abbildung 66 finden, indem ausschließlich Kunden des Quadranten 4 in diesem Kundenkubus eingeordnet werden, oder auf Basis von

---

1867 Vgl. zum Begriff der Geschäftsbeziehung auch Kapitel C.2.4.1.3 sowie ausführlich *Segbers, K.* (2007), S. 10 ff.

1868 Vgl. *Tomczak, T./Reinecke, S./Finsterwalder, J.* (2000), S. 413 ff.

1869 Vgl. *Fischer, T./von der Decken, T.* (2001), S. 304.

1870 Vgl. *Diller, H.* (1995a), Sp. 1368.

1871 Vgl. *Droll, M.* (2008), S. 1 ff.; *Günter, B./Helm, S.* (2006), S. 48.

1872 COLLINGS/BAXTER verweisen in diesem Kontext auf die negativen Imagewirkungen, die mit der anbieterseitigen Beendigung einer Kundenbeziehung einhergehen können. Vgl. *Collings, D./Baxter, N.* (2005), S. 27.

1873 Die regional begrenzten Geschäftsgebiete der Genossenschaftsbanken führen dazu, dass negativen Referenzwirkungen und der Möglichkeit, dass ausgegrenzte Kunden ggf. bei anderen Banken Kunden werden und deren Marktanteil steigern, besonderes Augenmerk zu schenken ist. Vgl. *Günter, B./Helm, S.* (2003), S. 56.

verhaltens- und merkmalsbasierten Kundensegmentierungen.[1874] Fixpunkte möglicher Strategien wären z. B. die Risiken zu reduzieren, den Bindungsgrad oder das Ertragspotenzial zu erhöhen. HOMBURG/STEINER/TOTZEK sprechen in diesem Zusammenhang von offensiven Marketingstrategien, deren Ziel in der Erhöhung der einzelnen Kundenwerte liegt.[1875] Diese werden auch als Ausschöpfungs- oder Entwicklungsstrategien bezeichnet. Ansatzpunkte, den anbieterspezifischen Kundenwert bei gegebenem Kundenwertpotenzial zu erhöhen, liegen in der Steigerung der Kundenbindung, der Erhöhung des Loyalitätsgrads, der Ausschöpfung von Preiselastizitäten, dem Absatz bisher nicht abgesetzter Leistungen, dem Senken der kundenbezogenen Prozesskosten oder der kundenbezogenen Risiken.[1876] Selten wird in diesem Zusammenhang auch genannt, dass sich der anbieterspezifische monetäre Kundenteilwert steigern lässt, wenn die Bankleistungen früher als prognostiziert abgesetzt werden können.[1877]

Bei Kunden im Quadranten 1 und 2 ist zu ermitteln, wie diese gegenüber wettbewerblichen Akquisitionsbemühungen abgesichert werden können. Insbesondere der Aufbau von Wechselbarrieren kann als Gegenstand einer entsprechenden Absicherungsstrategie genannt werden. Ziel dieser defensiven Marketingstrategien ist es, den anbieterspezifischen Kundenwert dieser Kundenbeziehungen zu stabilisieren.[1878]

Zeigt sich, dass mit diesen kundenstammbezogenen Vertriebsstrategien die strategischen Vertriebsziele nicht erreicht werden können, sind Strategien zu diskutieren, wie werthaltige Nichtkunden gewonnen werden können. Zu den Nichtkunden zählen nach dem numerischen Kundenpotenzial[1879] sowohl diejenigen Firmenkunden, die bisher keine Geschäftsbeziehung zur Kreditgenossenschaft hatten, als auch diejenigen Firmenkunden, die in der Vergangenheit mal eine Geschäftsbeziehung zur Kreditgenossenschaft unterhalten haben. Folglich sind Akquisitions- und/oder Rückgewinnungsstrategien zu formulieren. Für eine

---

1874 Vgl. dazu ausführlich *Duderstadt, S.* (2006), S. 74 ff. In der Literatur wird allenthalben argumentiert, die kundenwertbezogene Kundensegmentierung sei alleinig ausreichend, für die Ableitung von Vertriebsstrategien. Zu empfehlen ist jedoch ein zweistufiges Kundensegmentierungsmodell, das zunächst die Kunden nach adäquaten Kriterien zur Ansprache segmentiert und innerhalb dieser Segmente dann nach Kundenwerten klassifiziert. Für mögliche Kriterien der Segmentierung von Firmenkunden vgl. *Schulz, T. C.* (2005).

1875 Vgl. *Homburg, C./Steiner, V. V./Totzek, D.* (2009), S. 71 f.

1876 Ähnlich *Stüker, D.* (2008), S. 328; *Tewes, M.* (2003), S. 200 f.

1877 Dies begründet sich mit der positiven Zeitpräferenzrate der Entscheider, wonach heutige Zuflüsse höher gewertet werden als künftige. Vgl. dazu die Ausführungen in Kapitel D.5.2.2.3.

1878 Vgl. *Homburg, C./Steiner, V. V./Totzek, D.* (2009), S. 71 f.

1879 Vgl. dazu die Ausführungen in Kapitel D.1.2.2.

erfolgreiche Strategieentwicklung bei Nichtkunden stehen die Analyse der Bindung an die gegenwärtige(n) Bank(en) sowie die Möglichkeiten, die Bindung zu reduzieren, im Fokus der Betrachtung. Bei den Rückgewinnungsstrategien ist zusätzlich zu fragen, aus welchen Gründen die ehemals vorhandene Geschäftsbeziehung beendet wurde (Kündigungsgrundanalyse) und ob diese Gründe auch gegenwärtig ein Hemmnis für den Wiederaufbau der Geschäftsbeziehung darstellen. Uninteressant für die Rückgewinnungsabsichten sind nach STAUSS/ FRIEGE absichtlich vertriebene, sich entfernte und weggekaufte Kunden.[1880]

Wenngleich sich diese Normstrategien plausibel aus dem Kundenportfolio in Abbildung 65 ableiten lassen, gilt, dass „diese keinen dogmatischen Charakter haben. Sie geben nur die Richtung der zu treffenden Entscheidungen an."[1881] Letztlich kann hinsichtlich des Kundenwerts in Verbindung mit den formulierten Zielen für den „Potenzialwert" der Kreditgenossenschaft allgemein für die Strategiefindung nur postuliert werden, keine Geschäftsbeziehungen zu Kunden langfristig aufrecht zu erhalten, die einen negativen Kundenwert und keine Chancen der Änderung dieser Situation aufweisen. In diesem Postulat werden sowohl der monetäre als auch der nicht-monetäre Kundenwert gleichermaßen berücksichtigt, indem folgendes Entscheidungsmodell der Kundenbewertung und -selektion in der strategischen Planung berücksichtigt wird (siehe Abbildung 68).

[1880] Vgl. *Stauss, B./Friege, C.* (2006), S. 517.

[1881] *Götz, P./Diller, H.* (1991), S. 28.

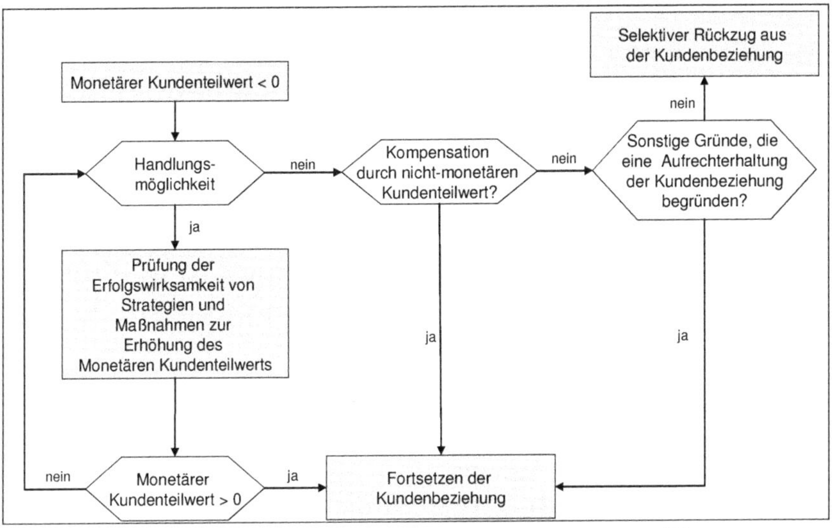

Abbildung 68: Entscheidungsmodell im Rahmen der strategischen Planung[1882]

2.1.4 Strategiebewertung

Nachdem verschiedene strategische Handlungsalternativen des Vertriebs auf Basis des Kundenwerts bei gegebenen Strategien und Plänen bestimmt worden sind, ist deren Erfolgswirksamkeit hinsichtlich des zu erreichenden Ziels zu bestimmen, um eine finale Auswahl der Vertriebsstrategie vorzunehmen. Dazu wird im Schrifttum eine aus der wertorientierten Unternehmensführung bekannte Vorgehensweise empfohlen.[1883] Im ersten Schritt wird der monetäre und nicht-monetäre Kundenteilwert zu Beginn der Planungsrunde auf Basis der bestehenden Planung bestimmt. Dieser Wert ist, wie schon in den bisherigen Ausführungen zur Analyse- und Planungsphase deutlich wurde, die Ausgangsbasis der Strategiebewertung und gilt als auch als Werthürde. Im zweiten Schritt sind die Kundenteilwerte und die Ausprägungen Kundenwerttreiber auf Basis der entwickelten strategischen Handlungsalternativen (Vertriebsstrategiealternativen) zu ermitteln. Die Differenz zwischen den Kundenwerten auf Basis der bestehenden Planung und auf Basis einer bestimmten Vertriebsstrategie stellt den Maßstab für die Erfolgswirksamkeit dieser einzelnen Vertriebsstrategiealternative dar. In diesem Zusammenhang ist es von Bedeutung, auch die mit der Strategie verbundenen Vertriebskosten in die Kundenbewertung einfließen zu lassen.

1882 In Anlehnung an *Tewes, M.* (2003), S. 205.

1883 Vgl. *Tewes, M.* (2003), S. 208; *Stüker, D.* (2008), S. 211.

Dieser Vergleich ist für alle formulierten Vertriebsstrategien vorzunehmen, um eine Entscheidungsgrundlage für die Auswahl einer tatsächlich umzusetzenden Vertriebsstrategie zu erhalten. Es sollte die Strategie ausgewählt werden, die die höchste strategieinduzierte Wertgenerierung impliziert.[1884]

Eines der zentralen Probleme dieser Vorgehensweise ist die Schätzung der Reagibilität des Kundenwerts oder Kundenwerttreiber auf die Vertriebsstrategie. Es ist begründet davon auszugehen, dass sich die einzelnen Strategien unterschiedlich auf die Höhe, die zeitliche Struktur und die Varianz der Kundenwerttreiber auswirken. Eine punktgenaue Schätzung dürfte beinahe unmöglich sein, während es realistisch erscheint, Wirkungsszenarien aufzubauen, die die Bandbreite der Strategiewirkung abbilden. Um dieser Problematik Rechnung zu tragen, lassen sich die bereits erwähnten Szenarioanalysen in der Kundenbewertung ergänzen um Sensitivitätsanalysen und Monte-Carlo-Simulationen.[1885] Während die Sensitivitätsanalyse nur einzelne Faktoren hinsichtlich ihrer Wirkung auf den Kundenwert betrachten kann, stellt die Monte-Carlo-Simulation ein Verfahren dar, das mehrere unsichere Einflussfaktoren gleichzeitig bezüglich ihrer Wirkung auf den Kundenwert berücksichtigen kann.[1886]

Da die Kundenwerte nicht nur durch die Strategien des Vertriebs, sondern auch durch die Strategien der Produktionsbank und der Steuerungsbank beeinflusst werden können, sind diese in einer übergeordneten Betrachtung auf Gesamtbankebene durch das zentrale Gesamtbankcontrolling zusammenzuführen, um den aus Gesamtbanksicht optimalen Strategiemix zu ermitteln.[1887] Somit kann der Kundenwert als Bewertungsmaßstab für die Koordination der strategischen Ziele der einzelnen Teilbanken verwendet werden:[1888]

$$\begin{aligned} KW_c(s_i^V, s_i^P, s_i^{St}) = {} & KW_c^{Vertrieb}\left(s_i^V\right) \\ & + KW_c^{Produktion}\left(s_i^P\right) \\ & + KW_c^{Steuerung}\left(s_i^{St}\right) \end{aligned} \qquad [E.4]$$

mit:

1884 Vgl. *Tewes, M.* (2003), S. 208; *Stüker, D.* (2008), S. 334.

1885 Vgl. hier und nachfolgend *Stüker, D.* (2008), S. 333 ff.

1886 Vgl. für einen beispielhaften Einsatz der Monte-Carlo-Simulation im Rahmen der Kundenbewertung *Gleißner, W.* (2005), S. 416 f.

1887 Vgl. *Blattberg, R. C./Deighton, J.* (1997), S. 25, die beispielweise fordern, dass die Beurteilung neuer Produkte, neuer Maßnahmen oder neuer Kundendienstangebote an der Frage gemessen werden, ob der Nettokapitalwert der Kunden dadurch wächst.

1888 Analog zur Formel [E.3] gilt, dass die additive Verknüpfung nur zu Darstellungszwecken gewählt wurde. Realiter ist von einer vergleichsweise komplexen und mehrfach rückgekoppelten Verknüpfung der einzelnen Strategiealternativen auszugehen.

s_i^V : Strategiealternative *i* der Vertriebsbank

s_i^P : Strategiealternative *i* der Produktionsbank

s_i^S : Strategiealternative *i* der Steuerungsbank

2.1.5 Strategieumsetzung

Die Phase der Strategieumsetzung stellt die sachlogische Verknüpfung zwischen der strategischen und der operativen Ebene der Vertriebssteuerung dar. Die strategische Vertriebsplanung schafft den organisatorischen Rahmen, so dass in der operativen Vertriebsplanung die langfristige Perspektive der strategischen Planung in kurz- und mittelfristige Bezugszeiträume dekomponiert wird, welchen operative Ziele und Maßnahmen zugeordnet werden.[1889] Die Herleitung von operativen Zielen und Maßnahmen ist Gegenstand der operativen Vertriebsplanung.

2.2 Operative Vertriebsplanung

Während die strategische Vertriebsplanung festlegt, welche Erfolgspotenziale, verstanden als Kombination aus Kundenwerten und bankinternen Vertriebsstrukturen und -prozessen, mit welchen Strategien abgesichert, ausgeschöpft oder entwickelt werden sollen, liegt der Fokus der operativen Vertriebsplanung in einer detaillierten und kurz- bis mittelfristigen Umsetzung dieser Strategien. Die operative Vertriebsplanung konkretisiert die Aufgaben des Ausführungssystems hinsichtlich der Ziele, Aktivitäten, Ressourcen, Aufgabenträger und Termine.[1890] Operative Vertriebspläne stellen die detaillierte gedankliche Antizipation der Handlungen im Vertriebsausführungssystem dar und dienen insbesondere der Kommunikation zwischen Vertriebsleitung und Vertriebsmitarbeitern.[1891] Im Gegensatz zur strategischen Vertriebsplanung zeichnet sich die operative Vertriebsplanung durch ein höheres Maß der Präzisierung aus und erfolgt auf Basis kurzfristig gegebener Strukturen und Kapazitäten.[1892] Die höhere Präzisierung zeigt sich insbesondere darin, dass kundengruppenbezogene und kundenindividuelle, differenzierte Ziele formuliert werden.[1893]

1889 Vgl. *Wille, K.* (2005), S. 246.

1890 Vgl. *Duderstadt, S.* (2006), S. 127; *Wille, K.* (2005), S. 250 f.

1891 Vgl. *Tewes, M.* (2003), S. 183.

1892 Vgl. *Althans, M.* (2001), S. 76.

1893 Vgl. *Reiners, J. O.* (2004), S. 124; *Hilgenstock, G. P.* (1995), S. 815.

Aufgrund der Zukunftsgerichtetheit und Langfristigkeit der Planung sollte die Vertriebsplanung als rollierende Planung ausgestaltet werden, so dass der Präzisionsgrad der Vertriebsplanung mit zunehmendem Zukunftsbezug abnimmt und auf Basis der aus der Überwachsungs- und Sicherungsphase gewonnen Informationen Planänderungen vorgenommen werden können.[1894]

Analog zur strategischen Planung sind für die weiteren Ausführungen zunächst die Anknüpfungspunkte an die operative Gesamtbankplanung herauszustellen, um Gestaltungsempfehlungen im Lichte des Kundenwerts ableiten zu können.

2.2.1 Anknüpfungspunkte an die operative Gesamtbankplanung

Der operative Planungsprozess in den Kreditgenossenschaften wird in VR-CONTROL konzeptionell als Gegenstromverfahren beschrieben, das in drei Phasen eingeteilt wird. Im Rahmen der strategischen Gesamtbankplanung werden Entwicklungspfade und Anspruchsniveaus für die Zielgrößen im und zum Ende des Planungszeitraums Top-Down festgelegt. Daraus werden qualitative Leitlinien und quantitative Vorgaben des Gesamtbankergebnisses für die kommende Planungsperiode abgeleitet. Qualitative Leitlinien und quantitative Vorgaben beziehen sich auf die einzelnen in Tabelle 29 aufgeführten Zielgrößen.[1895] Die jeweiligen Zielgrößen werden so operationalisiert, d. h. hinsichtlich des Ausmaßes und Zeitbezugs festgelegt, dass der gewünschte Ergebnispfad der Bankentwicklung im Sinne der strategischen Ziele erreicht wird. Dabei werden diese Größen zunächst auf Gesamtbankebene definiert und abschließend auf die einzelnen Teilportfolios disaggregiert, so dass diese in Summe die Gesamtbankziele ergeben.

1894 Vgl. zur rollierenden Planung *Horváth, P.* (2003), S. 204; *Küpper, H.-U.* (2001), S. 303.

1895 Vgl. hier und nachfolgend *BVR* (2002b), S. 122; *BVR* (2003b), Kapitel 1.2, S. 3 ff.

Tabelle 29: Beispielhafte kurzfristige (1-Jahres-)Gesamtbankplanung[1896]

Zielgröße	Qualitative Leitlinie	Quantitative Vorgabe	
Gesamtbankergebnis und Eigenkapitalrentabilität		Eigenkapitalrendite:	20 %
		Ergebnisanspruch:	25 Mio. EUR
Gesamtbankrisikolimit	Aus dem Firmenkundenportfolio ist ein ausgeglichenes Adressrisikoergebnis zu erwirtschaften. Im schlimmsten Fall dürfen Markt- und Adressrisiken das geplante Gesamtbankergebnis erreichen, so dass das Vermögen der Bank unangetastet bleibt.	Gesamtbankrisikolimit:	25 Mio. EUR
		Adressrisikolimit:	20 Mio. EUR
		Adressrisikoergebnis:	0 Mio. EUR
		Marktpreisrisikolimit:	5 Mio. EUR
		Marktrisikoergebnis:	2 Mio. EUR
Kundengeschäftsergebnis	Aus dem Kundengeschäft der Kreditgenossenschaft ist eine stabile Grundrentabilität zu erwirtschaften. Systematische Verbesserung der Risikoqualität des Firmenkundengeschäftes durch Beschränkung der Neugeschäfte auf die Kunden der besten Ratingkategorien.	Deckungsbeitrag III:	24 Mio. EUR
Produktivitätsergebnis	Es ist ein ausgeglichenes Produktivitätsergebnis anzustreben, so dass die vereinnahmten Standardstückkosten sämtliche dem Kundengeschäft zurechenbaren Kosten abdecken.	Produktivitätsergebnis:	0 Mio. EUR
Overheadergebnis	Die Overheadkosten dürfen nur in Höhe der Inflationsrate steigen.	Overheadergebnis:	-1 Mio. EUR

Die Skizze der operativen Gesamtbankplanung macht deutlich, dass die Schnittstelle zwischen Gesamtbankplanung und Vertriebsplanung das Kundengeschäftsergebnis ist. Die zentrale Zielgröße dabei ist der Deckungsbeitrag III, für

1896 Eigene Darstellung. Beispielhafte Zahlen modifiziert entnommen aus *BVR* (2003b), Kapitel 1.2, S. 5.

den ein eigenständiger Planungsprozess vorliegt.[1897] Ausgangspunkt dieses Planungsprozesses ist zunächst die Feststellung, auf welcher Ebene der Kundengeschäftserfolg geplant wird. Als Alternativen werden die Ebenen des Controllingwürfels genannt, der die Dimensionen Kunden, Produkte und Profit-Center hat. Anschließend wird in der Top-Down-Planung der Ertragsanspruch an das Kundengeschäft formuliert, der sich aus der strategischen Planung ergibt. Dazu werden die strategischen Ziele in periodische Ziele heruntergebrochen. Auf Basis der Ergebnisse der Top-Down-Planung führen die Ergebnisverantwortlichen die Planung auf Basis einzelkunden- oder kundengruppenbezogener Geschäftszahlen durch (Bottom-Up-Planung). Der höhere Detaillierungsgrad der Bottom-Up-Planung in Verbindung mit bankleistungsbezogenen Volumen- und Stückzahlen stellt die Ausgangsbasis für die Anforderungen hinsichtlich personeller und sachlicher Kapazitäten sowie der Koordinationstätigkeiten mit den Verbundunternehmen dar, die auf Basis der Planzahlen formuliert und vorgenommen werden. Zwar stellt der Deckungsbeitrag III die wichtigste und zentrale Größe dar, dennoch sind Hilfsgrößen in Form der Inputgrößen des Deckungsbeitrags III (Volumen, Marge, Risikoprämien und Standardstückkosten) ebenfalls Gegenstand der Zielplanungsüberlegungen. Basierend auf dem Planergebnis obliegt es den Führungskräften der jeweiligen strategischen Geschäftseinheiten und Funktionalbereiche, die Planrealisation vorzunehmen. Dazu gehört vornehmlich die Planung der Maßnahmen, mit denen die Ziele erreicht werden sollen.[1898] In der Abstimmungsphase werden die Top-Down- und Bottom-Up-Planung koordiniert, so dass ein finales operatives Planungsergebnis für den Deckungsbeitrag III für das Kundengeschäft der Firmenkundenbank resultiert.

Basierend auf dieser Vorgehensskizze lassen sich nun für die operative kundenwertorientierte Vertriebsplanung die Anknüpfungspunkte an die Gesamtbankplanung finden. In einer engen Perspektive stellt der monetäre Kundenteilwert eine Hilfsgröße für die Ableitung der Zielplanung in der Bottom-Up-Planung dar, da dieser Wert idiomatisch für die Abbildung des Markt- bzw. Kundenpotenzials in der Vertriebsplanung ist. In einer weiteren Betrachtung kann jedoch postuliert werden, die Top-Down-Planung – wie auch bei der strategischen Planung – um den Kundenwert als Zielgröße zu ergänzen. Dies erscheint auch insofern notwendig, als die reine auf dem Deckungsbeitrag III basierende Planung, die konsequent in der wertorientierten Vertriebssteuerung[1899] mündet, dysfunktionale Verhaltensanreize bei den Mitarbeitern setzt, wenn

---

1897 Vgl. hier und nachfolgend *BVR* (2002b), S. 8 und 73 ff. sowie grundlegend *Schierenbeck, H.* (2003), S. 387 ff.

1898 Vgl. *BVR* (2002b), S. 65.

1899 Vgl. zur wertorientierten Vertriebssteuerung die Ausführungen in Kapitel A.1.

kurzfristig Erfolge zu Lasten späterer Perioden realisiert werden. Weiterhin dient der Kundenwert als Bezugsrahmen für die Ableitung von operativen Maßnahmen im Rahmen der jeweiligen kundenwertbezogenen Strategien sowie der Abstimmung mit den Verbundunternehmen. Letzteres wird zwar nicht explizit in der operativen Planungsskizze genannt, ergibt sich jedoch zwingend aus den Ausführungen der Verbundorientierung in Kapitel C.1.4.4.

2.2.2 Kundenwertorientierte Balanced Scorecard als Bezugsrahmen der operativen Vertriebsplanung

Die Operationalisierung der strategischen Ziele und Strategien ist eine zentrale Herausforderung der operativen Planung.[1900] Für diese Aufgabe hat sich das Instrument der Balanced Scorecard als leistungsfähiger Ansatz in der Theorie und Praxis erwiesen. Zugleich ist die Balanced Scorecard ein multidimensionaler Ansatz, der quantitative und qualitative Aspekte vereint[1901] und daher kongruent zum dem dieser Arbeit zugrunde liegenden Kundenwert als zentraler Ziel- und Steuerungsgröße. Aus diesen Gründen soll die Balanced Scorecard als Bezugsrahmen der operativen Vertriebsplanung genutzt werden.

2.2.2.1 Konzept der Balanced Scorecard

In der Wissenschaft und Praxis erfreut sich die Balanced Scorecard als Bindeglied zwischen Strategieentwicklung und -umsetzung großer Aufmerksamkeit, da sie als konzeptionell einfach und flexibel gilt.[1902] Die Balanced Scorecard ist ein leistungsorientiertes und leistungssteuerndes Informations- und Managementsystem, das eine mehrfache Ausgewogenheit verschiedener Informations- und Steuerungsgrößen anstrebt.[1903] Die Ausgewogenheit bezieht sich zum einen auf interne und externe sowie auf finanzielle und nicht-finanzielle Steuerungsgrößen, zum zweiten auf Früh- und Spätindikatoren der Leistungsmessung sowie zum dritten auf objektive, leicht messbare und subjektive, abgeschätzte Maßgrößen. Grundlegende Idee der Balanced Scorecard ist die Übersetzung der Unternehmensstrategie in ein „übersichtliches System der Leistungsmessung, welches den Rahmen für ein strategisches Leistungsmessungs- und Managementsystem bildet."[1904]

1900 Vgl. *Köhler, A. G./Rolvering, H.-S./Germann, S.* (2005), S. 418.

1901 Vgl. *Klein, N.* (2009), S. 10.

1902 Vgl. *Kaufmann, L.* (1997), S. 428; *Schlüchtermann, J./Klöpfer, R./Braun, A.* (1999), S. 103.

1903 Vgl. hier und nachfolgend *Karlowitsch, M.* (1997), S. 1131.

1904 *Kaplan, R. S./Norton, D. P.* (1997), S. 2.

Die ausgewählten, ausgewogenen Steuerungs- und Maßgrößen fokussieren die Unternehmensleistung aus vier Perspektiven.[1905] KAPLAN/NORTON nutzen die finanzielle Perspektive, die Kundenperspektive, die interne Prozessperspektive sowie die Lern- und Entwicklungsperspektive zwar für die Vorstellung und Einführung der Balanced Scorecard in die wissenschaftliche und praktische Diskussion,[1906] führen aber aus, dass die „Scorecard als Schablone und nicht als Zwangsjacke gedacht ist."[1907]

Die ausgewählten Ziele und Kennzahlen der einzelnen Perspektiven stehen im Konzept der Balanced Scorecard nicht isoliert nebeneinander, sondern bilden ein inhaltlich zusammenhängendes Netzwerk, das durch die verfolgte Strategie determiniert wird. Die Verknüpfung der Ziele und Kennzahlen mit der Strategie kann über drei Prinzipien erfolgen. Erstens über Ursache-Wirkungsbeziehungen, die die Hypothesen der Strategieentwicklung wiedergeben, zweitens über Ergebnisse und Leistungstreiber, die vermitteln, wie die Ergebnisse erreicht werden sollen, und drittens mit der strikten Verbindung nicht-finanzieller Ziele mit den Finanzen.[1908]

Mit der Gestaltung und Nutzung der Balanced Scorecard als Managementsystem in Banken, insbesondere Kreditgenossenschaften und Genossenschaften im Allgemeinen haben sich KRING, HONOLD-REICHERT und BEUTHIEN/DIERKES/WEHRHEIM wissenschaftlich auseinandergesetzt.[1909] Vom BVR gibt es seit 2004 einen Leitfaden für die Einführung und Nutzung der Balanced Scorecard und einzelne Kreditgenossenschaften berichten von deren Anwendung.[1910] Eine Studie aus dem Jahr 2007 hat zum Ergebnis, dass die Einführung einer Balanced Scorecard in Regionalbanken zu den „prioritären Zielen der nächsten Jahre zu zählen ist"[1911]. Daher kann begründet angenommen werden, dass das Konzept der Balanced Scorecard hinreichend bekannt und etabliert ist.

1905 Vgl. *Kaplan, R. S./Norton, D. P.* (1997), S. 8.

1906 Vgl. dazu ausführlich *Kaplan, R. S./Norton, D. P.* (1997), S. 46 ff.

1907 *Kaplan, R. S./Norton, D. P.* (1997), S. 33.

1908 Vgl. *Kaplan, R. S./Norton, D. P.* (1997), S. 145. Die letztgenannte Möglichkeit stellen KAPLAN/NORTON als Forderung auf, um zu vermeiden, dass Maßnahmen und Programme zum Selbstzweck degradieren. Vgl. *Kaplan, R. S./Norton, D. P.* (1997), S. 145.

1909 Vgl. *Kring, T. I.* (2005); *Honold-Reichert, T.* (2002). Eine Kritik der beiden Autoren findet sich bei *Beuthien, V./Dierkes, S./Wehrheim, M.* (2008), S. 188 ff.

1910 Vgl. z. B. für die VOLKSBANK MITTELHESSEN EG *Bernhardt, H.-H./Dokter, M.* (2008), S. 38 ff. und für die VOLKSBANK FORCHHEIM EG *Homann-Wenig, S.* (2008), S. 217 ff.

1911 *Natter, N./Meck, A./Jodlbauer, R.* (2007), S. 120. In der Befragung wurden 80 Regionalbanken aus Österreich und Deutschland befragt, von denen 33 Prozent angaben, in den nächsten bis zu 5 Jahren eine Balanced Scorecard einzuführen.

Die Balanced Scorecard wird regelmäßig als Konzept definiert, das auf Ebene des Unternehmens oder auf Ebene der strategischen Geschäftseinheiten zu verwenden ist. Gleichwohl finden sich in Theorie und Praxis Beispiele für eine Balanced Scorecard auf Funktionalbereichsebene und insbesondere auf Ebene des Vertriebs.[1912] ROßNER/SEIFFERT sehen in der Balanced Scorecard eine Methode, die die einzelnen strategischen und operativen Elemente der Vertriebssteuerung zu einem Gesamtkonzept verbindet. Darüber hinaus nutzen sie die grundlegenden Prinzipien der Balanced Scorecard, um eine Kundenwertmanagement-Scorecard zu entwickeln, die die Vertriebssteuerung dazu befähigt, kundenorientierte Strategien konsequent umzusetzen.[1913] Auch PUFAHL empfiehlt die Vertriebsstrategie mittels der Balanced Scorecard zu operationalisieren.[1914] Gegensätzlich argumentiert KÖRNERT, der ein Herunterbrechen der Balanced Scorecard auf den Vertrieb als problematisch ansieht, da noch viele konzeptionelle Probleme zu bewältigen seien.[1915]

Vor dem Hintergrund dieser Überlegungen wird im nächsten Kapitel die Entwicklung einer kundenwertorientierten Balanced Scorecard als Bezugsrahmen der operativen Vertriebsplanung erörtert, die sich inhaltlich über den Kundenwert bzw. die vertrieblich relevanten Kundenwerttreiber und Bestimmungsfaktoren konkretisiert und auf den grundlegenden Prinzipien der Balanced Scorecard aufbaut.

2.2.2.2 Entwicklungsoptionen einer kundenwertbezogenen Balanced Scorecard

Ausgangspunkt der Entwicklung einer kundenwertbasierten Balanced Scorecard ist die Frage nach der Bezugsebene der Strategie. In Kapitel E.2.1.3.2 wurde erläutert, dass bei der Strategieformulierung zwischen der Ebene des Vertriebs und der Kundenebene zu unterscheiden ist und im Schrifttum empfohlen wird, als Bezugsgröße kundenwertorientierter Strategien die einzelne aktuelle oder potenzielle Kundenbeziehung zu nutzen.[1916] Konsequent ist es demnach, eine kundeninviduelle, kundenwertorientierte Balanced Scorecard zu formulieren, so dass auch die operative Vertriebsplanung auf Kundenebene erfolgt. HOMBURG ET AL. zeigen, dass eine so ausgeprägte Marktorientierung der Vertriebsplanung

---

1912 Vgl. z. B. *Deking, I./Meier, R.* (2000), S. 263 ff. So steuert z. B. die WGZ BANK ihren Vertrieb mit der Balanced Scorecard. Vgl. *Knappkötter, R./Völkel, H.-G.* (2002), S. 16. DOKTER/LÜDERS nutzen die Balanced Scorecard zur Steuerung von Profit-Centern. Vgl. *Dokter, M./Lüders, U.* (2003), S. 29.

1913 Vgl. *Roßner, R./Seiffert, S.* (2005), S. 508.

1914 Vgl. *Pufahl, M.* (2006), S. 162 ff.

1915 Vgl. *Körnert, J.* (2003), S. 41 f.

1916 Vgl. dazu die Ausführungen in Kapitel E.2.1.3.

den Planungserfolg erhöht.[1917] Eine kundenbezogene Balanced Scorecard hat sich jedoch in ein Scorecardnetzwerk zu integrieren, das auch eine Balanced Scorecard auf Gesamtbankebene, auf der Ebene der strategischen Geschäftseinheiten und der Funktionalbereichsebene umfasst.[1918] Die jeweils übergeordneten Scorecards sind als Orientierung für die Ausgestaltung der kundenwertorientierten Balanced Scorecard zu nutzen.[1919]

BEUTHIEN/HANRATH/WEBER konstruieren auf Gesamtbankebene eine Balanced Scorecard, die zwischen den Perspektiven Finanzen und Risiko, (Nur-)Kunden, Mitglieder, Lernen und Entwicklung und interne Prozesse unterscheidet.[1920] HOMANN-WENIG stellt für den Vertrieb im Firmenkundengeschäft eine Balanced Scorecard vor, die begründet die von KAPLAN/NORTON genannten Perspektiven nutzt, jedoch von der Unterscheidung der Kundenebene in Mitglieder und Nur-Kunden abstrahiert.[1921] Bemerkenswert ist, dass sie die Balanced Scorecard für das Geschäftsfeld Firmenkunden implizit der Balanced Scorecard für den Vertrieb gleichsetzt. Soll sich die kundenwertorientierte Balanced Scorecard auf Kundenebene an diesen überzuordnenden Scorecards orientieren, ist zu diskutieren, welchen Einfluss die explizite Berücksichtigung des Kundenwerts auf die Balanced Scorecard hat. Der Einfluss auf die Balanced Scorecard kann formell sein und die Architektur der Balanced Scorecard betreffen (z. B. die Art und Anzahl der relevanten Perspektiven) oder auch die Auswahl der Ziele und Kennzahlen berühren und somit materiell sein. Dazu ist ein Rückgriff auf in der Literatur vorgestellte kundenwertbezogene Scorecards sinnvoll.

Die Customer-Sorecard von RUDOLF-SIPÖTZ soll einen „umfassenden und schnellen Überblick über alle für das Kundenwertmanagement relevanten Informationen liefern."[1922] Diese Scorecard unterscheidet drei Wirkungsebenen, denen vier Dimensionen zugeordnet sind.[1923] In der interpersonellen Wirkungsebene wird der komplementäre Wert, bestehend z. B. aus dem Kooperations-

---

1917 Vgl. *Homburg, C. et al.* (2008), S. 655 und S. 659 f.

1918 Vgl. *Deking, I./Meier, R.* (2000), S. 263.

1919 Vgl. *Rudolf-Sipötz, E.* (2001), S. 199. In der VOLKSBANK MITTELHESSEN EG wurden aus der Balanced Scorecard auf Gesamtbankebene Scorecards für die Vertriebs-, die Steuerungs- und die Produktionsbank deduziert. Die Scorecard der Vertriebsbank konstituiert sich aus einzelnen Geschäftsfeldscorecards für Privat- und Firmenkunden. Vgl. *Bernhardt, H.-H./Dokter, M.* (2008), S. 41. So auch *Deking, I./Meier, R.* (2000), S. 263, die von einem top-down-Implementierungsprozess ausgehen, um die Strategiekonsistenz der einzelnen Scorecards zu gewährleisten.

1920 Vgl. *Beuthien, V./Hanrath, S./Weber, H. O.* (2008), S. 75.

1921 Vgl. *Homann-Wenig, S.* (2008), S. 222 ff.

1922 *Rudolf-Sipötz, E.* (2001), S. 200 f.

1923 Vgl. hier und nachfolgend *Rudolf-Sipötz, E.* (2001), S. 200 ff.

und dem Referenzpotenzial eines Kunden fokussiert. Der intrapersonellen Wirkungsebene sind kundenindividuell der gegenwärtige Erfolgsbeitrag und das künftige Erfolgspotenzial, zu dem z. B. das Cross-Selling-Potenzial gehört, zugeordnet. Die Dimension Kundenstamm ist der Makroebene, auf der z. B. Kundenabwanderungen betrachtet werden, zugeordnet. Es ist zu konstatieren, dass die kundenwertbezogene Ausgestaltung der Balanced Scorecard formell und materiell erfolgt. Die formelle Architektur dieser Customer-Scorecard orientiert sich eng an dem von RUDOLF-SIPÖTZ entwickelten Kundenwertmodell.[1924] Aus diesem Grund ist eine Übernahme auf den vorliegenden Anwendungskontext nicht möglich. Auch die Übernahme der verwendeten Verfahrensweise, d. h. dem Aufbau einer kundenbezogenen Scorecard, bei der sich die Dimensionen der Scorecard unmittelbar aus den Dimensionen des Kundenwertmodells ergeben, ist nicht sinnvoll, da damit ein Übergewicht externer Kennzahlen vorläge und die Innensicht der Vertriebssteuerung z. B. auf Mitarbeiter und Prozesse vernachlässigt werden würde.

In der Customer Equity Scorecard von WILLE sowie BARTH/WILLE wird die von KAPLAN/NORTON formulierte Struktur weitgehend beibehalten und inhaltlich an die kundenwertbezogenen Überlegungen angepasst.[1925] Der Finanzperspektive ordnen die Autoren z. B. die monetären Kundenwerte und die monetären Einflussgrößen und der Kundenperspektive die nicht-monetären Einflussgrößen (z. B. Kundennutzen, Kundenzufriedenheit) zu. Die Prozessperspektive umfasst Bestimmungsfaktoren des Kundenwerts, die dieser Ebene sachlogisch zugeordnet werden können (z. B. interne Strukturen zur Aufnahme von Kundeninformationen). Die Perspektive des Lernens und Wachsens umfasst in ähnlicher Weise diejenigen kundenwertbezogenen Bestimmungsfaktoren, die ihr sinnvoll zugeordnet werden können. Dazu zählen z. B. Maßnahmen zur Steigerung der Integration des Kunden in die Leistungserstellung. Im Gegensatz zur Scorecard von RUDOLF-SIPÖTZ hat die kundenwertorientierte Ausgestaltung der Scorecard keine formellen Auswirkungen, die sich in der Architektur der Scorecard niederschlagen, sondern lediglich materielle Auswirkungen, die sich in der Auswahl der Ziele und Maßnahmen in den einzelnen Perspektiven zeigen.

Auf der letztgenannten Vorgehensweise basiert auch die Entwicklung einer kundenwertbezogenen Balanced Scorecard bei FRIEDERICHS-SCHMIDT.[1926] Dabei werden auf der Finanz- und Kundenperspektive Kundenwerttreiber und kundenbezogene Bestimmungsfaktoren des Kundenwerts und der Kundenwerttreiber

---

[1924] Vgl. dazu Kapitel D.4.3.1.4.

[1925] Vgl. hier und nachfolgend *Wille, K.* (2005), S. 242 ff. sowie *Barth, K./Wille, K.* (2000), S. 77 ff.

[1926] Vgl. hier und nachfolgend *Friederichs-Schmidt, S.* (2006), S. 199 ff.

als Basis der Kennzahlen ausgewählt. Auf der Prozessperspektive und auf der Perspektive des Lernens und Entwickelns werden unternehmensinterne Einflussfaktoren herangezogen, um die Kennzahlen auszuwählen.

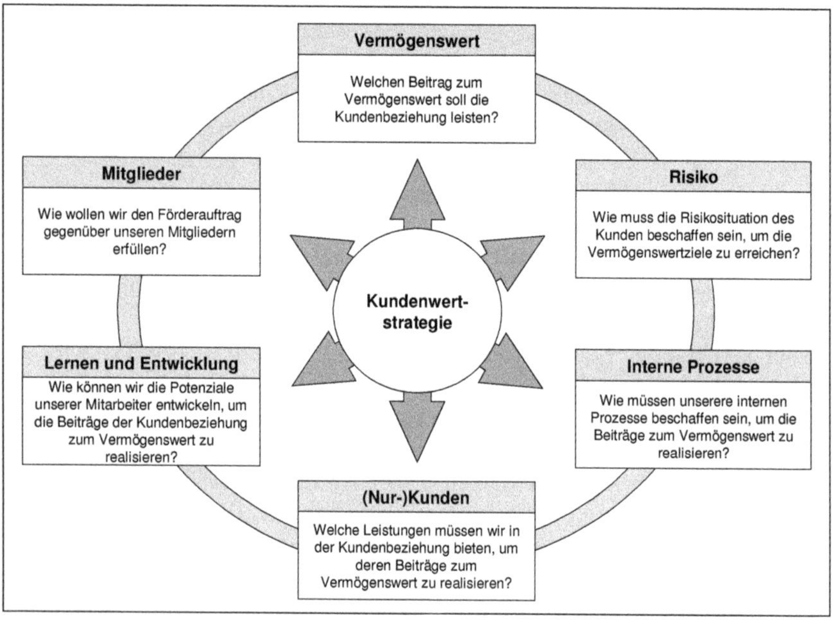

Abbildung 69: Grundlegende Architektur der kreditgenossenschaftlichen Balanced Scorecard[1927]

Dieses Vorgehen bietet eine größere Flexibilität und kann gewährleisten, dass die fundamentalen Prinzipien der Balanced Scorecard hinreichend berücksichtigt werden. Daher ist sie handlungsleitend für die Ableitung einer kundenwertbezogenen Balanced Scorecard des Vertriebs, die sich materiell am Kundenwert ausrichtet. Diese materielle Ausrichtung erfolgt über die Auswahl der Kennzahlen, für die Ziele definiert und Maßnahmen zur Zielerreichung geplant werden. Formelle Grundlage ist eine an die bisherigen Ergebnisse dieser Arbeit angepasste Balanced Scorecard von BEUTHIEN/HANRATH/WEBER für Kreditgenossenschaften.[1928] Abweichend von der Vorlage ist die oberste Ebene die Vermö-

[1927] In Anlehnung an *Beuthien, V./Hanrath, S./Weber, H.-O.* (2008), S. 75.

[1928] Vgl. hier und nachfolgend *Beuthien, V./Hanrath, S./Weber, H. O.* (2008), S. 75.

genswertperspektive.[1929] Zudem gibt es eine eigenständige Risikoperspektive. Die übrigen Perspektiven entsprechen der Vorlage der Autoren und umfassen die Perspektive des Lernens und Entwickelns, die interne Prozessperspektive und die (Nur-)Kunden Perspektive. Dies stellt die grundlegende Architektur der im Weiteren zugrunde gelegten kundenwertbezogenen Balanced Scorecard dar (siehe Abbildung 69).

2.2.2.3 Ziel- und Maßnahmenplanung auf Basis der Balanced Scorecard

2.2.2.3.1 Zielplanung

Die Zielplanung im Kontext der operativen Vertriebsplanung auf Basis der kundenbezogenen Balanced Scorecard umfasst die Auswahl relevanter Ziele, die Festlegung von Kennzahlen, mit denen die Ziele gemessen werden können, sowie die zu erreichenden Zielwerte und Bezugszeiträume. Bei der Ableitung der Ziele ist darauf zu achten, dass die Ziele sich – der Idee der Balanced Scorecard folgend – nicht widersprechen, die Strategie konkretisieren und in einem Netzwerk aus Kausalitätsbeziehungen eingeordnet werden.

In der Vermögenswertperspektive ist das aus der strategischen Planung auf Kundenebene formulierte Ziel hinsichtlich des anbieterspezifischen Kundenwerts aus Vertriebsperspektive maßgeblich.[1930] Im Gegensatz zur strategischen Planung werden für die operative Planung in der Balanced Scorecard auch dessen einzelne Bestimmungsgrößen näher betrachtet und ggf. als Ziele der Finanzperspektive berücksichtigt.[1931] Formal-analytisch lassen sich die Einflussgrößen aus der Formel [D.30] ableiten und als Zielgrößen definieren. Dazu gehören das Basisgeschäftspotenzial, das Entwicklungspotenzial bzw. dessen Ausprägungen, das Cross-Selling-Potenzial (alternativ bei Nichtkunden das Gesamtgeschäftspotenzial), die kundenbezogenen Prozesskosten und das Kooperationspotenzial. Eine mögliche Zielformulierung ist, den anbieterspezifischen Kundenwert in der laufenden Periode um einen zu bestimmenden Betrag zu erhöhen. Insgesamt stellen diese Größen Ergebnisgrößen im Sinne der Gestaltungsprinzipien der Balanced Scorecard dar. Hinsichtlich des Vermögenswertes und dessen Aufteilung in einem „Potenzialwert" und einen „Zerschlagungswert" würde mit diesen Zielgrößen bisher lediglich der „Potenzialwert" adressiert. Um auch die Ziele für den „Zerschlagungswert" zu berücksichtigen, sind periodenbezogene Ziele zu formulieren. Diese beziehen sich auf den in der betrachteten Periode zu reali-

[1929] Vgl. dazu die Ausführungen zum Zielsystem in Kapitel D.3.1.3.

[1930] Zu beachten ist, dass die Finanzperspektive hier als oberste Zielebene verstanden wird, die trotz ihrer Bezeichnung nicht zwingend nur Zielgrößen aufweisen muss, die in Geldeinheiten nominiert sind.

[1931] Vgl. *Stüker, D.* (2008), S. 347.

sierenden Kundenwert und dessen oben genannten Kundenwerttreiber. Eine mögliche Zielformulierung wäre, das Cross-Selling-Potenzial in Höhe eines bestimmten Betrags auszuschöpfen. Die vorhergehenden Überlegungen sollen anhand eines einfachen Beispiels erläutert werden.

Für einen Kunden wurden basierend auf den gegebenen Strategien und Maßnahmenprogrammen die in Tabelle 30 aufgeführten Werte für das Basisgeschäfts-, das Entwicklungs- und das Cross-Selling-Potenzial für vier Planungsperioden ermittelt. Unter Berücksichtigung der Kundenbindungsraten von 80 Prozent und der mittels des internen Ratingverfahrens ermittelten Ausfallwahrscheinlichkeit ergibt sich bei einem Diskontierungsfaktor ein anbieterspezifischer monetärer Kundenwert von 329 Geldeinheiten (GE).[1932]

Tabelle 30: Beispielhafte Berechnung des anbieterspezifischen monetären Kundenwerts[1933]

Kundenwertgrößen vor Maßnahmenplanung	**t=1**	**t=2**	**t=3**	**t=4**
Basisgeschäftspotenzial	100	120	125	130
Entwicklungspotenzial	30	35	35	40
Cross-Selling-Potenzial	20	25	28	25
Monetärer anbieterspezifischer Kundenwert vor Risiko je Periode t[1934]	**150**	180	188	195
Barwertiger monetärer anbieterspezifischer Kundenwert vor Kreditrisiko je Periode t (aus Sicht Beginn t=1) (Diskontierungsfaktor inkl. Risikozuschlag = 10%)	136	149	141	133
Kundenbindungsrate (80%)	80,0%	64,0%	51,2%	40,1%
Ausfallwahrscheinlichkeit	0,1%	0,2%	0,3%	0,4%
Barwertiger monetärere anbieterspezifischer Kundenwert nach Risiko je Periode t	109	95	72	53
Monetärer anbieterspezifischer Kundenwert (Beginn t=1)	329			
Monetärer anbieterspezifischer Kundenwert (Beginn t=2)		**242**		
Monetärer anbieterspezifischer Kundenwert (Beginn t=3)			151	
Monetärer anbieterspezifischer Kundenwert (Beginn t=4)				70

1932 Angenommen wird, dass die periodischen Zuflüsse jeweils am Periodenende realisiert werden und der Kundenwert für den Beginn der Periode t=1 ermittelt wird.

1933 Eigene Darstellung. Die Berechnung der monetären anbieterspezifischen Kundenwerte für die Zeitpunkte Beginn t=2, Beginn t=2 und Beginn t=3 erfolgt auf Basis der vorhandenen Daten. Das impliziert, dass die für die Berechnung aus Sicht des aktuellen Planungszeitpunktes keine Anpassung der Einflussgrößen vorliegt. Die Analyse etwaiger Abweichungen, die aus Veränderungen dieser Abweichungen resultieren, ist in der Vertriebskontrolle zu thematisieren.

1934 Vgl. auch Formel [D.25].

Für die operative Zielplanung kann darauf aufbauend im Sinne des „Zerschlagungswerts“ das Ziel formuliert werden, in der Periode t=1 einen Anteil von 80 Prozent des ermittelten anbieterspezifischen monetären Kundenwerts aus Vertriebssicht, mithin also einen Betrag von 120[1935] GE zu realisieren. Dies entspricht weitgehend der Planung mit barwertigen Deckungsbeitrags-III-Daten im Rahmen der oben skizzierten operativen Planung. Im Sinne des „Potenzialwerts“ kann das Ziel formuliert werden, den prognostizierten monetären anbieterspezifischen Kundenwert aus Vertriebssicht, zu Beginn der Periode t=2 von 242 GE um 10 Prozent auf etwa 264 GE zu steigern. Die Zielformulierung kann auch auf Ebene der einzelnen Kundenwerttreiber des monetären Kundenwerts erfolgen, die dieser Perspektive der Balanced Scorecard zugewiesen sind. Gleichfalls können auf dieser Ebene die Ziele für die nicht-monetären Kundenwerte formuliert werden, da diese einen mittelbaren Einfluss auf den „Potenzialwert“ der Kreditgenossenschaft haben. Ziele ließen sich z. B. für den Kunden Gesamtscorewert oder den Teilscorewerten für das Informations-, das Referenz- und das Member Value-Potenzial gem. Formel [D.32] formulieren.

Auf den übrigen Perspektiven der Balanced Scorecard sind nun im Sinne der oben genannten Prinzipien zur Verknüpfung von Zielen jeweils relevante Ziele auszuwählen. Die Kausalitätsbeziehungen können dabei sowohl Top-Down oder Bottom-Up als auch induktiv oder deduktiv bestimmt werden.[1936]

Die Mitgliederperspektive fokussiert die Fragen, wie der Förderauftrag den Mitgliedern gegenüber erfüllt werden soll bzw. wie der Member Value geschaffen werden soll. Kundenwertbezogene Ziele lassen sich auf zwei Arten einbinden. Erstens über den direkten Bezug zum Member Value-Potenzial, so dass Ziele formuliert werden, wie das Member Value-Potenzial des Firmenkunden entwickelt und genutzt werden kann. Und zweitens über die Mitgliedsbeziehung zum Firmenkunden, so dass Ziele formuliert werden, wie der Förderauftrag in Richtung des Firmenkunden als Mitglied auf der Einleger- und Leistungsbeziehungsebene erfüllt werden soll.[1937]

Die Perspektive des Lernens und Entwickelns orientiert sich an der Frage, wie die Potenziale der (Vertriebs-)Mitarbeiter entwickelt und genutzt werden können, um die Kundenwertziele zu erreichen. Diese Perspektive umfasst somit ebenfalls bankseitige Einflussfaktoren des Kundenwerts (z. B. die Qualifikation

---

1935 Der Wert 120 resultiert aus der Multiplikation des anbieterspezifischen monetären Kundenwerts der Periode t=1 in Höhe von 150 mit dem Zielanspruch von 80 Prozent.

1936 Vgl. *Klein, N.* (2009), S. 9 f.

1937 Beispiele möglicher Kennzahlen geben *Beuthien, V./Dierkes, S./Wehrheim, M.* (2008), S. 192.

und Zufriedenheit der Mitarbeiter[1938]). Auf dieser Perspektive finden sich Leistungstreiber und Ergebnisgrößen gleichermaßen.

Auf der (Nur-)Kundenperspektive finden sich Ziele, die die Fragen adressieren, welche kundenseitigen Leistungstreiber zu realisieren sind. Die Kundenperspektive sollte daher Ziele enthalten, die die kundenseitigen Bestimmungsfaktoren der Kundenwerttreiber betrachten. Dazu gehören z. B. die Kundenbindung, die Kundenzufriedenheit, die Cross-Buying-Bereitschaft, die Informationsbereitschaft oder die Bereitschaft zur Weitergabe von positiven Weiterempfehlungen. Kennzahlen auf dieser Ebene stellen für die Vermögenswertperspektive Frühindikatoren dar und werden selbst beeinflusst von der Prozessperspektive und der Perspektive des Lernens und Entwickelns.

Die Prozessperspektive sucht danach, strategiekonsistente Ziele zu formulieren, die eine Antwort auf die Fragen finden, welche organisatorischen, technologischen und prozessualen Voraussetzungen (im Vertrieb) vorliegen müssen, um die Kundenwertziele zu erreichen. Es sind also die bankseitigen Möglichkeiten in Zielen zu formulieren, die strategiekonsistent einen Beitrag dazu leisten, kundenseitige Einflussfaktoren des Kundenwerts zu fördern (z. B. die Kundenzufriedenheit durch kurze und einfache Kreditantragsprozesse zu steigern) und bankseitige Voraussetzungen des Kundenwerts zu schaffen und auszubauen (z. B. die Strukturen zur Beschaffung von kundenwertrelevanten Informationen).

Die Risikoperspektive umfasst Kennzahlen zu den beziehungsbezogenen Kundenwerttreibern der Risikodimension. Dazu gehören sowohl Ziele hinsichtlich des Adressrisikos (z. B. die Ratingnote oder die Besicherungsquote) als auch Ziele des Kundenplanungsrisikos[1939] (z. B. den Informationsstand zu einem Kunden und dessen Unternehmensplanung). Die Ziele dieser Perspektive wirken über die Ausfallwahrscheinlichkeit, den Risikozuschlag und den Ausfallschaden auf die Ziele der Vermögenswertperspektive positiv oder negativ ein. Gleichzeitig kann eine Abhängigkeit von internen Prozessen und den Fähigkeiten der Mitarbeiter, die über die Perspektive des Lernens und Entwickelns adressiert werden, unterstellt und abgebildet werden, z. B. bei der Fähigkeit Kundeninformationen aufzunehmen und in für die Prognose des Kundenwerts relevante Informationen zu transferieren.

Im Sinne der Umsetzung der Vertriebsstrategie sind für jede Zielgröße Kennzahlen zur Operationalisierung der Ziele sowie Soll-Werte und Maßnahmen zur Zielerreichung zu formulieren. Gegenstand des nächsten Kapitels ist die kundenwertorientierte Maßnahmenplanung im Kontext der operativen Vertriebsplanung.

1938 Vgl. *Beuthien, V./Hanrath, S./Weber, H.-O.* (2008), S. 70.

1939 Vgl. *Schöner, M./Graeve, A.* (2004), S. 156 f.

2.2.2.3.2 Maßnahmenplanung

Im Anschluss an die Zielplanung erfolgt die Maßnahmenplanung,[1940] in deren Fokus die Auswahl derjenigen Vertriebsmaßnahmen liegt, die dazu beitragen, die operativen und strategischen Ziele zu erreichen. Die Auswahl der operativen Maßnahmen erfolgt unter Beachtung der strategischen Zielvorgaben und Strategien.[1941] Zudem ist bei der Maßnahmenplanung deren Effektivität und Effizienz zu berücksichtigen.

Hinsichtlich der Effektivität, also der Auswahl der richtigen Maßnahmen aus Kundensicht, kann z. B. auf das KANO-Modell rekurriert werden, indem die Maßnahmen dahingehend unterschieden werden, ob sie Basisfaktoren, Leistungsfaktoren oder Begeisterungsfaktoren darstellen.[1942] Im personalisierten Vertrieb hat die Kundenkontaktstrategie eine besondere Bedeutung. Kundenkontaktstrategien sind generelle, einigermaßen zeitstabile Anspracheregeln, die eine Aussage darüber treffen, wie oft ein Kunde über welche Kanäle zu welchen Themen in einer Planungsperiode angesprochen werden soll.[1943] REINARTZ/THOMAS/KUMAR kommen in ihrer Studie im industriellen Kontext zu dem Ergebnis, dass persönliche Kontakte die größte Effektivität hinsichtlich der Ziele der Akquisition, der Bindung und der Profitabilität eines Kunden aufweisen.[1944] An zweiter Stelle stehen Telefon- und an dritter Stelle E-Mail-Kontakte. Sie weisen jedoch darauf hin: „It is also important to note that these modes of contacts are distinctive in terms of their costs.“[1945] Aus diesem Grund ist für die Kundenkontaktstrategie im Speziellen und die Maßnahmenplanung im Allgemeinen die Effizienz der Maßnahmen als Entscheidungskriterium zu berücksichtigen.

Bei der Umsetzung der Strategien sollen den Firmenkunden personelle und finanzielle Ressourcen in Abhängigkeit ihres erwarteten Kundenwertes differenziert zugewiesen werden.[1946] Regelmäßig und nahezu einmütig wird im Schrifttum postuliert, hohe Kundenwerte rechtfertigten ein hohes Maß an Ressourcenzuordnungen.[1947] Konkret hieße dies, dass ein Firmenkunde mit einem hohen Kundenwert intensiv betreut wird. „Die reine Kundenbewertung gibt alleine

---

[1940] Vgl. *Reiners, J. O.* (2004), S. 124.

[1941] Vgl. *Stüker, D.* (2008), S. 317.

[1942] Vgl. dazu ausführlich *Stahl, H. K. et al.* (2009), S. 247 ff.; *Matzler, K./ Sauerwein, E./Stark, C.* (2009), S. 324 ff. Für eine praxisbezogene Anwendung dieses Modells in Verbindung mit der Importance-Grid Methode vgl. *Raddao, N.* (2009), S. 61 ff.

[1943] Vgl. *Schöner, M./Graeve, A.* (2004), S. 150 f.

[1944] Vgl. *Reinartz, W. J./Thomas, J. S./Kumar, V.* (2005), S. 70.

[1945] *Reinartz, W. J./Thomas, J. S./Kumar, V.* (2005), S. 70.

[1946] Vgl. *Pelzl, U.* (2006), S. 12; *Buhl, H. U./Dzienziol, J./Heidemann, J.* (2010), S. 2.

[1947] Vgl. z. B. *Duderstadt, S.* (2006), S. 110 ff.

noch keine Indikation hinsichtlich optimaler Besuchsfrequenzen bzw. Ressourcenallokationen.“[1948] Da die betrieblichen und vertrieblichen Ressourcen knapp sind, ist eine zentrale Überlegung bei der Maßnahmenplanung die Frage nach der Beeinflussbarkeit des Firmenkunden durch Vertriebsmaßnahmen.[1949] Die zentrale Frage ist, welche Erfolge hinsichtlich der kundenwertbezogenen Ziele je eingesetzter Einheit an finanziellen und personellen Ressourcen erreicht werden kann.[1950] Daraus wiederum folgt die für die Maßnahmenplanung relevante Entscheidungsgrundlage, bei welchen Kundenbeziehungen vor dem Hintergrund der gegebenen Möglichkeiten mit welchem Aufwand welche Kundenwerte realisiert oder entwickelt werden können. Ein einfaches Modell zur Budgetallokation haben KRAFFT/ALBERS entwickelt.[1951] Im Zentrum ihres Modells steht die erfahrungsbasierte Ableitung von Budget-Elastizitäten, die sich aus dem Einfluss der Budgetveränderung auf die Veränderung einer bestimmten Zielgröße, die bei ihnen der Deckungsbeitrag ist, ergibt. Zwar ist mit der Anwendung eines solchen für den hier betrachteten Kundenwert als Zielgröße angepassten Modells noch nicht die Frage nach der Art der Maßnahme beantwortet, jedoch ergeben sich Hinweise auf die Budgethöhe, die für Maßnahmen zur Verfügung steht.[1952] Es ist somit zu konstatieren, dass eine optimale Vertriebspolitik am Kundenwert und an der Beeinflussbarkeit der Kunden durch Veränderungen im Marketing-Mix (im Sinne von Elastizitäten) auszurichten ist.[1953] Bei der Maßnahmenplanung geht es folglich darum, diejenigen Maßnahmen in den Vordergrund zu stellen, die erstens eine besonders starke Wirkung auf einen Kundenwerttreiber oder dessen Bestimmungsfaktoren ausüben und zweitens jene Kundenwerttreiber oder deren Bestimmungsfaktoren mit der höchsten Hebelwirkung auf den Kundenwert beeinflussen.[1954]

1948 *Reiners, J. O.* (2004), S. 157. In der kreditgenossenschaftlichen Praxis äußert sich diese These regelmäßig so, dass große Firmenkunden vergleichsweise öfter besucht werden. Problematisch daran ist jedoch, dass nicht nachgehalten wird, welchen periodischen oder überperiodischen Erfolg eine Betreuungsstunde des Vertriebsmitarbeiters generiert.

1949 Vgl. *Krafft, M./Rutsatz, U.* (2006), S. 282.

1950 Diese Aussage orientiert sich an den Überlegungen der engpassorientierten Planung. Vgl. dazu z. B. *Adam, D.* (1998), S. 224 ff. für die operative Produktionsprogrammplanung bei einem Engpass.

1951 Vgl. hier und nachfolgend *Krafft, M./Albers, S.* (2000), S. 523 ff.

1952 Vgl. für eine Übersicht von Modellen, die auch die Art der Maßnahme berücksichtigen z. B. *Fischer, M./Albers, S.* (2007), S. 177 ff.

1953 Vgl. *Krafft, M./Rutsatz, U.* (2006), S. 282.

1954 Vgl. *Tewes, M.* (2003), S. 212. So auch *Blattberg, R. C./Deighton, J.* (1997), S. 25.

Nach WEBER/LISSAUTZKI sind bei der Maßnahmenplanung zudem zwei weitere Aspekte von besonderer Bedeutung.[1955] Zum einen ist die Dynamik des Kundenwerts zu berücksichtigen, da es von enormer Bedeutung ist, ob Potenziale in der betrachteten Periode vorliegen oder erst in einer der nachfolgenden Perioden. So ist z. B. die Ansprache auf Finanzierungsbedarfe, die erwartungsgemäß erst in einigen Jahren auftreten, etwa aufgrund bekannter Fälligkeiten, in der kommenden Periode nicht zielführend. Zum zweiten ist es notwendig, einen hohen Detaillierungsgrad zu haben, so dass es nicht ausreichend ist, z. B. das Cross-Selling-Potenzial zu betrachten, sondern dessen Bestimmungsfaktoren in Form von konkreten Produkten des Cross-Selling-Potenzials. Die Anforderung eines hohen Detaillierungsgrads unterstreicht die Forderung einer kundenindividuellen Vorgehensweise in der Vertriebsplanung. Auf Ebene von Kundensegmenten festgelegte Maßnahmen gehen i. d. R. mit hohen Streuverlusten einher. Aus diesem Grund sieht STÜKER die Maßnahmenplanung auf Basis von Kundensegmenten nur als Notlösung an und postuliert eine kundenindividuelle Maßnahmenplanung, die sich an einer kundenindividuellen Entwicklungsstrategie orientiert.[1956]

Das zentrale Problem der Maßnahmenplanung ist nach TEWES die Verbindung zwischen einzelnen Maßnahmen und deren konkrete Wirkung auf das Kundenverhalten und damit die Kundenwerttreiber, da diese Beziehungen i. d. R. nicht als eindeutige Ursache-Wirkungs-Beziehung darzustellen und insbesondere nur schwierig zu quantifizieren sind.[1957] „Zudem werden Aussagen über Ursache-Wirkungs-Beziehungen erschwert, wenn - wie bei der überwiegenden Zahl der Maßnahmen zu erwarten ist - sachliche und zeitliche Wirkungsinterdependenzen auftreten, d. h. die Wirkung einer Maßnahme nicht auf eine Periode beschränkt bleibt und zudem mehrere Determinanten des Kundenwerts oder mehrere Kunden beeinflusst.“[1958] Dieses Problem dürfte im Kontext des Beziehungssystems der Balanced Scorecard noch stärker evident sein.

Wie bereits ausgeführt, stellt der Vertrieb die Kontaktfunktion zwischen Absatzmarkt und Bank dar. Aufgrund dieses Sachverhaltes und der systemorientierten Definition des Bankbetriebs, der sich bei den Kreditgenossenschaften im Drei-Banken-Modell und der Verbundstruktur der FinanzGruppe konkretisiert, sind interne und externer Koordinationsanforderungen in der Vertriebsplanung zu berücksichtigen.

1955 Vgl. hier und nachfolgend *Weber, J./Lissautzki, M.* (2004), S. 31 f.

1956 Vgl. *Stüker, D.* (2008), S. 299.

1957 Vgl. *Tewes, M.* (2003), S. 212 f.

1958 *Tewes, M.* (2003), S. 213.

2.3 Interne und externe Koordination des Vertriebsplans

Im Drei-Banken-Modell[1959] bestehen Interdependenzen zwischen der Planung der Vertriebsbank sowie der Steuerungs- und der Produktionsbank, die zu internen Koordinationsbedarfen führen. Darüber hinaus führt die Verbundorientierung zu externem Koordinationsbedarf. Diese Koordinationsbedarfe werden nachfolgend im Lichte des Kundenwerts betrachtet.

2.3.1 Interne Koordination mit der Steuerungsbank

Wie in Kapitel C.2.4.1.1 beschrieben, übernimmt die Steuerungsbank Aufgaben der Portfoliosteuerung für die einzelnen Anlageklassen und Geschäftsarten, des Risikomanagements sowie weitere zentralisierte Aufgaben mit Blick auf das Personal, das Bankmanagement, das Controlling und das Marketing, worin auch Entscheidungen zur Produkt- und Preispolitik enthalten sind[1960]. Nachfolgende Ausführungen fokussieren auf das für das Firmenkundengeschäft bedeutende Kreditrisikomanagement, da hieraus die größten Einflüsse auf die Vertriebsplanung resultieren und die Produktpolitik, die einen wesentlichen Einfluss auf die Kundenwertermittlung hat.

In Banken ist es notwendig, die Vertriebsplanung eng mit der Kreditrisikostrategie zu verknüpfen, da letztere einen limitierenden Faktor im Rahmen einer kundenfokussierten Planung darstellen kann. Bei produkt- oder rein wertmäßigen Vertriebsplanungen, die keinen unmittelbaren Bezug zu einzelnen Kundenbeziehungen herstellen, ist dieser Schritt nicht von Relevanz. In den MaRisk wird festlegt, dass jede Bank eine Kreditrisikostrategie zu erarbeiten und durch die Geschäftsleitung zu verabschieden hat, welche bspw. geplante Aktivitäten im Kreditgeschäft, angestrebte Marktsegmente, Branchen, regionale Schwerpunkte und Portfoliostrukturen umfassen sollte.[1961] Zudem sind nach MaRisk Limitsysteme zu implementieren, die auch auf Einzelgeschäftsebene Kreditlimite enthalten.[1962] Die operative Vertriebsplanung muss bedingt durch ihre Fristigkeit also von gegebenen Strukturen hinsichtlich der Kreditrisikostrategie ausgehen. Dies betrifft insbesondere auch die gegenwärtige Risikostruktur sowie die Ausstattung mit Risikokapital, die eine zentrale Ressource für die Realisierung von risikobezogenen Kundenwertpotenzialen ist (z. B. Kreditfinanzierungsvolumina). Aus diesem Grund muss die operative Vertriebsplanung zwingend mit

---

1959 Vgl. zum Drei-Banken-Modell die Ausführungen in Kapitel C.2.4.1.1.

1960 Vgl. *BVR et al.* (2006), S. 23.

1961 Vgl. *Bundesanstalt für Finanzdienstleistungsaufsicht* (2010) AT 4.2; *Russ, A.* (2003), S. 351.

1962 Vgl. *Bundesanstalt für Finanzdienstleistungsaufsicht* (2010) BTR 1.

der Kreditrisikostrategie koordiniert werden, was sich in den Abstimmungsfeldern der Adressrisikostruktur, der Adressrisikolimite und der Risikokapitalausstattung konkretisiert.

Hinsichtlich so genannter Problemkredite wird unterstellt, dass die betroffenen Kundenbeziehungen, die zu Beginn der Vertriebsplanungsaktivitäten bekannt sind und in spezifischen Abteilungen des Markts oder der Marktfolge betreut werden (Problemkreditbetreuung),[1963] nicht Bezugsobjekte der Vertriebsplanung sind. Daher reduziert sich die Anzahl der Kundenbeziehungen auf diejenigen Firmenkunden, die zum Zeitpunkt der operativen Planung nicht als Problemkredite qualifiziert sind. Gleichwohl ist zu fordern, diese Kunden hinsichtlich des hier vorgestellten Kundenwertmodells zu bewerten, um Entscheidungen hinsichtlich der Sanierungswürdigkeit im Rahmen des Problemkreditmanagements zu fundieren.[1964]

Anders als bei der strategischen Planung, in der alle Strukturen und Prozesse der Kreditgenossenschaft als gestaltbar angesehen werden und daher eine gegenseitige Abhängigkeit der Vertriebsplanung von der Adressrisikostruktur in der Planung berücksichtigt werden kann, ist bei der operativen Vertriebsplanung die Adressrisikostruktur sowie deren geplante Änderung in der betrachteten Periode ein Datum für die Vertriebsplanung.[1965] Daraus resultiert, dass Kundenbeziehungen, die nicht oder hinsichtlich der gewollten Veränderung nicht mehr kompatibel zur Risikostruktur, d. h. dem Soll-Portfolio der Kreditrisiken, der Kreditgenossenschaft sind, zu selektieren sind. Für diese Kunden wird, wie in Kapitel E.2.1.3.2 erläutert, ein vollständiger oder partieller Rückzug aus der Kundenbeziehung durch die Kreditgenossenschaft zu initiieren sein, sofern keine Maßnahmen der aktiven Kreditrisikosteuerung angewendet werden können.[1966] Werden Instrumente der aktiven Kreditrisikosteuerung angewendet, sind die anfallenden Kosten in die Kundenbewertung zu integrieren und der anbieterspezifische Kundenwert nach Durchführung der Steuerungsmaßnahme zu ermitteln. Auf Basis dieser Größe kann dann im Einzelfall entschieden werden, welche Vorgehensweise zielsetzungsgerecht ist. In diesem Kontext ist zu erwarten, dass die Kundenwertorientierung auch zu einem stärkeren Einsatz aktiver Steuerungsinstrumente der Kreditrisikosteuerung führt, da abgelehnte Kredite Fir-

1963 Vgl. zur organisatorischen Gestaltung des Problemkreditmanagement *Rübenacker, C.* (2009), S. 62 ff.

1964 Vgl. *Rübenacker, C.* (2009), S. 250 ff.

1965 Vgl. *Juncker, K.* (2006), S. 346.

1966 Charakteristisch für aktive Instrumente der Kreditrisikosteuerung ist, dass Kredite und/oder Kreditausfallrisiken an Dritte übertragen werden. Vgl. *Hartmann-Wendels, T./Pfingsten, A./Weber, M.* (2010), S. 594.

menkunden zum Wechsel oder Verlagerung der Bankbeziehung motivieren können.[1967]

Ein weiterer für die kundenwertorientierte Vertriebsplanung relevanter Abstimmungspunkt ist der Abgleich der prognostizierten Kundenwerte mit den für die jeweilige Kreditgenossenschaft zu berücksichtigenden Großkreditobergrenzen[1968] bzw. den in der Gesamtbankplanung festgelegten Adressrisikolimiten.[1969] Insbesondere bei Firmenkunden, bei denen in Zukunft hohe Finanzierungsbedarfe erwartet und der Kundenwertermittlung zugrunde gelegt werden, ist zu fragen, ob diese die vorgegebenen Limite auf Einzelengagementebene übersteigen werden. Ist dies der Fall, ist zu überlegen, wie mit dieser Situation umgegangen werden kann. In diesen Fällen können ebenfalls Instrumente der aktiven Kreditrisikosteuerung eingesetzt werden. Neben Lösungen, die über den Kreditrisikotransfer in der FinanzGruppe realisiert werden können, besteht auch die Möglichkeit mit Drittbanken zusammenzuarbeiten (jeweils z. B. in Form von Konsortialkrediten[1970]). Problematisch bei diesem Abstimmungspunkt ist, die zeitliche Kongruenz zwischen der Kundenwertermittlung und der Adressrisiko- bzw. Kreditportfolioplanung herzustellen. Gelingt es nicht, die Großkreditgrenzen durch geeignete Maßnahmen zu umgehen, so impliziert dies nicht, dass die Kundenbeziehung beendet werden muss. Für die Kundenwertermittlung muss jedoch der Wegfall dieses in der Zukunft liegenden Geschäfts sowie die damit einhergehenden Verbundeffekte berücksichtigt werden. Gelingt es, die Großkreditgrenze zu umgehen, sind die mit den zu treffenden Maßnahmen verbundenen

1967 So zeigt sich z. B., dass bei kleineren Firmenkunden, bei denen ein abgelehnter Kreditantrag Bestandteil der Historie der Geschäftsbeziehung ist, die Geschäftsbeziehung zur Bank und die Bank selbst anders wahrnehmen als Firmenkunden bei denen das nicht der Fall ist. Vgl. *KPMG* (2009), S. 31, S. 46 und S. 57.

1968 Gemäß § 13 Abs. 1 KWG ist ein Großkredit ein Kredit, der 10 Prozent des haftenden Eigenkapitals des kreditgebenden Kreditinstituts erreicht oder übersteigt. Dabei wird jedoch nicht nur ein einzelnes Kreditgeschäft betrachtet, sondern die Gesamtheit der Kreditgeschäfte an einen einzelnen Kreditnehmer im Sinne des § 19 Abs. 2 KWG. Einzelne Großkredite dürfen eine Höhe von 25 Prozent des haftenden Eigenkapitals nicht überschreiten. Zudem darf die Summe aller Großkredite einen Betrag des 8-fachen des haftenden Eigenkapitals nicht überschreiten. Vgl. dazu *Hartmann-Wendels, T./Pfingsten, A./Weber, M.* (2010), S. 631 ff. Die Großkreditgesamtobergrenze, die bei 800 Prozent des haftenden Eigenkapitals lag, soll im Zuge der Änderungen der EU-Bankenrichtlinie abgeschafft werden. Vgl. *Deutsche Bundesbank* (2009), S. 69.

1969 Vgl. zur Planung des Adressrisikos *BVR* (2003b), Kapitel 4.3 sowie zur Limitsystematik bezüglich des Adressrisikos *BVR* (2003b), Kapitel 4.6.

1970 Bei Konsortialkrediten treten mehrere Banken, die den Kredit vergeben, als Zweckgemeinschaft gegenüber dem kreditnehmenden Firmenkunden auf. Vgl. *Tolkmitt, V.* (2007), S. 195 ff.

wirtschaftlichen Konsequenzen in der Kundenwertermittlung zu berücksichtigen. Dies gilt sowohl dann, wenn das gesamte Kreditgeschäft vermittelt als auch dann, wenn es nur zum Teil von der betrachteten Kreditgenossenschaft kontrahiert wird.

Im Sinne der integrierten Rendite-Risiko-Steuerung ergeben sich für die kundenwertorientierte Vertriebsplanung Koordinationsbedarfe für den Fall, dass Risikokapital knapp ist. Während die Ausführungen zur Kundenwertorientierung bisher implizit von der auch realiter zu beobachtenden Situation ausgingen,[1971] dass das Risikokapital nicht knapp ist und somit absolute Größen zur Vertriebssteuerung ausreichend sind, ist bei knappem Risikokapital der Kundenwert im Verhältnis zum einzusetzenden Risikokapital in Beziehung zu setzen. Im Ergebnis resultiert ein modifizierter Kundenwert. Stellt neben den finanziellen und personellen Vertriebsressourcen damit die Risikokapitalausstattung einen weiteren Engpass dar, sind optimale Lösungen mit linearen Planungsmodellen möglich.[1972] Die damit verbundenen Aufgaben sind originär dem Vertriebscontrolling in der Steuerungsbank zuzuweisen und werden an dieser Stelle nicht weiter thematisiert.

Zusammenfassend lässt sich für die Abstimmung von Vertriebs- und Steuerungsbank der in Abbildung 70 skizzierte Ablaufplan herleiten, der den Abstimmungsprozess konzeptionell beschreibt.

1971 Als Indiz für diese Aussage können die Inhalte der Pressemitteilungen angeführt werden, die mehrheitlich von einer Vermeidung der Kreditrisikoklemme durch die VR-Banken und zusätzlichem Kreditvergabepotenzial berichten. Zudem weist der BVR im Jahresbericht 2010 daraufhin: „Mit der seit Langem weitestgehend konstant bleibenden Eigenkapitalquote – berechnet als bilanzielles Eigenkapital im Verhältnis zur Bilanzsumme – verfügen die Kreditgenossenschaften über ausreichende Wachstumsspielräume. […] Die Volksbanken und Raiffeisenbanken sind mit Eigenkapital und Liquidität gut ausgestattet, um ihr Kreditvolumen auch zukünftig weiter zu steigern. Sie haben gut 100 Milliarden Euro mehr Einlagen als Kredite im Bestand.“ *BVR* (2011a), S. 46.

1972 Vgl. dazu z. B. im Kontext der Produktionsprogrammplanung *Adam, D.* (1998), S. 242 ff.

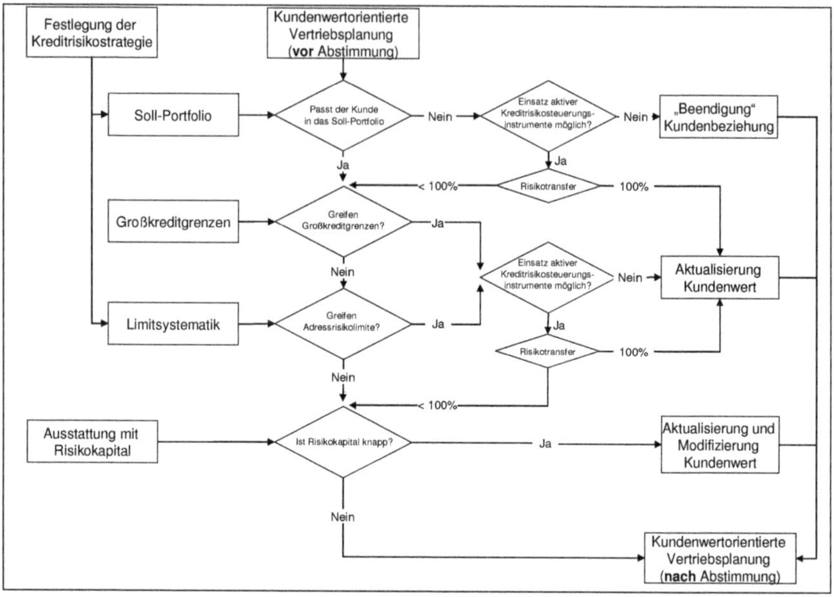

Abbildung 70: Ablaufplan der Abstimmung zwischen Vertriebs- und Steuerungsbank[1973]

Darüber hinaus besteht Abstimmungsbedarf zwischen der Vertriebs- und der Steuerungsbank mit Blick auf die Sortimentsplanung. Dieser Abstimmungsbedarf liegt in zwei Richtungen vor. Zum einen von der Steuerungsbank in die Vertriebsbank, sofern neue Bankleistungen im Planungszeitraum im Firmenkundenmarkt angeboten werden sollen. Denn diese neuen Produkte sind im Rahmen der Kundenwertermittlung und der darauf aufbauenden Vertriebsplanung zu berücksichtigen. In entgegengesetzter Richtung resultiert Abstimmungsbedarf, sofern sich aus der Kundenwertermittlung im Kontext des Cross-Selling-Potenzials zeigt, dass kundenwertinduzierende Produkte gegenwärtig nicht Bestandteil des Bankleistungssortiments sind, aber erfolgversprechend angeboten werden können.[1974] In diesem Zusammenhang sollte der Kundenwert als wesentliche Entscheidungsgrundlage der Produktpolitik verwendet werden.

2.3.2 Interne Koordination mit der Produktionsbank

Die Notwendigkeit zur Koordination der Planung der Vertriebs- und der Produktionsbank begründet sich mit der Aussage, dass „Aufgaben jenseits des direkten

[1973] Eigene Darstellung.

[1974] Vgl. Kapitel D.4.5.3.1.3. Dies betrifft in Abbildung 46 die „Würfel" 1b, 2b und 3b.

Kundenprozesses in den Bereich der Produktionsbank [verlagert werden, Erg. d. Verf.]“ [1975]. Vertriebs- und Produktionsbank stehen in einem unmittelbaren komplementären Verhältnis zueinander, das daraus resultiert, dass die Abwicklung eines abgeschlossenen Geschäfts in der Produktionsbank[1976] gebündelt ist, die entweder gleichzeitig auch Produktanbieter ist oder lediglich die Produktion im Sinne der weiteren Bearbeitung übernimmt.[1977] Aufgrund der Tatsache, dass der Absatz einer Bankleistung der Produktion ebendieser i. d. R. zeitlich vorgelagert ist, ist der Einfluss der Vertriebsplanung für die Produktionsbankplanung evident. Eine Vertriebstrategie und -planung, die die Akquisition von Kunden höher gewichtet als die Ausschöpfung bestehender Kundenbeziehungen, stellt ebenso andere Anforderungen an die Produktionsbank wie eine Strategie, deren Schwerpunkt in einer Intensivierung von großen Firmenkunden statt in einem Ausbau des kleinteiligen Geschäfts mit kleinen Firmenkunden liegt.

Für die vorzunehmende Koordination ist die gewählte Vorgehensweise der Ermittlung des Kundenwerts anhand der Bankleistungen, die dem Basisgeschäfts-, dem Entwicklungs- oder dem Cross-Selling-Potenzial zugeordnet werden, von besonders hoher Bedeutung. In Verbindung mit den Vertriebszielen und -maßnahmen können die operativ und strategisch fokussierten Bestandteile des bestehenden Bankleistungssortiments identifiziert werden, so dass sich darauf aufbauend relevante Informationen für die Kapazitätsplanung nach Art und Menge der Produktionsbank ableiten lassen. Wird bspw. das ressourcenintensive Kreditgeschäft zugunsten von Provisionsgeschäften mit Verbundpartnern reduziert, so bedingt dies mittel- bis langfristig kapazitive Änderungen in der Produktionsbank, da die produzierenden Tätigkeiten in diesen Fällen außerhalb der Kreditgenossenschaft anfallen. Irrelevant ist nach DAUSER die Frage nach restringierenden Einflüssen der Produktionsbank auf die Vertriebsplanung. “Anders als bei den meisten produzierenden Unternehmen unterliegt die Herstellung von Bankdienstleistungen nur geringen oder keinen Kapazitätsbeschränkungen. Somit ist, anders als bei den produzierenden Unternehmen, der Absatz am Markt der dominante Engpassbereich.“[1978] Gleichwohl kann es zu saisonalen Kapazitätsengpässen in der Produktionsbank kommen, die kaum durch den Vertrieb beeinflusst werden können. Bei der Kapazitätsplanung ist die Planung so vorzunehmen, dass die unzureichenden Kapazitäten bei Nachfragespitzen einen geringeren negativen Einfluss auf Realisierung gegenwärtiger oder Schaffung von

1975 *BVR et al.* (2006), S. 23.

1976 Synonyme Begriffe, die in der Literatur verwendet werden, sind Transaktionsbank, Back-Office-Fabrik oder Abwicklungsbank. Vgl. *Moormann, J./Schmidt, G.* (2007), S. 21.

1977 Vgl. *Segbers, K.* (2007), S. 114.

1978 *Dauser, A.* (1995), S. 14.

künftigen Kundenwerten haben als die Kosten der Unterbeschäftigung bei geringerer Nachfrage.

Ein weiteres Koordinationsfeld resultiert aus Überlegungen der Produktionsbank, Verfahren, Strukturen oder Prozesse zu innovieren, die Einfluss auf den Kundenwert bzw. die Beeinflussung des Kundenwerts durch den Vertrieb haben. Dies liegt bspw. vor, wenn Firmenkundenportale eingesetzt werden,[1979] die der Externalisierung von Aufgaben im Sinne der Isoleistungskurve dienen, und so Einfluss auf die Beziehung zwischen Firmenkundenberater und Firmenkunde haben (Verringerung der Kontakte zwischen Firmenkunde und Firmenkundenberater). Aus diesem Grund sollten in die Entscheidungsfindung von Verfahrensinnovationen in der Produktionsbank deren Auswirkungen auf die Kundenwerte berücksichtigt werden.

2.3.3 Externe Koordination mit Verbundunternehmen

Die Verbundorientierung der Kreditgenossenschaften sowie die Prinzipien der Subsidiarität und Kooperation[1980] machen es notwendig, dass sich die Kreditgenossenschaften hinsichtlich der operativen, aber auch strategischen Vertriebsplanung mit den Verbundunternehmen abstimmen. Dies hat zweierlei Konsequenz. Zum einen bedarf es seitens der Verbundpartner einer konkreten Planungsgrundlage für die eigene Vertriebsplanung und die damit verbundenen übrigen Unternehmensteilpläne. Zum anderen ist es in den Fällen einer gemeinsamen Betreuung von Firmenkunden durch Firmenkundenberater der Kreditgenossenschaft und Vertriebsmitarbeiter der Verbundunternehmen notwendig, die operativen Ansatzpunkte auf Basis der strategischen Überlegungen zu begründen. Das heißt konkret, dass die kreditgenossenschaftliche Bewertung eines Firmenkunden und die damit verbundene Vertriebsstrategie handlungsleitend auch für die Aktivitäten und Maßnahmen der Verbundunternehmen sein muss.

WIEDEMANN empfiehlt hinsichtlich des Planungsprozesses im Verbund, dem Prinzip der demokratischen Willensbildung „von unten nach oben" folgend, einen Bottom-Up-Planungsansatz.[1981] Das impliziert, dass die Kreditgenossenschaft in ihrer operativen Vertriebsplanung die Verbundprodukte berücksichtigt. Der bankleistungsbasierte Ansatz der Kundenwertermittlung unterstützt dieses Vorgehen. So lassen sich bei entsprechender Abgrenzung des relevanten Verbundprogramms die in den monetären Kundenwerten enthaltenen Verbundprodukte mengen- und volumenmäßig für die betrachteten Planperioden aggregie-

[1979] Vgl. zu Firmenkundenportalen, deren Nutzen und Akzeptanz durch die Firmenkunden *Schwencke, M.* (2008), S. 35 ff.

[1980] Vgl. dazu die Ausführungen in Kapitel C.1.4.4.

[1981] Vgl. *Wiedemann, A.* (1992), S. 106 f.

ren und als Planungsgrundlage den Verbundpartnern mitteilen. Übertragen auf den monetären Kundenwert bedeutet dies, dass sich der Kundenwert aus Sicht der Kreditgenossenschaft in einzelne Kundenwerte aus Sicht des jeweils relevanten Leistungsprogramms der beteiligten Unternehmen der FinanzGruppe zerlegen lässt (siehe Formel [E.5][1982]). Es sei jedoch darauf hingewiesen, dass der Kundenwert, der sich über das relevante Leistungsprogramm der beteiligten Verbundunternehmen der FinanzGruppe ergibt, nicht dem Kundenwert aus Sicht der betroffenen Verbundunternehmen selbst entsprechen muss. Dies begründet sich damit, dass die Kreditgenossenschaft in ihrer Kundenbewertung lediglich die ihr aus dem Absatz der Verbundprodukte zufließenden Erträge und Aufwendungen respektive Einnahmen und Ausgaben berücksichtigt. Diese können jedoch von den Einnahmen und Ausgaben, die das Verbundunternehmen aus der Kundenbeziehung erhält, abweichen.

$$KW_c^m = \sum_{u=1}^{U} KW_c^{m,u} \qquad \text{[E.5]}$$

mit:

$KW_c^{m,u}$:	Monetärer Kundenwert eines Kunden c aus Sicht des relevanten Leistungsprogramms des Unternehmens u der FinanzGruppe aus Sicht der Kreditgenossenschaft
u:	Index der beteiligten Unternehmen der FinanzGruppe[1983]

Auf Basis dieses Zusammenhangs lassen sich die verbundleistungsbezogenen Mengen- und Volumeninformationen generieren, die für die Verbundunternehmen die Vertriebsplanung fundieren. Weiterhin wird damit kommuniziert, welche Bedeutung die Leistungen der Verbundunternehmen aus Sicht der Kreditgenossenschaft im Kontext der jeweiligen Kundenbeziehung oder in aggregierter Form aus Sicht des gesamten Kundenstamms haben. Mit diesen Informationen können die Verbundunternehmen wiederum über die regionale Ebene bis zur Bundesebene die Plandaten aggregieren und ihren Planungen zugrunde legen.

Probleme resultieren, wenn größere Abweichungen zwischen den so ermittelten Plandaten der Kreditgenossenschaften und den Ansprüchen der Verbundpartner vorliegen. Der damit verbundene Anpassungs- und Korrekturprozess würde nämlich erst bei abgeschlossenem Vertriebsplanungsprozess der Kreditgenossenschaften einsetzen. Dies dürfte somit die Akzeptanz auf Seiten der Genossenschaftsbanken einschränken. Aus organisatorischer Sicht empfiehlt WIE-

---

1982 Die verwendete additive Verknüpfung ist eine vereinfachte Darstellung.

1983 In dieser formalen Darstellung gehört die Kreditgenossenschaft selbst auch zur Mengen der beteiligten Verbundunternehmen der FinanzGruppe.

DEMANN daher, eine frühzeitige Konkretisierung der Planung der Verbundunternehmen, so dass die Voraussetzungen einer rechtzeitigen Abstimmungsphase geschaffen sind.[1984] Für die Abstimmungsprozesse ist aus Sicht der Kundenorientierung und der Kundenwertorientierung zu fordern, dass Plananpassungen nicht dergestalt sein dürfen, dass die zwischen dem Verbundunternehmen und der Kreditgenossenschaft vereinbarten Ziele losgelöst von den Kundenbeziehungen und den damit verbundenen Kundenwerten erfolgen.[1985] Ansonsten würde nämlich die kundenwertorientierte Vertriebsplanung, die die Kundenbeziehungen individuell betrachtet ad absurdum geführt. Möglichkeiten der Anpassung liegen z. B. in der zeitlichen Verschiebung, so dass für spätere Perioden angesetzte Bedarfspotenziale früher vertrieblich angesprochen werden können oder in einer Vereinbarung, die die Kreditgenossenschaft teilweise aus dem Vertriebsprozess des Verbundunternehmens exkludiert. Dies wäre z. B. dann der Fall, wenn das Verbundunternehmen versucht, die Differenz zwischen Bottom-Up-Planwerten der Kreditgenossenschaft und dem eigenen Top-Down-Anspruch durch eigenständige Vertriebsmaßnahmen im Absatzmarkt der VR-Bank zu kompensieren.[1986]

Für diejeningen Verbundunternehmen, die zur Entlastung der Produktionsbank tätig werden, empfiehlt WIEDEMANN, dass diese ohne weiteren Abstimmungsbedarf von den von der Kreditgenossenschaft getroffenen Planungen ausgehen.[1987] Da sie als Service Center fungieren, die ihre Leistungen über Verrechnungspreise an die Kreditgenossenschaften abgeben, werden Leistungsverrechnungspreise auf Basis der so ermittelten Plankapazität bestimmt. Mengenabweichungen gehen dann zu Lasten der Kreditgenossenschaften und Preisabweichungen stellen von den Verbundunternehmen zu verantwortende Unwirtschaftlichkeiten dar.

„Die besondere Schwierigkeit der Kundenorientierung im Verbund liegt im Problem der Abstimmung der Aktivitäten der einzelnen Kreditgenossenschaften mit denen der Verbundpartner."[1988] Wie bereits angeklungen, ist es nicht zwingend, dass die Bewertung eines Kunden hinsichtlich des relevanten Leistungsprogramms aus Sicht der Kreditgenossenschaft und aus Sicht des Verbundunter-

---

1984 Vgl. *Wiedemann, A.* (1992), S. 107.

1985 Dies wäre beispielsweise der Fall, wenn Volumina und Stückzahlen auf Basis historischer Werte fortgeschrieben werden oder wenn der Planung Marktpotenzialdaten zugrunde gelegt werden, die sich nicht auf einzelne Kundenbeziehungen zurückführen lassen.

1986 Dies entspräche auch den Leitlinien aus dem Strategiepapier BÜNDELUNG DER KRÄFTE: DIE GEMEINSAME STRATEGIE. Vgl. *BVR* (2001b), S. 8 f.

1987 Vgl. hier und nachfolgend *Wiedemann, A.* (1992), S. 108.

1988 *Wiedemann, A.* (1992), S. 117.

nehmens identisch ist. Damit geht einher, dass die relative Bedeutung eines Kunden aus Sicht der Kreditgenossenschaft von der aus Sicht des Verbundunternehmens abweicht. Dies gilt auch, wenn aus Sicht des Verbundunternehmens nur das numerische Kundenpotenzial der Kreditgenossenschaft betrachtet wird. Für die Abstimmung der vertrieblichen Aktivitäten in der Firmenkundenbank ist daher konsistent zum subsidiären Antritt der Verbundunternehmen zu fordern, dass die gemeinsame Festlegung von Maßnahmen dominiert wird durch die kreditgenossenschaftliche Beurteilung des Werts der Kundenbeziehung. Nur dadurch ist gewährleistet, dass die Kreditgenossenschaft eine widerspruchsfreie Planung der vertrieblichen Maßnahmen vornehmen kann. Im alternativen Fall würden die ggf. inkonsistenten relativen Bedeutungen der einzelnen Kundenbeziehungen aus Sicht der beteiligten Verbundunternehmen zu widersprüchlichen und damit für die Kreditgenossenschaft nicht umsetzbaren Handlungsempfehlungen führen.

3 Kundenwertorientierte Vertriebsplanrealisation

Die operative Vertriebsplanung der Ziele und Maßnahmen stellt die Basis der Planrealisation dar. Die konkrete Umsetzung ist ein Gegenstand der Vertriebssteuerung im engeren Sinn.[1989] Gestaltungsparameter dieses Teilaspekts der Vertriebssteuerung im Lichte der Kundenwertorientierung werden im Folgenden erläutert.

3.1 Steuerungsadäquate aufbauorganisatorische Struktur

Die Kreditgenossenschaften weisen eine markt- und kundenorientierte Aufbauorganisation auf.[1990] Dies stellt die notwendige Bedingung einer kundenwertorientierten Vertriebssteuerung dar. Die inhaltliche Festlegung über den Umfang der Verantwortlichkeit eines so genannten Responsibility-Centers[1991] stellt die hinreichende Bedingung dar. Je nach Umfang der Verantwortlichkeit werden Cost-Center, Expense-Center, Revenue-Center, Profit-Center und Investment-Center unterschieden.[1992] Bei Cost -Centern umfasst die Verantwortlichkeit die Effizienz der Leistungserstellung und als Kontrollgrößen werden die Kosten betrachtet. Die Verantwortlichkeit von Expense-Centern umfasst die Ausgaben, die innerhalb einer Periode angefallen sind. Die Steuerung dieser Einheiten erfolgt über Ausgabenbudgets. Statt Größen des internen Ressourcenverbrauchs beinhaltet die Verantwortlichkeit von Revenue-Centern ausschließlich die Erlösseite, so dass vor allem Erlösanalysen Gegenstand der Steuerung sind. Zentrale Annahme dabei ist, dass Kosten zwar hier entstanden, jedoch nicht von diesen Bereichen verursacht worden sind. Bei Profit-Centern werden sowohl die Erlös- als auch die Kostenseite betrachtet. Um die Leistung eines Profit-Centers zu beurteilen, wird der Saldo aus Erlösen und Kosten einer Periode herangezogen. Dabei ist die tatsächlich relevante Gewinngröße zu konkretisieren. Investment-Center sind Bereiche der Renditeverantwortung, d. h. auch Entscheidungen über Investitionen und Desinvestitionen können vom jeweiligen Verantwortlichen getroffen werden. Dieser Art von Responsibility-Centern ist die periodenübergreifende Betrachtung von Einnahmen und Ausgaben inhärent.

Im Kontext der auf Basis des Einzelgeschäfts aufbauenden wertorientierten Steuerung werden Vertriebseinheiten als Profit-Center geführt. Die Größe der

---

[1989] Vgl. dazu die Ausführungen in Kapitel C.2.4.2.

[1990] Vgl. dazu die Ausführungen in Kapitel C.2.4.2.

[1991] Vgl. zum Begriff der Responsibility Center *Eigler, J.* (2004), S. 668 sowie *Hofmann, C./Daugart, J.* (2004), S. 194.

[1992] Vgl. hier und nachfolgend *Coenenberg, A. G.* (1999), S. 531 ff.

Profit-Center ist institutsindividuell zu bestimmen und ist abhängig von dem Differenzierungsgrad der vorliegenden Kostenrechnungsinformationen.[1993] Auf Ebene des Profit-Centers werden sowohl die Kundengeschäftsergebnisse der Vertriebsbank als auch die Produktivitätsergebnisse der Produktionsbank ermittelt, um die Leistung des Profit-Centers zu beurteilen.[1994] Diese Profit-Center-Struktur der Kreditgenossenschaften ist konzeptionell kongruent zur bisher ausschließlichen Betrachtung des „Zerschlagungswerts" als relevanten Bestandteil des „Vermögenswerts", der der Gesamtbanksteuerung zugrunde gelegt wird. Mit der in dieser Arbeit vorgenommenen Ausweitung der Betrachtungsperspektive auch auf den „Potenzialwert" ließe sich begründen, Investment-Center statt Profit-Center als Grundlage der steuerungsadäquaten aufbauorganisatorischen Struktur im Vertrieb zu nutzen. Aufgrund der dualen Steuerungskonzeption ist jedoch zu beachten, dass Investitions- und Desinvestitionsentscheidungen mit Blick auf Kundenbeziehungen nicht zwingend ausschließlich von den Vertriebseinheiten bestimmt werden können. Insbesondere die damit verbundenen adressrisikobezogenen Auswirkungen auf das Kreditrisikoportfolio der Gesamtbank, das auf Gesamtbankebene gesteuert wird, können als Gründe dagegen vorgebracht werden. Somit wären dezentrale Organisationsbereiche nicht als Investment-Center in reiner Form umsetzbar. Unter Berücksichtigung dieser Einschränkung sollten Vertriebseinheiten eher im Sinne des hier konzeptualisierten und operationalisierten Modells als Investment-Center geführt werden.

Unabhängig von ihrer konkreten Ausgestaltung bedingen dezentrale Entscheidungseinheiten, dass „ein dezentrales Selbstcontrolling durch Instrumente zur Planungs-, Steuerungs- und Ergebnisverantwortung unterstützt und gefördert wird."[1995] Kern dessen ist, die dezentralen Entscheidungen im Sinne der Gesamtbank- und Vertriebsziele zu koordinieren.

3.2 Auswahl eines geeigneten Koordinationsmechanismus

Wie die ökonomische Definition des Bankbegriffs im Allgemeinen und die Etablierung dezentraler Entscheidungsbereiche im Speziellen begründet, handelt es sich bei den Kreditgenossenschaften um arbeitsteilige Systeme mit Interdependenzen der relevanten Subsysteme bzw. Organisationsebenen und –ein-

---

[1993] So nennt ANACKER Regional- oder Teilbanken, d. h. die Privatkundenbank und die Firmenkundenbank, als Beispiele für Profit-Center. Vgl. *Anacker, R.* (2003), S. 25. Im Gegensatz zu den Kosten ist es bei den Einzelgeschäftsabschlüssen vergleichsweise einfach möglich, diese den jeweiligen Vertriebseinheiten zuzuordnen.

[1994] Vgl. *Kuß, H.* (2002), S. 36.

[1995] *Hilgenstock, G. P.* (1995), S. 802.

heiten.[1996] Es wurde bei der Deskription des institutionellen Vertriebs der Kreditgenossenschaften deutlich, dass der bedeutendste Vertriebskanal der persönliche Vertrieb durch die Firmenkundenberater ist.[1997] Daraus resultiert eine besonders hohe Bedeutung der Personalführung in der Vertriebssteuerung. Es besteht somit die evidente Notwendigkeit, die Handlungen der einzelnen Akteure, d. h. Firmenkundenberater zu koordinieren, um das oder die definierten Ziele auf Gesamtbank- und Vertriebsebene zu erreichen.[1998] Gleichwohl bemängelt DUDERSTADT, dass das „Thema der Mitarbeiterführung [.] von vielen Banken als Erfolgsfaktor unterschätzt [wird; Erg. d. Verf.]“[1999].

Als Koordination wird in diesem Kontext die Abstimmung zwischen interdependenten, aber durch die Aufgabenteilung organisatorisch getrennten Tatbeständen verstanden.[2000] Objekte der Koordination können Handlungsprozesse oder Handlungsergebnisse sein. In der organisationstheoretischen Literatur gibt es eine Reihe von Koordinationsmechanismen, die sich in unterschiedlicher Weise auf die Koordinationsobjekte beziehen. Aufbauend auf der Ordnungsstruktur der Koordinationsmechanismen nach KIESER/KUBICEK lassen sich persönliche und technokratische Koordinationsmechanismen unterscheiden (siehe Tabelle 31).

Tabelle 31: Gegenüberstellung von Koordinationsmechanismen und -objekten[2001]

Koordinationsobjekt / ***Koordinationsmechanismus***		**Handlungs-prozess**	**Handlungs-ergebnis**
Persönliche Koordination	durch Weisung	x	x
	durch Selbstabstimmung	x	x
Technokratische Koordination	durch Programme	x	
	durch Pläne	x	x

Die persönliche Koordination durch Weisung erfolgt situativ und ist somit weitgehend flexibel und leicht zu gestalten.[2002] Sie kann sich sowohl auf den Handlungsprozess als auch auf das Handlungsergebnis beziehen. Dieser Koordinati-

1996 Vgl. zur ökonomischen Definition von Banken die Ausführungen in Kapitel C.1.1.2.

1997 Vgl. dazu die Ausführungen in Kapitel C.2.4.2.

1998 Vgl. *Eigler, J.* (2004), S. 668.

1999 *Duderstadt, S.* (2006), S. 173.

2000 Vgl. *Weber, J.* (2004), S. 75.

2001 Eigene Darstellung auf Basis *Kieser, A./Kubicek, H.* (1992) S. 102 ff.; *Weber, J.* (2004), S. 75 ff.

2002 Vgl. hier und nachfolgend *Kieser, A./Kubicek, H.* (1992) S. 102 ff.; *Weber, J.* (2004), S. 75 ff.

onsmechanismus erfolgt vor allem bei vertikalen Interaktionen, wie sie in Unternehmenshierarchien vorliegen. Bei horizontalen Interaktionen auf einer Hierarchieebene kommt dagegen die Koordination durch Selbstabstimmung zum Tragen, bei der die Abstimmung durch Gruppenentscheidungen erfolgt.[2003] Auch dieser Koordinationsmechanismus kann sich auf Handlungsprozesse oder Handlungsergebnisse beziehen. Die technokratische Koordination ist dadurch gekennzeichnet, dass sie über unpersönliche Medien erfolgt. Bei der programmbasierten Koordination werden die Handlungen in Inhalt und Ablauf festgelegt und im Bedarfsfall ausgeführt. Die Koordination durch Pläne basiert auf den verbindlichen Vorgaben von Zielen für die Aufgabenerfüllung. „Führen über Ziele bedeutet [...], die einem Unternehmen von ihrem(n) Eigner(n) und anderen Vergleichsgruppen gesetzten Ziele auf alle Organisationsbereiche herunterzubrechen, dass jede dezentrale Einheit zu ihrer Erreichung beiträgt."[2004] Das Führen über Ziele wird auch unter dem Begriff des Management-by-Objectives (MbO) oder Management-by-Results (MbR) diskutiert.[2005] Diese Ansätze kommen häufig in Verbindung mit der Balanced Scorecard zur Anwendung,[2006] da diese über das Instrument der Ziele die Strategieformulierung operationalisiert.[2007]

Der Einsatz der persönlichen Koordinationsmechanismen erfolgt situativ und einzelfallabhängig, so dass eine intensive Auseinandersetzung mit diesen Koordinationsmechanismen vor dem Hintergrund der Kundenwertorientierung keine verallgemeinerbaren Aussagen erwarten lässt. Zu den antizipativen und somit vorab gestaltbaren Steuerungsmaßnahmen der Planrealisation gehören die technokratischen Koordinationsmechanismen. Aufgrund der hohen Bedeutung des persönlichen Vertriebs im Firmenkundengeschäft sowie der in Kapitel C.2.3 beschriebenen Besonderheiten des Vertriebs von Bankleistungen ist davon auszugehen, dass die damit verbundene Flexibilität und Individualität der einzelnen Interaktionen nur einen geringen Standardisierungsgrad zulassen,[2008] so dass die programmbasierte Koordination eher irrelevant für die Vertriebssteuerung ist. Besonderheiten resultieren aus der Kundenwertorientierung mit dem Kunden-

---

2003 Derartige Entscheidungen liegen z. B. bei Vorstandsbeschlüssen vor. Aufgrund der Anforderungen der MaRisk gibt es grds. in jeder Kreditgenossenschaft mindestens zwei Vorstandsstellen, so dass dadurch horizontaler Abstimmungsbedarf begründet wird.

2004 *Weber, J.* (2004), S. 78.

2005 Vgl. dazu ausführlich *Noll, J.* (2003), S. 899 f. Es handelt sich bei den MbR-Ansätzen im Wesentlichen um MbO-Ansätze mit dem Unterschied, dass die Organisationseinheiten, welche bestimmte Ziele (Ergebnisse) erreichen sollen, organisatorisch getrennt sind. Vgl. *Noll, J.* (2003), S. 900.

2006 Vgl. *Homburg, C./Schäfer, H./Schneider, J.* (2010), S. 148.

2007 Vgl. *Eigler, J.* (2004), S. 667.

2008 Vgl. *Tewes, M.* (2003), S. 302 f.

wert als zentrale Ziel- und Steuerungsgröße zudem vornehmlich in der technokratischen Koordination mittels Plänen. Aus diesem Grund wird dieser Koordinationsmechanismus ausgewählt und den weiteren Ausführungen zugrunde gelegt.

In den Kreditgenossenschaften hat in den vergangenen Jahren eine Intensivierung der technokratischen Koordination der arbeitsteiligen Prozesse über plan- bzw. zielbasierte Ansätze stattgefunden.[2009] In seiner Studie stellt KRAUß fest, dass über 80 Prozent der befragten Kreditgenossenschaften Zieldefinitionen im Rahmen der Vertriebssteuerung bzw. der Planrealisation nutzen (siehe Abbildung 71).[2010] Dies zeigt die insgesamt hohe praktische Bedeutung der Management-by-Objectives-Ansätze, wenngleich die verwendeten Zielgrößen zur kundenwertorientierten Vertriebssteuerung im Widerspruch stehen.

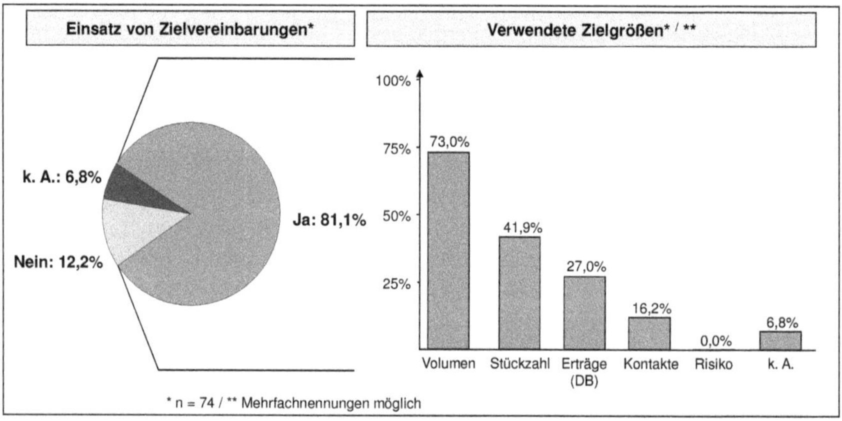

Abbildung 71: Einsatz von Zielvereinbarungen und Zielgrößen in Kreditgenossenschaften[2011]

Aus führungstheoretischer Sicht sollen Ziele Orientierungs- und Anhaltspunkte für die Mitarbeiter sein, die eine direkte persönliche Führung partiell substituieren.[2012] KÖHLER/ROLVERING/GERMANN argumentieren in diesem Zusammen-

2009 Vgl. *Kaiser, A./Kehr, J.* (2003), S. 31.

2010 Dass Zielvereinbarungen übergreifend bei Banken und Sparkassen das dominierende Führungsinstrument sind, indizieren auch die Ergebnisse der empirischen Studie von NIRSCHL/PAUL bei 85 Instituten mit Bezug zum Retailgeschäft. Sie geben an, dass bei 91,5 Prozent der Befragten die Zielvereinbarung das zentrale Führungsinstrument sei. Vgl. *Nirschl, M./Paul, V.* (2007), S. 200.

2011 Eigene Darstellung auf Basis der Daten in *Krauß, C.* (2005), S. 238.

2012 Vgl. *Schmoll, A.* (2006), S. 149.

hang gar, dass die „Voraussetzung für eine effektive Wahrnehmung der Steuerungsfunktion [durch die Vertriebssteuerung als Führungs- und Koordinationsinstrument, Erg. d. Verf.] [...] die Vorgabe von Zielen“[2013] sei.

Beim Festlegen dezentraler Ziele kann zwischen der Zielvorgabe und der Zielvereinbarung unterschieden werden. Während erstere einseitig erfolgt und stärker den Zielanspruch und damit den Zielinhalt der Bank berücksichtigt, liegt der Vorteil bei Zielvereinbarungen darin, dass der Mitarbeiter auf die Höhe der Ziele Einfluss nimmt und so keine demotivierenden Ziele zur Grundlage der Erfolgsmessung herangezogen werden. Konsistent zum Gegenstromverfahren der Vertriebsplanung und aufgrund tarifvertraglicher Vereinbarungen können nur Zielvereinbarungen die Grundlage eines leistungsorientierten Vergütungssystems in den Kreditgenossenschaften sein.[2014] Die leistungs- und erfolgsorientierte Vergütung, die als wesentlicher Bestandteil in der Vertriebssteuerung von Kreditinstituten angesehen wird,[2015] ist eine Ausprägungsform von Anreizsystemen. Der Einsatz von Anreizsystemen ergänzt die Koordinationsfunktion der Ziele um eine Motivationsfunktion, indem die Zielerreichungsgrade mit bestimmten positiven oder negativen Sanktionen verbunden sind.[2016] Soll eine kundenorientierte Unternehmenskultur etabliert werden, kommt der leistungs- und erfolgsorientierten Vergütung nach WEBSTER eine Schlüsselrolle zu.[2017] Die zielbasierte Führung stellt somit die notwendige Grundlage eines kundenwertorientierten Anreizsystems, das die tarifvertragliche Bedingung der leistungs- und erfolgsorientierten Vergütung in den Kreditgenossenschaften ist,[2018] dar.[2019] Die hinreichende Bedingung ist die Eignung des Kundenwerts oder kundenwertbasierter Ziele als Basis eines solchen Anreizsystems. Daher ist die Eignung des Kundenwerts als Grundlage eines vertriebliche Anreizsystems zu prüfen.

2013 *Köhler, A. G./Rolvering, H.-S./Germann, S.* (2005), S. 417.

2014 Vgl. *Kaiser, A./Kehr, J.* (2003), S. 31 f.

2015 Vgl. *Henke, K.* (1995), S. 758.

2016 Vgl. *Tewes, M.* (2003), S. 306 f. sowie die dort genannte Literatur.

2017 „The key to developing a market-driven, customer oriented business lies in how managers are evaluated and rewarded.“ *Webster, F. E.* (1988), S. 38. So auch *Hamel, W.* (2006), S. 462.

2018 Vgl. *AGVR* (2001), S. 6.

2019 Jedoch gilt: „Bei den meisten Kreditgenossenschaften werden die Kundenbetreuer [im Firmenkundengeschäft, Erg. d. Verf.] auch heute noch ohne Leistungsbezug bezahlt“. *Berhorst, U.* (2008), S. 36.

3.3 Kundenwertbasierte Anreizsysteme als Basis leistungs- und erfolgsorientierter Vergütung

3.3.1 Grundlagen von Anreizsystemen

Unter einem Anreizsystem wird die Gesamtheit von bewusst gestalteten und aufeinander abgestimmten Reizen verstanden, die dazu dienen, bestimmte Verhaltensweisen der Mitarbeiter durch positive Anreize oder Belohnungen auszulösen bzw. zu verstärken, um die Wahrscheinlichkeit des Auftretens unerwünschter Verhaltensweisen zu minimieren.[2020] In einem weit gefassten Verständnis sind die gesamten bankbetrieblichen Bedingungen als Anreize zu interpretieren. In einer engeren und hier relevanten Definition wird das Anreizsystem als Steuerungsinstrument verstanden, das durch zielgerichtete Gestaltung Einfluss auf das Verhalten der Mitarbeiter nimmt.[2021]

Der Einsatz von Anreizsystemen kann aus Sicht der Prinzipal-Agenten-Theorie und verhaltenswissenschaftlich begründet werden.[2022] Aus Sicht der Prinzipal-Agenten-Theorie wird mit einem Anreizsystem versucht, das Verhalten der Mitarbeiter unternehmenszielsetzungsgerecht zu steuern sowie partiell Risiken von der Unternehmensebene auf die Mitarbeiterebene zu verlagern, wenn es mit einem leistungs- und erfolgsorientierten Vergütungssystem verknüpft wird.[2023] Begründet wird dies u. a. mit der Gefahr opportunistischen Verhaltens des Agenten, die auf der einen Seite aus Informationsasymmetrien zwischen der Bank als Prinzipal und dem Firmenkundenberater als Agent und auf der anderen Seite aus der Tatsache, dass die Zielfunktion des Agenten von der des Prinzipals abweichen kann, resultiert.[2024] Verhaltenswissenschaftlich ist das vornehmliche Ziel von Anreizsystemen, die Motivation des Mitarbeiters zu erhalten oder zu erhöhen. Angelehnt an zwei unterschiedliche Motivationsquellen menschlichen Handelns, werden die einem Anreizsystem zugrunde liegende Anreize in Abhängigkeit ihrer Wirkungsweise in intrinsische Anreize mit langfristiger Motivationswirkung und extrinsische Anreize mit einer eher kurzfristigen Wirkungsdauer unterschieden.[2025] Anreizsysteme müssen so ausgestaltet sein, dass sie Zielkongruenz und Anreizkompatibilität gewährleisten.[2026] Zielkongruenz liegt vor, wenn der Firmenkundenberater über kundenbezogene Maß-

---

[2020] Vgl. *Büschgen, H. E.* (2001), S. 533 f.; *Eigler, J.* (2004), S. 679.

[2021] Vgl. ähnlich *Schmöller, P.* (2001), S. 203.

[2022] Vgl. *Homburg, C./Jensen, O.* (2000), S. 58; *Tewes, M.* (2003), S. 302.

[2023] Vgl. *Duderstadt, S.* (2006), S. 157 f.; *Eichelmann, T./Duderstadt, S.* (2005), S. 10.

[2024] Vgl. *Pfaff, D.* (2004), S. 169; *Buhl, H. U./Dzienziol, J./Heidemann, J.* (2010), S. 3.

[2025] Vgl. *Bastian, C.* (2000), S. 295.

[2026] Vgl. hier und nachfolgend *Buhl, H. U./Dzienziol, J./Heidemann, J.* (2010), S. 8.

nahmen so entscheidet wie die Bank entscheiden würde, so dass konsistente Verhaltensweisen vorliegen. Anreizkompatibel ist ein Anreizsystem dann, wenn der Firmenkundenberater dann einen Vorteil erfährt, wenn er sich so verhält, wie die Bank das machen würde.

Die zu gestaltenden Elemente eines Anreizsystems sind eine Menge von Anreizen, eine Menge von Bemessungsgrundlagen und die Kriterium-Anreiz-Relationen, die bestimmten Ausprägungen der jeweiligen Bemessungsgrundlage zu gewährende Anreize zuordnen.[2027] Die Ausgestaltung eines Anreizsystems umfasst viele einzelne Entscheidungstatbestände,[2028] die in weiten Teilen ohne Bezug zum Kundenwert festgelegt werden können und daher hier nicht weiter betrachtet werden. Unmittelbaren Bezug zum Kundenwert hat die Auswahl der relevanten Bemessungsgrundlage[2029] sowie der betroffenen Adressaten. Als Adressaten im Vertrieb der Firmenkundenbank sind sowohl die leitenden Angestellten auf der Leitungsebene als auch die einzelnen Firmenkundenberater auf der Ausführungsebene zu nennen. In Anlehnung an TEWES soll vereinfachend davon ausgegangen werden, dass auf Leitungsebene vor allem strategische Entscheidungen und auf Ausführungsebene vor allem operative Entscheidungen getroffen werden.[2030] Letztere stellen den Schwerpunkt der Vertriebssteuerung im engeren Sinn und somit den Schwerpunkt der weiteren Ausführungen dar.

KÖHLER/ROLVERING/GERMANN sehen aufgrund der im Firmenkundengeschäft vorherrschenden individuellen Kundenzuordnung die Möglichkeit, die kundenwertbezogenen Ziele der Vertriebsplanung in Ziele auf Mitarbeiterebene zu übersetzen.[2031] Die Vertriebsplanung wird in diesem Fall mit dem Anreizsystem verknüpft, indem das Ausmaß der Zielerreichung ein wesentlicher Maßstab der Leistungsbeurteilung und ggf. auch der Vergütung der betroffenen Mitarbeiter ist.[2032] Ziele im Sinne der technokratischen Koordination auf Basis von Plänen stellen die Grundlage für die Ableitung von Bemessungsgrundlagen dar, was als das zentrale Problem bei der Ausgestaltung von Anreizsystemen angesehen wird.[2033] Bezogen auf den Kundenwert ist daher zu fragen, welche Ziel-

---

2027 Vgl. *Homburg, C./Jensen, O.* (2000), S. 58.

2028 Vgl. dazu z. B. *Hamel, W.* (2006), S. 472 ff.; *Duderstadt, S.* (2006), S. 159 ff.; *Tewes, M.* (2003), S. 301 ff.

2029 Vgl. *Schmöller, P.* (2001), S. 204.

2030 Vgl. *Tewes, M.* (2003), S. 310.

2031 Vgl. *Köhler, A./Rolvering, H.-S./Germann, S.* (2005), S. 424. Zudem argumentieren die Autoren auf Basis ihrer Studienergebnisse, dass individuelle Zielvereinbarungen als eins von drei Kriterien sehr erfolgreiche Sparkassen von durchschnittlich erfolgreichen Sparkassen abgrenzen.

2032 Vgl. *Homburg, C. et al.* (2008), S. 645.

2033 Vgl. *Duderstadt, S.* (2006), S. 161 f.

größe sich als Bemessungsgrundlage für kundenwertorientierte Anreizsysteme eignet.[2034] Soll diese Frage beantwortet werden, sind die Anforderungen an eine Bemessungsgrundlage zu konkretisieren.

3.3.2 Anforderungen an Bemessungsgrundlagen

Bezogen auf den kreditgenossenschaftlichen Kontext sind die arbeitsrechtlichen Anforderungen an Ziele zu berücksichtigen. Ziele und die Kriterien der Zielvereinbarungen sollen sich auf das Aufgabenfeld des Einzelnen oder des Teams beziehen und durch den Vertriebsmitarbeiter beeinflussbar sein.[2035] Darüber hinaus existieren allgemeine Anforderungen. Bemessungsgrundlagen müssen den Zusammenhang zwischen den individuellen Zielen und Gesamtbankzielen herstellen, [2036] da nur so die postulierte Zielkongruenz hergestellt werden kann. Im Sinne der Anreizkompatibilität gilt, dass die Verbesserung der Ausprägung der Bemessungsgrundlage einen Beitrag zur Zielerreichung der Gesamtbank leistet.[2037] Zudem muss die Bemessungsgrundlage objektiv messbar und in ihrer Ausprägung durch den betroffenen Mitarbeiter und dessen Entscheidungen und Verhaltensweisen beeinflussbar sein.[2038] HAMEL schränkt diese Forderung durch sein Argument, es komme bei Anreizsystemen „nicht auf minutiöse Ursache-Wirkungsbeziehungen an, sondern vielmehr auf kategoriale Anerkennungen spezifischer Leistungen“[2039] leicht ein. Weiterhin wird gefordert, dass die Bemessungsgrundlage nicht durch den Berater manipuliert werden kann.[2040]

3.3.3 Würdigung kundenwertbasierter Bemessungsgrundlagen

Als mögliche Bemessungsgrundlagen bei kundenwertorientierten Anreizsystemen lassen sich der monetäre und der nicht-monetäre Kundenwert, die jeweiligen Kundenwerttreiber, die Steigerung der Kundenwerte oder die positive Veränderung der Kundenwerttreiber oder Bestimmungsfaktoren sowie die Implementierung kundenwertorientierten Verhaltens nennen.[2041]

Den kapitalwertbasierten Kundenwert gemäß Formel [D.30] als Bemessungsgrundlage zu nutzen, kann mittels verschiedener Zieloperationalisierungen

---

[2034] Vgl. *Tewes, M.* (2003), S. 311.

[2035] Vgl. *Kaiser, A./Kehr, J.* (2003), S. 31.

[2036] Vgl. *Duderstadt, S.* (2006), S. 162.

[2037] Vgl. *Eigler, J.* (2004), S. 672; *Homburg, C./Daugart, J.* (2004), S. 197; *Pfaff, D.* (2004), S. 179; *Schmöller, P.* (2001), S. 205.

[2038] Vgl. *Eigler, J.* (2004), S. 670 und S. 672.

[2039] *Hamel, W.* (2006), S. 470.

[2040] Vgl. *Buhl, H. U./Dzienziol, J./Heidemann, J.* (2010), S. 9.

[2041] Vgl. *Hamel, W.* (2006), S. 464.

erfolgen. Möglich wäre es, die absolute oder relative Veränderung zwischen zwei Bewertungszeitpunkten als Ziel für den Firmenkundenberater zu formulieren. Vorteilhaft ist, dass mit dieser Größe die postulierte Zielkongruenz und Anreizkompatibilität erreicht und der Zusammenhang zwischen Gesamtbank-, Vertriebs- und Mitarbeiterzielen leicht hergestellt werden kann. Jedoch ist zu konstatieren, dass der kapitalwertbasierte Kundenwert gemäß Formel [D.30] nicht ausschließlich durch Entscheidungen des einzelnen Firmenkundenberaters beeinflusst wird, da z. B. auch Entscheidungen der Steuerungs- und der Produktionsbank Einfluss nehmen. Würde nur derjenige Teil des kapitalwertbasierten Kundenwerts betrachtet, der durch den Vertrieb bzw. den Firmenkundenberater beeinflusst werden kann,[2042] würde dieses Defizit zwar behoben.[2043] Es würden implizit jedoch künftige, aber zum Zeitpunkt der Anreizgewährung noch nicht realisierte Verhaltensweisen und Maßnahmen in die Beurteilung einfließen. Zudem würde davon ausgegangen, dass das Kundenverhalten so eintritt, wie es bei der Ermittlung des Kundenwerts angenommen wird. Diese Annahme ist zwar obligatorisch für die Nutzung des Kundenwerts in der Vertriebssteuerung. Jedoch ist es problematisch, wenn die Anreizgewährung auf Basis dieser Größe erfolgen soll. Zwar kann argumentiert werden, dass das relevante Kundenverhalten gerade durch den Firmenkundenberater beeinflusst werden soll und kann. Gleichwohl müssten künftige negative Abweichungen, die den kapitalwertbasierten Kundenwert negativ beeinflussen und so auch einen Vermögenswertverlust für die Kreditgenossenschaft darstellen, dem Firmenkundenberater im Nachgang anzulasten sein.[2044] Daraus resultieren die aus der wertorientierten Vertriebssteuerung bekannten Probleme, dass die Bemessungsgrundlage der Anreizgewährung aufgrund der Dynamik des Kundenverhaltens oder nicht durch den Firmenkundenberater beeinflussbarer Störgrößen wie z. B. eine Insolvenz des Kunden, nicht stabil, sondern im Zeitablauf volatil ist. Damit verbundene und konsequenterweise durchzuführende negative Sanktionen dürften das Motivationsziel des Anreizsystems konterkarieren, da hier das so genannte Belohnungsrisiko virulent wird.[2045] Ähnliches gilt für den nicht-monetären Kunden-

---

[2042] Der durch die Vertriebsbank beeinflussbare Kundenwert wurde in Formel [E.3] als ein Summand des gesamten anbieterspezifischen Kundenwerts beschrieben.

[2043] So argumentiert TEWES, dass der Vertrieb nachhaltigen Einfluss auf den vom Kunden wahrgenommenen Nutzen hat, dieser wiederum den Kundenwert stark beeinflusst und daher der anbieterspezifische Kundenwert als Bemessungsgrundlage verwendet werden kann. Vgl. *Tewes, M.* (2003), S. 313.

[2044] Dies erfolgt in Analogie zu barwertigen Produktsteuerung, mithin der wertorientierten Steuerung auf Basis von Einzelgeschäftsbarwerten. Vgl. z. B. *Dokter, M./Lüders, U.* (2003), S. 29.

[2045] Vgl. *Schmöller, P.* (2001), S. 205.

wert, der aufgrund der Messmetrik noch stärker subjektiv geprägt ist und zudem von Einflussgrößen abhängt, die der Firmenkundenberater nicht bis kaum beeinflussen kann.

Weiterhin ist die Manipulationsfreiheit des monetären Kundenwerts zu diskutieren. Wie die Ausführungen in Kapitel D.5.2 verdeutlichen, hängt die Messung des Kundenwerts in beträchtlichem Umfang von den Informationen der Bank über die zukünftige Planung des Kunden ab. Diese Informationen werden in maßgeblicher Weise über den Firmenkundenberater durch den persönlichen Austausch mit dem Firmenkunden generiert. In dieser Situation liegt die Gefahr begründet, die Entwicklung des betrachteten Firmenkunden verzerrt darzustellen. Denn durch eine realistische Weitergabe der Informationen resultiert aus Sicht des Firmenkundenberaters die Gefahr, dass er hinsichtlich seiner Leistung realistischer und ggf. schlechter eingeschätzt wird.[2046]

Zusammenfassend ist zu konstatieren, dass die Verwendung des monetären oder nicht-monetären Kundenwerts als ausschließliche Bemessungsgrundlagen nicht geeignet ist. Die vorstehende Argumentation kann auf die Ebene der einzelnen Kundenwerttreiber, die in die Messung des Kundenwerts eingehen, übertragen werden und führt zum gleichen Ergebnis.

Alternativ zum monetären Kundenwert, der prospektiv ist, ist der in einer Periode realisierte Kundenwert als Bemessungsgrundlage zu diskutieren. Auch diese Größe weist den Vorteil der Zielkongruenz auf. Fraglich ist jedoch, ob auch die Anreizkompatibilität damit erreicht wird. Liegen periodische Ziele vor, kann dies dysfunktionales Verhalten bei den Vertriebsmitarbeitern induzieren,[2047] indem kurzfristige Erfolge zu Lasten langfristiger Erfolgspotenziale realisiert werden. Dies könnte etwa dann der Fall sein, wenn der Firmenkunden durch den Firmenkundenberater zu kurzfristigen Produktabsätzen gedrängt wird und dies dazu führt, dass ein daraus resultierend unzufriedener Kunde die Geschäftsbeziehung beendet. Zudem resultieren Probleme aus der Messung und der Manipulierbarkeit dieser Größe. Da die realisierten Erfolge im Rahmen der Einzelgeschäftskalkulation barwertig gemessen werden, liegen weiterhin Manipulationsspielräume vor.[2048] Aufgrund der periodenübergreifenden Dauer der Bankleistungserstellung können zudem die Probleme auftreten, die aus einer in der Zukunft abweichenden Ist-Situation von der erwarteten und der Messung zugrunde gelegten Situation resultieren. Eine isolierte Verwendung des in einer Periode realisierten Kundenwerts sollte daher auch nicht als Bemessungsgrundlage verwendet werden.

[2046] Vgl. *Buhl, H. U./Dzienziol, J./Heidemann, J.* (2010), S. 7.

[2047] Vgl. *Tewes, M.* (2003), S. 312.

[2048] Vgl. dazu z. B. *Batz, C.* (2004), S. 32 ff.; *Dokter, M./Lüders, U.* (2003), S. 29.

Als mögliche Bemessungsgrundlage können auch diejenigen Bestimmungsfaktoren des monetären und nicht-monetären Kundenwerts genutzt werden, die zum einen wichtige Einflussgrößen auf den Kundenwert sind und zum zweiten durch den einzelnen Mitarbeiter beeinflusst werden können.[2049] Für das Firmenkundengeschäft können dies bspw. die Kundenloyalität, die Kundenzufriedenheit, die Kundenbindung oder die vom Kunden aktiv aufgenommen Informationen sein. Dabei können sowohl Ergebnisse als auch Verhaltensweisen des einzelnen Vertriebsmitarbeiters in Betracht gezogen werden.[2050] Im Sinne eines inputorientierten Ansatzes, der sich auf Handlungsprozesse ausrichtet, kann auch eine auf am Kundenwert ausgerichtete Aktivitätensteuerung ergänzend eingesetzt werden. So stellt z. B. auch die Durchführung der in der Vertriebsplanung festgelegten Maßnahmen eine mögliche Bemessungsgrundlage dar. Problematisch ist dabei jedoch, dass damit die Ziele auf Gesamt-, Vertriebs- und Mitarbeiterebene nicht mehr unmittelbar miteinander verbunden werden können. Es ist daher begründet davon auszugehen, dass die Zielkongruenz geringer ausfällt. Zudem sind die einzelnen Bestimmungsfaktoren isolierte Größen, so dass die Gefahr vorliegt, den komplexen Gesamtzusammenhang der einzelnen Einflussfaktoren auf den Kundenwert nicht genügend zu berücksichtigen.

Wie die vorstehenden Ausführungen zeigen, ist auch bei Rekurs auf den Kundenwert als zentrale Ziel- und Steuerungsgröße im Vertrieb die Auswahl einer Bemessungsgrundlage eine bemerkenswerte Herausforderung für die Vertriebssteuerung. Zentrale Postulate der Kundenwertorientierung für die Auswahl von kundenwertbasierten Bemessungsgrundlagen sind, die Ausgewogenheit zwischen kurz- und langfristigen Zielen aufrechtzuerhalten sowie einen strikten Bezug zu den einzelnen Kundenbeziehungen aufrechtzuerhalten. Aus diesem Grund bietet es sich an, die kundenwertbezogene Balanced Scorecard als Orientierungspunkt für die Ziele und damit die Bemessungsgrundlage eines Firmenkundenberaters zu nutzen. Dies gewährleistet die notwendig Ausgewogenheit der Ziele und gleichzeitig deren Strategiekonsistenz, wodurch mittelbar auch die Zielkongruenz erreicht wird. Mit dem strikten Kundenwertbezug resultiert jedoch für die Gestaltung des Anreizsystems eine relativ höhere Komplexität. So würde die finale Höhe des Erfolgs als konkrete Ausprägung der Bemessungsgrundlage erst dann bekannt sein, wenn für alle von einem Firmenkundenberater betreuten Kunden die Ergebnisse vorliegen. Dies impliziert auch, dass eine unterjährige Positionsfeststellung aus Sicht des Firmenkundenberaters kaum möglich ist. Dies dürfte auch dann zu Problemen führen, wenn binnenperiodische

[2049] Vgl. hier und nachfolgend *Tewes, M.* (2003), S. 312.

[2050] Vgl. *Homburg, C./Schäfer, H./Schneider, J.* (2010), S. 150 f.

Wechsel in der Betreuung von Firmenkunden, d. h. in der Zuordnung von Firmenkunden zu Firmenkundenberater vorgenommen werden.

Abschließend sei noch auf die Möglichkeiten einer kundenwertorientierten Bemessungsgrundlage für die Leitungsebene eingegangen. Auf der strategischen Ebene sind Entscheidungen hinsichtlich der Selektion und Priorisierung von Kundenbeziehungen zu treffen. Daraus resultiert, dass geeignete Bemessungsgrundlagen auf dem anbieterspezifischen Kundenwert und das Kundenwertpotenzial sowie der Kombination aus diesen beiden Größen, mithin der Ausschöpfungsquote dieser Geschäftsbeziehungen im Einzelnen und in der Aggregation über den gesamten Kundenbestand basieren.[2051] Den dem prospektiven Kundenwert inhärenten Manipulationsspielräumen muss durch geeignete Kompensationsmaßnahmen bei der Ausgestaltung der übrigen Parameter des Anreizsystems Rechnung getragen werden.[2052] Mögliche Abweichungen, die durch externe Störgrößen bedingt werden, können unberücksichtigt bleiben, da es gerade die Aufgabe der Entscheidungsträger ist, auf externe Einflussgrößen so zu reagieren, dass die kundenwertbezogenen Ziele und damit die potenzialwertbezogenen Vermögenswertziele der Kreditgenossenschaft erreicht werden.[2053]

Neben diesen auf die Selbstkoordination ausgerichteten Aspekten der Vertriebssteuerung im engeren Sinn enthält VR-CONTROL mit der so genannten Vorsteuerung auch Instrumente der Fremdkoordination.[2054] Die Vorsteuerung umfasst Regelungen und Vorgaben, die zielsetzungsgerechte Entscheidungen beeinflussen sollen. Auf die Möglichkeiten den Kundenwert in der Vorsteuerung zu berücksichtigen, wird nachfolgend eingegangen.

3.4 Kundenwertorientierte Vorsteuerung

3.4.1 Grundlagen der Vorsteuerung nach VR-CONTROL

Ausgangspunkt der Vorsteuerung ist die Überlegung, dass die Einzelgeschäftskalkulation mit dem Deckungsbeitrag III keine abschließende Aussage über die Vorteilhaftigkeit eines Geschäfts zulässt.[2055] Diese Aussage gelingt nur, wenn dem ermittelten Deckungsbeitrag III ein Ergebnisanspruch gegenübergestellt

---

2051 Vgl. *Tewes; M.* (2003), S. 311 f.

2052 Z. B. durch eine Anreizgewährung bei einer leistungsorientierten Vergütung, die zeitlich gestreckt erfolgt. Vgl. *Tewes, M.* (2003), S. 322. Gestaltungsempfehlungen für ein monetäres Anreizsystem in diesem Zusammenhang zeigen *Buhl, H. U./Dzienziol, J./Heidemann, J.* (2010), S. 22 f. auf.

2053 Vgl. *Tewes, M.* (2003), S. 322.

2054 Vgl. *BVR* (2003b), Kapitel 2.3.

2055 Vgl. hier und nachfolgend *BVR* (2003b), Kapitel 2.3, S. 3 f.

wird, der als Mindestdeckungsbeitrag zu verstehen ist und diejenigen Kosten berücksichtigt, die noch nicht im Deckungsbeitrag III enthalten sind. Wesentlicher Bestandteil der Vorsteuerung ist daher die zentrale Vorkalkulation, die Höchst- und Mindestkonditionen vorgibt.[2056]

Vor diesem Hintergrund besteht die Aufgabe der Vorsteuerung darin, über die Preispolitik sicherzustellen, dass die betriebswirtschaftliche Vorteilhaftigkeit eines Einzelgeschäfts gewährleistet wird. Es wird unterschieden, ob es sich um endgültig festgelegte Konditionen im Standard- bzw. Massengeschäft oder um einzelgeschäftsbezogene Konditionsverhandlungen handelt. Im erstgenannten Fall wird ausgehend von einem Mindestdeckungsbeitrag ermittelt, wie die Konditionspolitik für die einzelnen Bankleistungen erfolgen soll. Im letztgenannten Fall dagegen wird auf Basis der individuell vereinbarten Kondition der resultierende Deckungsbeitrag III ermittelt. In diesem Fall wird ein kombiniertes Preisuntergrenzen- und Kompetenzsystem genutzt, um konditionspolitische Entscheidungen sach- und zielsetzungsgerecht durchführen zu können. Abbildung 72 zeigt den allgemeinen Aufbau eines Preisuntergrenzensystems in Anlehnung an die Einzelgeschäftskalkulation.[2057]

			Von der Geschäftskondition abgedeckte Positionen...				
			Kosten der Opportunität	Risikoprämie	Standard-stückkosten	Ertrags-anspruch	Überergebnis
Ausgangspunkt		DB-III + Ertragsanspruch > 0					
Preisuntergrenze	I	DB-III + Ertragsanspruch = 0					
	II	DB-III = 0					
	III	DB-II = 0					
	IV	DB-Ib = 0					
	V	DB-Ib < 0					

Abbildung 72: Preisuntergrenzensystem[2058]

Aufbauend auf diesem Preisuntergrenzsystem kann ein Kompetenzsystem in der Vertriebsteuerung implementiert werden. Dabei werden bspw. mit den jeweiligen Preisuntergrenzen in Kombination mit Aussagen über den Deckungsbeitrag

2056 Vgl. *Hilgenstock, G. P.* (1995), S. 808.

2057 Vgl. zur Einzelgeschäftskalkulation die Ausführungen in Kapitel C.4.3.

2058 *Hensel, V./Wild, C.* (2003), S. 37. Ähnlich auch *BVR* (2003b), Kapitel 2.3, S. 6. Die Preisuntergrenze IV gilt nur für Aktivgeschäfte. Die grau kolorierten Felder weisen zeilenweise diejeningen Kosten- und Ergebniskomponenten aus, die durch die Preisuntergrenze gedeckt sind.

der gesamten bisherigen Kundenbeziehung konkrete Kompetenzzuordnungen verknüpft, die vorgeben, welche Kompetenzstufe der Bank entscheidungsbefugt ist (siehe Tabelle 32). Unterschieden wird zwischen risikobehafteten Aktivgeschäften (AG) und sonstigen Geschäften (SG).

Tabelle 32: Beispielhafte Ausgestaltung eines Ergebniskompetenzsystems[2059]

Deckungsbeitrag des Einzelgeschäfts	**Deckungsbeitrag der Kundenverbindung insgesamt**											
	DB III+EA > 0		DB III+EA = 0		DB III = 0		DB II = 0		DB Ib = 0		DB Ib < 0	
	AG	SG	AG	SG	AG	SG	AG	SG	AG	SG	AG	SG
DB III+EA>0	1		1		1		1		1		1	
DB III+EA=0	1		1		1		1		1		1	
DB III=0	1		1		2	1	2	1	2	1	2	1
DB II = 0*	1		1		2		2		2		3	
DB Ib = 0	1		1		2		3	2	3	2	3	
DB Ib < 0	1		1		2		3		3		3	

*Preisuntergrenze gilt nur für Aktivgeschäfte; „Sonstige Geschäfte“ umfassen Passiv- und Dienstleistungsgeschäfte

1: 1. Kompetenzstufe (z. B. Berater)
2: 2. Kompetenzstufe (z. B. Bereichsleitung)
3: 3. Kompetenzstufe (z. B. Vorstand)

SG: Sonstige Geschäfte
DB: Deckungsbeitrag
EA: Ergebnisanspruch

Um in der wertorientierten Steuerung mögliches Margendumping bei individuell zu verhandelnden Geschäften zu vermeiden, wird empfohlen, Restriktionen in Form mindestens zu erzielender Nettomargen, die den barwertigen Deckungsbeitrag III in Prozent ausdrücken, festzulegen.[2060] Diese Nettomargen werden produktindividuell ermittelt. Das Kompetenzsystem kann in diesem Zusammenhang auch auf Basis der Nettomargen definiert werden (siehe Tabelle 33). Danach darf z. B. ein Firmenkundenberater mit einem Verhandlungsspielraum von 40 Prozent ausgehend von der mindestens zu erzielen Nettomarge in die Verhandlungen mit dem Kunden gehen. Reicht dieser Verhandlungsspielraum nicht aus, so muss die übergeordnete disziplinarische Stelle angerufen und in die Verhandlungen integriert werden.

2059 *BVR* (2003b), Kapitel 2.3, S. 8; *Hensel, V./Wild, C.* (2003), S. 38.

2060 Vgl. hier und nachfolgend *Batz, C.* (2004), S. 33 ff.

Tabelle 33: Kompetenzregelung auf Basis von Nettomindestmargen[2061]

	Privat-kundenberater	Firmen-kundenberater	Bereichsleiter	Vorstand Markt
Kompetenz Netto-marge (Reduktion bis...)	80%	60%	40%	Positives Ergebnis des Kunden

Im Standardgeschäft werden für einzelne Bankleistungen konkrete Konditionen vorgegeben,[2062] zu denen diese angeboten werden. Auch dieses Verfahren kann mit den vorgestellten Kompetenzregelungen verknüpft werden.

Aus prozessualer Sicht wird die Vorsteuerung in Form der Vorkalkulation der Produktionsphase zugeordnet und unterteilt in die Prozessschritte Konditionierung, die Ermittlung des Deckungsbeitrags III, die Ermittlung des Ergebnisanspruchs, die Ermittlung einer möglichen Über- oder Unterdeckung und in die Entscheidung über den Geschäftsabschluss gemäß des Ergebniskompetenzsystems (siehe Abbildung 73).[2063]

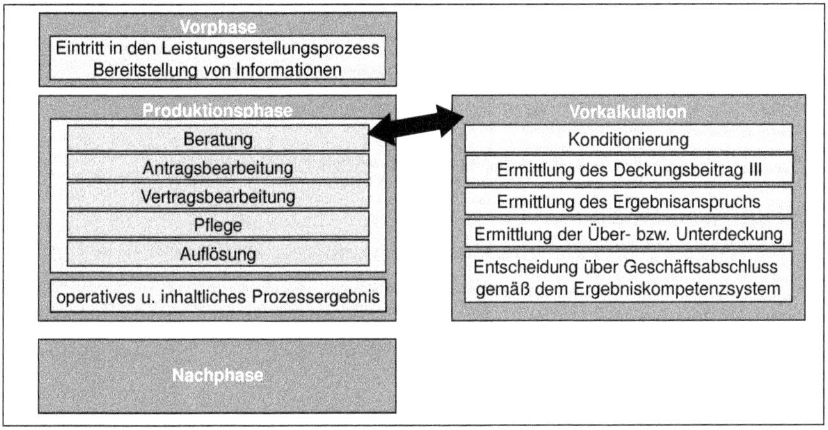

Abbildung 73: Vorkalkulation als integraler Bestandteil des Geschäftsprozesses[2064]

2061 In Anlehnung an *Batz, C.* (2004), S. 35.

2062 Dabei werden neben den Produktparametern wie Volumen und Laufzeit bei Aktivgeschäften auch kundenbezogene Merkmale wie Risikoklasse berücksichtigt. Vgl. *BVR* (2003b), Kapitel 2.3, S. 11 ff.

2063 Vgl. *BVR* (2003b), Kapitel 2.3, S. 14 ff.

2064 *BVR* (2003b), Kapitel 2.3, S. 15.

Zwar gilt nach Vorgabe von VR-CONTROL grundsätzlich, dass jeder Deckungsbeitrag positiv sein soll.[2065] Allerdings wird auch in dieser Steuerungskonzeption berücksichtigt, dass der Produktdeckungsbeitrag nicht für jedes Geschäft generell positiv sein muss. Vom Grundsatz abweichende Entscheidungen sollen die gesamte Geschäftsbeziehung mit dem Kunden ins Kalkül ziehen. Insbesondere sind die Deckungsbeiträge der sonstigen mit dem Kunden kontrahierten Bankleistungen zu beachten. BATZ weist daher darauf hin, „dass die einzelnen Kompetenzstufen in ihre Konditionsentscheidungen auch den gesamten Ertrag der Kundenverbindung integrieren müssen."[2066] Diese Forderungen zielen vornehmlich darauf ab, die zum Verhandlungszeitpunkt bereits mit einem Kunden bestehenden Geschäfte und deren monetären Wertbeiträge zu berücksichtigen. Sie berücksichtigen damit ausschließlich die bisherige Geschäftsbeziehung. Den Zukunftsbezug berücksichtigen HENSEL/WILD, die die Gefahr sehen, rentable Kunden aufgrund eines einzelnen nicht abgeschlossenen Einzelgeschäfts mit negativem Deckungsbeitrag III zu verlieren. Daher akzeptieren sie defizitäre Einzelgeschäfte, weisen aber darauf hin, dass langfristig die Erzielung positiver Deckungsbeiträge III im Vordergrund stehen muss.[2067] Und KIPKER/SCHÖNHALS fordern, dass die Firmenkundenberater einen gewissen Gestaltungsspielraum für Konditionen und Preise haben sollten, der abhängig von der aktuellen Rentabilität sowie dem weiteren Geschäftspotenzial des Kunden ist.[2068] Diese Forderungen können durch die Integration des Kundenwerts systematisch und methodisch in die Vorsteuerung integriert werden.

3.4.2 Integration des Kundenwerts in die Vorsteuerung

Die Integration des Kundenwerts kann vor dem Hintergrund der vorhergehenden Ausführungen auf drei kombinierbare Arten erfolgen. Erstens indem er Einfluss auf die Vorkalkulation nimmt, zweitens indem der Kundenwert entsprechend den genannten Anforderungen das oben skizzierte Preis- und Kompetenzsystem ergänzt und drittens indem auf Basis des Kundenwerts ein eigenständiges Kompetenzsystem entwickelt wird.

Im einfachsten Fall wird der Kundenwert insofern in die Vorsteuerung integriert, als die für die Ermittlung des monetären anbieterspezifischen Kundenwerts prognostizierten Einzelgeschäfte auch zur Grundlage der Vorkalkulation gemacht werden. Wie erinnerlich stellen die einzelnen Bankleistungen, deren Absatz im Rahmen einer Kundenbeziehung für die Zukunft prognostiziert und

2065 Vgl. hier und nachfolgend *BVR* (2003b), Kapitel 2.3, S. 4.
2066 *Batz, C.* (2004), S. 35.
2067 Vgl. *Hensel, V./Wild, C.* (2003), S. 37.
2068 Vgl. *Kipker, I./Schönhals, F.* (2002), S. 17.

geplant wird, die Grundlage dar, auf der der monetäre anbieterspezifische Kundenwert ermittelt wird. Die Kalkulation der Bankleistungen orientiert sich dabei an den bankseitigen Ansprüchen und den kundenseitigen Möglichkeiten und Grenzen. Zu nennen ist hier z. B. das auf der vermuteten Zahlungsbereitschaft des Kunden basierende Up-Selling-Potenzial. Mit der Prognose des monetären Kundenwerts ist verbunden, die Bankleistungen hinsichtlich der Produktparameter (z. B. Laufzeit, Volumen, Mengengerüste beim Zahlungsverkehr) und der Konditionengestaltung vorab zu definieren, da sonst eine Kalkulation der Bankdienstleistungen unmöglich ist. Die also im Kontext der Kundenwertermittlung genutzten Einzelgeschäftskalkulationen stellen somit die Grundlage des Kompetenzsystems im Sinne der Tabelle 33 dar. Abweichend zur Vorkalkulation im oben skizzierten Fall kann es jedoch sein, dass ein im Kontext des Kundenwerts kalkuliertes Geschäft bereits eine Preisuntergrenze unterschreitet und damit der Firmenkundenberater nicht mehr entscheidungsbefugt ist. In diesen Fällen könnten entsprechende ex ante Genehmigungen unter Vorbehalt an den Firmenkundenberater ausgestellt werden. Der Vorbehalt bezieht sich auf die Identität der tatsächlichen von der prognostizierten Ausführung sowie die Periode des Absatzes der Bankleistung.

Die im vorhergehenden Kapitel zitierten Forderungen machen deutlich, dass eine preispolitische Entscheidung bei einer konkreten Transaktion zulasten der Kreditgenossenschaft nur dann vorgenommen werden sollte, wenn ein Abwägen mit transaktionsexternen Vorteilen eine solche Entscheidung begründen kann. Ein solches Abwägen ist implizit in dem in Tabelle 32 dargestellten Kompetenzsystem enthalten. Problematisch daran ist jedoch, dass die betrachteten transaktionsexternen Vorteile vergangenheitsorientiert und somit steuerungstheoretisch inadäquat sind. Die Integration des Kundenwerts in die Vorsteuerung kann diese Entscheidungen intersubjektiv nachvollziehbar und steuerungstheoretisch besser begründen. In diesem Fall wäre die prozessuale Vorsteuerung um einen Schritt zu ergänzen. Und zwar wäre bei der Entscheidung über den Geschäftsabschluss eine Prüfung der Auswirkungen auf den monetären und/oder nicht-monetären Kundenwert und deren Kundenwerttreiber als Entscheidungsfundierung vorzunehmen. Bei der Prüfung dieses Einflusses ist zu unterscheiden, ob komplementäre oder kompensatorische Einflüsse vorliegen. Komplementäre Effekte liegen z. B. dann vor, wenn sich das Entwicklungspotenzial auf eine prognostizierte höhere Zahlungsbereitschaft für Leistungen des Basisgeschäftspotenzials bezieht. In diesem Fall kann eine Situation resultieren, in der die Verhandlung über eine Leistung den Wert des Basis- und des Entwicklungspotenzials beeinflusst. Kompensatorische Effekte können dagegen vorliegen, wenn eine geringere Kondition bei einer Leistung des Basisgeschäftspotenzials durch eine höhere Kondition bei einer Cross-Selling-Leistung aufgewogen werden kann.

Diese Überlegung sei an einem Beispiel verdeutlicht. Für den in Tabelle 30 betrachteten Firmenkunden wird ein monetärer Kundenwert von 329 GE ermittelt. Das Basisgeschäftspotenzial in der Periode t=1 weist einen Wert von 100 GE auf. Es sei angenommen, dass dieses unter anderem aus einer Anschlussfinanzierung für ein auslaufendes Investitionsdarlehen resultiere, wobei weiterhin angenommen werde, das Up-Selling-Potenzial von 30 GE begründe sich unter anderem in einem im Vergleich zur gegenwärtigen Finanzierung höheren Konditionsbeitrag. In der Periode t=1 verhandele nun der Firmenkundenberater mit dem Firmenkunden über die Kondition der Anschlussfinanzierung. Die verhandelte Kondition führe zu einem negativen Deckungsbeitrag III, so dass der Firmenkundenberater nicht mehr kompetent und entscheidungsbefugt ist. Der Bereichsleiter auf der nächsthöheren Kompetenzstufe muss nun über den Geschäftsabschluss entscheiden und dazu abwägen. Es ließen sich dazu die folgenden Überlegungen anstellen. Die verringerte Kondition reduziere c. p. das Basisgeschäftspotenzial auf 70 GE und das Up-Selling-Potenzial auf 15 GE. Daraus resultiere ein Kundenwert zu Beginn der Periode t=1 von 297 GE. Dieser Vermögenswertverlust in Höhe von 10 Prozent zum ursprünglich prognostizierten Kundenwert von 329 GE darf nur begründet in Kauf genommen werden. Es sei angenommen, die Anschlussfinanzierung habe Einfluss auf die Kundenbindungsrate der folgenden Perioden, da damit die Wechselkosten für den Kunden bedingt durch die vertragliche Bindungswirkung des Kreditvertrags steigen. Annahmegemäß steige die Kundenbindungsrate um 10 Prozent in den folgenden Perioden t=2 bis t=4. Es resultiert unter Berücksichtigung der gestiegenen Kundenbindungsrate ein monetärer anbieterspezifischer Kundenwert zu Beginn der Periode t=1 von 319 GE. Dieser liegt nur noch um 3 Prozent unter dem ursprünglich für diesen Zeitpunkt ermittelten Kundenwert. Damit steht der gegenwärtigen Vermögenswertreduzierung eine begründet anzunehmende künftige Vermögenswertsteigerung gegenüber, die diese teilweise kompensieren kann. Im alternativen Fall, dass die Anschlussfinanzierung nicht kontrahiert wird, möge das Basisgeschäftspotenzial einen Wert von 80 GE haben und die Kundenbindungsrate in den folgenden Perioden annahmegemäß um 15 Prozent sinken. Unter Berücksichtigung dieser Veränderungen resultiere c. p. ein monetärer anbieterspezifischer Kundenwert zu Beginn der Periode t=1 von 271 GE, der etwa 18 Prozent unter dem ursprünglich angenommenen Wert liegt. Ein analoges Vorgehen ließe sich über den nicht-monetären Kundenwert und dessen Kundenwerttreiber vornehmen. Auf Basis einer dergestalten Integration des Kundenwerts kann der Abwägungsprozess intersubjektiv nachvollziehbar auf Basis steuerungstheoretisch adäquater Bezugsgrößen vorgenommen werden.

Als dritte Möglichkeit könnte ein Kompetenzsystem direkt auf Basis des Kundenwerts konzeptualisiert werden. Dabei lässt sich eine breite Vielfalt an

Möglichkeiten vorstellen, von denen nachfolgend zwei skizziert werden sollen. Eine Möglichkeit wäre ein Kompetenzsystem zu implementieren, dass nicht auf Preisuntergrenzen eines einzelnen Geschäfts, sondern auf Werthürden für die Bezugsgröße des monetären Kundenwerts rekurriert. So könnten bspw. die anbieterspezifischen monetären Kundenwerte der laufenden Periode in Summe betrachtet werden. Alternativ zum Vorgehen in Tabelle 32 wird die Preisuntergrenze einer konkreten Transaktion nicht der Summe der Deckungsbeiträge der gesamten Geschäftsbeziehung, sondern dem Erfolg der laufenden Periode hinsichtlich der kundenwertbasierten Vertriebsziele gegenübergestellt. Zur Verdeutlichung dieser Überlegung sei auf das Beispiel in der Tabelle 30 verwiesen. Das Ziel für die relevante Kundenbeziehung sei auf 120 GE festgelegt.[2069] Im Rahmen der Verhandlung einer (oder mehrerer) Bankleistungen, die die Grundlage der Kalkulation des Kundenwerts in der Periode t=1 sind und dem Basis-, dem Entwicklungs- oder dem Cross-Selling-Potenzial zugeordnet wurden, ist nun zu prüfen, ob die verhandelte(n) Kondition(en) Einfluss auf die Zielerreichung haben kann (können). In Abhängigkeit der kundenindividuellen Konstellation der monetären Werttreiber impliziert dieses Vorgehen ein größeres Ausmaß an Kundensicht als das einzeltransaktionsbezogene Kompetenzsystem auf Basis von Preisuntergrenzen. Zudem sei darauf hingewiesen, dass dieses Vorgehen nicht zwingend auf einem bankleistungsbezogenem Mengengerüst basieren muss, sondern auch praktikabel ist, wenn die Ziele lediglich auf Basis wertmäßiger Informationen formuliert werden.

Eine alternative Möglichkeit dazu wäre es, den prognostizierten anbieterspezifischen monetären Kundenwert als periodenübergreifende Bezugsgröße zu nutzen. In diesem Fall rekurriert die Vorsteuerung auf die Vertriebsziele hinsichtlich des Kundenwerts für die künftigen Perioden. Dabei können zwei Fälle unterschieden werden. Im ersten Fall bezieht sich die Vorsteuerung auf den für die laufende Periode ursprünglich prognostizierten und der Vertriebsplanung zugrunde gelegten monetären anbieterspezifischen Kundenwert. Im Beispiel aus Tabelle 30 entspricht dies dem Wert von 329 GE zu Beginn der Periode t=1. Eine Abweichung der verhandelten von den prognostizierten und geplanten Deckungsbeiträgen oder Konditionen wirkt sich auf diesen Wert in zweierlei Weise aus. Erstens durch die Veränderung des periodischen Kundenwerts der Periode t=1 und zweitens durch die Auswirkungen auf die künftigen periodischen Kundenwerte der Perioden t=2 bis t=4. In Abhängigkeit der durch die konkrete Transaktion induzierten Veränderung des monetären anbieterspezifischen Kundenwerts lässt sich der Erfolg hinsichtlich der kundenwertbasierten Ziele für

[2069] Der Wert 120 resultiert aus der Multiplikation des anbieterspezifischen monetären Kundenwerts der Periode t=1 in Höhe von 150 mit dem Zielanspruch von 80 Prozent.

diese Zielgröße ermitteln. In Abhängigkeit des Ausmaßes der negativen Zielabweichung sind entsprechende Entscheidungskompetenzen zu vergeben.

Unabhängig davon auf welche Art der Kundenwert in die Vorsteuerung integriert wird, zeigen die vorhergehenden Ausführungen, dass der Kundenwert eine sinnvolle methodische Ergänzung der bestehenden Vorsteuerungsansätze ist. Die Integration des Kundenwerts gewährleistet, dass die Kundensicht, die der relationalen Vertriebsansätze zu Eigen sein soll, auch in der Vorsteuerung die produktzentrierte Sicht ergänzen oder gar substituieren kann. Die konkrete Art der Integration des Kundenwerts in die Vorsteuerung ist institutsindividuell vorzunehmen. Abschließend sei darauf hingewiesen, dass die Integration des Kundenwerts nach HOMBURG/DAUM auch als Argumentationshilfe in der Verhandlung mit dem Kunden verwendet werden kann.[2070]

4 Kundenwertorientierte Überwachung und Sicherung

Wie in grundlegenden Ausführungen in Kapitel B.2 erörtert wurde, umfasst der Planungs- und Steuerungsprozess in dem dieser Arbeit zugrunde gelegten Verständnis auch die Aufgaben der Überwachung und Sicherung. Für die Erfüllung dieser Aufgaben sind kundenwertorientierte Kontrollen notwendig. Dabei gilt, dass je unsicherer die Planungsinhalte sind, desto umfangreicher und vielfältiger die Kontrollen zur Wissensgenerierung in der Überwachungs- und Sicherungsphase sein sollten.[2071] Im Folgenden werden relevante Kontrollarten begründet ausgewählt sowie deren Beziehungen untereinander im Kontext der Wissensgenerierung beschrieben.

4.1 Auswahl relevanter Kontrollarten

In der kundenwertorientierten Überwachungs- und Kontrollphase sind Vertriebsplanabweichungen zu identifizieren, zu analysieren und ggf. als Basis zu verwenden, um Gegensteuerungsmaßnahmen einzuleiten. Es ist hierzu nicht ausreichend, den Kontrollumfang auf den vergangenheitsorientierten Soll-Ist-Abgleich zu beschränken.[2072]

In der Literatur werden im Kontext der Kundenwertorientierung die auf gegenwärtige und künftige interne und externe Entwicklungen ausgerichtete Prämissenkontrolle, die auf die Umsetzung der kundenwertbezogenen Maßnahmen

---

[2070] Vgl. *Homburg, C./Daum, D.* (1998), S. 141.

[2071] Vgl. *Stüker, D.* (2008), S. 353.

[2072] Vgl. *Eberling, G.* (2002), S. 304; *Küpper, H.-U.* (2001), S. 19.

ausgerichtete Durchführungskontrolle sowie die auf die Ergebnisse der Maßnahmen ausgerichtete Ergebniskontrolle als relevante Kontrollarten genannt.[2073] Diese Auswahl wird um die Verhaltenskontrolle und die strategische Überwachung ergänzt.

Die Verhaltenskontrolle ist zu berücksichtigen, da sie prüft, „ob der Ausführende unter den eingetretenen Situationsbedingungen die geeigneten Maßnahmen ergriffen hat."[2074] Im Gegensatz zur Ergebniskontrolle werden bei der Planumsetzung die geänderten Rahmenbedingungen berücksichtigt. Insbesondere aufgrund der Zukunftsorientierung des Kundenwerts und der damit verbundenen Unsicherheit ist diese Kontrollart daher eine sinnvolle Ergänzung.

Die bisher genannten Kontrollarten haben gemeinsam, dass sie so genannte gerichtete Kontrollen sind. Insbesondere bei strategischen Planungen ist es notwendig, auch ungerichtet Informationen zu sammeln. Dazu wird die strategische Überwachung eingesetzt. Diese ist zwar keine Kontrollart im wörtlichen Sinn, da sie aufgrund ihrer Ungerichtetheit keinen Vergleichsmaßstab hat.[2075] Sie stellt jedoch eine beachtenswerte Ergänzung der Prämissenkontrolle dar. Die Einordnung in die Überwachungs- und Sicherungsphase begründet sich damit, dass sie trotz ihrer inhaltlichen Ungerichtetheit einen Ausgangspunkt benötigt, der sich aus der Planungs- und Umsetzungsphase ergibt.

Diese fünf Arten von Kontrollen stehen in einer engen Beziehung zueinander, so dass die gemeinsame Analyse der Kontrollergebnisse eine höhere Qualität an Erkenntnissen für die Erfolgswirksamkeit des kundenwertorientierten Vertriebs im Firmenkundengeschäft liefert. Aufgaben der Vertriebskontrolle sind die Feed-back-Kontrolle sowie die Feed-forward-Überwachung. Ausgehend von identifizierten Vertriebsplanungsabweichungen auf Basis der Ergebniskontrolle im Rahmen der Feed-back-Kontrolle ist zu analysieren, worin etwaige Abweichungen begründet sind. Abweichungen können durch interne oder externe Einflussfaktoren verursacht werden. Typischerweise liegen intern begründete Ab-

---

2073 Vgl. *Wille, K.* (2005), S. 251; *Tewes, M.* (2003), S. 217 ff.; *Eberling, G.* (2002), S. 304 f. Für die allgemeine Wertorientierte Unternehmensführung werden diese Kontrollarten auch von *Ebeling, C.* (2007), S. 113 ff. als relevant beurteilt. Vgl. für eine Übersicht von Kontrollarten auch Abbildung 3. Alternative Bezeichnungen für die Durchführungskontrolle sind Maßnahmenkontrolle, Planfortschritts- und Verfahrenskontrolle. Die Ergebniskontrolle wird auch als Realisationskontrolle bezeichnet. Aus Controllingsicht können die Durchführungs- und die Ergebniskontrolle sich sowohl auf den Planungsprozess als auch auf den Umsetzungsprozess beziehen. Im ersten Fall ist es Gegenstand der Kontrolle, zu prüfen, ob die Planung wie gefordert umgesetzt wird. Für die Vertriebssteuerung ist die Kontrolle auf den Umsetzungsprozess zu richten.

2074 *Küpper, H.-U.* (2001), S. 173.

2075 Kontrollen beruhen stets auf einem Vergleich. Vgl. *Küpper, H.-U.* (2001), S. 169.

weichungen vor, wenn in der Planrealisationsphase Maßnahmen nicht oder nur unvollständig vorgenommen wurden. Dies zu prüfen, ist Gegenstand der Durchführungskontrolle. Extern begründete Abweichungen liegen vor, wenn grundlegende Prämissen der strategischen Planung nicht eintreten oder nicht eingetreten sind. Da die Prämissenkontrolle nur diejenigen Annahmen über die Umfeld- und Unternehmensentwicklung fokussiert, die in der strategischen Planung adressiert wurden, stellt sie eine selektive oder spezialisierte Kontrolle dar.[2076] Sie wird daher ergänzt um die strategische Überwachung, die die strategierelevanten Parameter sucht, die „weder als explizite Prämisse in den Prozess der Strategieformulierung eingegangen sind noch bei der Strategieumsetzung bereits wirksam geworden sind.“[2077] Bei fehlerhaften Prämissen dient die Verhaltenskontrolle dazu, Informationen darüber zu generieren, ob das Verhalten eines Vertriebsmitarbeiters an die geänderte Situation angepasst wurde und damit Ergebnisabweichungen erklärt werden können.

Nachfolgend werden die für eine kundenwertorientierte Vertriebssteuerung als relevant erachteten Kontrollarten vorgestellt und hinsichtlich ihrer Ausgestaltung erörtert.

4.2 Ergebniskontrolle

Gegenstand der Ergebniskontrolle ist es, das Ausmaß der Zielerreichung zu überprüfen.[2078] Aufgrund ihres unmittelbaren Zielbezugs ist es zwingend erforderlich, dass sich die Ergebniskontrolle auf die identischen Bezugsgrößen sowie deren Messverfahren stützt, die auch in der Planung als Basis für die Zielformulierung genutzt wurden. Im Kontext der kundenwertorientierten Vertriebssteuerung können der Ergebniskontrolle als Kontrollobjekte der monetäre und nichtmonetäre Kundenwert sowie deren Kundenwerttreiber und Bestimmungsfaktoren, die als Zielgrößen in der kundenwertbasierten Balanced Scorecard zugrunde gelegt wurden, zugerechnet werden. Die Ergebniskontrolle bezieht sich stets auf Handlungsergebnisse, so dass die in der kundenwertbasierten Balanced Scorecard definierten Maßnahmen nicht Kontrollobjekte der Ergebniskontrolle sein können.[2079] In Abhängigkeit der betrachteten Kontrollobjekte kann die Ergebniskontrolle der operativen oder strategischen Planung als Informationslieferant zugeordnet werden.

2076 Vgl. *Börner, C. J.* (2005), S. 37.

2077 *Börner, C. J.* (2005), S. 37.

2078 Vgl. *Küpper, H.-U.* (2001), S. 173.

2079 Die Handlungen selbst sind Gegenstand der Umsetzungs- und der Verhaltenskontrolle. Siehe dazu die Kapitel E.4.3 und E.4.4.

In der operativen Ausprägung resultiert aus der Ergebniskontrolle eine Aussage über den Erfolg von Entscheidungen und Maßnahmen, die auf Basis des Vergleichs einer normativen Soll-Größe und einer faktischen Ist-Größe getroffen wird.[2080] In diesem Fall ist die Ergebniskontrolle eine Feed-back-Kontrolle, die der Sicherstellung der Planerreichung dient.[2081] Die als Soll-Ist-Kontrolle konzipierte Ergebniskontrolle fokussiert vornehmlich auf die Frage, in weit es in der Kontrollperiode gelungen ist, die prognostizierten und geplanten Kundenwerte auszuschöpfen, mithin zu realisieren.[2082] Vor diesem Hintergrund ist der Bezugszeitraum der Ergebniskontrolle kurzfristig und periodisch, weswegen sie der operativen Planung gegenübergestellt wird.[2083] Die in der Ergebniskontrolle erkannten Mängel können für die nachfolgenden operativen Planungen korrigiert werden. Die Gestaltungsprinzipien der kundenwertbasierten Balanced Scorecard sollten für die Rhythmik der operativen Ergebniskontrolle handlungsleitend sein. So sollten die Frühindikatoren, die als Zielgrößen verwendet werden, in kürzeren Zeiträumen, und somit häufiger kontrolliert werden, als die Ergebnisgrößen. Durch dieses Vorgehen ist gewährleistet, dass die erstgenannten ihrer Indikatorfunktion gerecht werden können.[2084] Für strategische Planungen ist die Ergebniskontrolle auf Basis von Ist-Größen ineffektiv, da sie auf in der Vergangenheit realisierte und somit nicht mehr veränderliche Entscheidungen und Maßnahmen ausgerichtet ist.[2085] Gerade die Langfristigkeit und Zukunftsgerichtetheit der Kundenwertorientierung impliziert jedoch die Anforderung, wertverändernde Einflüsse frühzeitig zu identifizieren.

Die Ergebniskontrolle auf Basis des anbieterspezifischen monetären Kundenwerts ist der strategischen Vertriebsplanung gegenüberzustellen und als Soll-Wird-Analyse zu konzipieren. Soll-Wird-Analysen vergleichen die angestrebte Ausprägung einer Zielgröße mit dem für sie zum Zeitpunkt der Kontrolle prognostizierten Wert.[2086] Die kundenwertbasierte Wird-Größe zeigt, welcher anbieterspezifische monetäre Kundenwert ohne Anpassungsmaßnahmen bei fortgeführter Strategie erreicht wird.[2087] Liegen veränderte Informationen zu den Kun-

2080 Vgl. *Horváth, P.* (2003), S. 175.
2081 Vgl. *Weber, J.* (2004), S. 314.
2082 Vgl. *Tewes, M.* (2003), S. 220.
2083 Vgl. *Küpper, H.-U.* (2001), S. 173.
2084 Vgl. *Ebeling, C.* (2007), S. 111.
2085 Vgl. *Stüker, D.* (2008), S. 368.
2086 Vgl. *Küpper, H.-U.* (2001), S. 175.
2087 Vgl. *Stüker, D.* (2008), S. 370. Erfolgt die Soll-Wird-Analyse auf Basis angepasster Strategien ist keine sinnvolle Kontrolle möglich, da Abweichungen der Wird-Größe durch Planungen verändert werden, die gerade aus dem Ergebnis der Kontrolle abgeleitet werden sollen. Ähnlich auch *Ebeling, C.* (2007), S. 110.

denwerttreibern des anbieterspezifischen monetären Kundenwerts vor, so sind diese für die periodische Ergebniskontrolle der aktualisierten Prognose des Kundenwerts zugrunde zulegen. Zudem können darauf aufbauend Informationen hinsichtlich notwendiger Anpassungen kundenindividueller Strategien gewonnen werden, die bei der Überprüfung der definierten Strategien zu berücksichtigen sind. In diesen Überlegungen zeigt sich die Beziehung zwischen der Prämissen- und der strategischen Ergebniskontrolle, wonach die erstgenannte Informationen für die zweitgenannte Kontrollart liefert. Die strategische Ergebniskontrolle ist ihrem Zweck nach daher als Feed-Forward-Kontrolle zu verstehen. Die so ausgestaltete Ergebniskontrolle stellt eine wichtige Grundlage einer kundenwertorientierten rollierenden Planung dar, die die Verbindung zur strategischen Vertriebsplanung herstellt.[2088] Aufgrund der höheren Komplexität der Kontrollgröße ist eine Rhythmik mit längeren Zeitintervallen als bei der operativen Ergebniskontrolle sinnvoll.

Werden im Rahmen der Ergebniskontrolle Abweichungen zwischen der Zielgröße und den Ist- bzw. Wird-Größen festgestellt, umfasst die Ergebniskontrolle stets auch eine Abweichungsanalyse, die die Entscheidungsgrundlage für Sicherungs- und Anpassungsmaßnahmen darstellt. Grundlegende Aufgabe der Abweichungsanalyse ist es, Ursachen und Verantwortlichkeiten der festgestellten Abweichungen zu identifizieren, damit künftig möglichst keine Wiederholung von Zieluntererfüllungen auftreten.[2089] Dazu wird in der Abweichungsanalyse die festgestellte Gesamtabweichung in aussagekräftige Teilabweichungen disaggregiert. In dieser Aufspaltung der Gesamtabweichung liegt hinsichtlich des Kundenwerts die zentrale Herausforderung, da aufgrund des konzeptionell sachlogischen und operationell-funktionalen Zusammenhangs nennenswerte Abweichungsüberschneidungen (Sekundärabweichungen) bedingt durch die multiplikative Verknüpfung der Kundenwerttreiber und/oder Bestimmungsfaktoren auftreten können. Diese Situation impliziert, dass eine festgestellte Abweichung durch einen Faktor oder durch eine Kombination mehrerer Faktoren entstehen kann.[2090] Ein zweckmäßiger Umgang mit diesen Abweichungsüberschneidungen ist die zentrale Voraussetzung, um zielgerichtete Gegensteuerungsmaßnahmen abzuleiten. In der Literatur wird empfohlen, das Verfahren der kumulativen Abweichungsanalyse anzuwenden, um die Abweichungsüberschneidungen zu verrechnen.[2091] Bei der kumulativen Abweichungsanalyse wer-

2088 Vgl. zum Begriff der rollierenden Planung *Küpper, H.-U.* (2001), S. 303.

2089 Vgl. *Coenenberg, A. G.* (1999), S. 372.

2090 Vgl. *Tewes, M.* (2003), S. 230.

2091 Vgl. *Tewes, M.* (2003), S. 231; *Stüker, D.* (2008), S. 373. Vgl. zur Kritik an der kumulativen Abweichungsanalyse *Coenenberg, A. G.* (1999), S. 387.

den die Teilabweichungen bestimmt, indem sukzessive die Ist-Größen der relevanten Einflussfaktoren auf Plan-Größen oder korrigierte Plan-Größen umgestellt werden. Von besonderer Bedeutung für die Aussagefähigkeit der Abweichungsanalyse ist dabei die Reihenfolge, in denen die Einflussfaktoren des Kundenwerts von Ist auf Plan umgestellt werden. „Die Reihenfolge ist dabei so festzulegen, daß eine unter dem Aspekt der [.] [Kontrolle; Erg. d. Verf.] weniger aussagefähige Abweichung [...] zuerst errechnet und abgespalten wird, da hier der Anteil sekundärer bzw. höhergradiger Abweichungen am größten ist."[2092] Neben der hohen Praktikabilität dieses Vorgehens wird ein Vorteil darin gesehen, dass die Summe der Teilabweichungen der Gesamtabweichung entspricht.[2093] Die Notwendigkeit zur Festlegung einer Reihenfolge entfällt bei der differenziert-alternativen Abweichungsverrechnung, deren Ziel weniger in der Erklärung der Gesamtabweichung, sondern in der Demonstration der Auswirkungen der durch die verantwortlichen Mitarbeiter verursachten Abweichungen auf Abweichungen höheren Grades liegt.[2094] Der aus dem damit möglichen Lerneffekt resultierende Erkenntnisgewinn kann zur besseren Steuerung der Verantwortlichen bzw. deren Verhalten beitragen. Zudem kann dies die Akzeptanz des Kundenwerts als Ziel- und Steuerungsgröße erhöhen, wenn den Verantwortlichen differenziert verständlich gemacht werden kann, welcher Anteil der Abweichung durch ihr Verhalten und welcher Anteil nicht durch sie beeinflusst werden kann.[2095] PUFAHL unterscheidet in diesem Kontext zwischen endogenen und exogenen Abweichungen, wobei erstere vom Verantwortlichen und letztere nicht vom Verantwortlichen zu vertreten sind.[2096]

Unabhängig von der angewandten Methode der Abweichungsanalyse bietet es sich bei dem dieser Arbeit zugrunde gelegten Kundenwertmodell und dessen Operationalisierung an, zunächst die isoliert analysierbaren Kundenwerttreiber und Bestimmungsfaktoren zu betrachten. Im Anschluss daran sollten die Größen (wie z. B. die Kundenbindungsrate, die Ausfallwahrscheinlichkeit oder der Risikozuschlag auf den Diskontierungsfaktor), die auf die Kundenwerttreiber gemeinsam wirken, hinsichtlich ihres Einflusses auf die Abweichungen Eingang in die Analyse finden.

2092 *Coenenberg, A. G.* (1999), S. 381.

2093 Vgl. *Stüker, D.* (2008), S. 375.

2094 Vgl. hier und nachfolgend *Coenenberg, A. G.* (1999), S. 387 ff.

2095 Vgl. ähnlich *Herrmann, J./Lüders, U.* (2004), S. 52, die dies für den Deckungsbeitrag III fordern.

2096 Vgl. *Pufahl, M.* (2006), S. 206.

4.3 Umsetzungskontrolle

Die Umsetzungs- oder Durchführungskontrolle erfolgt wie die Ergebniskontrolle auf Basis realisierter Größen, jedoch bezieht sie sich nicht auf Handlungsergebnisse, sondern auf die zugrundeliegenden Handlungen selbst. Die Umsetzungskontrolle ist insbesondere im Kontext der Realisationsphase von strategischen Vertriebsplänen relevant, da sie die Umsetzung einzelner zeitlicher Planbestandteile überprüft, in die die langfristigen Pläne zerlegt werden (Meilensteine).[2097] In einer engen Betrachtung hat die Umsetzungskontrolle die im Rahmen der kundenwertorientierten Strategien definierten Maßnahmen aus der kundenwertbasierten Balanced Scorecard zum Inhalt und fragt, ob die geplanten Maßnahmen in Art und Anzahl umgesetzt wurden. In Verbindung mit der Ergebniskontrolle bezieht sich die Umsetzungskontrolle auch auf die Analyse der Wirkung der geplanten Maßnahmen.[2098]

Mittels der Umsetzungskontrolle lassen sich Erkenntnisse darüber gewinnen, warum Ergebnisse in einer abgelaufenen Planperiode nicht erreicht wurden und ob Ergebnisse in einer künftigen Planperiode erreicht werden können. Im letztgenannten Fall werden Soll- und Wird-Größen miteinander verglichen,[2099] um festzustellen, ob Planabweichungen aufgrund nicht umgesetzter Maßnahmen zu erwarten sind. Folglich prüft die Umsetzungskontrolle, ob der gewählte strategische Kurs gefährdet oder ob dieser noch gültig ist.[2100] So wie die Ergebniskontrolle kann die Umsetzungskontrolle daher auch als Feed-Back- oder Feed-Forward-Kontrolle verstanden werden. Teilweise wird argumentiert, die Umsetzungskontrolle liefere ausschließlich Feed-back-Informationen, da die Erkenntnisse nur ex-post nach Durchführung der Handlungen gewonnen werden könnten. Im Gegensatz zur reinen Feed-Back-Kontrolle jedoch „wird bei der [.] Durchführungskontrolle im Lichte vergangener Ereignisse die Strategie selbst in Frage gestellt.“[2101] Die strategische Umsetzungskontrolle kann und soll daher dazu dienen, aus den vorliegenden Informationen, die aus der Realisation der kundenwertbezogenen Strategie resultieren, die Notwendigkeit oder Chance einer Strategieanpassung zu identifizieren, um mittels dieser Änderung den Kun-

---

2097 Vgl. *Küpper, H.-U.* (2001), S. 173.

2098 Vgl. *Tewes, M.* (2003), S. 234.

2099 Vgl. *Horváth, P.* (2003), S. 175.

2100 Vgl. *Stüker, D.* (2008), S. 384.

2101 *Nuber, W.* (1995), S. 135.

denwert zu steigern.[2102] Die Umsetzungskontrolle wird aus diesem Grund der strategischen Planung gegenübergestellt.[2103]

In wie weit die realisierten Maßnahmen auch bei einer zum Zeitpunkt der Umsetzung der Maßnahmen geänderten Ausgangslage geeignet waren, wird in der Verhaltenskontrolle geprüft. Sie ergänzt somit die Ergebnis- und die Durchführungskontrolle.

4.4 Verhaltenskontrolle

Die Verhaltenskontrolle bezieht sich im Gegensatz zu den bisher genannten Kontrollarten auf die handelnden Akteure. In dieser Kontrolle wird geprüft, ob der Ausführende unter den tatsächlich eingetretenen Situationsbedingungen geeignete Maßnahmen ergriffen hat.[2104] Es würde jedoch der technokratischen Koordination über Pläne und Ziele widersprechen, wenn die Verhaltenskontrolle prüft, ob die geplanten, operativen Maßnahmen durchgeführt wurden. Vielmehr fragt die Verhaltenskontrolle, ob bei erkannten geänderten Rahmenbedingungen, die somit eben nicht Gegenstand der Planung waren, richtig gehandelt wurde. Gleichfalls kann die Verhaltenskontrolle eine Aussage darüber treffen, ob der Ausführende „trotz intensiven Bemühens kein besseres Ergebnis erreichen konnte“[2105] oder ob ein „planmäßig erreichtes Ergebnis nicht auf seinen Einsatz, sondern z. B. auf ungeplant günstigere Bedingungen im Umfeld zurückzuführen sind, und er bei entsprechender Anpassung ein wesentlich besseres Ergebnis hätte erzielen können.“[2106] Wird bspw. das Ziel verfolgt, das Potenzial eines Kunden aus steigenden Zahlungsverkehrstransaktionen zu steigern, so basiert dieses Ziel auf einer entsprechenden Annahme über die Entwicklung des gesamten Zahlungsverkehrsaufkommens des Kunden. Zeigt sich nun, dass der Kunde ein weit höheres Zahlungsverkehrsaufkommen aufweist, so ist zu fragen, ob bedingt durch diesen Sachverhalt nicht ein höherer Ergebnisbeitrag als das avisierte und ggf. erreichte Ziel möglich gewesen wäre.

Das Ziel der Verhaltenskontrolle liegt darin, die Entscheidungs- und Handlungsfähigkeit des Kontrollierten beurteilen zu können.[2107] Daraus lassen sich wiederum Impulse über den Status quo und über etwaige Veränderungsnotwen-

---

2102 Vgl. *Stüker, D.* (2008), S. 385.

2103 Die Umsetzungskontrolle darf daher nicht mit der in Kapitel A.1 genannten Aktivitätensteuerung gleichgesetzt werden. Diese ist der operativen Planung gegenüberzustellen.

2104 Vgl. *Küpper, H.-U.* (2001), S. 173.

2105 *Küpper, H.-U.* (2001), S. 173.

2106 *Küpper, H.-U.* (2001), S. 173.

2107 Vgl. *Küpper, H.-U.* (2001), S. 173.

digkeiten hinsichtlich der bankinternen Einflussfaktoren auf den Kundenwert gewinnen, die mit den involvierten Mitarbeitern verbunden sind.

Es ist evident, dass die Verhaltenskontrolle nur dann sinnvoll eingesetzt werden kann, wenn zum einen eine Prämissenkontrolle implementiert ist und diese zum zweiten nennenswerte Abweichungen zwischen den erwarteten und tatsächlichen Entwicklungen zum Ergebnis hat. Insofern stellt die Verhaltenskontrolle keine obligatorische, sondern eine fakultative Kontrollart dar. Was als nennenswerte Abweichungen zu bezeichnen ist, ist bankindividuell zu entscheiden.

4.5 Prämissenkontrolle

Die Ausführungen zum Kundenwert in Teil A der Arbeit machen deutlich, dass die Messung des Kundenwerts unter Unsicherheit und auf Basis geeigneter Annahmen erfolgt. Dies gilt vor allem für die Entwicklung des Firmenkunden hinsichtlich seines Bedarfspotenzials und seiner Risikosituation als auch für die Entwicklung der Geschäftsbeziehung. Diejenigen Annahmen, die nicht vollständig durch die Kreditgenossenschaft beeinflusst werden können, sind Gegenstand der Prämissenkontrolle.[2108] Prämissenkontrollen gewinnen dann an Bedeutung, wenn der Planungszeitraum oder die Zielgrößen der Planung durch Langfristigkeit gekennzeichnet[2109] und daher annahmenbedingte Fehlprognosen zu erwarten sind. Der Kundenwert als zentrale Ziel- und Steuerungsgröße einer darauf aufbauenden Vertriebssteuerung wurde in dieser Arbeit als zukunftsbezogene Größe konzeptualisiert und operationalisiert, die eben dieses Charakteristikum der Langfristigkeit und Prognosefehleranfälligkeit aufweist.[2110]

Die Prämissenkontrolle erfolgt kontinuierlich, d. h. sowohl während des Planungs- als auch des Umsetzungsprozesses. Ihre Aufgaben sind die Prüfung, ob die grundlegenden Planungsprämissen gegenseitig verträglich, d. h. widerspruchsfrei sind, und der kontinuierliche Abgleich zwischen den geplanten und erwarteten oder tatsächlichen Entwicklungen der für die Planung relevanten internen und externen Rahmenbedingungen.[2111] Während widersprüchliche Prämissen vor der vertrieblichen Ziel- und Maßnahmenplanung und der Realisationsphase der Vertriebsplanung korrigiert werden müssen, führen Abweichungen bei diesen Plandeterminanten zu unmittelbaren Änderungen der Ziele, Strategien und operativen Maßnahmenpläne in der laufenden Realisation.[2112] Ergebnis der

2108 Vgl. *Küpper, H.-U.* (2001), S. 175.
2109 Vgl. *Ebeling, C.* (2007), S. 117.
2110 Vgl. *Tewes, M.* (2003), S. 218.
2111 Vgl. *Horváth, P.* (2003), S. 175.
2112 Vgl. *Weber, J.* (2004), S. 318.

Prämissenkontrolle sind daher wesentliche Feed-forward-Informationen, die in die nachfolgenden Planungsaktivitäten eingehen.

Im Kontext der Ausgestaltung der Prämissenkontrolle ist festzulegen, auf Basis welcher Informationen die Kontrolle erfolgen und wie umfangreich die Prämissenkontrolle ausfallen soll, um hinreichend steuerungsrelevante Feed-forward-Informationen zu generieren. Aufgrund ihres Feed-forward-Charakters ist es nicht ausreichend, die Prämissenkontrolle ausschließlich in Form eines Wird-Ist-Vergleichs auf Basis faktischer Informationen vorzunehmen, da dann Abweichungen bereits vorliegen und somit die antizipativen Handlungsmöglichkeiten für die Kreditgenossenschaft eingeschränkt sind. Zu ergänzen ist eine Prämissenkontrolle in Form von Wird-Wird-Vergleichen, bei denen die Voraussage einer den Kundenwert beeinflussenden Annahme mittels neuer Informationen getroffen werden kann.[2113]

Hinsichtlich des Umfangs der Prämissenkontrolle ist zu konstatieren, dass eine vollständige Kontrolle aller Prämissen wegen ihrer zu vermutenden hohen Anzahl nicht realisierbar sein dürfte.[2114] Aus diesem Grund sind die zu überwachenden Prämissen zu priorisieren. Es sollten diejenigen Annahmen fokussiert werden, bei denen bereits sehr geringe Abweichungen die Ausprägung des Kundenwerts, erheblich beeinflussen.[2115] Die Identifikation solcher Prämissen kann z. B. auf Basis von Sensitivitätsanalysen erfolgen. Ist die Anzahl der identifizierten Prämissen immer noch zu hoch, ist die Auswahl auf diejenigen Prämissen zu beschränken, bei denen Abweichungen besonders wahrscheinlich auftreten. Dies ist vor allem bei denjenigen kundenwertkritischen Annahmen der Fall, die nur auf Basis schwacher Informationsgrundlagen getroffen wurden[2116] oder die von Natur aus im Laufe der Zeit erheblichen Instabilitäten unterworfen sind.[2117]

Auf Basis der Prämissenkontrolle lassen sich während der Planrealisation Soll-Wird-Kontrollen der kundenwertbezogenen Ziele durchführen. Bei diesen Soll-Wird-Kontrollen werden die angestrebten kundenwertbezogenen Ziele mit den auf Basis der neuen Informationen hinsichtlich der relevanten Prämissen prognostizierten Ausprägungen der Zielgrößen verglichen.[2118] Die Soll-Wird-Kontrollen sind sowohl auf der strategischen als auch auf der operativen Vertriebsplanungsebene zu implementieren.

2113 Vgl. *Küpper, H.-U.* (2001), S. 175.

2114 Vgl. *Ebeling, C.* (2007), S. 117 sowie die dort genannte Literatur.

2115 Vgl. *Ebeling, C.* (2007), S. 117; *Tewes, M.* (2003), S. 219.

2116 Vgl. *Ebeling, C.* (2007), S. 117.

2117 Vgl. *Stüker, D.* (2008), S. 383 sowie die dort genannte Literatur.

2118 Vgl. *Küpper, H.-U.* (2001), S. 175.

Voraussetzung der für die kundenwertorientierte Vertriebssteuerung wichtigen Prämissenkontrolle ist die differenzierte Dokumentation der zu überwachenden Prämissen zum Planungs- und Bewertungszeitpunkt.[2119]

Da die Prämissenkontrolle direkt an die explizit formulierten Planungsprämissen anknüpft, stellt sie eine „gerichtete Kontrollart" auf der strategischen Ebene dar. Aufgrund dieses Rahmens geht die Prämissenkontrolle selektiv vor, worin die Gefahr liegt, wesentliche kundenwertrelevante Änderungen interner oder externer Rahmenbedingungen zu übersehen. Kompensiert wird dieses konzeptionelle Defizit durch die Integration einer ihrem Wesen nach ungerichteten strategischen Überwachung.[2120]

4.6 Strategische Überwachung

Die strategische Überwachung soll jegliche Entwicklungen identifizieren, die für die Kreditgenossenschaft zum Risiko für den Erfolg der eingeschlagenen Strategie werden könnten. Insbesondere sollen solche Entwicklungen erkannt werden, die sonst bewusst oder unbewusst vernachlässigt werden.[2121] In diesem Zusammenhang wird jedoch das zentrale Problem der strategischen Überwachung offenbar, das sich darin zeigt, dass „die strategische Überwachung als ungerichtete Kontrolle theoretisch alles Erdenkliche überwachen müsste und konkrete Kontrollobjekte im Vorfeld nicht bekannt sind"[2122]. Daraus abzuleitende Hindernisse des praktischen Einsatzes der strategischen Überwachung sind die Informationsüberflutung sowie die knappen Ressourcen auf Seiten der Kreditgenossenschaften.

Dieses konzeptionelle Problem löst sich dadurch auf, indem die Strategie selbst zum Bezugspunkt der strategischen Überwachung wird. Durch die vorhergehende planerische Selektionsleistung und die Ableitung einer Strategie ist nach SCHREYÖGG/STEINMANN der Bezugspunkt gesetzt, an dem jeglichen neuen Informationen einen krisenhafter Sinn zugeschrieben werden kann. Problematisch an diesem Vorgehen ist jedoch, dass es das Ziel verfolgt, das Selektionsrisiko der strategischen Planung zu kompensieren.[2123] Dieses theoretische Ideal

[2119] Vgl. *Tewes, M.* (2007), S. 219. TEWES nennt als wichtigste Prämissen der Kundenwertorientierung die Kundenbindung und die auf diese wirkende Kundenzufriedenheit. Diesen beiden Größen unterstellt er zwar den größten Einfluss auf den Kundenwert, stellt jedoch fest, dass es schwierig sein dürfte, für diese Größen Schwellenwerte zu definieren, da der messbar funktionale Einfluss weitgehend unbekannt ist.

[2120] Vgl. *Stüker, D.* (2008), S. 384.

[2121] *Schreyögg, G./Steinmann, H.* (1985), S. 404.

[2122] *Stüker, D.* (2008), S. 386.

[2123] Vgl. *Schreyögg, G./Steinmann, H.* (1985), S. 405 f.

dürfte praktisch kaum umsetzbar sein,[2124] weshalb auf leichter anwendbare Ansätze zurückgegriffen wird. Zu diesen Ansätzen zählen Verfahren, deren Grundlage die Indikatorhypothese ist, nach der Veränderungen nicht abrupt auftreten, sondern durch Signale angekündigt werden.[2125] In diesem Kontext wird regelmäßig auf das Konzept der „Weak Signals" von ANSOFF verwiesen.[2126] ANSOFF unterteilt die Stärke von Signalen in Ungewissheitsgrade („States of Ignorance") mit den Ausprägungen "Sense of Threat/Opportunity", "Source of Threat/ Opportunity", "Threat/Opportunity", "Response Concrete" und "Outcome Concrete".[2127] Als schwache Signale werden schlecht-definierbare Informationen verstanden, die einen Empfänger in einem hohen Stadium der Ungewissheit belassen.[2128] Zudem sind ihre Wirkungen auf die strategische Planung nur unzureichend bekannt und müssen als mehrdeutig und spekulativ angesehen werden. Den jeweiligen Ungewissheitsgraden stellt ANSOFF Reaktionsstrategien gegenüber, die sich auf die Wahrnehmungsfähigkeit des Unternehmens, dessen Flexibilität oder unmittelbar auf dessen Handlungsfähigkeit beziehen.[2129]

Die Etablierung von Frühwarnsystemen, die auf dem Konzept der „Weak Signals" aufbauen, ist kein neues Thema der Gesamtbanksteuerung.[2130] Voraussetzung eines zielsetzungsgerechten Einsatzes ist die „systematische Sensibilisierung für schwache Signale"[2131] wobei jedoch gerade die Systematisierung der Signale aufgrund ihres Wesens problematisch ist. Auf Gesamtbankebene schlägt SCHIERENBECK vor, die Signale in politische und kulturelle, in markt- und wettbewerbspolitische und in geschäfts- und risikostrukturelle Faktoren zu unterscheiden.[2132] Weiterhin unterscheidet er zwischen bankexternen und -internen Wirkungsbereichen der schwachen Signale. Aufbauend auf dieser Systematisierung sind die schwachen Signale zu identifizieren, zu diagnostizieren und zu evaluieren. Die in der Identifikations- und Diagnosephase generierten und validierten Informationen werden in der Evaluationsphase aus Sicht der kundenwertbezogenen Strategie beurteilt.[2133] Dazu ist zu prüfen, welche Kundenwert-

2124 Vgl. *Stüker, D.* (2008), S. 389.

2125 Vgl. *Coenenberg, A. G./Baum, H.-G./Günther, T.* (2007), S. 329 ff.

2126 Vgl. z. B. *Schierenbeck, H.* (2003), S. 668 ff.; *Stüker, D.* (2008), S. 387 ff.

2127 Vgl. *Ansoff, H. I.* (1976), S. 135.

2128 Vgl. hier und nachfolgend *Schierenbeck, H.* (2003), S. 669 sowie die dort genannte Literatur.

2129 Vgl. *Ansoff, H. I.* (1976), S. 136 ff.

2130 Vgl. *Schierenbeck, H.* (2003), S. 669.

2131 *Schierenbeck, H.* (2003), S. 670.

2132 Vgl. hier und nachfolgend *Schierenbeck, H.* (2003), S. 670, der darauf hinweist, dass dies aufgrund der genannten Schwierigkeit nur eine „gewisse Hilfestellung" sein könne.

2133 Vgl. hier und nachfolgend *Schierenbeck, H.* (2003), S. 671 f.

treiber oder Bestimmungsfaktoren tendenziell durch die beobachteten Veränderungen tangiert werden (Relevanz), welche Auswirkungen zu erwarten sind (Ausmaß) und mit welcher vermuteten Geschwindigkeit bei den Änderungen zu rechnen ist (Geschwindigkeit). Aus den dadurch gewonnenen Erkenntnissen lässt sich ableiten, welche Auswirkungen auf die kundenwertbezogenen Ziele und Strategien zu erwarten sind und mit welchen Änderungen der Strategien als Ganzes oder einzelner Maßnahmen darauf in welcher Dringlichkeit reagiert werden kann.

Die strategische Überwachung ist wie auch die Prämissenkontrolle als kontinuierlicher Prozess, der permanent durchzuführen ist, zu implementieren.

5 Zusammenfassung der Ergebnisse und kritische Würdigung

In den vorangegangenen Kapiteln des Teils E wurden Gestaltungsempfehlungen für die kundenwertorientierte Vertriebssteuerung im Firmenkundengeschäft der Kreditgenossenschaften erörtert und erläutert. In diesem Zusammenhang wurde auf die in der kreditgenossenschaftlichen Umgebung bestehenden Konzepte und Verfahrensweisen (z. B. VR-CONTROL[2134], Leitfäden der strategischen Steuerung einer Primärbank[2135]) rekurriert, um dem praxeologischen Anspruch der Arbeit gerecht zu werden.

Basis einer kundenwertorientierten Vertriebssteuerung ist ein zweckmäßig ausgestaltetes Vertriebsinformationssystem. Dieses dient dazu, die für die Kundenwertermittlung relevanten Informationen zu beschaffen, aufzubereiten, zu speichern und die aufbereiteten Ergebnisse adressaten- und aufgabengerecht weiterzuleiten. Besondere Bedeutung hat die Beschaffung kundenwertrelevanter Informationen aus internen und externen Quellen. Der Firmenkunde selbst ist die zentrale Informationsquelle. Dem Firmenkundenberater obliegt es als gewichtige Vertriebsaufgabe, diese Information vom Kunden zu beschaffen. Gegenwärtig eingesetzte Verfahren können dazu ausgebaut und genutzt werden. Besonderes Augenmerk sollte darauf gelegt werden, die erhaltenen Kundeninformationen – analog zum risikopolitisch motivierten Vorgehen bei der Analyse der wirtschaftlichen Verhältnisse – aus Kundenwertsicht zu plausibilisieren. Dabei sollte auf den Erfahrungs- und Datenfundus der FinanzGruppe Bezug genommen werden.

Hinsichtlich der Vertriebsplanung wurden zwischen der strategischen und der operativen Planung unterschieden. Angelehnt an die etablierten und vom

[2134] Vgl. dazu *BVR* (2003b).
[2135] Vgl. dazu *BVR* (2007b).

BVR empfohlenen Konzepte und Verfahrensweisen wurde dargelegt, das Kundenwertmodell als Analyse- und Beurteilungsraster in der Analyse der strategischen Ausgangssituation sowohl für die Umfeld- als auch für die Kundenbeziehungs- sowie die Unternehmens- und Funktionalbereichsanalyse zu nutzen. Auch hier sollten Benchmarks über die gesamte FinanzGruppe ein Analysebaustein sein. Vertriebsziele beziehen sich auf den Zerschlagungs- und auf den Potenzialwert der Kreditgenossenschaft und betrachten die individuelle Kundenbeziehung. Im Rahmen dieses Schrittes bietet es sich an, die strategischen Ziele der Teilbanken mit Hilfe des Kundenwerts als Zielgröße zu koordinieren und zu optimieren. Vertriebsstrategien sollten sich an den kundenbeziehungsindividuellen Vertriebszielen ausrichten und die Erkenntnisse der Kundenbeziehungsanalyse nutzen. Für die Bewertung und Auswahl von Vertriebsstrategien ist ebenfalls der Kundenwert als zentrale Größe heranzuziehen.

Um in der operativen Vertriebsplanung die kundenbeziehungsbezogenen Strategien umzusetzen, wurde empfohlen, eine kundenwertbezogene Balanced Scorecard als konzeptionellen Bezugsrahmen zu verwenden. Diese ist konsequent zur Strategie einzelkundenbezogen. Anhand der Balanced Scorecard sollen die kurzfristigen Vertriebsziele und Maßnahmenpläne formuliert und vereinbart werden, da die dort hinterlegten Kennzahlen auch die Grundlage einer zielbasierten Personalführung im Vertrieb sein sollen.

Die Planung der Vertriebsbank ist mit der Steuerungs- und der Produktionsbank intern und mit den Verbundpartnern extern zu koordinieren. Hinsichtlich der Steuerungsbank wurde unterstellt, dass diese in der kurzen Sicht bestehende Risikostrukturen aufweist, die Einfluss auf die Möglichkeiten der Schaffung und Ausbeutung von Kundenwerten haben. Diese Grenzen sind in der Vertriebsplanung zu berücksichtigen. Darüber hinaus sind die produktpolitischen Entscheidungen der Steuerungsbank mit der Vertriebsbank abzustimmen, so dass neue Produkte in der Kundenwertermittlung zu berücksichtigen sind und ggf. Impulse an die Produktentwicklung aus der Kundenwertermittlung des Vertriebs resultieren. Mit Blick auf die Produktionsbank dient die Vertriebsplanung als Vorgabe für die Kapazitätsplanung. Dabei ist die Planung der Kapazität in Qualität und Quantität in Abhängigkeit der kundenwertbezogenen Strategien vorzunehmen. Bei der Einführung verfahrenstechnischer Innovationen in der Produktionsbank sind deren Auswirkungen auf den Kundenwert in die Entscheidungsfundierung einzubeziehen. Besondere Herausforderungen stellt die externe Koordination mit den Verbundpartnern dar. Hier sollte die Kreditgenossenschaft mit den Verbundunternehmen vereinbaren, dass die relative Bedeutung eines Kunden aus Sicht der Kreditgenossenschaft handlungsleitend für die gemeinsame Ansprache ist.

Die Vertriebssteuerung im engeren Sinn auf Basis des Kundenwerts fußt auf einer steuerungsadäquaten Aufbauorganisation, die sinnvollerweise den Vertrieb als Investment-Center aufbaut. Hinsichtlich der damit verbundenen Kompetenzen sind jedoch die Einschränkungen aus der Steuerungsbank als zentrale Vorgabe zu berücksichtigen. Wie bereits erwähnt, sollte die Personalführung im Vertrieb mittels der in der kundenwertbezogenen Balanced Scorecard hinterlegten Ziele erfolgen. Damit ist die Personalführung im Gegensatz zu gegenwärtigen Verfahrensweisen,[2136] konsistent zur Kunden(wert)orientierung ausgestaltet. Um den methodeninhärenten Schwächen der barwertigen Einzelgeschäftskalkulation entgegenzuwirken, sollte der Kundenwert in die Verfahren der Vorsteuerung nach VR-CONTROL integriert werden. Welche der erörterten Vorgehensweisen ausgewählt wird, ist institutsindividuell zu entscheiden.

Für die Überwachungs- und Sicherungsphase in der kundenwertorientierten Vertriebssteuerung wurden die Ergebniskontrolle, die Umsetzungskontrolle, die Verhaltenskontrolle, die Prämissenkontrolle und die strategische Überwachung vorgestellt. Mit der Integration dieser Kontrollarten in die kundenwertorientierte Vertriebssteuerung gelingt es, den Besonderheiten dieser zukunftsgerichteten Ziel- und Steuerungsgröße hinreichend gerecht zu werden.

In Teil der C Arbeit wurden kontextinduzierte Implikationen für die kundenwertorientierte Vertriebssteuerung im Firmenkundengeschäft der Kreditgenossenschaften hergeleitet. Werden die Gestaltungsempfehlungen vor dem Hintergrund der sich aus dem Kontext ergebenden Implikationen betrachtet, bildet dies die Klammer, um die Ergebnisse zu würdigen. Aus der Definition der Banken wurde hergeleitet, dass die aufsichtsrechtlichen Vorgaben und Zielunterschiede zwischen Bank und Bankmitarbeitern zu berücksichtigen seien. Erstgenannte Implikation findet sich mittelbar in der Gesamtheit der Ausgestaltung einer kundenwertorientierten Vertriebssteuerung wieder (z. B. durch die Abstimmung der Vertriebs- mit der Steuerungsbank sowie der Berücksichtigung der aufsichtsrechtlichen Vorgaben in der Phase der Prognose von Kundenwerten). Die Implementierung eines kundenwertbasierten Anreizsystems kann helfen, die Zieldifferenzen zwischen Bank und Bankmitarbeiter zugunsten der Bank zu reduzieren.

Die relative Kleinheit der Kreditgenossenschaften wurde berücksichtigt, indem als methodische Basis der Kundenbewertung die Einzelgeschäftskalkulation aus VR-CONTROL ausgewählt wurde. Zu konstatieren ist jedoch, dass die kundenorientierte Prozesskostenrechnung weder praktisch noch konzeptionell in der FinanzGruppe vorliegt. Entsprechende Anstrengungen, diese zu gestalten, sind zu wünschen und sicherlich nicht unmöglich. Insgesamt soll sich die kun-

2136 Vgl. dazu die Ausführungen in Kapitel A.1.

denwertbasierte Vertriebssteuerung durch einen pragmatischen Kern auszeichnen. Wenngleich davon ausgegangen wird, ist die Anforderungserfüllung erst mittels eines entsprechenden praktischen Einsatzes möglich.

Dem Wesenzweck der Kreditgenossenschaften, mithin der Mitgliederorientierung und dem Förderauftrag wurde breiter Raum in dieser Arbeit gegeben. Insgesamt ist daher davon auszugehen, dass den daraus resultierenden Implikationen hinreichend Rechnung getragen wurde.

Dem Regionalprinzip wurde insbesondere durch das differenzierte Kundenwertmodell Rechnung getragen. Aber auch die Überlegungen zum Referenzpotenzial und zu strategischen Entscheidungen im Hinblick auf die Beendigung von Kundenbeziehungen wurden stark durch das Regionalprinzip beeinflusst. Dies gilt gleichsam für die Berücksichtigung der Wettbewerbsintensität und der „vermuteten" Ertragsschwäche. Die Kundenwertmodellierung und –operationalisierung berücksichtigt die Besonderheiten von Bankdienstleistungen, die ebenfalls damit verbunden sind.

Den in der Verbundorientierung und dem Bankleistungsprogramm begründeten Implikationen wurde ebenfalls entsprochen. Ersterem insofern, als der Kundenwert der relevante Maßstab für die Koordination von Verbundunternehmen und Kreditgenossenschaften ist. Die Breite des Kundenwertmodells trägt dem Allfinanzangebot Rechnung.

Die Nutzung des Kundenwerts als Ziel- und Steuerungsgröße dürfte es auch erlauben, zwischen transaktionalen und relationalen Vertriebsansätzen zu unterscheiden. Zudem erlaubt die kundenwertbezogene Formulierung von Vertriebsstrategien die Entscheidung für den einen oder den anderen Vertriebsansatz zu fundieren. Gleichwohl ist die zentrale These im Bankgeschäft, dass die relationalen Vertriebsansätze das Geschäft weitestgehend dominieren.

Wie die Ausführungen zur strategischen Planung verdeutlichen, ist der Kundenwert als Ziel- und Steuerungsgröße hinreichend offen, um sich konsistent zur Gesamtbankstrategie gestalten zu lassen. Dies begründet sich damit, dass der Kundenwert sich per definitionem unmittelbar an den Zielen und nicht an den Wegen zur Zielerreichung orientiert.

In den vorliegenden Ausführungen wurde außerhalb der Betrachtung des Drei-Banken-Modells nicht expressis verbis auf die Frage eingegangen, in wie weit der Kundenwert die arbeitsteiligen Prozesse und gemeinschaftlichen Betreuungsaktivitäten unterstützt. Ausgangspunkt dieser Implikation ist, dass die gegenwärtig vorzufindende wertorientierte Steuerung zwingend eine mitarbeiterindividuelle Erfolgstrennung und -zurechnung benötigt, da sonst die Erfolgsmessung auf Basis der Ziele und ggf. die Ermittlung der Ausprägung der Bemessungsgrundlage im Kontext des Anreizsystems nicht erfolgen kann. Vor dem Hintergrund der Ausführungen zum Personalführungssystem auf Basis des

Kundenwerts ist jedoch zu fragen, ob die mitarbeiterindividuelle Erfolgstrennung und -zurechnung in gleicher Art und Stringenz sinnvoll ist. Dieser Frage sollte in weiteren Forschungen nachgegangen werden. Es wird aber angenommen, dass der Kundenwert in diesem Kontext keine methodenbedingte Restriktion darstellt.

F Schluss

1 Zusammenfassung

Die Motivation der vorliegenden Arbeit war, einen umfassenderen konzeptionellen Ansatz einer kundenwertorientierten Vertriebssteuerung im Firmenkundengeschäft der Kreditgenossenschaften zu entwickeln. Diese Motivation wurde über drei Forschungsziele konkretisiert.

Die Auseinandersetzung mit dem ersten Forschungsziel, ein Modell und eine Metrik zu entwickeln, anhand derer die Attraktivität von Kunden ermittelt und beurteilt werden kann, erfolgte in Teil A der Arbeit. Ausgehend von der begrifflichen Festlegung des Kundenwerts als Maßstab der Attraktivität wurden theoretische und praktische Modelle anhand eines formulierten Kriterienkatalogs gewürdigt. Es zeigte sich, dass keines der gegenwärtig vorhandenen Kundenwertmodelle hinreichend war, so dass ein eigenes Kundenwertmodell konzeptualisiert und operationalisiert wurde. In diesem Zusammenhang wurde auch das dritte Forschungsziel in den Blick gerückt. Als drittes Forschungsziel wurde formuliert, die ersten beiden Forschungsziele mit Blick auf die Besonderheiten der eingetragenen Genossenschaft zu verfolgen. Daher musste zwingend von der in bisherigen Forschungsbeiträgen gängigen Vorgehensweise abgewichen werden und das zugrunde zu legende Zielsystem der Kreditgenossenschaften analysiert werden. Gleichsam ist dem dritten Forschungsziel die Betrachtung des Untersuchungskontextes in Teil C, aus dem sich zentrale Hinweise auf die auf Plausibilitätsüberlegungen basierende Herleitung des Kundenwertmodells, zuzuordnen. Die Analyse der Erfolgsvoraussetzungen einer Kundenwertorientierung machte deutlich, dass diese erfolgversprechend für das Firmenkundengeschäft der Kreditgenossenschaften ist.

Auch das zweite Forschungsziel wurde mit dem Blick auf die Besonderheiten der Kreditgenossenschaften bearbeitet. Es wurden im Teil E der Arbeit Gestaltungsempfehlungen für eine kundenwertorientierte Vertriebssteuerung erläutert und erörtert, die in ihrer Gesamtheit einen Beitrag dazu leisten, die Kongruenz zwischen externer Ausrichtung am Markt in Form der relationalen Vertriebsansätze und interner Steuerung der marktgerichteten Aktivitäten, mithin also des Vertriebs, herzustellen.

2 Ausblick auf weiteren Forschungsbedarf

Mit der vorliegenden Untersuchung ist ein Fundament und gleichzeitig ein Rahmen für weitere Forschungsarbeiten gelegt worden. Künftige Forschungsbeiträge, die an diese Arbeit anschließen, lassen sich unterscheiden in Forschungsfragen, die sich auf die Fundierung, Würdigung und Messung des Kundenwertmodells auf der einen Seite und auf die Vertiefung und Würdigung der Gestaltungsempfehlungen auf der anderen Seite beziehen.

Auf Basis theoretisch und empirisch gestützter Plausibilitätsüberlegungen wurde das Kundenwertmodell konzeptualisiert. Dabei gelang es nur zum Teil anwendungskontextspezifische Ergebnisse früherer Arbeiten heranzuziehen. Unter der Annahme, dass Ergebnisse aus abweichenden Anwendungskontexten nur partiell auf das Firmenkundengeschäft der Kreditgenossenschaften übertragen werden können, kann konstatiert werden, dass in einigen Fällen weiterer Forschungsbedarf besteht, um die getroffenen Annahmen und damit die Konzeptualisierung des Kundenwertmodells zu bestätigen oder zu verwerfen.

Die Schätzung von Kundenpotenzialen mittels Potenzialmodellen ist seit vielen Jahren Bestandteil der Vertriebsplanung. Experten kritisieren jedoch, dass die angebotenen Verfahren sich primär an den erwerbswirtschaftlichen Interessen der Anbieter orientieren und eine empirische Prüfung der Leistungsfähigkeit der Analysen außen vor bleibt.[2137] Aus diesem Grund erscheint es zwingend, dass in dieser Arbeit theoretisch hergeleitete Kundenwertmodell empirisch zu überprüfen. Diese Überprüfung umfasst zwei Schwerpunkte. Der erste Schwerpunkt liegt in der Überprüfung des jeweiligen Erklärungsgehalts der hergeleiteten Kundenwerttreiber für den firmenkundenspezifischen Kundenwert aus Sicht der Kreditgenossenschaften. Der zweite Schwerpunkt liegt darauf, empirisch zu begründen, ob mittels dieses Kundenwertmodells statistisch signifikant werthaltige Kunden von nichtwerthaltigen Kunden diskriminiert werden können. Um diesen zweiten Schwerpunkt zum Gegenstand empirischer Forschungen zu machen, bedarf es jedoch einer wichtigen Voraussetzung in Form der Entwicklung von zukunftsbezogenen Prognosemodellen, die auf den Planungen des Unternehmens ansetzen und daraus Bedarfspotenziale für Bankleistungen ableiten.

In der vorliegenden Arbeit wurde als Beispiel das Prognosemodell von GLEIßNER/JAHN vorgestellt. Modelle dieser Art, die hinsichtlich der Inputdaten vergleichsweise einfach zu bestücken sind, müssen für alle relevanten Bankleistungen im Firmen- und Privatkundengeschäft aufgebaut und eingesetzt werden. Nur so gelingt es, den Kundenwert zukunftsbezogen zu schätzen und als Ziel- und Steuerungsgröße in der Vertriebssteuerung zu integrieren.

Die in der vorliegenden Arbeit dargestellten Gestaltungsempfehlungen sind aus Sicht des Vertriebs im Speziellen und aus Sicht der Gesamtbank im Allgemeinen nicht abschließend. Insbesondere die internen und externen Koordinationsaufgaben bedürfen einer tieferen Analyse. Anschließende Forschungsarbei-

---

2137 Vgl. *Kölker, A.* (2010), S. 395.

ten sollten daher prüfen, wie die Steuerungs- und die Produktionsbank im Lichte des Kundenwerts auszugestalten sind. Dies betrifft z. B. die Frage, mittels Kundenwerten die Bankleistungspolitik zu steuern. Liegen dazu Erkenntnisse vor, sind diese mit den vorliegenden zu verknüpfen, um ein kundenwertorientiertes Gesamtbanksteuerungsmodell zu entwerfen.

In der Problemstellung (Kapitel A.1) der vorliegenden Arbeit wurden drei Vertriebssteuerungsansätze kursorisch vorgestellt und knapp hinsichtlich ihrer allgemeinen und mit Blick auf das Beziehungsmanagement im Firmenkundengeschäft speziellen Schwächen skizziert. Diese Würdigung erfolgt auf Basis plausibler Überlegungen, die in der grundsätzlichen These dieser Arbeit mündeten, dass die genannten Vertriebssteuerungsansätze inferior gegenüber einer kundenwertorientierten Vertriebssteuerung seien. Die empirische Prüfung dieser These ist jedoch noch offen. So ist weiterer Forschungsbedarf insoweit zu konstatieren, als die verschiedenen Vertriebssteuerungsansätze hinsichtlich ihrer Bedeutung für den Vertrieb einer Kreditgenossenschaft zu evaluieren sind.

Literaturverzeichnis

A

Aaker, David A. (1991):
Managing Brand Equity: Capitalizing on the Value of a Brand Name, New York: Maxwell Macmillian International, 1991

Abell, Derek F. (1980):
Defining the Business: the Starting Point of Strategic Planning, 2nd Printing, New Jersey: Prentice Hall, 1980

Adam, Dietrich (1996):
Planung und Entscheidung: Modelle, Ziele, Methoden, 4., vollständig überarbeitete und wesentlich erweiterte Auflage, Wiesbaden: Gabler Verlag, 1996

Adam, Dietrich (1998):
Produktions-Management, 9., überarbeitete Auflage, Wiesbaden: Gabler Verlag, 1998

Adam, Dietrich (2000):
Investitionscontrolling, 3., völlig neu bearbeitete und wesentlich erweiterte Auflage, München/Wien: Oldenbourg Verlag, 2000

Agthe, Klaus (1959):
Stufenweise Fixkostendeckung im System des Direct Costing, erschienen in: Zeitschrift für Betriebswirtschaft (ZfB), 29. Jg., 1959, S. 404-418

AGVR (2001):
Mitarbeiterführung und leistungs- und erfolgsbezogene Vergütung, hrsg. von Arbeitgeberverband der Deutschen Volksbanken und Raiffeisenbanken e.V., Bonn, 2001

AGVR et al. (2008):
Marktforschung: Rahmendaten für das Firmenkundengeschäft, Oldenburg et al., 2009

Ahlert, Dietrich (1996):
Distributionspolitik, 3. Auflage, Stuttgart: Fischer Verlag, 1996

Ahrendt, Bernd (2007):
Die wertorientierte Bonitätsprüfung als Kernkompetenz im Firmenkundengeschäft, Nordhausen: Verlag Traugott Bautz, 2007

Algner, Markus R. (2006):
Neue Impulse für die Beteiligungsfinanzierung der Genossenschaft durch die Investorenmitgliedschaft?, erschienen in: Zeitschrift für das gesamte Genossenschaftswesen (ZfgG), 56. Jg., 2006, Heft 3, S. 186-199

Algner, Markus R./Bolsinger, Harald J. (2006):
Der Ansatz der derivativen Mitgliederförderung als Konzept der Neuinterpretation des Förderauftrags durch Kreditgenossenschaften im Firmenkundengeschäft, in: Zukunftsperspektiven für Genossenschaften: Bausteine für typgerechte Weiterentwicklung, hrsg. von Münkner, Hans-Hermann/Ringle, Günther, Bern/Stuttgart/Wien: Haupt Verlag, 2006, S. 180-202

Altenbäumker, Dirk/Andritzky, Ulrich (2007): Bestätigt, erschienen in: Bankinformation, 34. Jg., 2007, Heft 12, S. 26-29

Althans, Mark (2001):
Konzeption eines Vertriebscontrolling-Informationssystems für Unternehmen der liberalisierten Elektrizitätswirtschaft, Göttingen: Cuvillier Verlag, 2001

Anacker, Ralf (2003):
Wesentlicher Erfolgsfaktor, erschienen in: Bankinformation, 33. Jg., 2003, Heft 10, S. 23-27

Andon, Paul/Baxter, Jane/Bradley, Graham (2003):
Calculating Customer Lifetime Value (CLV): Theory and Practice, in: Kundenwert: Grundlagen, Innovative Konzepte, Praktische Umsetzungen, 2., überarbeitete und erweiterte Auflage, hrsg. von Günter, Bernd/Helm, Sabrina, Wiesbaden: Gabler Verlag, 2003, S. 299-315

Angelini, Paolo/Di Salvo, Roberto/Feri, Giovanni (1998):
Availability and Cost of Credit for Small Businesses: Customer Relationships and Credit Cooperatives, erschienen in: Journal of Banking and Financing, 22. Jg., 1998, Heft 6-8, S. 925-954

Ansoff, H. Igor (1976):
Managing Surprise and Discontinuity – Strategic Response to Weak Signals, erschienen in: Zeitschrift für betriebswirtschaftliche Forschung (ZfbF), 28. Jg., 1976, Heft 3, S. 129-152

Aoki, Masahiko/ Dinç, Serdar (1997):
Relational Financing as an Institution and its Viability under Competition, Working Paper 97011, Stanford University, Department of Economics, 1997

Armbruster, Holger/Hallberg, Inken (2006):
Optimal produzieren: Potenziale zur Leistungssteigerung auch in kleinen und mittleren Genossenschaftsbanken, erschienen in: Bankinformation, 33. Jg., 2006, Heft 1, S. 59-63

Arrow, Kenneth J.(1970):
Social Choice and Individual Values, 2nd ed., 4th print, New Haven: Yale University Press, 1970

Aubin, Peter (2011):
Null Gewinn aus dem Kundengeschäft…, erschienen in: Genograph, o. Jg., 2011, Heft 5, S. 42-46

Averbeck, Christian/Frére, Eric/Reuse, Svend (2008):
Wertorientierte Vertriebssteuerung in Banken, erschienen in: Controller Magazin, 33. Jg., 2008, Heft 6, S. 40-47

Averbeck, Christian/Frère, Eric/Reuse, Svend/Svoboda, Martin (2008):
Managemententscheidungen einer wertorientierten Vertriebssteuerung, erschienen in: BankPraktiker, o. Jg., 2008, Heft 9, S. 406-412

B

Baas, Timo/Schrooten, Mechthild (2006):
Relationship Banking and SMEs: A theoretical Analysis, erschienen in: Small Business Economics, 27. Jg., Heft 2/3, S. 127-137

Backhaus, Klaus (1999):
Industriegütermarketing, 6., überarbeitete Auflage, München: Verlag Franz Vahlen, 1999

Backhaus, Klaus/Aufderheide, Detlef/Späth, Georg-Michael (1994):
Marketing für Systemtechnologien: Entwicklung eines theoretisch-ökonomisch begründeten Geschäftstypenansatzes, Stuttgart: Schäffer-Poeschel, 1994

Backhaus, Klaus/Bauer, Mathias (2003):
Zufriedenheit in industriellen Geschäftsbeziehungen, in: Relationship Marketing: Standortbestimmung und Perspektiven, hrsg. von Rese, Mario/Söllner, Albrecht/Utzig, Peter, Berlin et al.: Springer Verlag, 2003, S. 105-138

Bakonyi, Emmerich (1983):
Mitgliedererwartungen in Bankgenossenschaften: Erwartungsstrukturen und Urteile der Mitglieder (Kunden) von Genossenschaftsbanken über den Förderungsauftrag, unter besonderer Berücksichtigung der Methoden zur Erfassung der Präferenzen und zur Messung des Zufriedenheitsgrades von Individuen und Gruppen, Stuttgart: Ulmer Verlag, 1983

Baran, Pavel (1990):
Werte, in: Europäische Enzyklopädie zu Philosophie und Wissenschaften, Band 4: R-Z, hrsg. von Sandkühler, Hans Jörg, Hamburg: Meiner Verlag, S. 805-815

Barth, Klaus/Wille, Kai (2000):
Customer Equity: Ein prozessorientierter Ansatz zur Kundenbewertung, Arbeitspapier Nr. 276, Duisburg: Gerhard-Mercator-Universität Duisburg, 2000

Barth, Wolfgang (2009):
Kundenwertmessung bei Kreditinstituten, WISU-Studienblatt Oktober 2009, erschienen in: WISU - Das Wirtschaftsstudium, 38. Jg., 2009, Heft 10, o. S.

Basler Ausschuss für Bankenaufsicht (2004):
Internationale Konvergenz der Kapitalmessung und Eigenkapitalanforderungen: Überarbeitete Rahmenanforderungen, Basel: Bank of International Settlement, 2004

Basler Ausschuss für Bankenaufsicht (2009):
Consultative Document: Strengthening the Resilience of the Banking Sector, Basel: Bank of International Settlement, 2009

Bastian, Christina (2000):
Mitarbeiterführung im Vertrieb: Anreizsysteme auf dem Prüfstand, in: Vertriebs-Management: Organisation, Technologieeinsatz, Personal, hrsg. von Reichwald, Ralf/Bullinger, Hans-Jörg, Stuttgart: Schäffer-Poeschel, 2000, S. 293-323

Batz, Christian (2004):
Auslegungssache, erschienen in: Bankinformation, 31. Jg., 2004, Heft 12, S. 32-35

Batz, Christian/Hensel, Volkmar (2010):
Kundengeschäftssteuerung mit VR-Control, in: Wertorientierte Vertriebssteuerung in Banken und Sparkassen: Potenzialermittlung, Deckungsbeitragsmessung, Multikanal-Steuerung, erfolgsabhängige Vergütung, 3. Auflage, hrsg. von Wimmer, Konrad, Heidelberg: FinanzColloquium, 2010, S. 102-131

Bauer, Hans H./Hammerschmidt, Maik (2005):
Customer-Based Corporate Valuation: Integrating the Concepts of Customer Equity and Shareholder Value, erschienen in: Management Decision, 43. Jg., 2005, Heft 3, S. 331-348

Bauer, Hans-H./Hammerschmidt, Maik/Brähler, Matthias (2002):
Kundenwertbasierte Unternehmensbewertung: Das Customer-Lifetime-Value-Konzept und sein Beitrag zu einer Marketingorientierten Unternehmensbewertung, in: Jahrbuch der Absatz und Verbrauchsforschung, 48. Jg., 2002, Nr. 4, S. 324-344

Bauer, Rudolf A. (2000):
Vertriebsorganisation: Kundenorientierung durch effektive Strukturen, in: Vertriebs-Management: Organisation, Technologieeinsatz, Personal, hrsg. von Reichwald, Ralf/ Bullinger, Hans-Jörg, Stuttgart: Schäffer-Poeschel, 2000, S. 33-83

Baum, Heinz-Georg/Coenenberg, Adolf G./Günther, Thomas (2007):
Strategisches Controlling, 4., überarbeitete Auflage, Stuttgart: Schäffer-Poeschel, 2007

Baumeister, Alexander (2003):
Risikoadjustierte Performancemaße, erschienen in: Die Betriebswirtschaft (DBW), 63. Jg., 2003, Heft 2, S. 221-226

Baumgärtler, Thomas (2000):
Neuere Ansätze zur Erklärung der Mitgliederpartizipation in Kreditgenossenschaften: eine empirische Analyse der Beteiligungsfaktoren, Nürnberg: Forschungsinstitut für Genossenschaftswesen, 2000

Bayón, Tomás/Gutsche, Jens/Bauer, Hans H. (2002):
Customer Equity Marketing: Touching the Intangible, erschienen in: European Management Journal, 20. Jg., 2002, Heft 3, S. 213-222

Bayón, Tomás/Herrmann, Andreas/Huber, Frank (Hrsg.) (2007):
Vielfalt und Einheit in der Marketingwissenschaft: Ein Spannungsverhältnis – Hans-H. Bauer zum 60. Geburtstag, Wiesbaden: Gabler Verlag, 2007

Bea, Franz-Xaver/Dichtl, Erwin/Schweitzer, Marcell (Hrsg.) (1985):
Allgemeine Betriebswirtschaftslehre, Band 2, Führung, 2. Auflage, Stuttgart/Jena: G. Fischer, 1985

Bea, Franz-Xaver/Dichtl, Erwin/Schweitzer, Marcell (Hrsg.) (2001):
Allgemeine Betriebswirtschaftslehre, Band 2: Führung, 8. Auflage, Stuttgart: Lucius & Lucius Verlag, 2001

Becker, Jörg (1994):
Strategisches Vertriebscontrolling, München: Verlag Franz Vahlen, 1994

Becker, Jörg/Kahn, Dietrich (2005):
Der Prozess im Fokus, in: Prozessmanagement: Ein Leitfaden zur prozessorientierten Organisationsgestaltung, 5., überarbeitete und erweiterte Auflage, hrsg. von Becker, Jörg/Kugeler, Martin/Rosemann, Michael, Berlin/Heidelberg/New York: Springer Verlag, 2005, S. 3-16

Becker, Jörg/Kugeler, Martin/Rosemann, Michael (Hrsg.) (2005):
Prozessmanagement: Ein Leitfaden zur prozessorientierten Organisationsgestaltung, 5., überarbeitete und erweiterte Auflage, Berlin/Heidelberg/New York: Springer Verlag, 2005

Beeser, Anemone Johanna (2003):
Kundenertragswert im Retail Banking: Konzeption und Operationalisierung, Dissertation, Frankfurt am Main: Universität, 2003

Bell, David/Deighton, John/Reinartz, Werner J./Rust, Roland T./Swartz, Gordon (2002):
Seven Barriers to Customer Equity Management, erschienen in: Journal of Service Research, 5. Jg., 2002, Heft 1, S. 77-86

Benölken, Heinz (2002):
Neue Strategien für das Firmenkundengeschäft in Banken und Sparkassen: Risikosteuerung, Marketing und Prozessmanagement, Wiesbaden: Gabler Verlag, 2002

Berens, Wolfgang/Bertelsmann, Rene (2002):
Controlling, in: Handwörterbuch Unternehmensrechnung und Controlling, hrsg. von Küpper, Hans-Ulrich/Wagenhofer, Alfred, 4., völlig neu gestaltete Auflage, Stuttgart: Schäffer-Poeschel, 2002, Sp. 280-288

Berens, Wolfgang/Brauner, Hans U./Frodermann, Jürgen (Hrsg.) (2005):
Unternehmensentwicklung mit Finanzinvestoren: Eigenkapitalstärkung, Wertsteigerung, Unternehmensverkauf, Stuttgart: Schäffer-Poeschel, 2005, S. 7-30

Berens, Wolfgang/Delfmann, Werner (2002):
Quantitative Planung, 3., überarbeitete Auflage, Stuttgart: Schäffer-Poeschel, 2002

Berens, Wolfgang/Högemann, Bernd/Segbers, Klaus (2005):
Das mittelständische Unternehmen - Status quo und Perspektiven der Finanzierung, in: Unternehmensentwicklung mit Finanzinvestoren: Eigenkapitalstärkung, Wertsteigerung, Unternehmensverkauf, hrsg. von Berens, Wolfgang/Brauner, Hans U./Frodermann, Jürgen, Stuttgart: Schäffer-Poeschel, 2005, S. 7-30

Berens, Wolfgang/Knappkötter, Ralf/Segbers, Klaus/Siemes, Andreas/Ullrich, Thomas (2005):
Der Firmenkundenkredit - Kreditprozess, Informationsbeschaffung und Kundenorientierung, erschienen in: Zeitschrift für das gesamte Kreditwesen (ZfgK), 58. Jg., 2005, Heft 15, S. 784-789

Berger, Allen N./Udell Gregory F. (1995):
Relationship Lending and Lines of Credit in Small Firms, erschienen in: Journal of Business, 68. Jg., 1995, Heft 3, S. 351-382

Berger, Paul D./Nasr, Nada I. (1998):
Customer Lifetime Value: Marketing Models and Applications, erschienen in: Journal of Interactive Marketing, 12. Jg., 1998, Heft 1, S. 17-30

Berghaus, Uwe (2008):
Es geht viel mehr, erschienen in: Bankinformation, 35. Jg., 2008, Heft 8, S. 40-42

Berghaus, Uwe (2009):
Genossenschaftsbanken: Ein Drittel des Kundenbedarfs bleibt unerkannt, erschienen in: Bank und Markt, o. Jg., 2009, Heft 2, S. 34-36

Berhorst, Ute (2008):
Ganzheitlich gedacht, erschienen in: Bankinformation, 35. Jg., 2008, Heft 6, S. 36-37

Berlin, Mitchell S. (1996):
For Better and for Worse: Three Lending Relationships, Business Review, Federal Reserve Bank of Philadelphia, December 1996, S. 3-12

Bernet, Beat/Westerfeld, Simone (2008):
KMU-Ratingmodelle und Ratingqualität: Auswirkungen der Ratingarchitektur auf die ex-ante Risikoklassifikation von KMU-Kreditkontrakten, erschienen in: Zeitschrift für Betriebswirtschaft (ZfB), 78. Jg., 2008, Heft 10, S. 1011-1060

Bernhardt, Hans-Heinrich/Dokter, Markus (2008):
Balanced Scorecard: Kein operatives Zielsystem für den Vertrieb, erschienen in: BankPraktiker, o. Jg., 2000, Heft 1, S. 38-42

Berthel, Jürgen (1975):
Betriebliche Informationssysteme, Stuttgart: Schäffer-Poeschel, 1975

Berthel, Jürgen (1992):
Informationsbedarf, in: Handwörterbuch der Organisation, hrsg. von Frese, Erich, 3., völlig neu gestaltete Auflage, Stuttgart: Schäffer-Poeschel, 1992, Sp. 872-886

Betsch, Oskar/Wiechers, Rüdiger (Hrsg.) (1995):
Handbuch Finanzvertrieb: Vertriebsstrategien, Vertriebswege, Vertriebsmanagement, Frankfurt am Main: Fritz Knapp Verlag, 1995

Beuthien, Volker (2006):
Wie kapitalistisch darf eine Genossenschaft sein?, erschienen in: Die Aktiengesellschaft, 51. Jg., 2006, Heft 3, S. 53-62

Beuthien, Volker/Dierkes, Stefan/Wehrheim, Michael (2008):
Die Genossenschaft - mit der europäischen Genossenschaft: Recht, Steuer, Betriebswirtschaft, Berlin: Erich Schmidt Verlag, 2008

Beuthien, Volker/Hanrath, Stefanie (2008): Den Förderauftrag prüfen – wie soll der Prüfer das machen?, erschienen in: Zeitschrift für das gesamte Genossenschaftswesen (ZfgG), 58. Jg., 2008, Heft 2, S. 85-97

Beuthien, Volker/Hanrath, Stefanie/Weber, Heinz-Otto (2008):
Mitglieder-Fördermanagement in Genossenschaftsbanken: Analysen, Erläuterungen und Gestaltungsempfehlungen aus ökonomischer, rechtlicher und steuerlicher Sicht, Marburger Schriften zum Genossenschaftswesen, Nr. 106, Göttingen: Verlag Vandenhoeck & Ruprecht, 2008

Beutin, Nikolas/Klenk, Peter (2005):
Erfolgreiches Cross Selling: Potenzialen auf der Spur, erschienen in: Die Bank, o. Jg., 2005, Heft 5, S. 51-54

Birkelbach, Ralf (1988):
Strategische Geschäftsfeldplanung im Versicherungssektor, München: Beck Verlag, 1988

Black, Joan S. (1982):
Opinion Leadership: Is anyone following?, erschienen in: Public Opinion Quaterly, 46. Jg., 1987, Heft 1, S. 169-176

Blackwell, David W./Winters, Drew B. (1997):
Banking Relationships and the Effect of Monitoring in Loan Pricing, erschienen in: Journal of Financial Research, 20. Jg., 1997, Heft 2, S. 275-289

Blattberg, Robert C./Deighton, John (1996):
Manage Marketing by the Customer Equity Test, erschienen in: Harvard Business Review, 74. Jg., 1996, Heft 4, S. 136-144

Blattberg, Robert C./Deighton, John (1997):
Aus rentablen Kunden vollen nutzen ziehen, erschienen in: Harvard Business Manager, 19. Jg., 1997, Heft 1, S. 24-34

Blattberg, Robert C./Getz, Gary/Thomas, Jacquelyn S. (2001):
Customer Equity: Building and Managing Relationships as Valuable Assets, Boston: Harvard Business School Press, 2001

Bleicher, Knut/Meyer, Erik (1976):
Führung in der Unternehmung: Formen und Modelle, Reinbek bei Hamburg: Rowohlt Verlag, 1976

Bleymüller, Josef/Gehlert, Günther/Gülicher, Herbert (1998):
Statistik für Wirtschaftswissenschaftler, 11., überarbeitete Auflage, München: Verlag Franz Vahlen, 1998

Bliemel, Friedhelm/Eggert, Andreas (1998):
Kundenbindung – die neue Sollstrategie, erschienen in: Marketing – ZFP, 20. Jg., 1998, Heft 1, S. 37-46

Blome-Drees, Johannes (2008):
Modelle einer Erfolgsorientierung von Genossenschaften, erschienen in: Zeitschrift für das gesamte Genossenschaftswesen (ZfgG), 58. Jg., 2008, Heft 1, S. 12-23

Blut, Markus (2008):
Der Einfluss von Wechselkosten auf die Kundenbindung: Verhaltenstheoretische Fundierung und empirische Analyse, Wiesbaden: Gabler Verlag, 2008

Boettcher, Erik (1979):
Die Problematik der Operationalisierung des Förderauftrags in Genossenschaften: Förderplan und Förderbericht, in: Zeitschrift für das gesamte Genossenschaftswesen (ZfgG), 29. Jg., 1979, Heft 3, S. 198-216

Boettcher, Erik/Großfeld, Bernhard/Wagner, Helmut (Hrsg.) (1984):
Die Genossenschaftsidee im Widerstreit der Meinungen, Tagungsbericht, Institut für Genossenschaftswesen der Universität Münster, Münster: Regensberg, 1984

Böhm, Anne (1999):
Geschäftsbeziehungen zwischen Bank und Existenzgründer; Lohmar et al.: Eul Verlag, 1999

Bohnenkamp, Peter (1995):
Prozessorientierte Standard-Kostenrechnung: Innovative Ansätze der Betriebskosten Ermittlung und dimensionalen Integration im Bank-Controlling, Berlin/Stuttgart/Wien: Verlag Paul Haupt, 1995

Böhrs, Sandra (2004):
Customer Value Management: Die Integration von Kundenwert und Kundennutzen als Marketingansatz im Verkehrsdienstleistungsbereich, Dissertation, Mannheim: Universität Mannheim, 2004

Bonus, Holger (1994):
Das Selbstverständnis moderner Genossenschaften: Rückbindung von Kreditgenossenschaften an ihre Mitglieder, Tübingen: Mohr Verlag, 1994

Bonus, Holger (1998):
Das Mitglied in Ihrer Bank: Zukunftsperspektiven der Genossenschaftsbanken, in: Freiheit und Bindung als Grundlage der marktwirtschaftlichen und demokratischen Ordnung, hrsg. von Jäger, Wilhelm, Münster: Institut für Genossenschaftswesen der Westfälischen Wilhelms-Universität Münster, 1998, S. 317-342

Bonus, Holger/Greve, Rolf/Kring, Thorn/Polster, Dirk (1999):
Der genossenschaftliche Finanzverbund als Strategisches Netzwerk: Neue Wege der Kleinheit, Arbeitspapier des Instituts für Genossenschaftswesen der Westfälischen Wilhelms-Universität Münster, Nr. 16, 1999

Bonus, Holger/Grossfeld, Bernhard/Jäger, Wilhelm (Hrsg.) (1989):
Die Genossenschaft im Spiegel des Rechts: Ansprachen auf dem Akademischen Festakt "100 Jahre Genossenschaftsgesetz" am 12. Juni 1989 im Schloß und im Rathaus zu Münster, Münster: Regensberg Verlag, 1989

Bookhagen, Bettina (2006):
Mitgliedschaft kommunizieren, erschienen in: Bankinformation, 32. Jg., 2006, Heft 11, S. 38-41

Boot, Arnoud W. A. (2000):
Relationship Banking: What do we know?, erschienen in: Journal of Financial Intermediation, 9. Jg., 2000, Heft 1, S. 7-25

Boot, Arnoud W. A./Thakor, Anjan V. (2000):
Can Relationship Banking survive Competition?, erschienen in: Journal of Finance, 55. Jg., 2000, Heft 2, S. 679-713.

Bornemann, Sven/Schaaff, Claudia/Pfingsten, Andreas (2009):
Genossenschaftliches Eigenkapital im HGB und den IFRS - ökonomische und regulatorische Implikationen für Kreditgenossenschaften, erschienen in: Zeitschrift für das gesamte Genossenschaftswesen (ZfgG), 59. Jg., 2009, Heft 1, S. 18-32

Börner, Christoph J. (2005):
Konzeptioneller Rahmen: Strategisches Management und Strategieparameter, in: Bankstrategien im Firmenkundengeschäft: Konzeption, Management, Dimensionen, hrsg. von Börner, Christoph J./Maser, Harald/Schulz, Thomas Christian, Wiesbaden: Gabler Verlag, 2005, S. 31-63

Börner, Christoph J./Maser, Harald/Schulz, Thomas Christian (Hrsg.) (2005): Bankstrategien im Firmenkundengeschäft: Konzeption, Management, Dimensionen, Wiesbaden: Gabler Verlag, 2005

Boston Consulting Group (2009a):
Creating Value in Banking: Living with New Realities, Boston: Boston Consulting Group, Inc., 2009

Boston Consulting Group (2009b):
Deutschland 2015 - Szenarien für den Industriestandort, Boston: Boston Consulting Group, Inc., 2009

Bradza, Johann/Kleer, Jerzy (Hrsg.) (1996):
Genossenschaften vor neuen Herausforderungen: Festschrift für Prof. Dr. Dr. Juhani Laruinkari, Augsburg: Maro-Verlag, 1996

Bramsemann, Rainer (1997):
Ein Erfolgssteuerungssystem für Kleinbetriebe, in: Management und Mittelstand: erfolgsorientierte Ansätze und Perspektiven, hrsg. von Bührens, Jürgen, Sternenfels/Berlin: Verlag Wissenschaft und Praxis, 1997, S. 29-54

Braun, Peter Alexander (1981):
Das Firmenkundengeschäft der Banken im Wandel: die Entwicklung eines neuen Leistungsverhaltens, Dissertation, Augsburg: Universität Augsburg, 1981

Breuer, Wolfgang (1997):
Die Wertadditivität von Marktbewertungsfunktionen, erschienen in: WISU - Das Wirtschaftsstudium, 26. Jg., 1997, Heft 12, S. 1148-1153

Brixner, Ulrich (2006):
Verbünde vs. Konzerne, in: Banken auf der Suche nach strategischem Profil, hrsg. von Tietmeyer, Hans/Rolfes, Bernd, Wiesbaden: Gabler Verlag, 2006b, S. 57-85

Brockmann, Günter M. (Hrsg.) (1992):
Erfolgreiches Verkaufs-Management, Landsberg/Lech: Verlag Moderne Industrie, 1992

Broschart, Jürgen/Conradt, Axel/Schlarb, Nicole/Wagner Petra (2002a):
Fluch oder Segen?: Die Bedeutung des Förderauftrags in der strategischen Planung einer ländlichen Genossenschaftsbank!, unveröffentlichte Studienarbeit, Trier: Bankakademie, 2002

Broschart, Jürgen/Conradt, Axel/Schlarb, Nicole/Wagner Petra (2002b):
Fluch oder Segen?: Die Bedeutung des Förderauftrags bei der strategischen Planung einer Genossenschaftsbank, erschienen in: Bankinformation, 29. Jg., 2002, Heft 7, S. 12-14

Brost, Heike/Dahmen, Andreas/Lippmann, Ingo (Hrsg.) (2006):
Corporate Banking: Zukunftsorientierte Strategien im Firmenkundengeschäft, 5., vollständig überarbeitete Auflage, Frankfurt am Main: Bankakademie Verlag, 2006

Brost, Heike/Neske, Rainer/Wrabetz, Wolfram (Hrsg.) (2008):
Vertriebssteuerung in der Finanzdienstleistungsindustrie, Frankfurt am Main: Frankfurt School Verlag, 2008

Bruhn, Manfred (1997):
Qualitätsmanagement bei Dienstleistungen: Grundlagen, Konzepte, Methoden, Berlin: Springer Verlag, 1997

Bruhn, Manfred (2009):
Relationship Marketing: Das Management von Kundenbeziehungen, München: Verlag Franz Vahlen, 2009

Bruhn, Manfred/Georgi, Dominik/Treyer, Mathias/Leumann, Simon (2000):
Wertorientiertes Relationship Marketing: Vom Kundenwert zum Customer Lifetime Value, erschienen in: Die Unternehmung, 54. Jg., 2000, Heft 3, S. 167-187

Bruhn, Manfred/Homburg, Christian (Hrsg.) (1999):
Handbuch Kundenbindungsmanagement: Grundlagen, Konzepte, Erfahrungen, 2. Auflage, Wiesbaden: Gabler Verlag, 1999

Bruhn, Manfred/Lusti, Markus/Müller, Werner R./Schierenbeck, Henner/Studer, Tobias (Hrsg.) (1998):
Wertorientierte Unternehmensführung: Perspektiven und Handlungsfelder für die Wertsteigerung von Unternehmen, Wiesbaden: Gabler Verlag, 1998

Bruhn, Manfred/Meffert, Heribert (Hrsg.) (1998):
Handbuch Dienstleistungsmanagement: von der strategischen Konzeption zur praktischen Umsetzung, Wiesbaden: Gabler Verlag, 1998

Bruhn, Manfred/Stauss, Bernd (Hrsg.) (2000):
Dienstleistungsmanagement Jahrbuch 2000, Wiesbaden: Gabler Verlag, 2000

Bruhn, Manfred/Stauss, Bernd (Hrsg.) (2000):
Dienstleistungsqualität: Konzepte, Methoden, Erfahrungen, 3., vollständig überarbeitete und erweiterte Auflage, Wiesbaden: Gabler Verlag, 2000

Bruhn, Manfred/Stauss, Bernd (Hrsg.) (2005):
Dienstleitungscontrolling, Wiesbaden: Gabler Verlag, 2005

Buhl, Hans-Ulrich/Dzienziol, Jochen/Heidemann, Julia (2010):
Vertriebssteuerung auf Basis des Customer Lifetime Value am Beispiel der Finanzmarktbranche, Diskussionspapier WI-69, Universität Augsburg, 2010

Buhl, Hans-Ulrich/Heinrich, Bernd (2008):
Valuing Customers Portfolios under Risk-Return-Aspects: a Model-based Approach and its Implication in the Financial Services Industry, erschienen in: Academy of Marketing Science Review, 12. Jg., 2008, Heft 5, S. 1-32

Bührens, Jürgen (Hrsg.) (1997):
Management und Mittelstand: Erfolgsorientierte Ansätze und Perspektiven, Sternenfels/Berlin: Verlag Wissenschaft und Praxis, 1997

Bundesanstalt für Finanzdienstleistungsaufsicht (2010):
Rundschreiben 11/2010 – Mindestanforderungen an das Risikomanagement – MaRisk, Online im Internet, URL: http://www.bafin.de/cln_179/nn_722758/SharedDocs/Veroeffentlichungen/DE/Service/Rundschreiben/2010/rs__1011__ba__marisk.html#doc2028716bodyText9 (Stand: 02.06.2011)

Bundesfinanzhof (2009):
Urteil vom 20.08.2009, Aktenzeichen: V R 32/08, Online im Internet, URL: http://www.bundesfinanzhof.de/pressemitteilungen (Stand: 03.09.2010)

Bungenstock, Christian (2002):
Kommentar – Kundenerfolgsrechnung – warum und wie funktioniert sie?, erschienen in: Kostenrechnungs-Praxis: krp; Zeitschrift für Controlling, Accounting & System-Anwendungen, o. Jg., 2002, Heft 1, S. 9-10

Burmann, Christoph:
„Customer Equity" als Steuerungsgröße für die Unternehmensführung; erschienen in: Zeitschrift für Betriebswirtschaft (ZfB), 73. Jg., 2003, Heft 2, S. 113-138

Busch, Rainer/Fuchs, Wolfgang/Unger, Fritz (2008):
Integriertes Marketing: Strategie, Organisation, Instrumente, 4., vollständig überarbeitete Auflage, Wiesbaden: Gabler Verlag, 2008

Büschgen, Hans E. (1998):
Bankbetriebslehre: Bankgeschäfte und Bankmanagement, 5., vollständig überarbeitete und erweiterte Auflage, Wiesbaden: Gabler Verlag, 1998

Büschgen, Hans E. (2001):
Leistungsorientierte Vergütungssysteme als Instrument der Synchronisation von Mitarbeiterinteressen und Gesamtbankzielen, in: Handbuch Bankcontrolling, 2., überarbeitete und erweiterte Auflage, hrsg. von Schierenbeck, Henner, Wiesbaden: Gabler Verlag, 2001, S. 531-546

Bussiek, Jürgen (1996):
Anwendungsorientierte Betriebswirtschaftslehre für Klein- und Mittelunternehmen, 2., durchgesehene Auflage, München/Wien: Oldenbourg Verlag, 1996

Buzziol, Steffi (2004):
Operationelle Risiken managen, erschienen in: Bankinformation, 31. Jg., 2004, Heft 9, S. 16-20

BVR (1999a):
Bündelung der Kräfte: Ein Verbund – eine Strategie, Band 1, Bonn: Bundesverband der deutschen Volksbanken und Raiffeisenbank e. V., 1999

BVR (1999b):
Bündelung der Kräfte: Ein Verbund – eine Strategie, Band 2, Bonn: Bundesverband der Deutschen Volksbanken und Raiffeisenbanken e. V., 1999

BVR (1999c):
Bericht und Zahlen '98: Jahresbericht 1998 des Bundesverbandes der Deutschen Volksbanken und Raiffeisenbanken, Bonn: Bundesverband der Deutschen Volksbanken und Raiffeisenbanken e. V., 1999

BVR (2000):
Bericht und Zahlen 1999: Jahresbericht 1999 des Bundesverbandes der Deutschen Volksbanken und Raiffeisenbanken, Bonn: Bundesverband der Deutschen Volksbanken und Raiffeisenbank e. V., 2000

BVR (2001a):
Wege 2000: Jahresbericht des Bundesverbandes der Deutschen Volksbanken und Raiffeisenbanken, Bonn: Bundesverband der Deutschen Volksbanken und Raiffeisenbanken e. V., 2001

BVR (2001b):
Bündelung der Kräfte: Die gemeinsame Strategie, Bonn: Bundesverband der Deutschen Volksbanken Raiffeisenbanken e. V., 2001

BVR (2001c):
Abschlussbericht Projekt 1: Aktive und effiziente Marktbearbeitung unter Einbeziehung der Neugestaltung und Differenzierung der Erlösströme im FinanzVerbund, Bonn: Bundesverband der Deutschen Volksbanken Raiffeisenbanken e. V., 2001

BVR (2002a):
Wir bündeln Kräfte: Jahresbericht 2001 des Bundesverbandes der Deutschen Volksbanken und Raiffeisenbanken, Bonn: Bundesverband der Deutschen Volksbanken und Raiffeisenbanken e. V., 2002

BVR (2002b):
Weißbuch der Banksteuerung, 2. Auflage, Bonn: Bundesverband der Deutschen Volksbanken und Raiffeisenbanken e. V., 2002

BVR (2003a):
Wege 2002: Wir bündeln Kräfte. Jahresbericht 2002 des Bundesverbandes der Deutschen Volksbanken und Raiffeisenbanken, Berlin: Bundesverband der Deutschen Volksbanken und Raiffeisenbanken e. V., 2003

BVR (2003b):
VR-Control Wissen: Die Zusammenfassung der Fachkonzepte zu VR-Control für Controller und Spezialisten, Wiesbaden: DG Verlag eG, 2003

BVR (2003c):
Geschäftsprozesse: Ermittlung der Standardbearbeitungszeiten und -kosten: Anwendungsleitfaden, Berlin: Bundesverband der Deutschen Volksbanken und Raiffeisenbanken e. V., 2003

BVR (2004a):
Wir bündeln Kräfte: Jahresbericht 2003 des Bundesverbandes der Deutschen Volksbanken und Raiffeisenbanken, Berlin: Bundesverband der Deutschen Volksbanken und Raiffeisenbanken e. V., 2004

BVR (2004b):
Genossenschaftsbanken beschließen Neupositionierung des BVR, Online im Internet, URL:
http://www.bvr.de/public.nsf/detail.html?ReadForm&main=6&sub=1&ParentUNID=FD62C7195FBA8126C1256FB6004AE122 (Stand: 12.08.2011)

BVR (2005a):
Im Netzwerk gestalten: Jahresbericht 2004 des Bundesverbandes der Deutschen Volksbanken und Raiffeisenbanken, Berlin: Bundesverband der Deutschen Volksbanken und Raiffeisenbanken e. V., 2005

BVR (2005b):
Ergebnisse der vierten quantitativen Auswirkungsstudie zu den neuen Eigenkapitalregelungen (QIS 4) für die Kreditgenossenschaften Az.:BIZ-QIS, Vermerk, Berlin, 2005

BVR (2006):
Dezentrales Unternehmertum: Jahresbericht 2005 des Bundesverbandes der Deutschen Volksbanken und Raiffeisenbanken, Berlin: Bundesverband der Deutschen Volksbanken Raiffeisenbanken e. V., 2006

BVR (2007a):
Früchte der Reformen: Jahresbericht 2006 des Bundesverbandes der Deutschen Volksbanken und Raiffeisenbanken, Berlin: Bundesverband der Deutschen Volksbanken und Raiffeisenbanken e. V., 2007

BVR (2007b):
Leitfaden "Strategische Steuerung einer Primärbank", Berlin: Bundesverband der Deutschen Volksbanken Raiffeisenbanken e. V., 2007

BVR (2008):
Wertorientiert und sicher: Jahresbericht 2007 des Bundesverbandes der Deutschen Volksbanken und Raiffeisenbanken, Berlin: Bundesverband der Deutschen Volksbanken und Raiffeisenbanken e. V., 2008

BVR (2009a):
Stabilität schafft Vertrauen: Jahresbericht 2008 des Bundesverbandes der Deutschen Volksbanken und Raiffeisenbanken, Berlin: Bundesverband der Deutschen Volksbanken und Raiffeisenbanken e. V., 2009

BVR (2009b):
Kompass 2010 . Analyse, Prognose, Bewertung und Maßnahmen, Berlin: Bundesverband der Deutschen Volksbanken und Raiffeisenbanken e. V., 2009

BVR (2010a):
Kompass 2011: Analyse, Prognose, Bewertung und Maßnahmen, Berlin: Bundesverband der Deutschen Volksbanken Raiffeisenbanken e. V., 2010

BVR (2010b):
Stark und zuverlässig: Jahresbericht 2009 des Bundesverbandes der Deutschen Volksbanken und Raiffeisenbanken, Berlin: Bundesverband der Deutschen Volkbanken und Raiffeisenbanken e. V., 2010

BVR (2011a):
Erfolgreich und eigenständig: Jahresbericht 2010 des Bundesverbandes der Deutschen Volksbanken und Raiffeisenbanken, Berlin: Bundesverband der Deutschen Volksbanken und Raiffeisenbanken e. V., 2011

BVR (2011b):
Entwicklung der Entwicklung der Volksbanken und Raiffeisenbanken ab 1970, Berlin: Bundesverband der Deutschen Volksbanken und Raiffeisenbanken e. V., 2011, Online im Internet, URL: http://www.bvr.de/p.nsf/5A4114C95B305875C12578530050F644/$FILE/Entwicklungseit1970Mappe.pdf (Stand: 04.06.2011)

BVR et al. (2006):
VR-FINANZPLAN MITTELSTAND – Projekthandbuch, Berlin: Bundesverband der Deutschen Volksbanken Raiffeisenbanken, 2006

BVR et al. (o. J.b):
Weißbuch der Unternehmenssteuerung, Bundesverband der Deutschen Volksbanken Raiffeisenbanken et al., o. O., o. J.

C

Canning, Gordon (1982):
Do a Value Analysis of your Customer Base, erschienen in: Industrial Marketing Management, 11. Jg., 1981, Heft 2, S. 89-93

Cario, Daniela (2005):
Aufnahme von Investorenmitgliedern in die SCE?: Gedanken zur Ausübung des Wahlrechts in Art. 14 Abs. 2 SCE-VO durch den deutschen Gesetzgeber, erschienen in: Zeitschrift für das gesamte Genossenschaftswesen (ZfgG), 55. Jg., 2005, Heft 2, S. 146-152

Cavallo, Laura/Rossi, Stefania P. S. (2001):
Scale and Scope Economies in the European Banking Systems, erschienen in: Journal of Multinational Financial Management, 11. Jg., 2001, Heft 4-5, S. 515-531

Choi, Duke Hyun/Kim, Chul Min/Kim, Sang-Il/Kim, Soung Hie (2006):
Customer Loyalty and Disloyalty in Internet Retail Stores - Its Antecedents and its Effect on Customer Price Sensitivity, erschienen in: International Journal of Management, 23. Jg., 2006, Heft 4, S. 925-941

Coenenberg, Adolf G. (1999):
Kostenrechnung und Kostenanalyse, 4., aktualisierte Auflage, Landsberg/Lech: Verlag Moderne Industrie, 1999

Collings, David/Baxter, Nicole (2005):
Valuing Customers, erschienen in: BT – Technology Journal, 23. Jg., 2005, Heft 3, S.24-29

Cornelsen, Jens (2000):
Kundenwertanalysen im Beziehungsmarketing: Theoretische Grundlagen und Ergebnisse einer empirischen Studie im Automobilbereich, Nürnberg: GIM Gesellschaft für innovatives Marketing, 2000

Cornelsen, Jens (2006):
Kundenbewertung mit Referenzwerten, in: Kundenwert: Grundlagen, Innovative Konzepte, Praktische Umsetzungen, 3., überarbeitete und erweiterte Auflage, hrsg. von Günter, Bernd/Helm, Sabrina, Wiesbaden: Gabler Verlag, 2006, S. 183-215

Corsten, Hans (2000):
Der Integrationsgrad des externen Faktors als Gestaltungsparameter in Dienstleistungsunternehmen: Voraussetzungen und Möglichkeiten der Externalisierung und Internalisierung, in: Dienstleistungsqualität: Konzepte, Methoden, Erfahrungen, 3., vollständig überarbeitete und erweiterte Auflage, hrsg. von Bruhn, Manfred/Stauss, Bernd, Wiesbaden: Gabler Verlag, 2000, S. 145-168

Cox Donald F. (Hrsg.) (1967):
Risk Taking and Information Handling in Consumer Behavior, Boston: Harvard University, 1967

Cromme, Gerhard (2008):
Gute Corporate Governance als Bindeglied zwischen Bank und Unternehmen, erschienen in: Zeitschrift für das gesamte Kreditwesen (ZfgK), 61. Jg., 2008, Heft 10, S. 431-433

Cunningham, Scott M. (1967):
Perceived Risk as a Factor in Informal Consumer Communications, in: Risk Taking and Information Handling in Consumer Behavior, hrsg. von Cox, Donald F., Boston: Harvard University, 1967, S. 265-288

D

Dahmen, Andreas (2006):
Finanzmanagement der Firmenkunden, in: Corporate Banking: Zukunftsorientierte Strategien im Firmenkundengeschäft, 5., vollständig überarbeitete Auflage, hrsg. von Brost, Heike/Dahmen, Andreas/Lippmann, Ingo, Frankfurt am Main: Bankakademie Verlag, 2006, S. 19-102

Darfener, Stefan/Lüders, Uwe (2002):
Ohne Schulung kein Erfolg, erschienen in: Bankinformation, 29. Jg., 2002, Sonderheft 1, S. 38-43

Daschmann, Hans-Achim (1993):
Erfolgsfaktoren mittelständischer Unternehmen: Ein Beitrag zur Erfolgsfaktorenforschung, Dissertation, München: Ludwig-Maximilians-Universität München, 1993

Dauber, Markus (2008):
Aktive Vertriebssteuerung im genossenschaftlichen Finanzverbund, erschienen in: Immobilien & Finanzierung, o. Jg., 2008, Heft 24, S. 890-892

Dauser, Achim (1995):
Vertriebs-Controlling in Versicherungsunternehmen: Theoretische Grundlagen und praktische Anwendungsmöglichkeiten im Rahmen der Lebens- und Kraftfahrtversicherung, Dissertation, Universität Ulm, 1995

Degryse, Hans/Ongena, Steven (2004):
The Impact of Competition on Bank Orientation and Specialization, Arbeitspapier des Centre for Economic Policy Research, London, 2004

Degryse, Hans/Van Cayseele, Patrick (2000):
Relationship Lending within a bank-based System: evidence from European Small Business Data, erschienen in: Journal of Financial Intermediation, 9. Jg., Heft 1, S. 90-109

Dehne, Thorsten (2008):
Der "VR-FinanzPlan Mittelstand": Die neue Vertriebskonzeption für das Firmenkundengeschäft der Volksbanken Raiffeisenbanken in Deutschland, in: Erfolgreiche Vertriebsstrategien im Firmenkundengeschäft: Konzepte, Praxisberichte, Lösungen, hrsg. von Schmoll, Anton, Köln: bank-verlag medien, 2008, S. 53-67

Deking, Ingo/Meier, Roland (2000):
Vertriebscontrolling: Grundlagen für ein innovatives, anwendungsorientiertes Verständnis, in: Vertriebs-Management: Organisation, Technologieeinsatz, Personal, hrsg. von Reichwald, Ralf/Bullinger, Hans-Jörg, Stuttgart: Schäffer-Poeschel, 2000, S. 249-268

Dellmann, Klaus (1992):
Eine Systematisierung der Grundlagen des Controlling, in: Controlling: Grundlagen – Informationssysteme – Anwendungen, hrsg. von Spremann, Klaus/Zur, Eberhard, Wiesbaden: Gabler Verlag, 1992, S. 113-140

Deppe, Hans Dieter (1969):
Bankbetriebliches Wachstum: Funktionalzusammenhänge und Operations Research in Kreditinstituten, Stuttgart: Poeschel Verlag, 1969

Deutsche Bundesbank (2001):
Monatsbericht September 20001, Frankfurt am Main, Deutsche Bundesbank, 2001

Deutsche Bundesbank (2002):
Monatsbericht September 2002, Frankfurt am Main: Deutsche Bundesbank, 2002

Deutsche Bundesbank (2003): Monatsbericht September 2003, Frankfurt am Main: Deutsche Bundesbank, 2003

Deutsche Bundesbank (2004): Monatsbericht September 2004, Frankfurt am Main: Deutsche Bundesbank, 2004

Deutsche Bundesbank (2005): Monatsbericht September 2005, Frankfurt am Main: Deutsche Bundesbank, 2005

Deutsche Bundesbank (2006a): Monatsbericht September 2006, Frankfurt am Main: Deutsche Bundesbank, 2006

Deutsche Bundesbank (2006b): Monatsbericht Juni 2006, Frankfurt am Main: Deutsche Bundesbank, 2006

Deutsche Bundesbank (2007): Monatsbericht September 2007, Frankfurt am Main: Deutsche Bundesbank, 2007

Deutsche Bundesbank (2008): Monatsbericht September 2008, Frankfurt am Main: Deutsche Bundesbank, 2008

Deutsche Bundesbank (2009): Monatsbericht September 2009, Frankfurt am Main: Deutsche Bundesbank, 2009

Deutsche Bundesbank (2010): Monatsbericht September 2010, Frankfurt am Main: Deutsche Bundesbank, 2010

Deutsche Bundesbank (2011a): Statistik – Zeitreihen – Banken, Online im Internet, URL: http://www.bundesbank.de/statistik/statistik_zeitreihen.php?lang=de&open=&open_node_id=153975, (Stand: 23.05.2011)

Deutsche Bundesbank (2011b): Bankenstatistik: Statistisches Beiheft zum Monatsbericht 1, Frankfurt am Main: Deutsche Bundesbank, 2011

DGRV (2005):
Die Prüfung der Geschäftsführung von Genossenschaften, DGRV-Schriftenreihe Band 10, 3., vollständig überarbeitete Auflage, Wiesbaden: Deutscher Genossenschaftsverlag, 2005

Dhar, Ravi/Glazer, Rashi:
Hedging Customers, erschienen in: Harvard Business Review, 81. Jg., 2003, Heft 5, S. 86-93

Dihlmann, Claus (1991):
Kennzahlen zur Vertriebssteuerung, erschienen in: Zeitschrift für betriebswirtschaftliche Forschung (ZfbF), 43. Jg., 1991, Heft 2, S. 157-171

Diller, Hermann (1975):
Produkt-Management und Marketing-Informationssysteme: Tätigkeitsbild und Informationsbedarf des Produktmanagers als Determinanten von Marketing-Informationssystemen, Berlin: Duncker & Humblot, 1975

Diller, Hermann (1992):
Vahlens großes Marketinglexikon, München: Beck Verlag, 1992

Diller, Hermann (1995a):
Kundenmanagement, in: Enzyklopädie der Betriebswirtschaftslehre, Band 4: Handwörterbuch des Marketing, 2., völlig neu gestaltete Auflage, hrsg. von Tietz, Bruno/Köhler, Richard/Zentes, Joachim, Stuttgart: Schäffer-Poeschel, 1995, Sp. 1363-1376

Diller, Hermann (1995b):
Beziehungs-Marketing, erschienen in: WiSt - Wirtschaftswissenschaftliches Studium, 24. Jg., 1995, Heft 9, S. 442-447

Diller, Hermann (1998):
Nutzwertanalysen, in: Marketingplanung, hrsg. von Diller, Hermann, 2., vollständig neu bearbeitete und ergänzte Auflage, München: Vahlen Verlag, 1998, 247-265

Diller, Hermann (2001):
Probleme des Kundenwerts als Steuerungsgröße im Kundenmanagement, Arbeitspapier des Lehrstuhl für BWL, insb. Marketing, Universität Erlangen-Nürnberg, Nürnberg: Universität Erlangen-Nürnberg, 2001

Diller, Hermann (2006):
Die Bedeutung des Beziehungsmarketings für den Unternehmenserfolg, in: Grundlagen des CRM: Konzepte und Gestaltung, 2., überarbeitete und erweiterte Auflage, hrsg. von Wilde, Klaus D./Hippner, Hajo, Wiesbaden: Gabler Verlag, S. 98-120

Diller, Hermann (Hrsg.) (1998):
Marketingplanung, 2., vollständig neu bearbeitete und ergänzte Auflage, München: Vahlen Verlag, 1998

Dittmar, Matthias (2000):
Profitabilität durch das Management von Kundentreue: Theoretische Diskussion, Methodik und empirische Ergebnisse am Beispiel der Automobilindustrie, Wiesbaden: Deutscher Universitäts-Verlag, 2000

Doberanzke, Volker (1993):
Exklusive Finanzierung mittelständischer Unternehmen, Schriftenreihe für Kreditwirtschaft und Finanzierung, Band 16, Wiesbaden: Gabler Verlag, 1993

Döhring, Jens/Hromadka, Jürgen (2002):
Die Ausfallrisikobepreisung: VR-Control Teil 4, erschienen in: Bankinformation, 29. Jg., 2002, Heft 6, S. 36-41

Dokter, Markus/Lüders, Uwe (2003):
Basis für aktive Vertriebssteuerung schaffen, erschienen in: Bankinformation, 30. Jg., 2003, Heft 10, S. 28-31

Dorsch, Michael J./Carlson, Les (1996):
A Transaction Approach to Understanding and Managing Customer Equity, erschienen in: Journal of Business Research, 35. Jg., 1996, Heft 3, S. 253-264

Dorsch, Michael J./Carlson, Les/Raymond, Mary Anne/Ranson, Robert (2001):
Customer Equity Management and Strategic Choices for Sales Managers, erschienen in: Journal of Personell Selling & Sales Management, 21. Jg., Heft 2, S. 157-166

Dreyer, Karl-Joachim (2000):
Kundensegmentierung im Mittelstand - Ansätze einer Großsparkasse, erschienen in: Zeitschrift für das gesamte Kreditwesen (ZfgK), 53. Jg., 2000, Heft 10, S. 516-518

Droll, Mathias (2008):
Kundenpriorisierung in der Marktbearbeitung: Gestaltung, Erfolgsauswirkungen und Implementierung, Wiesbaden: Gabler Verlag, 2008

DSGV (2009):
Sparkassen-Finanzgruppe in Zahlen, Berlin/Bonn: Deutscher Sparkassen- und Giroverband, 2009, Online im Internet, URL: http://www.dsgv.de/_download_gallery/Publikationen/SFG_in_Zahlen.pdf, (Stand: 04.06.2011)

DSGV (2011):
Sparkassenrangliste 2010, Berlin/Bonn: Deutscher Sparkassen- und Giroverband, 2008, Online im Internet, URL: http://www.dsgv.de/_download_gallery/statistik/Sparkassenrangliste_2010.p df (Stand: 04.06.2011)

Duderstadt, Stefan (2006):
Wertorientierte Vertriebssteuerung durch ganzheitliches Vertriebscontrolling: Konzeption für das Retailbanking, Wiesbaden: Deutscher Universitätsverlag, 2006

Duhnkrack, Thomas (2002):
Wertorientierte Steuerung im Firmenkundengeschäft, in: Handbuch Firmenkundengeschäft: Technologie, Rating und Risikosteuerung als Kernkompetenz der Banken, 2., völlig überarbeitete Auflage, hrsg. von Juncker, Klaus/Priewasser, Erich, Frankfurt am Main: Fritz Knapp Verlag, 2002, S. 153-162

Duhnkrack, Thomas (2006):
Investment Banking für den Mittelstand - Herausforderungen für den genossenschaftlichen FinanzVerbund, in: Die strukturelle Ertragsschwäche der Banken, hrsg. von Tietmeyer, Hans/Rolfes, Bernd, Wiesbaden: Gabler Verlag,2006a, S. 35-58

Duhnkrack, Thomas (2007):
Genossenschaftlicher Finanzverbund: geborener Partner des Mittelstands, erschienen in: Zeitschrift für das gesamte Kreditwesen (ZfgK), 60. Jg., 2007, Heft 22, S. 1220-1222

Dülfer, Eberhard (1980):
Welche Relevanz hat der "Förderungsauftrag" für die Pflichtprüfung der eingetragenen Genossenschaft?, erschienen in: Zeitschrift für das gesamte Genossenschaftswesen (ZfgG), 30. Jg., 1980, Heft 1, S. 47-61

Dülfer, Eberhard (1995):
Betriebswirtschaftslehre der Genossenschaften und vergleichbarer Kooperative, 2., überarbeitete und erweiterte Auflage, Göttingen: Verlag Vandenhoeck & Ruprecht, 1995

Duttenhöffer, Stephan (2004):
Identifikation von Cross-Selling- und Neukundenpotenzial, in: Handbuch Vertriebsmanagement Finanzdienstleistungen: Analyse, Umsetzung und Perspektiven bei Banken und Sparkassen, hrsg. von Duttenhöffer, Stephan/Keller, Bernhard, Frankfurt am Main: Fritz Knapp Verlag, 2005, S. 505-524

Duttenhöffer, Stephan/Keller, Bernhard (Hrsg.) (2004):
Handbuch Vertriebsmanagement Finanzdienstleistungen: Analyse, Umsetzung und Perspektiven bei Banken und Sparkassen, Frankfurt am Main: Fritz Knapp Verlag, 2004

Dwyer, F. Robert (1997):
Customer Lifetime Valuation to Support Decision Making, erschienen in: Journal of Direct Marketing, 11. Jg., 1997, Heft 4, S. 6-13

Dwyer, F. Robert/Schurr, Paul H./Oh, Sejo (1987):
Developing Buyer-Seller Relationships, erschienen in: Journal of Marketing, 51. Jg., 1987, Heft 2, S. 11-27

E

Ebeling, Cordula (2007):
Erfolgsfaktoren einer wertorientierten Unternehmensführung, Wiesbaden: Deutscher Universitäts-Verlag, 2007

Eberling, Gunther (2002):
Kundenwertmanagement: Konzeption zur wertorientierten Analyse und Gestaltung von Kundenbeziehungen, Wiesbaden: Deutscher Universitäts-Verlag, 2002

Eckstein, Bernd (1998):
Mit Mitgliedermarketing im Wettbewerb gewinnen, erschienen in: Genossenschaftsblatt, Nr. 2, 1998, S. 5-10

Eggert, Andreas (1999):
Kundenbindung aus Kundensicht: Konzeptualisierung, Operationalisierung, Verhaltenswirksamkeit, Wiesbaden: Deutscher Universitäts-Verlag, 1999

Eggert, Andreas (2006):
Die zwei Perspektiven des Kundenwerts: Darstellung und Versuch einer Integration, in: Kundenwert: Grundlagen, Innovative Konzepte, Praktische Umsetzungen, 3., überarbeitete und erweiterte Auflage, hrsg. von Günter, Bernd/Helm, Sabrina, Wiesbaden: Gabler Verlag, 2006, S. 41-59

Eichelmann, Thomas/Bufka, Jürgen/Lehmann, Marcus (2001):
Verbesserte Kundenzufriedenheit durch bedarfsorientierte Segmentierung, erschienen in: Die Bank, o. Jg., Heft 11, S. 778-781

Eichelmann, Thomas/Duderstadt, Stefan (2005):
An den Kundenpotenzialen ausgerichtetes Vertriebs-Controlling, in: Wertmanagement in Banken: Festschrift zum 65. Geburtstag von Guido Eilenberger, hrsg. von Lange, Thomas A./Schulze, Heiko, Wiesbaden: Gabler Verlag, 2005, S. 3-29

Eigler, Joachim (2004):
Controlling, Anreizsysteme und Verhaltenswissenschaften, in: Controlling: Theorie und Konzeptionen, hrsg. von Scherm, Ewald/Pietsch, Gotthard, München: Verlag Franz Vahlen, 2004, S. 665-690

Eilenberger, Guido (1996):
Bankbetriebswirtschaftslehre: Grundlagen, internationale Bankleistungen, Bank-Management, 6., völlig neu bearbeitete Auflage, München/Wien: Oldenbourg Verlag, 1996

Eim, Alexander (2004):
Das Drei-Säulen-System der deutschen Kreditwirtschaft unter besonderer Berücksichtigung des Genossenschaftlichen Finanzverbundes, Arbeitspapier des Instituts für Genossenschaftswesen der Westfälischen Wilhelms-Universität, Nr. 40, 2004

Eisenächer, Harald W./Backofen, Olaf, J./Hilverkus, Sascia (2006):
Zielkundenmanagement als Erfolgsfaktor für nachhaltigen Unternehmenserfolg am Beispiel der Lufthansa, in: Kundenwert: Grundlagen, Innovative Konzepte, Praktische Umsetzungen, 3., überarbeitete und erweiterte Auflage, hrsg. von Günter, Bernd/Helm, Sabrina, Wiesbaden: Gabler Verlag, 2006, S. 781-797

Eller, Roland (Hrsg.) (2001):
Handbuch Gesamtbanksteuerung: Integration von Markt-, Kredit und operationalen Risiken, Stuttgart: Schäffer-Poeschel, 2001

Elsas, Ralf (2001):
Die Bedeutung der Hausbank: Eine ökonomische Analyse, Wiesbaden: Deutscher Universitäts-Verlag, 2001

Elsas, Ralf (2005):
Empirical Determinants of Relationship Lending, erschienen in: Journal of Financial Intermediation, 14. Jg., 2005, Heft1, S. 32-57

Elsas, Ralf/Krahnen, Jan-Pieter (1998):
Is Relationship Lending special? Evidence from Credit-File Data in Germany, erschienen in: Journal of Banking and Financing, 22. Jg., 1998, Heft 10/11, S. 1283-1316

Emmerich, Norbert/Rossbach, Peter (Hrsg.) (2006):
Der Bankensektor im Wandel: Festschrift zum 65. Geburtstag von Professor Erich Priewasser, Frankfurt am Main: Fritz Knapp Verlag, 2006

Engelhard, Hans A. (1989):
Staat und Genossenschaften: Die ordnungspolitische Bedeutung des Genossenschaftsgesetzes, in: Die Genossenschaft im Spiegel des Rechts: Ansprachen auf dem Akademischen Festakt "100 Jahre Genossenschaftsgesetz" am 12. Juni 1989 im Schloß und im Rathaus zu Münster, hrsg. von Bonus, Holger/Grossfeld, Bernhard/Jäger, Wilhelm, Münster: Regensberg Verlag, 1989, S. 29-40

Engelhardt, Werner Hans/Kleinaltenkamp, Michael (1989):
Strategische Planung, Technischer Vertrieb (TV) Lehrbrief, Projektgruppe an der Freien Universität Berlin, Berlin, 1989

Engelhardt, Werner Hans/Kleinaltenkamp, Michael/Reckenfelderbäumer, Martin (1993):
Leistungsbündel als Absatzobjekte, erschienen in: Zeitschrift für betriebswirtschaftliche Forschung (ZfbF), 45. Jg., 1993, Heft 5, S. 395, 426

Engels, Michael (1997):
Verwässerung der Verfügungsrechte in Genossenschaften, erschienen in: Zeitschrift für betriebswirtschaftliche Forschung (ZfbF), 49. Jg., 1997, Heft 7/8, S. 674-684

Engels, Wolfram (1962):
Betriebswirtschaftliche Bewertungslehre im Licht der Entscheidungstheorie, Köln/Opladen: Westdeutscher Verlag, 1962

Engstler, Martin/Praeg, Peter/Vocke, Christian (2007):
Trendstudie: Bank & Zukunft 2007: Mit Prozessexzellenz und Vertriebsinnovationen die Bank der Zukunft gestalten, Stuttgart: Fraunhofer Institut für Arbeitswirtschaft und Organisation, 2007

Erben, Edgar (1970):
Genossenschaftsbanken im Wettbewerb: die Mitgliedschaft als ein Weg zur Differenzierung, erschienen in: Zeitschrift für das gesamte Kreditwesen (ZfgK), 23. Jg., 1970, Heft 11, S. 10-13

Erichson, Bernd/Hammann, Peter (1985):
Grundlagen der Informationsbeschaffung und -aufbereitung, in: Allgemeine Betriebswirtschaftslehre, Band 2, Führung, 2. Auflage hrsg. von Bea, Franz Xaver/Dichtl, Erwin/Schweitzer, Marcell, Stuttgart/Jena: G. Fischer, 1985, S. 137-171

Eschenbach, Rolf (Hrsg.) (1996):
Controlling, 2., überarbeitete und erweiterte Auflage, Stuttgart: Schäffer-Poeschel, 1996

Euler Hermes (2005):
FiKomm 2005: Finanzkommunikation im Mittelstand, Hamburg: Euler Hermes Kreditversicherung AG, 2005

F

Falk, Bernd/Wolf, Jakob (Hrsg.) (1982):
Das große Lexikon für Handel und Absatz, 2., überarbeitete und erweiterte Auflage, Landsberg/Lech: Verlag Moderne Industrie, 1982

Fayol, Henri (1929):
Allgemeine und industrielle Verwaltung, München/Berlin: Oldenbourg Verlag, 1929

Fehl, Ulrich/Kuhn, Norbert (2005):
Kreditgenossenschaften im Spannungsfeld von Wettbewerb und Regulierung, erschienen in: Zeitschrift für das gesamte Genossenschaftswesen (ZfgG), 55. Jg., 2005, Heft 3, S. 190-209

Fischer, Klaus (1990):
Hausbankbeziehungen als Instrument der Bindung zwischen Banken und Unternehmen: Eine theoretische und empirische Analyse, Dissertation, Bonn: Universität, 1990

Fischer, Marc/Albers, Sönke (2007):
Aktuelle Entwicklungen im Customer Equity Management, in: Vielfalt und Einheit in der Marketingwissenschaft: Ein Spannungsverhältnis, hrsg. von Bayón, Thomas/Herrmann, Andreas/Huber, Frank, Wiesbaden: Gabler Verlag, 2007, S. 168-186

Fischer, Thomas M./Schmöller, Petra (2006):
Kundenwert als Entscheidungskalkül für die Beendigung von Kundenbeziehungen, in: Kundenwert: Grundlagen, Innovative Konzepte, Praktische Umsetzungen, hrsg. von Günter, Bernd/Helm, Sabrina, Wiesbaden: Gabler Verlag, 2006, S. 483-507.

Fischer, Thomas M./von der Decken, Tim (2001):
Kundenprofitabilitätsrechnung in Dienstleistungsgeschäften - Konzeption und Umsetzung am Beispiel des Car Rental Business, erschienen in: Zeitschrift für betriebswirtschaftliche Forschung (ZfbF), 52. Jg., 2001, Heft 5, S. 294-323

Flacke, Klaus (2006):
Controlling in mittelständischen Unternehmen: Ausgestaltung, Einflussfaktoren der Instrumentennutzung und Einfluss auf die Bankkommunikation, Dissertation, Münster: Westfälische Wilhelms-Universität, 2006

Foehn, Pascal (2006):
Kundenwert im Private Banking, Dissertation, Zürich: Universität Zürich, 2006

Fornell, Claes/Johnson, Michael D./Anderson, Eugene W./Cha, Jaesung/Bryant, Barbara Everitt (1996):
The American Customer Satisfaction Index: Nature, Purpose, and Findings, erschienen in: Journal of Marketing, 60. Jg., 1996, Heft 4, S. 7-18

Freiberger, Thorsten/Kornatz, Christiane (2007):
Die Ertragskraft stärken, erschienen in: Die Bank, o. Jg., 2007, S. 54-59

Freiling, Jörg (2006):
Kundenwert aus ressourcentheoretischer Sicht, in: Kundenwert: Grundlagen, innovative Konzepte, Praktische Überlegungen, 3., überarbeitete und erweiterte Auflage, hrsg. von Günter, Bernd/Helm, Sabrina, Wiesbaden: Gabler Verlag , 2006, S. 83-102

Freiling, Jörg/Reckenfelderbäumer, Martin (2000):
Kundenerfolgsrechnung für industrielle Dienstleistungen: Probleme und Lösungsansätze, in: Dienstleistungscontrolling, hrsg. von Bruhn, Manfred/Stauss, Bernd, Wiesbaden: Gabler Verlag, 2000, S. 501-524

Frese, Bernd (2009):
Kundenwertorientierte Vertriebssteuerung bei Regionalbanken, nicht veröffentliche Diplomarbeit am Lehrstuhl für Betriebswirtschaftslehre insb. Controlling an der Westfälischen Wilhelms-Universität, Münster: Universität Münster, 2009

Frese, Erich (Hrsg.) (1992):
Handwörterbuch der Organisation, 3., völlig neu gestaltete Auflage, Stuttgart: Schäffer-Poeschel, 1992

Freter, Hermann (1983):
Marktsegmentierung, Stuttgart et al.: Kohlhammer Verlag, 1983

Freter, Hermann (1998):
Marktsegmentierung im Dienstleistungsbereich, in: Handbuch Dienstleistungsmanagement: von der strategischen Konzeption zur praktischen Umsetzung, hrsg. von Bruhn, Manfred/Meffert, Heribert, Wiesbaden: Gabler Verlag, 1998, S. 229-264

Friedag, Herwig R./Schmidt, Walter (2001):
Vom Trugbild der Ursache-Wirkungsketten zu einem ausgewogenen Management, erschienen in: Controller Magazin, 26. Jg., 2001, Heft 5, S. 494-497

Friederichs-Schmidt, Silke (2006):
Kundenwert aus Sicht von Versicherungsunternehmen: Analyse der Determinanten und Umsetzungsmöglichkeiten in anwendungsorientierten Modellen,Wiesbaden: Deutscher Universitäts-Verlag, 2006

G

Gabler Wirtschaftslexikon (2000):
Gabler Wirtschafts-Lexikon, 15., vollständig überarbeitete und aktualisierte Auflage, Wiesbaden: Gabler Verlag, 2000

Gadde, Lars-Erik/Mattson, Lars-Gunnar (1987):
Stability and Change in Network Relationships, erschienen in: International Journal of Marketing Research, 4. Jg., 1987, Heft 1, S. 29-41

Galasso, Gianpierro (1999):
Retention Marketing im Private Marketing: Theoretische und empirische Analyse des Kundenbeziehungsmarketings im schweizerischen Private Banking, Bern/Stuttgart/Wien: Haupt Verlag, 1999

Gälweiler, Aloys (1987):
Strategische Unternehmensführung, Frankfurt/New York: Campus Verlag, 1987

Gawlik, Helmut/Rathmann, Birgit (2008):
Regionalprinzip und Internet: Kein Widerspruch, erschienen in: Bank und Markt, o. Jg., 2008, Heft 9, S. 18-22

Geilmann-Ebbert, Andreas/Heine, Stefan (2006):
Kreditrisikotransfer im genossenschaftlichen Finanzverbund, erschienen in: Zeitschrift für das gesamte Kreditwesen (ZfgK), 59. Jg., Heft 21, S. 1147-1150

Gelbrich, Katja (2001):
Kundenwert: Wertorientierte Akquisition von Kunden im Automobilbereich, Göttingen: Cuvillier Verlag, 2001

Gelbrich, Katja/Müller, Stefan (2006):
Kundenwert: Hintergrund, Konzeptualisierung und Messmethoden, in: Wertorientiertes Management: Werterhaltung, Wertsteuerung, Wertsteigerung ganzheitlich gestalten, hrsg. von Schweickart, Nikolaus/Töpfer, Armin, Berlin/Heidelberg/New York: Springer Verlag, S. 448-487

Gemünden, Hans Georg (1993):
Information: Bedarf, Analyse, Verhalten, in: Handwörterbuch der Betriebswirtschaft, 5., völlig neu gestaltete Auflage, hrsg. von Wittmann, Waldemar/Kern, Werner/Köhler, Richard/Küpper, Hans-Ulrich/von Wysocki, Klaus, Schäffer-Poeschel, 1993, Sp. 1725-1735

Gensler, Sonja/Skiera, Bernd/Böhm, Martin (2007):
Einfluss der Nutzung des Online-Banking auf das Produktnutzungsverhalten und die Profitabilität von Bankkunden, erschienen in: Zeitschrift für Betriebswirtschaft (ZfB), 77. Jg., 2007, Heft 5, S. 675-695

Georgi, Dominik/Hadwich, Karsten (Hrsg.) (2010):
Management von Kundenbeziehungen: Perspektiven, Analysen, Strategien, Instrumente, Wiesbaden: Gabler Verlag, 2010

Gerstenmaier, Volker (1992):
Zukunftsperspektiven der Kreditgenossenschaften unter besonderer Berücksichtigung der Mitgliederförderung, Frankfurt am Main: Peter Lang Verlag, 1992

Geschwandtner, Marcus/Kuttler, Hans-Jörg (2009):
Mitgliederzweckgeschäfte: kein gesetzlich verordneter Ballast, sondern einzigartiger Unternehmenszweck, erschienen in: Zeitschrift für das gesamte Kreditwesen (ZfgK), 62. Jg., 2009, Heft 6, S. 264-268

Gierl, Heribert/Kurbel, Thomas (1997):
Möglichkeiten zur Ermittlung des Kundenwertes, in: Database Marketing, hrsg. von Link, Jörg/Brändli, Dieter/Schleuning, Christian/Kehl, Roger E., Handbuch Database Marketing, 2., korrigierte Auflage, Ettlingen: IM-Fachverlag Marketing-Forum, 1997, S. 175-189

Gleißner, Werner (2005):
Wertorientierte Unternehmensführung - Kundenmanagement und Vertrieb einer Versicherungsgesellschaft, in: Integriertes Risiko- und Ertragsmanagement: Kunden- und Unternehmenswert zwischen Risiko und Ertrag, hrsg. von Keuper, Frank/Roesing, Dirk/Schomann, Marc, Wiesbaden: Gabler Verlag, 2005, S. 407-437

Gleißner, Werner/Jahn, Clemens (2005):
Kundenwert als Maßstab für die Vertriebssteuerung, in: Bankstrategien im Firmenkundengeschäft: Konzeption, Management, Dimensionen hrsg. von Börner, Christoph J./Maser, Harald/Schulz, Thomas Christian, Wiesbaden: Gabler Verlag, 2005, S. 147-165

Gloger, Klaus/Lehmann, Stefan (2008):
Aufgedeckt, erschienen in: Bankinformation, 35. Jg., 2008, Heft 8, S. 64-66

Götz, Peter/Diller, Hermann (1991):
Die Kundenportfolio-Analyse: Ein Instrument zur Steuerung von Kundenbeziehungen, Arbeitspapier Nr. 1 des Lehrstuhl für Marketing an der Universität Erlangen-Nürnberg, 1991

Gouthier, Matthias H. J. (2003):
Kundenentwicklung im Dienstleistungsbereich: Wiesbaden: Deutscher Universitätsverlag, 2003

Grauer, Frank (1998):
Mitgliederinteressen und Managementverhalten: Ausgewählte Konfliktfelder im genossenschaftlichen Marketing- und Finanzierungsbereich - Helmut Wagner zum Gedächtnis, in: Freiheit und Bindung als Grundlage der marktwirtschaftlichen und demokratischen Ordnung: 50 Jahre Institut für Genossenschaftswesen der Westfälischen Wilhelms-Universität Münster, hrsg. von Jäger, Wilhelm, Münster: Regensberg Verlag, 1998, S. 191-209

Grob, Heinz Lothar (1989):
Investitionsrechnung mit vollständigen Finanzplänen, München: Verlag Franz Vahlen, 1987

Grob, Heinz-Lothar (1999):
Einführung in die Investitionsrechnung, 3., vollständig überarbeitete und erweiterte Auflage, München: Verlag Vahlen, 1999

Grochla, Erwin (Hrsg.):
Handwörterbuch der Organisation, 2. völlig neu gestaltete Auflage, Stuttgart: Schäffer-Poeschel, 1980

Grochla, Erwin/Wittmann, Waldemar (Hrsg.) (1976):
Handwörterbuch der Betriebswirtschaft, Band 3, 4., völlig neu gestaltete Auflage, Stuttgart: Schäffer-Poeschel, 1976

Groß, Thomas/Michaelis, Holger (2002):
Wertschöpfungskettenmanagement im Firmenkundengeschäft der Bank, in: Handbuch Firmenkundengeschäft: Technologie, Rating und Risikosteuerung als Kernkompetenz der Banken, 2., völlig überarbeitete Auflage, hrsg. von Juncker, Klaus /Priewasser, Erich, Frankfurt am Main: Fritz Knapp Verlag, 2002, S. 163-176

Großfeld, Bernhard/Jäger, Wilhelm/Lenfers, Guido (1989):
Tradition und Zukunft im Genossenschaftsrecht, in: Die Genossenschaft im Spiegel des Rechts: Ansprachen auf dem Akademischen Festakt "100 Jahre Genossenschaftsgesetz" am 12. Juni 1989 im Schloß und im Rathaus zu Münster, hrsg. von Bonus, Holger/Grossfeld, Bernhard/Jäger, Wilhelm, Münster: Regensberg Verlag, 1989, S. 73-101

Grosskopf, Werner (1990):
Der Förderungsauftrag moderner Genossenschaftsbanken und seine Umsetzung in der Praxis, Frankfurt am Main: Fritz Knapp Verlag, 1990

Grosskopf, Werner (2001):
Kapitalverfügbarkeit in genossenschaftlichen Unternehmen, in: Die Zukunft der Genossenschaft in der Europäischen Union an der Schwelle zum 21. Jahrhundert, hrsg. von Harbrecht, Wolfgang, Nürnberg, 2001, S. 239-250

Günter, Bernd (2006): Kundenwert – mehr als nur Erlös, in: Kundenwert: Grundlagen, Innovative Konzepte, Praktische Umsetzungen, 3., überarbeitete und erweiterte Auflage, hrsg. von Günter, Bernd/Helm, Sabrina, Wiesbaden: Gabler Verlag, 2006, S. 241-265

Günter, Bernd/Helm, Sabrina (2003):
Die Beendigung von Geschäftsbeziehungen aus Anbietersicht, in: Relationship Marketing: Standortbestimmung und Perspektiven, hrsg. von Rese, Mario/Söllner, Albrecht/Utzig, Peter, Berlin et al.: Springer Verlag, 2003, S. 45-70

Günter, Bernd/Helm, Sabrina (Hrsg.) (2003):
Kundenwert: Grundlagen, Innovative Konzepte, Praktische Umsetzungen, 2., überarbeitete und erweiterte Auflage, Wiesbaden: Gabler Verlag, 2003

Günter, Bernd/Helm, Sabrina (Hrsg.) (2006):
Kundenwert: Grundlagen, Innovative Konzepte, Praktische Umsetzungen, 3., überarbeitete und erweiterte Auflage, Wiesbaden: Gabler Verlag, 2006

Gupta, Sunil/Lehmann, Donald R. (2003):
Customers as Assets, erschienen in: Journal of Interactive Marketing, 17. Jg., 2003, Heft 1, S. 9-24

Gupta, Sunil/Lehmann, Donald R./Stuart, Jennifer Ames (2004):
Valuing Customers, erschienen in: Journal of Marketing Research, 41. Jg., 2004, Heft 1, S. 7-18

H

Hagemann, H. (1986):
Lebenszyklus-Management: Konzept zur Ausschöpfung des Finanzdienstleistungsbedarfs von Privatkunden, in: Planung und Prognose in Dienstleistungsunternehmen, hrsg. von Hammer, Gerald/Kremer, Egon/Schwebler, Robert/Zink, Achim, Karlsruhe: Verlag Versicherungswirtschaft, 1986, S. 1-21

Hagenloch, Thorsten (2008):
Effizienzberechnung mit der Data Envelopement Analysis (I), erschienen in: WISU - Das Wirtschaftsstudium, 37. Jg., 2008, Heft 10, S. 1372-1382

Hahn, Oswald (1980):
Die Unternehmensphilosophie einer Genossenschaftsbank: [stark erweiterte Fassung eines Vortrages vor dem Arbeitskreis Westfälischer Genossenschaften, Münster, 7. August 1979], Tübingen: Mohr Siebeck Verlag, 1980

Hamel, Winfried (2006):
Kundenwertorientierte Anreizsysteme, in: Kundenwert, Grundlagen, Innovative Konzepte, Praktische Umsetzungen, 3., überarbeitete und erweiterte Auflage, hrsg. von Günter, Bernd/Helm, Sabrina, Wiesbaden: Gabler Verlag, 2006, S. 461-482

Hamer, Eberhard (1987):
Das mittelständische Unternehmen: Eigenarten, Bedeutung, Risiken und Chancen, Stuttgart: Verlag Horst Poller, 1987

Hammer, Gerald/Kremer, Egon/Schwebler, Robert/Zink, Achim (Hrsg.) (1986):
Planung und Prognose in Dienstleistungsunternehmen, Karlsruhe: Verlag Versicherungswirtschaft, 1986

Hammerschmidt, Michael (2000):
Mitgliedschaft als Alleinstellungsmerkmal für Kreditgenossenschaften: Empirische Ergebnisse und Handlungsvorschläge, Arbeitspapier des Instituts für Genossenschaftswesen der Westfälischen Wilhelms-Universität Münster, Nr. 17, 2000

Hammerschmidt, Michael/Hellinger, Carsten (1998):
Mitgliedschaft als Instrument der Kundenbindung in Genossenschaftsbanken, Arbeitspapier des Instituts für Genossenschaftswesen der Westfälischen Wilhelms-Universität Münster, Nr. 15, 1998

Hampl, Herbert (1999):
Ausfallrisiken im mittelständischen Firmenkundenkreditgeschäft, erschienen in: Finanzierung Leasing Factoring, 46. Jg., 1999, Heft 1, S. 17-21

Hanrath, Stefanie/Weber, Heinz-Otto (2008):
Strategien zur Mitgliederförderung in Genossenschaftsbanken: eine Soll-Ist-Analyse, erschienen in: Zeitschrift für das gesamte Genossenschaftswesen (ZfgG), 58. Jg., 2008, Heft 4, S. 248-259

Hansen, Ursula (2006):
Beziehungslos im Dschungel des Beziehungsmarketing: Grenzen des Beziehungsmarketing aus Verbraucherperspektive, in: Grundlagen des CRM: Konzepte und Gestaltung. 2., überarbeitete und erweiterte Auflage, hrsg. von Wilde, Klaus D./Hippner, Hajo, Wiesbaden: Gabler Verlag, 2006, S. 146-166

Harbrecht, Wolfgang (2005):
Kundenorientierte Mitgliederförderung durch Wettbewerb im Genossenschaftssektor, erschienen in: Zeitschrift für das gesamte Genossenschaftswesen (ZfgG), 55. Jg., 2005, Heft 2, S. 99-113

Harbrecht, Wolfgang (Hrsg.) (2001):
Die Zukunft der Genossenschaft in der Europäischen Union an der Schwelle zum 21. Jahrhundert Bericht der 14. Internationalen Genossenschaftswissenschaftlichen Tagung vom 5. bis 7. September 2000 in Nürnberg, Nürnberg, 2001

Harhoff, Dietmar/Körting, Timm (1998):
Lending Relationships in Germany – Empirical Evidence from Survey Data, erschienen in: Journal of Banking and Financing, 22. Jg., 1998, Heft 10, S. 1317-1353

Harnischfeger, Uwe (1996):
Umziehen in ein House of Relations?, erschienen in: Absatzwirtschaft, o. Jg., 1996, Sondernummer Oktober 1996, S. 14-23

Hartmann-Wendels, Thomas/Pfingsten, Andreas/Weber, Martin (2010):
Bankbetriebslehre, 5., überarbeitete Auflage, Heidelberg et al.: Springer Verlag, 2010

Hax, Arnoldo C./Majluf, Nicolas S. (1996):
The Strategy Concept and Process, 2nd Edition, New Jersey: Prentice Hall, 1996

Heckl, Diana/Moormann, Jürgen (2005):
Modellierung von Geschäftsprozessen am Beispiel des mittelständischen Firmenkundengeschäfts, erschienen in: BIT – Banking and Information Technology, 6. Jg., Heft 3, S. 9-24

Heckt, Joachim G. (1977):
Fusionen von Primärgenossenschaften: Analyse zu einer Konfliktsituation und Ansätze zu ihrer Lösung, Dissertation, Hamburg, 1977

Heidemann, Julia/Kamprath, Nora/Görz, Quirin (2009):
Customer Lifetime Value: Entwicklungspfade, Einsatzpotenziale und Herausforderungen, erschienen in: Journal für Betriebswirtschaft, 59. Jg., 2009, Heft 4, S. 183-199

Heine, Jens Michael (2007):
zeb/-Firmenkundenstudie: Ergebnisse und Handlungsfelder, Münster: zeb/rolfes.schierenbeck.associates, 2007

Helm, Sabrina (2003):
Calculating the Value of Customers' Referrals, erschienen in: Managing Service Quality, 13. Jg., 2003, Heft 2, S. 124-133

Helm, Sabrina (2006):
Der Wert der Kundenbeziehungen aus der Perspektive des Transaktionskostenansatzes, in: Kundenwert, Grundlagen, Innovative Konzepte, Praktische Umsetzungen, 3., überarbeitete und erweiterte Auflage, hrsg. von Günter, Bernd/Helm, Sabrina, Wiesbaden: Gabler Verlag, 2006, S. 103-124

Helm, Sabrina/Günter, Bernd (2006):
Kundenwert - eine Einführung in die theoretischen und praktischen Herausforderungen der Bewertung von Kundenbeziehungen, in: Kundenwert, Grundlagen, Innovative Konzepte, Praktische Umsetzungen, 3., überarbeitete und erweiterte Auflage, hrsg. von Günter, Bernd/Helm, Sabrina, Wiesbaden: Gabler Verlag, 2006, S. 3-40

Hempell, Hanna S. (2002):
Testing for Competition Among German Banks, Discussion Paper 04/02: January 2002, Frankfurt am Main: Economic Research Center of the Deutsche Bundesbank, 2002

Hempelmann, Bernd/Lürwer, Martin (2003):
Der Customer Lifetime Value-Ansatz zur Bestimmung des Kundenwertes, erschienen in: WISU - Das Wirtschaftsstudium, 32. Jg., 2003, Heft 3, S. 336-341

Henke, Klaus (1995):
Vertriebssteuerung und Vertriebscontrolling in Kreditinstituten, in: Handbuch Finanzvertrieb: Vertriebsstrategien, Vertriebswege, Vertriebsmanagement, hrsg. von Betsch, Oskar/Wiechers, Rüdiger, Frankfurt am Main: Fritz Knapp Verlag, 1995, S. 757-774

Hensel, Volkmar/Wild, Christian (2003):
Vorsteuerung und Preisuntergrenzen in VR-Control, erschienen in: Bankinformation, 30. Jg., 2003, Heft 1, S. 35-40

Henseler, Jörg,/Hoffmann, Thorsten (2003):
Kundenwert als Baustein zum Unternehmenswert, Hamburg: Verlag Dr. Kovač, 2003

Hepp, Christina (2008):
Fehler- und Fehlerfolgekosten in Banken: Messung und Steuerung der internen Dienstleistungsqualität, Wiesbaden: Gabler Verlag, 2008

Heppelmann, Stefan/Middelmann, M. (2006):
Kundenorientiertes Vertriebsmanagement für Finanzdienstleister: Profitabel wachsen statt effizient schrumpfen, Stern Stewart & Co. in Practice No. 3, München: Stern Stewart & Co., 2006

Herrmann, Andreas/Homburg, Christian (Hrsg.) (2000):
Marktforschung, 2., aktualisierte Auflage, Wiesbaden: Gabler Verlag, 2000

Herrmann, Andreas/Jasny, Ralf/Vetter, Ingrid (Hrsg.) (1999):
Kundenorientierung von Banken: Strategie für Kundennähe und effektives Beziehungsmanagement, Frankfurt am Main: FAZ Verlagsbereich Buch, 1999

Herrmann, Andreas/Johnson, Michael D. (1999):
Die Kundenzufriedenheit als Bestimmungsfaktor der Kundenbindung, erschienen in: Zeitschrift für betriebswirtschaftliche Forschung (ZfbF), 51. Jg., Heft 6, S. 579-598

Herrmann, Jochen/Lüders, Uwe (2004):
Den Anforderungen an die integrierte Banksteuerung gerecht werden, erschienen in: Bankinformation, 31. Jg., 2004, Heft 4, S. 48-52

Heskett, James L./Jones, Thomas O./Loveman, Gary W./Sasser, W. Earl./Schlesinger, Leonard A. (1994):
Putting the Service-Profit Chain to Work, erschienen in: Harvard Business Review, 72. Jg., March-April 1994, S. 164-174

Hettrich, Eduard/Pöhlmann, Peter (2001):
Genossenschaftsgesetz: Kommentar zu dem Gesetz betreffend die Erwerbs- und Wirtschaftsgenossenschaften und zu umwandlungsrechtlichen Vorschriften für Genossenschaften, 2., neubearbeitete und erweiterte Auflage, München: Beck Verlag, 2001

Heyd, Reiner/Beyer, Michael (2007):
Bilanzielles Eigenkapital von Kreditgenossenschaften: Eine Bestandsaufnahme zur gegenwärtigen Kapitalabgrenzung sowie zu den laufenden Projekten, erschienen in: Zeitschrift für internationale Rechnungslegung (IRZ), 2. Jg., 2007, Heft 1, S. 53-59

Hildner, Rolf (2003):
Rede von Herrn Rolf Hildner (Vorsitzender des Verbandsrats) zur Verabschiedung Verbandspräsident Klaus Lambert, Online im Internet, URL: http://217.195.36.211/gvf_online/gvfdlc.nsf/0/F4B41BF79B6A167CC12573560023676B/$file/Re_RH.pdf (Stand: 24.08.2010)

Hilgenstock, G. Peter (1995):
Dezentrales Controlling als Voraussetzung für den Marketingerfolg im Bankbetrieb, in: Handbuch Finanzvertrieb: Vertriebsstrategien, Vertriebswege, Vertriebsmanagement, hrsg. von Betsch, Oskar/Wiechers, Rüdiger, Frankfurt am Main: Fritz Knapp Verlag, 1995, S. 800-822

Hinterhuber, Hans. H. (1997):
Strategische Unternehmensführung, Band 2: Strategisches Handeln: Direktiven, Organisation, Umsetzung, Unternehmungskultur, strategisches Controlling, strategische Führungskompetenz, 6., völlig neubearbeitete Auflage, Berlin/New York: de Gruyter, 1997

Hinterhuber/Hans H./Matzler, Kurt (Hrsg.) (2009):
Kundenorientierte Unternehmensführung: Kundenorientierung, Kundenzufriedenheit, Kundenbindung, 6., überarbeitete Auflage, Wiesbaden: Gabler Verlag, 2009

Hippner, Hajo (2004):
CRM: Grundlagen, Ziele und Konzepte, in: CRM: Grundlagen und Konzepte, hrsg. von Hippner, Hajo/Wilde, Klaus D., Wiesbaden: Gabler Verlag, 2004, S. 13-42

Hoberg, Peter (2007):
Kundenwert, erschienen in: WISU - Das Wirtschaftsstudium, 37. Jg., 2007, Heft 10, S. 1248

Hoekstra, Janny C./Huizingh, Eelko K. R. E. (1999):
The Lifetime Value Concept in Customer-Based Marketing, erschienen in: Journal of Market Focused Management, 3. Jg., 1999, Heft 3, S. 257-274

Hofbauer, Günter/Hellwig, Claudia (2009):
Professionelles Vertriebsmanagement: der prozessorientierte Ansatz aus Anbieter- und Beschaffersicht, 2., aktualisierte und erweiterte Auflage, Erlangen: Publicis Verlag, 2009

Hoffmann, Werner/Klien, Wolfgang/Unger, Martin (1996):
Strategieplanung, in: Controlling, 2., überarbeitete und erweiterte Auflage, hrsg. von Eschenbach, Rolf, Stuttgart: Schäffer-Poeschel, 1996, S. 211-213

Hofinger, Hans/ van Husen, Rainer (2004):
Eigenkapitalaufbringungsinstrumente bei Genossenschaften, erschienen in: Zeitschrift für das gesamte Genossenschaftswesen (ZfgG), 54. Jg., 2004, Heft 2, S. 90-99

Hofmann, Christian/Daugart, Jan (2004):
Bereichs- und unternehmensbezogene Performancemaße zur Koordination und Steuerung von Bereichsleitern – eine agencytheoretische Analyse, in: Controlling: Theorie und Konzeptionen, hrsg. von Scherm, Ewald/Pietsch, Gotthard, München: Verlag Franz Vahlen, 2004, S. 191-214

Hofmann, Gerhard (2010):
Neue Finanzarchitektur, Reform der Aufsichtsregeln und Aufsichtspraktiken in Europa: Sind wir auf dem richtigen Weg?, erschienen in: Zeitschrift für das gesamte Kreditwesen (ZfgK), 63. Jg., 2010, Heft 13, S. 664-667

Hogan, John E./Lehmann, Donald R./Merino, M./Srivastava, Rajendra K./Thomas, Jacquelyn S./Verhoef, Peter C. (2002):
Linking Customer Assets to Financial Performance, erschienen in: Journal of Service Research, 5. Jg., 2002, Heft 1, S. 26-38

Hogan, John E./Lemon, Katherine N./Libai, Barak (2003):
What ist the True Value of a Lost Customer?, erschienen in: Journal of Service Research, 5. Jg., 2003, Heft 3, S. 196-208

Hölzer, Klaus (2006):
Das mittelständische Firmenkundengeschäft der Banken: eine gesamteuropäische Perspektive, erschienen in: BIT – Banking and Information Technology, 7. Jg., 2006, Heft 1, S. 47-54

Hölzer, Klaus/Schnarr, Thomas (2010):
Neue Spielregeln im Firmenkundengeschäft, erschienen in: Die Bank, o. Jg., 2010, Heft 8, S. 54-57

Homann-Wenig, Sabine (2008):
Ansätze einer ertrags- und risikoorientierten Vertriebssteuerung für das Firmenkundengeschäft, in: Erfolgreiche Vertriebsstrategien im Firmenkundengeschäft: Konzepte, Praxisberichte, Lösungen, hrsg. von Schmoll, Anton, Köln: bank-verlag medien, 2008, S. 217–238

Homburg, Christian (2007):
Kundenprofitabilitätsrechnung als Aufgabe des Marketingcontrolling, in: Vielfalt und Einheit in der Marketingwissenschaft: Ein Spannungsverhältnis – Hans-H. Bauer zum 60. Geburtstag, hrsg. von Báyon, Tomás/Herrmann, Andreas/Huber, Frank, Wiesbaden: Gabler Verlag, 2007, S. 398-418

Homburg, Christian/Artz, Martin/Wieseke, Jan/Schenk, Bernhard (2008):
Gestaltung und Erfolgsauswirkungen der Absatzplanung: Eine branchenübergreifende empirische Analyse, erschienen in: Zeitschrift für betriebswirtschaftliche Forschung (ZfbF), 60. Jg., 2008, Heft November 2008, S. 634-670

Homburg, Christian/Beutin, Nikolas (2001):
Kundenstrukturmanagement als Controllingherausforderung, in: Marketing-Controlling: Marketing als Motor von Wachstum und Erfolg, hrsg. von Reinecke, Sven/Tomczak, Torsten/Geis, Gerold, Frankfurt/Wien: Wirtschaftsverlag Ueberreuter, 2001, S. 212-233

Homburg, Christian/Droll, Mathias/Totzek, Dirk (2008):
Customer Priorization: Does it pay off, and how should it be implemented?, erschienen in: Journal of Marketing, 72. Jg., 2008, Heft 5, S. 110-130

Homburg, Christian/Giering, Annette/Hentschel, Frederike (1999):
Der Zusammenhang zwischen Kundenzufriedenheit und Kundenbindung, in: Handbuch Kundenbindungsmanagement: Grundlagen, Konzepte, Erfahrungen, 2. Auflage, hrsg. Von Bruhn, Manfred/Homburg, Christian, Wiesbaden: Gabler Verlag, 2000, S. 81-112

Homburg, Christian/Hoyer, Wayne D./Koschate, Nicole (2005):
Customers' Reactions to Price Increase: Do Customer Satisfaction and Perceived Fairness matter?, erschienen in: Journal of the Academy of Marketing Science, 33. Jg., 2005, Heft 1, S. 36-49

Homburg, Christian/Jensen, Ove (2000):
Kundenorientierte Vergütungssysteme: Voraussetzungen, Verbreitung, Determinanten, erschienen in: Zeitschrift für Betriebswirtschaft (ZfB), 70. Jg., 2000, Heft 1, S. 55-74

Homburg, Christian/Pflesser, Christian (2000): Konfirmatorische Faktorenanalyse, in: Marktforschung, 2., aktualisierte Auflage, hrsg. von Herrmann, Andreas/Homburg, Christian, Wiesbaden: Gabler Verlag, 2000, S. 413-438

Homburg, Christian/Schäfer, Heiko (2000):
Cross-Selling: Aus der Kundenbeziehung mehr herausholen, erschienen in: Harvard Business Manager, 22. Jg., 2000, Heft 6, S. 35-44

Homburg, Christian/Schäfer, Heiko (2006):
Die Erschließung von Kundenwertpotenzialen durch Cross-Selling, in: Kundenwert: Grundlagen, Innovative Konzepte, Praktische Umsetzungen, 3., überarbeitete und erweiterte Auflage, hrsg. von Günter, Bernd/Helm, Sabrina, Wiesbaden: Gabler Verlag, 2006, S. 159-181

Homburg, Christian/Schäfer, Heiko/Schneider, Janna (2010):
Sales Excellence: Vertriebsmanagement mit System, 6., überarbeitete und erweiterte Auflage, Wiesbaden: Gabler Verlag, 2010

Homburg, Christian/Schnurr, Phoebe (1998):
Kundenwert als Instrument der wertorientierten Steuerung, in: Wertorientierte Unternehmensführung: Perspektiven und Handlungsfelder für die Wertsteigerung von Unternehmen, hrsg. von Bruhn, Manfred/Lusti, Markus/Müller, Werner R./Schierenbeck, Henner/Studer, Tobias, Wiesbaden: Gabler Verlag ,1998 ,S. 169-189

Homburg, Christian/Steiner, Viviana V./Totzek, Dirk (2009):
Managing Dynamics in a Customer Portfolio, erschienen in: Journal of Marketing, 73. Jg., 2009, Heft 5, S. 70-89

Homburg, Christian/Weber, Jürgen/Karlshaus, Jan Thido/Aust, René (2000):
Interne Kundenorientierung der Kostenrechnung, erschienen in: Die Betriebswirtschaft, 60. Jg., 2000, Heft 2, S. 241-256

Honold-Reichert, Thomas (2002):
Portfolio und Balanced Scorecard: Instrumente für das strategische Management in Genossenschaften. Arbeitspapiere der Forschungsstelle für Genossenschaftswesen an der Universität Hohenheim, Band 17, Stuttgart-Hohenheim 2002

Hopfenbeck, Waldemar (1997):
Allgemeine Betriebswirtschafts- und Managementlehre: das Unternehmen im Spannungsfeld zwischen ökonomischen, sozialen und ökologischen Interessen, 11. Auflage, Landsberg/Lech: Verlag Moderne Industrie, 1997

Hoppen, Dieter (1999):
Vertriebsmanagement: Steuerung des Firmenkundengeschäfts im Inland und im Export: Lehrbuch und Nachschlagewerk, München/Wien: Oldenbourg Verlag, 1999

Horváth, Peter (2003):
Controlling, 9., vollständig überarbeitete Auflage, München: Verlag Franz Vahlen, 2003

Howald, Bettina (2007):
Kundenwert im Private Banking: Eine Analyse der Einflussfaktoren und der Wirkungszusammenhänge, Bern/Stuttgart/Wien: Haupt Verlag, 2007

Huth, Hans-Olaf (1985):
Ertragsorientiertes Zielgruppenmarketing im mittelständischen Firmenkundenbereich von Universalbanken, Frankfurt am Main: Fritz Knapp Verlag, 1985

I

IfM (2009):
Schlüsselzahlen des Mittelstands in Deutschland 2007/2008, Bonn: Institut für Mittelstandsforschung, Online im Internet, URL: http://www.ifm-bonn.de/index.php?id=99 (Stand: 24.09.2009)

International Monetary Fund (2003):
IMF Country Report No. 03/343: Germany: Financial System Stability Assessment, Washington D.C., International Monetary Fund, 2003

J

Jacobs, Bernd/Krauß, Carsten (2005):
Vertrieb, Vertrieb und noch mal Vertrieb, erschienen in: Bankinformation, 32. Jg., 2005, Heft 7, S. 24-27

Jäger, Wilhelm (Hrsg.) (1998):
Freiheit und Bindung als Grundlage der marktwirtschaftlichen und demokratischen Ordnung: 50 Jahre Institut für Genossenschaftswesen der Westfälischen Wilhelms-Universität Münster, Münster: Regensberg Verlag, 1998

Jain, Diapak/Singh, Siddartha S. (2002):
Customer Lifetime Value Research in Marketing: A Review and Future Directions, erschienen in: Journal of Interactive Marketing, 16. Jg., 2002, Heft 2, S. 34-46

Jansen, Sven (2009):
Das mittelständische Firmenkundengeschäft deutscher Kreditinstitute im Spannungsfeld der Finanzmarktkrise: zeb/-Firmenkundenstudie 2009, Vortrag auf dem Düsseldorfer Finanz-Forum, Düsseldorf: zeb/rolfes.schierenbeck.associates, 2009

Juncker, Klaus (1979):
Marketing im Firmenkundengeschäft: Frankfurt am Main: Fritz Knapp Verlag, 1979

Juncker, Klaus (2006):
Kundenwertmanagement: Der Schlüssel zum Erfolg im Firmengeschäft, in: Der Bankensektor im Wandel: Festschrift zum 65. Geburtstag von Professor Erich Priewasser, hrsg. von Emmerich, Norbert/Roßbach, Peter, Frankfurt am Main: Fritz Knapp Verlag, S. 343-354

Juncker, Klaus/Priewasser, Erich (Hrsg.) (2002):
Handbuch Firmenkundengeschäft: Technologie, Rating und Risikosteuerung als Kernkompetenz der Banken, 2., völlig überarbeitete Auflage, Frankfurt am Main: Fritz Knapp Verlag, 2002

K

Kalefeld, Klaus (2002):
Ein Verbund - eine Strategie, erschienen in: Bankinformation, 29. Jg., 2002, Sonderheft 1, S. 44-45

Kaiser, Annette/Kehr, Joachim (2003):
Ziele setzen, Wandel fördern, erschienen in: Bankinformation, 30. Jg., 2003, Heft 9, S. 30-35

Kamakura, Wagner A. et al. (2005):
Choice Models and Customer Relationship Management, erschienen in: Marketing Letters, 16. Jg., 2005, Heft 3/4, S. 279-291

Kamakura, Wagner A./Ramaswami, Sridhar N./Srivastava Rajendra K. (1991):
Applying Trait Analysis in the Evaluation of Prospects for Cross-Selling of Financial Services, erschienen in: International Journal of Research in Marketing, 8. Jg., 1991, Heft 4, S. 329-349

Kaminsky, Stefan (1955):
Die Kosten- und Erfolgsrechnung der Kreditinstitute: eine theoretische, systematische und verfahrenstechnische Untersuchung, 2., verbesserte Auflage, Meisenheim am Glan: Hain, 1955

Kammlott, Christian/Schiereck, Dirk (2000):
Wachstum, Förderungsauftrag und Markterfolg von deutschen Kreditgenossenschaften, erschienen in: Zeitschrift für das gesamte Genossenschaftswesen (ZfgG), 50. Jg., 2000, Heft 4, S. 265-280

Kano, Noriaki (1984):
Attractive Quality and Must-be Quality, erschienen in: Hinshitsu: The Journal of the Japanese Society for Quality Control, 14. Jg., 1984, Heft 4, S. 39-48

Kaplan, Robert S./Narayan, V. G. (2001):
Measuring and Managing Customer Profitability, erschienen in: Journal of Cost Management, 15. Jg., 2001, Heft 5, S. 5-15

Kaplan, Robert S./Norton, David P. (1997):
Balanced Scorecard: Strategien erfolgreich umsetzen, aus dem Amerikanischen von Péter Horváth, Stuttgart: Schäffer-Poeschel, 1997

Karlowitsch, Martin (1997)
Balanced Scorecard, erschienen in: WISU - Das Wirtschaftsstudium, 26. Jg., 1997, Heft 12, S. 1131

Käser, Burkhard (2009):
Vorstellung der zeb/-Firmenkundenstudie, München: zeb/rolfes.schierenbeck.associates, 2009

Käser, Burkhard/Ritzenhofen, Andreas (2009):
Erfolgsfaktoren im Firmenkundengeschäft - Vorstellung der Ergebnisse der zeb/-Firmenkundenstudie 2009, Frankfurt am Main: zeb/rolfes.schierenbeck.associates, 2009

Kasselmann, Joachim/Melter, Oliver (2006):
Mehr Zeit für den Kunden - mehr Umsatz für die Bank, erschienen in: Bankmagazin, 55. Jg., 2006, Heft 9, S. 24-26

Kasten, Lars/Reents, Börchert (o. J.):
Strategische Geschäftsfeldanalyse: Empirische Analyse von Marktattraktivität und Wettbewerbsintensität im Privat- und Firmenkundengeschäft, Hannover: GRZ Genossenschaftsrechenzentrale Norddeutschland und GVBH Genossenschaftsverband Berlin-Hannover, o. J.

Kaufmann, Lutz (1997):
ZP-Stichwort: Balanced Scorecard, erschienen in: Zeitschrift für Planung, 8. Jg., 1997, S. 421-428

Keuper, Frank/Roesing, Dirk/Schomann, Marc (Hrsg.) (2005):
Integriertes Risiko- und Ertragsmanagement: Kunden- und Unternehmenswert zwischen Risiko und Ertrag, Wiesbaden: Gabler Verlag, 2005

KfW Bankengruppe (Hrsg.) (2007):
KfW-Research: Mittelstands- und Strukturpolitik, Ausgabe 38, Frankfurt am Main: KfW Bankengruppe, 2007

Kieser, Alfred/Kubicek, Herbert (1992):
Organisation, 3., völlig neubearbeitete Auflage, Berlin et al.: de Gruyter, 1992

Kipker, Ingo/Schönhals, Frank (2002):
Strukturwandel im Firmenkundengeschäft, erschienen in: Geldinstitute, o. J., 2002, Heft 11, S. 16-20

Kiwitz, Katharina/Scheffler, Constanze (2009):
Methoden zur Messung dynamischer Kundenwerte: ein State-of-the-Art-Review, Arbeitspapier des Betriebswirtschaftlichen Instituts an der Universität Erlangen-Nürnberg, Nr. 165, Erlangen, 2009

Klein, Nicole (2009):
Die Balanced Scorecard als Basis einer Customer Care Scorecard zur Konzeption einer systematischen Kundenpflege, Tübinger Diskussionsbeitrag Nr. 322, Tübingen: Wirtschaftswissenschaftliche Fakultät der Eberhard-Karls-Universität Tübingen, 2009

Kleinaltenkamp, Michael/Dahlke, Beate (2006):
Der Wert eines Kunden als Informant – auf dem Weg zu einem „knowledge based customer value", in: Kundenwert: Grundlagen, Innovative Konzepte, Praktische Umsetzungen, 3., überarbeitete und erweiterte Auflage, hrsg. von Günter, bernd/Helm, Sabrina, Wiesbaden: Gabler Verlag, 2006

Kleinaltenkamp, Michael/Plinke, Wulff (Hrsg.) (1997):
Geschäftsbeziehungsmanagement, Berlin et al.: Springer Verlag, 1997

Kleinaltenkamp, Michael/Schweikart, Jörg (1998):
Controlling der Kundenintegration, in: Marketingcontrolling, hrsg. von Reinecke, Sven/Tomczak, Torsten/Dittrich, Sabine, St. Gallen: Verlag Thexis, 1998, S. 110-124

Klemm, Volker/Mönke, Reinhard (2009):
Wichtige Ansatzpunkte, erschienen in: Bankinformation, 36. Jg., 2009, Heft 3,
S. 48-52

Kluge, Arnd Holger (1991):
Geschichte der deutschen Bankgenossenschaften: Zur Entwicklung mitgliederorientierter Unternehmen, Frankfurt am Main: Fritz Knapp Verlag, 1991

Knappkötter, Ralf/Völkel, Hans-Georg (2002):
Mehr als ein Auto mieten, erschienen in: Bankinformation, 29. Jg., 2002, Heft 3, S. 14-16

Knöbel, Ulf (1997):
Kundenwertmanagement im Retailbanking: Kundenprofitabilitätsanalyse und Customer-Life-Cycle Costing am Beispiel einer Universalbank, Dissertation, St. Gallen: Universität St. Gallen, 1997

Köhler, Annette G./Rolvering, Heinrich-Stefan/Germann, Stephan (2005):
Zusammenhänge zwischen Vertriebssteuerungsmaßnahmen und Vertriebserfolg – erste empirische Ergebnisse aus dem Sparkassensektor, in: Dienstleistungscontrolling, hrsg. von Bruhn, Manfred/Stauss, Bernd, Wiesbaden: Gabler Verlag, 2005, S. 416-430

Kölker, Andreas (2010):
Potenzialmodelle – Konstrukt und Wirklichkeit, in: Wertorientierte Vertriebssteuerung in Banken und Sparkassen: Potenzialermittlung, Deckungsbeitragsmessung, Multikanal-Steuerung, Erfolgsabhängige Vergütung, hrsg. von Wimmer, Konrad, Heidelberg: Finanz Colloquium, 2010, S. 395-415

Koneberg, Markus (2006):
Management von Firmenkreditportfolios in Genossenschaftsbanken: Der Einsatz von Kreditderivaten zur Portfoliooptimierung, Stuttgart-Hohenheim: Forschungsstelle für Genossenschaftswesen an der Universität Hohenheim, 2006

König, Wolfgang/Beimborn, Daniel/Martin, Sebastian/Blumenberg, Stefan/Homann, Ulrich (2007):
Mittelständler und Banken: Einflussfaktoren der Bereitschaft zur Kooperation auf der Basis eingebetteter Informationssysteme, in: Architekturen und Prozesse: Strukturen und Dynamik in Forschung und Unternehmen, hrsg. von Loos, Peter/Krcmar, Helmut, Berlin/Heidelberg/New York: Springer Verlag, 2007, S. 133-150

Körnert, Jan (2003):
Balanced Scorecard: Theoretische Grundlagen und Perspektivenwahl für Kreditinstitute, Berlin: BMV – Berliner Wissenschaftsverlag, 2003

Kotler, Philip/Bliemel, Friedhelm (1999):
Marketing-Management: Analyse, Planung, Umsetzung und Steuerung, 9., überarbeitete und aktualisierte Auflage, Stuttgart: Schäffer-Poeschel, 1999

KPMG (2009):
Mythos Beratung – was „kleine" mittelständische Unternehmer von ihrer Bankbeziehung erwarten!: Ergebnisse einer Kundenbefragung zu Wahrnehmungen und Perspektiven im Retail-Firmenkundengeschäft, Frankfurt am Main: KPMG AG, 2009

Krafft, Manfred (1997):
Kundenzufriedenheit und Kundenwert: Ergebnisse der gleichnamigen Studie der VDI-Gesellschaft „Entwicklung Konstruktion Vertrieb" VDI-EKV und von CEO GmbH Krefeld, Kiel: Lehrstuhl für Marketing an der Universität Kiel, 1997

Krafft, Manfred (2007):
Kundenbindung und Kundenwert, 2., überarbeitete und erweiterte Auflage, Heidelberg: Physica Verlag, 2007

Krafft, Manfred/Albers, Sönke (2000):
Ansätze zur Segmentierung von Kunden - Wie geeignet sind herkömmliche Konzepte?, erschienen in: Zeitschrift für betriebswirtschaftliche Forschung (ZfbF), 52. Jg., 2000, Heft 6, S. 515-536

Krafft, Manfred/Rutsatz, Uwe (2006):
Konzepte zur Messung des ökonomischen Kundenwerts, in: Kundenwert, Grundlagen, Innovative Konzepte, Praktische Umsetzungen, 3., überarbeitete und erweiterte Auflage, hrsg. von Günter, Bernd/Helm, Sabrina, Wiesbaden: Gabler Verlag, 2006, S. 269-291

Kramer, Friedrich (2006):
Vorbild Industrie, erschienen in: Bankinformation, 33. Jg., 2006, Heft 8, S. 58-61

Krämer, Werner (2003):
Mittelstandsökonomik: Grundzüge einer umfassenden Analyse kleiner und mittlerer Unternehmen, München: Vahlen Verlag, 2003

Krammer, Christian K./Sinn, Walter/Weiß, Hans H./Wieandt, Alexander (2001):
Wertmanagement in Banken, erschienen in: Zeitschrift für Betriebswirtschaft (ZfB), 71. Jg., 2001, Heft 12, S. 1445-1457

Krapfel, Robert E./Salmond, Deborah/Spekman, Robert (1991):
A Strategic Approach to Managing Buyer-Seller-Relationships, erschienen in: European Journal of Marketing, 25. Jg., 1991, Heft 9, S. 22-37

Krauß, Carsten (2005):
Mehr PS im Vertrieb: Erfolgreiches Change Management im Firmenkundengeschäft (ostdeutscher) Kreditgenossenschaften, Berlin: Logos Verlag, 2005

Krebber, Marita/Melter, Oliver/Sträter, Jens (2006):
Vertriebsaktivierung im mittelständischen Firmenkundengeschäft, erschienen in: BIT- Banking and Information Technology, 7. Jg., 2006, Heft 1, S. 40-46

Kring, Thorn Ingo (2002):
Neue Strategien - neue Managementmethoden: Eine empirische Analyse zum Strategischen Management von Genossenschaftsbanken in Deutschland, Arbeitspapier des Instituts für Genossenschaftswesen der Westfälischen Wilhelms-Universität, Nr. 24, 2002

Kring, Thorn Ingo (2005):
Die Balanced Scorecard als Managementsystem für Banken: Handlungsempfehlungen auf Basis einer institutionenökonomischen Analyse von Managementdefiziten in Genossenschaftsbanken, Aachen: Shaker Verlag, 2005

Krob, Bernhard (2007):
Die Bank marktorientiert steuern, erschienen in: Bankinformation, 34. Jg., 2007, Heft 11, S. 55-57

Krob, Bernhard/Zimmermann, Yvonne:
Dem Potenzial auf der Spur, erschienen in: Bankinformation, 34. Jg., 2007, Heft 4, S. 22-23

Krol, Florian (2009):
Wertorientierte Unternehmensführung im Mittelstand – Erste Ergebnisse einer empirischen Studie, Arbeitspapier des Lehrstuhls für Controlling an der Westfälischen Wilhelms-Universität Münster, Münster: Westfälische Wilhelms-Universität, 2009

Kroon, Gerhard (2009):
Messung und Steuerung von Kreditrisiken: Empirischer Befund und Handlungsempfehlungen, Wiesbaden: Gabler Verlag, 2009

Kroon, Gerhard/Pool, Frank (2002):
Professionelle Kundengeschäftssteuerung mit Hilfe der DB-III-Rechnung, erschienen in: Bankinformation, 29. Jg., 2002, Heft 4, S. 26-34

Krüger, Sabine Maria (1997):
Profitabilitätsorientierte Kundenbindung durch Zufriedenheitsmanagement: Kundenzufriedenheit und Kundenwert als Steuerungsgröße für die Kundenbindung in marktorientierten Dienstleistungsunternehmen, München: FGM-Verlag, 1997

Kumar, V./George, Morris (2007):
Measuring and maximizing Customer Equity: a Critical Analysis, erschienen in: Journal of the Academy of Marketing Science, 35. Jg., 2007, Heft 2, S. 151-171

Kumar, V./Ramani, Girish/Bohling, Timothy (2004):
Customer Lifetime Value Approaches and Best Practice Application, erschienen in: Journal of Interactive Marketing, 18. Jg., 2004, Heft 3, S. 60-72

Kunesch, Hermann (1996):
Besonderheiten des Controlling in Klein- und Mittelbetrieben, in: Controlling, 2., überarbeitete und erweiterte Auflage, hrsg. von Eschenbach Rolf, Stuttgart: Schäffer-Poeschel, 1996, S. 630-635

Küpker, Horst/Pudig, Andrea (2001):
Organisatorischer Aufbau der Gesamtbanksteuerung, in: Handbuch Gesamtbanksteuerung: Integration von Markt-, Kredit- und operationalen Risiken, hrsg. von Eller, Roland, Stuttgart: Schäffer-Poeschel, 2001, S. 571-590

Küpper, Hans-Ulrich (2001):
Controlling: Konzeption – Aufgaben – Instrumente, 3., überarbeitete und erweiterte Auflage, Stuttgart: Schäffer-Poeschel, 2003

Küpper, Hans-Ulrich/Wagenhofer, Alfred (Hrsg.) (2002):
Handwörterbuch Unternehmensrechnung und Controlling, 4., völlig neu gestaltete Auflage, Stuttgart: Schäffer-Poeschel, 2002

Kuß, Heike (2002):
VR-Control: der Soll-Ist-Vergleich, erschienen in: Bankinformation, 29. Jg., 2002, Heft 11, S. 34-36

L

Lam, Shun Yin/Shankar, Venkatesh/Erramilli, M. Krishna/Murthy, Bvsan (2004):
Customer Value, Satisfaction, Loyalty, and Switching Costs: An Illustration from a Business to Business Service Context, erschienen in: Journal of the Academy of Marketing Science, 32. Jg., 2004, Heft 3, S. 293-311

Lamprecht, Dirk/Donschen, Alexander (2006):
Der Nutzen des Member Value Reporting für Genossenschaftsbanken: eine ökonomische und juristische Analyse, Arbeitspapier des Instituts für Genossenschaftswesen der Westfälischen Wilhelms-Universität Münster, Nr. 58, 2005

Lang, Johannes/Weidmüller, Ludwig (2005):
Genossenschaftsgesetz: (Gesetz, betreffend die Erwerbs- und Wirtschaftsgenossenschaften) – mit Erläuterungen zum Umwandlungsgesetz: Kommentar, 34., völlig neu bearbeitete Auflage, Berlin: Verlag de Gruyter, 2005

Lang, Johannes/Weidmüller, Ludwig (Hrsg.) (2006):
Genossenschaftsgesetz: (Gesetz, betreffend die Erwerbs- und Wirtschaftsgenossenschaften): Kommentar, 35., neu bearbeitete Auflage, Berlin: Verlag de Gruyter, 2006

Lange, Thomas A./Schulze, Heiko (Hrsg.) (2005):
Wertmanagement in Banken: Festschrift zum 65. Geburtstag von Guido Eilenberger, Wiesbaden: Gabler Verlag, 2005

Langguth, Heike (1994):
Strategisches Controlling, Ludwigsburg/Berlin: Verlag Wissenschaft und Praxis, 1994

Lehmann, Stefan (2008):
Die "Richtigen" ansprechen, erschienen in: Bankinformation, 35. Jg., 2008, Heft 3, S. 64-65

Licht, Wolfgang (1980):
Die Beteiligungsfinanzierung der Kreditgenossenschaften, in: Veröffentlichungen des Forschungsinstituts für Genossenschaftswesen an der Universität Nürnberg, Erlangen, Nürnberg, 1980

Lichtsteiner, Hans/Schwarz, Peter (2008):
Genossenschaftsmanagement im Spannungsfeld zwischen Mission und Ökonomie, erschienen in: Zeitschrift für das gesamte Genossenschaftswesen (ZfgG), 58. Jg., 2008, Heft 3, S. 195-206

Link, Jörg/Brändli, Dieter/Schleuning, Christian/Kehr, Roger E. (Hrsg.) (1997):
Handbuch Database Marketing, 2., korrigierte Auflage, Ettlingen: IM-Fachverlag Marketing Forum, 1997

Link, Jörg/Hildebrand, Volker G. (1997):
Ausgewählte Konzepte der Kundenbewertung im Rahmen des Database Marketing, in: Handbuch Database Marketing, 2., korrigierte Auflage, hrsg. von Link, Jörg/Brändli, Dieter/Schleuning, Christian/Kehr, Roger E., Ettlingen: IM-Fachverlag Marketing Forum, 1997, S. 158-173

Lipfert, Helmut (1988):
Mitgliederförderndes Kooperations- und Konkurrenzmanagement in genossenschaftlichen Systemen, 2., unveränderte Auflage, Göttingen: Vandenhoeck & Ruprecht, 1988

Lippmann, H. (1992):
Kennzahlen und Steuerungssysteme, in: Erfolgreiches Verkaufs-Management, hrsg. von Brockmann, Günter M., Landsberg/Lech: Verlag Moderne Industrie, 1992,
S. 1-19

Lippmann, Ingo (2006):
Marekting, in: Corporate Banking: Zukunftsorientierte Strategien im Firmenkundengeschäft, 5., vollständig überarbeitete Auflage, hrsg. von Brost, Heike/Dahmen, Andreas/Lippmann, Ingo, Frankfurt am Main: Bankakademie Verlag, 2006

Lissautzki, Marius (2007):
Kundenwertorientierte Unternehmenssteuerung, Wiesbaden: Gabler Verlag, 2007

Loos, Peter/Krcmar, Helmut (Hrsg.) (2007):
Architekturen und Prozesse: Strukturen und Dynamik in Forschung und Unternehmen, Berlin/Heidelberg/New York: Springer, 2007

Lubitz, Karl-Joachim (1984):
Bankmarketing gegenüber mittelständischen Betrieben, Frankfurt am Main: Fritz Knapp Verlag, 1984

Lürig, Rolf (1985a):
Verkörperte Selbsthilfe - Einst und jetzt, in: Unternehmenspolitik von Genossenschaftsbanken: Neun Vorträge, hrsg. von Lürig, Rolf, Göttingen: Vandenhoeck & Ruprecht, 1985 ,S. 1-17

Lürig, Rolf (1985b):
Das Verhältnis der Volksbanken zu ihren Mitgliedern, in: Unternehmenspolitik von Genossenschaftsbanken: Neun Vorträge, hrsg. von Lürig, Rolf, Göttingen: Vandenhoeck & Ruprecht, 1985 ,S. 19-36

Lürig, Rolf (1985c):
Aktive Kreditpolitik genossenschaftlicher Banken, in: Unternehmenspolitik von Genossenschaftsbanken: Neun Vorträge, hrsg. von Lürig, Rolf, Göttingen: Vandenhoeck & Ruprecht, 1985 ,S. 37-52

Lürig, Rolf (Hrsg.) (1985):
Unternehmenspolitik von Genossenschaftsbanken: Neun Vorträge, Göttingen: Vandenhoeck & Ruprecht, 1985

Lütje, Sebastian (2009):
Kundenbeziehungsfähigkeit: Konzeptualisierung und Erfolgswirkung, Wiesbaden: Gabler Verlag, 2009

M

Macharzina, Klaus (2003):
Unternehmensführung: Das internationale Managementwissen: Konzepte, Methoden, Praxis, 4., grundlegend überarbeitete Auflage, Wiesbaden: Gabler Verlag, 2003

MacNeil, Ian. R. (1979):
The New Social Contract: An Inquiry into Modern Contractual Relations, New Haven/London: Yale Univ. Press, 1979

Malthouse, Edward C./Blattberg, Robert C. (2005):
Can we predict Customer Lifetime Value?, erschienen in: Journal of Interactive Marketing, 19. Jg., 2005, Heft 1, S. 2-16

Männel, Wolfgang (Hrsg.) (1992):
Handbuch Kostenrechnung, Wiesbaden: Gabler Verlag, 1992

Matzler, Kurt/Sauerwein, Elmar/Stark, Christian (2009):
Methoden zur Identifikation von Basis-, Leistungs- und Begeisterungsfaktoren, in: Kundenorientierte Unternehmensführung: Kundenorientierung, Kundenzufriedenheit, Kundenbindung, hrsg. von Hinterhuber, Hans H./Matzler, Kurt, Wiesbaden: Gabler Verlag, 2009, S. 320-344

Matzler, Kurt/Stahl, Heinz K./Hinterhuber, Hans H. (2009):
Die Customer-Based View der Unternehmung, in: Kundenorientierte Unternehmensführung: Kundenorientierung, Kundenzufriedenheit, Kundenbindung, hrsg. von Hinterhuber, Hans H./Matzler, Kurt, Wiesbaden: Gabler Verlag, 2009, S. 3-31

Meffert, H. (1998):
Marketing: Grundlagen marktorientierter Unternehmensführung: Konzepte – Instrumente – Praxisbeispiele, 8., vollständig neu bearbeitete und erweiterte Auflage, Wiesbaden: Gabler Verlag, 1998

Meffert, Heribert/Bruhn, Manfred (2009):
Dienstleistungsmarketing, 6., vollständig neu bearbeitete Auflage, Wiesbaden: Gabler Verlag, 2009

Meiler, Rudolf Carl (Hrsg.) (1999):
Mittelstand und Betriebswirtschaft: Beiträge aus Wissenschaft und Praxis, Wiesbaden: Gabler Verlag, 1999

Mengen, Andreas/Mettler, Alina (2008):
Kundenwertermittlung: wie viel Vertrieb ist uns der Kunde wert?, erschienen in: Zeitschrift für Controlling & Management (ZfCM), 52. Jg., 2008, Heft 1, S. 30-36

Mercedes Benz (2011):
Online im Internet, URL: http://www.mercedes-benz.de/content/germany/mpc/mpc_germany_website/de/home_mpc/van/home/financial_services/home.html (Stand: 10.08.2011)

Meyer, Anton/Dornach, Frank (1997):
Das Deutsche Kundenbarometer - Qualität und Zufriedenheit: Eine Studie zur Kundenzufriedenheit in Deutschland, Wiesbaden: Gabler Verlag, 1997

Meyer, Anton/Kantsperger, Roland/Wilkoszewski, Alexander (2005):
Strategien zur Messung und Ausschöpfung des Kundenwerts, in: Integriertes Risiko- und Ertragsmanagement: Kunden- und Unternehmenswert zwischen Risiko und Ertrag, hrsg. von Keuper, Frank/Roesing, Dirk/Schomann, Marc, Wiesbaden: Gabler Verlag, 2005, S. 297-320

Meyer, Anton/Oevermann, Dirk (1995):
Kundenbindung, in: Handwörterbuch des Marketing, 2., überarbeitete Auflage, hrsg. von Tietz, Bruno/Köhler, Richard/Zentes, Joachim, München: Vahlen Verlag, 1995, Sp. 1340-1351

Meyer, Matthias/Shaffu, Nina (2007):
Unternehmenswertorientiertes Kundenmanagement auf Basis des Customer Lifetime Value: Eine empirische Studie bei Mobilfunknutzern in Deutschland und Frankreich, erschienen in: Zeitschrift für Controlling & Management (ZfCM), 51. Jg., 2007, Heft 1, S. 54-62

Meyer, Paul Werner (1990):
Der integrative Marketingansatz und seine Konsequenzen für das Marketing, in: Integrierte Marketingfunktionen, 2., überarbeitete Auflage, hrsg. von Meyer, Paul Werner, Stuttgart: Kohlhammer Verlag, 1990, S. 13-30

Meyer, Paul Werner (Hrsg.) (1990):
Integrierte Marketingfunktionen, 2., überarbeitete Auflage, Stuttgart: Kohlhammer Verlag, 1990

Middermann, Frank (1987):
Darstellung und Analyse größenspezifischer Probleme für ein Controlling in mittleren Unternehmen: Erarbeitung von Lösungsansätzen unter besonderer Berücksichtigung organisatorischer Gestaltungsmöglichkeiten, Dissertation, Berlin: Technische Universität Berlin, 1987

Mödritscher, Gernot J. (2008):
Customer Value Controlling: Hintergründe, Herausforderung, Methode, Wiesbaden: Gabler Verlag, 2008

Monßen, Hans-Georg (1998):
Shareholder-Value und Förderauftrag, erschienen in: Zeitschrift für das gesamte Genossenschaftswesen (ZfgG), 48. Jg., 1998, Heft 4, S. 288-297

Moormann, Jürgen/Schmidt, Günter (2007):
IT in der Finanzbranche: Management und Methoden, Berlin/Heidelberg/New York: Springer Verlag, 2007

Morgan, Robert M./Hunt, Shelby D. (1994):
The Commitment-Trust Theory of Relationship Marketing, erschienen in: Journal of Marketing, 58. Jg., 1984, Heft 3, S. 20-38

Moriarty, Rowland T./Kimball, Ralph, C./Gay, J. H. (1983):
The Management of Corporate Banking Relationships, erschienen in: MIT – Sloan Management Review, 24. Jg., 1983, Heft 3, S. 3-15

Mugler, Josef (1995):
Betriebswirtschaftslehre der Klein- und Mittelbetriebe, 2., neubearbeitete Auflage, Wien/New York: Springer Verlag, 1995

Mulhern, Francis J. (1999):
Customer Profitability Analysis: Measurement, Concentration, and Research Directions, erschienen in: Journal of Interactive Marketing, 13. Jg., 1999, Heft 1, S. 25-40

Müller, Klaus (1998):
Kommentar zum Genossenschaftsgesetz GenG, 2. Aufl., Bielefeld: Gieseking Verlag, 1998

Münkner, Hans-Hermann (1990):
Forschungsprojekt 'Strukturfragen der deutschen Genossenschaften': Zusammenfassung der wichtigsten Ergebnisse einer Umfrage bei Genossenschaftsbanken, Marburg: o. V., 1990

Münkner, Hans-Hermann (Hrsg.) (2002):
"Nutzer-orientierte" versus "Investor-orientierte" Unternehmen, Göttingen: Vandenhoeck & Ruprecht, 2002

Münkner, Hans-Hermann/Ringle, Günther (Hrsg.) (2006): Zukunftsperspektiven für Genossenschaften: Bausteine für typgerechte Weiterentwicklung, Bern/Stuttgart/Wien: Haupt Verlag, 2006

N

Nader, Georg (1995):
Zufriedenheit mit Finanzdienstleistungen: Erfolgswirksamkeit, Messung und Modellierung, Wien/New York: Springer Verlag, 1995

Nagel, Dörte (2008):
Kundenwertorientierte Vertriebssteuerung für Firmen- und Gewerbeversicherungen, erschienen in: Zeitschrift für das Versicherungswesen, 59. Jg., 2008, Heft 6, S. 188-191

Natter, Nikolaus/Meck, Andreas/Jodlbauer, Richard (2007):
Studie: Vertriebssteuerung, Aktivitätensteuerung und „Mobiler Vertrieb“ in Regionalbanken auf dem Prüfstand, erschienen in: BankPraktiker, o. Jg., Heft 3, 2007, S. 120-121

Nieschlag, Robert/Dichtl, Erwin/Hörschgen, Hans (2002):
Marketing, 19. überarbeitete und ergänzte Auflage, Berlin: Duncker & Humblot, 2002

Nirschl, Marco/Paul, Verena (2007):
Vertriebsstärke in Banken und Sparkassen: Status quo und Gestaltungsmöglichkeiten, in: Wertorientierte Vertriebssteuerung in Banken und Sparkassen: Deckungsbeitragsmessung, Multikanal-Steuerung, erfolgsabhängige Vergütung, 2., überarbeitete und ergänzte Auflage, hrsg. von Wimmer, Konrad, Heidelberg: FinanzColloquium, 2007, S. 180-213

Noll, Jürgen (2003):
Unternehmensführung durch Management-by-Methoden, erschienen in: WISU - Das Wirtschaftsstudium, 32. Jg., 2003, Heft 7, S. 898-902

Nötzel, Rötger (1982):
Kunde, in: Das große Lexikon für Handel und Absatz, 2., überarbeitete und erweiterte Auflage, hrsg. von Falk, Bernd/Wolf, Jakob, Landsberg/Lech: Verlag Moderne Industrie, 1982, S. 453-454

Nowak, Helge (2002):
Ratings für alle Segmente, erschienen in: Bankinformation, 29. Jg., 2002, Heft 5, S. 22-28

Nuber, Wolfgang (1995):
Strategische Kontrolle: Konzeption, Organisation und kontextspezifische Differenzierung, Wiesbaden: Deutscher Universitäts-Verlag, 1995

O

Oellerking, Klaus/Holzgrabe, Manfred (1990):
Sparkassen und Genossenschaftsbanken im Spannungsverhältnis zwischen Moral und Ökonomie: Strukturelemente, Organisationsgrundsätze und Geschäftspolitik, Frankfurt am Main/Bern/New York et al.: Peter Lang Verlag, 1990

Oelrich, Jan-Frederik (2008):
Kundensegmentierung im Firmenkundengeschäft von Kreditgenossenschaften, nicht veröffentlichte Diplomhausarbeit, Münster: Westfälische Wilhelms-Universität, 2008

Olbrich, Rainer (2006):
Marketing: Ein Einführung in die marktorientierte Unternehmensführung. 2., überarbeitete und erweiterte Auflage, Berlin/Heidelberg/New York: Springer Verlag, 2006

Osman Yasmin (2009):
Der Erfolg ist bedroht, Handelsblatt vom 24.07.2009

Osman, Yasmin/Schrinner, Axel (2009):
Genossenschaftsbanken zahlen Umsatzsteuer, Handelsblatt vom 30.09.2009

P

Padberg, Thomas (2005):
Analyse der Jahresabschlüsse von Genossenschaftsbanken und Sparkassen, Frankfurt am Main: Bankakademie Verlag, 2005

Pangratz, Olaf (1979):
Vertriebssteuerung in der Industrie unter besonderer Berücksichtigung der Konzentration des Nachfragepotenzials im Handel, Dissertation, Berlin: Freie Universität Berlin, 1979

Panzar, John C./Rosse, James N. (1987):
Testing for "Monopoly" Equilibrium, erschienen in: Journal of Industrial Economics, 35. Jg., 1987, Heft 4, S. 443-456

Paul, Michael/Paul, Stephan/Stein, Stefan (2003):
Szenen einer Ehe - Mittelstand und Banken in der Beziehungskrise?, Berlin/Bochum/Wien, 2003

Paul, Stephan (2008):
Preispolitik deutscher Banken – Zentrale Lücken identifizieren und schließen, in: Vertriebssteuerung in der Finanzdienstleistungsindustrie, hrsg. von Brost, Heike/Neske, Rainer/Wrabetz, Wolfgang, Frankfurt am Main: Frankfurt School Verlag, 2008, S. 401-425

Paulick, Heinz (1956):
Das Recht der eingetragenen Genossenschaft: ein Lehr- und Handbuch, Karlsruhe: Müller Verlag, 1956

Payne, Adrian/Frow, Pennie (2005):
A Strategic Framework for Customer Relationship Management, erschienen in: Journal of Marketing, 69. Jg., 2005, Heft 4, S. 167-176

Pelzl, Ursula (2006):
Am Kundenwert wachsen, erschienen in: Bankenpartner, o. Jg., 2006, Heft 4,
S. 8-14

Pepels, Werner (2002):
Stellenwert des Vertriebs in Literatur und Praxis, in: Handbuch Vertrieb, hrsg. von Pepels, Werner, München/Wien: Hanser Verlag, 2002, S. 3-9

Pepels, Werner (Hrsg.) (2002):
Handbuch Vertrieb, München/Wien: Hanser Verlag, 2002

Perridon, Louis/Steiner, Manfred (1999):
Finanzwirtschaft der Unternehmung, 10., überarbeitet Auflage, München: Vahlen Verlag, 1999

Peter, Sibylle Isabelle (1999):
Kundenbindung als Marketingziel: Identifikation und Analyse zentraler Determinanten, 2., überarbeitete und aktualisierte Auflage, Wiesbaden: Gabler Verlag, 1999

Peters, Thomas J./Waterman, Robert H. (1991):
Auf der Suche nach Spitzenleistungen: Was man von den bestgeführten US-Unternehmen lernen kann, 14. Auflage, Landsberg/Lech: Verlag Moderne Industrie, 1991

Petersen, Mitchell. A./Rajan, Raghuram G. (1994):
The Benefits of Lending Relationships: Evidence form Small Business Data, erschienen in: Journal of Finance, 49 Jg., 1994, Heft 1, S. 3-37

Pfaff, Dieter (2005):
Performancemessung aus agencytheoretischer Sicht, in: Controlling: Theorie und Konzeptionen, hrsg. von Scherm, Ewald/Pietsch, Gotthard, München: Verlag Franz Vahlen, 2005, S. 167-189

Pfaffenberger, Kay (2009):
Demokratie und Partizipation in Kreditgenossenschaften - erlebte Teilhabe oder Illusion?, erschienen in: Zeitschrift für das gesamte Kreditwesen (ZfgK), 62. Jg., 2009, Heft 6, S. 279-284

Pfeifer, Hans (2009):
Mitgliedschaft - der Schatz der Kreditgenossenschaften, erschienen in: Zeitschrift für das gesamte Kreditwesen (ZfgK), 62. Jg., 2009, Heft 6, S. 269-270

Pfeiffer, Marion/Pirngruber, Norman/Mährle, Christine (2003):
Kunden- und Risikomanagement: Strategien im Firmenkundengeschäft der Kreditinstitute (Kurzversion), Frankfurt am Main: IBM Business Consulting Services, Strategic Change Solutions, 2003

Picot, Arnold/Neuburger, Rahild (2000):
Banken und das Firmenkundengeschäft im Internet-Zeitalter: der Einfluss der Informations- und Kommunikationstechnik auf Bankgeschäft und Bankorganisation unter besonderer Berücksichtigung des Firmenkundengeschäftes privater Banken, Köln: Bank-Verlag, 2000

Pieper, Bernhard (2006):
Das Drei-Säulen-System: Erfolgsbremse der deutschen Banken?, erschienen in: Wirtschaftsdienst, 86. Jg., Heft 7, S. 452-461

Pigou, Arthur Cecil (1960):
The Economics of Welfare, 4. Auflage, London: Macmillan, 1960

Pleister, Christopher (2002):
VR-CONTROL: Das Zukunftsprojekt des genossenschaftlichen FinanzVerbunds, erschienen in: Bankinformation, 29. Jg., 2002, Sonderheft 1, S. 4-5

Pleister, Christopher (2006):
Der genossenschaftliche FinanzVerbund - Ein Netzwerk für die Zukunft, in: Der Bankensektor im Wandel: Festschrift zum 65. Geburtstag von Professor Erich Priewasser, hrsg. von Emmerich, Norbert/Roßbach, Peter, Frankfurt am Main: Fritz Knapp Verlag, 2006, S. 113-122

Plinke, Wulff (1989):
Die Geschäftsbeziehung als Investition, in: Marketing-Schnittstellen: Herausforderungen für das Management, hrsg. von Specht, Günter/Abel, Bodo, Stuttgart: Poeschel Verlag, 1989, S. 305-325

Plinke, Wulff (1997):
Bedeutende Kunden, in: Geschäftsbeziehungsmanagement, hrsg. von Kleinaltenkamp, Michael/Plinke, Wulff, Berlin et al.: Springer Verlag, S. 113-159

Pohle, Andreas/Schlütken, Robert (2009):
Geschäftsgrundlagen, erschienen in: Bankinformation, 36. Jg., 2009, Heft 10, S. 40-44

Pohlkamp, Andre (2009):
Identifikation und Abschöpfung von Up-Selling-Potenzialen, Wiesbaden: Gabler Verlag, 2009

Polak, Benjamin/Herrmann, Andreas/Heitmann, Mark/Einhorn, Martin (2008):
Die Macht des Defaults: Wirkung von Empfehlungen und Vorgaben auf das individuelle Entscheidungsverhalten, erschienen in: Zeitschrift für Betriebswirtschaft (ZfB), 78. Jg., 2008, Heft 10, S. 1033-1060

Polster, Dirk (2001):
Finanzintermediation und institutioneller Wandel: Finanzsysteme, Universalbanken, Kreditgenossenschaften, Aachen: Shaker Verlag, 2001

Porter, Michael E. (1997):
Wettbewerbsstrategie: Methoden zur Analyse von Branchen und Konkurrenten, 9. Auflage, Frankfurt am Main: Campus Verlag, 1997

Praxmarer, Marc Andreas (1993):
Allfinanzstrategien aus Sicht der Banken: Rahmenbedingungen und Gestaltungsansätze unter besonderer Berücksichtigung der Universalbanken in der Schweiz, Bern et al.: Haupt Verlag, 1993

Preißner, Andreas (2003):
Kunden-Controlling: erfolgreiche Steuerung der Kundenbeziehung, München/Wien: Hanser Verlag, 2003

Priewasser, Erich (2001):
Bankbetriebslehre, 7., erweiterte und vollständig überarbeitete Auflage, München/Wien: Oldenbourg Verlag, 2001

Proeller, Michael (1996):
Konzipierung und Einführung eines Vertriebs-Controlling-Systems, Dissertation, St. Gallen: Universität St. Gallen, 1996

Pufahl, Mario (2003):
Implementierungsaspekte eines Vertriebscontrollings, erschienen in: Zeitschrift für Controlling und Management (ZfCM), 47. Jg., 2003, Heft 5, S. 339-343

Pufahl, Mario (2006):
Vertriebscontrolling: so steuern Sie Absatz, Umsatz und Gewinn, 2., erweiterter Auflage, Wiesbaden: Gabler Verlag, 2006

Püthe, Tina (2009):
Mittelständische Unternehmen und Genossenschaftsbanken: Eine empirische Analyse der Wirkung ökonomischer und verhaltenswissenschaftlicher Faktoren, Frankfurt am Main: Peter Lang Verlag, 2009

R

Raddao, Nino (2009):
Potenzialorientierte Kundensegmentierung zur Optimierung des Leistungsportfolios in der Firmenkundenbank, Frankfurt am Main: Peter Lang Verlag, 2008

Rappaport, Alfred (1999):
Shareholder value: ein Handbuch für Manager und Investoren, 2., vollständig überarbeitete und aktualisierte Auflage, Stuttgart: Schäffer-Poeschel, 1999

Recht, Andreas/Holm, Rüdiger (2005):
Grundelemente der Vertriebssteuerung im Firmenkundengeschäft, in: Bankstrategien im Firmenkundengeschäft: Konzeption – Management – Dimensionen, hrsg. von Börner, Christoph J./Maser, Harald/Schulz, Thomas Christian, Wiesbaden: Gabler Verlag, 2005, S. 133-146

Reckenfelderbäumer, Martin (2005):
Konzeptionelle Grundlagen des Dienstleistungscontrolling: Kritische Bestandsaufnahme und Perspektiven der Weiterentwicklung zu einem Controlling der Kundenintegration, in: Dienstleistungscontrolling, hrsg. von Bruhn, Manfred/Stauss, Bernd, Wiesbaden: Gabler Verlag, 2005 ,S. 32-51

Reckenfelderbäumer, Martin/Welling, Martin (2006):
Der Beitrag einer relativen Einzel-, Prozesskosten- und Deckungsbeitragsrechnung zur Ermittlung von Kundenwerten - konzeptionelle Überlegungen und Gestaltungsempfehlungen, in: Kundenwert: Grundlagen, innovative Konzepte, Praktische Umsetzungen, 3., überarbeitete und erweiterte Auflage, hrsg. von Günter, Bernd/Helm, Sabrina, Wiesbaden: Gabler Verlag, 2006, S. 337-368

Rehbach, Stefan (2003):
Kundenwert und Unternehmenswert: Eine Analyse am Beispiel des E-Commerce, Wiesbaden: Deutscher Universitäts-Verlag, 2003

Rehberg, Michael (2004):
Die Mischung macht's, erschienen in: Bankinformation, 31. Jg., 2004, Heft 8, S. 10-11

Rehbock, Tobias (2007):
Unternehmenskreditzinsen und Ertragslage der Banken im europäischen Vergleich, in: KfW-Research: Mittelstands- und Strukturpolitik, Ausgabe 38, Frankfurt am Main: KfW Bankengruppe, 2007

Rehkugler, Heinz (1989):
Erfolgsfaktoren der mittelständischen Unternehmen, erschienen in: WiSt - Wirtschaftswissenschaftliches Studium, 18. Jg., 1981, Heft 11, S. 626-632

Reichel, Richard (2005):
Zinsspanne und Marktmacht im genossenschaftlichen Bankensektor: Theoretische Überlegungen und empirische Befunde, erschienen in: Zeitschrift für das gesamte Genossenschaftswesen (ZfgG), 55. Jg., 2005, Heft 4, S. 277-288

Reichheld, Frederick F./Sasser, W. Earl (1990):
Zero Defections: Quality comes to services, erschienen in: Harvard Business Review, 68. Jg., 1990, Heft 5, S. 105-111

Reichwald, Ralf/Bastian, Christina/Lohse, Christoph (2000):
Vertriebsmanagement im Wandel: Neue Anforderungen an die Gestaltung der Kundenschnittstelle, in: Vertriebs-Management: Organisation, Technologieeinsatz, Personal, hrsg. von Reichwald, Ralf/ Bullinger, Hans-Jörg, Stuttgart: Schäffer-Poeschel, 2000, S. 3-31

Reichwald, Ralf/Bullinger, Hans-Jörg (Hrsg.) (2000):
Vertriebs-Management: Organisation, Technologieeinsatz, Personal, Stuttgart: Schäffer-Poeschel, 2000

Reinartz, Werner J./Krafft, Manfred (2001):
Überprüfung des Zusammenhangs von Kundenbindungsdauer und Kundenertragswert, erschienen in: Zeitschrift für Betriebswirtschaft (ZfB), 71. Jg., 2001, Heft 11, S. 1263-1281

Reinartz, Werner J./Krafft, Manfred/Hoyer, Wayne D. (2004):
The Customer Relationship Management Process: Its Measurement and Impact on Performance, erschienen in. Journal of Marketing Research, 41. Jg., 2004, Heft 3, S. 293-305

Reinartz, Werner J./Kumar, V. (2000):
On the Profitability of long-life Customers in a noncontractual setting: an empirical investigation and implications for Marketing, erschienen in: Journal of Marketing, 64. Jg., 2000, Heft October 2000, S. 17-35

Reinartz, Werner J./Thomas, Jaquelyn S./Kumar, V. (2005):
Balancing Acquisition and Rentention Resources to Maximize Customer Profitability, erschienen in: Journal of Marketing, 69. Jg., 2005, Heft 1, S. 63-79

Reinecke, Sven/Reibstein, David J. (2002):
Performance Measurement in Marketing und Verkauf – Einsatz von Marketingkennzahlen in den USA und in Kontinentaleuropa, erschienen in: Kostenrechnungspraxis: krp, 46. Jg., 2002, Heft 1, S. 18-25

Reinecke, Sven/Tomczak, Torsten (2001):
Einsatz von Instrumenten und Verfahren des Marketingcontrollings in der Praxis, in: Marketing-Controlling: Marketing als Motor von Wachstum und Erfolg, hrsg. von Reinecke, Sven/Tomczak, Torsten/Geis, Gerold, Frankfurt/Wien: Wirtschaftsverlag Ueberreuter, 2001, S. 76-89

Reinecke, Sven/Tomczak, Torsten/Dittrich, Sabine (Hrsg.) (1998):
Marketingcontrolling, St. Gallen: Verlag Thexis, 1998

Reinecke, Sven/Tomczak, Torsten/Geis, Gerold (Hrsg.) (2001):
Marketing-Controlling: Marketing als Motor von Wachstum und Erfolg, Frankfurt/Wien: Wirtschaftsverlag Ueberreuter, 2001

Reiners, Jens Oliver (2004):
Kundenwertsteigerung und Außendienst: Organisation, Personal- und Informationsmanagement im persönlichen Vertrieb, Berlin: Erich Schmidt Verlag, 2004

Reith, Daniel (1997):
Innergenossenschaftlicher Wettbewerb im Bankensektor, Sternenfels: Verlag Wissenschaft und Praxis, Studienreihe der Stiftung Kreditwirtschaft, 1997

Renker, Clemens (2004):
Vertriebsmanagement: Konzept - Prozess – Kultur: Grundlagen ganzheitlich vertikalen Marketings, Görlitz: Neisse Verlag, 2004

Renker, Clemens (2005):
Relationship Marketing im Firmenkundengeschäft: Konzepte, Erfolgsfaktoren, Umsetzung, 2., überarbeitete Auflage, Stuttgart: Deutscher Sparkassenverlag, 2005

Rese, Mario/Söllner, Albrecht/Utzig, Peter (Hrsg.) (2003):
Relationship Marketing: Standortbestimmung und Perspektiven, Berlin et al.: Springer Verlag, 2003

Richter, Klaus/Sterk, Thomas (2004):
Stellhebel kennen, Wirkungen entfalten, erschienen in: Bankinformation, 31. Jg., 2004, Heft 5, S. 36-39

Ridder, Thomas (2002):
Portfoliosteuerung der Adressrisiken, erschienen in: Bankinformation, 29. Jg., 2002, Heft 7, S. 17-24

Riebel, Paul (1992):
Einzelerlös-, Einzelkosten- und Deckungsbeitragsrechnung als Kern einer ganzheitlichen Ergebnisrechnung, in: Handbuch Kostenrechnung, hrsg. von Männel, Wolfgang, Wiesbaden: Gabler Verlag, 1992, S. 247-299

Rieker, Stephen (1995):
Bedeutende Kunden: Analyse und Gestaltung von langfristigen Anbieter-Nachfrager-Beziehungen auf industriellen Märkten, Wiesbaden: Deutscher Universitäts-Verlag, 1995

Ringle, Günther (1996):
Genossenschaftliches Identitätsprinzip: Zur Frage der Vermeidbarkeit von Abweichungen, in: Genossenschaften vor neuen Herausforderungen: Festschrift für Prof. Dr. Dr. Juhani Laurinkari, hrsg. von Brazda, Johann/Kleer, Jerzy, Augsburg: Maro-Verlag, 1996, S. 245-258

Ringle, Günther (2006):
"Erfolg" aus genossenschaftsspezifischer Sicht, erschienen in: Zeitschrift für das gesamte Genossenschaftswesen (ZfgG), 56. Jg., 2006, Heft 3, S. 207-220

Rinn, Thomas (1991):
Vertriebs-Controlling: Ausgestaltungs- und Anwendungsmöglichkeiten – dargestellt am Beispiel einer Unternehmung der bundesdeutschen Automobilindustrie, Dissertation, Gießen: Justus-Liebig-Universität Gießen, 1991

Ritter, Dwight S. (1993):
Relationship Banking: Cross-Selling the Bank's Products & Services to meet your Customer's every Financial Need, Chicago, Illinois: Bankers Publishing, 1993

Rolfes, Bernd (1993):
Marktzinsorientierte Investitionsrechnung, erschienen in: Zeitschrift für Betriebswirtschaft (ZfB), 63. Jg., 1993, Heft 7, S. 691-713

Rolfes, Bernd (1999):
Gesamtbanksteuerung, Stuttgart: Schäffer-Poeschel, 1999

Rolfes, Bernd (2002):
Das Firmenkundengeschäft – ein "Wertvernichter"?, in: Handbuch Firmenkundengeschäft: Technologie, Rating und Risikosteuerung als Kernkompetenz der Banken, 2., völlig überarbeitete Auflage, hrsg. von Juncker, Klaus /Priewasser, Erich, Frankfurt am Main: Fritz Knapp Verlag, 2002, S. 140-152

Rolfes, Bernd (2006):
Wachstumsgrenzen im Bankgeschäft und ihre Folgen, in: Die strukturelle Ertragsschwäche der Banken, hrsg. von Tietmeyer, Hans/Rolfes, Bernd, Wiesbaden: Gabler Verlag, 2006

Rolfes, Bernd (2008):
Integration, Effektivität, Effizienz: Neue Entwicklungen in der Banksteuerung, erschienen in: Bankinformation, 35. Jg., 2008, Heft 9, S. 22-25

Rolfes, Bernd/Dartsch, Andreas (1997):
Marge und Barwert im Konzept der Marktzinsmethode, erschienen in: WISU – Das Wirtschaftsstudium, 26. Jg., 1997, Heft 10, S. 926-940

Rometsch, Sieghardt (1999):
Die Zukunft des Firmenkundengeschäfts der Banken, in: Kundenorientierung von Banken: Strategie für Kundennähe und effektives Beziehungsmanagement, hrsg. von Herrmann, Andreas/Jasny, Ralf/Vetter, Ingrid, Frankfurt am Main: FAZ Verlagsbereich Buch, 1999, S. 157-178

Rosset, Saharon/Neumann, Einat/Eick, Uri/Vatnik, Nurik (2003):
Customer Lifetime Value Models for Decision Support, erschienen in: Data Mining and Knowledge Discovery, 7. Jg., 2003, Heft 3, S. 321-339

Roßner, Reik/Seifert, Sabine (2005):
Kundenorientiertes Vertriebscontrolling, erschienen in: Controller-Magazin, 30. Jg., 2005, Heft 6, S. 507-511

Rübenacker, Christoph (2009):
Die Bedeutung der organisatorischen Gestaltung für die Effektivität und Effizienz des Problemkreditmanagements: Analysen und Empfehlungen für Sparkassen und Kreditgenossenschaften auf der Grundlage einer empirischen Untersuchung, Dresden: TUDpress, 2009

Rudolf-Sipötz, Elisabeth (2001):
Kundenwert: Konzeption, Determinanten, Management, St. Gallen: Thexis Verlag, 2001

Rudolf-Sipötz, Elisabeth/Tomczak, Torsten (2001):
Kundenwert in Forschung und Praxis, Fachbericht für Marketing 2001/2, Institut für Marketing und Handel, St. Gallen: Thexis, 2001

Russ, Andreas (2003):
Die MAK: eine schöne neue Welt?, erschienen in: WiSt – Wirtschaftswissenschaftliches Studium, 32. Jg., 2003, Heft 6, S. 351-353

Rust, Roland T./Lemon, Katherine N/Zeithaml, Valerie A. (2004):
Return on Marketing: Using Customer Equity to focus Marketing Strategy, erschienen in: Journal of Marketing, 68. Jg., 2004, Heft 1, S. 109-127

Rust, Roland T./Zeithaml, Valerie A./Lemon, Katherine N. (2000):
Driving Customer Equity: How Customer Lifetime Value is Reshaping Corporate Strategy, New York et al.: Free Press, 2000

Ryals, Lynette (2003):
Making Customers Pay: Measuring and Managing Customer Risk and Returns, erschienen in: Journal of Strategic Marketing, 11. Jg., 2003, Heft 3, S. 165-175

Ryals, Lynette (2005):
Making Customer Relationship Management work: the measurement and profitable management of Customer Relationships, erschienen in: Journal of Marketing, 69. Jg., 2005, Heft 4, S. 252-261

S

Sachenbacher, Hans-Ulrich (1990):
Prospektive Lebensdauerkalkulation von Kundenverbindungen in Kreditinstituten: ein anwendungsorientiertes, empirisch fundiertes Konzept für das Privatkundengeschäft unter Berücksichtigung der Konsequenzen in Marketing und Controlling, München: VVF, 1990

Sandkühler, Hans Jörg (Hrsg.) (1990):
Europäische Enzyklopädie zu Philosophie und Wissenschaften, Bd. 4: R-Z, Hamburg: Meiner Verlag, 1990

Sauerbrey, Christa/Henning, Rolf (2000):
Kunden-Rückgewinnung: erfolgreiches Management für Dienstleister, München: Vahlen Verlag, 2000

Schäfer, Heiko (2002):
Die Erschließung von Kundenpotenzialen durch Cross-Selling: Erfolgsfaktoren für ein produktübergreifendes Beziehungsmanagement, Wiesbaden: Gabler Verlag, 2002

Schäfer, Stefan/Weistroffer, Christian (2007):
Determinanten des deutschen Kreditzinsniveaus, erschienen in: Wirtschaftsdienst, 87. Jg., 2007, Heft 7, S. 600-606

Schäffer, Utz/Steiners, Daniel (2005):
Wie nutzen Geschäftsführung und Vorstände in deutschen Industrieunternehmen ihre Kostenrechnung?, erschienen in: Controlling: Zeitschrift für erfolgsorientierte Unternehmenssteuerung, 17. Jg., 2005, Heft 6, S. 321-325

Schax, Elke (2007):
Strategieorientierte Personalentwicklung in Genossenschaftsbanken: Eine empirische Untersuchung zur betrieblichen Weiterbildung, Wiesbaden: Deutscher Universitätsverlag, 2007

Scheer, Olaf (1989):
Potentialorientierte Analyse und Steuerung des Privatkundengeschäfts von Banken, Kiel: Vauk Verlag, 1989

Schemuth, Jan (1996):
Möglichkeiten und Grenzen der Bestimmung des Werts eines Kunden für ein Unternehmen der Automobilindustrie: dargestellt am Beispiel eines Käufers der C-Klasse von Mercedes Benz, München: Verlag der FGM Fördergesellschaft Marketing in München e.V., 1996

Scherm, Ewald/Pietsch, Gotthard (Hrsg.) (2004):
Controlling: Theorie und Konzeptionen, München: Verlag Franz Vahlen, 2004

Schewe, Gerhard (1999):
Unternehmensstrategie und Organisationsstruktur: ein systematischer Überblick zum Stand der Forschung, erschienen in: Die Betriebswirtschaft, 59. Jg., 1999, Heft 1, S. 61-75

Schierenbeck, Henner (2003):
Ertragsorientiertes Bankmanagement, Band 1: Grundlagen, Marktzinsmethode und Rentabilitäts-Controlling, 8., überarbeitete und erweiterte Auflage, Wiesbaden: Gabler Verlag, 2003

Schierenbeck, Henner (Hrsg.) (2001):
Handbuch Bankcontrolling, 2., überarbeitete und erweiterte Auflage, Wiesbaden: Gabler Verlag, 2001

Schierenbeck, Henner/Lister, Michael/Kirmße, Stefan (2008):
Ertragsorientiertes Bankmanagement, Band 2: Risiko-Controlling und integrierte Rendite-/Risikosteuerung, 9., aktualisierte und überarbeite Auflage, Wiesbaden: Gabler Verlag, 2008

Schierenbeck, Henner/Pohl, Michael (2010):
Sicherstellung nachhaltiger Kundenbeziehungen als Herausforderung für bankbetriebliche Steuerungssysteme, in: Management von Kundenbeziehungen: Perspektiven, Analysen, Strategien, Instrumente, hrsg. von Georgi, Dominik/Hadwich, Karsten, Wiesbaden: Gabler Verlag, 2010, S. 277-294

Schildbach, Thomas (1992):
Begriff und Grundprobleme des Controlling aus betriebswirtschaftlicher Sicht, in: Controlling: Grundlagen – Informationssysteme – Anwendungen, hrsg. von Spremann, Klaus/Zur, Eberhard, Wiesbaden: Gabler Verlag, 1992, S. 21-36

Schleuning, Christian (1994):
Dialogmarketing: Theoretische Fundierung, Leistungsmerkmale und Gestaltungsansätze, Ettlingen: IM-Marketing-Forum, 1994

Schleuning, Christian (1997):
Die Analyse und Bewertung der einzelnen Interessenten und Kunden als Grundlage für die Ausgestaltung des Database Marketing, in: Handbuch Database Marketing, 2., korrigierte Auflage, hrsg. von Link, Jörg/Brändli, Dieter/Schleuning, Christian/Kehl, Roger E., Ettlingen: IM-Fachverlag Marketing Forum, 1997, S. 142-157

Schlüchtermann, Jörg/Klöpfer, Ralf/Braun, Andrea (1999):
Balanced Scorecard: Ein neues Konzept zur kennzahlengestützten Strategieumsetzung – Einsatzmöglichkeiten für den Mittelstand, in: Mittelstand und Betriebswirtschaft: Beiträge aus Wissenschaft und Praxis, hrsg. von Meiler, Rudolf Carl, Wiesbaden: Gabler Verlag, 1999, S. 103-121

Schmalen, Helmut (1999):
Grundlagen und Probleme der Betriebswirtschaft, 11., vollständig überarbeitete und erweiterte Auflage, Köln: Wirtschaftsverlag Bachem, 1999

Schmid, Stefan/Kutschker, Michael (2002):
Zentrale Grundbegriffe des strategischen Managements, erschienen in: WISU - Das Wirtschaftsstudium, 31. Jg., 2002, Heft 10, S. 1238-1246

Schmidt, Thomas (2001):
Die neue Rolle des Firmenkundenbetreuers im mittelständischen Firmenkundengeschäft: eine Analyse etablierter Firmenkundenbetreuungsansätze und Entwicklung eines Modells zur Erfolgspartnerschaft, Aachen: Shaker Verlag, 2001

Schmidt-Sudhoff, Ulrich (1967):
Unternehmensziele und unternehmerisches Zielsystem, Wiesbaden: Gabler Verlag, 1967

Schmoll, Anton (2004):
Firmenkundengeschäft mit System: Bausteine zur Optimierung des Vertriebserfolgs, erschienen in: Bankinformation, 31. Jg., 2004, Heft 6, S. 24-29

Schmoll, Anton (2006):
Vertriebsoptimierung im Firmenkundengeschäft: Lösungen für nachhaltige Ertragssteigerung, Wiesbaden: Deutscher Genossenschafts-Verlag, 2006

Schmoll, Anton (2008):
Transparenz schaffen: Konsequentes Aktivitätencontrolling macht Vertriebsaktivitäten sichtbar, erschienen in: Bankinformation, 35. Jg., 2008, Heft 2, S. 52-55

Schmoll, Anton (Hrsg.) (1997):
Bankmarketing für Firmenkunden: Tendenzen, Strategien, Neue Wege, Wien: Manz Verlag, 1997

Schmoll, Anton (Hrsg.) (2008):
Erfolgreiche Vertriebsstrategien im Firmenkundengeschäft: Konzepte, Praxisberichte, Lösungen, Köln: bank-verlag medien, 2008

Schmöller, Petra (2001):
Kunden-Controlling: Theoretische Fundierung und empirische Erkenntnisse, Wiesbaden: Gabler Verlag, 2001

Schneider, Nicole C. (2007):
Kundenwertbasierte Effizienzmessung: Der Beitrag von Marketingmaßnahmen zur Unternehmenswerterhöhung, Wiesbaden: Gabler Verlag, 2007

Schneider, Yvonne (2007):
Die wertorientierte Planung von Tangibles: Eine Untersuchung am Beispiel des Kundenwerts, Hamburg: Verlag Dr. Kovač, 2007

Schöffling, Bernd (1991):
Der Wert einer Genossenschaft für ihre Mitglieder: Ein mehrdimensionales Bewertungsmodell, Arbeitspapier der Forschungsstelle für Genossenschaftswesen an der Universität Hohenheim, Nr. 8, Stuttgart-Hohenheim: Universität Hohenheim, 1992

Schöner, Manfred/Graeve, Andrea (2004):
Kundenkontaktmanagement – Herzstück der Vertriebsstrategie einer Retailbank, in: Handbuch Vertriebsmanagement Finanzdienstleistungen: Analyse, Umsetzung und Perspektiven bei Banken und Sparkassen, hrsg. von Duttenhöfer, Stephan/Keller, Bernhard, Frankfurt am Main: Fritz Knapp Verlag, 2004, S. 141-170

Schöning, Stefan (2003):
Herausforderungen an das Bankmarketing, erschienen in: WISU - Das Wirtschaftsstudium, 32. Jg., 2003, Heft 8/9, S. 1043-1047.

Schramm, Bernhard (1982):
Die Volksbanken und Raiffeisenbanken, 2. überarbeitete und aktualisierte Auflage, Frankfurt am Main: Fritz Knapp Verlag, 1982

Schreiber, Horst (2007):
Neue Vertriebserfolge im Privat- und Firmenkundengeschäft am Beispiel der Volksbank Trier eG (Lektion 6), Schriftlicher Management-Lehrgang in 8 Lektionen: Vertriebssteuerung in Banken und Sparkassen, Düsseldorf: Euroforum Verlag, 2007

Schreyögg, Georg/Geiger, Daniel (2005):
Zur Konvertierbarkeit von Wissen – Wege und Irrwege im Wissensmanagement, erschienen in: Zeitschrift für Betriebswirtschaft (ZfB), 75. Jg., 2005, Heft 5, S. 433-454

Schreyögg, Georg/Koch, Jochen (2007):
Grundlagen des Managements: Basiswissen für Studium und Praxis, Wiesbaden: Gabler Verlag, 2007

Schreyögg, Georg/Steinmann, Horst (1985):
Strategische Kontrolle, erschienen in: Zeitschrift für betriebswirtschaftliche Forschung (ZfbF), 37. Jg., 1985, Heft 5, S. 391-410

Schroeder, Nina (2006):
Kundenwert als zentrale Größe zur wertorientierten Unternehmenssteuerung, Hamburg: Verlag Dr. Kovač, 2006

Schüller, Stephan/Schneider, Mike (2006):
Zwischen Kostendegression und Kundennähe: welche Unternehmenskonzepte haben Zukunft?, in: Banken auf der Suche nach strategischem Profil, hrsg. von Tietmeyer, Hans/Rolfes, Bernd, Wiesbaden: Gabler Verlag, 2006, S. 1-30

Schulte, Günther (2006):
Errichtung der Genossenschaft, in: Genossenschaftsgesetz (Gesetz betreffend die Erwerbs- und Wirtschaftsgenossenschaften): Kommentar, 35., neu bearbeitete Auflage, hrsg. von Lang, Johannes/Weidmüller, Ludwig, Berlin: de Gruyter, 2006, S. 53-233

Schulz, Bernhard (1995):
Kundenpotentialanalyse im Kundenstamm von Unternehmen, Frankfurt am Main: Peter Lang Verlag, 1995

Schulz, Thomas Christian (2005):
Strategische Segmentbildung im Firmenkundengeschäft der Banken, in: Bankstrategien im Firmenkundengeschäft: Konzeption, Management, Dimension, hrsg. von Börner, Christoph, J./Maser, Harald/Schulz, Thomas Christian, Wiesbaden: Gabler Verlag, 2005, S. 65-92

Schuster, Peter (1991):
Erfolgsorientierte Steuerung kleiner und mittlerer Unternehmen: funktionale, instrumentelle und organisatorische Aspekte eines größengerechten Controlling-Systems: Berlin et al.: Springer Verlag, 1991

Schweickart, Nikolaus/Töpfer, Armin (Hrsg.) (2006):
Wertorientiertes Management: Werterhaltung, Wertsteuerung, Wertsteigerung ganzheitlich gestalten, Berlin/Heidelberg/New York, 2006

Schweitzer, Marcell (2001):
Planung und Steuerung, in: Allgemeine Betriebswirtschaftslehre, Band 2: Führung, 8. Auflage, hrsg. von Bea, Franz-Xaver/Dichtl, Erwin/Schweitzer, Marcell, Stuttgart: Lucius & Lucius Verlag, 2001, S. 16-126

Schweitzer, Marcell/Friedl, Birgit (1992):
Beitrag zu einer umfassenden Controlling-Konzeption, in: Controlling: Grundlagen – Informationssysteme – Anwendungen, hrsg. von Spremann, Klaus/Zur, Eberhard, Wiesbaden: Gabler Verlag, 1992, S. 141-167

Schweitzer, Marcell/Küpper, Hans-Ulrich (2008):
Systeme der Kosten- und Erlösrechnung, 9., überarbeitete und erweiterte Auflage, München: Verlag Franz Vahlen, 2008

Schwencke, Moritz (2008):
Die Akzeptanz von Firmenkundenportalen: Eine empirische Studie der Einflussfaktoren, Wiesbaden: Gabler Verlag, 2008

Segbers, Klaus (2007):
Die Geschäftsbeziehung zwischen mittelständischen Unternehmen und ihrer Hausbank: Eine ökonomische und verhaltenswissenschaftliche Analyse, Frankfurt am Main: Peter Lang Verlag, 2007

Segbers, Klaus/Wirtz, Harald (2008):
Schlank und effektiv: Der VR-FinanzPlan Mittelstand bei der Volksbank Bitburg, erschienen in: Bankinformation, 35. Jg., 2008, Heft 1, S. 47-49

Sheth, Jagdish N./Parvatiyar, Atul (1995):
Relationship Marketing in Consumer Markets: Antecedents and Consequences, erschienen in: Journal of the Academy of Marketing, 23. Jg., 1995, Heft 4, S. 255-271

Siebert, Johannes (2009):
Modellierung komplementärer Interaktionen: Konzeption eines innovativen Ansatzes in der multiattributiven Theorie, Lohmar-Köln: Eul-Verlag, 2009

Siemes, Andreas (2002):
Marktorientierte Kreditrisikobewertung: Eine empirische Untersuchung mittels Künstlicher Neuronaler Netze, Frankfurt am Main: Peter Lang Verlag, 2002

Simon, Hermann (1992):
Lessons from German's Midsize Giants, erschienen in: Harvard Business Review, 70. Jg., 1992, Heft 3, S. 115-123

Söllner, Fritz (2001):
Die Geschichte des ökonomischen Denkens, 2., verbesserte Auflage, Berlin et al.: Springer Verlag, 2001

Specht, Günter/Abel, Bodo (Hrsg.) (1989):
Marketing-Schnittstellen: Herausforderungen für das Management, Stuttgart: Poeschel Verlag, 1989

Sperling, Dietrich (1984):
Grundlagenbeitrag, in: Die Genossenschaftsidee im Widerstreit der Meinungen: Tagungsbericht, hrsg. von Boettcher, Erik/Grossfeld, Bernhard/Wagner, Helmut, Münster: Regensberg, 1984, S. 33-42

Spremann, Klaus/Zur, Eberhard (Hrsg.) (1992):
Controlling: Grundlagen – Informationssysteme – Anwendungen, Wiesbaden: Gabler Verlag, 1992

Srivastava, Rajendra K./Shervani, Tasadduq A./Fahey, Liam (1997):
Driving Shareholder Value: The Role of Marketing in Reducing Vulnerability and Volatility of Cash Flows, erschienen in: Journal of Market Focused Management, 2. Jg., 1997, Heft 2, S. 49-64

Srivastava, Rajendra K./Shervani, Tasadduq A./Fahey, Liam (1998):
Market-Based Assets and Shareholder Value - A Framework for Analysis, erschienen in: Journal of Marketing, 62. Jg., 1998, Heft 1. S. 2-18

Srivastava, Rajendra K./Shervani, Tasadduq A./Fahey, Liam (1999):
Marketing, Business Processes, and Shareholder Value: An Organizational Embedded View of Marketing Activities and the Discipline of Marketing, erschienen in: Journal of Marketing, 63. Jg., 1999, Heft 4, S. 168-179

Stahl, Heinz K. (1996):
Zero-Migration: ein kundenorientiertes Konzept der strategischen Unternehmensführung, Wiesbaden: Gabler Verlag, 1996

Stahl, Heinz K./Hinterhuber, Hans-H./von den Eichen, Stephan A. Friedrich/Matzler, Kurt (2009):
Kundenzufriedenheit und Kundenwert, in: Kundenorientierte Unternehmensführung: Kundenorientierung, Kundenzufriedenheit, Kundenbindung, 6., überarbeitete Auflage, hrsg. von Hinterhuber, Hans H./Matzler, Kurt, Wiesbaden: Gabler Verlag, S. 247-266

Stahl, Heinz K./Matzler, Kurt/Hinterhuber, Hans-H. (2006):
Kundenbewertung und Shareholder-Value: Versuch einer Synthese, in: Kundenwert: Grundlagen, Innovative Konzepte, Praktische Umsetzungen, 3., überarbeitete und erweiterte Auflage, hrsg. von Günter, Bernd/Helm, Sabrina, Wiesbaden: Gabler Verlag, 2006, S. 425-445

Stanger, Martin (2007):
Customer Value Management: Kundenwert aus Anbieter- und Nachfragersicht, Saarbrücken : VDM Verlag Dr. Müller, 2007

Stanley, Scott M./Markman, Howard J. (1992):
Assessing Commitment in Personal Relationships, erschienen in: Journal of Marriage and the Family, 54. Jg., 1992, Heft 2, S. 595-608

Stauss, Bernd/Friege, Christian (1999):
Regaining Service Customers: Costs and Benefits of Regain Management, erschienen in: Journal of Service Research, 1. Jg., 1999, Heft 4, S. 347-361

Stauss, Bernd/Friege, Christian (2006):
Kundenwertorientiertes Rückgewinnungsmanagement, in: Kundenwert: Grundlagen, Innovative Konzepte, Praktische Umsetzungen, 3., überarbeitete und erweiterte Auflage, hrsg. von Günter, Bernd/Helm, Sabrina, Wiesbaden: Gabler Verlag, 2006, S. 509-530

Steck, Werner/Kundisch, Dennis (2005):
Langfristig ertragreiche Kundenbeziehungen im Finanzdienstleistungsbereich – Einflußfaktoren und Handlungsempfehlungen, in: Integriertes Risiko- und Ertragsmanagement: Kunden- und Unternehmenswert zwischen Risiko und Ertrag, hrsg. von Keuper, Frank/Roesing, Dirk/Schomann, Marc, Wiesbaden: Gabler Verlag, 2005, S. 441-464

Stein, Stefan/Kaltofen, Daniel/Paul, Stephan (2009):
Weitere 10 Jahre Einsatz von Ausfallprognoseverfahren im Firmenkreditgeschäft: Eine Vermessung der Neuen (Basler) Welt, erschienen in: Die Betriebswirtschaft (DBW), 69. Jg., 2009, Heft 3, S. 351-372

Steiner, Joachim (1980):
Die personelle Führungsstruktur in mittelständischen Betrieben, Göttingen: Otto Schwartz, 1980

Steiner, Jürgen (1995):
Der genossenschaftliche Bankenverbund im Spannungsfeld von Organisationszwang und Wettbewerbsdruck, in: Stabilität und Effizienz hybrider Organisationsformen: die Kooperation im Lichte der neuen Institutionenökonomik, hrsg. von Wagner, Helmut/Jäger, Wilhelm, Münster: Regensberg, 1995, S. 15-38

Steinmann, Horst/Schreyögg, Georg (2005):
Management: Grundlagen der Unternehmensführung: Konzepte – Funktionen – Fallstudien, 6., vollständig überarbeitete Auflage, Wiesbaden: Gabler Verlag, 2005

Steria Mummert Consulting (2007):
Branchenkompass 2007 Kreditinstitute, Frankfurt am Main/Hamburg: Steria Mummert Consulting AG; F.A.Z.-Institut für Management-, Markt- und Medieninformationen GmbH, 2007

Stock, Ruth (2003):
Der Einfluss der Kundenzufriedenheit auf die Preissensitivität von Firmenkunden: Transaktionstheoretische Betrachtung und empirische Analyse, erschienen in: Die Betriebswirtschaft (DBW), 63. Jg., 2003, Heft 3, S. 333-348

Storbacka, Kaj (1993):
Customer Relationship Profitability in Retail Banking, GTF: Forskningsrapporter, Nr. 29, Helsingfors, 1993

Sträter, Jens (2005):
Der Customer-Lifetime-Value von Privatkunden im Finanzdienstleistungssektor: Modellaufbau und Anwendung an der Kundengruppe der Hochschulabsolventen wirtschaftswissenschaftlicher Fakultäten, Frankfurt am Main: Fritz Knapp Verlag, 2005

Strauß, Marc R. (2006):
Erfolgsfaktoren von Banken im mittelständischen Firmenkundengeschäft: Empirische Analyse und konzeptionelle Anwendung, Wiesbaden: Deutscher Universitäts-Verlag, 2006

Stricker, Per/Leßmann, Johannes (2008):
Portfoliosteuerung von Immobilienkrediten unter Berücksichtigung des demographischen Wandels, erschienen in: Kredit & Rating Praxis, 34. Jg., 2008, Heft 1, S. 11-14

Strieder, Thomas (2000):
Eigenkapitalbeschaffung bei genossenschaftlichen Unternehmen, erschienen in: Zeitschrift für das gesamte Genossenschaftswesen (ZfgG), 50. Jg., 2000, Heft 3, S. 214-229

Stüker, David (2008):
Evaluierung und Steuerung von Kundenbeziehungen aus Sicht des unternehmenswertorientierten Controlling, Wiesbaden: Gabler Verlag, 2008

Stützel, Wolfgang (1976):
Wert und Preis, in: Handwörterbuch der Betriebswirtschaft, Band 3, 4., völlig neu gestaltete Auflage, hrsg. von Grochla, Erwin/Wittmann, Waldemar, Stuttgart: Schäffer-Poeschel, 1976, Sp. 4404-4426

Szyperski, Norbert (1980):
Informationsbedarf, in: Handwörterbuch der Organisation, 2., völlig neu gestaltete Auflage, hrsg. von Grochla, Erwin, Stuttgart: Schäffer-Poeschel, 1980, Sp. 904-913

T

Tebroke, Hermann-Josef (1998):
Kontrolle der Managementleistung in Kreditgenossenschaften, erschienen in: Zeitschrift für Organisation und Management (ZfO), 67 Jg., 1998, Heft 6, S. 332-339

Tewes, Matthias (2003):
Der Kundenwert im Marketing: Theoretische Hintergründe und Umsetzungsmöglichkeiten einer wert- und marktorientierten Unternehmensführung, Wiesbaden: Deutscher Universitäts-Verlag, 2003

Theurl, Theresia (2002):
"Shareholder Value" und "genossenschaftlicher Förderauftrag" - Zwei unvereinbare strategische Ausrichtungen?, in: Vom Modell zur Umsetzung: Strategische Herausforderungen für Genossenschaften, hrsg. von Theurl, Theresia/Greve, Rolf, Aachen: Shaker Verlag, 2002, S. 51-91

Theurl, Theresia (2003):
Dabei sein ist nicht alles, erschienen in: Bankinformation, 30. Jg., 2003, Heft 2, S. 12-15

Theurl, Theresia (2004):
Genossenschaftsbanken: Vertriebsautomaten oder Unternehmen?, erschienen in: Genossenschafts-Kurier, o. Jg., 2004, Heft 4, S. 1-2

Theurl, Theresia (2005):
Genossenschaftliche Mitgliedschaft und Member Value Reporting als Konzepte für die Zukunft, erschienen in: Zeitschrift für das gesamte Genossenschaftswesen (ZfgG), 55. Jg., 2005, Heft 2, S. 136-145

Theurl, Theresia/Böttiger, Jan-Matthias (2007):
Stakeholderorientierte Berichterstattung von Genossenschaften: Einordnung, Zielsetzung und Grundsätze des Member Value-Reportings, Arbeitspapier des Instituts für Genossenschaftswesen an der Westfälischen Wilhelms-Universität Münster, Nr. 70, Münster, 2007

Theurl, Theresia/Greve, Rolf (Hrsg.) (2002):
Vom Modell zur Umsetzung: Strategische Herausforderungen für Genossenschaften, Aachen: Shaker Verlag, 2002

Tietmeyer, Hans/Rolfes, Bernd (Hrsg.) (2006a):
Die strukturelle Ertragsschwäche der Banken, Wiesbaden: Gabler Verlag, 2006

Tietmeyer, Hans/Rolfes, Bernd (Hrsg.) (2006b):
Banken auf der Suche nach strategischem Profil, Wiesbaden: Gabler Verlag, 2006

Tietz, Bruno/Köhler, Richard/Zentes, Joachim (Hrsg.) (1995):
Enzyklopädie der Betriebswirtschaftslehre, Band 4: Handwörterbuch des Marketing, 2., völlig neu gestaltete Auflage, Stuttgart: Schäffer-Poeschel, 1995

tns infratest (2008):
Firmen-FMDS 2007: Analyse des deutschen Firmenkundenmarktes, Köln: tns infratest, 2008

tns infratest (2010):
Firmen-FMDS 2010: Analyse des deutschen Firmenkundenmarktes, Köln: tns infratest, 2010

Töfferl, Siegwald (1997):
Consulting-Banking: Unternehmensberatung für den Mittelstand, in: Bankmarketing für Firmenkunden: Tendenzen, Strategien, Neue Wege, hrsg. von Schmoll, Anton, Wien: Manz Verlag, 1997, S. 109-124

Tolkmitt, Volker (2007):
Neue Bankbetriebslehre: Basiswissen zu Finanzprodukten und Finanzdienstleistungen, 2., überarbeitete Auflage, Wiesbaden: Gabler Verlag, 2007

Tomczak, Torsten/Dittrich, Sabine (1997):
Erfolgreich Kunden binden: eine kompakte Einführung, Zürich: Werd-Verlag, 1997

Tomczak, Torsten/Reinecke, Sven/Finsterwalder, Jörg (2000):
Kundenausgrenzung: Umgang mit unerwünschten Dienstleistungskunden, in: Dienstleistungsmanagement Jahrbuch 2000, Wiesbaden: Gabler Verlag, 2000, S. 399-421

Tomczak, Torsten/Rudolf-Sipötz, Elisabeth (2006):
Bestimmungsfaktoren des Kundenwertes: Ergebnisse einer branchenübergreifenden Studie, in: Kundenwert: Grundlagen, Innovative Konzepte, Praktische Umsetzungen, 3., überarbeitete und erweiterte Auflage, hrsg. von Günter, Bernd/Helm, Sabrina, Wiesbaden: Gabler Verlag, 2006 ,S. 127-155

Topp, Stefan (1999):
Die Pre-Fusionsphase von Kreditinstituten: Eine Untersuchung der Entscheidungsprozesse und ihrer Strukturen, Sternenfels: Verlag Wissenschaft & Praxis, 1999

Treiblmaier, Horst (2007):
Beziehungsmarketing aus Kundensicht, erschienen in: Wirtschaftsinformatik, 49. Jg., 2007, Heft 1, S. 42-48

Tschöpel, Michael (2010):
Operationalisierungsversuche des Förderauftrags – Ereignisse und Implikationen einer Literaturstudie, Arbeitspapier Nr. 99, Institut für Genossenschaftswesen an der Westfälischen Wilhelms-Universität: 2010

V

Varmaz, Armin (2006):
Rentabilität im Bankensektor: Identifizierung, Quantifizierung und Operationalisierung werttreibender Faktoren, Wiesbaden: Deutscher Universitätsverlag, 2006

Venkatesan, Rajkumar/Kumar, V. (2004):
A Customer Lifetime Value Framework for Customer Selection and Resource Allocation Strategy, erschienen in: Journal of Marketing, 68. Jg., 2004, Heft 4, S. 106-125

Viehoff, Felix (1979):
Möglichkeiten der Zielkonkretisierung des gesetzlichen Förderungsauftrags bei Kreditgenossenschaften, in: Zeitschrift für das gesamte Genossenschaftswesen (ZfgG), Sonderheft: Erwartungen der Genossenschaftspraxis an die Wissenschaft: Tagungsbericht der IX. Internationalen Genossenschaftswissenschaftlichen Tagung 1978 in Freiburg/Schweiz, Göttingen: Vandenhoeck & Ruprecht, 1979

Vogel, Verena (2006):
Kundenbindung und Kundenwert: Der Einfluss von Einstellungen auf das Kaufverhalten, Wiesbaden: Gabler Verlag, 2006

Vogel, Verena/Evanschitzky, Heiner/Ramaseshan, B. (2008):
Customer Equity Drivers and Future Sales, erschienen in: Journal of Marketing, 72. Jg., 2008, Heft 4, S. 98-108

Vogel, Wolfgang (2002):
Zielbildung in Unternehmen mit aktiver Beteiligung der Mitglieder/Nutzer, in: "Nutzer-orientierte" versus "Investor-orientierte" Unternehmen, hrsg. von Münkner, Hans-Hermann, Göttingen: Vandenhoeck & Ruprecht, 2002, S. 59-85

Völckner, Frankziska (2006):
Methoden zur Messung individueller Zahlungsbereitschaften: Ein Überblick zum State of the Art, erschienen in: Journal für Betriebswirtschaft, 56. Jg., 2006, Heft 1, S. 33-60

Volk, K. O./ Volk, Günter (1989):
Genossenschaftsbanken zwischen Principal-Agent-Realität und dem Ideal der Wirtschaftsdemokratie, in: Genossenschaften und genossenschaftliche Forschung, hrsg. von Zerche, Jürgen/Herder-Dorneich, Philipp, Regensburg: Transfer Verlag, 1989, S. 139-156

Vomhoff, Stephan/Richter, Klaus (2006):
Weit entfernt, erschienen in: Bankinformation, 33. Jg., 2006, Heft 4, S. 22-25

von Boehm-Bezing, Carl-Ludwig (2002):
Zukunftsstrategien im europäischen mittelständischen Firmenkundengeschäft, in: Handbuch Firmenkundengeschäft: Technologie, Rating und Risikosteuerung als Kernkompetenz der Banken, 2., völlig überarbeitete Auflage, hrsg. von Juncker, Klaus/Priewasser, Erich, Frankfurt am Main: Fritz Knapp Verlag, 2002, S. 81-95

von der Oelsnitz, Dietrich/Nirsberger, Ina (2007):
Marktaustritt – Gründe und Barrieren, erschienen in: WISU – Das Wirtschaftsstudium, 36. Jg., 2007, Heft 10, S. 1288-1296

von Hippel, Eric (1986):
Lead Users: A Source of Novel Product Concepts, erschienen in: Management Science, 32. Jg., 1986, Heft 7, S. 791-805

von Usslar, Ludolf (2007):
Der genossenschaftliche Förderauftrag in der Problemkreditbetreuung, erschienen in: Zeitschrift für das gesamte Genossenschaftswesen (ZfgG), 57. Jg., 2007, Heft 1, S. 41-51

von Wangenheim, Florian (2003):
Weiterempfehlung und Kundenwert.: Ein Ansatz zur persönlichen Kommunikation, Wiesbaden: Deutscher Universitäts-Verlag, 2003

Von Wangenheim, Florian/Bayón, Tomás/Herrmann, Andreas (2006):
Die Abgabe von Weiterempfehlungen – Determinanten und ökonometrische Modellierung, erschienen in: Zeitschrift für betriebswirtschaftliche Forschung (ZfbF), 58. Jg., 2006, Heft 5, S. 304-336

VR-Bank Westmünsterland eG (2010):
Jahresabschluss zum 31. Dezember 2009, online im Internet, URL: https://www.ebundesanzeiger.de/ebanzwww/wexsservlet?session.sessionid=dff39ab68921e0609dcd2a42809d4cbe&page.navid=detailsearchlisttodetailsearchdetail&fts_search_list.selected=aa2f29bc2e92dd4e&fts_search_list.destHistoryId=55918 (Stand: 07.07.2011)

W

Wagner, Helmut/Jäger, Wilhelm (Hrsg.) (1995):
Stabilität und Effizienz hybrider Organisationsformen: die Kooperation im Lichte der neuen Institutionenökonomik, Münster: Regensberg Verlag, 1995

Wald, Matthias (2002):
Kreditderivate und modernes Kreditportfoliomanagement: Innovationspotenzial und Anwendungsbarrieren, Dissertation, Universität Mannheim. 2002

Walter, Georg (1996):
Beurteilung der Kundenattraktivität durch Kreditinstitute: Anforderungen an die Database und Umsetzungsmöglichkeiten für ein Gesamtkonzept, München: FGM Verlag, 1996

Wayland, Robert E./Cole, Paul M. (1997):
Customer Connections : New Strategies for Growth, Boston: Harvard Business School Press, 1997

Weber, Jürgen (2004):
Einführung in das Controlling, 10., überarbeitete und aktualisierte Auflage, Stuttgart: Schäffer-Poeschel, 2004

Weber, Jürgen/Lissautzki, Marius (2004):
Kundenwert-Controlling, Vallendar: WHU-Otto-Beisheim-Hochschule, Lehrstuhl für Betriebswirtschaftslehre, 2004

Weber, Jürgen/Lissautzki, Marius (2005):
Kundenwert-Controlling: Dienstleistungsunternehmen kundenorientiert steuern, in: Dienstleistungscontrolling, hrsg. von Bruhn, Manfred/Stauss, Bernd, Wiesbaden: Gabler Verlag, 2005, S. 304-323

Weber, Jürgen/Schäffer, Utz (2008):
Einführung in das Controlling, 12., überarbeitete und aktualisierte Auflage, Stuttgart: Schäffer-Poeschel, 2008

Weber, Werner (1981):
Strategien zur Verbesserung des Managements in kleinen und mittleren Unternehmen, München: Minerva Publikation, 1981

Webster, Frederick E. (1988):
The Rediscovery of the Marketing Concept, erschienen in: Business Horizons, 31. Jg., 1988, Heft 3, S. 29-39

Weegen, Marc (2006):
Mitgliedschaft auf sechs Säulen, erschienen in: Bankinformation, 33. Jg., 2006, Heft 11, S. 26-30

Weiber, Rolf/Adler, Jost (1995a):
Informationsökonomisch begründete Typologisierung von Kaufprozessen, erschienen in: Zeitschrift für betriebswirtschaftliche Forschung (ZfbF), 47. Jg., 1995, Heft 1, S. 43-65

Weiber, Rolf/Adler, Jost (1995b):
Der Einsatz von Unsicherheitsreduktionsstrategien im Kaufprozess: eine informationsökonomische Analyse, erschienen in; Zeitschrift für betriebswirtschaftliche Forschung (ZfbF), 47. Jg., 1995, Sonderheft 35, S. 61-77

Welsh, John A./White, Jerry F. (1981):
A small business is not a little big business, erschienen in: Harvard Business Review, 59. Jg., 1981, Heft 4, S. 18-32

WGZ BANK (2008):
Ergebnisse der Studie zur Potenzialanalyse im Firmenkundengeschäft, nicht veröffentlicht, Düsseldorf: WGZ BANK AG, 2008

WGZ Initiativkapital (2005):
Intelligente Finanzierungsalternativen, erschienen in: Initiativbanking, o. Jg., 2005, Heft 3, S. 10-11

Wiedemann, Arnd (1992):
Verbundstrategien für Kreditgenossenschaften, Bern/Stuttgart: Verlag Paul Haupt, 1992

Wiedmann, Klaus Peter/Hennigs, Nadine/In der Stroth, Anne-Berit (2006):
Wertorientiertes Kundenmanagement im Premiumsegment: Der Kunde als zentrale Steuerungsgröße in der Finanzdienstleistungsbranche, Arbeitspapier des Instituts Marketing & Management, Hannover: Universität Hannover, 2006

Wiemeyer, Joachim/Carstensen, Christiane (1984):
Zusammenfassendes Protokoll der Podiums- und Plenumsdiskussion, in: Die Genossenschaftsidee im Widerstreit der Meinungen, Tagungsbericht, hrsg. von Boettcher, Erik/Großfeld, Bernhard/Wagner, Helmut, Institut für Genossenschaftswesen der Universität Münster, Münster: Regensberg Verlag, 1984, S. 60-67

Wiesel, Thorsten/Skiera, Bernd (2007):
Unternehmensbewertung auf der Basis von Kundenlebenswerten, erschienen in: Zeitschrift für betriebswirtschaftliche Forschung (ZfbF), 59. Jg., 2007, Heft 6, S. 706-731

Wild, Jürgen (1974):
Betriebswirtschaftliche Führungslehre und Führungsmodelle, in: Unternehmensführung - Festschrift für Erich Kosiol zu seinem 75. Geburtstag, hrsg. von Wild, Jürgen, Berlin: Duncker & Humblot, S. 141-179

Wild, Jürgen (Hrsg.) (1974):
Unternehmensführung - Festschrift für Erich Kosiol zu seinem 75. Geburtstag, Berlin: Duncker & Humblot, 1974

Wilde, Klaus D./Hippner, Hajo (Hrsg.) (2006):
Grundlagen des CRM: Konzepte und Gestaltung, 2., überarbeitete und erweiterte Auflage, Wiesbaden: Gabler Verlag, 2006

Wille, Kai (2005):
Customer Equity: Grundlagen einer kundenwertorientierten Unternehmensführung, Wiesbaden: Deutscher Universitäts-Verlag, 2005

Wimmer, Konrad (2009):
Aktuelle Herausforderungen für die Vertriebssteuerung, erschienen in: Betriebswirtschaftliche Blätter, 58. Jg., 2009, Heft 4, S. 210-214

Wimmer, Konrad (2010):
Grundlagen der wertorientierten Vertriebssteuerung, in: Wertorientierte Vertriebssteuerung in Banken und Sparkassen: Potenzialermittlung, Deckungsbeitragsmessung, Multikanal-Steuerung, Erfolgsabhängige Vergütung, 3. Auflage, Heidelberg: FinanzColloquium, 2010, S. 13-43

Wimmer, Konrad (Hrsg.) (2007):
Wertorientierte Vertriebssteuerung in Banken und Sparkassen: Deckungsbeitragsmessung, Multikanal-Steuerung, erfolgsabhängige Vergütung, 2., überarbeitete und ergänzte Auflage, Heidelberg: FinanzColloquium, 2007

Wimmer, Konrad (Hrsg.) (2010):
Wertorientierte Vertriebssteuerung in Banken und Sparkassen: Potenzialermittlung, Deckungsbeitragsmessung, Multikanal-Steuerung, Erfolgsabhängige Vergütung, 3. Auflage, Heidelberg: FinanzColloquium, 2010

Wimmer, Konrad/Schilcher, Norbert/Windbichler, Christian (2010):
Bestandteile und Konzeptionen der wertorientierten Vertriebssteuerung, in: Wertorientierte Vertriebssteuerung in Banken und Sparkassen: Potenzialermittlung, Deckungsbeitragsmessung, Multikanal-Steuerung, Erfolgsabhängige Vergütung, 3. Auflage, Heidelberg: FinanzColloquium, 2010, S. 47-101

Wings, Heinz/Benölken, Heinz (2004):
Konsequent Kurs halten, erschienen in: Bankinformation, 31. Jg., 2004, Heft 4, S. 37-42

Winkelmann, Peter (2003):
Vertriebskonzeption und Vertriebssteuerung: Die Instrumente des integrierten Kundenmanagements (CRM), 2., völlig überarbeitete und erweiterte Auflage, München: Verlag Franz Vahlen, 2003

Wittmann, Georg (2006):
Steigerung des Cross-Selling-Erfolgs im mittelständischen Firmenkundengeschäft durch Produktbündelung, erschienen in: BIT – Banking and Information Technology, 7. Jg., 2006, Heft 1, S. 9-24

Wittmann, Georg/Drexler, Christian/Breitschaft, Markus/Krabichler, Thomas/Stahl, Ernst (2007):
Cross-Selling bei Banken und Sparkassen: Empirische Analyse zu Status quo, Trends und zukünftigen Anforderungen, Regensburg: ibi Research an der Universität Regensburg, 2007

Wittmann, Waldemar (1959):
Unternehmung und unvollkommene Information: Unternehmerische Voraussicht, Ungewissheit und Planung, Köln/Opladen: Westdeutscher Verlag, 1959

Wittmann, Waldemar/Kern, Werner/Köhler, Richard/Küpper, Hans-Ulrich/von Wysocki, Klaus (Hrsg.) (1993):
Handwörterbuch der Betriebswirtschaft, 5., völlig neu gestaltete Auflage, Stuttgart: Schäffer-Poeschel, 1993

Wöhe, Günter (1996):
Einführung in die Allgemeine Betriebswirtschaftslehre, 19., überarbeitete und erweiterte Auflage, München: Verlag Franz Vahlen, 1996

Wölfel, Julia/Denzler, Michaela (2008):
Der Umgang mit Kundendaten im Marketing aus Sicht des Datenschutzes: Eine Metaanalyse empirischer Studien, Arbeitspapier Nr. 161, Lehrstuhl für Marketing, Universität Erlangen-Nürnberg, 2008

Woodruff, Robert B. (1997):
Customer Value: the Next Source for Competitive Advantage, erschienen in: Journal of the Academy of Marketing Science, 25. Jg., 1997, Heft 2, S. 139-153

Woratschek, Herbert/Roth, Stefan (2006):
Die Ermittlung von Preisbereitschaften und Preisreaktionen als Basis zur Bestimmung des Kundenwerts, in: Kundenwert: Grundlagen, Innovative Konzepte, Praktische Umsetzungen, 3., überarbeitete und erweiterte Auflage, hrsg. von Günter, Bernd/Helm, Sabrina, Wiesbaden: Gabler Verlag, 2006, S. 369-395

Wossidlo, Peter Rüttger (1993):
Stichwort: Mittelständische Unternehmen, in: Handwörterbuch der Betriebswirtschaft, 5., völlig neu gestaltete Auflage, hrsg. von Wittmann, Waldemar/Kern, Werner/Köhler, Richard/Küpper, Hans-Ulrich/von Wysocki, Klaus, Schäffer-Poeschel, 1993, Sp. 2888-2898

Wübben, Markus/von Wangenheim, Florian (2008):
Instant Customer Base Analysis: Managerial Heuristics often "get it right", erschienen in: Journal of Marketing, 72. Jg., 2008, Heft 3, S. 82-93

Wübker, Georg/Baumgarten, Jens (2004):
Der Markenwert von Banken: Messung komplexer Sachverhalte mit modernen Verfahren, erschienen in: Zeitschrift für betriebswirtschaftliche Forschung (ZfbF), 56. Jg., 2004, Heft 9, S. 577-592

Wünschmann, Stefan (2007):
Beschwerdeverhalten und Kundenwert, Wiesbaden: Gabler Verlag, 2007

Z

Zerche, Jürgen/Herder-Dorneich, Philipp (Hrsg.) (1989):
Genossenschaften und genossenschaftliche Forschung, Regensburg: Transfer Verlag, 1989

Zieger, Thomas (2007):
Die Novellierung des deutschen Genossenschaftsrechts, erschienen in: WiSt-Wirtschaftswissenschaftliches Studium, 36. Jg., 2007, Heft 6, S. 313-320

Zollner, Georg (1995):
Kundennähe in Dienstleistungsunternehmen: empirische Analyse von Banken, Wiesbaden: Deutscher Universitätsverlag, 1995

Beiträge zum Controlling

Herausgegeben von Wolfgang Berens

Band 1 Wolfgang Berens / Joachim Strauch: Due Diligence bei Unternehmensakquisitionen – eine empirische Untersuchung. Unter Mitarbeit von Thorsten Behrens und Julia Lescher. 2002.

Band 2 Andreas Siemes: Marktorientierte Kreditrisikobewertung. Eine empirische Untersuchung mittels Künstlicher Neuronaler Netze. 2002.

Band 3 Karl Christoph Heinen: Die Berücksichtigung von Kosten in der Konkurrenzanalyse. 2002.

Band 4 Thomas Mosiek: Interne Kundenorientierung des Controlling. 2002.

Band 5 Vera Südmeyer: Wettbewerbsvorteile durch strategisches Betriebsformenmanagement. Ein dynamischer Bezugsrahmen für Einzelhandelsunternehmen. 2003.

Band 6 Wolfgang Berens / Walter Schmitting (Hrsg.): Controlling im E-Business. Rückkehr zur Rationalität. 2004.

Band 7 René Bertelsmann: Entwicklung einer Controlling-Konzeption im verallgemeinerten Neuen Steuerungsmodell für Trägerorganisationen der gesetzlichen Unfallversicherung. 2005.

Band 8 Mirko Tillmann: Risikokapitalbasierte Steuerung in der Schaden- und Unfallversicherung. Konzeption einer modellgestützten Risikoanalyse. 2005.

Band 9 Andreas Hoffjan: Risikorechnung bei industrieller Auftragsfertigung. Theoretische Konzeption und Anwendung für die Bauwirtschaft. 2006.

Band 10 Pascal Nevries: Die Marketingwirkungen von Börsengängen. Eine konzeptionelle Analyse. 2006.

Band 11 Klaus Segbers: Die Geschäftsbeziehung zwischen mittelständischen Unternehmen und ihrer Hausbank. Eine ökonomische und verhaltenswissenschaftliche Analyse. 2007.

Band 12 Andreas Röhrig: Wirkungsorientiertes Controlling im politisch-administrativen System. Unter besonderer Berücksichtigung der Gestaltungsmöglichkeiten von öffentlichen Verwaltungen. 2008.

Band 13 Nino Raddao: Potenzialorientierte Kundensegmentierung zur Optimierung des Leistungsportfolios in der Firmenkundenbank. Konzeption und Implementierung einer efficient customization am Beispiel von Genossenschaftsbanken. 2009.

Band 14 Thorsten Pieper: Wirkungsorientiertes Controlling staatlichen Handelns. Systematische Identifikation und Bewertung der gesamtgesellschaftlichen Wirkungen staatlichen Handelns. 2009.

Band 15 Tina Püthe: Mittelständische Unternehmen und Genossenschaftsbanken. Eine empirische Analyse der Wirkung ökonomischer und verhaltenswissenschaftlicher Faktoren. 2009.

Band 16 Christian Buschhoff: Evaluation von Verwaltungsmodernisierung. Empirische Erkenntnisse auf Grundlage der Binnenmodernisierung in einer Landesverwaltung. 2009.

Band 17 Florian Wüller: Kundenwertorientierte Vertriebssteuerung im Firmenkundengeschäft der Kreditgenossenschaften. Kundenwertmodellierung, -operationalisierung und konzeptionelle Gestaltungsempfehlungen. 2012.

www.peterlang.de